CERAMIC MATRIX COMPOSITES

CERAMIC MATRIX COMPOSITES

MATERIALS, MODELING AND TECHNOLOGY

Edited by

Narottam P. Bansal
Jacques Lamon

The American Ceramic Society

WILEY

Published by John Wiley & Sons, Inc., Hoboken, New Jersey.
Published simultaneously in Canada.

Library of Congress Cataloging-in-Publication Data:

Ceramic matrix composites (Bansal)
Ceramic matrix composites : materials, modeling and technology / edited by Narottam P. Bansal, Jacques Lamon.
pages cm
Includes bibliographical references and index.
ISBN 978-1-118-23116-6 (hardback)
1. Ceramic-matrix composites. I. Bansal, Narottam P., editor. II. Lamon, Jacques, editor. III. Title.
TA418.9.C6.C47 2014
620.1′4–dc23

2014018426

10 9 8 7 6 5 4 3 2 1

Contents

Preface *xv*
Contributors *xvii*

Part I Fibers: Interface and Architecture 1

1 **Reinforcement of Ceramic Matrix Composites: Properties of SiC-Based Filaments and Tows** 3
Jacques Lamon, Stéphane Mazerat, and Mohamed R'Mili

1.1 Introduction 3
1.2 Processing of SiC-Based Filaments 4
1.3 Fracture Characteristics of Single Filaments 6
1.3.1 Statistical Strength Distributions 6
1.3.2 Weibull Distribution of Failure Strengths 6
1.3.3 Determination of Weibull Statistical Parameters 8
1.3.4 Normal Distribution 9
1.4 Multifilament Tows 11
1.4.1 The Bundle Model 13
1.4.2 Filaments–Tows Relations: Tow-Based Testing Methods for Determination of Single Filament Properties 14
1.5 Mechanical Behavior at High Temperatures 16
1.5.1 Strength Degradation and Oxidation at High Temperature 16
1.5.2 Static Fatigue Under Constant Load at Intermediate Temperatures: Subcritical Crack Growth 16
1.6 Summary 23
References 23

2 **Carbon Fibers** 27
Herwig Peterlik

2.1 Introduction/Production Routes 27
2.2 Structure of Carbon Fibers 28
2.2.1 Levels 1 and 2, Atomic level 28
2.2.2 Level 3, Lower Nanometer Range 28
2.2.3 Level 4, Upper Nanometer Range 31
2.2.4 Level 5, 10-μm Range 32
2.3 Stiffness and Strength of Carbon Fibers 32
2.4 Concluding Remarks and Future Directions 36
Acknowledgments 37
References 37

3 **Influence of Interfaces and Interphases on the Mechanical Behavior of Fiber-Reinforced Ceramic Matrix Composites** 40
Jacques Lamon

3.1 Introduction 40

3.2 Role of Interfacial Domain in CMCs 41
3.2.1 Crack Initiation at Interfaces 41
3.2.2 Crack Deflection at Interfaces 41
3.2.3 Approaches to Crack Deflection at Interfaces 42
3.2.4 Deflection Criteria Based on the Cook and Gordon Mechanism 44
3.2.5 Influence of Material Elastic Properties on Crack Deflection 45
3.3 Influence of Deflected Cracks 49
3.4 Strengthened Interfaces and Interphases 51
3.5 Various Concepts of Weak Interfaces/Interphases 56
3.6 Determination of Interfacial Properties 56
3.6.1 The Interfacial Tensile Strength 57
3.6.2 Interfacial Shear Strength or Stress 57
3.7 Interface Selection 60
3.8 Conclusions 60
References 61

4 **Textile Reinforcements: Architectures, Mechanical Behavior, and Forming** **65**
Philippe Boisse

4.1 Introduction 65
4.2 Textile Composite Reinforcements 65
4.2.1 Multiscale Materials: Fibers, Tows, Fabrics 65
4.2.2 Architecture and Geometry of the Unit Woven Cell 67
4.2.3 Experimental Analysis of the Mechanical Behavior 67
4.2.4 Mechanical Behavior Modeling 73
4.3 Reinforcements of Ceramic Composites 74
4.3.1 Silicon Carbide Fibers 74
4.3.2 Textile Reinforcement 75
4.3.3 Infiltration of the Textile Preform 75
4.4 Preforming Simulation 76
4.4.1 Fishnet Algorithm 76
4.4.2 Continuous FE Approaches 76
4.4.3 Hypoelastic Behavior: Simulation of a Double-Dome Forming 77
4.4.4 Composite Reinforcement Forming Using a Semidiscrete Approach 77
4.5 Conclusion 81
References 82

Part II Composite Materials 85

5 **Carbon/Carbons and Their Industrial Applications** **87**
Hiroshi Hatta, Roland Weiss, and Patrick David

5.1 Introduction 87
5.2 Manufacturing of Carbon/Carbons 87
5.2.1 Carbon Fiber Reinforcements 88
5.2.2 Matrix Systems 90
5.2.3 Redensification/Recarbonization Cycles 95
5.2.4 Final Heat Treatment 95
5.3 Strengths 97
5.3.1 Introduction 97
5.3.2 Fiber/Matrix Interface 98
5.3.3 Tensile Strength 101
5.3.4 Shear Strength 104
5.3.5 Compressive Strength 106

5.3.6 Fatigue Behavior 108
5.3.7 Concluding Remarks 109
5.4 Thermal Properties of Carbon/Carbon Composites 109
5.4.1 Introduction 109
5.4.2 Thermophysical Properties of Monolithic Carbons 110
5.4.3 Thermal Conductivity of Carbon/Carbons 110
5.4.4 Electrical Properties 115
5.4.5 Thermal Expansion 116
5.4.6 Specific Heat 117
5.4.7 Thermal Shock Resistance 118
5.4.8 Concluding Remarks and Future Directions 118
5.5 Oxidation Protection of Carbon/Carbon 118
5.5.1 Bulk Protection Systems for Carbon/Carbons 119
5.5.2 Outer Multilayer Coatings 122
5.5.3 Outer Glass Sealing Layers 124
5.6 Industrial Applications of Carbon/Carbons 126
5.6.1 Carbon/Carbons for High-Temperature Furnaces 129
5.6.2 Application for Thermal Treatments of Metals 134
5.6.3 Application of Carbon/Carbon in the Solar Energy Market 137
References 140

6 C/SiC and C/C-SiC Composites 147
Bernhard Heidenreich

6.1 Introduction 147
6.2 Manufacturing Methods 149
6.2.1 Chemical Vapor Infiltration 151
6.2.2 Polymer Infiltration and Pyrolysis 157
6.2.3 Melt Infiltration 160
6.3 Properties 174
6.3.1 General Properties 174
6.3.2 Material Composition and Microstructure 176
6.3.3 Mechanical Properties 178
6.3.4 Thermal Properties 187
6.3.5 Oxidation 190
6.3.6 Tribological Properties 191
6.4 Applications 191
6.4.1 Space Applications 192
6.4.2 Applications for Aeronautics 200
6.4.3 Applications for Friction Systems 202
6.4.4 Applications for High Temperature Treatment of Metals 207
6.5 Summary 209
Acknowledgments 209
Abbreviations 210
References 211

7 Advances in SiC/SiC Composites for Aero-Propulsion 217
James A. DiCarlo

7.1 Introduction 217
7.2 Materials and Process Requirements for Structurally Reliable High Temperature SiC/SiC Components 218
7.3 Current Fabrication Routes for SiC/SiC Engine Components 219
7.4 Recent NASA Advancements in SiC/SiC Materials and Processes 220
7.4.1 Advances in SiC-Based Fibers 220
7.4.2 Advances in Interfacial Fiber Coatings 224

7.4.3 Advances in SiC Fiber Architectures 225
7.4.4 Advances in SiC-Based Matrices 227
7.4.5 Advances in SiC/SiC Microstructural Design Methods 229
7.5 Current Microstructural Design Guidelines and Potential Service Issues for Higher Temperature SiC/SiC Components 232
7.6 Concluding Remarks 233
Acknowledgments 233
References 233

8 **Oxide–Oxide Composites** 236
Kristin A. Keller, George Jefferson, and Ronald J. Kerans

8.1 Introduction 236
8.2 Composite Design for Tough Behavior 237
8.2.1 Porous Matrices 237
8.2.2 Interface Control 238
8.3 Fibers and Fiber Architecture 240
8.4 Processing Methods 241
8.4.1 Processing of Interface Coatings 242
8.4.2 Matrix Infiltration 243
8.4.3 Consolidation 247
8.4.4 Metal Oxidation Processing 248
8.5 Porous Matrix Composite Systems 248
8.6 Properties 250
8.6.1 Basic Physical Characteristics 250
8.6.2 Room Temperature Uniaxial Mechanical Properties 251
8.6.3 Long-Term Thermal Exposure 253
8.6.4 Durability 254
8.6.5 Notch Sensitivity and Toughness 255
8.6.6 Off-Axis Properties 256
8.7 Composites with Interface Coatings 257
8.7.1 Weak Oxide–Oxide Phase Boundaries/Weak Oxides 257
8.7.2 Porous Coatings 260
8.7.3 Fugitive Coatings 260
8.7.4 Other Coatings 261
8.8 Technology Development 261
8.9 Potential Future for Oxide–Oxide Composites 263
Acknowledgments 264
References 264

9 **Ultrahigh Temperature Ceramic-Based Composites** 273
Yutaka Kagawa and Shuqi Guo

9.1 Introduction 273
9.2 Ultrahigh Temperature Ceramic-Based Composites with Particulates 273
9.2.1 Fabrication Methods 273
9.2.2 Physical Properties 278
9.2.3 Mechanical Properties 282
9.3 Ultrahigh Temperature Ceramic-Based Composites with Short Fibers 285
9.3.1 Carbon Fiber-Reinforced ZrB_2- or HfB_2-Based Ceramics Matrix Composites 286
9.3.2 Silicon Carbide Fiber-Reinforced ZrB_2-Based Ceramics Matrix Composites 288
9.4 Summary Remarks and Future Outlook 288
References 290

Part III Environmental Effects and Coatings 293

10 Environmental Effects on Oxide/Oxide Composites 295
Marina B. Ruggles-Wrenn

10.1 Introduction/Background 295
10.2 Mechanical Behavior—Effects of Environment 296
10.2.1 Tensile Stress–Strain Behavior 296
10.2.2 Tensile Creep 298
10.2.3 Tensile Creep and Recovery 304
10.2.4 Tensile Creep—N720/A and N720/AM Composites with ±45 Fiber Orientation 308
10.2.5 Compression and Compression Creep 313
10.2.6 Creep in Interlaminar Shear 317
10.2.7 Tension–Tension Fatigue 321
10.2.8 Tension–Tension Fatigue with Hold Times 324
10.3 Concluding Remarks and Future Directions 330
References 331

11 Stress-Environmental Effects on Fiber-Reinforced SiC-Based Composites 334
Gregory N. Morscher

11.1 Introduction/Background 334
11.2 Mechanisms 334
11.2.1 Surface Recession 336
11.2.2 Interior Oxidation and "Oxidation Embrittlement" (Via Cracks or Exposed Volatile Pathways) 336
11.3 Composite Systems 337
11.3.1 Carbon-Fiber-Reinforced SiC Composites (C_f/SiC_m) 337
11.3.2 SiC-Fiber-Reinforced SiC Matrix Composites with Carbon Interphases ($SiC_f/C_i/SiC_m$) 338
11.3.3 SiC-Fiber-Reinforced SiC Matrix Composites with BN Interphases ($SiC_f/BN_i/SiC_m$) 341
11.4 Modeling and Design for Stress-Oxidation Degradation 345
11.4.1 Modeling Stressed-Oxidation Degradation in C/SiC Composites—C Fiber Removal 345
11.4.2 Modeling Stressed-Oxidation Degradation in SiC/C/SiC Composites: Interphase Recession and Fiber Flaw Growth 346
11.4.3 Modeling Stressed-Oxidation Degradation in SiC Fiber Composites—Unbridged Crack Growth Due to Fiber Weakening and Stress Concentration at Unbridged Crack Tip 347
11.4.4 Modeling Stressed-Oxidation Degradation in SiC/BN/SiC Composites: Unbridged Crack Growth Due to Strongly Bonded Fibers and Local Fiber Failure 347
11.4.5 Mechanism Map: Henegar and Jones 348
11.4.6 Simple Design Approach: Matrix Cracking Stress 348
11.5 Concluding Remarks and Future Directions 350
Acknowledgments 350
References 350

12 Environmental Effects: Ablation of C/C Materials—Surface Dynamics and Effective Reactivity 353
Gerard L. Vignoles, Jean Lachaud, and Yvan Aspa

12.1 Introduction/Background 353
12.1.1 Materials Description 355

12.1.2 Materials Tests 357
12.1.3 Observation of Roughness Features 359
12.2 Materials Observation: Recession Rate 365
12.2.1 Theory 368
12.2.2 Results and Discussions 374
12.3 Concluding Remarks and Future Directions 383
Acknowledgments 384
References 384

13 **Radiation Effects** **389**
Yutai Katoh

13.1 Introduction 389
13.2 Theory of Radiation Damage 389
13.2.1 What Is Radiation Damage? 389
13.2.2 Radiation Defect Production 390
13.2.3 Defect Migration and Evolutions 391
13.2.4 Resultant Macroscopic Effects 392
13.3 Radiation Effects on Ceramics 392
13.3.1 Silicon Carbide 392
13.3.2 Radiation Damage Evolution in SiC 392
13.3.3 Radiation Effects in SiC 393
13.3.4 Carbon Materials 394
13.4 Radiation Effects in Ceramic Matrix Composites 394
13.4.1 SiC/SiC Composites 394
13.4.2 Interphase in SiC/SiC Composites 398
13.4.3 Carbon Fiber Composites 399
13.5 Concluding Remarks and Future Directions 401
Acknowledgment 402
References 402

14 **Foreign Object Damage in Ceramic Matrix Composites** **405**
Sung R. Choi

14.1 Introduction/Background 405
14.2 Experimental Techniques 406
14.2.1 FOD Test Rig 406
14.2.2 Materials: CMC and Projectile Materials 406
14.2.3 Configuration and Support of CMC Targets 407
14.2.4 Impact Damage Assessments 408
14.3 Phenomena of Foreign Object Damage in CMCs 409
14.3.1 Residual Strength 409
14.3.2 Impact Damage Morphology 410
14.3.3 Prediction of Impact Force 417
14.3.4 Other Effects in FOD 420
14.4 FOD Response of Environmental Barrier Coatings 422
14.5 Comparison of CMCs and Silicon Nitrides 424
14.6 Consideration Factors of FOD in CMCs 425
14.7 Concluding Remarks 426
Acknowledgments 426
References 426

15 **Environmental Barrier Coatings for SiC_f/SiC** **430**
Kang N. Lee

15.1 Introduction 430

15.2 Background 431
15.2.1 Silica Volatility 431
15.2.2 Key Requirements for EBC 432
15.3 Evolution of EBCs 437
15.3.1 Mullite Coating 437
15.3.2 First Generation Environmental Coatings 438
15.3.3 Second Generation Environmental Coatings 441
15.3.4 Next Generation Environmental Coatings 442
15.4 Processing, Testing, and Lifing 442
15.4.1 Processing 442
15.4.2 Testing 443
15.4.3 Lifing 444
15.5 Concluding Remarks and Future Directions 448
References 448

16 Oxidation Protective Coatings for Ultrahigh Temperature Composites 452
Qiangang Fu and Yiguang Wang

16.1 Introduction 452
16.2 Basic Requirements of Anti-Oxidation Coating for C/C and C/SiC Composites 453
16.2.1 Inhibiting Ability of Oxygen Diffusion 453
16.2.2 Good Match of Thermal Expansion with C/C and C/SiC Composites 453
16.2.3 Low Volatility during Service Process 453
16.2.4 Compatible Stability with C/C or C/SiC Composites 453
16.2.5 Good Interfacial Bonding with C/C or C/SiC Composites 454
16.3 Preparation Methods of Anti-Oxidation Coatings 454
16.3.1 Pack Cementation 454
16.3.2 Chemical Vapor Deposition 454
16.3.3 Liquid Phase Reaction 454
16.3.4 Plasma Spraying 455
16.3.5 Sol-Gel Method 455
16.3.6 Supercritical Fluid Technique 455
16.3.7 Slurry Method 455
16.3.8 *In Situ* Reaction 456
16.4 Oxidation-Resistant Coating Systems 456
16.4.1 Glass Coatings 456
16.4.2 Metal Coatings 456
16.4.3 Ceramic Coatings 457
16.5 Composite Coating 460
16.6 Summary 460
References 461

Part IV Modeling 465

17 Damage and Lifetime Modeling for Structure Computations 467
Pierre Ladevèze, Emmanuel Baranger, Martin Genet, and Christophe Cluzel

17.1 Introduction 467
17.2 Damage Modeling Based on an Anisotropic Damage Theory Including Closure Effects 468
17.2.1 General Notations 468
17.2.2 Thermodynamic Framework 469
17.2.3 A First CMC Damage Model 471

17.2.4 An Advanced CMC Damage Model 472
17.2.5 Some tools for the Implementation of the Anisotropic Framework 479
17.3 Multiscale Modeling of the Oxidation/Damage Coupling and the Self-Healing Effects 481
17.3.1 Modeling of Fatigue and of the Transverse Intra-yarn Crack Opening 482
17.3.2 Modeling of the Healing Process 490
17.3.3 Modeling of the SubCritical Cracking of Fibers 496
17.3.4 Simulation of Degradation Mechanisms and Prediction of Lifetime 497
17.4 Prediction Capabilities 503
17.4.1 Calculation of the Local Behavior 503
17.4.2 Lifetime Predictions and Application to Damage Tolerance Analysis 506
17.4.3 Control of the Damage Localization and Static Failure Predictions 513
References 515

18 | Approach to Microstructure–Behavior Relationships for Ceramic Matrix Composites Reinforced by Continuous Fibers 520
Jacques Lamon

18.1 Introduction 520
18.2 Composite Mechanical Behavior 521
18.2.1 Tensile Stress–Strain Behavior of Composites Reinforced by Continuous Fibers 521
18.2.2 Ultimate Failure 524
18.3 Constituent Properties and Length Scales 526
18.3.1 Length Scales: Micro and Minicomposites 527
18.3.2 Fracture Strength of Matrix 527
18.3.3 Flaw Strength Distributions 528
18.3.4 Damage Tolerance and Influence of Macroscopic Flaws 528
18.3.5 Interface Strength and Influence of Interface Cracks on the Mechanical Behavior 530
18.4 Modeling of Stress–Strain Behavior 531
18.4.1 Stochastic Model of Matrix Fragmentation in 1D Composite and Minicomposite 531
18.4.2 Ultimate Failure in 1D Composite and Minicomposite 535
18.4.3 Toward a Probability-Based General Model of Composite Behavior 536
18.4.4 The Stress–Strain Behavior of 1D Composites and Minicomposites 536
18.4.5 Alternate Approach 537
18.5 Virtual Testing: Computational Approach for Woven Composites 539
18.5.1 Multiple Cracking in Transverse Tows 539
18.5.2 Matrix Damage in 2D Woven Composites 541
18.6 Predictions of Rupture Time 542
18.7 Conclusions 545
References 546

Part V Joining 549

19 | Integration and Joining of Ceramic Matrix Composites 551
Monica Ferraris and Valentina Casalegno

19.1 Introduction/Background 551
19.2 Mechanical Joining and Integration of CMC 552
19.3 Adhesive Joining of CMC 553
19.4 Brazing of CMC 553
19.5 Liquid Silicon Infiltration 554
19.6 ArcJoinT 554

19.7 "Exotic" Techniques for Integration And Joining of CMC 555
19.7.1 Transient-Liquid-Phase Bonding 555
19.7.2 Nanopowder Infiltration and Transient Eutectic Phase 555
19.7.3 Spark Plasma Sintering 556
19.7.4 Micro-Wave-Assisted Joining 557
19.7.5 Laser-Assisted Joining 557
19.7.6 Glass and Glass-Ceramic as Joining Materials for CMC 557
19.7.7 Solid State Displacement Reactions 557
19.7.8 Preceramic-Polymer Joints 558
19.8 Back to Basic: Joints for CMC Like in Wood-Based Products 558
19.9 Special Issues 560
19.9.1 Joining of CMC for Nuclear Applications 560
19.9.2 Joining of CMC for Ultrastable Structures 561
19.10 Mechanical Tests on Joined CMC 561
19.10.1 Shear Strength Tests 561
19.10.2 Nondestructive Tests 562
19.11 Concluding Remarks and Future Directions 562
Acknowledgments 563
References 563

Part VI Nondestructive Evaluation 569

20 Use of Acoustic Emission for Ceramic Matrix Composites 571
Gregory N. Morscher and Nathalie Godin

20.1 Introduction/Background 571
20.2 AE Principles and Practice 572
20.3 Event-Based AE Monitoring of CMCs 575
20.3.1 Event-Based AE of CMC Stress–Strain Behavior 576
20.3.2 Event-Based AE of CMC-Elevated Temperature Stress-Rupture 579
20.3.3 Event-Based AE of CMC C-Coupon Testing 580
20.4 AE Signal Analysis Using Pattern Recognition Techniques 580
20.4.1 Unsupervised Clustering Methodology Applied to AE Signals 583
20.4.2 Supervised Pattern Recognition Method 584
20.5 High Temperature Testing and AE Monitoring 584
20.5.1 Identification of Damage Mechanism on SiC_f/[Si-B-C] Composite at Intermediate Temperatures (450–750°C) 585
20.5.2 Identification of Damage Mechanism on C_f/[Si-B-C] Composite at High Temperature (700–1200°C) 586
20.6 Acoustic Emission and Lifetime Prediction During Static Fatigue Tests 586
20.6.1 Detection of Energy Release Acceleration 587
20.6.2 Application of the Benioff Law to Assess Lifetime of the CMCs 588
20.7 Concluding Remarks and Future Directions 588
References 589

Part VII Applications 591

21 CMC Applications to Gas Turbines 593
Patrick Spriet

21.1 Introduction 593
21.2 CMC Developments for Military Engines 594
21.2.1 Demonstration and Developments for Exhaust Section 594
21.2.2 Demonstration of CMC Afterburner Components 598
21.2.3 Turbine Component Demonstrations on Military Engines 599

21.3 CMC R&D for Commercial Engines 600
21.3.1 CMC R&D for Combustion Chamber 601
21.3.2 R&D on CMC Turbine Components 602
21.3.3 Development Activities in the Exhaust Section 604
21.3.4 CMC Program Demonstrations on Industrial Gas Turbine 606
21.4 Summary and Insertion Issues 607
References 608

22 **Ceramic Matrix Composites: Nuclear Applications** **609**
Cédric Sauder

22.1 Introduction 609
22.2 CMC Fusion Applications 610
22.2.1 Introduction 610
22.2.2 The Divertor 612
22.2.3 First Wall and Blanket 614
22.3 CMC Fission Applications 616
22.3.1 Introduction 616
22.3.2 CMCs for Structural Materials in VHTRs and SFRs 618
22.3.3 CMCs for Pin Cladding or Structural Materials in GFRs and LWRs 621
22.4 Processing of C/C Composites for Nuclear Applications 624
22.4.1 Introduction 624
22.4.2 C/C Composites for Fusion Applications 624
22.4.3 C/C Composites for Fission Reactor Applications 627
22.5 Processing of SiC/SiC Composites for Nuclear Applications 627
22.5.1 Introduction 627
22.5.2 Selection of SiC Fibers 628
22.5.3 Selection of Architecture 630
22.5.4 Choice of Interphase 630
22.5.5 Densification Processes and Associated SiC/SiC Composite Properties 631
22.5.6 Designs for Hermetic Composites Adapted for Nuclear Applications 637
22.6 Conclusions and Perspectives 641
Acknowledgment 642
References 642

23 **Ceramic Matrix Composites for Friction Applications** **647**
Walter Krenkel and Jacques Georges Thébault

23.1 Introduction 647
23.2 Carbon/Carbon for Friction Applications 647
23.2.1 Aircraft Braking Particularities 647
23.2.2 History and Materials Choice 648
23.2.3 C/C Tribological Behavior and Constituents 651
23.2.4 Other C/C Tribological Applications 655
23.2.5 C/C Limitations and Trends 657
23.3 Carbon/Ceramic for Friction Applications 657
23.3.1 History and Milestones of the C/SiC Development 659
23.3.2 C/SiC Brake Disks for Passenger Cars 664
23.3.3 C/SiC Materials for Industrial Applications 666
23.3.4 Future Challenges for C/SiC Materials 668
23.4 Conclusions 668
Acknowledgments 669
References 669

Index 673

Preface

The concept of composites covers a wide variety of materials of which one can tailor the properties with respect to end-use applications, through a clever combination of constituents. A large variety of composite properties are theoretically foreseeable. A sound combination of continuous brittle fibers and brittle matrix ends up in damage-tolerant ceramic matrix composites (CMCs) with good mechanical properties in a broad range of temperatures. CMCs are lightweight and exhibit much greater resistance to high temperatures and aggressive environments than metals or other conventional engineering materials. They consist of ceramic or carbon matrix reinforced with continuous ceramic or carbon fibers. Their mechanical behavior displays several typical features that differentiate them from the other composites (such as polymer matrix composites and metal matrix composites) and from the homogeneous (monolithic) materials. These features depend on fiber and matrix properties, composite microstructure, interfaces/interphases, and so on.

CMCs are of great interest as thermostructural materials and were developed initially for military and aerospace applications. Now they are being introduced into many other new fields and their range of applications will grow when production cost is drastically reduced. Potential applications include heat exchangers, heat engines, gas turbines, structural components for the aerospace industry, nuclear reactors, chemical industries, automotive industry, biological implants, and so on.

Several approaches are available for processing of CMCs, using liquid, gaseous, or solid precursors. The chemical vapor infiltration (CVI) technique fostered the development of CMCs. The continuous fiber-reinforced CMCs followed the development of C/C composites and the CVI manufacturing technique. Development of CVI SiC/SiC composites began in the 1980s when SEP (now SNECMA), Amercorm, Refractory Composites, and others began to develop equipment and processes for producing CMC components for aerospace, defense, and other applications. SNECMA was at the forefront of this technology and demonstrated satisfactory performance of various CVI SiC/SiC components in turbine engine and flight tests. A number of CVI SiC/SiC components have performed successfully in turbine engines or full-scale tests.

The CVI technique has been studied since the 1960s, as an extension of chemical vapor deposition (CVD) technology. CVD involves the deposition of a solid on a heated substrate, from gaseous precursors. It has been used for many years to produce wear resistant coatings, coatings for nuclear fuels, thin films for electronic circuits, ceramic fibers, and so on. When the CVD technique is used to impregnate rather large amounts of matrix materials in fibrous preforms, it is called chemical vapor impregnation or infiltration. CVI was first used for the fabrication of carbon–carbon composites via pyrolysis of CH_4 at 1000–2000°C. Carbon–carbon (C/C) composites display several advantageous properties (such as low density and good mechanical properties at high temperatures). However, it was realized ca. 1973 that applications of C/C composites would be limited because of their poor oxidation resistance at temperatures higher than 450°C. SiC matrix composites were considered to be a solution to overcome the above shortcoming of C/C composites for long service life at elevated temperatures in oxidative environments. The feasibility of CVI SiC matrix composites was established in 1977 and confirmed independently in 1978. CVI SiC matrix composites reinforced by SiC fibers were manufactured in 1980. The current CVI SiC/SiC composites exhibit excellent performance in extreme environments at elevated temperatures. In the meantime, alternative manufacturing methods have been devised such as melt infiltration (MI), polymer infiltration and pyrolysis (PIP), and hybrid approaches which are a combination of CVI and other methods.

During the past 30 years, tremendous progress has been made in CMC developments. A large amount of knowledge has been acquired, and numerous processing/microstructure/property data have been produced. Results and state-of-the-art at a certain point in time were discussed in published conference proceedings, journal articles, and a few books. However, since we have entered the new millennium, the CMCs have reached maturity for use in systems working at high temperatures. They have now reached a high level of technological development. The issues have moved from processing methods and basic characteristics to stability at high temperatures in aggressive environments, life of material and components, predictive models, and simulation.

Among the features of long fiber-reinforced composites that have to be capitalized, versatility is quite important. For this purpose, characteristics and behavior of constituents have to be considered. The performance of primary constituents requires further and full consideration. From this point of view, the fibers, the interfaces/interphases, the reinforcing preforms, and the matrices are key elements. Some efforts have been devoted to multiscale approaches to thermostructural behavior, including (i) bottom-up approaches to establish microstructure/property relations with a view to a complete approach integrating processing/microstructure/property relations and (ii) top-down approaches for component simulations.

In this book, opportunity was given to international specialists to produce a comprehensive review of recent developments in their field of expertise. They were encouraged to discuss state-of-the-art knowledge on the following topics related to CMCs:

- Behavior and properties of constituents: fibers (including single filaments and tows), fiber/matrix interfaces and interphases, and preforms. (***Chapters 1–4***)
- Carbon/carbon, C/SiC, C/C-SiC, SiC/SiC, oxide/oxide, and ultrahigh temperature ceramic-based composites: manufacturing methods, properties, technology development, and recent advances for major continuous fiber-reinforced composite systems. (***Chapters 5–9***)
- Environmental effects, including effects of steam on high temperature performances of oxide/oxide composites, stress-oxidation degradation in SiC-based composites, thermomechanical ablation, radiation effects on SiC-based and carbon fiber composites, and foreign object damage. (***Chapters 10–14***)
- Protective coatings against surface recession in SiC/SiC and oxidation of ultrahigh temperature composites. (***Chapters 15 and 16***)
- Multiscale modeling of material behavior and computational simulation of life of engineering structures. (***Chapters 17 and 18***)
- Integration and joining of CMCs and mechanical testing of joined structures. (***Chapter 19***)
- Acoustic emission–based detection and quantification of damage with a view to life prediction. (***Chapter 20***)
- Applications of CMCs in key sectors including aeronautics, space, and nuclear plants. (***Chapters 21–23; also Chapters 5, 6, 7, and 8***)

The editors hope that this book will spark interest in the ceramic composite design and development communities and within other organizations that relate to these activities. Universities that educate future engineers who will be working with these complex materials will find much that is valuable and intellectually challenging.

The authors are fully responsible for the material presented in their chapters and they deserve the credit also. The editors are greatly indebted to them for their hard work and dedication in presenting their material in a timely fashion. Thanks are due to Ms. Anita Lekhwani, Senior Acquisitions Editor for Chemistry, Biotechnology, and Materials Science, John Wiley & Sons, Inc., for her cooperation and understanding during the publication of this book.

NAROTTAM BANSAL
JACQUES LAMON

Contributors

Yvan Aspa, Laboratoire des Composites Thermostructuraux, Université de Bordeaux, Pessac, France

Narottam P. Bansal, Materials & Structures Division, NASA Glenn Research Center, Cleveland, OH, USA

Emmanuel Baranger, Laboratoire de Mécanique et Technologie (LMT), CNRS (Centre National de la Recherche Scientifique), ENS Cachan (Ecole Normale Supérieure), UPMC (Univesité Pierre et Marie Curie), Cachan, France

Philippe Boisse, Laboratoire LaMCoS, INSA de Lyon, Université de Lyon, Villeurbanne Cedex, France

Valentina Casalegno, Dipartimento di Scienza dei Materiali e Ingegneria Chimica (DISAT), Politecnico di Torino, Torino, Italy

Sung R. Choi, Naval Air Systems Command, Patuxent River, MD, USA

Christophe Cluzel, Laboratoire de Mécanique et Technologie (LMT), CNRS (Centre National de la Recherche Scientifique), ENS Cachan (Ecole Normale Supérieure), UPMC (Univesité Pierre et Marie Curie), Cachan, France

Patrick David, CEA Le Ripault, Monts, France

James A. DiCarlo, Materials and Structures Division, NASA Glenn Research Center, Cleveland, OH, USA

Monica Ferraris, Dipartimento di Scienza dei Materiali e Ingegneria Chimica (DISAT), Politecnico di Torino, Torino, Italy

Qiangang Fu, National Key Laboratory for Thermostructure Composite Materials, Northwestern Polytechnical University, Xi'an, Shaanxi, China

Martin Genet, Laboratoire de Mécanique et Technologie (LMT), CNRS (Centre National de la Recherche Scientifique), ENS Cachan (Ecole Normale Supérieure), UPMC (Univesité Pierre et Marie Curie), Cachan, France

Nathalie Godin, Laboratoire MATEIS, CNRS, INSA de Lyon, Universit de Lyon, Villeurbanne, France

Shuqi Guo, Hybrid Materials Unit, National Institute for Materials Science, Tsukuba, Ibaraki, Japan

Hiroshi Hatta, Division of Space Propulsion, Institute of Space and Astronautical Science, Japan Aerospace Exploration Agency, Sagamihara, Kanagawa, Japan

Bernhard Heidenreich, German Aerospace Center (DLR), Institute of Structures and Design, Department of Ceramic Composites and Structures, Stuttgart, Germany

George Jefferson, Materials and Manufacturing Directorate, Air Force Research Laboratory, Wright-Patterson Air Force Base, OH, USA

Yutaka Kagawa, Research Center for Advanced Science and Technology, The University of Tokyo, Meguro-ku, Tokyo, Japan

Yutai Katoh, Materials Science and Technology Division, Oak Ridge National Laboratory, Oak Ridge, TN, USA

Kristin A. Keller, Materials and Manufacturing Directorate, Air Force Research Laboratory, Wright-Patterson Air Force Base, OH, USA

Ronald J. Kerans, Materials and Manufacturing Directorate, Air Force Research Laboratory, Wright-Patterson Air Force Base, OH, USA

Walter Krenkel, Ceramic Materials Engineering, University of Bayreuth, Bayreuth, Germany

Jean Lachaud, Laboratoire des Composites Thermostructuraux, Université de Bordeaux, Pessac, France

Pierre Ladèveze, Laboratoire de Mécanique et Technologie (LMT), CNRS (Centre National de la Recherche Scientifique), ENS Cachan (Ecole Normale Supérieure), UPMC (Univesité Pierre et Marie Curie), Cachan, France

Jacques Lamon, Laboratoire de Mécanique et Technologie (LMT), CNRS (Centre National de la Recherche Scientifique), ENS Cachan (Ecole Normale Supérieure), UPMC (Univesité Pierre et Marie Curie), Cachan, France

Kang N. Lee, Rolls-Royce Corporation, Indianapolis, IN, USA

Stéphane Mazerat, Laboratoire des Composites Thermostructuraux, Université de Bordeaux, Pessac, France

Gregory N. Morscher, Department of Mechanical Engineering, The University of Akron, Akron, OH, USA

Herwig Peterlik, Faculty of Physics, University of Vienna, Vienna, Austria

Mohamed R'Mili, Laboratoire MATEIS, CNRS, INSA de Lyon, Université de Lyon, Villeurbanne, France

Marina B. Ruggles-Wrenn, Department of Aeronautics & Astronautics, Air Force Institute of Technology, Wright-Patterson Air Force Base, OH, USA

Cédric Sauder, CEA, DEN, DMN, SRMA, F-91191 Gif Sur Yvette, France

Patrick Spriet, Herakles, Safran Group (Retired), Aeronautics and Hi-Temp Composites B.U., Les Cinq Chemins, Le Haillan, France

Jacques Georges Thébault, Thermostructural composites expert, SAFRAN group (Retired), Bordeaux, France

Gerard L. Vignoles, Laboratoire des Composites Thermostructuraux, Université de Bordeaux, Pessac, France

Yiguang Wang, National Key Laboratory for Thermostructure Composite Materials, Northwestern Polytechnical University, Xi'an, Shaanxi, China

Roland Weiss, Schunk Kohlenstofftechnik GmbH, Abteilung Composites, Giessen, Germany

PART I

FIBERS: INTERFACE AND ARCHITECTURE

CHAPTER 1

REINFORCEMENT OF CERAMIC MATRIX COMPOSITES: PROPERTIES OF SiC-BASED FILAMENTS AND TOWS

JACQUES LAMON,[1] STÉPHANE MAZERAT,[2] AND MOHAMED R'MILI[3]

[1]Laboratoire de Mécanique et Technologie (LMT), CNRS (Centre National de la Recherche Scientifique), ENS Cachan (Ecole Normale Supérieure), UPMC (Univesité Pierre et Marie Curie), Cachan, France
[2]Laboratoire des Composites Thermostructuraux, Université de Bordeaux, Pessac, France
[3]Laboratoire MATEIS, CNRS, INSA de Lyon, Université de Lyon, Villeurbanne, France

1.1 INTRODUCTION

Proper reinforcement of ceramics is aimed at increasing the resistance to crack propagation by introducing elements that can arrest the cracks. Only continuous fibers are able to arrest the cracks through deflection at fiber/matrix interfaces. Composite damage tolerance requires strong fibers and appropriate interfaces. Composite strength requires strong fibers and damage tolerant multifilament tows.

In continuous-fiber-reinforced ceramics, only those fibers that can withstand the high temperatures required by matrix processing (above 1000°C) can be used. Other high temperature requirements to be met include long-term stability, creep resistance, and oxidation resistance. A wide spectrum of continuous fiber–reinforced ceramic matrix composites (CMCs) can be foreseen owing to a wide variety of matrices and fibers. Non-oxide CMCs reinforced by non-oxide fibers have been the most studied. The reason for this is that carbon and silicon carbide fibers display the highest properties for use at high temperature. Second, for compatibility reasons, non-oxide fibers can be combined essentially to non-oxide matrices. However, carbon fibers degrade in oxidizing atmosphere at temperatures as low as 450°C, and they must be protected. SiC-based fibers are much more resistant to oxidation. Oxide fibers are inherently resistant to oxidation, but they have limited creep resistance and undergo grain growth at high temperatures, which causes strength degradation. Further, they display much higher densities than carbon and SiC-based fibers. Despite these drawbacks, alumina-based CMCs have been extensively studied.

The literature abounds in papers, book chapters, and books on microstructure/properties relationships for oxide and non-oxide fibers [1–4]. Carbon fibers are discussed in another chapter of this book. The present chapter focuses on a recent important issue for continuous SiC-based reinforcing ceramic fibers that has been ignored because potential use of CMCs was driven by their superior temperature resistance over metals. Use at very high temperatures above 1000°C was essentially foreseen. Recently, a growing interest was fostered by lightweight features of CMCs for use in the range of allowable temperatures for metals (below 1000°C in aeronautical engines).

Ceramic Matrix Composites: Materials, Modeling and Technology, First Edition. Edited by Narottam P. Bansal and Jacques Lamon.

The Nextel oxide fibers are the most widely used reinforcements for continuous fiber oxide–oxide composites [5–7]. Nextel 610 fiber has the highest strength and elastic modulus (3.1 GPa and 380 GPa, respectively), but it is limited by creep to temperatures <1000°C. Nextel 720 fiber has lower room temperature strength (2.1 GPa), but higher creep resistance, which allows use at higher temperatures (up to 1200°C). Sapphire (single crystal Al_2O_3) fibers are no longer available, their cost and diameter (>50 μm) limit their use in composites.

Non-oxide fibers exhibit superior tensile strength and creep resistance to the oxides. They possess comparable Young's moduli and diameters. Table 1.1 lists the main properties of SiC-based ceramic fibers [1, 2, 8–29]. Lower creep rates are observed at temperatures >1200°C, even under high stresses, whereas the oxide fibers can barely exceed 1000°C [1, 3, 30–35]. For example, Sylramic SiC fibers show less than 1% creep strain after 1000 hours at 1350°C and 100 MPa stress, or Hi-Nicalon type S fibers show less than 0.5% creep strain after 60 hours at 1350°C and 850 MPa stress. Creep strain of 10^{-8} per second is obtained at 1000°C and 100 MPa stress on the most creep-resistant oxide fiber (Nextel 720), at 1400°C and 300 MPa stress on Tyranno SA3, and at 1350°C and 850 MPa on Hi-Nicalon type S. Tyranno SA3 and Hi-Nicalon type S exhibit higher resistance to creep than Hi-Nicalon. The creep resistance is commensurate with low oxygen content, SiC grain size, and small amount of amorphous phase. It has been shown to be improved after high temperature treatment under various atmospheres. Creep behavior of Hi-Nicalon can be improved by using high temperature treatment that eliminates amorphous phase and organizes better carbon structure [3].

1.2 PROCESSING OF SiC-BASED FILAMENTS

Three generations of SiC-based fibers were manufactured essentially by two companies: Nippon Carbon Co. Ltd. and UBE INDUSTRIES Ltd. They differ by composition and properties as a result of the organosiliced precursors and manufacturing processes used (Table 1.1).

Fibers of the first generation (NicalonTM (Nicalon NLP101, NLM202, NLM207) and Tyranno (Tyranno ZMI, LOX-M, S, AM) Si-O-C fibers) were spun from a molten polycarbosilane (PCS) based ceramic precursor, cured under oxygen around 180°C during about 1 hour, and then submitted to a pyrolysis treatment under an inert gas, giving rise to a fine diameter SiC-based fiber [36, 37]. The first ones NLM 101 and NLM 200 have been commercialized since the late 1970s by Nippon Carbon Co., Yokohama, Japan (NC).

All these fibers contain a large amount of oxygen combined into amorphous silicon oxycarbide phases that are thermodynamically instable and decompose beyond 1100°C with a SiO(g) and CO(g) evolution and SiC crystal growth. This decomposition yields significant strength degradation [23, 29, 38–41].

For the Tyranno family fibers, a precursor resulting from the reaction of PCS with an alkoxide (M = Ti, Zr, or Al) was used [18, 19]. These Si-M-C-O Tyranno fibers containing metal additives show improved thermal stability and chemical corrosion resistance, and the degradation occurs at higher temperature (>1300°C) when compared with the Si-C-O Nicalon fibers [17, 42].

To avoid thermal instability due to presence of the oxycarbide phase, a nearly oxygen-free SiC fiber (Hi-Nicalon fiber from Nippon Carbon) was developed in 1990 by melt spinning, electron beam curing, and pyrolysis of PCS precursor under anaerobic conditions [15, 43]. This fiber exhibits much better thermal stability than the standard Nicalon fiber; however, it also contains a large excess of turbostratic carbon, which affects resistance to oxidation and creep [3, 4, 8, 44].

To reduce the free carbon content and to improve the high temperature properties of the fibers, near pure SiC fibers were produced [1, 14, 45]. Those third-generation fibers include the Hi-Nicalon type S fibers [42], Tyranno SA3 fibers [43], and Sylramic fibers [16].

The Hi-Nicalon S fiber process is similar to that of Hi-Nicalon one except the thermal treatment at high temperature under hydrogen [45]. The Hi-Nicalon S fiber consists of β-SiC sub-micrometer crystals with traces of carbon and oxygen [46]. This fiber exhibits a high Young's modulus, a high creep and oxidation resistance, and an excellent thermal stability up to 1600°C [1, 14, 47].

The Tyranno SA3 fiber was obtained by heat treatment of the amorphous AM fiber at about 1800°C. It shows no significant degradation in strength or change in composition on heating up to 1800°C in argon (1 hour) and up to 1000 hours in air at 1000°C [1, 14, 48].

The Sylramic fiber is no longer produced but it is similar to the Tyranno SA3 fiber. It was manufactured from Lox-M fiber heat-treated in boron oxide environment at high temperature in order to help sintering, to decrease the oxygen content, and to obtain a finely crystallized pure SiC fiber [1, 13, 49].

TABLE 1.1 Physical Properties and Composition of Various SiC-Based Fibers [1, 2, 8–29]

Fiber	Diameter (μm)	Fibers/fil	Curing	$T_{pyrolyse}$ (°C)	Composition (at. %) Si	C	O	Hetero-element	C_{libre} (at. %)	<SiC-β> (nm)	σ (S/m)	$\alpha \times 10^{-6}$	Density (g/cm^3)
NL101	15 [23]	500	ox	1200	36.4 41.1	46.8 43.3	16.7 15.6	–[22] –[1]	18.8 [22] 10.0 [1]	1.2 [22] 1.1 [23]	0.02 [28]	3.1 [29]	2.57 [21]
NL102	15 [23]	500	ox	1200	33.9 35.0	48.8 51.4	17.3 13.6	–[22] –[1]	23.6 [22] 23.3 [1]	1.7 [23]		3.1 [29]	2.61 [21]
NL207	14 [11]	500	ox	>1200 [1, 8, 10]	38.6 37.4	48.6 49	12.8 13.6	–[22] –[1, 10]	16.4 [22] 18.4 [1, 10]	1.9 [22] ~2 [10, 11, 13]	0.006 [28] 0.1–0.01 [15, 25, 27] 0.005 [17]	3.2 [1, 15]	2.55 [1, 10, 11, 13, 15, 17] 2.58 [21]
Hi-Ni	14 [8] 12 [1]	500	e⁻	1300 [1] 1500 [12]	39 41.6 41.9	60.4 57.8 57.2	0.6 0.6 0.9	–[1, 8, 13] –[21]	21.7 [1, 8, 13] 15.8 [22]	5 [8, 12, 13]	23 [28] 71 [25] 80 [27]	3.5 [1]	2.74 [1, 10, 11, 13, 15]
Hi-Ni-S	13 [14]	500	e⁻	>1500	48.2 48.7	50.8 51.0	1.0 0.3	–[22] –[1, 9, 13, 14]	3.1 [22] 2.6 [1, 9, 13]	50 [11–13]	0.14 [2, 8, 27]		3.05 [1, 13, 14] 3.10 [9]
T-S	11 [13] 8.5 [13]	800 1600	ox		31.6 31.8	41.7 48.1	26.0 19.4	(Ti) 0.7 [21] (Ti) 0.7 [22]	23.0 [21] 26.0 [22]	1.2 [22]	0.09 [28]	3.1 [24]	2.35 [21, 24]
Lox-M	8.5 [13] 11 [13]	800	ox	1200 [1] 1300 [12]	35.4 35.8 36.9	53.3 49.0 50.4	10.4 14.4 11.9	(Ti) 0.9 [22] 0.8 [1, 10, 13] 0.8 [17]	23.1 [22] 20.4 [1, 10, 13] 19.5 [17]	1.9 [22] 1.4 [18]	2.6 [28] 3.3 [17]	3.1 [1]	2.37 [10, 11, 13] 2.48 [1, 9, 17, 19, 24]
Lox-E	11 [1]	800	e⁻	1300 [12]	35.7 36.4	57.0 57.8	6.6 5.8	(Ti) 0.7 [1, 10, 17] –[9]	24.6 [1, 10] 24.3 [9]	2 [10]	125 [17] 40 [27]		2.39 [10, 11, 13] 2.55 [9, 17]
ZMI	11 [24]	800	ox	1300	35.5 37.3	54.3 53.7	10.0 8.8	(Zr) 0.2 [22] 0.2 [10]	23.9 [22] 20.7 [10]	2.2 [22] 2.0 [18]	16 [28] 50 [25]	4.0 [19, 24]	2.48 [10, 18, 21, 24]
ZM	11 [1]	800	ox	1300 [1] 1550 [17]	34.7 37.7	56.5 53.4	8.6 8.7	(Zr) 0.2 [21] 0.2 [1, 17]	26.0 [21] 20.5 [1]	2.48 [22] ~2 [17]	47 [28] 50 [17]		2.48 [1, 13, 17]
ZE	11 [1]	800	e⁻	1300 [1] 1580 [17]	36.3 38.7	61.9 59.1	1.7 2.0	(Zr) 0.1 [21] 0.2 [1, 13, 17]	26.4 [21] 21.4 [1, 10]	3.5 [22]	380 [28] 330 [17, 25]		2.55 [1, 10, 17]
ZX	11 [19]	800	ox	1500 [19]	37.0 34.8	53.7 52.9	9.1 12.1	(Zr) 0.2 [19] 0.2 [19]	21.2 [19] 24.1 [19]	~2 [19]			2.48 [19] 2.42 [19]
AM	11 [18] 8.5	800 1600	ox	1500 [18]	32.9 34.6	52.5 51.3	14.2 13.7	(Al) 0.4 [22] 0.4 [9, 18]	26.7 [22] 23.5 [9, 18]	2.2 [22] 1.8 [18]	0.15 [28]		2.42 [9, 18]
SA3	10 [11] 7.5 [24]	800/1600	ox	AM+1700 + >1800 [9, 12, 13, 18]	47.8	52.4	0.3	(Al) 0.5 [1, 9, 13, 14]	4.75 [1, 9, 13]	300 [13] >200 [12]	80 [28] 660 [26]	4.5 [24]	3.02 [9, 13] 3.1 [1, 24]
Sylramic	10 [1]	800	ox	Lox-M + 1600 [13] or >1700 [1]	46.4	47	1.2	(Ti/B/N) 0.8/4.0/0.6 [1, 13, 14]	1.2 [1, 13]	200 [13] 60–90 [20]	100 [26]	5.4 [1] 5.3 [20]	3.05 [1, 11, 13, 16]

ox: cured by heating in air; e⁻: cured by means of electron beams

1.3 FRACTURE CHARACTERISTICS OF SINGLE FILAMENTS

Fibers are brittle ceramics. The tensile behavior is completely described by Young's modulus and stress and strain at failure.

Fracture toughness, which measures the resistance to crack propagation, is difficult to measure using notched specimen-based conventional fracture mechanics techniques. However, Morishita and coworkers created a notch on a SiC Tyranno-SA3 fiber using a focused-ion beam (FIB) [50]. In most cases, available fracture toughness data are derived from diameters of fracture-inducing flaws identified by fractography in the fracture surface, using the conventional fracture mechanics equation: $K_{IC} = \sigma_r Y \sqrt{a_c}$, where Y is a geometrical factor characteristic of flaw shape and a_c is the critical flaw size. Fiber failure stress σ_r is either measured during the test or estimated from fracture mirror size. SiC-based fibers exhibit high resistances to failure and very low toughness (Table 1.2) [8–12, 14–21, 24]. Toughness estimates are in the range 1–2 MPa $m^{0.5}$ for most SiC fibers. Values close to 3 MPa $m^{0.5}$ have been measured on Sylramic fibers and on notched SiC Tyranno SA3 fibers. Tensile strengths and strains-to-failure reported in the literature range, respectively, between 1500 and 3500 MPa and between 1% and 1.9%. Values measured on significant batches of filaments with identical gauge length of 25 mm span comparable range (Table 1.3). Low toughness and high strengths are indicative of high sensitivity to fracture-inducing flaws and the presence of very small fracture-inducing flaws (submicron size). These features inherent to ceramic fibers reflect a scale effect. Flaw size is commensurate with fiber small dimensions and microstructure: a few microns large diameter (typically around 10 μm, Table 1.1) and fine grain size (a few nanometers).

However, it is worth emphasizing that the high strengths reported in Tables 1.2 and 1.3 are average values. They are approximate estimates depending on the number of data they average. Strength and strain data exhibit a wide variability as a result of inherent variability in failure origins, and also of uncertainties introduced by practical difficulties in the determination of fiber diameter (enhanced by diameter variation along fiber) and in the direct measurement of deformations. During tensile tests, deformations are derived from the displacement of load train. Correction for load train compliance is necessary. It is based on a calibration technique [51]. The estimation of strength from failure load is affected by uncertainty in fiber cross-sectional area or in strain. Practically much effort is done to improve accuracy in the determination of strengths, using up-to-date measurement techniques or by testing multifilament tow specimens, as discussed in section 1.4. It is thus reasonable to consider that strength variability is dominated by the distribution of fracture-inducing flaws. The scatter is generally very wide: from 500 to 3000 MPa and 0.2–1.2% for Nicalon fibers.

1.3.1 Statistical Strength Distributions

As usual with brittle ceramics, fracture data of single filaments exhibit a significant scatter, since inherent flaws with a random distribution induce fracture. An important consequence is that the fracture stress is not an intrinsic characteristic. It is instead a statistical variable, which depends on several factors including the stress state, the size of test specimens, and the number of test specimens [52]. Therefore, a unique reference value of fracture strength cannot be recommended.

1.3.2 Weibull Distribution of Failure Strengths

It is widely accepted that the Weibull model satisfactorily describes the statistical distribution of failure strengths of single filaments under tensile loads:

$$P = 1 - \exp\left\{ - \int (\sigma/\sigma_0)^m \, \mathrm{d}V/V_0 \right\} \tag{1.1}$$

where P is the probability of failure, σ is the stress, σ_0 is the scale factor, m is the Weibull modulus, V is the volume of specimen, V_0 is a reference volume ($V_0 = 1\ \mathrm{m}^3$ should be preferred), and σ_0 is related to the mean value of strength. The Weibull modulus reflects the scatter in strength data.

The strength for given size and stress state can be determined using Equation (1.1) when proper m, σ_0, and associated V_0 are known. If V_0 is not specified, the estimate of σ_0 is meaningless; the strength cannot be determined safely.

TABLE 1.2 Elastic and Failure Characteristics of Various SiC-Based Fibers

Fiber	σ_r (MPa)	E (GPa)	ε_r (%)	Weibull modulus	σ_l (MPa)	K_{IC} (MPa m$^{1/2}$)
NL101	1840 [23] 2200 [21]	134 [24] 180 [21]	1.4 [24] 1.2 [21]	2.3 [24] 4.3 [21]	2400 [21]	1.2 [21]
NL102	1490 [24] 2600 [21]	144 [24] 180 [21]	1.0 1247 1.4 [21]	2.7 [24] 7.6 [21]	2700 [21]	1.2 [21]
NL207	2800 [21] 2700 [8] 3000 [1, 13, 15]	210 [21] 220 [13, 15] 200 [1]	1.3 [21] 1.0 [13] 1.5 [1, 15]	5.0 [21] 4.3 [21]	3100 [21] 2900 [21]	1.2 [21] 0.9 [8] 2 [1]
Hi-Ni	3000 [21] 2800 [1, 13–15, 20]	300 [21] 270 [1, 9–11, 13, 15, 20]	1.0 [8, 10, 11, 13, 15, 19–21]	9.8 [21] 5.0 [25]	3300 [21] 2880 [25]	1.7 [21]
Hi-Ni-S	2900 [21] 2600 [9] 2500 [13, 14]	420 [9] 400 [1, 13, 14]	0.6 [9, 13, 14, 21]	8.4 [21] 4.5 [25]	3000 [21] 1330 [25]	1.9 [21]
TS	3200 [21] 3300 [24]	180 [21] 170 [24]	1.8 [21] 1.9 [24]	5.9 [21]	3400 [21]	1.7 [21]
TS11	2800 [21]	180 [21]	1.5 [21]	4.8 [21]	3000 [21]	1.1 [21]
Lox-M	3000 [21] 3300 [9, 17, 24] 2500 [13] 3200 [20]	200 [21] 187 [1, 9, 13, 17, 24] 180 [20]	1.5 [21] 1.8 [9, 17, 20, 24] 1.4 [13]	4.6 [21]	3300 [21]	1.1 [21]
Lox-E	3000 [21] 3400 [9, 17] 2900 [1, 13]	200 [21] 206 [9 17] 200 [1, 13]	1.5 [21] 1.7 [9, 17] 1.4 [1, 11]	5.5 [21]	3200 [21]	1.0 [21]
ZMI	3100 [21] 3400 [9, 10, 18, 24]	210 [21] 200 [9, 10, 18, 24]	1.5 [21] 1.7 [9, 10, 18, 24]	4.8 [21]	3400 [21]	0.7 [8] 1.0 [21]
ZM	3400 [1 17 21]	220 [21] 200 [1, 17]	1.6 [21] 1.7 [1, 17]	5.3 [21]	3700 [21]	1.0 [21]
ZE	3300 [21] 3500 [9, 10, 17]	230 [1, 9, 10, 17, 21]	1.4 [21] 1.5 [9, 10, 17]	5.7 [21]	3500 [21]	1.3 [21]
ZX	3000 [19]	195 [19]	1.5 [19]			
AM	2800 [8, 21]	180 [9, 18, 21]	1.6 [9, 18, 21]	6.2 [21]	3000 [21]	1.2 [21]
SA3	2800 [9, 13, 14, 24] 2500 [11]	420 [9] 330 [11] 375 [1, 13, 24] 303 [13]	0.7 [9, 24] 0.75 [11] 1.45 [13] 0.9 [13]			
Sylramic	3000 [11, 13] 3200 [10, 13] 2800 [20]	390 [11] 400 [1, 13, 20] 372 [9, 10, 13]	0.8 [1, 13, 14] 1.7 [20]	4.6 [20]		3–3.5 [20]

It is important to stress that the estimate of σ_0 depends on dimension units [52]. For instance, the σ_0 estimates are substantially different if V_0 is set to 1 m^3 or 1 mm^3 as shown by Equation (1.2) derived from Equation (1.4) given in the next section for a uniform uniaxial stress state:

$$\sigma_0(\mathrm{m}^3) = \sigma_0(\mathrm{mm}^3)\{V_0(\mathrm{mm}^3)/V_0(\mathrm{m}^3)\}^{1/m} = \sigma_0(\mathrm{mm}^3)10^{-9/m} \tag{1.2}$$

where σ_0 (m^3) and σ_0 (mm^3) are estimates corresponding, respectively, to V_0 (m^3) and V_0 (mm^3).

TABLE 1.3 Failure Characteristics Measured on Various SiC-Based Filaments Under Identical Conditions (Gauge Length 25 mm) [21]

Fibers	Nb of specimens	$\sigma_r \pm$ std dev (MPa)	m	σ_l (MPa)	σ_0 (MPa)
NL101	50	2166 ± 559	4.28	2391	5.5
NL102	50	2563 ± 422	7.64	2720	87.2
NL202			5.5	2270	19
NL207	100	2839 ± 667	5.05	3090	16.7
Hi-Ni	49	3136 ± 368	9.84	3295	225.5
Hi-Ni-S	44	2877 ± 414	8.42	3062	129.2
TS	72	3152 ± 624	5.90	3413	35.4
TS11	40	2776 ± 647	4.79	3034	11.5
Lox-M	44	2984 ± 695	4.63	3314	9.9
Lox-E	41	2939 ± 582	5.54	3238	25
ZMI	49	3076 ± 742	4.81	3372	12.8
ZM	46	3439 ± 751	5.29	3688	23.6
ZE	50	3250 ± 683	5.69	3521	32.16
AM	91	2816 ± 534	6.15	3031	40.4

Except for the Nicalon NL202 Fiber, σ_l was Estimated Since Fiber Diameters were not Measured Before Tests, and σ_0 ($V_0 = 1$ m^3) was Derived from σ_l using Equation (1.3) and Average Diameters given in Table 1.1. For the NL202 Fiber, σ_0 was Estimated from Test Data, Fiber Diameters were Measured Before Tests using Laser Diffractometry [53]. σ_l was Derived from σ_0.

Authors assume implicitly that the sectional area is identical in all the fibers when they use instead a characteristic strength (σ_l) that is associated to a characteristic length (Equation 1.5). σ_0 is related to the characteristic strength by the following equation derived from Equations (1.4) and (1.5):

$$\sigma_0(\mathrm{m}^3) = \sigma(\mathrm{mm})\{V(\mathrm{mm}^3)L_0(\mathrm{mm})/V_0(\mathrm{m}^3)\}^{(1/m)} \tag{1.3}$$

σ_l is the characteristic strength for gauge length L, V is the volume for L, and $L_0 = 1$ mm. The values of Weibull modulus of most SiC-based fibers are rather small (Tables 1.2 and 1.3). They span a range from 2.3 to 9.9. A significant scatter in m estimates is generally observed for each fiber: for instance, m $\cong$ 2.3–7.1 is reported for Nicalon fibers [53–55]. The variation is inherent to the use of an estimator to construct the Weibull plot, to the use of limited sample size, and to the selection of sample [56]. Tables 1.2 and 1.3 also report the scale factors estimated for various SiC fibers. Because of lack of uniformity in treatment of test data in the literature, only those scale factors obtained in comparable conditions were kept. Table 1.3 reports σ_0 values that were derived from characteristic strengths σ_l for the average diameters given in Table 1.1. For the Nicalon NL202 fibers, σ_0 value was determined on single filaments for which diameters were measured before the tests. σ_l was calculated from σ_0. These data are given for information. Potential users should be reminded that when the diameter is neglected, the scale factor might be affected by variability in fiber diameter. It is clear that the σ_0 estimates obtained for diameters measured on each fiber are the most pertinent.

1.3.3 Determination of Weibull Statistical Parameters

For a single gauge length and uniform tensile stress, Equation (1.1) reduces to

$$P = 1 - \exp - \left[\frac{V}{V_0}\left(\frac{\sigma}{\sigma_0}\right)^m\right] \tag{1.4}$$

or to Equation (1.5), when the sectional area is neglected to be constant:

$$P = 1 - \exp - [(\sigma/\sigma_0)^m 1/L_0] \tag{1.5}$$

The Weibull modulus m is obtained by fitting Equation (1.4) to the distribution of experimental strength data or as the slope of the "Weibull plot" of ln (ln(1/1−P)) vs. ln σ. The scale factor σ_0 is determined from the ordinate at origin k:

$$\ln(\ln(1/1 - P)) = m \ln \sigma + k \tag{1.6}$$

Either graphical or curve-fitting methods (like linear regression analysis, least squares method, or maximum likelihood estimator) can be used [52].

The Weibull plot is constructed, using an estimator for estimation of failure probabilities associated to failure data ranked in ascending order. Various estimators have been devised in the literature [52]. The estimator $P_j = j/N$ can be used on large sample sizes. The estimator $P_j = (j - 0.5)/N$ is recommended for limited sample sizes (>30 specimens); j is the rank of filament strength, N is the total number of data.

Another method of determining Weibull modulus is to measure fiber strength as a function of gauge length. With increasing gauge length, the chance of finding a large flaw increases, so fiber strength decreases. The effect of gauge length is given by the equation:

$$\ln \sigma = -1/m \ln L + k' \tag{1.7}$$

where L is gauge length and k' is a constant. Thus, the slope of a log–log plot of strength vs. gauge length is $-1/m$.

1.3.4 Normal Distribution

The normal distribution should be a natural solution. It is considered the most prominent probability distribution in statistics. It indicates the probability of occurrence of a characteristic in a population of infinite size. Then, certain distributions can be approximated by the normal distribution when the sample size is large (e.g., the binomial distribution, the Poisson distribution, the chi-squared distribution, the Student's t-distribution). It is reported that this trend is also observed with the Weibull distribution when the shape parameter $3 \leq m \leq 4$ [57]. The Normal distribution is not used in fracture statistics. It has been assumed by a few researchers for the distribution of strengths in the locality of flaws [58,59]. It has been demonstrated recently on SiC Nicalon fibers that flaw strengths follow normal distribution [56]. For the demonstration, large sets of 500 and 1000 strength data were used. Effort was devoted to the elimination of sources of variability. The analysis was based on strain-to-failure, which allowed elimination of the influence of fiber diameter on fiber strength. Strains were measured using an extensometer clamped to test specimens, which consisted of tows of 500 or 1000 filaments. A lubricant was introduced to avoid interfiber friction. Filament failures were detected from acoustic emission.

Equations of Gaussian probability density function f(ε) and normal distribution $P_N(E < \varepsilon)$ are

$$f(\varepsilon) = \frac{1}{S\sqrt{2\pi}} \exp\left[-\frac{(\varepsilon - \mu)^2}{2S^2}\right] \tag{1.8}$$

$$P_N(E \leq \varepsilon) = \int_0^{\varepsilon} f(\varepsilon) d\varepsilon \tag{1.9}$$

where ε is the strain to failure, μ is the mean, and S is the standard deviation.

The probability density function derived from acoustic emission monitoring is a bell curve described by a Gaussian function, symmetric about its mean (Figure 1.1). Quite identical parameters μ and S were estimated on three bundles of Nicalon filaments (Table 1.4).

The corresponding normal distribution coincides with the Weibull distribution function (Equation 1.4) for the statistical parameters reported in Table 1.4 (Figure 1.2). The scatter in Weibull scale factors is quite negligible. The shape parameter values (5.23–5.43; Table 1.4) show a very small variation when comparing to the data reported in the literature: 2.3–7.1 [53 and 55]. Referring to the limited scatter in Weibull parameters that was obtained, and to the large size of the data samples that were analyzed, it is considered that these data approach the true statistical parameters.

The Gaussian probability density function shows the presence of a single population of fracture-inducing flaws, while volume- and surface-located flaws have been observed by Scanning Electron Microscope (SEM) fractography [53]. It confirms previous results, which showed that the contributions of the populations of surface- and volume-located flaws cannot be separated [53].

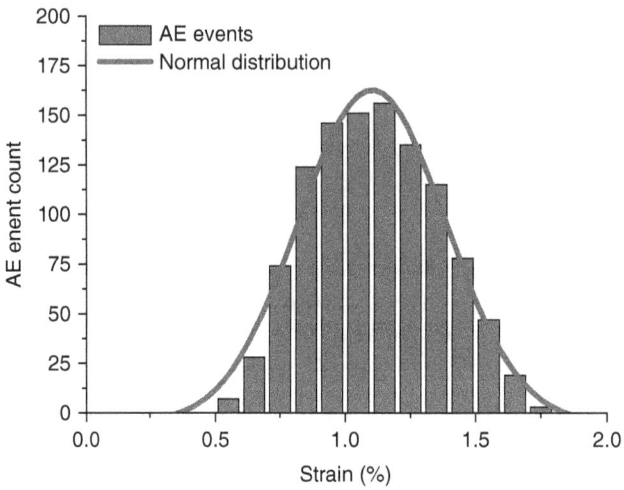

FIGURE 1.1
Histogram of filament strengths described by the distribution of AE events during a tensile test on a bundle of 1000 Nicalon SiC filaments. Reprinted with permission from APS.

TABLE 1.4 Statistical Parameters of Normal and Weibull Distributions of Flaw Strengths for Nicalon Filaments

		Normal distribution		Weibull distribution			
Test specimen	Number of filaments	μ(%)	S(%)	m	ε_l(%)	ε_0(%)	σ_0(MPa)
1	487	1.16	0.25	5.30	1.25	0.01	23.4
2	986	1.11	0.24	5.23	1.20	0.01	21.1
3	924	1.15	0.24	5.43	1.23	0.01	25.7

$V_0 = 1\ m^3$.

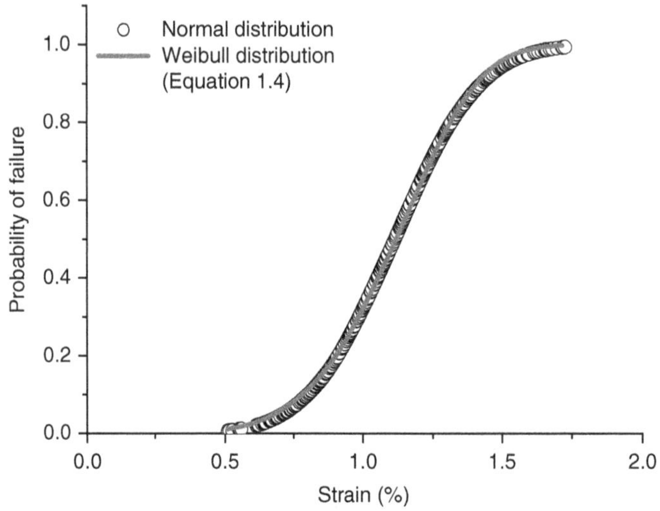

FIGURE 1.2
Cumulative distribution functions of filament failure strains obtained on a Nicalon SiC fiber bundle: Normal distribution vs. Weibull distribution (Equation 1.4). Reprinted with permission from APS.

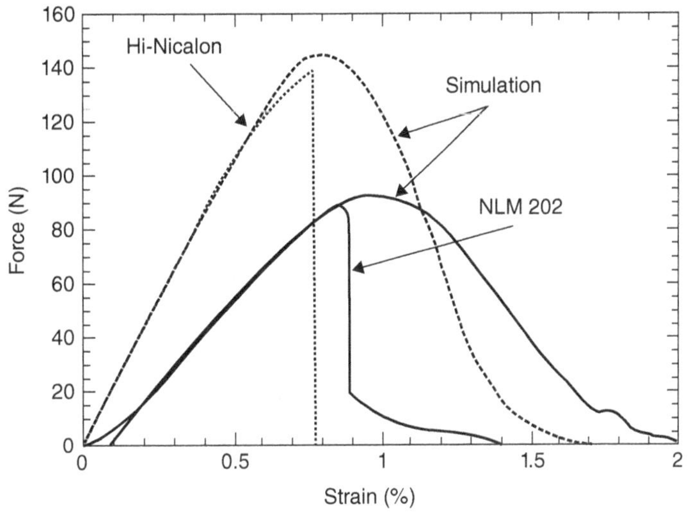

FIGURE 1.3
Ideal tensile behavior of SiC tows obtained under conditions of equal load sharing and parallel and independent filaments, and behavior obtained experimentally. Reprinted with permission from Elsevier.

1.4 MULTIFILAMENT TOWS

Multifilament tows represent a fundamental entity in composites reinforced with fabrics. They comprise several hundreds or thousands of single filaments. They progressively carry the load as the matrix is damaged, and they control ultimate failure [60]. Multifilament tows are elastic and damage tolerant, whereas single fibers are brittle.

Two different tensile behaviors are observed depending upon loading conditions and fiber (Figure 1.3) [61]:

- either a nonlinear force–strain relation resulting from successive individual fiber breaks under controlled deformation rate. This ideal behavior is obtained on carbon fiber tows or on certain SiC-based fiber tows with a lubricant (like SiC Nicalon). A typical force–strain curve together with locations of acoustic emission events in the gauge length is shown in Figure 1.4. The curve displays the conventional features of bundle tensile behavior, that is, initial elastic deformations for strains <0.5%, and then nonlinear deformations and stable failure as a result of individual successive fiber breaks as indicated by acoustic emission events.
- or a two-step nonlinear force–strain relation (Figure 1.4) under controlled deformation or load rate. After the initial elastic step, stable failure ends at maximum load. Instability results from the catastrophic failure of the filaments unbroken during the stable step. It is favored by interfiber friction. It is observed even in the presence of a lubricant on those fibers with a rough surface, associated to a large grain microstructure, like the Hi-Nicalon fibers. By contrast, the Nicalon fiber tows with fine grain microstructure (a few nanometers) exhibit a stable behavior when interfiber friction is reduced with a lubricant.

The fraction of fiber fractures at maximum force in both types of tow response controls the failure of composites. It takes the critical value $\alpha_c = N_c/N_0$, with N_0 the initial number of intact fibers, N_c the number of broken fibers at maximum load. The particular fiber that fails at maximum load is referred to as the "critical fiber." It has rank N_c, when filament strengths are ordered from smallest to largest. The fraction α_c can be satisfactorily estimated using the following relationship:

$$\alpha_c = 1 - \exp(-1/m) \tag{1.10}$$

α_c = 12–17% for Nicalon and Hi-Nicalon fiber tows, and, more generally, for those fibers with Weibull modulus $5 < m < 8$.

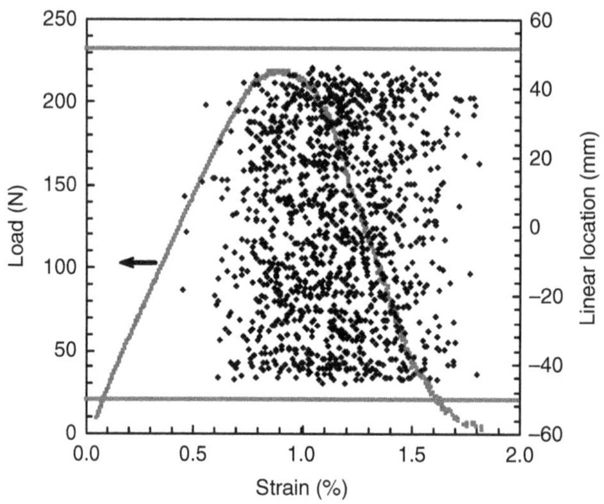

FIGURE 1.4
Load–strain curve and linear location of AE events for a Nicalon fiber bundle with a lubricant. Solid parallel lines delineate the gauge length (115 mm). Reprinted with permission from APS.

The initial number of intact filaments needs to be determined because it generally differs from the number provided by vendors, since some filaments are broken during specimen preparation. The number N_0 is derived from the initial tow compliance:

$$N_0 = N_t L_0 / E_f C_0 S_t \tag{1.11}$$

Where N_t is the total number of filaments in tow including those broken during handling or specimen preparation, L_0 is the initial gauge length, E_f is filament's Young's modulus, C_0 is the initial tow compliance, and S_t is total tow cross-sectional area.

Table 1.5 provides failure characteristics measured under monotonous tensile loading on tows of various SiC fibers. Tow strength is derived from the maximum force using the following equation:

$$\sigma_F = F_{max} / S_0 (1 - \gamma)(1 - \alpha) \tag{1.12}$$

where γ is the fraction of fibers broken prior to testing, $\gamma = 1 - N_0/N_t$; cross-sectional area is derived from tow mass $S_0 = m_0 / L_0 \rho$, ρ is density, m_0 is mass, and L_o is tow length.

TABLE 1.5 Failure Characteristics Measured on Tows for Various SiC Fibers (Gauge Length = 25 mm)

	$\sigma_F(\gamma, \alpha_c)$ (MPa)	F_{max} (N)	$\bar{\gamma}$(%)	$\bar{\alpha}_c$(%)	ε_r(%)	C_0(μm/N)
NL101	970	57	15 ± 13	10	0.84	1.6
NL207	940	60	20 ± 6.0	9.0	0.60	0.60
Hi-Ni	2100	120	14 ± 12	12	0.93	1.6
Hi-Ni-S	1800	82	25 ± 16	10	0.80	1.3
TS	1400	100	16 ± 11.2	7.3	1.1	0.90
TS11	1300	87	10 ± 8.5	10	0.98	1.6
Lox-M	1500	95	25 ± 8.1	4.5	1.0	0.80
ZMI	1400	110	14 ± 8.5	4.3	0.90	1.2
ZMId	1000	57	25 ± 6.5	3.4	0.71	0.68
AM	850	60	13 ± 11	4.9	0.71	1.1

$\sigma_F()$ is Tow Strength, F_{max} is the Maximum Force, γ is the Fraction of Filaments Broken Prior to Testing, C_0 is the Initial Compliance.

Several authors have modeled relationships between tow tensile behavior and fiber properties. H. E. Daniels [62] considered tows containing parallel and noncontacting fibers. He demonstrated that tow strengths are described by a normal distribution for large numbers of fibers. Coleman [63] proposed a relationship between tow and fiber strengths and evidenced a significant drop in tow strength when comparing to mean fiber strength. S. L. Phoenix and H. M. Taylor [64, 65] introduced effects of nonuniform loading resulting from fiber misalignment and scatter in fiber lengths within tows. V. Calard and J. Lamon [61] introduced random load sharing. They predicted tow strength drops caused by filament interactions.

1.4.1 The Bundle Model

The bundle models are based upon the following hypotheses [62, 63]: the bundle contains $N_t(= N_0)$ identical and parallel fibers (radius R_f, length l), and the statistical distribution of fiber strengths is described by the Weibull model (Equation 1.4).

When a fiber fails, equal load sharing is assumed. This means that the load is carried equally by all the surviving fibers, whereas the broken fiber no longer carries any load.

Under load-controlled conditions, the load that was carried by the broken fiber is shared equally by the surviving fibers, which experience overloading by an increment $\Delta\sigma_i$:

$$\Delta\sigma_i = \frac{\sigma_i}{N - N_i} \tag{1.13}$$

where i designates the fiber that failed, σ_i is the stress that was operating on this fiber before failure, and N_i is the number of broken fibers. Ultimate failure occurs when $\Delta\sigma_i > \sigma_{i+1} - \sigma_i$, where σ_{i+1} is the strength of the fiber having rank $i + 1$, the strengths being in ascending order. At this stage, the surviving fibers are generally unable to withstand the load increment $\Delta\sigma_i$. Most of them fail catastrophically. Failure becomes unstable at maximum load.

Under strain-controlled conditions, there is no overloading of surviving fibers when a fiber fails: $\Delta\sigma_i = 0$. As a consequence, failure is a stable phenomenon.

The ratio $\alpha(\sigma)$ of the number of broken fibers N to the initial number of fibers $N_t(= N_0)$ is approximately equal to failure probability when N_t is large and when fiber failures are equally probable events:

$$\alpha(\sigma) = \frac{N}{N_t} = P(\sigma) \tag{1.14}$$

Consequently, the total force $F(\sigma)$ applied to the bundle is

$$F(\sigma) = N_t(1 - \alpha(\sigma))S_f\sigma = N_f S_f \exp - \left[\frac{V}{V_0}\left(\frac{\sigma}{\sigma_0}\right)^m\right] \tag{1.15}$$

with S_f is the fiber cross-sectional area.

The maximum force F_{max} is given by one of the following conditions, depending on the loading mode:

$\frac{dF}{d\sigma} = 0$ (stable failure under deformation-controlled conditions)

$\alpha = \alpha_c$ (unstable failure under load-controlled conditions)

The first condition of stable failure yields

$$\sigma_{max} = \sigma_0 \left(m\frac{V}{V_0}\right)^{-\frac{1}{m}} \tag{1.16}$$

$$\alpha(\sigma_{max}) = 1 - \exp\left(-\frac{1}{m}\right) \tag{1.17}$$

$$F_{max} = F(\sigma_{max}) = N_t S_f \sigma_0 \left(m\frac{V}{V_0}\right)^{-\frac{1}{m}} \exp -\frac{1}{m} \tag{1.18}$$

The second condition of unstable failure yields

$$F_{\max} = F(\alpha_c) = N_t S_f \sigma_F (1 - \alpha_c) \tag{1.19}$$

where σ_F is the strength of tow at instability. F_{max} corresponds now to tow ultimate strength. It can be shown that α_c is given by Equation (1.10), which implies that $\sigma_F = \sigma_{max}$. As a consequence, F_{max} is also given by Equation (1.15) in the presence of unstable failure. The only difference between both tow behaviors under load- or strain-controlled conditions lies in the post maximum force responses.

To estimate scattering of the maximum force F_{max}, the binomial function can be considered $B(N_t, \alpha(\sigma_{max}))$ (respectively, α_c), which is naturally equal to the critical number of broken fibers N_c (statistical definition):

$$N_c = B(N_t, \alpha(\sigma_{\max})) \tag{1.20}$$

Thus, the expectation $E()$ and the variance $S^2()$ of the critical number of broken fibers N_c are deduced from the expectation and the variance of the binomial function, which finally yields the expectation and the variance of the maximum force F_{max}:

$$E(N_c) = N_t(\sigma_{\max}) \Rightarrow E(F_{\max}) = N_t(1 - \alpha(\sigma_{\max}))S_f\sigma_{\max} \tag{1.21}$$

$$S^2(N_c) = N_t(1 - \alpha(\sigma_{\max}))\alpha(\sigma_{\max}) \Rightarrow S^2(F_{\max}) = (S_f\sigma_{\max})^2 N_t(1 - \alpha(\sigma_{\max}))\alpha(\sigma_{\max}) \tag{1.22}$$

This allows calculation of the coefficient of variation C_v, given by Daniels [62], McCartney and Smith [66], and Gurvich and Pipes [67]:

$$C_V(F_{\max}) = \frac{S(F_{\max})}{E(F_{\max})} = \sqrt{\frac{\alpha(\sigma_{\max})}{N_t(1 - \alpha(\sigma_{\max}))}} \tag{1.23}$$

An important consequence of Equation (1.23) is that the coefficient of variation is small when the total number of fibers N_t is high: $N_t \geq 500$ in most SiC fibers. Therefore, according to theory, the maximum force (tow strength) should not show variation. Experimental results are at variance with theory: an unpredicted scatter in SiC Nicalon and Hi-Nicalon tow strengths was observed (Figure 1.5). It was attributed to imperfect local load sharing caused by fiber interactions (such as friction) or dynamic effects due to individual fiber breaks. Imperfect load sharing induces a drop in tow strengths (Figure 1.6). Fiber stress redistribution depends on the degree of interaction of the broken fiber with its neighbors. Taking into account random local load sharing allowed sound predictions of tow strengths and of associated scatter [61].

Although tow strength data are influenced by extrinsic factors, testing of tows is an interesting technique to determine fiber properties [68–72].

1.4.2 Filaments–Tows Relations: Tow-Based Testing Methods for Determination of Single Filament Properties

It is worth pointing out that the critical number of fibers broken individually at maximum force determines tow strength:

$$\sigma(F_{\max}) = \sigma(\alpha = \alpha_c) \tag{1.24}$$

whereas the average filament strength corresponds to the particular value $\alpha = 0.5$.

$$\overline{\sigma} = \sigma(\alpha = 0.5) \tag{1.25}$$

Since $\alpha_c < 0.5$, $\sigma(F_{max}) < \overline{\sigma}$ (Tables 1.3 and 1.5). Both strengths can be calculated using Equation (1.4) when m, σ_0, and V_0, are available. α_c is derived from m using Equation (1.10). When theoretical α_c is identical to the experimental value at F_{max}, this indicates that there was no fiber interaction during the tests (equal load sharing).

The failure characteristics of single filaments can be extracted from the tensile stress–strain curves determined on tows [68–72]. For this purpose, filament strengths are derived from the force and the effective bundle section,

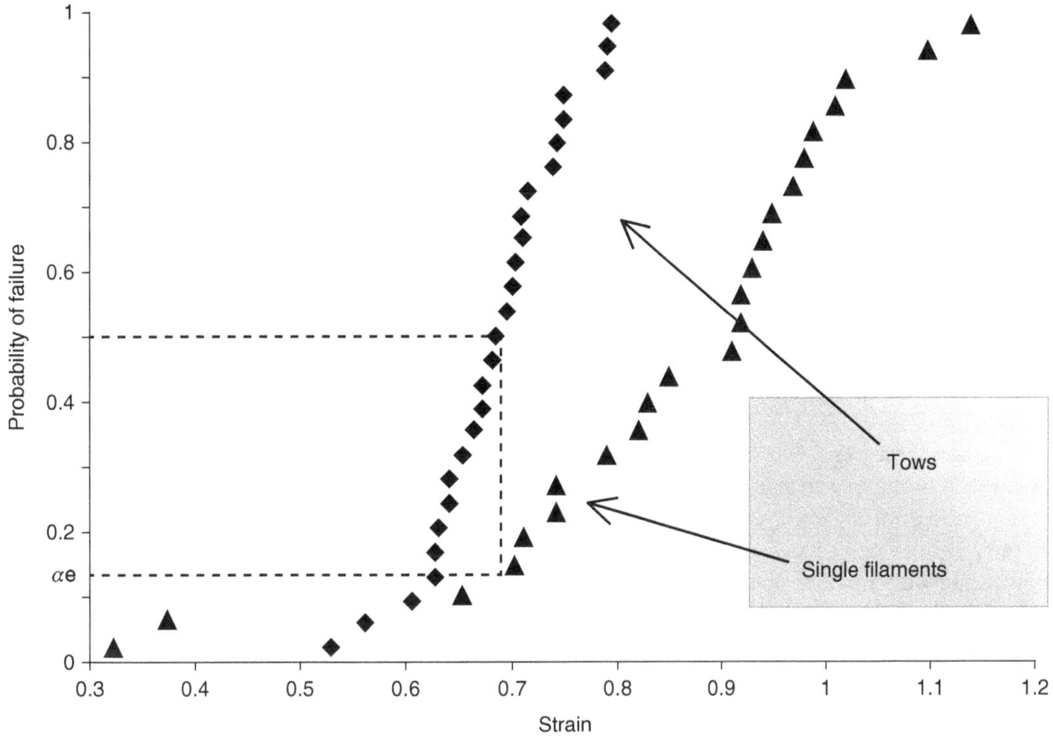

FIGURE 1.5
Statistical distributions of strains-to-failure for Nicalon SiC-based single filaments and multifilament tows.

taking into account the number of fibers broken individually. The number (j) of fibers broken at load F_j is derived from compliance C_j:

$$N_j = N_0(1 - C_0/C_j) \tag{1.26}$$

where N_0 is the initial number of intact fibers in the tow and C_0 is the corresponding compliance.

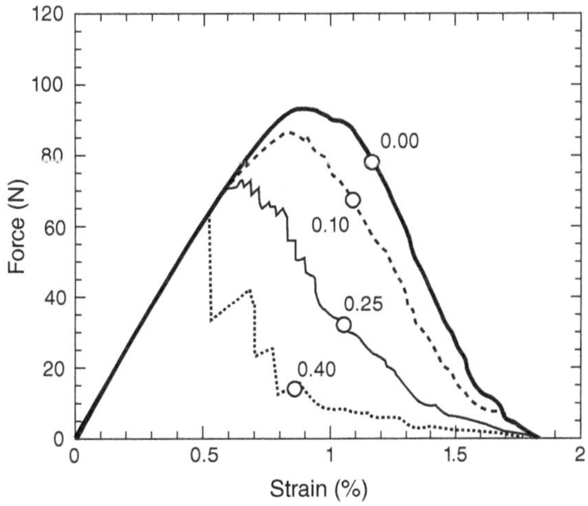

FIGURE 1.6
Influence of fiber interactions by random load sharing on the tensile load strain behavior of multifilament tows (predictions).

The estimator $P = N_j/N_0 = 1 - C_0/C_j$, can be used since the sample size is generally large, owing to the number of filaments present in a tow.

Deformation-based analysis should be preferred since measurement of fiber sections is not necessary (Figure 1.2). However, a sound method of strain measurement is required [71]. Then, deriving stresses on filaments from strains is straightforward when Young's modulus is available.

The tow testing technique is interesting because a significant sample size can be obtained using a single test. However, an artifact may result from fiber misalignment or fiber fractures during test specimen preparation. Specimen preparation and testing require much care. The degree of fiber interaction can be checked from the comparison of theoretical and experimental values of α_c. α_c data are reported in Table 1.5.

1.5 MECHANICAL BEHAVIOR AT HIGH TEMPERATURES

At intermediate temperatures <1000°C, most SiC-based fibers are sensitive to subcritical crack growth. Strength degradation of SiC single filaments starts at temperatures >1000°C [73,74], and creep at temperatures >1100°C [8]. It was also found that growth of a silica layer at fiber surface as well as oxygen diffusion is enhanced under load [75]. The intermediate temperature range has been investigated essentially on low oxygen content SiC fibers and on Nicalon fibers. Recent unpublished works on SiC fibers of first generation confirmed the phenomenon of subcritical crack growth. Oxide fibers have not been investigated whereas carbon fibers are consumed in the presence of oxygen at temperatures >400°C.

1.5.1 Strength Degradation and Oxidation at High Temperature

Non-oxide fibers experience oxidation-related effects in the high temperature range (above 900°C). However, this temperature may be lower for fibers that contain larger amounts of oxygen. Quite all the results found in the literature have been determined on single filaments. The authors were essentially interested in creep and strength retention after heat treatment or oxidation at high temperatures. The strength retention at room temperature generally decreases with increasing the heat treatment temperature [1,76,77] and the load [77]. At temperatures >1000°C, the oxidation was found to cause the growth of SiC grains, drop of resistivity, and degradation of strength for Nicalon and Hi-Nicalon fibers [76]. At all temperatures, the formation of a uniform silicon oxide layer at the surface of fibers has been observed. A very thin oxide film was formed at the temperatures between 650°C and 730°C [78].

Microstructure observations after heat treatments at 1300°C revealed that oxidation and loading accelerated new flaw nucleation and growth resulting in stress corrosion cracks in Hi-Nicalon fibers [77]. Oxidation in air at 800–1000°C only slightly decreased the strength of Nicalon fibers, whereas a decrease of strength and no noticeable change of the Young's modulus were reported for Tyranno fibers at lower temperatures (650–730°C) [78]. This strength decrease was related to flaw size increase after oxidation [78]. Hi-Nicalon single filaments retain >90% of their room temperature strength at 1300°C. By contrast, the strength data measured on multifilament strands show a steep decrease from 400°C. Yun and DiCarlo also reported strength data measured on Hi-Nicalon tows at high temperatures in air [79]. They found that tensile strength degradation started at 300°C. This phenomenon results from fiber bonding by a SiO_2 layer that grows at the surface of fibers.

E. Lara Curzio modeled the effect of oxidation on the stress–rupture time behavior of fiber bundles and composites [80,81]. The delayed failure of fiber bundles was attributed to formation of a silica layer on the surface of fibers; the thickness of this layer was introduced in place of critical flaw size in the linear fracture mechanics equation of strength leading to a fiber bundle strength decrease with time as $t^{-1/4}$ [80,81], which corresponds to a fiber independent stress exponent $n = 4$.

1.5.2 Static Fatigue Under Constant Load at Intermediate Temperatures: Subcritical Crack Growth

Several researchers have demonstrated since the 1960s the sensitivity of refractory materials to slow (subcritical) crack growth. Slow crack growth leads to failure, which implies that the strength is also time dependent.

1.5.2.1 Stress–Rupture Time Diagrams

Delayed failure is indicated by finite rupture times under stresses much smaller than the failure stress (Figures 1.7–1.9) [82–84]. The rupture times generally decrease when the stresses are

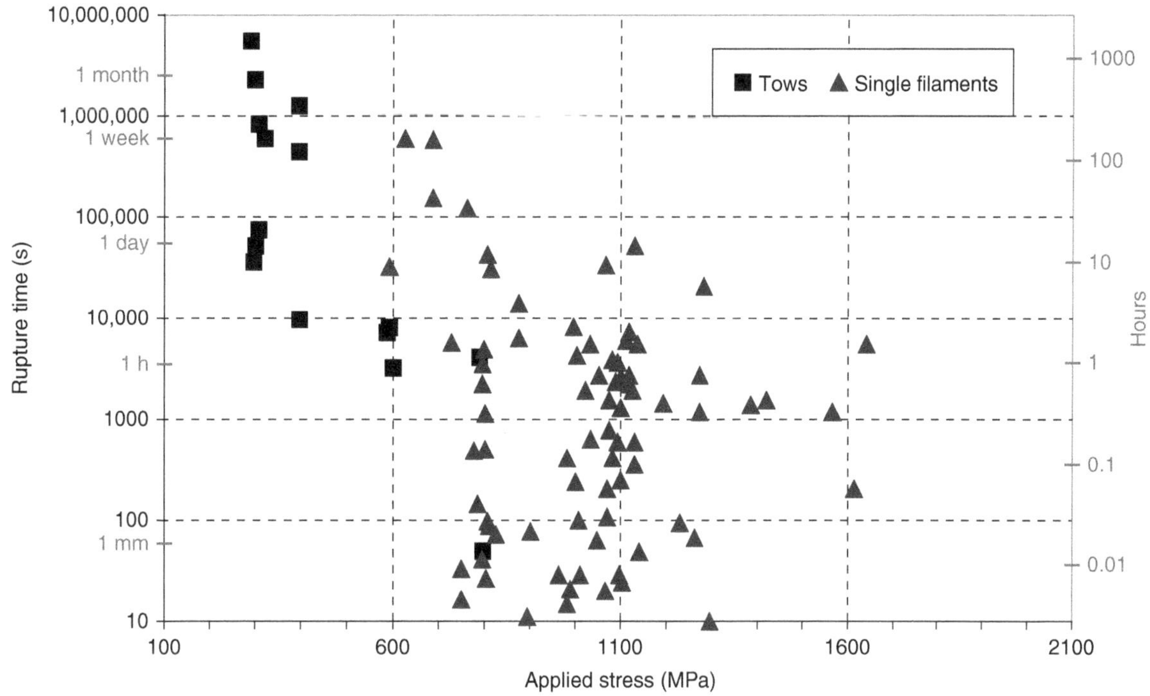

FIGURE 1.7

Stress–rupture time data determined on single SiC-based filaments (Hi-Nicalon) at 800°C. From Journal of the American Ceramic Society, March 2009.

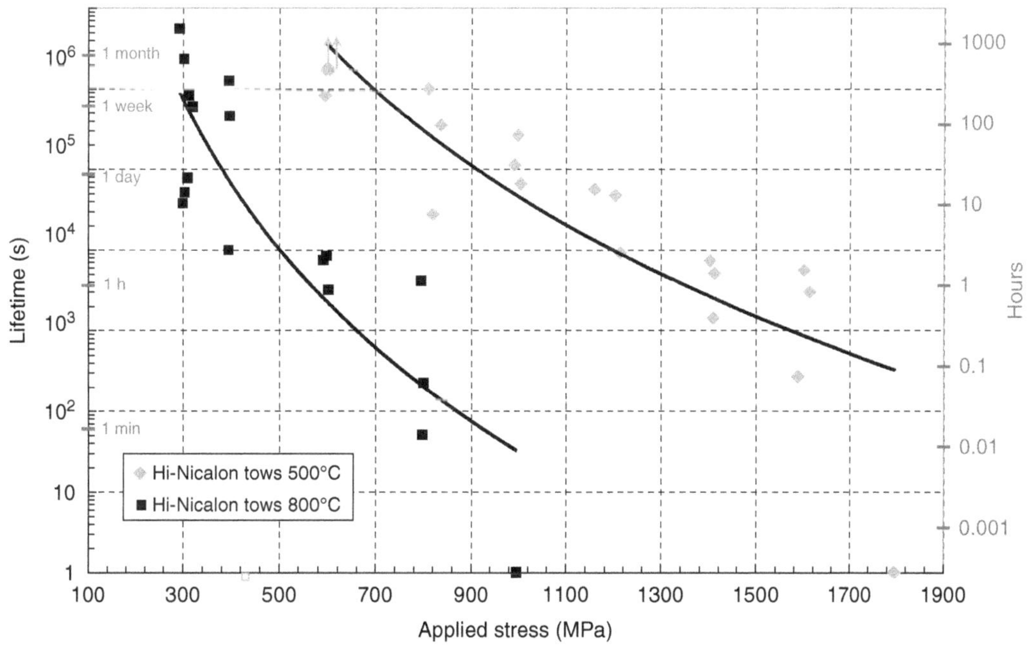

FIGURE 1.8

Stress–rupture time diagrams for Hi-Nicalon multifilament tows tows at 500°C and 800°C. From Journal of the American Ceramic Society, March 2009.

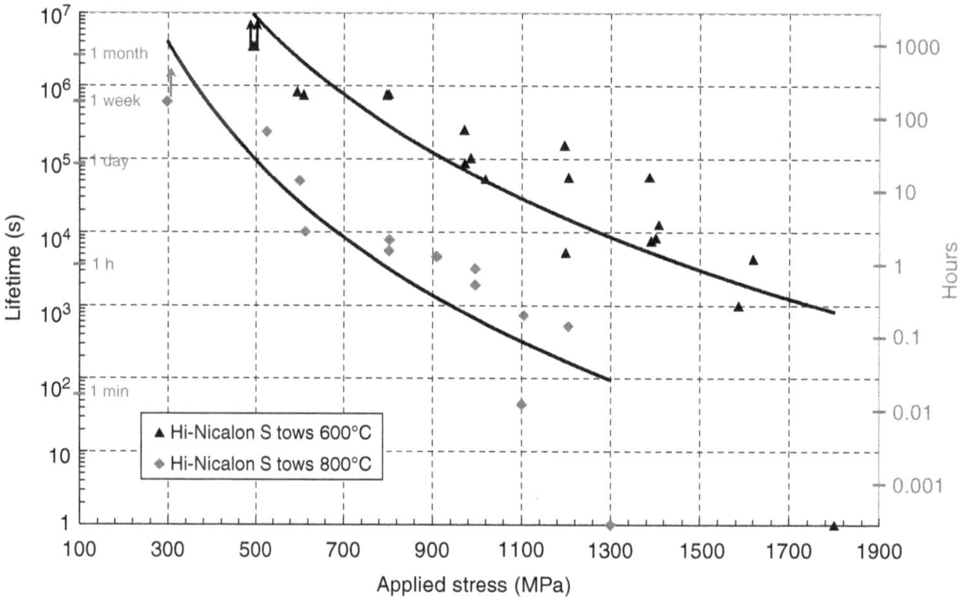

FIGURE 1.9

Stress–rupture time data for Hi-Nicalon S fiber bundles at 500°C and 800°C. From Journal of the American Ceramic Society, March 2009.

increased. Lifetimes of single filaments exhibit a significant scatter (Figure 1.7). The statistical variability at constant stress is similar to that one observed with ceramics when various specimens are tested. It results from the presence of flaws with different sizes.

The stress–rupture time data obtained for tows align on a curve that is fitted by the following power law (Figures 1.8 and 1.9):

$$t\sigma^n = A \tag{1.27}$$

where t is the rupture time, σ is the applied stress, n and A are constants depending, respectively, on material and environment.

This power law often describes nonlinear time-dependent responses such as that one dictated by slow crack growth activated by the environment under low stresses in ceramics [85]. The stress–rupture time behavior of tows must be regarded as a remarkable feature since it is at variance with that dispersed one of ceramics and single fibers. The constant A depends on the initial flaw size. It takes the same value for those specimens that possess flaws with the same initial size [85]. The response of tows can be related to the critical fiber-dictated failure mode, as discussed in section 1.4.2. When α_c is a constant, the critical fibers correspond to identical initial flaw sizes.

A and n are estimated from the stress–rupture time diagram using a regression technique (Tables 1.6 and 1.7). Unlike A, the stress exponent n does not display significant temperature dependence. It seems to depend on fiber. Close values of n were estimated for a large spectrum of commercial SiC-based fibers (Tables 1.6 and 1.7).

TABLE 1.6 Main Static Fatigue Constants Estimated on Tows for Nicalon-Type Silicon Carbide-Based Fibers [21,75]

	Nicalon	Hi-Nicalon		Hi-Nicalon S	
T (°C)	600	500	800	600	800
n	7.3	8.4	8.3	7.2	7.2
A (MPa^n sec)	2.10^{24}	$1.05.10^{30}$	$3.36\ 10^{26}$	$3.15.10^{26}$	$3.33.10^{24}$
E_a (KJ/mol)	140	185	185	177	177

TABLE 1.7 Main Static Fatigue Constants Estimated on Tows for Ube Silicon Carbide-Based Fibers at Various Temperatures [21]

	NL207		TS		TSI1		Lox-M		ZMI	
T (°C)	n	$A \times 10^{25}$ (s $MPa^{7.3}$)	n	$A \times 10^{19}$ (s $MPa^{5.0}$)	n	$A \times 10^{21}$ (s $MPa^{5.8}$)	n	$A \times 10^{19}$ (s $MPa^{4.9}$)	n	$A \times 10^{31}$ (s $MPa^{9.1}$)
350	–	–	–	600	–	–	–	–	–	–
400	–	–	5.59	200	–	–	–	1 200	9.99	10 000
450	–	1 000	5.59	80	6.87	400	–	20	9.99	2 000
500	6.92	400	4.81	38	5.34	80	4.67	80	11.35	300
550	7.66	40	5.26	17	4.68	30	5.49	20	9.24	60
600	–	60	4.84	6.0	5.48	20	5.17	1.8	8.37	6.0
650	7.05	15	4.34	0.80	6.81	4.0	4.25	0.40	8.71	0.70
750	7.63	1.0	5.29	0.50	–	2.0	–	0.10	8.39	0.30
850	7.25	0.30	–	0.20	–	1.0	–	0.10	9.16	0.12
$\bar{n}$	**7.3**		**5.0**		**5.8**		**4.9**		**9.1**	

1.5.2.2 Mechanism of Slow Crack Growth Figure 1.10 demonstrates that oxygen is responsible for the delayed failure of fibers at high temperature under a stress far smaller than the failure stress. In pure nitrogen atmosphere, failure did not occur (the tests were interrupted after 1 month) whereas rupture time decreased to a plateau with increasing oxygen concentration. Above 15–20%, oxygen was in excess for the chemical reactions.

The following typical features were identified by SEM on the fracture surface of fibers (Figures 1.11 and 1.12):

- successive marks, indicative of slow crack growth;
- and a thin film of oxide on the external surface. The thickness of this film was generally quite small after tests lasting not more than 2 or 3 months. This duration is reasonable for practical reasons.

The delayed failure of SiC-based fiber bundles at temperatures below 800°C results from the subcritical crack growth of the surface defects by the oxidation of free carbon at grain boundaries and of the SiC nanograins or silicon oxicarbide at crack tip (Figure 1.13) [75]. Both phenomena may contribute simultaneously (Nicalon fibers) or sequentially (Hi-Nicalon S and SA3 fibers, which contain little free carbon not connected). Growth of a silica layer at fiber surface is not responsible for fracture. In the presence of free carbon, the crack length increase is attributed to the consumption

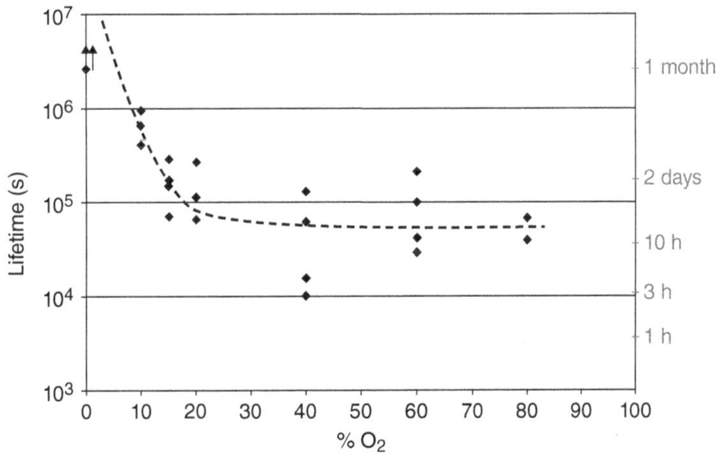

FIGURE 1.10

Influence of oxygen content on the rupture time of Hi-Nicalon fiber bundlles at 500°C under 1000 MPa. From Journal of the American Ceramic Society, July 2009.

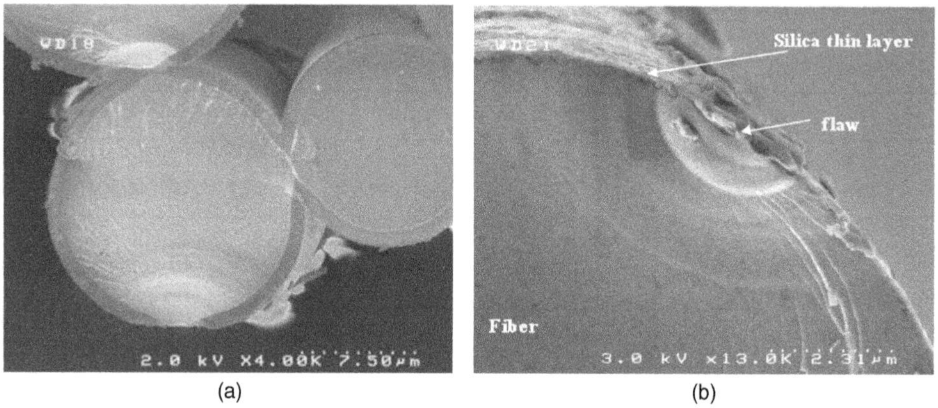

FIGURE 1.11

SEM micrographs showing evidence of slow crack growth and oxide layer after static fatigue on (a) Hi-Nicalon fibers under 250 MPa at 1000°C during 13 days. (b) Nicalon fibers under 540 MPa at 700°C during 4 hours. From Journal of the American Ceramic Society, July 2009.

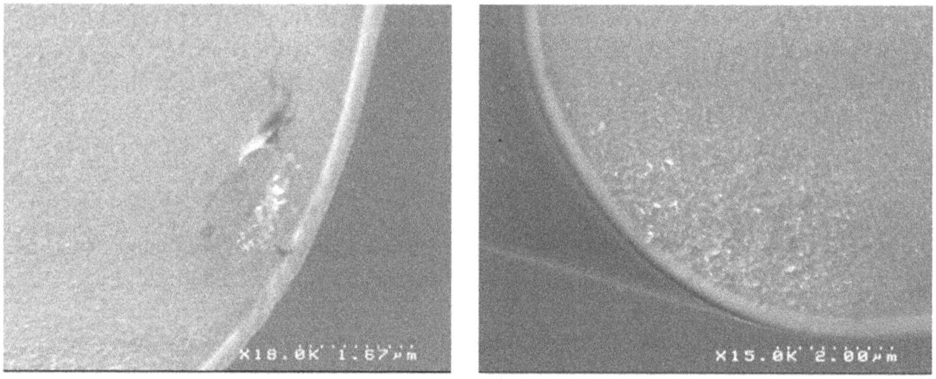

FIGURE 1.12

Typical examples of oxide layers observed at the surface of Nicalon fibers after static fatigue at 800°C under 300 MPa during ~2 months. From Journal of the American Ceramic Society, July 2009.

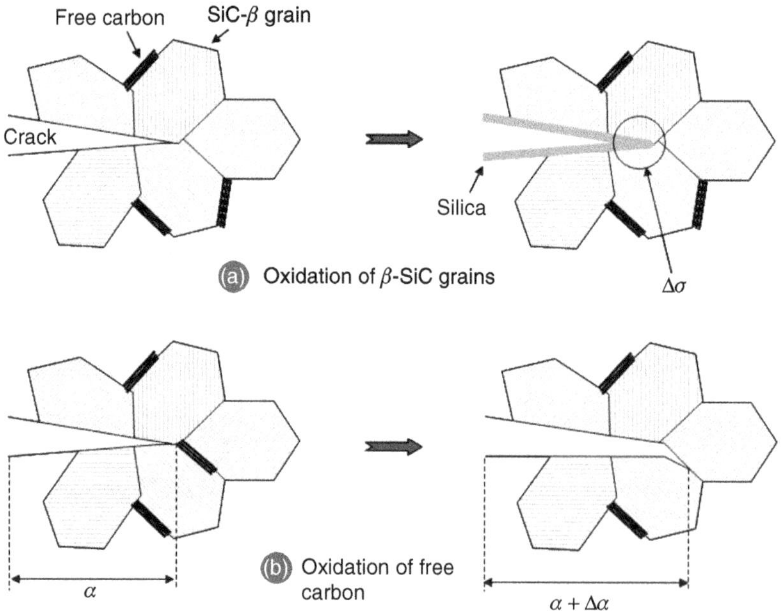

FIGURE 1.13

Schematic diagrams showing the mechanisms of slow crack growth at high temperature in air. From Journal of the American Ceramic Society, July 2009.

of free carbon. In the presence of silicon carbide or silicon oxicarbide, the silicon oxide formed at the defect/crack tip causes local volume increase (by 2.1 for SiC and less for SiO_xC_y), which induces a tensile stress $\Delta\sigma$. Those fibers with the largest fractions of free carbon are the least resistant to static fatigue:

Tyranno SA3 ≈ Hi-Nicalon S < Hi-Nicalon ≈ Nicalon
(≈ 2 at.%) (≈ 3 at.%) < (≈ 17 at.%) (≈ 15at.%): atomic concentrations of free carbon.

- Nicalon fiber contains large amounts of very reactive free carbon (≈15 at.%), and very small β-SiC nanograins (~5 nm) embedded in silicon oxicarbide. These elements are very sensitive to oxidation due to their structure (free carbon) or their size; Valhas and Laanani demonstrated that smaller are the grains, the faster they oxidize [86]. In consequence both mechanisms can operate simultaneously and this is consistent with the small resistance to fatigue of this fiber.
- Hi-Nicalon fiber is essentially made of bigger β-SiC nanograins (~10 nm) and a free carbon network (the electrical conductivity is high: about 70 S/m, compared to the Hi-Nicalon S and Nicalon fibers: 0.1 S/m). The oxidation of this network from the surface defects should be the predominant mechanism of slow crack growth. As a consequence, longer rupture times would be expected when compared to the Nicalon fibers.
- Hi-Nicalon S fiber contains much bigger β-SiC nanograins (~20 nm) without the silicon oxicarbide, and a little free carbon located at grain boundaries and not connected. The oxidation of the β-SiC nanograins would contribute to slow crack growth. The resistance to static fatigue is greater than that of Hi-Nicalon fiber because free carbon is not connected and the SiC grains are bigger.
- Tyranno SA3 fiber possesses the same amount of free carbon in the surface as the Hi-Nicalon S fiber, but the difference lies in the SiC grain size: 50–100 nm vs. 20 nm for the Hi-Nicalon S fiber. The longer rupture times exhibited by the SA3 fiber can be attributed to the presence of much bigger grains, which oxidize more slowly than the smaller ones present in the other SiC-based fibers [51]. In this case the oxidation rate of the SiC grains would determine slow crack growth. Both phenomena of oxidation of SiC grains and free carbon may contribute sequentially to slow crack growth in the Hi-Nicalon S and Tyranno SA3 fibers.

Finally, the lifetime microstructure relationships can be highlighted by comparing rankings of resistances to static fatigue and microscopic features, which control the reactivity of constitutive elements:

Lifetime: Tyranno SA3 > Hi-Nicalon S > Hi-Nicalon > Nicalon
Free carbon: (≈ 2 at.%) (≈ 3 at.%) < (≈ 17 at.%) (≈ 15 at.%)
Grain size: (≈50–100 nm) (≈20 nm) > (≈10 nm) (≈5 nm)

In order to improve the fatigue resistance of these SiC-based fibers in air at intermediate temperatures between 400°C and 800°C, the manufacturers will have to reduce drastically the amounts of oxygen and free carbon near fiber surface and also to eliminate the surface defects, in order to limit the oxidation.

1.5.2.3 Model of Slow Crack Growth in Filaments Under a Constant Stress

Models of slow crack growth have been proposed for engineering ceramics [85] and for tows [83]. The delayed failure of a single fiber is caused by the slow growth of a crack from an initial flaw size to a critical length. Crack growth rate is assumed to obey the following equation, which is usually employed to describe the slow propagation of cracks in fatigue in ceramics or metallic materials:

$$v = \frac{da}{dt} = A_1 K_I^n \tag{1.28}$$

where v is crack velocity, a is crack length, t is time, K_I is the stress intensity factor, and A_1 and n are constants depending, respectively, on environment and material.

Under constant stress σ, the lifetime of a single filament is the time period required for the particular flaw responsible for filament failure to grow from the initial size C_j to the critical length a_c:

$$t = \int_{C_j}^{a_c} \frac{da}{V} = \frac{2}{\sigma^n A_1 (n-2)} \left[\frac{C_j^{2-n/2}}{Y^n} - \frac{K_{IC}^{2-n} \sigma^{n-2}}{Y^2} \right] \tag{1.29}$$

where K_{IC} is the critical value of K_I (fracture toughness) and $Y = 2/\pi^{1/2}$ the geometrical parameter for a penny-shaped crack.

The initial flaw size C_j can be characterized by the tensile strength of the fiber, σ_f, in absence of environmental effects:

$$C_j = \frac{K_{IC}^2}{\sigma_f^2 Y^2} \tag{1.30}$$

Moreover, the filament failure strength distribution is described by the Weibull equation:

$$P(\sigma) = \alpha = 1 - \exp\left[-\frac{V}{V_0}\left(\frac{\sigma_f}{\sigma_{0f}}\right)^{m_f}\right] \tag{1.31}$$

where V_0 is the reference volume ($V_0 = 1\ m^3$) and m_f and σ_{0f} are the statistical parameters.

Inserting Equations (1.30) and (1.31) into (1.29) gives the following stress–rupture time relationship:

$$t = \frac{2}{Y^2 A_1 (n-2)\sigma^n}\left[\frac{\sigma_{0f}^{n-2}}{K_{IC}^{n-2}}\left[\frac{V_0}{V}\ln\frac{1}{1-\alpha}\right]^{\frac{n-2}{m_f}} - \frac{\sigma^{n-2}}{K_{IC}^{n-2}}\right] \tag{1.32}$$

A_I can be expressed in terms of temperature T and the activation energy E_a related to the chemical reaction at crack tip:

$$A_1 = A_{10}\exp\left[-\frac{E_a}{RT}\right] \tag{1.33}$$

where A_{10} is constant and $R = 8.314$ J/K/mol.

Within the domain of low stresses, where the contribution of environment is effective, Equation (1.32) reduces to

$$t\sigma^n = A_0(\alpha)\exp\left[\frac{E_a}{RT}\right] = A(\alpha, T, E_a) \tag{1.34}$$

where

$$A_0(\alpha) = \frac{2}{Y^2(n-2)A_{10}}\left[\frac{\sigma_{0f}^{n-2}}{K_{IC}^{n-2}}\left[\frac{V_0}{V}\ln\frac{1}{1-\alpha}\right]^{\frac{n-2}{m_f}}\right] \tag{1.35}$$

Equations (1.34) and (1.35) show that A depends on fiber rank, temperature, and environment. Furthermore, it is worth emphasizing that the power law form of time to failure $t\sigma^n = A$ results from the use of power form for crack velocity (Equation 1.28).

1.5.2.4 Application to Static Fatigue of Tows Equations (1.32), (1.34), and (1.35) assume that fibers are subjected to stresses that remain constant during time. Lifetime of a tow is thus obtained for the particular value of α that corresponds to the critical fiber. The value of α may be equal to α_c determined under inert conditions or it may be smaller since all the filaments in a tow experience slow crack growth. But, this issue has not been examined yet.

For tows subjected to a constant load, the condition of constant stresses on filaments is not fulfilled because of reloading of surviving filaments at each filament failure from slow crack growth. As a consequence, the stresses on filaments increase step by step. The successive stresses and time steps can be determined to adapt Equations (1.32), (1.34), and (1.35) to the condition of constant load. In a first step, Equation (1.27) was used to extract slow critical crack growth constants from stress–rupture time diagrams, assuming that the stress on fibers was equal to the initial applied stress. It appeared that Equation (1.27) fitted satisfactorily the experimental stress–rupture time curves. It can be considered that the obtained slow crack growth data are not intrinsic values pertinent to filaments, but instead they characterize the behavior of tows.

1.6 SUMMARY

Using continuous refractory fibers is the most efficient way to obtain strong and tough CMCs at high temperature. Owing to a wide variety of potential matrices, the fibers must meet several requirements. They must withstand the high temperatures required by composite processing above 1000°C. Their thermoelastic and fracture characteristics must match the matrix. Fiber resistance to fast or delayed fracture determines composite ultimate strength under monotonous or fatigue loading. Fiber strength degrades at high temperatures in air, as a result of various temperature-dependent phenomena like subcritical crack growth, creep, and oxidation.

Fast fracture data for filaments exhibit a statistical distribution that can be approximated by Weibull equation. However, they follow normal distribution, which can be determined safely on large sample size. The statistical parameters depend tremendously on the selection of samples. Tensile tests on tows with large amounts of filaments are appropriate to estimate reliable filament characteristics, provided filaments interactions during tests are not significant. Multifilament tows are fundamental entities in textile composites. They control ultimate failure and they determine scatter in composite strength. They exhibit a damage-tolerant behavior, which depends on statistical distribution of filament strengths. A critical filament having 15–20% failure probability in the distribution of reference strengths dictates their ultimate strength.

REFERENCES

1. A. R. Bunsell and A. Piant. A review of the development of three generations of small diameter silicon carbide fibers. *J. Mater. Sci.*, 41, 823–839 (2006).
2. B. Clauss. Fibers for ceramic matrix composites [Chapter 1], in *Ceramic Matrix Composites: Fibre-Reinforced Ceramics and Their Applications*, edited by W. Krenkel (Wiley-VCH, Weinheim, Germany, 2008), pp. 1–19.
3. G. Chollon, R. Pailler, R. Naslain, and P. Olry. Correlation between microstructure and mechanical behaviour at high temperatures of a SiC fibre with a low oxygen content (Hi-Nicalon). *J. Mater. Sci.*, 32, 1133–1147 (1997).
4. R. Pailler, J. Lamon, A. Guette, and I. Martin-Litas. Non-oxide ceramic fibres: relationships between nanostructure or composition and properties. *Ann. Chim. Sci. Mat.*, 30(6), 565–578 (2005).
5. K. A. Keller, G. Jefferson, and R. J. Kerans. Progress in oxide composites, *Ann. Chim. Sci. Mat.*, 30, 659–671 (2005).
6. K. A. Keller, G. Jefferson, and R. J. Kerans. Oxide–oxide composites, in *Handbook of Ceramics and Glasses*, edited by N. P. Bansal (Kluwer Academic Publishers, New York, 2005), pp. 377–421.
7. A. Bunsell. Oxide fibers, in *Handbook of Ceramics and Glasses*, edited by N. P. Bansal (Kluwer Academic Publishers, New York, 2005), pp. 3–31.
8. R. Bodet, X. Bourrat, J. Lamon, and R. Naslain. Tensile creep behaviour of a silicon carbide-based fibre with a low oxygen content. *J. Mat. Sci.*, 30, 661–677 (1995).
9. G. A. Budnitskii and N. N. Machalaba. Some research trends at the institute. *Fibre Chem.*, 33(2), 85–95 (2001).
10. A. R. Bunsell and M. H. Berger. *Fine Ceramic Fibres.* (CRC Press, 1999).
11. A. R. Bunsell and M.-H. Berger. Fine diameter ceramic fibres. *J. Eur. Ceram. Soc.*, 20, 2249–2260 (2000).
12. A. R. Bunsell. Fracture processes in fine silicon carbide fibers, in *Fiber Fracture*, edited by M. Elices and J. Llorca (2002), pp. 75–87.
13. A. R. Bunsell and J. Renard. *Fundamentals of Fibre Reinforced Composite Materials.* Series in Material Science and Engineering, Taylor & Francis Group (IOP Publishing Ltd., Bristol and Philadelphia, 2005).
14. S. M. Dong, G. Chollon, C. Labrugère, M. Lahaye, A. Guette, J. L. Bruneel, M. Couzi, R. Naslain, and D. L. Jiang. Characterization of nearly stoichiometric SiC ceramic fibres. *J. Mat. Sci.*, 36, 2371–2381 (2001).
15. T. Ishikawa. Recent developments of the SiC fiber Nicalon and its composites, including properties of the SiC fiber Hi-Nicalon for ultra-high temperature. *Compos. Sci. Technol.*, 51, 135–144 (1994).
16. R. E. Jones, D. Petrak, J. Rabe, and A. Szweda. Sylramic SiC fibers for CMC reinforcement. *J. Nucl. Mater.*, 283–287, 556–559 (2000).

17. K. Kumagawa, H. Yamaoka, M. Shibuya, and T. Yamamura. Thermal stability and chemical corrosion resistance of newly developed continuous Si-Zr-C-O Tyranno fiber. *Ceram. Eng. Sci. Proc.*, 18, 113–118 (1997).
18. K. Kumagawa, H. Yamaoka, M. Shibuya, and T. Yamamura. Fabrication and mechanical properties of new improved Si-M-C-(O) Tyranno fiber. *Ceram. Eng. Sci. Proc.*, 19, 65–72 (1998).
19. K. Kumagawa, H. Yamaoka, M. Shibuya, M. Suzuki, and T. Yamamura, Fabrication and mechanical properties of Si-M-C-(O) Tyranno fibers, *Ceram. Eng. Sci. Proc.*, 21(4), 291–298 (2000).
20. J. Lipowitz, J. A. Rabe, K. T. Nguyen, L. D. Orr, and R. R. Androl. Structure and properties of polymer-derived stoichiometric SiC fiber. *Ceram. Eng. Sci. Proc.*, 16(4), 55–62 (1995).
21. S. Mazerat. Fibres de renfort pour composites SiC/SiC: amélioration et corrélation de la durée de vie sous air à T<900°C avec la réactivité chimique (Fibres for reinforcement of SiC/SiC composite: improvement of lifetime and correlation with chemical reactivity). Ph.D. Thesis, No. 4554 (University of Bordeaux I, France, July 20, 2012).
22. G. Puyoo. Augmentation de la durée de vie de composites à matrice céramique: rôle des fibres Si-O-C et des interphases B-C-N (Improvement of lifetime for ceramic matrix composites: role of Si-O-C fibres and B-C-N interphases). Ph.D. Thesis, No. 4497, (University of Bordeaux I, France, March 9, 2012).
23. G. Simon and A. R. Bunsell, Mechanical and structural characterisation of the Nicalon silicon carbide fibre. *J. Mater. Sci.*, 19, 3649–3657 (1984).
24. Tyranno Fiber, Ube Industries, 1978-10, Kogushi, Ube City, Japan.
25. A. Hasegawa, A. Kohyama, R. H. Jones, L. L. Snead, B. Riccardi, and P. Fenici. Critical issues and current status of SiC/SiC composites for fusion. *J. Nucl. Mater.*, 283–287, 128–137 (2000).
26. C. Sauder. Effet de l'irradiation sur le comportement des fibres de carbure de silicium. Internal report LCTS, SEMES-050001 (2004).
27. G. Chollon, R. Pailler, R. Canet, and P. Delhaes. Correlation between microstructure and electrical properties of SiC-based fibres derived from organosilicon precursors. *J. Eur. Ceram. Soc.*, 18, 725–733 (1998).
28. R. Sarrazin. Phénomène d'endommagement progressif des fibres SiC en fatigue statique aux températures intermédiaires. Internal report LCTS (2011).
29. T. Ishikawa, H. Ishikawa, and H. Teranishi. Strength and structure of SiC fiber after exposure to high temperature. *J. Mater. Sci.*, 36, 205–217 (1987).
30. J. A. DiCarlo, and H.-M. Yun. Non-oxide (silicon carbide fibers, in *Handbook of Ceramics and Glasses*, edited by N. P. Bansal (Kluwer Academic Publishers, New York, 2005), pp. 32–53.
31. H. M. Yun, J. C. Goldsby, and J. A. DiCarlo. Tensile creep and stress–rupture behavior of polymer derived SiC fibers, in *Advances in Ceramic-Matrix Composites II*, edited by J. P. Singh and N. P. Bansal (American Ceramic Society, Westerville, OH, 1995), pp. 17–28.
32. H. M. Yun and J. A. DiCarlo. Comparison of the tensile, creep, and rupture strength properties of stoichiometric SiC fibers. Report NASA/TM-1999-209284 (1999).
33. R. Bodet, X. Bourrat, J. Lamon, and R. Naslain Tensile creep behavior of a silicon carbide-based fiber with a low oxygen content. *J. Mater. Sci.*, 30 661–677 (1995).
34. J. A. DiCarlo and H. M. Yun. Microstructural factors affecting creep–rupture failure of ceramic fibers and composites. *Ceram. Trans.*, 99, 119–134 (1998).
35. C. Sauder and J. Lamon, The tensile creep behavior of SiC-based fibers with low oxygen content. *J. Am. Ceram. Soc.*, 90(4), 1146–1156 (2007).
36. J. Hayashi, C. F. Liaw, T. Matzuzawat, K. Okamura, I. M. Omori, and S. Yajima. Simple synthesis of the continuous SiC fiber with high tensile strength. *Chem. Lett.*, 551, (1976).
37. Y. Hasegawa, M. Iimura, and S. Yajima. Synthesis of continuous silicon carbide fibre. Part 2: conversion of polycarbosilane fibre into silicon carbide fibres. *J. Mater. Sci.*, 15, 720–728 (1980).
38. P. Le Costumer, M. Monthioux, and A. Oberlin. Mécanisme de dégradation d'une fibre Nicalon Série 200, in *Comptes Rendus des Journées AMAC-CODEMAC*, edited by R. Naslain, J. Lamalle, and J. L. Zulian (Bordeaux, France, 1990), pp. 43–53.
39. O. Greck, J. P. Viricelle, D. Bahloul-Hourlier, P. Goursat, M. Dalbin, S. Thomin, and A. M. Flank. SiC-based ceramic fibres: thermal stability and oxidation behaviour. *Key Eng. Mater.*, 132–136, 1950–1953 (1997).
40. H.-E. Kim, and A. J. Moorhead. Strength of Nicalon silicon carbide fibers exposed to high-temperature gaseous environments. *J. Am. Ceram. Soc.*, 74(3), 666–669 (1991).
41. R. Bodet, J. Lamon, N. Jia, and R. E. Tressler. Microstructural stability and creep behavior of Si-O-C (Nicalon) fibers in carbon monoxide and argon environments. *J. Mater. Sci.* 30, 2241–2253 (1995).
42. K. Igashira. Decomposition and oxidation resistance of Si-M(Ti, Zr)-C-O fibers at elevated temperature. *J. Am. Soc.*, 255–260 (2003).

43. C. D. Batich, G. J. Choi, M. D. Sacks, M. Saleem, and W. M. Toreki. Polymer-derived silicon carbide fibers with low oxygen content. *Ceram. Eng. Sci. Proc.*, 198–208 (1992).
44. G. Chollon, R. Pailler, R. Naslain, F. Laanani, M. Monthioux, and P. Orly. Thermal stability of a PCS-derived SiC fibre with a low oxygen content (Hi-Nicalon). *J. Mater. Sci.*, 32, 327–347 (1997).
45. K. Okamura, T. Shimoo, W. Takano, and Y. Katase. Carbon elimination by heat-treatment in hydrogen and its effect on thermal stability of polycarbosilane-derived silicon carbide fibers. *J. Mater. Sci.*, 204, 6243–6251 (2004).
46. M. Takeda, J. Sakamoto, A. Saeki, and H. Ichikawa. Mechanical and structural analysis of silicon carbide fiber Hi-Nicalon type S. *Ceram. Eng. Sci. Proc.*, 17(4–5) 35–42 (1996).
47. M. Takeda, A. Urano, J. I. Sakamoto, and Y. Imai. Microstructure and oxidative degradation behavior of silicon carbide fiber Hi-Nicalon type S. *J. Nucl. Mater.*, 258–263, 1594–1599 (1998).
48. K. Morishita, S. Ochiai, H. Okuda, T. Ichikawa, and M. Sato. Fracture toughness of a crystalline silicon carbide fiber (Tyranno-SA3®). *J. Am. Ceram. Soc.*, 89(8) 2571–2576 (2006).
49. H. M. Yun, J. A. DiCarlo, R. T. Bhatt, and J. B. Hurst. Processing and structural advantages of the Sylramic-iBN SiC fiber for SiC/SiC components. *Ceram. Eng. Sci. Proc.*, 24(4) 247–253 (2003).
50. K. Morishita, S. Ochiai, H. Okuda, T. Ishikawa, M. Sato, and T. Inoue. Fracture toughness of a crystalline silicon carbide fiber (Tyranno-SA3). *J. Am. Ceram. Soc.*, 89(8) 2571–2576 (2006).
51. Association Française de Normalisation [AFNOR] NF EN 1007-4: Advanced Technical Ceramics—Ceramic Composites—Methods of Test for Reinforcement—Part 4: Determination of Tensile Properties of Filaments at Ambient Temperature (2004).
52. J. Lamon. *Mechanics of Brittle Fracture and Damage: Statistical Probabilistic Approaches* (In French "Mécanique de la rupture fragile et de l'endommagement: Approches Statistiques-Probabilistes") (Editions Hermès-Lavoisier, Paris, France, 2007).
53. N. Lissart, and J. Lamon, Statistical analysis of failure of SiC fibres in the presence of bimodal flaw populations. *J. Mater. Sci.* 32, 6107–6117 (1997).
54. A. J. Eckel, and R. C. Bradt. Statistical analysis of failure of SiC fibres in the presence of bimodal flaw populations. *J. Am. Ceram. Soc.* 72(3), 455–458 (1989).
55. K. Goda, and H. Fukunaga. The evaluation of the strength distribution of silicon carbide and alumina fibres by a multi-modal Weibull distribution. *J. Mater. Sci.* 21, 4475–4480 (1986).
56. M. R'Mili, N. Godin, and J. Lamon. Flaw strength distributions and statistical parameters for ceramic fibers: the normal distribution. *Phys. Rev. E*, 85, 051106 (2012).
57. E. G. Stoner, D. D. Edie, and S. D. Durham. An end-effect model for the single-filament tensile test. *J. Mater. Sci.* 29, 6561–6574 (1994).
58. F. T. Peirce. Journal of the textile institute. *Transactions*, 17, 355 (1926).
59. B. Epstein. Application of the theory of extreme values in fracture problems. *J. Am. Stat. Assoc.*, 43(243), 403–412 (1948).
60. J. Lamon. A micromechanics-based approach to the mechanical behavior of brittle-matrix composites. *Comp. Sci. Technol.*, 61, 2259–2272 (2001).
61. V. Calard and J. Lamon. Failure of fibres bundles. *Comp. Sci. Technol.*, 64, 701–710 (2004).
62. H. E. Daniels. The statistical theory of the strength of bundles of threads I. *Proc. R. Soc. A*, 183, 405–435 (1945).
63. B. D. Coleman. On the strength of classical fibers and fibers bundle. *J. Mech. Phys. Solid*, 7, 60–70 (1958).
64. S. L. Phoenix and H. M. Taylor. The asymptotic strength distribution of a general fiber bundle. *Adv. Appl. Prob.*, 5, 200–216 (1973).
65. S. L. Phoenix. Probabilistic strength analysis of fiber bundles structures. *Fiber Sci. Technol.*, 7, 15–31 (1974).
66. L. N. McCartney and R. L. Smith. Statistical theory of the strength of fiber bundles. *ASME J. Appl. Mech.* 105, 601–608 (1983).
67. M. Gurvich and R. Pipes. Strength size effect of laminated composites. *Comp. Sci. Technol.*, 55, 93–105 (1995).
68. Z. Chi, T. W. Chou, and G. Shen. Determination of single fiber strength distribution from fiber bundle testing. *J. Mater. Sci.*, 19, 3319–3324 (1984).
69. N. Lissart and J. Lamon. Evaluation des propriétés de monofilaments à partir d'essais de traction sur mèches, in *Comptes-rendus des 9° Journées Nationales sur les Composites (JNC9)*, vol. 2, edited by J. P. Favre and A. Vautrin (AMAC, 1994), pp. 589–598.
70. M. R'Mili and M. Murat. Caractérisation des fibres par amélioration de l'essai sur mèche avec mesure directe de la déformation. *C. R. Acad. Sci.*, 324(6), 355–364 (1997).
71. M. R'Mili, T. Bouchaour, and P. Merle. Estimation of Weibull parameters from loose bundle tests. *Comp. Sci. Technol.*, 56, 831–834 (1996).
72. ENV1007-5: Advanced Technical Ceramics—Ceramic Composites—Methods of Test for Reinforcements—Part 5: Determination of Distribution of Tensile Strength and Tensile Strain to Failure of Filaments within a Multifilament Tow at Ambient Temperature (European Committee for Standardization, CEN TC 184 SC1, Brussels, Belgium, 1997).

73. M. Berger, N. Hochet, and A. R. Bunsell. Small diameter SiC-based fibers, in *Fine Ceramic Fibers*, edited by A. R. Bunsell and M. H. Berger (Marcel Dekker, New York, 1999), p. 265.
74. H. M. Yun and J. A. Di Carlo. Time/temperature-dependent tensile strength of SiC and Al_2O_3-based fibers. *Ceram. Trans.*, 74, 17–26 (1996).
75. W. Gauthier, F. Pailler, J. Lamon, and R. Pailler. Oxidation of silicon carbide fibers during static fatigue in air at intermediate temperatures. *J. Am. Ceram. Soc.*, 92(9), 2067–2073 (2009).
76. T. Shimoo, K. Okamura, and W. Mutoh. Oxidation behavior and mechanical properties of low-oxygen SiC fibers prepared by vacuum heat-treatment of electron-beam-cured poly(carbosilane) precursor. *J. Mater. Sci.*, 38, 1653–1660 (2003).
77. J. J. Sha, J. S. Park, T. Hinoki, and A. Kohyama. Tensile properties and microstructure characterization of Hi-Nicalon SiC fibers after loading at high temperature. *Int. J. Fract.*, 142, 1–8 (2006).
78. Y. Gogotsi and M. Yoshimura. Oxidation and properties degradation of SiC fibers below 850°C. *J. Mat. Sci. Lett.*, 13, 680–683 (1994).
79. H. M. Yun and J. A. Di Carlo. Thermomechanical behavior of advanced SiC fiber multifilament tows. *Ceram. Eng. Sci. Proc.*, 17(4), 61–67 (1996).
80. E. Lara-Curzio. Stress–rupture of Nicalon/SiC continuous fiber ceramic composites in air at 950°C. *J. Am. Ceram. Soc.*, 80 (12) 3268–3272 (1997).
81. E. Lara-Cuzio. Oxidation induced stress-rupture of fiber bundles. *J. Eng. Mater. Technol.*, 120, 105–109 (1998).
82. S. Bertrand, R. Pailler, and J. Lamon. Influence of strong fibre/coating interfaces on the mechanical behaviour and lifetime of Hi-Nicalon/$(PyC/SiC)_n$/SiC minicomposites. *J. Am. Ceram. Soc.*, 84(4), 787–794 (2001).
83. P. Forio, F. Lavaire, and J. Lamon, Delayed failure at intermediate temperatures (600°–700°C) in air in silicon carbide multifilament tows. *J. Am. Ceram. Soc.*, 87(5), 888–893 (2004).
84. W. Gauthier, and J. Lamon. Delayed failure of Hi-Nicalon and Hi-Nicalon S multifilament tows and single filaments at intermediate temperatures (500–800°C). *J. Am. Ceram. Soc.*, 92, 3, 702–709 (2009).
85. R. W. Davidge, J. R. McLaren, and G. Tappin. Strength-probability-time (SPT) relationships in ceramics. *J. Mater. Sci.*, 8, 1699–1705 (1973).
86. C. Valhas and F. Laanani. Thermodynamic study of the thermal degradation of SiC-based fibres: influence of grain size, *J. Mater. Sci. Lett.*, 14(22), 1558–1561 (1995).

CHAPTER 2

CARBON FIBERS

HERWIG PETERLIK

Faculty of Physics, University of Vienna, Vienna, Austria

2.1 INTRODUCTION/PRODUCTION ROUTES

The particular advantage of carbon fibers is the extremely high specific strength. The huge strength-to-weight ratio is based on two peculiar properties of carbon. It is on the one hand one of the lightest elements and on the other hand it is able to form strong covalent bonds. In carbon fibers, the sp^2 hybridization leads to an extremely high stiffness with an in-plane Young's modulus in the range of 1 TPa. Though the first carbon fibers were produced for lamp filaments by carbonized cotton and bamboo by Thomas Edison in the late nineteenth century, a key step was the development of carbon fibers on the basis of the synthetic polymer polyacrylonitrile (PAN) [1, 2] in Japan. The fibers are wet spun from a precursor, which consists mainly of acrylonitrile, together with some other monomers as copolymers, which are added to obtain a better control of the oxidative stabilization process in a temperature range of 200–300°C [3]. The stabilization leads to a thermally stable structure with an arrangement of carbon hexagons in a preferably one-dimensional (1D) order, which is also described as cyclized ladder structure [3]. It is followed by carbonization, a heat treatment at a temperature of about 1000–1500°C in inert atmosphere. This results in carbon fibers with a turbostratic structure and a preferred orientation of the carbon crystallites with respect to the fiber axis, but still a considerable amount of impurity atoms (mainly nitrogen). Only at higher temperatures above 1400–1500°C it is possible to get rid of residual nitrogen, which could lead to instability due to the formation of flaws and a reduction of strength with a minimum heat treatment temperature (HTT) of 1800°C [3]. Molecular dynamic simulations were performed to get a clearer image of the formation mechanisms of five-, six-, and seven-membered carbon-only ring structures and the release of gaseous species (nitrogen and hydrogen compounds) [4].

The second main production route is the use of mesophase pitch precursor (MPP). This approach was also originally developed in Japan [5] and is nowadays used for fibers with special properties such as extremely high stiffness (Young's modulus of more than 900 GPa and thus approaching the Young's modulus of diamond [6]) or high thermal conductivity, up to nearly 1000 W/m K for fibers grafted with carbon nanotubes (CNT) [7]. In this case, the precursor is based on discotic nematic liquid crystalline material obtained, for example, by thermal polymerization of petroleum- or coal-tar-based pitches. The fibers are spun and the shear stresses induced by the flow lead to a 1D-preferred orientation of the structure, with an alignment of the originally disc-shaped molecules. After stabilization and heat treatment at high temperatures, pure carbon fibers are obtained. Details of the chemistry of MPP fibers and their properties are found in the overview article by Edie [8].

There are other techniques to produce carbon fibers for example, from rayon fibers [9] but the procedure how a preferred orientation develops and can be controlled is rather unclear [10]. Therefore, these fibers are mainly a field of research in the scientific community, but do not have a wide application. The situation is similar for vapor-grown carbon fibers or carbon nanofibers [11], where the idea is to use them for battery applications [12]. More information on the details on the fiber production and their properties can be found in textbooks [3, 13, 14] and overview articles [8, 15–17].

Ceramic Matrix Composites: Materials, Modeling and Technology, First Edition. Edited by Narottam P. Bansal and Jacques Lamon.

2.2 STRUCTURE OF CARBON FIBERS

Carbon fibers exhibit a hierarchical structure with a specific texture on different length scales, from atomic dimensions to the micrometer range. The structural basis is the graphene layer, which consists of planar hexagonal cells. Different to perfect nanostructures—like zero-dimensional fullerenes [18] or 1D CNT [19]—the structure of carbon fibers is imperfect and their graphene layers typically contain a high number of defects. They could be vacancies, interstitials, Stone–Wales defects or impurities. This led to the statement that the basic unit of carbon fibers is the imperfect graphene sheet [20]. In the literature, three different hierarchical levels were defined for carbon fibers, with the focus on the size of each level [20]. The first hierarchical level is the atomic one, with the arrangement of carbon atoms in a hexagonal lattice and the stacking of these sheets. The second hierarchical level is the nanometer level, consisting of basic structural units (BSU), the crystallites, with a lateral extension (a coherently scattering region) L_a and a stacking height L_c in the range of few to some tenths of nanometers. As carbon fibers are usually porous and comprise elongated and needle-like pores with a size in the range of typically 10 nm [21], they are also attributed to this structural level. The third structural level is in the size of 100 nm and describes large inhomogeneities such as a skin-core structure or a cross-sectional texture [22].

In this work, we complement the structural levels by the level of the whole fiber (typically in the 10-μm range) and by splitting the atomic level into two, one being the in-plane distance of atoms of about 0.14 nm and the other the out-of plane distance of graphene sheets with a distance of approximately 0.34 nm, depending on the degree of order. More precisely, the true interlayer spacing was determined for disordered carbons in an early work of Rosalind Franklin [23] by $d_{002} = 0.3440–0.0086\ (1 - p^2)$, where p is the proportion of disoriented layers.

In Figure 2.1, the five structural levels as defined above are depicted.

2.2.1 Levels 1 and 2, Atomic level

The imperfectness of the structure is one of the most striking features of carbon fibers, whereas graphitic carbon has an extreme anisotropy, an extraordinary high in-plane modulus and in-plane strength together with a very weak bonding perpendicular to the graphene sheets, the imperfectness of carbon fibers allows to modify and to control the anisotropy to a wide degree. This is the case for the structure as well as mechanical, electrical, or thermal properties. The main techniques, which allow obtaining information on the structure at the atomic level, are X-ray diffraction (XRD) or electron diffraction. Figure 2.2a shows a typical 2D-scattering pattern of a highly oriented MPP carbon fiber, where the lateral distance of the graphene planes is calculated from the 00l reflections (visible up to 006 in Figure 2.2a) and the in-plane distance of the carbon atoms from the *hk*0 reflections (usually from the 100 reflection). The turbostratic structure of carbon fibers is visible in the scattering pattern as no 3D crystal structure is formed. Thus, one usually uses only two indices for the in-plane *hk*0 reflections, which are frequently denoted as *hk*-bands. A consequence of the turbostratic structure, the 1D stacking of distorted 2D-graphene sheets, is that the typical distance between two sheets is larger than in the ABAB-stacking of perfect graphite of 0.3354 nm (in the limit of ordered crystals described by Franklin's formula [23]). The in-plane lattice constant a of the carbon atoms can be calculated from the maximum of the intensity of the *hk*-band in reciprocal space q_{hk} by $d_{hk} = 2\pi/q_{hx}$ and $a = d_{hk}(4(h^2 + k^2 + hk)/3)^{1/2}$. The absolute value of the scattering vector q in reciprocal space is related to the scattering angle θ by $q = 4\pi \sin\theta/\lambda$, where λ is the radiation wavelength and 2θ is the scattering angle. A peculiarity is the occurrence of layer lines [24] in the diffraction pattern (Figure 2.2a) at the positions of the *hk* reflections, which are due to the more less periodic arrangement of the graphene layers [20]. These layer lines lead to a characteristic asymmetric shape of the integrated intensities with a steep flank toward the low q-side and a broad flank toward the high one [20].

2.2.2 Level 3, Lower Nanometer Range

In the low nanometer range (denoted as mesoscale in chemistry) between 2 and 50 nm, carbon fibers are built up of BSUs or crystallites as well as of pores between these BSUs. The size and orientation of these crystallites is accessible by techniques such as XRD, transmission electron microscopy (TEM), Raman scattering, or atomic force microscopy (AFM). The advantage of XRD in this case is that parameters describing the size of the BSUs are easily accessible by applying Scherrer's formula. The size of a BSU is $L = 2\pi K/\Delta q$, where Δq is the full width at half maximum of the respective peak intensity width and K is a constant depending on the specific shape of the crystallite (frequently a value of $K = 0.92$ is taken for the size of the stacked graphene layers, L_c, and $K = 1.84$ for the in-plane size of the BSU, L_a).

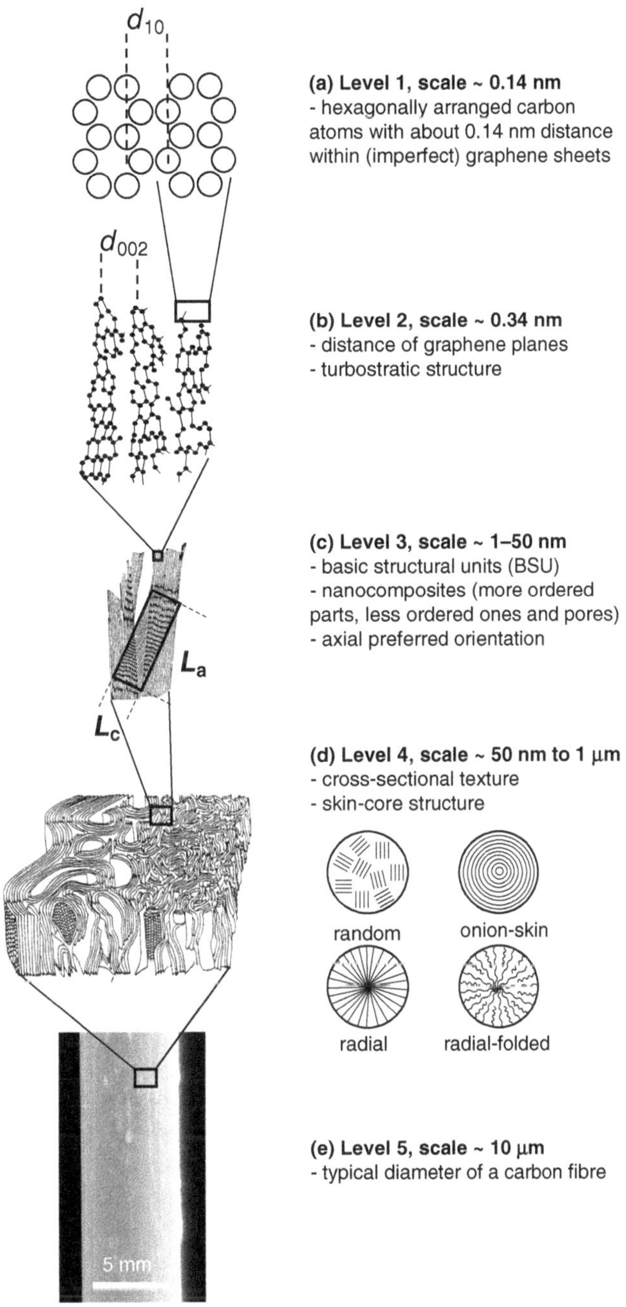

FIGURE 2.1

Five structural levels of a carbon fiber. Atomic levels: In-plane atomic distances d_{10}, distance of graphene planes d_{002}. Nanometer level, coherent scattering length (crystallite size) L_a (in-plane), L_c (out-of-plane), and pores between the crystallites. Submicrometer range, skin-core structure, and cross-sectional texture. Micrometer range, diameter of fibers.

More advanced techniques for evaluation are available, if more than one reflection can be measured [24,25]. A concurrent technique is to derive the size of the BSUs by the measurement of lattice fringe images in TEM. The in-plane layer size L_a from TEM is typically larger than in XRD, as the coherence length in XRD measures primarily the straight portion of the planes [26]. In a recent work, different carbon fibers were compared and L_a from TEM succeeded the value of XRD by a factor up to three for highly oriented MPP fibers, whereas TEM and XRD gave nearly identical values for some PAN fibers [27]. The authors were not aware of the previous paper from Johnson [26], but their results support

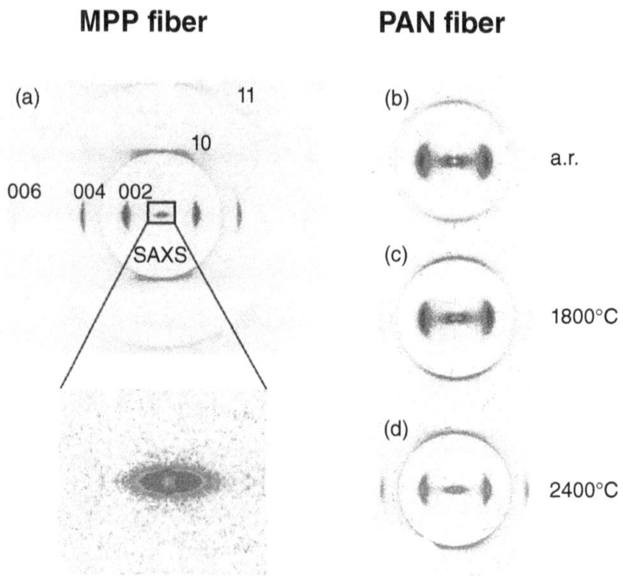

FIGURE 2.2

Typical XRD pattern of carbon fibers: (a) MPP fiber with 00l reflections and *hk*-bands, (b)–(d) PAN fiber HTA5131 as received and with a heat treatment of 1800°C and 2400°C, respectively. The increasing visibility of higher order reflections visualizes the increasing order and the smaller spread in azimuthal direction of the intensity of the 002 reflection, the increased alignment of the structure with respect to the fiber axis.

his arguments concerning the size difference between TEM and XRD. If data from the literature [27,28] are summarized as in Figure 2.3, the significantly higher values, in particular, for the in-plane crystallite size from TEM (triangles) in comparison to XRD (circles and squares) are obvious. This effect is also more pronounced for PAN (Figure 2.3, gray symbols) than for MPP fibers (Figure 2.3, black symbols).

A completely different approach is the determination of the crystallite size from Raman scattering. The Raman G-band is attributed to the crystalline portion of the fiber and the D-band to the disordered portion, as the G-band is very pronounced and the D-band is absent in single crystal graphite. From the intensity ratio of these bands, one can calculate the crystallite size [29]. There is a very good coincidence of the crystallite size from Raman measurements with XRD, though this relation seems to break down for a very low crystallite size less than 2 nm [30].

A measure for the orientation of the BSUs with respect to the fiber axis is directly obtained from the intensity spread of the d_{002} reflection in azimuthal direction. As the pores usually have a needle-like shape, the orientation distribution in real space can be calculated by an integral transformation [31]. An example for an increasing amount of order and orientation is seen in Figure 2.2b–d, where XRD patterns of a high tenacity (HT-) PAN fiber (HTA5131 from Tenax) are shown. This PAN fiber was subjected to a different HTT, whereas for the as-received fiber (Figure 2.2b) it is even difficult or nearly impossible to identify the 004 reflection. With increasing the HTT to 1800°C (Figure 2.2c) and to 2400°C (Figure 2.2d), the order of the PAN fiber increases, indicated by the appearance of a distinct 004 reflection (due to the increase in the stacking height of the crystallites L_c) and the decrease of the orientational disorder, that is, the higher orientation of the crystallites (visible by the lower spread of the 002 reflection in azimuthal direction).

PAN fibers have a rather low density, however, this density increases with increasing heat treatment [28]. Typically, the density is only in the range of about two-thirds of the density of graphite, thus a considerable porosity exists. TEM was used to characterize the porosity of carbon fibers [26] and a model of the structure of porous carbons as well as its transformation by stretching in one direction for high tensile strength PAN-based carbon fibers is found in [32]. The effect of crumpling of graphene sheets (the number of folds per BSU) on the pore texture, the mechanical properties, and the electrical resistance were investigated by the same group [33]. In a recent work, the temperature dependence of the microstructure of a T700 carbon fiber—pores and crystallites—was determined using TEM, HRTEM, and selective area electron diffraction (SAED) to determine the mean misorientation angle from the 002 reflection [34]. With increasing temperature, many pores start to develop in the skin region, and after a high HTT of 2800°C they could also be observed in the core region. Obtaining statistically relevant information on the pore size and pore orientation with a dimension in the low nanometer range across the cross section of a carbon fiber with a size of 10 μm is, however, a

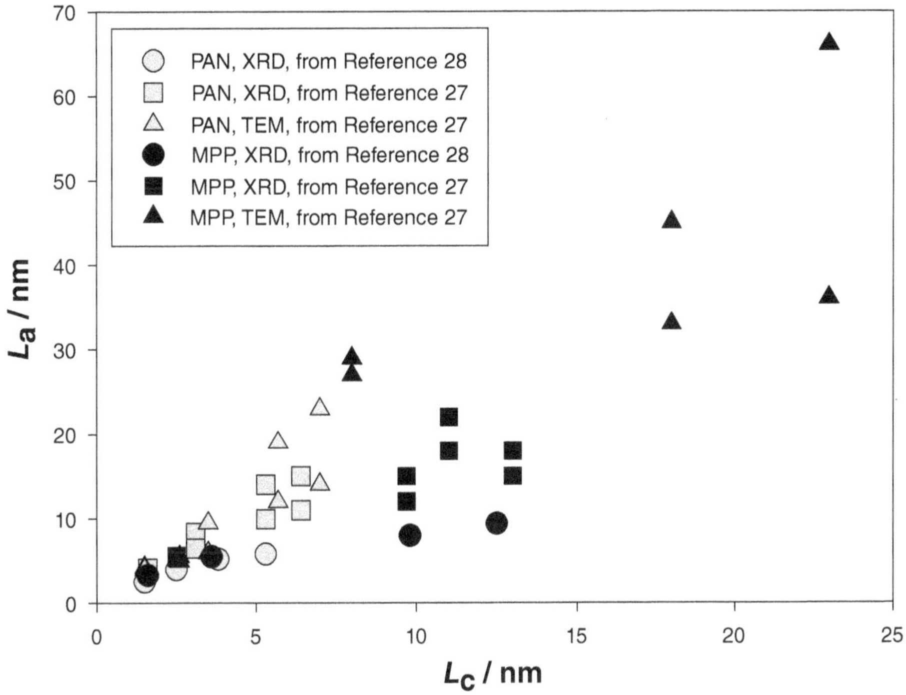

FIGURE 2.3

The in-plane layer size L_a from TEM (triangles) is typically larger than in XRD (circles and squares), as the coherence length in XRD measures primarily the straight portion of the planes [26]. This effect is also more pronounced for PAN than for MPP fibers. Data are taken from References 27 and 28.

rather tedious work. A very suitable technique for the determination of pore size and orientation is small-angle X-ray scattering (SAXS). Only a very low effort in preparation as well as a short measurement time in the range of minutes is required to collect data on the pore structure. Different to XRD, SAXS detects the density fluctuations within a material. For needle-like pores, either the radius of gyration or a mean chord length can be evaluated [21,31]. The latter is related to the small dimension of the pores. Using microbeam SAXS across the fiber cross section, different types of pores could be distinguished in as-received PAN fibers. In an isotropic outer ring along the fiber axis microvoids were observed, and in the center less oriented porosity was found [35]. The different appearances of the pore structure in the skin and in the core carry the topic to the next level, the upper nanometer range (hundred to some hundreds of nanometers).

2.2.3 Level 4, Upper Nanometer Range

In the high nanometer range, two structural levels have been observed: a skin-core structure, in particular, in PAN carbon fibers [36] and a cross-sectional texture, which is not random but could be radial, radial-folded, or onion-like, in particular, for MPP fibers [8]. TEM was used for the investigation of the skin-core structure and a higher orientation and larger crystallites were found in the skin region of the carbon fibers in comparison to the less oriented core of PAN fibers with a HTT of 2500°C by Bennett and Johnson [36]. They did not found any large structural heterogeneity between skin and core by varying the time of oxidation or at low HTT (below 1500°C). A precise numerical value for the size of the skin is obtained from scanning across the cross section of a carbon fiber with an X-ray microbeam from a synchrotron radiation source [37]. First experiments used a microbeam from a glass capillary with a width of about 2 μm [37] and in a later work, a 100-nm wide microbeam from a waveguide structure was used [28]. The small beam size and therefore the high resolution in real space allowed the precise determination of the size of the skin region by scanning across the fiber. Figure 2.4 shows data from Reference 28 for a PAN HTA5131 fiber with 2100°C HTT (Figure 2.4, circles), with a higher crystallite orientation (lower mean misorientation angle) in the skin. For a fiber, which is subjected to bending (Figure 2.4, triangles), the misorientation angle decreases in the tension part and increases in the compression part of the fiber, but the higher orientation in the skin is still visible.

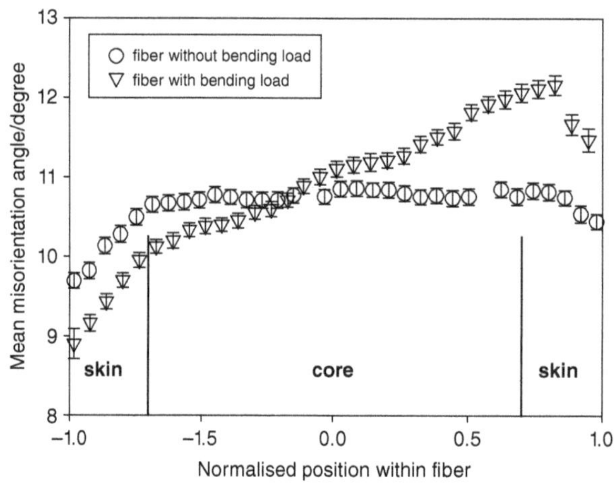

FIGURE 2.4

Mean misorientation angle with respect to the fiber axis, evaluated from the intensity spread of the 002 reflection in azimuthal direction, in dependence on the position within the fiber (PAN fiber HTA5131, 2100°C HTT, data are taken from Reference 28). Fiber without bending load (circles) with a lower misorientation angle in the skin, fiber subjected to bending load (triangles). The crystallite orientation increases in the tension part and decreases in the compression part of the fiber, but the difference between skin and core is still visible.

For MPP fibers, the cross-sectional texture led to a different intensity of the 002 reflection and the 10-band. By fitting with an appropriate two-phase model with a radially folded ordered and an amorphous disordered phase together with deconvolution with the fiber shape function and the beam profile [37], numerical values on the amount of both phases and the degree of folding were obtained [37]. Quantitative values for the amount of the phases and the magnitudes of the undulations were also obtained in Reference 22, where a correlation between the axial preferred orientation and the cross-sectional texture in MPP fibers was found. As the size of the skin is in the upper nanometer range, many research groups used scanning electron microscope (SEM) of fracture surfaces to investigate the cross-sectional texture [38–41]. Figure 2.5 shows images from optical microscopy (left column) and SEM images of polished (center column) as well as from fracture surfaces (right column) for four different fibers, two PAN fibers with a HTT of 1300°C and of 2400°C, where the cross-sectional texture is amorphous and no specific features are visible, and two MPP fibers, where in the optical microscopy image an interference pattern viewed under convergent light and crossed polars indicates a cross-sectional texture. For the MPP fibers, a radially folded texture is also visible in the SEM image from the fracture surface. One should, however, be cautious with the interpretation of fracture surfaces, as even in an amorphous material like silica the fracture pattern exhibits features (e.g., mirror-mist-hackle), but they depend more on crack propagation than on a specific structure. The origin of the skin-core structure is still a matter of discussion; it is related either to the stabilization process [42], where the overpreoxidized skin leads to more defects and a different structure during heating [43], at higher temperatures the development of a skin-core morphology was observed and attributed to the release of nitrogen [44].

2.2.4 Level 5, 10-μm Range

Carbon fibers typically have a diameter between 5 and 15 μm. Though carbon fibers could have been produced in a wide size range, this dimension is a compromise between high fracture strength (which increases with decreasing fiber diameter), low health risk (which increases toward lower fiber diameters, as concluded in analogy to asbestos fibers), and good processability in the industrial manufacturing.

2.3 STIFFNESS AND STRENGTH OF CARBON FIBERS

The high stiffness is one of the most striking properties of carbon fibers, as structural components in technical applications are often not strength but stiffness limited—the reason is that Euler buckling of struts in lightweight constructions

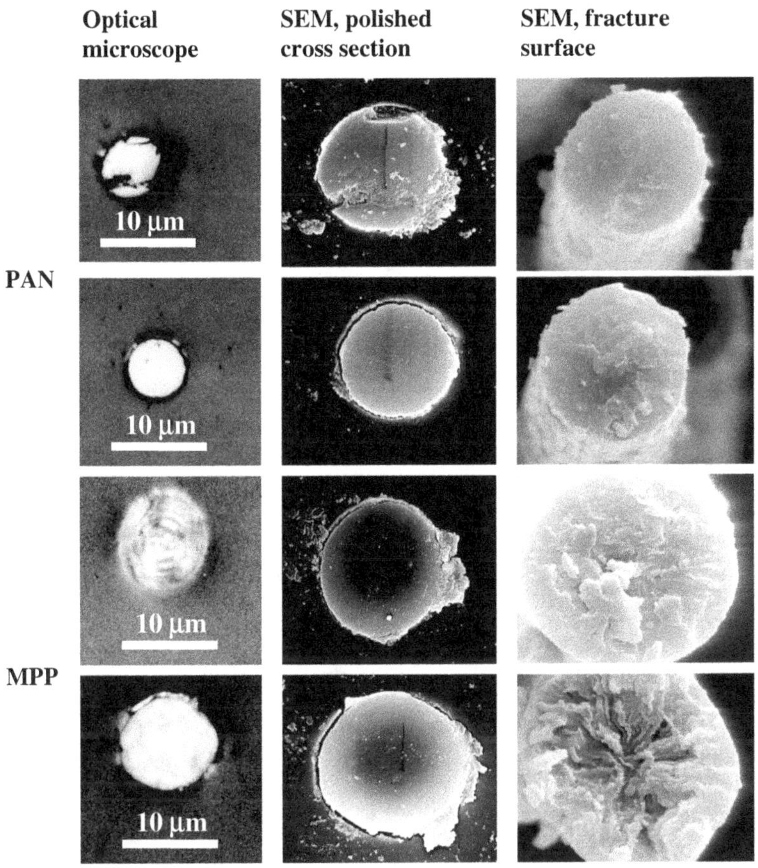

FIGURE 2.5
Images from optical microscopy (left), SEM, polished cross sections (middle) and SEM, fracture surfaces (right). For the PAN fibers (upper two rows) the cross-sectional texture is random, different to the MPP fibers, where an optical anisotropy is visible in the optical micrographs as well as a radially folded structure in the fracture surface.

is linearly dependent on the Young's modulus. Carbon fibers are nowadays available with a Young's modulus of more than 500 GPa for PAN fibers and more than 900 GPa for MPP fibers [6]. On the other hand, the fracture strain of these fibers with an extremely high modulus is rather low. The Young's modulus of carbon fibers can be measured by fiber bundle tests, where with one test the statistical average of thousands of fibers is obtained, or by a certain number of single fiber tests. In the latter case, the cross-sectional area of each fiber has to be determined either in the SEM or by laser diffraction, whereas in the first case a polished cross section in the SEM or optical microscope results in a numerical value for the mean cross-sectional area of a typical fiber. The advantage of the single fiber test for the determination of the stress–strain curve and calculation of the Young's modulus there from is obvious (Figure 2.6). In Figure 2.6a, a typical force–strain diagram of a fiber bundle test is depicted together with the slope from the initial part of the curve, from which the Young's modulus is evaluated (here up to about 0.2% strain). In comparison, single fibers exhibit a nonlinear stress–strain curve (Figure 2.6b), and the insert in Figure 2.6b shows the increase of the Young's modulus. The corresponding values from the fiber bundle test are drawn as thick black line into the insert, and it is clearly visible that the nonlinearity can only be identified if the whole stress–strain curve is evaluated. In fiber bundle tests, fiber failure from weak fibers starts very early and masks the increase in the Young's modulus. Some carbon fibers have an extraordinary pronounced nonlinearity and a strong increase of the Young's modulus up to 50%with increasing load was observed [45]. The structural origin of the increase follows in the uniform stress model [46,47] from crystallite stretching and crystallite rotation during loading [48].

The high tensile Young's modulus of the whole carbon fiber is the consequence of the extreme stiff sp^2 bonding of the graphene planes within the crystallites. A value of 1.02 ± 0.02 TPa was measured for the in-plane modulus by dynamic and static methods for compression-annealed pyrolytic graphite [49] and 1.14 ± 0.01 TPa for the nanometer-large crystallites within carbon fibers [48]. Similar results were obtained for CNT (1.1 TPa by static test methods [50] and 1.25 TPa by dynamic ones [51]) and for graphene monolayers (1.0 ± 0.1 TPa [52]). However, for nanotubes and graphene sheets the published values scatter enormously, because they are based on continuum models and it is rather difficult to define a cross section for a row of atoms.

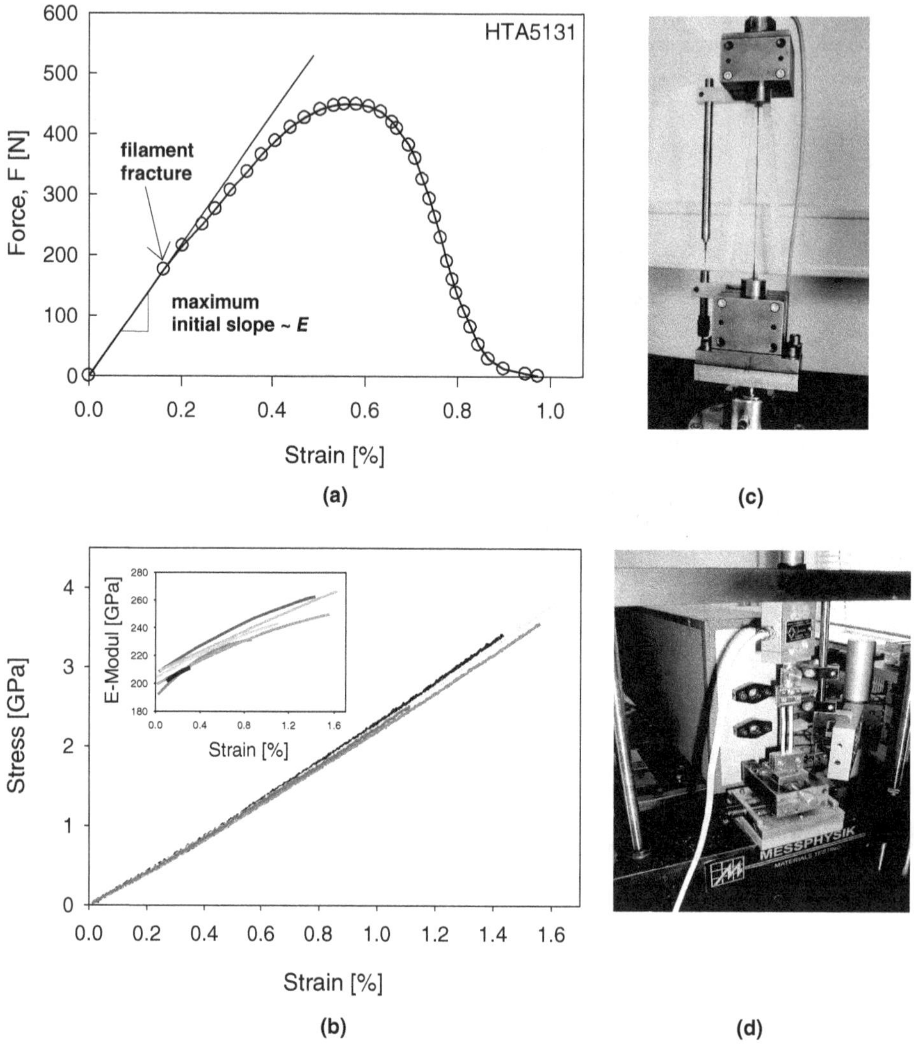

FIGURE 2.6

(a) Typical force–strain diagram of a fiber bundle test, from which Young's modulus is calculated from the initial (linear) slope, here up to about 0.2% strain; (b) experimental equipment for fiber bundle tests with 200-mm gauge length; (c) typical stress–strain diagram from a single fiber test, the insert shows the increase of Young's modulus with increasing strain and the thick black line shows the corresponding regime, from which Young's modulus is evaluated from a fiber bundle test; (d) experimental equipment for single fiber tests with 50-mm gauge length and a direct optical strain measurement.

As the high modulus of graphene planes is the basis of the high modulus of carbon fibers, it is only natural that a misorientation of graphene planes leads to a decreasing modulus and vice versa. The first quantitative relation between crystallite orientation and the numerical value of the Young's modulus was developed by Fischer and Ruland [53]. They found experimentally a distinct relation between the Young's modulus and an orientation parameter obtained from XRD. The model was denoted "elastic unwrinkling model" and was based on the strain of straight parts and a reversible increase of distorted BSUs [20, 53]. Northolt et al. [46] compared the two limiting cases: the uniform stress model, where the fiber is seen as built up of identical fibrils exposed to the same stress (a fiber as a bundle of fibrils with no interaction in between) and the uniform strain model, where every part of the identical fibrils is subjected to the same axial strain [20]. The uniform stress model was favored by Northolt [46], supported by experimental data for the Young's moduli and the argument that the structure consists of long elongated ribbons, which are separated by long needle-like voids [53]. This might also be the case for different structural models, but the situation is much less clear, if one follows the view that carbon fibers consist of crumpled and folded sheets of entangled layer planes, which are interlinked by their boundaries [32]. A comparison of the deformation of crystals and the effect on the tangent modulus is given by Shioya et al. [54] for different models such as the undulating ribbon, the zigzag ribbon, the series elements, the series rotable elements, the parallel elements, and the mosaic model.

Two approaches have been followed in the past decade to relate the structure and the structural development to the nonlinear stress–strain curve of carbon fibers. The first was *in situ* XRD of single carbon fibers with a microbeam under tensile [48] and bending [28] load. From the tensile experiment, a value of 1.14 TPa for the in-plane Young's modulus of graphene planes was obtained, and the increase of the modulus with increasing strain was related to the influence of the shear modulus and the deformation and rotation of the crystals [48] similar as proposed in References 46, 53, and 54. The second approach is *in situ* Raman scattering, which was also performed in tension [47, 55] and compression [56, 57]. Four-point-bending tests of graphene on a polymeric support led to a splitting of the G-band into two peaks, similar to the effect of the curvature of the CNT on the G-peak [58]. This could be used to measure the amount of uniaxial and biaxial strains [58], and the method was proposed as universal stress sensor for graphene and carbon fibers [59].

In uniaxial loading, the modulus hardening and softening was followed by the nonlinear shift of the Raman G-band [56, 57]. However, the continuous change of the elastic properties during tensile or compressive loading is still more complicated in bending. Additionally, a rearrangement of the tension and compression zone was observed by bending of single carbon fibers and *in situ* measuring the in-plane deformation by XRD [28]. Based on elastic analysis of the shift of the neutral zone in bending [60], a ratio of the effective Young's moduli in compression to the one in tension can be calculated [28], which is depicted in Figure 2.7 in dependence on the in-plane length of the crystallites L_a. The diagram shows that in compression PAN fibers exhibit a much higher modulus than MPP fibers.

The high temperature behavior of the Young's modulus is of particular interest, as aerospace industries are one of the world-wide most important producers using carbon fibers for reinforcement of components. For many of their applications, excellent high temperature behavior is crucial. Carbon fibers here are unrivalled, as they are thermally very stable up to temperatures of more than 2000°C [17]. Surprisingly, the tensile strength increases with increasing test temperature, being attributed to the improved alignment of crystallites and possibly the release of residual shear strains in the fibers [20, 61]. The Young's modulus was observed either to increase [62] or decrease [63] with increasing test

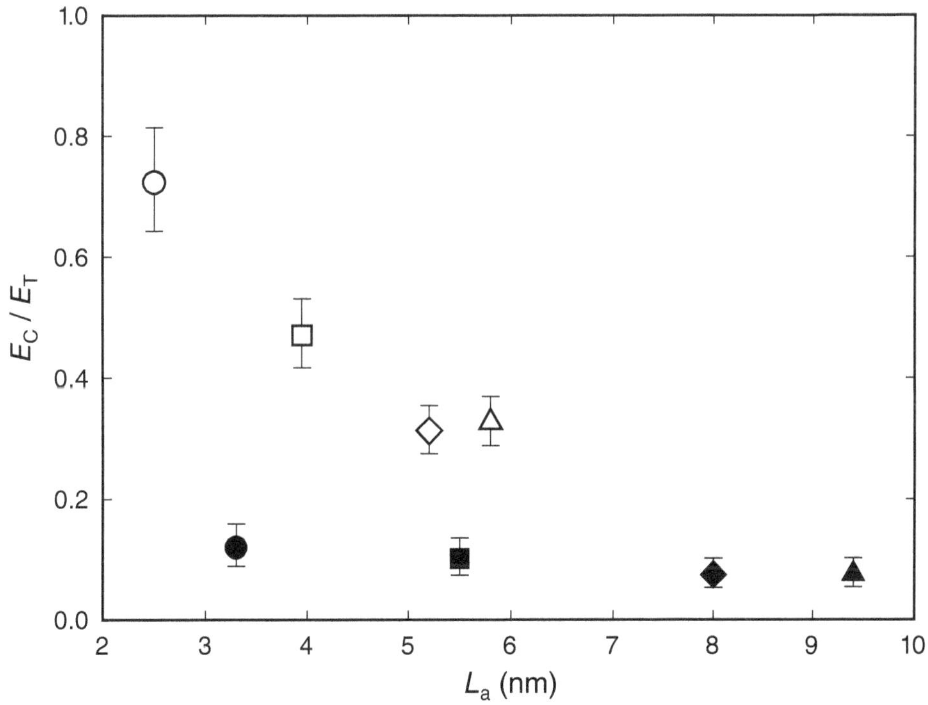

FIGURE 2.7

Ratio of effective Young's modulus in compression to the one in tension, E_C/E_T, in dependence on crystallite size L_a, obtained from the shift of the neutral zone during bending and *in situ* XRD. Data are taken from Reference 28.

temperature. At high temperatures above 1400°C, carbon fibers show viscoelastic behavior, and above 1800°C, they might even behave viscoplastic [63]. This was explained by molecular deformations [63], which contribute to a shear strain and lead to viscoelasticity. Initiated by these results, *in situ* XRD of carbon fiber bundles was performed at high temperatures under load [64]. The authors suggested that three different mechanisms might occur, which lead to an increase of the crystallite size L_a and L_c: a rearrangement of crystallites with graphene layers facing either their edges (increase of L_a), with graphene layers facing the planes (increase of L_c), and a process similar to dislocation glide, which leads to a simultaneous increase of both crystallite sizes (L_a as well as L_c) [64]. This was further supported by mechanical creep experiments, from which an activation energy of 1.7 ± 0.3 eV and a microscopic activation energy of 2.0 ± 0.3 eV was calculated [65]. A heat treatment without load leads to a complete stabilization of the structure and no further creep is observed up to the HTT [65].

Considering tension, compression, and bending strength of carbon fibers, it is clear that tension strength is the most important one, as most of the fibers are used as reinforcement of structural parts and are therefore loaded primarily in tension. Due to weaving, bending and compression cannot be excluded and mechanical properties depend strongly on the crimp angle [66]. This is, however, strongly dependent on the interface between the carbon fiber and the matrix in the composites and not a theme of this section. Different to the Young's modulus, which is primarily dependent on the orientation of the BSUs, the tensile strength is much more difficult to interpret. In general, for brittle materials, Weibull [67] theory is used, which describes the strength distribution of a material with two parameters, the scale parameter (for the mean strength) and the Weibull modulus (for the width of the distribution). It was observed that the Weibull modulus had the lowest numerical value for ultrahigh modulus MPP fibers, and the Weibull modulus decreases with an increase in the tensile modulus and the mean tensile strength [6]. However, the Weibull theory is based on a power-law dependence of the size of flaws, a unique relation of strength and flaw size and noninteracting flaws. Concerning carbon fibers, from a structural point of view, the evaluation according to the Weibull theory was seen more as a technical approach. The strength is nearly unaffected by an increase of the crystallite size of a factor of five because the fiber strength is not affected by an increase of the crystallite size [44] or an increase of the pore size by a factor of two [21]. As the size of these structural parameters does not seem to have a decisive effect on the strength properties, the general view is that essentially the interlinking and misalignment of crystallites is responsible for failure [61]. Therefore, a bimodality in the strength distribution, which is often observed, if a significant large number of carbon fibers is tested, is attributed more to the skin-core structure than to different stress intensity factors from flaws on the surface and in the bulk [44]. This assumption was supported by the observation that the strength distribution is much more uniform, if PAN fibers are subjected to a HTT of 2400°C, and by concomitant XRD measurements, which showed that the skin-core structure nearly completely vanishes at this HTT [44].

2.4 CONCLUDING REMARKS AND FUTURE DIRECTIONS

Carbon fibers are nowadays available covering thermal and mechanical properties in an extremely wide range, with a Young's modulus up to more than 900 GPa [6], a thermal conductivity of nearly 1000 W/m K [7], and a tensile strength of more than 5 GPa [6]. A lot of information on the structure of carbon fibers is available and many processing parameters have been investigating extensively [43, 68–70].

Still an open question is the understanding of the precise defect structure on the atomic scale, the role of cross-links between graphene sheets on the mechanical properties, and the way how impurities such as noncarbon atoms influence the structural development, in particular, in dependence on temperature. Though theoretical work suggested a number of defect types being possible candidates for cross-links, it is still a matter of discussion, if covalent cross-links arise from defects within the graphene sheets, from the boundaries of crystallites, or if a more or less amorphous phase links more or less ordered crystallites. Therefore, even from a basic science view, still more research in this field is required. A better understanding of the nature of defects also opens the field for further applications as functional materials with specific thermal and electrical properties.

From a technical point of view, obviously there is an increasing demand for carbon fibers at the world-wide market. This is driven by the requirement to reduce weight and energy. No other material until now can rival with composites being reinforced with carbon fibers. As example, the use of composites as structural parts increased from about 10% for the Boeing 777 to more than 20% for the Airbus A380 and should reach a part of 50% for the future generation of airplanes such as the Airbus A350 and the Boeing Dreamliner. Materials with lower specific weight would fulfill the ecological requirement of the reduction of CO_2 emissions, and large programs in the European Union have been started, for example, to develop new technologies and materials to tackle the problems connected to the climate change. However, research has also to take into account the demand from industry that the production of new materials should be possible at reasonable costs.

ACKNOWLEDGMENTS

Oskar Paris (University of Leoben, Austria) and Roland Weiß (Schunk Kohlenstofftechnik, Germany) are gratefully acknowledged for many years of constructive collaboration in the field of carbon fibers, concerning structure as well as mechanical properties. I am also indebted to Dieter Loidl (PDTS GmbH), Harald Rennhofer (University of Natural Resources and Applied Life Sciences, Vienna, Austria), and Stephan Puchegger (University of Vienna, Austria), who laid the basis for the research presented in this chapter during their PhD theses in my group. For their support in performing successful microbeam diffraction experiments and for many discussions, Martin Müller (University of Kiel, Germany), Manfred Burghammer, and Christian Riekel (ESRF Grenoble, France) are acknowledged.

REFERENCES

1. Patent announcement Japan, Shouwa-37-4405 [in Japanese] (1959).
2. G. Shindo. Studies on graphite fibre, Rep. 317 (Govt. Indust. Res. Inst. Osaka) (1961).
3. E. Fitzer and L. M. Manocha. *Carbon Reinforcements and Carbon/Carbon Composites* (Springer, Berlin-Heidelberg, 1998).
4. B. Saha and G. C. Schatz. Carbonization in polyacrylonitrile (PAN) based carbon fibers studied by ReaxFF molecular dynamics simulations. *J. Phys. Chem. B*, 116(15), 4684–4692 (2012).
5. S. Otani, K. Yamada, T. Koitabashi, and A. Yokoyama. On the raw materials of MP carbon fiber. *Carbon*, 4(3), 425–432 (1966).
6. K. Naito, Y. Tanaka, J.M. Yang, and Y. Kagawa. Tensile properties of ultrahigh strength PAN-based, ultrahigh modulus pitch-based and high ductility pitch-based carbon fibers. *Carbon*, 46(2), 189–195 (2008).
7. K. Naito, J. M. Yang, Y. Xu, and Y. Kagawa. Enhancing the thermal conductivity of polyacrylonitrile and pitch-based carbon fibers by grafting carbon nanotubes on them. *Carbon*, 48(6), 1849–1857 (2010).
8. D. D. Edie. The effect of processing on the structure and properties of carbon fibers. *Carbon*, 36(4), 345–362 (1998).
9. R. Bacon and M. M. Tang. Carbonization of cellulose fibers-II. Physical property study. *Carbon*, 2(3), 221–225 (1964).
10. O. Paris, C. Zollfrank, and G. A. Zickler. Decomposition and carbonisation of wood biopolymers—a microstructural study of softwood pyrolysis. *Carbon*, 43(1), 53–66 (2005).
11. M. Endo and H. W. Kroto. Formation of carbon nanofibers. *J. Phys. Chem.*, 96(17), 6941–6944 (1992).
12. M. Endo, Y. A. Kim, T. Hayashi, K. Nishimura, T. Matusita, K. Miyashita, and M. S. Dresselhaus. Vapor-grown carbon fibers (VGCFs)—basic properties and their battery applications. *Carbon*, 39(9), 1287–1297 (2001).
13. D. Edie and J. McHugh, High performance carbon fibers, in *Carbon Materials for Advanced Technologies*, edited by T. Burchell (Pergamon, Elsevier, Amsterdam, 1999)
14. M. Dresselhaus and M. Endo. Relation of carbon nanotubes to other carbon materials, in *Carbon Nanotubes*, edited by T. Burchell M. Dresselhaus, G. Dresselhaus and P. Avouris, (Springer, Berlin, 2001)
15. R. J. Diefendorf and E. Tokarsky. High-performance carbon fibers. *Polymer Eng. Sci.*, 15(3), 150–159 (1975).
16. A. Oberlin. Carbonization and graphitization. *Carbon*, 22(6), 521–541 (1984).
17. E. Fitzer. Carbon fibres—the miracle material for temperatures between 5 and 3000 K. *High Temp.–High Press.*, 18(5), 479–508 (1986).(10 ETPC Proceedings)
18. H. W. Kroto, J. R. Heath, S. C. Obrien, R. F. Curl, and R. E. Smalley. C-60 Buckminsterfullerene. *Nature*, 318(6042), 162–163 (1985).
19. S. Iijima. Helical microtubules of graphitic carbon. *Nature*, 354(6348), 56–58 (1991).
20. O. Paris and H. Peterlik. The structure of carbon fibers, in *Handbook of Textile Fibre Structure Volume 2*, edited by S. J. Eichhorn, J. W. S. Hearle, M. Jaffe and T. Kikutani (Woodhead Publishing Ltd, Oxford, 2009).
21. H. Peterlik, P. Fratzl, and K. Kromp. Pore structure of carbon/carbon composites studied by small-angle X-ray scattering. *Carbon*, 32(5), 939–945 (1994).
22. O. Paris, D. Loidl, and H. Peterlik. Texture of PAN- and pitch-based carbon fibers. *Carbon*, 40(4), 551–555 (2002).

23. R. E. Franklin. The structure of graphitic carbons. *Acta. Cryst.*, 4(3), 253–261 (1951).

24. R. Perret and W. Ruland. Profile analysis of random-layer lines. *J. Appl. Cryst.*, 1, 257–262 (1968).

25. W. Ruland and B. Smarsly. X-ray scattering of non-graphitic carbon: an improved method of evaluation. *J. Appl. Cryst.*, 35, 624–633 (2002).

26. D. J. Johnson and C. Frank. Recent advances in studies of carbon fiber structure. *Phil. Trans. Royal Soc. Lond.on A – Math. Phys. Eng. Sci.*, 294(1411), 443–449 (1980).

27. X. Qin, Y. Lu, H. Xiao, Y. Wen, and T. Yu. A comparison of the effect of graphitization on microstructures and properties of polyacrylonitrile and mesophase pitch-based carbon fibers. *Carbon*, 50(12), 4459–4469 (2012).

28. D. Loidl, O. Paris, M. Burghammer, C. Riekel, and H. Peterlik. Direct observation of nanocrystallite buckling in carbon fibers under bending load. *Phys. Rev. Lett.*, 95, 225501 (2005).

29. F. Tuinstra and J. L. Koenig. Raman spectrum of graphite. *J. Chem. Phys.*, 53(3), 1126–1130 (1970).

30. G. A. Zickler, B. Smarsly, N. Gierlinger, H. Peterlik, and O. Paris. A reconsideration of the relationship between the crystallite size L_a of carbons determined by X-ray diffraction and Raman spectroscopy. *Carbon*, 44(15), 3239–3246 (2006).

31. R. Perret and W. Ruland. Single and multiple X-ray small-angle scattering of carbon fibres. *J. Appl. Cryst.*, 2, 209–218 (1969).

32. M. Guigon, A. Oberlin, and G. Desarmot. Microtexture and structure of some high tensile strength, PAN-based carbon fibres. *Fibre Sci. Technol.*, 20(1), 55–72 (1984).

33. M. Guigon, A. Oberlin, and G. Desarmot. Microtexture and structure of some high-modulus, PAN-based carbon fibres. *Fibre Sci. Technol.*, 20(3), 177–198 (1984).

34. G. Zhou, Y. Liu, L. He, Q. Guo, and H. Ye. Microstructure difference between core and skin of T700 carbon fibers in heat-treated carbon/carbon composites. *Carbon*, 49(9), 2883–2892 (2011).

35. D. Lozano-Castelló, J. A. Macia-Agulló, D. Cazorla-Amorós, A. Linares-Solano, M. Müller, M. Burghammer, and C. Riekel. Isotropic and anisotropic microporosity development upon chemical activation of carbon fibers, revealed by microbeam small-angle X-ray scattering. *Carbon*, 44(7), 1121–1129 (2006).

36. S. C. Bennett and D. J. Johnson. Electron-microscope studies of structural heterogeneity in Pan-based carbon fibres. *Carbon*, 17(1), 25–39 (1979).

37. O. Paris, D. Loidl, M. Müller, H. Lichtenegger, and H. Peterlik. Cross-sectional texture of carbon fibres analysed by scanning microbeam X-ray diffraction. *J. Appl. Cryst.*, 34, 473–479 (2001).

38. L. P. Kobets and I. S. Deev. Carbon fibres: structure and mechanical properties. *Comp. Sci. Technol.*, 57(12), 1571–1580 (1997).

39. S. Hong, Y. Korai, and I. Mochida. Development of mesoscopic textures in transverse cross-section of mesophase pitch-based carbon fibers. *Carbon*, 37(6), 917–930 (1999).

40. S. Hong, Y. Korai, and I. Mochida. Microscopic texture at the skin area of mesophase pitch-based carbon fiber. *Carbon*, 38(6), 805–815 (2000).

41. F. Watanabe, Y. Korai, I. Mochida, and Y. Nishimura. Structure of melt-blown mesophase pitch-based carbon fiber. *Carbon*, 38(5), 741–747 (2000).

42. M. J. Yu, C. G. Wang, Y. J. Bai, M. X. Ji, and Y. Xu. SEM and OM Study on the microstructure of oxidative stabilized polyacrylonitrile fibers. *Polymer Bull.*, 58(5–6), 933–940 (2007).

43. J. Liu and W. X. Zhang. Structural changes during the thermal stabilization of modified and original polyacrylonitrile precursors. *J. Appl. Polymer Sci.*, 97(5), 2047–2053 (2005).

44. D. Loidl, O. Paris, M. Rennhofer, M. Müller, and H. Peterlik. Skin-core structure and bimodal weibull distribution of the strength of carbon fibers. *Carbon*, 45(14), 2801–2805 (2007).

45. C. Reder, D. Loidl, S. Puchegger, D. Gitschthaler, H. Peterlik, K. Kromp, G. Khatibi, A. Betzwar-Kotas, P. Zimprich, and B. Weiss. Non-contacting strain measurements of ceramic and carbon single fibres by using the laser-speckle method. *Composites Part A – Appl. Sci. Manufact.*, 34(11), 1029–1033 (2003).

46. M. G. Northolt, L. H. Veldhuizen, and H. Jansen. Tensile deformation of carbon fibers and the relationship with the modulus for shear between the basal planes. *Carbon*, 29(8), 1267–1279 (1991).

47. Y. Huang and R. J. Young. Effect of fibre microstructure upon the modulus of PAN- and pitch-based carbon fibres. *Carbon*, 33(2), 97–107 (1995).

48. D. Loidl, H. Peterlik, M. Müller, C. Riekel, and O. Paris. Elastic moduli of nanocrystallites in carbon fibers measured by in-situ X-ray microbeam diffraction. *Carbon*, 41(3), 563–570 (2003).

49. O. L. Blakslee, D. G. Proctor, E. J. Seldin, G. B. Spence, and T. J. Weng. Elastic constants of compression-annealed pyrolytic graphite. *J. Appl. Phys.*, 41(8), 3373–3382 (1970).

50. S. Akita, H. Nishijima, Y. Nakayama, F. Tokumasu, and K. Takeysu. Carbon nanotube tips for a scanning probe microscope: their fabrication and properties. *J. Phys. D (Appl. Phys.)*, 32(9), 1044–1048 (1999).

51. A. Krishnan, E. Dujardin, T. W. Ebbesen, P. N. Yianilos, and M. M. J. Treacy. Young's modulus of single-walled nanotubes. *Phys. Rev. B*, 58(20), 14013–14019 (1998).
52. C. Lee, X. Wei, J. W. Kysar, and J. Hone. Measurement of the elastic properties and intrinsic strength of monolayer graphene. *Science*, 321(5887), 385–388 (2008).
53. L. Fischer and W. Ruland. The influence of graphitization on the mechanical properties of carbon fibers. *Coll. Polym. Sci.*, 258(8), 917–922 (1980).
54. M. Shioya, E. Hayakawa, and A. Takaku. Non-Hookean stress–strain response and changes in crystallite orientation of carbon fibres. *J. Mater. Sci.*, 31(17), 4521–4532 (1996).
55. I. M. Robinson, M. Zakikhani, R. J. Day, R. J. Young, and C. Galiotis. Strain dependence of the Raman frequencies for different types of carbon-fibres. *J. Mater. Sci. Lett.*, 6(10), 1212–1214 (1987).
56. N. Melanitis and C. Galiotis. Compressional behaviour of carbon fibers Part 1. A Raman spectroscopic study. *J. Mater. Sci.*, 25(12), 5081–5090 (1990).
57. N. Melanitis, P. L. Tetlow, C. Galiotis, and S. B. Smith. Compressional behavior of carbon-fibers 2. Modulus softening. *J. Mater. Sci.*, 29(3), 786–799 (1994).
58. T. M. G. Mohiuddin, A. Lombardo, R. R. Nair, A. Bonetti, G. Savini, R. Jalil, N. Bonini, D. M. Basko, C. Galiotis, N. Marzari, et al. Uniaxial strain in graphene by Raman spectroscopy: G peak splitting, gruneisen parameters, and sample orientation. *Phys. Rev. B*, 79(20), 205433 (2009).
59. O. Frank, G. Tsoukleri, I. Riaz, K. Papagelis, J. Parthenios, A. C. Ferrari, A. K. Geim, K. S. Novoselov, and C. Galiotis. Development of a universal stress sensor for graphene and carbon fibres. *Nat. Commun.*, 2, 255 (2011).
60. J. M. Gere and S. P. Timoshenko. *Mechanics of Materials* (Chapman and Hall, London, 1987).
61. W. N. Reynolds and J. V. Sharp. Crystal shear limit to carbon fibre strength. *Carbon*, 12(2), 103–110 (1974).
62. Y. Tanabe, Y. E. Yasuda, A. R. Bunsell, Y. Favry, M. Inagaki, and M. Sakai. The strength of pitch-based carbon-fibre at high-temperature. *J. Mater. Sci.*, 26(6), 1601–1604 (1991).
63. C. Sauder, J. Lamon, and R. Pailler. The tensile behavior of carbon fibers at high temperatures up to 2400 degrees C. *Carbon*, 42(4), 715–725 (2004).
64. H. Rennhofer, D. Loidl, S. Puchegger, and H. Peterlik. Structural development of PAN-based carbon fibers studied by in situ X-ray scattering at high temperatures under load. *Carbon*, 48(4), 964–971 (2010).
65. H. Rennhofer, D. Loidl, J. Brandstetter, K. Kromp, R. Weiss, and H. Peterlik. Structural change of carbon-fibres at high temperatures under load, fatigue. *Fract. Eng. Mater.*, 29(2), 167–172 (2006).
66. P. B. Pollock. Tensile failure in 2-D carbon-carbon composites. *Carbon*, 28(5), 717–732 (1990).
67. W. Weibull. A statistical distribution function of wide applicability. *J. Appl. Mech.-Trans. ASME*, 18(3), 293–297 (1951).
68. Y. X. Wang and Q. Wang. Evaluation of carbonization tar in making high performance polyacrylonitrile-based carbon fibers. *J. Appl. Polymer Sci.*, 104(2), 1255–1259 (2007).
69. Y. X. Wang, C. G. Wang, J. W. Wu, and M. Jing. High-temperature DSC study of polyacrylonitrile precursors during their conversion to carbon fibers. *J. Appl. Polymer Sci.*, 106(3), 1787–1792 (2007).
70. M. Jing, C. G. Wang, B. Zhu, Y.-X. Wang, X. P. Gao, and W. N. Chen. Effects of preoxidation and carbonization technologies on tensile strength of PAN-based carbon fiber. *J. Appl. Polymer Sci.*, 108(2), 1259–1264 (2008).

CHAPTER 3

INFLUENCE OF INTERFACES AND INTERPHASES ON THE MECHANICAL BEHAVIOR OF FIBER-REINFORCED CERAMIC MATRIX COMPOSITES

JACQUES LAMON

Laboratoire de Mécanique et Technologie (LMT), CNRS (Centre National de la Recherche Scientifique), ENS Cachan (Ecole Normale Supérieure), UPMC (Univesité Pierre et Marie Curie), Cachan, France

3.1 INTRODUCTION

The fiber/matrix interfacial domain is a decisive constituent of fiber-reinforced ceramic matrices. Depending on characteristics of this domain, the composite will be either a brittle ceramic or a damage-tolerant composite. The key role of interfacial domain is to protect the reinforcing fibers against fracture. Several requirements, that may appear opposite, have to be met:

- The fibers have to be bonded to the matrix, in order to ensure material integrity, and to end up with a continuous medium.
- Fiber failures have to be prevented when matrix cracks. This is achieved by crack deflection.
- After deflection the loads have to be still transferred efficiently through the interfacial domain, to limit fiber overloading induced by matrix cracking.
- In aggressive environments, the fibers should not be exposed to aggression from species that diffuse through the matrix.

The fiber/matrix interfacial domain may consist of an interface or an interphase. An interface is a surface between two phases, or between the fiber and the matrix. An interphase is a film of one or several layers bonded to the fiber and the matrix. An interphase implies the presence of at least two interfaces: one with the matrix, and one with the fiber and more when the interphase consists of a multilayer.

There can be two main types of bonding at an interface: mechanical or chemical bonding. Mechanical bonding results from residual stresses. For those ceramic matrix composites manufactured at high temperature, when the matrix radially shrinks more than the fiber on cooling down from the processing temperature, this will lead to gripping of the fiber by the matrix. Radial gripping can be enhanced when the interface is rough.

Ceramic Matrix Composites: Materials, Modeling and Technology, First Edition. Edited by Narottam P. Bansal and Jacques Lamon.

An interfacial reaction zone with a certain thickness is involved in chemical bonding. The strength of a mechanical bond is lower than that of a chemical bond.

Interface properties depend on the fiber and the matrix since bonding results from chemical reactions during processing or thermal shrinkage during cooling. Therefore, the number of routes that permit to meet the above-mentioned requirements for the number of fibers and matrices that are compatible is limited. Furthermore, the prevalent assumption is that good composite behavior requires weak fiber/matrix interfaces that debond upon impingement of a matrix crack. Interphases allow these limitations to be overcome, and the interfacial characteristics to be tailored with respect to composite properties. A certain amount of effort has been directed toward either the optimization of interfacial domain in CVI SiC/SiC composites obtained via chemical vapor infiltration (CVI), and more recently in C/C composites, or the development of oxide interphases for oxide/oxide composites. The concept of *strengthened interfaces* has been established on CVI SiC/SiC composites with pyrocarbon (PyC) or multilayered (PyC/SiC) fiber coatings. Improved properties have been obtained using treated fibers that are more strongly bonded to coating. In this chapter, interfaces characteristics are examined with respect to crack deflection and composite mechanical behavior: that is, conditions of crack deflection, mechanisms controlling the behavior of debonded interfaces, influence of interface characteristics on composite mechanical behavior.

3.2 ROLE OF INTERFACIAL DOMAIN IN CMCs

Bonding fibers to matrix to keep composite material integrity is the first role assigned to interfaces. Any debonding affects a lot of properties including mechanical properties, thermal properties, and resistance to environment. Under stresses, fiber/matrix interfaces can initiate cracks as weakest links or arrest the cracks.

3.2.1 Crack Initiation at Interfaces

In certain loading conditions versus fiber orientation, fiber/matrix interfaces are the weakest elements. This is essentially encountered under the following loading conditions:

- Tensile stresses operating in mode I with respect to fiber axis: interface opening stresses. Such stresses are either thermally induced stresses due to thermal expansion mismatch and generated during cooling down from the processing temperature in carbon-reinforced SiC or C matrices [1] or stresses operating on transverse tows in multidirectionally reinforced composites (Figure 3.1) [2].
- Shear stresses generated by shear loads applied parallel to fiber plies [1]. This may be observed on brake discs.

The interfaces act as flaws. In the first case, they dictate failure of 2D composites loaded perpendicular to plies and the limit of linearity of 2D or 3D composites loaded to a certain angle with respect to fiber axis. Their resistance is measured by the opening failure stress perpendicular to fiber axis. It may be characterized by a single value or by a statistical distribution.

3.2.2 Crack Deflection at Interfaces

The interfaces associated to those fibers parallel to the load direction are not the weakest elements when the matrix stiffness is sufficiently large, so that the matrix carries large stresses and breaks first. The matrix cracks are perpendicular to fiber axis (transverse cracks: Figure 3.1). Crack arrest or deflection at interfaces is the mechanism that governs damage tolerance. Crack deflection requires interfaces that can be debonded.

When matrix stiffness is small compared to that of fibers, the matrix is subjected to low stresses whereas the fibers carry most of the load (carbon/carbon composites) [2]. The initiation of cracks in the matrix is more difficult or nonexistent, and thus, the crack deflection phenomenon is less critical. However, as it will be discussed in a subsequent section, some combinations such as carbon fibers and a carbon matrix are favorable to debonding.

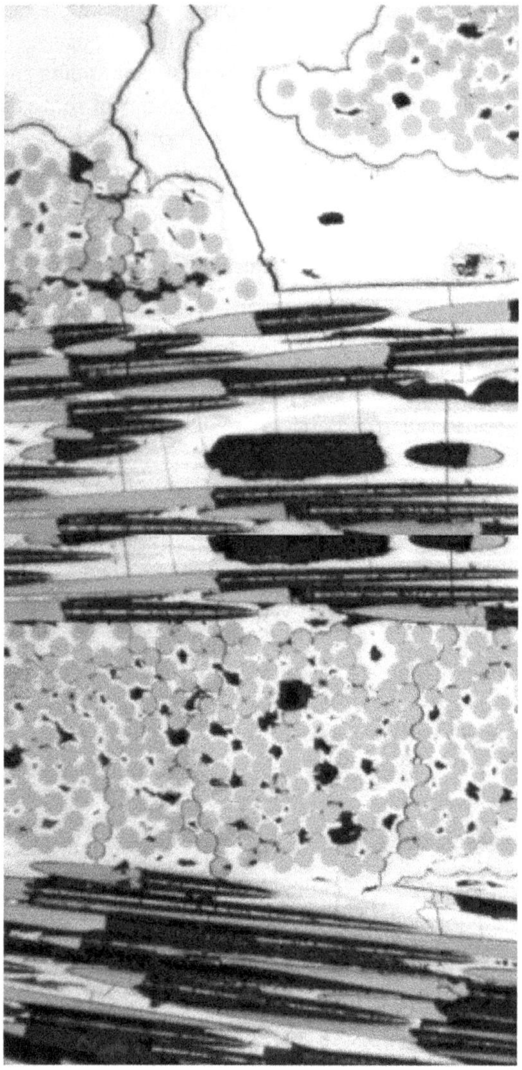

FIGURE 3.1
Micrograph of a 2D woven SiC/SiC composite showing debonding in transverse tows perpendicular to loading direction and transverse cracks in the matrix of longitudinal tows oriented in the loading direction (fiber diameter $\approx 15\ \mu m$). The former cracks appeared under low stresses, whereas the latter cracks appear under high stresses. Reproduced with permission from Wiley-VCH Verlag GmbH & Co. KGaA.

3.2.3 Approaches to Crack Deflection at Interfaces

Two main approaches to crack deflection have been proposed in the literature, depending on the mechanism considered: either the matrix crack is supposed to reach the interface and to be lying stationary [3–7], or a crack is nucleated at the interface ahead of the matrix crack (Cook and Gordon mechanism [8]) (Figure 3.2).

In the first approach, further crack extension is the result of a competition between two phenomena: propagation along fiber/matrix interface and propagation through the fiber. The strength or toughness of interface and fiber is compared to a ratio of stresses or strain energy release rates. In an energy-based approach, the interfacial crack growth is controlled by an interfacial fracture energy G_i. Debonding will occur only if G_i is sufficiently low compared to the fracture energy G_0 that is required for continuation in the plane of the original crack. The transition between debonding and penetration occurs at a critical value of G_i/G_0; He and Hutchinson have developed a fracture mechanics model for the dependence of the critical ratio versus the elastic constants of the constituent materials. For the fiber/matrix debonding to occur, arrest and deflect the matrix micro cracks, G_i should be low compared to G_0, such that $G_i/G_0 < 1/4$ [4].

These models may be appropriate for computational purposes, but they present several shortcomings from a physical point of view. First, they are not based on the true debonding mechanism. Indeed, the crack is assumed to propagate through the matrix, to stop at the interface, and then to extend further in one direction. An arbitrary short crack must be thus introduced in the interface or in the fiber to express the conditions of extension in the analysis based on an energy balance. Initial crack size is arbitrary which may affect the results. Finally, these models cannot be used to predict the behavior of a crack in the presence of an interphase that is observed with certain SiC/SiC composites as

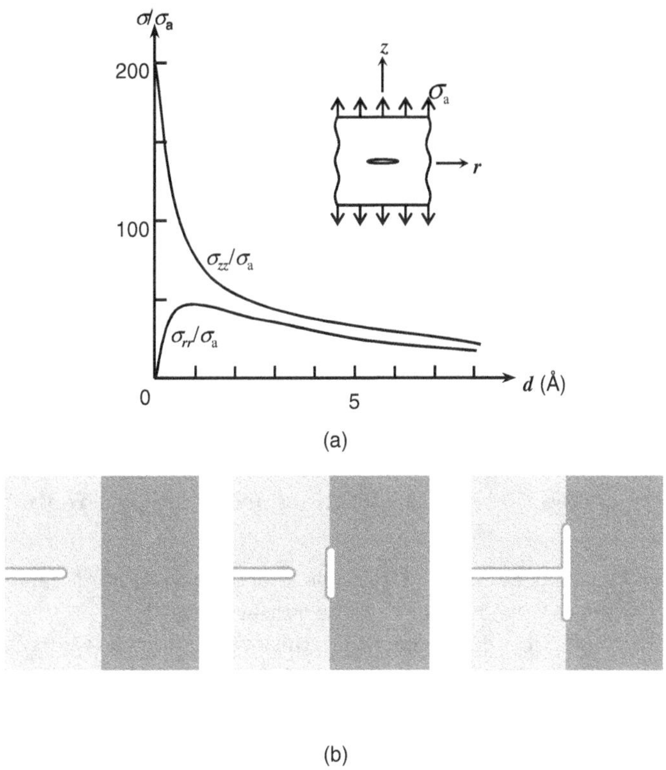

FIGURE 3.2

(a) Stress profile at crack tip in a cell subjected to uniaxial tension; (b) schematic diagram showing crack deviation by debonding at an interface and coalescence with the advancing crack. Reproduced with permission from Elsevier Limited, Composites Science and Technology, vol 67, iss. 10.

mentioned above. A previous work has shown that an energy-based approach is unable to account for the influence of interphase/fiber interface strength on crack deviation [9, 10]. It appeared that a criterion for initiation of a debond crack is required.

The crack deflection mechanism at a weak interface between two identical homogeneous materials was proposed by Cook and Gordon [8]. Cook and Gordon have computed the stress field induced by a semielliptical crack placed in a single homogeneous material subjected to uniaxial tension. They have shown that the stress state at crack tip was polyaxial, although a uniaxial tensile stress was applied. In particular, they evidenced that the stress component σ_{rr} parallel to crack plane (r-axis) reaches a maximum $\sigma_{rr}^{\max}$ in the crack plane at a distance l* from crack tip. Figure 3.2 shows schematically the stress distribution at the crack tip. If an interface is perpendicular to the main advancing crack, it can crack in front of the crack tip if the interfacial tensile strength is less than $\sigma_{rr}^{\max}$. From the comparison of maximum stress components $\sigma_{zz}^{\max}$ and $\sigma_{rr}^{\max}$, respectively, with fiber and interface strengths, Cook and Gordon estimated that an interfacial strength of one-fifth of fiber strength or less is required for debonding at interface in front of the crack tip.

An analytical expression of stresses at the tip of an elliptical crack in a homogeneous material has been established by Maugis [11]:

$$(a-b)^2\left(\frac{\sigma_{rr}}{\sigma_a}\right)_{z=0} = -a^2 + \frac{a\,|r|}{\sqrt{r^2-a^2+b^2}}\left[a - \frac{b^2\,(a-b)}{r^2-a^2+b^2}\right] \tag{3.1}$$

$$(a-b)^2\left(\frac{\sigma_{zz}}{\sigma_a}\right)_{z=0} = b^2 + \frac{a\,|r|}{\sqrt{r^2-a^2+b^2}}\left[a - 2b + \frac{b^2\,(a-b)}{r^2-a^2+b^2}\right] \tag{3.2}$$

where σ_a is the applied stress, a and b are respectively the major and the minor axes of ellipse. For the same conditions as above ($a = 1$ μm, $b = 0.01$ μm, $\rho = b^2/a = 0.1$ nm), Equation (3.1) indicates that σ_{rr} reaches a maximum at a distance in the order of ρ. Furthermore, the magnitude of $\sigma_{rr}^{\max}$ is identical to that one computed by Cook and Gordon.

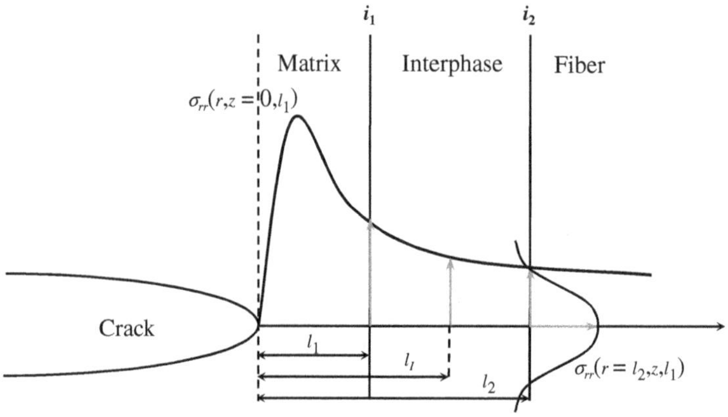

FIGURE 3.3
Stress state at crack tip (σ_{rr} component) in the presence of an interphase and two interfaces i_1 and i_2 [9]. Reproduced with permission from Elsevier Limited, Composites Science and Technology, vol 67, iss. 10.

Computations, as well as Equations (3.1) and (3.2), show that location of σ_{rr}^{max} depends on tip radius ρ. Distance to crack tip increases with ρ, whereas magnitude of σ_{rr}^{max} decreases slightly [11].

Deflection of matrix crack results from coalescence with the interfacial crack (Figure 3.2). This mechanism has been viewed on several systems [12–18]. Figure 3.3 shows the stress state at crack tip in the presence of an interphase in a fiber-reinforced ceramic [9]. Location of debonding is dictated by interface and interphase strengths (perpendicular to matrix crack direction) respective to the magnitude of σ_{rr}.

3.2.4 Deflection Criteria Based on the Cook and Gordon Mechanism

The above deflection criterion is based on maximum stress components. The criterion can be refined considering the stress components at location of interface, at distance l from crack tip. It can be extended to interface domains involving various interfaces and interphases. In the following, subscript 1 refers to the cracked material (matrix in CMC), material 2 to the uncracked material (fiber in CMC) [9].

Failure of interface occurs when the stress at distance l from crack tip exceeds opening strength (Equation 3.3) while failure of material 2 does not occur (at a distance $>l$ the stresses are smaller than material 2 strength). These conditions are described by the following expressions, for the position of the main crack defined by l:

$$\sigma_{rr}(r = l, l) = \sigma_i^c \tag{3.3}$$

$$\max_r \sigma_{zz}(r > l, l) < \sigma_2^c \tag{3.4}$$

The stress components at any point depend on l: $\sigma(r, z, l)$. Moreover, the highest stresses operate along the r-axis ($z = 0$). They are denoted as $\sigma(r, z = 0, l) = \sigma(r, l)$. $\sigma_i{}^c$ is interface opening strength and $\sigma_2{}^c$ is material 2 strength. Condition (3.4) corresponds to a simple deterministic failure criterion. Statistical aspects of failure can be introduced.

A debonding criterion can be summarized by the following condition derived from (3.3) and (3.4):

$$\frac{\sigma_i^c}{\sigma_2^c} \le \frac{\sigma_{rr}(r = l, l)}{\max_r \sigma_{zz}(r > l, l)} \tag{3.5}$$

Use of (3.5) requires computations of the stress state for various positions of crack, that is, various l.

Equation (3.5) can be applied in the presence of various interfaces and interphases. For instance, let us consider a composite with an interphase (Figure 3.3). A crack is located in the matrix (material 1) at a distance l_1 from the first interface (interphase/matrix interface). Debonding within the interphase, at a distance l_1 from the crack tip, corresponds to the following conditions:

no debonding at interface 1 (interphase/matrix interface):

$$\frac{\sigma_{i_1}^c}{\sigma_1^c} > \frac{\sigma_{rr}(r = l_1, l_1)}{\max_r \sigma_{zz}(r > l_1, l_1)} \tag{3.6}$$

debonding within the interphase:

$$\frac{\sigma_{i_1}^c}{\sigma_1^c} \leq \frac{\sigma_{rr}(r = l_1, l_1)}{\max\limits_r \sigma_{zz}(r > l_1, l_1)} \tag{3.7}$$

no debonding at interface 2 (fiber/interphase interface):

$$\frac{\sigma_{i_2}^c}{\sigma_2^c} > \frac{\sigma_{rr}(r = l_2, l_1)}{\max\limits_r \sigma_{zz}(r > l_2, l_1)} \tag{3.8}$$

Subscripts i_1 and i_2 designate, respectively, the interphase/matrix interface and the fiber/interphase interface. Subscript i_I refers to debonding location within the interphase.

When the opening strength of the interface is not available, the critical ligament length l_0 cannot be determined. An alternative condition for debonding can be established, by considering the peak stress σ_{rr}^{max}, that is, the highest value that can be reached by $\sigma_{rr}(r, z, l)$:

$$\sigma_{rr}^{\max} = \max_l \sigma_{rr}(r = l, l) = \sigma_{rr}(r = l^*, l^*) \tag{3.9}$$

Thus from (3.3) and (3.4) it comes as follows:

$$\sigma_{rr}^{\max} = \sigma_i^c \tag{3.10}$$

$$\sigma_{zz}^{\max} < \sigma_2^c \tag{3.11}$$

$$\frac{\sigma_i^c}{\sigma_2^c} \leq \frac{\sigma_{rr}^{\max}}{\sigma_{zz}^{\max}} \tag{3.12}$$

where

$$\sigma_{zz}^{\max} = \max_r \sigma_{zz}(r > l^*, l^*) \geq \max_r \sigma_{zz}(r > l_0, l_0) \tag{3.13}$$

Thus, when condition (3.10) is fulfilled, it is obvious that condition (3.3) is also satisfied and vice-versa. Note that according to Equation (3.9), for a cell of similar materials, the peak stress σ_{rr}^{max} is observed for $r = \rho$, $z = 0$, $l = \rho$ [7, 9, 11]:

$$\sigma_{rr}^{\max} = \sigma_{rr}(r = \rho, \rho) \tag{3.14}$$

Determination of $\sigma_{rr}^{max} = \sigma_{rr}(r = \rho, \rho)$ and $\sigma_{zz}^{max} = \max\limits_r \sigma_{zz}(r > \rho, \rho)$ does not require a lot of computations. Note that Equation (3.12) provides a necessary but not sufficient condition. When (3.12) is not satisfied, one of the three following situations can arise:

- Conditions (3.10) and (3.11) are not satisfied: no crack deflection and failure of material 2 (brittle fracture).
- Condition (3.10) is satisfied whereas condition (3.11) is not: debonding and failure of material 2 (brittle failure).
- Condition (3.10) is not satisfied whereas condition (3.11) is: no debonding nor failure of material 2 (crack arrest).

3.2.5 Influence of Material Elastic Properties on Crack Deflection

Relationships between the deflection criterion and Young's moduli of bonded materials were established for the stress state in a cell with an elliptical crack (tip radius ρ) located at a short distance $l^* = \rho$ from the interface (Figure 3.2b), such that:

$$\sigma_{rr}^{\max} = \max_r \sigma_{rr}(r, z = 0, l) = \sigma_{rr}(r, z = 0, l^*) \tag{3.15}$$

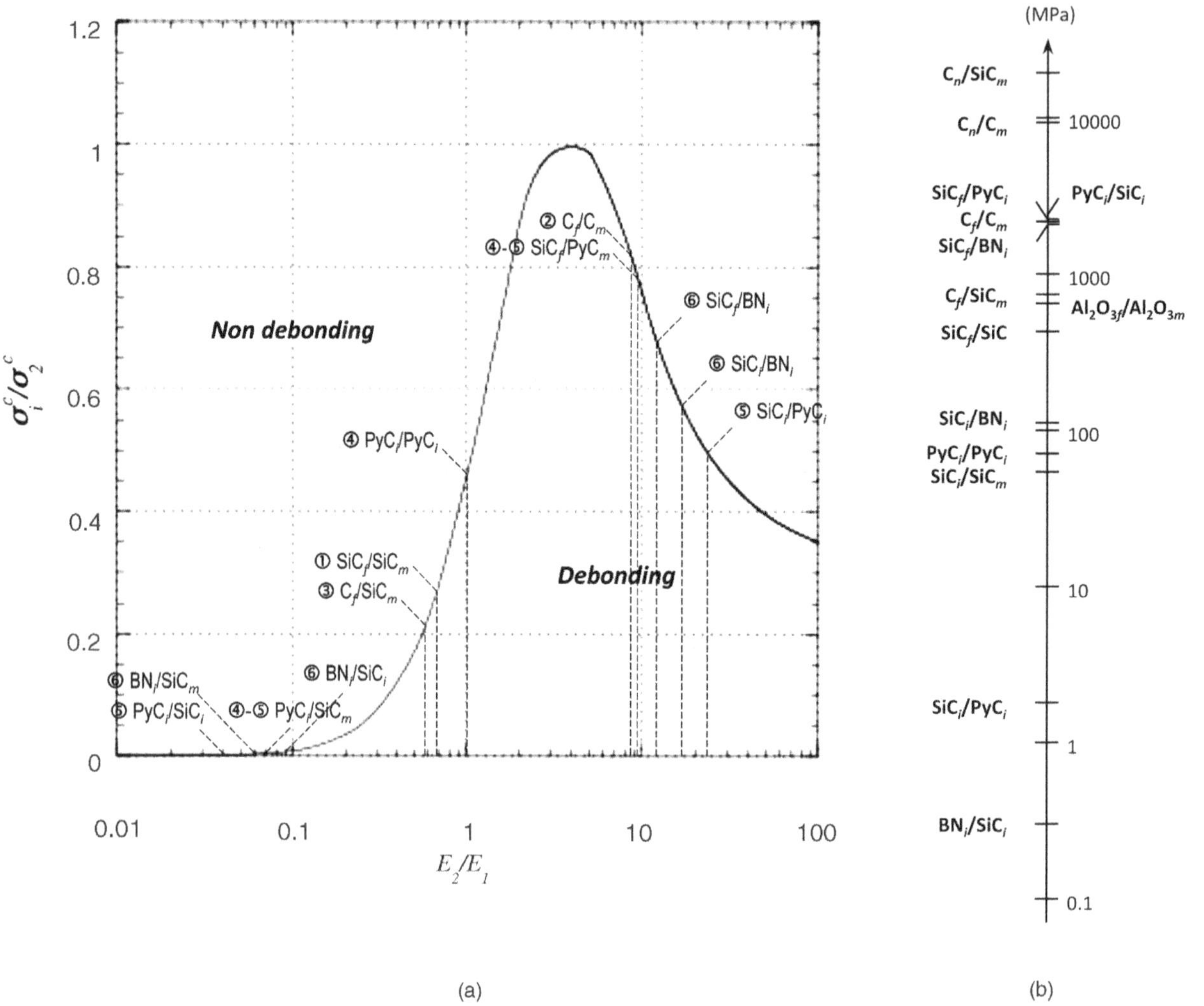

FIGURE 3.4

Master curve (a) and scale of debonding potentials (b) for various couples of materials: the cracked material (material 1) is cited second. Subscripts: m refers to matrix, f to fiber, I to interphase, n to nanotube. Reproduced with permission from Elsevier Limited, Composites Science and Technology, vol 67, iss. 10.

This is a critical configuration [7, 11, 19]. A uniaxial tensile stress σ_a was applied uniformly and parallel to fiber axis to both constituents. Computations were run for Young's modulus ratios E_2/E_1 ranging from 0.01 to 100, Poisson's ratios $\nu_1 = \nu_2 = 0.2$, and various crack tip radii. Crack tip radius had no significant influence on results [9].

The master curve $\frac{\sigma_{rr}^{\max}}{\sigma_{zz}^{\max}} = f(\frac{E_2}{E_1})$ plotted in Figure 3.4 delineates the domain where condition (3.12) is satisfied. The domain under the curve thus contains the appropriate values of constituent properties $(E_2/E_1, \sigma_i^c/\sigma_2^c)$ for which debonding occurs. Prediction of behavior of material 2 requires condition (3.11) or (3.12) to be met, as mentioned above. The master curve exhibits a few interesting features.

1. First, the maximum is smaller than 1 ($\sigma_i^c/\sigma_2^c = 0.99$). This result is in agreement with logical expectation: debonding requires that the interface opening strength does not exceed material 2 strength. The corresponding ratio of elastic moduli is $E_2/E_1 = 3.86$. Debonding capability is maximum for this typical material combination.
2. Second, when the elastic moduli ratio decreases to zero, debonding becomes quite impossible. This is in agreement with logical expectation. Material 2 can deform easily as a result of very small elastic modulus. As a consequence, the interface is subjected to a low stress.

TABLE 3.1 Mechanical Properties of Constituents

Material	Young's modulus (GPa)	Strength σ_2^c (MPa)	References
BN interphase	24	115	[20]
C fiber	349 ± 20	2845 ± 558	[21]
C matrix	40		[22]
PyC interphase ⊥	12		[23–25]
PyC interphase //	30	151	[23–25]
SiC fiber (Hi-Nicalon)	280	1500	[26, 27]
SiC interphase	280	124–266	[28]
SiC matrix	415		[29, 30]

3. Third, the master curve is not symmetrical. Debonding is favored when the uncracked constituent (material 2) is stiffer than material 1.
4. Finally, it is interesting to examine the typical case of a homogeneous isotropic material, without interface: $E_2/E_1 = 1$, $\sigma_i^c/\sigma_2^c = 1$. It can be noticed that the (1,1) point is located in the no debonding domain (Figure 3.4). This result is in agreement with common sense. It is well known that a homogeneous and isotropic material under tension cannot fail parallel to the load direction.

When σ_i^c is available, interface behavior can be predicted, since E_1, E_2, and σ_2^c are available for most materials. When σ_i^c is unknown, the master curve along with images of experimental interface behavior obtained by fractography can be used to estimate the debonding potential defined as follows:

$$\sigma_i^c \leq \sigma_2^c \frac{\sigma_{rr}^{\max}}{\sigma_{zz}^{\max}} = \sigma_{ic}^{+} \tag{3.16}$$

When debonding occurs, $\sigma_i^{c\dagger}$ is an upper bound of the current interface strength. When debonding does not occur, this is a lower bound. Thus $\sigma_i^{c\dagger}$ reflects the debonding potential of interface: the chances of interface debonding increase with $\sigma_i^{c\dagger}$.

Debonding potentials estimated for carbon or ceramic fiber/matrix combinations are summarized in Table 3.2, for the properties given in Table 3.1.

The failure of SiC_f/SiC_m composites is essentially brittle in the absence of interphase. Therefore, $\sigma_i^c > \sigma_i^{c\dagger} = 420$ MPa (Table 3.2). Debonding requires either smaller interface strength $\sigma_i^c < 420$ MPa, or larger potential $\sigma_i^{c\dagger}$, or both. Introduction of a compliant interphase, such as PyC, causes an increase in $\sigma_i^{c\dagger}$ (Table 3.2). Table 3.2 shows that, in SiC_f/PyC/SiC_m composites, $\sigma_i^{c\dagger}$ is much larger at the SiC_m/PyC interface ($\sigma_i^{c\dagger} \approx 2200$ MPa), which appears to be the most probable location for debonding. The PyC/SiC_m interface is the less probable location, since σ_i^c should be as small as 0.3 MPa. However, data show that the interior of PyC interphase is a possible place for debonding: σ_i^c is as small as 69.5 MPa as indicated above. Fractography and microscopy evidenced crack deviation within the PyC interphase in SiC_f/PyC/SiC_m composites when the interphase/fiber bond has been modified via appropriate fiber treatments [31].

Note that $\sigma_i^{c\dagger}$ estimated for the PyC interphase, in a direction perpendicular to graphitic planes, is reasonable (Table 3.2). It is smaller than the strength in a direction parallel to graphitic planes (Table 3.2). Orientation of graphitic planes perpendicular to crack plane is the most favorable deviation (Figure 3.5). Application of the debonding condition when graphitic planes are parallel to crack plane shows that, as expected, debonding within the interphase is impossible. The intermediate orientation (inclined to crack plane) can also be examined using stress components parallel to graphitic planes [9].

Multilayered interphases such as $(PyC/SiC)_n$ are comparable to PyC. Table 3.2 shows that similar $\sigma_i^{c\dagger}$ were estimated. The most favorable debonding location is still the fiber/PyC interface (high $\sigma_i^{c\dagger}$). Note that the PyC/SiC_i interface $\sigma_i^{c\dagger}$ is of the same order of magnitude that the $PyC_\perp$ one, when the crack is located within the PyC layer and the graphitic

TABLE 3.2 Interface Debond Stresses $\sigma_i^{c\dagger}$ Determined for Various Materials Combinations and Results of Fractography Analysis of Current Materials

System	Uncracked material (2) versus cracked material (1)	Young's modulus ratio E_2/E_1	$\sigma_i^{c\dagger}$ (MPa)	Experimental data
SiC_f/SiC_m	SiC_f/SiC_m	0.68	420	Brittle failure of composite observed [32]
C_f/C_m	C_f/C_m	8.73	2142	No observation available under stress-controlled conditions
C_f/SiC_m	C_f/SiC_m	0.56	730	Debonding at fiber/matrix interface
$SiC_f/PyC/SiC_m$	$PyC_\perp/SiC_m$	0.07	0.3	Debonding at fiber/PyC interface (as-received fibers) or within the interface (with treated fibers) [31]
	$PyC_\perp/PyC_\perp$	1	69.5	
	$SiC_f/PyC_\perp$	9.33	2200	
$SiC_f/(PyC/SiC)_n/SiC_m$	$PyC_\perp/SiC_m$	0.07	0.3	Debonding within interphase or along SiC_i/PyC interface, depending of fiber surface [26, 27]
	$SiC_i/PyC_\perp$	23.4	62–133	
	$PyC_\perp/SiC_i$	0.04	1	
	$SiC_f/PyC_\perp$	9.33	2200	
$SiC_f/(BN/SiC)_n/SiC_m$	BN/SiC_m	0.06	1.15	Debonding within interface [26]
	SiC_i/BN	17.3	69.4–149	
	BN/SiC_i	0.09	2	
	SiC_f/BN	11.67	2070	

Indexes f, i, and m refer respectively to fiber, interphase, and matrix.

planes are oriented perpendicular to crack plane. This implies that the fiber/PyC bond can also influence the location of debonding.

$SiC_f/BN/SiC_m$ was not examined, because the properties of $BN_\perp$ in an anisotropic BN interphase were not available ($BN_\perp$ designates the direction perpendicular to crack plane, by analogy with the orientation of crystallites in PyC). Analysis of $SiC_f/(BN/SiC)_n/SiC_m$ containing a multilayered (BN/SiC) interphase leads to results comparable to those obtained with PyC. Deflection is highly improbable at the BN_i/SiC_m interface ($\sigma_i^{c\dagger}$ is rather small), and it is much easier at the SiC_f/BN interface ($\sigma_i^{c\dagger}$ is very large; Table 3.2). As previously, it can occur at the SiC_i/BN_i interface when the crack is located in BN ($\sigma_i^{c\dagger}$ is moderate). It is quite impossible at BN_i/SiC_i interface when the crack is located in the SiC_i layer ($\sigma_i^{c\dagger}$ is rather small). Figure 3.6 illustrates the above analysis for $SiC_f/(BN/SiC)_n/SiC_m$ composites.

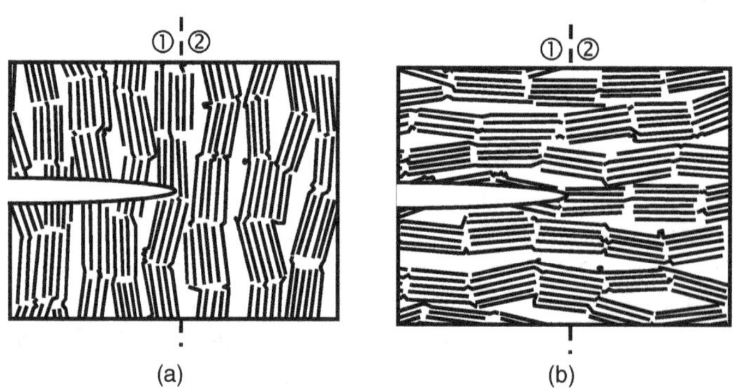

FIGURE 3.5

Schematic diagrams showing the orientation of graphitic planes with respect to crack plane in a PyC interphase: (a) perpendicular, (b) parallel to crack plane. Reproduced with permission from Elsevier Limited, Composites Science and Technology, vol 67, iss. 10.

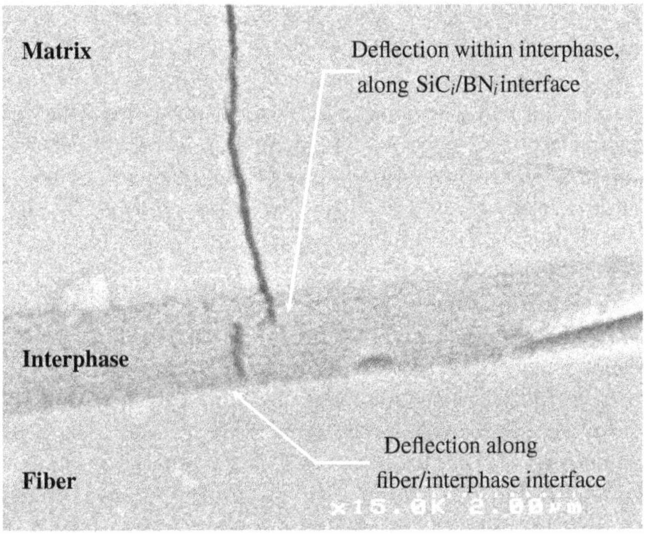

FIGURE 3.6
Micrograph showing crack path through a $(SiC/BN)_n$ multilayered interphase in a SiC/SiC composite. Reproduced with permission from Elsevier Limited, Composites Science and Technology, vol 67, iss. 10.

C_f/C_m is also an appropriate combination with respect to debonding, since $\sigma_i^{c\dagger}$ is very large (Table 3.2). Under strain-controlled boundary conditions, stresses in the carbon matrix would be much smaller, and thus cracking of matrix would be more difficult.

3.3 INFLUENCE OF DEFLECTED CRACKS

The discontinuities created by matrix and interface cracks induce disturbances in the stress state: overstressing of fibers and a stress lag in the matrix and fibers. A quantitative description of these phenomena is necessary with a view to a better control of composite mechanical behavior.

Opening of a transverse matrix crack under further loading may cause fiber/matrix sliding along the debonded interface and extension of the interface crack (Figure 3.7). The significance of sliding depends on several factors, first, the presence of residual contact between the interface surfaces, and, then, debond length, surface roughness, and

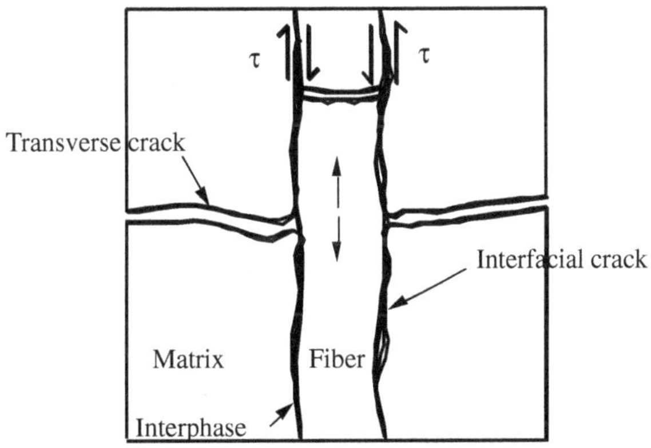

FIGURE 3.7
Local phenomena associated to matrix cracking and deviation at interfaces. Reproduced with permission from Wiley-VCH Verlag GmbH & Co. KGaA.

compressive residual stresses in the matrix acting across the interface. Surface roughness and compressive residual stresses impede relative slip of the matrix and fiber [33–35].

Surface roughness was found to be critical for Tyranno SA3 and Sylramic polycrystalline SiC fibers [36, 37]. Like in the Nicalon, Hi-Nicalon, and Hi-Nicalon S-reinforced SiC matrix minicomposites, crack deflection was shown to occur at the SA3 fiber/pyrocarbon interface; but unlike the former minicomposites (strain-to-failure $\approx$ 0.8%), the SA3-reinforced ones exhibited very small strain-to-failure (<0.2%) although bonding of SA3 was not significantly stronger. This behavior was attributed to the roughness of debonding crack surface formed by the SA3 fiber surface that exhibits a very significant surface roughness; amplitude $h = 60$ nm measured using AFM was found to be four times larger when compared to Hi-Nicalon S [36].

The strain-to-failure of Sylramic/BN interphase/SiC matrix minicomposites was also found to be quite small (<0.27%) when compared to Nicalon or Hi-Nicalon/BN interphase/SiC matrix minicomposites (0.8–0.9%) [37]. This effect was due to the rough surface of Sylramic fiber; peak amplitude $h = 90$ nm has been measured by AFM [37].

By contrast, the topography of the sliding surface may differ from the surface of fiber, and roughness of the corresponding debonding crack may contribute to improved properties of composite. This interesting effect has been observed on Nicalon/C/SiC composites that consist of pyrolitic carbon-coated SiC fibers in a matrix of chemical vapor-deposited silicon carbide [26, 27, 31, 38, 39]. TEM analysis of tensile specimens reinforced with treated fibers has indicated that the debonding fracture consists of fine, diffuse multiple cracking of the carbon layer itself, as opposed to the single crack at or very near the coating/fiber interface that is common in the untreated-fiber material. The topography of the sliding surface was also influenced by the presence of a pyrocarbon interphase in SA3/C/SiC composites; increasing thickness of fiber coating reduced the negative influence of fiber surface roughness [36].

Debonding crack topography complicates analysis of interfacial stress. Difficulties have motivated several approximations that permit characterization of interfaces and determination of relationships between interface characteristics and composite properties. Thus, the stress across the interface crack is assumed to be a constant value τ. This approximation has been found to be valid a posteriori. Nonconstant interfacial stress does not appear necessary to predict the mechanical behavior of ceramic matrix composites. The interfacial shear stress τ is a useful characteristic of interface efficiency in the post-deflection behavior of composites. It varies inversely with interface crack size (l_d), according to the following classical equation:

$$l_d = \frac{\sigma a R_f}{2V_f(1+a)\tau} \quad (3.17)$$

where σ is the applied stress, $a = \frac{E_m V_m}{E_f V_f}$, E_m is the matrix Young's modulus, E_f is the fiber one, V_m and V_f are the volume fractions of fiber and matrix, and R_f is the fiber radius.

It is considered that interface weakness is commensurate with debond length, and that τ is a convenient characteristic of interfacial domain weakness.

Depending on the magnitude of fiber/matrix interaction through the interface crack, stress on fiber increases more or less abruptly (Figure 3.8). Load sharing is maximum value outside the debonded portion. Pull out of broken fibers causes further sliding along the debonded interface.

The magnitude of these phenomena depends strongly on the size of interface crack. In the presence of long interface cracks, the portion of fibers that carries high stresses is larger, whereas the domain of stressed matrix is smaller and the contribution of matrix to load sharing is reduced when compared to short interface cracks. This has several implications on the mechanical properties and the shape of the tensile stress strain curves. In the presence of long debond cracks, composite is weakened as a result of fiber overloading and size effects on strength, whereas energy dissipation associated to sliding is enhanced. In the presence of short debond cracks, the composite can carry higher loads because of better contribution of matrix to load sharing, and matrix crack density is enhanced as a result of a larger stressed volume of matrix.

Energy-absorbing phenomena like sliding and multiple matrix cracking lead to an enhanced fracture toughness, whereas contribution of matrix to load sharing leads to an enhanced load carrying capacity and ultimate strength. Long interface cracks are detrimental to ultimate strength, whereas short interface cracks are more favorable because they allow both a high toughness and a high ultimate strength to be obtained.

Figure 3.9 illustrates the influence of debonding (characterized by τ) on the tensile stress–strain behavior of SiC/SiC minicomposites [40]. Interface crack length is characterized through τ. Note that the carried force increases with τ, that is, when the debond length decreases.

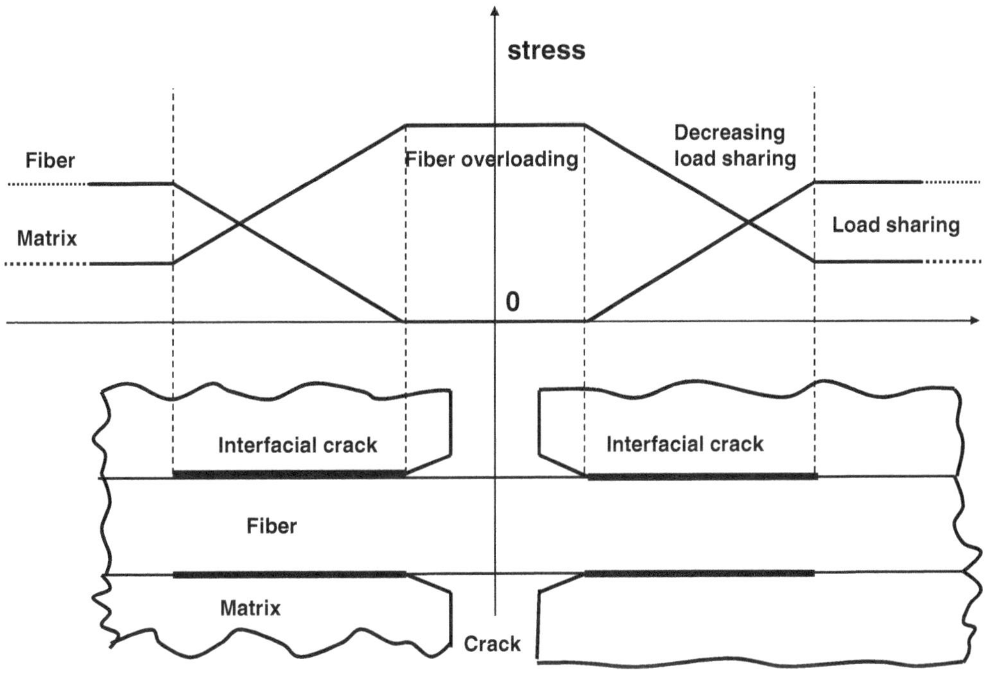

FIGURE 3.8

Disturbances in the stress state caused by matrix crack deviation at an interface between fiber and matrix. Reproduced with permission from Wiley-VCH Verlag GmbH & Co. KGaA.

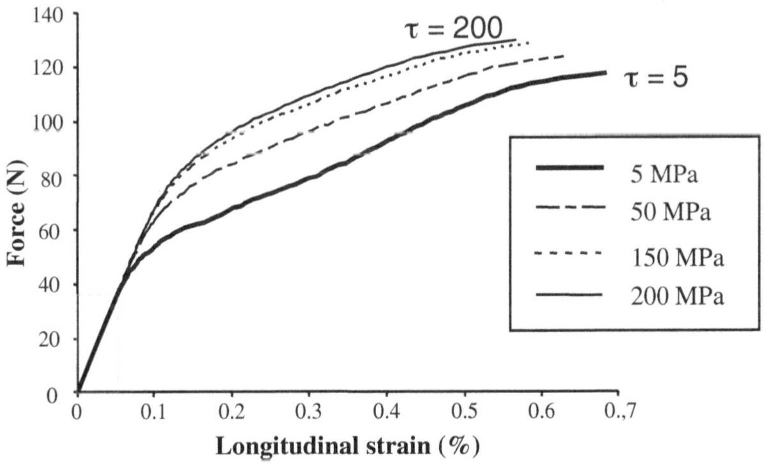

FIGURE 3.9

Influence of interfacial shear stress on the stress–strain behavior of 1D SiC/SiC composites: predictions [40]. Reproduced with permission from Wiley-VCH Verlag GmbH & Co. KGaA.

3.4 STRENGTHENED INTERFACES AND INTERPHASES

The fiber/matrix interfacial domain is a critical part of composites, not only for crack arrest and damage tolerance but also for further contribution of matrix to load sharing in the presence of debonding cracks. Most authors focus on crack deflection from a qualitative point of view in the sense that the condition for crack deflection is interface weakness. Magnitude of weakness with respect to mechanical properties is not specified. The prevalent assumption is that good composite behavior requires that the debond crack toughness and interfacial sliding friction should both be quite small. A significant effort has been directed toward the control of interface properties in SiC/SiC composites made via CVI technique in order to improve the very favorable properties of these composites. The control of the interfacial domain involves appropriate interphase and fiber/interphase interface. The concept of composite with a tailored interfacial domain has been established on CVI SiC/SiC composites, using anisotropic interphases made of pyrocarbon (PyC) or a SiC/PyC multilayer, and fibers that had been previously treated with a view to increase the fiber/interphase bond [26,27, 31, 38, 39]. The proprietary treatment is reported to clean the fiber surface and remove any native oxide layer. The composites made with treated fibers demonstrated 30% higher strength at the same ≈1% strain-to-failure value, much finer matrix crack spacing, and different stress–strain behavior.

Less interesting results have been obtained with BN [41]. Table 3.3 gives various values of interfacial shear stresses measured on CVI SiC/SiC composites. It can be noticed that high interfacial shear stresses in the 100–300 MPa range have been measured for the strengthened interfacial domains.

TABLE 3.3 Available Interfacial Shear Stresses (MPa) Measured Using Various Methods

	τ (MPa)	Specimens	Methods	References
C/PyC/SiC	13–30	Minicomposites	Equation (3.20)	[42]
Nicalon/BN/SiC	40–100	Microcomposites	Push-in/push-out	[41]
	40–140	2D composites	Push-out	
Nicalon/BN/SiC	15–35	Minicomposites	Hysteresis loops	[43]
Hi-Nicalon/BN/SiC	5–25	–	–	
Sylramic/BN/SiC	70	–	–	
Nicalon/PyC/SiC	12	2D composites	Equation (3.20)	[44]
Untreated fibers	8	–	Equation (3.21)	
	0.7–4	–	Hysteresis loops	
	14–16	–	Push-out	
	3	–	Hysteresis loops	
	4–20	–	Model of tensile behavior	
	21–115	Minicomposites		
	40–80	–	Hysteresis loops	
Hi-Nicalon/PyC/SiC	50–12	–	Model	
Untreated fibers	34–81	–	Hysteresis loops	
	29–71	–	Equation (3.20)	
	100	–	Equation (3.21)	
			Model	
Nicalon/$(PyC/SiC)_4$/SiC	9	2D composite	Hysteresis loops	
Untreated fibers	12–28	–	Push-out	
Nicalon/$(PyC/SiC)_{10}$/SiC	35–100	Minicomposite	Hysteresis loops	
Untreated fibers	25–89	–	Equation (3.20)	
	22–78	–	Equation (3.21)	
	100	–	Model	
Nicalon/PyC/SiC	203	2D composite	Equation (3.21)	
Treated fibers	140	–	Equation (3.20)	
	190–370	–	Hysteresis loops	
	100–270	–	Push-out	
Nicalon/$(PyC/SiC)_2$/SiC	150	2D composite	Hysteresis loops	
Treated fibers	133	–	Push-out	
Nicalon/$(PyC/SiC)_4$/SiC	90	2D composite	Hysteresis loops	
Treated fibers	90	–	Push-out	
Nicalon/PyC/SiC	46–127	Minicomposites	Hysteresis loops	[27]
Treated fibers	54–160	–	Equation (3.20)	
	50–140	–	Equation (3.21)	
	150	–	Model	
Nicalon/$(PyC/SiC)_{10}$/SiC	113–216	Minicomposites	Hysteresis loops	[27]
Treated fibers	61–117	–	Equation (3.20)	
	54–103	–	Equation (3.21)	
	350	–	Model	

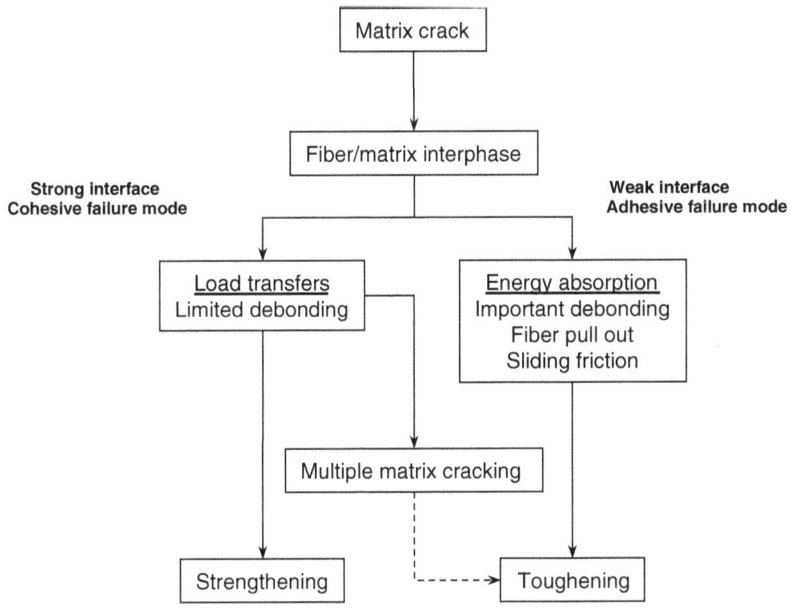

FIGURE 3.10

Schematic diagram which summarizes the features associated to interfacial domain strength: weak fiber/matrix interface versus strong fiber coating bond. Reproduced with permission from Wiley-VCH Verlag GmbH & Co. KGaA.

The main features that have been observed on SiC/SiC composites may be summarized as follows. In the presence of weak fiber/coating bonds, the matrix cracks generate a single long debond at the surface of fibers (adhesive failure type, Figures 3.10 and 3.11); the associated interface shear stresses are low (Table 3.3); multiple matrix cracking is limited by the presence of long debonds; the matrix cracks are widely opened (Figure 3.12); the crack spacing distance at saturation as well as the pull-out length tends to be rather long (>100 μm). The corresponding tensile stress–strain curve exhibits a narrow curved domain limited by a stress at matrix saturation which is distinct of ultimate strength (Figure 3.13).

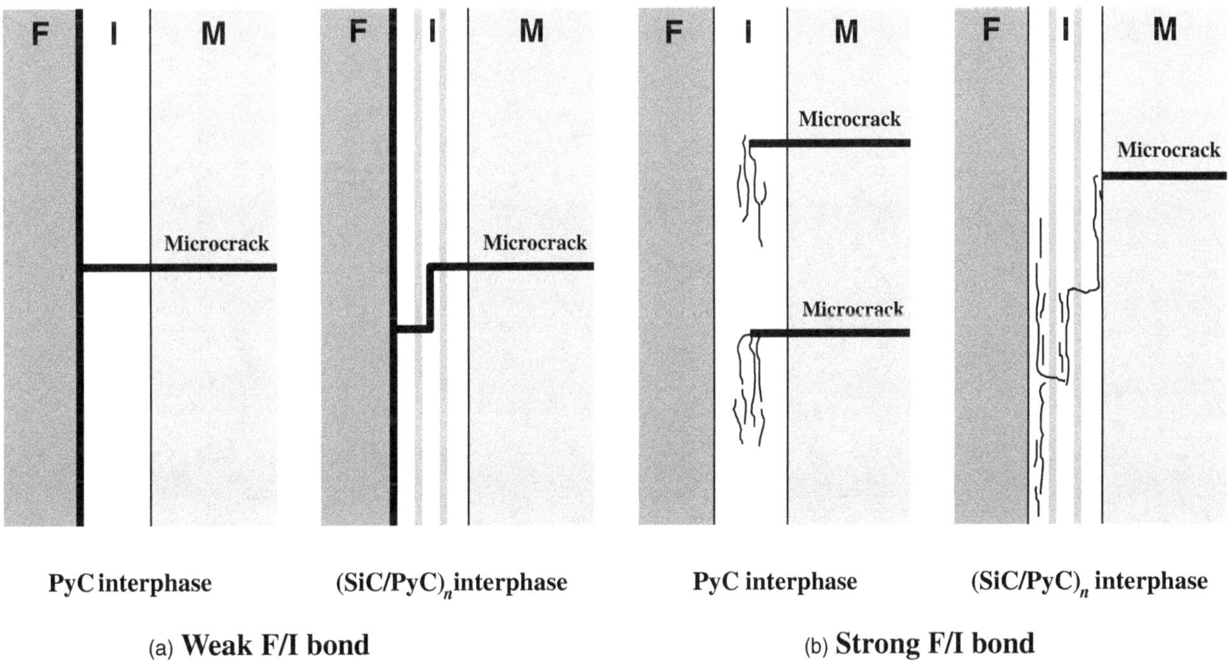

FIGURE 3.11

Schematic diagram showing matrix interfacial crack patterns when the fiber coating/interface is weak (a) or strong (b). Reproduced with permission from Wiley-VCH Verlag GmbH & Co. KGaA.

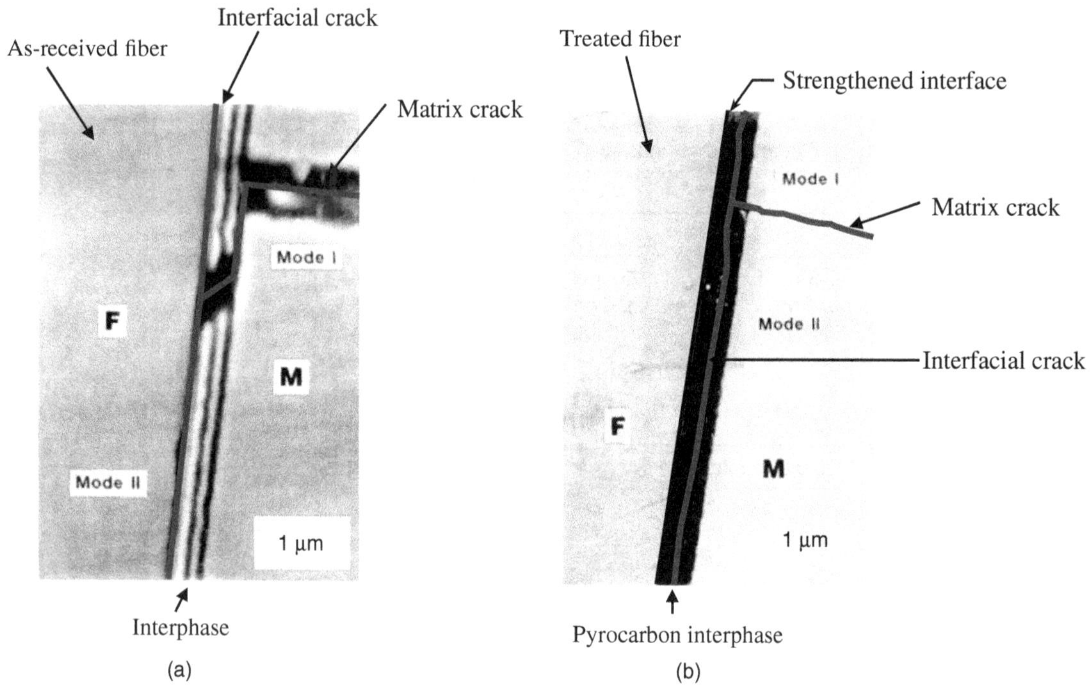

FIGURE 3.12

Micrographs showing crack deviation in a SiC/SiC composite with a weak fiber/interphase interface (a) or in a SiC/SiC composite with a strengthened fiber/interphase interface (b) [31]. Reproduced with permission from Wiley-VCH Verlag GmbH & Co. KGaA.

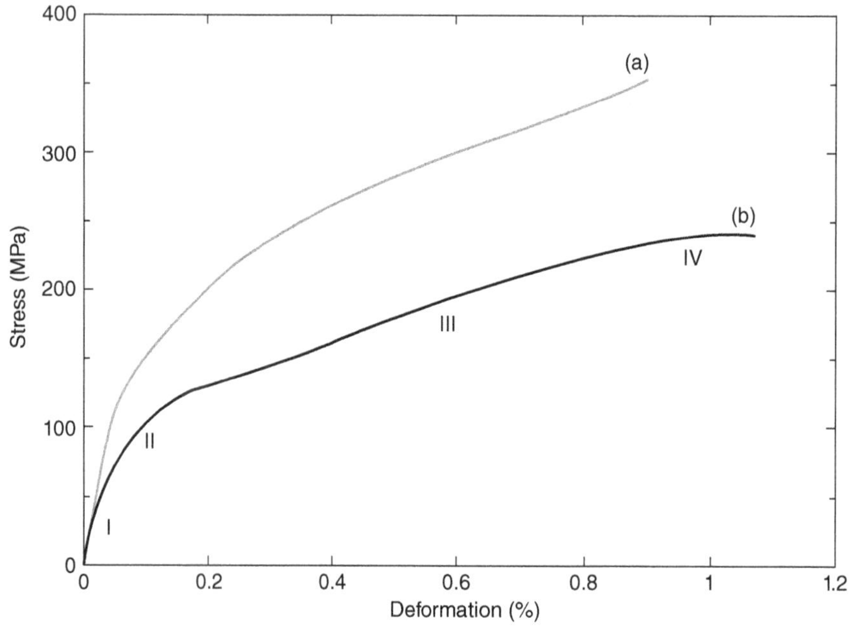

FIGURE 3.13

Typical tensile stress–strain curves determined on 2D woven SiC/SiC composites with a PyC interphase and with (a) a tailored interfacial domain (treated fibers) or (b) a weak fiber/coating interface (as-received fibers). I: elastic deformation, II: matrix multiple cracking, III: fiber deformation, IV: fiber failures. Reproduced with permission from Wiley-VCH Verlag GmbH & Co. KGaA.

In the presence of stronger fiber/coating bonds, the matrix cracks are deflected within the coating (cohesive failure type, Figures 3.10 and 3.11) into short and branched multiple cracks. Short debonds as well as associated improved load transfers allow further cracking of the matrix leading to a higher density of matrix cracks; the crack spacing distance may be as small as 10–20 μm. The matrix cracks are slightly opened (Figure 3.13). Sliding friction within the coating as well as multiple cracking of the matrix increase energy absorption, leading to toughening. Limited debonding and improved load transfers reduce the load carried by the fibers, leading to strengthening (Figure 3.13). The associated tensile stress–strain curve exhibits a wide curved domain, and the stress at matrix cracking saturation is close to ultimate failure (Figure 3.13).

Fracture toughness values (strain energy release rate) increasing from 3 to 8 kJ/m^2 have been determined using compact tension specimens of CVI SiC/SiC composites respectively with weak interfaces or high-strength-tailored interfacial domains [31]. A process zone of matrix microcracks forms at notch tip or at the tip of a preexisting main macroscopic crack. Microcrack density is commensurate with τ. Experiments have shown that a tailored interfacial domain is also beneficial to lifetime and creep resistance [27, 45, 46]. High-strength-tailored interphases are based on PyC. Examples of PyC-based interphases are shown in Figures 3.12 and 3.14. Interfacial shear stresses are reported in Table 3.3. Although PyC-based interphases are subject to degradation and elimination by oxidation at temperatures above 500°C, they are a viable concept. It has been shown [27] that SiC/SiC minicomposites with strong tailored interphases made of a single layer of PyC or a (PyC/SiC) multilayer exhibited lifetimes which compared favorably with those of minicomposites with BN interphases produced via CVD/CVI. A self-healing matrix is able to protect interphases against oxidation [27].

(a)

(b)

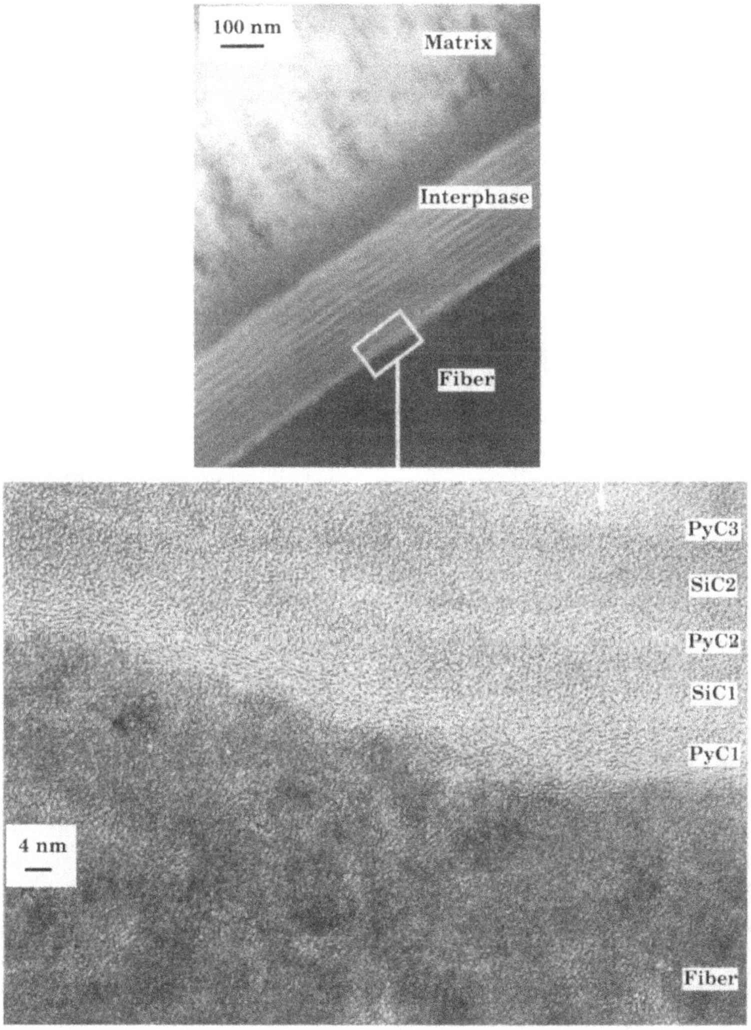

FIGURE 3.14
TEM micrographs of a multilayered interphase: (a) interfacial zone, (b) HRTEM micrograph of the interfacial bond between the fiber and the interphase [27]. Reproduced with permission from Wiley-VCH Verlag GmbH & Co. KGaA.

Lifetimes as long as a few thousands of hours have been observed on CVI SiC/SiBC composites with a self-healing matrix and a PyC interphase.

3.5 VARIOUS CONCEPTS OF WEAK INTERFACES/INTERPHASES

Various types of interphases have been proposed. Unlike for the high-strength-tailored interphases, research was motivated mainly by crack deflection, and, secondarily by resistance to oxidation. A big effort was spent on processing of oxide systems. Interface characteristics are generally not available.

NASA developed an *in situ* grown BN interphase on Sylramic SiC fibers. A thermal treatment in a controlled nitrogen environment allows mobile boron sintering aids in the Sylramic fiber to diffuse out of the fiber and to form a thin *in situ* grown BN layer on fiber surface [43]. This Sylramic-iBN SiC fiber reinforces SiC/SiC composites with a CVI SiC matrix with porosity filled via infiltration approaches, such as melt infiltration of silicon metal. These composites have been reported to exhibit a higher oxidation resistance when compared to those with a CVD/CVI-produced BN coating [47]. An interfacial shear strength $\approx$70 MPa has been reported [43] (Table 3.3).

Several *oxide coatings* have been developed for oxide fibers-reinforced oxide matrix composites: $LaPO_4$ (monazite), $CaWO_4$ (Scheelite), $NdPO_4$. Of these, monazite has been the most widely studied. Indentation testing and push out testing have shown that these coatings provide the weak interface function in a dense oxide matrix [48]. Furthermore, the following interesting feature was evidenced during push out of oxide fibers coated with monazite. The monazite coating exhibited a permanent deformation. The contribution of monazite plastic deformation to composite mechanical properties has not been investigated. Significant composite lifetime improvements (by a factor of several hundreds) have been observed on alumina composites with a monazite interphase. Results of tensile tests or 4-point bending tests suggest that oxide coatings are viable interphase candidates. But the strain-to-failure of these composites is rather modest (<0.2%), indicating a limited damage tolerance capability [49].

Porous fiber coatings are expected to cause matrix crack deflection. It is postulated that cracks will meander to connect pores and hence arrive at fibers in pores, imposing lesser stress concentration on the fibers than in the dense matrix system. Various oxide-based porous coatings have been examined including zircon [50], zirconia [51], and rare earth aluminates [52]. The porous coating approach does not seem viable for small-diameter filaments, because the corresponding scale of porosity required cannot be expected to be stable [49].

The term "fugitive" indicates that the coating material, such as molybdenum or carbon, can be removed after composite fabrication, for instance through oxidation giving a gap at the fiber/matrix interface. *Fugitive coatings* have been demonstrated in several composite systems [53] such as sapphire-reinforced YAG and Nextel 720/CAS composites. Two-layer coatings of fugitive carbon + zirconia have also been shown to provide a weak interface in a sapphire/alumina composite [54]. Molybdenum was not found superior to fugitive carbon coatings [49]. In composites with fugitive coatings there is an obvious sacrifice in strength, stiffness, and thermal conductivity in directions perpendicular to the fiber axes. Furthermore, since the interfacial domain is extremely weak, the contribution of matrix to load carrying is quite small. This may eliminate fugitive-interface composites from consideration for many applications.

Research work has been reported on other interfacial coating concepts. They are briefly mentioned here: easy-cleaving coatings including phyllosilicates [55], hexaluminates [56] and layered perovskites [57], ductile coatings [58], segregation-weakened interfaces, and reactive coatings [59].

3.6 DETERMINATION OF INTERFACIAL PROPERTIES

Among the properties that characterize the behavior of interfacial domain, the following ones are of importance for interface design or prediction of composite mechanical behavior: the interfacial strength (tensile or shear strength), the coefficient of friction, the interfacial shear stress, and the interfacial debond energy. The interfacial tensile strength is defined as the mode I resistance to debonding under a tensile stress perpendicular to fiber axis. The interfacial shear strength is the stress required to initiate an interfacial crack. The interfacial shear stress is the stress required for fiber sliding in the debonded interface. The tensile properties characterize initiation of debonding, whereas the shear properties refer to load transfers. The tests used for the determination of these properties are aimed at producing tensile or shear loading conditions onto the interfacial domain.

3.6.1 The Interfacial Tensile Strength

Data on interfacial tensile strength are generally not available in the literature. A few techniques have been proposed to generate stresses perpendicular to interfaces:

- *Transverse bend test:* A three-point bend test with fibers aligned perpendicular to the specimen length [60].
- *Transverse tensile test:* A tensile test with a fiber direction aligned perpendicular to the specimen length [61].
- *Brazilian test:* Diametral compression of disks generates tensile stresses which operate perpendicular to fiber direction when plies are parallel to loading direction and perpendicular to disc faces.
- *Flexural test on bilayered specimens:* This technique is appropriate to layered systems [62].

These tests give an estimate of the interfacial tensile strength. However, appropriate models are still missing for the relationships between interfacial stresses and force-strain data for tensile, Brazilian, and bending tests on composite materials. Existing models need to be upgraded to account for composite features.

3.6.2 Interfacial Shear Strength or Stress

Most of the test techniques were developed first for polymer matrix composites. They consist in pushing or pulling one or several fibers out of the matrix. They have been extended to ceramic matrix composites. Interfacial properties are derived from the force–displacement curve.

Many methods involving the pressing of an indenter on a fiber cross section have been devised. The push-in and the push-out tests have been the subject of an abundant literature [63–71]. During a push-in test, fiber debonding length and sliding distance are much smaller than specimen thickness. This behavior is characterized by a nonlinear load displacement curve. It is observed on specimens with significant thickness relative to interface strength. It is also observed on push-out tests during stable interface crack growth prior to fiber sliding (Figure 3.15). During push-out tests the fiber slides through the specimen. Figure 3.15 shows a typical single push-out curve measured under constant displacement rate of indenter. The common response of untreated fiber materials, involves a nonlinear domain of stable interface crack growth, a load drop under constant deformation at instability, and fiber sliding under constant load. This behavior is well described by various push-out models [64–69].

An estimate of interfacial shear stress is often derived from the force at plateau using the following equation:

$$\tau = F/2\pi R_f t \tag{3.18}$$

where F is the force at plateau, R_f is fiber radius, t is the embedded fiber length.

When the test is carried out under a constant load rate, the instability occurs under constant load [68], which exceeds that required for fiber steady sliding. Figure 3.16 shows the main features of the load displacement curve. Tests must be preferentially performed at constant displacement rate.

Figure 3.15 also shows a typical push-out curve measured on SiC/SiC composites with a strengthened interfacial domain [69]. The curve differs from those expected from commonly used models and observed for conventional composites with untreated fibers: the upward curvature was attributed to increasing compression in the fiber due to very limited propagation of the debond, with the presence of rough crack surfaces preventing sliding, that resolves into a single crack at some critical displacement [71].

Rough-interface formalism permitted an interpretation of this push-out behavior [71]. Two stages of cracking were assumed; each stage has the same fracture energy (G) but different roughness parameters that describe the two types of surfaces. The roughness parameters have been adjusted to generate curves that fit the respective portions of the experimental curve. It is interesting to note the relevant parameters for the initial crack (roughness amplitude $h = 50$ nm, period $d = 50$ nm, $G = 28$ J/m^2) and the secondary crack ($h = 90$ nm, $d = 2.2$ µm, $G = 28$ J/m^2). The value of 28 J/m^2 for the fracture energy of pyrocarbon is in good agreement with literature mode I values, but it is much higher than the 0–3 J/m^2 value typically inferred from composites with carbon coatings, often considered to be a necessary condition.

Alternative methods are based on the analysis of tensile behavior of a model unidirectional composite (microcomposite or minicomposite reinforced by one single filament or one multifilament tow) or of 2D composite [31, 40, 72–75]. Interfacial characteristics are determined from features induced by the presence of interfacial crack: hysteresis loops during unloading–reloading cycles, matrix cracking, or nonlinear deformations.

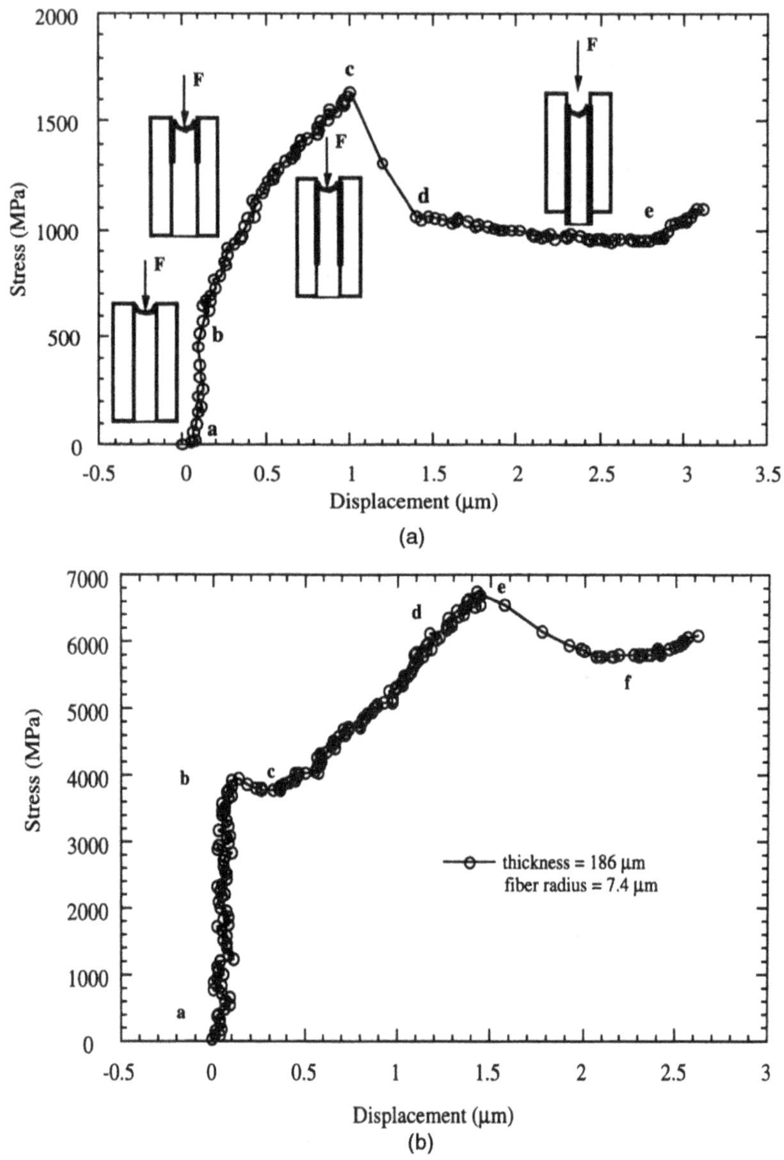

FIGURE 3.15
Single-fiber push-out curves measured on SiC/C/SiC composite samples (constant indenter displacement rate) reinforced with as-received fibers (a) (ab: elastic deformation of fiber, bc: stable fiber debonding, cd: unstable debonding, de: fiber sliding) or treated fibers (b) (ab: elastic deformation of fiber, bc: fiber debonding, cd: fiber loading, de: stable debonding, ef: unstable debonding, f: fiber sliding) [68]. Elastic deformations were removed. Reproduced with permission from Wiley-VCH Verlag GmbH & Co. KGaA.

The interfacial shear stress is related to the width of hysteresis loops ($\delta\Delta$) measured during unloading–reloading cycles, by the following equation [73]:

$$\tau = \frac{b_2 N(1 - a_1 V_{\mathrm{f}})^2 R_{\mathrm{f}}}{2V_{\mathrm{f}}^2 E_{\mathrm{m}}} \left(\frac{\sigma_{\mathrm{p}}^2}{\delta\Delta}\right)\left(\frac{\sigma}{\sigma_{\mathrm{p}}}\right)\left(\frac{1-\sigma}{\sigma_{\mathrm{p}}}\right) \tag{3.19}$$

with

$$a_1 = \frac{E_{\mathrm{f}}}{E_{\mathrm{c}}}$$

$$b_2 = \frac{(1+\nu)E_{\mathrm{m}}[E_f + (1+2\nu)E_{\mathrm{c}}]}{E_{\mathrm{f}}[(1+\nu)E_{\mathrm{f}} + (1-\nu)E_{\mathrm{c}}]}$$

where σ is the applied stress in the unloading–reloading sequence that corresponds to $\delta\Delta$, σ_{p} the initial stress level at unloading, E_{c} the Young's modulus of the minicomposite, R_{f} the fiber radius, the Poisson's ratio ($\nu = \nu_{\mathrm{m}} = \nu_{\mathrm{f}}$), E_{m}

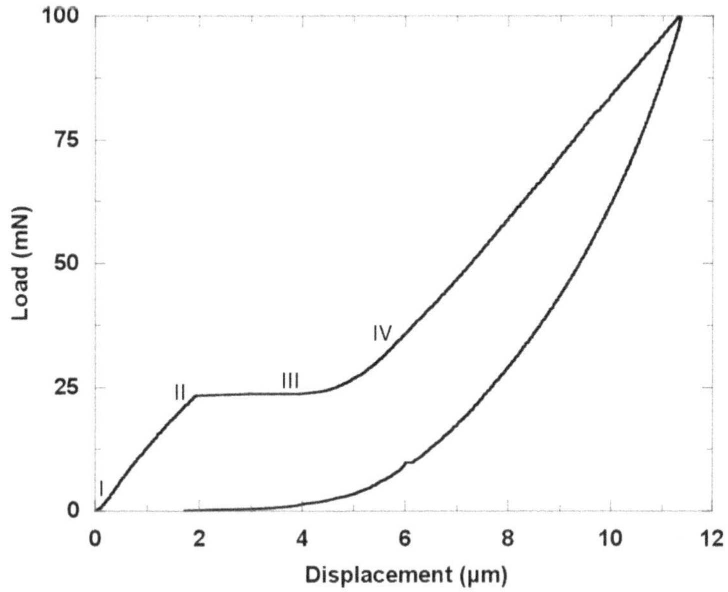

FIGURE 3.16
Single-fiber push-out curves measured on C/C composite samples (constant load rate). I: elastic deformation of fiber, II: stable fiber debonding, III: fiber is pushed out, IV: the indenter hits the matrix.

the Young's modulus of the matrix, E_f that of fiber, and V_f the fiber volume fraction. τ is derived from the $\delta\Delta$–σ data measured during the last unloading–reloading sequence before ultimate failure. The number of matrix cracks in gauge length, N, is determined from scanning electron microscopy (SEM) inspection of the specimen after ultimate failure.

The interfacial shear stress is also related to the spacing distance of the matrix cracks at saturation (l_s) and to the corresponding stress σ_s by the following equations [42, 76, 77]:

$$\tau = \frac{\sigma_s R_f}{2V_f l_s \left(1 + \dfrac{E_f V_f}{E_m V_m}\right)} \tag{3.20}$$

$$\tau = \frac{\sigma_s R_f V_m}{2V_f l_s} \tag{3.21}$$

The interfacial shear stress can also be extracted from the tensile stress–strain curve of micro- or minicomposites using a model that has been detailed and validated in previous works [26, 40]. The input data are thermoelastic properties and flaw strength parameters of the matrix and fibers. τ is estimated by fitting the model to experimental stress–strain curve.

Interfacial shear stresses available in the literature are given in Table 3.3. The data exhibit a certain scatter and dependence on the measurement method. Push-in and push-out tests provide values for a single fiber loaded over a short distance (less than a few hundreds of microns), whereas data from tensile tests pertain to an upper length scale. Note that the largest τ estimates for those conventional untreated fibers composites were obtained using the push-out test method which may be attributed to enhanced friction as a result of loading conditions. Anyway, an unambiguous trend can be identified from the interfacial shear stresses summarized in Table 3.3: τ is less than 100 MPa for SiC/SiC composites reinforced by untreated SiC fibers, and it can be as high as 350 MPa for strengthened interfaces obtained with treated SiC fibers.

Several limitations to the push-in and push-out tests can be found. First, the thickness of specimens must be very small for strong interfaces or for samples prepared from composites with woven reinforcement. Typically less than 100 µm may be required owing to fiber curvature. Such small size is not representative of composite at larger length scales. Then, interfacial phenomena depend on loading conditions; for instance, the roughness effects may be enhanced during push-out tests, and they must be properly taken into account in the models. Finally, the inherent wide variation in behavior can be increased at micrometrical length scale. The test methods based on tensile tests on minicomposites or composite test specimens seem to be more appropriate since a significant volume of material is concerned.

3.7 INTERFACE SELECTION

Since the interfacial domain determines composite damage tolerance and mechanical properties, composite design should integrate the selection of interfacial domain with respect to crack deflection and further load sharing. Design of composites is generally based on an empirical approach involving both processing and testing. The interface characteristics result from the interaction between the matrix and the fiber during processing. Fundamental, intrinsic, or generic data on interface resistance are generally not available.

Some general guidelines to designing the interfacial domain can be proposed. One of the requirements most often cited is that the ratio of fracture energies of the interface and the fiber be less than ≈0.25. But this criterion cannot be used for composite design since data on fracture energy of the interface are generally not available. The material has to be made first. Debond energy can be determined using the previously mentioned testing methods. Then, it is worth mentioning that data on debond energy are obtained with uncertainty and scatter.

On the basis of debonding potential approach, those materials that maximize σ_{ic}^{+} should be selected. Figure 3.14 shows a debonding potential chart for various combinations of materials. Four groups of debonding potentials σ_{ic}^{+} can be identified:

- Very high values >10 000 MPa obtained with C nanotubes. Debonding is highly probable. But it is worth pointing out that efficiency of nanotubes requires continuous nanotubes.
- High values >1000 MPa obtained with PyC or BN coatings on SiC or C fibers: crack deviation is expected at the coating/fiber interface. High interface strengths are permissible.
- Intermediate values ≈100 MPa. Debonding is less easy, low interface/interphase strength or low interphase transverse strength is required. PyC_i/PyC_i refers to deflection within pyrocarbon interphase.
- Low values <2 MPa: debonding requires very weak interfaces. It is not probable. If it occurs, the composite will exhibit very poor properties.

Then, the interfacial tensile strength should be close to debonding potential σ_{ic}^{+}. For this requirement to be met, interface engineering may be required. The relationship between interfacial tensile strength and debond length has not been established yet. However, it can be conjectured that such relationship exists, and that debond length is shorter when interfacial tensile strength is larger. However, the resulting interfacial shear stress should not be enhanced by surface roughness. The interfacial shear stress should not exceed a critical value, in order to obtain large strains-to-failure. This limit has not been determined yet.

3.8 CONCLUSIONS

Interphases/interfaces are critical constituents of fiber-reinforced ceramic matrix composites. They are assigned several functions which look opposite from a qualitative point of view. These functions depend upon several factors including loading conditions, environment, fiber orientations. The importance of interfacial domain depends on respective properties of fibers and matrix. It is less significant when matrix stiffness is small compared to fibers. It is crucial when the matrix is able to carry a substantial part of applied loads (stiff matrix).

Quantitative criteria for interfacial domain characteristics are required in order to develop composites with excellent properties. Looking only for weak interfaces is not sufficient. A significant effort has been devoted to the development of SiC/SiC composites with an interfacial domain tailored with respect to composite mechanical behavior. Interfacial domain characteristics can be adapted via an anisotropic interphase (such as a carbon coating or a multilayer) and a strengthened interphase/fiber bond. The behavior of the interfacial domain can be controlled also via a sound selection of constituents. Combinations of certain materials favor debonding. For this purpose, the elastic modulus of the uncracked material (material 2 or fiber) must be larger than that of the cracked material (material 1 or matrix) and the strength of material 2 or fiber must be large.

In those composites with a matrix stiffer than fibers, the contribution of the matrix to load sharing after matrix crack deflection must be favored. This phenomenon requires short interfacial cracks, which implies the presence of a tailored interface. Strengthened interfaces were developed on SiC/SiC composites using an anisotropic pyrocarbon interphase and treated fibers. Similar effect has not been reproduced with BN interphases.

The pyrocarbon-based tailored interphase demonstrated to be a viable concept. Various weak interface concepts have been examined in the literature. For oxide systems, a big effort was put on processing. Interface characteristics are not documented.

Owing to the importance of interfacial domain in composite performances, relations between interfaces characteristics and composite mechanical behavior must be established. For this purpose, appropriate characteristics must be measured, so that interfacial domain selection can be made possible, and then integrated into the composite design process.

REFERENCES

1. O. Siron and J. Lamon. Damage and failure mechanisms of a 3D carbon–carbon composite under uniaxial tensile and shear loads. *Acta Mater.*, 46(18), 6631–6643 (1998).
2. J. Lamon. Micromechanics-based approach to the mechanical behavior of brittle matrix composites. *Compos. Sci. Technol.*, 61, 2259–2272 (2001).
3. K. Kendall. Transition between cohesive and interfacial failure in a laminate. *Proc. R. Soc.*, A344, 287–302 (1975).
4. M. Y. He and J. W. Hutchinson. Crack deflection at an interface between dissimilar elastic materials. *Int. J. Solids Struct.*, 25(9), 1053–1067 (1989).
5. C. Lacroix, D. Leguillon, and E. Martin. The influence of the interphase on the deflection of a matrix crack in a ceramic–matrix composite. *Compos. Sci. Technol.*, 62, 519–523 (2002).
6. A. K. Kaw and N. J. Pagano. Axisymmetric thermoelastic response of a composite cylinder containing an annular matrix crack. *J. Compos. Mater.*, 27(6), 540–571 (1993).
7. D. O. Swenson and C. A. Rau Jr. The stress distribution around a crack perpendicular to an interface between materials. *Int. J. Fracture Mech.*, 6, 357–365 (1970).
8. J. Cook and J. E. Gordon. A mechanism for the control of crack propagation in all-brittle systems. *Proc. Roy. Soc.*, 28A, 508–520 (1964).
9. S. Pompidou and J. Lamon. Micromechanics-based evaluation of interfaces in ceramic matrix composites. *Compos. Sci. Technol.*, 67, 2052–2060 (2007).
10. N. Carrère, E. Martin, and J. Lamon. The influence of the interphase and associated interfaces on the deflection of matrix cracks in ceramic matrix composites. *Composites, Part A*, 31, 1179–1190 (2000).
11. D. Maugis. Contact, adhesion and rupture of elastic solids. *Springer Ser. Solid-State Sci.*, 130 (2000).
12. P. S. Theocaris and J. Milios. The disruption of a longitudinal interface by a moving transverse crack. *J. Reinf. Plast. Comp.*, 2, 18–28 (1983).
13. W. Lee, S. J. Howard, and W. J. Clegg. Growth of interface defects and its effect on crack deflection and toughening criteria. *Acta Mater.*, 44(10), 3905–3922 (1996).
14. W. J. Clegg, K. S. Blanks, J. B. Davis, and F. Lanckmans. Porous interfaces as crack deflecting interlayers in ceramic laminates. *Key Eng. Mater.*, 132–136, 1866–1869 (1997).
15. J. Zhang and J. J. Lewandowski. Delamination study using four-point bending of bilayers. *J. Mater. Sci. Lett.*, 32, 3851–3856 (1997).
16. S. G. Warrier, B. S. Majumdar, and D. B. Miracle. Interface effects on crack deflection and bridging during fatigue crack growth of titanium matrix composites. *Acta Mater.*, 45(12), 4969–4980 (1997).
17. Y. Kagawa and K. Goto. Direct observation and modelling of the crack-fibre interaction process in continuous fibre-reinforced ceramics: model experiments. *Mater. Sci. Eng., A.*, 250, 285–290 (1998).
18. B. S. Majumdar, D. B. Gundel, R. E. Dutton, S. G. Warrier, and N. J. Pagano. Evaluation of the tensile strength in brittle-matrix composite systems. *J. Am. Ceram. Soc.*, 81(6), 1600–1610 (1998).
19. J. M. Dorlot, J. P. Baïlon, and J. Masounave. Des matériaux, deuxième édition (Editions de l'Ecole Polytechnique de Montréal, 1986).

20. A. Lipp, K. A. Swchetz, and K. Hunold. Hexagonal boron nitride: fabrication, properties and applications. *J. Europ. Ceram. Soc.*, 5, 3–9 (1989).
21. A. Mauchin. Comportement mécanique à haute température de composites carbone/carbone. Ph.D. Thesis, No. 1728 (University of Bordeaux, 1997).
22. B. Rand. Matrix precursor for carbon–carbon composites, in *Essentials of Carbon–Carbon Composites* (Chap. 3), edited by C. R. Thomas (Royal Society of Chemistry, 1993), pp. 67–102.
23. J. Rappeneau, M. Yvars, M. Bocquet, C. Davis, and A. Auriol. Dépôt et propriétés d'un carbone pyrolitique peu orienté. *Carbon*, 2, 149–161 (1964).
24. J. L. Bobet and J. Lamon. Thermal residual stresses in ceramic matrix composites – I: axisymmetrical model and finite element analysis. *Acta. Metall. Mater.*, 43(6), 2241–2253 (1995).
25. G. A. Lebrun. Comportement thermomécanique et durée de vie des composites à matrice céramique: théorie et expérience. Ph.D. Thesis, No. 1606 (University of Bordeaux, 1996).
26. S. Bertrand, P. Forio, R. Pailler, and J. Lamon. Hi-Nicalon/SiC minicomposites with (PyC/SiC)n nanoscale-multilayered interphases. *J. Am. Ceram. Soc.*, 82(9), 2465–2473 (1999).
27. S. Bertrand, R. Pailler, and J. Lamon. Influence of strong fibre-coating interfaces on the mechanical behaviour and lifetime of Hi-Nicalon/(PyC/SiC)n/SiC minicomposites. *J. Am. Ceram. Soc.*, 84(4), 787–794 (2001).
28. F. Heurtevent. Matériaux multicouches nanoséquencés (PyC/SiC)n. Application en tant qu'interphases dans les composites thermostructuraux. Ph.D. Thesis, No. 1476 (University of Bordeaux, 1996).
29. J. Lamon, N. Lissart, C. Rechiniac, D. H. Roach, and J. M. Jouin. Micromechanical and statistical approach to the behaviour of CMCs. *Am. Ceram. Soc.*, 1115–1124 (1993).
30. A. Dalmaz. Etude du comportement en fatigue cyclique à hautes températures du composite tissé fibres de carbone/matrice carbure de silicium 2,5D-C/SiC. Ph.D. Thesis, No. 97ISAL 0085 (INSA Lyon, 1997).
31. C. Droillard and J. Lamon. Fracture toughness of 2-D woven SiC/SiC CVI-composites with multilayered interphases. *J. Am. Ceram. Soc.*, 79, 849–858 (1996).
32. H. W. Carpenter and J. W. Bolhen. Fibre coatings for ceramic matrix composites, in *Proceedings of the 16th Annual Conference on Composites and Advanced Ceramic Materials, Part 1 of 2: Ceramic Engineering and Science Proceedings*, Vol. 13, Issue 7/8, edited by J. B. Wachtman (John Wiley & Sons, Inc., Hoboken, NJ, 2008), pp. 238–256. doi:10.1002/9780470313954.ch26 (1992).
33. P. D. Jero and R. J. Kerans. The contribution of interfacial roughness to sliding friction of ceramic fibers in a glass matrix. *Scr. Metall. Mater.*, 24, 2315–2318 (1991).
34. T. A. Parthasarathy, D. B. Marshall, and R. J. Kerans. Analysis of the effect of interfacial roughness on fiber debonding and sliding in brittle composites. *Acta Metall. Mater.*, 42(11), 3773–3784 (1994).
35. T. A. Parthasarathy and R. J. Kerans. Predicted effects of interfacial roughness on the behavior of selected ceramic composites. *J. Am. Ceram. Soc.*, 80(8), 2043–2055 (1997).
36. C. Sauder, A. Brusson, and J. Lamon. Influence of interface characteristics on the mechanical properties of Hi-NicalonS and SA3 fiber reinforced SiC/SiC minicomposites. *Int. J. Appl. Ceram. Tec.*, 7(3), 291–303 (2010).
37. G. N. Morscher and J. Martinez-Fernandez. Fiber effects on minicomposite mechanical properties for several silicon carbide fiber-chemically vapor-infiltrated silicon carbide matrix systems. *J. Am. Ceram. Soc.*, 82(1), 145–155 (1999).
38. R. Naslain. Fibre-matrix interphases and interfaces in ceramic matrix composites processed by CVI. *Compos. Interface.*, 1(3), 253–286 (1993).
39. J. M. Jouin, J. Cotteret, F. Christin. SiC/SiC interphase: case history, in *Proceedings of 2nd European Colloquium "Designing Ceramic Interfaces"*, edited by S. D. Peteves (Office for Official Publications of the European Communities, Luxembourg, 1993), pp. 191–202.
40. N. Lissart, and J. Lamon. Damage and failure in ceramic matrix minicomposites: experimental study and model. *Acta Metall.*, 45, 1025 (1997).
41. F. Rebillat, J. Lamon, and A. Guette. The concept of strong interface applied to SiC/BN/SiC composites with BN interphases. *Acta Mater.*, 48, 4609–4618 (2000).
42. P. Dupel, J. L. Bobet, R. Pailler, and J. Lamon. Influence d'interphases pyrocarbone déposées par CVI pulsée sur les caractéristiques mécaniques de matériaux composites unidirectionnels. *J. Phys. III France*, 5, 937–951 (1995).
43. J. A. DiCarlo, H-M. Yun, G. N. Morscher, and R. T. Bhatt. SiC/SiC composites for 1200°C and above, in *Handbook of Ceramic Composites*, edited by N. Bansal (Kluwer Academic Publishers, 2005), pp. 77–98.
44. J. Lamon. Chemical vapor infiltrated SiC/SiC composites, in *Handbook of Ceramic Composites*, edited by N. Bansal (Kluwer Academic Publishers, 2005), pp. 55–76.
45. S. Pasquier, J. Lamon, and R. Naslain. Tensile static fatigue of 2D SiC/SiC composites with multilayered (PyC-SiC)n interphases at high temperatures in oxidizing atmosphere. *Composites – Part A*. 29A, 1157–1164 (1998).

46. K. L. Rugg, R. E. Tressler, and J. Lamon. Interfacial behavior during creep of microcomposites at elevated temperatures. *J. Eur. Ceram. Soc.*, 9, 2297–2303 (1999).
47. H. M. Yun, J. Z. Gyekenyesi, Y. L. Chen, D. R. Wheeler, and J. A. DiCarlo. Tensile behavior of SiC/SiC composites reinforced by treated Sylramic SiC fibers. *Cer. Eng. Sci. Proc.*, 22(3), 521–531 (2001).
48. P. E. D. Morgan and D. B. Marshall. Ceramic composites of monazite and alumina. *J. Am. Ceram. Soc.*, 78(6), 1553–1563 (1995).
49. K. A. Keller, G. Jefferson and R. J. Kerans. Oxide-oxide composites, in *Handbook of Ceramic Composites*, edited by N. Bansal (Kluwer Academic Publishers, 2005), pp. 377–421.
50. E. Boakye, R. S. Hay, M. D. Petry, and T. A. Parthasarathy. Sol-gel synthesis of zircon–carbon precursors and coatings of Nextel 720 fiber tows. *Cer. Eng. Sci. Proc. A*, 20(3), 165–172 (1999).
51. M. Holmquist, R. Lundberg, O. Sudre, A. G. Razzell, L. Molliex, J. Benoit, and J. Alderborn. Alumina/alumina composite with a porous zirconia interphase-processing, properties and component testing. *J. Eur. Ceram. Soc.*, 20, 599–606 (2000).
52. M. K. Cinibulk, T. A. Parthasarathy, K. A. Keller, and T. Mah. Porous rare-hearth aluminate fiber coatings for oxide–oxide composites. *Ceram. Eng. Sci. Proc.*, 21(4), 219–228 (2000).
53. K. A. Keller, T. Mah, C. Cooke, and T. A. Parthasarathy. Fugitive interfacial carbon coatings for oxide/oxide composites. *J. Am. Ceram. Soc.*, 83(2), 329–336 (2000).
54. O. Sudre, A. G. Razzell, L. Molliex, and M. Holmquist. Alumina single-crystal fiber reinforced alumina matrix for combustor tiles. *Ceram. Eng. Sci. Proc.*, 19(4), 273–280 (1998).
55. G. Demazeau. New synthetic mica-like materials for controlling fracture in ceramic matrix composites. *Mat. Tech.*, 10, 43–58 (1995).
56. M. K. Cinibulk and R. S. Hay. Textured magnetoplumbite fiber/matrix interphase derived from sol-gel fiber coatings. *J. Am. Ceram. Soc.*, 79(5), 1233–1246 (1996).
57. G. Fair, M. Shenkunas, W. T. Petuskey, and S. Sambasivan. Layered perovskites as "soft-ceramics". *J. Eur. Ceram. Soc.*, 19, 2437–2447 (1999).
58. J. Wendorff, R. Janssen, and N. Claussen. Platinum as a weak interphase for fiber-reinforced oxide-matrix composites. *J. Am. Ceram. Soc.*, 81(10), 2738–2740 (1998).
59. R. S. Hay. The use of solid-state reactions with volume loss to engineer stress and porosity into the fiber matrix interface of a ceramic composite. *Acta. Metall.*, 43(9), 333–3348 (1995).
60. K. K. Chawla. *Ceramic Matrix Composites* (Chapman & Hall, London, 1993).
61. M. Rollin, J. Lamon, and R. Pailler. Evaluation of fibre/matrix interfaces in carbon/carbon composites. Proceedings of the 12th European Conference on Composite Materials, Biarritz, France, August 29–September 1, 2006, CD ROM, edited by J. Lamon and A. T. Marques.
62. S. Pompidou and J. Lamon. Determination of interface opening strength. *Ceram. Eng. Sci. Proc.*, 27(2), 207–216 (2007).
63. A. A. Wereszczak, M. K. Ferber, and R. A. Lowden. Development of an interfacial test system for the determination of interfacial properties in fiber reinforced ceramic composites. *Ceram. Eng. Sci. Proc.*, 14(7–8), 156–167 (1993).
64. C. H. Hsueh. Evaluation of interfacial properties of fiber-reinforced ceramic composites using a mechanical properties microprobe. *J. Am. Ceram. Soc.*, 76(12), 3041–3050 (1993).
65. R. J. Kerans and T. A. Parthasarathy. Theoretical analysis of the fiber pullout and pushout tests. *J. Am. Ceram. Soc.*, 74(7), 1585–1596 (1991).
66. D. B. Marshall. Analysis of fiber debonding and sliding experiments in brittle matrix composites. *Acta Metall. Mater.*, 40(7), 1585–1596 (1992).
67. C. Liang and J. W. Hutchinson. Mechanics of the fiber push-out test. *Mech. Mater.*, 14, 207–221 (1993).
68. F. Rebillat, J. Lamon, R. Naslain, E. Lara-Curzio, M. K. Ferber, and T. M. Besmann. Interfacial bond strength in SiC/C/SiC composite materials as studied by single fiber push-out tests. *J. Am. Ceram. Soc.*, 81(4), 965–978 (1998).
69. F. Rebillat, J. Lamon, R. Naslain, E. Lara-Curzio, M. K. Ferber, and T. M. Besmann. Properties of multilayered interphases in SiC/SiC CVI-composites with "weak" and "strong" interface. *J. Am. Ceram. Soc.*, 81(9), 2315–2326 (1998).
70. M. Rollin, S. Jouannigot, J. Lamon, and R. Pailler. Characterization of fiber/matrix interfaces in carbon/carbon composites. *Compos. Sci. Technol.*, 69, 1442–1446 (2009).
71. R. J. Kerans, T. Parthasarathy, F. Rebillat, and J. Lamon. Interface properties in high strength NICALON/C/SiC composites as determined by rough surface analysis of fiber pushout tests. *J. Am. Ceram. Soc.*, 81(7), 1881–1887 (1998).
72. L. Guillaumat and J. Lamon. Fracture statistics applied to modelling the nonlinear stress-strain behavior in microcomposites: influence of interfacial parameters. *Int. J. Fracture*, 82, 297–316 (1996).
73. J. Lamon, F. Rebillat, and A. Evans. Assessment of a microcomposite test procedure for evaluating constituent properties of ceramic matrix composites. *J. Am. Ceram. Soc.*, 78(2), 401–405 (1995).

74. E. Vagaggini, J. M. Domergue, and A. G. Evans. Methodology for relating the tensile constitutive behavior of ceramic-matrix composites to constituent properties. *J. Am. Ceram. Soc.*, 77(6), 1425–1435 (1994).

75. E. Vagaggini, J. M. Domergue, and A. G. Evans. Relations between hysteresis measurements and the constitutive properties of ceramic matrix composites. *J. Am. Ceram. Soc.*, 78(10), 2709–2720 (1995).

76. D. B. Marshall, B. N. Cox, and A. G. Evans. The mechanics of matrix cracking in brittle-matrix fiber composites. *Acta Metall.*, 33(11), 2013–2021 (1985).

77. J. Aveston, G. A. Cooper, and A. Kelly. *Conference Proceedings of the National Physical Laboratory*, Vol. 15 (IPC Science and Technology Press Ltd., Surrey, UK, 1971).

CHAPTER 4

TEXTILE REINFORCEMENTS: ARCHITECTURES, MECHANICAL BEHAVIOR, AND FORMING

PHILIPPE BOISSE

Université de Lyon, LaMCoS, INSA-Lyon, Villeurbanne, France

4.1 INTRODUCTION

The mechanical properties of ceramic materials are high and retained at high temperature. Nevertheless they need reinforcements because of their brittleness. Therefore, in order to keep the good mechanical properties at high temperature, ceramic matrix composites (CMCs) are reinforced by textile reinforcements made of continuous fibers, generally a carbon (C) fiber fabric or a silicon carbide (SiC) fiber fabric. The mechanical properties of the matrix of the CMCs are higher than those of the fibers. This is quite specific to these composites. In most composites with organic matrix, the strength and stiffness of the composites are because of the fibers.

The objectives of this chapter concern the analysis of the architecture and mechanical properties of the textile reinforcement. In particular, the forming process is analyzed and modeled. This forming will give a preform in which the ceramic matrix is introduced by chemical vapor infiltration (CVI). This manufacturing process corresponds to liquid composite molding (LCM) in case of organic matrix composites. The principal experimental tests for the determination of tensile, in-plane shear, bending, and compaction properties are described. Then a 3D hypoelastic model for textile reinforcement is presented. The main point in such a model lies in the follow-up of the fiber directions. Finally, two approaches for the simulation of textile reinforcement forming are described. The first is based on the hypoelastic model and the second uses semidiscrete shell finite elements.

4.2 TEXTILE COMPOSITE REINFORCEMENTS

4.2.1 Multiscale Materials: Fibers, Tows, Fabrics

The internal structure of a textile reinforcement mainly influences its mechanical behavior. Textiles are made up of woven warp and weft yarns. These yarns are themselves made up of thousands of fibers (Figure 4.1). This fibrous structure makes relative motion possible between yarns and between fibers. The resulting mechanical behavior is very specific. The tensile stiffness in the fiber direction is high while the other rigidities (shear, bending, compaction) are much weaker, often close to zero. Three scales can be clearly distinguished in a woven fabric (Figure 4.1). These scales are simultaneously present, but depending on the specific analysis, one or more scales of the woven fabric have to be

Ceramic Matrix Composites: Materials, Modeling and Technology, First Edition. Edited by Narottam P. Bansal and Jacques Lamon.

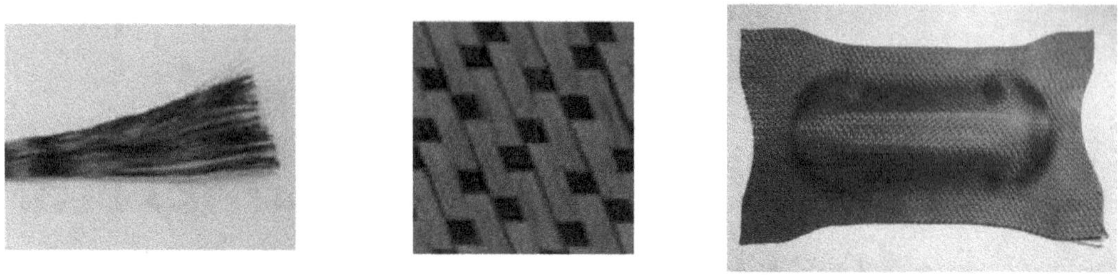

Microscopic scale (a) Mesoscopic scale (b) Macroscopique scale (c)

FIGURE 4.1
The three scales of the fiber reinforcement.

explored in the modeling. The microscopic scale is the scale of the fiber (one to several micrometers). Each yarn is made up of thousands of fibers which interact (Figure 4.1a). These interactions can be analyzed at the microscopic scale. Within the fiber the material is actually continuous. It is the only scale at which the material is continuous. At the mesoscopic scale, the woven reinforcement is seen as a set of yarns. Generally, in case of a woven fabric, there are two yarn directions, the warp and the weft yarns (Figure 4.1b). The corresponding working scale is the one of the yarn section, (some millimeters). For periodic materials, mesoscopic models consider representative unit cell (RUC) (Figure 4.2), that is, the smallest elementary pattern which can represent the whole fabric by translations. The macroscopic scale refers to the whole part level, with dimensions in the order of some centimeters to several meters (Figure 4.1c).

Some analyses of the textile reinforcement are made at microscopic scale. The fibers are generally considered as 3D beams in contact and friction with their neighbors [1–4]. In practice the simulations at microscopic scale concern one or a few woven unit cells (Figure 4.3). The information obtained from such simulations is interesting especially regarding fiber arrangements during and after deformation. Nevertheless, the number of fibers within a yarn (500–48 000 in most composite reinforcements) is currently too high for the microscopic analyses. Those are performed on a smaller number of fibers. This reduces the conclusions that can be obtained from such simulations. It is not possible to simulate the forming of a whole preform at this scale.

In mesoscopic simulation, the yarn is modeled as a continuous material. These analyses are used as virtual tests to determine the mechanical properties of a textile, in particular before manufacture [5–9]. Mesoscopic simulations are also used to estimate the permeability tensor of the preformed fabric, taking into account the local geometry of the deformed reinforcement [10–12]. The mesoscopic simulations need an accurate description of the geometry of the RUC. This is developed in Section 4.2.2. Some approaches have been proposed to simulate the forming of composite reinforcement at mesoscale [13, 14]. Some more investigation are, however, necessary to use this approach in industrial cases.

At macroscopic scale, the mechanical model must take into account the textile nature of the reinforcement. A large number of continuous mechanical models have been proposed for textile reinforcements [15–22]. There is no widely accepted model that accurately describes all aspects of the mechanical behavior of fabrics. Nevertheless, the macroscopic approach is the most popular (and today, quite unique) approach for forming simulations. An example of the model (based on a hypoelastic approach) is presented in Section 4.2.4.

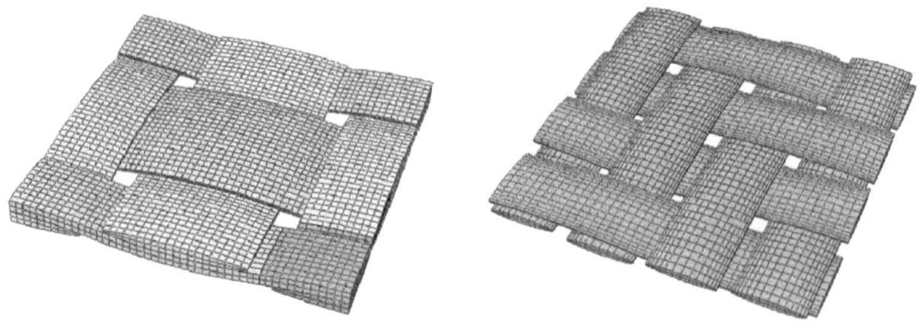

FIGURE 4.2
Representative unit cell (RUC): (a) a plain weave; (b) a 2 × 2 twill weave.

FIGURE 4.3
Analysis of a sheared plain weave at the microscopic scale [3].

4.2.2 Architecture and Geometry of the Unit Woven Cell

The modeling of the architecture of textile composite reinforcement is important for several applications. It is used to define an initial mesh to compute the deformation of the RUC at mesoscopic scale [5–9]. It is also used to determine the homogenized properties of the composite [23] and to compute damage initiation [24]. Several models and software programs have been developed [25–28]. A consistent 3D geometrical model for the fabric unit cell is presented below. It guarantees that there is no penetration between the warp and weft yarns [29, 30]. The cross section along the yarn is not constant and the reorganization of the fibers in the yarn is an important phenomenon. In this consistent model the cross section is dissymmetric due to the boundary conditions prescribed by other yarns and has to change along the trajectory. This aspect is not present in most of the models. Some control sections at control points are used to take into account the variations of section shape along the yarn (Figure 4.4). The 3D model of the yarn is obtained through a smooth interpolation between the control sections. CAD software are used for the interpolation [31]. They use a "swept blend" feature that is able to build volumes using control sections and trajectories. The standard complete 3D model of a fabric need seven parameters for unbalanced fabrics: yarn width, yarn density, crimp of each direction, and thickness of the fabric. They are quite easy to measure. In case of balanced fabrics the number of parameters amounts to three. The consistent CAD model (i.e., without penetration) obtained for a 4 × 3 twill and the corresponding finite element mesh are shown in Figure 4.5.

4.2.3 Experimental Analysis of the Mechanical Behavior

Because the mechanical behavior of textile reinforcements is very particular due to the fibrous composition, specific mechanical tests have been developed to measure the mechanical properties and to identify the constitutive law of those fabric.

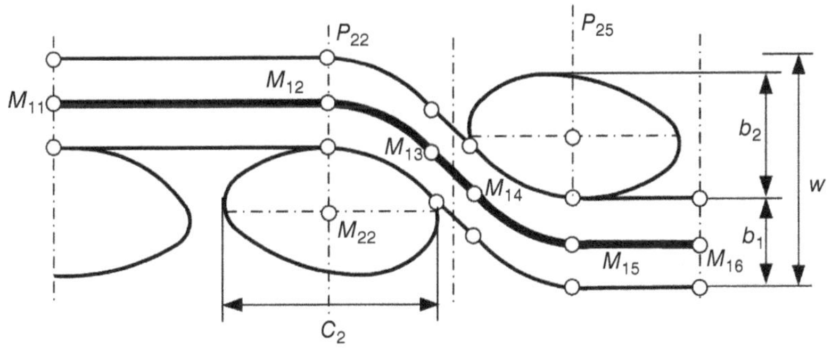

FIGURE 4.4
Control points for the trajectory of a twill $m*n$ [29]. Reproduced with permission from Elsevier, 2004.

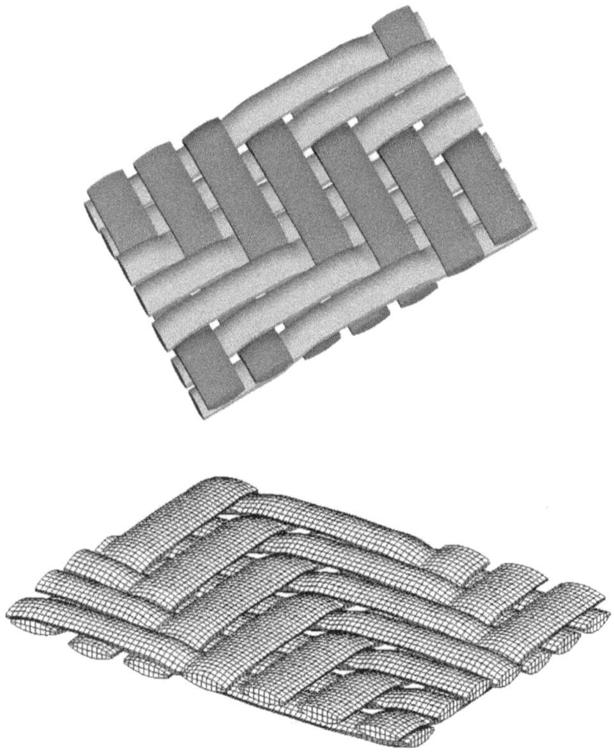

FIGURE 4.5

3D consistent geometry and mesh of a 4 × 3 twill.

4.2.3.1 In-Plane Shear Behavior A great literature is dedicated to the experimental tests used to characterize the in-plane shear behavior of textile composite reinforcements [32–39]. The picture frame is presented in Figure 4.6 and the bias-extension tests in Figure 4.7. The in-plane shear is the most dominant deformation mode in woven composite forming. The shear angle can reach 50° when the manufactured part is strongly doubly curved. This in-plane shear can lead to wrinkling depending on the process parameters and on the material properties [40]. An international benchmark has been launched recently concerning these two tests. Eight laboratories of six different countries have performed picture frame and bias-extension tests on the same textile composite reinforcements [39].

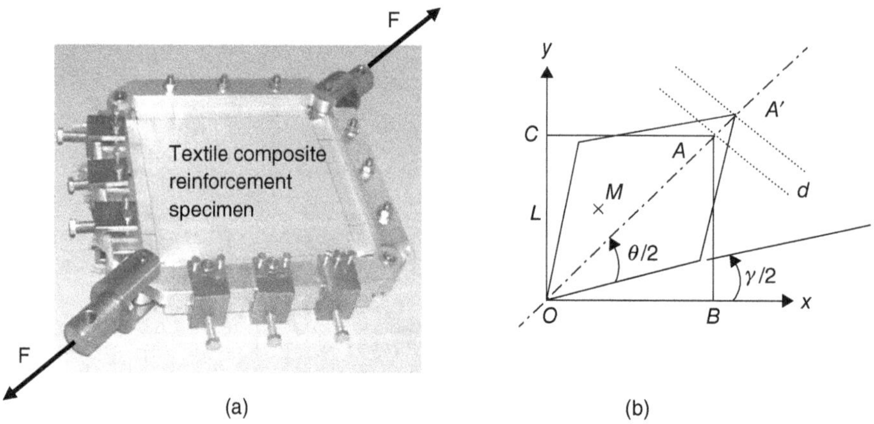

FIGURE 4.6

Picture frame test: (a) experimental device; (b) kinematics of the test.

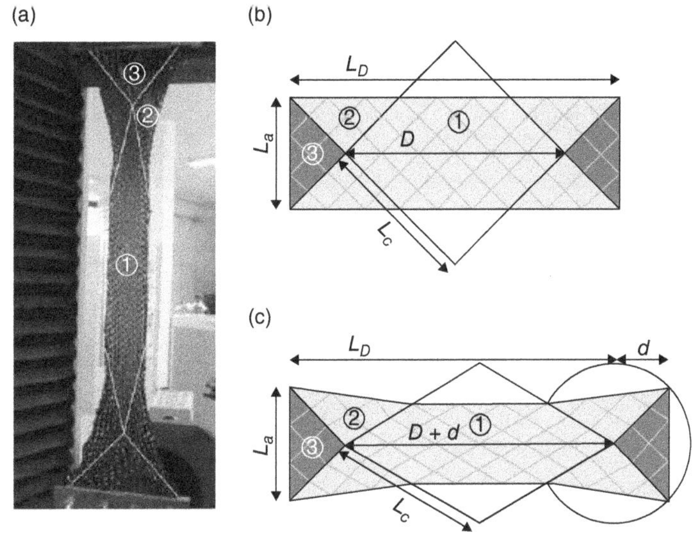

FIGURE 4.7
Bias extension test with three zones: (a) the device; (b) initial rectangular specimen with yarns oriented at ±45°; (c) deformed specimen.

In the picture frame, a hinged frame with four bars of equal length is assembled in a tensile testing machine. A tensile force is applied across diagonally opposing corners of the picture frame. Consequently, it is theoretically subjected to pure in-plane shear angle γ, that is, function of the displacement d of the mobile clamp:

$$\gamma = \frac{\pi}{2} - 2\arccos\left(\frac{1}{\sqrt{2}} + \frac{d}{2L}\right) \quad M_s(\gamma) = F\frac{S_c}{\sqrt{2}\,L}\left(\cos\frac{\gamma}{2} - \sin\frac{\gamma}{2}\right) \tag{4.1}$$

The in-plane shear moment on a unit shell M_s is related to the force on the frame F. S_c is the surface of the unit woven cell in the initial state.

In the bias-extension test, a rectangular specimen of woven fabric with warp and weft directions initially oriented at 45° is submitted to a tensile load applied by a tensile machine (Figure 4.7). The initial length of the specimen L must be larger than twice the width ℓ. The zone 1 in the center of the specimen is submitted to a (quasi) pure shear γ if the yarns are assumed to be inextensible. That is a correct assumption for the yarns used as composite reinforcements. This inextensibility imposes that the shear angle in the zone 2 is half the value in the central region 1. The shear angle γ is related to the specimen elongation d, and the in-plane shear moment M_s for a shear angle γ is related to the force F on the tensile machine:

$$\gamma = \frac{\pi}{2} - 2\arccos\left(\frac{D+d}{\sqrt{2}D}\right) \quad M_s(\gamma) = \frac{FDS_c}{\ell\,(2D-\ell)}\left(\cos\frac{\gamma}{2} - \sin\frac{\gamma}{2}\right) - \frac{\ell}{2D-\ell}M_s\left(\frac{\gamma}{2}\right) \tag{4.2}$$

Figure 4.8 shows a shear curve $M_s(\gamma)$ for a glass plain weave. For small shear angles, the in-plane shear stiffness is weak. This rigidity increases for larger shear angles and becomes significant. The relative displacement within a yarn is measured by an optical method [38]. These displacement fields show that the tows rotate in a rigid body motion during the first part of the loading (Figure 4.8b). Lateral contacts between the yarns occur when the shear angle becomes larger. The tows are transversally compressed and the shear rigidity increases significantly. This increase of shear stiffness leads to wrinkling onset [40]. The corresponding shear angle is called locking angle [32, 39]. This wrinkling is not only depending on the shear angle but also from the other loadings [40]. If there is no tension, the locking angle is in order of 40–45° for textile composite reinforcements.

4.2.3.2 *Tensile Behavior* When a woven material is stretched, the tension leads to a decrimping of tows. The resulting tensile behavior presents specific nonlinearities. The tensile stiffness of the fabric is much lower than those of

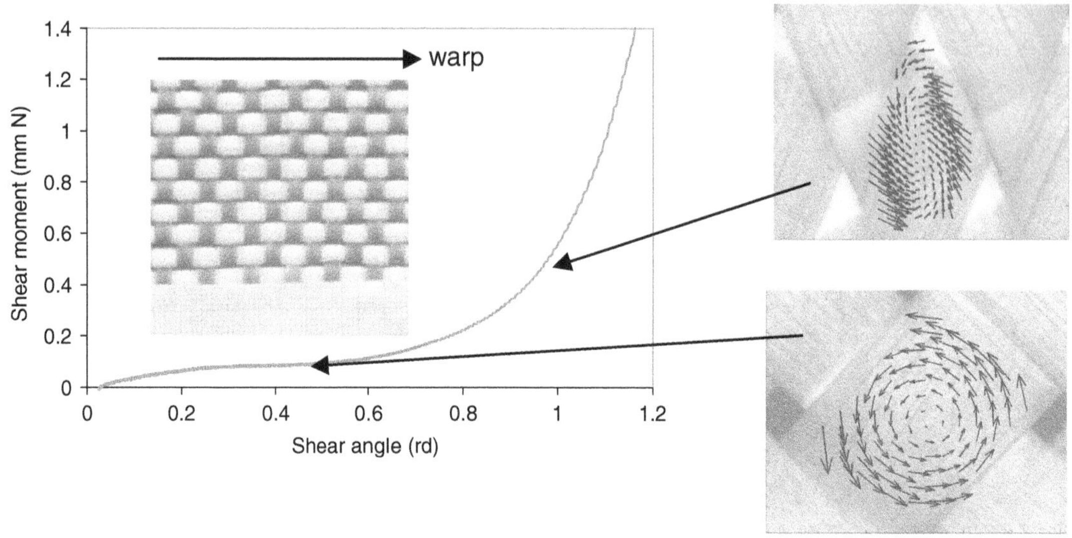

FIGURE 4.8
(a) Shear curve for a glass plain weave; (b) relative displacement field within a yarn.

the tow for small axial strains. The decrimping phenomena in warp and weft directions are interdependent because of the weaving. The tensile behavior is biaxial [41–44]. The cross-shaped specimen is submitted to a biaxial tension by a device such as shown in Figure 4.9 [42] (Figure 4.10). The tensions in warp and weft directions $T^1(\varepsilon_{11}, \varepsilon_{22})$ and $T^2(\varepsilon_{11}, \varepsilon_{22})$ are measured for different ration between warp and weft strains (Figure 4.11). The influence of the shear angle on the tensile behavior can usually be neglected [42].

4.2.3.3 Bending Behavior The possible motion between fibers making up a fibrous reinforcement leads to a low bending stiffness. In simulations, this stiffness is often neglected and membrane approaches are often used. This membrane assumption is, however, not sufficient to obtain accurate wrinkle shape. For this type of computation, the bending stiffness is important [40]. Specific tests have been developed to measure the stiffness of textile materials and especially of composites reinforcements [45–47]. Figure 4.12 shows a cantilever bending test. The own weight of the fabric defines the bending moments for a given length of the specimen.

FIGURE 4.9
Biaxial tensile machine.

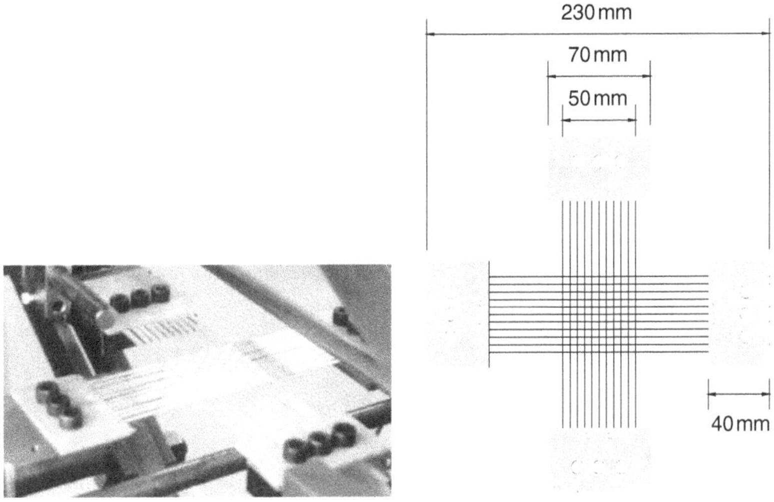

FIGURE 4.10
Cross-shape specimen. Reproduced with permission from Springer, Experimental Mechanics, Vol. 1, 1961–Vol. 53, 2013.

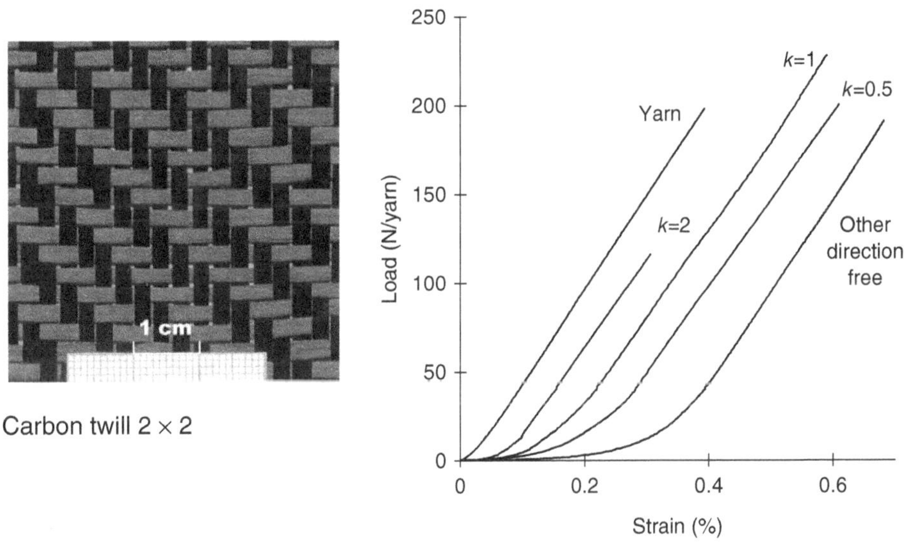

FIGURE 4.11
Load–strain curves for carbon twill weave at different ratios $k = \varepsilon_{warp}/\varepsilon_{weft}$. Reproduced with permission from Springer, Experimental Mechanics, Vol. 1, 1961–Vol. 53, 2013.

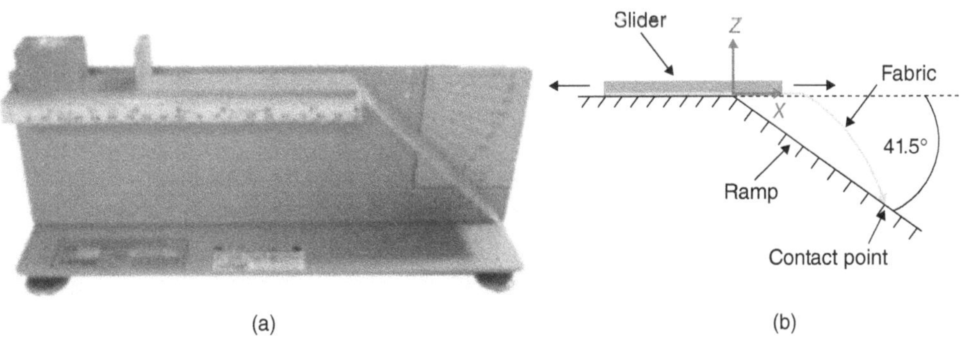

FIGURE 4.12
Device for the measurement of the bending stiffness of a textile composite reinforcement.

FIGURE 4.13

Transverse compression test.

4.2.3.4 ***Transverse Compaction Behavior*** Transverse compaction is an important property. In particular, in the case of LCM processes or CVI processes, the fibrous reinforcement is compacted in the mold before the resin is injected through the textile reinforcement [9, 48, 49]. The transverse compression behavior of the textile reinforcement determines the quality of the process and the thickness of the composite part. A transverse compaction device is shown in Figure 4.13. The compaction experiments are performed on a single fabric layer or on several fabric layers. In this case the layers can have different orientations. Figure 4.14 shows transverse compaction curves for one or three layers with different orientations. In order to avoid the difficulty of the determination of the start of the test, the curves give the transverse pressure on the fabric versus the fiber volume fraction. It can be seen from Figure 4.14 that the transverse stiffness increases much when the compaction raises. A single layer is stiffer than a three-layer fabric [9]. This one is less stiff when the three layers are in the same direction because of nesting.

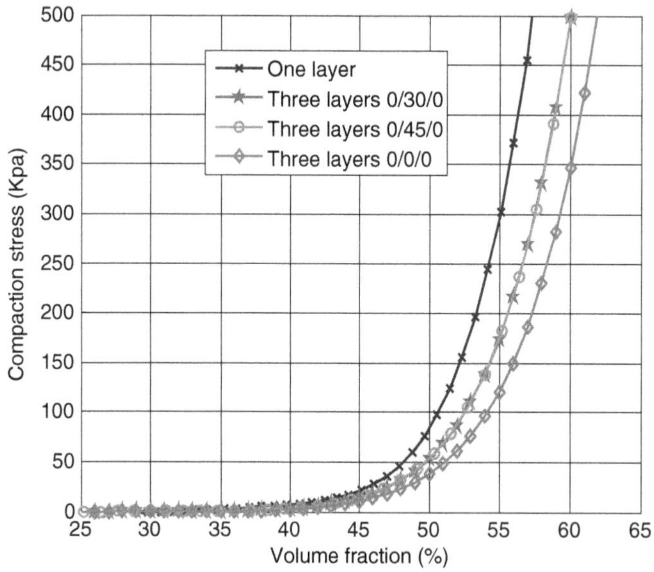

FIGURE 4.14

Compaction stress versus volume fraction for three plies with different orientations of the central ply (experiment) [9].

4.2.4 Mechanical Behavior Modeling

The analyses of the deformation of woven fabric can be made at the three scales presented in Section 4.2.1. A large part of the forming simulation is made at the macroscopic scale for numerical efficiency. At this scale the mechanical model must take into account the specific behavior of the reinforcement that is made up of yarns and fibers that can slide relatively. In particular the direction of the fibers must be strictly followed. Several models have been proposed [15–22]. A hypoelastic model for the large deformation of textile membrane is described below. In these models (also called rate constitutive models) [50,51], the strain rate tensor $\underline{\underline{\mathbf{D}}}$ is related to the objective rate of the stress tensor $\underline{\underline{\sigma}}^{\nabla}$ by a constitutive tensor $\underline{\underline{\underline{\underline{\mathbf{C}}}}}$:

$$\underline{\underline{\sigma}}^{\nabla} = \underline{\underline{\underline{\underline{\mathbf{C}}}}}:\underline{\underline{\mathbf{D}}} \quad \text{with} \quad \underline{\underline{\sigma}}^{\nabla} = \underline{\underline{\mathbf{Q}}}\cdot\left(\frac{\mathrm{d}}{\mathrm{d}t}\left(\underline{\underline{\mathbf{Q}}}^{\mathrm{T}}\cdot\underline{\underline{\sigma}}\cdot\underline{\underline{\mathbf{Q}}}\right)\right)\cdot\underline{\underline{\mathbf{Q}}}^{\mathrm{T}} \tag{4.3}$$

The objective derivative $\underline{\underline{\sigma}}^{\nabla}$ is the time derivative for an observer who is fixed with respect to the material. $\underline{\underline{\mathbf{Q}}}$ is the rotation of the so-called rotating frame where the time objective derivative is made. The objective derivatives of Jaumann and Green–Naghdi are most commonly used. They are based on the rotation of the corotational frame and on the polar rotation. These are much used for the simulation of metals forming [51].

In the case of a material with one fiber direction, the proper objective rotational derivative is based on the rotation of the fiber [7]. In the current configuration, the unit vectors $(\underline{\mathbf{e}}_1, \underline{\mathbf{e}}_2)$ of the Green–Naghdi's (GN) frame and the unit vectors in the warp and weft fiber directions $(\underline{f}_1, \underline{f}_2)$ are obtained from the initial orientation axes $(\underline{\mathbf{e}}_1^0, \underline{\mathbf{e}}_2^0)$ and from the initial warp and weft directions $(\underline{\mathbf{f}}_1^0, \underline{\mathbf{f}}_2^0)$, respectively:

$$\underline{\mathbf{e}}_1 = \underline{\underline{\mathbf{R}}}\cdot\underline{\mathbf{e}}_1^0 \quad \underline{\mathbf{e}}_2 = \underline{\underline{\mathbf{R}}}\cdot\underline{\mathbf{e}}_2^0 \quad \underline{\mathbf{f}}_1 = \frac{\underline{\underline{\mathbf{F}}}\cdot\underline{\mathbf{f}}_1^0}{\left\|\underline{\underline{\mathbf{F}}}\cdot\underline{\mathbf{f}}_1^0\right\|} \quad \underline{\mathbf{f}}_2 = \frac{\underline{\underline{\mathbf{F}}}\cdot\underline{\mathbf{f}}_2^0}{\left\|\underline{\underline{\mathbf{F}}}\cdot\underline{\mathbf{f}}_2^0\right\|} \tag{4.4}$$

where $\underline{\underline{\mathbf{F}}} = \underline{\underline{\mathbf{R}}}\cdot\underline{\underline{\mathbf{U}}}$ is the deformation gradient tensor and $\underline{\underline{\mathbf{R}}}$, the proper rotation.

$(\underline{\mathbf{e}}_1^0, \underline{\mathbf{e}}_2^0)$ and $(\underline{\mathbf{f}}_1^0, \underline{\mathbf{f}}_2^0)$ are assumed to coincide initially (Figure 4.15). The two orthonormal frames $g(\underline{\mathbf{g}}_1, \underline{\mathbf{g}}_2)$ and $h(\underline{\mathbf{h}}_1, \underline{\mathbf{h}}_2)$ are based on the warp and weft directions: with $\underline{\mathbf{g}}_1 = \underline{\mathbf{f}}_1$ and with $\underline{\mathbf{h}}_2 = \underline{\mathbf{f}}_2$ (Figure 4.15).

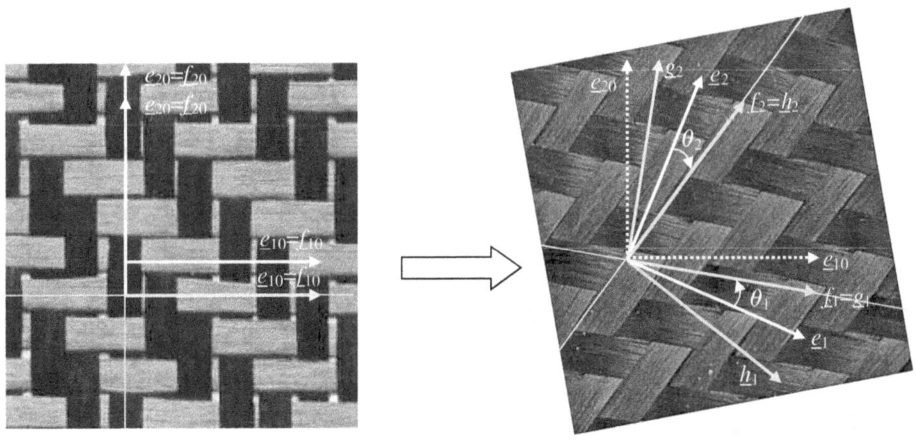

FIGURE 4.15

Initial and current GN and fiber axes. Reproduced with permission from Springer, International Journal of Material Forming.

The constitutive law will give the stress state at time t^{n+1} from the strain increment $\underline{\underline{d\varepsilon}}$ on the time increment (t^n, t^{n+1}). The components of this strain increment are usually given in the GN frame. The components of the strain increment in the two frames g and h are calculated:

$$\underline{\underline{d\varepsilon}} = d\varepsilon^{g}_{\alpha\beta}\, \underline{g}_\alpha \otimes \underline{g}_\beta = d\varepsilon^{h}_{\alpha\beta}\, \underline{h}_\alpha \otimes \underline{h}_\beta \tag{4.5}$$

$$d\varepsilon^{g}_{11} = \underline{g}_1 \cdot \underline{\underline{d\varepsilon}} \cdot \underline{g}_1 \quad d\varepsilon^{g}_{12} = \underline{g}_1 \cdot \underline{\underline{d\varepsilon}} \cdot \underline{g}_2 \quad d\varepsilon^{h}_{22} = \underline{h}_2 \cdot \underline{\underline{d\varepsilon}} \cdot \underline{h}_2 \quad d\varepsilon^{h}_{12} = \underline{h}_1 \cdot \underline{\underline{d\varepsilon}} \cdot \underline{h}_2 \tag{4.6}$$

where α and β are indexes taking value 1 or 2 and where $\varepsilon_{\alpha\alpha}$ are direct strain and ε_{12} shear strains. The axial stress component and shear stress components are calculated in frame g and h:

$$d\sigma^{g}_{11} = E^{g} d\varepsilon^{g}_{11} \quad d\sigma^{g}_{12} = G d\varepsilon^{g}_{12} \quad d\sigma^{h}_{22} = E^{h} d\varepsilon^{h}_{22} \quad d\sigma^{h}_{12} = G d\varepsilon^{h}_{12} \tag{4.7}$$

where E^g and E^h are the tensile stiffness in the warp and weft fiber directions, respectively, and G is the in-plane shear stiffness of the fabric. These moduli are usually not constant. In particular, G strongly depends on the in-plane shear (see Figure 4.8). The stresses are then integrated on the time increment (t^n, t^{n+1}) using the Hughes and Winget scheme [52]:

$$\left(\sigma^{g}_{11}\right)^{n+1} = \left(\sigma^{g}_{11}\right)^{n} + d\sigma^{g^{n+1/2}}_{11} \quad \left(\sigma^{g}_{12}\right)^{n+1} = \left(\sigma^{g}_{12}\right)^{n} + d\sigma^{g^{n+1/2}}_{12} \tag{4.8}$$

$$\left(\sigma^{h}_{11}\right)^{n+1} = \left(\sigma^{h}_{11}\right)^{n} + d\sigma^{h^{n+1/2}}_{11} \quad \left(\sigma^{h}_{12}\right)^{n+1} = \left(\sigma^{h}_{12}\right)^{n} + d\sigma^{h^{n+1/2}}_{12} \tag{4.9}$$

The addition of the stresses in warp and weft directions gives the stress in the fabric at time t^{n+1}:

$$\underline{\underline{\sigma}}^{n+1} = \left(\underline{\underline{\sigma}}^{g}\right)^{n+1} + \left(\underline{\underline{\sigma}}^{h}\right)^{n+1} \tag{4.10}$$

The components of the Cauchy stress tensor are then computed in any frame by change-of-basis. More detail on this approach can be found in References 13 and 53. An example of use of this approach to simulate an international forming benchmark is given in Section 4.4.3.

4.3 REINFORCEMENTS OF CERAMIC COMPOSITES

Ceramic materials exhibit high mechanical properties. These characteristics are kept at high temperature (up to 1500°C). Consequently, ceramic materials can be used for parts submitted to high loads and high temperatures. Nevertheless, these ceramic materials have low fracture toughness. Cracks initiate and propagate easily in the material. Some reinforcements are necessary to increase the resistance to cracks of monolithic ceramics. If different elements such as short fibers and whiskers are used in this goal, only continuous fiber reinforcements are actually efficient to stop the cracks [54]. With these continuous fibers, in case of local matrix failure, the strength is insured by the textile reinforcement. Figure 4.16 shows a C/SiC composite with transverse cracks [55]. The textile reinforcements of the ceramic matrix are mainly insured by carbon fibers (C/SiC composites) or SiC fibers (SiC/SiC composites). Basically these CMCs are very different from organic matrix composite. For the last ones, the strength and stiffness are insured by the fibers; the matrix is weak and its role is to prevent the relative motion of the fibers and yarns. In the case of ceramic composite, the stiffness of the matrix is very high. The role of the fiber is to stop the crack development.

4.3.1 Silicon Carbide Fibers

Silicon carbide fibers can be used up to 1000°C and up to 1400°C for the last generation [56]. Silicon carbide in bulk can be used up to 1600°C. The first silicon carbide fibers have been developed in the 1980s in Japan [57]. A second and a third generation of the fibers have improved their properties. At room temperature, the young modulus of the

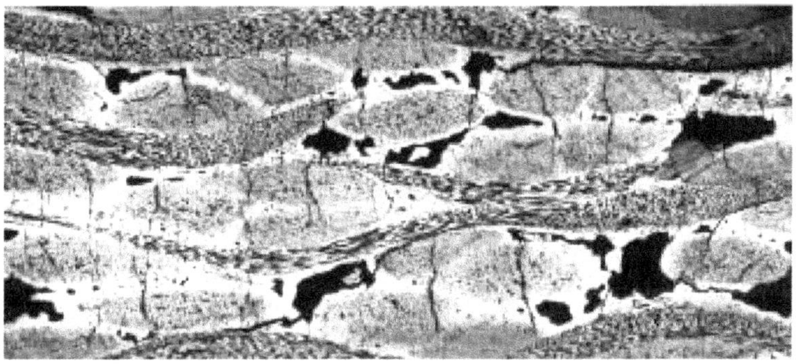

FIGURE 4.16
Transverse cracks in a C/SiC [55].

fibers is in the range 200–400 GPa, the strength is in the range 2.5–3 GPa, and the diameter of the fibers is in the range 7.5–14 μm [56].

4.3.2 Textile Reinforcement

The textile reinforcement is necessary to stop cracks in the ceramic matrix. It is generally made of carbon fibers or SiC fibers. Both of these fibers can withstand high temperature. The architecture of the reinforcements is that presented in Section 4.2. They can be made of 2D fabrics. For instance the ceramic composite presented in Figure 4.17 is made up of SiC fibers woven into eight balanced 5-hardness satin layers [58]. The matrix is introduced by CVI.

The textile reinforcement can also have a 2.5D architecture. Figure 4.18 presents a C/SiC composite where the textile reinforcement is made with a 2.5D fabric [59]. The weaving insures a joint between the yarns through the thickness. This architecture overcomes the delamination between the layers that can happen in laminate composites.

4.3.3 Infiltration of the Textile Preform

The LCM process is efficient to manufacture organic matrix composites [60]. After manufacture of the preform, the thermoset liquid resin is injected within this fiber preform. These processes permit to manufacture composite parts with complex shapes and good mechanical properties. CMCs with continuous fibers are also be produced by infiltration. Silicon carbide composites are produced by CVI of a carbon or SiC preform [61]. The infiltration of the gaseous precursor, at a temperature larger than 850°C, into the continuous fiber structure is driven by diffusion process or external pressure. The deposition fills the space between yarns and fibers and becomes the matrix of the composite material. The CVI process has several advantages; in particular, it leads to good mechanical properties. It increases creep and oxidation resistance. Nevertheless, the process is slow and costly and there is high residual porosity [62].

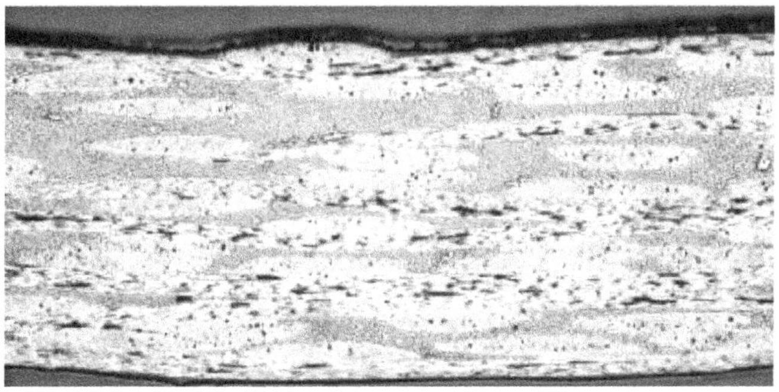

FIGURE 4.17
SiC/SiC composite [58]. Reproduced with permission from Elsevier.

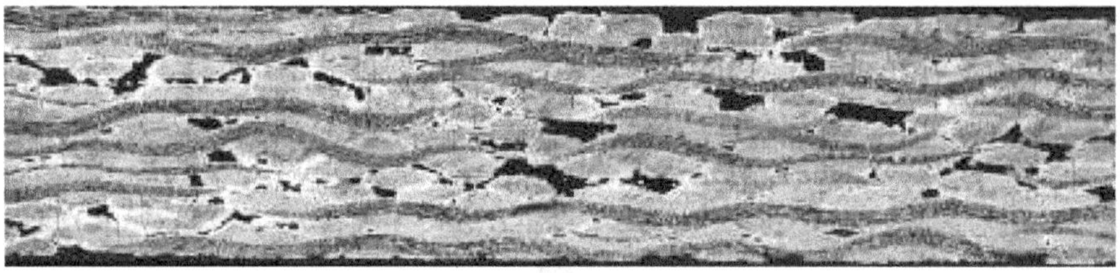

FIGURE 4.18
2.5D C/SiC composite [59].

4.4 PREFORMING SIMULATION

4.4.1 Fishnet Algorithm

Fishnet algorithms, also called "kinematic models," are used in industry for analyzing draping in woven composite reinforcements [63–66]. The approach is based on the following assumptions: (i) there is no sliding at the intersection between warp and weft yarns; (ii) the yarns are inextensible; (iii) the rotations between warp and weft yarns are free; (iv) there is no sliding between the fabric and the tool. The position of the node C on Figure 4.19 can be determined from the position of its neighbors A and C. The lengths of AC and BC are fixed. C is calculated as the intersection of two geodesics segments of given length coming from A and B and that intersect in C. This constitutes a small nonlinear scalar problem. It can be solved very fast. The surface of the tool must be defined analytically by curved elements. To initiate the draping it is necessary to position a first node and to prescribe two initial draping directions. These directions are the symmetry axes if they exist. The result of the draping depends on this initial point and on these directions.

This algorithm is fast and efficient in particular for hand draping in thermoset prepreg fabrics. However, the approach does not account for mechanical behavior of the textile reinforcement for load boundary conditions and for possible sliding of the fabric in relation to the tools.

4.4.2 Continuous FE Approaches

In order to avoid these drawbacks of the kinematics approach, simulations of draping are based on the Finite Element (FE) analyses of the composite reinforcement forming. These approaches take into account the mechanical behavior of the textile reinforcement described in Section 4.2. In the continuous approaches, the woven fabric is seen as a continuous material with high anisotropy and the ability to exhibit very large shearing and bending deformations. The continuous approach is commonly used for composite reinforcement forming simulations. It permits using standard finite elements.

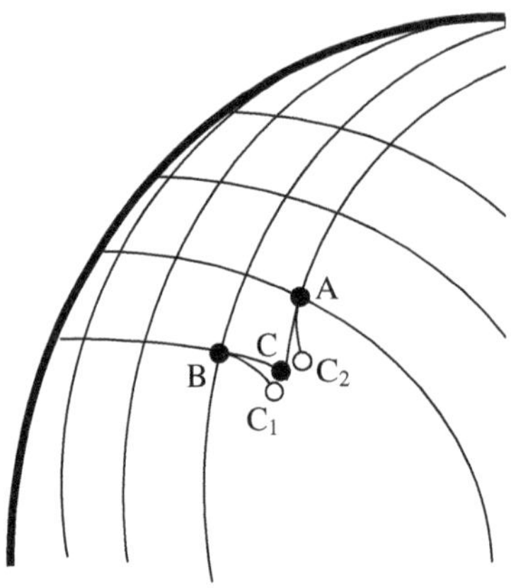

FIGURE 4.19
Kinematics approach: determination of the position of point C knowing those of its neighbors A and B.

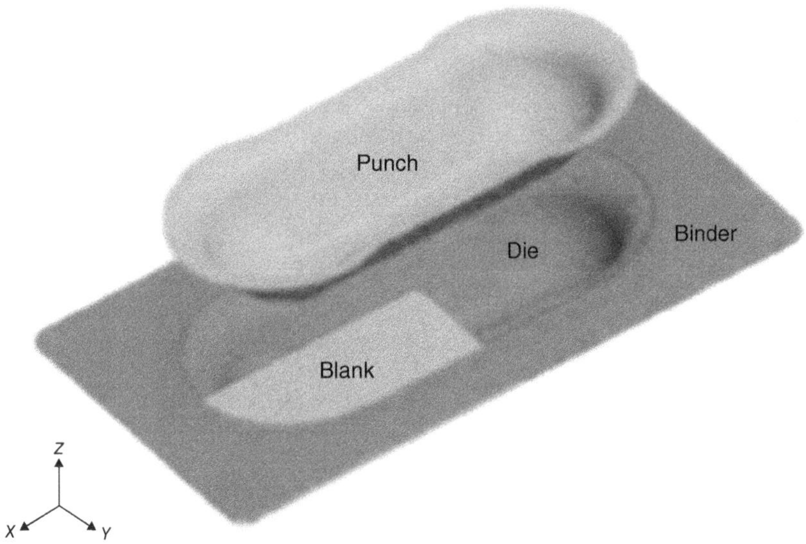

FIGURE 4.20
Double-dome punch, die, and binder [69].

The description of the internal textile material structure is not directly done but must be taken into account in the constitutive model. A large number of continuous mechanical models have been proposed for textile reinforcements [15–22]. A simulation made using the hypoelastic model presented in Section 4.2.4 is presented in section 4.4.3.

4.4.3 Hypoelastic Behavior: Simulation of a Double-Dome Forming

The "double-dome" benchmark has been performed by several labs in order to validate and compare different approaches [53, 67–69]. The tool selected for the benchmark program is shown in Figure 4.20 [69]. The simulation is based on a dynamic explicit approach [70] using the hypoelastic approach presented in Section 4.2.4. The computed deformed shape is compared (Figure 4.21) with the shape obtained by an experimental forming [53]. The plain weave fabric is initially oriented at 0°/90°. To compare experiment and simulation, the shear angles are measured in 10 points of the preform and the draw-in are measured in 10 points of the boundary. The comparisons are presented in Figures 4.22 and 4.23. They show a good accuracy of the forming simulation based on the hypoelastic law.

4.4.4 Composite Reinforcement Forming Using a Semidiscrete Approach

In this approach the textile composite reinforcement is seen as a set of unit woven cells submitted to biaxial tensions, in-plane shear, and bending (Figure 4.24) [71] The woven fabric reinforcement internal work in a virtual displacement field $\underline{\eta}$ is separated in the internal virtual work of tension, in-plane shear, and bending:

$$W_{\text{int}}(\underline{\eta}) = W_{\text{int}}^{\text{t}}(\underline{\eta}) + W_{\text{int}}^{\text{s}}(\underline{\eta}) + W_{\text{int}}^{\text{b}}(\underline{\eta}) \tag{4.11}$$

with:

$$W_{\text{int}}^{\text{t}}(\underline{\eta}) = \sum_{p=1}^{\text{ncell}} {}^{p}\varepsilon_{11}(\underline{\eta})\, {}^{p}T^{11}\, {}^{p}L_1 + {}^{p}\varepsilon_{22}(\underline{\eta})\, {}^{p}T^{22}\, {}^{p}L_2 \tag{4.12}$$

$$W_{\text{int}}^{\text{s}}(\underline{\eta}) = \sum_{p=1}^{\text{ncell}} {}^{p}\gamma(\underline{\eta})\, {}^{p}M^{\text{s}} \tag{4.13}$$

$$W_{\text{int}}^{\text{b}}(\underline{\eta}) = \sum_{p=1}^{\text{ncell}} {}^{p}\chi_{11}(\underline{\eta})\, {}^{p}M^{11}\, {}^{p}L_1 + {}^{p}\chi_{22}(\underline{\eta})\, {}^{p}M^{22}\, {}^{p}L_2 \tag{4.14}$$

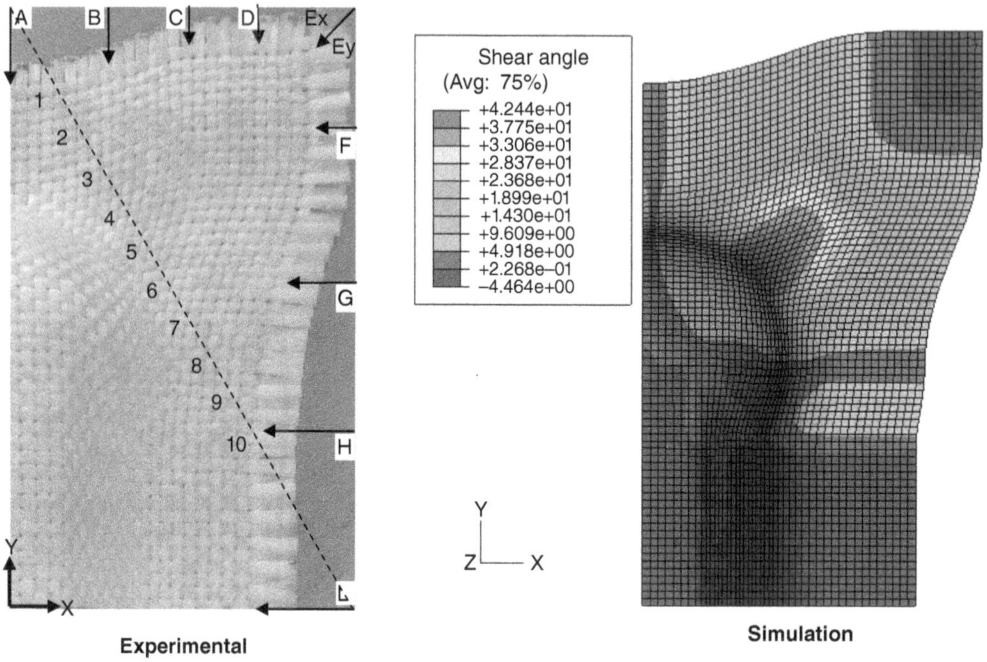

FIGURE 4.21

Double-dome forming: experimental and numerical deformed shape. Reproduced with permission from Elsevier, Journal of Materials Processing Technology.

where $\varepsilon_{11}(\underline{\eta})$ and $\varepsilon_{22}(\underline{\eta})$ are the virtual axial strains in the warp and weft directions. $\gamma(\underline{\eta})$ is the virtual angle between warp and weft directions. $\chi_{11}(\underline{\eta})$ and $\chi_{22}(\underline{\eta})$ are the virtual curvatures of warp and weft directions. ncell is the number of woven cells. ${}^{p}A$ means that the quantity A is considered for the woven cell number p. L_1 and L_2 are the lengths of unit woven cell in warp and weft directions. T^{11} and T^{22} are the tensions on the unit woven cell in warp and weft directions. M^{11} and M^{22} are the bending moments on the woven cell respectively in warp and weft directions. M^s is the in-plane shear moment (Figure 4.25). The mechanical behavior of the textile reinforcement defines a relation between the loads $T^{\alpha\alpha}$, M^s, and $M^{\alpha\alpha}$ and the strain field. The experimental tests specific to textile composite reinforcements that have been presented in Section 4.2 are used to obtain T^{11}, T^{22}, M^s, M^{11}, and M^{22} in function of ε_{11}, ε_{22}, γ, χ_{11}, and χ_{22}.

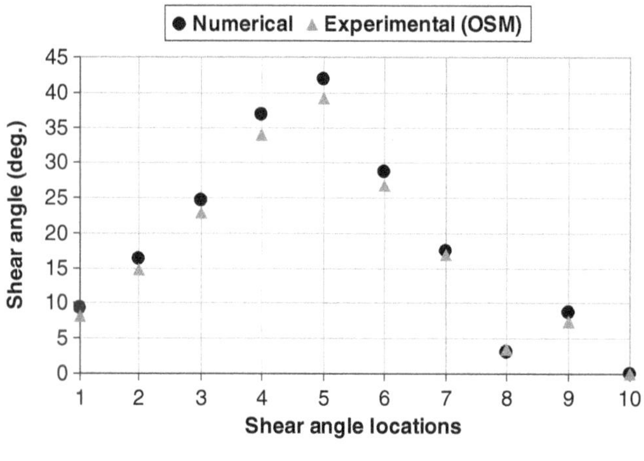

FIGURE 4.22

Double-dome forming: shear angle comparison. Reproduced with permission from Elsevier, Journal of Materials Processing Technology.

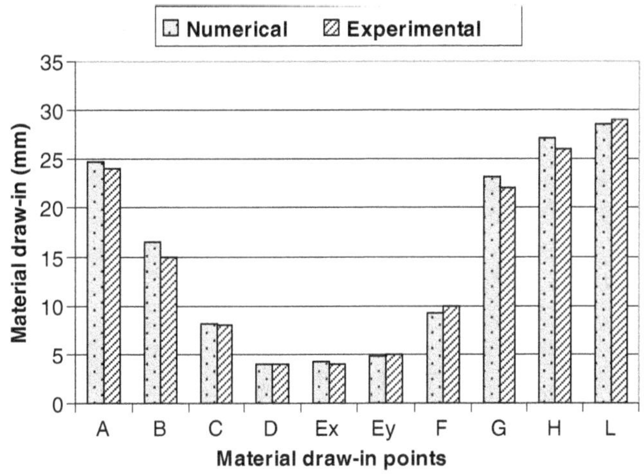

FIGURE 4.23

Double-dome forming: shear angle comparison. Material draw-in comparison.

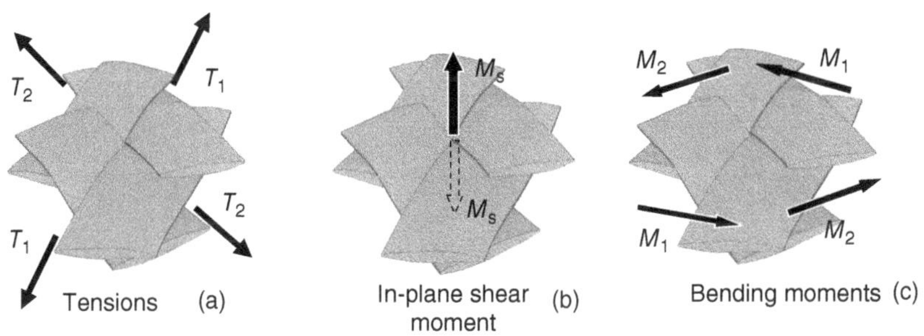

FIGURE 4.24

Load resultants on a unit woven cell.

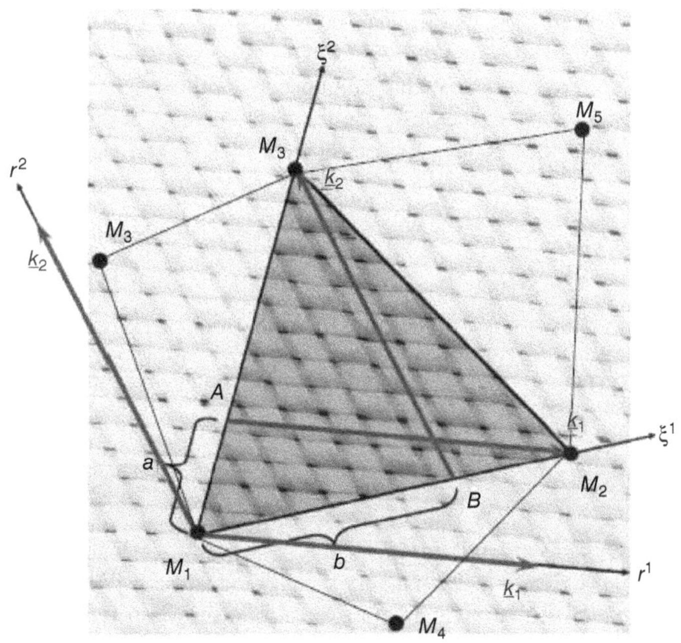

FIGURE 4.25

Three-node finite element made of woven cell.

Shell finite elements can be derived from Equation (4.11)–(4.14). The displacement $\underline{u}$ and virtual displacement $\underline{\eta}$ of any point within an element are interpolated:

$$\underline{u} = \mathbf{N}\mathbf{u}^{\mathrm{e}} \quad \text{and} \quad \underline{\eta} = \mathbf{N}\boldsymbol{\eta}^{\mathrm{e}} \tag{4.15}$$

where $\mathbf{u}^{\mathrm{e}}$ and $\boldsymbol{\eta}^{\mathrm{e}}$ are the matrices of the nodal displacements and virtual displacements, respectively. $\mathbf{N}$ is the interpolation matrix of the element.The three-node shell finite element $M_1M_2M_3$ made up of ncelle woven cells is considered (Figure 4.25). The vectors $\underline{k}_1 = \underline{AM_2}$ and $\underline{k}_2 = \underline{BM_3}$ in warp and weft directions are defined. The internal virtual work of tension on the element defines the element nodal tensile internal forces $\mathbf{F}^{\mathrm{te}}_{\mathrm{int}}$:

$$\sum_{p=1}^{\text{ncelle}} {}^{p}\varepsilon_{11}(\underline{\eta})\, {}^{p}T_1\, {}^{p}L_1 + {}^{p}\varepsilon_{22}(\underline{\eta})\, {}^{p}T_2\, {}^{p}\mathrm{L}_2 = \boldsymbol{\eta}^{eT}\mathbf{F}^{\mathrm{te}}_{\mathrm{int}} \tag{4.16}$$

The internal tensile force components are calculated from the tensions T_1 and T_2:

$$\left(\mathbf{F}^{\mathrm{te}}_{\mathrm{int}}\right)_{ij} = \text{ncelle}\left(B_{1ij}T_1\frac{L_1}{\|\underline{k}_1\|^2} + B_{2ij}T_2\frac{L_2}{\|\underline{k}_2\|^2}\right) \tag{4.17}$$

i is the index of the direction ($i = 1$ to 3), j is the index of the node ($j = 1$ to 3). B_{1ij} and B_{2ij} are strain interpolation components. The internal virtual work of in-plane shear on the element defines the element nodal tensile internal forces $\mathbf{F}^{\mathrm{se}}_{\mathrm{int}}$:

$$\sum_{p=1}^{\text{ncelle}} {}^{p}\gamma(\underline{\eta})\, {}^{p}M_{\mathrm{s}} = \boldsymbol{\eta}^{\mathrm{eT}}\mathbf{F}^{\mathrm{se}}_{\mathrm{int}} \tag{4.18}$$

The internal in-plane shear force components are calculated from the in-plane shear moment:

$$\left(\mathbf{F}^{\mathrm{se}}_{\mathrm{int}}\right)_{ij} = \text{ncelle}B_{\gamma ij}M^{\mathrm{s}}(\gamma) \tag{4.19}$$

The bending stiffness is taken into account within an approach without rotational degree of freedom [72]. The curvatures of the element are computed from the positions and displacements of the nodes of the neighboring elements (Figure 4.25). The internal virtual work of bending on the element defines the element nodal bending internal forces $\mathbf{F}^{\mathrm{be}}_{\mathrm{int}}$:

$$\sum_{p=1}^{\text{ncelle}} {}^{p}\chi_{11}(\underline{\eta})\, {}^{p}M_1\, {}^{p}L_1 + {}^{p}\chi_{22}(\underline{\eta})\, {}^{p}M_2\, {}^{p}L_2 = \boldsymbol{\eta}^{\mathrm{eT}}\mathbf{F}^{\mathrm{be}}_{\mathrm{int}} \tag{4.20}$$

The internal bending force components are calculated from the bending moments M_1 and M_2:

$$\left(\mathbf{F}^{\mathrm{be}}_{\mathrm{int}}\right)_{km} = \text{ncelle}\left(Bb_{1km}M_1\frac{L_1}{\|\underline{k}_1\|^2} + Bb_{2km}M_2\frac{L_2}{\|\underline{k}_2\|^2}\right) \tag{4.21}$$

Details of the formulation can be found in Reference 71.

The simulation of a hemispherical forming with a very unbalanced fabric is performed. The blank-holder is a 6-kg ring (Figure 4.26a). The experimental shape obtained after forming is shown in Figure 4.26b and the computed shape after forming is shown in Figure 4.26c. It is in good agreement with experiments. Especially, the extension ratio at the center of the hemisphere ($l_{\mathrm{weft}}/l_{\mathrm{warp}} = 1.8$) is correctly computed. The shape of the many wrinkles in the flat part of the textile is also properly simulated. Some other composite reinforcement forming simulations based on this semidiscrete approach can be found in References 40, 71, 73, and 74.

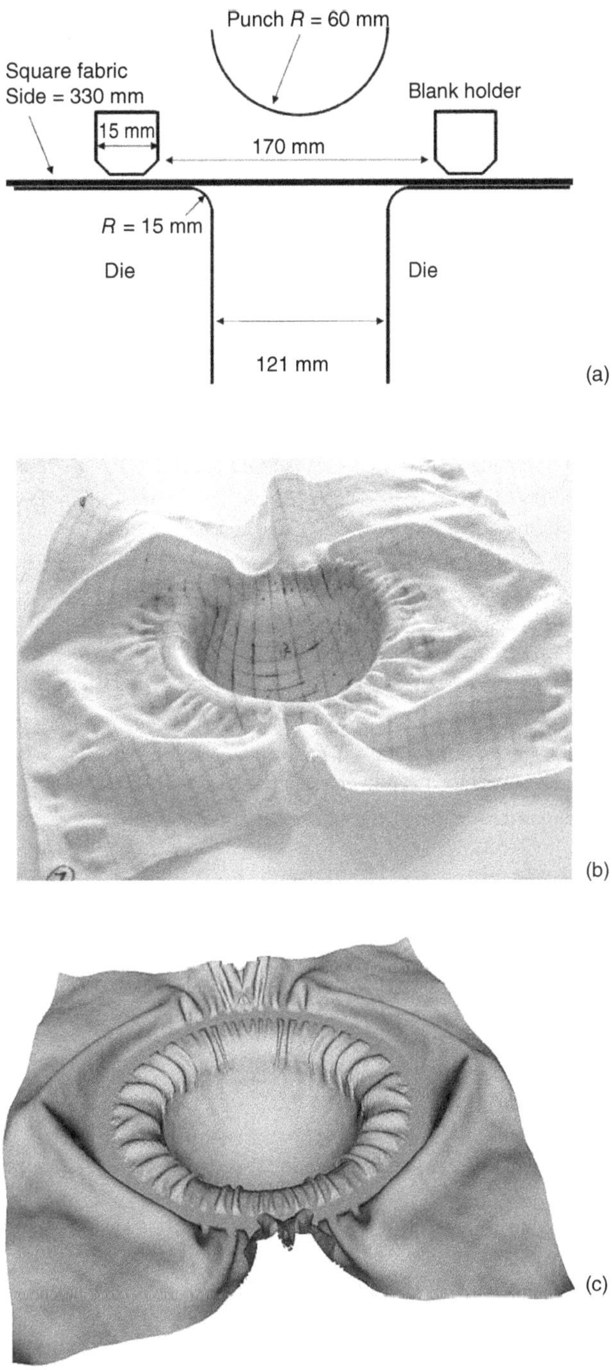

FIGURE 4.26

Forming of an unbalanced textile reinforcement.

4.5 CONCLUSION

Carbon or silicon carbide fiber reinforcements are necessary for the use of ceramic materials because of their brittleness. These reinforcements are shaped in a preform before the infiltration of the ceramic matrix by CVI. The mechanical properties of the textile reinforcement during forming are specific because of the possible motion between the fibers. Specific mechanical tests have been developed for textile reinforcements. In particular, the in-plane shear, that is, the angle variation between warp and weft yarns, can be very large and is the main deformation mode during forming. The bending stiffness is very low in comparison to the tensile rigidity. The textile reinforcements do not follow the standard plate theory. Hypoelastic and hyperelastic models have been proposed to describe at a macroscopic scale the large deformation due to forming. The simulations of the forming process are possible by different approaches. They permit to determine the process parameters to avoid defects such as wrinkles of the fabric. They also give the density and the direction of the fibers in the preform. This information is important for a fine analysis of the CVI process. The textile reinforcement forming simulations can also determine the residual stresses in the preform. The analysis of their consequences on the final ceramic composite part will probably be developed in the next future.

REFERENCES

1. G. Zhou, X. Sun, and Y. Wang. Multi-chain digital element analysis in textile mechanics. *Compos. Sci. Technol.*, 64, 239–244 (2004).
2. D. Durville. Numerical simulation of entangled materials mechanical properties. *J. Mater. Sci.*, 40, 5941–5948 (2005).
3. D. Durville. Finite element simulation of textile materials at mesoscopic scale, in *Finite Element Modelling of Textiles and Textile Composites* (Saint-Petersbourg: Russian Federation. Hal-00274046, 2008).
4. Y. Miao, E. Zhou, Y. Wang, and B. A. Cheeseman. Mechanics of textile composites: micro-geometry. *Compos. Sci. Technol.*, 68, 1671–1678 (2008).
5. A. Gasser, P. Boisse, and S. Hanklar. Analysis of the mechanical behaviour of dry fabric reinforcements. 3D simulations versus biaxial tests. *Compos. Mater. Sci.*, 17(1), 7–20 (2000).
6. P. Badel, E. Vidal-Sallé, E. Maire, and P. Boisse. Simulation and tomography analysis of textile composite reinforcement deformation at the mesoscopic scale. *Compos. Sci. Technol.*, 68(12), 2433–2440 (2008).
7. P. Badel, S. Gauthier, E. Vidal-Sallé, and P. Boisse. Rate constitutive equations for computational analyses of textile composite reinforcement mechanical behaviour during forming. *Compos. Part A-Appl. Sci.*, 40(8), 997–1007 (2009).
8. P. Vandeurzen, J. Ivens, and I. Verpoest. A three-dimensional micromechanical analysis of woven-fabric composites: II. Elastic analysis. *Compos. Sci. Technol.*, 56, 1317–1327 (1996).
9. Q. T. Nguyen, E. Vidal-Sallé, P. Boisse, C. H. Park, A. Saouab, J. Breard, and G. Hivet. Mesoscopic scale analyses of textile composite reinforcement compaction. *Compos. Part B-Eng.*, 44(1), 231–241 (2013).
10. S. Bickerton, P. Simacek, S. E. Guglielmi, and S. G. Advani. Investigation of draping and its effects on the mold filling process during manufacturing of a compound curved composite part. *Compos. Part A-Appl. Sci.*, 28A, 801–816 (1997).
11. E. B. Belov, S. V. Lomov, I. Verpoest, T. Peters, D. Roose, R. S. Parnas, K. Hoes, and H. Sol. Modelling of permeability of textile reinforcements: lattice Boltzmann method. *Compos. Sci. Technol.*, 64, 1069–1080 (2004).
12. F. Loix, P. Badel, L. Orgéas, C. Geindreau, and P. Boisse. Woven fabric permeability: From textile deformation to fluid flow mesoscale simulations. *Compos. Sci. Technol.*, 68, 1624–1630 (2008).
13. P. Boisse, Y. Aimène, A. Dogui, S. Dridi, S. Gatouillat, N. Hamila, M. A. Khan, T. Mabrouki, F. Morestin, and E. Vidal-Sallé. Hypoelastic, hyperelastic, discrete and semi-discrete approaches for textile composite reinforcement forming. *Int. J. Mater. Forming*, 3, 1229–1240 (2010).
14. S. V. Lomov, D. S. Ivanov, I. Verpoest, M. Zako, T. Kurashiki, H. Nakai, and S. Hirosawa. Meso-FE modelling of textile composites: road map, data flow, and algorithms. *Compos. Sci. Technol.*, 67, 1870–1891 (2007).
15. T. G. Rogers. Rheological characterisation of anisotropic materials. *Composites*, 20, 21–27 (1989).
16. A. J. M. Spencer. Theory of fabric-reinforced viscous fluids. *Compos. Part A-Appl. Sci.*, 31, 1311–1321 (2000).
17. W. R. Yu, F. Pourboghrata, K. Chungb, M. Zampaloni, and T. J. Kang. Non-orthogonal constitutive equation for woven fabric reinforced thermoplastic composites. *Compos. Part A-Appl. Sci.*, 33, 1095–1105 (2002).
18. L. Dong, C. Lekakou, and M. G. Bader. Processing of composites: simulations of the draping of fabrics with updated material behaviour law. *J. Compos. Mater.*, 35, 138–163 (2001).
19. P. Boisse, A. Gasser, B. Hagège, and J. L. Billoet Analysis of the mechanical behavior of woven fibrous material using virtual tests at the unit cell level. *J. Mater Sci*, 40, 5955–5962 (2005).
20. B. Hagège, P. Boisse, and J. L. Billoët. Finite element analyses of knitted composite reinforcement at large strain. *Eur. J. Comput. Mech.*, 14, 767–776 (2005).
21. X. Peng and J. Cao. A continuum mechanics-based non orthogonal constitutive model for woven composite fabrics. *Compos. Part A-Appl. Sci.*, 36, 859–874 (2005).
22. R. H. W. ten Thije, R. Akkerman, and J. Huétink. Large deformation simulation of anisotropic material using an updated Lagrangian finite element method. *Comput. Methods Appl. Mech. Eng.*, 196, 3141–3150 (2007).

23. S. V. Lomov, D. S. Ivanov, I. Verpoest, M. Zako, T. Kurashiki, H. Nakai, and S. Hirosawa. Meso-FE modelling of textile composites: road map, data flow and algorithms. *Compos. Sci. Technol.*, 67, 1870–1891 (2007).
24. S. Daggumati, W. Van Paepegem, J. Degrieck, J. Xu, S. V. Lomov, and I. Verpoest. Local damage in a 5-harness satin weave composite under static tension: part II—meso-FE modelling. *Compos. Sci. Technol.*, 70, 1934–1941 (2010).
25. S. V. Lomov. Modelling the geometry of textile reinforcements for composites: WiseTex, in *Composite Reinforcements for Optimum Performance* (Woodhead Publishing Limited, 2011), pp. 200–238.
26. A. C. Long and L. P. Brown. Modelling the geometry of textile reinforcements for composites, in *Composite Reinforcements for Optimum Performance* (Woodhead Publishing Limited, 2011), pp. 239–266.
27. WiseTex http://www.mtm.kuleuven.ac.be/Research/C2/poly/software.html, last accessed on 27 July 2014.
28. TexGen http://www.nottingham.ac.uk/~emxmns/texgen.htm, last accessed on 27 July 2014.
29. G. Hivet and P. Boisse. Consistent 3D geometrical model of fabric elementary cell. Application to a meshing preprocessor for 3D finite element analysis. *Finite Elem. Anal. Des.*, 42, 25–49 (2005).
30. G. Hivet and P. Boisse. Consistent mesoscopic mechanical behaviour model for woven composite reinforcements in biaxial tension. *Compos. Part B-Eng.*, 39, 345–361 (2008).
31. PROEngineer, http://www.ptc.com/product/creo/parametric, last accessed on 27 July 2014.
32. A. G. Prodromou and J. Chen. On the relationship between shear angle and wrinkling of textile composite preforms. *Compos. Part A-Appl. Sci.*, 28A, 491–503 (1997).
33. K. Potter. Bias extension measurements on cross-plied unidirectional prepreg. *Compos. Part A-Appl. Sci*, 33, 63–73 (2002).
34. G. Lebrun, M. N. Bureau, and J. Denault. Evaluation of bias-extension and picture-frame test methods for the measurement of intraply shear properties of PP/glass commingled fabrics. *Compos. Struct.*, 61, 341–352 (2003).
35. X. Peng and J. Cao. A continuum mechanics-based non-orthogonal constitutive model for woven composite fabrics. *Compos. Part A-Appl. Sci*, 36, 859–874 (2005).
36. S. V. Lomov, A. Willems, I. Verpoest, Y. Zhu, M. Barburski, and T. Z. Stoilova. Picture frame test of woven composite reinforcements with a full-field strain registration. *Text. Res. J.*, 76(3), 243–252 (2006).
37. J. Launay, G. Hivet, A. V. Duong, and P. Boisse. Experimental analysis of the influence of tensions on in plane shear behaviour of woven composite reinforcements. *Compos. Sci. Technol.*, 68, 506–515 (2008).
38. S. Lomov, P. Boisse, E. Deluycker, F. Morestin, K. Vanclooster, D. Vandepitte, I. Verpoest, and A. Willems. Full field strain measurements in textile deformability studies. *Compos. Part A-Appl. Sci.*, 39, 1232–1244 (2008).
39. J. Cao, R. Akkerman, P. Boisse, J. Chen, H. S. Cheng, E. F. de Graaf, J. L. Gorczyca, P. Harrison, and G. Hivet. Characterization of mechanical behavior of woven fabrics: experimental methods and benchmark results. *Compos. Part A-Appl. Sci.*, 39, 1037–1053 (2008).
40. P. Boisse, N. Hamila, E. Vidal-Sallé, and F. Dumont. Simulation of wrinkling during textile composite reinforcement forming. Influence of tensile, in-plane shear and bending stiffnesses. *Compos. Sci. Technol.*, 71, 683–692 (2011).
41. S. Kawabata, M. Niwa, and H. Kawai. The Finite deformation theory of plain weave fabrics part i: the biaxial deformation theory. *J. Text. Inst.*, 64(1), 21—46 (1973).
42. K. Buet-Gautier and P. Boisse. Experimental analysis and modeling of biaxial mechanical behavior of woven composite reinforcements. *Exp. Mech.*, 41(3), 260–269 (2001).
43. V. Carvelli, C. Corazza, and C. Poggi. Mechanical modelling of monofilament technical textiles. *Comp. Mater. Sci.*, 42, 679–691 (2008).
44. A. Willems, S. V. Lomov, I. Verpoest, and D. Vandepitte. Optical strain fields in shear and tensile testing of textile reinforcements. *Compos. Sci. Technol.*, 68, 807–819 (2008).
45. S. Kawabata. *The Standardization and Analysis of Hand Evaluation* (The Textile Machinery Society of Japan, Osaka, 1986).
46. T. J. Lahey and G. R. Heppler. Mechanical modeling of fabrics in bending. *ASME J. Appl. Mech.*, 71, 32–40 (2004).
47. E. de Bilbao, D. Soulat, G. Hivet, and A. Gasser. Experimental study of bending behaviour of reinforcements. *Exp. Mech.*, 50, 333–351 (2010).
48. S. Comas-Cardona, P. Le Grognec, C. Binetruy, and P. Krawczak. Unidirectional compression of fibre reinforcements. Part 1: a non-linear elastic-plastic behaviour. *Compos. Sci. Technol.*, 67, 507–514 (2007).
49. P. A. Kelly. Transverse compression properties of composite reinforcements, in *Composite Reinforcements for Optimum Performance*. (Woodhead Publishing Limited, 2011), pp. 333–366.
50. J. K. Dienes. On the analysis of rotation and stress rate in deforming bodies. *Acta Mech.*, 32, 217–232 (1979).
51. T. Belytschko, W. K. Liu, and B. Moran. *Nonlinear Finite Elements for Continua and Structures* (John Wiley & Sons, Chichester, 2000).
52. T. J. R. Hughes and J. Winget. Finite rotation effects in numerical integration of rate constitutive equations arising in large deformation analysis. *Int. J. Numer. Meth. Eng.*, 15, 1862–1867 (1980).

53. M. A. Khan., T. Mabrouki, E. Vidal-Sallé, and P. Boisse. Numerical and experimental analyses of woven composite reinforcement forming using a hypoelastic behaviour. Application to the double dome benchmark. *J. Mater. Process. Tech.*, 210(2), 378–388 (2010).
54. J. Lamon. Ceramic reinforcements for composites, in *Composite Reinforcements for Optimum Performance* (Woodhead Publishing Limited, 2011), pp. 51–86.
55. P. Pineau. Essais virtuels et modèle statistique de multifissuration transverse des fils dans les composites tissés à matrice céramique, Ph.D. thesis (University of Bordeaux France, 2010)
56. A. R. Bunsell. Fibres for composite reinforcement: properties and microstructures, in: *Composite Reinforcements for Optimum Performance* (Woodhead Publishing Limited, 2011), pp. 3–31.
57. S. Yajima, K. Okamura, J. Hayashi, and M. Omori. Synthesis of continuous SiC fibers with high tensile strength. *J. Am. Ceram. Soc.*, 59, 324–327 (1976).
58. Y. Gowayed, G. Ojard, R. Miller, U. Santhosh, J. Ahmad, R. John. Correlation of elastic properties of melt infiltrated SiC/SiC composites to in situ properties of constituent phases. *Compos. Sci. Technol.*, 70 435–441 (2010).
59. G. Couegnat. Approche multiéchelle du comportement mécanique de matériaux composites à renfort tissé, Ph.D. thesis (University of Bordeaux France, 2018).
60. S. G. Advani. *Flow and Rheology in Polymeric Composites Manufacturing* (Elsevier, Amsterdam, 1994).
61. S. Samal and M. Lanka. Ceramic matrix composites: carbon fiber reinforced, in *Encyclopaedia of Composites* (John Wiley & Sons, 2012), pp. 251–259.
62. M. Leuchs. *Ceramic Matrix Composites: CVI*, (John Wiley & Sons, 2012), pp. 259–271.
63. C. Mark and H. M. Taylor. The fitting of woven cloth to surfaces. *J. Text. Inst.*, 47, 477–488 (1956).
64. F. Van Der Ween. Algorithms for draping fabrics on doubly curved surfaces. *Int. J. Numer. Meth. Eng.*, 31, 1414–1426 (1991).
65. A. C. Long and C. D. Rudd. A simulation of reinforcement deformation during the production of preform for liquid moulding processes. *I. Mech. E. J. Eng. Manuf.*, 208, 269–278 (1994).
66. H. Borouchaki and A. Cherouat. Une nouvelle approche pour le drapage des structures composites. *Rev. Comp. Mat. Avanc.*, 32, 407–422 (2002).
67. P. Harrison, R. Gomes, N. Correia, F. Abdiwi, and W. R. Yu. Press forming the double-dome benchmark geometry using a0/90uniaxial cross-ply advanced thermoplastic composite. Proceedings ECCM15 - Venice, Italy, June 24–28, 2012.
68. J. Sargent, J. Chen, J. Sherwood, J. Cao, P. Boisse, A. Willem, K. Vanclooster, S. Lomov, M. Khan, T. Mabrouki, et al. Benchmark study of finite element models for simulating the thermostamping of Woven-fabric reinforced composites. *Int. J. Mater. Form.*, 3, 683–686 (2010).
69. Woven Composites Benchmark Forum, 2008, http://www.wovencomposites. org/., last accessed on 2011-05-09.
70. T. Belytschko. An overview of semidiscretisation and time integration procedures, in *Computation Methods for Transient Analysis*, edited by T. Belytschko and T. J. R. Hughes (Elsevier, Amsterdam, 1983).
71. N. Hamila, P. Boisse, F. Sabourin, and M. Brunet. A semi-discrete shell finite element for textile composite reinforcement forming simulation. *Int. J. Numer. Meth. Eng.*, 79, 1443–1466 (2009).
72. F. Sabourin and M. Brunet. Detailed formulation of the rotation-free triangular element "S3" for general purpose shell analysis. *Eng. Computation*, 23(5), 469–502 (2006).
73. S. Allaoui, P. Boisse, S. Chatel, N. Hamila, G. Hivet, D. Soulat, and E. Vidal-Salle. Experimental and numerical analyses of textile reinforcement forming of a tetrahedral shape. *Compos. Part A-Appl. Sci.*, 42, 612–622 (2011).
74. S. Bel, N. Hamila, P. Boisse, and F. Dumont. Finite element model for NCF composite reinforcement preforming, importance of inter-ply sliding. *Compos. Part A-Appl. Sci.*, 43(12), 2269–2277 (2012).

PART II COMPOSITE MATERIALS

CHAPTER 5

CARBON/CARBONS AND THEIR INDUSTRIAL APPLICATIONS

HIROSHI HATTA,[1] ROLAND WEISS,[2] AND PATRICK DAVID[3]

[1] *Division of Space Propulsion, Institute of Space and Astronautical Science, Japan Aerospace Exploration Agency, Kanagawa, Japan*
[2] *Schunk Kohlenstofftechnik GmbH, Abteilung Composites, Giessen, Germany*
[3] *CEA Le Ripault, Monts, France*

5.1 INTRODUCTION

The lack of suitable high-temperature materials resulted in the development of carbon/carbon materials in the late 1960s [1,2].

First developments in the United States and Europe were tailored to military applications like rocket nozzles and re-entry parts for missiles due to the high costs of these materials [3–8].

Carbons, graphites, and their modifications are typical high temperature materials. Their intrinsic high thermal stability (above 3000°C) and very low density (below 2.2 g/cm^3) make carbon-based materials one of the most promising candidates for high temperature applications.

The requirements for higher mechanical properties at high temperatures from Air Force and NASA research programs resulted in the reinforcement of carbon and the development of carbon/carbons.

The carbon/carbons used in rocket nozzles [3–8] and nose tips for re-entry components were reinforced with carbon fabrics made from low-modulus rayon precursor fibers.

Thereby, those carbon/carbons were low performance qualities compared to the carbon/carbon grades, which are available nowadays.

The influence of fibers, reinforcement patterns, matrix systems, and manufacturing parameters will be discussed and demonstrated in the following sections.

Since the early days of the development of carbon/carbon, the manufacturing technique for industrial applications is mainly based on the reinforcement of polymers and their conversion to carbon/carbon via heat treatments. The first matrix systems were selected in respect to high carbon yields, which is also valid for all manufacturing techniques until now.

5.2 MANUFACTURING OF CARBON/CARBONS

The manufacturing technique, which will be finally used, depends on the geometry, the size, and the number of parts and the mechanical and thermal requirements of the components. Figure 5.1 shows the parameters which can be modified for the production of carbon/carbon (C/C) or carbon/silicon carbide (C/SiC) materials.

Ceramic Matrix Composites: Materials, Modeling and Technology, First Edition. Edited by Narottam P. Bansal and Jacques Lamon.

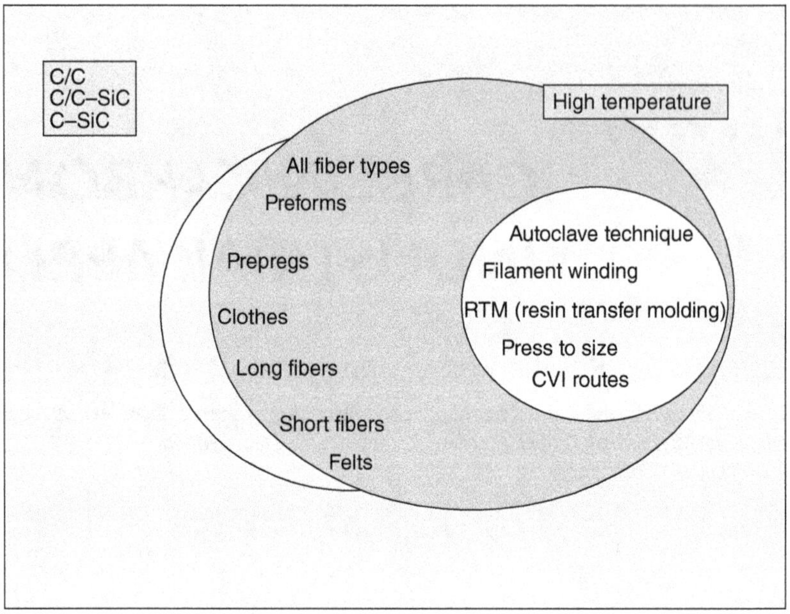

FIGURE 5.1
Parameters for carbon/carbon and ceramic matrix composite (CMC) production. Reproduced with permission from Pergamon, Composites Science and Technology.

Figure 5.2 shows a typical manufacturing cycle for carbon/carbons.

The carbon fibers (C-fibers) and their orientation in the composites are dominating the mechanical properties. The matrix is responsible for the load transfer in a composite and determines the physical and chemical properties of the carbon/carbon material.

Prepregs (pre-impregnated C-fibers) are mainly used as a commercially available semi-product in composite manufacturing for press molding or autoclave procedures.

5.2.1 Carbon Fiber Reinforcements

High performance carbon/carbon materials are tailored to the mechanical requirements of the final composite structure.

Three different types of C-fibers are commercially available, such as high tensile (HT), high modulus (HM), and intermediate modulus (IM) fibers. These fibers are based either on pitch or polyacrylonitride (PAN), whereas rayon-based C-fibers are only used for some few specified applications in military or space structures.

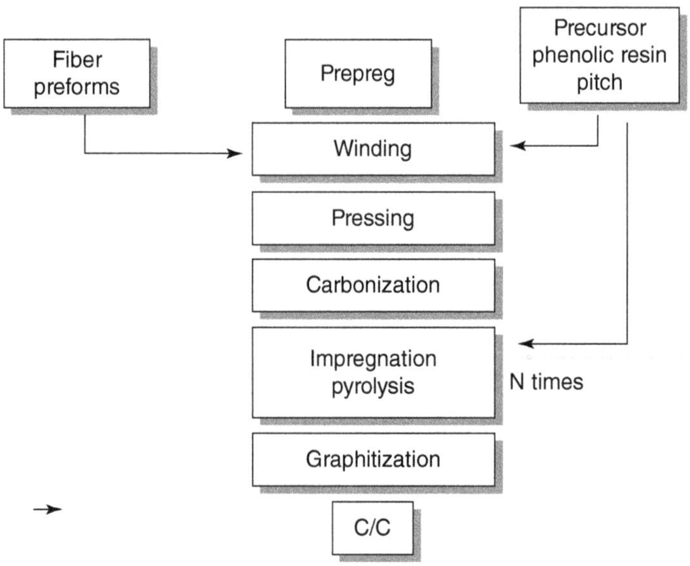

FIGURE 5.2
Schematic manufacturing cycle of carbon/carbon [9]. Reproduced with permission from Pergamon, Composites Science and Technology.

TABLE 5.1 Properties of Commercially Available C-fibers

Manufacturer	Product Name	Precursor	Filament Count	Density (g/cm^3)	Tensile Strength (MPa)	Tensile Modulus (GPa)	Strain-to-Failure (%)
Amoco (USA)	Thornel 75	Rayon	10K	1.9	2520	517	1.5
	T300	PAN	1, 3, b, 15K	1.75	3310	228	1.4
	P55	Pitch	1, 2, 4K	2.0	1730	379	0.5
	P75	Pitch	0.5, 1, 2K	2.0	2070	517	0.4
	P100	Pitch	0.5, 1, 2K	2.15	2240	724	0.31
HEXEL (USA)	AS-4	PAN	6, 12K	1.78	4000	235	1.6
	IM-6	PAN	6, 12K	1.74	4880	296	1.73
	IM-7	PAN	12K	1.77	5300	276	1.81
	UHMS	PAN	3, 6, 12K	1.87	3447	441	0.81
Toray (Japan)	T300	PAN	1, 3, 6, 12K	1.76	3530	230	1.5
	T800H	PAN	6, 12K	1.81	5490	294	1.9
	T1000G	PAN	12K	1.80	6370	294	2.1
	T1000	PAN	12K	1.82	7060	294	2.4
	M46J	PAN	6, 12K	1.84	4210	436	1.0
	M40	PAN	1, 3, 6, 12K	1.81	2740	392	0.6
	M55J	PAN	6K	1.93	3920	540	0.7
	M60J	PAN	3, 6K	1.94	3920	588	0.7
	T700	PAN	6, 12K	1.82	4800	230	2.1
Toho Rayon (Japan)	Besfight HTA	PAN	3, 6, 12, 24K	1.77	3800	235	1.6
	Besfight IM 60	PAN	12K, 24K	1.80	5790	285	2.0
Nippon Graphite Fiber Corp. (Japan)	CN60	Pitch	3K, 6K	2.12	3430	620	0.6
	CN90	Pitch	6K	2.19	3430	860	0.4
	XN15	Pitch	3K	1.85	2400	155	1.5

Table 5.1 gives an incomplete overview of some commercially available C-fibers.

The broad variety of the mechanical properties of the C-fibers allows either composites with optimized strength or with maximum stiffness. The best utilization of these mechanical properties can be realized in unidirectional (UD) C/C composites as shown in Table 5.2.

The mechanical properties of structural components can be varied in a broad range and be adapted to the requirements gained by design criteria from finite element calculations. Therefore, load-carrying structures are modeled via finite element analysis to minimize the structural weight and to guarantee a high service life time. The C-fibers can be orientated according to the mechanical load requirements.

However, for cost effective manufacturing cycles, C-fabrics are used, which are commercially available including all fiber types as well as the desired fiber orientations. Some typical fabric types are given in Figure 5.3.

Although the transverse strength properties are low and thereby a limiting factor in the load capability of carbon/carbons, 3D-reinforcement architectures or 2.5D reinforcements are only used for some few critical applications in space or military structures. The first automated 3D-fiber preforms were manufactured by Brochier in France. The technology was transferred to the US defense company, Textron Specialty Materials, in 1983/1986 [11]. Nowadays, 3D- and 2.5D-fiber preforms are manufactured by various companies worldwide, such as FMI, Brochier, Aerospatiale, EADS,

TABLE 5.2 Mechanical Properties of Unidirectional Carbon/Carbon Composites [10]

	Flexural Strength (MPa)	Flexural Modulus (GPa)	Strain-to-Failure (%)	Tensile Strength (MPa)	Tensile Modulus (GPa)
C/C with HT-C-fibers	1200	220	0.55	1100	250
C/C with HM-C-fibers	600	480	0.15	700	480

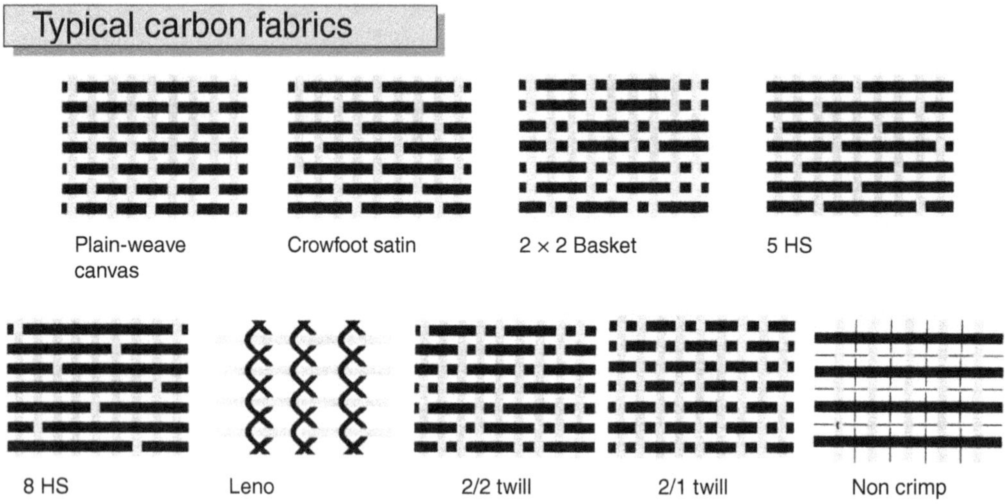

FIGURE 5.3
Typical carbon fabrics for manufacturing of 2D carbon/carbons. Reproduced with permission from Pergamon, Composites Science and Technology.

SNECMA, Carbone Industries, and others. However, these preforms are very often restricted for internal use and are not completely commercially available. Furthermore, the preform costs increase with an increasing number of desired fiber reinforcement directions in multiple-directional reinforced preforms. Therefore, industrially used carbon/carbons are mostly based on a 2D reinforcement which itself is based on C-fabrics or on noncrimped lay-ups as shown in Figure 5.3.

5.2.2 Matrix Systems

The carbon matrix in carbon/carbon composites has to transfer the mechanical loads to the reinforcing fibers even if the mechanical strength of the matrix is negligible. However, it is known from the literature that the matrix is multifunctional and as important as the reinforcement.

The matrix contributes to the physical, chemical, and mechanical properties [12–15]. In particular, the failure behavior strongly depends on the microstructure of the matrix system [12].

The most important criteria for the selection of the matrix (carbon) precursors are:

1. High carbon yield
2. Minimal shrinkage during pyrolysis
3. Easiest processability for all types of manufacturing routes like resin transfer molding (RTM), prepreg manufacturing, filament winding
4. Low solvent content
5. High degree of pre-polymerization with lowest viscosity
6. Availability from multiple sources
7. Low in costs
8. High pot and storage life times

The high carbon yield is required in order to obtain a completely densified carbon/carbon without additional reimpregnation/recarbonization cycles, as shown in Figure 5.2. Fully densified means in principle carbon/carbons with an open remaining porosity below 10%. This goal can be reached by using matrix precursors with carbon yields beyond 84%. Higher carbon yields are always combined with a reduced shrinkage during pyrolysis and thereby with a minimized damage of the carbon/carbons [12]. Solvents as well as a low degree of prepolymerization reduce the carbon yield. An easy processability is an important requirement because the manufacturing of the green bodies (carbon-fiber-reinforced polymers (CFRP)) has to be performed more or less perfectly. All damages occurring during the production of the semiproduct CFRP are detectable later on in the final carbon/carbon component.

Three types of matrix systems are used for the industrial manufacturing processes for carbon/carbons: thermosetting resins, thermoplastic precursors, and, as shown in Figure 5.2, gas-phase-derived carbons or ceramics.

5.2.2.1 Thermosetting Resins as Matrix Precursors

Phenolics, phenolic furfuryl alcohols, furnaces, polyimides, polyphenylenes, and polyarylacetylenes have been used as thermosetting resins [1, 11, 16].

Phenolics and their modifications were already available at the early beginnings of C/C developments in 1960 in the United States. The resins were cheap due to the application for glass reinforcements, easy to process, and they possessed a reasonable char yield. Nowadays, phenolic resins are available with carbon yields up to 74% (resols).

In addition, other types of precursor systems like isocyanates are used now, due to a high carbon yield and due to a better pyrolysis behavior, because these resins are more or less thermally stable up to 400°C. Therefore, the pyrolysis starts later in comparison to standard thermosets and it leaves the possibility of manufacturing a more perfect green body with lower damage and reduced shrinkage.

These improvements of thermal stability and increased carbon yield resulted in the development of special types of polyimides with a carbon yield above 84% (in the late 1970s). Another activity was a thermosetting resin used by the Aerospace Corporation [11] and in research activities in Europe [17] called polyarylacetylene. Both thermosets enabled a carbon/carbon within a single pyrolysis cycle, a time- and cost-saving criteria. However, on the one hand, the high price for the polymers (up to some hundred $/kg) was inappropriate and even more strikingly, was the poor handling qualities of the polymers.

Initially, all thermosetting resins form a glassy carbon after pyrolysis, which cannot be fully graphitized. The influence of the microstructure of the carbon will be discussed later; see chapter "Final Heat Treatment."

5.2.2.2 Thermoplastics as Matrix Precursors

Low-cost precursors are isotropic pitches. All pitches belong to the second type of matrix precursors, the thermoplastics. However, high carbon yields are only feasible in case of high-pressure carbonization (Figure 5.4) with the disadvantage of higher carbonization costs and very often a limitation in size and geometry. Pressureless carbonization of isotropic pitches results in carbon yields comparable to low-cost phenolics.

The development of mesophase pitches and their industrialization by Ashland, Mitsubishi, and Osaka has offered a new class of matrix systems. Their polyaromatic structure with a very high carbon/hydrogen ratio is responsible for their superior carbon yield (Figure 5.4). However, the higher the carbon yield is, the higher will be its viscosity and consequently the melting temperature. Therefore, in the early beginning, mesophase-derived pitches were only used as matrix for carbon/carbons for hot pressing techniques a further limitation for low-cost industrial applications. A second

Matrix precursors

	Carbon yield		Processibility
Phenolics:	Novolacs:	50–55%	Good
	Resoles:	50–55%	Good
Furanes:		70–75%	Medium
Polyarylacetylene:		85–90%	Medium
Pitches:	MPP + MPP-cokes	85–95%	Bad
	Isotropic pitch (with high pressure carbonization)	> 85%	Medium
	Isotropic pitch (carbonization without pressure)	50–55%	Good

FIGURE 5.4

Polymer precursors, their carbon yield, and processability. Reproduced with permission from Pergamon, Composites Science and Technology.

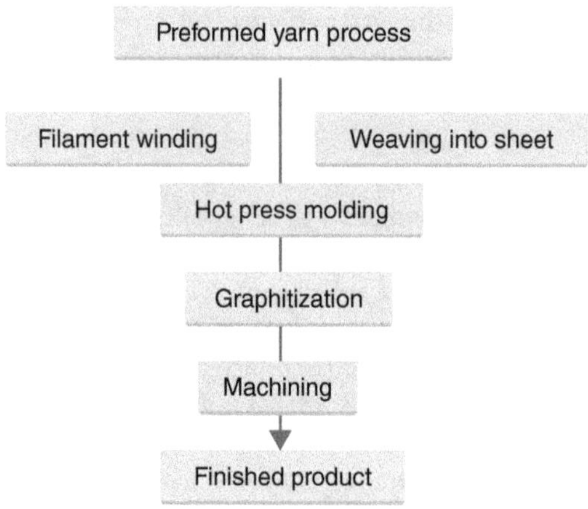

FIGURE 5.5
Yarn prepreg process of Accross [19–21]. Reproduced with permission from Pergamon, Composites Science and Technology.

disadvantage was the thermoplastic behavior. Cross-linking of these pitches was only feasible by oxidation or treatments with sulfur.

Thereby, a manufacturing of prepregs with pure mesophase pitches was actually not feasible. Nowadays, such MPP are already commercially available in a pre-oxidized type and thereby suitable for prepregging via powder technology to be used in a one-step pyrolysis cycle for carbon/carbon manufacturing (Figure 5.2).

Industrial applications of such a one-step-cycle have been put into reality by Schunk [18] and Accross [19, 20]. Schunk formed the bulk mesophase during the molding step up to 400°C and added binders like sulfur or binder pitches during impregnation to the pitches responsible for bulk mesophase. A standard method is the blending of MPP with other carbon precursors. A similar procedure is patented by Accross [19, 20], (Figure 5.5).

In principle, the C-fiber bundle is coated with a blend of pitch and coke. A thermoplastic coating (sleeve) surrounds the C-fiber bundle to encase the powder blend. The C-fiber bundle can be directly used to make all textile preforms as shown in Figure 5.3 and can afterwards be processed to carbon/carbon according to the manufacturing schemes of Figures 5.2 and 5.5. The patents mentioned of Schunk and Accross are open for use throughout the carbon/carbon community.

5.2.2.3 Gas-phase-derived Carbon Matrices As shown in Figure 5.2, carbon/carbons can also be manufactured via chemical vapor infiltration (CVI) techniques.

The CVI process is based on a gas-phase pyrolysis of gaseous species deposited on the surface (pore walls) of the fibrous preform. The process itself is controlled by the deposition parameters and the in-pore diffusion rates. The microstructure of the deposits is mainly determined by the surface of the substrates and the deposition parameters. The CVI process for carbon/carbon and other CMC composites is extensively described by Savage [11, 16]. The CVI parameters as in pore diffusion and deposition kinetics are described by Diefendorf et al. [22]. The homogeneity of pore filling with a minimum of closed porosity is the key factor for fast CVI densification procedures. The most common CVI process is the isothermal impregnation method.

There are four basic CVI processes like:

- Isothermal CVI
- Thermal gradient CVI
- Pressure gradient CVI
- Rapid CVI

All four methods have advantages and disadvantages. All methods are used, modified with each other, or combined with polymer impregnation techniques depending on the requirements of the composite part.

The four basic methods are described in principle in the following.

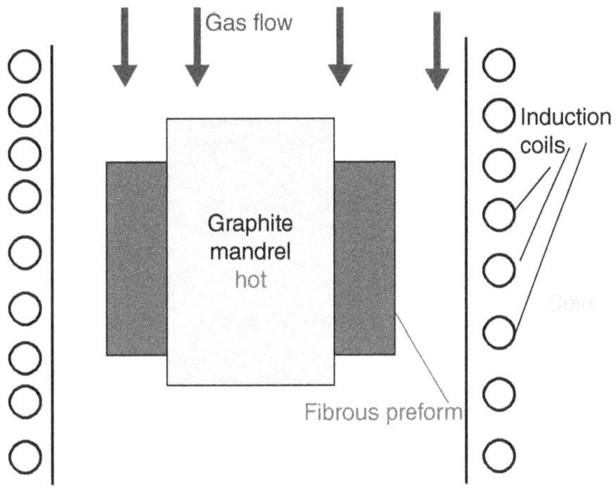

FIGURE 5.6
Principle of thermal gradient chemical vapor infiltration. Reproduced with permission from Pergamon, Composites Science and Technology.

5.2.2.3.1 Isothermal CVI The fiber preforms are placed in a furnace and heated up to the CVI temperature which can be chosen in the range between 800°C and 1200°C. The furnace temperature and pressure (typically below 100 mbar) are kept constant throughout the complete deposition process. Typical resident times for CVI impregnation of C/C brakes for aircraft are 120–200 hours. C/C brake manufacturers like Dunlop have a furnace capacity of about 10 tons of C/C brakes per process cycle and furnace. The used precursor gas is methane, which can be taken in the Dunlop process simultaneously for CVI impregnation and furnace heating.

The pyrocarbon (PyC) closes very often the open porosity on the outer surface before the final densification is carried out. Therefore, the C/C parts have to be removed from the furnace as an intermediate step and machined to open the closed porosity. The parts have to be reheated and the CVI process can be continued up to the required final densification. The number of intermediate machining steps depends on the thickness of the C/C component which has to be densified via CVI. The machining step can be performed up to three times in case of C/C components with a thickness of 40 mm.

The isothermal CVI process has an excellent reproducibility. The resulting PyC matrix possesses a high density and modulus and a good graphitizability. The process is time consuming but cost competitive as proven by the suppliers of C/C brake disks such as Dunlop, SEP, Hitco, Bendix, Allied Signal, Carbone Lorraine, and others.

5.2.2.3.2 Thermal Gradient CVI The thermal gradient method is normally performed as a cold wall CVI procedure. The fibrous preform is placed on a graphite tool (mandrel). The graphite tool is inductively heated (Figure 5.6).

The highest temperature in the fibrous preform is highest at the interface graphite mandrel/fibrous preform and the temperature decreases in direction to the cold wall. The deposition rate increases with increasing temperature. Therefore, densification starts at the interface. The deposited PyC increases the thermal conductivity of the fibrous preform and the highest temperature region moves toward the cold wall. The deposition front follows the thermal front.

Therefore, only large individual parts are densified via thermal gradient methods like rocket nozzle. The method is described in detail by Kotlensky [23].

5.2.2.3.3 Pressure Gradient CVI The pressure gradient CVI method is a modification of the isothermal process. The fiber preform is sealed into a gas-tight tool and heated up in the furnace to the deposition temperature. The deposition gas, usually methane for carbon/carbon, is forced to flow through the fiber preform. The pressure drop in the flow direction, caused by the porous fiber preform, increases with increasing deposition. In contrast to isothermal CVI processes, the deposition rate is independent from diffusion effects due to the forced flow and it increases with increasing pressure drops.

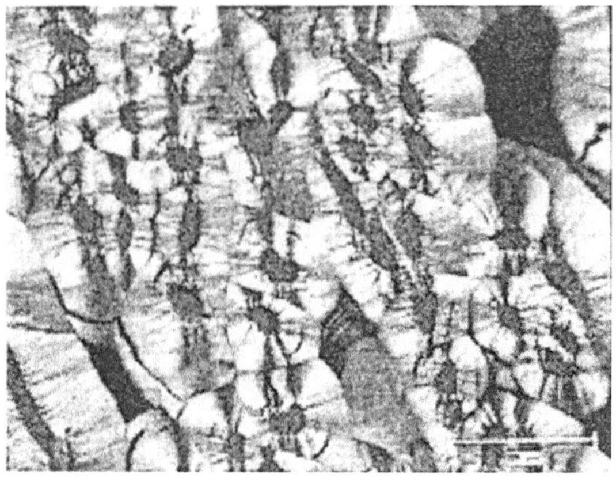

FIGURE 5.7
Microstructure of the carbon deposited (optical microscopy with polarized light) [26]. Reproduced with permission from Pergamon, Composites Science and Technology.

The main drawback of this method is the limitation to single item processing.

Nevertheless, this method and a combination of thermal and pressure gradient CVI was licensed by MAN in Germany from Oak Ridge National Laboratory, TN, USA.

5.2.2.3.4 Rapid CVI Methods Two different methods of rapid CVI densification for carbon/carbons have been developed and can be applied industrially for full densifications of C-fiber preforms.

The first method, the so-called film-boiling method, was developed by CEA in France [24]. The porous C-fiber preform acts as a carbon susceptor and is fully immersed in a hydrocarbon liquid, for example, cyclohexane or toluene. The carbon yield is higher with toluene. The liquid boils during inductive heating and the vapor penetrates into the porous structure. The thermal decomposition of these vapors of hydrocarbons forms a carbon deposit on the inner surface of the porous structure and densifies via the CVI preform. The vapor which is not decomposited is cooled back and can be used again [25]. A complete densification of C/C brakes with a 2.5D C-fiber preform can be done within 10 hours [26].

The carbon matrix is highly oriented and can easily be graphitized [26, Figure 5.7].

The CVI reactor is very simple and made of glass, because the C-fiber preform is cooled over the complete densification cycle by the surrounding liquid precursor. A second method was developed by Huettinger et al. [27, 28]. The porous C-fiber preform, often carbon felt, is placed in an isothermally heated furnace with an outer graphite tool which only leaves a very small gap for the forced gas flow of the precursor gas between graphite tool and porous C-fiber preform. The relation of inner and outer surface volume dominates the deposition chemistry and kinetics [27–30].

The rapid CVI method is controlled by kinetics. The gaseous species can diffuse into the porous preform. However, reactions inside the preform structure can occur by deposition even before the diffusion back to the gas flow has come to an end. A complete overview of carbon/carbon manufacturing by CVI was given by Diefendorf, a pioneer of CVD and CVI techniques [31].

As explained here, the selected matrix precursors have a dominant influence on the manufacturing process. CVI methods can be used, if carbon/carbon parts have to be manufactured near-net shape, because no shrinkage occurs during PyC deposition in the porous structure of the fiber preform.

In case of matrix precursors, which have to be carbonized, like pitches or thermosetting polymers, the shrinkage can be reduced by increased carbon yields of the precursor.

Industrial manufacturing processes are selected from the viewpoint of cost effectiveness. Therefore, polymer precursors are widely used. Polymer precursors can be applied for all technologies known from fiber-reinforced polymers, as is depicted in Figure 5.1. These composite technologies are available at all carbon/carbon manufacturers.

The special knowledge of the manufacturers centers mainly on the precursors that are used, how the different manufacturing techniques are combined, and, last but not least, the parameters for the final heat treatment of the carbon/carbon.

TABLE 5.3 Influence of Densification on the Mechanical Properties of 2D Carbon/Carbons

Grade	Density (g/cm^3)	Porosity (%)	Flexural Strength (MPa)	Tensile Strength (MPa)
CF222	1.55	<8	240	200
CF222/2	1.4	20	140	160
CF226	1.5	<8	150	180
CF226/2	1.4	20	120	120

5.2.3 Redensification/Recarbonization Cycles

The mechanical properties of carbon/carbons increase with an increasing degree of densification (Table 5.3).

Therefore, it is very important to control the pore-filling behavior during reimpregnation and recarbonization to avoid the formation of a high amount of closed pores. The pore-filling mechanisms have been well described in the literature (Figures 5.9, 5.32).

In case of CVI densification the inner walls of the pores are directly coated with carbon. The advantage of CVI impregnations is their capability of filling up large pores. The disadvantages are the small pore-mouth diameters that can be closed resulting in closed porosity (Figure 5.9).

In case of liquid reimpregnations the pore filling depends on different parameters:

- Impregnation pressure
- Viscosity of the polymer
- Carbonization pressure
- Type of polymer

Low impregnation pressures in combination with low viscosities of the polymer are filling larger pores easier than smaller ones. However, during the curing process, or carbonization, the polymers flow out from the pores. In case of pitch impregnation it can be prevented by the application of high pressures (up to 100 bars) during carbonization, whereas for impregnations with thermosets the curing of the polymers is preformed under pressure (up to 10 bars).

The disadvantage is the encapsulation of the condensates (in case of thermosets) or pyrolysis gases (in case of pitch) which can result in a damage of the composite itself.

In principle, reimpregnations with pitches are better for composites with large pores, because the carbon shrinks during pyrolysis onto the surface of the inner wall of the pore (Figure 5.9).

Carbonized and graphitized resins (thermosets) show a shrink off from the inner wall surface, influencing thereby the graphitization behavior during the final heat treatment (Figure 5.8).

5.2.4 Final Heat Treatment

The final heat treatment (HTT) determines the chemical and physical properties of carbon/carbon.

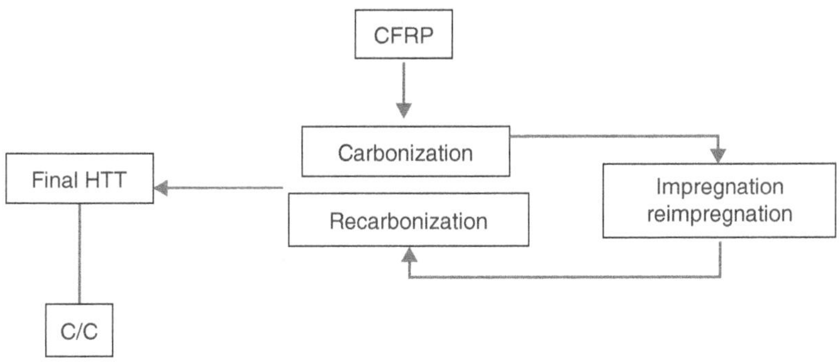

FIGURE 5.8
Recarbonization/reimpregnation cycles. Reproduced with permission from Pergamon, Composites Science and Technology.

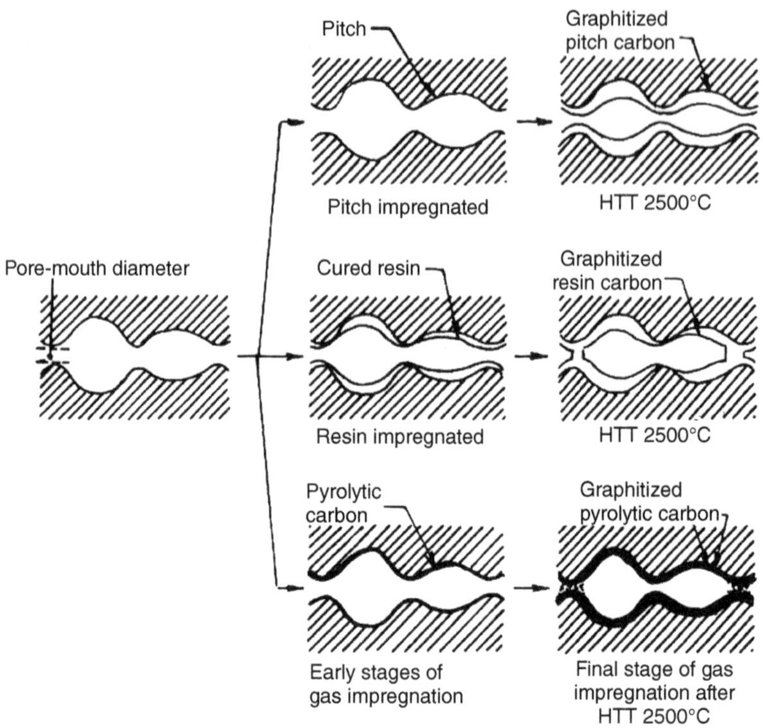

FIGURE 5.9
Carbon pore-filling mechanisms [32]. Reproduced with permission from Pergamon, Composites Science and Technology.

The influence of the HTT on the microstructure is well described in the literature [33–35]. The HTT is performed in between 1700°C and more than 3000°C. Increasing temperatures result in increasing graphitization of the fibers and the matrix in carbon/carbon, which can be investigated by TEM [33]. Pitch- or CVI-derived matrix carbon shows a better graphitization than carbons derived from phenolic resins.

Although phenolic resins form a glass-like carbon, it can be graphitized by *in situ* mechanical stress during HTT (Figure 5.10a, b) [36].

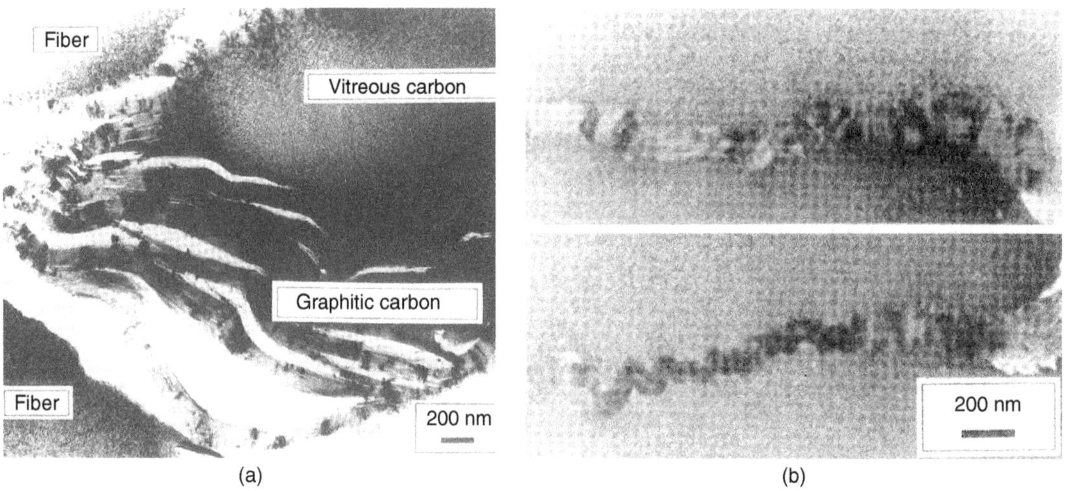

FIGURE 5.10
(a) Transverse cross-section of unidirectional carbon/carbon. The fibers are surrounded by stress-oriented graphite which split into smaller crystallites easily. Thus cracks are deflected around fibers and vitreous carbon. (b) Two examples of closely packed fibers with equidistant surfaces. Reproduced with permission from Pergamon, Composites Science and Technology.

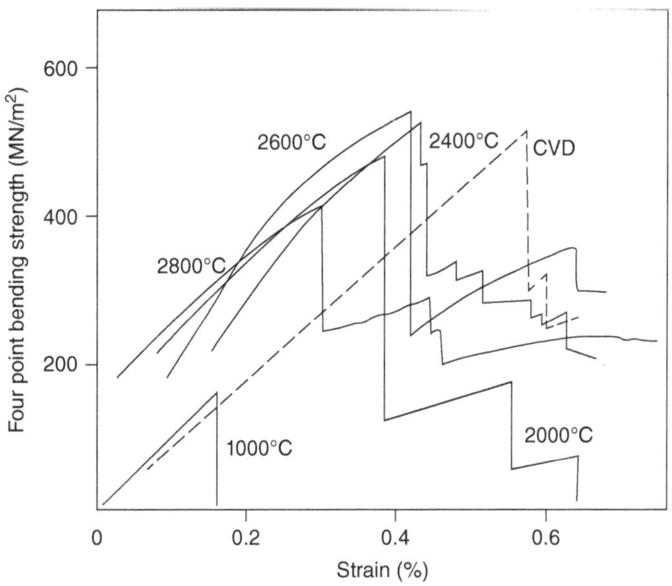

FIGURE 5.11

Fracture behavior of carbon/carbon composites made from a furan precursor, with respect to HTT [37]. Reproduced with permission from Pergamon, Composites Science and Technology.

The degree of graphitization determines electrical and thermal resistivity. Increasing graphitization improves the electrical and thermal conductivity which is important for heating elements or first wall materials in fusion reactors. However, the failure behavior is affected, too. High degrees of graphitized matrix carbon result in a pure shear failure of the matrix with loss of ductility of the carbon/carbon. However, the creep resistance increases with increasing final heat treatment temperature (HTT).

In the case of carbon/carbons with thermosets as matrix precursor, the final HTT dominates the mechanical properties as well as the fracture behavior of the carbon/carbon composites [35–39].

The failure behavior can be modified and tailored from pure brittle failure [37] with an HTT of 1000°C to a pure shear failure (Figure 5.11), [37].

At intermediate temperatures, between 1800°C and 2600°C, a mixed mode of shear failure of the matrix and delamination of the fiber/matrix interface [35] is observed, resulting in a high energy of fracture [38, 39]. The fracture energy is increased from 0.3 kJ/m^2 (HTT 1000°C) to 4 kJ/m^2 (HTT 2700°C) as reported by Zhao et al. [38]. However, the maximum of fracture energy depends on the type of reinforcing C-fibers as well as the polymer precursor [39]. Therefore, the best ductility (pseudo-ductility) of carbon/carbons can be tailored to more convenient temperatures in order to reduce the manufacturing costs.

The optimization of the mechanical and physical properties depends on the final applications.

5.3 STRENGTHS

5.3.1 Introduction

carbon/carbons exhibit superior thermomechanical properties at elevated temperatures above 2000 K. In light of this advantage, carbon/carbons are being expected to be applicable for use in high temperature structures. However, the mechanical behavior of carbon/carbons has not been fully elucidated to the extent that their mechanical properties can be tailored to certain purpose in contrast to the cases with fiber-reinforced plastics and ceramics. Thus, carbon/carbon structures have been designed on a trial-and-error basis, and therefore lack sufficient reliability for use in primary load-bearing structures. As a result, the application of carbon/carbons has been restricted to structures in which high strength is not required, but rather in which only high-temperature capabilities are necessary. Recently, the mechanical

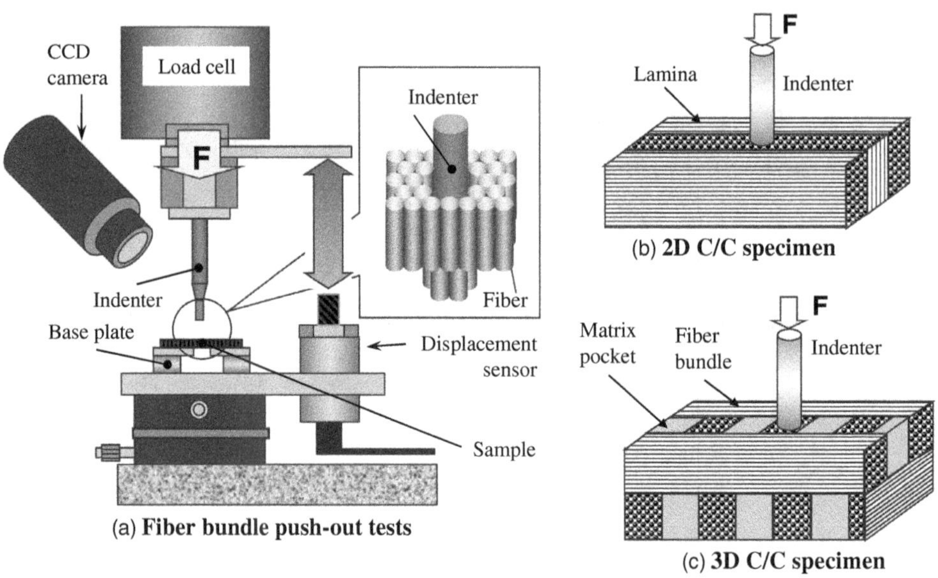

FIGURE 5.12

Schematic drawings of fiber bundle push-out test: (a) test fixture; (b) specimen arrangement for 2D carbon/carbon; (c) 3D carbon/carbon. Reproduced with permission from Pergamon, Composites Science and Technology.

properties of carbon/carbons have found to be profoundly affected by their fiber/matrix interfacial properties [40–45]. This section deals with such recent progress regarding the strengths of carbon/carbons, including studies of their tensile, shear, compressive, and fatigue strengths.

5.3.2 Fiber/Matrix Interface

5.3.2.1 Interfacial Strength at Room Temperature Several attempts have been made to measure the fiber/matrix interfacial shear strength of carbon/carbons using methods often adopted for other composites, that is, polymer matrix and ceramic matrix composites [46–50], including fiber push-in, push-out, and pull-out tests [51–54]. The difficulty associated with the determination of the interfacial strength of carbon/carbons is twofold. First, C-fiber is prone to compressive failure upon compressive axial loading. This tendency is especially noticeable in the case of pitch-based fibers. Accordingly, a specimen must be prepared such that interfacial debonding readily occurs at a sufficiently low load. This means that the specimen should be extremely thin, and less than approximately 100 μm. However, in such a case, the fiber/matrix interface is often damaged during preparation. The second difficulty in this context is the thickness of the C-fibers, which are around 5–10 μm in diameter. Hence, the application of load precisely at the center of fiber cross-sections and confirmation of damage at such interfaces remain rather difficult [53]. Thus, the utilization of the single fiber was limited to PAN-based C-fibers, and to specimens with weak interfaces.

In order to avoid such difficulties, we attempted a fiber bundle push-out test. The procedure of this test is illustrated in Figure 5.12. In this test, an indenter measuring 50 μm in diameter was successfully employed [44,55]. Thus, the diameter of the pushed-out bundle is large and the thickness of the specimen can be set at about 100–300 μm. Observation of pushed-out bundles assured that fiber/matrix interfacial fracture dominated the fractured area. This result guaranteed reliability of this test method. However, the single and bundle push-out tests resulted in different interfacial strengths [53]. Accordingly, the present methods are useful in comparative purpose.

Typical results obtained by the fiber bundle push-out tests are shown in Figure 5.13 [43] and 3 [56] for carbon/carbons reinforced with PAN-based fiber IM-600 and pitch-based fibers K321 and K633, respectively. In Figure 5.13, 2D-RC-2573K and 2D-HIP-2573K represent carbon/carbons with cross-ply lamination (CP) densified by the resin char method (RC) or the hot isostatic pressurization method (HIP), and heat treated at 2573 K. The fiber volume fraction (V_f) of the precursor (carbon-fiber-reinforced phenolic resin) of these carbon/carbons was 60%. As this figure shows, the interfacial sliding stress τ_s and the debonding stress τ_d were enhanced with increases in density ρ, but the ultimate tensile fracture strain ε_b was degraded with ρ. By densification, interfacial debonding is gradually eliminated,

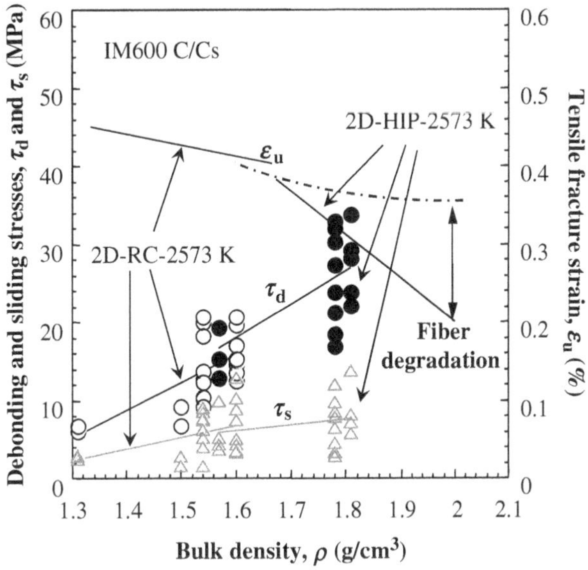

FIGURE 5.13

Tensile fracture strain, ε_u; interfacial debonding stress, τ_d; and interfacial sliding stress, τ_s of a cross-ply-laminated carbon/carbon reinforced with PAN-based fiber IM600 as a function of bulk density, ρ, densified by repeating resin char method and hot isostatic pressuring method. Reproduced with permission from Pergamon, Composites Science and Technology.

as is clear from a comparison of the photos in Figure 5.14a, b. In Figure 5.15, K321 and K633 represent pitch-based reinforcing fibers produced by Mitsubishi Chemical Co., Japan. For these carbon/carbons, the τ_ds are shown as functions of the HTT. The τ_ds of these carbon/carbons clearly decrease with increases in HTT. The decrease in the τ_ds was partly due to the extension of fiber/matrix interfacial debonding with increasing HTT, as is shown in Figure 5.14c and d. This development of debonding was facilitated primarily by the development of graphitization in the matrix [56]. The other source of the reduction in τ_d was found to be a reduction in the bonding strength of truly bonded region due

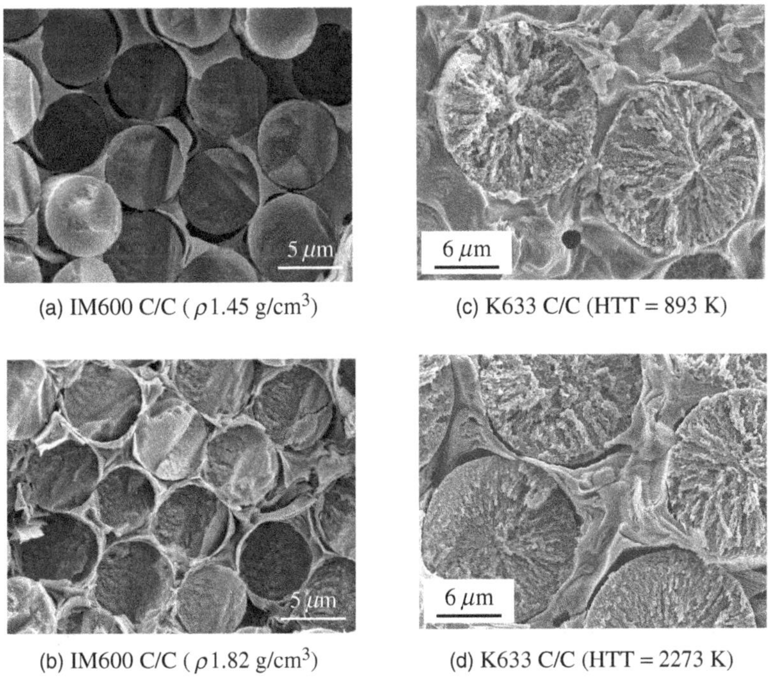

(a) IM600 C/C (ρ1.45 g/cm^3)

(c) K633 C/C (HTT = 893 K)

(b) IM600 C/C (ρ1.82 g/cm^3)

(d) K633 C/C (HTT = 2273 K)

FIGURE 5.14

(a) and (b) Cross-sections of IM600 carbon/carbons, with different densities, ρs; (c) and (d) K633 carbon/carbons, heat treated at different temperatures. Reproduced with permission from Pergamon, Composites Science and Technology.

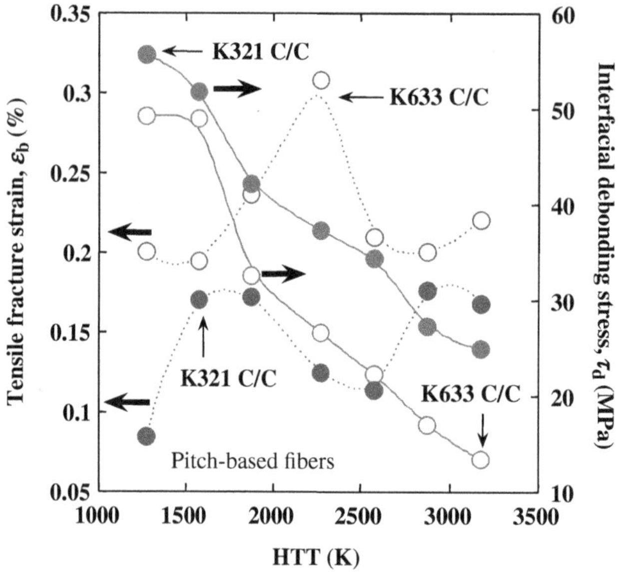

FIGURE 5.15

Tensile fracture strain, ε_u and interfacial debonding stress of cross-ply-laminated carbon/carbons reinforced with pitch-based fiber K321 and K633 as a function of heat treatment. Temperature, HTT densified by five cycles of hot isostatic pressuring, HIP. Reproduced with permission from Pergamon, Composites Science and Technology.

to graphitization, by which the crystalline units increased in size and the (0 0 2) surface was aligned in a more parallel manner with the interfaces [56].

5.3.2.2 *Interfacial Strength at Elevated Temperatures*

In order to examine the interfacial debonding stress at elevated temperatures, a specially arranged fiber bundle push-out test was conducted [45,54]. Figure 5.16 [54] summarizes the temperature dependence of the fiber bundle interfacial debonding stress τ_d of 3D carbon/carbon as a function of the test temperature, T. As is clearly seen in this figure, τ_d monotonically increases with increasing T. The matrix in carbon/carbons shrinks to a great extent during the processing of carbon/carbon, at the carbonization stage, due to the conversion of resin into carbon, and the cooling stage from the HTT to room temperature by reversed thermal expansion. Hence, significant tensile thermal stresses accumulate in the matrix of carbon/carbons [57] in the directions parallel

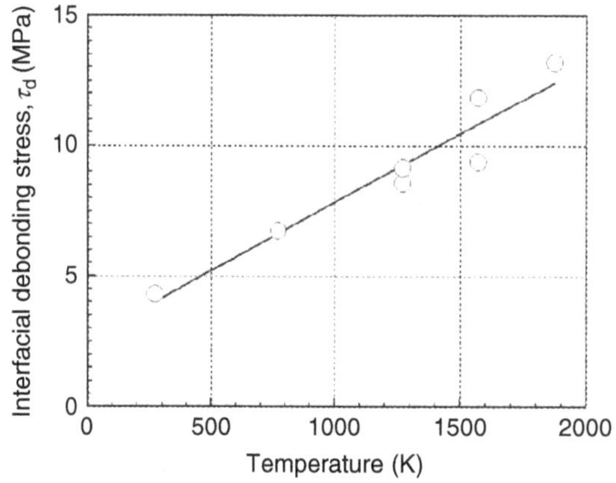

FIGURE 5.16

Temperature dependence of fiber/matrix interfacial debonding stress of the 3D carbon/carbon. Reproduced with permission from Pergamon, Composites Science and Technology.

and perpendicular to the fiber/matrix interface. These tensile stresses reduce the τ_d. Thus, the dominant mechanism leading to increases in τ_d at elevated temperatures is thought to be the relief of thermal stresses in carbon matrices at the fiber bundle interfaces.

5.3.3 Tensile Strength

5.3.3.1 Parameters Determinative of Tensile Fracture The tensile strength of carbon/carbons is typically much weaker than that expected from the rule of mixture. In order to identify the mechanisms yielding such low strength, the effects of various potential factors including the HTT [58,59], the fiber/matrix interfacial strength [60,61], and defects in the carbon/carbons [62–64] have been investigated. Continuous fiber-reinforced ceramic matrix composites (CFCCs) are also known to elongate to a much lesser degree than do reinforcing fibers [65, 66]. Statistical models have been proposed [67–69] to account for the tensile strength of CFCCs, in which the fiber/matrix interfacial sliding stress τ_s has been assumed to support a portion of the tensile load. Recently a model proposed by Curtin et al. [67] has attracted attention for its successful prediction of the tensile fracture stress of CFCCs. In their model, tensile fracture stress is slightly enhanced with increases in τ_s. C/C usually exhibit the opposite tendency [70, 71]; the tensile strength of carbon/carbons was found to improve with decreases in interfacial bonding, τ_d. Thus, a new model is required in order to simulate the tendency of carbon/carbons toward tensile fracture.

Hatta et al. recently examined the effects of density on the tensile strength of various carbon/carbons, and they reconfirmed that the tensile strength of the carbon/carbons decreased with the enhancement of matrix/fiber interfacial strength, τ_d, [44] as shown in Figure 5.13. The 2D-HIP-2573 K and 2D-RC-2573 K in the figure represent that the C/Cs are a lamination type (reinforced with PAN-based fiber in the direction of 0°/90°), produced by HIP and RC, respectively, and heat treated 2573 K. The 2D-HIP-2573 K samples shown in this figure degraded more rapidly than 2D-RC-2573 K specimens. This was found to be due to fiber degradation, which occurred only in the case of 2D-HIP-2573 K samples by simultaneous and repeated applications of high pressure and high temperature, HIP treatment [44]. In this figure, the tensile fracture strain was diminished with density, ρ. However, this result was not entirely in agreement with findings reported as results in other studies of the densification effect on the tensile strength of carbon/carbons; the tensile strength of carbon/carbons has often been reported to increase with ρ when carbon/carbon density is rather low [72]. However, if one considers the data 2D-RC-2573 K in Figure 5.13, in which the tensile fracture stress is plotted instead of strain, then the revised data demonstrates that the fracture stress increases with ρ [44]. This behavior can be explained in terms of the Young modulus E; the Es of low density carbon/carbons were much lower than the predictions obtained by the rule of mixture. Such low Es were shown to be obtained by the tests yielding nonuniform stress distributions. At a low density, the tensile strain was not uniform across a cross-section of a tensile specimen due to the extremely low interfacial strength of carbon/carbons. The load from the tensile test fixture of a tensile loading machine to the surfaces of a specimen was not transferred to the core (center region) of the specimen. Thus, low density carbon/carbon tended to be loaded only in near-surface regions [44].

Figure 5.15 shows the tensile fracture strain ε_us and interfacial shear strength τ_δs of pitch-fiber-reinforced carbon/carbons as functions of HTT [56]. In this figure, the relationship between the ε_us and τ_δs appears rather complicated; although the τ_δs monotonically decrease with HTT, the ε_us show more complex behaviors. The peculiar behavior of the ε_us could be accounted for as a result of fiber degradation [63]. Thus, Figures 5.13 and 5.15 suggest that the tensile fracture strains of carbon/carbons can be determined on the basis of fiber strength and interfacial strength, provided the statistical distributions of both are included [64]. The tensile fracture surfaces of K633 carbon/carbons heat treated at various temperatures are shown in Figure 5.17. As this figure shows, the tensile fracture of carbon/carbons occurred by intermittent fractures by fiber bundle units; the thicknesses of the fractured fiber bundles consistently decreased with decrease in τ_δ. This tendency was also observed in the cases of the K321 carbon/carbons and IM600 carbon/carbons (the 2D-HIP-2573 K and 2D-RC-2573 K in Figure 5.13a). These findings as well as the previously demonstrated low interfacial strength of carbon/carbons led to the development of the following tensile fracture model [56].

1. From a single-fiber fracture, a crack propagates until the fracture tip meets interfaces possessing an extremely low τ_d. At such locations, the crack front is deflected. Thus, the fracture bundle is determined by the probability of encountering an extremely weak interface. As shown in Figure 5.13, interfacial debonding stress was scattered very widely. These findings indicate that when the mean τ_d is high, then the fracture bundles are thick.
2. The type of bundle-unit fractures described above intermittently continues to proceed.

(a) 1273 K, $\tau_d = 49$ MPa
(b) 2273 K, $\tau_d = 25$ MPa
(c) 2873 K, $\tau_d = 17$ MPa
(d) 3173 K, $\tau_d = 13$ MPa

FIGURE 5.17
Comparison of tensile fracture surfaces of K633 carbon/carbons heat treated at designated different temperatures with different interfacial shear debonding stresses, τ_d. Reproduced with permission from Pergamon, Composites Science and Technology.

3. The load, which is sustained by broken bundles, is nearly uniformly transferred to surviving fibers, because the τ_d is extremely low in the carbon/carbons.
4. When the remaining fibers cannot sustain the load, then the ultimate fracture occurs. Thus, the ultimate fracture occurs earlier when fracture bundle thicker, because the transferred load per fractured bundle is higher.

Thus, the tensile fracture stress of carbon/carbons can be predicted if we can determine the fracture fiber bundle by a simulation of fracture process of each fiber bundle using, for example, a Monte Carlo method. For a specially arranged mini-bundle carbon/carbon, fracture bundle could be easily determined and a Monte Carlo simulation resulted in excellent agreement with experimental results [73].

5.3.3.2 Tensile Properties at Elevated Temperatures Only a small number of reports have been published concerning high temperature behaviors [70,71,74,75], and the results have been contradictory, even with respect to the temperature dependence of tensile or flexural strength. Figure 5.18 shows the typical tensile stress–strain, σ–ε, curves of a CP-laminate carbon/carbon loaded in the 0° (fiber axis) direction at room temperature and 2273 K, and obtained at a crosshead speed of 0.1 mm/min or 5.0 mm/min [76]. As shown in this figure, the σ–ε curve of the C/C at room temperature was linear up to the point of total fracture, and the same line was traced during the repetition of loading–unloading. In contrast, a nonlinear deformation and permanent strain were observed at 2273 K at a test speed of 0.1 mm/min. However, the σ–ε curve at a test speed of 5.0 mm/min resulted in a nearly straight line up to the total fracture, with a slight decrease in the Young modulus, compared with the results obtained at room temperature shown as a broken line in Figure 5.18. These findings indicate that, at elevated temperatures, the σ–ε curves of carbon/carbons are similar to those at room temperature, and apparent nonlinear behavior appears due to a strain-rate-dependent factor, that is, creep deformation.

The reported results regarding the temperature dependence of tensile and flexural strength have been contradictory; while some researchers have reported a weak dependence on temperature [70,71,74], others have shown a significant enhancement of strength [71,76]. To elucidate the mechanisms governing the enhancement of the strength of carbon/carbons at high temperatures, tensile tests of a 2D-laminate carbon/carbon parallel to the fiber axis were conducted at various temperatures of up to 2773 K [76]. The tensile strength of the carbon/carbon was monotonically enhanced with increases in temperature, as shown in Figure 5.19. This enhancement became rapid from 1773 K. One of the mechanisms responsible for this enhancement in strength was the degassing of absorbed water [77], which exerted a dominant influence on the enhancement in tensile strength observed at temperatures of up to 1773 K. The second mechanism was the enhancement of the strengths of the fibers and matrix due to stretching graphitization that took

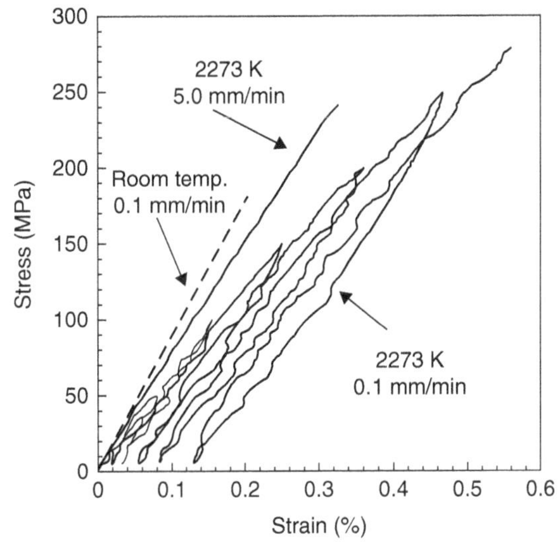

FIGURE 5.18

Stress–strain curves of the CP at room temperature and at 2273 K with different crosshead speeds of 0.1 mm/min and 5.0 mm/min. Reproduced with permission from Pergamon, Composites Science and Technology.

place during creep deformation. This phenomenon was especially notable at temperatures exceeding 1773 K, and it produced much greater levels of enhancement than did degassing.

5.3.3.3 Creep Behavior Creep deformation in a carbon/carbon has been reported at temperatures exceeding 1773 K [76,78], as shown in Figure 5.20 in the case of a 2D-laminate carbon/carbon reinforced with M40 C-fiber (M40 carbon/carbon). In this experiment, the tensile creep behavior was observed until a creep strain of 10%, but no creep rupture was observed. More rapid creep behavior was observed for a carbon/carbon reinforced with high-strength type C-fiber T300, Toray [76]. Generally, creep rates have been compared in terms of steady-state creep, where the creep

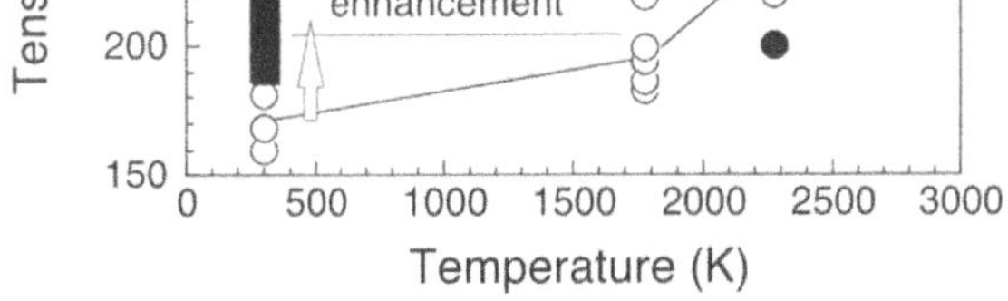

FIGURE 5.19

Temperature dependence of tensile strength in a cross-ply-laminated carbon/carbon (reinforced with high-modulus carbon fiber M40 Toray) obtained under a vacuum up to 2273 K and an Ar atmosphere over 2273 K. The difference between carbon/carbon strengths with and without de-gas treatment represents the effect of absorbed water. The de-gas treatment was made at 1473 K under a vacuum. The strength improvement up to1773 K was caused by the evaporation of absorbed gas, while the improvement over 1773 K was caused by creep deformation. Reproduced with permission from Pergamon, Composites Science and Technology.

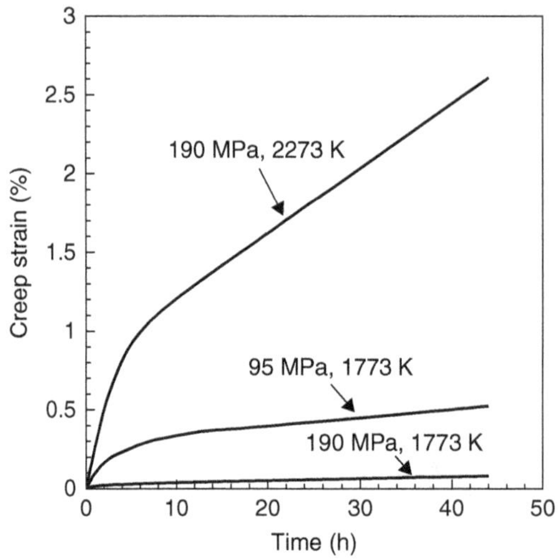

FIGURE 5.20

Creep curves of cross-ply-laminated carbon/carbon reinforced with high-modulus fiber M40. Reproduced with permission from Pergamon, Composites Science and Technology.

rate is constant. As regards steady-state creep, the Dorn equation [79], given by Equation (5.1), has frequently been adopted. The Dorn equation is characterized by the stress exponent, n, and the activation energy, Q, as follows:

$$\dot{\varepsilon} = A\sigma^n \exp(-Q/RT) \tag{5.1}$$

where $\dot{\varepsilon}$ is the steady-state creep rate, A is a constant, R is the gas constant, and T is the test temperature. From Figure 5.20, n and Q were determined to be 2.0 and 1100 kJ/mol, respectively, and those of a carbon/carbon reinforced with T300 were 6.0 and 730 kJ/mol, respectively; here, the Q value of these two carbon/carbons appears nearly the same, but n is different. A similar Q value has been reported for other carbon/carbons [78] and glassy carbons [80], that is, from 1000 to 1200 kJ/mol. As regards the value of n in the case of carbon materials, values of 6.0 and 8.0 [78,80] have been reported. This figure indicates that differences in fibers result in different values of n, but similar Q values. However, the underlying mechanisms for these results have not been elucidated.

5.3.4 Shear Strength

5.3.4.1 Stress–Strain Curve Although carbon/carbons exhibit high strength and an almost linear σ–ε curve in terms of tension parallel to a fiber axis, a significant nonlinearity in the stress–strain curve followed by extremely low strength appears in cases involving shear loading [81,82]. In particular, 3D-reinforced carbon/carbons have been shown to possess remarkably low shear strength and yet high strain capability [40]. For example, a tensile strength of 250 MPa and a shear strength of 20 MPa were reported for an orthogonal 3D C/C. As a result, the poor shear properties and the significant shear nonlinear behavior are often critical parameters in the context of designing carbon/carbon structures.

The evaluation of shear behavior at elevated temperatures remains a difficult task. Thermal expansion mismatch between a specimen and test fixtures prevents the utilization of test fixtures often employed for room-temperature shear tests, for example, the Iosipescu shear-test configuration. Aoki et al. [44] used an off-axis compressive tests method in which fibers were oriented at ±45° to the loading direction. This method enabled the use of self-standing specimens, and thus the use of a supporting fixture at elevated temperatures can be avoided. For example, Aoki et al. used dog bone specimens possessing a cross-section of 10 mm × 10 mm, as shown in Figure 5.21. The shear stress τ and the shear strain γ were calculated according to the following equations [83].

$$\tau = 0.5 \times P/A,\ \gamma = 2 \times \varepsilon_L \tag{5.2}$$

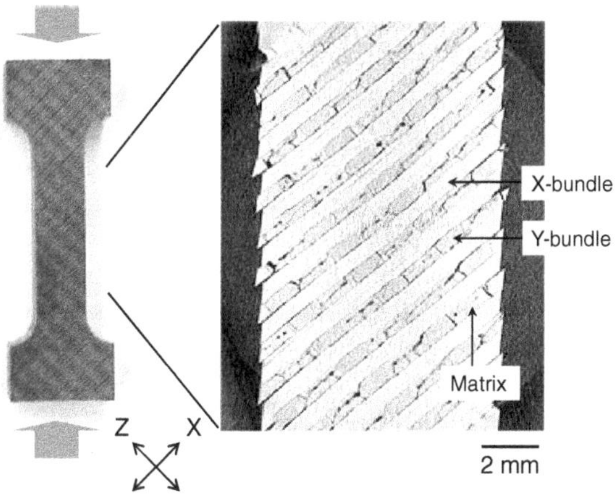

FIGURE 5.21

Specimen geometry of off-axis compressive test and shear damage observed after a shear strain γ of 10%. Reproduced with permission from Pergamon, Composites Science and Technology.

where P, A, and ε_L denote the applied load, the cross-sectional area of the gauge sections, and the strain in longitudinal directions of specimens, respectively. The problem associated with this test method is that not only pure shear stress but also normal stress are applied to the fracture surface. The corresponding fracture shear stress might be affected by the normal stress.

Figure 5.22 shows shear stress–strain (τ–γ) curves of an orthogonal 3D carbon/carbon obtained at room temperature and also at elevated temperatures obtained using the off-axis compressive method. The shear behavior represented by the τ–γ curves was divided roughly into three regions: (i) initial elastic, (ii) progressive nonlinear, and (iii) nearly plateau stress regions. The initial shear modulus obtained in region (i) was as low as 2 GPa, and loading capability did not degrade even at $\gamma > 10\%$. The onset stress of nonlinearity and maximum shear stress regions significantly increased with increases in the test temperature. To investigate the underlying mechanisms yielding substantial shear deformation and strength improvement at elevated temperatures, the evolution of shear damage was observed at different loading stages. At approximately the transition from region (i) to region (ii), debonding at fiber bundle surfaces started to propagate; this led to nonlinearity in the τ–γ curve. In region (ii), the debonding of fiber bundles occurred progressively. When the debonding in the gauge section was completed, a nearly constant shear stress continued to be observed (region (iii)). In region (iii), the 3D carbon/carbon was deformed by the sliding at fiber bundle interfaces. Figure 5.10 shows the shear damage observed in a specimen deformed up to region (iii) at room temperature. This photograph clearly

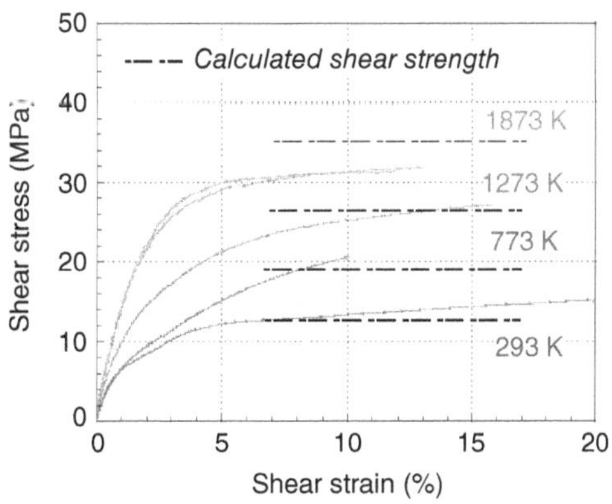

FIGURE 5.22

Shear stress–strain curves of the orthogonal 3D carbon/carbon from room temperature to elevated temperatures. Reproduced with permission from Pergamon, Composites Science and Technology.

demonstrates the shear damage at fiber bundle interfaces. This finding indicates that the shear strength enhancement at elevated temperatures can be attributed to the increase in the interfacial shear strength of fiber bundles.

5.3.4.2 Shear Fracture Model

Based on the above results, an analytical model was proposed by Aoki et al. for simulating the shear damage process of a 3D carbon/carbon [54]. This model was derived as a modification of Keith's model, which was originally developed for laminated composites [54]. In the present model, the shear damage of the 3D carbon/carbon was simplified. The applied load in the plateau stress region is sustained by two frictional sliding stresses, acting in the directions parallel and perpendicular to (rotational) fiber bundles. The sliding stresses were determined by the fiber bundle push-out tests shown in Figure 5.16. Thus, the shear strength τ_{max} of the 3D carbon/carbon can be expressed as [54]:

$$\tau_{\max} = \frac{\pi}{2}\tau_d + 4\sum \left[\frac{\pi}{2}\frac{\tau_d}{L_z d^2}\int\int r^2\, dr\, d\phi\right] \tag{5.3}$$

where L_z and d denote the width of the z bundles and the size of the unit cell used for the calculations, respectively. The first and second terms represent the respective influences of axial bundle sliding and rotational sliding. Equation (5.3) predicts high shear strength results for composites with strong interfaces. The calculated τ_{max}s of the 3D C/C are also indicated in Figure 5.22 by the dotted lines. As can be seen in Figure 5.22, reasonable agreement was observed between the calculated and the experimental shear strengths.

5.3.5 Compressive Strength

The difficulty of evaluating the compressive strength of continuous fiber-reinforced composites is well known [85]. Compressive tests of composite materials have generally been carried out using a supporting jig in order to prevent premature bending deformation. Because carbon/carbons have extremely low shear modulus and strength [40, 86], prevention of such bending deformation during compression tests is of primary concern. However, the supporting jig approach is difficult to apply at elevated temperatures, due to the difficulty of avoiding a thermal expansion mismatch between the specimen and the test fixtures.

The compressive strength of fiber-reinforced composites are classified into three categories, that is, compressive failure of the fiber, shear failure, and kinking [85, 87]. Among these categories, kinking is generally the most frequently observed. This type of fracture tends to be preceded by the microbuckling of the fiber. The microbuckling stress σ_c has been discussed by a number of groups [88, 89]; the following (or a similar) criterion has been given for the onset of fiber buckling:

$$\sigma_c = \frac{G_{\mathrm{m}}}{1 - V_{\mathrm{f}}} \tag{5.4}$$

where G_m is the shear modulus of the matrix. This equation is known to be quantitatively incorrect, but is nonetheless employed to indicate qualitative tendencies [89].

As regards the compressive fracture behavior of carbon/carbons, there is little available data. In particular, very little is known about high temperature results. Fitzer [90] and Monacha [91] first examined the compressive strength of carbon/carbons, and more recently Gupta et al. [92] and Kishi et al. [93] have reported more detailed results. Gupta et al. examined the compressive fracture behavior of laminated carbon/carbon reinforced with eight harness satin weave cloths using rectangular parallelepiped block specimens. They concluded that the compressive failure of the carbon/carbon finally results in interlaminar shear fracture, but the initial failure is due to kinks originating from internal defects such as interfacial debonding between the fiber and the matrix. Kishi et al. also observed similar results in an initial report [93], but when an inclined load was applied against the fiber direction in a UD-reinforced carbon/carbon, they observed that the dominating type of fracture was in-plane shear fracture. This result was apparently caused by the low shear strength of the C/Cs as noted above.

Hatta et al. also investigated the compressive strength of laminated [94] and 3D-reinforced carbon/carbons [95]. As regards the laminated carbon/carbon, they observed interlaminar fracture under a compressive load, but the kink type of fracture was observed in the case of the 3D carbon/carbon. Moreover, they examined the compressive strength of 3D carbon/carbon at temperatures from room temperature to 3000 K. The 3D carbon/carbon had the tendency to induce kinks at the ends of the specimen, due to stress concentrations at the specimen ends, as well as due to their

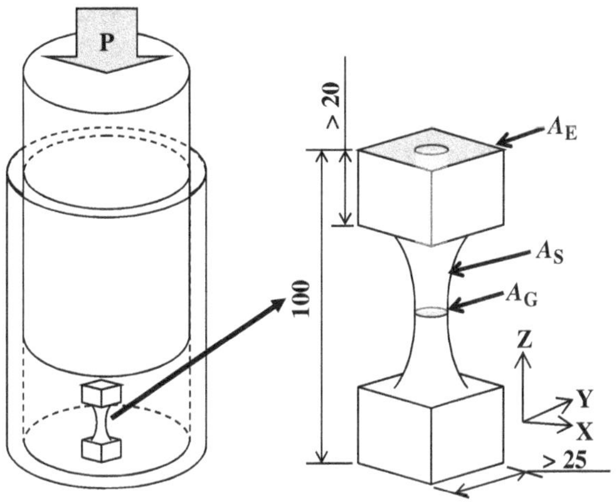

FIGURE 5.23
The configuration of compressive test adopted for high temperature test of 3D carbon/carbon, in which $A_E/A_G > 10$ and $A_S/A_G > 20$ for reduction of shear deformation effect. Reproduced with permission from Pergamon, Composites Science and Technology.

extremely low shear strength of the specimen. In order to avoid this type of premature fracture and to conduct high temperature tests, the specimen geometry and testing procedure were considered in more detail. The configuration illustrated in Figure 5.23 was found to be effective. This self-standing specimen was adopted in order to avoid thermal mismatch strain between the specimen and the test fixture at elevated temperatures; a sufficient loading area, A_E, and a shear supporting area, A_S, were arranged in order to alleviate the stress concentrations at specimen ends and to avoid shear deformation, respectively. Figure 5.24 summarizes the results of high-temperature tensile and compressive tests of the 3D carbon/carbon. As this figure depicts, the compressive strength of the 3D carbon/carbon increased up to a temperature of 2273 K and decreased at temperatures beyond 2273 K.

The experimentally observed kinks (i.e., fiber fracture by bending) for the 3D carbon/carbon were found to be the result of microbuckling. When Equation (5.4) is used to account for the compressive fracture in the present cases, V_f should include only those fiber bundles oriented in the loading direction, and G_m should be taken as the shear modulus of the composite, including the matrix and fiber bundles oriented in directions other than in the loading direction. In Equation (5.4), only G_m depends on temperature. Direct evaluation of G_m is not easy. However, as mentioned earlier, the shear behavior is closely related to the interfacial strength between the fiber bundle and matrix, τ_d. As shown in

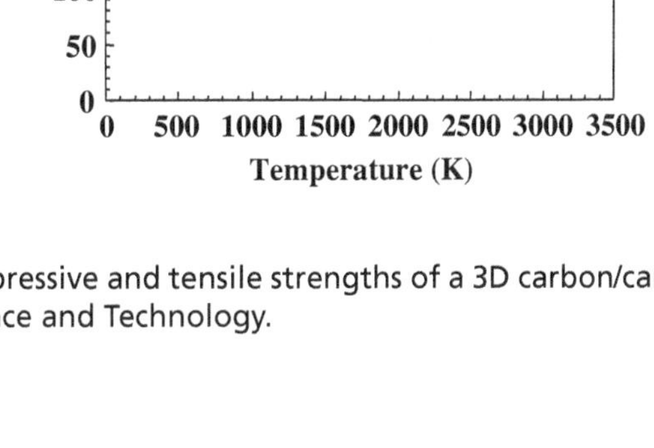

FIGURE 5.24
Temperature dependence of compressive and tensile strengths of a 3D carbon/carbon. Reproduced with permission from Pergamon, Composites Science and Technology.

Figure 5.16, τ_d is monotonically increased at temperatures of up to 2273 K. Thus, the improvement in σ_c with increases in temperature of up to 2300 K can be explained in terms of improvements in interfacial bonding. Another factor affecting G_m is the shear modulus of the matrix of the 3D carbon/carbon. Generally, the elastic modulus of the carbon material degrades at temperatures higher than 2000 K [71]. Thus, the temperature dependence of σ_c in the region from 2000 to 2800 K is thought to be the result of a trade-off between the advantageous factor of τ_d and the disadvantageous factor of the shear modulus degradation of the matrix. The rapid decreases in the compressive strength at temperatures exceeding 2800 K might be due to the degradation of the T300 fibers [44].

5.3.6 Fatigue Behavior

Fatigue behavior is one of the most important design properties for primary load-bearing structures intended for long-term use. However, only a few studies on the fatigue behavior of C/Cs have been reported [41,96–101].

5.3.6.1 S–N Curves for Tensile Fatigue The fracture strain versus the number of cycles-to-failure, that is, the S–N curves of UD-reinforced, CP-laminated, and quasi-isotropically laminated (QI) carbon/carbons fatigue-tested at a frequency of 10 Hz, are shown in Figure 5.25 [100]. The ultimate tensile failure strains obtained by static tensile tests were plotted as the data at a single fatigue cycle, $N = 1$. From Figure 5.25, it is clear that the S–N curves of all of the carbon/carbons tested here were similar. The fracture strains gradually diminished from the static fracture strains to fatigue limit strains of approximately 0.2 ± 0.01%, and those limit strains were 85–92% of the static fracture strains. The fatigue limit cycle was ca. 10^4 cycles, and no fatigue fracture was observed beyond this cycle. This finding implies that the fatigue limits of the carbon/carbons were not affected by the fracture strain or fatigue limit of the matrix as is also the case with CFRPs [101].

5.3.6.2 Residual Strength The fracture strains of the fatigue-loaded but surviving specimens were larger than the fatigue limit strains (≈ 0.2%), as shown in Figure 5.26a and b. This finding accounts for the observation that no fatigue fractures were observed at over 10^4 cycles. This enhancement of the tensile strength was facilitated by increases in the number of fatigue cycles. In contrast, the carbon/carbons show slight increases in fracture strains with increases in the maximum applied fatigue loading strain. Thus, the factor determining the strength enhancement in cases involving fatigue loading after 10^4 cycles is not the maximum applied strain, but rather the number of fatigue cycles.

In order to characterize the microfractures that occur during the fatigue loading in the principal load-bearing (0°) plies, a comparison of samples with and without clear strength enhancement caused by fatigue loading was carried out; the number of interfacial fractures and matrix cracks was compared between the samples [41]. Observation of cross-sections by SEM indicated that the strength enhancement mechanism required to recover or strengthen the

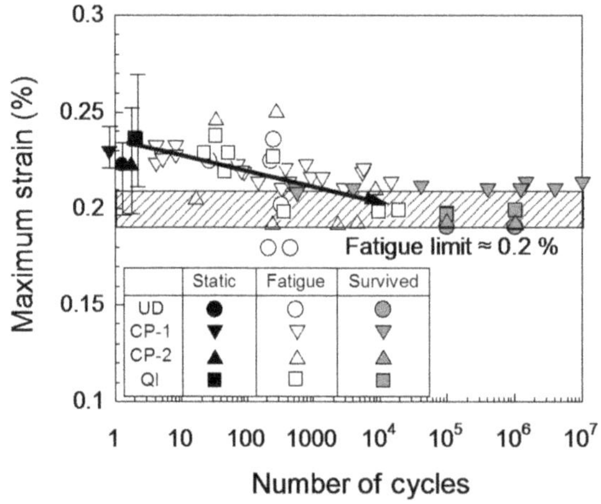

FIGURE 5.25

Relation between maximum strain and cycles-to-failure of 2D carbon/carbons during fatigue loading. Reproduced with permission from Pergamon, Composites Science and Technology.

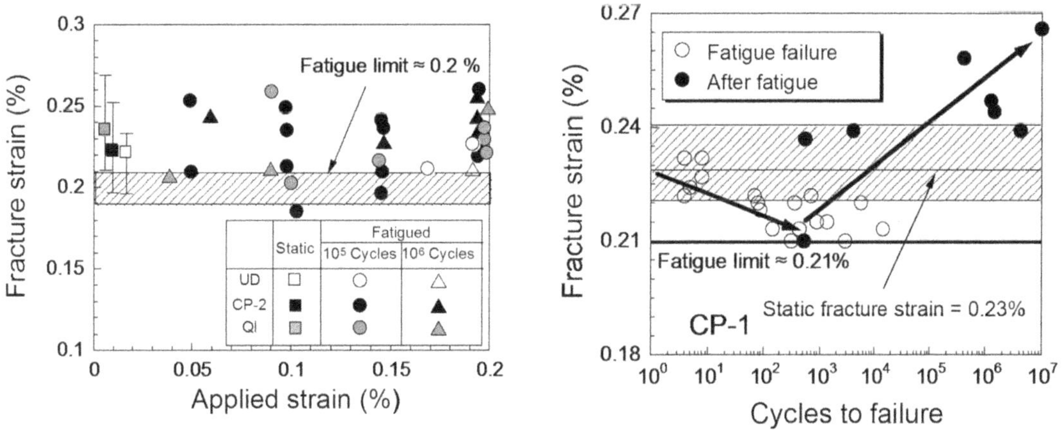

(a) (b)

FIGURE 5.26

Residual fracture strains after fatigue loading as functions of applied maximum fatigue strain (a) and applied cycles (b). Reproduced with permission from Pergamon, Composites Science and Technology.

carbon/carbons after fatigue was closely related to increases in the number of cracks near the fiber/matrix interfaces, that is, the degradation of the fiber/matrix interface in 0° plies.

The S–N curves and residual failure strain after fatigue loading indicated that two types of damage occur during cyclic loading, namely degradation and enhancement of fracture strains of the 0° plies of the carbon/carbons. The first type of damage, that is, the strength degradation mechanism, operated above the fatigue limit, and was most likely caused by fiber damage induced by wear due to sliding at interfaces, which is similar to the process that occurs in CFCCs [102–104]. The mechanism responsible for the other type of damage, that is, enhancement of fracture strains, appeared to be the evolution of the fiber/matrix interface debonding, which had accumulated in the carbon/carbons during fatigue loading, since carbon/carbons have been reported to be strengthened by weakened fiber/matrix interface bonding [44, 61, 105].

5.3.7 Concluding Remarks

The fiber/matrix interfacial strength of carbon/carbons is generally weak, such that tensile, shear, compressive, and fatigue fractures accompany fiber/matrix interfacial fractures. Hence, analytical models that can be used to predict the abovementioned types of fracture should take into account the contributions made by the interfacial damage. The low interfacial strength (τ_d) of carbon/carbons also affects other fracture phenomena. For example, the notch sensitivity of carbon/carbons (not included in this review) is an important property for designing load-bearing structures. In fact, carbon/carbons are known to possess substantial toughness and notch insensitivity [44, 105–107]. In particular, high toughness has been observed in 3D carbon/carbons. Such high toughness behaviors are principally the result of extremely low τ_d values [40]. The low coefficient of the thermal expansion of carbon/carbons is in part due to low τ_d values, as well as partially debonded fiber interfaces [57]. The fiber/matrix interface is thus an important research issue. However, investigations of the carbon/carbon interfaces remain at a primitive stage. For example, the interfacial shear strength shown in Figures 5.13, 5.15, and 5.16 significantly varied when the test geometry changed. Hence, the determination of the principal parameters that truly characterize interfacial fractures of carbon/carbons will be necessary.

5.4 THERMAL PROPERTIES OF CARBON/CARBON COMPOSITES

5.4.1 Introduction

Thermal properties of C/C composites, such as the thermal conductivity and thermal coefficient expansion, are very important characteristics for the main applications of these materials, such as brakes, re-entry shields, rocket nozzles, first wall tokamaks, and heat exchangers (thermal radiators, electronic heat sinks). The literature on the thermal properties of C/C composites is quite abundant, with some extensive reviews [108, 109] and important studies driven by the need for highly conductive materials for plasma-facing components of ITER [110, 111] as well as for aircraft brakes [112, 113]. The thermal properties, as the mechanical ones, depend on several parameters including the characteristics of the constituents (fibers, matrix, and interphase) and geometry (spatial arrangement) of carbon and voids (porosity, cracks). The field is complex because:

- The constituents, fibers and matrix, can exhibit a variability of about three orders of magnitude (for conductivity) due to the anisotropy of graphite (i.e., its hexagonal structure),
- The matrix homogeneity and distribution of voids are quite difficult to control, since they are dependent on the fiber architecture, precursor, and process,
- Fiber reinforcement solutions (architecture and fiber types), to tailor the properties, are numerous.

5.4.2 Thermophysical Properties of Monolithic Carbons

The thermophysical properties of monolithic carbons have been mainly studied in the 1960s and 1970s, with a very detailed literature [114, 115], due to the important industrial development of the material at that time. It is interesting to recall some important results, since they are applicable to matrices, fibers, and finally to C/C materials.

Table 5.4 [116–128] summarizes the properties of various carbon structures, ranging from graphite to vitreous carbon. The graphite structure possesses very anisotropic properties with very high thermal and electrical conductivities in the hexagonal plane (*a* axis), and very low ones in the transverse direction (*c* axis). Synthetic carbons, obtained through a gaseous (CVD or CVI) or liquid approach (pitch and polymer pyrolysis), give materials with a more or less ordered structure, depending on the synthesis conditions and post-treatment temperature (graphitization).

Carbons can be characterized through the determination of their structural parameters, including the interlayer spacing (d_{002}), crystallite size (L_c, along the *c* axis; L_a, in the plane), and the degree of preferred orientation (*R* anisotropy factor). The lower the value of d_{002}, and the higher those of *R*, L_c, and L_a, the more the properties resemble those of graphite. This has been shown through numerous correlations (Table 5.5) [108, 118, 129–132], and also through modeling of the thermal conductivity of fibers in relation to their structure [133]. Carbon materials, which are graphitizable, can be treated at high temperature to adjust several properties, such as thermal and electrical conductivities, thermal expansion, etc. The tendencies for C/C composites are the same as for monolithic carbons, but with the opportunity to tune the properties, both by the choice of fibers and the type of matrix.

5.4.3 Thermal Conductivity of Carbon/Carbons

It is possible to obtain carbon/carbon materials with very different values of thermal conductivity, sometimes as high as 600 W/(m·K) (Table 5.6) [110, 112, 127, 134–139]. They can either present similar values in the different directions or be very anisotropic (Figure 5.27). The thermal conductivity is mainly governed by the fiber part (Figure 5.28) and can be estimated, in the direction parallel to the fibers, with a mixture law [108]:

$$K_c = K_f V_f + K_m(1 - V_f - V_p) \tag{5.5}$$

Here, K_c, K_f, and K_m are the thermal conductivities of the composite, fibers, and matrix, respectively, and V_f and V_p are the volume fractions of the fibers and the porosity, respectively.

This formula gives a relatively good approximation when the content of fiber is large and when the fiber's conductivity is high with respect to that of the matrix. Models have been developed and are either analytical, for laminated composites [140], or based on finite element modeling, for 1D composites. Nevertheless, they are quite limited, as they are not coupled with deformation and cannot be used for predictions at high temperatures.

The matrix can also bring an important contribution to the thermal conductivity, since values of 100–150 W/(m·K) can be reached in directions with relatively low fiber fractions (8% for 3D DRDL and 4% for SEP NB31, Table 5.6). One way to obtain such matrices is through densification by hot isostatic pressure of pitch, with small amounts of quinoline insoluble, which can lead to matrices that are very conductive in several directions [141, 142]. Nevertheless, the development of the process requires a specific optimization due to the texture of the ex-pitch carbon depending

TABLE 5.4 Characteristics and Properties of Different Carbon Materials

Material Characteristics	Material References	Density	Young Modulus (Gpa)	Thermal Exp. Coeff. (10^{-6}/K)	Thermal Conductivity (W/(m·K))	Thermal Anisotropy	Electrical Resistivity (μΩ·m)	Heat Capacity (25°C) (J/kg·K)	Data reference
Perfect crystal					3000 (//) 1 (⊥)	2000			[116]
Pseudo-crystal	Mersen PGCCL	2.2	1000	0.4 (//, 525°C) 29 (⊥, 525°C)	2000 (//) 7 (⊥)	300	0.35–0.5 (//) 2.5–5·10^5 (⊥)	710	[117]
Carbon nanotube					3000–6000 (//)				[118]
C pyrolytic (CVD High T)	Mersen G21	2.2	30–40	1 (//, 500°C) 25 (⊥, 500°C)	375 (//) 2–8 (⊥)	50–200	2–4 (//) 100–1000 (⊥)		[117]
C pyrolytic (CVI 1000°C)		2.05			107 (//) 4 (⊥)	27			[119]
Carbon fibers	Thornel P100 (pitch)	2.1	760	−1.6	590		2.5		[120, 121]
	T300 graphitized				170 (//)	10			[122]
	T300 HR (ex-Pan)		230	0.3 (//) 36 (⊥)	6–11 (//) 0.7–2.8 (⊥)	4–9	18		[123, 124]
	XN05 (isotrope pitch)	1.65	54	1.4	7.4		28		[125]
	SPS TC2 (rayon)	2.0	35		3		50		[126]
Isotropic graphite	Mersen 6503	1.74		3.3	200	1	8		[127]
	Mersen 2020	1.77		4.3	85	1	15.5		
	Toyo Tanso IG56	1.77	10.3	4.7	104	1	12	700	[127]
Vitreous carbon	GC-10 (1300°C)	1.47–1.51	31	2.0–2.2	3.7–4.6	1	45–65		[128]
	GC-30 (3000°C)	1.43–1.47	23	2.0–2.2	15–18	1	35–40		

TABLE 5.5 Relationship between Physical Properties and Microstructural Characteristics for Different Carbon Materials

Physical Property	Microstructural Characteristics	Material	Reference
Thermal conductivity	Crystallite size, L_c	Isotropic and laminar	[130]
	Crystallite size, L_a	Petroleum coke	[129]
	Interlayer space d_{002}, L_a, L_c	Ex-pan and pitch fibers	[131]
Thermal conductivity temperature dependence	Crystallite size	Polycrystalline carbons	[130]
Dilatation under neutron irradiation	Preferred orientation, R	Polycrystalline carbons	[130]
Thermal expansion	Preferred orientation, R	Pyrolytic carbons	[130]
	Interlayer space d_{002}, L_c, Preferred orientation, R	Carbon fibers	[132]
Electric resistivity	Crystallite size, L_a	Carbon fibers (ex-pitch)	[118]
Young modulus	Preferred orientation, R	Carbon fibers (ex-pitch)	[108]

on several factors such as the composition of the pitch and its pre-treatment, the heating cycle, the surface properties of the fibers, and the effect of particle addition [143, 144].

Another way to achieve very conductive matrices is through CVI. The pyrolytic carbon deposit is oriented in the same direction as that of the fibers. Three types of structures, which can be differentiated through optical microscopy (with polarized light), can be observed: isotropic, smooth laminar, and rough laminar. Only the latter can be graphitized and gives a relatively high thermal conductivity of up to about 100 W/(m·K). This has been determined locally in a composite, using photothermal microscopy [122].

The addition of carbon nanotubes or nanofibers in the matrix or on the surface of the fibers also represents a solution to increase the conductivity having undergone much investigation [145–147]. Concerning matrices issued from resin pyrolysis, they possess a quite poor organization, and thus a low thermal conductivity of a few W/(m·K), and less than 20 W/(m·K) after a high-temperature treatment (Table 5.7). Nevertheless, for certain applications such as insulation, it can be interesting to realize low conductivity–low density carbon/carbon materials, known as carbon-bonded carbon fibers (CBCF). In this case, either a pyrolyzed resin or a CVI matrix can be used to optimize the properties [148, 149].

High-temperature treatments represent a simple way to improve the thermal conductivity of C/C composites, as demonstrated in Figure 5.29 [111, 134, 150–154]. For materials with rough laminar CVI or pitch matrices, there is generally a significant increase of the conductivity when treating the material at a temperature between 2200°C and 3000°C. The higher the temperature, the more significant the increase in conductivity. This evolution is caused by the graphitization and can be related, as previously mentioned, with the L_a and L_c parameters (Table 5.5) [111] of the material. This increase is mainly parallel to the fiber direction, since CVI or pitch layer deposits follow this orientation.

The texture formation from pitch precursors is a quite complex phenomenon that depends on numerous parameters (e.g., pitch characteristics, fiber–mesophase interfacial energy, fiber compaction) [144] and the conductivity increase can thus differ depending on the process. For example, the addition of coke filler particles to the pitch leads to the disruption of the matrix orientation and to a preferential increase of the transverse conductivity [155].

The conductivity improvement can also partially come from the graphitization of the fibers, as it is the case for the T300 sample (Figure 5.29), with a conductivity increase from 6 to 170 W/(m·K). Much higher values have been measured with methane- or benzene-derived fibers, with an increase from 40 up to 2000 W/(m·K) (equivalent to that of graphite crystals) after a treatment at 3000°C [156, 157]. Nevertheless, heat treatments must be carefully optimized, taking into account not only the thermal conductivity but also other properties. If the temperature is too high, heat treatments can give rise to an increased crack density, shrinkage of the matrix and fibers, pore enlargement [158], and a decrease in the Young modulus [153].

Concerning the evolution of the thermal conductivity with temperature, there is a maximal value, within the range 180 K (for pure graphite) and 500 K (for low ordered carbons). Above this maximum, the thermal conductivity is mainly limited by phonon–phonon interactions of "Umklapp" type and decreases quite quickly. The more the material is ordered, the more significant is this decrease, and the values should tend, for all carbon and graphite, toward the same value [159]. This behavior has been modeled for thin perfect graphite, in the basal plane, and along the c axis, using a Debye model [160] and has been found to be in quite good agreement with experimental measurements [120]. Figure 5.30 presents the evolution of the thermal conductivity as a function of temperature, for graphite [120, 161], an ex-pitch

TABLE 5.6 Characteristics and Properties of Several Carbon/Carbon Materials

Material Characteristics	Material Reference	Density	Young Modulus (Gpa)	Thermal Exp. Coeff. (10^{-6}/K)	Thermal Conductivity (W/(m·K))	Thermal Anisotropy	Electrical Resistivity (μΩ·m)	Heat Capacity (25°C) (J/kg·K)	Data Reference
2D									
Ex-Pan cloth/C_{CVI}	Dunlop DMS 678	1.72		0.69^{*}	$110^{x,y}$, 23^{z}	4.8		669	[110, 134]
Ex-Pan cloth/C_{CVI}	Schunk CF222	1.55	80	$0.8^{x,y}$, 7^{z}	$40^{x,y}$, 10^{z}	4	$34.1^{x,y}$, 18.6^{z}		[136]
Ex-Pan cloth/C	Toyo Tanso CX761	1.51		$0^{x,y}$, 8.4^{z}	$44^{x,y}$, 9^{z}	5	20	670	[127]
Ex-Pan felt/C_{CVI}	Toyo Tanso 2002U	1.65	11^{x},8.1^{y},3.4^{z}	1.7^{x}, 2.3^{y}, 5.3^{z*}	390^{x}, 320^{y}, 190^{z}	2	2.7^{x}, 3.4^{y}, 5.1^{z}		[127, 137]
Felt/C_{Resin}	ABS (B767 brake)			2.88 (1173 K)	$66^{x,y}$, 13^{z}	5			[114]
2.5D (needled z)									
Ex-Pan/(C_{CVI} + H.P Pitch)	SEP N112	2.01		0.55^{*}	220^{z}, $166^{x,y}$	1.4		669	[110, 134]
Ex-Pan/C_{CVI}	SEP N11	1.79		1.34^{*}	248^{z}			673	[110]
P55(z, 27%) Pan (4%)/C_{CVI}	SEP NB31	1.9	15^{x}, 12^{y}, 107^{z}	1^{x}, 2.1^{y}, 0.4^{z}	$117^{x,y}$, 323^{z}	2.8		720	[135, 138]
P130/C_{Pitch}	FMI A27-130	2.03		1.17^{*}	309^{z}			668	[110]
3D									
Granoc (z, 43%; x, 18%; y, 6%)/C	Nisseki NIC-01			0.6^{x}, 0.9^{y}, -0.6^{z}	537^{z}			717	[135]
Ex-Rayon (z, 28%; x = y = 8.1)/C	DRDL	1.8	$25^{x,y}$, 89^{z}		$130^{x,y}$, 148^{z}	1.1			[139]
4D									
Ex-Pan (z, 14%; u = v = w = 8)/C	DRDL	1.8	$31^{u,v,w}$, 38^{z}		$54^{u,v,w}$, 70^{z}	1.3			[139]

*: average linear TCE 293-1273 °K; ^{x}: longitudinal direction; ^{y}: cross direction; ^{z}: perpendicular direction.

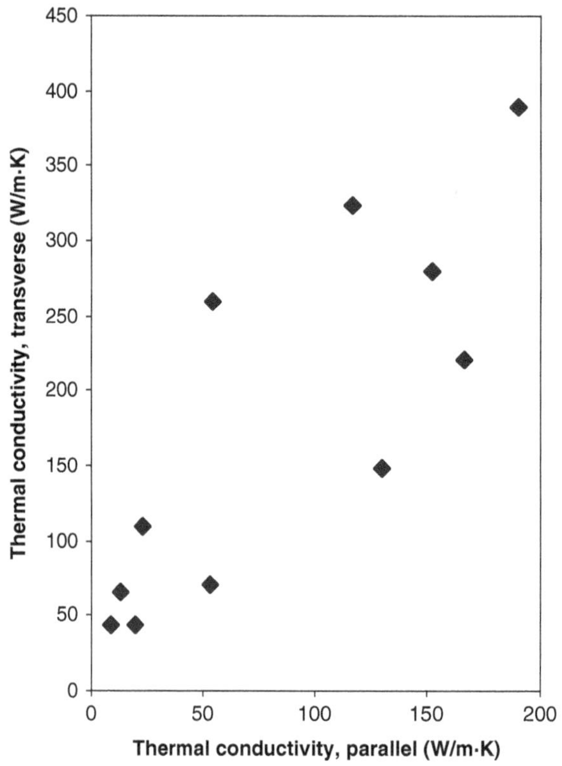

FIGURE 5.27

Values of transversal and parallel thermal conductivity for various carbon/carbon materials (values issued from Table 5.3, [108, 141]). Reproduced with permission from Pergamon, Composites Science and Technology.

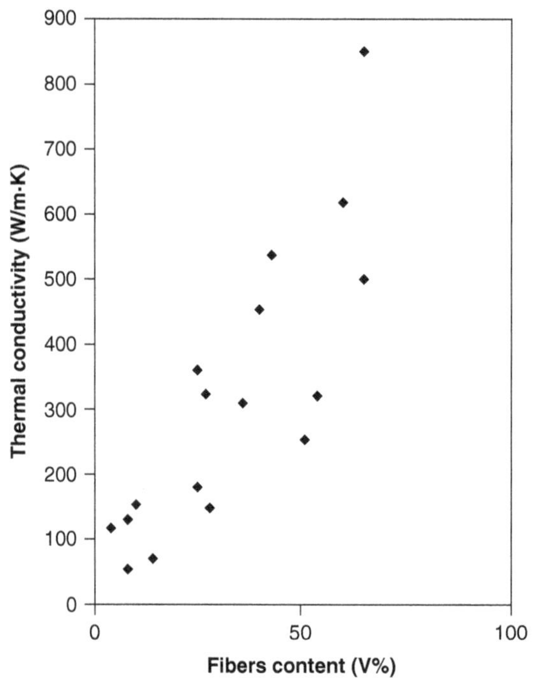

FIGURE 5.28

Thermal conductivity as a function of fiber content (values issued from Table 5.3 [108, 110, 141]). Reproduced with permission from Pergamon, Composites Science and Technology.

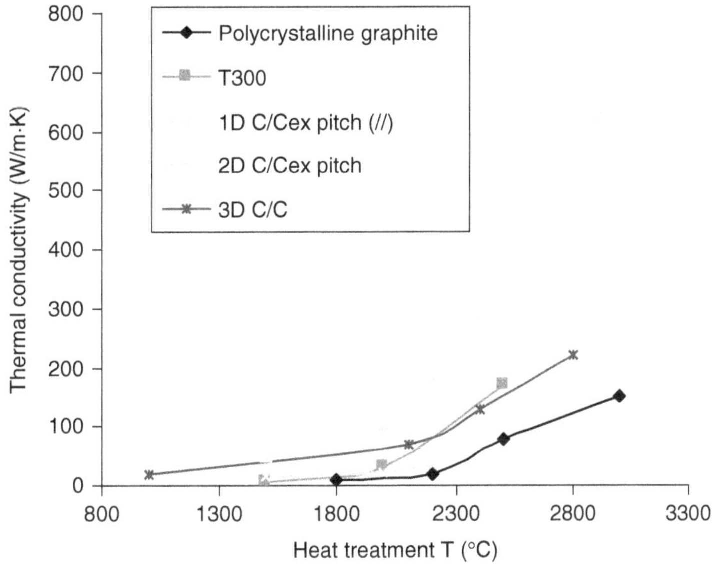

FIGURE 5.29

Effect of heat treatment on the thermal conductivity of carbon materials and carbon/carbon composites (values issued from [111, 134, 150–154]). Reproduced with permission from Pergamon, Composites Science and Technology.

C-fiber (P100) [162], and two C/C composites studied for fusion applications [135], whose behavior is similar to that of other C/C materials [121, 153].

5.4.4 Electrical Properties

The literature comprises much less data on electrical properties than on thermal conductivity of C/C composites, since this characteristic is important only in very few applications, including heating elements [163, 164], current collection [166], and grids for ion engines [166]. Nevertheless, the electrical resistivity is sometimes determined for other uses such as nuclear fusion. The anisotropy of the graphite structure leads, for very well-organized and oriented graphites, to values ranging from 0.5 $\mu\Omega\cdot m$ (*c* axis) to $5\cdot 10^5$ $\mu\Omega\cdot m$ (*a* axis) (Table 5.4).

The same is true for C/C composites, with literature values generally ranging from a few $\mu\Omega\cdot m$ [24] to 50 $\mu\Omega\cdot m$ [30], but some higher values, around $16\cdot 10^3$ $\mu\Omega\cdot m$, are also mentioned [166]. The lowest values correspond to the materials

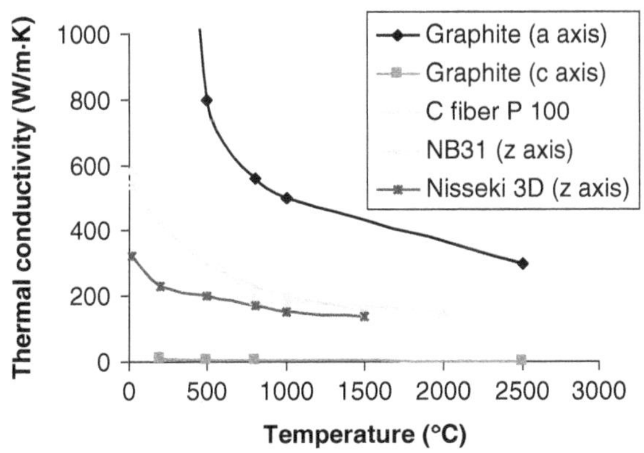

FIGURE 5.30

Evolution of thermal conductivity with temperature for different carbon materials and two carbon composites (values issued from [116, 133, 135, 161, 162]. Reproduced with permission from Pergamon, Composites Science and Technology.

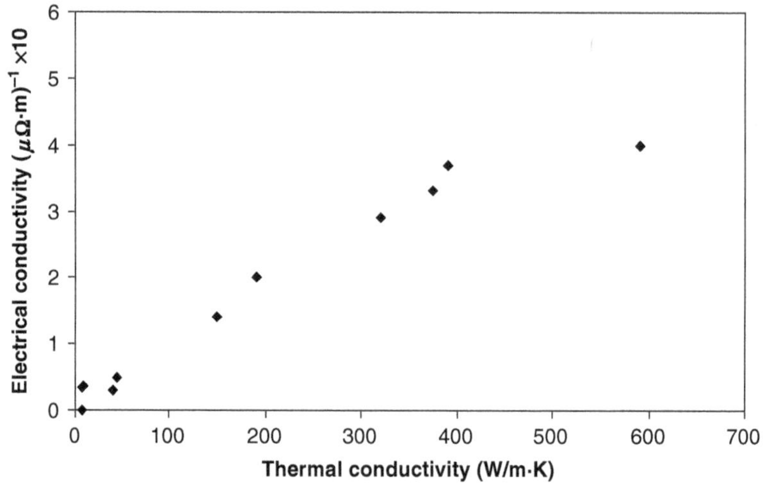

FIGURE 5.31
Electrical and thermal conductivity values for several carbon fibers and carbon/carbon materials (values issued from Tables 5.1 and 5.3 [167]). Reproduced with permission from Pergamon, Composites Science and Technology.

with the highest organization and with an orientation of the planes in the direction of the electrical measurements. For well-organized C/C composites, at room temperature, there seems to exist a good correlation between thermal conductivity and electrical conductivity [60], as can been seen in Figure 5.31.

The electrical resistivity generally decreases slowly when going from room temperature to 1500°C, as for other carbon materials such as fibers (Figure 5.32) [136, 137, 151]. The evolution at higher temperature has been little studied, but it seems that the electrical temperature may slowly increase linearly [168], which is in agreement with the similar behavior of high-temperature ($T_{Synthesis} > 2200°C$) polycrystalline carbons [169].

5.4.5 Thermal Expansion

Thermal expansion, characterized by the coefficient of thermal expansion (CTE), is an important property to take into account, both for materials processing, the conception of parts, and for control of dimension variations. As C/C materials are mainly employed for high-temperature applications, this property has been extensively studied [108]. The CTE is a characteristic that, once again, can possess varying features and evolves differently as a function of direction and temperature, in relation with the anisotropic microstructure of the material. The CTE for graphite crystal at room temperature is about 30×10^{-6} K^{-1} in the *c* direction, with a weak increase as a function of temperature. In the *a* direction, it is slightly negative, -1.8×10^{-6} K^{-1}, and also exhibits a weak increase with temperature, leading to a zero value at about 400°C [116] [121].

As in the case for the thermal conductivity, the thermal expansion, for high contents of fibers, is mainly controlled by the corresponding value for the fibers, which is negative in the fiber direction (–0.6 to -1.5×10^{-6} K^{-1}), and positive in the transverse direction ($17–20 \times 10^{-6}$ K^{-1}) [121]. This can be seen in Table 5.2, with the UD C/C (K321) as well as

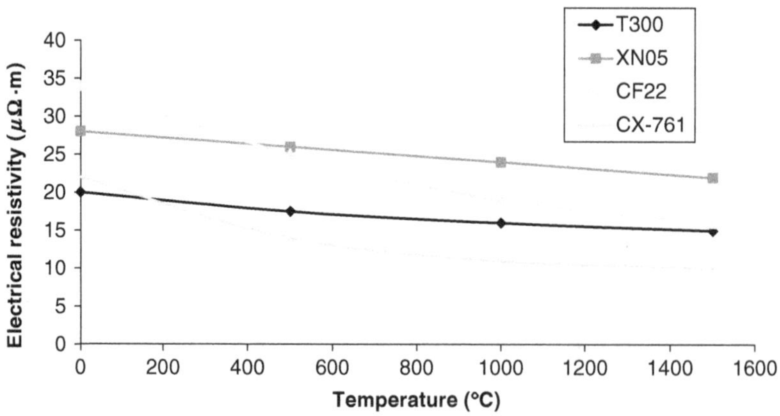

FIGURE 5.32
Evolution of the electrical resistivity with temperature for two carbon fibers (T300 and XN05) [151] and two industrial carbon/carbon materials (CX-761 [127] and CF222 [136]). Reproduced with permission from Pergamon, Composites Science and Technology.

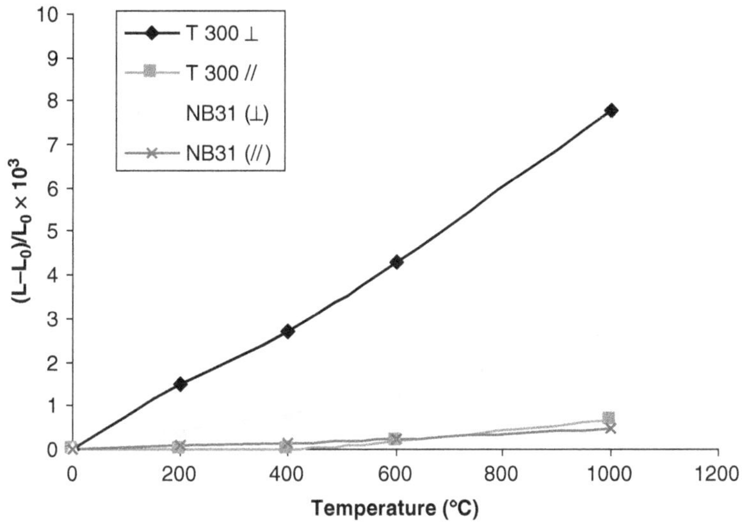

FIGURE 5.33
Effect of the temperature on the dimensional changes (parallel and perpendicular to the fibers axis) of a carbon fiber [132] and of a carbon/carbon material [135]. Reproduced with permission from Pergamon, Composites Science and Technology.

the 3D Nic-01 (with a high percentage of fiber in the z direction), constituted of ex-pitch fibers, which possess negative CTE values in the z direction. To illustrate the thermal expansion, Figure 5.33 represents the evolution of the variation in dimension as a function of temperature, both for T300 fibers (heat treated at 2500°C) [132] and a 2.5D C/C material [135].

Logically, the matrix and porosity also participate to the global behavior, but to a lesser extent. Several approaches exist to obtain low CTE values which are required for several applications such as thermal management. One way to achieve this is to optimize the conception with the choice of fibers, reinforcement, matrix type, residual porosity, and process [108, 148, 170–173]. Another way, for use at either moderate or high temperature, is to substitute a part of the carbon matrix by a metal, such as aluminum [174, 175].

5.4.6 Specific Heat

Specific heat is a less important property than the others, but it is often characterized and given as part of the basic information in the material data sheet. Figure 5.34 presents the evolution of the specific heat as a function of temperature for a graphite crystal [116], a 2D C/C material [136] and a 3D C/C material [153]. As can be seen, the values are quite close, but this is not always the case: differences may be as high as 10% or 20%, depending on the carbon type and heat treatment, as mentioned by several authors [102, 139, 141, 142].

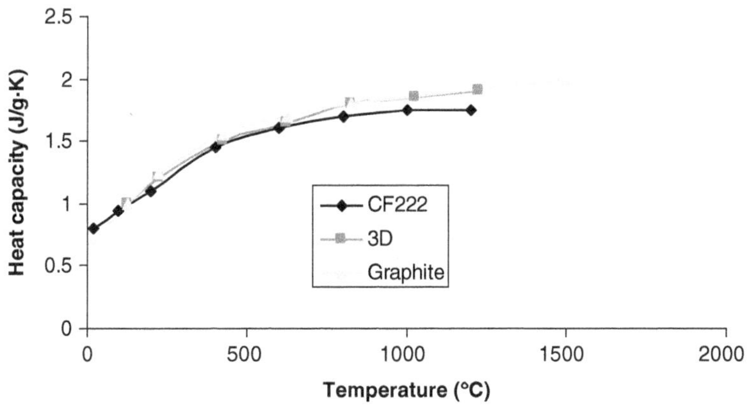

FIGURE 5.34
Evolution of the heat capacity for two carbon/carbon materials [136, 153] compared with graphite [116]. Reproduced with permission from Pergamon, Composites Science and Technology.

5.4.7 Thermal Shock Resistance

A high thermal shock resistance is often cited as an important characteristic that has rendered possible the development and use of C/C composites as brakes, heat shields for re-entry vehicles, and rocket nozzles [109], but there are very little data and studies available [69, 70, 176, 177]. Nevertheless, for a simple comparative approach, two merit indices [121, 178], for cold or hot shock, can be used:

$\sigma/\alpha \cdot E$, in the case of ideal surface heat transfer (in K)

$K \cdot \sigma/\alpha \cdot E$, in the case of poor surface heat transfer (in W/m)

K, σ, α, and E represent the thermal conductivity, tensile strength, linear CTE and the Young modulus, respectively. The higher the value of the indices, the higher is the material's shock resistance.

5.4.8 Concluding Remarks and Future Directions

C/C materials can be used under extreme conditions (temperature, thermal shocks, nuclear irradiation) and possess exceptional thermophysical properties (extremely high thermal and electrical conductivities, dimensional stability) which can be tailored through the composite constituents and architecture. The main fields that need to be investigated include thermomechanical modeling for conception optimization, the addition of carbon nanotubes and nanofibers for improvement of properties, and cost reduction for access to new applications.

5.5 OXIDATION PROTECTION OF CARBON/CARBON

As mentioned beforehand, the application of carbon/carbon started with rocket nozzles and re-entry parts for military or space projects in about 1960. The first industrial products were carbon/carbon brakes for civil aircraft developed by Super Temp Division of B. F. Goodrich and were licensed by Dunlop. Today, carbon/carbon brake disks are used for most civil aircraft and manufactured worldwide by companies like Dunlop, SEP, Hitco, Bendix, Carbone, and others. The matrix is PyC (CVI) and the materials are finally heat treated to achieve an optimum ductile failure behavior.

The market volume is estimated with some 50% of the total carbon/carbon used worldwide.

The brake disks have to be protected against oxidation. This can be done by design and construction of the braking system itself, that is, by avoiding more or less oxygen contact with the carbon/carbon surface or by minimizing the contact with air during braking. In such an optimized design the carbon must be highly graphitized up to approximately 2500°C in order to get the best oxidation resistance together with a reasonable ductility of the material. Oxidation protection of carbon/carbon has to be applied for all materials which are used in an oxidative atmosphere above 420°C. The principles of oxidation protection are described in the following.

All oxidation protections for carbon/carbons basically include three components that help to protect against an oxidation attack for long-term applications (Figure 5.35).

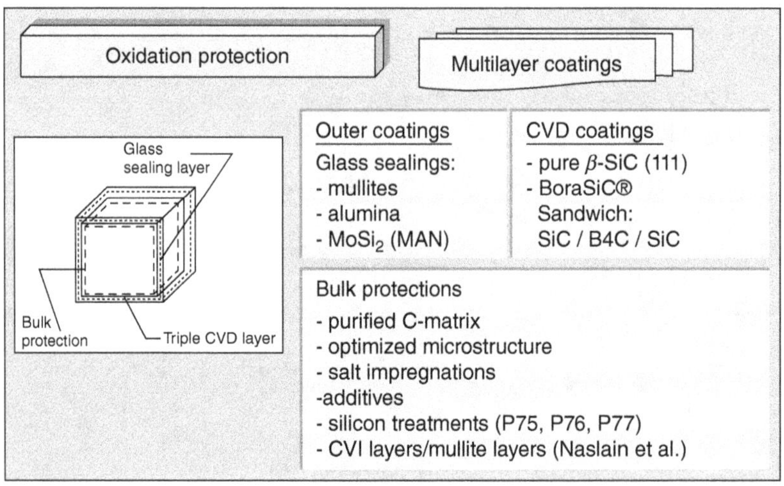

FIGURE 5.35

Multilayer coated carbon/carbon. Reproduced with permission from Pergamon, Composites Science and Technology.

These three protection systems are an inner bulk protection, an outer multilayer CVD coating and finally a glass sealing layer serving as a surface coating.

The first patent on oxidation protection of carbons was given to the National Carbon Company in 1934 [179]. Already then, first oxidation protection systems included a bulk protection by the formation of SiC and a sealing layer of B_2O_3. CVD coatings were unknown at this early stage of oxidation protection of carbons. However, the principles of protecting carbon materials are still valid even today.

5.5.1 Bulk Protection Systems for Carbon/Carbons

The oxidation protection systems depend on the oxidation conditions, like temperature, atmospheres, architecture of reinforcement, matrix systems, and required life times.

Carbon/carbons which are applied in oxygen-containing atmospheres below 650°C can be protected by pure bulk treatments. Carbon/carbons are oxidized according to the following chemical reactions:

1. $C_{(s)} + O_{2(g)} \rightarrow CO_{2(g)}$
2. $2\ C_{(s)} + O_{2(g)} \rightarrow 2\ CO_{(g)}$

At higher temperatures, the second chemical reaction is preferred (Boudouard's reaction), whereas at lower temperatures, CO_2 is formed. The oxidation attack starts at the edge atoms of the carbon material, which means at crystalline defects of the carbon structure.

Therefore, the carbon matrix is more sensitive against oxidation than the reinforcing C-fibers, which have been studied in detail by Kimura and Yasuda [180]. Oxidation rates at low temperatures can be reduced by increased crystallinity and purification of the carbon matrix by higher final HTT as can be seen from Figure 5.36 [16, 180].

Furthermore, it can be concluded from Figure 5.36, which shows the Arrhenius plots, that the oxidation mechanism depends on temperature. At lower oxidation temperatures (below 800°C) the oxidation rate is controlled by the chemical oxidation attack (kinetics), whereas at higher temperatures (low 1/T values) the burn-off rate is limited by diffusion effects, which was also confirmed by oxidation studies on graphite materials [181].

The active sites in the carbon matrix can be reduced by increased final HTT or blocked by contaminations applied as salt impregnations.

Industrially available are those grades possessing phosphoric salt impregnations, which are applied as standard carbon/carbon materials in manufacturing hollow glasses at application temperatures of approximately 700°C without any further oxidation protection. The mechanical properties of these grades are given in Table 5.7.

The influence of the salt impregnation can be shown by comparison of the oxidation behavior of CF264 (Figure 5.37, without salt) and CF264Q (Figure 5.38, with salt).

The salt impregnation inhibits the oxidation rate measured as weight loss via thermogravimetrical analysis. Furthermore, it can be concluded from these results, that salt impregnations decrease the influence of higher oxygen flow rates

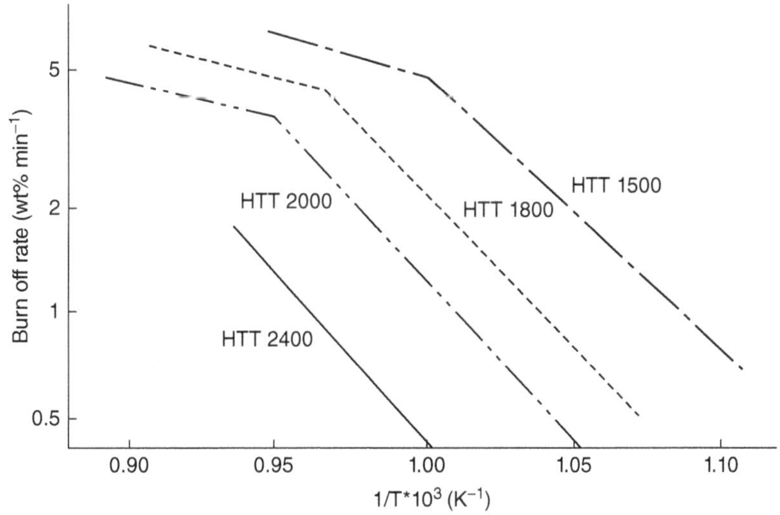

FIGURE 5.36
Arrhenius plots of the oxidation of carbon/carbon composites [16, 180]. Reproduced with permission from Pergamon, Composites Science and Technology.

TABLE 5.7 Properties of C/C Grades with Bulk Protection Based on Impregnations with Phosphoric Salts

Grade	Reinforcement	X_F (%)	γ (g/cm^3)	P (%)	Flexural Strength (MPa)	ILSS (MPa)	HTT (°C)
CF260Q	2D	55	1.40	8	80	7	2100
CF264Q	2D	55	1.45	8	80	8	2400

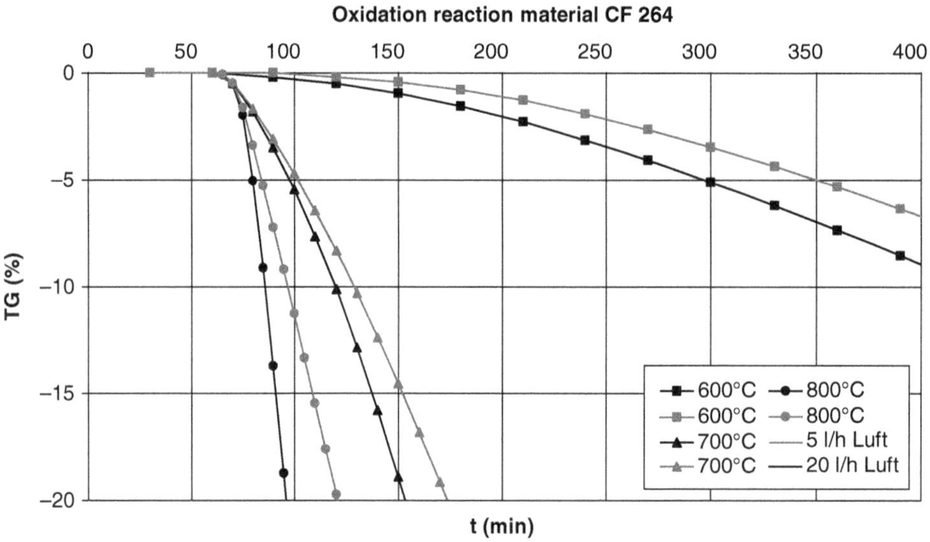

FIGURE 5.37
Oxidation behavior of CFC 264. Reproduced with permission from Pergamon, Composites Science and Technology.

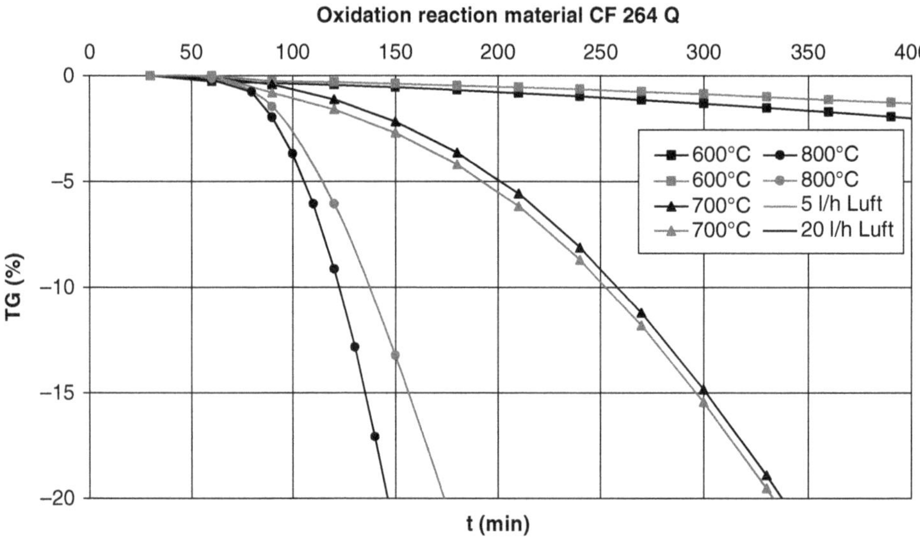

FIGURE 5.38
Oxidation behavior of CFC without salt impregnation 264Q with phosphoric salt impregnation. Reproduced with permission from Pergamon, Composites Science and Technology.

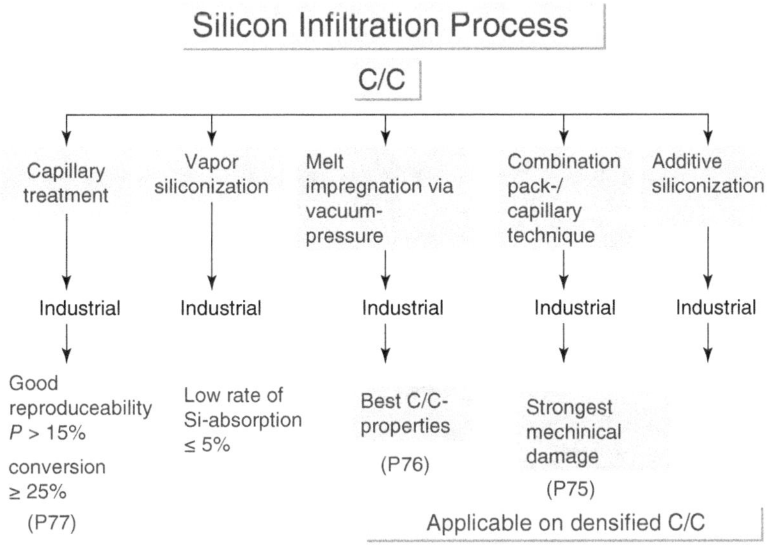

FIGURE 5.39

Bulk protection methods via silicon treatments. Reproduced with permission from Pergamon, Composites Science and Technology.

and thereby the oxidation attack. CF264Q is used in hollow glass manufacturing up to 2000 hours, if the contact times and contact areas are sufficiently low.

The oxidation rate depends on the temperature, time, and surface area attacked by oxygen.

As already mentioned and shown in Figure 5.35, bulk treatments are very often performed with silicon to form SiC, which possesses superior oxidation protection behavior when compared to salt impregnations.

Silicon treatments of the bulk material are well known and described in the literature [10, 182–185]. Different methods are industrially available as shown in Figure 5.39.

The advantages and disadvantages of the different methods are given in Figure 5.39.

Silicon treatments can be performed with Si vapor, by melt impregnation under vacuum and pressure (P76 process), by capillary impregnations via liquid-phase reaction (P77, Figure 5.39) called liquid silicon impregnation (LSI) process, by pack cementation and by a combination of cementation and liquid-phase impregnation (P75 process; Figure 5.39).

The advantages and disadvantages of the different methods are given in Figure 5.39.

The silicon treatment via P75 is the only procedure which can be used for fully densified carbon/carbons without any dimensional change of the treated carbon/carbon [10]. The degree of matrix conversion from carbon to SiC depends on the reactivity of the carbon, its open porosity and density. A broad range of C/C over C/C–SiC up to C/SiC materials can be realized according to the final requirements.

The higher the amount of formed SiC is, the better the results for oxidation resistance. However, increasing degrees of conversion are always combined with increasing embrittlement of the composites including a fiber damage by chemical reactions between C-fibers and silicon.

Silicon treatments of C/C are used not only for bulk protection, but also for improvement of the wear behavior for tribological applications. The best bulk protection is obtained by multilayer CVI coatings on the inner surface of bulk materials. Such methods have been described and used [186–188] in particular for C/SiC and SiC/SiC components applied in space or aircraft structures. The combination of nanoscaled multilayers with PyC, SiC, BN, or B_4C, directly applied on the internal surface of fiber preforms improves their oxidation resistance as well as the ductile failure behavior of the high-temperature composites. Boron-containing intermediate layers are able to form boron oxide (B_2O_3) even at very low oxidation temperatures resulting in a self-healing behavior of the matrix during an oxygen attack. However, the same oxidation protection of the bulk material can be obtained by using additives with boron (Figure 5.35).

A further method for bulk protection is powder immersion reaction assisted coating (PIRAC), an *in situ* CVI method developed by Prof. E. Gutmanas, Technion, Haifa.

An overview on bulk protection of C/C with chromium powder is given in the thesis of R. Eliezer [189].

PIRAC coatings with titanium can also be used as bulk protection for carbon/carbons (Figure 5.40).

The TiC formed filled the intra-laminar cracks in the composites and thereby prevented a strong oxidation attack by closing possible diffusion paths for the oxygen.

However, for long-term oxidation resistance outer CVD coatings must be applied additionally. The PIRAC method is a very simple and cost-efficient CVI technique. The samples or components, which have to be protected, are packed

FIGURE 5.40
SEM mircograph of the cross-section of carbon/carbon PIRAC treated showing the inter-fiber cracks filled by titanium carbide. Reproduced with permission from Pergamon, Composites Science and Technology.

in a powder mixture of metal and powder of the same metal halogenide or a pure halogenide like jodium. The package is vacuum tight closed by, for example, a metal foil and heated up in a conventional furnace to the CVI reaction temperature, preferably below 1100°C. The furnace itself can be used under inert gas. Enclosed in the metal foil under low pressure a metal subhalogenide is formed which diffuses as a CVI reaction inside the carbon/carbon. The halogenide acts as a catalyst. The bulk protection itself shows a gradient layer and can be modified from composition as shown in Figure 5.41 [190].

A SiC-protected carbon/carbon was treated with a mixture of Cr and J_2 and oxidized after the PIRAC treatment for 2 hours at 1000°C. As can be seen, the PIRAC treatment can also be applied for outer coatings.

5.5.2 Outer Multilayer Coatings

CVD-SiC coatings are normally used as outer multilayer coatings of carbon/carbons (Figures 5.2 and 5.41). The mismatch in CTE of the bulk carbon/carbon and the CVD-SiC coating results in thermally induced crack formations during cooling down from CVD-coating temperature to room temperature (Figures 5.35 and 5.42).

FIGURE 5.41
SEM micrograph showing the oxidation scale on the surface of coating obtained by oxidizing coated SiC at 1000°C for 2 hours in static dry air [190]. Reproduced with permission from Pergamon, Composites Science and Technology.

FIGURE 5.42

(a) Equidistant crack formation on a CVD-SiC-coated carbon/carbon perpendicular carbon/carbon to the reinforcement [191]. (b) Irregular crack formation on the surface of a CVD-SiC coated. Reproduced with permission from Pergamon, Composites Science and Technology.

An increase in the CVD-SiC layer thickness results in a decrease of the crack distance, that means increased crack densities due to the induced higher thermal stresses [191]. The cracks formed are diffusion paths for oxygen in case of oxidation. However, the thermally induced cracks can be nearly completely closed by heating up the samples to the CVD coating temperatures.

Therefore, the oxidation rate of multilayer coated carbon/carbon depends not only on physical parameters like reaction kinetics and diffusion rates, but also on crack opening and crack closing effects, which are influenced by the coating parameters and the temperature.

It is well known from oxidation kinetics, that the maximum oxidation rate for multilayer coated C/C is observed at about 800°C [190]. This maximum is caused by the ratio of internal surface area available for oxidation (by crack opening) and highest oxidation kinetics.

However, for long-term applications such cracks are a limiting life time factor. Therefore, self-healing effects have to be used in order to close these cracks. The self-healing capacity can be tailored according to the requirements from the viewpoint of oxidation temperatures and atmospheres. Although the thermodynamic and kinetic stability of the different oxidation protections will not be discussed in this contribution, some general remarks are necessary to understand the principles of oxidation protection.

Silicon carbide (SiC) is applied as the standard CVD-SiC coating, as shown in Figure 5.12. SiC forms a silica layer (SiO_2) in oxidizing atmospheres, with a parabolic growth of the layer thickness. SiC is selected as the most favorite oxidation protection, because the SiO_2 layers possess the lowest diffusion rate of oxygen. Therefore, the formation of silica during oxidation is an additional self-healing layer (glass sealing layer). The disadvantage of such a self-healing layer is the fact, that a complete covered SiO_2 layer can be obtained at temperatures above 1000°C, whereas at lower oxidation temperatures (700–1000°C) only isles of SiO_2 are formed on the attacked carbon surface accessible by the thermally induced cracks in the multilayer coating. Glass formers are needed for lower temperatures like boron-derived additives. Therefore, the bulk can be protected additionally as already mentioned, with borides as additives. Another possibility is to include a boron oxide former, like B_4C, in the multilayer coating (Figure 5.43) [9, 193].

Boron carbide is able to form, even at very low temperatures, a boron oxide, which can close the cracks in the CVD multilayer as shown in Figure 5.44 [193].

The cracks on the top surface of the multiplayer BoraSiC® coating are healed like cuts in a skin as can be seen in the micrograph of Figure 5.21. The crack is closed under the formation of B_2O_3; boron silicate, and SiO_2 as proved by EDX.

In principle, borides can be used up to 1600°C as self-healing glasses. However, the boron oxide is neither stable under water vapor flow, nor thermally stable at temperatures above 1200–1250°C due to evaporation of boron silicates. Therefore, additional outer glass sealing layers are necessary, with superior high temperature behavior.

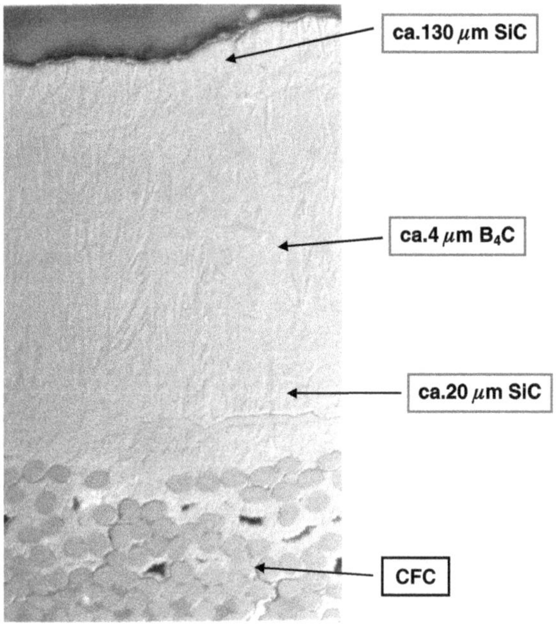

FIGURE 5.43
BoraSiC® coating with self-healing capacity [9, 193]. Reproduced with permission from Pergamon, Composites Science and Technology.

5.5.3 Outer Glass Sealing Layers

As mentioned above, multilayer CVD-SiC coatings (Figure 5.35) form *in situ*, during oxidation, a dense silica layer at temperatures above 1000°C. The diffusion rate of oxygen through the SiO_2 layer depends on the layer thickness, which increases with increasing oxidation time, and the oxidation temperature. The diffusivity of oxygen in silica is extremely low compared to the diffusivity in B_2O_3. The diffusion rate in SiO_2 of oxygen is seven orders of magnitude lower [194, 195] resulting in a superior oxidation protection.

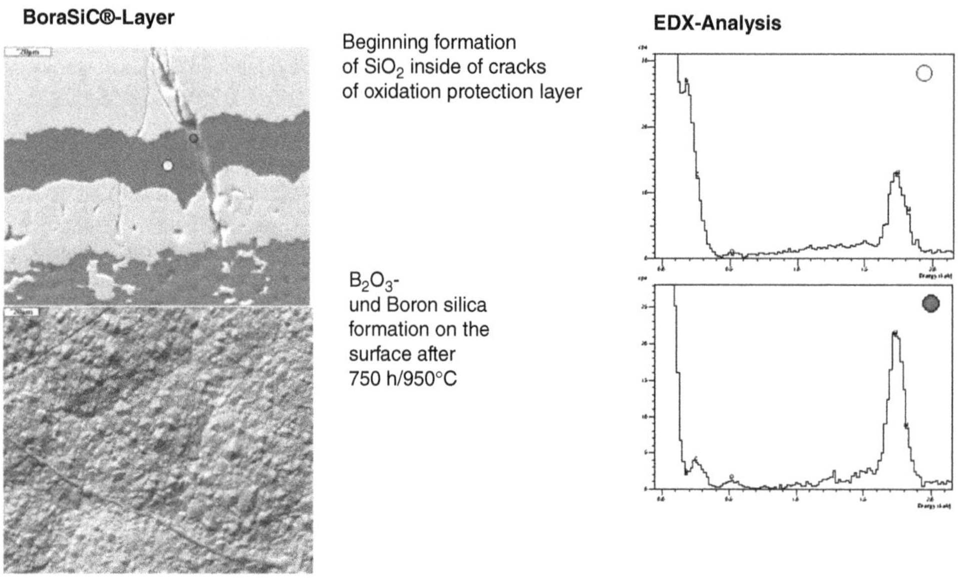

FIGURE 5.44
Self-healing capacity of BoraSiC® multilayer coatings [195]. Reproduced with permission from Pergamon, Composites Science and Technology.

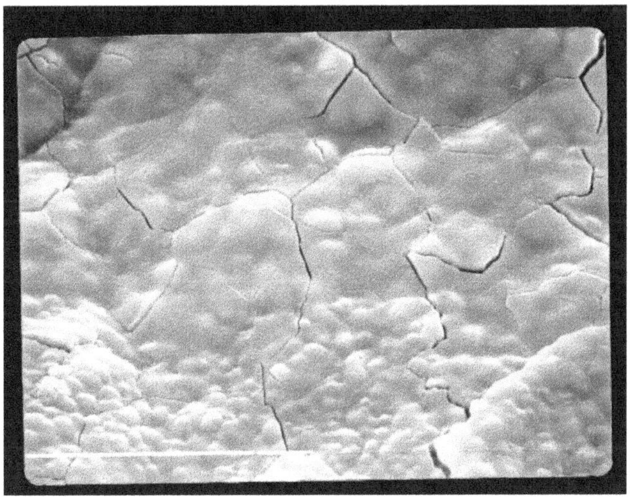

FIGURE 5.45

SEM of the protected carbon/carbon sample after 1000 hours at 1530 K in 80 L/h O_2; sealing thickness 3–4 μm (Magnification: 10 mm = 5 μm). Reproduced with permission from Pergamon, Composites Science and Technology.

Figure 5.45 shows a multilayer coated carbon/carbon with an *in situ* formed silica layer.

The cracks visible at a.T. are closed during oxidation. The sample did not show any weight loss after 1000 hours at 1530 K [183].

Nevertheless, glass sealings based on silica have their limitations.

The first one is the compatibility with the CVD-SiC layer underneath. The thermal shock resistance of these glass layers is limited by their thickness. Spallations can be observed in cases of layer thicknesses exceeding 8 μm. Furthermore, crystallinization occurs, which reduces the life time of the glass sealing. The viscosity of SiO_2 is one order of magnitude or more higher than B_2O_3 and thereby the sealing efficiency is extremely reduced. A nonstoichiometric CVD-SiC coating with an excess amount of Si shows a formation of pores in the SiO_2 glass sealing due to the following chemical reaction (Figure 5.46):

$$Si_{(s)} + SiO_{2(s)} \rightarrow SiO_{(g)}$$

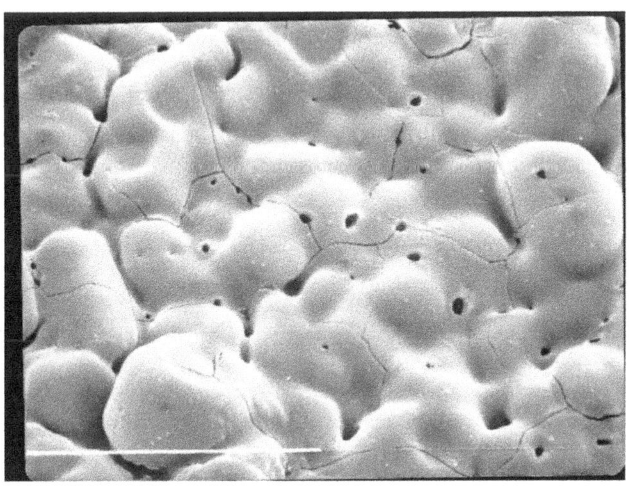

FIGURE 5.46

Perforation of SiO_2 glass sealings on multiplayer CVD-SiC-coated carbon/carbon after 1000 hours at 1530 K in air. Reproduced with permission from Pergamon, Composites Science and Technology.

Silica glass sealings can be replaced by glass sealings with $MoSi_2$ and their modifications, which have a lower viscosity and thereby a better sealing effect.

At temperatures above 1800°C the SiO_2 glass sealings are thermally unstable. Therefore, different high-temperature refractory oxides have to be applied, like ZrO_2, Y_2O_3, or HJO_2. However, the oxygen diffusivity is some orders of magnitude higher than in the case of SiO_2 glass sealings. A further disadvantage of silica is the insufficient corrosion resistance against water vapor with high flow rates as required for application in gas turbines. Therefore, environmental barrier coatings (EBC) are under development to overcome these problems.

For most of the industrial applications oxidation protection is not required, because carbon/carbons are normally applied in inert gas or under vacuum.

5.6 INDUSTRIAL APPLICATIONS OF CARBON/CARBONS

As already mentioned, the driving force for industrial applications of carbon/carbons have been the developments for space and military applications followed by carbon/carbons for military and civil aircraft brakes (1971 for Concorde, 1970 for F-15 fighters). Subsequently, the use of carbon/carbon brakes has been exploited to formula I racing cars by Hitco (1980), SEP 1981), AP Racing, Le Carbon Lorraine, and others. These carbon/carbon brake disks are sensitive against oxidation and can only be used for one single race. The formula I disks are reinforced with chopped fibers, carbon felts, or carbon cloth with a CVI-derived PyC matrix. The same is valid for carbon/carbon clutches in race cars.

However, these carbon/carbons could not be used for automotive applications due to their limited life time. Therefore, developments were started in Europe in 1979 to replace asbestos brake pads by metal-impregnated carbon/carbons to improve their wear and oxidation resistance [196–198]. Although these developments were successful from a technical viewpoint, a realization never occurred due to the high costs of these materials for mass applications. These materials were already optimized from the viewpoint of improved wear resistance and thermal conductivity by silicon treatments and combinations with copper impregnations. However, an industrial application took an additional 20 years. The restrictions for use only in luxury cars and the requirements of improved safety and life time concepts enabled the first CMC brake disks used in Porsche cars in the beginning of the year 2000.

Today, CMC brake disks for different luxury cars are commercially available from SGL carbon brakes for Porsche and Audi and from Brembo for Ferrari and Mercedes.

Approximately 50 000–70 000 CMC brake disks were manufactured in 2006. The broad application on cars with lower costs is still limited due to the strong competition with brake disks made of steel. The manufacturing costs of steel disks are below 10 €/disk, whereas CMC disks with a weight up to 5 kg/disk are above 500 €/disk.

Although CMC disks are superior from the viewpoint of performance, life time, friction behavior and safety aspects, their application seems to be feasible only in luxury cars within the next few years.

The manufacturing technologies are in an excellent progress (Figure 5.47) [10]

The different manufacturing steps are already, or rather, can be automated to a certain degree.

Compounding is executed batch-wise in a mixing device or continuously in an extruder. Forming and molding can be made semi-continuously with an automated molding device. The requested molding times depend on the process and can be reduced to times below 15 minutes/disk. Near-net-shape techniques minimize the machining costs, in particular for the final machining steps.

Carbonization can be performed continuously, whereas continuous silicon treatments are still under development. However, the price target given by metallic disks, can never be reached due to the costs of raw material and the processing costs.

The complex shape caused by internal vented disks can be achieved by different manufacturing techniques as demonstrated in Figure 5.48 [10]

The complete machining of the cooling channels after or before the silicon treatment step is too expensive for a production process and was only used for the development phase in order to be as flexible as possible. Industrially usable techniques for making cooling channels are given in Figure 5.48.

The most flexible method is the joining of three parts with a cover and bottom plate and a middle sector which contains all cooling channels. The middle part can be water jet cut after the molding step. This technique allows the usage of very simple molding tools, and due to the reduced thickness of a single part with approximately 13–14 mm, extreme short molding cycles. The disadvantage is the waste of the water jet cut pieces resulting in higher material

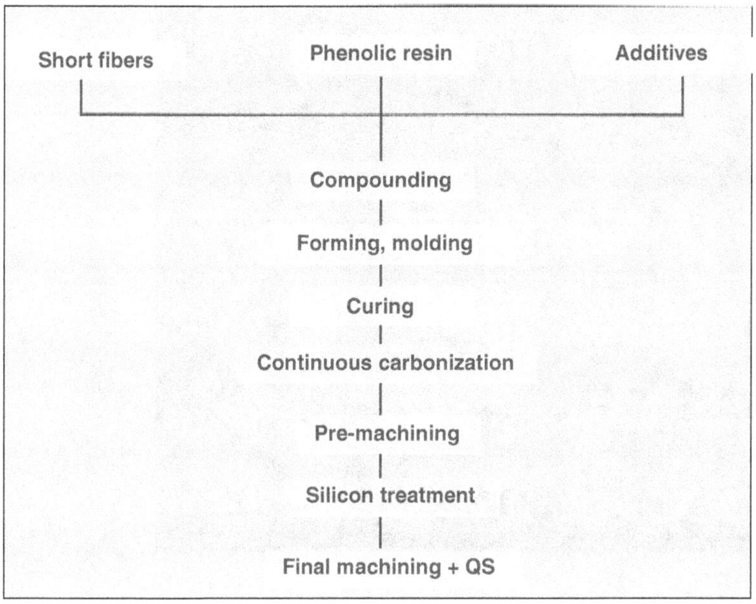

FIGURE 5.47
Manufacturing scheme of CMC brake disks [10]. Reproduced with permission from Pergamon, Composites Science and Technology.

costs. The molding process as well as the water jet cutting step can be fully automated. Water jet cutting can be avoided by net shape pressing of the middle section which requires an additional pressing tool.

In case of minimized material waste, two pieces can be net-shape-molded and afterward mechanically joined. Chemical joining (welding) takes place during the silicon treatment step (Figure 5.47).

Another possibility is net-shape molding of a single part by using dead molds or movable tools (Figure 5.25). The disadvantages are that movable tools are more complex and have to be cleaned after each molding cycle, or in case of dead molds, they must be replaced. Furthermore, thicker green bodies require longer molding and curing cycles, a disadvantage from the viewpoint of cost effectiveness.

As can be derived from Figure 5.48, the tribological surface can be adapted according to the requirements by using a surface coating, which is combined with increased costs.

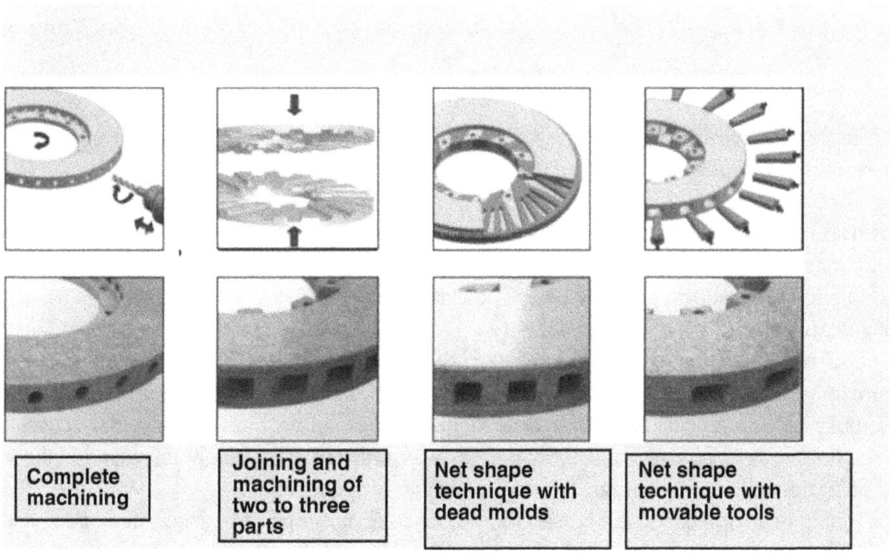

FIGURE 5.48
Manufacturing of complex vented brake disks [10]. Reproduced with permission from Pergamon, Composites Science and Technology.

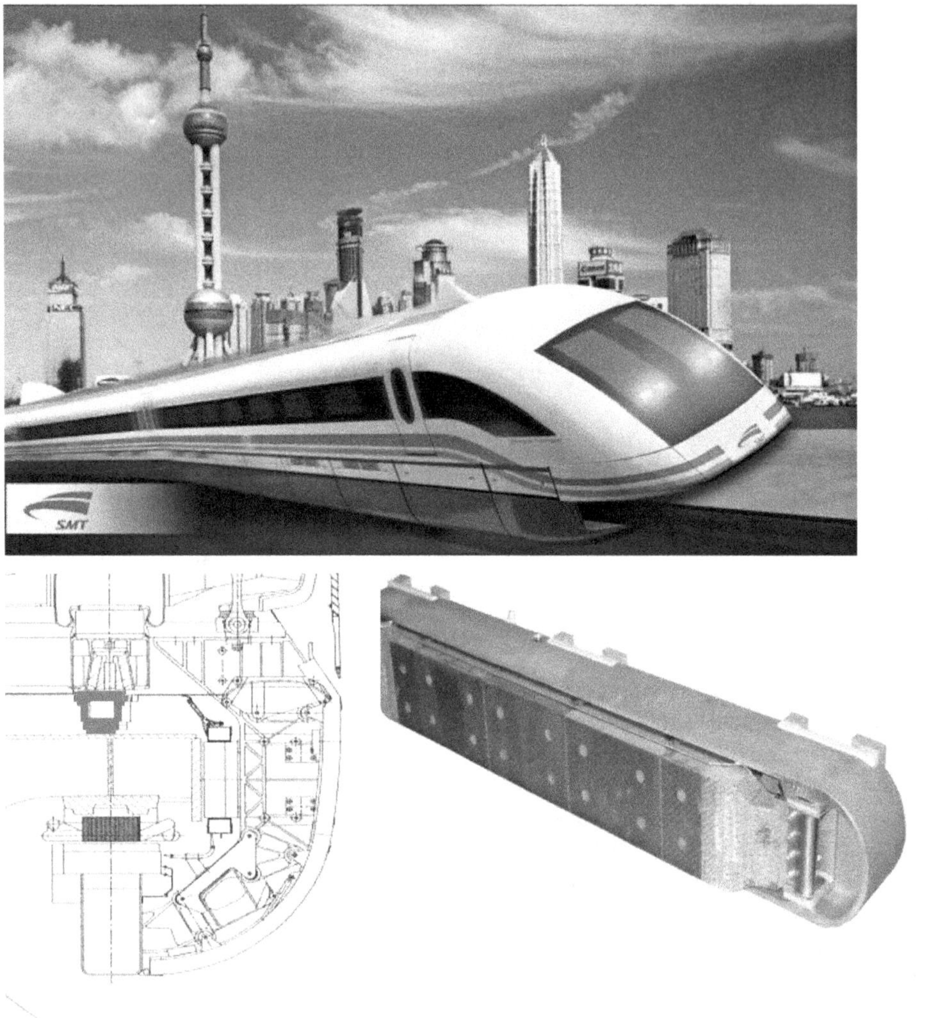

FIGURE 5.49

Sliding elements for the Transrapid [192]. Reproduced with permission from Pergamon, Composites Science and Technology.

The excellent friction behavior of C/C–SiC, the silicon-treated carbon/carbon is used for emergency braking systems like high-speed elevators, high-speed turning and milling machines, and amusement rides.

However, the silicon treatment contributes not only to an increased friction coefficient. C/C–SiC grades can also be tailored for a superior sliding behavior due to the extreme low wear rates and the possibility to combine hard carbide with a soft graphitic matrix.

A typical example for low friction and low wear applications are C/C–SiC sliding elements for the high-speed train Transrapid running in Shanghai (Figure 5.49) [192].

The requirements are wear rates below 0.1 mm/km sliding length with a friction factor below 0.3 against an iron track at a maximum speed of 500 km/h. The C/C–SiC sliding elements possess a high mechanical strength, low thermal conductivity, and must be biocompatible and environmentally compatible due to the wear losses. The good sliding behavior in combination with good mechanical properties was realized by a newly developed silicon treatment (P75, Figure 5.39), a combination of pack cementation and capillary impregnation.

The excellent sliding behavior can also be used in petrochemical pumps as bearings and sliding rings as can be seen in Figure 5.50.

The advantage of these parts for pumps is their superior failure behavior (pseudo-ductile) in comparison to monolithic ceramic elements.

The industrial application of carbon/carbons is based on the specific properties of these light-weight materials.

FIGURE 5.50

Bearing and sliding ring made from C/C–SiC for petrochemical pumps. Reproduced with permission from Pergamon, Composites Science and Technology.

Carbon/carbons possess the best specific mechanical high-temperature properties of all materials (Figure 5.51).

The strength and stiffness of carbon/carbons increases with increasing temperature. The improvement in strength and stiffness from a. T to 2000°C is about 25%. Therefore, the mechanical finite element analysis based on room temperature properties includes an additional safety factor for high-temperature applications.

Carbon/carbons are predestined for all high-temperature applications under inert gas or vacuum atmospheres like furnace materials, hot pressing dies, handling tools for hollow glass manufacturing, for semiconductor industry, and photovoltaic applications.

5.6.1 Carbon/Carbons for High-Temperature Furnaces

High-performance vacuum furnaces are fully equipped with carbon/carbon materials. Thermal insulation based on hard felts, CVI, or resin-impregnated soft felts, are used as standards in most of the furnaces (Figure 5.52).

The hard felt boards are protected at the edges with U- and L-shaped profiles (carbon-cloth reinforced) with a standard length of 1, 1.5, or 2 m. The profiles are used for protection against handling damages, against gas erosion effects during gas quenching, and for assembling of the thermal insulation with carbon/carbon screws and nuts.

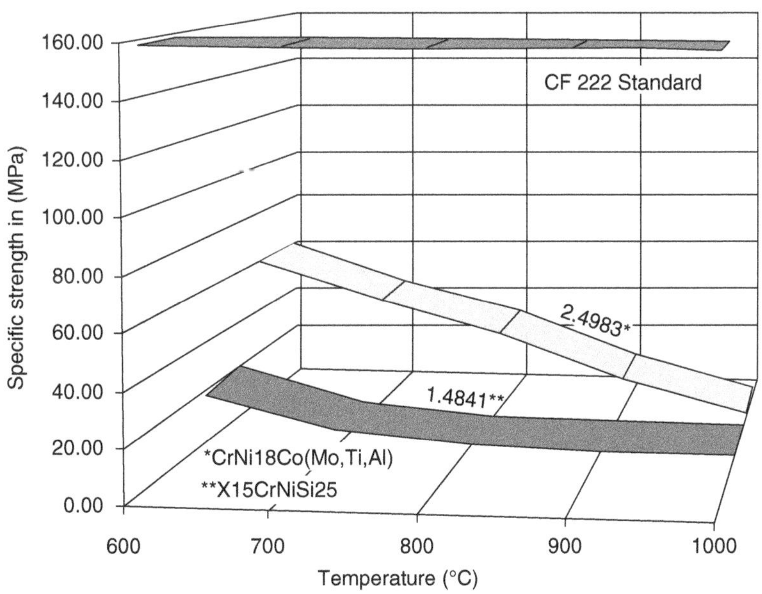

FIGURE 5.51

Specific strength of carbon/carbon in comparison to steel. Reproduced with permission from Pergamon, Composites Science and Technology.

FIGURE 5.52
Interior of an industrial furnace. Reproduced with permission from Pergamon, Composites Science and Technology.

Screws, bolts, and nuts are used for fixation and joining of mechanical loaded bars and pillars within the furnace. A broad data base on the mechanical properties of carbon/carbon joining elements exists for tensile as well as for shear load introductions (Figure 5.53) [199].

The furnace chamber itself is very often equipped with a carbon/carbon chamber (Figure 5.54).

The carbon/carbon furnace chamber shown in Figure 5.31 has a size of $2000 \times 1000 \times 1000\ mm^3$ at a total weight of 30 kg. The advantage of a furnace chamber is the improved homogeneity with respect to the temperature distribution. The maximum temperature deviation within a furnace chamber can be limited to about 10° over the full range of required temperatures. However, the used heater system has to be adapted in order to avoid heat sink effects. A further advantage of a carbon/carbon chamber in comparison to a graphite one is the required lower thickness, which results in a reduced dead mass and better heating and cooling efficiency. The disadvantage of the stiffness reduction due to the reduced wall thickness can be overcome by stiffening the whole component with profiles (Figure 5.54).

Heating elements for vacuum and inert gas furnaces can be tailored from electrical properties by using carbon/carbon materials as resistivity heaters. Their specific electrical resistivity is dominated by the resistivity of the fibers and their thermal post treatments.

Figure 5.55 shows the electrical resistivity of two different C/C grades with a 2D reinforcement in dependence of the application temperature.

Furthermore, the resistivity can be tailored by the design of the heating elements (Figure 5.56)

The total resistivity of the heating element can be calculated according to the following equation:

$$R_{\text{total}} = \frac{p \times L}{A}$$

where P is the specific electrical resistivity, L is the length of the heating element, and A is the cross-section.

The length and the cross-section can be optimized by the design of the heating elements within the given limitations of the geometry of the furnace facilities. The specific electrical resistivity depends on the fiber orientation, fiber volume fraction, the fiber type, and the final heat treatment of carbon/carbon. Thereby, a broad variety of total resistivity can be made according to the requirements of the customer. However, the higher the final application temperature, and thereby the final heat treatment has to be, the lower the specific resistivity.

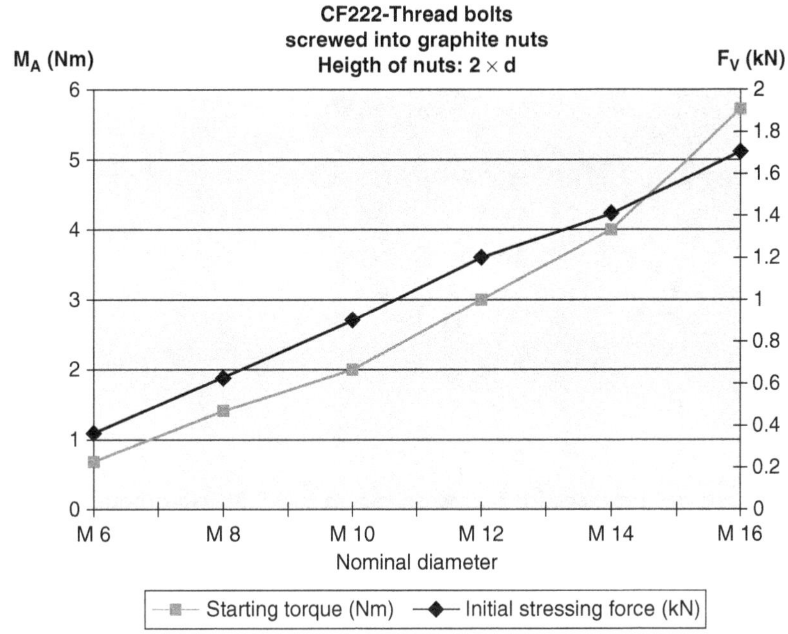

(a)

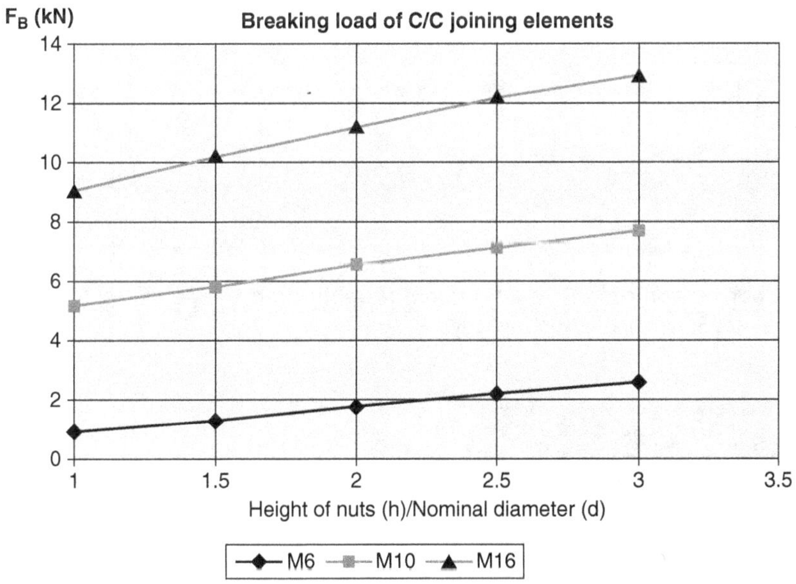

(b)

FIGURE 5.53

Mechanical properties of carbon/carbon screws [199]. (a) Starting torque and initial stressing force. (b) Breaking loads of C/C bolts of C/C bolts in dependence of the nominal diameter as a function of the ratio height of nut to nominal diameter. Reproduced with permission from Pergamon, Composites Science and Technology.

As mentioned above, an important target in the high-performance vacuum and inert gas furnaces is their homogeneity of the temperature distribution. Therefore, in inert gas furnaces a forced ventilation is required additionally, which can be realized by ventilators made of carbon/carbon applicable up to extreme temperatures above 1500°C (Figure 5.57) [200].

The superior mechanical behavior like high tensile strength in fiber direction is used in chains made of carbon for sinter furnaces as shown in Figure 5.58.

The whole chain is made of carbon/carbon. The maximum load can be overtaken by the UD-reinforced chain links with a specific tensile strength beyond 800 MPa. The maximum load is limited by the carbon/carbon axes and their cross-section.

FIGURE 5.54

Carbon/carbon chamber for vacuum and inert gas furnaces for temperatures up to 2400°C. Reproduced with permission from Pergamon, Composites Science and Technology.

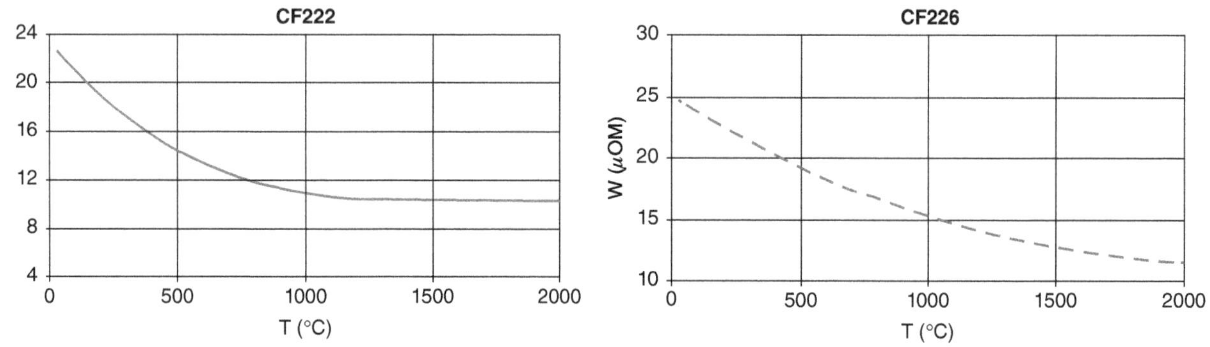

FIGURE 5.55

Electrical resistivity of two different carbon/carbon grades as a function of application temperature. Reproduced with permission from Pergamon, Composites Science and Technology.

FIGURE 5.56

Meander-shaped tubular and flat heating elements. Reproduced with permission from Pergamon, Composites Science and Technology.

FIGURE 5.57
Carbon/carbon ventilators for forced inert gas conversion. Reproduced with permission from Pergamon, Composites Science and Technology.

The chain links, the axes, and the distance holders are filament winded and near-net-shape manufactured. The parts have only to be machined according to the required length or width, respectively.

Carbon/carbon chains can be applied for vertical as well as for horizontal transportation in a furnace. The main disadvantage is their sensitivity against wear losses and different gas atmospheres.

As already mentioned, carbon/carbon materials are attacked by oxygen at temperatures beyond 420°C. In sinter furnaces such oxidations are suppressed by using inert gases and a mixture of nitrogen (N_2) and hydrogen (H_2) as endo or exo gases. In case of sintering, iron alloys or thermal treatments of steel, a tailored carburization of the metal alloy is required, which very often forms water as an intermediate product. Furthermore, open and closed furnaces sometimes contain within the applied insulation materials water due to the uptake at ambient temperature, if the furnace is cooled down. This low amount of water cannot always be completely removed during the heating-up phase in the temperature

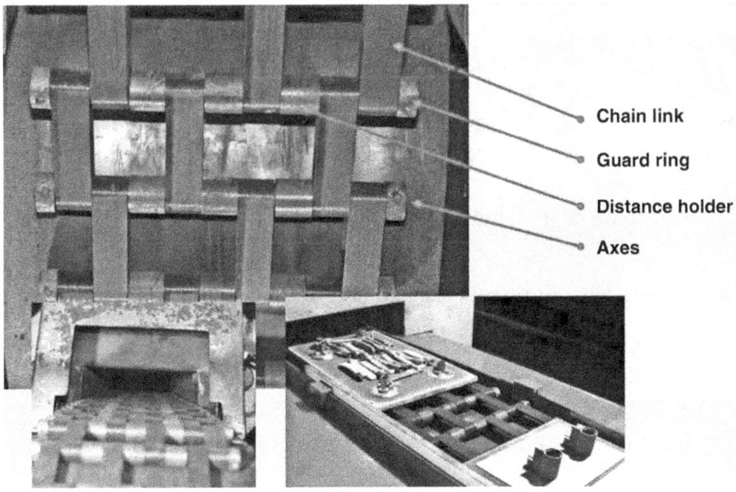

FIGURE 5.58
Chain made of carbon/carbon with a total length of 38 m applied in a sinter furnace. Reproduced with permission from Pergamon, Composites Science and Technology.

TABLE 5.8 Overview of Maximum Application Temperatures for Thermal Treatments of Iron Alloys in Furnaces with Carbon/Carbon Materials

Atmosphere	Maximum Application Temp. (°C)	Applicability of Carbon/Carbon
N_2	1080 (1100)	+++
H_2	1200	+++/o
Exo gas	1050	+++
Endo gas	1050	+++
Vacuum	1050	o
O_2	<420	+++

+++, applicable without restraint; o, applicable with limitations.

range below 420°C. Thereby, two typical chemical attacks can occur and affect the carbon/carbon materials inside the furnace.

1. $C + H_2O \leftrightarrows CO + H_2$ $\Delta H = -131.38$ kJ/Md at $T \approx 800–1000°C$
2. $C + 2H_2 \leftrightarrows CH_4$ $\Delta H = -74.86$ kJ/Md.

The most severe reaction is the formation of methane (reaction 2), because the oxygen attack or water attack can be controlled by *in situ* measurements of the water concentration in the furnace and an improved preheating to remove the remaining humidity from the thermal insulation materials.

Table 5.8 gives an overview of the maximum application temperature of carbon/carbon materials for thermal treatments of iron alloys with respect to chemical resistance including uncontrolled carburization effects of the metallic alloys.

In case of hydrogen, methanization reduces the life time of the carbon/carbon parts. This can be suppressed by using a gas mixture of hydrogen and nitrogen with low hydrogen content. An alternative method is the treatment of carbon/carbons with Si to form a C/C–SiC material, which reduces the attack of hydrogen for some orders of magnitude. The disadvantage is the formation of silicides with compounds within the metallic alloy, which has to be prevented. Therefore, the silicon treatment is limited to selected alloys and to a maximum temperature of approximately 1100°C for thermal treatment of metals.

The limitations in vacuum are mainly caused by carburization effects of the metallic alloy due to the increased diffusion activity of the carbon under vacuum. This behavior can be improved and shifted to higher temperatures by using an improved crystallinity of the matrix, which can be realized by CVI/CVD coatings with highly orientated PyC (pyrolithic graphite). Such coatings (impregnations) enable maximum application temperatures up to 1200°C.

5.6.2 Application for Thermal Treatments of Metals

Load-carrying carbon/carbon systems are industrially used as fixtures for thermal treatments of metallic parts like vacuum brazing of steel or stainless steel components, for annealing of high-quality steels for gas turbine compounds, low-pressure carburizing, gas quenching of automotive parts, and case hardening of steel parts with subsequent oil quenching (e.g., Figure 5.59).

The industrial motivation to use C/C fixtures instead of metallic ones is based on the following advantages:

- No distortion of C/C trays resulting in handling by roboting (increased efficiency, reduced man power costs)
- No embrittlement even after some years of application
- Increased life time of the fixtures
- Improved energy efficiency by reduced weight in comparison to steel fixtures
- Increased number of trays by reduced thickness of the C/C fixtures
- Increased number of rods by optimized design.

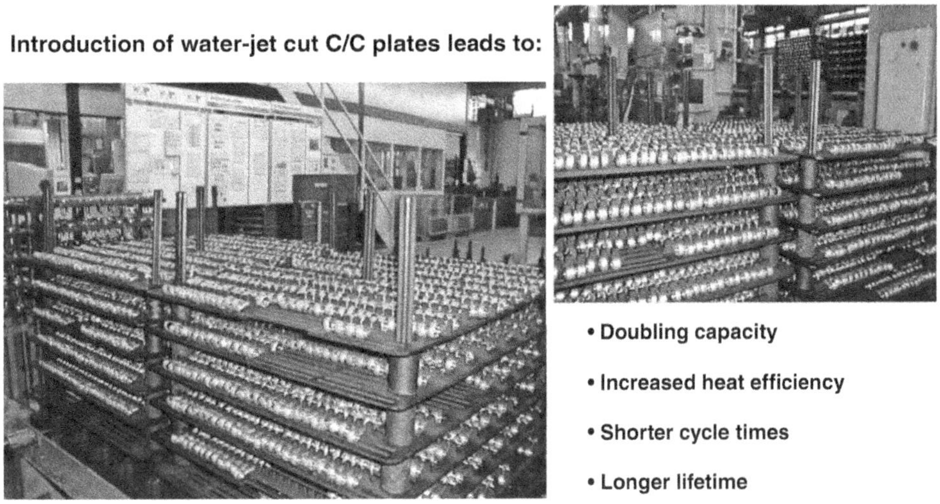

FIGURE 5.59

Carbon/carbon fixture with water jet cut carbon/carbon plates for case hardening of rods with subsequent oil quenching. Reproduced with permission from Pergamon, Composites Science and Technology.

All these advantages contribute to an improved efficiency and reduced costs in thermal treatments of metallic components. The return of investment for the more costly C/C fixtures in comparison to metallic ones is below 2 years. Therefore, this is one of the fast-growing markets for industrial C/C applications.

A comparison of a cast metal tray and a carbon/carbon tray is given in Figure 5.60.

As can be seen, even after 8 years in service, no distortion can be detected on the C/C tray. The weight saving is 28 kg per tray (Figure 5.60) and 336 kg per batch as shown in Figure 5.36. This illustrates the advantage of reduced dead mass per batch and the enormous potential of energy saving over a service period of 8 years with up to six or seven batches per day.

With increasing energy costs the use of C/C fixtures will be even more attractive and superior from the viewpoint of life cycle and manufacturing costs.

The trays shown in Figure 5.60 are used for the same metallic parts for case hardening. The reduced height of the C/C trays enabled additional stacks in comparison to the metallic ones (Figure 5.36) Furthermore, the number of connecting rods per tier could be increased by the thinner distance holder made from carbon/carbon. The overall efficiency improvement resulted in doubling of connection rods per batch. This demonstrates the economical and ecological superiority of C/C fixtures in comparison to metallic ones.

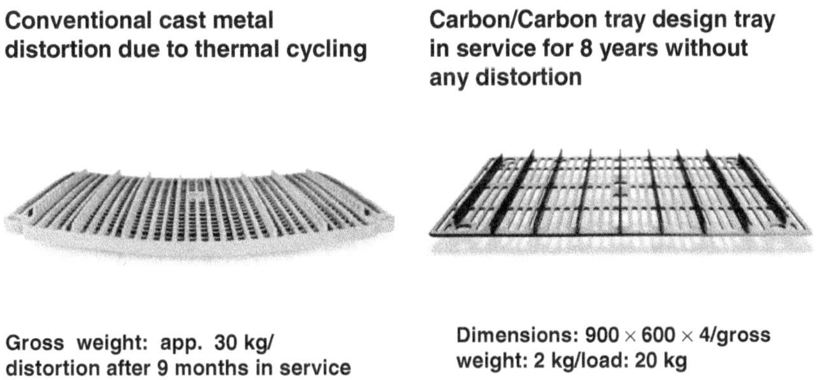

FIGURE 5.60

Comparison of carbon/carbon and cast metal trays. Reproduced with permission from Pergamon, Composites Science and Technology.

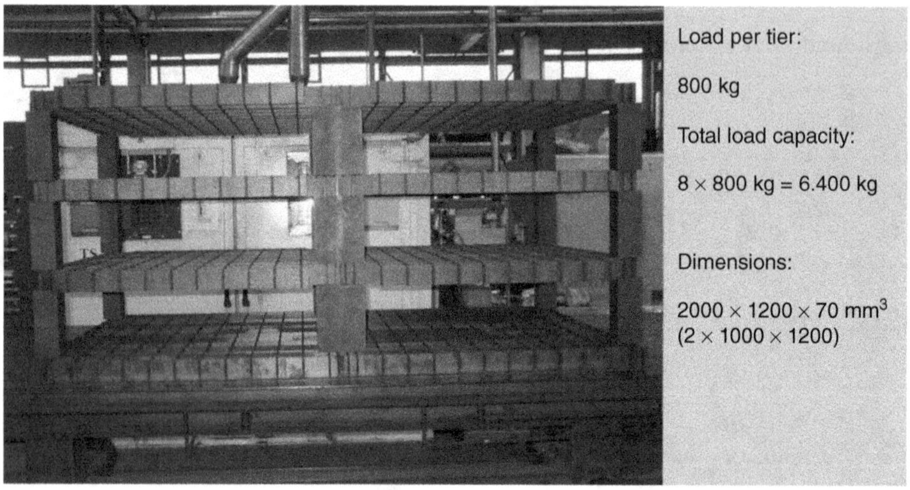

FIGURE 5.61

Carbon/carbon fixtures for heavy loads, for example, brazing of heat exchangers for cars. Reproduced with permission from Pergamon, Composites Science and Technology.

The C/C fixtures can be tailored to the required load capacity as shown in Figure 5.61.

The carbon/carbon trays are designed according to the requirements of the customer and can be used in their already existing facilities and furnaces.

A new development for vacuum brazing of turbine components (Figure 5.62) is the Unigrid™ system of Schunk. This light-weight grid, approximately 2 kg, is based on a textile preform made via tailored fiber placement (TFP) without any joinings manufactured near-net-shape. The manufacturing process can be performed via RTM or resin infusion methods (RI).

The direct contact of carbon and gas turbines has to be avoided to prevent an uncontrolled carburization. Therefore, CarboGard™ protective tiles based on alumina are available for the UniGrids™ system. The high flexibility of the UniGrids™ is shown in Figure 5.63.

The load capacity can be brought into line with these requirements by using graphite supports which connect the trays. The graphite supports can be fixed wherever it is necessary (in the middle, at the corners). The load capacity

FIGURE 5.62

Schunk UniGrid™ system with CarboGard™ protective tiles. Reproduced with permission from Pergamon, Composites Science and Technology.

FIGURE 5.63
Assembling of UniGrids™ with graphite supports [201, 202]. Reproduced with permission from Pergamon, Composites Science and Technology.

increases with reduction of the free length between two supporting points. The system can be assembled by the final end user and thereby be modified for a variety of different applications.

The use of C/C fixtures for thermal treatments of metals will be widespread in the future due to the high cost efficiency for automotive suppliers. The second fast-growing market is the field of photovoltaic industry.

5.6.3 Application of Carbon/Carbon in the Solar Energy Market

The growth of the global solar energy market is shown in Figure 5.64. [192, 203]

This market is mainly politically driven with a worldwide lead of Germany. Carbon/carbon structural components are necessary for the high-temperature processes in manufacturing photovoltaic cells like re-crystallization and CVD coatings, and for making polysilicon as one of the most important raw materials. The shortage in polysilicon on the market resulted in tremendous investments worldwide of some billions of US dollars. The planned increase in the worldwide capacity for polysilicon is given in Figure 5.65.

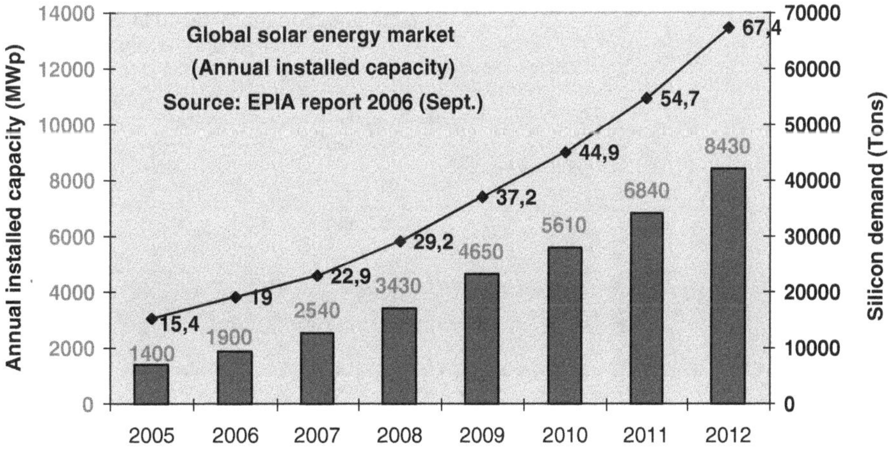

FIGURE 5.64
Growth of the Global Solar Energy Market. Reproduced with permission from Pergamon, Composites Science and Technology.

Capacity:	2007	2010
Hemlock	7700	36 000
Wacker	6500	21 500
REC	5300	12 500
Tokuyama	5200	5400
MEMC	3800	8000
Mitsubishi	2850	3500
Sumitomo	800	1600
Total China	130	6000
Elkem	-	10 000
Others	-	4500

Announced capacity
In 2010: 109 000 tons

Demand: PV: 45 000 tons
Demand SC: ??

Source: Flynn & Bradford Prometheus Institute "Publications from semiconductor companies"

FIGURE 5.65
Planned capacity of polysilicon. Reproduced with permission from Pergamon, Composites Science and Technology.

The announced capacity covers up to 2012 the demand of semiconductor industry as well as photovoltaic applications. The need of carbon/carbons is planned to increase from 2007 to 2012 up to more than 600%. Therefore, most of the investments of European carbon/carbon manufacturers are concentrated on these applications:

Typical requirements are:

- High purity levels
- CVI/CVD-coated carbon/carbons in order to avoid carbon particle release
- Coatings for improved corrosion resistance against SiO attacks
- Ductile
- Cost efficient
- Low methanization rates

These requirements can only be fulfilled with carbon/carbon with inner and outer protection systems. Typical structural components are heaters, thermal insulations, heat shields, crucibles, handling tools, and carrier systems for solar cells.

Most of the carbon/carbons for these applications are based on mixed matrix systems, including CVI and resin densifications [192]. Hard felt boards as thermal insulations possess only a limited life time in case of hydrogen or SiO environments of some months, whereas thermal insulations with mixed matrix systems survive more than 2 years under the same conditions. The methanization as well as the oxidation rate of such high-performance carbon/carbon insulations is superior to standard grades as demonstrated in Figure 5.66.

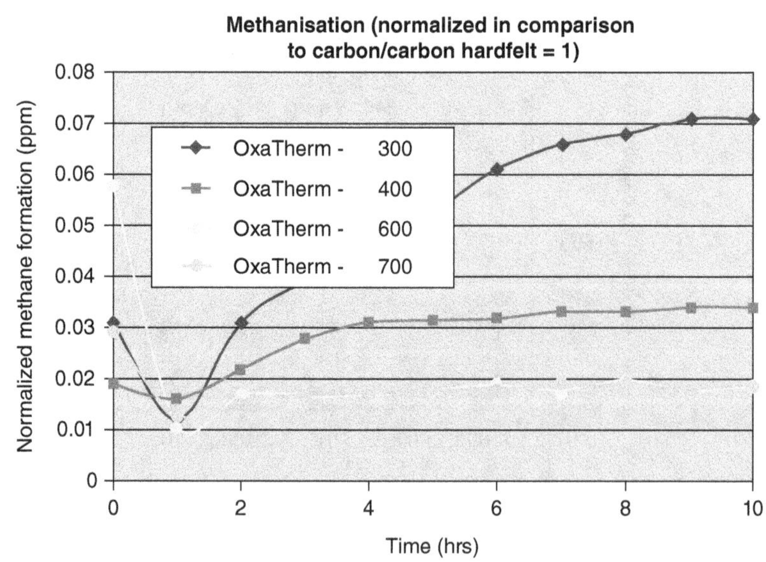

FIGURE 5.66
Methanization rate of thermal felt insulations normalized in comparison to standard carbon/carbon hard felt boards. Reproduced with permission from Pergamon, Composites Science and Technology.

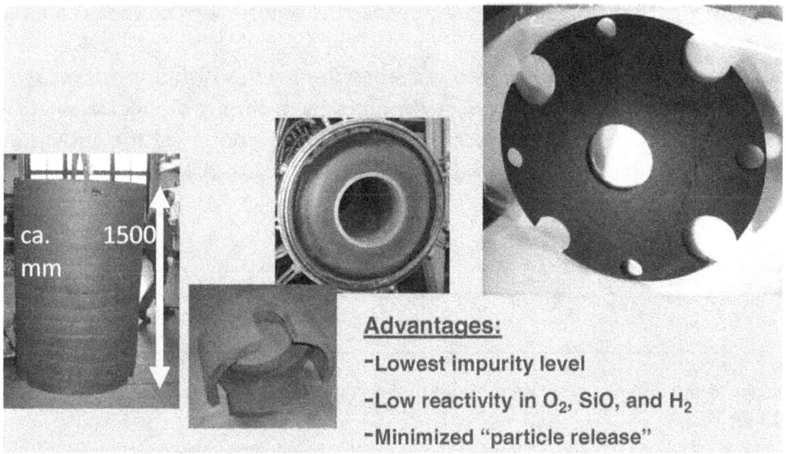

FIGURE 5.67
Carbon/carbon thermal insulations with mixed matrix systems. Reproduced with permission from Pergamon, Composites Science and Technology.

FIGURE 5.68
Crucibles made of carbon/carbon with mixed matrix systems for pulling of single crystals. Reproduced with permission from Pergamon, Composites Science and Technology.

The methanization rate can be reduced to 2% in comparison to standard hard felt boards. Some typical thermal insulations are shown in Figure 5.67.

Crucibles for pulling of single crystals are made of carbon/carbons with mixed matrix systems as well as heating elements in order to reduce their oxidation by SiO attacks (Figure 5.68).

Comparable chemical requirements are valid for etching processes of wafers by using felt susceptors [204], (Figure 5.69)

Felt susceptors are necessary for oxidative etching as well as for the doping processes of wafers. Therefore, the susceptor must be highly purified, oxidation-resistant, and should not be contaminated by the doping species. The gas

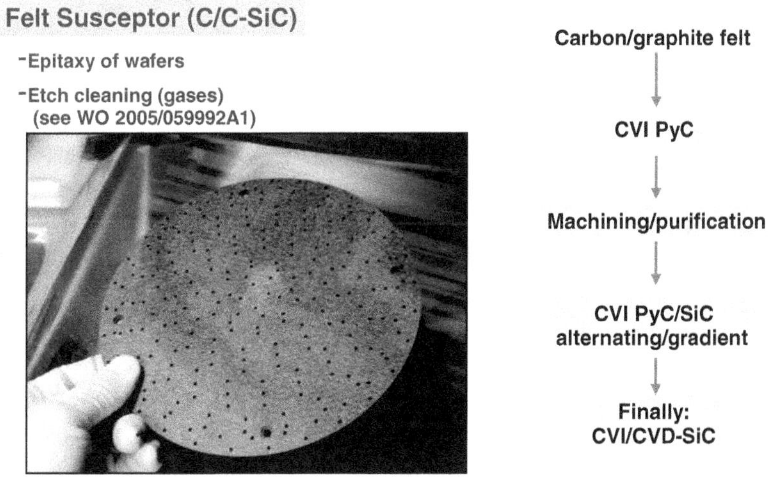

FIGURE 5.69
Carbon/carbon felt susceptor with mixed matrix systems. Reproduced with permission from Pergamon, Composites Science and Technology.

flow characteristics have to be homogeneous in order to guarantee an excellent homogeneity of the treated wafer surface.

For future applications in fixation and fusion reactors carbon/carbons are required with a high thermal conductivity combined with a low erosion and sublimation rate. Therefore, new development activities are undertaken to dope carbon/carbon with carbides and to maximize their thermal conductivity. Although the potential for such modified carbon/carbons is high, an industrial application will not be feasible within the next 5–10 years.

REFERENCES

1. C. R. Thomas and E. J. Walker. *Proceedings of the Fifth London International Carbon and Graphite Conference*, Vol. I (Society of Chemical Industry, London, 1978) p. 520 ff.
2. C. R. Thomas. *Essentials of Carbon/Carbon Composites* (The Royal Society of Chemistry, Cambridge, 1993) pp. 1–35.
3. E. Fitzer and A. Burger (1971) *International Conference on Carbon Fibres, Their Composites and Applications,* London paper no. 36.
4. L. E. McAllister and A. R. Taverna. *Proceedings International Conference on Composite Materials*, Vol. I (Metallurgical Society of AIME, New York, 1976) p. 307.
5. P. J. Lamieq. (1977), *Proceedings of the AIAA/SAE 13th Propulsion Conference, Paper no.* 77–882, Orlando.
6. E. Fitzer, K. H. Geigl, and W. Hüttner. *Proceedings of the Fifth London International Carbon and Graphite Conference*, Vol. I (Society of Chemical Industry, London, 1978), p. 493.
7. H. Girard. *Proceedings of the Fifth London International Carbon and Graphite Conference*, Vol. I (Society of Chemical Industry, London, 1978), p. 483.
8. C. R. Thomas and E. J. Walker. *Proceedings of the Fifth London International Carbon and Graphite Conference*, Vol. I (Society of Chemical Industry, London, 1978), p. 520.
9. R. Weiss. Carbon fibre reinforced CMCs: manufacture, properties, oxidation protections, in *High Temperature Ceramic Matrix Composites (HTCMC)*, edited by W. Krenkel, R. Naslain, and H. Schneider (VCH Verlag GmbH & Co. KGaA, Weinheim, 2001), pp. 440–456.
10. R. Weiss. *CMC and C/C-SiC-Fabrication* (CIMTEC, Sicily, 2006).
11. D. L. Schmidt. (1996), Wright Laboratory WL-TR-96-4107.
12. P. W. M. Peters, G. Lüdenbach, R. Plesger, and R. Weiss. *JECS*, 13, 561–569 (1994).
13. S. Takano, T. Kinjo, T. Uruno, T. Tlomak, and C. P. Ju. Investigation of process-structure-performance relationship of unidirectional reinforced carbon-carbon composites. *Ceram. Eng. Proc.*, 12(9–10), 1914–1930 (1991).
14. R. J. Zaldivar, G. S. Rellick, and J. M. Yang. Studies of fiber strength utilization in carbon/carbon composites, in *Extended Abstracts, 22nd Carbon Conference* (1991), pp. 400–401.
15. P. W. M. Peters, J. Schmauch, R. Weiß, and M. G. Baser. The strength of carbon fibre bundles, loose and embedded in a polymer and carbon matrix, in *Proceedings of the Fifth European Conference on Computational Mechanics (ECCM V),* Bordeaux, April 7–10, 1992, pp. 157–163.
16. G. Savage. *Carbon/Carbon Composites* (Chapmann & Hall, London, 1993).
17. Hüttner, W. (1980), Doctor thesis; University of Karlsruhe.
18. DE 2714364 (1977) Verfahren zur Herstellung von Kohlenstofffaserverstärkten Kohlenstoffkörpern.
19. JP 1986 – 182483.
20. JP 1985 – 276440.
21. Brochures of Accross, 2007, http://www.acrosscfc.com/aboutcfc.html
22. R. J. Diefendorf and Y. Sohda. *Extended Abstracts Program 17th Biennial Conference on Carbon*, 17, 1983, p. 31.
23. W. V. Kotlensky. *Chem. Phys. Carbon*, 9, 173 ff (1973).
24. FP (1981) (von CEA).
25. D. Rovillain, M. Trinquecoste, E. Bruneton, A. Derré, P. David, and P. Delhaes. Film boining chemical vapor infiltration—an experimental study on C/C composite materials. *Carbon*, 39, 1355–1365 (2001).
26. P. C. Davin, J. Blein, C. Gachet, and C. Robin-Brosse (2007), ICCM 16, Abstracts, pp. 736–737.
27. A. Becker and K. J. Hüttinger. Chemistry and kinetics of CVD of PyC II, III. *Carbon*, 36, 177–211 (1996).
28. W. Benzinger and K. J. Hüttinger. Chemistry and kinetics of chemical vapour infiltration of pyrocarbon of carbon felt. *Carbon*, 37(6), 941–946 (1999).

29. G. L. Doug and K. J. Hüttinger. Consideration of different mechanisms leading to pyrolytic carbon of different textures. *Carbon*, 40, 2515–2528 (2002).

30. Z. J. Hu and K. J. Hüttinger. Mechanisms of carbon deposition—kinetic approach. *Carbon*, 40(4), 624–628 (2002).

31. R. J. Diefendorf. *The Conference; Carbon/Carbon Composites Produced by Chemical Vapor Deposition* (Cocoa Beach, FL, 2006).

32. G. M. Jenkins and K. Kawamura. *Polymeric Carbons, Carbon Fibre, Glass and Char* (Cambridge University Press, 1976).

33. R. Pleger. *Vom monolithischen kohlenstoff zum verbundwerkstoff: Analyse der Strukturvariation und Gefuegeentwicklung mittels hochaufloesender Durchstrahlungselektronenmikroskopie* (Gund-und Werkstoffe, Reihe 5, 1994).

34. A. Wanner. (1991). Doctor Thesis *Structure and Properties of CMCs*, University of Stuttgart.

35. R. Weiss. *Plenary Lecture, Tsukuba Carbon Conference*, (Matsumoto, 1990).

36. P. W. M. Peters, G. Lüdenbach, R. Pleger, and R. Weiss. *JECS*, 13, 561–569 (1994).

37. D. F. Adams. *J. Comp. Mats.*, 8, 320 (1974).

38. J. X. Zhao, R. C. Bradt, and P. L. Walker Jr. *Extended Abstracts Program 15th Biennial Conference on Carbon*, (American Carbon Society, Washington, DC, 1981), p. 274.

39. K. Eckert. (1992), Diploma Thesis: Einfluss unterschiedlicher Endglühbehandlungstemperaturen auf die Duktilität von HT- , IM- und HM-faserverstärkten C/C-Werkstoffen.

40. M. S. Aly-Hassan, H. Hatta, S. Wakayama, M. Watanabe, and K. Miyagawa. Comparison of 2D and 3D carbon/carbon composites with respect to damage and fracture resistance. *Carbon*, 41, 1069–1078 (2003).

41. K. Goto, H. Hatta, D. Katsu, and T. Machida. Tensile fatigue of a laminated carbon-carbon composite at room temperature. *Carbon*, 41, 1249–1255 (2003).

42. Y. Kogo, R. Sumiya, H. Hatta, and Y. Sawada. Examination of strength-controlling factors in C/C composites using bundle composites. *Adv. Compos. Mater.*, 12(2–3), 139–154 (2003).

43. K. Goto, H. Hatta, M. Oe, and T. Koizumi. Tensile strength and deformation of a 2D carbon-carbon composite at elevated temperatures. *J. Am. Ceram. Soc.*, 86(12), 2129–2135 (2003).

44. H. Hatta, T. Aoi, I. Kawahara, Y. Kogo, and I. Shiota. Tensile strength of carbon/carbon composites. *J. Compos. Mater.*, 38(19) 1667–1689 (2004).

45. H. Hatta, K. Taniguchi, and Y. Kogo. Compressive strength of three-dimensionally reinforced carbon/carbon composite. *Carbon*, 43, 351–358 (2005).

46. V. Bianchi, P. Goursat, W. Sinkler, M. Monthioux, and E. Mensessier. Carbon fiber-reinforced (YMAS) glass-ceramic matrix composites. III. Interfacial aspects. *J. Eur. Ceram. Soc.*, 19, 317–327 (1984).

47. D. B. Marshall and W. C. Oliver. An indentation method for measuring residual stresses in fiber-reinfoced ceramics. *Mater. Sci. Eng.*, A126, 95–103 (1990).

48. D. K. Shetty. Shear-lag analysis of fiber push-out (indentation) tests for estimating interfacial friction stress in ceramic-matrix composites. *J. Am. Ceram. Soc.*, 71(2), C107–C109 (1988).

49. M. N. Kallas, D. A. Koss, H. T. Hahn, and J. R. Hellmann. Interfacial stress state present in a "thin-slice" fiber push-out test. *J. Mater. Sci.*, 27, 3821–3826 (1992).

50. X.-F. Zhou, H. D. Wagner, and S. R. Nutt. Interfacial propertied of polymer composites measured by push-out and fragmentation tests. *Composites*, A32, 1543–1551 (2001).

51. H. Serizawa. The study about the physical properties of fiber-reinforced ceramics composites (in Japanese), PhD thesis, (Tokyo University, 1997).

52. K. Fujita, H. Sakai, N. Iwashita, and Y. Sawada. Influence of heat treatment temperature on Interfacial Shear Strength of C/C. *Proceeding of the Fifth Japan International SAMPE Symposium* (Tokyo, 1997), pp. 1557–1562.

53. Y. Furukawa, H. Hatta, and Y. Kogo. Interfacial shear strength of C/C composites. *Carbon*, 41, 1819–1826 (2003).

54. T. Aoki, T. Ogasawara, and T. Ishikawa. Shear behavior and fiber bundle interfacial properties of 3D-C/C composites. *Proceedings of the Eighth International SAMPE Japan* (Tokyo, 2003), pp. 845–848.

55. H. Hatta, K. Goto, S. Ikegaki, I. Kawahara, S. A. H. Mohamed, and H. Hamada. Tensile and fiber/matrix interfacial properties of 2D- and 3D-carbon/carbon composites. *J. Euro. Ceram. Soc.*, 25(4), 535–542 (2005).

56. I. Kawahara, H. Hatta, and I. Siota. Effects of interfacial strength on tensile fracture mechanism of C/C composites. *Proceedings of the Eighth Japan International SAMPE Symposium* (Tokyo, 2003) 18–21.

57. H. Hatta, Y. Kogo, Y. Yoshihara, Y. Sawada, K. Takahashi, K. Hoshono, and T. Dozono. Thermal expansion behavior of C/C composites. *Mater. Syst.*, 14, 15–24 (1995).

58. W. C. Chang, C. C. M. Ma, N. H. Tai, and C. B. Chen. Effects of processing methods and parameters on the mechanical properties and microstructure of carbon/carbon composites. *J. Mater. Sci.*, 29, 5859–5867 (1994).

59. A. Domnanovich, H. Peterlik, A. Wanner, and K. Kromp. Elastic moduli and interlaminar shear strength of a bidirectional carbon/carbon composite after heat treatment. *Comps. Sci. Tech.*, 53, 7–12 (1995).

60. L. M. Manocha. The effect of heat treatment temperature on the properties of polyfurfuryl alcohol based carbon/carbon composite. *Carbon*, 32, 213–223 (1994).

61. B. Trouvat, X. Bourrat, and R. Naslain. Toughening mechanisms in C/C minicomposites with interface control, in *Extended Abstract, 23rd Biennial Conference on Carbon* (American Carbon Society, Strasbourg, 1997), pp. 536–537.

62. H. Weisshaus, S. Kenig, and A. Siegmann. Effect of materials and processing on the mechanical properties of C/C composites. *Carbon*, 29, 1203–1220 (1991).

63. G. Rellick. Densification efficiency of carbon-carbon composites. *Carbon*, 28(4), 589–594 (1990).

64. H. Hatta, K. Sizuki, T. Shigei, S. Somiya, and Y. Sawada. Strength improvement by densification of C/C composites. *Carbon*, 39, 83–90 (2001).

65. J. Aveston, G. A. Cooper, and A. Kelly. Paper 2 Single and multiple fracture, in *Proceedings of National Physical Laboratory: Properties of Fiber Composites* (IPC Science and Technology Press, Surrey, 1973), pp. 15–26.

66. A. G Evans and F. W. Zok. The physics and mechanics of fiber-reinforced brittle matrix composites (Review). *J. Mater. Sci.*, 29, 3857–3896 (1994).

67. W. A. Curtin and N. Takeda. Tensile strength of fiber-reinforced composites. *J. Compos. Mater.*, 32(22), 2042–2059 (1998); 32(22), 2060–2081 (2000); 34(15), 1301–1332, *J. Am. Ceram. Soc.*, 74(11), 2837–2845 (1991).

68. A. G. Evans, J. M. Domergue, and E. Vagaggini. Methodology for relating the tensile constituent properties. *J. Am. Ceram. Soc.*, 77(6), 1425–1435 (1994).

69. J. Aveston and A. Kelly. Theory of multiple fracture of fibrous composites. *J. Mater. Sci.*, 8, 352–362 (1973).

70. G. Savage. *Carbon-Carbon Composites* (Chapman & Hall, London, 1993).

71. C. R. Thomas. *Essentials of Carbon-Carbon Composites* (The Royal Society of Chemistry, Cambridge, 1993).

72. E. Fitzer and L. M. Manocha. *Carbon Reinforcements and Carbon/Carbon Composites*, Chap. 6 (Springer, 1998), pp. 190–236.

73. Y. Kogo, A. Kikkawa, W. Saito, and H. Hatta. Comparative study on tensile fracture behavior of monofilament and bundle C/C composites, in *Proceeding of 11th US-Japan Conference on Composite Materials* (Yonezawa, 2004–2009) C/C1.1–1.4.

74. H. Hatta, Y. Kogo, and A. Okura. Research Report for New Energy and Industrial Technology Development Organization, NEDO-ITK-9209, 1993, (in Japanese).

75. S. Sato, A. Kurumada, H. Iwaki, and Y. Komatsu. Tensile properties and fracture toughness of carbon-fiber felt reinforced carbon composites at high temperature. *Carbon*, 27, 791–801 (1989).

76. K. Goto, H. Hatta, M. Oe, and T Koizumi. Tensile strength and deformation of a two-dimensional carbon-carbon composite at elevated temperatures. *J. Am. Ceram. Soc.*, 86(12), 2129–2135 (2003).

77. G. W. Rowe. High temperature strength of clean graphite. *Nuclear Eng.*, March, 102–103 (1962).

78. G. Sines, Z. Yang, and B. D. Vickers. Creep of carbon yarn and a carbon-carbon composite at high temperature and high stress. *Carbon*, 27(3), 403–415 (1989).

79. J. E. Dorn. Some fundamental experiments on high temperature creep. Creep and fracture of metals at high temperature. *National Physical Laboratory Symposium* (HMSO, London, 1956) 89–138.

80. H. M. Hawthone. The mechanics of stretch-graphitization of glassy carbon fibers. *J. Mater. Sci.*, 11, 97–110 (1976).

81. K. Goto, H. Hatta, H. Takahashi, and H. Kawada. Effects of shear damage on the fracture behavior of carbon-carbon composites. *J. Am. Ceram. Soc.*, 84(6), 1327–1333 (2001).

82. L. Denk, H. Hatta, A. Misawa, and S. Somiya. Shear fracture of C/C composites with variable stacking sequence. *Carbon*, 39, 1505–1513 (2001).

83. H. Ho, M. Y. Tsai, and J. Morton. A comparison of three popular test methods for determining the shear modulus of composite materials, *Compos. Eng.*, 3(1), 69–81 (1993).

84. W. P. Keith and K. T. Kedward. Shear damage mechanisms in a woven, Nicalon-reinforced ceramic-matrix composite. *J. Am. Ceram. Soc.*, 80(2), 357–364 (1997).

85. D. Hull. *An Introduction to Composite Materials* (Cambridge University Press, Cambridge, 1981), pp. 154–163.

86. A. G. Evans and F. W. Zok. Review: the physics and mechanics of fiber-reinforced brittle matrix composites. *J. Mater. Sci.*, (29), 3857–3896 (1994).

87. M. R. Piggot. A theoretical framework for the compressive properties of aligned fiber composites. *J. Mater. Sci.*, 15, 2523–2538 (1981).

88. W. B. Rosen, N. F. Dow, and Z. Hashin. Mechanical properties fibrous composites. NASA CR-31, April, 1964, 2288.

89. H. T. Hahn, M. Soni, and S. Moon. Compression failure mechanisms of composite structures. NASA CR-3988, June, 1986.

90. F. Fitzer. The future of carbon–carbon composites. *Carbon*, 25, 163 (1987).

91. L. M. Manocha and O. P. Bahl. Influence of carbon fiber type and weave pattern on the development of 2D carbon-carbon composites. *Carbon*, 26, 13 (1988).
92. V. Gupta, K. Anand, and M. Kryska. Failure mechanisms of laminated carbon-carbon composites-I. Under uniaxial compression. *Acta. Meta.*, 42(3), 781–795 (1994); J. A. Grape, and V. Gupta. Failure mechanisms in laminated carbon-carbon composites under biaxial compression. *Acta Mater.*, 43(7), 2657–2665 (1995).
93. N. Tohyama, B. N. Kim, M. Enoki, and T. Kishi. Compressive strength and failure mechanisms of laminated C/C composites by using acoustic emission method. *Mater. Trans. JIM.*, 37(5), 1156–1160 (1996).
94. Y. Kogo, H. Hatta, and A. Misawa. Compressive behavior of carbon-carbon composites. Proceeding of the Fourth Japan International SAMPE Symposium, September 1995, pp. 1159–1164.
95. H. Hatta, K. Taniguchi, and Y. Kogo. Compressive strength of three-dimensionally reinforced carbon/carbon composite. *Carbon*, 43, 351–358 (2005).
96. H. Hatta, Y. Kogo, T. Tanimoto, and T. Morii. Static and fatigue fracture behavior of C/C composites. Proceeding of Fourth Japan International SAMPE Symposium, 1995, pp. 368–373.
97. D. Rouby and P. Reynaud. Fatigue behavior related to interface modification during load cycling in ceramic-matrix fibre composites. *Comp. Sci. Tech.*, 48, 109–118 (1993).
98. C. Tallaron, D. Rouby P. Reynaud, and G. Fantozzi. Improvement of cyclic fatigue analysis by the use of a tensile master curve in carbon/carbon composites. *Key Eng. Mater.*, 164–165, 329–332 (1999).
99. A. Ozturk and R. E. Moore. Tensile fatigue behavior of tightly woven carbon/carbon composites. *Composites*, 23, 39–46 (1992).
100. K. Goto, Y. Furukawa, H. Hatta, and Y. Kogo. Fatigue behavior of 2D laminate C/C composites at room temperature. *Compos. Sci. Tech.*, 65, 1044–1051 (2005).
101. R. Talreja. Structure and properties of composites, Chap 13, in *Fatigue of Fiber Composites*, edited by TW Chou (VCW, Tokyo, 1993), pp. 583–608.
102. B. F. Sorensen, J. W. Holmes, and E. L. Vanswijgenhoven. Rate of strength decrease of fiber-reinforced ceramic-matrix composites during fatigue. *J. Am. Ceram. Soc.*, 83(6), 1469–1475 (2000).
103. N, Chawla, Y. K. Tur, J. W. Holmes, J. R. Barber, and A. Szweda. High-frequency fatigue behavior of woven-fiber-fabric-reinforced polymer-derived ceramic-matrix composites. *J. Am. Ceram. Soc.*, 8(5), 1221–1230 (1998).
104. M. Mizuno, S. Zhu, Y. Nagano, Y. Sakaida, Y. Kagawa, and M. Watanabe. Cyclic-fatigue behavior of SiC/SiC composites at room and high temperatures. *J. Am. Ceram. Soc.*, 79(12), 3065–3077 (1996).
105. Y. Kogo, H. Hatta, H. Kawada, and T. Machida. Effect of stress concentration on tensile fracture behavior of carbon-carbon composites. *J. Comp. Matcr.*, 32(13), 1273–1291 (1998).
106. L. Denk, H. Hatta, S. Somiya, and H. Misawa. Fracture behavior of multi-holed C/C composites. *Adv. Compos. Mater.*, 12(2–3), 107–122 (2003).
107. M. S. Aly-Hassan, H. Hatta, and S. Wakayama. Effect of zigzag damage extension mechanism on fracture toughness of cross-ply laminated carbon/carbon composites. *Adv. Compos. Mater.*, 12(2–3), 223–236 (2003).
108. E. Fitzer and L. M. Manocha. *Carbon Reinforcement and C/C Composites* (Springer, Berlin, 1998).
109. G. Savage. *Carbon-Carbon Composites* (Chapman & Hall, London, 1993).
110. J. P. Bonal. Neutron irradiation effects on the thermal and dimensional stability of carbon fibre composites at divertor conditions. *J. Nucl. Mater.*, 228, 155–161 (1996).
111. J. Michalowski. Thermal conductivity of 2D C–C composites with pyrolytic and glass-like carbon matrices. *J. Nucl. Mater.*, 393, 47–53 (2009).
112. J. M. Su. Structure and properties of carbon/carbon composite materials for aircraft brake discs. *New Carbon Mater.*, 21(1), 81–89 (2006).
113. W. Krenkel. C/C–SiC composites for space applications and advanced friction systems, *Mater. Sci. Eng. A*, 412, 177–181 (2005).
114. C. L. Mantel, *Carbon and Graphite Handbook* (Krieger Publishing, New York, 1968).
115. P. L. Walker. *Chemistry and Physics of Carbon*. Vol. 9 (Marcel Dekker, 1973).
116. W. N. Reynolds. *Physical Properties of Graphite* (Elsevier, London, 1968), p. 75.
117. Carbone Lorraine. Data sheet: carbon pyrolytic material "G21."
118. P. Delhaes. *Solids and Carbonated Materials* (John Wiley & Sons, 2010).
119. J. Jumel. Micro thermal characterization of carbon/carbon composite matrix by the photoreflectance method. *AIP Conf. Proc.*, 557, 1039 (2001).
120. Amoco. Data sheet: carbon fiber "Thornel P100."
121. C. R. Thomas. *Essentials of C/C Composites* (The Royal Society of Chemistry, Cambridge, 1993).

122. J. Jumel. Micro thermal characterization of carbon/carbon composite matrix by the photoreflectance method. *AIP Conf. Proc.*, 557, 1045 (2001).
123. Torayca. Technical data sheet no. CFA-001: carbon fiber "T300."
124. D. Rochais. Microscale thermal characterization at temperatures up to 1000°C by photoreflectance microscopy. Application to the characterization of carbon fibres. *J. Phys. D: Appl. Phys.*, 38, 1498–1503 (2005).
125. Nippon Graphite. Data sheet: carbon fiber "XN05," http://www.ngfworld.com/en/en_product/en_yarn.html (accessed September 3, 2014).
126. P. Delhaes. *L'Actualité Chimique*, 295–296, 42–46 (2006).
127. Toyo Tanso. Data sheet: graphite "IG 56." Ref. TSE-04-01 2011-12.
128. Tokai Carbon. Data sheet: vitreous carbon "GC-10" and "GC-30."
129. P. L. Walker. *Chemistry and Physics of Carbon*. Vol. 7 (Marcel Dekker, 1971).
130. P. L. Walker. *Chemistry and Physics of Carbon*, vol. 5 (Marcel Dekker, New York, 1969), p. 84, 96, 99, 101.
131. P. M. Adams. Characterization of high thermal conductivity carbon fibres and a self-reinforced graphite panel. *Carbon*, 36, 233–245 (1998).
132. C. Pradere. Transverse and longitudinal coefficient of thermal expansion of carbon fibers at high temperatures (300–2500 K). *Carbon*, 46, 1874–1884 (2008).
133. N. C. Gallego, Edie D. D., Ntasin L. N., Ervin V. J. Modeling the thermal conductivity of carbon fibres. Extended Abstracts of the International Conference on Carbon, Carbon '01, Lexington, KY, July 14–19, 2001).
134. J. P. Bonal. Neutron induced thermal properties changes in carbon fibre composites irradiated from 600 to 1000°C. *J. Nucl. Mater.*, 230, 271–279 (1996).
135. V. Barabash. Carbon fibre composites application in ITER plasma facing components. *J. Nucl. Mater.* 258–263, 149–159 (1998).
136. Schunk. Data sheet: graphite materials properties. Ref. 21.02e/1.500/2009.
137. J. L. Chen, J. G. Li, H. Li, and Q. G. Quo. Development of carbon based plasma facing components for steady state operation of the fusion devices in China. *Phys. Scr.*, T111, 173–180 (2004).
138. M. Rodig. Effect of neutrons on carbon based materials. 10th International Workshop on Carbon Materials for Fusion Applications, Jülich, September 17–19, 2003.
139. R. Devi. Carbon-carbon composites—an overview. *Def. Sci. J.*, 43(4), 369–383 (1993).
140. M. R. Kulkarny. A model of global thermal conductivity in laminated C/C composites. *Sci. Technol.*, 57, 277–285 (1997).
141. L. M. Manocha. Thermophysical properties of densified pitch based carbon/carbon materials—I. Unidirectional composites. *Carbon*, 44, 480–487 (2006).
142. L. M. Manocha. Thermophysical properties of densified pitch based carbon/carbon materials—II. Bidirectional composites, *Carbon*, 44, 488–495 (2006).
143. L. M. Manocha. Carbon/carbon composites with heat-treated pitches II. Development of porosity in composites. *Carbon*, 39, 673–683 (2001).
144. G. Gupta. Texture modeling in carbon–carbon composites based on mesophase precursor matrices. *Carbon*, 43, 1400–1406 (2005).
145. C. Jie. The effect of carbon nanotube growing on carbon fibers on the microstructure of the pyrolytic carbon and the thermal conductivity of carbon/carbon composites. *Mater. Chem. Phy.*, 116, 57–61 (2009).
146. K. Naito. Enhancing the thermal conductivity of polyacrylonitrile- and pitch-based carbon fibres by grafting carbon nanotubes on them. *Carbon*, 48, 1849–1857 (2010).
147. J. Li, Effect of carbon nanofibers on the infiltration and thermal conductivity of carbon/carbon composites. *Mater. Res. Bull.*, 46, 1437–1442 (2011).
148. R. I. Baxter. Effect of chemical vapor infiltration on erosion and thermal properties of porous carbon/carbon composite thermal insulation. *Carbon*, 38, 441–449 (2000).
149. G. Marban. Dry formation of low-density NomexE rejects-based activated carbon fibre composites. *Carbon*, 38, 2167–2170 (2000).
150. P. L. Walker. *Chemistry and Physics of Carbon*, vol. 5 (Marcel Dekker, New York, 1969), p. 130.
151. E. Ruellan. Characterization of carbon fibers, CEA Internal Report, p. 64 (2000).
152. JAXA. Data (Kakuda).
153. S. Iqbal, R. Dinwiddie, W. Porter, M. Lance, and P. Filip. Effect of heat treatment on thermal properties of pitch-based carbon fibre and Pan based carbon fibre C/C composites. *Mechanical Properties and Performance of Engineering Ceramics and Composites, VI: Ceramic Engineering and Science Proceedings, Vol. 32,* edited by D. Singh, J. Salem, and S. Widjaja (John Wiley & Sons, Inc., Hoboken, NJ, 2011), pp. 245–255.

154. M. Montaudon. Novoltex textures for thermostructural materials. 27th Joint Propulsion Conference, Sacramento, CA, June 24–26, 1991, AIAA 91-1848.
155. S. P. Appleyard. Control of thermal properties of C/C composites by optimisation of structure. *Proc. 24th Biennial Conf. Carbon*, 46–47 (1999).
156. J. Heremans. Thermal conductivity and thermopower of vapor grown graphite fibers. *Phys. Rev. B*, 32(4), 1981–1986 (1985).
157. Heremens J, Thermal conductivity and Raman spectra of carbon fibers. *Phys. Rev. B*, 32(10), 6742–6747 (1985).
158. S. Ozcan. Microstructure and elastic properties of individual components of C/C composites. *Carbon*, 47, 3403–3414 (2009).
159. B. T. Kelly. *Physics of Graphite* (Applied Science Publishers, London, 1981).
160. K. Sun. Graphite C-axis thermal conductivity. *Superlattices Microstruct.*, 45, 60–64 (2009).
161. P. Pascal. Nouveau Traité de Chimie Minérale, Masson and Cie, Tome 8, p. 859 (1968).
162. C. Pradere. Thermal properties of carbon fibers at very high temperature. *Carbon*, 47, 737–743 (2009).
163. H. Fuzzelier, D. Guerard, R. Jaume, C. Oberlin. Etude du vieillissement d'éléments chauffants à haut-flux engraphite ou en composite carbone-carbone. *J. Phys. IV*, 3, Colloque C9–889 (1993).
164. J. P. Maumus. Electrical heating resistance using resistive elements made of C/C, Patent 523165 (1993).
165. B. Shangguan, Y. Zhang, J. Xing, L. Sun, and Y. Chen. Comparative study on wear behaviors of metal-impregnated carbon material and C/C composite under electrical sliding. *Tribol. Trans.*, 53, 933–938 (2010).
166. Y. Shimizu, H. Kuninaka, K. Nishiyama, and K. Toki. Evaluation of a new C/C composite material for a 20 cm ion engine. International Electric Propulsion Conference, paper 97 (2003).
167. P. David. Thermal properties of C/C composites. Proceedings Journées scientifiques du Groupe Français d'Etude des Carbones, May 12–15 (2014).
168. A. Cezairliyan. Specific heat and electrical resistivity of a C/C composite in the range 1500–3000 K by a pulse heating method. *Int. J. Thermophys.*, 1(3), 317 (1980).
169. P. Pascal. *Nouveau traité de chimie minérale. Carbone*, vol. 8 (Masson, 1968), p. 470.
170. I. Golecki. Properties of high thermal conductivity carbon–carbon composites for thermal management applications. *Proc. 24th Biennial Conf. Carbon*, 186–187 (1999).
171. X. Liao. Study on the thermal expansion properties of C/C composites. *J. Mater. Sci.*, 42, 3435–3439 (2007).
172. H. J. Joo. Effect of carbon preform architecture on the mechanical and thermal properties of 4D carbon/carbon composites. *Proceedings of Carbon Conference* (Seattle, 2007), p. 152.
173. M. Meftah, S. Lee, A. Irbah, and S. Ostergren. Carbon/Carbon for satellite applications. *Proc. SPIE*, 8044, 80440Y-1 (2011).
174. W. Shih, Carbon composites fort thermal management applications. *AIP Conf. Proc.* 361, 857 (1996).
175. H. Yamauchi. Development of three dimensional composite with extremely low thermal expansion properties. *IHI Eng. Rev.*, 40(1), 27–30 (2007).
176. S. Amada. Thermal shock resistance of carbon±carbon (C/C) composite by laser irradiation technique. *Ceram. Int.*, 25, 61–67 (1999).
177. H. W. Li. Evaluation of the thermal shock shear strength of carbon–carbon (C/C) composite by a laser irradiation technique. *J. Mater. Process. Technol.*, 138, 423–428 (2003).
178. T. J. Lu. The thermal shock resistance of solids. *ActaMater.*, 46(13), 4755–4768 (1998).
179. H. V. Johnson (1934), US Patent 1.948.382.
180. E. Yasuda, S. Kimura, and Y. Shilsusa. *Trans. ISCM.*, 6(1), 14 (1980).
181. P. L. Walker, F. Rusinko, and L. G. Austin. *Adv. Catal.*, 11, 164 (1959).
182. J. Chown, R. F. Dencon, N. Singer, and A. E. S. White. Refractory coatings on graphite, in *Special Ceramics*, edited by P. Popper (Academic Press, 1963), p. 81.
183. W. Huettner, R. Weiss, G. Dietrich, and R. Meistring. Space applications of advanced structural materials. Proceedings of an International Symposium Organised by the European Space Agency, ESTEC, Noordwijk, The Netherlands, March 21–23, 1990, pp. 91–95.
184. R. Weiss. *High Temp. Mater. Process.*, 3, 351–356 (1994).
185. W. Huettner, R. Weiss, and T. Scheibel. Oxidation resistance of 2D-C/C at 1530 K. Abstract 19. *Biennal Conference on Carbon,* Penastate University, PA, USA, June 25–30, 1986.
186. J. C. Cavalier, I. Berdoyes, and E. Bouillon. *Adv. Sci. Technol.*, 50(2006), 153–162 (2006).
187. R. R. Naslain. Processing of non-oxide ceramic composites. *Adv. Sci. Technol.*, 50, 64–74 (2006).
188. R. R. Naslain, R. Pailler, X. Bourrat, S. Bertrand, F. Heurtevent, P. Dupel, and F. Lamouroux. Synthesis of highly tailored ceramic matrix composites by pressure-pulsed CVI. *Solid State Ionics*, 141/142, 541–548 (2001).

189. R. Eliezer. Coating of Graphite and C/C Composites via Reaction with Cr-powder, M.Sc. Thesis Technion, Haifa.
190. R. Weiss and A. Lauer. Final Report, on the Nationally Funded Project: "Gradierte CVD- und PIRAC- Multibeschichtungen auf C/C als Korrosions- und Oxidationsschutz durch innovative Hochtemperaturprozesse", FKZ 03N5039B (2004).
191. R. Weiss. Final Report: RESTAND, EC Contract No: SMT4-CT97-2200 (2001).
192. R. Weiss. HTCMC6, 04–07, September, 2007, New Delhi.
193. R. Weiss. CMC and C/C-SiC fabrication. International Workshop on Nanostructured HT/High Performance Materials, Sardar Patel University, Vidyanagar, India, September 10, 2007.
194. E. W. Sucor. *JACS*, 46, 14 (1963).
195. A. L. Grigorev and D. J. Polishchuk. *Fiz. Aerodisp. Sist.*, 8, 87 (1973).
196. D. Kehr, BMBF-Projekt 01VM039, from August 1979–July 1983; BMBF.No.FB-T85-014, Entwicklung von asbestfreien Reibbelägen auf Kohlenstoffbasis in Form von C/C und Kohlenstoff-Metallcarbidverbundwerkstoffen (1983).
197. E. Fitzer, W. Fritz, A. Gkogkidis, and K. D. Moergenthaler. *Proceedings on Carbon 86 Conference* (Baden-Baden, 1986).
198. Brite-Euram Project, MA1E-0068, Entwicklung von CFC-Werkstoffen mitpartiell reduziertem Gehalt an C-Fasern für Anwendungen im PKW-Bremssystem (1990).
199. R. Weiss. Final Report, "Development of oxidation resistant C/C", BMBF Reference No.: 03M1019 BO (1992)
200. R. Weiss. DGM-Fortbildungsseminar, Universität Bayreuth (2007).
201. EP 1106334, (2001).
202. US-Patent 7,175, 787 (2007).
203. EPIA-Report, European Photovoltaic Industry Association (EPIA) (2006).
204. WO 2005/059992A1, (2005) Felt Susceptor.

CHAPTER 6

C/SiC AND C/C-SiC COMPOSITES

BERNHARD HEIDENREICH

German Aerospace Center (DLR), Institute of Structures and Design, Department of Ceramic Composites and Structures, Stuttgart, Germany

6.1 INTRODUCTION

Due to their high temperature and thermal shock resistance, low density, and non-brittle, quasi-ductile fracture behavior, carbon-fiber-reinforced carbon (C/C) materials pioneered the development of reliable lightweight structures in high temperature applications. C/C materials have been developed since the 1960s, originally for thermal protection systems (TPSs) of reusable spacecraft, such as the US space shuttle, and components for rocket motors. The application area could be extended to high performance brake discs in aircraft (1970 for F-15, 1971 for Concorde [1,2]) and racing cars (late 1970s), heating elements, charging rigs, and molds in high temperature furnaces for ceramic sintering and glass production, crucibles and charging devices in semiconductor industry and linings for nuclear technology [3]. However, C/C materials offer low oxidation stability at temperatures above 400°C. Therefore, service time at high temperatures is limited in oxidative atmosphere, restricting the use of C/C to short time applications, for example, in nozzles for space propulsion, or to nonoxidative environments, for example, for components in high temperature furnaces, operated with inert gas atmospheres.

In order to overcome this main drawback of C/C materials and to increase the oxidation stability, graphitized C fibers, such as high modulus (HM) or ultra-high modulus (UHM) fibers, and highly graphitized C matrices, obtained by thermal treatment of pitch or chemical vapor infiltration (CVI) derived C matrices at high temperatures, are used. Additionally, C/C materials are infiltrated with salts, acting as oxidation inhibitors up to temperatures of about 900°C for short-term applications. Furthermore, increased service times could be achieved by external coatings with oxidation protection layers, for example, based on pure SiC or in combination with a SiO_2 sealing layer or $SiC/B_4C/SiC$ multilayers, applied via chemical vapor deposition (CVD), pack cementation, or slurry coating [4,5]. However, these coatings typically are characterized by microcracks, caused by the combination of high process temperatures ($T > 1000°C$) and the significant differences in the coefficient of thermal expansion (CTE) of the C/C substrate and the ceramic coating layer, leading to tensile stresses in the coating during cooling down. At service temperatures below the coating temperature, the cracks are open, leading to an almost free access of oxygen to the substrate. At service temperatures above the coating temperature, compression stresses are built up in the coating layer, increasing the risk of spallation. Therefore, oxidation of C/C cannot be prevented completely in long-term applications, especially at low service temperatures or thermal cycling. Additionally, coating thicknesses generally are very low ($t_{coating, CVD} = 100–200\ \mu m$) and therefore not suitable for mechanically and abrasively loaded areas, such as friction surfaces of brake discs.

In another approach, the carbon matrix has been replaced partly or completely by more stable ceramic matrices, leading to a novel class of materials, so called ceramic matrix composites (CMCs), characterized by an internal oxidation protection of the carbon fibers.

Different matrix types, such as SiCN and Al_2O_3 [6] have been investigated. However, until now, all known industrial applications are based on SiC or SiSiC matrices and the current research activities are mainly focused on this class of materials. Depending on the manufacturing method, the C matrix is either replaced completely by a SiC matrix, leading

Ceramic Matrix Composites: Materials, Modeling and Technology, First Edition. Edited by Narottam P. Bansal and Jacques Lamon.

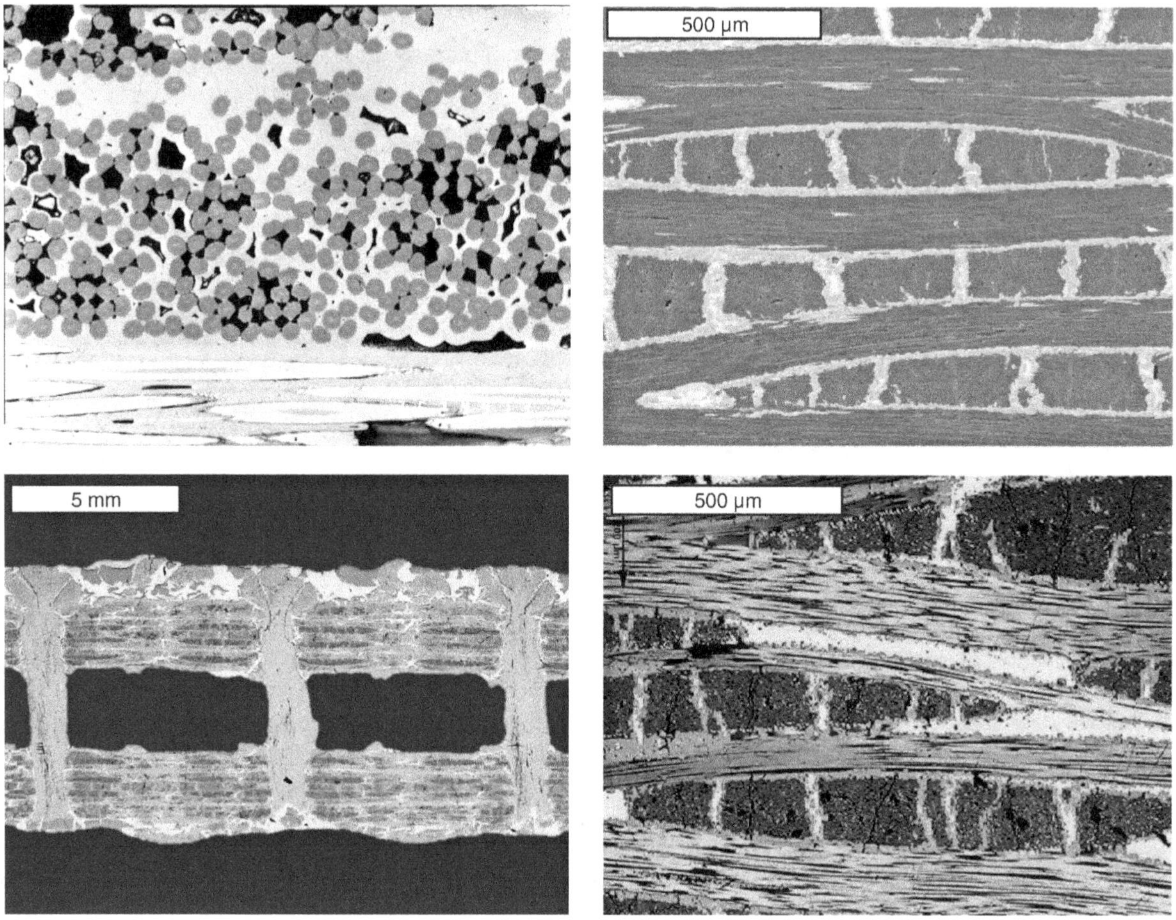

FIGURE 6.1

SEM views showing typical microstructures of carbon-fiber-reinforced SiC. C/SiC material (Keraman®, top left, MT) manufactured via chemical vapor infiltration (CVI) with single C fiber filaments (dark gray) embedded in the SiC matrix (light gray). C/C-SiC material (top right, DLR) with dense C/C bundles (dark gray) embedded in SiC matrix (gray) derived by liquid silicon infiltration (LSI). C/C-SiC (CARBOTEX®SI, bottom left, Astrium ST/EADS IW) material based on a dense C/C substrate material, obtained via CVI, and a SiSiC surface coating, subsequently built up by Si melt infiltration of a porous C coating. C/SiC material Sigrasic® 1500 J (bottom right, SGL), showing dense C/C bundles (dark gray) embedded in Si and SiC matrix (light gray).

to C/SiC materials, or tailored C/C preforms are densified with an additional SiC or SiSiC matrix, leading to C/C-SiC materials. Typical microstructures are characterized by either single C fibers (C/SiC) or dense bundles of C/C (C/C-SiC) embedded in SiC matrix. However, this nomenclature is not used systematically in the literature and the term C/C-SiC is also used for a dense C/C material, coated with a SiSiC surface layer derived by Si melt infiltration (CARBOTEX®SI and SICTEX®SI, respectively [7]). On the other hand, CMC materials characterized by dense C/C bundles, embedded in SiC matrix are often called C/SiC materials [8] (Figure 6.1). In this chapter, C/SiC is used for all types of CMC materials based on C fibers and SiC or SiSiC matrices in general, meaning C/SiC as well as C/C-SiC materials, and C/C-SiC is used if only this special type of CMC materials is addressed and if this is the official term of the material.

Starting in the late 1970s, the development of C-fiber-reinforced SiC materials was mainly driven by space applications. First components were used in TPSs for reusable spacecraft, such as the nose cap and wing-leading edges of the Buran (1974–1993) [9]. Due to the development of low cost materials and manufacturing methods, new industrial application areas outside aerospace could be opened successfully for these materials and the introduction of C/SiC brake discs in automobiles was an important milestone and breakthrough for this new material class.

Currently, C/SiC materials are only used in niche applications and worldwide production of C/SiC ($\approx$100 tons/year) is still low, compared to C/C ($\approx$1000 tons/year) and to fiber-reinforced plastics (FRP >6 Mio tons/year). However, a continuous increase of the manufacture of automobile brake discs is observed and new application areas in rocket propulsion and for friction components could be opened.

6.2 MANUFACTURING METHODS

For the manufacture of C/SiC materials, the C fibers have to be embedded in a SiC matrix. In practice, the SiC matrix is built up inside a fiber preform. Therefore, different methods are used, which are mainly derived from the industrial manufacture of SiC materials and components. Natural SiC, also called Moissanite is a very rare material on earth and can be found in some species of meteorites as well as in volcanic kimberlite [10]. In 1891, Edward Goodrich Acheson discovered SiC, he named it carborundum, during his work on preparing artificial diamonds, even before it was detected in nature by Henry Moissan in 1904 [11]. The first industrial application of this new, artificial material was as abrasives, which were produced by the Carborundum Company starting in 1893. The Acheson process, characterized by the reaction of silica sand and coke is used for the industrial production of bulk α-SiC to date. Higher quality SiC is obtained via CVD and thermal decomposition of preceramic polymers. SiC components are commonly manufactured by pressureless sintering (SSiC) and recrystallization (RSiC) or hot (isostatic) pressing (H(i)PSiC) of fine SiC powder as well as by liquid phase sintering (LPSSiC). For reaction bonded SiC (RBSiC), porous preforms based on SiC particulates and a carbon binder are infiltrated with molten silicon, leading to highly dense SiSiC materials and components.

For the manufacture of C/SiC materials, the processes based on CVD, pyrolysis of preceramic polymers, and reaction bonding have been adapted for the buildup of SiC matrices in fibrous preforms, leading to three different manufacturing methods:

CVI (Chemical Vapor Infiltration) = Deposition of gaseous SiC precursors
PIP (Polymer Infiltration and Pyrolysis) = Pyrolysis of Si polymers
MI (Melt Infiltration) = Reaction of molten Si with C

The Acheson process as well as sintering and hot pressing are not used for the manufacture of C/SiC materials, to date. However, hot pressing is well known for the realization of highly dense SiC-fiber-reinforced SiC materials [12, 13].

In the CVI method, the SiC matrix is built up in one process step, using a gaseous precursor. Thereby, a dry fiber preform is infiltrated with the precursor and SiC crystals are deposited on the hot C fiber filaments, simultaneously. In contrast, using polymer precursors, several separated processes are needed to create the SiC matrix in the so called PIP or LPI process (liquid polymer infiltration). In the first step, a CFRP preform (carbon-fiber-reinforced plastic) is manufactured by infiltrating the liquid polymer precursor into the fiber or fiber preform before curing to a thermoset polymer matrix. In the second step, the CFRP preform is pyrolyzed, converting the polymer matrix to a porous SiC matrix. Whereas in the PIP process, the SiC matrix is directly derived by the pyrolysis of the preceramic polymer matrix, two separate process steps are needed for the SiC matrix buildup in the MI or LSI (liquid silicon infiltration) process. Thereby, a C precursor polymer is used, which is transformed to an intermediate carbon matrix during pyrolysis in a first step, resulting in a porous C/C preform. In the second step, molten Si is infiltrated in the C/C preform, in order to build up the SiC matrix by chemical reaction of Si with a certain amount of C. Figure 6.2 gives a schematic overview of the different methods for matrix buildup. Additionally, combined processes, for example, CVI + PIP, CVI + LSI, and PIP + LSI, have been investigated and developed in order to overcome the drawbacks and to benefit from the advantages of the specific manufacturing methods.

In principal, all the different manufacturing methods for C/SiC materials are based on three main process steps:

1. Manufacture of a C fiber preform or CFRP preform.
2. Buildup of a weak fiber/matrix interphase.
3. Buildup of a SiC matrix.

For the manufacture of the fiber preform, generally all types of carbon fibers can be used. Due to economic aspects, high tenacity (HT) fibers, such as Tenax HTA, are most common in the LSI process. Due to the fact that C/SiC composites are fiber-dominant materials, improved material properties can be obtained by the use of intermediate modulus (IM, e.g.,

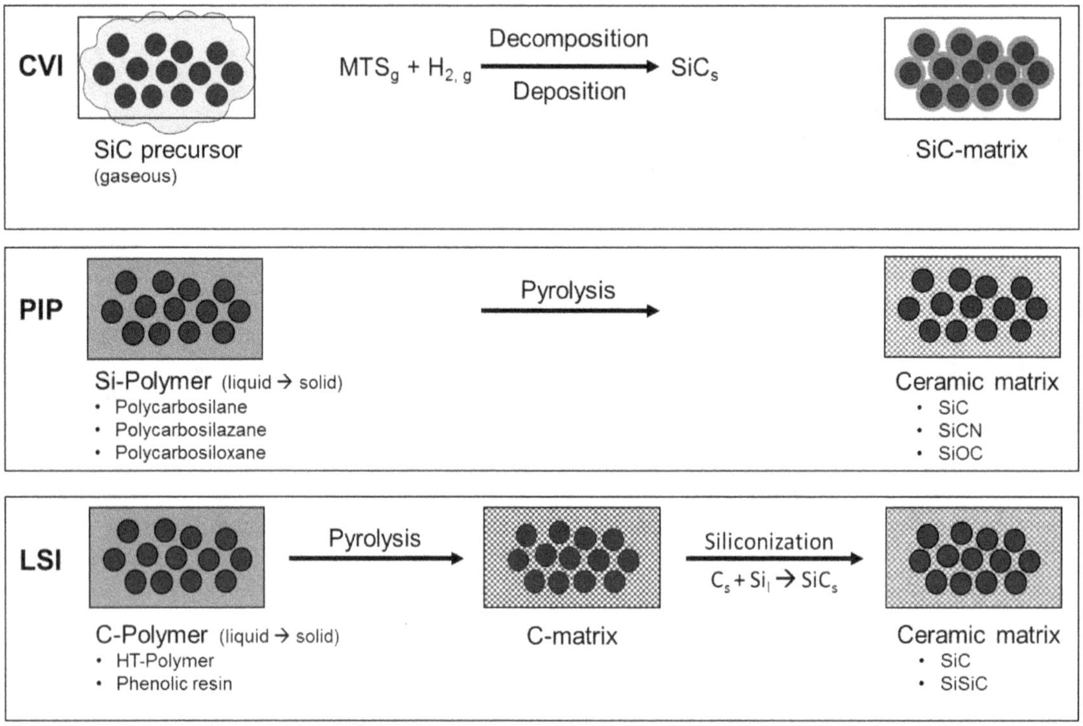

FIGURE 6.2
Schematic overview of the different methods used for the buildup of SiC matrix in C/SiC materials (DLR).

Torayca T 800) and high modulus (HM, e.g., Torayca M40) fibers, offering highest tensile strength ($\sigma_{t,T\,800} = 5490$ MPa, $\sigma_{t,\,HTA} = 3950$ MPa) and Young's modulus (M40: 392 GPa, HTA: 238 GPa). Therefore, IM fibers are widely used in PIP- and CVI-based C/SiC. At present, ulta high modulus (UHM) fibers, offering highest modulus of up to 880 GPa are not generally used, due to their high cost above 700 €/kg (HTA: 20–50 €/kg). Additionally, it was found that UHM fibers are critical in LSI processes due to an insufficient fiber embedding and protection, leading to excessive Si attack and conversion of the fibers to SiC.

In order to obtain a homogeneous distribution of the SiC matrix and the C fibers, as well as to take advantage of the fiber properties as much as possible, usually fine rovings based on 1000–6000 single filaments (1–6 K) are preferred.

Whereas in the CVI process, the SiC matrix is built up directly in the dry fiber preform, CFRP preforms also called "green bodies" have to be manufactured for PIP and LSI. Depending on the manufacturing method of the CFRP preform, dry fiber preforms are formed and infiltrated with the polymer, for example, via resin transfer molding (RTM), or polymer infiltrated fibers are used in the autoclave technique, warm pressing, and wet filament winding. For the latter, the fiber preform and the CFRP preform are built up simultaneously in one process step.

Dry fiber preforms are usually manufactured by stacking cut layers of two-dimensional (2D) woven fabrics, resulting in orthotropic (0°/90°) or quasi-isotropic (0°/90°, ± 45°) fiber architectures. Unidirectional or 1D fiber layers can be used as cross-ply laminates or in combination with fabric layers. Multiaxial fiber architectures are achieved by filament winding or braiding (Figure 6.3), which can be combined with 3D weaving, stitching, or needling, leading to 2.5D and 3D preforms. Thereby, even highly complex structures with integrated cooling channels can be realized (Figure 6.4) [14]. Additionally, 2.5D and 3D preforms have been developed, for example, by Herakles (Novoltex®, Skinex®, Guipex®), especially for the use in CMC materials. Guipex fiber preforms are commonly used for C/SiC materials and are characterized by planar layers of 2D-oriented fibers, linked together by additional fibers, running in the thickness direction. Available wall thicknesses are in the range of 2–7 mm [15]. Generally, 2.5D and 3D preforms are advantageous to form free standing near net shape preforms, ideal for gas phase infiltration. Even more important, interlaminar shear strength can be increased, compared to 2D fabrics, especially for CVI- and PIP-based C/SiC materials. Randomly oriented fiber reinforcements can be achieved using so called short fibers, representing cut fibers with typical lengths in the range of 2 mm to about 40 mm and above.

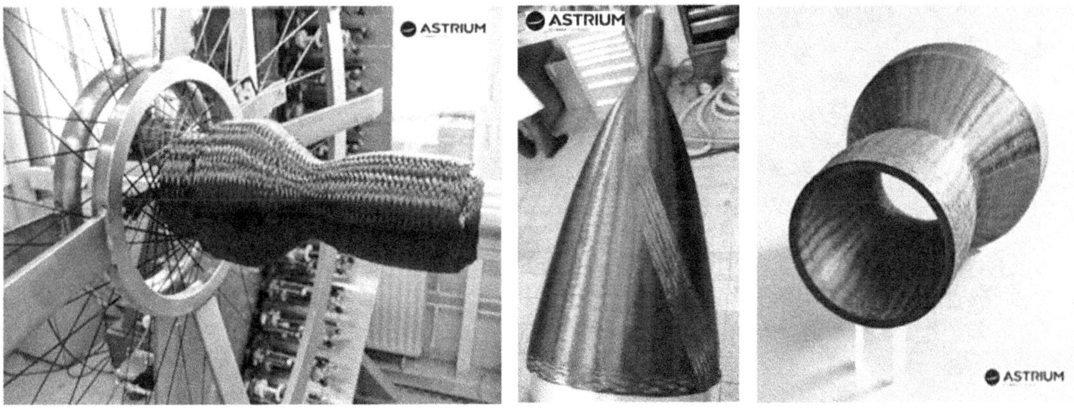

FIGURE 6.3
Manufacture of dry fiber preforms via robot-assisted braiding and fiber placement onto contoured mandrels (left and middle), resulting in satellite thruster and nozzle demonstrator (middle) and near net shape CFRP preform for a combustion chamber demonstrator (scale 1:5, right) for liquid rocket engines application (all figures Astrium ST/EADS IW) [14].

In common with all fiber-reinforced materials, the fiber and CFRP preforms are made in near net shape geometry, in order to ensure a lightweight design with fiber architecture and fiber orientation well fitted to the load paths. Additionally, waste of costly raw materials as well as costs for final machining can be reduced.

In order to take advantage of the high strength and elongation of the C fibers, as well as to obtain a C/SiC material offering high fracture toughness and damage tolerance, a weak embedding of the C fibers in the highly brittle SiC matrix is essential. In almost all manufacturing processes, the fibers are separated from the SiC matrix by coating the fibers with an interphase material, providing low shear strength of the coating material itself, between the interface of fiber and coating or between the coating and the SiC matrix [16, 17]. Typically, a thin layer of pyrolitic carbon (pyC), with a thickness of 0.1–1 μm, is applied on the fibers via CVD or CVI, prior to the buildup of the matrix [18, 19] (Figure 6.5). This leads to C/SiC materials, characterized by coated carbon fiber filaments embedded individually in SiC matrix. Alternatively, the C fibers are embedded in a dense C matrix via PIP. After the buildup of the SiC matrix, the final CMC materials are characterized by dense C/C bundles, embedded in the SiC matrix and consequently are called C/C-SiC materials.

6.2.1 Chemical Vapor Infiltration

The CVI process is characterized by the deposition of a ceramic matrix out of a gaseous precursor, infiltrated in a porous fiber preform. The CVI method was derived from CVD, a well-established technique for applying thin ceramic coatings [20]. Originally developed for the manufacture of C/C materials, the CVI process could be adapted to the deposition of SiC matrices using a defined mixture of methyltrichlorosilane (MTS) and H_2 as process gas.

$$CH_3SiCL_{3,gaseous} + H_{2,gaseous} \rightarrow SiC_{solid} + 3HCl_{gaseous} + H_{2,gaseous}$$

The CVI technique is used for the manufacture of simple plates as well as for very large and complex structures. One of the largest C/SiC components worldwide was manufactured via CVI. That is the body flap for X38, a thin-walled, lightweight structure with outer dimensions of $1.5 \times 1.5 \times 0.15\ m^3$ (see Figure 6.51) [21]. For the industrial manufacture of C/SiC components, CVI is used by Herakles, France, MT Aerospace (MT), Germany, General Electric (GE), Goodrich (now UTC), Hyper-Therm High-Temperature Composites Inc. (Hyper-Therm), all USA, and Surface Transforms PLC, UK.

The CVI process is subdivided into three main process steps (Figure 6.6):

1. Manufacture and stabilizing of a carbon fiber preform.
2. Deposition of fiber coating (interphase).
3. Deposition of the SiC matrix.

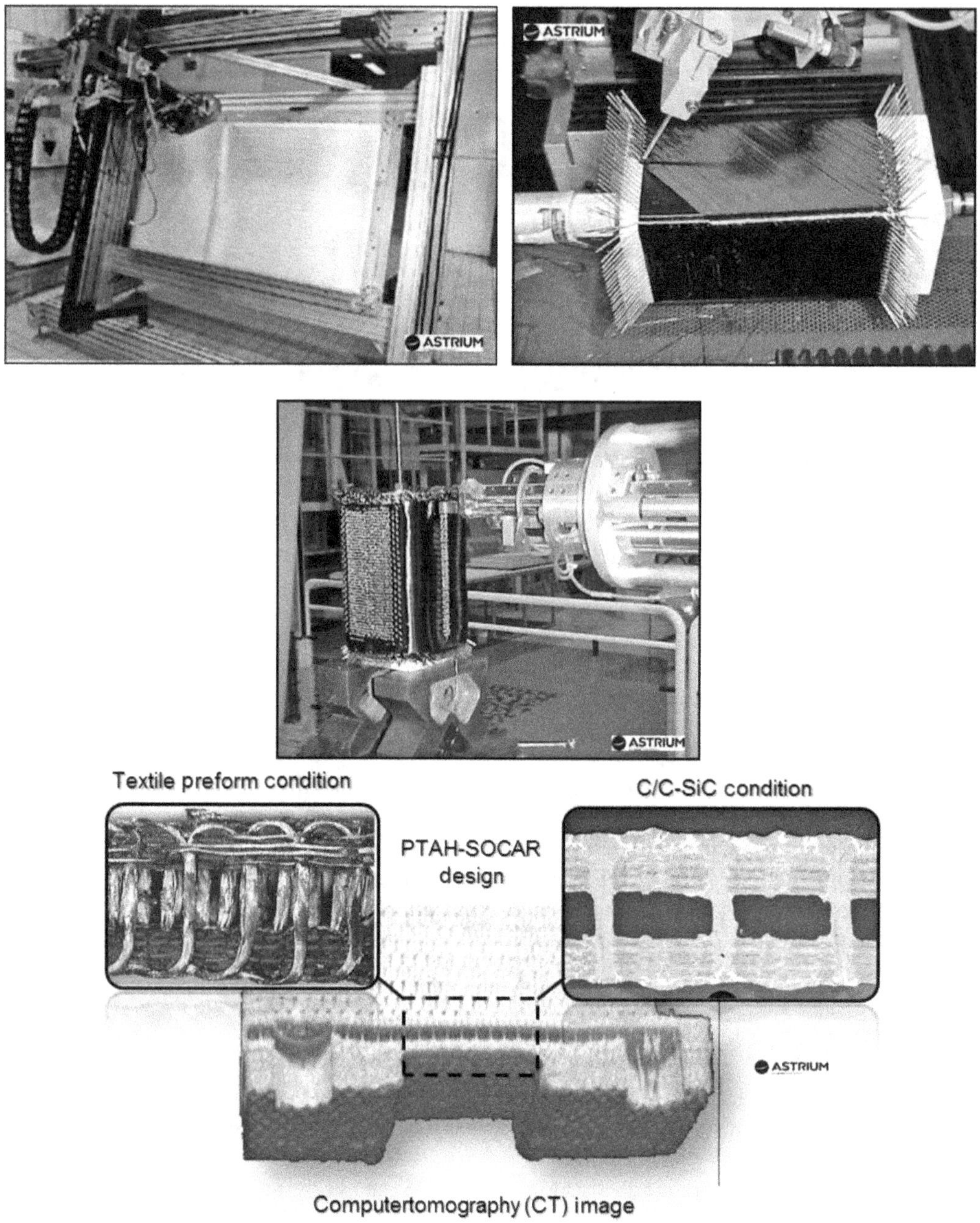

FIGURE 6.4

Manufacture of near net shape fiber preforms combining 3D weaving (upper left), filament winding (upper right), and stitching (middle), leading to highly complex structures with integrated cooling channels (bottom) for combustion chambers (all figures Astrium ST/EADS IW) [7].

In the first step, a porous preform made of carbon fibers is made in near net shape geometry. High fiber contents between 40 and 60 vol.% can be achieved by densifying dry fabrics in a heat-resistant graphite mold, during the first CVI densification step. For complex-shaped components, graphite molds can be avoided by using geometrically stable, dry preforms obtained via 3D textile techniques or by liquid-route hardening of the preform with a polymer binder, leading to so called stabilized preforms. Dry preforms are built up by special weaving and braiding methods (3D-woven fabrics) or by stitching of 2D fabric laminates (2.5D fabrics), as well as by a combination of dry filament winding or braiding and stitching. For stabilized preforms, 2D fabrics or cut fibers are coated with a C or SiC polymer precursor, for example,

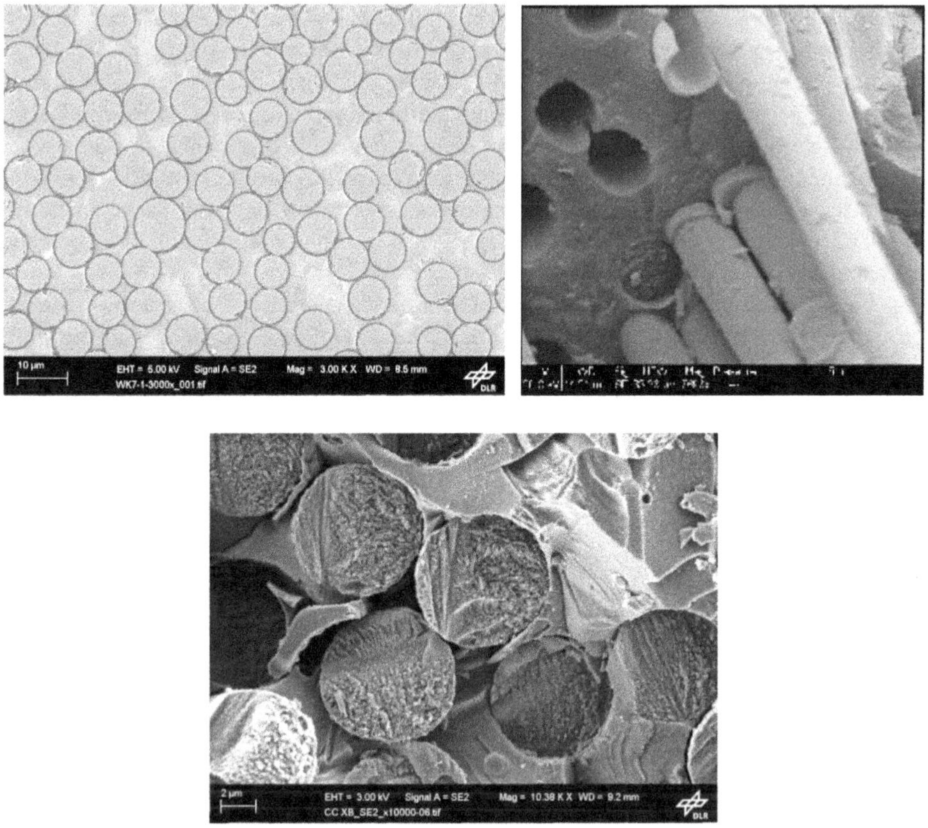

FIGURE 6.5

CVI-pyC fiber coating clearly visible on SiC fibers in a SiC/SiCN material derived from PIP (top left, DLR) and fiber pull out in a PIP-based C/SiC material (top right, COIC), showing fragments of the pyC fiber coating. C fibers, embedded in dense, PIP-based C matrix (bottom, DLR).

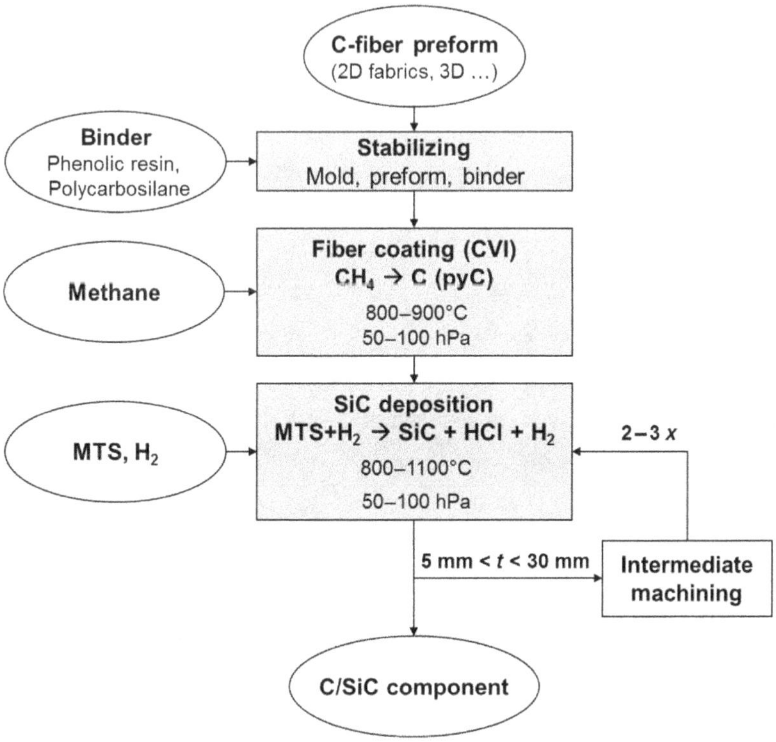

FIGURE 6.6

Schematic overview of the manufacture of C/SiC materials via I-CVI (DLR).

phenolic resin or polycarbosilane, and laminated or filled in a mold, where they are formed to near net shape geometry and densified to a certain fiber volume content in the range of 15–25% (short fibers) up to 30–50% (fabrics). After curing the precursor, the component is demolded and pyrolyzed, leading to a highly porous but stable preform.

The structure of the preforms porosity is an all-important parameter for the CVI process. A translaminar open pore system of evenly distributed, interconnected pores or microchannels with minimum diameters of 10–500 μm is necessary for a homogeneous densification. Suitable pore structures can be obtained by multidimensional fiber structures and average fiber contents of about 40–50 vol.%. Unidirectional or cross-ply laminates based on 1D fiber preforms are critical, due to small pores that cannot be filled easily, especially if the fiber volume content is high, and therefore such preforms are not common for CVI.

Fiber coating is essential to obtain a weak interphase between the fibers and the ceramic matrix. Typically, a thin layer of pyrolytic carbon with a thickness of about 0.1 up to 0.3 μm is applied by cracking methane in a first CVI process step, prior to the deposition of the SiC matrix.

In the third step of the CVI process, the porous fiber preform is set inside an infiltration chamber, and heated to a temperature typically between 800°C and 1100°C. The gaseous precursor passes through the infiltration chamber and penetrates into the porous fiber preform. Parallel to the infiltration, the chemical reaction of the process gas is activated at the hot surface of the porous fiber preform and β-SiC is deposited.

One main advantage of the CVI process is the highly pure and fine grained SiC matrix obtained, leading to excellent mechanical properties in the final C/SiC material, especially at high operating temperatures. However, to achieve this, the molar ratio $\alpha = [H_2]/[CH_3SiCl_3] = 10$ has to be controlled thoroughly and a homogeneous buildup of the matrix has to be ensured.

Homogeneity and deposition rate are determined by the process parameters, that is, temperature T and gas pressure p. High T and p leads to high deposition rates but also to an increased gradient of the SiC deposition, especially in thick-walled components. Thereby, the precursor is mainly reacting at the outer surface of the component, leaving impoverished, less reactive process gas for the center area. This leads to a higher deposition rate on the surface, and finally to a sealing of the surface before the matrix buildup in the center is completed. Figure 6.7 shows the calculated influence of temperature and pressure on the infiltration depth in a cylindrical pore channel [22]. The higher the temperature and the pressure, the lower is the infiltration depth. High infiltration depth, indicated by a constant thickness of the SiC deposited along the pore, can be obtained by low temperatures and pressures, due to an increased mean free path of the gas molecules, leading to a reduced reactivity of the precursor and to an increased penetration of the pore. However, the time needed to fill the pore with SiC matrix is increasing, too, resulting in longer process time.

Two main process variations are industrially used:

- Isothermal/Isobaric CVI (I-CVI) [23].
- Thermal gradient CVI with or without forced flow.

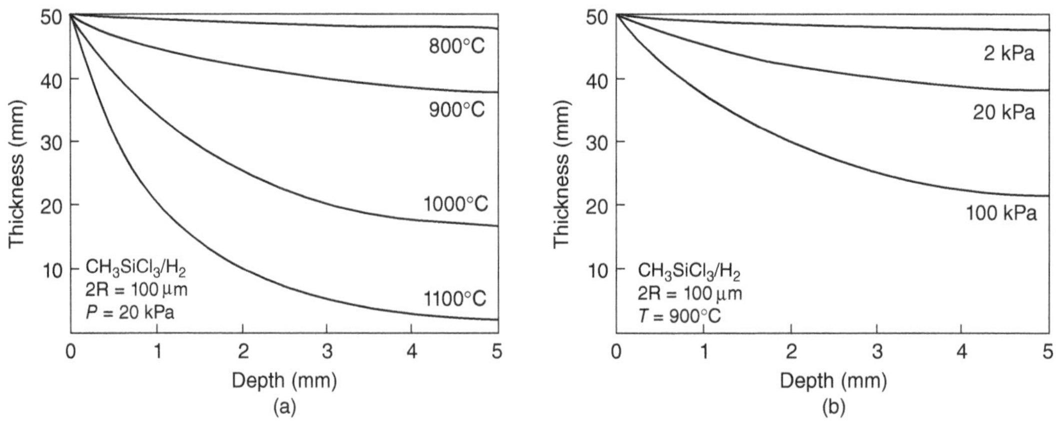

FIGURE 6.7

Calculated SiC deposition thickness dependent on the depth of a straight, cylindrical pore channel (Ø 100 μm) at different process temperatures (left) and pressures (right) [22] (with kind permission from Springer Science+Business Media B.V.).

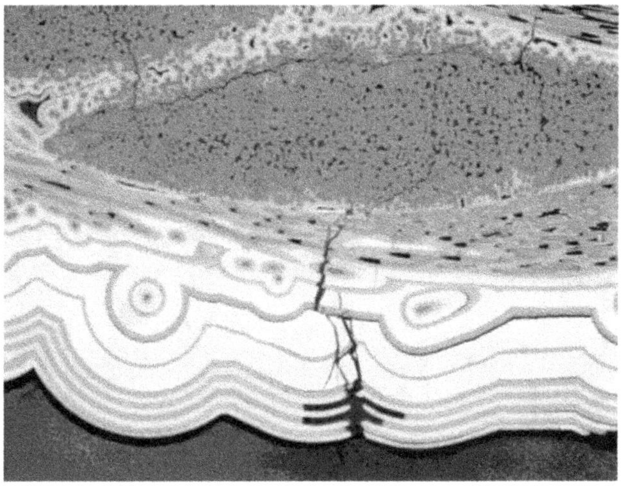

FIGURE 6.8

SEM image of the microstructure of SEPCARBINOX A500, characterized by a self-sealing matrix with alternating layers in the system Si, C, B (Herakles) [29].

For the latter the temperature/pressure gradient CVI (G-CVI) [24] is very common. Using I-CVI, the preforms are heated uniformly (T_{max} = 800–1100°C), and the process gas infiltrates the preforms solely by diffusion at a constant pressure of about 50–100 hPa. Thereby, very large components and complex-shaped structures can be manufactured. Furthermore, using I-CVI, multilayer interphases [25,26] as well as multilayer matrices can be deposited homogeneously in a submicron scale. For the latter, multiphase matrices are built up, characterized by alternating layers of different matrix materials, for example, SiC and C (pyC). One advantage of multilayered matrices is higher damage tolerance, due to increased crack deflection and crack stopping at the numerous interfaces of the matrix layers. Even more important, the oxidation stability at high temperatures can be increased by the introduction of so-called "self-sealing" matrices, originally developed by Herakles for SiC/SiC materials [27–29]. Thus, boron-bearing materials are introduced into the SiC matrix, by multilayer matrix systems typically based on Si, B, C system. Due to oxidation, liquid glassy phases of B_2O_3 and SiO_2 are formed at temperatures above 450°C, entrapping the penetrating oxygen in the whole volume of the material. Microcracks in the C/SiC material, induced during processing and caused by service loads, can be filled with the glassy phases and further attack of oxygen can be reduced, leading to increased oxidation stability (Figure 6.8) [30].

The main drawback of the I-CVI process is the comparatively slow matrix deposition rate, leading to long cycle times for the densification process. Additionally, the maximum wall thickness, which can be manufactured in one infiltration step, is limited to about 5 mm. Thicker components can be realized up to a maximum of about 30 mm and are obtained by using multiple (typically 2, up to 3) CVI densification steps with specific process parameters. After each intermediate densification step, the preform has to be machined in order to reopen the pores on the components surface. This leads to inhomogeneous matrix distribution, characterized by higher porosity and less density in the center area, compared to the skin area.

Overall manufacturing times, including preform manufacture, fiber coating and preform stabilizing, matrix buildup (loading of furnace, heating, densification, cooling, and unloading), intermediate and final machining, are several days or weeks for thin-walled components up to some weeks or months for thick-walled structures. However, very different components, from flat plates up to large, complex-shaped structures can be densified in one batch by using generic process parameters. For industrial production, large furnaces up to Ø 2.5 m × 2.5 m (Herakles) are used, in which high quantities of components can be densified in parallel, leading to low furnace utilization time per component and economic manufacturing processes.

In contrast to I-CVI, where the penetration of the precursor gas into the porous fiber preform is controlled only by diffusion, in the G-CVI process, the gaseous precursor is forced to flow through the fiber preform, well controlled by an applied pressure gradient, providing a sufficient amount of precursor gas to be cracked and deposited inside the fiber preform. Additionally, a temperature gradient is imposed by heating one surface of the preform to a temperature between 1100°C and 1200°C and cooling the opposite surface to lower temperatures of 800–900°C. Thereby, matrix

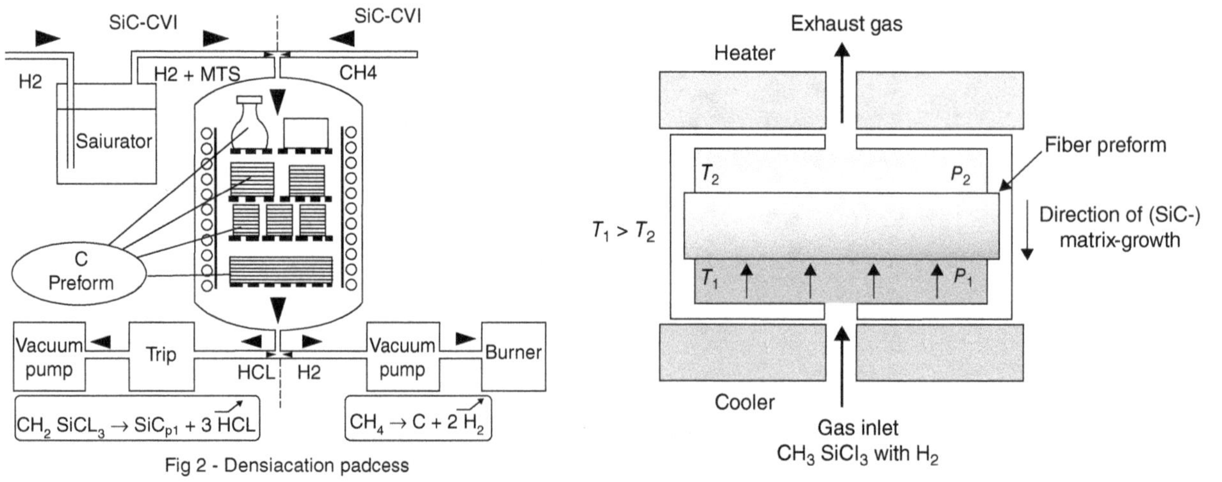

FIGURE 6.9

Schematic overview of a I-CVI facility (left, with kind permission from Springer Science+Business Media B.V.) [32] and experimental setup of the temperature/pressure gradient CVI (G-CVI) process (right, MT).

deposition can be controlled locally and starts on the hot side of the fiber preform. As a result of the matrix buildup, the thermal conductivity of the infiltrated material is increased, leading to an increased temperature inside the preform and a moving of the reaction zone from the hot side to the cooled side of the preform. Thereby, a homogeneously distributed SiC matrix is built up in one process step, from one side of the preform to the other, without sealing the surface. Deposition rates are higher compared to I-CVI, enabling short process times of about 5 days for thin plates of about 5 mm wall thickness and about 9 days for thick-walled components up to 30 mm [31]. However, to keep the process efficient, the densification is stopped at a residual open porosity of about 10–15%, similar to I–CVI. To ensure the temperature and pressure gradients, each preform has to be mounted in a specific graphite tool with integrated cooling system (Figure 6.9), and therefore the process generally is limited to simple components such as tubular structures or flat plates with constant wall thicknesses.

In order to increase material homogeneity and decrease densification time, several modifications of the I-CVI process, such as pressure-pulsed CVI (P-CVI), have been studied for the buildup of C as well as SiC matrices. However, to date, only I-CVI and G-CVI processes are industrially used.

Due to the fact that CVI is very favorable for filling the microporosity by densifying the fiber bundles, but not very effective at filling the macroporosity in between the fiber bundles and in the fiber crossover regions, process combinations of CVI and subsequent PIP are used for a fast densification of C/SiC materials. In order to simplify shaping and tooling, the matrix can also be built up in reverse order, starting with PIP and a subsequent densifying via CVI.

The main advantages of the CVI process are the relatively low process temperatures, uncritical for the carbon fibers and the high purity of the obtained SiC matrix, avoiding free Si and leading to high performance C/SiC materials. During matrix buildup, no contraction occurs, minimizing the risk of delaminations and finally enabling a safe manufacture even of large components with highly complex geometries. Additionally, complex-shaped lightweight structures can be built up by *in situ* joining of simple substructures, manufactured separately and bonded together with SiC at an intermediate step of matrix densification [32]. Whereas I-CVI offers high flexibility for the manufacture of different shaped components in very high quantities, G-CVI methods enable low process times. The main drawbacks of the CVI methods are the high investment cost for the facilities and the demanding process control as well as long process times and material inhomogeneities, limiting the maximum wall thickness, especially for I-CVI. In industrial production, the comparatively slow matrix deposition rate can be compensated by the manufacture of large quantities of components in one batch. On the other hand, the faster G-CVI methods are restricted to flat plates or simple-shaped components. Other disadvantages of CVI are the limited fiber content and the remaining open porosity. 3D-woven fiber performs have been developed and are a good fit with CVI process in order to increase interlaminar strength and to reduce the sensitivity to in service delaminations, experienced by materials exclusively based on 2D fabric preforms.

6.2.2 Polymer Infiltration and Pyrolysis

In a similar manner to the CVI process, PIP is also well established for the manufacture of C/C materials. For the production of C/SiC materials, however, the process had to be adapted to the chemistry of novel, preceramic precursor systems, quite different to carbon precursors [33, 34]. Silicon-based preceramic polymers have been developed since the 1960s and are used for the formation of bulk ceramics, ceramic matrices, and fibers as well as for coatings [35]. The characteristic of the PIP process is the matrix buildup via thermal decomposition of preceramic polymers in multiple densification cycles. Thereby, the fibers or fiber preforms are infiltrated by the liquid precursor in a first step. After curing of the polymer, the resulting thermoset matrix is converted into a ceramic matrix via heat treatment in a non-oxidizing atmosphere. SiC matrices are usually derived from polycarbosilanes with high yields of SiC, such as polymethylsilane (PMS) and allylhydridopolycarbosilane, characterized by a chain of alternating C and Si atoms [36,37]. Thus an amorphous SiC matrix is formed during pyrolysis at relatively low temperatures of about 850–1050°C. In order to increase matrix density and temperature stability, the matrix can be transformed to crystalline β-SiC by a subsequent heat treatment at temperatures of 1500–1600°C in inert gas (Figure 6.10).

Alternatively, amorphous SiCN matrices based on polysilazane precursors can be built up at pyrolysis temperatures of about 1000–1400°C, and they can be transformed to a crystalline nanocomposite matrix of Si_3N_4/SiC/C after annealing at temperatures above 1400°C (Figure 6.11). In cost critical applications also polysiloxane precursors are used for C/SiOC materials and components for lower service temperatures up to 1200°C. Furthermore, matrix properties, processing and costs can be tailored by combining polysilazane and polysiloxane precursors.

SiC forming polycarbosilane polymers are produced by Starfire Systems (SMP 10) and EEMS (Extreme Environment Materials Solutions (EEMS®), LLC), both USA, with an annual output of commercially available precursor of about 5 tons. Polycarbosilanes are also used for the production of SiC fibers by Nippon Carbon and UBE. For C/SiC manufacture via PIP, SMP 10 from Starfire Systems is widely used, due to its high purity and high ceramic yield of up to 90 wt.%. The main disadvantage of polycarbosilanes is the high cost of 1300 \$/kg up to 1800 \$/kg for SMP 10, caused by expensive raw materials, elaborate synthesis processes, and low volume production. Therefore, C/SiC materials based on polycarbosilanes typically are limited to aerospace and military applications.

Cost for polycarbosilazane matrix precursors are on a significantly lower level in the range of 50–200 €/kg and are sold by Clariant (Switzerland), AZ Electronic Materials (USA), and KiON Defense Technologies, Inc. (KDT, USA). Worldwide production of polycarbosilazanes was about 100 tons in 2012. However, only a very small amount of some tons is going into CMC manufacture, whereas most of the production is used for the coating market.

Due to mass production, the costs for polycarbosiloxane precursors are in the same range or even lower, compared to polycarbosilazanes and they are commercially available from well-known, large suppliers, such as Wacker Chemie

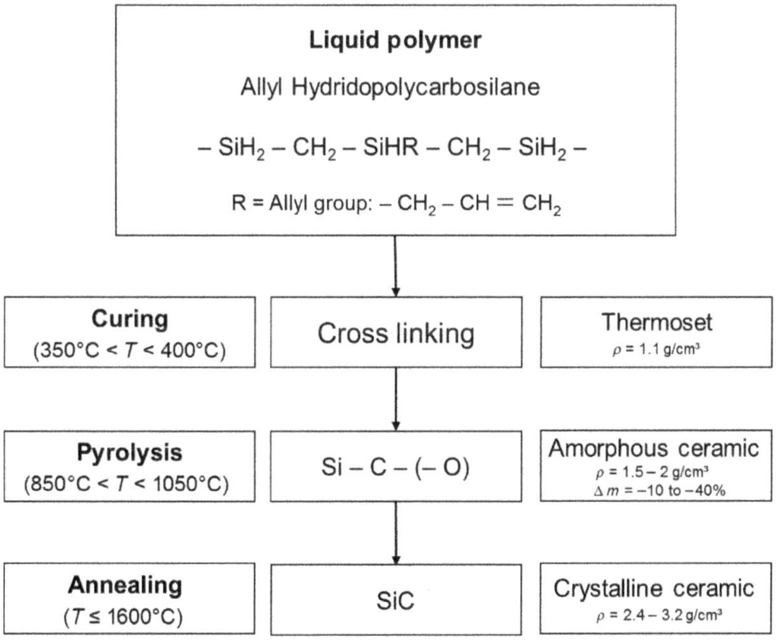

FIGURE 6.10

Schematic overview of the transformation of a polycarbosilane precursor (Starfire SMP 10) used in C/SiC materials based on PIP (DLR).

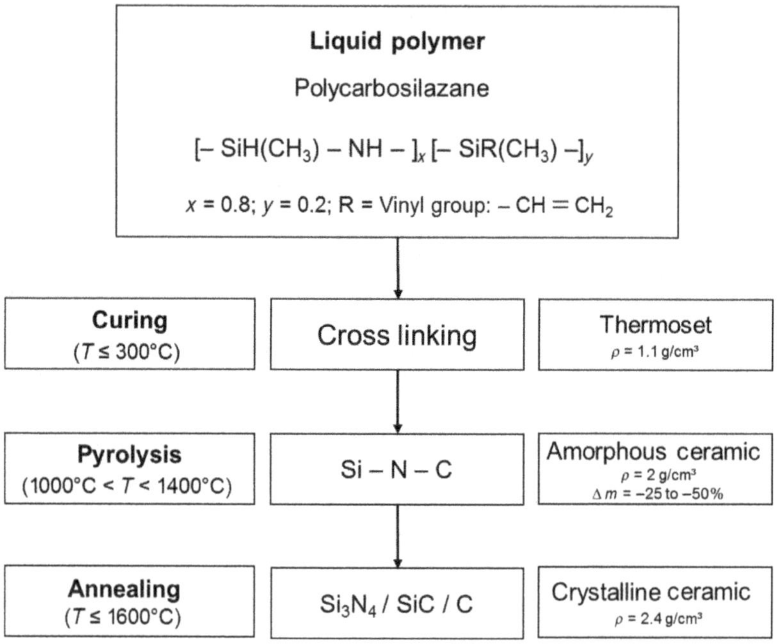

FIGURE 6.11 Schematic overview of the transformation of a typical polycarbosilazane precursor (Clariant Ceraset HTT 1800) for C/SiCN materials based on PIP (DLR).

AG (Germany), Dow Corning and Momentive Performance Chemicals (USA), and Shin-Etsu Chemicals (JP), but also from smaller companies such as EEMS, Starfire Systems (both USA), MS Production (Slovenia), and others.

The PIP technique is one of the most advanced methods for the manufacture of large and complex-shaped C/SiC structures in the aerospace industry [38]. As an example, nozzle extensions with a maximum diameter of 1.33 m and a height of 1.36 m could be realized by Astrium ST/EADS IW (see Figure 6.54).

Today, PIP-derived C/SiC materials are manufactured by Astrium ST/EADS IW (Germany), Herakles (France), COI Ceramics Inc. (COIC, USA), MS Production (Slovenia), and others.

The PIP process can be subdivided into four basic process steps (Figure 6.12):

1. Deposition of fiber coating (interphase).
2. Manufacture of a CFRP preform.
3. Pyrolysis of the CFRP preform.
4. Densification via repeated polymer infiltration an pyrolysis.

In the first process step, the carbon fiber or fiber preforms are coated with an interphase layer via continuous or discontinuous CVD or CVI processes. Usually a thin (0.1–0.3 μm) layer of pyC is deposited onto the fibers, to ensure a weak embedding of the fibers in the final SiC matrix.

For the manufacture of the CFRP preforms, several methods, well known from the manufacture of FRPs, such as wet filament winding and fiber placement, vacuum-assisted polymer (VAP) infiltration (developed by Astrium ST/EADS IW) or RTM as well as autoclave technique and warm pressing are used. However, in comparison to FRP, where epoxy or polyester resin systems are common, these techniques have to be adapted to the use of preceramic precursors, which require higher curing temperatures and careful handling to avoid oxidation. In order to increase the ceramic yield after curing and pyrolysis, submicron SiC particles or reactive fillers can be added to the liquid precursors [39]. The resulting slurry is infiltrated in the fiber bundle or roving by using well-known wet filament-winding techniques. By the use of nanoparticles, the slurries can also be applied to RTM and VAP, because filtration effects of the fiber preform can be avoided due to the small particle size, ensuring a homogeneous distribution of the nanoparticles, even inside the fiber bundles. However, in order to avoid fiber damage by the particles, the maximum fiber volume content is limited to about 50 vol.%, and due to high cost, reactive fillers, as well as nanosized particles are not currently used for industrial production.

Wet filament winding and fiber placement is used to manufacture CFRP preforms directly as well as for the manufacture of pre-impregnated plies (prepregs) with unidirectional or 0°/90° fiber orientation. These prepregs are favorably

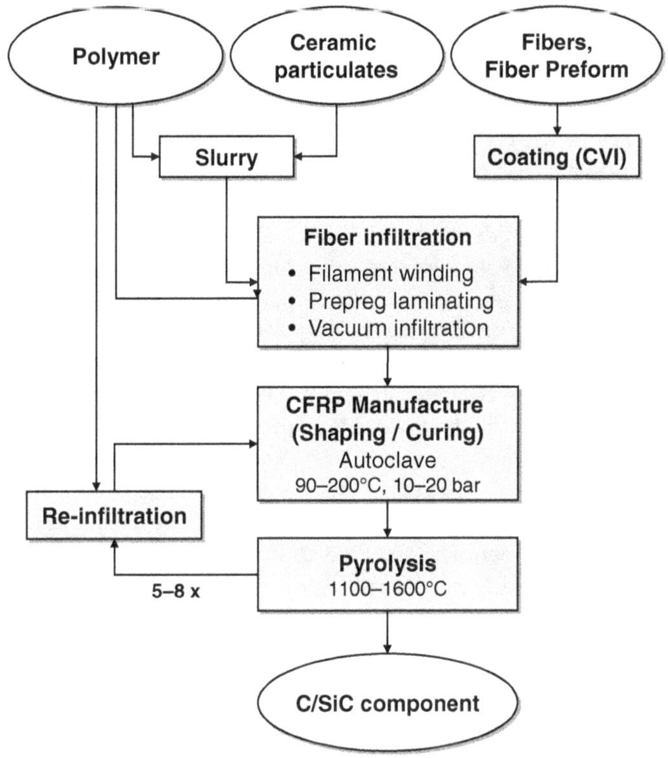

FIGURE 6.12

Schematic overview of the manufacture of C/SiC materials via PIP (DLR).

used to build up multiaxial, fiber-reinforced laminates for flat plates or complex-shaped structures, very similar to the manufacture of CFRP components. Shaping, consolidation, and curing of the filament wound or laminated structures takes place in an autoclave at maximum temperatures between 90°C and 200°C and at pressures in the range of 1–2 MPa.

In the third process step, the CFRP preform is pyrolyzed in inert gas atmosphere or vacuum. During pyrolysis, the chemical bonds within the polymer break, the organic side chains separate, and the polymer matrix is converted into the final ceramic matrix. Gaseous products, oligomers, and solvents as well as hydrocarbons and hydrogen evaporate, leading to a mass loss of 10–40 wt.% and porous ceramic matrices. The morphology of the resulting ceramic matrix is strongly influenced by the process temperature. Generally, low heating rates are applied in order to obtain dense SiC matrices with low porosity. Low maximum process temperatures of up to 1100°C lead to porous, amorphous SiC matrices, whereas low porous, crystalline matrices are obtained at maximum process temperatures of up to 1600°C.

Due to the mass loss and the significant increase of density from the polymer ($\rho \approx 1$–1.2 g/cm^3) to the ceramic matrix ($\rho \approx 2.4$–2.7 up to 3.2 g/cm^3), a high volume contraction of 70–80 vol.% is observed for the polymer matrix during pyrolysis. This results in C/SiC materials with open porosities between 20 and 40 vol.%, after the first pyrolysis step. To achieve sufficient mechanical properties in the final C/SiC, the porosity has to be reduced by multiple densifying cycles, each consisting of a polymer infiltration and a pyrolysis step. To obtain C/SiC materials with open porosities below 10% or even 5%, typically five to eight densification steps are necessary. In contrast to the manufacture of the CFRP preform, where the fibers or fabric layers can be infiltrated with the slurry matrix precursor separately and prior to the shaping and consolidation process, no fillers, or more exactly, only nanoscaled particles can be used for the densification steps, where the C/SiC component has to be infiltrated as a whole.

Process time is highly dependent on the ceramic yield of the precursor and the resulting ceramic matrix density, which determines the amount of densification steps. In order to avoid delaminations, pyrolysis time is up to 5 days for thick-walled components or complex-shaped, large structures, but can be reduced to 1–2 days for thin-walled ($t = 3$–10 mm), flat plates. Due to the lower risk of delaminations, 3D fiber architectures based on short fibers or braided and stitched preforms offer shorter pyrolysis cycles, compared to 2D fabric-based materials. Additionally, process time can be reduced significantly by the use of continuous pyrolysis processes, leading to overall manufacturing time of about 14 days (M. Zornik, MS Production, personal communication, 2012), for example, for brake discs.

The main advantage of the PIP process is the controllable matrix buildup, enabling stoichiometric SiC and avoiding free silicon. Process temperatures for the conversion of the cross-linked polymer to the amorphous SiC ceramic matrix are low, and well-established manufacturing methods from the plastics industry can be used in principle. Similar to the CVI process, the fibers are not influenced or damaged during the SiC matrix buildup, and, in combination with a single fiber embedment, high mechanical properties can be achieved. Additionally, unidirectional (1D) fiber reinforcement is possible, resulting in C/SiC materials with very high strength values in parallel to the fiber orientation. Due to a homogeneous matrix buildup, wall thickness is generally not limited, and plates with a maximum wall thickness above 25.4 mm (1 inch) up to 30 mm have been produced to date. Using filament wound preforms, very large and precise structures for space propulsion systems have already been realized in near net shape technique. *In situ* joining technologies are available for the manufacture of highly complex structures. Thereby, easily manufactured substructures of simple geometries are manufactured separately in the first process steps and joined together in the course of the densification process.

The disadvantages of the PIP process are high costs for the preceramic precursors and the long manufacturing times of several weeks up to months, especially for large, complex-shaped structures, mainly caused by the multiple infiltration and pyrolysis steps, essential for material densification. Due to the relatively high residual porosity, interlaminar shear strength usually is low, especially for 2D fiber-reinforced materials, but can be increased by the use of multiaxial fiber preforms for high performance components.

6.2.3 Melt Infiltration

The melt infiltration (MI) process is derived from the C/C manufacture via PIP as well as from the reaction bonding of SiSiC materials. These two processes have been combined and further developed since the 1970s [40] with the main focus on suitable C/C preforms, ensuring both, minimal fiber damage during siliconization and complete densification with a homogeneously distributed SiC matrix. These two objectives could be achieved by tailored microstructures, which enable a fast infiltration with molten silicon and a uniform buildup of the SiC matrix in the whole volume of the C/C preform. From the outset, the development of the MI process was driven by economic considerations, in order to provide a low cost alternative to the highly sophisticated CVI and PIP methods.

The characteristics of the MI method are the infiltration of a porous C/C preform with molten Si and the buildup of the SiC matrix by an exothermic chemical reaction of the liquid silicon with the solid carbon:

$$\mathrm{Si_{liquid}} + \mathrm{C_{solid}} \rightarrow \mathrm{SiC_{solid}}[\Delta H = -68\ \mathrm{kJ/mol}]$$

The MI technique is used for the manufacture of C/C-SiC as well as of C/SiC materials. Flat plates as well as complex-shaped, large structures are feasible. Due to economical and some specific technical advantages, such as high thermal conductivity and wide variability in materials composition and properties, MI is the method of choice for the manufacture of brake discs for automobiles and of other friction components in industrial applications.

Today, MI-derived C/C-SiC and C/SiC materials are industrially manufactured by Brembo SGL Carbon Ceramic Brakes (BSCCB), in Germany and Italy, SGL Group (SGL), Schunk Kohlenstofftechnik GmbH (SKT), both in Germany, M cubed and Ultramet, both USA, and others, as well as by FCT Ingenierkeramik (FCT), Germany, and Nammo Raufoss AS (Nammo), Norway, both of which have licensed the LSI process from DLR, Germany, to produce C/C-SiC friction pads and rocket motor components, respectively. Further activities have been reported by Surface transforms, UK. Additionally, the MI process is used as an optional manufacturing route, or in combination with CVI and PIP, as by Astrium ST/EADS IW, Germany, and Herakles, France.

The MI process can be subdivided in four main process steps (Figure 6.13):

1. Buildup of fiber/matrix interphase, for example, by fiber coating.
2. Manufacture of a CFRP preform.
3. Pyrolysis of the CFRP preform to a C/C preform.
4. Siliconization of the C/C preform.

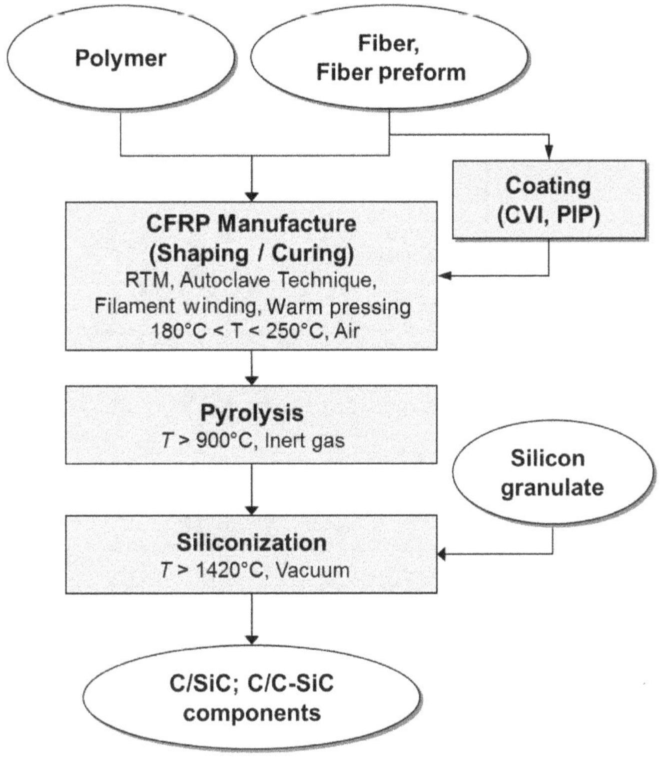

FIGURE 6.13

Schematic overview of the manufacture of C/SiC materials via MI (DLR).

For the siliconization, the LSI process, characterized by a capillary infiltration of molten Si is the most common method. Other MI methods used for industrial production are a combination of pack cementation and capillary infiltration (PCI) and pressurized melt infiltration (PMI), also called vacuum pressure infiltration.

In contrast to CVI and PIP, where the SiC matrix buildup does not influence or damage the carbon fibers, in MI, the molten silicon is highly reactive to the C matrix as well as to the C fibers. Therefore, a direct contact of Si melt and C fibers has to be avoided carefully. Additionally, a weak embedding of the brittle fibers in the brittle matrix is mandatory to obtain characteristic CMC properties, such as high strength, fracture toughness, and thermal shock resistance. To ensure both, fiber protection and weak fiber–matrix interface, three different methods are currently used for the industrial production of MI-based C/SiC and C/C-SiC materials, namely fiber coating with pyrolitic carbon (pyC) via CVI/CVD, fiber embedding in carbon matrix via multiple PIP, and *in situ* fiber embedding in carbon matrix (Figure 6.14).

Using CVI for fiber coating, a thin layer, (~0.1 to 0.3 μm) of pyC is deposited on each fiber filament, ideally, finally resulting in a C/SiC material with the filaments mainly embedded individually in the SiC matrix [7]. Fiber protection via PIP is widely used for the manufacture of short fiber-based C/SiC brake discs [41,42]. Thereby, the fiber bundles are impregnated with phenolic resin, which is cured and pyrolyzed. After repeated cycles of resin infiltration and pyrolysis, the fiber filaments are embedded in a dense carbon matrix, leading to a C/C-like raw material. The coated fiber bundles are cut to defined lengths and the resulting, short C/C sticks are mixed with phenolic resin and a near net shape CFRP preform, or more precisely, a C/C stick-reinforced polymer preform, is manufactured via warm pressing and subsequently pyrolyzed. During siliconization, the fiber filaments are well protected against Si contact, finally leading to a C/C-SiC material, characterized by dense C/C bundles embedded in SiSiC matrix. However, these materials usually are called C/SiC by the manufacturer. Time consuming and costly fiber coatings are not necessary at all, if suitable precursors are used for the manufacture of the CFRP preform. Thus, a strong fiber–matrix bonding (FMB) can be achieved in the CFRP preform, which leads to a segmentation of each fiber bundle into dense C/C bundles during pyrolysis, and again, C/C bundles embedded in SiSiC matrix are obtained after siliconization. However, the fibers at the outer surface of the C/C bundles are not protected, and therefore, a small amount of C fibers get in contact with the molten Si and are converted to SiC.

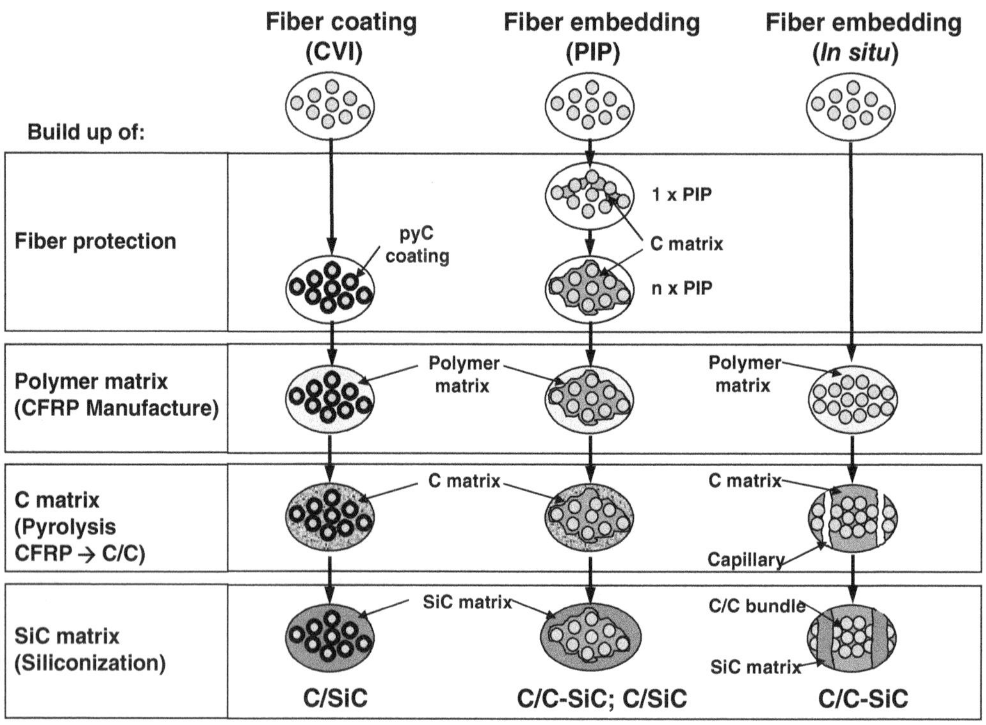

FIGURE 6.14
Schematic overview of industrially used methods for providing fiber protection and weak fiber–matrix interfaces in the MI process (DLR). Fiber coating with pyC interphase via CVI (left). Embedding C fibers in C matrix by multiple PIP (middle). *In situ* fiber embedding in C matrix (right).

In the following, the manufacture of C/C-SiC materials and components based on *in situ* fiber embedding and LSI is described in detail.

Manufacture of CFRP Preforms For the manufacture of CFRP preforms, commercially available C-fibers and polymer precursors are used. In order to ensure a single cycle process for pyrolysis and siliconization, the carbon yield of the polymer precursors has to be high, to ensure a low mass loss (<40 wt.%) and volume shrinkage (<60 vol.%) during pyrolysis. Typically, low cost phenolic resins (about 5–20 €/kg) or other high temperature resins are used. Almost all C-fiber types, from low cost HT to IM and HM fibers, with fiber counts of 1 K–24 K (1 K = 1000 fiber filaments) and above can be used. For high performance, lightweight structures, typically 3 K and 6 K fibers and 2D fabrics or 1D layers are favored, whereas low cost 12 K or 24 K and up to 240 K and 360 K fibers are common for short fiber materials as used for the manufacture of automobile brake disc [43]. For the near net shape manufacture of CFRP preforms, different technologies, well known from the plastics industry, are common:

- Resin transfer molding (RTM).
- Autoclave technique.
- Warm pressing.
- Wet filament winding.

In the RTM process, dry preforms based on stacked 2D fabrics or 1D cross plies as well as filament wound or braided preforms are pressure infiltrated with the liquid precursor. In order to avoid fiber displacement and to ensure a complete infiltration of the fiber preform, the viscosity of the precursor and the infiltration pressure should be low. Usually, the precursor is heated before infiltration, to obtain a viscosity below 100 mPas. Typical process parameters are infiltration pressures of 0.03–0.08 MPa and maximum pressures and temperatures of about 2.2 MPa and 200°C for the curing of the precursor. RTM is used favorably for the manufacture of very complex-shaped, integral structures

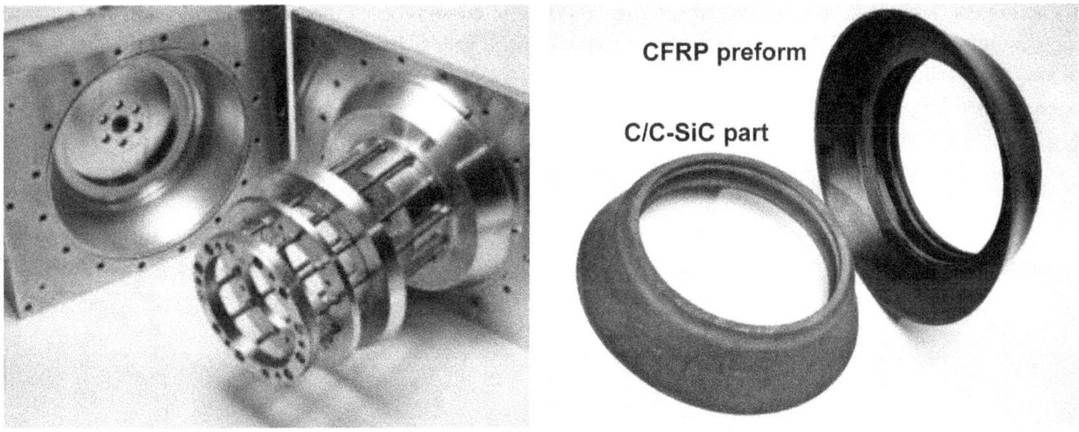

FIGURE 6.15
Manufacture of a C/C-SiC hot gas liner (design study) with integrated stiffeners in near net shape technique. RTM mold with multiple core elements, made of steel (left, DLR). CFRP preform made as integral component and C/C-SiC raw structure (right, DLR).

(Figure 6.15) but also for flat plates, especially with high wall thicknesses. As an example, CFRP preforms for brake discs with a diameter of up to 740 mm and a thickness of 70 mm could be achieved to date. Due to the use of a defined mold cavity, the fiber content can be well controlled and high fiber contents of up to 60 vol.% and above can be realized. The disadvantages of RTM are the high cost for the molds and the long manufacturing time of up to 30 hours for complex-shaped components, including heating up and cooling down.

In contrast to RTM, the autoclave technique is typically used for the manufacture of large, complex-shaped CFRP preforms, for example, for nose caps of spacecraft (Figure 6.16) [44]. As a raw material, prepregs based on phenolic precursors and 2D fabrics are common. The pre-cut prepreg plies are laminated on or into a mold with predefined fiber

FIGURE 6.16
CFRP preform (Ø 730 mm × 450 mm, DLR) for the Expert nose cap. The CFRP preform (center, lifted by a crane) was manufactured by laminating cut prepreg plies in a steel mold (bottom) and by curing the multiaxial laminate in an autoclave (background) [44].

orientations for each single layer. Subsequently, the laminate is covered by an airtight, flexible foil and is consolidated by applying vacuum ($p \approx -0.02$ MPa) between the mold and the foil. Curing takes place in an autoclave at a maximum pressure and temperature of about 1 MPa and 200°C, respectively. Process time for curing is about 6 hours. However, total process time for CFRP manufacture, including a post-curing step, varies from about 10 hours for thin-walled, flat plates, to up to 5 days, for complex structures, due to the time consuming, laminating process. As an example, for the nose cap of the Expert TPS, more than 1400 single prepreg plies had to be manually positioned precisely in the mold.

Warm pressing is used for the manufacture of very complex-shaped CFRP preforms, based on randomly oriented short fibers, such as internally ventilated brake discs (Figure 6.17) [45], as well as for profiles (Figure 6.18) [44] or flat

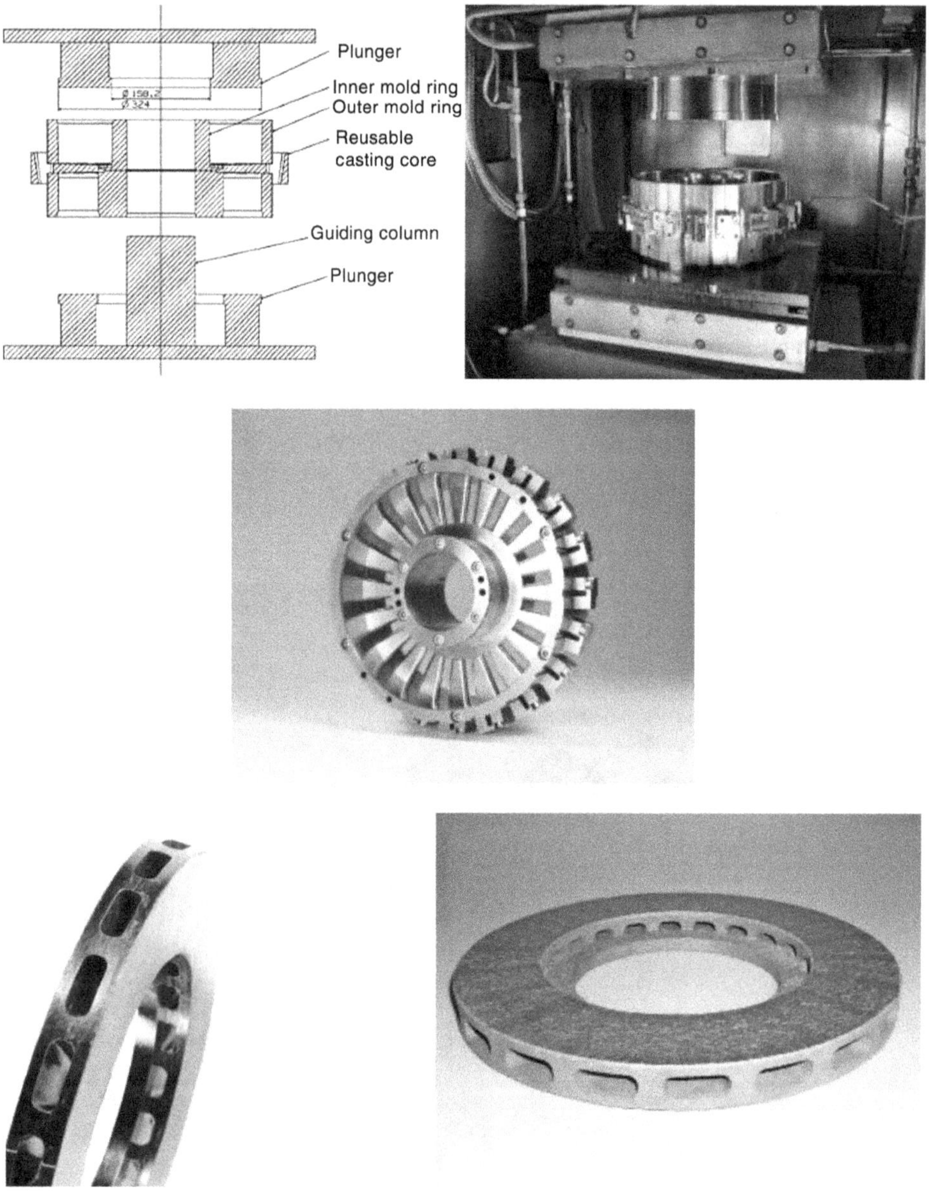

FIGURE 6.17

Warm pressing of complex-shaped CFRP preforms for inner ventilated C/C-SiC brake discs. Schematic view of the press mold made of steel (top left) and mold, integrated in the press (top right). Center part of the mold, showing the reusable cores for forming radial cooling channels (middle). Near net shape CFRP preform (bottom left) and final C/C-SiC brake disc after machining (bottom right, all figures DLR) [45].

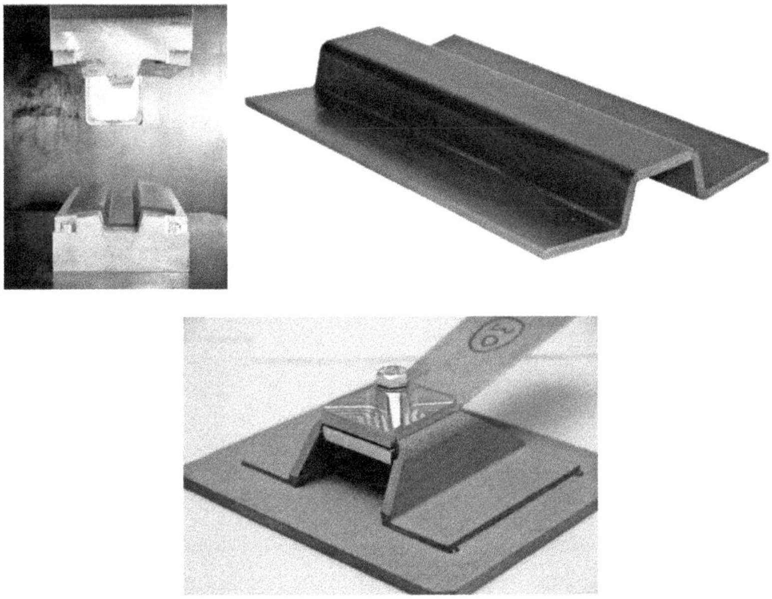

FIGURE 6.18

Warm pressing of a CFRP profile based on 2D fabric prepregs. Pressing mold made of aluminum (top left) with near net shape CFRP profile (top right). Test sample of C/C-SiC load introduction with *in situ* joined profile element (bottom, all figures DLR) [44].

plates, based on 2D fabrics. Thus very large plates of up to 2 m × 3 m × 10 mm are available from industrial production. For short fiber-based preforms, liquid as well as powder phenolic resins are common, whereas prepregs are used for 2D fabric-based components, similar to the autoclave technique. Typical process parameters are maximum pressures of 2–5 MPa and curing temperatures in the range of 200°C up to 250°C, both depending on the fiber architecture and the precursor. The main advantage of warm pressing are the low cycle times in the range of 3–7 hours down to less than 1 hour, especially for short fiber-based press masses, which can be filled in the mold automatically. Therefore, warm pressing is the process of choice for the mass production of automotive brake discs.

Similar to PIP, wet filament winding has been adapted for the manufacture of rotational symmetric C/C-SiC components via LSI, for example, tubes or integral combustion chambers (Figure 6.19) and nozzle extensions [46,47]. Thereby, the fiber is infiltrated with the phenolic precursor and wound on a turning mandrel with predefined angle or fiber orientation. After finishing the filament-winding process, the laminate is cured in an oven at ambient pressure or consolidated by a vacuum bag and cured in an autoclave, applying a defined pressure. Typical curing temperatures are 180–200°C, depending on the used type of precursor. Due to the computer controlled, automatized-winding process, a well-defined fiber architecture with multiaxial in-plane fiber orientations can be built up with high reproducibility. Fiber content is

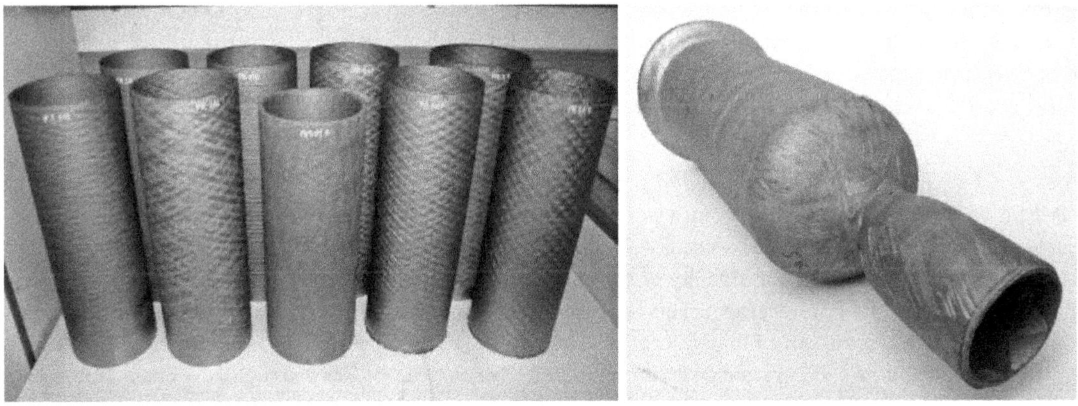

FIGURE 6.19

Wet filament wound C/C-SiC tubes (Ø 160 mm × 400 mm, $t = 6$ mm, left, DLR) and nozzle structure (feasibility study, Ø 140 mm × 550 mm, right, DLR). The figures show raw components after siliconization.

TABLE 6.1 Manufacturing Methods for CFRP Preforms Used in the LSI Process

Method	Preform fiber architecture	Application/feasibility						Fiber content (vol.%)	Manufacturing time for plate (250 × 250 × 3 mm³) (h)
		Plates			Structures				
		Thin walled	Thick walled	Large sizes	Complex	Large and complex	Rotational symmetric		
RTM	• Dry fiber preforms (1D, 2D, 2.5D, 3D) • Filament wound preforms (multiaxial)	++	++	**0**	++	–	+	40–60	20–30 (uncooled mold)
Autoclave technique	• prepregs (1D, 2D), • Wet filament wound laminates (multiaxial)	++	++	+	+	++ (nose caps)	**0**	50–60	10
Warm pressing	• prepregs (1D, 2D) • Wet, filament wound laminates (multiaxial)	++	+	++	+	–	**0**	50–60	6
	• Short fibers • Short fiber prepregs (randomly oriented)	–	++	++	++ (ventilated brake discs)	–	–	20–50	–
Wet filament winding	• Fiber rovings (multiaxial)	+	–	–	**0**	++ (nozzles)	++	40–60	–

in the range of 40–60 vol.%, depending on fiber architecture. Furthermore, complex-shaped, tubular structures with changing diameters can be realized without cutting the fibers, which is not possible if 2D fabrics are used. However, process time is usually longer, compared to fabric-based preform manufacture, due to the use of a single or a limited number of fiber tows, resulting in a lower turnover of fiber volume. In Table 6.1, an overview of the different CFRP manufacturing processes is given.

Pyrolysis of CFRP Preforms In the second process step, the CFRP preform is pyrolyzed at temperatures above 900°C and up to a maximum of 2000°C in a non-oxidizing atmosphere at ambient pressure. Thereby, the polymer matrix is converted to a glassy C matrix, resulting in a mass loss of about 36–40% and a volume contraction of about 50–60%. However, due to the thermal stability of the carbon fibers, the geometry of the CFRP preform remains almost unchanged, depending on the fiber preform. This leads to internal stresses in the component and tension stresses in the matrix, and finally to the formation of matrix cracks. The resulting C/C preform is characterized by an open porosity of about 10–30%, formed by a 3D system of interconnected microcracks or pore structures, but is stable enough for a safe handling and machining. The microstructure of the C/C preform is an all-important factor for the subsequent silicon infiltration and the properties of the final CMC material, and is influenced by the fiber preform and fiber orientations, but mainly determined by the strength of the FMB (fiber matrix bonding). Thereby, high FMB leads to the formation

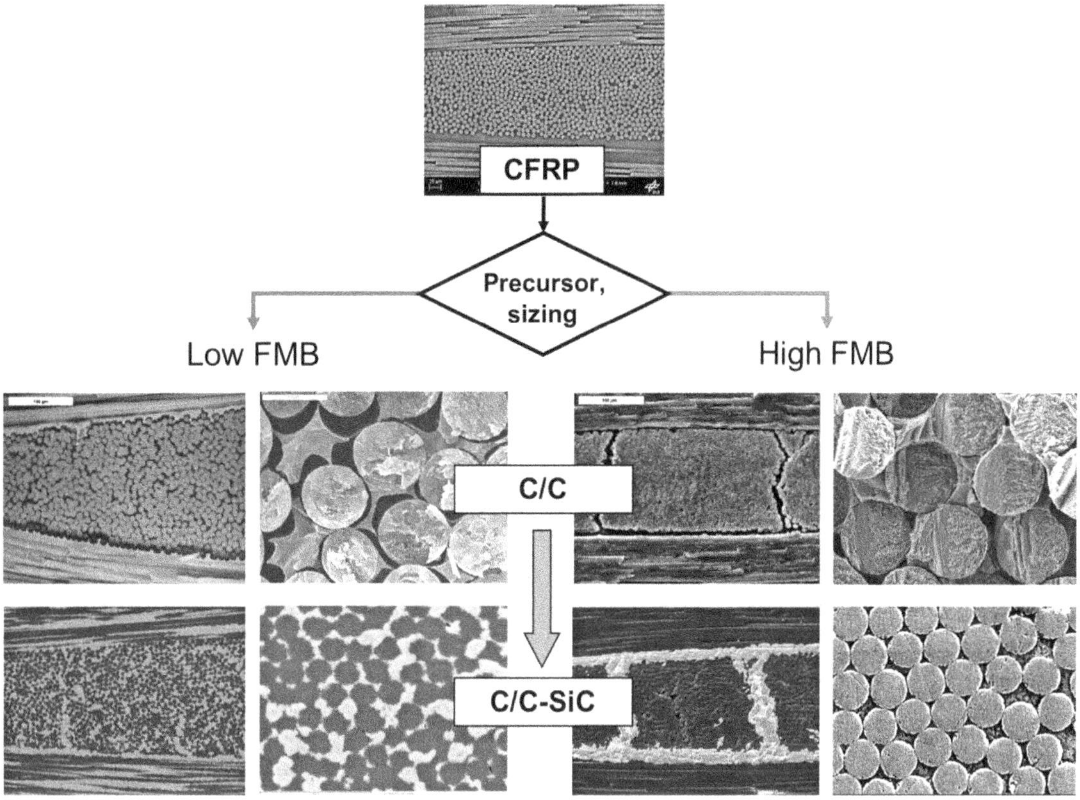

FIGURE 6.20

SEM figures showing the influence of the FMB strength in the CFRP preform on the microstructure of the C/C preform and the final C/C-SiC material. Low FMB with the matrix peeled off from the single fibers in the C/C preform and completely infiltrated fiber bundle after siliconization, with C fibers partially converted to SiC (left from top to bottom, DLR). High FMB, leading to a segmentation of the fiber bundle by microcracks, forming dense C/C bundles with the fibers embedded in C matrix, and C/C-SiC microstructure showing the microcracks filled with SiSiC matrix, but no infiltration of Si into the C/C bundle (right from top to bottom, DLR).

of distinctive microcracks and a fragmentation of the fiber bundles into several dense C/C bundles, which are able to protect the fibers from Si-attack during the subsequent siliconization step, and finally lead to a quasi-ductile, damage tolerant CMC material. In contrast, if the FMB is low, the matrix peels off from every single fiber filament and the whole fiber bundle is infiltrated with Si during siliconization. Thereby, all fibers are attacked by the Si and partially converted to SiC, leading to a SiC-rich material with high hardness but brittle fracture behavior, similar to monolithic ceramics (Figure 6.20). The FMB itself can be tailored by using suitable matrix precursors as well as by a thermal treatment of the fibers, modifying the fiber seizing or the chemically active groups (e.g., OH, COOH) on the fiber surface.

The pyrolysis of the CFRP preform is the most critical step in the LSI process. During the conversion of the polymer to the carbon matrix, high volumes of gaseous products are developed, similar to the PIP process. These gases have to be released safely out of the component, in order to avoid the buildup of excessive internal pressure, which can cause local delaminations and a complete destruction of the component. Due to the volume contraction of the matrix on the one hand and the geometrical stability of the fiber preform on the other hand, internal stress is built up, leading to microcracks or microchannels, substantially needed for the subsequent Si infiltration. However, these stresses can also cause delaminations and distortions, especially for curved or complex-shaped components. Laminates based on 2D fiber fabrics are critical, because of their anisotropic contraction behavior, characterized by a linear shrinkage of about 8–10% in the wall thickness, perpendicular to the laminate, but only about 0.5% in the in-plane direction, parallel to the fibers. Whereas in flat plates, the thickness contraction is not restricted and the single layers are able to move closer together, in curved or, even more critical, in closed tubular structures, the movement of the layers is blocked

FIGURE 6.21
Delaminations in tubular C/C preform (Ø 51/ 40 mm, DLR) caused by circumferential fiber orientation and blocked contraction during pyrolysis.

by the fibers oriented circumferentially. As a result, interlaminar stresses perpendicular to the laminate are induced, which cannot be carried by the matrix connecting the layers, finally leading to delaminations and local separation of the layers (Figure 6.21). To avoid delaminations and distortions, low heating rates are applied during pyrolysis, and high temperature stable, well-fitting supporting elements or structures made of graphite or C/C are used. Additionally, internal stresses during pyrolysis can be decreased by the use of quasi-isotropic lay-ups and by reducing the fiber length in critical areas, for example, by the use of small-sized fabric plies. In another approach, the effective fiber length in the fabrics can be reduced by a systematic local cutting of the fibers via a punching process [44].

Siliconization In the third and last process step, molten silicon is infiltrated in the porous C/C preform at temperatures of about 1650°C, well above the melting point of Si ($T_{melt, Si} = 1410°C$). Depending on the application, Si granulates of different qualities are used, from high purity solar grade Si (99.99% Si, 50–100 €/kg) for hot structures in space, to low cost, metallurgical raw Si (98–99.7% Si, 10–12 €/kg) for applications with lower service temperatures, such as brake discs.

For the Si-infiltration of the C/C preform, only capillary forces are used, and no pressure is applied. However, in order to support the infiltration and to avoid oxidation, the siliconization is performed in vacuum ($P \leq 2$ mbar). Thereby, almost all the open porosity can be filled with Si, from very fine capillaries with a width of about 1 μm, up to broad capillaries and pores of about 200 μm width or diameter, respectively.

Due to the low viscosity of the molten Si, comparable to water ($\eta_{Si, melt} = 0.77–0.46$ mPa s at 1560°C, $\eta_{H_2O} = 1$ mPa s), the good wettability to carbon and SiC (wetting angle $\Phi = 0–22°$ to C and 30–41° to SiC) and the high surface tension (0.72–0.75 N/m at 1550°C in vacuum), a fast infiltration even of large C/C preforms is possible. Analytical calculation models based on Navier stokes equation [48, 49] show that infiltration time is in the range of some minutes for an infiltration height up to 500 mm (Figure 6.22). The Si usually is supplied to the edge of a plate or structure, leading to an infiltration in parallel to the fibers and the surface of the component. Due to the simultaneous penetration of the whole cross section, even thick-walled components can be infiltrated homogeneously. In another method, the Si is supplied to the surface of the component. Thereby, the Si is initially infiltrated perpendicular to the fabric layers and subsequently is distributed in the whole volume of the C/C preform. Both methods can be combined, too, in order to ensure a safe and homogeneous infiltration of large components.

In parallel to the infiltration, the silicon reacts with the carbon in the contact areas and forms the SiC matrix. If the fibers are well protected, for example, via CVI-pyC coating or fiber embedding via PIP, only the C matrix is converted to SiC. Using *in situ* fiber embedding for the protection of the fibers, a small amount of both the carbon matrix and the carbon fibers are converted to SiC. Due to the fact that the molar volume of the formed SiC ($V_{SiC} = 12.45$ cm^3/mol) is 30% lower compared to the molar volume of the reactants, C + Si ($V_C + V_{Si} = 6.53 + 11.11$ cm^3/mol $= 17.64$ cm^3/mol) a single filling of the pore volume with Si is not sufficient, but further Si has to be infiltrated to obtain a dense SiC matrix and a complete filling of the pores. On the other hand, the volume of the formed SiC matrix is almost twice as high (191%) than the volume of the C, reacting with the Si, and therefore the pores and microchannels can be closed completely, leading to almost dense C/SiC and C/C-SiC materials. Depending on the microstructure of the C/C preform,

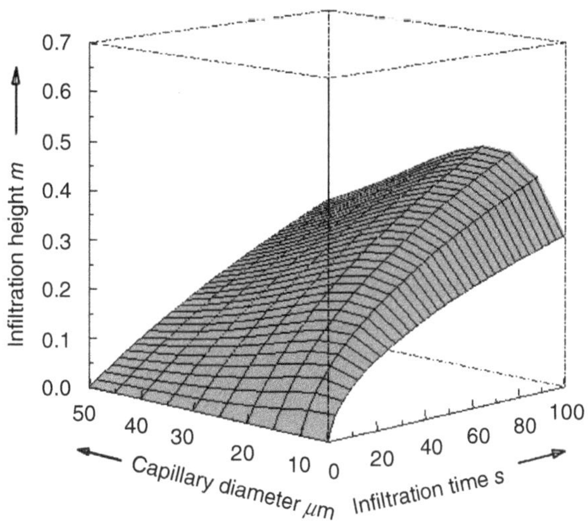

FIGURE 6.22

Calculated infiltration height of a single capillary in a C/C preform in dependence of infiltration time and capillary diameter (DLR) [49].

the infiltrated Si is not converted completely, but unreacted, free Si remains (ϕ_{Si} < 4 vol.% for fabric-based C/C-SiC), especially in large pores or wide microchannels, resulting in a dense SiSiC matrix.

In addition to the solely capillary infiltration of the LSI method, two other process variants are used industrially for the melt infiltration of Si in C/C preforms, a combination of PCI and PMI, also called vacuum pressure infiltration [4, 50, 51].

Capillary infiltration is mainly used in order to obtain C/SiC and C/C-SiC materials, characterized by minimal residual porosity as well as by a high conversion rate and SiC content. In contrast, PCI and PMI are used to manufacture materials with high C/C content and minimized fiber damage, achieved by a controlled and limited conversion of the C/C to SiC. In PCI, the C/C preform is spray coated by a mixture of Si particles and a binder. After curing the binder, the coated component is heated to temperatures above the melting point of Si and the molten Si is infiltrated in the porosity of the C/C preform, building up the SiC matrix. The amount of Si, supplied to the C/C preform can be controlled very well by the thickness of the coating. The resulting C/C-SiC material is characterized by a low content of SiC matrix and the absence of free Si as well as by minimized fiber damage, leading to high mechanical performance, similar to C/C. Compared to C/C, oxidation resistance is increased by a homogeneous buildup of a SiC coating, sealing the outer surfaces of the component as well as the inner surfaces of the pores. However, PCI is limited to highly densified C/C preforms with low porosity volumes and low Si uptake. Whereas a homogeneous SiC coating of the inner pore surface can be achieved in components with low wall thickness, high wall thicknesses above 10 mm lead to gradient materials, characterized by a dense C/C-SiC material at the outer surface and a C/C material in the center [50]. After PCI, no free Si remains on the surface and the component can be coated subsequently by CVD-SiC without the need of intermediate machining.

In contrast to PCI, where the conversion of C/C to SiC is controlled by a defined amount of Si supply, in PMI, developed at SKT in the 1970s [50], the buildup of the SiC matrix is varied by the time given for Si infiltration and chemical reaction. Thereby, the C/C preform is heated to temperatures up to 2000°C and transferred into an autoclave containing a crucible filled with molten Si. The autoclave is evacuated and the C/C preform is dipped into the Si. After the C/C preform is fully submerged, the autoclave is floated with inert gas and a pressure above 2 MPa is applied. Thereby, the C/C preform is fully infiltrated in seconds and, due to the pressure applied, even the finest micropores and microchannels are filled with Si. After a short dwell time of about 5 minutes, the component is taken out of the Si and the autoclave and cools down in a separate facility. The resulting C/C-SiC material is dense and the complete porosity of the C/C preform is filled with a SiSiC matrix, characterized by a homogeneous SiC coating layer, sealing the inner and outer surfaces of the material. The wall thickness of the SiC layer is constant in the whole volume and can be controlled by the dwell time for Si infiltration. Due to the short termed contact to Si, only a small amount of C/C is converted to SiC and the mechanical performance of the C/C material can be maintained, similar to PCI. If necessary, free silicon is extracted by a subsequent chemical or thermal treatment. The main advantages of the PMI process are the gradient-free infiltration and homogeneous SiC matrix buildup as well as the low cycle times for Si infiltration and the low overall process duration, caused by the quasi-continuous process with separated heating up, Si infiltration and cooling down in different facilities. However, investment costs for the facilities are high, especially for the transfer of the C/C preform. Additionally, the safe handling of large amounts (>100 kg) of molten Si is critical, and therefore the PMI process is limited to small components with maximum diameters of about 320 mm.

Material composition and morphology of LSI-based C/SiC materials can be varied in a wide range by the use of different fibers and fiber preforms as well as by the matrix precursor and the way, the SiC matrix is built up. Additionally, the FMB strength in the CFRP preform and during pyrolysis can be varied by a thermal treatment of the fibers leading to strongly different material variants. Thereby, the variants can be manufactured as homogeneous materials or different material variants can be combined in one single component, leading to gradient materials. As an example, for friction applications, SiC-rich and highly abrasive-resistant friction layers on the surfaces can be combined with a C/C-rich material in the center area, providing high strength and ductility. Therefore, single fabric layers are thermally treated at different temperatures in the range of 600–1100°C and stacked in a defined sequence. The higher the temperature, the lower the FMB and the higher the SiC content after siliconization. The transition from the C-rich to the SiC-rich material can be designed layer by layer, to give a seamless graded structure or as discontinuous graded structures (Figure 6.23) [52, 53]. Additionally, surface layers based on highly porous C felts or pyrolyzed, wooden preforms can be joined *in situ* with C/SiC substrates, leading to SiSiC surface layers. These so-called "SiCralee" coatings offer high abrasive and oxidation resistance, due to their high SiC content and the absence of C (Figure 6.24).

Due to low and predictable contraction rates during pyrolysis, almost no geometrical changes during siliconization and the fast and homogeneous Si infiltration process and SiC matrix buildup, structural components can be realized in near net shape technique with almost no restrictions to size and geometry. To date, C/C-SiC components with wall thicknesses below 1 mm up to 70 mm (Figure 6.25) and diameters of up to 740 mm have been manufactured successfully.

Additionally, complex-shaped, thin-walled structures can be realized via *in situ* joining. Thereby, the structure is divided into several subcomponents, which can be manufactured easily in near net shape technique. After pyrolysis, the subcomponents are assembled by using a slight pressure fit or by adhesive bonding the components with a joining paste, based on phenolic resin and carbon powder [44, 54, 55]. To assure a tight fit, the joint areas of the components are usually machined before assembly. The joining itself is performed *in situ*, during the siliconization of the assembled C/C

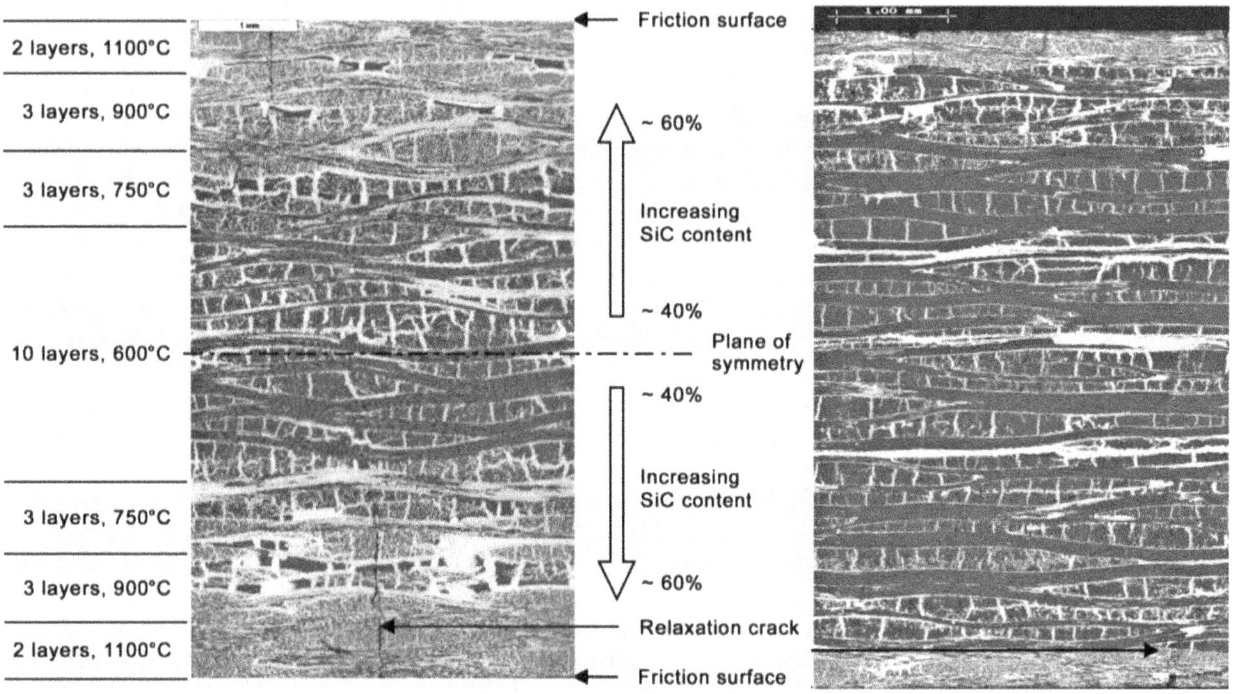

FIGURE 6.23

Graded C/C-SiC friction materials combining SiC-rich surface layers and C/C-rich material in the center by increasing the SiC content gradually (left, DLR) [53] and by a discontinuous lay-up (right, DLR).

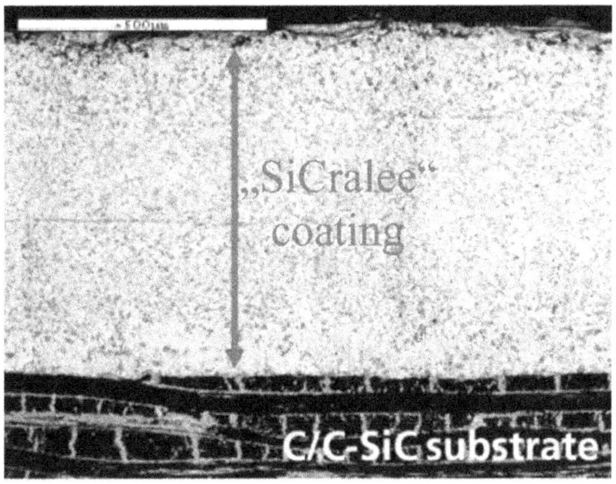

FIGURE 6.24
Thick-walled SiSiC surface layer, realized by joining of a pyrolyzed, wooden-based preform onto a C/C substrate and subsequent siliconization (DLR).

structure. Thereby, SiC matrix is built up in the joint, bonding together the components permanently to an integral structure (Figure 6.26).

Process time for LSI is mainly determined by the complexity of the component and by the pyrolysis process. CFRP preform manufacture takes some hours to 1 day for simple plates, and up to 1 week for large, complex-shaped structures, such as nose caps for spacecraft TPSs. For the siliconization, typically 1–2 days are needed. Process times for pyrolysis are determined by the geometry (wall thickness, volume, shape) of the component as well as by its fiber architecture and the open porosity in the CFRP preform. For components based on 2D fiber fabrics, the cycle time varies from about 3 to 5 days for thin-walled plates, up to 10 days for thick plates or large and complex-shaped structures. Due to the random fiber orientation in short fiber-reinforced preforms, the risk of delamination is reduced, and the pyrolysis

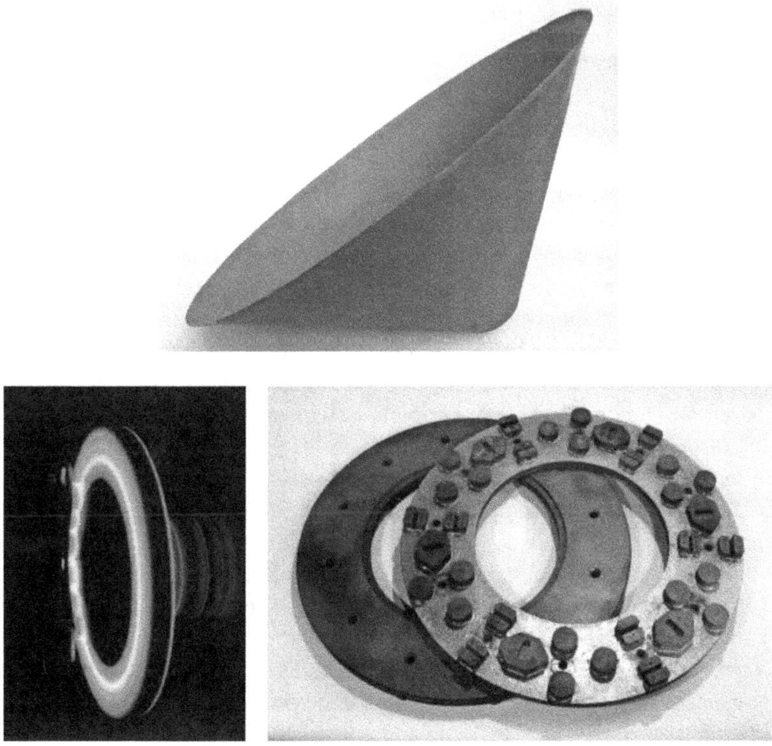

FIGURE 6.25
Exemplary C/C-SiC structures, manufactured via LSI process. Thin-walled antenna substructure (t = 1–2 mm, Ø 800 mm, h = 360 mm, top, DLR). Brake disc demonstrators for high speed trains (t = 60 mm, Ø 650 mm, bottom left, Knorr-Bremse AG; t = 35/70 mm, Ø 735 mm, bottom right, DLR).

FIGURE 6.26

In situ joining of load introduction elements in the Expert nose cap. C/C preform of the nose cap showing machined joining areas and one C/C profile in Ω shape, bonded to the nose cap with joining paste (top left). C/C-SiC nose cap with *in situ* joined Ω-profiles after siliconization (top right). SEM figure of a typical *in situ* joining, showing a homogeneous morphology of substrate and joining area (bottom, all figures DLR).

can be reduced to 1–3 days. Shortest process times are achieved with highly porous CFRP preforms, enabling a fast outgassing during matrix conversion.

For a further reduction of the cycle time, continuous processes for pyrolysis are already introduced in serial production. In batch processes, the furnaces are charged with a large number of components, for economic reasons. However, high masses of CFRP preforms, supporting structures, charging devices, and furnace equipment have to be heated up and cooled down. Additionally, the preforms usually are heated by radiation, through heating elements positioned at the interior furnace walls. However, in large staples of preforms, radiation heat transfer is limited to small areas of the preforms surface, usually to the edges, and the preforms volume have to be heated by thermal conduction. In order to avoid critical thermal gradients and inner stresses in the preforms, the heating has to be done carefully, leading to low heating rates and long process times. In continuous pyrolysis, the component is moved separately through the furnace with different zones of constant temperature and only the preform itself has to be heated and cooled, enabling significantly increased heating rates and, even more important, a fast cooling down. For industrial series production, the geometry of the furnaces can be well adapted to the preform geometry, ensuring an effective heat transfer and a homogeneous temperature distribution in the preform, finally avoiding inner stresses. In summary, due to a homogeneous temperature distribution and fast cooling rates, the temperature profile of the process can be well adapted to the individual limits of the preforms, enabling a fast and safe pyrolysis of typically less than 1 day for porous, short fiber-reinforced CFRP preforms. For fabric-based materials, process time for CFRP manufacture and pyrolysis can be reduced dramatically by integrating the curing of the polymer in the pyrolysis process. Thereby, cut prepreg layers or manually infiltrated wet fabrics are stacked to a laminate and pyrolyzed directly in about 24 hours, leading to highly porous C/C preforms and final C/C-SiC materials characterized by high Si contents. First investigations of new heating methods based on microwaves for CFRP manufacture and pyrolysis [56] as well as direct sintering for siliconization [57] showed a high potential for significant reductions of overall process time and energy consumption.

The main advantages of LSI are the short manufacturing times and the variability of the process. Thereby, the material properties can be adjusted over a wide range to meet specific requirements in very different application areas. Due to the fact that every process step is performed only once, process time is lower, compared to the other CMC manufacturing processes, especially for the manufacture of thick-walled structures and large, complex-shaped components. Additionally, low cost raw materials such as phenolic polymer precursors and Si alloys can be used and costly fiber coating can be avoided. The main disadvantages are the high process temperatures for siliconization and the loss of C fibers due to Si attack and conversion to SiC as well as the anisotropic contraction during pyrolysis, resulting in high internal stresses and risk of delaminations, especially for curved or tubular structures.

Comparison of Manufacturing Methods All manufacturing methods for C/SiC materials described above are feasible for the manufacture of simple components as well as for large, complex-shaped structures, which are realized in near net shape technology. Integral structures can be built up by the manufacture of subcomponents and *in situ* joining at an intermediate process step. In CVI, no contraction of the matrix takes place during processing and only the mismatch in the thermal expansion of C fiber and SiC matrix leads to internal stresses and microcracks during cooling down after matrix buildup. Therefore, the manufacture of curved structures is less critical by CVI, compared with PIP and LSI in particular, where matrix contraction during pyrolysis results in anisotropic geometry changes and interlaminar stresses, increasing the risk of delaminations. Additionally, 3D-woven fabrics are commonly used in CVI, which reduces delamination risk during processing and service. With CVI, component wall thickness is limited to about 30 mm, especially if I-CVI is used, whereas to date for PIP and LSI, no process-related restrictions for the component thickness have been observed.

For all manufacturing methods, process time is strongly dependent on component geometry and complexity. Large and complex-shaped components generally lead to long process times up to several weeks or months. The higher the wall thickness, the lower the heating rates for pyrolysis in PIP and LSI, resulting in long process times, especially for fiber architectures based on 2D fiber orientations (Figure 6.27). Additionally, time-consuming multiple densification

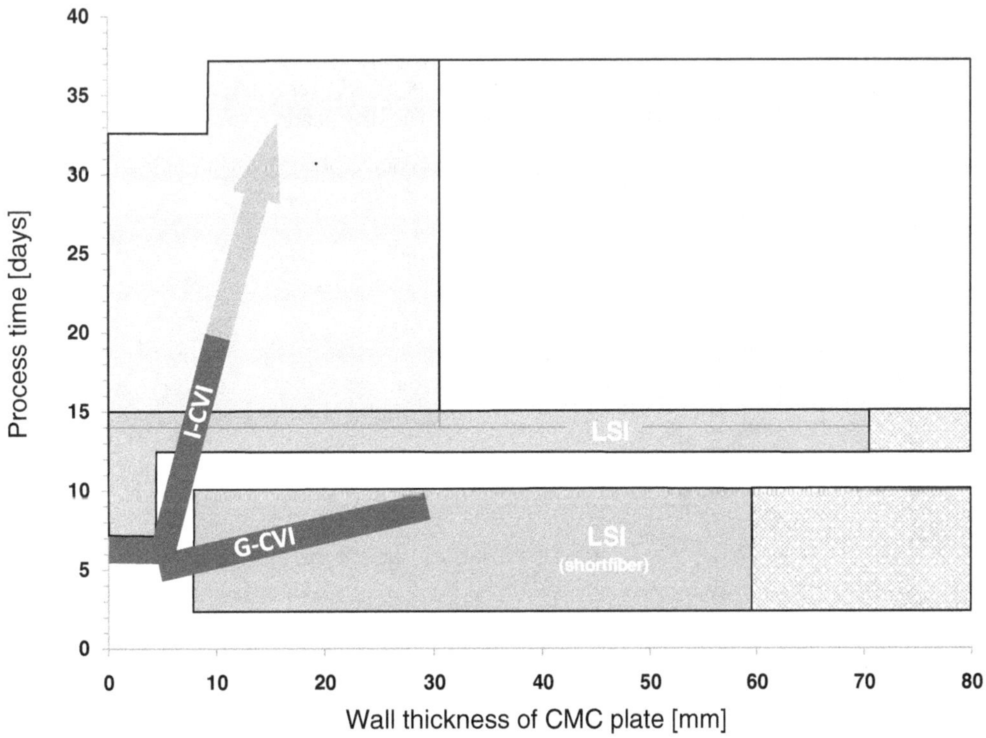

FIGURE 6.27

Typical process times for the manufacture of 2D fabric-based C/SiC plates (about 300 mm × 300 mm) dependent on the manufacturing method and wall thickness (DLR). The diagram is based on net process times, excluding dead times between the process steps and final machining. Maximum wall thicknesses manufactured to date are shown in dark color, whereas light color areas indicate prognosticated feasibilities.

cycles lead to generally longer process times for PIP, depending on the used precursor system. Very short process times can be obtained by randomly oriented fiber architectures, for example, based on short fibers, even for complex-shaped components and high wall thicknesses above 30 mm. In I-CVI, wall thicknesses above 6 mm generally tend to an inhomogeneous SiC matrix buildup with a gradient from the outer surface to the center, and therefore require multiple infiltration steps and intermediate surface machining. Using thermal and pressure gradient G-CVI, cycle times for thick-walled components can be reduced significantly to days and weeks, due to the densification in a single infiltration cycle.

Using CVI, a high temperature stable, stoichiometric and dense SiC matrix is achieved. PIP-derived matrices are reported to be less stable due to their characteristic inner porosity. SiSiC matrices, typically for LSI-derived materials, are critical in long-term applications and service temperatures above 1400°C, due to the resumption of diffusion reactions and the melting of free Si.

Production cost of a C/SiC component is mainly determined by the process and investment costs, which are strongly dependent on both the manufacturing method and the overall quantity and lot size of the component to be produced [58, 59]. Another important aspect is the machining, which can strongly influence the overall costs, especially for components requiring extensive machining in the final ceramic stage. In practice, the selection of the manufacturing method itself is not only determined by the costs but by the components geometry, the mechanical and thermal properties, required in each specific application and the available manufacturing technology.

6.3 PROPERTIES

6.3.1 General Properties

Among all available engineering materials, C and SiC fiber-reinforced ceramics offer the highest specific strength (Figure 6.28) at temperatures above 900°C and therefore C/SiC are the most promising materials for lightweight structures and components submitted to high thermal and mechanical loads.

Ceramics usually are well known for their high thermal and chemical stability, hardness, and abrasive wear resistance. Their main drawback is their brittleness and low damage tolerance. In bulk ceramics, cracks induced by local overloads cannot be stopped, but propagate rapidly through the entire structure, leading to catastrophic failure. Additionally,

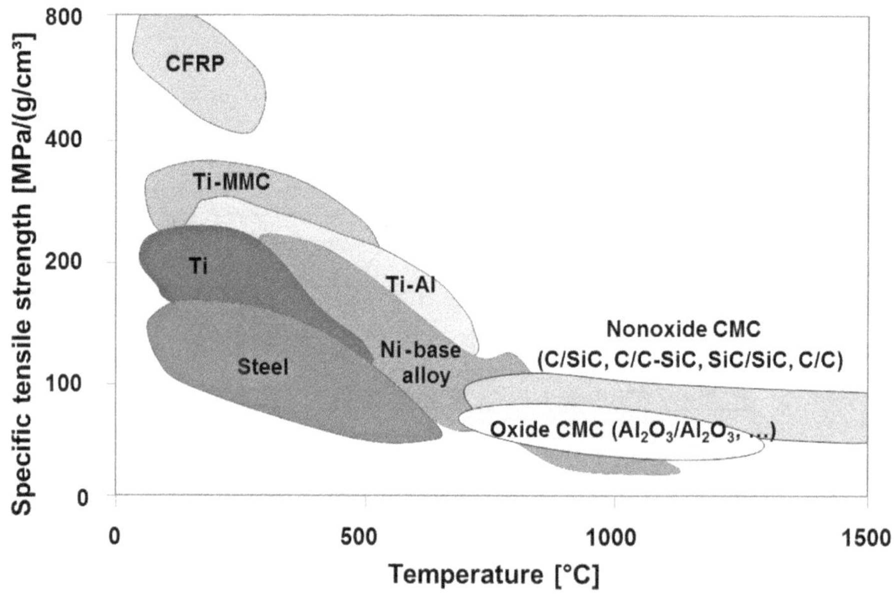

FIGURE 6.28

Weight-specific strength of materials as a function of temperature (DLR).

cracks are starting at material imperfections such as pores, impurities, microcracks, or surface flaws. The strength of bulk ceramics is determined by the quantity, size, and distribution of these imperfections, which are governed by Weibull statistics. Therefore, no universal strength values can be used for the design of monolithic ceramics. In contrast to the deterministic design of metal components, a probabilistic approach based on the principle of the weakest bond is mandatory. This leads to a failure probability, determined by a Weibull failure hypothesis. Thus large structures generally require high wall thicknesses in order to obtain a low stress level and high survivability and reliability, leading to heavy components.

The most promising method to improve the reliability and fracture behavior of ceramics and to increase the acceptable stress level of structural components is the embedment of continuous fibers. First studies in the late 1970s showed that the integration of C or SiC fibers into ceramic matrices led to a new class of ceramic materials with unique properties. These CMCs combine the outstanding properties of ceramics with novel qualities, extremely unusual for bulk ceramics:

- quasi-ductile fracture behavior;
- high fracture toughness;
- low Young's modulus;
- very low CTE;
- extreme thermal shock stability;
- low density.

In CFRP materials, the polymer matrix is reinforced by the fibers resulting in significantly higher strength and stiffness of the composite material compared to the bulk polymer. In contrast, in C/SiC materials, the integration of C fibers usually leads to only a slight increase of strength, and in many cases even to decreased strength, especially for MI materials, and generally results in a reduction of Young's modulus, compared to the bulk SiC ceramic. However, the integration of fibers is significantly increasing toughness and reliability by energy dissipating mechanisms such as crack deflection, crack splitting, and crack stopping at the fiber–matrix interfaces and at microcracks, as well as by crack bridging (Figure 6.29) of the fibers [60, 61]. To achieve this, a weak embedding of the fibers in the matrix is essential. In the case of a strong FMB, crack bridging cannot be realized due to a very low effective fiber length. Thereby, a matrix crack, arriving at the fiber, induces a theoretically infinite strain rate in the fiber, resulting in fiber rupture and brittle fracture behavior. If the FMB is weak, the fiber is able to debond from the matrix, increasing its effective length and decreasing the strain rate of the fiber. Thus advancing cracks can be bridged by the fibers, leading to a less brittle and damage tolerant material behavior (Figures 6.30 and 6.31) and the CMC material can capitalize on the high tensile strength ($\sigma_t > 3500$ MPa) and elongation at failure ($\varepsilon > 1.5\%$) of the fibers, which are significantly higher than that of bulk ceramics (SiC: $\sigma_t = 400$ MPa, $\varepsilon < 0.05\%$).

As a result, strength values of C/SiC structures are mainly determined by fiber content and fiber orientation and are usually independent from size or volume. This leads to thin-walled, lightweight structures, which can be dimensioned using a deterministic approach, similar to polymer-based composites.

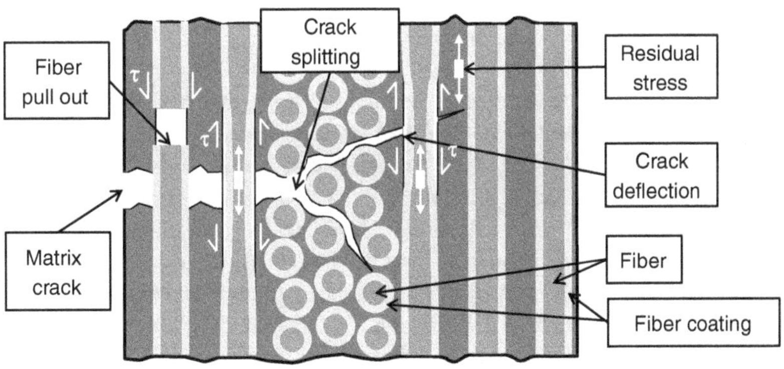

FIGURE 6.29
Schematic illustration of energy absorbing phenomena in CMC (DLR) [61].

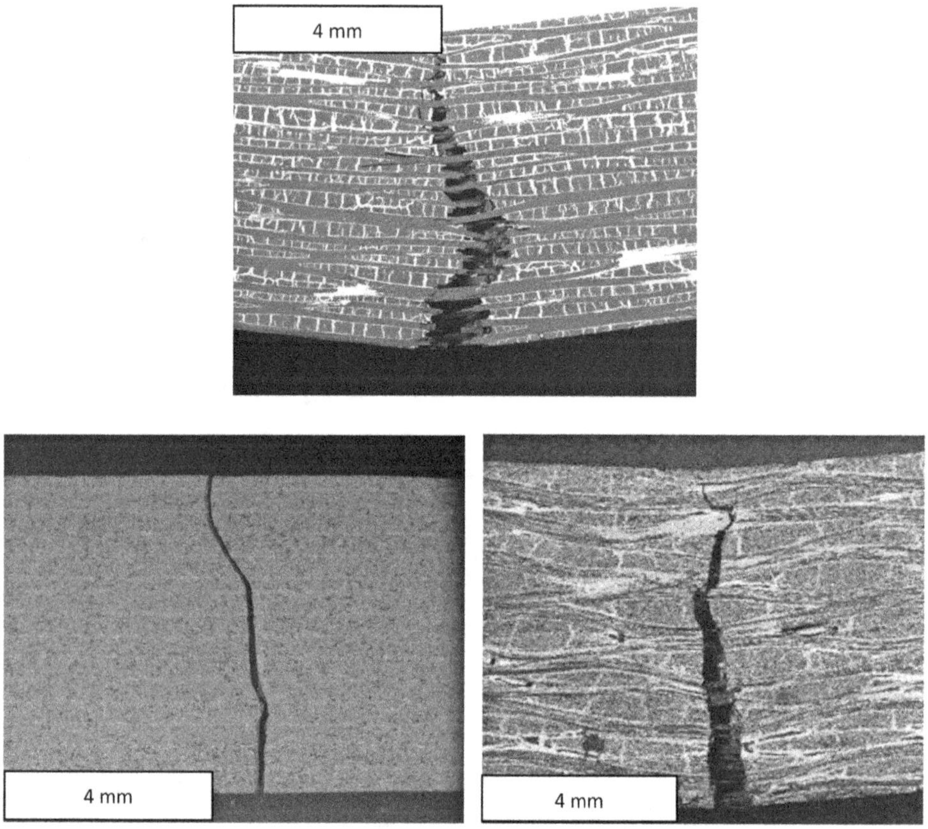

FIGURE 6.30

SEM of bending samples, showing the typical, quasi-ductile fracture behavior of C/C-SiC XB (top), characterized by crack deflection, fiber bridging and pull out of fibers and C/C bundles. Catastrophic failure of bulk SiSiC (bottom left) and C/C-SiC XD with high SiC content and almost all the C fibers converted to SiC (bottom right, all figures DLR).

6.3.2 Material Composition and Microstructure

C/SiC and C/C-SiC materials are multiphase materials consisting of C fibers, C fiber coating (in most cases), and β-SiC matrix and, in the case of LSI derived materials, additional C matrix as well as some residual Si. Using CVI, generally a stoichiometric SiC matrix can be obtained, whereas in the PIP process, the matrix composition depends on the nature of the preceramic precursor used. Moreover, the CVI process is capable of diversified matrix composition in the C/SiC/Si system.

Material composition and microstructure are mainly determined by the manufacturing method as well as by the used fiber preform and its architecture. CVI and PIP generally lead to so-called C/SiC materials, characterized by single C fiber filaments embedded in SiC matrix with a weak C interphase based on pyC fiber coatings, whereas in LSI-derived C/C-SiC materials, the carbon fibers primarily are embedded in C matrix, leading to dense C/C bundles, which again are embedded in SiC or SiSiC matrix (Figure 6.32).

Generally, C/SiC materials obtained by CVI and PIP offer a moderate overall C content, that is, C fiber and pyC fiber coating, and high SiC contents ($\phi_C \approx 42$ to 58 vol.%, $\phi_{SiC} \approx 35$ to– 50 vol.%), depending on the fiber content, usually in the range of 40–50 vol.% and up to 55 vol.%. In contrast, C/C-SiC materials based on LSI and high fiber contents are characterized by a high C content (C fiber and C matrix) of up to about 75 vol.% and a significantly lower SiC content ($\phi_{SiC} \approx 15$ to 20 vol.%) as well as some residual Si ($\phi_{Si} < 4$ vol.%). This characteristic difference of CVI/PIP- and LSI-derived materials is caused by the way the SiC matrix is built up, and the interphase between C fibers and SiC matrix is obtained. In CVI/PIP, the weak fiber interface is achieved by a thin pyC fiber coating, increasing the overall C content of the final C/SiC by only a small amount of $\phi_{pyC} \approx 2$ to 10 vol.%, and the SiC matrix is built up independently with no protection of the fibers needed. In the MI processes, a C matrix is built up in a first step in order to embed and protect the fibers, and subsequently, the C matrix is only partly converted to SiC matrix, resulting in a large amount of

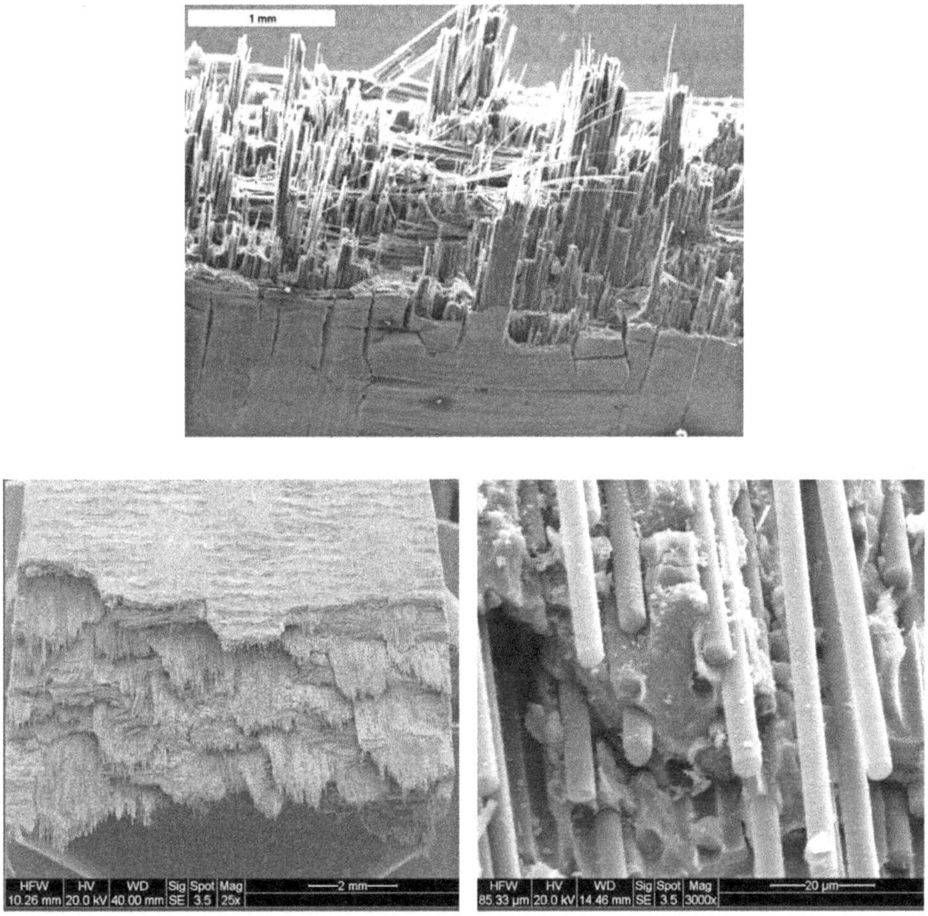

FIGURE 6.31

SEM figures of typical fracture surfaces of MI-based C/C-SiC XB, showing pull out of C/C bundles (top, DLR) and PIP-based C/SiC materials (bottom left, COIC) with single fiber pull out (bottom right, COIC).

residual C matrix in the final C/C-SiC or C/SiC material ($\phi_{C\ matrix} \approx$ 15 to 30 vol.%). However, the MI processes offer the possibility to adjust the material composition in a wide range. For commonly used 2D fabric-based C/SiC materials, the SiC content is varied from about 3 to 10 vol.%, using PCI or PMI processes and highly densified C/C preforms, up to a maximum of about 50 vol.% for LSI materials based on either highly porous C/C preforms or on a high conversion rate of the C fibers to SiC (Figure 6.33). For short fiber-reinforced materials (Figure 6.34), low overall C contents of about 43 vol.% and maximum contents of SiC ($\approx$60 wt.%, equivalent to $\approx$44 vol.%) and Si ($\approx$10 wt.%, equivalent to $\approx$11 vol.%) can be achieved by, again, using highly porous C/C preforms and by converting a defined amount of the C/C bundles to SiC [62].

In CVI-derived C/SiC materials, the residual open porosity is generally 8–12% up to 15%, as a result from the competition between surface deposition and in-depth infiltration mechanisms. Due to economic reasons, the matrix buildup in PIP is stopped at a residual open porosity of 6–10%, whereas the MI processes lead to relatively dense C/SiC and C/C-SiC materials with open porosities of 1–4% in one process cycle. In Figure 6.35, an overview of the typical phase compositions of C/SiC materials is given.

C/SiC materials are lightweight materials, with a typical density of 1.8–2.4 g/cm^3, slightly higher than that of CFRP ($\rho = 1.5$ g/cm^3) and magnesium ($\rho = 1.74$ g/cm^3) but about 10–30% lower than aluminum ($\rho = 2.7$ g/cm^3). Due to the usually high C fiber content of about 60 vol.% and the residual C matrix, 2D-reinforced LSI materials offer low densities of about 1.9 g/cm^3, despite their generally low open porosity. In comparison, C/SiC materials manufactured via CVI are limited to fiber volume contents of about 50 vol.% at most, enabling high SiC contents and typical densities in the range of 1.9–2.2 g/cm^3. Highest densities of up to 2.4 g/cm^3 can be obtained with short fiber-based LSI materials by increasing the conversion rate, leading to high contents of SiC and Si matrix.

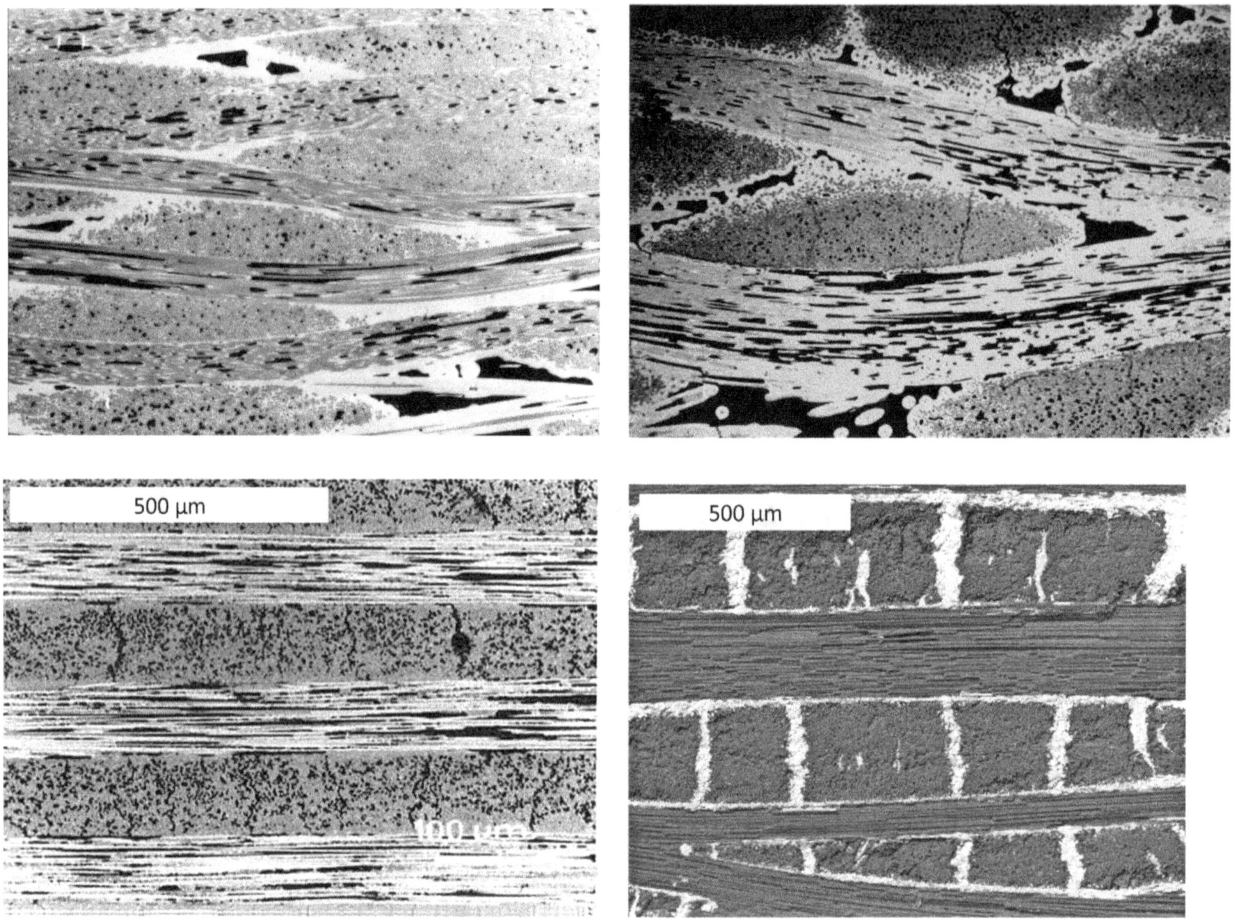

FIGURE 6.32

Typical microstructures of C fiber-reinforced SiC based on 2D fiber preforms. C/SiC materials manufactured via CVI (Keraman®, top left, MT; Sepcarbinox®, top right, Herakles) with C fiber filaments (dark gray) embedded in the SiC matrix (light gray). Porosity is shown in black. C/SiC material based on PIP and 0°/90° cross-ply laminate (SICARBON®, bottom left, Astrium ST/EADS IW), showing single C fibers (dark gray) embedded in SiC matrix (light gray). LSI-based C/C-SiC XB (bottom right, DLR) showing dense C/C bundles (dark gray) embedded in SiSiC matrix (light gray).

6.3.3 Mechanical Properties

Mechanical and thermal properties of C/SiC materials are strongly influenced by the properties of the basic materials (Table 6.2), the composite's phase composition and microstructure. However, single properties of the composite, such as strength, thermal conductivity or CTE, are controlled by either the fiber or the matrix properties and therefore cannot be calculated by applying a simple rule of mixtures. Depending on the fiber architecture, the composite materials are highly anisotropic, especially if laminated 2D fabrics or 1D fiber layers are used, leading to different properties in the in-plane directions, in parallel to the fibers, and the direction perpendicular to the fiber layers.

The mechanical properties of C fiber-reinforced SiC materials are dominated by the C fibers and therefore are strongly influenced by fiber type, fiber orientation, and fiber volume content as well as by the way the fibers are integrated in the composite material, which again is mainly determined by the manufacturing method.

In Table 6.3 an overview is given of the mechanical and thermal properties of C/SiC and C/C-SiC materials with 2D and 3D fiber architectures, based on woven fabrics or UD cross-ply laminates.

Although the values cannot be compared directly, due to different or unknown evaluation methods used, some characteristic properties of CVI, PIP, and LSI materials become apparent. Highest mechanical properties with tensile strengths and elongations at failure up to $\sigma_t = 340$ MPa and $\varepsilon = 0.7\%$, respectively, can be obtained via CVI or PIP. The mechanical strength of LSI materials based on *in situ* fiber embedding and dense C/C bundles is usually lower than that of CVI- and PIP-based materials, despite the higher fiber content in the CFRP and C/C preforms used for LSI ($\phi_{f,\ CFRP\ preform} = 50$–$60$ vol.%, $\phi_{f,\ CVI,\ PIP} = 45$–$50$ vol.%). This can be explained by the homogeneous pyC fiber coating

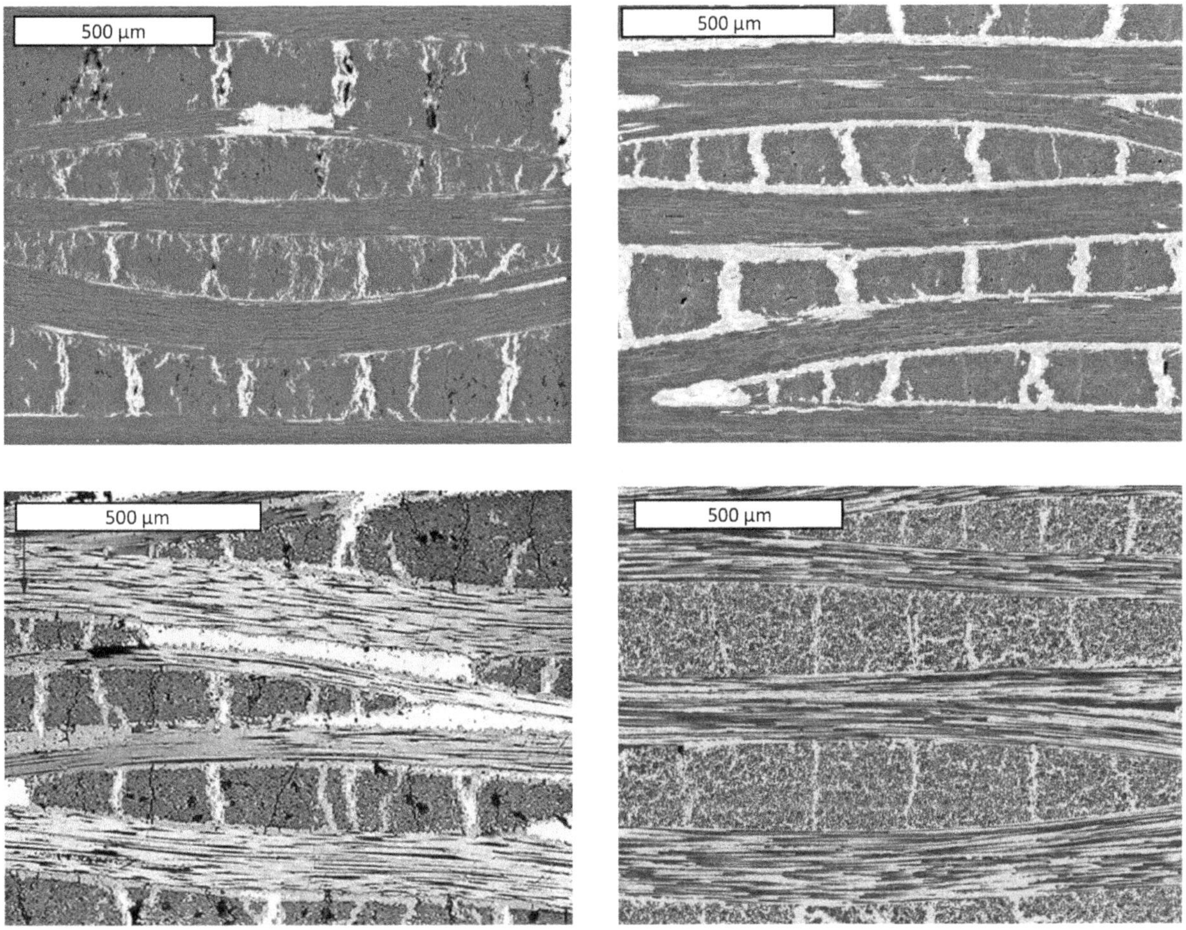

FIGURE 6.33

Typical microstructures of 2D fabric-based C/C-SiC and C/SiC materials based on Si melt infiltration, showing different phase compositions, achieved by different manufacturing methods. C/C-SiC material CF222 P75 manufactured via PCI of a highly densified C/C preform, showing low SiC content of about 5–10 vol.% (top left, SKT). LSI-based C/C-SiC XB with $\phi_{SiSiC} = 20$ vol.% (top right, DLR). C/SiC material (SIGRASIC® 1500 J) based on PIP fiber coating and LSI with increased SiSiC content of 52 wt.% (about 37 vol.%, bottom left, SGL) [63]. LSI-based C/C-SiC XD characterized by low fiber–matrix bonding, showing a high conversion rate of the C fibers and a high SiC content of $\phi_{SiC} = 40$ vol.% (bottom right, DLR).

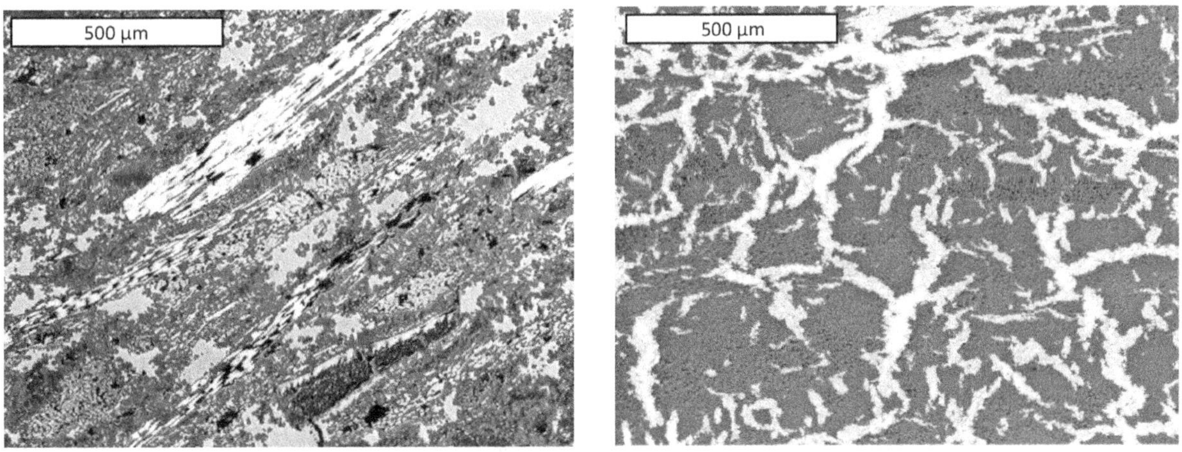

FIGURE 6.34

Microstructures of LSI materials based on random-oriented, short C fibers, showing dense C/C bundles (black) embedded in SiC matrix (dark gay) with residual Si (light gray). SIGRASIC® 6010 GNJ (left, SGL) based on PIP fiber coating, showing high amount of SiC and Si matrix ($\phi_{SiC} \approx 44$ vol.%, $\phi_{Si} \approx 11$ vol.%, left according to [62]) and C/C-SiC SF (right, DLR) based on high fiber content and *in situ* fiber embedding, showing low SiSiC content ($\phi_{SiC+Si} \approx 27$ vol.%).

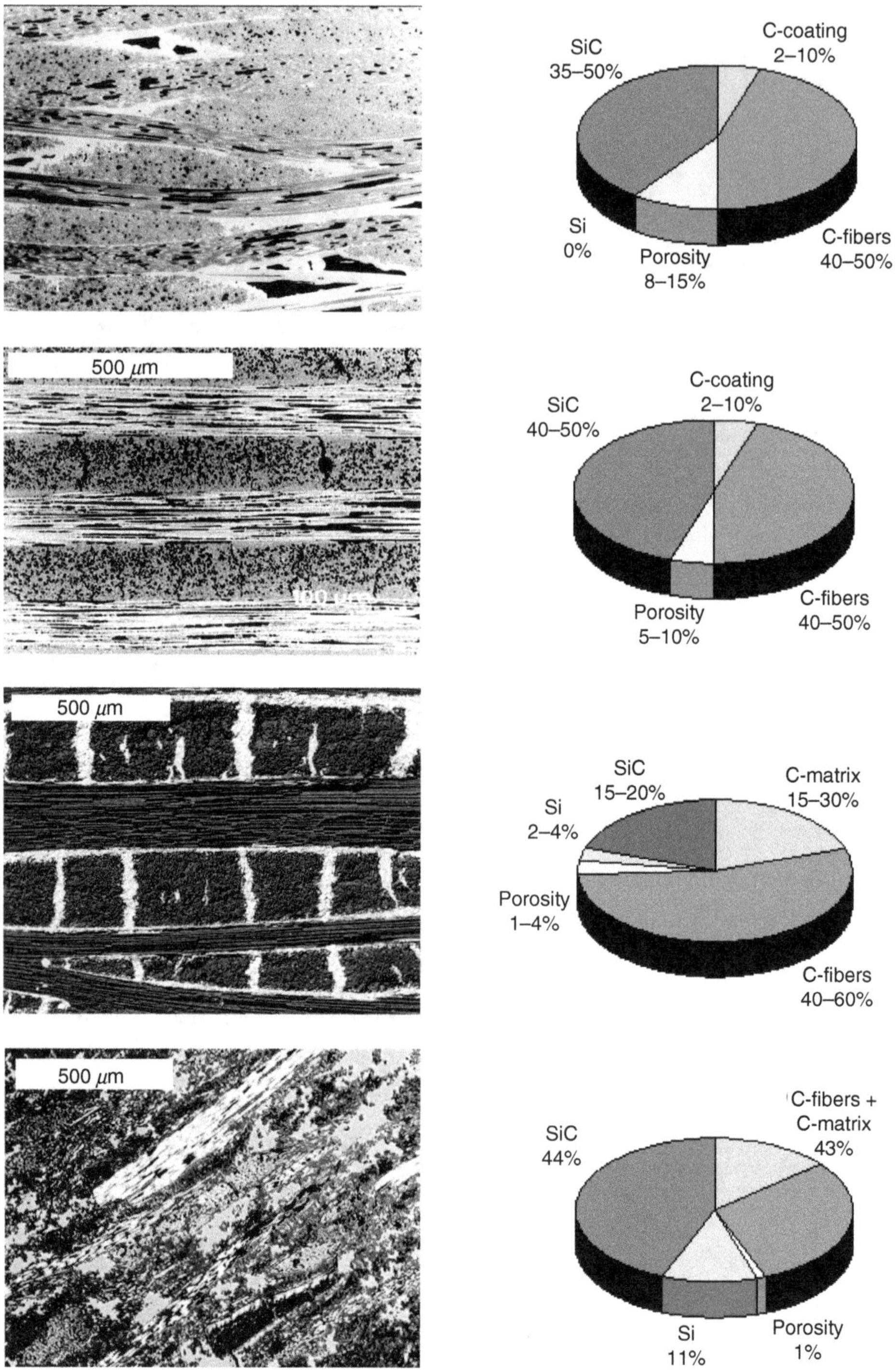

FIGURE 6.35

Typical microstructures and phase compositions (in vol.%) of C/SiC materials based on different manufacturing methods. From top to bottom: CVI-derived C/SiC materials based on 2D fiber fabrics (Keraman®, MT), PIP-derived SICARBON® (Astrium ST/EADS IW) based on UD cross-ply laminate, LSI-derived C/C-SiC based on 2D fiber fabrics (C/C-SiC XB, DLR) and typical LSI-derived C/SiC (SIGRASIC® 6010 GNJ, SGL) based on randomly oriented, short fibers.

TABLE 6.2 Typical Properties of C Fibers, Sintered and Si-Infiltrated SiC Materials and Silicon [64–68]

		C Fibers						
		HT	IM	HM	UHM	SiC (CVD)[a]	SiSiC[b]	Si
		HTA[c]	T 800[d]	M40J[d]	YS-90-A[e]	CVD Silicon Carbide™	CarSIK-NT	
Density	g/cm³	1.76	1.81	1.77	2.18	3.21	3,09	2.33/2.53[f]
Tensile strength	MPa	3950	5490	4410	3530	–	–	113
Strain to failure	%	1.7	3.5	1.2	0.3	–	–	–
Young's modulus	GPa	238	294	377	880	461	360	112
Compression strength	MPa	–	–	–	–	–	3000	–
Flexural strength	MPa	–	–	–	–	415/575[g]	280	300
CTE	10^{-6} K^{-1}	−0.1	−0.56	−0.83	−1.5	2.2/4.0[h]	3,9[i]	2.6
Thermal conductivity	W/(m K)	10	35	69	500	300/63[j]	120	156
Specific heat capacity	J/kg K	710	753	711	–	640/1295[j]	–	627
T_{melt}/T_{max}	°C	–	–	–	–	2700	1380	1414

All properties are at room temperature (RT) unless otherwise noted.
[a]The Dow Chemical Company.
[b]Schunk Ingenieur Keramik GmbH.
[c]TohoTenax.
[d]Toray.
[e]Nippon Graphite Fiber Corporation.
[f]Solid/melt.
[g]RT/1400°C.
[h]RT–1000°C.
[i]20–400°C.
[j]RT/1200°C.

and the individually embedded fibers in the SiC matrix that enhance the fiber reinforcement effect in CVI and PIP materials compared to the embedding of the C/C bundles in the LSI material. Additionally the fibers are not influenced or damaged at all during the matrix buildup in CVI and PIP, whereas in LSI a certain amount of fibers can be converted to SiC. This leads to a decrease of the load carrying fibers and the effective fiber content in the final C/C-SiC material, resulting in lower strength and failure strain values of $\sigma_t = 80$–190 MPa and $\varepsilon = 0.15$–0.35%, depending on the used C fiber type. However, due to the low open porosity of LSI-based materials, the interlaminar shear strength of C/C-SiC is at a similar level to CVI materials but is usually higher than that of PIP materials.

Fiber-reinforced materials generally offer higher fatigue strengths than unreinforced materials. Carbon-fiber-reinforced SiC materials can survive cyclic dynamic loads without a reduction of strength. Cyclic tension tests of CVI-based C/SiC with a maximum load of 150 MPa, which is one-third of the initial fracture strength, showed identical fracture strength even after 10^6 load cycles (Figure 6.36).

In MI materials, the content of load carrying C fibers and C/C bundles can be influenced by varying the fiber volume content in the CFRP preform or by densifying the C/C preform via repeated cycles of polymer infiltration and pyrolysis, like it is performed in the PCI and PMI process variations. The latter leads to low open porosities in the C/C preform resulting in a low Si uptake and a low conversion rate of the C fibers to SiC. The resulting C/C-SiC materials generally show higher strength compared to materials with high Si uptake and fiber conversion (Table 6.4).

In short fiber-based materials, material composition and texture can be varied over a wider range than in fabric-based materials. Table 6.5 gives an overview of different standard materials manufactured via LSI, which is currently the most common method used for the industrial manufacture of short fiber-based C/SiC and C/C-SiC materials.

Due to the lack of relevant testing standards for this novel class of materials, the values shown are determined by different test methods and cannot be compared directly. Furthermore, they are restricted to sample plates and the applicability of the given values for the design of components has to be checked thoroughly.

Strength values of short fiber-based C/SiC, determined by flexural or tensile testing of coupons, show a high scatter mainly due to inhomogeneous fiber orientation in the coupons [69, 70]. Depending on the material composition and manufacturing method, the sample geometry can influence the results for strength and Young's modulus. Particularly for the tensile strength of LSI-based C/C-SiC, an increase by a factor of 2.7 was observed, caused by an increase of the coupon cross section from 30 to 200 mm² [71]. Short fiber-based C/SiC is notch insensitive, which is explained by

TABLE 6.3 Properties of Industrially Produced C/SiC Materials Based on 2D and 3D Fiber Architectures [3, 4, 29, 32, 35, 63, 72]

		CVI C/SiC			PIP C/SiC			MI C/C-SiC	MI C/SiC
Manufacturer		Herakles	Herakles	MT	EADS	COIC	MSP	SKT	SGL
Manufacture method		I-CVI	I-CVI	G-CVI	PIP	PIP	PIP	LSI	LSI
Trade name		SEPCARBINOX® A262	Keraman® A500	SICARBON®	S 400	DMC 09/09	CF 226/1	SIGRASIC P77	1500J®
Fiber architecture		3D fabrics	3D fabrics	2D fabrics	UD crossply	2D fabrics	2D fabrics	2D fabrics	2D fabrics
Fiber content (vol.%)		45	40	42–47	50[a]	40[b]	≥55	60	–
Density (g/cm³)		2.1	1.9–2.1	2.1–2.2	1.8	2.1	1.7–1.95	>1.8	2
Porosity (%)		10	12–14	10–15	8	–	2.5–5	–	2
Tensile strength (MPa)		350[d]	230	300–320	260	290	–	–	110
Strain to failure (%)		0.9[d]	0.8	0.6–0.9	0.5	0.53	1–1.5	0.25–0.3	0.3
Young's modulus (GPa)		90/100[d]	65	90–100	80–90	110	–	50–55[c]	65
Compressive strength (MPa)		580/600[d]	–	450–550	590	–	61–85	–	470
Flexural strength (MPa)		500/700[d]	–	450–500	500	241	290–370	140–160	190
ILSS (MPa)		35[d]	–	45–48	11–20	28	52	15–17	–
CTE (10^{-6} K^{-1})	‖	3[d]	2.5	3	2[e]	3.5[f]	0.4[g]	1.0–1.5[g]	−0.3/1.36[h]
	⊥	5[d]	–	5	5[e]	–	3.9[g]	6.0–6.5[g]	–
Thermal conductivity W/(m K)	‖	14.3/20.6[d]	10	14	14[e]	–	11–13	28–33	–
	⊥	6.5/5.9[d]	3	7	6[e]	–	6.5	12–15	23/12[i]
Specific heat capacity kJ/(kg K)		0.62/1.4[d]	–	–	0.6[e]	–	1.4–2.2[d]	–	–

All properties are at room temperature unless otherwise noted.
‖ and ⊥ are fiber orientations.
[a]Toray T 800 [65].
[b]Hexcel HexTow® IM7.
[c]Flexural modulus.
[d]RT/1000°C.
[e]RT–1500°C.
[f]20–1200°C.
[g]25–800°C.
[h]25/1200°C.
[i]20/1200°C.

inelastic mechanisms such as initiation or extension of matrix microcracks and local load sharing mechanisms through load bridging by the fibers [73]. Fatigue effects are low as long as the applied load is below the inelastic region [70].

The mechanical strength of short fiber-reinforced C/SiC or C/C-SiC materials is strongly influenced by fiber orientation, fiber content, and fiber length. Generally, the longer the fibers and the higher the fiber content the higher the strength (Figures 6.37 and 6.38). Fiber length of about 6 mm, leads to low tensile strength of $\sigma_t = 20$–30 MPa but high failure strains of $\varepsilon = 0.3\%$ in a typical C/SiC material used for the manufacture of brake discs [62, 74]. In C/C-SiC

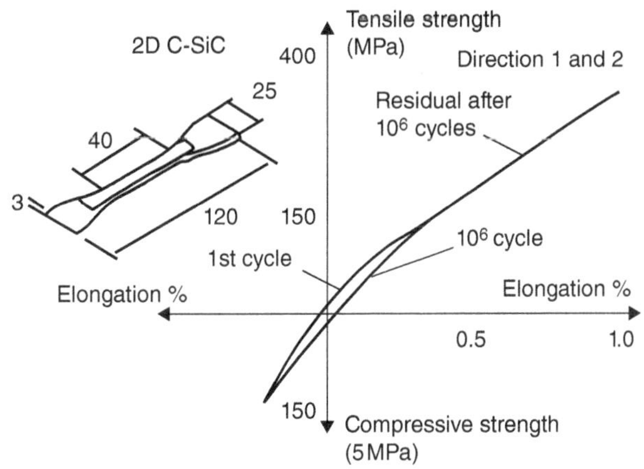

FIGURE 6.36

Stress/strain behavior of 2D-reinforced C/SiC materials after dynamic loading (±150 MPa, 10^6 cycles) in tension test at ambient temperature (Herakles) [32].

TABLE 6.4 Typical Properties of MI-Derived C/C-SiC and C/SiC Materials Based on 2D Fabrics [3, 4, 52, 63, 75]

		C/C-SiC XB	C/C-SiC XT	C/C-SiC XD	SIGRASIC 1500J®	CF 226 P75	CF 226 P76	CF 226/2 P77
Manufacturer		DLR	DLR	DLR	SGL	SKT	SKT	SKT
Manufacture method		LSI	LSI	LSI	LSI	PCI	PMI	LSI
Fiber content[a] (vol.%)		60	60	60	–	60	60	60
Fiber type		HT[b]	T 800[c]	HT[b]	HT	HT	HT	HT
Si uptake (wt.%)		35–40	35–40	50–60	–	10–20	>20	>45
SiC content (vol.%)		15–20	15–20	35–40	20–25[d]	3–10	5–10	–
Si content (vol.%)		2–4	2–4	1–2	15–17[d]	0	0	–
C content (vol.%)		70–75	70–75	60–65	50–60[d]	85–95	85–93	–
Density (g/cm^3)		1.9–2.0	1.9–2.0	2.2–2.3	1.8–2.2	>1.8	>1.8	>1.95
Porosity (%)		1–4	1–4	1	2	<6	<5	<5
Tensile strength (MPa)		80	190	30	110	–	–	–
Strain to failure (%)		0.15	0.35	0.04	0.3	0.3–0.35	0.25–0.30	0.23–0.27
Young's modulus (GPa)		60	60	100	65	55–60[e]	65–70[e]	55–60[e]
Flexural strength (MPa)		160	300	80	190	180–200	220–240	130–140
ILSS (MPa)		28	33	–	–	14–18	17–20	14–17
CTE (10^{-6} K^{-1})	∥	−1/2.4[f]	−1/2.0[f]	1.1/4.4[f]	−0.3/1.36[g]	1.0–1.5[h]	1.0–1.5[h]	0.8–1.3[h]
	⊥	2.1/7[f]	2.5/7[f]	4/8[i]		5.5–6.0[h]	5.5–6.0[h]	5.5–6.0[h]
Thermal conductivity	∥	17/17[j]	21/21[j]	33.6/23.3[j]	–	30–35	30–35	30–35
W/(m K)	⊥	9/7.5[j]	10.3/8.9[j]	18.1/12.6[j]	23/12[g]	12–15	12–15	18–22
Specific heat capacity J/(kg K)		750/1550[k]	690/1540[k]	720/1450[k]	–	–	–	–

All properties at room temperature unless otherwise noted.
∥ and ⊥ are fiber orientations.
[a] In CFRP preform.
[b] Tenax HTA [64].
[c] Torayca T 800 [65].
[d] Calculated from mass contents.
[e] Flexural Modulus.
[f] 100/1400°C.
[g] 25/1200°C.
[h] 25–800°C.
[i] 100/1000°C.
[j] 200/1600°C.
[k] 20/1400°C.

TABLE 6.5 Properties of LSI-Derived C/SiC Materials Based on Short Fibers Compared to a C/C-SiC Material Based on 2D Fabrics [4,52,62,69,74,75]

		C/C-SiC XB	C/C-SiC SF	Sigrasic 6010 GNJ	FU 2952 P77
Manufacturer		DLR	DLR	SGL	SKT
Fiber architecture		2D fabric	Random short fiber	Random short fiber	Random short fiber
Manufacture method		LSI	LSI	LSI	LSI
Fiber type		HTA	HTA	HT	HT
Fiber content (vol.%)		60[a]	50[a]	–	50
Density (g/cm^3)		1.9–2.0	2.0–2.1	2.45	>2
Open porosity (%)		1–4	3	<1	–
SiC content (vol.%)		15–20	24–28	45[b]	–
Si content (vol.%)		2–4	1–3	11[b]	–
C content (vol.%)		70–75	65–70	43[b]	–
Flexural strength (MPa)		160	–	50–80	60–80
Tensile strength (MPa)		80	–	20–60	
Young's Modulus (GPa)		60	50–70[c]	30	23–27[c]
Strain to failure (%)		0.15	0.15–0.25[c]	0.3	0.2–0.26
Thermal conductivity W/(m K)	‖	17/17[d]	–	–	18–23
	⊥	9.0/7.5[d]	25–30[e]	40–60	28–33
Specific heat capacity J/(kg K)		750	750	800	–
Thermal shock coefficient R_2 (W/m)		9 070	–	>27 000	–
CTE (Ref. 25 °C) (10^{-6} K^{-1})	‖	−1/2.4[f]	0.5/3.5[f]	–	1.2–1.6[g]
	⊥	2.1/7[f]	1.0/4.0[f]	2.6–3.0	5.2–5.6[g]

All properties at room temperature unless otherwise noted.
‖ and ⊥ are fiber orientations.
[a]In CFRP preform.
[b]Calculated from mass contents according to [62].
[c]Flexural test.
[d]200/1600°C.
[e]50°C.
[f]100/1400°C.
[g]25–800°C.

materials, flexural strength is increased with fiber length l_f and fiber volume content from $\sigma_b = 60$ MPa ($l_f = 10$ mm) to $\sigma_b = 120$ MPa ($l_f = 40$ mm) and from $\sigma_b = 70$ MPa ($\phi_f = 30$ vol.%) to $\sigma_b = 140$ MPa ($\phi_f = 55$ vol.%) [69]. Again, a high scatter of the results is observed, caused by varying fiber orientation in the test coupons.

Depending on the manufacturing method, the properties of the C fibers are transferred to the CMC material at a significantly lower level than that known for CFRP materials. The tensile strength of the CMC materials is about 3–10 times lower than that expected from a rough rule of mixtures calculation (Figure 6.39), taking into account the fiber

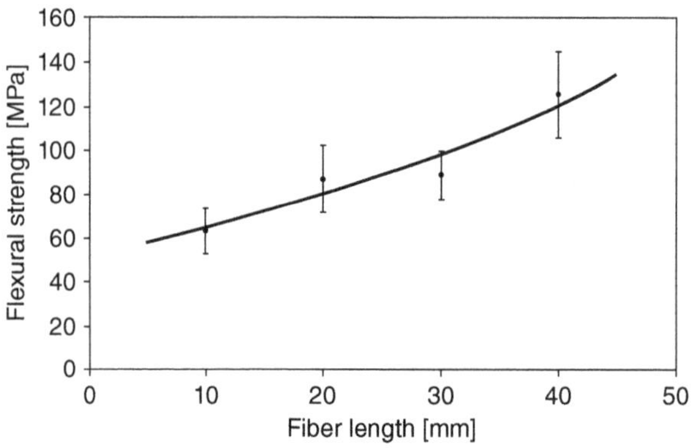

FIGURE 6.37

Flexural strength (3-point bending, DIN-EN 658-3) in dependence of fiber length of short fiber-based C/C-SiC (ϕ_f = 50 vol.%, DLR) [69].

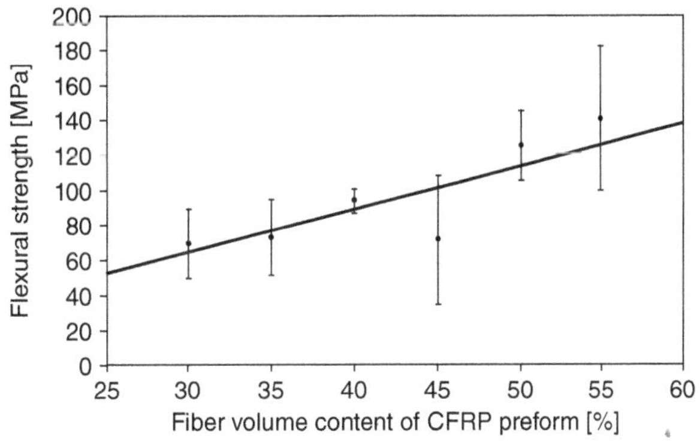

FIGURE 6.38

Influence of the fiber volume content on the flexural strength (3-point bending, DIN-EN 658-3) of short fiber-reinforced C/C-SiC ($l_f = 40$ mm, DLR) [69].

content and the content of load-carrying fibers (50% for 2D fiber architectures). Additionally, LSI-based materials take less advantage of the fiber strength, compared to CVI materials. In contrast to strength, the Young's modulus fits well to the fiber properties, but is about 3–6 times lower compared to bulk SiC. However, strength and Young's modulus of C/SiC materials can be tailored to a particular range by using appropriate fiber types (Figure 6.40).

As is the case for CFRP, fiber orientation strongly influences the mechanical properties of CMC materials. The more the fibers are oriented in the load direction, the higher the strength and stiffness. This becomes apparent in filament wound C/C-SiC materials with varying fiber orientations α relative to the load direction (Figure 6.41) [46,76]. Thereby, tensile strength values decrease from 120 MPa ($\alpha = \pm 15°$) down to 20 MPa ($\alpha = \pm 75°$).

For special applications with mainly unidirectional loads, the properties of 2D fabric-based materials can be increased by adding UD fiber layers in between the 2D fabric layers. Depending on the content of UD fibers, strength and Young's modulus can be increased by 50% and about 90%, respectively, compared to materials based on 2D fabrics

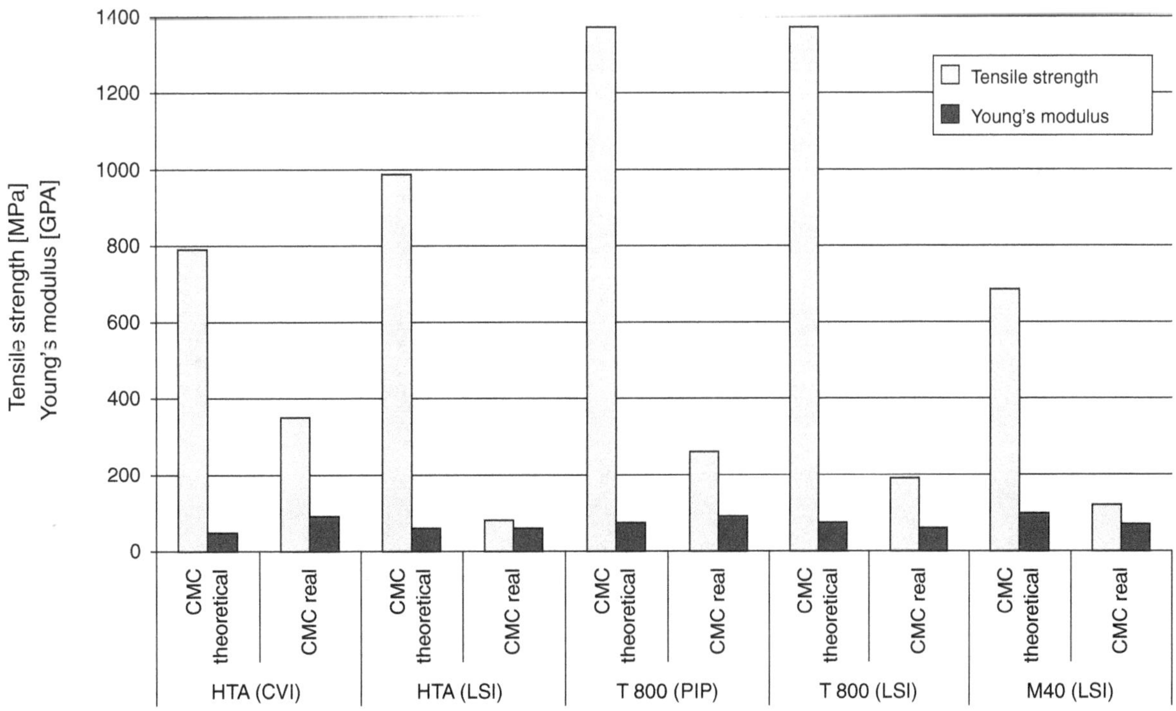

FIGURE 6.39

C/SiC properties calculated on the basis of fiber properties and fiber content in load direction, compared to typically C/SiC properties, obtained by different manufacturing methods (DLR).

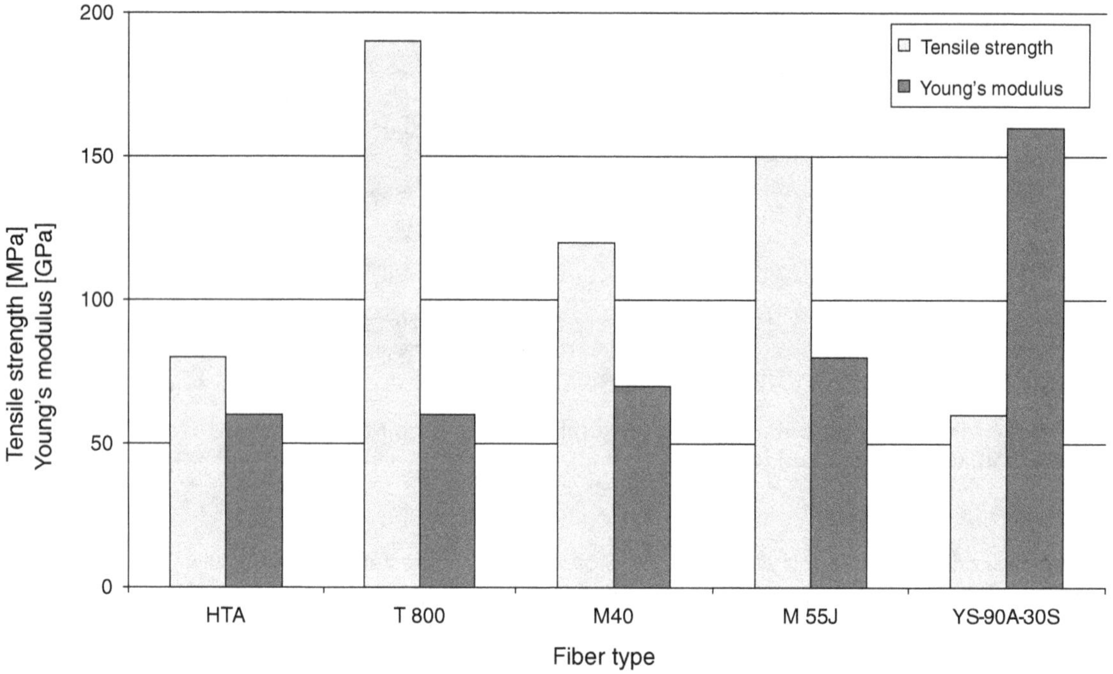

FIGURE 6.40

Influence of the fiber type on the tensile strength and Young's modulus of 2D fabric-based C/C-SiC materials, manufactured via LSI (DLR).

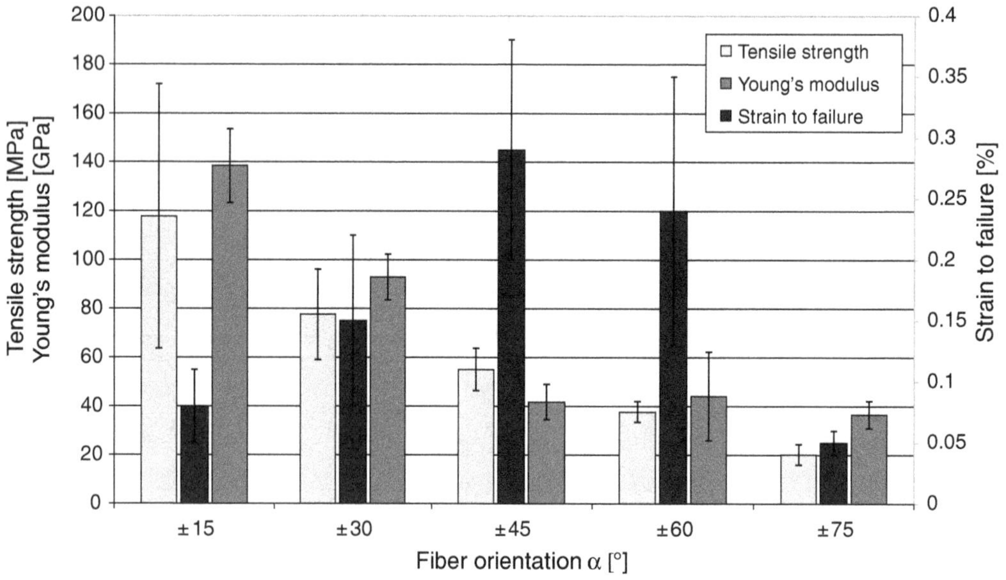

FIGURE 6.41

Influence of the fiber orientation to the tensile strength (DIN-EN 658-1) Young's modulus and strain to failure of C/C-SiC materials based on wet filament wound and warm pressed CFRP preforms (DLR) [76].

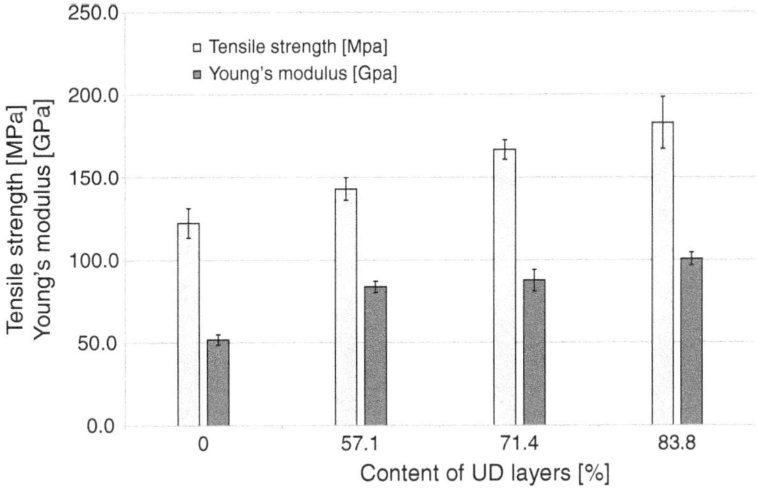

FIGURE 6.42

Tensile strength and Young's modulus (DIN-EN 658-1) of C/C-SiC materials based on 2D fiber fabrics and different contents of UD fiber layers added in between the fabric layers. The fibers of the UD layers are oriented parallel to the load direction (DLR).

only (Figure 6.42). However, as a consequence of the highly anisotropic fiber architecture, the in-plane properties perpendicular to the UD fiber layers are decreased by a similar extent (Figure 6.43). The manufacture of C/C-SiC materials, based entirely on unidirectional oriented fibers is critical, due to inhomogeneous buildup of microcracks and SiC matrix during pyrolysis and siliconization. Using the PIP process, C/SiC materials based on unidirectional (0°) fiber preforms have been developed, again leading to significantly higher tensile strength σ_t, and compression strength σ_c compared to UD cross-ply (0°/90°) laminates ($\sigma_{t,\,0^\circ} = 470$ MPa, $\sigma_{t,\,0^\circ/90^\circ} = 260$ MPa and $\sigma_{c,\,0^\circ} = 1080$ MPa, $\sigma_{c,\,0^\circ/90^\circ} = 590$ MPa) [35].

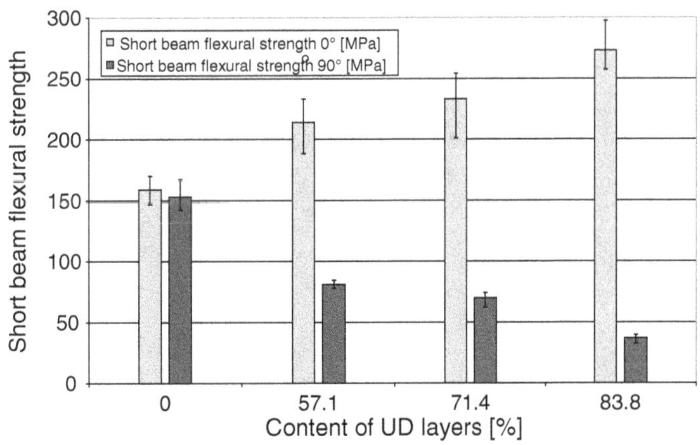

FIGURE 6.43

Short beam flexural strength (DIN-EN 658-3) of C/C-SiC materials based on 2D fiber fabrics and different contents of UD fiber layers added in between the 2D fabric layers. Strength values are determined on samples with the fibers of the UD layers oriented parallel (0°) and perpendicular (90°) to the load direction (DLR).

6.3.4 Thermal Properties

One main advantage of C fiber-reinforced SiC materials is the general stability of the mechanical properties at high temperatures, comparable to C/C. In contrast to metals, where strength values decrease significantly with increasing temperatures, C/SiC materials show either almost no decrease of strength or even a certain increase at higher temperatures up to 1200°C [32, 77]. As a result, hot structures are typically designed on the basis of room temperature properties, leading to a certain safety margin when loaded at high temperatures.

Thermal expansion of C/SiC is mainly determined by the C fiber, offering a very low CTE ($\alpha_{f,\,\parallel,\,RT} = -0.83$ to $-0.1 \times 10^{-6}\,K^{-1}$) compared to the SiC matrix ($\alpha_{SiC,\,RT} = 4\text{–}5 \times 10^{-6}\,K^{-1}$), and therefore is generally low ($\alpha_{C/SiC,\,\parallel,\,RT} = -1$ to $+3.5 \times 10^{-6}\,K^{-1}$). In C/C-SiC materials with 2D fiber architectures, the in-plane CTE can be tailored in the range of

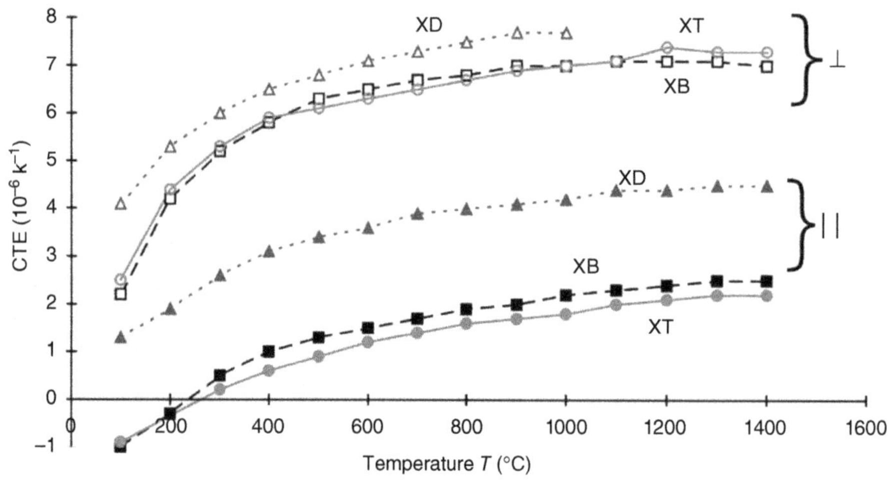

FIGURE 6.44
Typical CTE of 2D-reinforced C/C-SiC materials parallel and perpendicular to the fabric layers (DLR).

$\alpha_{\|,\, RT} = -0.5$ to $+\,0.8 \times 10^{-6}$ K^{-1} by varying the ratio of fiber and matrix content as well as the fiber orientation [78]. Thus, geometrically stable C/C-SiC materials and structures with a CTE of $\alpha_{\|} = 0 \pm 0.1 \times 10^{-6}$ K^{-1} in a temperature range of −30°C to +50°C could be realized [79, 80]. In the perpendicular direction, the SiC and C matrix control properties, leading to a CTE in the range of $\alpha_{\perp,\, RT} = 2$ to 5.5×10^{-6} K^{-1} (Figure 6.44). Short fiber-based C/SiC materials also show an anisotropic CTE with typical values of $\alpha_{\|,RT} = 0.5$ to 1.2×10^{-6} K^{-1} in the in-plane direction (perpendicular to press direction) and $\alpha_{\perp,\, RT} = 1$ to 5.2×10^{-6} K^{-1} in the perpendicular or thickness direction (parallel to press direction). This can be explained by the characteristic fiber orientation, perpendicular to the pressing direction, obtained during warm pressing of the CFRP preform.

Thermal conductivity of C/SiC-based on 2D fiber architectures as well as on short fibers is also anisotropic. For 2D fabric-based C/SiC materials typical values are in the range of $\lambda_{\|}$ = 10–35 W/(m K) and $\lambda_{\perp}$ = 3–12 W/(m K) up to 23 W/(m K) at ambient temperature. In-plane thermal conductivity is mainly determined by the fibers. In LSI-based C/C-SiC materials, it can be increased from $\lambda_{\|,\, 200°C}$ = 18.5 W/(m K) to $\lambda_{\|,\, RT}$ = 42.2 W/(m K) and up to $\lambda_{\|,\, RT}$ = 99 W/(m K) by

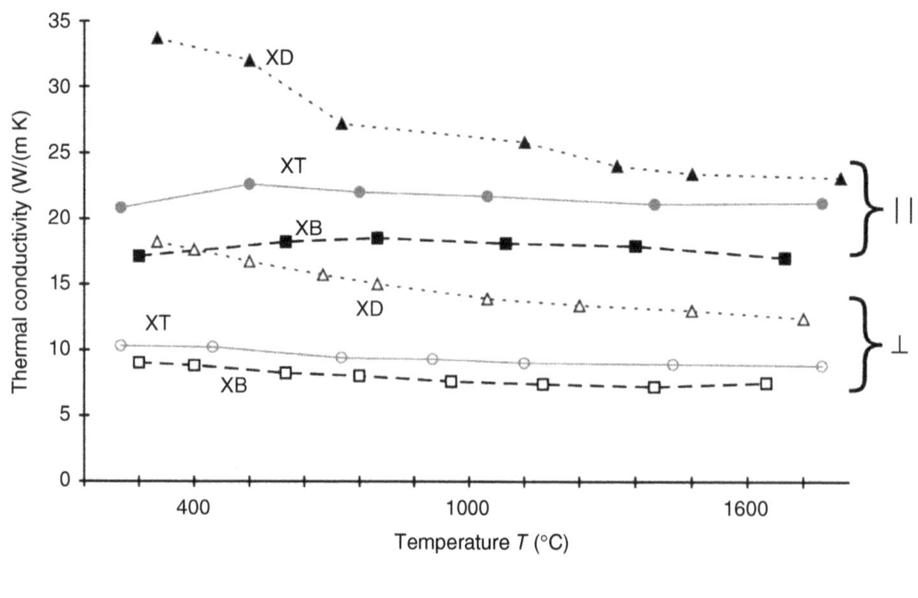

FIGURE 6.45
Typical thermal conductivity of 2D-reinforced C/C-SiC materials (DLR) parallel and perpendicular to the fabric layers.

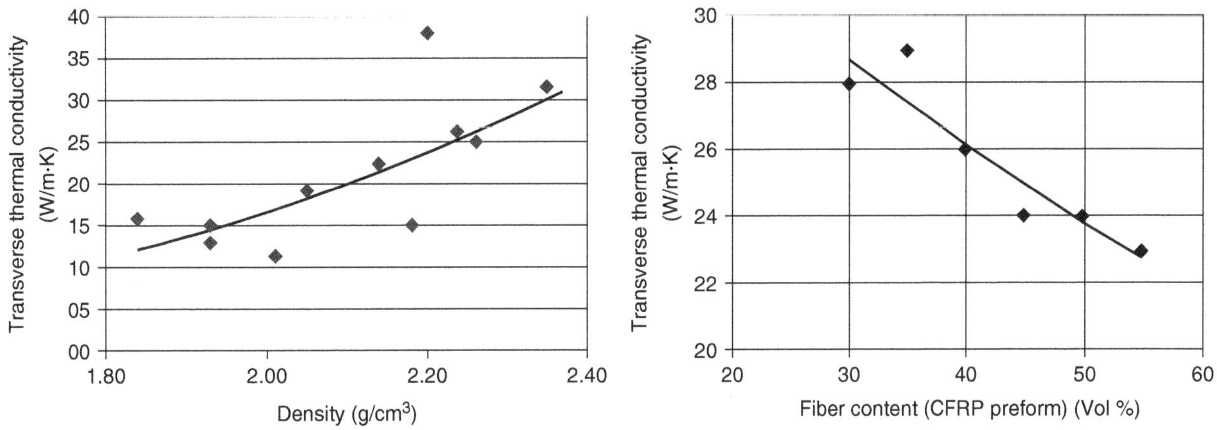

FIGURE 6.46
Thermal conductivity of short fiber-based C/C-SiC materials (DLR) in dependence of density and fiber content.

using highly graphitized C fiber types, such as HM (e.g., Torayca M40: $\lambda_{\parallel, RT} = 80$ W/(m K)) or UHM fibers (e.g., NGF-YS 15: $\lambda_{\parallel, RT} = 590$ W/(m K)) instead of standard HT fibers (e.g., Tenax HTA: $\lambda_{\parallel, RT} = 10$ W/(m K)). In another method, the C/C preform is graphitized by a heat treatment at temperatures up to 2400°C before siliconization, resulting in an increase of thermal conductivity up to $\lambda_{\parallel, RT} = 31$ W/(m K) [81]. This is explained by the graphitization of the C/C bundles and, more importantly, by the increase of the content of SiSiC matrix caused by an additional contraction of the C matrix in the C/C preform, again resulting in an increased porosity of the C/C preform and a higher Si uptake during siliconization. On the other hand, thermal conductivity is generally increased by increasing the ceramic matrix content and decreasing the C fiber content (Figures 6.45 and 6.46), leading to highest values for 2D fabric-based C/C-SiC XD ($\lambda_{\parallel, RT} = 33.7$ W/(m K), $\lambda_{\perp, RT} = 18.2$ W/(m K), $\phi_{SiC} = 40$ vol.%) and up to $\lambda_{\perp, RT} = 40$–60 W/(m K) for short fiber-based C/SiC ($\phi_{SiC + Si} = 56$ vol.%) [62].

The extremely high thermal and thermal shock stability was the main reason for the development of C/SiC materials for use in TPSs of spacecraft and for components in solid propulsion rocket motors. Here, heating rates of several thousand Kelvin per second are applied and have to be withstood. This can be achieved by the very low CTE, which minimizes local material expansion in hot spots or due to temperature gradients. Thereby internal stresses can be avoided, even in large structures. Additionally, Young's modulus is low compared to monolithic ceramics or metals, resulting in high thermal shock coefficients R_1 and R_2, respectively (Table 6.6):

$$R_1 = \frac{R_m}{E \cdot \alpha}\,[K] \qquad R_2 = \frac{R_m \cdot \lambda}{E \cdot \alpha}\left[\frac{W}{m}\right]$$

R_m = tensile strength [MPa], E = Young's modulus [GPa], α = CTE [10 ^{6} K ^{1}], λ = thermal conductivity [W/(m K)].

TABLE 6.6 Thermal Shock Coefficients R_1 and R_2 of Different C/C-SiC Materials Compared to Monolithic Ceramics and Gray Cast Iron [75]

	C/C-SiC			Monolithic ceramics			Gray cast iron	
	XB	XD	XT	Al_2O_3	SiSiC	RBSN	GG15	GG25 leg.
R_1 (K)	530	80	1 440	90	400	500–730	200	200
R_2 (W/m)	9070	1820	30 230	1000	18 000–40 000	7000–9000	12 000	8000

6.3.5 Oxidation

In inert atmosphere, C and C/C offers extreme thermal stability and is used for service temperatures up to 2700°C in charging components of graphitization furnaces. In oxidative atmospheres, oxidation of the C matrix is observed at temperatures above 450°C, whereas oxidizing of the C fibers starts at slightly higher temperatures above 500°C, due to their higher degree of graphitization [82]. Methanization of C is starting at about 700°C in hydrogen containing atmospheres. In SiC matrix materials, the reaction with hydrogen can be reduced by some orders of magnitude leading to significantly longer lifetimes of C/SiC compared to C/C, for example, in charging devices, used for the thermal treatment of metals [3].

In open systems, SiC decomposes at temperatures above 2300°C. In dry air, SiC offers high temperature stability as well as oxidation resistance. Thereby, a surface layer of SiO_2 is formed at temperatures above about 600–800°C, which is effective in preventing further oxidation (passive oxidation). At temperatures above 1000°C and in reducing atmospheres, gaseous SiO is formed in an active oxidation, leading to a decomposition of SiC. The main disadvantage of SiC is its limited corrosion resistance in water vapor atmosphere at high partial pressures, leading to the formation of SiO_2 and gaseous $SiOH_4$ and a rapid degradation at temperatures above 1000°C, especially in alkali containing atmospheres and at high gas velocities.

In LSI-based C/SiC and C/C-SiC materials, residual silicon oxidizes at higher temperatures, again forming SiO_2. At service temperatures above about 1400°C Si is melting and discharging, leading to increased porosity and reduced mechanical properties of the composite. Additionally, discharging Si melt can further react with free C, welding together C/SiC components in close contact, especially if Si-rich materials are used.

The oxidation behavior of C/SiC materials is mainly determined by the fiber architecture. Oxidation rates are highest parallel to the C fibers or C/C bundles, due to the unhindered access of oxygen. Therefore, 2D fabric-based materials as well as warm-pressed short fiber-based C/SiC show significantly lower degradation perpendicular to the fabric layers than parallel to the fibers (Figure 6.47). Due to a distinct fiber orientation during warm pressing, even short fiber-based C/SiC materials show twice as high oxidation rates perpendicular to the pressing direction (parallel to the fibers), compared to the direction parallel to the pressing direction (perpendicular to the fibers) [82]. Generally, oxidation rates strongly increase with temperature. However, weight loss rates of C/SiC materials are lower, compared to C/C materials or C fibers. This can be explained by the protective effect of the SiC matrix, which reduces the surface of the C, accessible to oxygen. Additionally, at higher temperatures above 700°C, the oxidation rate decreases with oxidation time. This is explained by the elongated diffusion path and reduced access of oxygen to the reaction front due to increased formation of CO and CO_2 in the capillaries [83].

Oxidation can be reduced by surface impregnation, for example, with phosphoric salt or with phosphate or boron-carrying solutions [84], and by applying oxidation protection coatings such as CVD-SiC combined with glassy layers or by multilayer coating systems based on boron carbide, silicon carbide, and self-sealing glass-forming layers [4]. In another approach, self-sealing matrices have been developed by Herakles for C/SiC (SEPCARBINOX® A 500) and SiC/SiC composites manufactured via CVI. Thereby, multilayer matrices consisting of alternating matrix layers in the system Si, C, B are built up in a defined sequence. At elevated temperatures, incoming oxygen is consumed by forming glassy phases, which again are able to seal microcracks and to prevent further access of oxygen to the carbon interphases and fibers [85]. Lifetime in tensile low cycle fatigue could be increased to more than 500 hours at 1000°C, at stress

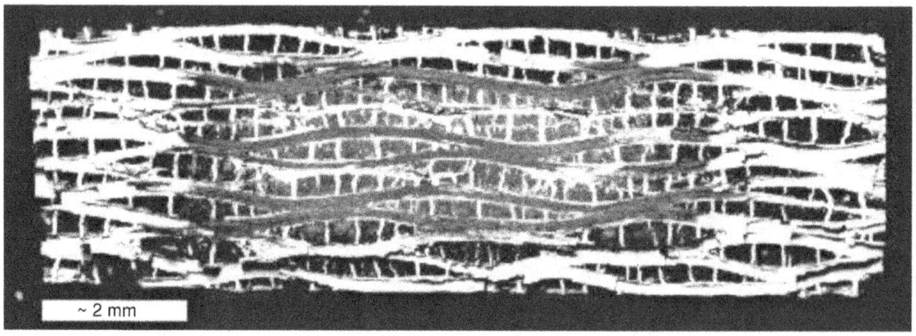

FIGURE 6.47

Cross section of a C/C-SiC specimen after thermal treatment in air (900°C, 1 hour), showing anisotropic burn out of C/C bundles and not oxidized core area. Oxidation depth parallel to the fabric layers is approximately three times higher than in the perpendicular direction (DLR) [75].

levels of 90 and 130 MPa and more than 100 hours at 1200°C and a stress level of 160 MPa [29, 30]. Tensile tests on samples cut out of nozzle flaps showed a decrease in tensile strength of only 6% after 1300 hours ground testing at realistic conditions. Compared to C/SiC materials based on a sole SiC matrix only, lifetime was two orders of magnitude higher [30].

The self-sealing matrix of the CVI-derived Sepcarbinox® A 500 material has been a way to extend to several thousands of hours the life endurance of components submitted to high thermal and mechanical loads typical of military engines applications. Despite the significant progress made in oxidation protection, C/SiC materials are generally not applicable for long-term use in oxidative atmospheres at high dynamic loads and temperatures, for example, in the combustion chamber or turbine area of jet engines and stationary gas turbines.

6.3.6 Tribological Properties

C/SiC and C/-SiC materials show unique properties for high performance friction applications, which include high wear resistance, thermal and thermal shock resistance, and a coefficient of friction (COF), which can be tailored over a wide range for the successful use in brakes as well as in gliding systems. These properties are presented in detail in chapter 23 "Ceramic Matrix Composites for Friction Applications".

6.4 APPLICATIONS

Carbon-fiber-reinforced SiC materials offer a unique combination of favorable thermal and mechanical properties with low density, and therefore are able to overcome technical barriers in numerous application areas by replacing conventional metals or monolithic ceramics. The application spectrum is generally determined by the combination of environmental conditions, temperature and atmosphere, with service time. Due to the excellent temperature stability of the carbon fibers and the SiC matrix, C/SiC and C/C-SiC materials currently are used at extreme temperatures up to 2700°C. However, service time is limited, mainly due to the oxidation of C at temperatures above 450°C, and to active oxidation of SiC above 1000–1900°C, depending on atmosphere. Therefore, possible applications vary over a wide range, comprising thermally highly loaded structures for short-term operation in rocket propulsion systems, as well as long-term stable components, such as brake discs or satellite structures, used in a moderate or low temperature regime (Figure 6.48).

The development of these materials was initiated and driven by space applications, especially by the development of lightweight hot structures in TPSs of reusable spacecraft. As in space structures, the high temperature and thermal shock stability at low structural weight were the main reasons for replacing metallic hot gas liners in aeronautic jet engines for military aircraft and in rocket propulsion. For the latter, the high abrasive resistance at high temperatures is an additional key advantage of C/SiC materials compared to metals. In the lower temperature regime, the abrasive resistance along with the favorable friction behavior is again the determining reason for the successful introduction of C/C-SiC materials in high performance brake systems, the predominant and most promising application area beyond aerospace. Besides the high temperature applications, C/SiC is also used for high precision satellite structures. Here, the low and controllable CTE as well as the non-brittle fracture behavior, the low porosity, and long-term stability are favorable for use in telescope structures, providing an alternative to CFRP, C/C, and monolithic ceramics such as Zerodur®.

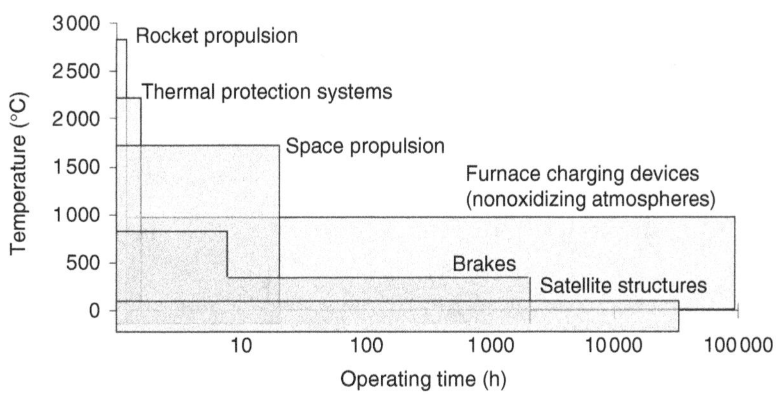

FIGURE 6.48

Typical application areas for C/SiC and C/C-SiC materials dependent on service temperature and time (DLR).

Due to economic reasons, C/SiC materials are typically used when there are no alternative materials, such as conventional metals or FRPs that meet the requirements, or when significant technical improvements or marketing advantages can be obtained that justify their use despite higher cost. Therefore, high performance materials such as continuous fiber-reinforced C/SiC based on CVI or PIP are mainly used for mechanically highly loaded aerospace applications, whereas low cost C/C-SiC materials based on short fibers and the LSI process opened up new industrial application areas, such as brake discs in automobiles.

Currently, the applications for C/SiC materials are still limited to niche markets. Components are manufactured as single items or in small batches for space research and demonstration programs as well as in small-scale series of several hundred components per year in aeronautics. In the area of friction materials, brake discs or pads are produced in some small- and medium-scale series of several thousand up to almost hundred thousand components per year.

6.4.1 Space Applications

The development of carbon-fiber-reinforced SiC materials started in the 1970s, initially for thin-walled, lightweight structures in TPSs of reusable spacecraft, which remains one of the main application areas for C/SiC materials. The other important areas are propulsion systems for rocket launchers, future spacecraft, and satellites. Here, the focus is on high temperature stable C/SiC nozzle extensions and combustion chambers. Other promising applications are highly precise, lightweight satellite structures for communication systems.

Thermal Protection Systems In TPSs of spacecraft, maximum temperatures of 1600°C up to 2000°C, high heating rates of several hundred Kelvin per second, and high thermal gradients are generated locally during the critical reentry phase, lasting up to 30 minutes. Conventional composite materials based on glass FRP, cannot withstand these extreme thermal and mechanical loads without severe damage and therefore are used as ablator systems, leading to thick-walled, heavy liners. In order to obtain lightweight, aerodynamically stable structures, C/C materials with oxidation protective coatings were introduced and were used for the extremely hot areas of the TPS structure of the US space shuttle, such as the nose cap and the leading edges of the wings. To further increase oxidation stability, bulk protected C/C has been developed and was first introduced in the form of LSI-based C/SiC composites in the TPS of the Buran (Figure 6.49), a spacecraft from the former Soviet Union [9].

FIGURE 6.49

Nose section of the Buran with nose cap made of C/SiC (top). Subscale flight model BOR 5 (bottom left) with wing-leading edges (bottom right) made of C/SiC (right structure) and ablator material (left structure, all figures DLR).

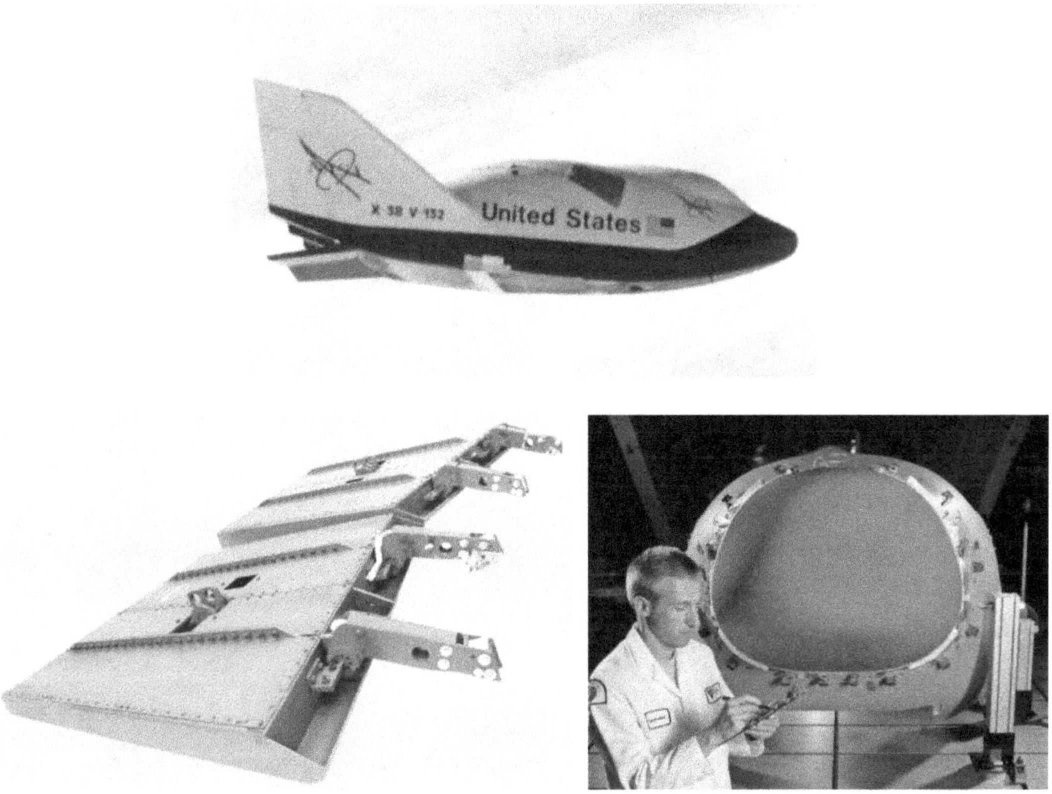

FIGURE 6.50

C/SiC and C/C-SiC structures for X38 spacecraft (top, NASA). C/SiC (G-CVI) body flap assembly (2 flaps, each 1600 mm × 1400 mm × 150 mm, t = 5–15 mm, m = 68 kg, bottom left, MT) based on substructures joined with C/SiC (CVI) screws. C/C-SiC (LSI) nose cap (about 740 mm × 640 mm × 170 mm, t = 6 mm, m = 7 kg, DLR, bottom right, DLR).

Further milestones were large TPS and structural components for the European Hermes program and the X-38 demonstrator (NASA), which was planned to be the future crew return vehicle for the international space station (ISS). Here, highly loaded, thin-walled body flaps for the flight control of the spacecraft as well as nose cap systems and leading edges for the protection of aerodynamic stagnation points were developed. This was carried out by SEP (now Herakles) and Dassault Aviation in the Hermes program, starting in the 1980s [32], and by a German consortium (MT, Astrium ST, DLR) for X38 in the TETRA program (1998–2002; [86–88]) leading to the largest C/SiC and C/C-SiC structures, worldwide, with maximum dimensions up to 1.6 m × 1.4 m (Figure 6.50). The CMC components were tested successfully and a TRL of 8 (Technology Readiness Level, NASA) could be achieved for the material. Currently, nose cap and body flap assemblies will be used in the ESA flight experiments EXPERT [44,89], (Figure 6.51) and the IXV [90,91] (Figure 6.52), an unmanned shuttle (about 2 × 2 × 4 m^3) based on an all composite design with CFRP internal substructure and C/SiC TPS, which will be launched in 2014.

In order to improve supersonic aerodynamic performance and to reduce manufacturing costs, novel TPSs based on facetted structures have been developed at DLR since 2002 in the SHEFEX (Sharp Edge Flight Experiment) projects I [92] and II [93,94]. In contrast to commonly used, curved shell structures, which require expensive molds for green body manufacture and ceramics conversion, these systems are built up using easy to manufacture flat plates or tiles (Figure 6.53). However, during ballistic reentry at maximum velocities of up to 10 000 km/h (≈Mach 10), the thermal load at the tip and at the edges of the tiles is higher than in curved structures, leading to high thermal shock and maximum temperatures above 1700 °C, which could be withstood by uncoated C/C-SiC.

Space Propulsion Whereas TPS and aerodynamic control structures of experimental spacecraft typically are manufactured as single items, small series production is planned for C/SiC components in space propulsion systems. Subscale as well as full scale C/SiC nozzle extension demonstrators were developed and tested for the main engine "Vulcain" of Ariane 5 (Figure 6.55, Astrium ST/EADS-IW) and the upper stage engines HM 7 (Herakles) and AESTUS (Astrium

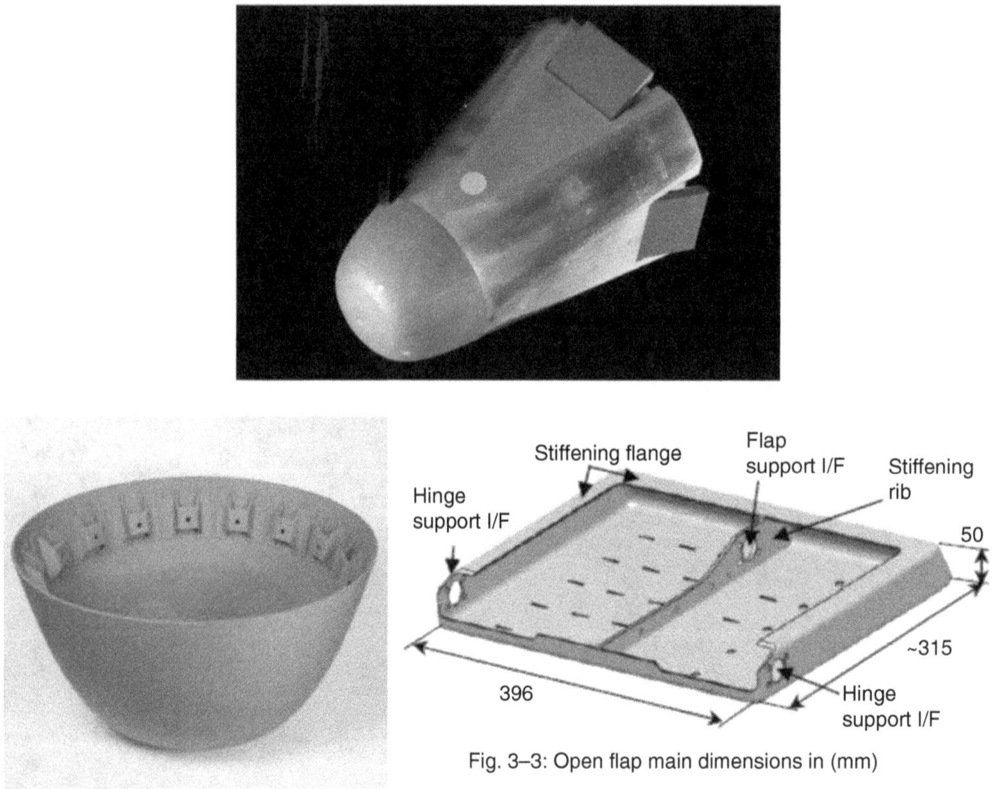

FIGURE 6.51

Reentry capsule (≈Ø 1200 mm × 1600 mm) of the EXPERT experiment during assembly, showing C/SiC (G-CVI) body flaps (396 mm × 315 mm × 50 mm, $t = 2.5$ mm, $m \approx 2$ kg, bottom right, MT) and C/C-SiC (LSI) nose cap (Ø 716 mm × 413 mm $t = 6$ mm, $m = 12.24$ kg, bottom left, DLR) with *in situ* joined flanges inside, used for assembly to the metallic substructure.

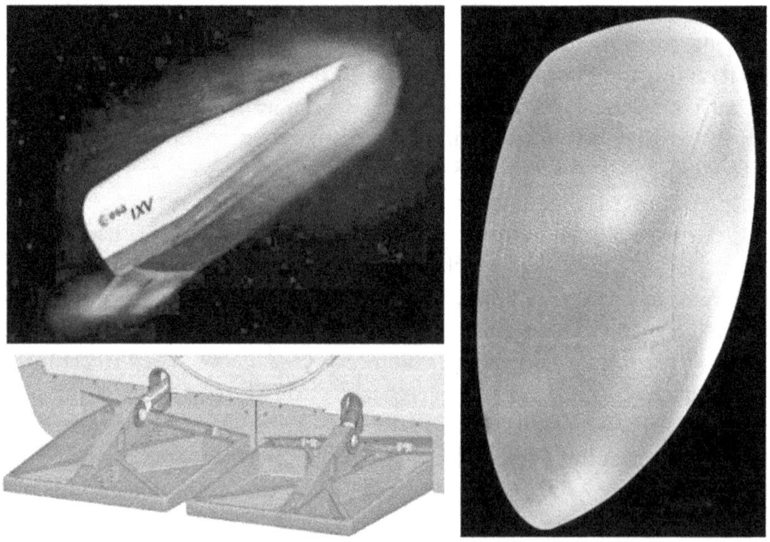

FIGURE 6.52

Artists view of the IXV experiment (top left, ESA) based on C/SiC front structure (≈1320 mm × 910 mm × 410 mm, m ≈ 15 kg, right, Herakles) and C/SiC body flap assembly (2 flaps each about 1000 mm × 1000 mm × 100–50 mm, $t =$ 1.5–4 mm: $m = 40$ kg, bottom left, MT), manufactured via I-CVI and G-CVI, respectively.

FIGURE 6.53

Launch of SHEFEX 2 from Andøya, Norway (top). Facetted TPS structures based on flat C/C-SiC plates for the Sharp Edge Flight Experiments SHEFEX 1 (Ø 370 mm × 800 mm, *t* = 3 mm, bottom left) and SHEFEX 2 (Ø 500 mm × 1 500 mm, *t* = 3 mm, bottom right, all figures DLR).

ST/EADS IW) of Ariane 4 and 5, respectively. The Aestus nozzle extension, successfully tested in the year 2000 with a total burn time of 150 seconds [95], was manufactured via filament winding and PIP and had a length of 1360 mm and a maximum diameter of 1330 mm (Figure 6.54). The wall thickness varied from 3.6 mm in the upper area to only 1.4 mm at maximum diameter. The total weight was 16 kg, which is only 40% compared to the conventional metal structure based on Haynes 25 super alloy.

The C/SiC exit cone for HM 7 was manufactured via I-CVI and extensively tested in several firings with a cumulated duration of more than 1500 seconds [29], which demonstrated a TRL of 8. Thereby, maximum temperatures above 1700°C were applied locally. The investigation of samples, cut from the exit cone after testing, showed no degradation of mechanical properties and no oxidation could be observed. The weight of the uncooled C/SiC exit cone was 24 kg. Compared to 84 kg for the cooled, metallic exit cone, a weight reduction of 70% was obtained. Based on these experiences, a C/SiC nozzle extension has been introduced in the upper stage Vinci® engine, intended to be used in a future Ariane 5 version [29,96]. This deployable nozzle extension is separated into two segments. Whereas C/C is used for the lower and therefore cooler component, the upper segment is made of CVI-based C/SiC in order to withstand the higher temperatures (Figure 6.56).

Satellite thrusters are used for altitude control and orbit raising (Figure 6.57). For liquid propellant systems, thrust levels in the range of 10–500 N are typical. In geostationary applications up to 60% of the satellite mass is propellant, mainly consumed by the apogee motor in order to lift the satellite from its transfer orbit to a geostationary orbit [97]. Typical burning times are above 8–15 hours, accumulated in 15–20 thermal cycles. Increasing the efficiency of apogee motors and control thrusters has a high impact on satellite weight and cost, and is the focus of several development projects worldwide [98–100], (Figures 6.58–6.60). To improve the performance of these propulsion systems, the operating temperature has to be increased. Currently the maximum temperature is limited to about 1600°C due to the materials used for the combustion chamber and nozzle. High temperature-resistant alloys based on refractory metals such as rhenium and niobium, which both require oxidation protection coatings usually based on iridium, as well as

FIGURE 6.54
C/SiC Vulcain subscale (1:5) nozzle extensions manufactured via PIP (bottom left). C/SiC nozzle on test bench (top) and during 40 bar hot-firing test (bottom right, all figures Astrium ST/EADS IW).

uncoated platinum are state of the art. By introducing C/SiC materials, lower costs for material and component manufacture and higher operating temperatures above 1600°C are in the focus. Further advantages expected are weight savings of 30–50% as well as a simplification of the design, achieved by the combination of combustion chamber and nozzle in one single, integral structure.

At Astrium, radiation-cooled combustion chambers for orbital satellite thrusters in the 400 N and 500 N class have been manufactured and tested in hot firings since 1998. Here, maximum temperatures up to 1980°C (short term) at combustion chamber pressures of 1 MPa and accumulated test times of up to 3.9 hours could be achieved with C/SiC structures, manufactured via filament winding and PIP (Figures 6.59 and 6.60) [95, 98, 101].

Radiation-cooled C/SiC integral structures, combining combustion chamber and nozzle extension in a single component (Figure 6.60) have been developed and tested successfully [97, 99]. The structures must withstand maximum chamber pressures up to 1 MPa at temperatures well above 1300°C in chemically aggressive environments caused by the hypergolic propellants (e.g., MMH/N_2O_4) and their combustion products.

C/SiC materials are promising candidates for future launch systems for space and are in the focus of various development programs, worldwide. For hypersonic cruise vehicles as well as for rockets, lightweight TPSs and actively cooled structures for flow-path components and combustion chambers in air-breathing propulsion systems are identified as key technologies. Thereby, CMC materials offer highest temperature stability at significantly reduced weight, compared to metals, enabling advanced concepts as well as increased payload and range.

For dual-mode ramjets (DMRs) in future hypersonic vehicles, propellant-cooled, high performance combustion chambers have been developed by MBDA, France and Astrium ST, Germany [14, 102, 103], using the so-called PTAH-SOCAR technology (Paroi Tissee Application Hypersonique—Simple Operational Composites for Advanced Ramjet technology). Thereby, a near net shape fiber preform with integrated cooling channels is manufactured via 3D weaving, filament winding, and stitching (see Figure 6.4). For densifying, C and SiC matrix is built up by a combination of CVI and LSI. The resulting C/C-SiC (CARBOTEX®SI) structure is characterized by dense C/C wall structures, columnar

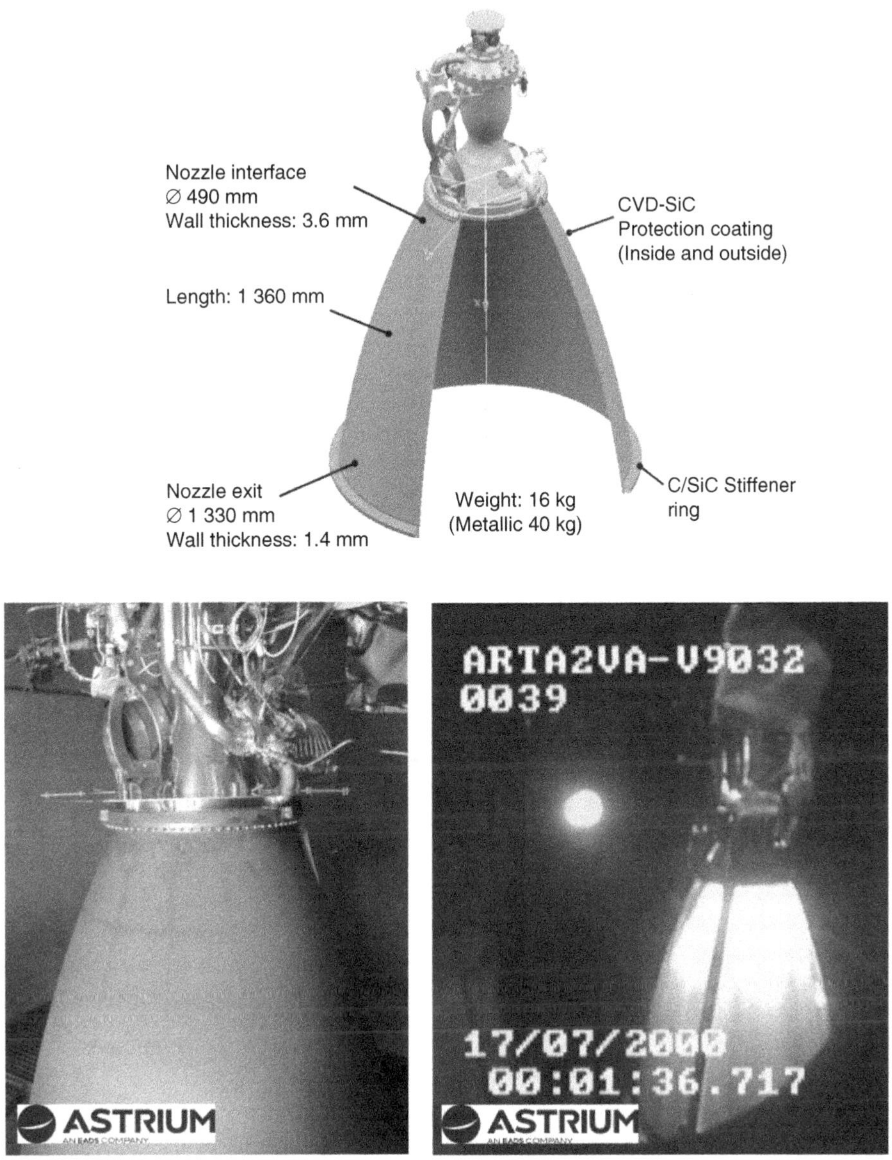

FIGURE 6.55

C/SiC nozzle extension technology demonstrator (Ø1330 /Ø490 mm, l = 1360 mm) for the upper stage engine AESTUS (top). SICARBON® structure in test bench (bottom left) and during hot-firing test (bottom right, all figures Astrium ST/EADS IW) [95].

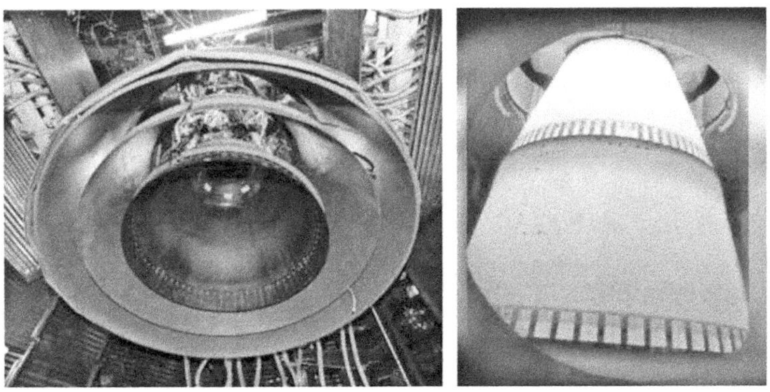

FIGURE 6.56

Nozzle extension made of C/SiC (Ø 2.2 m upper part) and C/C (lower part) for the deployable, divergent nozzle of the Vinci engine (180 kN), for the upper stage of the future version of Ariane 5, on test bench (left, Herakles) and during hot-firing test (right, DLR).

FIGURE 6.57

Satellite propulsion system with 400 N engine (bottom) and propellant tanks (Astrium).

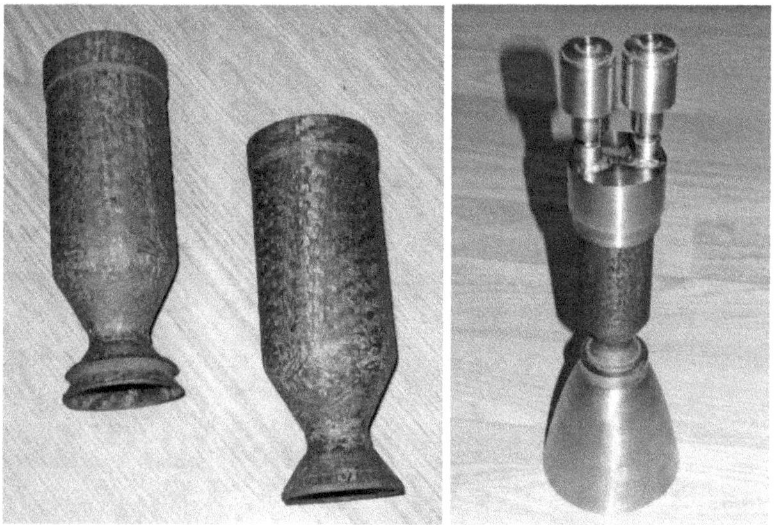

FIGURE 6.58

C/SiC (CVI) combustion chambers (Ø 42 mm, length 120 mm, left) and model for 200 N propulsion system MAI-202-200-OK (right, MAI/OAO "Kompozit") [100].

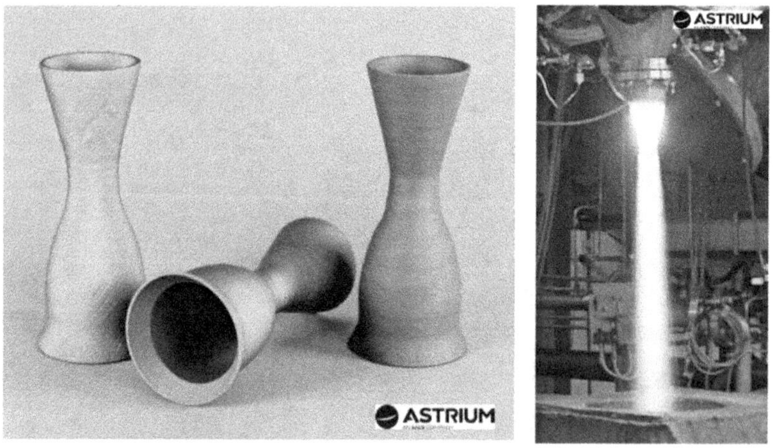

FIGURE 6.59

Coated C/SiC (SICARBON®, PIP) 400N combustion chambers (cylinder Ø 60 mm, throat Ø 16 mm, length 180 mm), hot-firing test at test bench P1.5 (Astrium, Lampoldshausen).

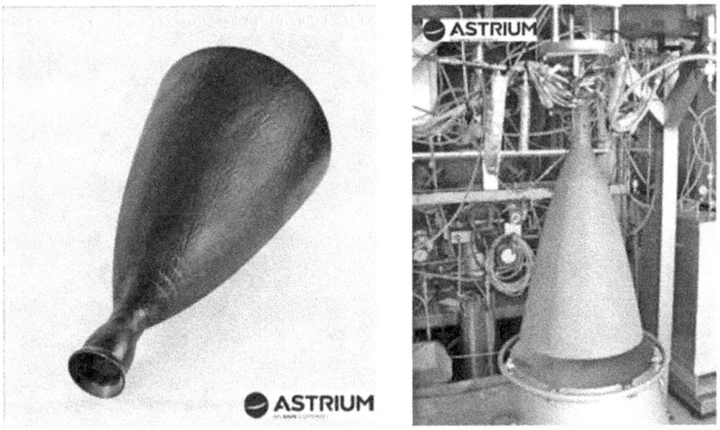

FIGURE 6.60
Integral structures of combustion chambers and nozzle extensions for satellite propulsion systems. Technology demonstrator (400 N, chamber: $\approx \varnothing$ 60 mm/l $\approx$ 100 mm, nozzle: $\approx \varnothing$ 300 mm, l $\approx$ 500 mm, left) and European Apogee Motor (EAM, 500 N, chamber: $\approx \varnothing$ 40 mm/l $\approx$ 170 mm, nozzle: $\approx \varnothing$ 400 mm, l $\approx$ 600 mm, right), both based on coated C/SiC (SICARBON®), manufactured via filament winding and PIP (all figures Astrium ST/EADS IW) [7].

C/C spacers, and by a gas-tight SiSiC layer at the outer surface (Figure 6.61) [14, 102, 103]. The combustion chamber has been tested successfully under hypersonic conditions at MBDA, France. Thereby, maximum temperatures of up to 1500°C and pressures of 1 MPa were obtained. Compared to metallic structures, weight savings above 30% are expected.

Satellite Structures The low thermal expansion of C/SiC materials opens up new application areas for very precise satellite components, such as telescope structures in laser communication terminals (LCTs). In these optical systems, laser beams are used for satellite communication, providing data transmission rates about three orders of magnitude higher than in conventional microwave technologies. Simultaneously, energy consumption is reduced enabling lightweight compact systems. Further advantages are operational security and immunity to interference sources [104]. To achieve this, extremely precise structures are required for an accurate positioning of the optical mirrors responsible for focusing and guiding the laser beams. Furthermore, these structures have to be stable in the μm-scale for long operation times above 5 years in space environment [79]. In a laser link experiment in 2008, data transmission rates of 5.5 Gigabit per second (Gbps) could be demonstrated between two satellites, the German TerraSAR-X and the US NFIRE, flying at a distance of 5000 km and a relative velocity of 25 000 km/h in low earth orbit. This equals a transfer of about 200 000 DIN A4 pages per second or 400 DVDs per hour and is a world record to date.

Figure 6.62 shows the optical unit used for the LCT in each of the two satellites with the telescope structure built up by an *in situ* joined C/C-SiC tube. The main task of the telescope tube is the accurate positioning of the mirrors mounted inside, in order to minimize angular deviations of the receiving and transmitting laser beam, and finally to ensure a reliable data transfer. Main sources for possible deviations are deformation of the telescope, caused by thermal expansion due to solar irradiation, or by vibrations induced by different sources in the satellite. For the telescope tube, a tailored C/C-SiC material could be developed characterized by a well-defined fiber architecture and material composition and providing a very low CTE ($0 \pm 0.1 \times 10^{-6}$ K^{-1}) in axial direction [80]. Thereby, a constant distance between the primary and secondary mirror in the temperature range of -50°C and $+70$°C could be achieved. The telescope tube was manufactured via LSI. For the CFRP preform, prepreg layers were cut to size, wound on an aluminum mandrel and warm pressed in a steel mold. Due to well-adapted graphite cores, used during pyrolysis for controlling the shrinkage of the CFRP tubes, a near net shape manufacture of the C/C-SiC tubes was achieved and extensive final machining of the external surface could be avoided. Two telescope structures were manufactured at DLR, qualified at Carl Zeiss Optronics (now Cassidian Optronics), and finally integrated in the LCT systems of the satellites by TESAT Spacecom.

Compared to other low expansion materials such as Invar, Zerodur, or SiC, C/C-SiC offers lower density and high specific stiffness. In contrast to the highly brittle monolithic ceramics (Zerodur, SiC), C/C-SiC provides fracture toughness and damage tolerance, enabling thin-walled lightweight structures and a near net shape manufacture of complex geometries at reasonable cost. Unlike CFRP, no outgasing in vacuum or hygroscopic swelling was observed. Due to its

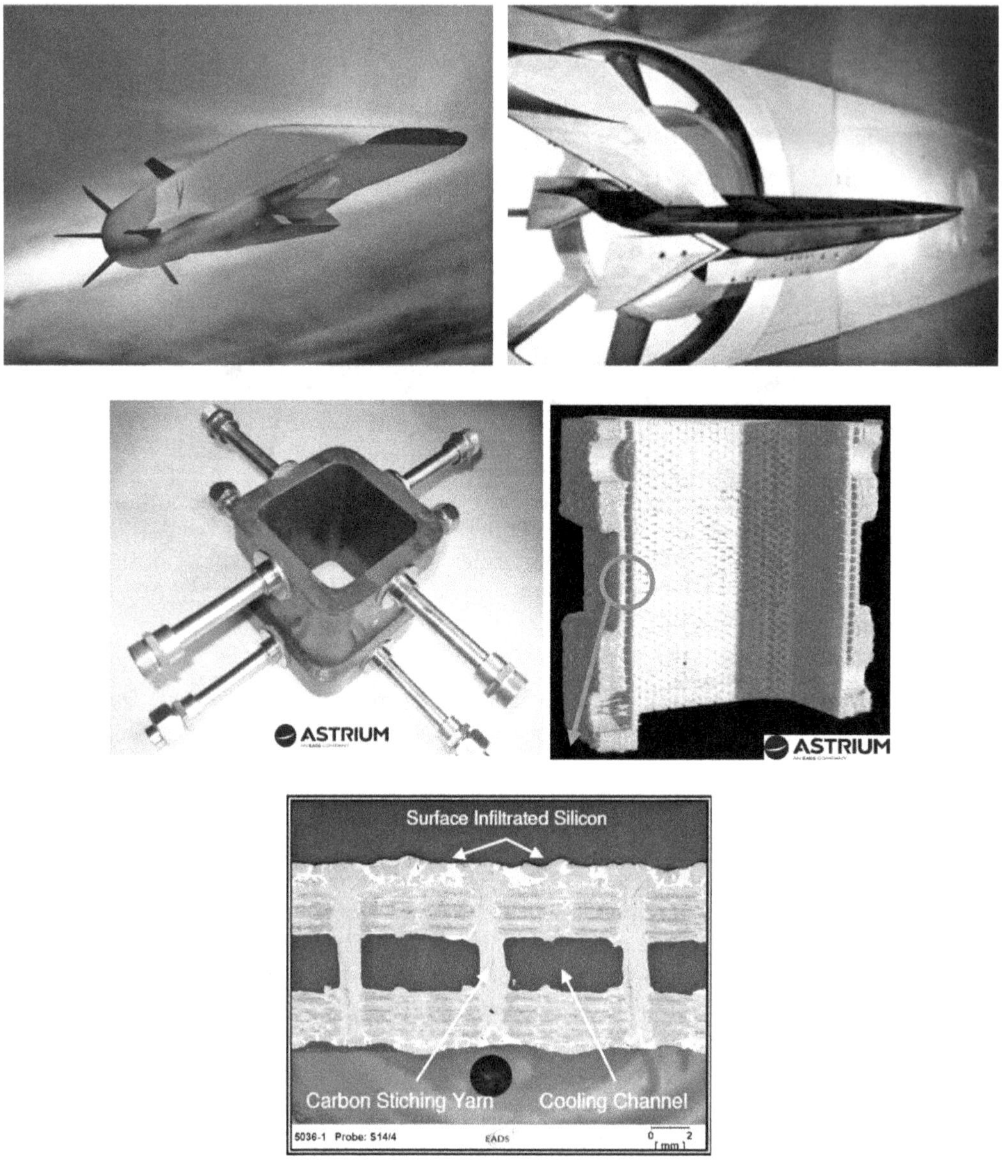

FIGURE 6.61

Hypersonic demonstration aircrafts X-51 WaveRider (artists view, top left, Boeing) and LEA (wind tunnel model, top right, MBDA). Test module of PTAH-SOCAR combustion chamber duct 135 mm × 135 mm × 150 mm, middle left, Astrium ST) developed for dual mode ramjet propulsion systems in future spacecraft. CT image of the cross section of the C/C-SiC monobloc structure, showing the internal cooling channel between inner and outer liners (middle right, Astrium ST). SEM image of the cross section of the cooled wall structure, showing the microstructure of the CARBOTEX$^{®SI}$ (CVI + LSI) material (bottom, Astrium ST/EADS IW).

low porosity, particle release and cleaning processes are reported to be less critical for C/C-SiC materials, compared to porous C/C.

As a future application, C/SiC antenna substructures will be used for the Bepi Colombo experiment, a joint mission of ESA and JAXA to the planet mercury [105].

6.4.2 Applications for Aeronautics

In 1996, serial production of CMC materials started with the manufacture of C/SiC nozzle flaps for the SNECMA M88-2 jet engine, used in the Rafale military aircraft (Figure 6.63) [96, 106]. To date more than 5000 C/SiC flaps were produced by Herakles via I-CVI. Compared to the previously used superalloy (Inconel 718), the so-called Sepcarbinox®

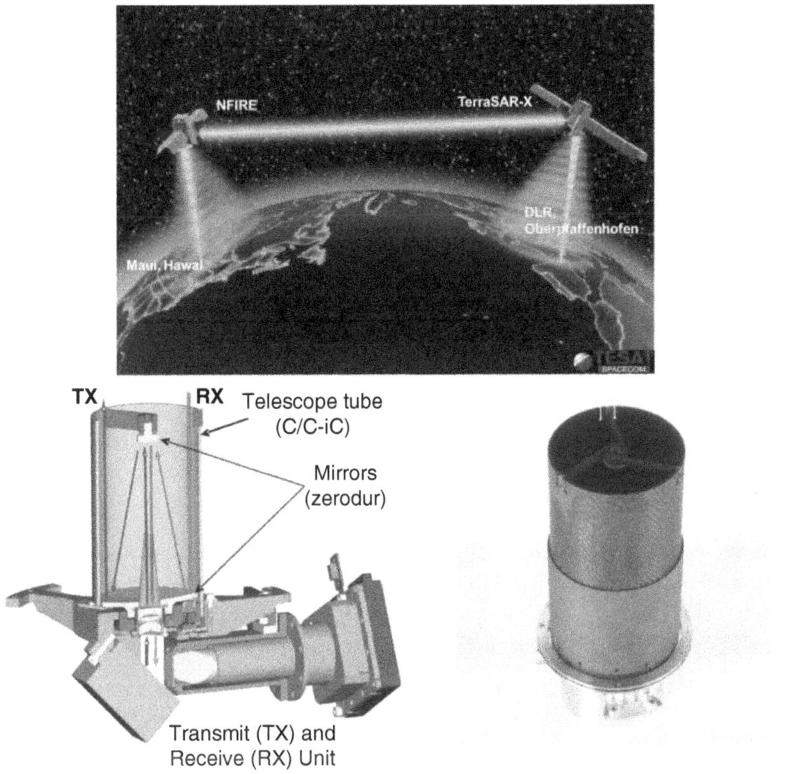

FIGURE 6.62

Artists view of the laser link experiment on the satellites TerraSAR-X and NFIRE (top, TESAT SPACECOM). *In situ* joined C/C-SiC telescope tube (Ø 140 mm, $l = 160$ mm, $t = 3$ mm, bottom right, Cassidian Optronics GmbH). CAD model of the LCT, showing the telescope assembly (bottom left, TESAT SPACECOM).

FIGURE 6.63

C/SiC outer flap ($430 \times 300 \times 2$ mm^3, bottom), manufactured via I-CVI, for M 88-2 jet engine in the military aircraft Rafale (top, all figures Herakles).

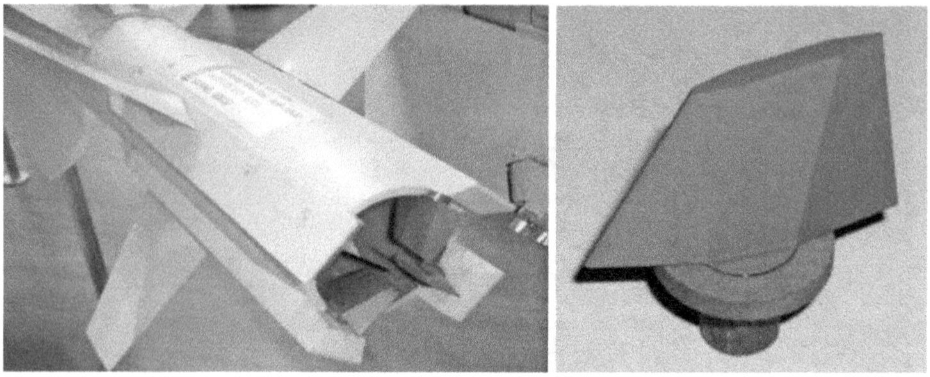

FIGURE 6.64
Thrust vector control (TVC) system of solid propellant rocket motors (left, DLR) and C/C-SiC jet vane (≈60 mm × 60 mm × 10 mm, right, DLR).

flaps provide weight savings of about 40% and typical drawbacks of the metallic pendant, such as cracking and excessive creep, could be overcome, resulting in extended lifetime. Further applications have been investigated and evaluated on engine divergent seals for the Pratt & Whitney F 100-PW-229 engine, powering F-16 and F-15 fighters as well as for novel systems with moveable vector nozzles [107]. A detailed description of this application area is given in chapter 21 "CMC Applications for Gas Turbines".

In military rocket motors, the first serial applications of C/SiC materials were nozzle throats and jet vanes (Figure 6.64). These materials have thus proven their high potential for short-term applications at highest thermal, mechanical, and abrasive loads.

Jet vanes are used for thrust vector control (TVC) systems of rockets since 2004, especially in solid propellant rocket motors enabling extreme maneuverability. The moveable jet vanes are positioned in the exhaust jet stream at the end of the nozzle extension and provide high steering forces immediately after takeoff, when the rocket velocity is still low and the exterior aerodynamic steering-devices are not yet effective [108–111]. Thereby, the jet vanes have to withstand high bending loads in the shaft as well as very high temperatures above 2800°C in the leading edge. Furthermore, these loads are applied in milliseconds, resulting in heating rates of several thousand Kelvin per second in the leading edge, as well as in local temperature gradients up to 200 K/mm in the blade area, leading to high thermally induced internal stresses. To increase thrust and specific impulse, aluminum particles are usually added to the propellant. During combustion, the particles are transformed to liquid Al_2O_3 droplets that impact the leading edge of the jet vane with a velocity of up to 2000 m/s, resulting in high abrasive loads. However, the operational time is very short, usually in the range of a few seconds up to some minutes, and therefore oxidation of the carbon fibers is not critical. Due to their high strength and fracture toughness at high temperatures, extreme thermal shock stability, and abrasive resistance, C/SiC materials could replace refractory metals such as tungsten, leading to weight savings of about 90% for the jet vanes. To date CMC jet vanes of various sizes in the range 60 mm × 60 mm × 10 mm up to about 150 mm × 150 mm × 40 mm are used in modern, highly agile air-to-air and ground-to-air rockets, for example, IRIS-T and the Exocet Block 3 booster [112]. In the future, almost all air-to-air rockets and some selected medium range missiles are expected to be equipped with CMC jet vanes. Currently, C/SiC and C/C-SiC jet vanes are produced by Herakles/Roxel (C/SiC via CVI) and Nammo (C/C-SiC via LSI). More than 4000 IRIS-T missiles initially ordered by the countries of the development consortium are scheduled to be delivered by the end of 2012 [113]. Annual production of C/SiC and C/C-SiC jet vanes is estimated to be in the range of several hundred kilograms.

6.4.3 Applications for Friction Systems

As for C/C, the main application area for C/SiC materials are high performance friction systems. C/C friction materials are state of the art in aircraft brakes as well as in brake discs and clutches for racing cars, as used in Formula 1 and the 24 Hours of Le Mans racing series as well as in rally racing [114]. Main drawbacks of C/C are high manufacturing cost as well as some unfavorable tribological properties, such as high wear, caused by abrasion and oxidation, and a significant decrease of the COF at low temperature and in humid environments, known as "early morning disease." These drawbacks could be overcome by replacing the C matrix by SiC as well as by the development of economical

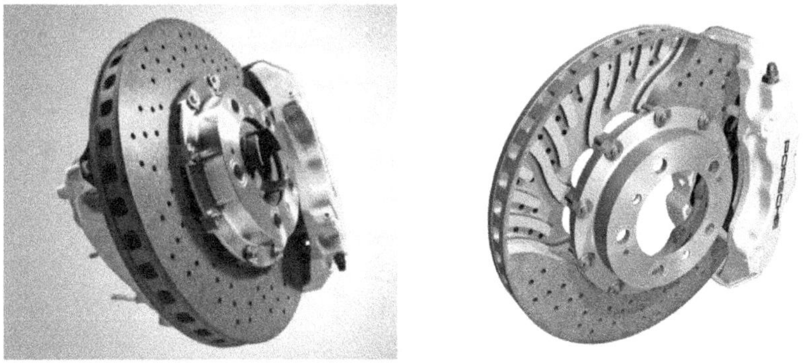

FIGURE 6.65
Porsche Ceramic Brake (PCCB) with short fiber-based C/SiC brake disc rotor, manufactured via LSI, with well-adapted aluminum brake caliper and brake pads as well as a metallic bell (all figures Porsche).

manufacturing processes, characterized by the use of low cost raw materials, such as phenolic resin precursors, cut C fibers, and a rapid buildup of the ceramic matrix via LSI. In this way, a wide and very promising application area beyond aerospace could be opened up for CMC materials for the first time.

In 1999, C/SiC brake discs for the automobile market were presented by DaimlerChrysler (now Daimler Benz) and Porsche at the International Motor Show (IAA) in Frankfurt, Germany, and were used for the first time for the Mercedes CL 55 AMG F1 Limited Edition in 2000. Serial production started in 2001 with the Porsche 911 GT2 (Figure 6.65) [115]. Currently, the so-called PCCB (Porsche Ceramic Composite Brake) is available for all Porsche models as an option and is a standard fitment for special edition cars. Furthermore, so-called Carbon Ceramic Brakes are available from all Original Equipment Manufacturers (OEM) of sports cars (e.g., Ferrari, Lamborghini, Bugatti, Corvette) as well as of premium cars and Sport Utility Vehicles (SUV), such as Audi (A8-W12, Q7, RS4, RS6, R8, etc.), AMG, Aston Martin, Bentley, and BMW (M5, M6 Coupe and Convertible).

Compared to gray cast iron, C/SiC brake discs offer weight savings up to 50%, reducing the unsprung mass by almost 5 kg per wheel and finally leading to improved agility and driving comfort. The friction behavior is characterized by increased braking response and a smooth and well-controllable braking process. Wear and corrosion stability is significantly increased, leading to service times of up to 300 000 km at standard road use. Even more important, the tribological behavior of C/SiC materials is almost independent of temperature and humidity, leading to high and stable COF even at low temperatures and almost no fading is observed even after repeated emergency stops. Furthermore, another argument for ceramic brake discs is the improved visual appearance, because rusty metal discs and wheels, heavily soiled with wear particles, are avoided.

For the serial production of CMC brake discs, new factories were built up by SGL Brakes GmbH in Germany and a joint venture of DaimlerChrysler and Brembo SpA in Italy. The start of serial production at SGL Brakes in 2001 was an important milestone and breakthrough for this new class of materials. For the first time ever, the CMC technology was transferred from the manufacture of prototypes, single components or very small-scale series to a reliable large-scale production of a safety-relevant component. Since 2009, CMC automotive brake discs are produced and further developed by Brembo SGL Carbon Ceramic Brakes GmbH (BSCCB) a joint venture of Brembo and SGL Brakes, bringing together two of the most experienced experts in brake and CMC technology. Production rate could be increased from several thousand discs in 2002 to about 90 000 discs per annum and about 240 tons per annum in 2011. Hence, production facilities at Brembo in Stezzano (Italy) and at SGL in Meitingen (Germany) are running at almost full capacity. For the manufacture of this safety-critical component, a quality assurance system according to VDA 6.1 and ISO 9001:2000 was introduced, documenting about 600 data from processing and quality testing for every single brake disc.

Additionally, CMC brake discs and systems have been developed by AP Racing Ltd., a subsidiary of Brembo SpA, Stillen (Great Britain), MS Production (Slovenia), MOV'IT (Germany), Surface Transforms (Great Britain), and others and are supplied to the tuning aftermarket for selected motorbikes and premium cars such as BMW S1000 RR and Nissan R35 GT-R, respectively, as well as for the use in racing motorbikes (Figure 6.66) and cars, especially for rally racing [116]. In addition to LSI, also PIP processes, characterized by continuous pyrolysis and short process cycles, as well as combined processes (Reaction bonding + PIP, CVI + PIP, CVI + LSI) are used for the manufacture of C/SiC and C/C-SiC brake discs based on chopped fibers and needled 3D fiber preforms [117–119].

Due to economic reasons, almost all of the CMC brake discs for automobiles are manufactured via LSI, using near net shape CFRP preforms based on short fibers and warm pressing. Typical geometries are in the range of 350–420 mm in diameter and 28–40 mm in thickness. In order to limit the temperature level of the disc and the overall brake system, C/SiC materials with high thermal conductivity of up to 40–60 W/(m K), obtained by high contents of SiSiC matrix,

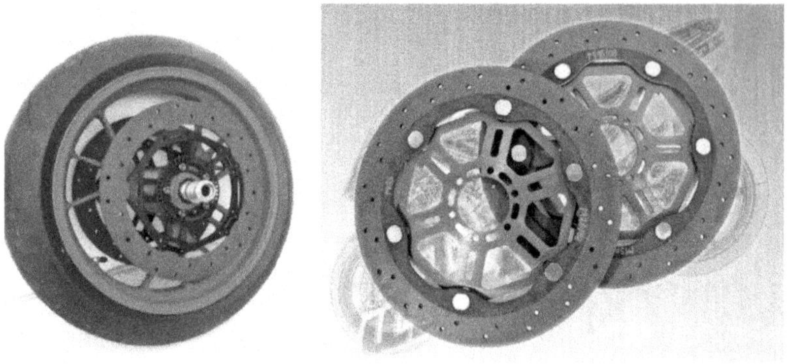

FIGURE 6.66
C/C-SiC aftermarket tuning brake discs for motorcycles. Front wheel with C/C-SiC brake discs (T-drive, Ø 330 mm, left) for Ducati 1199 Panigale and brake disc (Ø 320 mm, $m < 0.8$ kg) for BMW S1000 RR (right), manufactured via PIP (all figures MS Production).

as well as inner ventilated discs are used. To minimize oxidative and abrasive wear, the friction surface is usually protected by a thin, SiC-rich layer, built up via *in situ* joining during siliconization.

Prices for optionally available CMC brake systems are still very high and vary between 8000–12 000 Euros per car, depending on car type and manufacturer. Replacement cost for Ferrari sports cars are reported to be above €20 000 for four discs. This is mainly caused by a high diversity of brake disc geometries, leading to low production numbers of each disc type and finally to high manufacturing costs. For the same reason, tuning kits for the aftermarket are even more expensive with prices well above 14 000 Euros per car [117] or about €400–500 for one motorbike brake disc. Despite the high costs, production volume of C/SiC brake discs for the aftermarket, refit, and racing segments is several thousand brake discs per year.

In another automotive application, C/SiC friction rings have been used in clutches for high end sports and racing cars of Porsche Carrera GT and Audi R10 TDI, and are currently supplied to rally racing cars such as the Volkswagen Race Touareg 3 by ZF Sachs. In all these systems, C/SiC friction rings from BSCCB based on LSI are used. The main advantages of C/SiC friction rings compared to conventional metals and organic or sintermetallic materials are their higher temperature stability and strength, leading to significantly increased torque capacity and energy input and finally to a lightweight and compact design. Compared to C/C, the C/SiC materials offer lower wear, which is favorable for long distance racing series such as the 24 Hours of Le Mans or the Rally Dakar. In order to minimize mass and volume, continuous fiber reinforced materials based on 2D fabrics are used, leading to high in-plane strength and enabling low wall thicknesses down to 4.5 mm. Compared to a conventional clutch in a Porsche 911, the diameter and weight of the Porsche ceramic composite clutch (PCCC) in the Carrera GT (450 kW, maximum torque >1000 Nm) could be reduced by 30% from 240 to 169 mm and by 50% from 7.6 to 3.5 kg, respectively (Figure 6.67) [120].

In large cargo aircraft, propeller brakes are used for a fast stopping of the propellers after landing as well as for blocking the propeller to avoid uncontrolled movement due to heavy winds while parking or during maintenance. For

FIGURE 6.67
Porsche Ceramic Composite Clutch (PCCC) based on carbon-fiber fabric-reinforced C/SiC, manufactured via LSI (Porsche).

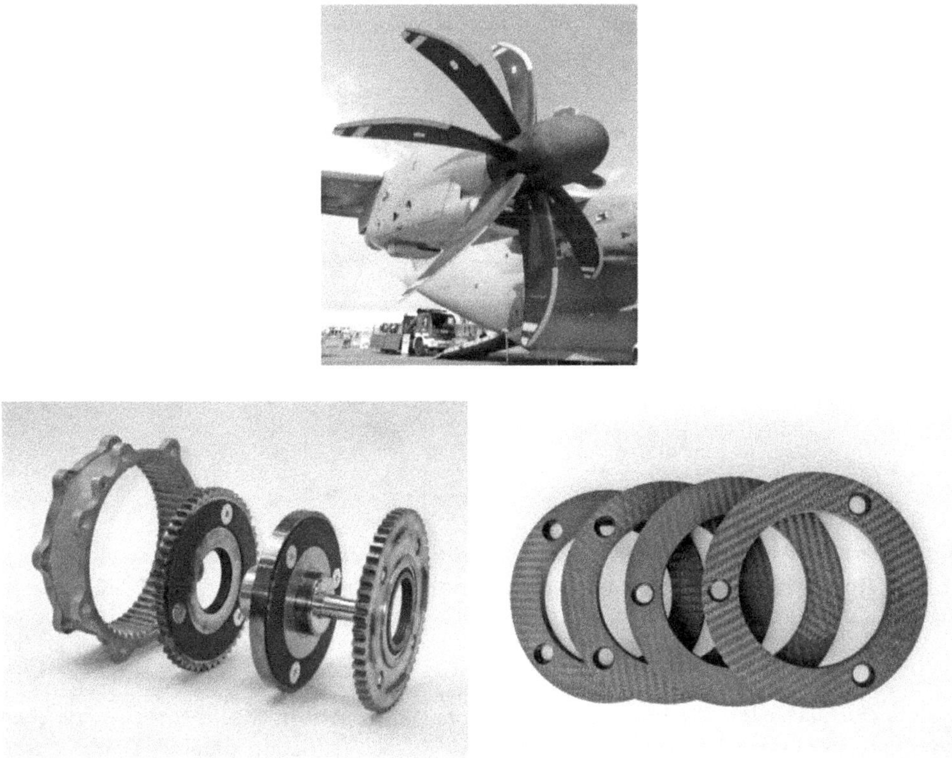

FIGURE 6.68
Multidisc brake system (bottom left) for the propeller of the A 400M (top), based on C/C-SiC friction pads (Ø120 mm, $t = 6$ mm, bottom right, all figures DLR).

the propellers of the A 400M, a highly compact multidisc propeller brake has been developed by Umbra Cuscinetti S.p.A., Italy. Here, four C/C-SiC friction pad rings with an outer diameter of only 120 mm (Figure 6.68) are used to decelerate the eight-blade propeller with a diameter of 5.3 m from 650 rpm to zero in less than 8 seconds. Thereby, a maximum kinetic energy of about 120 kJ has to be transformed into thermal energy in the fully enclosed brake unit. This leads to high temperatures that either could not be withstood by organic or sintermetallic materials or caused unacceptable high wear rates. As in the clutches described above, a 2D fabric-based C/C-SiC material is used in order to minimize weight and volume. The LSI-based friction rings have been developed at DLR and are produced in a small-scale series by SKT, since 2014.

Emergency brakes in high performance elevators have to stop the elevator from free falling in the very unlikely event of failure in the transporting cables. If the speed of the car exceeds a predefined nominal value, a controlled emergency stop is automatically started. Thereby, metal springs press the friction pads against the guidance rails, made of steel. Due to high carrier masses of up to 45 tons and high velocities up to 10 m/s, temperatures above 1200°C are obtained in the brake pads. This cannot be withstood by organic or metallic friction pads unless they are very large, thus they have been replaced by thermally stable and lighter C/C-SiC materials [121].

To ensure a specified constant deceleration graded C/C-SiC materials, characterized by a high ceramic content on the friction surface and a ductile core material, have been developed at DLR [122]. Thereby, brake pad wear could be reduced to a very low level below 0.5 μm/km braking distance. Furthermore, the C/C-SiC pads offer a high and stable COF in the whole temperature range [123]. The ceramic brake pads (Figure 6.69) improve braking performance by more than 33%, with a 65% reduction in volume, and 35% less weight, compared to conventional brake systems. As another advantage, service life is increased and the C/C-SiC brake pads do not cause wear or damage to the installation or elevator system when the brakes are tested periodically [121]. Since 2004 several thousands of C/C-SiC brake pads based on 2D-woven fabrics and LSI has been produced by FCT (Germany).

C/SiC friction pads are also used in niche markets such as in brake systems for sled or crash test facilities for automotive industry and in emergency brakes for fun rides in amusement parks. For the latter, controlled deceleration achieved by constant friction behavior together with long service live is the main advantage of C/SiC brake pads. Compared to conventional friction pads, C/SiC materials offer high abrasive resistance and can withstand the required daily safety test brakings for a complete season.

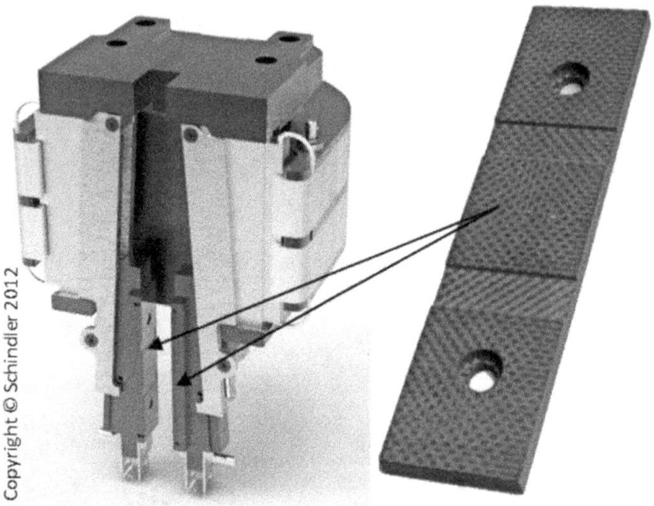

FIGURE 6.69

Emergency brake system of high speed elevators (left, Schindler Elevator Limited) and C/C-SiC brake pads (142 mm × 34 mm × 6 mm), manufactured via LSI (right, DLR).

For testing the crash behavior of automotive components, a high performance test facility has been developed by MESSRING Systembau MSG GmbH, Germany (Messring), using C/SiC as friction material. Thereby, test items, such as car seats are mounted on a sled ($m_{max} = 3$ tons) equipped with a brake wedge. The sled is decelerated from a predefined velocity of up to 80 km/h by pushing the brake wedge into a hydraulic brake system (Figure 6.70) [124]. Thereby, the brake wedge enters the gap between a top and bottom brake shoe and moves the top brake shoe against a hydraulic

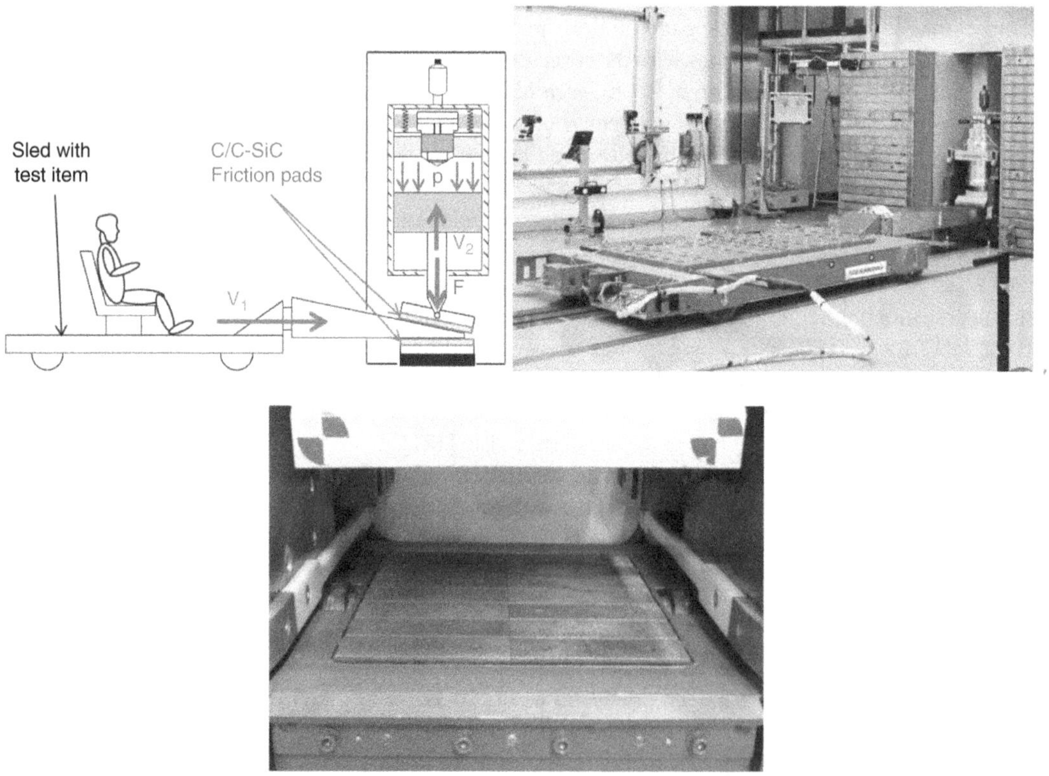

FIGURE 6.70

Schematic side view (top left) of a crash test facility (top right) and lower brake shoe, equipped with 14 C/C-SiC brake pads (200 mm × 50 mm × 8 mm each, bottom, all figures Messring).

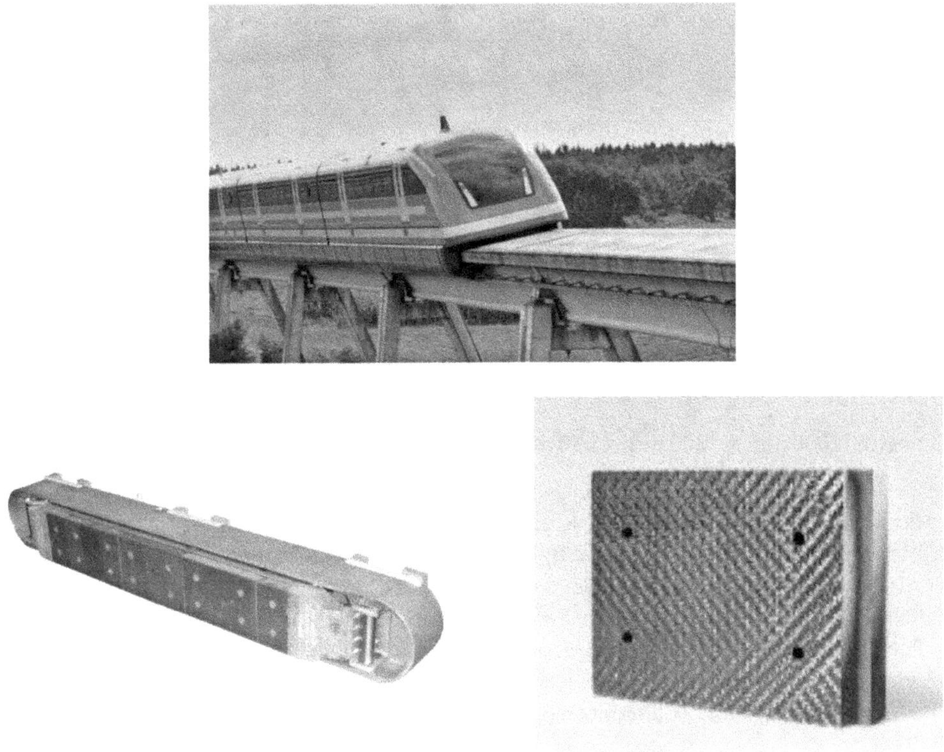

FIGURE 6.71

Emergency runner system for the Transrapid (top, ThyssenKrupp Transrapid GmbH © Fritz Stoiber Productions GmbH) in Shanghai. Sliding unit (bottom left, SKT), equipped with 5 C/C-SiC (PCI) pads (600 mm × 110 mm × 22 mm in total) and single C/C-SiC pad (159 mm × 110 mm × 28 mm, bottom right, SKT).

cylinder. In order to simulate realistic crash conditions, the deceleration can be controlled by varying the braking force generated by the hydraulic cylinder. Thus, maximum braking forces of 2–3.2 MN and contact pressures of 6–9 MPa are achieved, enabling maximum decelerations of 70 g. C/SiC friction pads mounted on the brake shoes offer sufficient mechanical strength to withstand the high impact and friction forces. Due to their high abrasive resistance, wear can be minimized leading to a reliable system and low maintenance cost.

Due to the variability of raw materials and manufacturing processes, C/C-SiC materials can be tailored in a wide range to match the specifications of different applications. This could be demonstrated by the development of C/C-SiC sliding elements. In contrast to brake applications where usually a high COF is welcome, low COF is required for the emergency runners of the Transrapid. In the case of malfunction of the electromagnetic hovering system, the train has to glide on the concrete driveway at a maximum starting velocity of up to 500 km/h. Thereby, high temperature-resistant C/C-SiC sliding elements manufactured by SKT provide both low wear (<0.1 mm/km) and low COF (<0.3) to ensure a safe emergency gliding and transport to the service station (Figure 6.71) [3]. Since 2002, several thousand pads have been produced for experimental trains as well as for the Shanghai MAGLEV Train.

6.4.4 Applications for High Temperature Treatment of Metals

For high temperature treatment of metallic components, C/SiC materials were introduced in the late 1990s for the design of lightweight and long-term stable charging devices (Figure 6.72). Due to their lower mass and heat capacity compared to metals, higher heating up and cooling down rates are possible, leading to faster processes and lower energy consumption. Additionally, due to the low wall thickness of the C/SiC structures, the number of components treated in one batch can be increased. Another main advantage of C/SiC devices is their high thermal stability. Due to the low CTE of carbon-fiber-reinforced SiC, even unsymmetrical structures with inhomogeneous wall thicknesses are geometrically stable at varying temperatures and do not show distortions even after years in service and several thousand heat treatment cycles. Additionally, high precision charging devices are possible, which can be handled and

FIGURE 6.72

CMC charging device (2000 mm × 1200 mm × 650 mm) for thermal treatment of metals after 8 years in service (SKT).

loaded automatically by robotic systems. In contrast to metallic structures, where lifetime is limited usually to less than 1 year due to distortions and cracks caused by thermally induced internal stresses, C/SiC structures offer almost unlimited operational life if used correctly. Compared to C/C-charging devices, C/SiC materials are more stable in hydrogen containing atmospheres due to reduced methanization. Thereby lifetime can be increased by some orders of magnitude [3]. Further advantages are higher abrasive resistance and lower oil take-up during quenching, due to higher hardness and lower porosity compared to C/C. However, silicides can be formed in the components that have to be heat treated, limiting the application of Si-containing C/SiC devices to selected alloys and maximum temperatures of about 1000°C [3].

The C/C-SiC handling device, shown in Figure 6.73, is used in the automotive industry for hardening of gear box components. Here, metal rings are taken out of the heating furnace (T_{max} = 900°C, carburization atmosphere) and

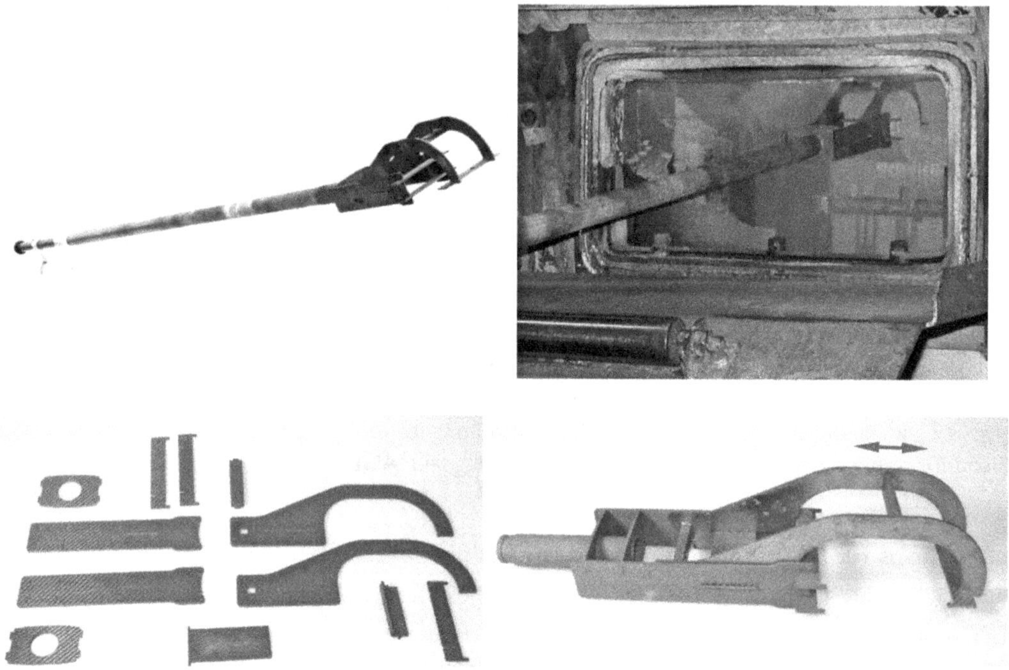

FIGURE 6.73

C/C-SiC (LSI) handling device (l = 2.1 m, top left, TECHNOKON®) during manual discharging of metal components from hardening furnace (top right, TECHNOKON®). Single components of the gripper, cut out of C/C plates (bottom left, DLR) and *in situ* joined gripper (two structures with moveable bow, ≈340 mm × 110 mm × 100 mm, bottom right, DLR) after siliconization and assembling. Permission from ThyssenKrupp Transrapid.

transported to the oil quenching bath, manually. Due to its low mass and heat capacity, the gripper is heated up and cooled down quickly in the furnace and during the transport in air at ambient temperature. In contrast to metallic structures, the C/C-SiC gripper stays stable during this thermal cycling (≈400 cycles per day) and no heat distortion is observed. This ensures a reliable operation of the moving clamp mechanism, integrated in the hot head section. Despite the high material temperatures, no severe oxidation is observed leading to a service life well above thousand hours to date. With the thin-walled, lightweight structure, weight can be minimized, and productivity can be increased.

6.5 SUMMARY

The industrial use of C/SiC materials is still focused on niche markets. However, existing application areas have been expanded and novel application areas, such as rocket propulsion and aircraft brakes, could be opened in recent years. This was mainly possible by continuously improving the manufacturing processes. Thereby, not only the increase of material performance, like strength or environmental stability, has been in the focus, but also process reliability and reproducibility as well as the reduction of manufacturing cost.

Highly sophisticated processes are available for the manufacture of high performance components in aerospace applications, which are produced as single parts or in very small series. For the production of C/SiC components in medium scale, the processes have been improved along the whole manufacturing chain. As an example, for the LSI process, cost efficient and reproducible methods for the buildup of CFRP preforms, especially for shaping and matrix curing has been investigated. Cycle times of the high temperature process steps, pyrolysis and siliconization, could be reduced significantly by the introduction of continuous processes, replacing time consuming, discontinuous batch processes. New technologies have been investigated successfully, offering very fast and controlled heating rates, and a high potential for further reductions of cycle times. Last but not least, cost efficient tools and strategies for the machining of C/SiC components have been in the focus of research projects. All these development activities are ongoing, in order to obtain significant reductions of production cost and increase of process reliability, which is still necessary to introduce C/SiC materials to novel, economically critical application areas or to expand present applications to mass markets.

In order to reduce development cost as well as to introduce C/SiC components in demanding applications, it is widely recognized and accepted that reliable tools for design and simulation are needed. Here, the main challenges are the high variability of the material properties, which are in many cases defined along with component manufacture, the anisotropic thermomechanical properties, the complex, inhomogeneous material composition and microstructure, and the complex failure mechanism. Other challenges are inhomogeneities and localized imperfections of the materials, which cannot be completely avoided during manufacture and service. In order to calculate the real performance of the component in its specific load scenario, these defects have to be detected by reliable nondestructive evaluation and implemented in simulation models. Simulation tools based on these realistic material and component models require further development in order to provide a reliable lifetime prediction, which is mandatory for the introduction of C/SiC components in safety critical applications.

ACKNOWLEDGMENTS

The author would like to acknowledge the valuable support received from:

- R. Weiß, A. Lauer, and B. Groos from Schunk Kohlenstofftechnik, and A. Kienzle from SGL Group for enhancing the contents regarding MI-based materials and their applications in friction systems and high temperature furnaces.
- K. Handrick and F. Motschmann from MT Aerospace as well as P. Spriet, J. Thebault, E. Bouillon, and T. Pichon from Herakles, for updating the data for CVI technology and aerospace applications.
- C. Wilhelmi from EADS-IW, S. Beyer from Astrium ST, A. Szweda from COIC, M. Zornik from MS Production, A. Müller and D. Decker from Clariant, H. Beurich from HB Consultancy, and M. Frieß from DLR for their additions to PIP technology and related space applications.
- K. Osl from Umbra Cuscinetti As, R. Renz from Dr. Ing. h.c. F. Porsche AG, R. Weber from MESSRING Systembau MSG GmbH, K. Berroth from FCT Ingenieurkeramik GmbH, E. Orbekk from Nammo AS, P. Kohler from

Inventio, F. Heine from Tesat-Spacecom GmbH & Co.KG, A. Vorobiev from Moscow Aviation Institute, and D. Kemmerle from Technokon® for detailed information regarding C/SiC applications.

- The colleagues from DLR, Institute of Structures and Design, Stuttgart with special thanks to T. Reimer, H. Böhrk, H. Weihs, C. Zuber, and H. Hald for background information and figures regarding TPSs for spacecraft.
- Many thanks to A. Johnson for his very valuable advices on correct English usage.

ABBREVIATIONS

Astrium ST	ASTRIUM Space Transportation, Munich, Germany
AZ electronics	AZ Electronic Materials Ltd.
BSCCB	Brembo SGL Carbon Ceramic Brakes, Stezzano, Italy
C	Carbon
Carbotex®	C/C material (CVI) of EADS-ST / EADS-IW
Carbotex®SI	Carbotex Surface infiltrated; C/C-SiC material of EADS-ST / EADS-IW
C/C	Carbon fiber reinforced Carbon
C/C- SiC	Carbon fiber reinforced Carbon with Silicon Carbide matrix
C/SiC	Carbon fiber reinforced Silicon Carbide
CFRP	Carbon Fiber Reinforced Plastic
Clariant	Clariant International Ltd., Muttenz, Switzerland
CMC	Ceramic Matrix Composites
COF	Coefficient of Friction
COIC	COI Ceramics, Inc., USA
CT	Computer Tomography
CTE	Coefficients of Thermal Expansion
CVI	Chemical Vapor Infiltration
CVD	Chemical Vapor Deposition
DMC	Double Matrix Composite; C/SiC material (PIP) of MS Production
DMR	Dual-Mode Ramjets
EADS-IW	EADS Deutschland GmbH, EADS Innovation Works, Munich, Germany
EEMS	Extreme Environment Materials Solutions (EEMS®, LLC), USA
ESA	European Space Agency
FCT	FCT Ingenieurkeramik GmbH, Rauenstein, Germany
F-CVI	Forced-Chemical Vapour Infiltration
FMB	Fiber–Matrix Bonding
FRP	Fiber-Reinforced Plastics
G-CVI	Gradient-Chemical Vapor Infiltration
GE	General Electric Ceramic Composite Products, Newark, USA
Goodrich	Goodrich's High Temperature Composites, Santa Fe Springs, USA; part of UTC
Herakles	Herakles, Safran Group, France
HT	High Tenacity (carbon fiber type)
HM	High Modulus (carbon fiber type)
Hyper-Therm	Hyper-Therm HTC (High Temperature Composites), USA
I-CVI	Isothermal Isobaric Chemical Vapor Infiltration
IM	Intermediate Modulus (carbon fiber type)
JAXA	Japan Aerospace Exploration Agency, Japan
K	Filament count of fiber (1K = 1000 filaments)
Keraman®	CMC materials (CVI) of MT Aerospace
Kion Defense	Kion Defense Technologies, Inc., USA
LCT	Laser Communication Terminals
LPI	Liquid Polymer Infiltration
LSI	Liquid Silicon Infiltration

MAI	Moscow Aviation Institute, Moscow, Russia
MAGLEV Train	Magnetic Levitation Train
Messring	MESSRING Systembau MSG GmbH, Krailling, Germany
MI	Melt Infiltration
MMH/ N_2O_4	Monomethylhydrazine / Di-Nitrogen-Tetroxide
MOV'IT	MOV'IT GmbH, Ottersweier, Germany
MSP	MS Production, Bled, Slovenia
MT	MT Aerospace, Augsburg, Germany
Nammo	Nammo AS, Norway
NGF	Nippon Graphite Fiber Corporation, Japan
OAO "Kompozit"	OAO "Kompozit," Moscow, Russia
PCCB	Porsche Ceramic Composite Brake
PCCC	Porsche Ceramic Composite Clutch
P-CVI	Pressure Pulsed-Chemical Vapor Infiltration
PCI	Pack Cementation and Capillary Infiltration
PMI	Pressurized Melt Infiltration
PMS	Polymethylsilane
PIP	Polymer Infiltration and Pyrolysis
prepreg	Resin Pre-impregnated Fibers/Fabrics
PTAH	Paroi Tissee Application Hypersonique
PTAH-SOCAR	Paroi Tissee Application Hypersonique (Simple Operational Composites for Advanced Ramjet)
pyC	pyrolitic Carbon
Roxel	ROXEL S.A.S., La Boursidière, France
RT	Room Temperature
RTM	Resin Transfer Molding
SEPCARBINOX®	C/SiC material (CVI) of Herakles
Si	Silicon
SiC	Silicon Carbide
SICARBON®	C/SiC Material (PIP) of EADS-ST / EADS-IW
SICTEX®	C/SiC Material (LSI) of EADS-ST / EADS-IW
SICTEX®SI	Surface Infiltrated CARBOTEX®; C/C-SiC material of EADS ST / EADS IW
SIGRASIC®	C/SiC Materials (LSI) of SGL
SGL	SGL Group, Meitingen, Germany
SSiC	Sintered Silicon Carbide
SiSiC	Silicon Infiltrated Silicon Carbide
SKT	Schunk Kohlenstofftechnik GmbH, Gießen, Germany
Surface Transforms	Surface Transforms PLC, UK
UHM	Ultra High Modulus (carbon fibers type)
UTC	UTC Aerospace Systems, USA
Starfire Systems	Starfire Systems, Malta, USA
Tenax	Toho Tenax Co. Ltd., Japan
Toray	Toray Industries, Inc., Japan
TPS	Thermal Protection Systems
TRL	Technology Readiness Level (NASA)
VAP	Vacuum Assisted Polymer Infiltration

REFERENCES

1. I. L. Stimson and R. Fisher. Design and engineering of carbon brakes. *Philos. Trans. R. Soc.*, 294(1411), 583–590 (1980).
2. Carbon brakes for Concorde. Flight International (December 1971), p. 1031, (1971).
3. R. Weiß. Carbon/carbons and their industrial applications, in *Ceramic Matrix Composites*, edited by W. Krenkel (Wiley-VCH Verlag, Weinheim, 2008), pp. 69–108.

4. R. Weiß. Carbon fibre reinforced CMCs: manufacture, properties, oxidation protection, in *High Temperature Ceramic Matrix Composites*, edited by W. Krenkel, R. Naslain, and H. Schneider (Wiley-VCH, Weinheim, Germany, 2001), pp. 440–456.
5. E. Cordano, C. Mao, M. Labanti, G. Martignani, G. L. Minoccari, L. Pilotti, R. Weiss, and J. Schmidt. Oxidation testing and mechanical characterisation of BoraSiC coated C/C-SiC composites. Materials Week, International Congress on Advanced Materials, Their Processes and Applications, Munich, Germany, September 30–October 2, 2002, published on CD.
6. A. S. Fareed. Silicon carbide and oxide fiber reinforced alumina matrix composites fabricated via direct metal oxidation, in *Handbook of Ceramic Composites*, edited by N. P. Bansal (Kluwer Academic Publishers, USA, 2005), pp. 277–307.
7. S. Beyer, S. Schmidt, F. Maidl, R. Meistring, M. Bouchez, and P. Peres. Advanced composite materials for current and future propulsion and industrial applications. *Adv.Sci. Technol.*, 50, 174–181 (Trans Tech Publications, Switzerland, 2006). doi:10.4028/www.scientific.net/AST.50.174
8. SGL Carbon Group. Industrial composites, SIGRASIC®/TAVCOR®, products in carbon fiber-reinforced silicon carbide, Company Brochure, Meitingen (2005).
9. V. I. Trefilov. *Ceramic- and Carbon-Matrix Composites* (Chapman & Hall, 1995).
10. H. Moissan. Nouvelles recherches sur la météorite de canon diabolo. *Comptes rendus*, 139, 773–786 (1904).
11. A. W. Weimer. *Carbide, Nitride and Boride Materials Synthesis and Processing* (Chapman & Hall, London, 1997), pp. 115–122, ISBN 0-412-54060-6.
12. Y. Katoh and S. M. Dong. Thermomechanical properties and microstructure of silicon carbide composites fabricated by nano-infiltration and transient eutectic processes. *Fusion Eng. Des.*, 61–62, 723–731 (2002).
13. A. Kohyama. CMC for nuclear applications, in *Ceramic Matrix Composites*, edited by W. Krenkel (Wiley-VCH Verlag, Weinheim, 2008), pp. 357–359.
14. S. Beyer, S. Schmidt, P. Peres, and M. Bouchez. PTAH-SOCAR fuel-cooled composite material structure for dual-mode ramjet and liquid rocket engines [AIAA 2005-3644]. 41st AIAA/ASME/SAE/ASEE Joint Propulsion Conference & Exhibit, Tucson, AZ, July 10–13, 2005.
15. I. Berdoyes, J. Thebault, and E. Bouillon. Improved SiC/SiC and C/C materials applications parts. Proceedings of the European Congress on Advanced Materials and Processes [EUROMAT 2005], Prague, Czech Republic, September 5–8, 2005.
16. G. Evans and R. Naslain. *High Temperature Ceramic Matrix Composites*. Vols 57 and 58: Ceramic Transactions (American Ceramic Society, Westerville, OH, 1995).
17. M. H. Lewis. Interphase formation, microstructure and performance, in *High Temperature Ceramic Matrix Composites*, edited by W. Krenkel, R. Naslain, and H. Schneider (Wiley-VCH, Weinheim, Germany, 2001), pp. 136–148.
18. J. Thebault, inventor; Societe Europeenne de Propulsion, assignee. Composite materials constituted by a matrix in resin coke carbon, reinforced with pyrolitic carbon-coated refractory fibers. U.S. Patent 4,748,079 [Filed 1985, Priority France 1983] (May 31, 1988).
19. J. Thebault, inventor; Societe Europeenne de Propulsion, assignee. Composite material with refractory fibrous reinforcement and ceramic matrix. U.S. Patent 5,026,604 [Filed 1988, Priority France 1984] (June 25, 1991).
20. D. P. Stinton, T. M. Besman, and R. A. Lowden. Advanced ceramics by chemical vapour deposition techniques. *Am. Ceram. Soc. Bull.*, 67, 350–355 (1988).
21. A. Mühlratzer and M. Leuchs. Applications of non-oxide CMC, in *High Temperature Ceramic Matrix Composites*, edited by W. Krenkel, R. Naslain, and H. Schneider (Wiley-VCH, Weinheim, Germany, 2001), pp. 288–289.
22. R. Naslain. CVI composites, in *Ceramic-Matrix Composites*, edited by R. Warren (Chapman & Hall, New York, 1992), pp. 199–244.
23. R. Naslain and F. Langlais. CVD-processing of ceramic–ceramic composite materials, in *Tailoring Multiphase and Composite Ceramics*, edited by R. E. Tressler, G. L. Messing, C. G. Pantano, and R. E. Newnham. Vol. 20: Materials Science Research, (Springer, 1986), pp. 145–164.
24. D. P. Stinton, A. J. Caputo, and R. A. Lowden. Synthesis of fibre-reinforced SiC composites by chemical vapour infiltration. *J. Am. Ceram. Soc. Bull.*, 65(2), 347–350 (1986).
25. R. Naslain. The concept of layered interphases in SiC/SiC. Proceedings of the Second International Conference on High Temperature Ceramic Matrix Composites (HTCMC II): Manufacturing and Materials Development, Santa Barbara, CA, August 21–24, 1995. Vol. 58: Ceramic Transactions, edited by A. G. Evans and R. Naslain (American Ceramic Society, 1995), pp. 23–26.
26. F. Lamouroux, S. Bertrand, R. Pailler, R. Naslain, and M. Cataldi. Oxidation-resistant carbon-fiber-reinforced ceramic-matrix composites. *Compos. Sci. Technol.*, 59(7), 1073–1085 (1999).
27. P. Bouillon, G. Habarou, P. C. Spriet, J. L. Lecordix, G. C. Ojard, G. D. Linsey, and D. T. Feindel. Characterization and nozzle test experience of a self sealing ceramic matrix composite for gas turbine applications [Paper GT-2002-30458]. ASME Turbo Expo Land, Sea and Air, Amsterdam, The Netherlands, June 3–6 (ASME, 2002).

28. R. Naslain, A. Guette, F. Rebillat, R. Pailler, F. Langlais, and X. Bourrat. Boron bearing species in ceramic matrix composites for long term aerospace applications. *J. Solid State Chem.*, 177, 449–456 (2004).

29. F. Christin. CMC materials for space and aeronautical application, in *Ceramic Matrix Composites*, edited by W. Krenkel (Wiley-VCH Verlag, Weinheim, 2008), pp. 357–359.

30. E. Bouillon, P. C. Spriet, G. Habarou, C. Louchet, and T. Arnold. Engine test and post engine test characterization of self sealing ceramic matrix composites for nozzle applications in gas turbine engines. Proceedings of ASME Turbo Expo 2004, Power for Land Sea and Air, Vienna, Austria, June 14–17 (ASME, 2004).

31. M. Leuchs and A. Mühlratzer. CVI-Verfahren zur Herstellung faserverstärkter Keramik—Herstellung, Eigenschaften, Anwendungen, in *Keramische Verbundwerkstoffe* edited by W. Krenkel (Wiley-VCH, Weinheim, Germany, 2003), pp. 163–173.

32. D. Desnoyer, A. Lacombe and J. M. Rouges. Large thin composite structural parts. Proceedings of the International Conference on Spacecraft Structures and Mechanical Testing, Nordwijk, The Netherlands, April 24–26 (ESA, 1991).

33. E. Fitzer and R. Gadow. Fibre-reinforced silicon carbide. *Am. Ceram. Soc. Bull.*, 65, 368–372 (1986).

34. R. P. Boisvert and R. J. Diefendorf. Polymeric precursor SiC matrix composites. *Ceram. Eng. Sci. Proc.*, 9, 873–880 (1988).

35. G. Motz, S. Schmidt, and S. Beyer. The PIP-process: Precursor properties and applications, in *Ceramic Matrix Composites*, edited by W. Krenkel (Wiley-VCH Verlag, Weinheim, 2008), pp. 357–359.

36. W. Schäfer and W. D. Vogel. Faserverstärkte Keramiken hergestellt durch Polymerinfiltration, in *Keramische Verbundwerkstoffe*, edited by W. Krenkel (Wiley-VCH, Weinheim, Germany, 2003), pp. 76–94.

37. W. J. Sherwood. CMCs come down to earth. *Am. Ceram. Soc. Bull.*, 82(8) (2003).

38. S. Schmidt, S. Beyer, H. Knabe, H. Immich, R. Meistring, and A. Gessler. Ceramic matrix composites: a challenge in space propulsion technology applications. *J. Appl. Ceram. Technol.*, 2(2), 85–96 (2005).

39. P. Greil. Polymer derived engineering materials. *Adv. Eng. Mater.*, 2, 339–348 (2000).

40. C. C. Evans, A. C. Parmee, and R. W. Rainbow. Silicon treatment of carbon fiber-carbon composites. Proceedings of 4th London International Carbon and Graphite Conference, London, UK, September 23–27, 1974, pp. 231–235.

41. Winnacker, Küchler, Chemische Technik: Prozesse und Produkte, vol. 8, edited by R. Dittmeyer, W. Keim, G. Kreysa, and A. Oberholz (Wiley-VCH, Weinheim, Germany, 2005), pp. 1166–1173.

42. U. Gruber and M. Heine, inventor; SGL Technik GmbH, assignee. Mit Graphitkurzfasern verstärkter Siliciumcarbidkörper. Patent DE 19710105 A1 [Applied December 3, 1997] (September 17, 1998).

43. D. Neudeck and A. Wüllner. *"Bremsen mit nichtmetallischen Bremsscheiben," Bremsenhandbuch*, edited by B. Breuer and K. H. Bill (Vieweg & Teubner/GWV Fachverlage, Wiesbaden, Germany, 2006), pp. 420–426.

44. C. Zuber, T. Reimer, K. Stubicar, B. Heidenreich, and H. Hald. Manufacturing of the CMC nose cap for the EXPERT spacecraft. 34th International Conference and Exposition on Advanced Ceramics and Composites, Daytona Beach, FL, January 24–29 (American Ceramic Society, 2010).

45. C. Zuber and B. Heidenreich. Development of a net shape manufacturing method for ventilated brake discs in single piece design, in *Ceramic Forum International*, Vol. 13, edited by W. Krenkel (DKG, DGM, Göller Verlag GmbH, Baden-Baden, 2005), pp. 16–22.

46. M. Frieß, M. Scheiffele, W. Zankl and F. Hofmann. Development, manufacture and characterization of C/C-SiC components based on filament winding, in *High Temperature Ceramic Matrix Composites*, edited by W. Krenkel, J. Lamon (Aviso Verlagsges, Berlin, 2010), pp. 92–97.

47. F. Breede, D. Koch, and M. Frieß. Development and characterisation of novel C/C-SiC material using LSI method and effective fibre preform techniques for rocket nozzle extensions. DFG SFB/TRR 40, Annual report 2012, (2012).

48. F. Gern. Kapillarität und Infiltrationsverhalten bei der Flüssigsilicierung von Carbon/Carbon-Bauteilen, Doctoral Thesis, University of Stuttgart, DLR-FB 95-26, (1995).

49. F. Gern and W. Krenkel. Morphology and infiltration dynamics of liquid siliconized carbon/carbon. *Proc. 10th Int. Conf.Compos. Mater.* [ICCM-10], 3, 61–68 (Whistler, BC, 1995).

50. R. Weiß. Oxidation behaviour of carbon/carbon composites. 2nd European Workshop on High Temperature Behaviour of Advanced Materials, Béziers, France, September 29–October 1, 1993.

51. U. Papenburg, K. K. O. Bär, R. Heidenreich, and R. Weiß. Die Herstellung und Charakterisierung von oxidationsgeschützten C/SiC-Verbundwerkstoffen. *Fortschrittsberichte der DKG*, 9(4), 141–186 (1994).

52. M. Frieß, W. Krenkel, R. Kochendörfer, R. Brandt, G. Neuer, and H.-P. Maier. High temperature materials and hot structures, in *Basic Research and Technologies for Two-Stage-to-Orbit Vehicles*, edited by D. Jakob, G. Sachs, and S. Wagner (Wiley-VCH, Weinheim, Germany, 2005), pp. 599–526.

53. M. Frieß, R. Renz, and W. Krenkel. Graded ceramic matrix composites by LSI processing, *Advances in Science and Technology*, 40, *Advanced Inorganic structural fiber composites IV*, edited by P. Vincenzini and C. Badini (Techna Srl, 2003), ISBN: 88-86538-42-1, 2003.

54. R. Kochendörfer and W. Krenkel. CMC intake ramp for hypersonic propulsion systems, in *Ceramic Matrix Composites I: Design Durability and Performance*, edited by A. G. Evans and R. Naslain. Vol. 57: Ceramic Transactions (American Ceramic Society, 1995), pp. 13–22.
55. W. Krenkel, T. Henke, and N. Mason. In situ joined CMC components. International Conference on Ceramic and Metal Matrix Composites CMMC, San Sebastián, Spain, September 9–12, 1996.
56. H.-S. Park, M. Willert-Porada, and W. Krenkel. Rapid pyrolysis of CFRP using microwaves for the production of C/SiC composite, in *High Temperature Ceramic Materials and Composites*, edited by W. Krenkel and J. Lamon (Aviso Verlagsgesellschaft, Berlin, Germany, 2010), pp. 132–137.
57. A. Konschak, R. Schulte, W. Krenkel, J. Hennicke, and H. U. Kessel. Fast Siliconization of C/C-Preforms via LSI, in *High Temperature Ceramic Materials and Composites*, edited by W. Krenkel and J. Lamon (Aviso Verlagsgesellschaft, Berlin, Germany, 2010), pp. 113–118.
58. H. Abu El-Hiya and W. Krenkel. Cost analysis for the manufacture of C/C-SiC structural parts, edited by W. Krenkel, R. Naslain, and H. Schneider (Wiley-VCH, Weinheim, Germany, 2001), pp. 440–456.
59. W. Krenkel. Kosten- und Qualitätssicherungs-Aspekte bei der Herstellung von CMC-Bauteilen, in *Keramische Verbundwerkstoffe*, edited by W. Krenkel (Wiley-VCH, Weinheim, Germany, 2003), pp. 173–192.
60. J. W. Hutchinson and H. M. Jensen. Models of fiber debonding and pullout in brittle composites with friction. *Mech. Mater.* 9, 139–163 (1990), Elsevier, (1990).
61. A. G. Evans, F. W. Zok, and T. J. Mackin. *High Temperature Mechanical Behavior of Ceramic Composites* (Butterworth-Heinemann, Boston, 1995) pp. 3–84.
62. SGL Carbon Group. SIGRASIC 6010 GNJ—Short fibre-reinforced ceramics. Data sheet, 11 2005/0.5, (2005).
63. SGL Carbon Group. SIGRASIC 1500 J – Fabric-reinforced ceramics, Data sheet, 11 2005/0.5, (2005).
64. Toho Tenax, Data sheet April 2011, www.tohotenax-eu.com, http://www.tohotenax-eu.com/fileadmin/tohotenax/downloads/Produkte/Technische%20Datenblaetter/en/HTA_en_2011-04.pdf, August 21st 2014.
65. Toray Carbon Fibers America, Inc., Technical data sheet No.CFA-007, 2012, http://www.toraycfa.com/pdfs/T800HDataSheet.pdf, last access August 21, 2014.
66. Toray Carbon Fibers America, Inc., Technical data sheet No.CFA-014, 2012, http://www.toraycfa.com/pdfs/M40JDataSheet.pdf, last access August 21, 2014.
67. Nippon Graphite Fiber Corporation. Technical Data YS-A, Granoc Yarn YS-A Series, http://www.ngfworld.com/dcms_media/other/skill_skill04.pdf, last access August 21, 2014.
68. Schunk Ingenieurkeramik GmbH, Wekstoffdatenblatt CarSiC, 2009, www.tribo@schunkgroup.com, last access August 21, 2014.
69. B. Heidenreich, R. Renz, and W. Krenkel. Short fibre reinforced CMC materials for high performance brakes, in *High Temperature Ceramic Matrix Composites*, edited by W. Krenkel, R. Naslain, and H. Schneider (Wiley-VCH, Weinheim, Germany, 2001), pp. 809–815.
70. A. Neubrand, B. Thielicke, and A. Kienzle. Four point bending tests of a C/SiC material for industrial applications under quasistactic and cyclic loading. Proceedings of the 5th International Conference on High Temperature Ceramic Matrix Composites, pp. 205–210 (2004).
71. M. Henrich, D. Regener, and R. Weiß. Querschnittseinfluss beim Zugversuch von siliziertem kohlenstoffkurzfaserverstärktem Kohlenstoff (C/C-SiC). *Materialwissenschaft und Werkstofftechnik*, 35(1), 45–54 (2004) Wiley-VCH, ISSN: 0933-5137.
72. COI Ceramics, Inc., Silicon Carbide Ceramic Matrix Composites, Typical Characteristics of S400, 2012, http://www.coiceramics.com/nonoxidepg
73. A. Neubrand, B. Thielicke, and E. Hirsch. Investigation of the notch sensitivity of a short fibre reinforced C/SiC. Proceedings of the 5th International Conference on High Temperature Ceramic Matrix Composites, (2004), September 12–16, Seattle, USA.
74. SGL Carbon Group, 2012. Carbon-ceramic-brake disks. Company Brochure, www.sglgroup.com/cms/international/products/product-groups/bd/carbon-ceramic-brake-disks. Last access July 28, 2012.
75. W. Krenkel, 2000. Entwicklung eines kostengünstigen Verfahrens zur Herstellung von Bauteilen aus keramischen Verbundwerkstoffen. *Development of a cost efficient process for the manufacture of CMC components*, Doctoral Thesis. University of Stuttgart, DLR Forschungsbericht 2000-4, 2004, ISSN 1434-8454, ISRN DLR-FB–2000-4.
76. F. Breede, E. Klatt, M. Frieß, and H. Voggenreiter. Mechanical properties and microstructures of C/C-SiC composite plates by wet filament winding technique, in *High Temperature Ceramic Matrix Composites*, edited by W. Krenkel and J. Lamon (Aviso Verlagsges, Berlin, 2010), pp. 276–282.
77. H.-P. Maier and K. Maile. Thermomechanisches Verhalten von C/C-SiC, Materialkennwertermittlung. Final report SFB 259, Teilprojekt B2b (2001).
78. R. Renz, B. Heidenreich, W. Krenkel, A. Schöppach, and F. Richter. CMC Materials for Lightweight and Low CTE Applications, 4th International Conference on High Temperature Ceramic Matrix Composites (HT-CMC 4), Munich, 1–3 October 2001, in

High Temperature Ceramic Matrix Composites, edited by W. Krenkel, R. Naslain, and H. Schneider (Wiley-VCH Verlag, Weinheim, 2001), pp. 839–845.

79. A. Schöppach, T. Petasch, B. Heidenreich, R. Renz, and W. Krenkel. Use of ceramic matrix composites in high precision laser communication optics. European Conference on Spacecraft Structures, Materials and Mechanical Testing, ESTEC, Nordwijk, The Netherlands (2000).
80. B. Heidenreich, M. Scheiffele, M. Tausendfreund, and H.-U. Wieland. C/C-SiC telescope structure for the laser communication terminal in TerraSAR-X, in *High Temperature Ceramic Matrix Composites*, edited by W. Krenkel and J. Lamon (Aviso Verlagsges, Berlin, 2010), pp. 505–512.
81. M. Frieß, W. Krenkel, R. Brandt, and G. Neuer. Influence of process parameters on the thermophysical properties of C/C-SiC, in *High Temperature Ceramic Matrix Composites*, edited by W. Krenkel, R. Naslain, and H. Schneider (Wiley-VCH, Weinheim, Germany, 2001), pp. 328–333.
82. J. Meinhardt, T. Woyke, F. Raether, and A. Kienzle. Measurement and simulation of the oxidation of carbon fibers and C/SiC ceramic. *Adv. Sci. Technol.*, 45, 1489–1494 (2006).
83. F. Raether, J. Meinhardt, and A. Kienzle. Oxidation behaviour of carbon short fibre reinforced C/SiC composites. *J. Eur. Ceram. Soc.*, 27, 1217–1221 (2007, Elsevier Ltd. (2007).
84. SGL Carbon AG, assignee. Verfahren zum Oxidationsschutz faserverstärkter kohlentoffhaltiger Verbundwerkstoffe und Verwendung eines nach dem Verfahren hergestellten Verbundwerkstoffes, Patent DE 10161218 B4 (June 3, 2003).
85. F. Christin, A global approach to fibre nD architectures and self sealing matrices: from research to production, in *High Temperature Ceramic Matrix Composites 5*, edited by M. Singh, R. J. Kerans, and E. Lara-Curzio (The American Ceramic Society, Westerville, Ohio, USA, 2004), pp. 477–484.
86. A. Mühlratzer and M. Leuchs. Applications of non-oxide CMC, in *High Temperature Ceramic Matrix Composites*, edited by W. Krenkel, R. Naslain, and H. Schneider (Wiley-VCH, Weinheim, Germany, 2001), pp. 288–289.
87. H. Hald, et al. Development of a Nose Cap System for X 38. Proceedings of International Symposium Atmospheric Reentry Vehicles and Systems, Arcachon, France (1999).
88. U. Trabandt and W. Fischer. International congress on environmental systems, paper 01-ICES-184, Orlando, USA (2001).
89. H. Lange, K. Handrick, and S. Weiland. Status of CMC open flap development for EXPERT, in *Thermal Protection Systems and Hot Structures*, Proceedings of the 5th European Workshop, ESTEC, Noordwijk, The Netherlands, 17–19 May, 2006, ESA SP.631E.41L, published on CD ROM, p.41.1, edited by K. Fletcher, (2006).
90. H. Lange, A. Steinacher, K. Handrick, S. Weiland, D. Sygulla, S. Guedron, and T. Salmon. *Status of Flap Development for Future Re-Entry Vehicles (Pre-X)* 5th European Workshop on Thermal Protection Systems and Hot Structures, Noordwijk, The Netherlands, May 17–19 (2006).
91. F. Infed, K. Handrick, H. Lange, A. Steinacher, S. Weiland, and C. Wegmann. Development of thermal protective seal for hot structure control surface actuator rod. *Acta Astronaut.*, 70, 122–138, Elsevier (2012).
92. A. Gülhan, F. Siebe, H. Weihs, T. Laux, J. Longo, T. Eggers, J. Turner, and M. Hörschgen. The sharp edge flight experiment—a mission overview. Conference proceedings of 5th European workshop on Thermal Protection Systems & Hot Structures, Nordwijk, The Netherlands, SP 631 (2006).
93. H. Böhrk, H. Elsäßer, and H. Weihs. *The SHEFEX II Thermal Protection System*, 7th Symposium on Aerothermodynamics for Space Vehicles, Brügge, Belgium, 9–12 May 2011, (2011).
94. R. Knoche, E. Werth, S. Goek, and M. Weth. Thermo-mechanical design of the Astrium C/SiC experiment on SHEFEX II, in *High Temperature Ceramic Matrix Composites*, edited by W. Krenkel and J. Lamon (Aviso Verlagsges, Berlin, 2010), pp. 543–552.
95. S. Schmidt, S. Beyer, H. Knabe, H. Immich, R. Meistring, and A. Gessler. Advanced ceramic matrix composite materials for current and future propulsion technology applications. *Acta Astronaut.*, 55, 409–420 (2004).
96. F. Christin. Design, fabrication and application of C/C, C/SiC and SiC/SiC composites, in *High Temperature Ceramic Matrix Composites*, edited by W. Krenkel, R. Naslain, and H. Schneider (Wiley-VCH, Weinheim, Germany, 2001), pp. 731–743.
97. D. Ehmann, U. Gotzig, G. Schulte, S. Beyer, and S. Schmidt. New European Apogee Motor—advanced propulsion technology. First Symposium on Potentially Disruptive Technologies and Their Impact in Space Programs, Marseille, France, July 4–6, 2005.
98. S. Schmidt, S. Beyer, G. GahuZac, R. Meistring, and M. Bouchez. Advanced ceramic matrix composite materials for current and future propulsion system applications [AIAA 2004-4019]. 40th AIAA/ASME/SAE/ASEE Joint Propulsion Conference, Fort Lauderdale, FL, July 11–14, 2004.
99. C. Liu, J. Chen, H. Han, Y. Wang, and Z. Zhang. A long duration and high reliability liquid apogee engine for satellites. *Acta Astronaut.* 55, 401–408 (2004).
100. A. A. Kozlov, A. G. Vorobiev, I. N. Borovik, I. S. Kazennov, A. V. Lahin, E. A. Bogachev, and A. N. Timofeev. Development liquid rocket engine of small thrust with combustion chamber from carbon–ceramic composite material, in *Nanocomposites with Unique Properties and Applications in Medicine and Industry*, edited by J. Cuppoletti (2011), ISBN: 978-953-307-351-4.

101. S. Beyer, S. Schmidt, F. Maidl, R. Meistring, M. Bouchez, and P. Peres. Advanced ceramic matrix composite materials for current and future propulsion and industrial applications, 11th Ceramics Congress (international) and 4th Forum on New Materials. Acireale, Italy, 2006.
102. M. Bouchez, and S. Beyer. Ptah-socar fuel-cooled composite materials structure. *Prog. Propulsion Phys.*, 1, 627–644 (2009), http://dx.doi.org/10.1051/eucass/200901627, last access August 21, 2014.
103. S. Beyer, S. Schmidt-Wimmer, K. Quering, C. Wilhelmi, and M. Steinhilber. Technology status of fuel cooled ceramic matrix composites for Dual-Mode Ramjet (DMR) and Liquid Rocket Engine (LRE) applications. 18th AIAA International Space Planes and Hypersonic Systems and Technologies Conference, Tours, France, September 24–28, 2012).
104. Tesat-Spacecom. Laser products, p. 5, Company Brochure (2012).
105. W. A. Imbriale, S. Gao, and L. Boccia. *Space Antenna Handbook* [Google eBook] (John Wiley & Sons, 2012), p. 126.
106. E. P. Bouillon, P. C. Spriet, G. Habarou, T. Arnold, G. C. Ojard, D. T. Feindel, C. P. Logan, K. Rogers, G. Doppes, R. J. Miller, Z. W. Grabowski, and D. P. Stetson. Engine test experience and characterization of self sealing ceramic matrix composites for nozzle applications in gas turbine engines. Proceedings of ASME Turbo Expo 2003, Power for Land Sea and Air, Atlanta, GA, June 16–19, (ASME, 2003).
107. L. Zawada, G. Richardson, and P. Spriet. Ceramic matrix composites for aerospace turbine engine exhaust nozzles, in *High Temperature Ceramic Matrix Composites 5*, edited by M. Singh, R. J. Kerans, and E. Lara-Curzio (The American Ceramic Society, Westerville, Ohio, USA, 2004), pp. 499–506.
108. M. Frieß, C. Zuber, B. Heidenreich, H. Ciezki, A. Feinauer, and J. V. Kampen. Keramische Verbundstrukturen für hochagile Flugkörper. Proceedings des Deutschen Luft und Raumfahrtkongress, Friedrichs-hafen, Germany (2005).
109. M. Frieß, C. Zuber, S. Hofmann, M. Crippa, and B. Heidenreich. CMC-Bauteile für Heißgasanwendungen: Von der Entwicklung des Prototypen bis hin zum Serienbauteil, in *Verbundwerkstoffe und Werkstoffverbunde*, edited by W. Krenkel (Wiley-VCH, 2009), pp. 257–264, ISBN 978-3-527-32615-0.
110. S. Kumara, A. Kumara, K. Sampathb, V. V. Bhanu Prasad, J. C. Chaudharya, A. K. Gupta, and G. Rohini Devi. Fabrication and erosion studies of C–SiC composite jet vanes in solid rocket motor exhaust. *J. Eur. Ceram. Soci.*, 31(13), 2425–2431 (Elsevier Science, 2011), ISSN: 0955-2219.
111. E. Orbekk. Novel TVC system utilizing guide vanes with jet flaps into a high efficiency compact nozzle. 41st AIAA/ASME/SAE/ASEE Joint Propulsion Conference & Exhibit, Tucson, AZ, July 10–13, 2005.
112. O. Verberne. Jet vane production facilities operational. Nammo Bulletin, 2006, http://de.scribd.com/doc/46524420/Nammo-Technical-Bulletin-2006, last access August 21 2014.
113. Net Resources International, Airforce-technology.com, Iris-t-air-air-guided-missile-germany, 2012, http://www.airforce-technology.com/projects/iris-t-air-air-guided-missile-germany/, last access August 21, 2014.
114. G. Savage. *C/C Composites* (Chapman & Hall, 1993), pp. 323–346.
115. R. J. Koehler. Manuscript of the presentation at the shareholders meeting of the SGL Carbon Group 2001, Germany (2001).
116. M. Zornik, inventors. Brake/clutch disc, such as for a vehicle. U.S. Patent 6,077,607 (June 20, 2000).
117. STILLEN Nissan GT-R R35 Carbon-Ceramic Matrix Brake Upgrade, 2012, http://www.stillen.com/advanced_search_result.php?keywords=Nissan+GT-R+R35+Carbon-Ceramic+Matrix&x=18&y=21, last access August 21, 2014.
118. Z. Stadler, M. Kermc, T. Kosmac, and A. Dakskobler. A motorcycle brake system with C/C–SiC composite brake discs. 28th International Conference on Advanced Ceramics and Composites, edited by E. Lara-Curzio and M. Readey. Cocoa Beach, FL, January 25–30, 2004. Vol. 25(4): Ceramic Engineering & Science Proceedings (American Ceramic Society, Westerville, OH, 2004), pp. 179–184.
119. K. Krnel, Z. Stadler, and T. Kosmač, Carbon/carbon–silicon-carbide dual-matrix composites for brake discs. *Mater. Manuf. Process.*, 23(6), 587–590 (2008). doi:10.1080/10426910802160437
120. C. Becker. Klein, stark, schwarz. *Christophorus—Das Porsche Magazin*, Christopherus, 306, 40–43 (2004).
121. R. Biffi, and B. Zäch. Schindler 700 – the Journey to the Top, Schindler Elevators, 2002, http://www.schindler.com/com/internet/en/media/press-releases-english/press-releases-2006-2000/schindler-700-the-journey-to-the-top.html, last access August 21, 2014.
122. H. Abu El-Hija, W. Krenkel, and S. Hugel. Development of C/C-SiC brake pads for high-performance elevators. *Int. J. Appl. Ceram. Technol.*, 2(2), 105–113 (2005).
123. H. Abu El-Hija, M. Kriescher, and B. Heidenreich. Herstellung von faserkeramischen Friktionswerkstoffen mit verschleißfester Reibschicht. CFI Ceramic Forum International, *Berichte der Deutschen Keramischen Gesellschaft, DKG*, 82(13), 23–30 (2005), ISSN 01739913.
124. Messring Systembau MSG GmbH. Servo Hydraulic Brake System for Test Sled. Code 2MC, Company Brochure (2010).

CHAPTER 7

ADVANCES IN SiC/SiC COMPOSITES FOR AERO-PROPULSION

JAMES A. DICARLO

Materials and Structures Division, NASA Glenn Research Center, Cleveland, OH, USA

7.1 INTRODUCTION

Silicon carbide (SiC) ceramic matrix composites (CMCs) reinforced by continuous-length, high performance, small-diameter (~10 to 15 μm), polycrystalline SiC fibers are considered enabling materials for a variety of advanced applications where lightweight reusable structural materials are required to operate for long time periods within extreme high temperature environments. Today the first generations of SiC/SiC CMCs with thermo-structural capability to ~1250°C are being introduced into the hot-section components of military and commercial gas turbine engines [1,2]. In comparison to superalloy metallic components, which at best operate at as high as ~1100°C, these CMC components will not only offer reduced engine weight, but also reduced component cooling-air requirements. Reduction in cooling air would then have the additional engine benefits of improved thrust-to-weight ratio, reduced fuel burn, and reduced harmful exhaust emissions. In order to seek further increases in these engine benefits, research has been ongoing within NASA aimed at developing SiC/SiC CMCs with even higher structural reliability and temperature capability.

The objective of this chapter is to detail, both fundamentally and practically, the key advances from recent SiC/SiC CMC research activities within NASA. Although these advances are primarily still under development and optimization, they do show various degrees of improvement over the materials and processes currently being implemented for SiC/SiC engine components. To this end, Section 7.2 presents information on the general constituent materials and process requirements for structurally reliable high temperature SiC/SiC engine components. Section 7.3 then discusses the primary fabrication routes currently being employed for production-ready SiC/SiC components in terms of their benefits and limitations. Section 7.4 then details recent advancements in the materials and processes for the SiC fiber, the fiber interfacial coating, the fiber geometry or architecture, the SiC-based matrix, and in the design methods for assembling these microstructural constituents into higher temperature and higher performance SiC/SiC composites and components. Where possible, CMC property data will be presented and analyzed that directly compares the materials and processes of the advanced approaches against those of current approaches. This property comparison will focus primarily on such key CMC properties as in-plane and thru-thickness tensile strength, thermal conductivity, creep resistance, and rupture behavior since it is these properties that directly control the CMC structural and upper temperature capability. Finally, Section 7.5 summarizes these results into current design guidelines for achieving SiC/SiC microstructures with improved intrinsic thermo-structural capability, and then discusses potential issues, such as thermo-mechanical shock, environmental attack, and creep-related residual stress development, which still require further research to understand whether they will be areas of concern for real engine components.

Ceramic Matrix Composites: Materials, Modeling and Technology, First Edition. Edited by Narottam P. Bansal and Jacques Lamon.

7.2 MATERIALS AND PROCESS REQUIREMENTS FOR STRUCTURALLY RELIABLE HIGH TEMPERATURE SiC/SiC COMPONENTS

As with any structural ceramic material, the first key performance requirement for a SiC/SiC CMC is the ability to display three-dimensional (3D) strength properties or allowables that are high enough to permit a component to achieve its desired service life. Like metallic hot-section components, ceramic components for aero-propulsion gas turbine engines will typically be designed to be thin-walled and cooled using conventional approaches such as film cooling (vanes, blades) and backside convective cooling (combustion liners, shrouds). Thus, component stresses typically arise from a combination of mechanical, aerodynamic, and thermal-gradient loads that produce both in-plane and thru-thickness stress–temperature distributions that can vary with time within the component walls and attachment areas. Under these stress conditions, over a time period generally equivalent to the component inspection cycle, the SiC/SiC composite's in-plane matrix cracking strength (MCS) should be sufficiently high enough to reliably avoid cracks that propagate into and across the component wall, thereby preventing aggressive application environments from reaching and degrading the crack-bridging fibers. Of additional importance is the ability of the CMC to avoid the occurrence of delamination cracks within the wall since these can significantly reduce the thru-thickness thermal conductivity, further increasing the thermal-gradient stresses and the probability of wall spallation. Thus, the constituents and geometries of the as-produced CMC should be optimized to ensure that the composite's MCSs, both in-plane and thru-thickness, sufficiently exceed the expected component service stresses in these directions.

In terms of maintaining these 3D strength characteristics *at high temperatures*, the key requirements of the SiC/SiC fiber and matrix constituents are (1) to be thermally stable at the expected application temperatures and (2) to avoid fracture under the maximum time, temperature, and stress conditions that they will experience during component service. Today these constituents will most always have a polycrystalline SiC microstructure in part due to the fact that single crystal microstructures are too costly to produce. Also, as seen for single crystal oxide fibers in oxide matrices, single crystal microstructures may not offer any significant performance advantage over polycrystalline materials due to rapid atomic diffusion on the fiber surface and pitting-induced fiber strength loss caused by irregular contact points with the multi-constituents in the composite environment [3]. Furthermore, if the SiC fiber and matrix contain a pure stoichiometric composition and grain sizes near ~500 nm, their intrinsic microstructural and strength stabilities under zero stress conditions are not an issue for thousands of hours at temperatures up to at least 1600°C. However, as soon as stress is applied to a SiC/SiC component, the grains in the constituent microstructures will begin to slide past each other, giving rise to measurable time-dependent deformation or creep that could start at as low as ~1100°C depending on the impurity phases in the grain boundaries. Two practical issues with constituent creep are undesirable dimensional changes for the CMC component and the growth of cavitation flaws at the grain triple points within the constituent microstructures. Eventually these flaws grow large enough to cause time-dependent constituent fracture or rupture, which in the worst case could cause rupture of the total SiC/SiC component. Unlike metals that can rupture at creep strains well over 1% so that undesirable dimensional changes are typically their life-limiting criteria, SiC-based constituents will typically rupture at strains below 1%, so that their use lives are primarily limited by rupture rather than by dimensional stability issues. Thus for structurally reliable high temperature operation for long times, the constituents of SiC/SiC composites should have microstructures and compositions that display the optimum in creep and rupture resistance under the stress and temperature conditions that they will encounter within the final application.

Another life-limiting and creep-related mechanism for high temperature SiC/SiC components is that the constituents almost always differ in creep behavior due to different compositions and/or to different stresses related to their geometries within the composite. To appreciate this issue, one can consider the situation where a constant and uniform mechanical load is applied to a composite and the fiber and matrix within the composite display significantly different creep behavior. For Case A, where the matrix is more creep prone than the fiber, the matrix will relax its initial stress and shed more of the CMC load onto the fibers, so that the composite upper-use temperature will then typically depend on such life-limiting issues as excessive fiber deformation, fiber time-dependent fracture or rupture, and/or adverse residual stress development within the composite. Regarding this last issue, after creep under uniform CMC tensile loads, the internal residual stress on the matrix at lower temperatures will be compressive, so that the CMC in-plane MCS will beneficially increase in this temperature regime [4]. However, after CMC creep under compressive loads, tensile residual stresses on the matrix will develop at lower temperatures, thereby decreasing the CMC MCS at these lower temperatures and increasing the chance of thru-thickness cracking if the local CMC load should become tensile. This reduction in MCS has been shown to be a likely scenario for components with nonuniform stresses in their walls, as, for example, when subject to thermal gradients across the walls [5]. As will be discussed, this Case A scenario is the situation with current SiC/SiC composites containing free silicon in the matrix. For Case B, where the matrix is more creep resistant than the fiber, the fiber in its primary creep stage will quickly shed its load to the matrix, so that initial composite tensile loads must be kept low to avoid the matrix stresses rising over time above their cracking limit. Bottom line is that, if possible, both the fiber and matrix should be highly creep resistant under the component service conditions, but if this is not the case, the matrix creep should be slightly greater than that of the fiber, so that during the component service life, the matrix carries some of the composite load and remains protective of the fiber without cracking.

7.3 CURRENT FABRICATION ROUTES FOR SiC/SiC ENGINE COMPONENTS

Today two primary SiC/SiC fabrication routes have achieved to a large degree the material and process requirements needed to produce complex-shaped engine components with microstructures capable of reliable structural operation for long times at temperatures greater than those of metallic superalloys. The first route was developed under the NASA Enabling Propulsion Materials (EPM) Program and was demonstrated by the production of small and large combustor liners [6]. Here continuous-length multi-fiber tows of a small-diameter, high performance SiC fiber type are woven into continuous fabric pieces, typically with a 0/90 balanced five-harness satin (5HS) fiber architecture. These 2D fabric pieces are then cut and judiciously stacked into component-shaped 3D preforms with fiber volume fractions up to 40%. Chemical vapor infiltration (CVI) is employed to first apply an interfacial coating or interphase on the fibers within the tows, and then to apply a thin protective SiC matrix layer over the interphase. To adequately achieve the key interfacial coating requirements of high environmental durability and high compliance for matrix crack deflection, the interphase composition is typically based on silicon-doped boron nitride (BN). The coating thicknesses can vary from ~0.1 to 0.5 μm depending on such factors as the fiber surface roughness and the degree of protection needed to avoid coating degradation during matrix formation and/or component service. The thin CVI SiC matrix coating is important in that it not only protects the interphase and fibers from subsequent matrix infiltration processes and from external environments, but it also provides a highly thermally conductive and creep-resistant matrix constituent. As will be discussed, NASA has determined that these latter two matrix properties, as well as the protective nature of the interface coating, can be further improved prior to the final matrix infiltration steps by a high temperature treatment of the CVI SiC-coated preform [7]. In the final steps of matrix fabrication, SiC particulate is slurry-cast into the component preform near room temperature, followed by a finishing step in which a molten metallic silicon alloy is infiltrated near 1400°C into any open remaining matrix porosity. The silicon bonds the SiC particulates together with little chemical reaction. This last step of this nonreactive melt infiltration (NRMI) route is important in that it results in a SiC-based matrix with low residual porosity, high thermal conductivity, and good environmental protection for the coated fibers. In addition, it minimizes the existence of large matrix flaws, thereby raising the stress levels required for formation of detrimental thru-thickness and delamination cracks in the composite. In following discussions, the matrices formed from the NRMI approach of CVI, slurry, and silicon melt infiltration (MI) will often be referred to as the CVI-MI matrices for simplicity.

In the second fabrication route, which has been developed and demonstrated by General Electric for a variety of turbine hot-section components, multi-fiber tows of a small-diameter, high performance SiC fiber type are first spread and coated by chemical vapor deposition (CVD) with a layered interfacial coating consisting of BN and silicon nitride (Si_3N_4) compositions [8]. The coated tows are then pre-pregged into unidirectional 2D tapes using a polymer-based binder containing SiC and carbon particulates. The tapes are then cut, oriented, and stacked into 3D preforms with total fiber volume content less than 30%. To achieve the component structural and shape requirements, the 3D preforms are assembled using methods much like those used for tape lay-up of polymer matrix composites. In the final steps of this "pre-preg" approach, the 3D preform is heated to a high temperature to decompose the polymer and to form the final matrix composition by infiltration with molten silicon. Unlike the NRMI route, the SiC-based matrix is formed *in situ* by the reaction of the silicon with the residual carbon in the preform. During this reactive MI (RMI) process, there is an exothermic reaction between silicon and carbon, resulting in an expansion of the matrix that puts the fibers under residual tension and the final matrix under residual compression, which is beneficial in increasing the composite in-plane cracking strength at low temperatures. However, the RMI route also causes a rise in local temperatures, which if not controlled could cause strength loss in those SiC fiber types that are produced at maximum temperatures near 1400°C. In addition, unless the interphase coating is thick enough (~1 μm) with the proper compositions, these temperature increases can increase the risk of fiber attack by the thermally induced diffusion of excess free silicon across the coating. Thus, SiC fibers with the highest production temperatures are generally preferred for strength retention during both the NRMI and RMI matrix fabrication routes.

In terms of temperature capability, both current CMC fabrication routes always result in some excess silicon in the matrix, thereby allowing SiC/SiC components under their application loads to display undesirable matrix creep that begins at ~1100°C, and becomes excessive above ~1300°C. As described in Case A above, if the SiC fiber is more creep resistant than these matrices, effectively, all the composite load will be shifted to the fiber, resulting in composite creep and rupture at the higher temperatures controlled by the fiber creep–rupture characteristics and its volume fractions. In addition, at temperatures approaching ~1400°C, not only does the excess silicon melt and tend to exude from the composite, but the beneficial residual stresses in the RMI route tend to disappear. Thus, for current production SiC/SiC composites fabricated by both MI routes and with high temperature SiC fibers, it is the final silicon infiltration step and not the reinforcing SiC fiber that is the prime mechanism that limits the intrinsic CMC upper-use temperature to only ~1250°C.

7.4 RECENT NASA ADVANCEMENTS IN SiC/SiC MATERIALS AND PROCESSES

Although the current CMC fabrication routes described above are adequate for the introduction of SiC/SiC components into the hot sections of aero-propulsion turbine engines, NASA has determined that these production approaches do not offer the optimum in SiC/SiC composite structural reliability or in high temperature service capability. The following sections will detail the various deficiencies of these routes and the advancement approaches recently developed by NASA and demonstrated with industry collaboration to address these shortcomings.

7.4.1 Advances in SiC-Based Fibers

First-level property requirements for a SiC fiber in high performance, high temperature SiC/SiC components are (1) to be produced in continuous lengths with a small diameter; (2) to display in its as-produced condition high tensile strength (~3 GPa) and high stiffness or modulus (~400 GPa); and (3) to have the ability to retain as much of this strength and stiffness as possible at high temperatures for long times under high stresses. Today the best approach for achieving these requirements is by the spinning, curing, and pyrolysis of multifilament tows of polymer-derived fibers [9]. For high strength and stiffness in their as-produced condition, key fiber microstructural requirements are, respectively, average grain sizes below ~500 nm and a near-stoichiometric SiC composition with high density (>3.0 g/cm^3) or low porosity [10]. Although decreasing grain size below 500 nm should theoretically increase tensile strength of polymer-derived SiC fibers, this is not observed because inherent in the polymer preparation and spinning processes, flaw-related particulates with sizes of the order of ~200 nm are allowed to pass into the green fibers. Thus for the finer-grained SiC fibers, these particulates rather than the fiber grains are limiting the strength to ~3 GPa. For high strength and stiffness retention (creep resistance) during service at high temperatures, key microstructural requirements are a pure near-stoichiometric SiC composition and grain sizes as large as possible without exceeding ~500 nm. These properties in turn will also provide a SiC fiber with high intrinsic thermal conductivity (>50 W/(m·K)), a property very important for cooled components. Finally, on the more practical side, the SiC fiber should be commercially available in large quantities and with low acquisition or life cycle cost.

Today the commercial SiC fiber type that is primarily being employed for both the NRMI and RMI SiC/SiC component fabrication approaches is the "Hi-Nicalon Type S" fiber (hereafter HNS fiber) that is produced by Nippon Carbon in Japan. This fiber meets most of the fiber's practical requirements, particularly commercial availability in large quantities [2], but still lacks some important microstructural requirements. The HNS fiber, like its "Hi-Nicalon" precursor fiber, is first spun from an organometallic "pre-ceramic" polymer precursor, followed by cross-linking or curing under electron irradiation. It is then subjected to high temperature pyrolysis and heat treatment steps to remove the excess carbon present in the Hi-Nicalon fiber (C/Si ~1.3) and to form a nearly stoichiometric (C/Si ~1.05) β-phase SiC fiber [11]. Some key chemical and physical properties for the as-produced HNS fiber are shown in Table 7.1, and compared with other SiC-based fibers with potential for high temperature CMCs [9, 11–13] (R. T. Bhatt, personal communication). Even though irradiation rather than oxidative curing is employed, the HNS fiber still contains residual oxygen (~0.7 wt.%), which typically is in the form of a silicon oxycarbide impurity phase. When exposed for short times at temperatures greater than ~1600°C, this phase decomposes into SiO and CO gases, which in turn creates pores in

TABLE 7.1 Some Key Properties of SiC-Based Fibers for High Temperature SiC/SiC CMC

Fiber type (trade name)	Vendor	Max temp.[a], °C	Key non-SiC elements in bulk, wt.%	Grain size[b], nm	Key surface phase	Tensile strength[c], GPa	R.T. thermal cond., W/(m K)	Ref.
Hi-Nicalon	Nippon Carbon	1450	0.5 Ox	5[d]	C	3.0	8	9, 11
Hi-Nicalon Type-S (NOX grade)	Nippon Carbon	1650	0.7 Ox	20[d]	C	2.8	18	9, 12
Hi-Nicalon Type-S (OX grade)	Nippon Carbon	1650	0.7 Ox	20[d]	SiC	2.6	18	12
Sylramic	ATK-COI Ceramics	1850	<0.1 Ox, 1.2 B, 2.4 Ti	100	B	3.2	46	13
Sylramic-iBN	ATK-COI Ceramics	1800	<0.1 Ox, ~0 B, 2.4 Ti	200	BN	3.1	>50	13
Super Sylramic-iBN	NASA	1800	<0.1 Ox, ~0 B, 2.4 Ti	200	BN	3.0	>50	14
Tyranno SA3	UBE Industries	1900	0.2 Ox, 0.6 Al	400	SiC	2.8	65	9

[a]Approximate maximum temperature for loss of >10% of as-produced strength after 10-hour inert exposure.
[b]Approximate average grain size measured by TEM.
[c]Average as-produced single fiber strength at room temperature and 25 mm gauge length.
[d]Spinning-related strength controlling flaws are ~200 nm in size.

the fiber, thereby reducing fiber strength [14]. In addition, the as-produced HNS (NOX grade) fiber has a carbon-rich surface, which can be detrimentally removed within the CMC when exposed to combustion environments [15]. Thus, for turbine applications, a judicious practice is to utilize HNS (OX grade) fibers where the carbon surface has been removed by the fiber vendor or by the fiber end user prior to CMC fabrication. However, as shown in Table 7.1, this will reduce fiber strength by ~10% by exposing flaws on the as-produced fiber surface. Adding a compliant interfacial coating to the fiber, such as BN, may act to blunt these flaws again, provided the coating is applied to all areas of the fiber surface.

Regarding structural performance at high temperatures, because of its thermal limit of ~1600°C for strength retention, the average grain size within the as-produced HNS fiber does not grow beyond ~20 nm [9] and the fiber still retains the oxide-based impurity phase in its grain boundaries [12]. Recent studies at NASA [16] suggest that these two factors are the prime contributors to undesirable fiber creep and a structural limit for this fiber type of less than ~1450°C. To better understand this limitation, Figure 7.1 compares the creep behavior of various polymer-derived SiC fiber types when subjected as single fibers to a constant stress σ of 270 MPa at a constant high temperature T of 1400°C in laboratory air [16]. One can see that all polycrystalline SiC fibers display a time-dependent transient or primary creep strain ε_p followed by a nearly steady-state secondary creep strain ε_s. NASA [10, 16] has identified the underlying creep mechanism for the secondary stage of these fine-grained SiC fibers As interface-controlled diffusional (ICD) creep in which the concentration of vacancies supporting fiber creep by grain-boundary diffusion is limited by the fine-sized grains [17]. In this case, the second-stage creep strain rate ε_s can be described analytically as

$$\dot{\varepsilon}_s = (C/d)\sigma^3 \exp[-Q/RT] \tag{7.1}$$

Here C is a composition-dependent diffusion constant; d is average fiber grain size; Q is the activation energy controlling steady-state creep; and R is the universal gas constant (8.314 J/(mol·K)). Thus, for a given applied stress and temperature, Equation (7.1) indicates that the HNS fiber secondary creep strain could possibly be reduced by an order of magnitude simply by increasing its grain size from 20 to 200 nm. But due to decomposition of the silicon oxycarbide phase preexistent in this fiber, heat treatments above ~1600°C, although increasing the HNS fiber grain size and improving its creep resistance, cause an undesirable loss in fiber strength [14].

Besides limiting grain size, the oxide-based phase in the HNS fiber also gives rise to an undesirable, large primary creep stage for this fiber type [16]. A convenient method for seeing this effect is by examining Table 7.2, which compares

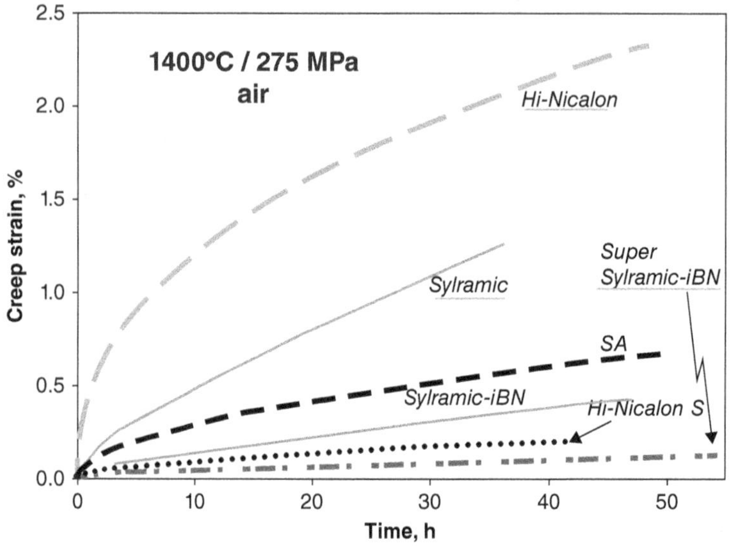

FIGURE 7.1 Creep strain behavior of various SiC fiber types in single fiber form [16].

the parameter m measured in a bend stress relaxation (BSR) test [18] after 1 hour exposure at various temperatures in argon. The listed m data are not only for the HNS fiber, but also for two other high performance SiC fiber types that possess less oxygen content after fabrication (G. N. Morscher, personal communication). Since stress relaxation is caused by creep, the following equation for m values >0.5 is a good approximation for a fiber's primary creep strain (ε_p) [16]:

$$\varepsilon_p = \varepsilon_e[(1/m) - 1)] \tag{7.2}$$

Here $\varepsilon_e = \sigma/E$ = elastic strain on the fiber. Thus, one can see in Table 7.2 that the as-produced HNS fiber shows primary creep as low as 1100°C where its m value begins to fall below unity after only 1 hour; whereas the other two fiber types do not show this behavior until higher temperatures. It is believed that this creep is caused by grain-boundary sliding accommodated by the low viscosity silicon oxycarbide phase within the HNS boundaries [19]. Furthermore, on the practical side, one can see by Equation (7.2) and Table 7.2 that the HNS primary strain is approximately equal to its elastic strain after only 1 hour at ~1450°C ($m = 0.5$). This implies a 50% loss in fiber stiffness or modulus in just that 1 hour period which, depending on the matrix type, could cause a stress overload on the matrix, causing it to crack. Thus based on the observed limitations of the HNS fiber, in order to achieve SiC/SiC composites with improved structural performance and temperature capability, *it is important to seek SiC fibers with little or no low viscosity phases in the grain boundaries and with average grain sizes up to ~500 nm*. The guideline for little or no low viscosity phases in the grain boundaries is also important for creep of the SiC matrix.

One proven approach for obtaining SiC fibers with greater temperature capability than the HNS fiber is to allow by thermal treatment the silicon oxycarbide phases that are typically created during fiber fabrication to decompose as gases, resulting in a weak SiC fiber with fine-sized pores. Then by incorporating sintering aids in the fiber pre-ceramic polymer [9, 20] or by gaseous infiltration of the pores with sintering aids after decomposition of a SiCO fiber [21, 22], the pores are then removed by sintering treatments to 1800°C and above. This approach typically results in high density, near-stoichiometric SiC fibers with little remaining oxide phases and with average grain sizes greater than 100 nm,

TABLE 7.2 Average BSR (m) Values for Single Near-Stoichiometric SiC-Based Fibers Measured After 1-hour Exposure in Argon at Various Temperatures [18] (G. N. Morscher, personal communication)

Fiber type	1100°C	1200°C	1300°C	1400°C	1500°C
Tyranno SA3		0.90	0.74	0.45	
Hi-Nicalon Type-S	0.96	0.89	0.83	0.64	0.40
Sylramic-iBN	1.00	0.97	0.91	0.84	0.74

two important microstructural conditions not available in the HNS fiber. Also, this sintering approach has the major advantage of producing strong fibers at very high production temperatures, which can significantly open the temperature window for CMC fabrication and service without experiencing microstructural instabilities in the fiber reinforcement.

Today two commercial SiC fiber types are using this sintering approach: the "Tyranno SA3" fiber from UBE Industries where an aluminum-based sintering aid is incorporated into a SiC-yielding pre-ceramic polymer [9]; and the "Sylramic" fiber originally produced by Dow Corning and now by ATK-COI Ceramics in San Diego, CA, where a boron-based sintering aid is infiltrated into the pores of a decomposed SiCO precursor fiber [22]. Table 7.1 lists some key as-produced properties of these two fiber types, and Figure 7.1 compares the tensile creep behavior of these fibers against HNS under the same test conditions of 1400°C, 270 MPa, and laboratory air. As can be seen, although the Tyranno SA and Sylramic fibers are near stoichiometric with high strength and high thermal conductivity, their high temperature tensile creep behavior is worse than that of the HNS fiber. This less than optimum creep behavior can be attributed to the aluminum and boron-based sintering aids that still remain in the grain boundaries of the as-produced fibers, thereby enhancing the rate of grain boundary sliding. Realizing this issue, NASA developed high temperature thermal treatments in selected environments to remove or alter the sintering aids in the as-produced fibers. It was determined that whereas the grains grew and creep-resistance improved, the low tensile strength of the as-produced Tyranno SA fibers degraded further with high temperature thermal treatment. On the other hand, the boron in the Sylramic fiber sintering aids could be removed in both nitrogen and argon atmospheres with retention of high tensile strength ($\sim$3 GPa), and with significant improvement in fiber creep resistance [23]. The fiber resulting from treatment at 1 atm pressure of highly pure nitrogen is labeled "Sylramic-iBN" because as the boron is removed from the as-produced Sylramic fiber, a thin protective *in situ* grown BN (iBN) coating is formed with uniform thickness on the circumference of each fiber, even those in tight contact with its neighbors within a woven tow [6]. When treated in highly pure nitrogen at pressures greater than 1 atm, fibers with even more creep resistance have been produced and have been labeled "Super Sylramic-iBN" [13]. Key properties and the improved tensile creep behavior of the Sylramic-iBN SiC fiber (hereafter iBN fiber) and the Super Sylramic-iBN fiber over their precursor Sylramic fiber are shown in Table 7.1 and Figure 7.1, respectively.

It is interesting to note in Figure 7.1 that under creep testing at 1400°C, the iBN fiber is slightly less creep resistant than the HNS fiber; whereas the Super iBN fiber is slightly more creep resistant. By Equation (7.1), if grain size were the only difference, this result would be unexpected since the average grain size of both iBN fibers is at least an order of magnitude larger than that of the HNS fiber. On the other hand, the BSR results of Table 7.2 show a significant improvement in creep resistance in the primary stage for the iBN fiber over the HNS fiber. Recent NASA studies attribute these apparently contradicting results to the fact that after sintering, the Sylramic as well as the Tyranno SA fibers display a cross-sectional microstructure consisting of large grains in a thin shell at the fiber surface and a core region consisting of smaller grains and fine-size pores [24]. It would appear that the shell region is created prior to or during sintering by the concentration near the fiber surfaces of the aids responsible for grain growth. Thus, similar to the fiber–matrix creep conditions described for Case A in Section 7.2, at high temperatures the more creep-prone inner region within the fibers will shed its load to the more creep-resistant shell region, which only occupies $\sim$30% of the total cross-sectional area. As such, the stress on the shell can be as much as $\sim$3 times larger than that on the HNS fiber, which contains a more uniform cross section [24]. Then by Equation (7.1), the tensile steady-state creep rate for the iBN would be $\sim 3^3$ times greater than the HNS fiber at the same grain size. Thus, the perceived tensile creep advantage of a larger grain size for the iBN fiber is effectively lost by its nonuniform cross section. Also, it follows that since the BSR test primarily samples the creep-resistant shell region, the BSR data for the iBN fiber would be expected to show much better creep-resistant behavior than that of the HNS fiber.

Because it can be more conveniently produced without the need for high pressure nitrogen, the iBN fiber in contrast to the Super Sylramic-iBN fiber has gone beyond the developmental stage to commercial availability as multifilament tows. It also has been available for nearly a decade to be employed in SiC/SiC CMC panels as 2D-woven fabric for the NRMI fabrication route and for 3D-woven preforms using CVI-MI and other matrix infiltration approaches. This has allowed a direct property comparison between CMC reinforced with the HNS and iBN fibers. For example, during creep–rupture tests at 1315°C in laboratory air [25], the HNS CMC showed a significant initial increase in creep strain in comparison to the iBN CMC when the applied on-axis stress increased from 103 to 138 MPa. Each CMC type was fabricated by the NRMI route with 2D-woven balanced fabric and $\sim$18% fiber content in the loading direction. It would appear that this behavior is caused by a lower MCS for the HNS CMC and/or by the larger primary creep stage of the HNS fiber. This later mechanism is supported by the BSR data in Table 7.2 that shows after 1 hour at 1300°C, the creep of the iBN fiber is much less than that of the HNS fiber. As a result, on initial CMC loading, the primary creep for the HNS fiber allowed the NRMI matrix to exceed its micro-cracking strain limit of $\sim$0.05% (CMC modulus $\sim$265 GPa) before it could stress relax and shift most of its load to the fiber. Thus, it would appear that the higher HNS CMC creep behavior at 138 MPa is indicative of local matrix cracking, leaving the crack-bridging fibers to carry much higher stress levels they would experience in an uncracked CMC [4]. On the other hand, matrix cracking was probably not occurring in the iBN CMC, allowing these fibers to creep at lower stress levels, and thus, providing a CMC with reduced creep.

For MI CMC specimens at low stresses with no matrix cracking, steady-state creep behavior can be predicted well by NASA's analytical creep models determined from the creep behavior of single HNS and iBN fibers by assuming that all the fibers after long times are carrying the entire CMC load, [16]. This is not only true for CMC fabricated by the NRMI route, but also for CMC fabricated with HNS fibers by the RMI route [26]. This implies that at least in laboratory air, the rupture life of these CMCs should be predictable by understanding the *intrinsic rupture characteristics* of these two fiber types. However, although these characteristics have been measured for single HNS and iBN fibers as a function of stress and temperature [27], one should realize that the fiber conditions within a cracked CMC during high temperature application could be considerably different than those employed during simple creep–rupture testing of single fibers. For example, the condition of tensile stress being held constant both in time and along the fiber length will probably not exist within a CMC due to fiber curvature induced by the fiber architecture and due to the typical occurrence of CMC matrix micro-cracking, where fiber stresses are highest within the matrix cracks and then drop off within the intact segments of the matrix. Also under these CMC conditions, not only is fiber gauge length considerably smaller than the 25 or 100 mm length typically used for single-fiber creep–rupture testing, but also rupture is controlled by bundles of fibers rather than single fibers. In addition, if the testing environment plays a role on fiber rupture, testing in laboratory air may give different results than testing within a given composite application environment. For this reason, NASA has found it more practically useful to understand and empirically model the general rupture characteristics of various fiber types by examining the rupture characteristics of CMC specimens in which the (1) reinforcing fibers are loaded on axis with the primary fiber direction at a known fiber content, (2) reinforcing fibers are essentially straight in the loading direction, and (3) CMC matrix carries little or no load because it is more creep prone than the fiber. The results of this approach are discussed in Section 7.4.5.

In summary, when compared to the HNS fiber, the iBN SiC fiber is more thermally stable due to its higher production temperature, thereby allowing higher SiC/SiC fabrication and service temperatures without fiber tensile strength degradation, Its higher production temperature also results in lower oxygen content for a reduced primary creep stage, plus larger grains for much improved thermal conductivity and the potential for improved creep if its microstructure can be made more uniform in grain size distribution. In addition, the *in situ* grown BN layer provides more environmental resistance than the carbon- or non-carbon-rich surfaces of the HNS fiber as well as the CVI produced BN-based interfacial coatings that cannot cover the full fiber surface in high fiber fraction CMC. However, beyond its nonuniform microstructure, certain other issues are currently limiting optimum application of the iBN fiber. For example, its larger grain sizes in comparison to those of the HNS fiber typically result in rougher fiber surfaces, which although beneficial for high interfacial stress and tight crack opening, can be detrimental in that they can cause fiber–fiber abrasion and fiber fracture when multi-fiber tows are woven or braided into complex preforms. Also, as will be discussed in Section 7.4.5, the larger grains appear to reduce the ultimate creep strains that the fiber can reach without rupture. In addition, there is the general increase in the acquisition costs for sintered fibers that typically accompanies their higher production temperatures. Finally, the Sylramic precursor fibers for the iBN SiC fibers are currently produced commercially by a combination of batch and continuous processing, which limits their availability in sufficient quantities for production-ready components.

7.4.2 Advances in Interfacial Fiber Coatings

Although the SiC fiber controls many of the key structural properties of SiC/SiC composite systems, the other key composite constituents also need to be optimized in composition and geometry for durable component use. For example, in early SiC/SiC composites, the preferred interfacial fiber coating composition was carbon as formed by the CVI process. However, under certain conditions, it is likely that these coatings will be exposed to aggressive environmental conditions, for example, within a matrix crack, but also at free or cut CMC edges. For this latter condition, NASA has shown that under the conditions of high velocity combustion gases, these carbon-based coatings can be quickly removed by oxidation, and that this removal could occur throughout the composite because of fiber–fiber contact [15, 28]. Working within the EPM program, NASA has demonstrated that a more stable coating would be based on a CVI-produced, silicon-doped BN composition. This is the case because BN is not only structurally weak for crack deflection, but instead of forming a volatile gas upon oxidation, it also forms in the presence of silicon a non-fugitive product (borosilicate glass), thereby preventing further oxygen ingress into the composite.

Although silicon-doped BN coatings have proven useful for CMC with free or cut CMC edges, if stress-induced matrix crack deflection occurs at the fiber surface rather than on the outside of the interfacial coating, the environment entering the matrix crack could reach and degrade the SiC fiber. To address this concern, NASA developed CMC thermal treatment processes that would allow the BN coating to shrink toward the fiber surface and debond from the matrix, thereby ensuring crack deflection to occur on the outside of the fiber-protective BN coating [29]. Even though the interphase coating remains attached to the fiber and debonded from the matrix, load transfer between the fibers and the matrix is still maintained within complex-shaped fiber architectures that allow the interfacial coating to mechanically slide against the matrix during the application of stress. In comparison to multilayer concepts for interfacial coatings, this "outside debonding" approach avoids the fabrication of complex interphase compositions and structures, does not rely on uncertain microstructural conditions for matrix crack deflection outside of the interphase, and provides a more reliable approach for retention of the total interphase on the fiber surface. As mentioned earlier, this oxidation protective scheme is further enhanced by use of the Sylramic-iBN fibers that also have a thin *in situ* grown environmentally resistant BN coating on their surfaces. As a result of outside debonding, a SiC/SiC system with 2D fiber architectures will typically display a slightly lower thru-thickness thermal conductivity due to partial loss of matrix thermal contact with the fibers. However, NASA has shown that this effect can be mitigated by the use of 3D architectures as discussed in Section 7.4.3.

7.4.3 Advances in SiC Fiber Architectures

As with any composite component reinforced by high performance continuous fibers, one of the primary requirements is to design the fiber architectures to provide sufficient fiber content in those directions that are structurally loaded during component service in order that the fiber-controlled structural properties of the component, such as MCS, will be sufficient to allow it to survive without failure during its required lifetime. For SiC/SiC components, life-limiting failure mechanisms are generally identified with (a) thru-thickness matrix cracking under in-plane tensile loading and (b) delamination cracking under thru-thickness tensile loading. In the first case, thru-thickness cracking will allow the service environment to reach and degrade the crack-bridging fibers within time periods typically less than the required component lifetime. For aero-propulsion components, this situation would apply, for example, to the walls of pressure-cooled airfoils and to centrifugally loaded blades [30]. In the second case, delamination cracking can increase the thermal stresses and wall temperatures for internally cooled turbine components, increasing the risk of wall spallation and other component damage. Thus, the general fiber architecture goals for SiC/SiC components are to achieve a starting preform with (1) a shape as close as possible to the final component shape, (2) a total fiber content as high as possible, and (3) a judicious splitting of this total fiber content between in-plane, thru-thickness, and locally to allow the CMC directional MCS properties to sufficiently exceed the component service stresses. Furthermore, if the fiber is thermally conductive, it will aid in increasing the CMC directional thermal conductivities, thereby reducing service-related thermal stresses in these directions.

When the current 2D fiber architecture designs of the NRMI and RMI fabrication routes are compared to the above goals, certain deficiencies can be observed. For example, regarding CMC in-plane properties, although the 2D-woven fabric approach of the NRMI fabrication route (and other routes that use 2D fabric) have reached total fiber contents as high as 50%, their in-plane fibers are not straight. Depending on the weave pattern, the effective on-axis fiber modulus and strength will be less than those of straight fibers, which in turn will debit CMC thru-thickness MCS, ultimate tensile strength (UTS), and rupture strength (to be discussed). In the case of the RMI or pre-preg fabrication route, although the in-plane fibers are effectively straight, total fiber contents are not high (<30%) due to the need to individually coat almost every fiber with a thick interphase. Thus, all fiber-controlled properties are less than optimum for the RMI route.

Regarding CMC thru-thickness properties, it is found that as the multi-ply 2D-woven fabric of for the NRMI route is compressed in tooling prior to interphase deposition, the resulting CMC thru-thickness tensile strength and thermal conductivity will generally increase. This effect is primarily due to increased fiber–fiber contact and intermingling or "nesting" of the fiber tows. But as might be expected, this nesting results in less opportunity to achieve uniform CVI-produced interfacial coatings on all the fiber surfaces in a tow or through the thickness, plus increased mechanical degradation of the fibers and CMC in-plane strength properties. Whereas the NRMI route relies on increasing fabric nesting to improve the CMC thru-thickness or Z-direction properties, the RMI route relies on its low total fiber content to allow the dense SiC-based matrix between the fibers to provide monolithic-like thru-thickness tensile strength. For both the NRMI and RMI routes, it is difficult to obtain reproducible CMC thru-thickness properties due to local microstructural variances. This results in large scatter, for example, in the thru-thickness or Z-direction properties,

TABLE 7.3 Typical Directional Strength and Thermal Conductivity Properties of SiC/SiC Panels with Various Woven Fiber Architectures

SiC/SiC fiber architecture		Fiber content, %			20°C CMC directional strengths, MPa			CMC directional thermal conductivity, W/(m K)			Ref.
					Y		Z	X	Z	Z	
		X	Y	Z	MCS	UTS	UTS	20°C	20°C	1400°C	
A	2D fabric	18	18	0	180	460	15	50	25	18	6, 32
B	2.5D	17	19	3	140	360	28		50	25	35, 36
C	3D Orthogonal	18	17	< 5	120	320					37
D	3D Orth-Unbal	10	25	< 5	240	600					37

X, Y stuffer fibers = Sylramic-iBN fiber; Matrix fabrication = CVI-MI.
2.5D Z weaver = Sylramic-iBN fiber; 3D Z weaver = ZMI SiC Fiber.

which in turn results in significantly low thru-thickness tensile strength allowables (<20 MPa) for current SiC/SiC components. Finally, both fabrication routes rely on the lay-up of finite-sized 2D plies, which means that for components with continuous surfaces, such as combustor liners and airfoils, there must be regions where the plies terminate and fiber reinforcement is absent. Thus unless the ply ends are intermingled in some judicious manner, or the ply ends are located in low stress regions, there can be a significant knockdown in component strength properties at these locations [31].

Realizing the general architecture goals and the deficiencies of the two MI fabrication routes, NASA in recent years has been examining SiC/SiC CMC panels primarily consisting of multiple types of 3D-woven preforms with thru-thickness fiber reinforcement. These types range from simple 2.5D-woven to more complex 3D-orthogonal architectures [32–35]. In most cases, iBN SiC fiber tows have been used as the in-plane warp and fill stuffers, and the interfacial coatings and matrices were formed by those used in the NRMI fabrication route. For the warp weaver or thru-thickness fibers, due to their high stiffness and tendency to fracture under high bend angles, iBN fibers were only used for preforms with low angle thru-thickness tows; whereas low stiffness, low performance SiC-based "ZMI" fibers were used for the high angle 3D-orthogonal preforms. Table 7.3 lists the key room temperature in-plane and thru-thickness properties for some 2.5D and 3D SiC/SiC panels and compares them with panels fabricated by the standard NRMI route using the lay-up of 2D-woven iBN fabric (NASA N24A SiC/SiC). For the Table 7.3 panels, the in-plane iBN stuffer tows were nearly straight with total fiber contents of ~36%, and were nearly balanced at ~18% in each of the warp and fill directions or were unbalanced at ~10% in the warp and ~25% in the fill direction. The indicated test data were measured in-plane in the fill or *Y* direction, as well as in the thru-thickness or *Z* direction.

Table 7.3 shows that for the NRMI fabrication route, significant advances can be achieved in SiC/SiC CMC structural and thermal properties by the use of 3D-woven preforms rather than by the current 2D-woven fabric lay-up approach. For example, when the 2D-woven Panel A is compared against the 2.5D-woven Panel B, although a small knockdown occurred for the in-plane strengths, one can see not only a doubling in the *Z*-direction properties of thru-thickness tensile strength (Z-UTS) and thermal conductivity (Z-TC) at room temperature, but also a significant increase in Z-TC at 1400°C (extrapolated). Indeed this latter Z-TC value of 25 W/(m·K) may be the highest ever measured at high temperature for SiC/SiC panels. These improvements were achieved with only ~3% iBN fibers in the *Z* direction, so that further improvements should be expected with higher *Z*-fiber content. Whereas the Z-UTS improvement can be explained by the use of strong iBN fibers, the improvement in Z-TC was probably due in part to the high conductivity of these fibers and in part to the fact that weaver fibers probably allowed better infiltration channels for the NRMI matrix.

Then when one compares Panel A with the balanced 3D-orthogonal Panel C, one finds a further decrease in in-plane strengths, even though the *Y*-stuffer fibers in Panel C are straighter than those in Panel A. This degradation in the warp and fill stuffers of Panel C may have been caused by excessive abrasion with the ZMI weaver fibers and/or due to reaction with these fibers during high temperature CMC fabrication. Finally, when one compares the balanced 3D-orthogonal Panel C with the unbalanced 3D-orthogonal Panel D, one sees an approximate doubling of the in-plane *Y*-direction strengths, even though the fiber content in this direction only increased from ~18% to 25%. Thus, although these results are preliminary, one might conclude that the demonstrated *Z*-direction property benefits of thru-thickness fibers need not necessarily exclude improvement at the same time of the in-plane properties, provided one makes a judicious choice of the SiC fiber and its 3D architecture.

Beyond improved thru-thickness strength and conductivity, NASA has shown that the 2.5D-woven architectures with iBN fibers also provide SiC/SiC CMC with improved delamination resistance and impact resistance in comparison to 2D architectures [36]. As discussed at the beginning of the section, another potential advantage for 3D architectures can be enhanced matrix infiltration down the weaver fibers, resulting in reduced matrix porosity. Furthermore, the possibility may exist that the automation aspects of 3D weaving or braiding will result in reduced manufacturing costs and reduced scatter in composite properties by eliminating the human element that is typically involved in 2D fabric and tape lay-up techniques. However, for the highest temperature SiC/SiC components, sintered SiC fibers, such as the iBN and Tyranno SA fibers, are near stoichiometric in composition with large nano-sized grains, resulting in high bending stiffness and rough fiber surfaces leading to poor abrasion resistance. As described for the panels in Table 7.3, these issues can enhance the probability of fiber fracture during 3D preforming and thus limit the available 3D-woven architectural designs and thru-thickness fiber fractions. Nevertheless, NASA has recently demonstrated with a commercial textile processing vendor that complex 3D-woven preforms for SiC/SiC panels as well as for SiC/SiC turbine airfoils can be fabricated using polymer-based "serving" fiber tows that are wound around sintered SiC fiber tows prior to weaving, thereby significantly reducing their abrasion issues (T.E.A.M: Textile Engineering and Manufacturing, Woonsocket, RI, personal communication). In subsequent CMC processing, these fibers will decompose leaving a negligible carbon char, which should be of little issue for the key CMC properties. Another potential approach developed at NASA for complex-shaped components is to first form 3D preforms with simpler shapes that are close to those required by the component, but do not contain the extreme bends that can cause fiber fracture. Then one can heat treat the architecture while applying continuous or discrete levels of pressure to creep-form the fibers without fracturing them into the tighter bends required by the component shape. NASA has demonstrated that this can be done at the same time as the iBN treatment with minimal loss in fiber strength [23].

7.4.4 Advances in SiC-Based Matrices

Since the free silicon in the NRMI and RMI matrices limits their long-term SiC/SiC CMC upper-use temperatures to less than 1300°C, NASA and the U.S. Air Force have recently been researching and developing SiC/SiC systems by using alternate matrix infiltration processes into stacked 2D-fabric preforms that result in SiC compositions that are effectively silicon free. To date, three such infiltration processes have been extensively evaluated [32]: (1) CVI processing where SiC deposition is continued as long as possible to fill the porosity within an interface-coated 3D preform (Full CVI approach); (2) polymer infiltration and pyrolysis (Full PIP approach) in which multiple infiltration cycles are performed, each consisting of infiltration of a SiC-yielding polymer into the remaining 3D preform porosity, followed by polymer decomposition at high temperatures into crystalline SiC particulates; and (3) a hybrid of these two processes, where typically the CVI approach is first used for partial filling of the preform porosity, and then the remaining porosity is filled as much as possible by subsequent multiple PIP cycles (CVI-PIP approach). Although achieving the goal of an effectively silicon-free SiC matrix, each of these three processes have various advantages and issues that may or may not limit their application for higher temperature SiC/SiC components.

In terms of advantages, the Full CVI method for matrix infiltration is amenable to the fabrication of SiC/SiC components having irregular, large, and complex shapes with near-net shape capability. High temperature properties are generally good because high purity matrices with controlled microstructures can be obtained. Aside from trapped porosity (10–15 vol.%), this approach typically results in a SiC matrix material with high thermal conductivity and high creep resistance in multiple directions. Also CVI is carried out at relatively low temperatures (<1000°C), so that there is little risk of fiber damage. In addition, the fiber–matrix interphase can be tailored by pre-coating the fiber preform during an initial CVI step using different reagents, but the *same* reactor. However, one of the major limitations of the Full CVI method is the long deposition times needed for the densification process (typically greater than 100 hours). This raises the cost of the component fabrication in comparison to processes that require much shorter times to complete the matrix infiltration. Furthermore, although the CVI matrix initially forms fairly uniformly within the fiber tows and then on the surface of the tows, as deposition continues nonuniform regions of matrix with large and sharp pores typically form between the tows due to their separation and due to their depth within the CMC thickness. Because of their size and irregular shapes, these trapped pores reduce the CMC MCS in multiple directions. Thus it is not possible to obtain fully dense matrices using the Full CVI approach. In addition, the typical presence of residual closed porosity may shorten the useful life of the CMC component particularly after matrix cracking in an oxidizing environment. Finally, in relation to fully dense CMC, the trapped or closed porosity will not allow CMC with maximum thermal conductivity.

Regarding the Full PIP approach, its main advantage is its use of methods typically employed and well established in the fabrication of polymer matrix composites. Thus in contrast to the Full CVI approach, the size of CMC components is not limited by the size of the CVI reactor. However, one of its main drawbacks is the long processing times and repeated infiltration/pyrolysis cycles in order to obtain sufficiently high matrix density. Due to polymer shrinkage during pyrolysis, another drawback is that matrix micro-cracking generally results within the CMC in the form of fine cracks between the crystallized particulates. Re-infiltration fills these cracks to some degree, but MCS and thermal conductivity in multiple directions generally remain low in comparison with the Full CVI approach. Finally, due to oxygen in the polymer and also multiple cycles from room to high temperature, oxygen is typically picked up within the fine cracks, which then can attack the interphase and fibers unless oxygen getters are also introduced into the matrix [32].

For recent research and development studies aimed at silicon-free matrices, NASA has been emphasizing the CVI-PIP hybrid approach because it avoids to a large degree the drawbacks of the Full CVI and Full PIP approaches while incorporating some of their advantages. For example, just as in the NRMI fabrication route, only partial CVI deposition is utilized, thereby affording much shorter deposition times and reduced risk of large inter-tow pores, while retaining a partial SiC matrix with high strength, high creep resistance, and high thermal conductivity in multiple directions. In addition, the partial CVI SiC matrix provides environmental protection for the interphase and fibers during subsequent matrix infiltration processes. Filling the remaining matrix by the PIP approach allows the introduction of high conductivity SiC particulates without the concern of oxygen pickup during multiple PIP cycles. Although fine cracks will still exist between the particulates, they are much finer than those that form during cracking of the CVI matrix so that they can seal much more rapidly in high-temperature oxidizing environment. Nevertheless, although the hybrid approach has these multiple advantages, it like the other approaches does not as yet yield matrices with zero or near-zero porosity.

Table 7.4 compares some key in-plane and thru-thickness properties at 20°C and 1450°C for SiC/SiC CMC panels fabricated with matrices infiltrated by the CVI-MI, Full CVI, Full PIP, and hybrid CVI-PIP approaches. The first set of panels employed 2D-woven iBN fabric; while the second set employed 2.5D-woven iBN tows. To achieve maximum performance in terms of reduced matrix defects and increased matrix creep resistance and thermal conductivity, all the panels with silicon-free matrices were post-processed at NASA by annealing treatments above 1500°C [6, 33]. Examining Table 7.4, one can see that currently all the indicated processes for Si-free matrices do not achieve the low ~5% porosity of the CVI-MI approach. The high porosity levels of ~15% typically result in lower MCSs depending on fiber content and the character of the porosity, and in variable ultimate strengths, also depending on fiber content and possibly on the resistance of the iBN fibers to attack during CMC processing and post treatment. One approach to acquire better information about the matrix structural quality is to test the CMC in an off-axis direction where the matrix carries more of the CMC load. Thus, when tested at 45° to the balanced X and Y directions, Table 7.4 shows a significant knockdown in the X/Y-UTS for the Full PIP panel, indicating that this matrix approach does not have the structural quality provided by the partial CVI or Full CVI approaches. On the other hand, the Full PIP displays a relatively high Z-UTS. As previously described, this is caused primarily by increased fabric compression and nesting

TABLE 7.4 Typical Effects of Various Matrix Infiltration Approaches on Directional Strength and Thermal Conductivity Properties of SiC/SiC Panels

SiC/SiC matrix approach	*Y* Fiber, %	Matrix: ~ Free silicon, %	Matrix: ~ Total porosity, %	20°C CMC Directional strengths, MPa: *Y* MCS	*Y* UTS	*XY* UTS	*Z* UTS	*Z* Thermal Conductivity, W/(m K): 20°C	1400°C	Ref.
				2D-woven Sylramic-iBN fabric						
CVI-MI	18	13	5	180	460	240	15	25	18	6, 32
Full CVI	18	0	15	120	350		7	28	8	34, 35
Full PIP	21	0	<15	140	400	<100	23	8	5	34, 32
CVI-PIP	18	0	14	150	360		10	28	10	34, 35
				2.5D-woven Sylramic-iBN tow						
CVI-MI	18	13	5	140	360		28	50	25	36
CVI-PIP	18	0	14	130	380		24	45	22	36

All silicon-free SiC/SiC CMC were post treated above 1500°C.

of the fiber tows between fabric plies. Furthermore, due to the fine cracks, the Full PIP panel does not compare in thermal conductivity with the other matrix approaches. This low Z-TC property is probably due in part to the need for boron-based oxygen getters purposely incorporated in the matrix to avoid the effects of oxygen pickup during the multiple PIP cycles [32].

Regarding the hybrid CVI-PIP approach, although it can provide SiC/SiC that can operate at higher temperatures, Table 7.4 shows that with conventional 2D-woven architectures, it does not as yet provide comparable properties with CMC fabricated with the CVI-MI matrix approach. As might be expected from their high residual porosity, the 2D-woven CVI-PIP panels show a small knockdown in Y-MCS and Z-UTS, as well as high temperature Z-TC, but their values are still higher than those of the Full CVI panels. Nevertheless, as also shown in Table 7.4, when the hybrid matrix approach is combined with a 2.5D-woven architecture, the resulting panels, while still retaining a small deficiency in in-plane properties, display significant improvement in Z-UTS and Z-TC in comparison to those fabricated by the 2D-woven route and CVI-MI matrix,. Thus, if the porosity can be reduced and advanced 3D-architectures with iBN fibers are employed, the hybrid CVI-PIP matrix approach has the potential of not only displaying better structural and thermal properties than the current NRMI or CVI-MI route for SiC/SiC components, but also higher temperature structural capability as discussed in the Section 7.4.5.

7.4.5 Advances in SiC/SiC Microstructural Design Methods

As discussed in Section 7.2, of key importance for SiC/SiC component implementation in high temperature applications is the ability to select the optimum constituents and their geometries so that the component microstructure has the intrinsic structural capability to achieve the required component life under its projected mechanical, aerodynamic, and thermal loads. For this selection process, a method often used for metals and ceramics is to perform stress–rupture tests in which a constant stress is applied to a material at a constant high temperature under laboratory conditions. The time for total material rupture is then recorded for various combinations of stress and temperature. One simple and convenient empirical approach for representing the data from these tests is by use of "Larson–Miller" (LM) plots. In these plots, master curves are generated for various materials, which describe the applied stress at rupture (material rupture strength) versus the time–temperature dependent parameter q given by

$$q \equiv T\,(\log t_r + D) \tag{7.3}$$

Here T (Kelvin) is the absolute temperature for the rupture test, t_r (hours) is the material rupture time, and D is a constant that best allows the rupture data at different temperatures to fall nearly on one master curve. The designer can then compare the rupture test results for prospective materials for an initial down-select of those materials that can best meet the component's service requirements. NASA has determined that this LM approach with $D = 22$ can be effective method for SiC-based fibers [27] and also for SiC/SiC CMC [37], not only for showing the effects of various constituents and architectures on the structural life of various SiC/SiC CMC, but also for demonstrating the benefits of the recent constituent advances discussed in Sections 7.4.1 through 7.4.4.

In Figure 7.2, the LM approach has been used to plot the in-plane rupture strength data for various SiC/SiC composites versus the time–temperature at which they failed by complete rupture [37]. All CMC were fabricated as panels by the NRMI route and tested on-axis with one of the primary fiber directions (content of ~18%), and were reinforced with balanced 2D-woven 0/90-balanced 5HS fabrics of different SiC fiber types as indicated by the different curves. Also included is the predicted curve if an NRMI CMC were fabricated in the same manner using the developmental Super Sylramic-iBN fiber. Thus, all CMC contained free silicon, so that their long-term creep and rupture behaviors were effectively controlled by the reinforcing fiber tows under the total CMC load. In Figure 7.2, one can see that the LM approach appears to be a good empirical method for predicting SiC/SiC rupture life in that the rupture strength data for various fiber types not only fall consistently near one best-fit curve for each type, but also the curves closely follow the creep ranking of the different SiC fiber types as measured on single fibers in Figure 7.1. Although the data for each fiber type show some scatter from its best-fit curve, the curves can still be clearly distinguished from one another. It is envisioned that some of the data scatter for a specific fiber type may be due to whether or not matrix cracking occurred during initial CMC loading. Also it should be realized that with the LM approach, the scatter in rupture-time data is greatly reduced by the time insensitivity of the LM (q) parameter. Thus the curves of Figure 7.2 are practically useful for understanding the projected rupture life of an NRMI-fabricated SiC/SiC component that needs to experience a maximum stress at a maximum temperature. For example, if the maximum on-axis stress for an uncracked woven iBN/SiC component was 100 MPa, Figure 7.2 predicts a maximum q value of ~39 400, which

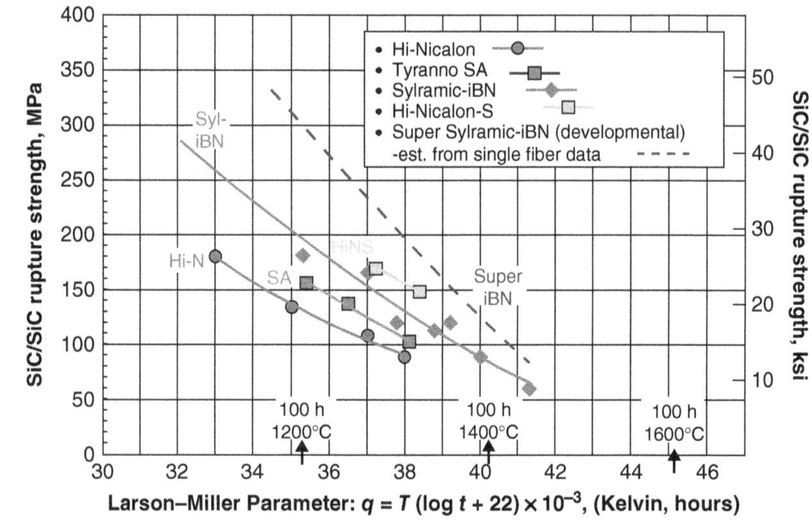

FIGURE 7.2

Effects of SiC fiber type on on-axis rupture strength in air for SiC/SiC CMC with 2D-woven 0/90-balanced 5HS fabric, on-axis fiber content of ~18%, and CVI-MI matrix [37].

in turn would allow an intrinsic rupture life for the component of 100 hours at 1642 K (1369°C), or 1000 hours at 1576 K (1303°C).

Besides its practical implications, there are also various mechanistic conclusions one can draw from the Figure 7.2 results. For example, some of the data were taken at a high enough CMC stress to initially crack the matrix, and yet the rupture strength data fall very close to the data where the CMC stresses did not cause initial matrix cracking. Thus, at least in laboratory air, the rupture mechanism for NRMI CMC appears to not be related to environmental attack, but to the fiber tows reaching a certain ultimate rupture strain during creep. For example, no matter the test conditions, the HNS CMC typically ruptures at ~0.6% total strain; whereas the iBN CMC ruptures at ~0.4% [38]. Thus in comparison to the iBN CMC, the slightly better rupture behavior of the HNS CMC in Figure 7.2 can be explained by its enhanced rupture strain, which in turn is probably related to its smaller grain size. It is also interesting to note that the rupture strains of the two fiber types are effectively equivalent to the ultimate fracture strains for their CMC in a room temperature fast-fracture test [16].

The LM approach has also proved useful in understanding fiber architecture effects. For example, Figure 7.3 plots rupture strength data vs. q data for a variety of different iBN-reinforced architectures within SiC/SiC panels fabricated

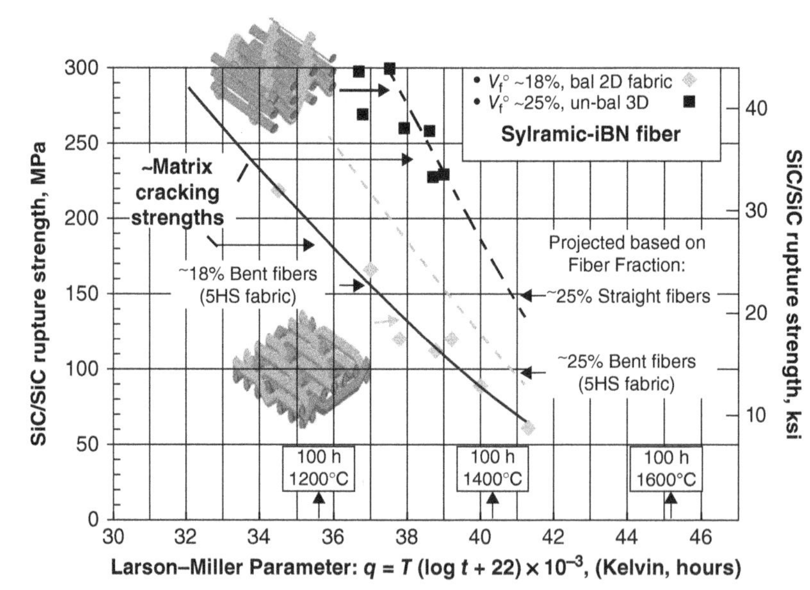

FIGURE 7.3

Effects of fiber architecture on on-axis matrix cracking and rupture strength in air for SiC/SiC CMC with Sylramic-iBN woven tows and CVI-MI matrix [37].

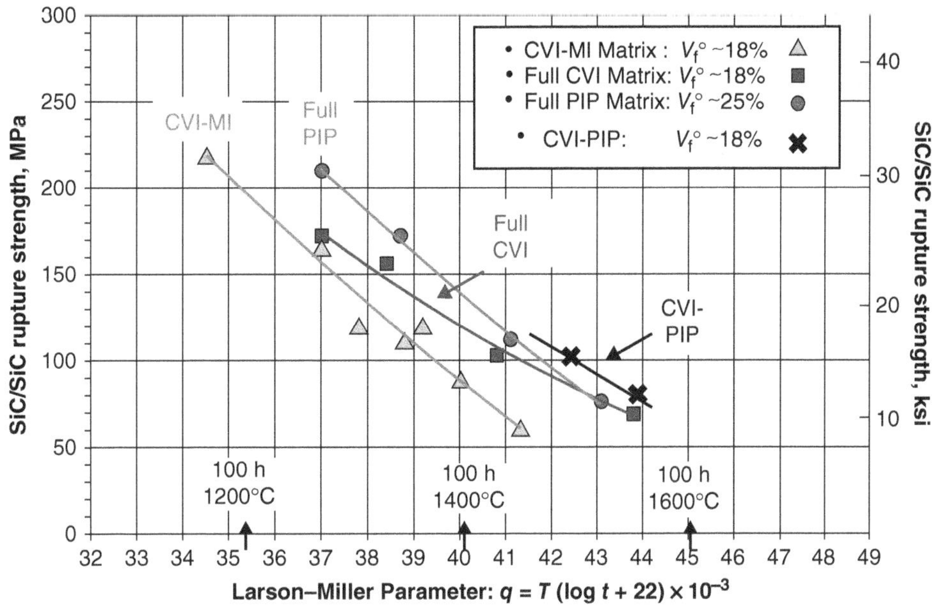

FIGURE 7.4

Effects of matrix infiltration approach on on-axis rupture strength in air for SiC/SiC CMC with 2D-woven 0/90-balanced Sylramic-iBN fabric [37, 39].

with CVI-MI matrices [37]. The first set of data on the left reproduces the data in Figure 7.2 for NRMI-fabricated CMC reinforced by 2D-woven 5HS fabric of iBN tows with on-axis content of ~18%. This curve is then projected to the right by assuming that if the same woven architecture could be produced with ~25% on-axis iBN fiber content, the final stress on the fiber would be reduced by a factor of 18/25 for the same CMC applied stress. Finally, in the upper center of Figure 7.3 are the CMC rupture data [38] measured in the fill direction for some of the 2.5D and 3D architectures described in Table 7.3. As previously indicated, in the fill direction the iBN tows were straight with an average fiber content of ~25%. When one compares this dataset with the first set, one can see a significant improvement in CMC rupture strength for the same time–temperature test conditions. As indicated, this improvement is still much better than the middle curve that has the same fiber content, but incorporates bent fibers as a result of the woven fabric. Bottom line from Figure 7.3 is that the structural life of SiC/SiC components can be significantly improved by (1) increasing the fiber content in the key component directions and (2) ensuring as much as possible that the fibers are in their as-produced straight condition.

Finally, one can use the LM approach in Figure 7.4 to understand the effects of different matrix infiltration approaches on SiC/SiC rupture behavior [37, 39]. Here all CMC are again fabricated with 2D-woven 5HS fabric of iBN tows, and the first dataset on the left was fabricated by CVI-MI or the NRMI route. The next curve at higher q values is for CMC with Full PIP matrices, but with ~25% fiber content due to enhanced fabric compression after lay-up. This dataset is almost exactly 25/18 times the CVI-MI dataset indicating that although containing no free silicon, the PIP matrix carries no long-term load, leaving the fibers to again carry the entire CMC load. This effect can be attributed to little bonding between the PIP-generated SiC particulates caused by particle shrinkage during pyrolysis. Thus, as previously discussed, the Full PIP matrix approach picks up its main advantage of higher fiber content in-plane, but suffers in other key CMC properties. On the other hand, the post-annealed CMC with Full CVI matrix and ~18% fiber content does show improved behavior over the silicon-containing CVI-MI matrix with the same fiber content, suggesting that this creep-resistant matrix can carry some of the CMC load at temperatures where the CVI-MI and Full PIP matrices cannot. Finally, Figure 7.4 shows that although the post-annealed hybrid CVI-PIP approach does not have the CVI SiC content of the Full PIP, this could be an advantage by allowing more load to be put on the CVI SiC, causing it to creep at a rate closer to that of the fiber. In effect, this would be similar to increasing the fiber content. Thus, for balanced 2D fiber architectures and post-annealed Si-free matrices, the hybrid CVI-PIP matrix approach currently yields the best high temperature SiC/SiC rupture behavior when compared to any other current approach [39].

7.5 CURRENT MICROSTRUCTURAL DESIGN GUIDELINES AND POTENTIAL SERVICE ISSUES FOR HIGHER TEMPERATURE SiC/SiC COMPONENTS

The advancements described in this chapter offer a better theoretical and experimental understanding of the effects of various SiC-based constituent compositions and geometries on the key mechanical and thermal properties of SiC/SiC CMC. As such, they suggest the following general microstructural design guidelines for achieving a SiC/SiC component with high structural reliability and higher temperature capability than current SiC/SiC fabrication routes.

- Select (or develop) a polycrystalline small-diameter, near-stoichiometric SiC fiber with average grain size between 200 and 500 nm, with grain boundaries free of such impurity elements as free silicon, free carbon, oxygen, boron, aluminum, and other elements that act as SiC sintering aids, and with a cross-sectional microstructure as uniform as possible in grain size and composition.
- Select (or develop) a fiber interfacial coating system that is compliant enough to deflect thru-thickness matrix cracks before reaching the fiber, is based on a composition that does not volatilize in an oxidative environment like carbon, and fully covers every fiber surface such as the iBN coating.
- Select, if possible, a fiber architecture that allows the final component shape to be textile formed, maximizes total fiber content in the component, allows sufficient fiber content in the key component stress directions, and eliminates or minimizes fiber curvature (straight fibers) in these directions.
- Select (or develop) a matrix process that yields a near-dense, Si-free, tough, and SiC-based matrix that has high thermal conductivity and creeps slightly more than the fiber.

Although these guidelines may provide a SiC/SiC component with constituents that are intrinsically able to withstand quasi-static structural loads under uniform stress conditions, there are certain situations that can arise during component service that may still limit its performance and life and thus require further investigation. As discussed in Section 7.2, one such situation is related to creep in that, even if the creep mismatch between the fiber and matrix can be minimized, there will certainly be stress and temperature gradients within the component leading to local amounts of differential creep strain, which in turn will result in local residual stress development that is not relieved at lower temperatures where the CMC behaves elastically [5]. These residual stresses can in turn reduce the local MCS, so that during some follow-up situation, the weakened area could be unpredictably subject to a tensile stress that could exceed the local cracking strength. Hopefully, the resulting cracking or damage will not affect the component's performance or allow the service environment to enter and degrade the CMC microstructure. Clearly one approach for avoiding this situation is to only employ the component at temperatures and times where the CMC exhibits little or no creep. Thus, if a certain high temperature performance is required for a SiC/SiC component, one should seek SiC fibers and matrices that display minimum long-term creep at this temperature.

Another possible life-limiting issue is flaw or crack creation in the SiC/SiC component generated by unexpected mechanical and/or thermal shock. Mechanical shock, for example, by foreign object impact, can not only occur during component service, but also during general handling and assembly of the component. Currently, without sufficient engine experience, it is very difficult to predict the flaw severity that can arise from this effect or its degrading influence on component residual strength and structural life. For this reason, the general design approach should be to only seek SiC-based matrices that in their as-produced condition display high fracture toughness or high resistance to crack propagation from surface flaws, so that the as-produced MCS of the as-produced component will be minimally compromised during service. Similar to mechanical shock, thermal shock can also lead to cracking, for example, in a component wall as the colder side experiences an unexpectedly high tensile stress that is proportional to the CMC thru-thickness thermal conductivity. If cracking should occur under either shock situations, the reinforcing fibers will bridge the cracks provided they have sufficient tensile strength and volume content to withstand the mechanical overload. Thus, for each engine component, it is important to understand the probability of shock, and where it might occur in order to ensure that the CMC will have sufficient overload capability in these locations to maintain component integrity and allow continuous engine operation. As discussed in Section 7.4.3, the use of 3D fiber architectures not only improves the mechanical shock or impact resistance of SiC/SiC panels, but also can improve the panel thru-thickness thermal conductivity, thereby reducing the stresses from any thermal shock situation.

Finally, a predictable and very important life-limiting issue for SiC/SiC components is erosion and/or corrosion of the component surface during high temperature gas-turbine service. This effect, which can be severe for all silicon-based ceramics, is typically due to the water vapor formed in the combustion process, which reacts with the oxides, such as silica, that continually form on the ceramic surface, causing them to volatilize. This in turn results in surface recession of the ceramic at a rate that increases with water vapor pressure, combustion gas velocity, and most importantly with CMC surface temperature [40]. To minimize this effect, environmental barrier coatings (EBCs) have been developed for both Si-based monoliths and composites [41,42]. For industrial gas turbines, these EBCs have been very instrumental in extending the life of SiC/SiC CMC combustor liners and shrouds at temperatures up to ~1250°C [43], but little experience exists for using EBCs under more aggressive engine conditions. Because EBCs themselves are typically produced from oxide compositions, such as barium aluminosilicate and mullite, they generally display very low thermal conductivity, thereby also acting as thermal barrier coatings (TBCs). Thus, EBCs are not only important for maintaining ceramic component cross section, they can also aid in reducing the surface temperature of a cooled ceramic component by providing another source of thermal resistance between the combustion gas temperature and the component surface. Generally EBCs are required for Si-based ceramic components that will require long-term service at surface temperatures above ~1000°C. As efforts are ongoing to increase the upper-use temperatures of SiC/SiC components beyond 1250°C, recession becomes the primary concern for long component life. Thus, high temperature EBC development is a key objective of current studies at NASA [44].

7.6 CONCLUDING REMARKS

This chapter has detailed the results of various studies conducted primarily at the NASA Glenn Research Center aimed at advancing the current state of experimental and mechanistic knowledge concerning the constituent materials and processes for SiC/SiC composites. When compared to current approaches for SiC/SiC aero-propulsion components, this knowledge has the potential of generating components with enhanced structural reliability and higher temperature performance. It should be realized that these advancements have been demonstrated primarily using single SiC fibers, woven SiC/SiC preforms, and SiC/SiC test coupons removed from panels that contain the improved approaches. These specimens were then tested under simple conditions such as laboratory air, constant temperature, and constant stress along the primary fiber direction. Thus, considerably more effort is required before these advancements can be comfortably implemented into real SiC/SiC engine components. Nevertheless, when compared to current approaches for SiC/SiC components, they clearly show substantial improvements, particularly, for achieving the multiple engine benefits that can be accrued by components with durable higher temperature performance. Because of these benefits, efforts at NASA are ongoing to further understand and demonstrate the practical viability of these advancements.

ACKNOWLEDGMENTS

Regarding the SiC/SiC advancements made over the last decade, the author gratefully acknowledges the funding support of the NASA Ultra Efficient Engine Technology program, the NASA Glenn Director's Discretionary Fund, and the NASA Fundamental Aeronautics program. As major contributors to these advancements, the author wishes to acknowledge the exceptional professional efforts of H. M. Yun, R. T. Bhatt, and G. N. Morscher, plus the excellent technical support of R. Phillips, R. Babuder, and R. Angus.

REFERENCES

1. D. A. Lewis, M. T. Hogan, J. McMahon, and S. Kinney. Application of uncooled ceramic matrix composite power turbine blades for performance improvement of advanced turboshaft engines. Proceedings of the 64th American Helicopter Society International Annual Forum, vol. 2, Montreal, QC, Canada, April 29–May 1, 2008 (Curran Associates, Inc., 2011), pp. 961–966.
2. M. Mecham. Out of the Lab. *Aviat. Week*, 30, 13 (2012)
3. M. H. Jaskowiak and J. A. Setlock, Processing and properties of sapphire reinforced alumina composites. Proceedings of the 20th Annual Conference on Ceramic, Metal, and Carbon Composites, Cocoa Beach, FL, January 23–25, 1996 (DoD Ceramics Information Analysis Center), pp. 843-862.

4. G. N. Morscher and V. Pujar. Creep and stress-strain behavior after creep for SiC fiber- reinforced melt-infiltrated SiC matrix composites. *J. Am. Ceram. Soc.*, 89, 1652–1658 (2006).
5. J. Lang and J. A. DiCarlo. Modeling creep-induced stress relaxation at the leading edge of SiC/SiC airfoils. Proceedings of the 31st Annual Conference on Ceramics, Metal &Carbon Composites, Materials and Structures, Restricted Sessions, Cape Canaveral, FL (2007), Daytona Beach, Fl, January 22–25, 2007 (ZAI Inc.).
6. J. A. DiCarlo, H. M. Yun, G. N. Morscher, and R. T. Bhatt. SiC/SiC composites for 1200°C and above, in *Handbook of Ceramic Composites*, edited by N. P. Bansal (Kluwer Academic Publishers, Boston, 2005), pp. 77–98.
7. R. T. Bhatt and J. A. DiCarlo. Method of Improving the Thermo-Mechanical Properties of Fiber-Reinforced Silicon Carbide Matrix Composites. U.S. Patent Application, 13/018,611.
8. G. S. Corman and K. L. Luthra. Silicon melt-infiltrated ceramic composites (HiPerComp). in *Handbook of Ceramic Composites*, edited by N. P. Bansal (Kluwer Academic Publishers, Boston, 2005), pp. 99–115.
9. T. Ishikawa. Advances in inorganic fibers. *Adv. Polym. Sci.*, 178, 109–144 (2005).
10. J. A. DiCarlo and H. M. Yun. Creep of ceramic fibers: mechanisms, models, and composite implications, in *Creep Deformation: Fundamentals and Applications*, edited by R. S. Mishra, J. C. Earthman, and S. V. Raj (The Minerals, Metals, and Materials Society, Warrendale, PA, 2002), pp. 195–208.
11. M. Takeda, J. Sakamoto, A. Saeki, Y. Imai, and H. Ichikawa. High performance silicon carbide fiber Hi-Nicalon for ceramic matrix composites. *Cer. Eng. Sci. Proc.*, 16(4), 37–44 (1995).
12. H. Ichikawa. High performance SiC fibers from polycarbosilane for high temperature applications. *Key Engineering Materials*, 352, 59–64 (2007).
13. H. M. Yun, D. Wheeler, Y. Chen, and J.A. DiCarlo. Thermo-mechanical properties of super Sylramic sic fibers. *Cer. Eng. Sci. Proc.*, 26(2), 59–65 (2005).
14. M. Takeda, J. Sakamoto, and H. Ichikawa. Mechanical and structural analysis of silicon carbide fiber Hi-Nicalon type S. *Cer. Eng. Sci. Proc.*, 17(4), 35–42 (1996).
15. H. M. Yun and J. A. DiCarlo. Tensile behavior of as-fabricated and burner-rig exposed SiC/SiC composites with Hi-Nicalon type S fibers. *Cer. Eng. Sci. Proc.*, 23(3), 571 (2002).
16. J. A. DiCarlo. SiC fiber creep and rupture models for understanding CMC behavior above 1400°C. NASA TM (in preparation).
17. R. M. Cannon, W. H. Rhodes, and A .H. Heuer. Plastic deformation of fine-grained alumina: I, interface-controlled diffusional creep. *J. Am. Ceram. Soc.*, 63(1–2) 46–53 (1979).
18. G. N. Morscher and J. A. DiCarlo. A simple test for thermo-mechanical evaluation of ceramic fibers. *J. Am. Ceram. Soc.*, 75, 136–140 (1992)
19. G. Pezzotti, H. J. Kleebe, H. Nishimura, and K. Ota. Grain-boundary viscosity of preoxidized and nitrogen-annealed silicon carbides. *J. Am. Ceram. Soc.*, 84(10), 2371–2376 (2001).
20. M. D. Sacks, W. Toreki, C. D. Batish, and G. J. Choi, inventors; University of Florida, assignee. Preparation of boron-doped silicon carbide sic fibers. U.S. Patent 5,851,942 (December 22, 1998).
21. J. Lipowitz, J. A. Rabe, A. Zangvil, and Y. Xu. Structure and properties of Sylramic^{TM} silicon carbide fiber—a polycrystalline, stoichiometric β–sic composition. *Ceram. Eng. Sci. Proc.*, 18(3), 147–157 (1997).
22. J. Lipowitz and J. A. Rabe, inventors; Dow Corning Corporation, assignee. Preparation of polycrystalline ceramic fibers. U.S. Patent 5,279,780 (January 18, 1994).
23. J. A. DiCarlo and H. M. Yun, inventors; The United States of America as represented by the Administrator of National Aeronautics and Space Administration, assignee. Methods for producing silicon carbide architectural preforms. U.S. Patent 7,687,016 (March 30, 2010).
24. S. M. Dong, G. Chollon, C. Labrugere, M. Lahaye, A. Guette, J. L. Bruneel, M. Couzi, R. Naslain, and D. L. Jiang. Characterization of nearly stoichiometric SiC ceramic fibers. *J. Mat. Sci.*, 36, 2371–2381 (2001).
25. G. N. Morscher and V. Pujar. Design guidelines for in-plane mechanical properties of SiC fiber-reinforced melt-infiltrated SiC composites. *Int. J. Appl. Ceram. Technol.*, 6, 151–163 (2009).
26. D. Dunn. The effect of fiber volume fraction in HiPerComp SiC-SiC composites. PhD Thesis. Alfred, New York (2010).
27. J. A. DiCarlo and H. M. Yun. Non-oxide (Silicon Carbide) fibers, in *Handbook of Ceramic Composites*, edited by N. Bansal (Kluwer Academic Publishers, Boston, 2005), pp. 33–52.
28. L. Thomas-Ogbuji. A pervasive mode of oxidation degradation in a SiC/SiC composite, *J. Am. Ceramic Soc.*, 81, 2777–2784 (1998).
29. G. N. Morscher, H. M. Yun, J. A. DiCarlo, and L. Thomas-Ogbuji. Effect of a BN interphase that debonds between the interphase and the matrix in SiC/SiC composites. *J. Am. Ceramic Soc.*, 87, 104–112 (2004). Also U.S. Patent 7,427,428 (September 23, 2008).
30. J. A. DiCarlo and M. van Roode. Ceramic composite development for gas turbine engine hot section components. Proceedings of ASME Turbo Expo 2006: Power for Land, Sea and Air, Paper GT2006–90151, Barcelona, Spain, May 8–11, 2006.

31. M. J. Verrilli, J. A. DiCarlo, H. M. Yun, and T. R. Barnett. Hoop tensile characterization of SiC/SiC cylinders fabricated from spliced 2D fabric. *Cer. Eng. Sci. Proc.*, 24(4), 407–413 (2003).
32. H. M. Yun, J. A. DiCarlo, and R. T Bhatt. Advanced SiC/SiC ceramic composites for air-breathing and rocket propulsion engine components. Proceedings of JANNAF Conference, Charleston, SC, June 15, 2005.
33. H. M. Yun, J. A. DiCarlo, R. T Bhatt, and T. Easler. Advanced SiC/SiC ceramic composites for hot-section aerospace components. Proceedings of National Space & Missile Materials Symposium, Summerlin, NV, June 27–July 1, 2005.
34. H. M. Yun, J. A. DiCarlo, R. T. Bhatt, and M. H. Jaskowiak. Thermo-structural properties of SiC/SiC panels with 2.5D and 3D fiber architectures. Proceedings of the 29th Annual Conference on Ceramics, Metal &Carbon Composites, Materials and Structures, Restricted Sessions, Cape Canaveral, FL, January 24–27, 2005 (ZAI Inc.).
35. G. N. Morscher, J. A. DiCarlo, J. D. Kiser, and H. M. Yun. Effects of fiber architecture on matrix cracking for melt-infiltrated SiC/SiC composites. *Int. J. Appl. Ceram. Technol.*, 7(3) 276–290 (2010).
36. R. T. Bhatt, L. M. Cosgriff, and D. S. Fox. Influence of Fiber Architecture on Impact Resistance of Uncoated SiC/SiC Composites, Proceedings of 8th Pacific Rim Conference on Ceramic and Glass Technology, Vancouver, BC, Canada, May 31–June 5, 2009.
37. J. A. DiCarlo and R. T. Bhatt. Modeling SiC/SiC Creep-Rupture Behavior from 2400 to 3000°F. Proceedings of 35th Annual Conference on Composites, Materials, and Structures, Restricted Sessions, Cocoa Beach/Cape Canaveral, FL, January 24–27, 2011 (US Advanced Ceramics Association).
38. G. N. Morscher. Tensile creep of melt-infiltrated SiC/SiC composites with unbalanced Sylramic-iBN fiber architectures. *Int. J. Appl. Ceram. Technol.*, 8(2), 239–250 (2011).
39. R. T. Bhatt, J. A. DiCarlo, and J. D. Kiser. Tensile and Creep Properties of Hybrid CVI-PIP SiC/SiC Composites at High Temperatures in Air. Proceedings of 36th Annual Conference on Composites, Materials, and Structures, Restricted Sessions, Cocoa Beach/Cape Canaveral, FL, January 23–26, 2012 (US Advanced Ceramics Association).
40. J. L. Smialek, R. C. Robinson, E. J. Opila, D. S. Fox, and N. S. Jacobson. SiC and Si_3N_4 recession due to SiO_2 scale volatility under combustor conditions, *Adv. Composite Mater*, 8(1), 33–45 (1999).
41. K. N. Lee, D. S. Fox, J. I. Eldridge, D. Zhu, R. C. Robinson, N. P. Bansal, and R. A. Miller. Upper temperature limit of environmental barrier coatings based on mullite and BSAS. *J. Am. Ceram. Soc.*, 86(8) 1299–1306 (2003).
42. K. N. Lee, D. S. Fox, and N. P. Bansal. Rare earth silicate environmental barrier coatings for SiC/SiC composites and Si3N4 ceramics. *J. Eur. Ceram. Soc.*, 25(10), 1705–1715 (2005).
43. M. van Roode, J. R. Price, J., Kimmel, N. Miriyala, D. Leroux, A. Fahme, and K. Smith. Ceramic matrix composite combustor liners: a summary of field evaluations. *J. Eng. Gas Turbines Power* [Transactions of the ASME], 129, 21–30 (2007).
44. D. Zhu, R. A. Miller, and D. S. Fox. *Thermal and Environmental Barrier Coating Development for Advanced Propulsion Engine Systems*, AIAA-2007–2130. American Institute of Aeronautics and Astronautics. Also NASA/TM-2008–215040.

CHAPTER 8

OXIDE–OXIDE COMPOSITES

KRISTIN A. KELLER,[1,2] GEORGE JEFFERSON,[1] AND RONALD J. KERANS[3]

[1] *Materials and Manufacturing Directorate, Air Force Research Laboratory, Wright-Patterson Air Force Base, OH, USA*
[2] *UES Inc., Dayton, OH, USA*
[3] *Emertius Member of Materials and Manufacturing Directorate, Air Force Research Laboratory, Wright-Patterson Air Force Base, OH, USA*

8.1 INTRODUCTION

Aerospace technology advancement is vitally dependent upon the availability of structural materials with ever increasing thermal capabilities. For example, by enhancing thermal capabilities, turbine engines can operate with reduced cooling requirements and at potentially higher combustion temperatures, leading to lower fuel costs and decreased gas emissions. Similarly, improved thermal protection systems (TPS) that offer both thermal and structural functions will significantly impact future space/reentry vehicle designs. Fiber reinforced ceramics (terminology includes "ceramic matrix composites" (CMC), "continuous fiber ceramic composites" (CFCC), and "ceramic fiber–matrix composites" (CFMC)) offer thermal performance beyond the service limits of high temperature metals, with damage tolerance superior to monolithic ceramics [1–4].

Currently, the most advanced ceramic composites are based on silicon carbide fibers; although extremely robust materials, their lifetimes are ultimately limited by susceptibility to oxidation and more rapid degradation in combustion environments [5, 6]. Superior environmental stability can be attained, in exchange for some reduction in mechanical performance, by using oxides for all constituents: fibers, matrices, and interface-control coatings. In addition, oxide fibers and some oxide composites also have a significant cost advantage over SiC-based composites. These two composite types have different strong points, leading to the preferential use of each in given applications.

The superior environmental stability of oxide composites generally comes at the expense of somewhat reduced mechanical properties when compared to non-oxide ceramic composites. Creep resistance is of particular concern, since this is the life limiting property in many high temperature applications. Further, oxides do not have the very low thermal expansivities of silicon-based ceramics, and their thermal conductivities are low. Consequently, thermal strains are larger than in non-oxide composites. To some degree, lower elastic moduli moderate the resulting stresses, but typically strengths are also lower. Hence, the design of components to be used in thermal gradients and transients will need to pay special attention to these issues. However, there is significant room for improvement in all constituents and improved properties can be expected in future materials.

This chapter is focused on providing an updated overview of high temperature (>1000°C) oxide–oxide composite technology, potential applications, current research issues, and what might be expected in future derivatives. Glass and glass–ceramic matrices and/or fibers will not be considered in this chapter. Section 8.2 discusses the fundamental requirements for tough behavior in oxide–oxide CMCs, Section 8.3 describes the ceramic fibers and fiber architectures typically used in oxide CMCs, and Section 8.4 deals with the various processing methods used to produce these composites. Section 8.5 presents the current porous matrix composite systems, while Section 8.6 provides property information on these materials. Section 8.7 discusses composites containing interface coatings. Section 8.8 provides details on the current efforts in technology development for oxide–oxide composites and Section 8.9 presents a view of the future of oxide–oxide CMCs. The reader will find additional insight from other chapters in this volume, along with reviews of oxide CMCs given in References 1, 2, and 7–9.

Ceramic Matrix Composites: Materials, Modeling and Technology, First Edition. Edited by Narottam P. Bansal and Jacques Lamon.

8.2 COMPOSITE DESIGN FOR TOUGH BEHAVIOR

The basic requirement for a tough ceramic composite is crack deflection at or near the fiber/matrix interface [3]. As the crack is deflected along the fiber, the primary matrix crack eventually moves on around the fiber, leaving it bridging the matrix crack and reducing the stress intensity at the matrix crack tip. The fiber ultimately fractures at a point away from the crack plane and begins to pull out from the matrix upon continued loading of the composite. The toughening mechanism of fiber debonding and non-coplanar fracture of the constituents is often called the distributed damage mechanism. The pull-out of the fiber requires some amount of work to overcome the roughness and frictional effects. This provides further toughening to the composite. Each fiber acts more or less independently; the fracture of one imposes only a modest increase in stress on its neighbors [10].

In contrast, matrix cracks in a composite with excessive fiber/matrix bonding propagate directly through fibers, and the composite behaves as if it were a monolithic ceramic. The load-deflection behavior of a good composite exhibits nonlinear, pseudo-ductile behavior, sometimes to a considerable extent, while a bad one, with excessive fiber/matrix bonding, is perfectly elastic and typically not very strong. This load-deflection behavior has led to fairly common use of the terms "non-brittle" and "brittle" to describe the two cases. This can be confusing because in both cases every fracture event of every constituent is brittle. The terms are understood to refer to the (observed or assumed) presence or lack of the desired nonlinear load-deflection behavior.

The two basic mechanisms utilized for crack deflection in CMCs are the use of a sufficiently weak (porous) matrix or the use of a weak engineered interface, which typically involves a fiber coating. The term "sufficiently weak," in this context, is in comparison to the fiber strength in the final composite; essentially all fabrication methods have some effect on the fiber strength. Crack behavior results from a competition between fracture in the fiber, coating (if present) and the matrix. The discussion of this section borrows from Reference 3, which provides a comprehensive treatment.

8.2.1 Porous Matrices

The most mature oxide–oxide composites utilize a sufficiently weak, porous matrix [11–22]. In porous matrix composites, a matrix crack is deflected within the matrix in the region of the fiber/matrix interface. Although the exact mechanism is not fully understood, it is taken to be some combination of deflection of cracks due to pores near the fiber surface, irrespective of potentially strong bonding between the fiber and the matrix particles, and intrinsic fracture of matrix particles from the fiber surface [3]. Matrix cracks essentially connect pores and arrive at the fiber surface in pores, imposing a lower stress concentration on the fiber as compared to the sharp crack tip of a dense matrix composite. The matrix is unable to impose sufficient stress concentrations to fracture the fibers until the fiber stress is quite high. Fiber failure occurs in a statistical manner at various points/planes within the fiber tow and final failure occurs when the differing crack planes unite to form a continual crack across the sample and comminution of the matrix has occurred [22]. A macroscopic comparison of (a) a good porous matrix composite showing distributed damage and nonlinear load-deflection behavior and (b) a bad brittle composite is shown in Figure 8.1. The brittle sample was initially porous, but was reinfiltrated multiple times, creating strong bonding between the fiber and the matrix.

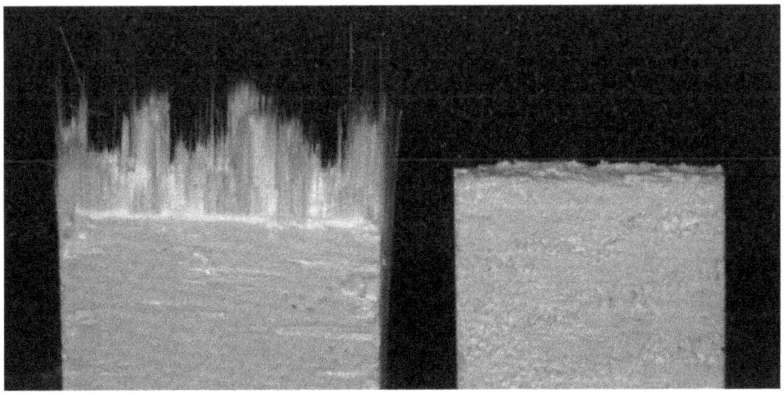

FIGURE 8.1
Comparison between brushy fracture surface of a tough ceramic composite (porous matrix, left side) and a brittle fracture surface (right side).

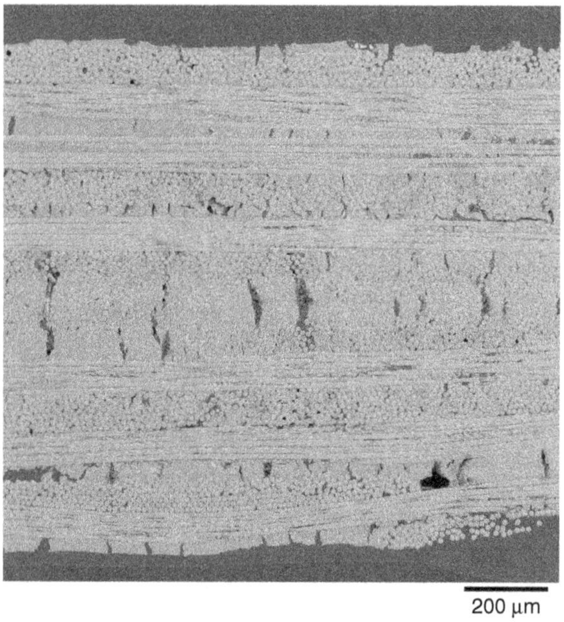

FIGURE 8.2
Typical microstructure of a porous matrix oxide–oxide composite.

One should note that the terms "porous matrix composite" and "dense matrix composite" are often used to differentiate between families of composites (without and with fiber coatings, respectively) and are used loosely with regard to the actual density of the material. Current technology composites of the porous matrix type are not fully sintered and have approximately 35–50% matrix porosity/25–40% total porosity (see Table 8.2 for examples and references). The composite consists of a volume fraction of fibers, f. The remainder of the volume, $(1 - f)$ is considered to be matrix. The term matrix porosity (p_m) refers to the pore or void fraction of the matrix, so that, as the fibers are assumed to be fully dense, the composite or "total" porosity is $p_c = p_m(1 - f)$. For example, if $f = 0.4$, a 50% matrix porosity would correspond to 30% total porosity. Matrices tend to develop microcracks during drying and/or sintering due to the constraint induced by the fibers. A typical microstructure of a porous matrix composite is shown in Figure 8.2. The weak matrix necessary for crack deflection compromises matrix-dominated properties to some degree, for example, shear properties, and transverse and interlaminar properties in 2D lay-ups.

The mechanical behavior of porous matrix composites has been modeled, but the compromises between matrix properties and toughness are not yet fully quantified. Initial estimates indicated that a matrix porosity of approximately 30% is needed for crack deflection within a porous matrix [29,30]. Subsequent efforts have attempted to better quantify the transition between "porous" and "dense" CMC failure behavior [23]; however, further investigation is essential, especially considering the temperature dependent nature of the porous microstructure.

While mechanistic details remain a subject of investigation, the time (t)–temperature (T) use limits of porous matrix composites can be ultimately limited by matrix sintering. At high temperature, the matrix continues to sinter, gaining strength (and stiffness) until it is no longer effective in deflecting cracks around fibers. The current generation of polycrystalline fibers have t–T capabilities similar to the matrix materials, so the matrix-related limits are somewhat masked. While it is not coincidental that the matrix and fiber limits are similar, there are apparent paths to improved fibers that can be pursued when it is clear that the fibers are the limiting factor. Moreover, recent results on composites using fiber coatings for interface control indicate that greater improvements than anticipated can be attained even with current fibers.

8.2.2 Interface Control

Engineered weak fiber–matrix interfaces have been the traditional mechanism for enabling tough behavior in ceramic composites. In this case, as a matrix crack approaches a coated fiber, the crack is deflected either at the matrix/coating

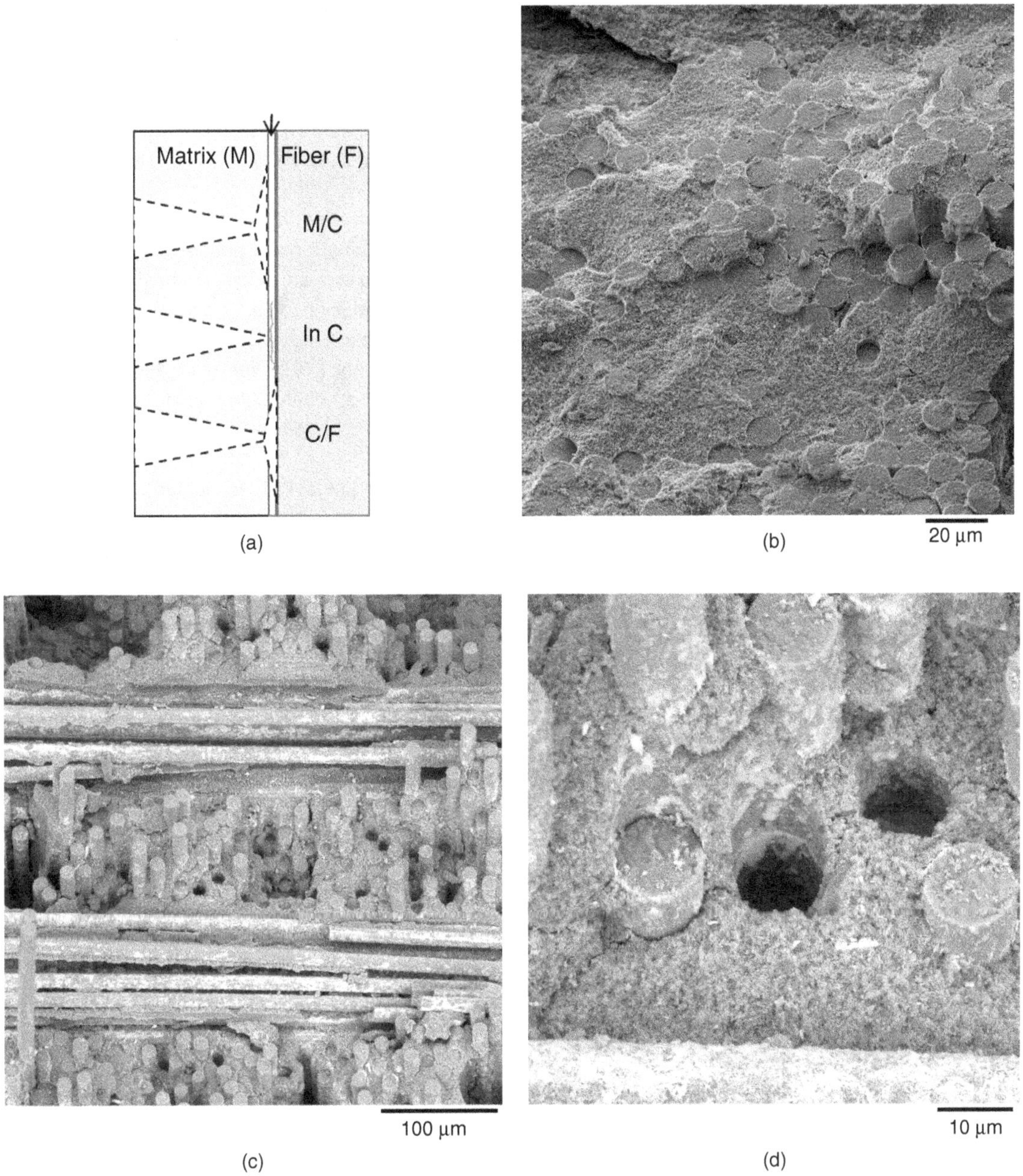

FIGURE 8.3

(a) Schematic showing potential crack-deflection areas in an oxide composite containing an engineered interface; (b) fracture surface of Nextel 610/alumina composite showing brittle failure due to the absence of an engineered interface; (c) fracture surface of Nextel 610/monazite/alumina composite showing fiber pull-out due to weak fiber/matrix interface; (d) higher magnification view showing monazite coating in trough and on fiber, indicating deflection within the coating.

interface, within the coating layer or at the coating/fiber interface (Figure 8.3a). The fiber ultimately fractures at a point away from the crack plane and begins to pull out from the matrix upon continued loading of the composite. The pull-out of the fiber requires some amount of load to overcome the roughness and frictional effects. For comparison, an example of a brittle fracture surface, with virtually no fiber pull-out, is shown in Figure 8.3b, while a typical fracture surface showing fiber pull-out of coated oxide fibers is shown in Figure 8.3c. A higher magnification view, shown in Figure 8.3d, reveals coating on the fiber and in the trough, indicating deflection within the coating.

The optimal interface properties, such as coating thickness, frictional stresses, debond lengths, etc., are dependent upon the particular composite constituents and must be evaluated on an individual basis. There are some general guidelines associated with choosing an interface coating with appropriate properties; one of the most commonly used requirements is that the ratio of the fracture energies of the interface and the fiber be less than a particular value dependent upon the elastic properties of the fiber and matrix (0.25 for like values) (He and Hutchinson criterion) [31]. A thorough treatment of this topic is beyond the scope of this chapter and the reader is referred to References 1, 3, 32, and 33 for a more in-depth discussion.

The design and evaluation of oxide fiber coating substitutes for controlled interface properties is a multifaceted problem. Although carbon (C) and boron nitride (BN) have unusual combinations of properties that make them ideal choices for fiber coatings, their poor oxidation resistance precludes their use in all-oxide composites. Our accumulated understanding of composites based on C and BN coatings provides a good foundation, however composites based on oxide coatings can be expected to behave in very different ways. For example, oxide coatings are typically less compliant; hence the optimum coating thicknesses are likely to be substantially different [34]. Even with thicker coatings, the low levels of friction achievable with C are probably unattainable. This will put an even higher priority on strong fibers and the retention of that strength throughout processing and service in order to attain sufficiently long debond lengths for high levels of distributed damage.

Despite the significant challenges, several coatings, $LaPO_4$ (monazite) and $CaWO_4$ (scheelite), have been definitively demonstrated to provide not only the crack-deflection function, but, in the former case, substantial improvement in alumina-based composite properties [35, 36]. For example, the use of monazite coatings in alumina reinforced alumina composites was found to increase the composite life as compared to a coatingless control composite by a factor of several hundred [36]. Such dramatic improvements are rare in structural materials and portend significant improvements in future composite systems. Further discussion of research on composites containing interface coatings is addressed in Section 8.8.

8.3 FIBERS AND FIBER ARCHITECTURE

The reinforcement of choice for oxide–oxide composites is typically the fine diameter (~10–12 μm) Nextel™ fibers (3M Corporation, St. Paul, MN), due to their maturity and the limited number of other available oxide fibers. Of the Nextel fibers (312, 440, 550, 610, and 720), Nextel 610 (Alumina, Al_2O_3) and 720 (Mullite ($3Al_2O_3$-$2SiO_2$)-Al_2O_3) fibers are used for high temperature applications (>1000°C), while the 312 and 440 fibers (Al_2O_3-SiO_2-B_2O_3) are generally used for lower temperature (<1000°C) applications, such as thermal insulation. Nextel 550 fiber contains only γ-Al_2O_3 and amorphous SiO_2 and its use temperature is limited by the onset of crystallization in the fiber (<1200°C). Nextel 312 and 440 fibers contain a glassy phase that dramatically affects their creep properties, which, in turn, limits their maximum use temperature [37]. This chapter focuses principally on composites based on Nextel 610 and 720 fibers.

For the higher temperature fibers, Nextel 610 has a higher tensile strength than the 720, while the 720 has superior creep resistance at high temperature. Selected properties of the Nextel fibers are given in Table 8.1. 3M has also presented information on other improved fibers, such as Nextel 650 (Al_2O_3 + 10% ZrO_2 + 1% Y_2O_3) and a Nextel 720 fiber that displays a slight shrinkage at 1200°C under creep conditions, versus the elongation seen in the current Nextel 720 fiber [48]. However, these fibers are not commercially available.

Sapphire (single crystal alumina) fibers have also been utilized in oxide–oxide composites in the past (see, e.g., [49, 50]); however, their current typical use is as an optical material. Various suppliers produce optical grade sapphire fiber, including MicroMaterials Inc. (Tampa, FL) and Photran LLC (Poway, CA). Although sapphire fibers can be used at higher temperatures than the polycrystalline Nextel fibers [51], they generally have a substantially larger diameter (>100 μm), which limits their handleability in comparison to the fine diameter tows. These fibers cannot be woven into fabrics or easily wound due to their large diameter. Further, an intermediate temperature strength drop and susceptibility to slow crack growth have been reported for sapphire [52].

Other oxide fiber types have been studied, including zirconia (ZrO_2), YAG ($Y_3Al_5O_{12}$—yttrium aluminum garnet), as well as various eutectic (e.g., [Al_2O_3-YAG]) compositions. Of these, YAG is arguably the most creep resistant oxide and is therefore an excellent material choice [53]. Fine diameter YAG fibers have been produced [38, 54, 55] and continued commercial development has been reported [56]. However, recent research activity in this area has focused on transparent polycrystalline YAG fibers for high energy laser (HEL) applications [57]. Work on the functional

TABLE 8.1 Constituent Properties

	Composition, wt%			Strength, MPa			Modulus	Density	Diameter	Thermal limits, °C		Expansion
	Al_2O_3	SiO_2	Other	Filament[d]	Roving[e]	Tow[d]	GPa	g/cc	μm	Creep[f]	Strength[g]	ppm/°C
Fibers												
Nextel 312	63	25[b]	B_2O_3	1700	850	–	150	2.7	10–12	600	850	3.0
Nextel 610	>99	<0.3	Fe_2O_3	3100	1130	1600	380	3.9	10–12	1000	1225	8.0
Nextel 720	85	15[c]	–	2100	540	800	260	3.4	10–12	1200	1350	6.0
Sapphire[a]	100	0	–	3100	–	2250	435	3.8	70–250	–	–	–
Alumina	100	0	0	–	–	–	380	3.96	–	–	–	8.1
Mullite	75.5	24[c]	<.5	–	–	–	220	3.16	–	–	–	5.0

Source: From References: Fibers: 4, 37–44; mullite/alumina: 30, 45–47.
[a]Saphikon inc., c-axis single-crystal.
[b]Partly amorphous silica.
[c]In crystalline mullite phase.
[d]2.5 mm gage length.
[e]150 mm gage length.
[f]1% strain after 1000 hours at 70 MPa, single filament.
[g]90% tow strength retention after 1.5 minute exposure.

polycrystalline HEL YAG fibers has resulted in improved tensile properties [58] and the dual structural/functional application provides further impetus for YAG fiber development.

In terms of fiber architecture, the most mature composites utilize Nextel fibers in the woven form and generally in an 8 harness-satin weave, although other weaving patterns are available. The woven fabric enables easy handling of the fiber and is formable to more complex shapes. Alternately, fiber tows can be utilized by forming tapes of the fibers and then stacking them for consolidation into a green body. These tapes can be formed through filament winding [59] or tape casting [60], although the latter is more applicable to larger diameter fibers (~100 μm) than to the small diameter (10–12 μm) Nextel fibers. In a few cases, 3D woven oxide–oxide composites have been produced and evaluated, although these have been primarily with the lower temperature Nextel 312 and 440 fibers [61–63]. The startup costs associated with the manufacturing of 3D preforms will necessitate either substantial production volumes or significant cost penalties.

8.4 PROCESSING METHODS

Processing of oxide-based fibrous composites is complicated by the reinforcing fibers, whose presence constrains the densification of the matrix material [64,65] and also limits the temperatures, pressures, and chemistries that can be used during consolidation. If the processing parameters are not adequately controlled, considerable damage can be done to the fiber. Creating an oxide matrix is therefore a balance between avoiding microstructural and morphological changes in one refractory constituent while affecting large changes in another nearly equally refractory constituent. It is noted that some efforts have been made toward taking advantage of extreme parameters; for example, the grains within the Nextel 610 fibers have been intentionally grown from submicron size to tens of microns in size in order to improve the creep resistance of the fibers [66]. Such actions have a substantial corresponding reduction in other fiber properties, such as tensile strength. This discussion will focus on the more benign processing conditions typically encountered in oxide–oxide composite production.

The constraining effect of fiber reinforcement is further complicated by the use of complex fiber architectures, such as 3D reinforcements. While 2D cloth/tape reinforced composites are constrained from shrinking within the fiber plane (*x*–*y*), there is some freedom to shrink in the third direction (z). In the case of a 3D reinforcement, the external bounds of the composite are determined by the fiber preform, with virtually no shrinkage in the *x*–*y* or z directions. Processing options are confined to filling the inter-fiber/inter-tow space and most processes are inherently unable to fill it completely.

Since some porosity is virtually unavoidable in oxide–oxide composites, the prudent course of action for composite designers is to decide on how to best distribute the void volume. It is presumed that uniformly distributed, fine, closed porosity is preferable to having the void volume present as large cracks in the matrix. These large cracks are commonly seen in 2D reinforced composites, due to the constraint imposed on the matrix by the fibers during the sintering process; this issue is further magnified in the 3D reinforced composites. Some research has been devoted to eliminating such cracks and work in this area is expected to increase [63, 67]. While the constraint provided by the fibers is a significant hindrance to obtaining dense(r) matrices, it makes crack deflecting porous matrices easy to achieve and maintain to relatively high temperatures.

The processing of a composite typically involves the following separate steps: (1) coating fibers (optional), (2) arranging fibers in the desired architecture, (3) infiltrating fibers with the matrix precursor, (4) consolidating the matrix, including sintering and (5) final machining. This section focuses primarily on steps 1, 3, and 4; steps 2 and 5 are beyond the scope of this work and are mentioned only briefly in this text. It is noted that the methods described in this section have been used for both porous matrix and dense matrix composite production.

8.4.1 Processing of Interface Coatings

For composites utilizing a weak engineered interface for crack deflection, there have been various methods used to deposit coatings onto ceramic fibers, including solution/sol/slurry precursors [68–70], chemical vapor deposition (CVD) [71–74] and electrophoretic deposition (EPD) [75,76]; Reference 3 provides a review of coating technology. The choice of coating process is often dictated by the material being deposited; for example, carbon (C) and boron nitride (BN) are commonly deposited by CVD (see, for example, [77–80]). This process, however, is not efficient for applying coatings of crystalline high temperature oxides, such as alumina, and particularly mixed oxides. Both solution/sol/slurry precursors and EPD have been utilized for the deposition of oxide coatings onto fibers.

Coating large single filaments, >100 μm diameter, is generally easier than coating fine diameter fiber tows ($\sim$10–12 μm diameter, 400–800+ fibers/tow). Attaining effective, discrete coatings on fibers within a tow and further, on tows within a fabric, was initially a challenge, due to the tendency toward bridging of the coating as the coating thickness increased. The fiber tow essentially became encrusted at a critical point, preventing further matrix infiltration.

This issue has been successfully addressed using the liquid precursor (solution/sol/slurry) coating techniques for the discrete deposition of oxides, such as monazite ($LaPO_4$) and porous oxides (carbon–oxide mixtures) onto oxide fiber tows [68,69, 81,82] without significant fiber coating bridging. A typical fiber coater is shown in Figure 8.4a [83]. In this coating process, a lighter, immiscible liquid is floated on the surface of the coating precursor, which acts to remove excess sol from the coated tow. This in turn allows for the coating of individual filaments with minimal bridging. Figure 8.4b shows monazite coated Nextel 720 fiber in an alumina matrix. Some variation in the coating thickness can be seen; however, each fiber is separately coated, with no apparent bridging between the fibers. Note that the stoichiometry of the coating precursor has been shown to be critical in the retention of fiber strength during coating [84]. However, washing of the coating precursor solution has been effective for retaining the fiber strength [85].

Although coated fiber tows can be woven into a fabric [86], it is much more efficient to deposit a coating onto the already-woven fabric. The deposition of a discrete fabric coating, (i.e., each fiber individually coated without excessive coating bridging), had been elusive; however, recent work by Fair et al. [87–89] demonstrated the deposition of discrete coatings on oxide fabrics without fiber degradation. Figure 8.5 shows a schematic of the fabric coating process, along with a micrograph showing a cross-section of coated Nextel fabric. In this process, the woven fabric is completely saturated with a dilute, cooled ($\sim$5°C) coating precursor solution. For monazite ($LaPO_4$) coatings, a mixture of phosphoric acid (H_3PO_4) and lanthanum citrate are used. The cool saturated fabric is then transferred to a warm ($\sim$50°C) deionized water bath, which causes heterogeneous precipitation of the monazite particles. The concentration of the precursor controls the amount of material precipitated; a lower concentration results in coatings on the individual filaments without crusting of the tows. The coated fabric is then heat treated at high temperature to drive off the organics and to strengthen the coating. Multiple infiltration cycles are used to build up a coating of sufficient thickness ($>\sim$100 nm). This advancement in coating technology will be critical as denser matrices and/or more complex fiber architectures are utilized.

The deposited coatings are typically characterized in terms of coating uniformity, continuity, thickness, porosity, and composition using scanning electron microscopy (SEM). The coating composition can also be confirmed using either x-ray powder diffraction and/or transmission electron microscopy [90]. Control over the coating characteristics is critical for imparting the needed crack-deflection behavior in dense matrix oxide–oxide composites. This is particularly true if a hermetic coating is needed for environmental protection of the fiber.

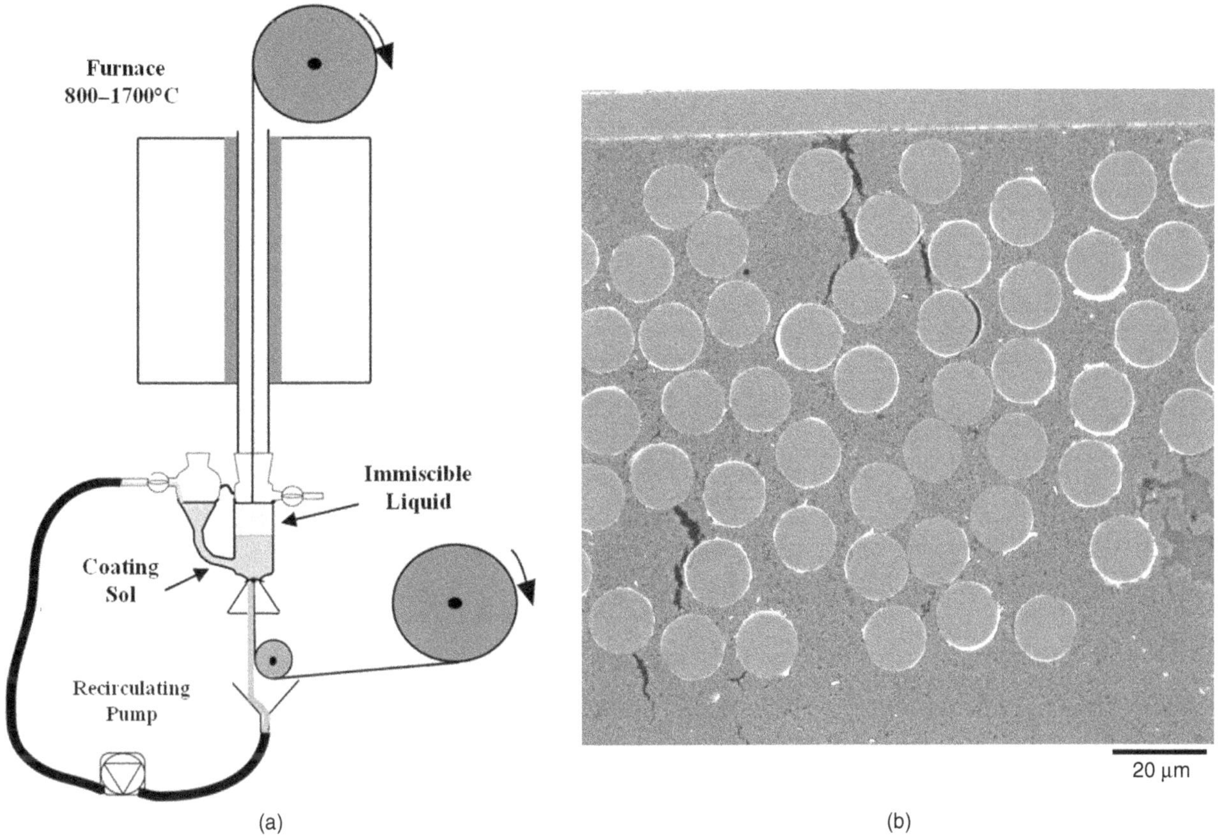

FIGURE 8.4
(a) Schematic of fiber tow coater, showing immiscible liquid addition; (b) SEM cross-section of Nextel 720/$LaPO_4$/Al_2O_3 composite fabricated with Nextel 720 tow coated using the fiber coater.

8.4.2 Matrix Infiltration

Oxide fiber tapes, woven fabric, or a more complex (2.5D, 3D) woven fiber structure can be infiltrated with oxide matrices. In the first case, single filaments or fiber roving/tows are incorporated into an easily handled form, such as a tape. This is commonly achieved through filament winding of the fiber onto a mandrel [59]. During the winding process, the tow is passed through a controlled ceramic slurry to infiltrate the matrix material within the tow. The resultant tape has some level of green strength, which enables handling of the fiber tows. An alternate method is to first filament wind the fiber and then tape cast an oxide layer within/on the fiber tape; this process is practical for the larger (>100 μm) diameter fibers. Hand lay-up of fibers can also be used, although this is labor intensive and used only with the larger diameter (>100 μm) fibers. The matrix infiltration processes outlined in this section can be used directly with woven fabric or with the more complex fiber architectures.

One of the most critical aspects in attaining a good quality oxide–oxide composite is control over the solution/slurry used to fabricate the CMC. Dispersed powder slurries are most commonly used and the formulation of the slurry is extremely important for attaining a homogeneous green composite and an optimal final product. Oxide powder is generally dispersed in a carrier, such as water, along with various additives, such as binders, dispersants, plasticizers, etc. The powder particle size is limited in relation to the fiber diameter. For example, it is difficult to infiltrate a tow containing 10 μm diameter fibers with 10 μm size powder. A review of particle–particle interactions and the stabilization of slurries is given in Reference 91.

8.4.2.1 Prepreg Processing The most commonly used process for fabricating oxide–oxide composites is the use of prepreg processing; this approach is akin to the techniques used in polymer matrix composite fabrication. For example, ATK-COIC (San Diego, CA) produces their Nextel 720/aluminosilicate composites by submersing the desized

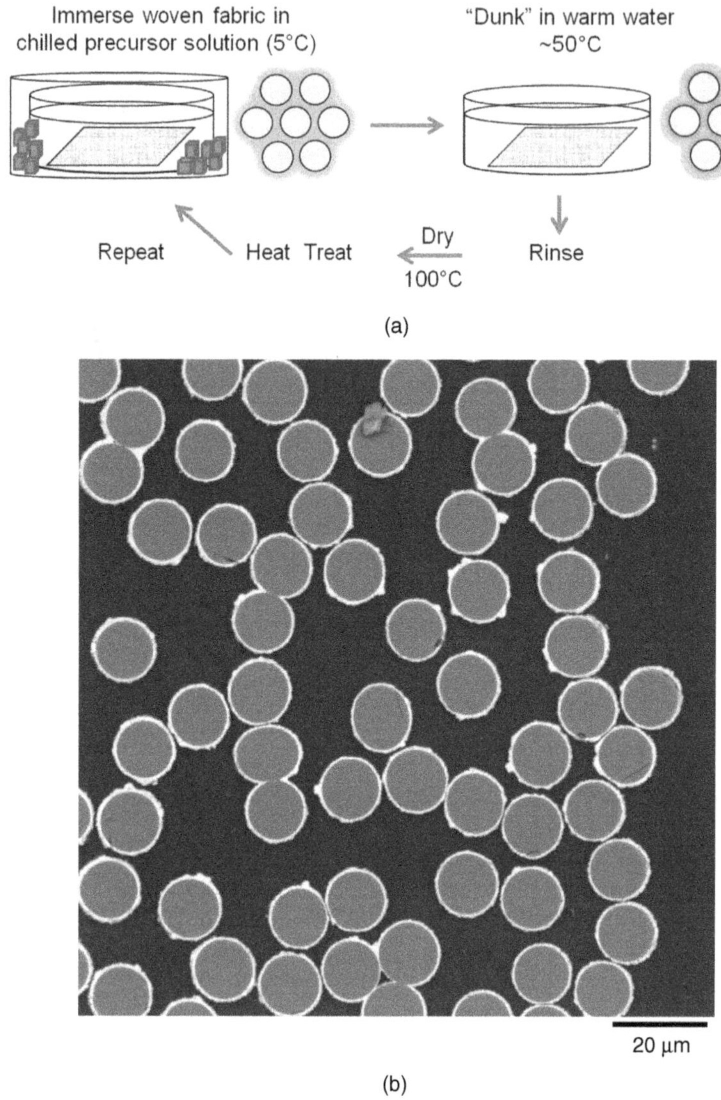

FIGURE 8.5
(a) Schematic of fabric coating process; (b) SEM micrograph showing cross-section of coated Nextel fabric.

fabric within a slurry ("pre-impregnating") to form a matrix-infiltrated fabric known as a prepreg. These prepregs are then stacked onto tooling (Figure 8.6) or within a mold and consolidated using vacuum bagging and warm (<150°C) temperatures, followed by pressureless sintering at high temperature (1000–1200°C) [24]. This is akin to the stacking of polymer-impregnated tapes into a mold and then autoclaving, which involves pulling a vacuum on the material, applying exterior gas pressure in the autoclave, and heating to cure the polymer resin. In the case of oxide–oxide composites, the warm temperatures are used to drive off any water or remaining solvents. The primary advantage of the prepregging process is the ability to form near-net shape, complex parts with relatively inexpensive tooling.

8.4.2.2 Pressure Infiltration This processing technique is used primarily with fabric-based samples, although it can be used for fiber tow samples [92] as well as complex 3D preforms. Layers of desized fabric are stacked into a mold in the desired orientation. The fabric is constrained during the infiltration process to avoid "floating," which would lead to undesirably thick matrix regions between layers and consequently a low fiber volume fraction. Depending upon the individual set-up, the mold can either be pressurized itself or the mold can be placed within a pressure vessel.

Prior to pressurization, a dispersed slurry is poured into the mold. The amount of slurry is dictated by the volume fraction of solids in the slurry and the optimal green matrix density, which is typically approximately 65% of full density (corresponding to close packing of powders). The slurry is produced by dispersing the oxide powder within a solvent, with powder-specific additives to ensure the stability of the slurry. Without stabilization, sedimentation of the slurry

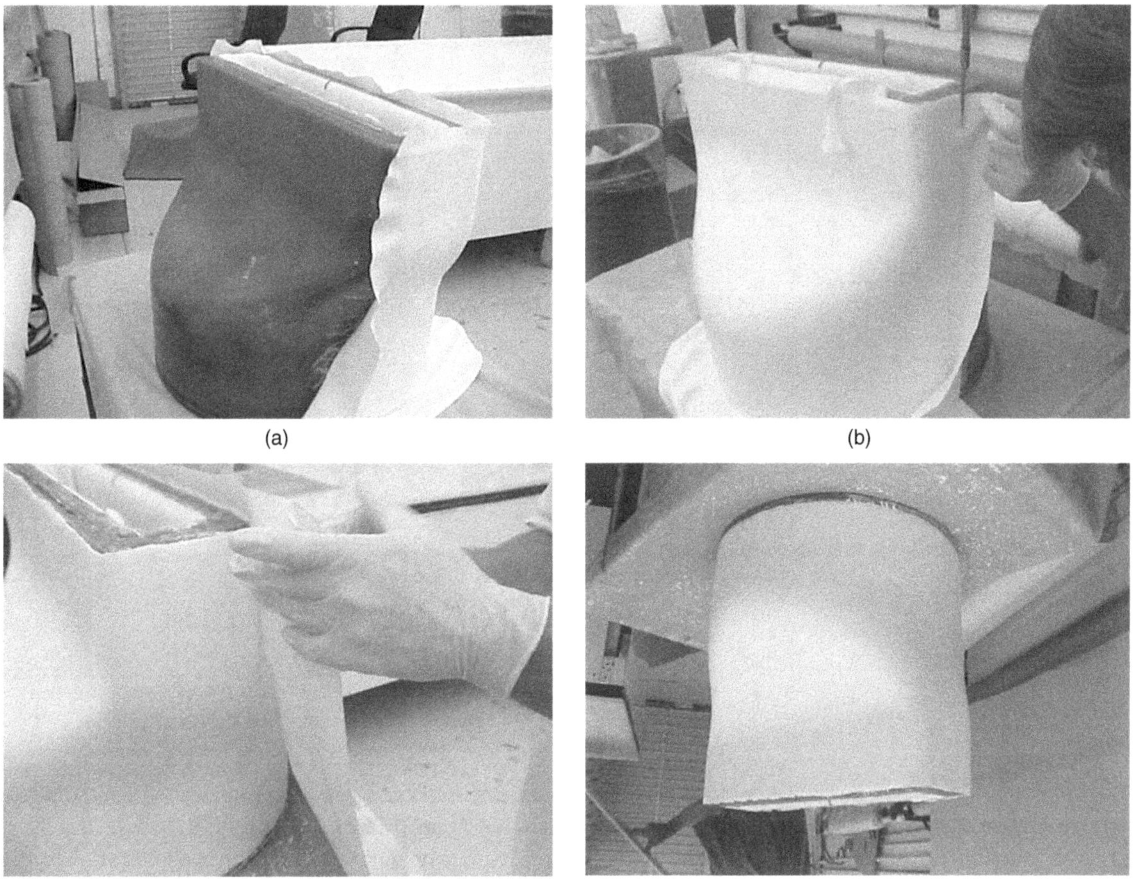

(a) (b) (c) (d)

FIGURE 8.6

Hand lay-up of oxide–oxide CMC onto component mold and resultant green body. Photos courtesy of ATK-COIC.

during the infiltration process could occur, leading to matrix gradients within the sample. This can be a particular problem with non-uniform powder size distributions and with increasing variations between the particle sizes. Pressure is applied to the unit to force the slurry through the fabric preform. The fabric essentially acts as a filter, building the matrix through the thickness of the sample. Vacuum may also be applied to aid in the filtration.

Once the slurry has filtered through the sample, the green body is removed from the unit. Care must be taken during this step, depending upon the amount of pressure applied to the sample. If excessive pressure is used, there can be a "springback" action of the green body upon release of the pressure, as discussed by Lange et al. [92]. Proper control over processing parameters can reduce this effect. The green body can then be dried and pressurelessly sintered to form the final porous oxide–oxide composite. Subsequent infiltrations with liquid precursors, such as aqueous colloidal sols, can increase the matrix density significantly, although a dense matrix is difficult to form in this manner. The density increases rapidly during the first few infiltrations; however, the effect is greatly diminished as the interior channels begin to fill with material and/or the outer pores fill up, preventing the flow of liquid into the interior of the body. Moreover, volatiles must escape during each infiltration cycle, which requires a minimum level of connected porosity.

8.4.2.3 Vibration-Assisted Vacuum Infiltration Vacuum infiltration is very similar to pressure infiltration, except that no additional pressure is added. The fiber preform is loaded into the chamber and an appropriate amount of slurry is poured onto the preform, as discussed for pressure infiltration. The chamber is evacuated from both the top and bottom to eliminate any air bubbles within the sample and the slurry. The chamber is then vibrated during the initial portion of the vacuum infiltration process, which aids in packing the powder uniformly within the fiber preform. The top of the chamber can be opened after the initial deairing/vibration process, to provide atmospheric pressure to the composite.

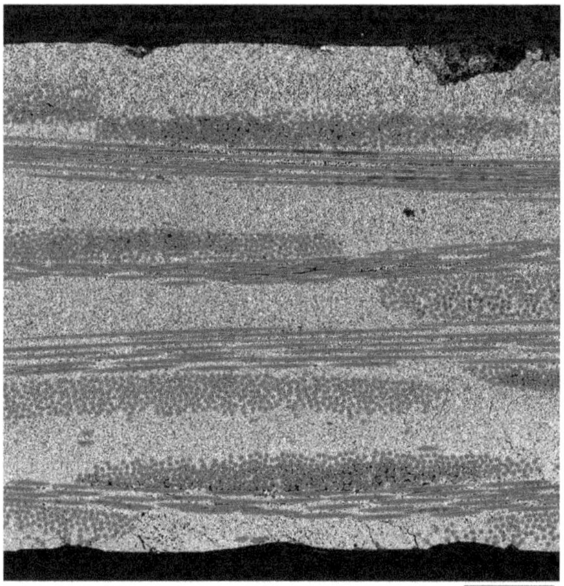

FIGURE 8.7
Virtually crack-free ceramic composite fabricated using freeze forming.

Yang et al. [63] looked at producing crack-free oxide fiber reinforced composites by infiltrating controlled particle sizes within a fiber preform using this vibration-assisted vacuum infiltration technique. In this process, large (>10 μm) SiC particles, combined with finer mullite powder (~1 μm) and submicron alumina (0.2 μm), are infiltrated into 2D and 3D Nextel 720 samples [63]. The large particles affect the shrinkage strain (t/R) that occurs in the matrix during drying, where t is the thickness of the liquid layer and R is the particle radius. With increasing particle radius, the strain will decrease, ultimately leading to the elimination of drying shrinkage cracks. A bi- or tri-modal particle distribution is used to ensure matrix infiltration within the fiber tows. The initial results for the N720/SiC-mullite-alumina samples look very promising and it is noted that a large oxide particle should provide the same effect as SiC; this material was used due to availability.

8.4.2.4 Freeze Forming/Freeze Drying/Freeze Gelation The primary advantage of freeze forming of composites is the ability to produce a virtually crack-free green composite (Figure 8.7). While freeze forming/drying has been increasingly applied to ceramic monolith formation, very limited studies have been conducted on freeze forming oxide–oxide composites [93, 94].

In freeze forming composites, a ceramic powder is dispersed in a sublimable solvent, such as water, along with binders and dispersants to impart strength to the green body. Alternately, camphene is an attractive solvent, since it exhibits approximately 3% shrinkage upon solidification (whereas water *expands* ~9% on freezing) and it is sublimable under ambient conditions [67]. The slurry is infiltrated into the oxide fiber tow either through filament winding or by pouring the slurry onto the oxide fabric, followed by rolling of the fabric to force the ceramic slurry within the fiber tows. The fiber tow/fabric layers are then stacked and warm molded to form the green body. This body is next cooled to below the freezing point of the solvent, wherein the solidifying solvent rejects the oxide powder ahead of the solidification interface and essentially pushes the fine (<2–3 μm) particles. Water and camphene most commonly solidify in a dendritic manner, therefore the oxide powder is forced into the interdendritic regions.

The composite is then freeze dried under vacuum in the case of water, or sublimated in air at room temperature, in the case of camphene. Eliminating the solvent through sublimation prevents the buildup of capillary pressure that is associated with drying of a liquid from a body. This capillary pressure leads to tension in the body and subsequent cracking of the ceramic matrix. The resultant composite contains a porous matrix, where the pores replicate the frozen solvent structure. The green body is then sintered; the pore structure is usually retained during this process.

Freeze forming of oxide composites was first reported by R. Russell-Floyd et al. [93]. In this case, the Nextel fiber was wound with an aqueous-based slurry that also contained a colloidal silica sol. During freezing of the composite, the colloidal silica particles were forced into close physical proximity, leading to gelling of the sol. The silica gel that formed was sufficiently strong to withstand the tensile stresses introduced by the drying liquid. This technique has been investigated recently for the formation of Nextel 610/SiO_2-mullite composites with promising initial results; however, continued study of process controls is needed [94].

8.4.2.5 Electrophoretic Deposition EPD is a process that utilizes the behavior of charged particles under an applied electric field to infiltrate ceramic fibers/fabric (for further information on this topic, see References 95, 96, and 97). In this method, colloidal sols containing fine charged particles are subjected to an electric field, which drives the charged particles toward a deposition electrode. The non-conducting oxide fabric/tape must be either coated with a conductive coating or held against a conductive material to allow for infiltration of the colloidal particles. A strong field must also be applied, which gives rise to concerns over the stability of the EPD coating solution under such conditions.

The process is initially quite rapid; however, the rate decreases as the thickness of the infiltrated layer increases. EPD is generally limited in terms of the size and thickness of the material that can be formed. Thinner layers, such as a single fabric layer, can be infiltrated reasonably well; however, the process is not as applicable for thicker samples, due to the decreased deposition rate with increasing distance from the electrode. EPD can therefore be used as a "prepreg" process, where single layer fabric mats are infiltrated with this process and then taken through other processing steps, such as pressure infiltration [98]. Results of work completed at the University of Birmingham [99] indicated that approximately 81–90% composite densities could be attained by using EPD combined with either pressure filtration or with consolidation of the stacked layers [100]. Alternately, Bao and Nicholson demonstrated the potential of infiltrating thicker sections using electrophoretic infiltration deposition (EPID). In this case, the repulsive forces between the fibers and the particles aid in the infiltration of the oxide fiber preform [101].

8.4.3 Consolidation

8.4.3.1 Pressureless Sintering Pressureless sintering comprises heating a green composite, attained through the various methods described in this text, at high temperature in air with virtually no external pressure on the material to form a porous matrix CMC. For these materials, the goal in the sintering step is not necessarily to densify the matrix, but to strengthen the necks between the particles to improve the matrix strength. Significant densification of the porous matrix would result in the further opening of microcracks within the matrix. Aside from air, pressureless sintering in an HCl-containing atmosphere has been shown to provide strengthening of mullite/zirconia matrices through evaporation/condensation reactions [102]. Basic discussions of particle sintering can be found in References 103 and 65.

8.4.3.2 Hot Pressing/Hot Isostatic Pressing/Field-Assisted Sintering (FAST/SPS) The three listed processes, hot pressing, hot isostatic pressing, and field-assisted sintering (FAST/SPS), all involve the use of high temperatures (>1000°C) combined with high(er) pressures to consolidate the composites. For oxide–oxide composites, there has been limited use of these processes for non-glassy matrices, due to the potential damage to the fibers caused by the high temperatures needed to consolidate the oxide matrices, along with the possible deformation of the fiber, depending upon the applied pressure.

In these three processes, the prepreg material or the individual components (i.e., fibers placed by hand within a powder) are loaded within a graphite die or metallic mold (HIP). In hot pressing and SPS processing, uniaxial pressure is applied to the die as it is heated to high temperature (>1000°C) in an inert atmosphere. High(er) pressures are applied to the material during heating or at a temperature to facilitate the consolidation of the powder [104]. For hot isostatic pressing, gas pressure is utilized for consolidation versus the uniaxial pressure applied in hot pressing [105]. This allows for much higher, isostatic pressure on the sample. The primary drawback is the need for a sealed (or canned) rigid mold to prevent gas intrusion within the porous composite sample and to prevent distortion of the final product due to the anisotropy introduced by the reinforcing fibers, along with the potential of fiber deformation during the consolidation process.

High densities (>90%) can be attained with these methods, particularly for glass and glass–ceramic matrices, which soften and flow during heating. Additives can also be added to the starting material to enhance densification. However, the potential for damage to the fiber is significant.

8.4.3.3 Chemical Vapor Deposition/Infiltration Although CVD/chemical vapor deposition/infiltration (CVI) holds some promise for producing dense oxide matrix composites, it has not been widely used to form these composites and has been used primarily in SiC- and C-based composites [72, 106, 107]. This process involves forming a fiber preform through stacking filament wound tapes, woven fabric or utilizing a 3D woven structure. The preform is placed within the deposition chamber, and the desired product (oxide) is deposited on and within the heated preform from gaseous precursors. The deposition process is slow (up to days depending upon sample size) and is further hindered due to the filling of the outer portions of the preform, which inhibits infiltration within the preform. The primary drawback for oxide materials is the difficulty in forming multicomponent oxides, and often even single component oxides, along with the increased cost due to the slow process. Reference 107 provides a more in-depth discussion of this topic.

8.4.4 Metal Oxidation Processing

The processes of directed metal oxidation (DIMOX), melt infiltration, and reaction bonding are similar in that each seeks to fill the open porosity within an oxide matrix composite through oxidation of a metallic precursor. The primary advantage of metal oxidation is the typical increase in volume in going from the metallic to the oxidized state; this helps to increase the density of the final oxide matrix composite.

DIMOX [108] involves coating an oxide preform on five sides with a barrier material, while the sixth side is held adjacent to a molten metal, generally aluminum. Additives can be used to prevent the formation of a protective scale and to enhance the wetting capability of the aluminum. Metal continues to wick through channels in the matrix through capillary action. The resultant composite contains a matrix of aluminum + alumina; the majority of the aluminum can be removed through oxidation. A small amount of metal (<5%) is typically left in the composite and this amount can be reasonably controlled.

Similarly, melt infiltration involves infiltrating a ceramic preform with a molten metal [109, 110] using pressure, vacuum, wetting agents, etc., to force the metal into the preform. The wetting characteristics of the metal and preform are critical and special care should be taken to properly prepare the sample. The metal is oxidized after infiltration resulting in a pore-filling volume expansion; however, some residual metal may remain.

Reaction bonding has been increasingly investigated due in part to its cost-effectiveness, along with the ability to oxidize virtually all of the metallic material. It has been successfully demonstrated with sapphire fiber reinforcements [111, 112], with Nextel 610 fibers [113] and Nextel 720 fibers [114]. This process involves combining a metal powder, such as aluminum, with an oxide powder, such as alumina. The powders are milled together to obtain the optimal size and shape of the metal powders; the milling parameters are critical to obtaining a metallic phase that completely oxidizes. Undermilling results in larger metal particles that may not oxidize completely; overmilling produces hydroxides on the surface of the powder that may not be removed until very high temperature, resulting in bloating of the composite due to gas formation.

The oxidation of the metal must be fully understood to obtain optimal results, including obtaining a virtually crack-free composite material. Metastable or low temperature phases, formed upon oxidation, may convert upon exposure to higher temperatures to less desirable phases that lead to a lower density than predicted. There has been minimal application of this process to composite fabrication; however, with proper control, it appears to be a viable method worthy of further exploration.

8.5 POROUS MATRIX COMPOSITE SYSTEMS

While there are abundant reasons for academic consideration of porous matrix oxide–oxide composites, much of the practical interest, and opportunity to pursue them, is owed to two factors: the economical availability of numerous types of fibers, and the absence of the necessity to coat fibers. Both of these factors significantly affect the ease and cost of experimentation, and, more importantly, the ultimate cost of the materials. An additional significant advantage is that in many cases fabrication can be very similar to that employed for polymeric composites.

The fibers discussed in Section 8.3 are combined with various matrix materials to produce porous oxide–oxide CMCs. Coatingless, porous matrix CMCs based on fabric reinforcement are the most fully developed and examples of these materials are given in Table 8.2. General Electric's (GE, Cincinnati, OH) GEN-IV was one of the earliest porous matrix, all-oxide fiber composites, while ATK-COIC is currently the most prominent commercial supplier of

TABLE 8.2 Physical Properties of Selected Woven Cloth CMC Laminates

Designation	Source	Fiber	Matrix	Fiber volume, %	Matrix porosity, %	Composite porosity, %	Density, g/cc
COI-312/AS	COI[a]	Nextel 312	Aluminosilicate	48	42	22	2.30
COI-610/AS	COI	Nextel 610	Aluminosilicate	51	51	25	2.82
GE-610/GEN-IV	GE[b]	Nextel 610	Aluminosilicate	30	35	25	2.90
UCSB-610/M	UCSB[c]	Nextel 610	Mullite-alumina	40	40	24	–
UCSB-720/M	UCSB	Nextel 720	Mullite-alumina	40	29–35[d]	17–21	–
COI-720/AS	COI	Nextel 720	Aluminosilicate	48	48	25	2.60
COI-720/A	COI	Nextel 720	Alumina	46	46	25	2.71

Source: From References 12, 23–28.
[a]ATK-COI Ceramics Inc.
[b]General Electric.
[c]University of California, Santa Barbara.
[d]Variation in properties with porosity adjustment examined.

oxide–oxide composites. The matrix materials are primarily aluminosilicate (e.g., manufacturers: GE, ATK-COIC), alumina-mullite (University of California, Santa Barbara [UCSB]), and/or alumina (ATK-COIC).

For the aluminosilicate matrices, the main constituent is alumina, which is held together by a silica binder. The silica may be present as a highly porous filler between the alumina particles, as in the GE-based materials, or it may be a continuous SiO_2 film on the alumina particles, as in the ATK-COIC-based materials. The porosity in the former case is much finer than the porosity seen in the ATK-COIC material.

Alternately, an alumina-mullite matrix was developed in an effort to improve long-term thermal stability of porous matrix CMCs. This matrix utilizes a bimodal mixture of mullite (~1 μm) and alumina (~0.2 μm) powders [30]. The mullite exhibits relatively slow sintering that inhibits shrinkage and maintains the overall porous structure, while the fine alumina sinters more readily, cementing the mullite particles and providing strength to the porous matrix structure [45, 46]. Zirconia has also been studied as the primary phase, with mullite used to stabilize the matrix [115].

Another relatively mature porous matrix composite, wound highly porous oxide ceramic composite (WHIPOX) has been developed by the German Aerospace Center (DLR) [116]. These materials utilize either Nextel 720 or 610 fibers filament wound into a highly porous (60–80% porosity) mullite or alumina matrix [116–118]. These composites are pressurelessly sintered at temperatures up to 1300°C and the resulting composites contain between 25–50 vol% fiber. Various shapes have been produced using this process, including flat plates and more complex structures such as cylinders and grids. The composites exhibit good thermal shock resistance and fatigue behavior, along with high in-plane strengths; however, the absolute property values depend upon the specific fiber/matrix/processing combination. Sintering at temperatures up to 1300°C can be detrimental to the Nextel 610 fibers due to excessive grain growth in the fiber; however, recent work from DLR has shown that this effect can be minimized through the addition of silica to the matrix [119, 120]. This reduces the driving force for silica diffusion out of the fiber, which is present to inhibit grain growth. Continued research in this area may provide increased temperature capabilities for these materials.

Both single-phase mullite and two-phase mullite-alumina mixtures can be termed "silica-free." However, due to the high silica activity in mullite and the possibility of residual glassy phase material, it may be subject to degradation in environments that are detrimental to silica [121]. For example, vapor-deposited mullite coatings exhibited vaporization of the SiO_2 in a high pressure, high temperature H_2O environment, leading to cracking and coating recession [122, 123]. Conversely, CVD mullite coatings, characterized as high purity and fully crystalline, appeared to be stable in similar environments [124].

For high temperature applications, there is an effort toward eliminating all silica in the matrix, even in mullite form, due to concerns over degradation in high temperature water and fuel rich environments [37, 125]. Therefore, matrices such as all-alumina and alumina-YAG have been investigated for use in oxide–oxide composites. ATK-COIC has utilized the bimodal concept to increase the thermal stability of an all alumina matrix [25] by using a colloidal alumina (nanometer size) as the filler for the larger alumina powder. In work on unidirectional porous matrix composites based on alumina-YAG matrices reinforced with Nextel 610 fiber, the addition of even a small amount of YAG or Y^+ was shown to significantly slow the sintering of fine (~0.2 μm) alumina in the composite [126, 127]. Previous work on alumina monoliths showed similar results with the addition of Y^+ [128]. Higher volume fractions of YAG (~50%) have been used in a bimodal morphology similar to mullite-alumina, where the YAG phase is much larger (~1–2 μm) than the finer alumina phase (~0.2 μm) [127].

Applied Thin Films, Inc. (ATFI, Evanston, IL) has also developed microporous oxide/oxide CMCs with a novel aluminophosphate-based matrix (Cerablak™) that contains no silica. This matrix system offers unique benefits including non-reactivity with fibers, minimal sintering or shrinkage, and thermal stability up to 1315°C (2400°F). Flat, cylindrical, and 3D CMCs have been fabricated with a variety of fiber types (Nextel 312, Nextel 720, Nextel 610) and fiber architectures (2D layup, 3D woven, braiding, filament winding). Cerablak™-based CMCs are ultralightweight with densities ranging from 1.8–2.2 g/cc (for 40% fiber volume Cerablak™/N720 configuration). Composite behavior under mechanical loading suggests a relatively weak fiber–matrix interface compared to other porous oxide CMCs.

8.6 PROPERTIES

The physical properties of composites are dependent upon the constituent properties and the details of their assembly. Because target service conditions for ceramic composites are in thermal and environmental regimes where fibers, matrices and interfaces evolve microstructurally, and potentially react and degrade, it is important to develop an understanding of the time–temperature dependence and environmental performance of the materials. However, as these are emerging materials, mostly under active development, property characterization is limited, and it is not unusual to encounter significant variations in important properties resulting from small refinements in the processing. The best possible understanding of the dependence of time–temperature performance on microstructural details, such as matrix composition and porosity, must be developed so that the limited available data may be meaningfully interpreted. Most of the available data has been generated on the porous matrix oxide–oxide composites produced by GE, UCSB, and ATK-COIC, only the latter of which is a current CMC producer. These will be the primary materials considered in this section.

This section is organized as follows: the basic physical characteristics are introduced, followed by discussions of uniaxial room temperature mechanical properties, the effect of long-term thermal exposure, composite durability (creep behavior/fatigue performance), and notch sensitivity. The significantly different off-axis, matrix-dominated properties are described in the final section.

8.6.1 Basic Physical Characteristics

A number of characteristics are common to most fibrous oxide–oxide composites. Nearly all have a large matrix void volume (30–50%). Furthermore, all of the appreciably characterized materials depend on that void volume for crack deflection and toughness. The void volume often includes shrinkage cracks and microcracks, as well as literal pores. The relative amounts of these forms of porosity can be expected to significantly affect properties; however, the relative distribution is rarely known, hence the topic is essentially uninvestigated. In any case, the matrices are typically very compliant and very weak compared to the fibers. Consequently, the fiber-dominated properties are very much so, and the matrix-dominated properties are quite modest. The stiffness and strength in loading directions that put fibers in tension are nearly those of the loaded fibers. The properties in unreinforced directions are those of porous, and sometimes poorly consolidated, oxides.

Selected composite materials and their basic physical properties are summarized in Table 8.2. The GE GEN-IV all-oxide fiber composites have a relatively low volume fraction (30%) of Nextel™ 610 fiber and a matrix composed of alumina and silica with approximately 35% porosity [12, 16, 129]. ATK-COIC produces a number of aluminosilicate matrix materials with a selection of different Nextel™ oxide fibers, but with higher fiber loading and matrix porosity than the GE materials [24, 26]. Two examples of nominally silica-free porous composites (Table 8.2) with improved thermal stabilities are the mullite-alumina formulation developed at UCSB [23, 27, 30, 130] and the all-alumina ATK-COIC N720/A [24–26, 28].

As shown in Table 8.2, for all systems, matrix porosity is of the order of 40% with the overall porosity near 25%, resulting in relatively low composite densities. Although the low density is beneficial in weight-driven applications, such high, interconnected matrix porosity is undesirable where hermetic fluid containment is required and also leads to concerns with use in erosive environments [12, 125].

The thermal and dielectric properties of several composite systems are shown in Table 8.3. Note that the thermal expansion, conductivity, and specific heat all vary by approximately a factor of two over the range of temperature from 23°C to 1000°C.

TABLE 8.3 Thermal and Electrical Characteristics of All Oxide CFCCs

	GE-610/GEN-IV	COI-720/A-1
Thermal expansion coefficient[a] at RT, ppm/°C	4.5	3.5
Thermal expansion coefficient[a] at 1000°C, ppm/°C	8.2	6.2
Thermal conductivity[a], RT, W/m K	4.5	4.25
Thermal conductivity[a], 1000°C, W/m K	2.5	2.3
Specific heat at RT, W-s/gK	–	0.75
Specific heat at 1000°C, W-s/gK	–	1.30
Dielectric constant (loss tangent)	–	5.74 (<1)

Source: From References 28 and 129.
[a]Principle fiber direction.

8.6.2 Room Temperature Uniaxial Mechanical Properties

The as-fabricated room temperature tensile properties of the composites are summarized in Table 8.4. It is important to recognize that, due to their high porosity, the matrices in all of these systems have elastic moduli and strengths an order of magnitude less than those of the fibers [132], hence the composites are highly orthotropic [27, 131]. See [133] for an extensive review of the mechanics of brittle matrix composites. The properties in Table 8.4 represent the response to loading in the fiber direction. The fiber direction orientation, denoted as 0°/90°, is the most favorable orientation; the strength and stiffness in any other orientation, denoted as "off-axis," is substantially lower.

Due to the extreme difference between the matrix and fiber properties, the in-plane modulus is dominated by the fiber modulus and volume fraction. For example, the GE GEN-IV material has a modulus of 70 GPa. With 15% fiber in the loading direction and a fiber elastic modulus of 380 GPa, the stiffness due to the fibers alone is expected to be 57 GPa. The UCSB Nextel 610-based material, with a 33% higher fiber content, likewise has roughly a 35% higher modulus. The small contribution of the matrix makes it particularly difficult to infer the elastic properties of the matrix from the composite test results. Although the porous matrix materials are difficult to produce without any fiber reinforcement, several studies have found the intrinsic matrix properties to be consistent with composite behavior [132, 134].

Figure 8.8 shows a typical 0°/90° orientation stress–strain plot. The response is initially linear and is somewhat stiffer than would be expected due to the fibers alone; however it becomes nonlinear at relatively low strain. This behavior is consistent with the matrix contribution to the modulus decreasing with stress due to cracking [18]. The total matrix contribution to load carrying is sufficiently small that a proportional limit can be difficult to identify. However, this inelastic behavior contributes to the composite's ability to tolerate stress concentrations, as will be discussed. As matrix damage develops fully, the tangent modulus decreases to nearly the value that would be expected by considering the fiber contribution alone [133].

Unlike the tensile modulus, the ultimate tensile strength can be substantially impacted by the matrix formulation. In the idealized case, as the ultimate strength of the composite is approached, damage accumulates in the matrix, without

TABLE 8.4 Mechanical Properties—Room Temperature 0°/90° Fast Fracture Data

	Tension					Flexure			Compression		
Designation	E, GPa	v12	prop.limit MPa	UTS, MPa	ε	E, GPa	Strength, MPa	ε	E, GPa	Strength, MPa	ε
COI-312/AS	31.0	–	–	125	0.42	48.3	159	0.33	–	–	–
COI-610/AS	124	–	–	366	0.35	141	352	0.35	141	270	0.24
GE-610/GEN-IV	70	–	100	205	0.33	–	–	–	–	–	–
UCSB-610/M	95	0.05	–	215	0.22	–	255	0.29	–	–	–
UCSB-720/M	60	–	80	150	–	–	215	–	–	–	–
COI-720/AS	75.6	0.09	–	180/220[a]	0.30	98.6	216	0.22	80.0	186	0.22
COI-720/A-1	75.1	–	50	177	0.31	90.9	218	0.23	–	253	–

Source: From References 12, 23, 24, 26–28, 130, and 131.
Improved generation material recently reported by the manufacturer.

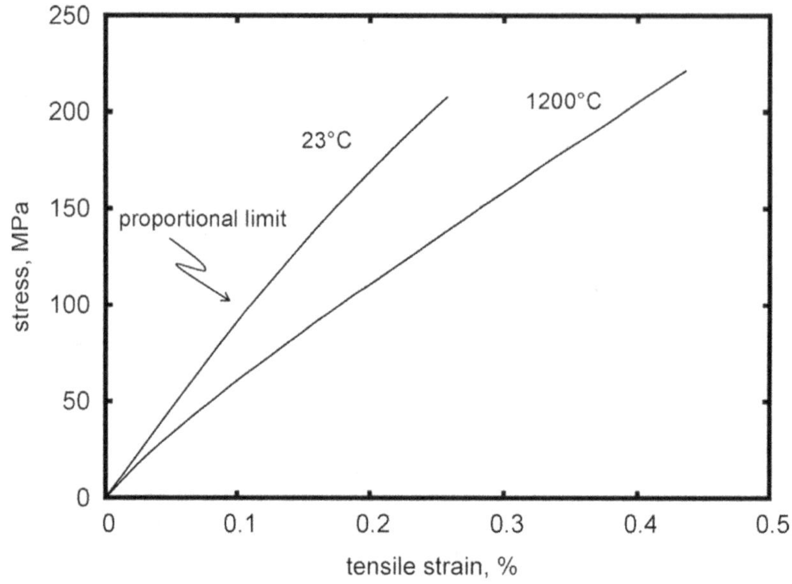

FIGURE 8.8

Typical stress–strain plot for 0°/90° oxide–oxide CMC at room and elevated temperatures.

damaging the fibers, until the matrix is not load bearing at all. Failure thus occurs as individual fibers reach their statistically distributed breaking load, each transferring their load to the remaining fibers [10]. However, if the matrix and/or fiber interface is not sufficiently weak, the failure of one fiber results in localized load transfer to its immediate neighbors; hence the composite's ultimate strength is dominated by the weakest fibers [135].

In a good quality composite, many fibers fail some distance away from the fracture surface. In dense matrix composites with a weak fiber/matrix interface, they continue to offer some load bearing capability due to frictional sliding as they pull out of the matrix [3], and the fracture surfaces feature significant bare fiber, along with distinct holes resulting from the fiber pull-out (see e.g., Figure 8.3c). Typical fracture surfaces of porous matrix composites also feature considerable lengths of bare fiber (Figure 8.1), however as the porous matrix is intrinsically weak, this pull-out likely provides minimal additional load bearing capability.

In any case, the ultimate strength of the composite approaches the tow strength of the constituent fiber multiplied by the loaded fiber fraction to a reasonable approximation. For example, the ATK-COIC 610/AS composite has 25% fibers in the loading direction with a tow strength of 1600 MPa; the composite strength of 366 MPa is within 10% of the 400 MPa value calculated based on the fiber tow strength. See [136] for discussion of the relation between fiber filament strength and tow strength. In the opposite extreme, the matrix offers no fiber damage protection mechanism and fracture of the composite is observed to occur on a single plane with no fiber pull out. The composite strength in that case is that of a poor quality monolithic ceramic.

Table 8.4 also shows flexure strengths for comparison; often, these are the only strength values reported for ceramic composites. Flexure tests can be very useful for several reasons: they better mimic the actual loading conditions in many applications, and they sometimes more clearly reveal high damage tolerance. With that said, the reported values should be viewed with caution. Flexure strengths are generally higher than tensile strengths derived from uniaxial tension tests by as much as 20–30%. This is a consequence of the tension–compression asymmetry in the failure of the composite, which strictly invalidates the simple beam theory stress measure as well as the localization of the peak stress over a small volume. CMC strength volume dependence is a consequence of the statistical nature of the fracture of brittle materials, and may play an important role in the notch tolerant behavior of ceramic composites [135, 137, 138].

The strengths of the composites in compression are also listed in Table 8.4. It is notable that the compressive strength is comparable to the tensile strength. This is in contrast with monolithic ceramics, where the compressive strength can be substantially higher than the tensile strength. In the fibrous composite case, compressive failure occurs by fiber buckling, and thus the strength is limited because of the low strength and stiffness of the matrix [139, 140].

The effect of matrix porosity has been examined extensively for the UCSB 720/mullite-alumina composites. The baseline matrix consists of an 80/20 vol% mixture of mullite and alumina particles, with approximately 35% matrix porosity by volume. The matrix is then infiltrated with an alumina precursor solution, resulting in a small density increase with each infiltration [23]. As the matrix density increases, the fiber-dominated flexure strength decreases and the fracture mechanism changes from fibrous to monolithic, with only a 2–4% change in matrix density. As noted by the authors, the possibility of progressive fiber degradation seriously complicates interpretation of results of this kind. In contrast to the fiber-dominated in-plane strength, the interlaminar shear strength of the composite improves, to a point, as the matrix density is increased. Hence there is a trade-off to be considered when tailoring the matrix design [23]. Perhaps more importantly, however, recognizing this strong sensitivity of room temperature fracture behavior to matrix properties provides general insight into the behavior of the porous composites under thermal exposure, to be discussed in the following sections.

8.6.3 Long-Term Thermal Exposure

The composite service temperature/time limit for porous matrix oxide composites is bounded by the combination of two inherent temperature/time limits: degradation of the fibers themselves and degradation of the crack-deflection mechanism via matrix evolution. The details of the degradation of the crack-deflection mechanism are not fully understood and probably vary somewhat from system to system. Nevertheless, it is some combination of sintering and strengthening within the matrix, matrix stiffening due to the associated densification, and sintering of the matrix to the fiber. At some point, the crack-deflection mechanism ceases to operate and the properties in all loading directions become those of a poor quality monolithic ceramic. This is particularly a problem for the aluminosilicate matrix systems [24]. The ATK-COIC 720/AS composite strength drops dramatically when exposed to temperatures above 1000°C for 1000 hours (Figure 8.9). At 1200°C the degradation occurs quickly, well within 100 hours. The aluminosilicate matrix therefore may be serviceable for short-term exposures as high as 1000°C [24]. Alternately, the GE GEN-IV material suffers only 15% strength degradation for short-term exposures up to 1100°C [12].

The UCSB matrix is composed primarily of a network of relatively coarse mullite. Mullite is resistant to sintering up to 1300°C, which increases the temperature at which shrinkage and degradation would be problematic to above the degradation temperature of the fiber. The alumina in the matrix serves to bond the mullite particles together, but because the mullite forms a continuous network, there is no global shrinkage for long duration high temperature exposure [30, 130]. The efficacy of this concept is shown in Figure 8.9 as the mullite-alumina composite strength remains essentially unchanged following 1000-hour heat treatments up to 1200°C. The ATK-COIC all-alumina matrix performs comparably well by employing a high initial matrix porosity as well as a heterogeneous alumina particle formulation to enable long-term stability against densification.

As discussed previously, an alumina-YAG matrix behaves in a manner similar to the UCSB material (d_{YAG} ~12 μm, $d_{alumina}$ ~0.2 μm) and has shown long-term stability against densification at temperatures up to 1200°C [127, 141]. It has the additional advantage of being entirely silica-free, which may be important in some atmospheres.

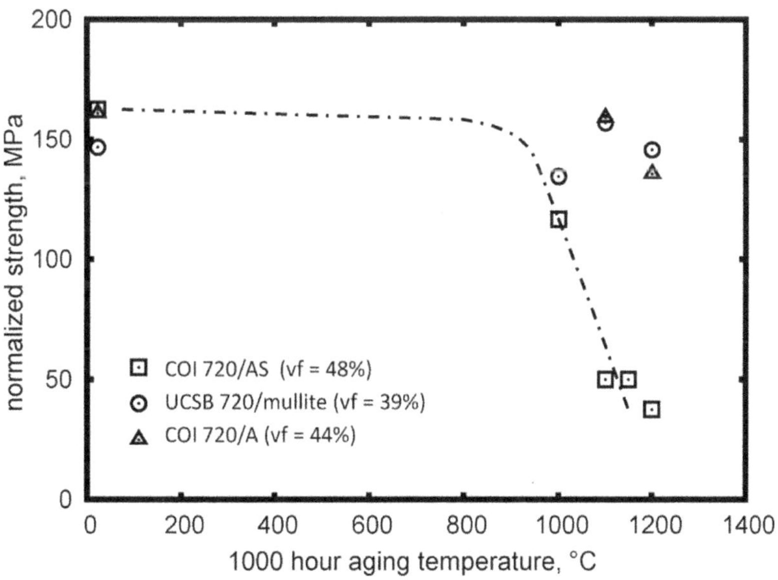

FIGURE 8.9

Retained strength after 1000 hour aging. The dashed line indicates the trend for the silica-containing matrix, which is presumed to be stable up to 800°C.

8.6.4 Durability

Oxide composites are typically expected to survive for extended times in harsh combined thermal, chemical, and cyclic load environments. It is therefore critical to understand their creep, cyclic fatigue, and chemical degradation characteristics. As the damage mechanisms associated with each can significantly interact and the environmental conditions invariably occur simultaneously, testing for the influence of the isolated mechanisms is quite challenging.

Oxide composites have shown excellent room temperature fatigue resistance, surviving up to 10^5 cycles at approximately 85% of their room temperature strength [25, 142–144]. Such room temperature performance is, however, of minimal practical interest and the majority of recently published characterization efforts have focused on cyclic loading combined with thermochemical exposure.

Under sustained load at high temperature, oxide composites are susceptible to creep rupture [145–148]. For load in the fiber direction, creep deformation is dominated by the creep behavior of the fibers, with negligible environmental degradation associated with the matrix cracking. The selection of structural ceramic fibers is a compromise between strength and creep resistance. The fine grain size necessary for the very high strengths of quality polycrystalline fibers also makes them prone to creep. Fibers such as Nextel 312 have especially low creep resistance due to the viscous behavior of the glass-phase silica, while the fully crystalline 610 and 720 fibers offer improved creep characteristics. At 1100°C and a stress of 100 MPa, Nextel 610 fibers creep at a rate of the order of 10^{-6} per second, due to its fine grain size. Under the same conditions, the creep rate of Nextel 720 is four orders of magnitude lower, with a rate of 10^{-10} per second, owing to its coarser-grained mullite morphology [38–40].

The creep characteristics of the fibers translate directly to the creep behavior of the composites in the principle fiber directions. The deformation rate of both fibers and composites has a stress dependence given by $\dot{\varepsilon} = A\sigma^n$, where the creep exponent n is approximately 3 for both Nextel 610 and 720 fibers [38]. Similar stress exponents are found for porous matrix CMCs. For example, the GE GEN-IV 610-based composite shows an exponent of 3.06. The strain rate at 1000°C under loads from 75 to 135 MPa is nearly constant until failure; there is no primary or tertiary creep. The 610/aluminosilicate composite creep rate is $\sim 2.7 \times 10^{-8}$ per second at 1000°C and 75 MPa stress, and retains a residual strength of 205 MPa after 100 hours under those conditions. At 1100°C, the creep rate jumps two orders of magnitude and the specimen ruptures in less than 1 hour, clearly limiting the 610 system to 1000°C applications even under moderate load [12, 142, 149].

The Nextel 720-based composites show substantially improved creep performance in correlation with the superior fiber properties. A creep limit of 145 MPa, which is 80% of the ultimate strength, was identified by John et al. as the level where the ATK-COIC 720/AS composite could survive 100 hours at 1100°C. The linear creep rate is $\sim 1 \times 10^{-8}$ per second under these conditions; however, the creep response at this high load is significantly greater during the initial 10 hours before decreasing to a steady rate [150]. The ATK-COIC 720/A material with a silica-free matrix has been examined extensively for creep resistance in laboratory air. At 1100°C, the alumina matrix composite survived 100 hours at 150 MPa [151] with the 100 hour run-out stress dropping to 80 MPa at 1200°C and less than half of that at 1330°C [152].

Although oxide composites have been developed for their durability in oxidizing environments, steam has been found to have a detrimental effect on high temperature creep performance, which is a significant concern for combustion applications. Among the oxide CMCs, ATK-COIC's 720/A has been specifically developed to address moisture degradation through elimination of any silica in the matrix. Steam, however, still degrades the fiber by reacting with mullite to form $Si(OH)_4$ [153]. Consequently, the creep of ATK-COIC's 720/A composite is accelerated by nearly an order of magnitude in the presence of steam, with a rupture time of 46 hours at 80 MPa [151, 152]. Under compressive loading, the creep under steam lifetime was even further degraded, with failure occurring in under 1 hour at 60 MPa due to fiber microbuckling [154].

As observed at room temperature, oxide composites are also generally fatigue tolerant when cyclic loading is combined with elevated temperatures and oxidizing environments. In temperature regimes where the static strength retention is significantly degraded, the fatigue resistance also severely declines [142, 143]. For example, the ATK-COIC 720/AS material survives only 10^4 cycles at 60% of its ultimate strength at 1200°C [25]. However, under conditions where the composite demonstrates significant creep rupture life, cyclic load application does not further shorten the time to rupture and indeed may even improve it.

The combined creep/oxidation/fatigue performance has been extensively studied for the 720/A system [151, 152, 154–158]. In a series of high temperature fatigue tests where the peak loads were sustained for times exceeding 10 seconds per cycle, it was found that the composite strain paralleled the static creep behavior; this indicates that time-at-temperature-and-stress is more significant than the cyclic loading [142]. Figure 8.10 shows the time to rupture at

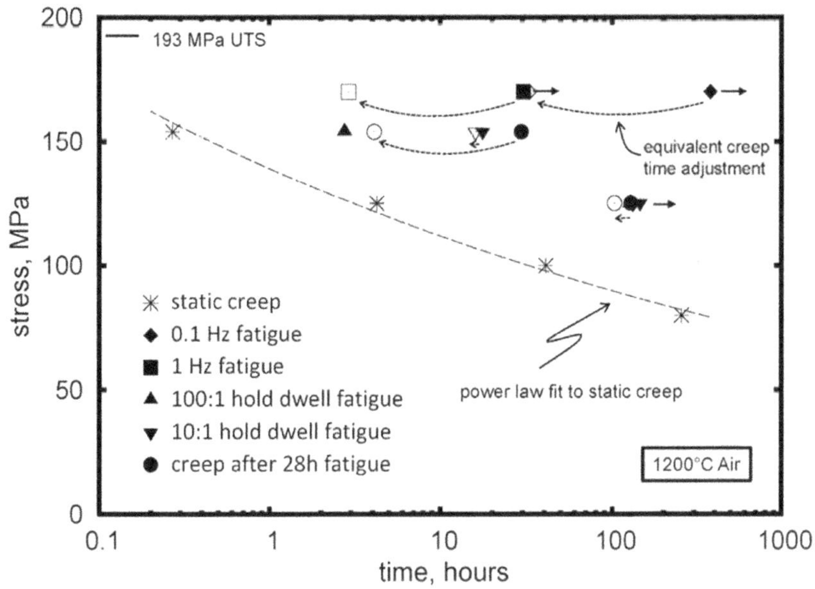

FIGURE 8.10

Time to rupture at 1200°C under static creep as well as various cyclic load conditions. Filled markers indicate actual test time, open markers denote times adjusted for time-at-stress [156].

1200°C under static creep as well as various cyclic load conditions. The fatigue rupture times being considerably longer than the static creep rupture times may be explained in part because the material is only under load for a portion of the cycle. However, even after calculating an equivalent time assuming power law creep [155], indicated by filled markers on Figure 8.10, the cyclically loaded specimens still exceeded the rupture life of the statically loaded specimens.

8.6.5 Notch Sensitivity and Toughness

Notch sensitivity is evaluated by considering a tensile specimen with a stress concentrating feature such as a hole. For an isotropic elastic tensile specimen with a circular hole of one half of the section width, as an example, linear elasticity theory predicts a maximum stress at the edge of the hole in the plate to be 2.15 times the average section stress [159], while for the 0°/90° layup orthotropic materials under consideration here, the concentration is somewhat higher, having a value of ~2.5 [160]. A notch sensitive material will fail as a result of that peak stress; hence the hole effectively reduces the strength of the part by a factor of more than 2 compared with the ultimate strength measured in simple tension.

The inelastic behavior of ceramic composites that results from the formation of microcracks and fiber–matrix debonding at high stress levels serves to substantially reduce the peak stress, thereby improving the material's notch sensitivity and damage tolerance [27]. This peak stress reduction is notably independent of the size of the stress concentrating feature, that is, geometrically self-similar specimens would be expected to fail at the same stress regardless of their absolute size. However, experiments show that the notch sensitivity of fibrous composites is strongly affected by size scale—notch sensitivity generally improves with smaller defect sizes. The UCSB 610/M composite shows a 30% effective 0°/90° tensile strength reduction due to the presence of a 10-mm diameter hole, but less than a 10% reduction due to a 3-mm hole; a highly notch sensitive material would show a 65% strength reduction in both cases. The 610 notched tensile specimens exhibit a stable progressive failure following fracture initiation, thus permitting the calculation of a fracture energy of 6000 J/m^2. Notably, this is an order of magnitude higher than the fracture energy typical of monolithic ceramics [27]. The UCSB 720 composite shows a fracture energy of about 3000 J/m^2 [23]. The ATK-COIC 720 material is likewise shown to be notch insensitive, with negligible strength reduction in the presence of sharp or semicircular edge notches in small size tensile specimens [161], however significant notch sensitivity develops at temperatures above 1100°C [162].

The scale dependence on strength can be explained by utilizing the *point stress* failure criterion, that is used with polymer matrix composites [163, 164], as well as by considering the statistics of fiber strength distribution [135]. These models suggest a characteristic length scale of ~0.5 mm over which the stress must exceed the strength of the material in order for failure to initiate.

8.6.6 Off-Axis Properties

As noted, these composites are highly orthotropic. Although there is a small difference between the warp and fill directions of the fiber fabric itself [37], the composites for which data have been reported were fabricated with the warp directions of alternating layers of cloth at 0° and ± 90°; the resulting laminate has two essentially identical principal fiber direction orientations. Although it is well understood that the off-axis properties are strongly influenced by the porous matrix properties—the strength and stiffness in any load direction not aligned with the fiber directions is substantially lower—there is relatively little published test data for off-axis properties. The most complete set of test data for a single material is that of the ATK-COIC 720/AS, as shown in Table 8.5 [131]. Subscripted quantities in the table are the elastic tensor components in the coordinate system aligned with the fiber directions, that is, 1,2 refer to the in-plane principle fiber directions with 3 being the through-thickness direction. The elastic moduli in the through-thickness direction (E_{33}) and in the ±45° in-plane orientation (E^{45}) are nearly equal and approximately 60% of the 0°/90° value. The 0°/90° and ±45° moduli, along with the value of 0.09 for the principle direction Poisson's ratio, have been used to compute the shear modulus values in Table 8.5 [2, 159].

The UCSB Nextel™ 720 reinforced material, with its lower stiffness mullite matrix, has a ± 45° tensile elastic modulus of only 23 MPa, or less than 40% of its fiber direction stiffness [130].

The thermal expansion coefficient and thermal conductivity of the ATK-COIC 720 material are orthotropic, but with lower anisotropy than the elastic properties, as Table 8.5 indicates. The thermoelastic properties are reasonably computed by micromechanical models that incorporate both porosity and microcracking in the matrix [131].

The anisotropy of the composite strength is even greater than the anisotropy of the thermoelastic properties (Table 8.6). The ultimate strength of specimens loaded in a ±45° tensile configuration is approximately 20% of the fiber direction strength. Failure in these specimens typically occurs in shear bands at 45° , that is, parallel to one set of fibers and perpendicular to the other, hence the resolved shear stress on that plane may be interpreted as the shear strength of the material [12]. However, there is also a tensile stress component on the 45° plane, so that the measured apparent shear strength is typically somewhat lower than the strength measured in pure shear [138, 166]. The aforementioned UCSB study of matrix density found that the ±45° tensile fracture mode transitions from matrix failure and fiber scissoring to tensile fiber failure in the higher density, stronger, matrices [130]. (Again, some caution in interpretation is warranted as this result could be a consequence of fiber degradation.) Interlaminar shear and transthickness tensile strengths are even lower, as the failure occurs by delamination and is purely a matrix failure [27, 165].

Off-axis properties are difficult to interpret and additional test techniques and dedicated modeling will be necessary to clarify the design implications. Notably, strengths derived from ±45° testing show a strong dependence on specimen geometry, owing to the fiber's ability to rotate and stiffen the response of wider specimens following the initial shear failure of the matrix [167]. This is just one of the many issues that complicate characterization of composite failure under multiaxial stress states [168, 169]. Hence, the need to measure basic properties and develop a thorough understanding of materials' design for actual application requirements is reinforced.

TABLE 8.5 Anisotropic Elasticity and Thermal Properties of COI 720/AS

$E_{11} = E_{22}$	80 GPa
E_{33}	50^a GPa
ν_{12}	0.09
$\nu_{13} = \nu_{23}$	0.12
G_{12}	17^b GPa
E^{45}	50 GPa
$G_{13} = G_{23}$	40^c GPa
Thermal expansion in plane	6 ppm/K
Thermal expansion through thickness	7 ppm/K
Thermal conductivity in plane	2 W/m K
Thermal conductivity through thickness	1.6 W/m K

Source: From Reference 131.

[a] Ultrasonic estimate.

[b] Calculated.

[c] Model-based estimate.

TABLE 8.6 Off-Axis Strength.

	GE GEN-IV	COI 610/AS	COI 720/AS	UCSB 720 (N = 2)
0/90 strength, MPa	205	366	195	150
±45° strength, MPa	54	–	–	28
In-plane shear, MPa	27[a]	48.3[b]	31.0[b]	–
Inter-laminar shear, MPa	–	15.2	11.7	10
Transthickness	7.1	–	2.7	–

Source: From References 12, 23, 26, 130, and 165.
[a]±45° test.
[b]Iosipescu (V-notch) shear test.

8.7 COMPOSITES WITH INTERFACE COATINGS

Although there are distinct advantages to porous matrix, coatingless composites, to fully optimize the matrix-dominated mechanical properties and ultimate use temperature, a dense matrix composite is needed. As discussed previously for porous matrix composites, when the matrix strength exceeds a critical level, the crack-deflection function ceases to operate. Moreover, apparently this level can be exceeded by over-sintering without any total densification, that is, even though the matrix porosity is maintained, crack deflection is lost. Consequently, achieving the best balance of properties and highest temperature capabilities will require oxidation resistant, crack deflecting coatings and the densest possible matrices. (The following discussion borrows heavily from Reference 3.)

The results of interface control studies are complicated by various factors and should therefore be approached keeping several facts in mind. As discussed elsewhere in this chapter, crack deflection is a competition between fracture of the interface/coating and fracture of the fiber and many processes are aggressive to fibers. It is very difficult to distinguish between degradation of a crack-deflection mechanism and degradation of the fibers [170]. On the other hand, due to the problematic nature of making high density matrices, it can be equally difficult to distinguish between a successful coating and the porous matrix debonding mechanism. The transition point between the crack deflecting behavior of the porous matrix composite and the brittle behavior of a dense(r) matrix composite is critical and very limited work has focused on identifying this point. It is likely that the transition point will change based on the composite constituents, along with the processing parameters [1].

Research on interface control has also been performed on composites with a wide range of fiber fractions, from single fibers with very little matrix (high fiber volume fraction, >50%), to single or multiple tow minicomposites, unidirectional and 0°/90° fabric composites (moderate volume fractions, ~20–50%), to hand-assembled mono-filament reinforced composites (very low fiber volume fractions). Since the tensile strengths of composites are highly dependent upon the volume fraction of fiber in the loading direction, this needs to be considered when comparing results.

Finally, these factors make it especially important to carefully choose appropriate control specimens (with no fiber coatings) and control tests to properly evaluate the effect of the coatings on the composite behavior. All of these issues should be taken into consideration when comparing strengths of composites and the success of interface control mechanisms.

8.7.1 Weak Oxide–Oxide Phase Boundaries/Weak Oxides

Monazites and xenotimes are two classes of weak oxide coatings that satisfy the criteria for crack deflection for the currently available fibers [171, 172]. Of these, monazite ($LaPO_4$) has been the most widely studied, since it is desirable as an interfacial material due to its high melting point (>2000°C) and its high temperature stability with oxides such as alumina [173]. This material also bonds weakly to oxides, particularly alumina. Morgan and Marshall first demonstrated that monazite is a promising weak interface material by testing monazite coated sapphire fiber hot pressed in an alumina matrix [174]. Cracks formed from indentation testing did not penetrate into the fiber; instead the crack was deflected at the monazite/fiber interface. This demonstration provided a catalyst for concentrated research on monazite coatings.

Various research efforts on dense oxide fiber/monazite/alumina composites with low fiber volume fractions (5–10%) verified the efficacy of monazite as a weak interface. [41, 175–177]. Indentation testing, push-out testing [178] and three-point bend testing were utilized to demonstrate the low strength of the fiber/matrix interface as well as to

TABLE 8.7 Oxide–Oxide Composites with Weak Oxide (ABO_4) Interface Coatings

Fiber/interface/matrix	Fiber Arch.	V_f	Test type[a]	Test condition[a]	UTS (MPa)	%ε	Reference
N610/LaPO4/Al_2O_3	0°	~20%[b]	Tensile	RT	198 ± 12	~0.1	36
				RT, after HT at 1200°C/100 h	143 ± 7		
N610/Al_2O_3	0°	~20%[b]		RT	45 ± 20	~0.03	
N610/$LaPO_4$/Al_2O_3	0°	~20%[b]	Tensile	RT	168.4 ± 10.5	~0.1	182
			(10-min soak at T)	1200°C	167.2 ± 2.5		
N610/Al_2O_3	0°	~20%[b]		RT	93.15 ± 6.4		
				1200°C	99.6 ± 26	~0.05	
N610/$LaPO_4$/Al_2O_3	8HS	~33%	Tensile[d]	1100°C	88.5	0.11	183
N610/Al_2O_3	8HS	~33%	(20-min soak at T)	1100°C	66.6	0.08	
N610/$LaPO_4$/Al_2O_3	0°		Tensile[d]	900°C	180	0.31	184
			(20-min soak at T)	1000°C	162	0.28	
				1100°C	157	0.34	
				1200°C	130	1.42	
N610/Al_2O_3	0°			1100°C	105	0.11	
				1200°C	95	0.46	
N610/$LaPO_4$/AS[c]	8HS	–	Tensile	RT	~140	~0.23	185
N610/AS[c]	8HS		Tensile	RT	~210	~0.45	
N610/$LaPO_{4\,in\,situ}$/ Al_2O_3-$LaPO_4$	8HS	40%	Tensile	RT	200–250	–	84, 186
N610/$CaWO_4$/ Al_2O_3-$CaWO_4$	0°	–	Tensile	RT	~350	0.25%	35, 171
N720/$NdPO_4$/Mullite + 5wt.% ZrO_2	8HS	35%	4-point bend	RT	235 ± 32	–	187
			(1 h at HT)	1000°C	234 ± 17		
				1200°C	233 ± 33		
				1300°C	230 ± 41		
N720/$NdPO_4$/Al_2O_3	8HS	40%	4-point bend	RT	279	–	100
			(1 h at HT)	1300°C	266		
			Tensile[e]	RT	142		
N720/$AlPO_4$/RBM (reaction bonded mullite)	0°	25%	4-point bend	RT	175 ± 20	–	114
			(1 h at HT)	1100°C	170 ± 25		

[a]RT ≡ Room temperature test; HT ≡ High temperature test.
[b]Composite strengths normalized to a V_f = 20%. Actual V_f ranged from 18–35%, although most ~18–25%.
[c]Aluminosilicate matrix from COI.
[d]Only one sample tested in each condition.
[e]Unknown number of samples tested per condition.

determine the effect of fiber roughness on the interfacial properties [179, 180]. In the latter case, results showed that the monazite coating was deformed during push-out in relation to the roughness of the interface, which further validates the effectiveness of monazite as a weak interfacial layer [181].

The production of composites containing higher volume fractions of monazite-coated fibers has been somewhat limited and a summary of the relevant work is listed in Table 8.7. These composites were subjected to mechanical testing, either tensile or flexural, to evaluate the effectiveness of the coatings.

Discrete monazite coatings were shown to provide increased strengths and use temperatures for Nextel 610/(porous) alumina composites, as compared to similar composites containing no interfacial coatings [36] (Table 8.7). In these composites, an initial strength loss (~28%) was seen after thermal exposure at 1200°C for 100 hours, while control samples containing uncoated fiber displayed >70% strength losses after only 5 hours at 1200°C. After the initial

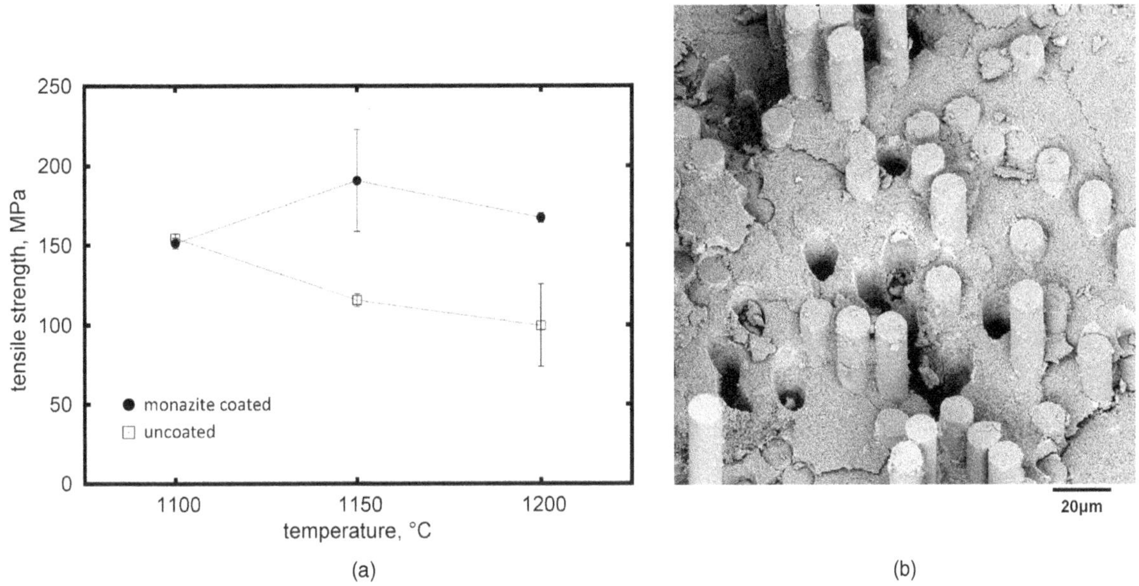

FIGURE 8.11

(a) Average tensile strength values of Nextel 610/monazite/alumina composites versus control samples (Nextel 610/alumina) tensile tested at high temperature; (b) typical fracture surface of Nextel 610/monazite/alumina composite showing portions of monazite coatings attached to pulled-out fibers and to fiber troughs.

strength loss with short-term thermal exposure at 1200°C, the remaining composite strength (>60%) was retained for up to 1000 hours at 1200°C [36]. High temperature tensile test results for Nextel 610/$LaPO_4$/alumina composites also indicated no difference between samples tested at room temperature or at temperatures up to 1200°C (Figure 8.11a) [188]. The fracture surfaces of the tested specimens revealed coated fibers that had pulled out from the alumina matrix, with portions of the monazite coatings remaining on the pulled-out fibers and in the fiber troughs (Figure 8.11b). Control samples containing uncoated fibers showed brittle behavior, with no fiber pull-out.

In situ monazite coatings, developed by Davis et al. for the production of porous matrix Nextel 610-based composites [84, 186], were also shown to provide improved composite properties. The *in situ* monazite coating was formed by mixing the monazite precursor with the alumina matrix powder during composite fabrication (~10–20 vol% monazite). The final product displayed higher damage tolerance and a reduced notch sensitivity compared to similar porous matrix composites without the monazite phase. This process is beneficial in that it precludes an individual coating step for the fiber. Either monazite alone or monazite mixed with a refractory oxide is machinable, which further enhances its use [189].

Another monazite coating method, developed by Weaver et al. utilized fugitive carbon coatings during the composite processing [190]. After sintering the composite under vacuum, the carbon coating was oxidized by heat treating in air at 1100°C. This oxidative heat treatment left a gap where the carbon had been; multiple infiltrations with a monazite precursor partially filled this gap, ultimately forming a monazite coating. Note that the monazite was not confined to only the interface region; some monazite was present within the matrix. Composites containing either the monazite coatings or the fugitive carbon coatings displayed higher notch strengths and fracture energies than composites with no coatings.

Several efforts examined the creep behavior of Nextel 610/$LaPO_4$/alumina composites at high temperatures; each effort utilized composites produced with varying processing methods and fiber architectures (filament wound vs. 8HS) but was tested with the same protocol [183, 184]. Uncoated samples of Nextel 610/alumina were also tested as a control. The monotonic tensile strength values varied depending upon the composite processing; irrespective of this, the monazite coatings significantly improved the strength of the composites in comparison to control samples. It is noted that only one sample was tested in each condition and for creep testing, run-out was defined to be 100 hours at the given creep stress.

N610/$LaPO_4$/alumina composites produced using filament winding were creep tested at temperatures of 900°C and 1100°C [184]. Creep run-out was attained at 900°C, at stresses up to 120 MPa, with a creep rate of approximately 1×10^{-8} per second. The material retained >90% of its monotonic strength after creep exposure. Control samples also showed run-out at 900°C with stresses up to 73 MPa and a similar creep rate; however, a slight increase in creep stress to 80 MPa resulted in a failure after only 5.5 hours. At 1100°C, the creep response of the N610/$LaPO_4$/alumina composite was significantly increased, with creep rates from 1.4×10^{-6} per second to 7.9×10^{-5} per second. No run-out was achieved at this temperature.

Alternately, in Reference 183, the N610/$LaPO_4$/alumina composites, tested at 1100°C, showed creep run-out at stresses up to 72 MPa, with creep rates ranging from 1.1×10^{-9} per second to 2.0×10^{-8} per second. In comparison, none of the control samples containing uncoated fiber attained run-out during testing at 1100°C. For both sample types, creep rupture testing in steam significantly accelerated the creep properties. For example, the monazite-containing samples tested at 72 MPa, which showed run-out during testing in air, failed after only 2.25 hours in steam. Evaluation of the composite fracture surfaces did not provide a clear answer as to the degradation mechanism.

The results of these efforts verified that monazite coatings improve both the tensile and creep properties of Nextel 610-based composites. However, these studies also highlighted the variability in results based on the composite processing. Further process optimization is needed to obtain reproducible property values.

Other ABO_4 coatings, such as scheelite ($CaWO_4$), neodymium phosphate ($NdPO_4$), and aluminum phosphate ($AlPO_4$) have been found to provide a weak interface in Nextel-based composites [172, 187]. Goettler et al. [35, 171], reported that relatively dense (>80% dense) unidirectional N610/$CaWO_4$/alumina composites were produced with tensile strengths approaching 350 MPa and strains-to-failure of the order of 0.25%. Debonding at the fiber/matrix interface was evident through microstructural evaluation (SEM, TEM). As with monazite, the stoichiometry of the scheelite coating precursor was found to be critical in maintaining the strength of the fibers.

Kaya et al. [187] produced Nextel 720/$NdPO_4$/mullite composites and the resultant samples were tested using 4-point flexural tests at high temperature. The composites were reasonably dense (~86%), with some porosity within the fiber tows. The strengths of these materials were ~235 MPa at room temperature; this strength was retained up to 1300°C. The strength decreased <15% after 300 thermal cycles to 1150°C. The results indicated that $NdPO_4$ is also a viable coating candidate. Similar results were found for N720/$NdPO_4$/Al_2O_3 composites, where damage tolerant behavior was exhibited from room temperature up to 1300°C during flexural testing [100].

Bao and Nicholson evaluated Nextel 720/$AlPO_4$/Al_2O_3 composites, hot pressed at 1250°C/20 MPa/1 hour, that contained ~14% porosity. Mechanical testing revealed distinct fiber pull-out on the fracture surface, indicating that $AlPO_4$ may be useful as a weak interface [191]. As previously noted, Applied Thin Films, Inc., (Evanston, IL) has also successfully used $AlPO_4$ as a matrix material for oxide–oxide composites.

Research into composites containing weak oxide interfaces is on-going, including evaluation of other rare-earth orthophosphates [192], and will likely continue until a commercial product is developed and the full advantages of these coatings are realized.

8.7.2 Porous Coatings

Porous coatings can provide a similar toughening function as the porous matrices; in this case, the porosity is localized around the fiber within a coating. Various porous coatings have been examined, including zircon [193], ZrO_2 [50], and rare-earth aluminates [194].

Research on sapphire/porous ZrO_2/alumina (~30 vol% fiber) composites was conducted as part of a project focused on developing oxide–oxide composites for high temperature (~1400°C) combustor applications [50]. Porous zirconia (ZrO_2) coatings were deposited onto sapphire fibers using a mixed carbon + ZrO_2 slurry; the carbon was removed by oxidation after incorporation of the coated fibers in an alumina matrix. Flat combustor tiles were fabricated and exposed in a combustor test rig for times up to 1.5 hours. Examination of the tested samples showed microcracking of the matrix, but the fibers remained intact and held the matrix material together. These initial results confirmed that the porous ZrO_2 coating was effective.

8.7.3 Fugitive Coatings

The term fugitive is used for coatings that are (or can be) removed after composite fabrication. Various materials have been examined for this function, such as molybdenum [195], but carbon is the obvious choice, based on cost and commercial availability. Carbon coatings can be incorporated into oxide–oxide composites and then removed through oxidation either prior to or during use, leaving a gap at the fiber/matrix interface. Oxide fibers are oxidation resistant, so the intermediate temperature embrittlement encountered for SiC-based composites is not a concern [196].

A review of this concept is given in Reference 197. Fugitive carbon coatings were shown to provide a weak interface for sapphire reinforced YAG and for Nextel 720/CAS (calcium aluminosilicate, glass–ceramic) composites. Push-out testing was utilized for the sapphire/C/YAG composites, while tensile testing was used with the Nextel 720/C/CAS composites. In the latter case, higher strengths were found for the fugitive carbon-containing samples (92 ± 42 MPa) versus control samples with no interfacial coating (22 ± 3 MPa). The thickness of the carbon was found to be very important, as also indicated by other researchers for SiC-based composites [198].

Recent work by Weaver et al. has further demonstrated the viability of very thin (~10 nm) fugitive carbon coatings, deposited using a 5% PVA solution onto Nextel 720 fabric [199]. The coated fabric was consolidated in a mullite-alumina matrix, with multiple reinfiltration cycles to increase the composite density. Composites with fugitive carbon coatings showed higher retained strengths versus control samples after exposure at 1200°C for 1000 hours, for both unnotched and notched samples. Extensive fiber pull-out, of the order of several millimeters, was observed in the as-processed condition; fiber pull-out was still observed after the aging heat treatment, although the length was reduced. The results confirm that fugitive carbon coatings are effective as a weak interface; however, the thickness of the carbon coating requires further optimization.

Two-layer coatings of fugitive carbon + zirconia (ZrO_2) have also been shown to provide a weak interface in a sapphire/alumina composite [200]. In this work, the carbon coating alone did not improve the composite properties; however, it is possible that the carbon coating was not thick enough or was discontinuous. The fugitive layer in combination with ZrO_2 was successful, however, providing flexural strengths of ~600 MPa after oxidation of the carbon. Other works on duplex fugitive carbon + ZrO_2 coatings can be found in References 74 and 201. Fugitive carbon coatings were also examined for the small diameter Nextel 610 fibers in a mullite matrix [202]. These coatings were not successful due to incomplete coverage of the fibers. These examples further illustrate the need for strict control over the fugitive coating characteristics.

8.7.4 Other Coatings

Research has been conducted on other interfacial coating concepts; however, these will only be briefly mentioned in this text. Reference 3 gives an in-depth description of these classes: easy-cleaving coatings, including phyllosilicates, hexaluminates, and layered perovskites, ductile coatings such as platinum, segregation weakened interfaces, multilayered coatings, and reactive coatings. Aside from these classes, various other interface coatings, including TiO_2 [203] and ZrO_2-SiO_2 [204], have been proposed and studied.

8.8 TECHNOLOGY DEVELOPMENT

Significant progress has been made over the past decade in the transitioning of oxide CMCs to industrial and aerospace applications through a number of technology demonstration programs. These introductory applications have primarily been modestly loaded static structures such as hot gas structures, exhaust structures, and TPS elements.

The use of oxide–oxide CMCs as combustor liners in gas turbine engines has been actively pursued, since the oxide–oxide composite allows for higher temperature, more efficient burning of the fuel, which translates into lower NO_x emissions. The potential use of an oxide–oxide CMC in such an application was demonstrated through field tests of a combustor liner, that survived for over 25000 hours and 109 starts in a *Centaur*® 50S (Solar Turbines) gas turbine engine [205–207]. These liners were exposed to through-thickness thermal gradients during use and it is noted that an oxide-based FGI layer (friable graded insulation [208]) was added to the hot surface of the porous matrix CMC. The FGI layer provided thermal protection to the composite, which is temperature limited (~1200°C) due to creep of the fibers [209, 210].

Porous matrix oxide–oxide CMCs have also been evaluated for use in exhaust components, including nozzle flaps for a fighter aircraft [129], and lightweight helicopter exhaust ducts that insulate and protect the surrounding structure [211]. An example of the helicopter exhaust component is shown in Figure 8.12. Other components include the seal/casing shroud and stationary vanes [212] in a turbine engine. It is noted that rotating components have generally not been considered due to the modest high temperature strength and creep resistance of the available oxide fibers.

Several US programs are currently underway that utilize oxide–oxide composites as exhaust components. The use of oxide–oxide CMC exhaust nozzles in subsonic jet engines to improve component durability (in comparison with titanium) and avoid weight increases associated with switching to heavier metallic alloys such as Inconel is being evaluated by several companies, including Rolls-Royce Liberty Works (RRLW) and Boeing [213–216].

The NASA Glenn Research Center (GRC) and RRLW teamed with ATK-COIC in the NASA ERA (Environmentally Responsible Aviation) project [213] with the goal of advancing CMC mixer nozzle technology to full scale engine testing [214]. As a result of this collaboration, a subscale oxide fiber/oxide matrix composite mixer nozzle (~8″ diameter, as

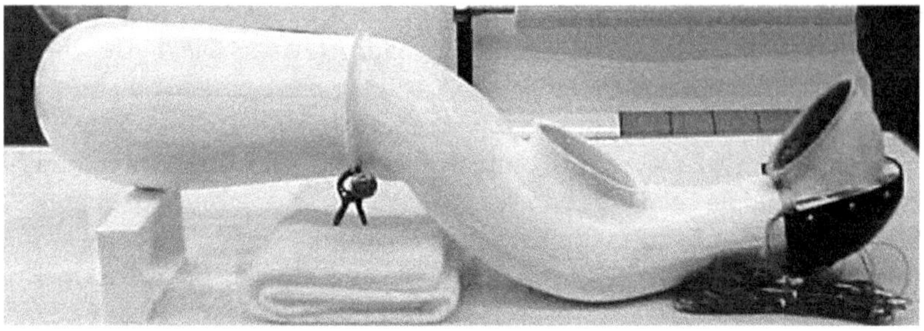

FIGURE 8.12

Oxide–oxide exhaust system. Photo courtesy of ATK-COIC.

shown in Figure 8.13) was fabricated by ATK-COIC. Aero-performance of the nozzle assembly was ground tested in a Hot/Cold Dual Stream Static Nozzle Test Facility and acoustic performance was evaluated in the NASA GRC Aero-Acoustic Propulsion Laboratory. Various combinations of temperature and pressure ratios were used during the tests in both facilities. The testing yielded positive results and nondestructive evaluation (using x-ray computed radiography) of the nozzle assembly following these tests did not show any damage. Full scale oxide/oxide CMC mixer nozzles are being fabricated for vibration testing and engine testing [214].

The Boeing Company, in collaboration with the FAA under the Continuous Lower Energy, Emissions and Noise (CLEEN) Technologies Development Program, is working to produce an acoustic Nextel 610/aluminosilicate composite centerbody and exhaust nozzle (Figure 8.14) for commercial aircraft [215, 216, 217]. The addition of this nozzle will ultimately allow for weight savings, enhanced acoustic attenuation, and improved lifetime. Subelement/subcomponent testing was successfully completed and the program is expected to progress to ground testing followed by prototype flight testing. Boeing is being supported by suppliers ATK/COIC and Albany Engineered Composites to fabricate the Boeing-designed exhaust system for a Rolls-Royce Trent 1000 engine.

GE Aviation has significantly invested in oxide–oxide composites and their "Ox–Ox" material was first used as a divergent exhaust seal for the F414 engine (2011). More recently, GE has announced the incorporation of their Ox–Ox material in multiple components in the Passport engine for the Bombardier Global 7000 and 8000 aircrafts, including the exhaust mixer, centerbody, and core cowls. The Passport engine is currently undergoing certification testing, with the final certification expected by 2015 [218].

Internationally, a significant amount of work on porous matrix oxide–oxide composites has been completed in Germany. A recent program, High Performance Oxide Composites (HiPOC), which began in February 2009, focused on the development of various oxide-based CMCs (WHIPOX, UMOX, OXIPOL) for use in hot section applications for either land-based or aerospace turbine engines [219].

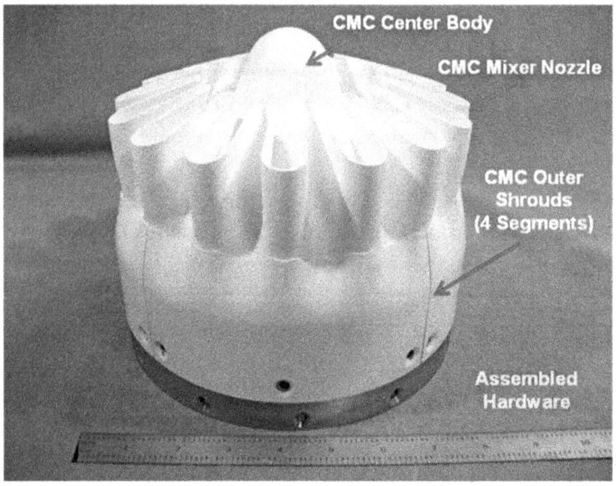

FIGURE 8.13

Subscale oxide/oxide CMC mixer nozzle assembly. Photo courtesy of NASA.

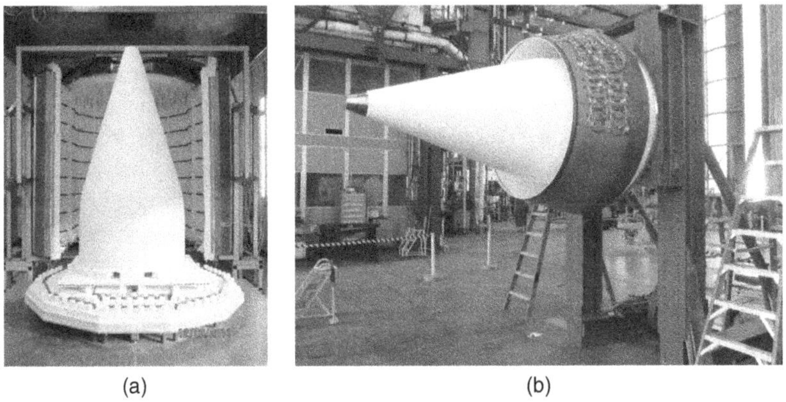

FIGURE 8.14
Boeing oxide CMC acoustic nozzle. (a) Centerbody; (b) CMC Nozzle assembly instrumented for testing. Photos courtesy of Boeing and ATK-COIC.

Alternative uses include semistructural applications, such as TPS, which typically combine various elements, including insulation, an outer skin and some provision for attachment, that protect the underlying structure from exposure to high temperature and hot gases [220, 221]. An example of such a system is the silica-based composites used as thermal blankets for the space shuttles (maximum temperature ~650°C) [222]. The durability of such systems is enhanced if the outer surface is robust against abuse such as impact damage and erosion. For example, Nextel™ 440/monazite composites allow for over a 50% increase in the maximum temperature capability of the current TPS system (SiO_2 based), while also being much more durable [223].

There has been a continued industrial effort to improve oxide–oxide TPS systems through innovative design and fabrication; several examples of this effort are given here. Boeing has developed and improved shuttle-type tile insulation, where an oxide–oxide CMC face sheet is bonded to the surface of a lightweight ceramic foam insulation tile. The CMC provides wear protection, impact damage resistance, improved thermal resistance and a smooth surface [224]. ATK-COIC patented an advanced non-tile-based TPS component that uses a bonded mortise and tenon system to fabricate honeycomb-like CMC panels that are mechanically fastened to the substructure (Figure 8.15) [225].

The potential benefits of oxide–oxide CMCs in the various applications discussed here will continue to drive the research, development, and technology transition of these materials.

8.9 POTENTIAL FUTURE FOR OXIDE–OXIDE COMPOSITES

The applications discussed in the previous section highlight the increased focus on porous matrix oxide–oxide composites. The key limitations of this class of composites are inferior interlaminar properties and low creep resistance. While there have been studies that attempt to improve interlaminar fracture resistance, there have been none to improve the use temperature limit. The most straightforward way to improve the creep resistance of these materials is through the introduction of higher temperature capable fibers. There has been no funded work in this area for the past decade,

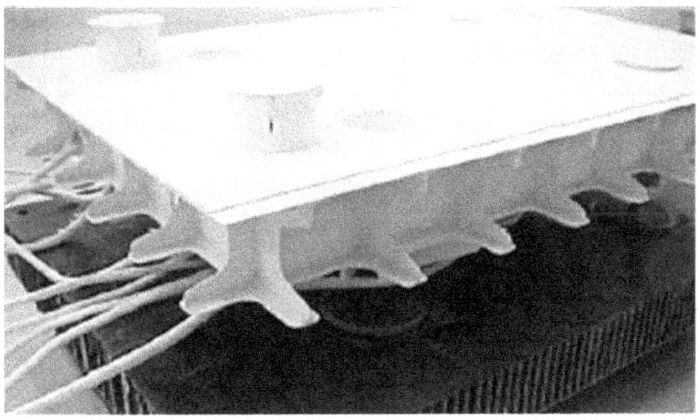

FIGURE 8.15
All-oxide standoff thermal protection system. Photo courtesy of ATK-COIC.

presumably due to lack of insertion. Over the decade, oxide composites have found niche applications, and it is likely the right time to invest the needed capital to develop these materials.

A positive note is that there are clear approaches and in some cases, incentives to the development of higher temperature, fine diameter oxide fibers. As mentioned previously, YAG is the most creep resistant oxide reported [53] and therefore a good candidate for a fiber. Polycrystalline YAG fibers are also currently being developed for high energy laser applications, which provides an additional motivation for development [226]. Mullite, spinel, and zircon are other possible compositions worth exploring. There may also be potential in other double or complex oxides, which a basic research study may help identify. Further, there are microstructural approaches to improving creep resistance. It is hoped that the success of porous matrix oxide composites in practical use applications will provide the required impetus to produce improved fibers.

The other aspect of oxide matrix composites that has been overlooked is improvements in matrix-dominated properties such as resistance to wear, abrasion, and uneven thermal loads. This will require a significant increase in matrix density, which will necessitate an oxide- or oxidation-resistant interface that debonds readily. Thus the full potential of dense matrix oxide composites has not yet been realized. Further, it has been difficult to form dense matrices in these composites. There will also need to be a sufficient payoff in increasing the operating temperatures of current technology composites to justify the expense associated with the adoption of fiber coatings. This exists to some extent in land-based turbine engines, which require higher operating temperatures for efficiency and lower pollution, and long operating times, which preclude non-oxide composites.

Aside from material development, modeling and design of structural components will require careful allowance for the matrix-dominated properties and careful placement of fibers in all loading directions. This implies a high degree of sophistication in component design and in many cases, designing and producing weavable 3D fiber architectures. Processing composites with the additional fiber constraint implied by complex architectures will require a corresponding enhancement of processing sophistication. Ultimately these improvements will need to be accomplished without substantial cost penalties. That, in turn implies that these infrastructural needs may require significant commercial markets to provide the production volumes necessary to make them economically viable.

Finally, tough oxide–oxide composites useful at high temperatures, in some cases up to ~1200°C, have been produced, are commercially available and are bridging the "valley of death" commonly seen between laboratory development/initial demonstration and the tangible implementation of a material within an actual use application. With continued research and development, particularly on improved fibers, the optimal use temperature of oxide–oxide composites is expected to increase, along with their use in practical applications.

ACKNOWLEDGMENTS

The authors would like to thank Douglas Kiser of NASA-Glenn, Mitch Petervary of the Boeing Company, Dr. Andy Szweda of ATK-COIC, Inc., Dr. Benjamin Mangrich of ATFI, Inc., Drs. L. P. Zawada and G. Fair of AFRL/RXCC, and Drs. Tap Parthasarathy, Thomas Key, and Emmanuel Boakye of UES, Inc. for contributions to this chapter. This work was performed by one of the authors (K. Keller) under Air Force Contract # FA8650-10-D-5226.

REFERENCES

1. F. W. Zok. Developments in oxide fiber composites. *J. Am. Ceram. Soc.*, 89, 3309–3324, (2006).
2. K. A. Keller, G. Jefferson, and R. J. Kerans. Oxide-oxide composites, in *Handbook of Ceramic Composites*, edited by N. A. Bansal (Kluwer Academic, 2005), pp. 377–421.
3. R. J. Kerans, R. S. Hay, T. A. Parthasarathy, and M. K. Cinibulk. Interface design for oxidation-resistant ceramic composites. *J. Am. Ceram. Soc.*, 85, 2599–2632 (2002).
4. D. W. Richerson. Ceramic matrix composites, in *Composites Engineering Handbook*, edited by P. K. Mallick (Marcel Dekker, New York, 1997), pp. 983–1038.
5. R. C. Robinson and J. L. Smialek. SiC recession caused by SiO_2 scale volatility under combustion conditions: i, experimental results and empirical model. *J. Am. Ceram. Soc.*, 82, 1817–1825 (1999).
6. N. S. Jacobson. Corrosion of silicon-based ceramics in combustion environments. *J. Am. Ceram. Soc.*, 76, 3–28 (1993).

7. D. B. Marshall and J. B. Davis. Ceramics for future power generation technology: fiber reinforced oxide composites. *Curr. Opin. Solid State Mater. Sci.*, 5, 283–289 (2001).
8. A. G. Evans, D. B. Marshall, F. Zok, and C. Levi. Recent advances in oxide-oxide composite technology. *Adv. Comp. Mater.*, 8, 17–23 (1999).
9. R. J. Kerans, R. S. Hay, and T. A. Parthasarathy. Structural ceramic composites. *Curr. Opin. Solid State Mater. Sci.*, 4, 445–451 (1999).
10. W. A. Curtin. Fiber pull-out and strain localization in ceramic matrix composites. *J. Mech. Physics Solids*, 41, 35–53 (1993).
11. G. H. Cullum. Ceramic matrix composite fabrication and processing: sol-gel infiltration, in *The Handbook on Continuous Fiber-Reinforced Ceramic Matrix Composites* edited by R. Lehman, S. El-Rahaiby, and J. Wachtman (Purdue Research Foundation, 1995).
12. L. P. Zawada, R. S. Hay, S. S. Lee, and J. Staehler. Characterization and high-temperature mechanical behavior of an oxide/oxide composite. *J. Am. Ceram. Soc.*, 86, 981–990 (2003).
13. T. J. Dunyak, D. R. Chang, and M. L. Millard. Thermal aging effects in oxide/oxide ceramic matrix composites. *Proceedings of 17th Conference on Metal Matrix, Carbon, and Ceramic Matrix Composites*, edited by J. D. Buckley, Part 2 (NASA Conference Publication 3235, NASA, 1993) pp. 675–690.
14. T. J. Lu. Crack branching in all-oxide ceramic composites. *J. Am. Ceram. Soc.*, 79, 266–274 (1996).
15. W.-C. Tu, F. F. Lange, and A. G. Evans. Concept for a damage-tolerant ceramic composite with "strong" interfaces. *J. Am. Ceram. Soc.*, 79, 417–424 (1996).
16. A. Szweda, M. L. Millard, and M. G. Harrison. Fiber reinforced ceramic matrix composite member and method for making. U.S. Patent No. 5,601,674. Feb 11 (1997).
17. L. P. Zawada. Longitudinal and transthickness tensile behavior of several oxide/oxide composites. *Cer. Eng. Sci. Proc.*, 19, 327–339 (1998).
18. C. G. Levi, J. Y. Yang, B. J. Dalgleish, F. W. Zok, and A. G. Evans. Processing and performance of an all-oxide ceramic composite. *J. Am. Ceram. Soc.*, 81, 2077–2086 (1998).
19. F. F. Lange, W. C. Tu, and A. G. Evans. Processing of damage tolerant, oxidation resistant ceramic-matrix composites. *Mater. Sci. Eng.*, A195, 145–150 (1995).
20. B. N. Cox and F. W. Zok. Advances in ceramic composites reinforced by continuous fibers. *Curr. Opin. Solid State Mater. Sci.*, 1, 666–673 (1996).
21. A. G. Evans. The mechanical properties of reinforced ceramic, metal and intermetallic matrix composites. *Mater. Sci. Eng. A*, 143, 63–76 (1991).
22. C. G. Levi, F. W. Zok, J. Y. Yang, M. Mattoni, and J. P. A. Lofvander. Microstructural design of stable porous matrices for all-oxide ceramic composites. *Z. Metallk.*, 90, 1037–1047 (1999).
23. M. A. Mattoni, J. Y. Yang, C. G. Levi, and F. W. Zok. Effects of matrix porosity on the mechanical properties of a porous-matrix, all-oxide ceramic composite. *J. Am. Ceram. Soc.*, 84, 2594–2602 (2001).
24. R. A. Jurf and S. C. Butner. Advances in all-oxide CMC. *J. Eng. Gas Turbines Power*, 122, 202–205 (2000).
25. S. G. Steel, L. P. Zawada, and S. Mall. Fatigue behavior of a Nextel(tm) 720/alumina (N720/A) composite at room temperature. *Ceram. Eng. Sci. Proc.*, 22, 695–702 (2001).
26. *Oxide-Oxide CMC Data Sheets* (COI Ceramics, Inc., San Diego, CA, 2003).
27. J. A. Heathcote, X. Gong, J. Y. Yang, U. Ramamurty, and F. W. Zok. In-plane mechanical properties of an all-oxide ceramic composite. *J. Am. Ceram. Soc.*, 82, 2721–2730 (1999).
28. S. C. Butner, personal communication.
29. A. G. Evans, F. W. Zok, and J. B. Davis. The role of interfaces in fiber-reinforced brittle matrix composites. *Compos. Sci. Technol.*, 42, 3–24 (1991).
30. C. G. Levi, F. W. Zok, J. Y. Yang, M. Mattoni, and J. P. A. Lofvander. Microstructural design of stable porous matrices for all-oxide ceramic composites. *Z. Metallk.*, 90, 1037–1047 (1999).
31. M.-Y. He and J. W. Hutchinson. Crack deflection at an interface between dissimilar elastic materials. *Int. J. Solids Struct.*, 25, 1053–1067 (1989).
32. R. J. Kerans and T. A. Parthasarathy. Crack deflection in ceramic composites and fiber coating design criteria. *Comp. Part A*, 30A, 521–524 (1999).
33. K. T. Faber. Ceramic composite interfaces: properties and design. *Annu. Rev. Mater. Sci.*, 27, 499–524 (1997).
34. R. J. Kerans. The role of coating compliance and fiber/matrix interfacial topography on debonding in ceramic composites. *Scri. Metall. Mater.*, 32, 505–509 (1995).
35. R. W. Goettler, S. Sambasivan, V. Dravid, and S. Kim. Interfaces in oxide fiber-oxide matrix ceramic composites, in *Computer Aided Design of High Temperature Materials*, edited by A. Pechenik, R. Kalia, and P. Vashishta (Oxford University Press, 1999), pp. 333–349.

36. K. A. Keller, T.-I. Mah, T. A. Parthasarathy, E. E. Boakye, P. Mogilevsky, and M. K. Cinibulk. Effectiveness of monazite coatings in oxide/oxide composites after long-term exposure at high temperature. *J. Am. Ceram. Soc.*, 86, 325–332 (2003).
37. 3M Nextel Ceramic Textiles Technical Notebook. 2001.
38. D. M. Wilson and L. R. Visser. High performance oxide fibers for metal and ceramic composites. *Comp. Part A*, 32, 1143–1153 (2001).
39. D. M. Wilson, S. L. Lieder, and D. C. Lueneburg. Microstructure and high temperature properties of Nextel 720 fibers. *Ceram. Eng. Sci. Proc.*, 16, 1005–1014 (1995).
40. D. M. Wilson, D. C. Lueneburg, and S. L. Lieder. High temperature properties of nextel 610 and alumina-based nanocomposite fibers. *Ceram. Eng. Sci. Proc.*, 14, 609–621 (1993).
41. T. A. Parthasarathy, E. Boakye, M. K. Cinibulk, and M. D. Petry. Fabrication and testing of oxide/oxide microcomposites with monazite and hibonite as interlayers. *J. Am. Ceram. Soc.*, 82, 3575–3583 (1999).
42. G. B. Merrill and J. A. Morrison. High temperature erosion resistant coating and material containing compacted hollow geometric shapes. US Patent No. 6,641,907 (2003).
43. M. K. Cinibulk, T. A. Parthasarathy, K. A. Keller, and T.-I. Mah. Porous yttrium aluminum garnet fiber coatings for oxide composites. *J. Am. Ceram. Soc.*, 85, 2703–2710 (2002).
44. T. A. Parthasarathy, E. Boakye, K. A. Keller, and R. S. Hay. Evaluation of porous zirconia silica and monazite coatings using nextel 720 fiber reinforced blackglas minicomposites. *J. Am. Ceram. Soc.*, 84, 1526–1532 (2001).
45. A. I. Kingdon, R. F. Davis, and M. M. Thackeray. Engineering properties of multicomponent and multiphase oxides, in *Engineered Materials Handbook, Vol. 4, Ceramics and Glasses* (ASM International, Metals City, OH, 1991), pp. 758–774.
46. M. Miyayama, K. Koumoto, and H. Yanagida. Engineering properties of single oxides, in *Engineered Materials Handbook, Vol. 4, Ceramics and Glasses* (ASM International, Metals City, OH, 1991), pp. 748–757.
47. M. D. Petry and T.-I. Mah. Effect of thermal exposures on the strengths of nextel 550 and 720 filaments. *J. Am. Ceram. Soc.*, 82, 2810–2807 (1999).
48. D. M. Wilson. High temperature oxide fibers. 105th Annual Meeting of the American Ceramic Society, Nashville, TN (2003).
49. M. Koopman, S. Duncan, K. K. Chawla, and C. Coffin. Processing and characterization of barium zirconate coated alumina fibers/alumina matrix composites. *Comp. Part A*, 32, 1039–1044 (2001).
50. M. Holmquist, R. Lundberg, O. Sudre, A. G. Razzell, L. Molliex, J. Benoit, and J. Adlerborn. Alumina/alumina composite with a porous zirconia interphase - Processing, properties and component testing. *J. Eur. Ceram. Soc.*, 20, 599–606 (2000).
51. National Research Council, *Ceramic Fibers and Coatings: Advanced Materials for the Twenty-First Century* (The National Academic Press, Washington, DC, 1998).
52. S. A. Newcomb and R. E. Tressler. Slow crack growth in sapphire fibers at 800C to 1500C. *J. Am. Ceram. Soc.*, 76, 2505–2512 (1993).
53. G. S. Corman. Creep of yttrium aluminum garnet single crystals. *J. Mater. Sci. Let.*, 12, 379–382 (1993).
54. G. N. Morscher, K. C. Chen, and K. S. Mazdiyasni. Creep resistance of developmental polycrystalline yttrium-aluminum garnet fibers. *Cer. Eng. Sci. Proc.*, 15, 181–188 (1994).
55. B. H. King and J. W. Halloran. Polycrystalline yttrium aluminum garnet fibers from colloidal sols. *J. Am. Ceram. Soc.*, 78, 2141–2148 (1995).
56. E. J. A. Pope. (2012, Aug 7). *The All American Composite* [press release]. www.matechgsm.com. Accessed 5 October 2012.
57. G. E. Fair, H. J. Kim, H. Lee, T. A. Parthasarathy, K. A. Keller, and Z. D. Miller. Development of ceramic fibers for high energy laser applications. *Proc. SPIE* 8039, Laser Technology for Defense and Security VII (8039)0X (May 13. 2011).
58. K. Hyun Jun, G. E. Fair, A. M. Hart, S. A. Potticary, and N. G. Usechak. Influence of processing variables on the properties of polycrystalline YAG fibers. *Proc. SPIE* 8381, Laser Technology for Defense and Security VIII (8381)11 (April 23—25, 2012).
59. S. T. Peters and Y. M. Tarnopol'skii. Filament winding, in *Composites Engineering Handbook*, edited by P. K. Mallick ((Marcel Dekker, New York, 1997), pp. 515–548.
60. A. R. Bhatti and P. M. Farries. Preparation of long-fiber-reinforced dense glass and ceramic matrix composites, in *Carbon/Carbon, Cement and Ceramic Matrix Composites*. vol. 4, edited by R. Warren (Elsevier Science Ltd., Oxford, 2000), pp. 645–667.
61. E. H. Moore, C. A. Folsom, K. A. Keller, and T. Mah. 3D Composite fabrication through matrix slurry pressure infiltration. *Ceram. Eng. Sci. Proc.*, 15, 113–120 (1994).
62. S. M. Sim and R. J. Kerans. Slurry infiltration of 3-D woven composites, *Ceram. Eng. Sci. Proc.*, 13, 632–641 (1992).
63. J. Y. Yang, J. H. Weaver, F. W. Zok, and J. J. Mack. Processing of oxide composites with three-dimensional fiber architectures. *J. Am. Ceram. Soc.*, 92, 1087–1092 (2009).
64. D. C. C. Lam and F. F. Lange. Microstructural observations on constrained densification of alumina powder containing a periodic array of sapphire fibers. *J. Am. Ceram. Soc.*, 77, 1976–1978 (1994).

65. M. N. Rahaman, *Ceramic Processing and Sintering*, 1st edition (Marcel-Dekker, Inc., New York, 1995).
66. A. Krell, H. Klemm, T. Hutzler, and A. Bales. Approach to the fabrication of oxide/oxide single-crystalline fiber composites from polycrystalline fiber bundles, in *High Temperature Ceramic Matrix Composites 5* (American Ceramic Society, 2005), pp. 155–160.
67. T. Mah, K. A. Keller, M. K. Cinibulk, and R. J. Kerans. Towards crack-free oxide-oxide CMC development, in *High Temperature Ceramic Matrix Composites 6* (American Ceramic Society, 2005).
68. D. B. Gundel, P. J. Taylor, and F. E. Wawner. Fabrication of thin oxide coatings on ceramic fibres by a sol-gel technique. *J. Mater. Sci.*, 29, 1795 (1994).
69. R. S. Hay and E. E. Hermes. Sol-Gel coatings on continuous ceramic fibers. *Ceram. Eng. Sci. Proc.*, 11, 1526–1532 (1990).
70. H. Dislich and E. Hussmann. Amorphous and crystalline dip coatings obtained from organometallic solutions: procedures, chemical processes, and products. *Thin Solid Films*, 77, 129–139 (1981).
71. P. D. Jero, F.Rebillat, D. J. Kent, and J. G. Jones. Crystallization of lanthanum hexaluminate from MOCVD precursors. *Ceram. Eng. Sci. Proc.*, 19, 359–360 (1998).
72. J. A. Haynes, K. M. Cooley, D. P. Stinton, and R. A. Lowden. Corrosion-resistant CVD mullite coatings for Si_3N_4. *Ceram. Eng. Sci. Proc.*, 20, 355–362 (1999).
73. P. V. Chayka. Liquid MOCVD precursors and their application to fiber interface coatings. *Ceram. Eng. Sci. Proc.*, 18, 287–294 (1997).
74. K. Nubian, B. Saruhan, B. Kanka, M. Schmucker, H. Schneider, and G. Wahl. Chemical vapor deposition of ZrO_2 and C/ZrO_2 on mullite fibers for interfaces in mullite/aluminosilicate fiber-reinforced composites. *J. Eur. Ceram. Soc.*, 20, 537–544 1 (2000).
75. P. W. Brown. Electrophoretic deposition of mullite in a continuous fashion utilizing non-aqueous polymeric sols, in *Ceramic Transactions, Vol. 56*, edited by K. V. Logan (American Ceramic Society, Columbus, OH, 1995), pp. 369–376.
76. T. J. Illston, C. B. Ponton, P. M. Marquis, and E. G. Butler. Electrophoretic deposition of silica/alumina colloids for the manufacture of CMC's. *Ceram. Eng. Sci. Proc.*, 15, 1052–1059 (1994).
77. R. W. Rice. BN Coating of Ceramic Fibers for Ceramic Fiber Composites. US Patent No. 4,642,271 (1987).
78. R. Naslain, O. Dugne, A. Guette, J. Sevely, C. R. Brosse, J. P. Rocher, and J. Cotteret. Boron nitride interphase in ceramic matrix composites. *J. Am. Cer. Soc.*, 74, 2482–2488 (1991).
79. C. J. Griffin and R. R. Kieschke. CVD processing of fiber coatings for CMCs. *Ceram. Eng. Sci. Proc.*, 16, 425–432 (1995).
80. M. H. Lewis. Interfaces in ceramic matrix composites, in *Carbon/Carbon, Cement and Ceramic Matrix Composites*. Vol. 4, edited by R. Warren (Elsevier, Oxford, 2000), pp. 289–322.
81. R. S. Hay, M. D. Petry, K. A. Keller, M. K. Cinibulk, and J. R. Welch. Carbon and oxide coatings on continuous ceramic fibers, in *Ceramic Matrix Composites - Advanced High Temperature Structural Materials*. Vol. 365, edited by R. A. Lowden, M. K. Ferber, J. R. Hellmann, and S. G. DiPetro (Materials Research Society, 1995), pp. 377–382.
82. E. Boakye, R. S. Hay, and M. D. Petry. Continuous coating of oxide fiber tows using liquid precursors: monazite coatings on Nextel 720. *J. Am. Ceram. Soc.*, 82, 2321–2331 (1999).
83. R. S. Hay and E. E. Hermes. Coating Apparatus for Continuous Fibers. US Patent No. 5,217,533 (1993).
84. J. B. Davis, D. B. Marshall, and P. E. D. Morgan. Monazite containing oxide-oxide composites. *J. Eur. Ceram. Soc.*, 20, 583–587 (2000).
85. E. E. Boakye, R. S. Hay, P. Mogilevsky, and L. M. Douglas. Monazite coatings on fibers: II, coating without strength degradation. *J. Am. Ceram. Soc.*, 84, 2793–2801 (2001).
86. E. E. Boakye, T. Mah, C. M. Cooke, and K. A. Keller. Initial assessment of the weavability of fiber tows coated with monazite. *J. Am. Ceram. Soc.*, 87, 1775–1778 (2004).
87. G. E. Fair, R. S. Hay, and E. E. Boakye. Precipitation coating of monazite on woven ceramic fibers: I, feasibility. *J. Am. Ceram. Soc.*, 90, 448–455 (2007).
88. G. E. Fair, R. S. Hay, and E. E. Boakye. Precipitation coating of monazite on woven ceramic fibers: II. Effect of processing conditions on coating morphology and strength retention of Nextel 610 and 720 fibers. *J. Am. Ceram. Soc.*, 91, 1508–1516 (2008).
89. G. E. Fair, R. S. Hay, E. E. Boakye, P. E. D. Morgan, R. F. Marzke, and R. Sharma. Precipitation coating of monazite on woven ceramic fibers: III-Coating without strength degradation using a phytic acid precursor. *J. Am. Ceram. Soc.*, 93, 420–428 (2010).
90. R. S. Hay, J. R. Welch, and M. K. Cinibulk. TEM specimen preparation and characterization of ceramic coatings on fiber tows. *Thin Solid Films*, 308–309, 389–392 (1997).
91. F. F. Lange. Shape forming of ceramic powders by manipulating the interparticle pair potential. *Chem. Eng. Sci.*, 56, 3011–3020 (2001).
92. F. F. Lange, C. G. Levi, and F. W. Zok. Processing fiber reinforced ceramics with porous matrices, in *Carbon/Carbon, Cement and Ceramic Matrix Composites*. Vol. 4, edited by R. Warren (Elsevier Science Ltd., Oxford, 2000), pp. 427–447.

93. R. S. Russell-Floyd, B. Harris, R. G. Cooke, J. Laurie, F. W. Hammett, R. W. Jones, and T. Wang. Application of sol-gel processing techniques for the manufacture of fiber-reinforced ceramics. *J. Am. Ceram. Soc.*, 76, 2635–2643 (1993).
94. T. Machry, C. Wilhelmi, and D. Koch. *Novel High Temperature Wound Oxide Ceramic Matrix Composites Manufactured Via Freeze Gelation* (John Wiley & Sons, Inc., 2011), 201–212.
95. P. Nicholson, P. Sarkar, and X. Huang. Electrophoretic deposition and its use to synthesize ZrO_2/Al_2O_3 micro-laminate ceramic/ceramic composites. *J. Mat. Sci.*, 29, 6274–6278 (1993).
96. A. R. Boccaccini, C. Kaya, and K. K. Chawla. Use of electrophoretic deposition in the processing of fibre reinforced ceramic and glass matrix composites: a review. *Comp. Part A*, 32, 997–1006 (2001).
97. S. Kooner, W. S. Westby, C. M. A. Watson, and P. M. Farries. Processing of Nextel 720/mullite composition composite using electrophoretic deposition. *J. Eur. Ceram. Soc.*, 20, 631–638 (2000).
98. C. Kaya, X. Gu, I. A. H. Al-Dawery, and E. G. Butler. Microstructural development of woven mullite fibre-reinforced mullite ceramic matrix composites by infiltration processing. *Sci. Technol. Adv. Mater.*, 3, 35–44 (2002).
99. I. A. H. Al-Dawery and E. G. Butler. Fabrication of high-temperature resistant oxide ceramic matrix composites. *Comp. Part A*, 32, 1007–1012 (2001).
100. C. Kaya, F. Kaya, E. G. Butler, A. R. Boccaccini, and K. K. Chawla. Development and characterisation of high-density oxide fibre-reinforced oxide ceramic matrix composites with improved mechanical properties. *J. Eur. Ceram. Soc.*, 29, 1631–1639 (2009).
101. Y. Bao and P. S. Nicholson. Constant current electrophoretic infiltration deposition of fiber-reinforced ceramic composites. *J. Am. Ceram. Soc.*, 90, 1063–1070 (2007).
102. J. J. Haslam and F. F. Lange. Strengthening of porous mullite and zirconia CMC matrices by evaporation/condensation. *J. Am. Ceram. Soc.*, 89, 2043–2050 (2006).
103. W. D. Kingery, H. K. Bowen, and D. R. Uhlmann, *Introduction to Ceramics* (John Wiley & Sons, New York, 1976).
104. P.Y. Lee, M. Imai, and T. Yano. Effects of sintering condition on mechanical properties of monazite-coated alumina-fiber/alumina-matrix composites fabricated by hot-pressing. *J. Ceram. Soc. Jpn.*, 112, 29–34 (2004).
105. H. T. Larker and R. Lundberg. Near net shape production of monolithic and composite high temperature ceramics by hot isostatic pressing (HIP). *J. Eur. Ceram. Soc.*, 19, 2367–2373 (1999).
106. F. Langlais. Chemical vapor infiltration processing of ceramic matrix composites, in *Carbon/Carbon, Cement, and Ceramic Matrix Composites*. Vol. 4, edited by R. Warren (Elsevier, Oxford, 2000), pp. 611–644.
107. J. Lamon. Chemical vapor infiltrated SiC/SiC composites (CVI SiC/SiC), in *Handbook of Ceramic Composites*, edited by N. P. Bansal (Kluwer Academic Publishers, New York, NY, 2005).
108. M. S. Newkirk, A. W. Urquhart, H. R. Zwicker, and E. Breval. Formation of lanxide ceramic composite materials. *J. Mater. Sci.*, 81–89 (1987).
109. W. B. Hillig. Making ceramic composites by melt infiltration. *Am. Cer. Soc. Bull.*, 73, 56–62 (1995).
110. W. B. Hillig and H. C. McGuigan. An exploratory study of producing non-silicate all-oxide composites by melt infiltration. *Mat. Sci. Eng.*, A196, 183–190 (1995).
111. J. Wendorff, R. Janssen, and N. Claussen. Platinum as a weak interphase for fiber-reinforced oxide-matrix composites. *J. Am. Ceram. Soc.*, 81, 2738–2740 (1998).
112. J. Wendorff, R. Janssen, and N. Claussen. Model experiments on pure oxide composites. *Mat. Sci. Eng.*, A250, 186–193 (1998).
113. K. A. Keller, T.Mah, E.E.Boakye, and T.A.Parthasarathy. Gel-casting and reaction bonding of oxide-oxide minicomposites with monazite interphase. *Ceram. Eng. Sci. Proc.*, 21, 525–534 (2000).
114. Y. Bao and P. S. Nicholson. $AlPO_4$-coated mullite/alumina fiber reinforced reaction-bonded mullite composites. *J. Eur. Ceram. Soc.*, 28, 3041–3048 (2008).
115. J. J. Haslam, K. E. Berroth, and F. F. Lange. Processing and properties of an all-oxide composite with a porous matrix. *J. Eur. Ceram. Soc.*, 20, 607–618 (2000).
116. M. Schmücker and H. Schneider. WHIPOX all oxide ceramic matrix composites, in *Handbook of Ceramic Composites*, edited by N. A. Bansal (Kluwer Academic, 2005), pp. 423–435
117. M. Schmucker, A. Grafmuller, and H. Schneider. Mesostructure of WHIPOX all oxide CMCs. *Comp. Part A*, 34, 613–622 (2003).
118. H. Schneider, J. Goring, B. Kanka, and M. Schmucker. WHIPOX: a new oxide fibre/oxide matrix composite for high-temperature applications. *Keram. Zeit.*, 53, 788–791 (2001).
119. M. Schmucker and P. Mechnich. Microstructural coarsening of nextel (Tm) 610 fibers embedded in alumina-based matrices. *J. Am. Ceram. Soc.*, 91, 1306–1308 (2008).
120. M. Schmucker and P. Mechnich. Improving the microstructural stability of nextel (Tm) 610 alumina fibers embedded in a porous alumina matrix. *J. Am. Ceram. Soc.*, 93, 1888–1890 (2010).

121. A. Souto and F. Guitian. Purification of mullite by reduction and volatilization of impurities. *J. Am. Ceram. Soc.*, 82, 2660–2664 (1999).

122. N. S. Jacobson, E. J. Opila, and K. N. Lee. Oxidation and corrosion of ceramics and ceramic matrix composites. *Curr. Opin.Solid State Mater. Sci.*, 5, 301–309 (2001).

123. K. N. Lee, N. S. Jacobson, and R. A. Miller. Refractory oxide coatings on SiC ceramics. *MRS Bull.*, 19, 35–38 (1994).

124. J. A. Haynes, M. J. Lance, K. M. Cooley, M. K. Ferber, R. A. Lowden, and D. P. Stinton. CVD mullite coatings in high-temperature, high-pressure air-H_2O. *J. Am. Ceram. Soc.*, 83, 657–659 (2000).

125. L. P. Zawada, J. Staehler, and S. G. Steel. Consequence of intermittent exposure to moisture and salt fog on the high-temperature fatigue durability of several ceramic matrix composites. *J. Am. Ceram. Soc.*, 86, 1282–1291 (2003).

126. M. K. Cinibulk, K. A. Keller, and T. Mah. Effect of yttrium aluminum garnet additions on alumina-fiber-reinforced porous-alumina-matrix composites. *J. Am. Ceram. Soc.*, 87, 881–887 (2004).

127. M. K. Cinibulk. Effect of yttria and yttrium-aluminum garnet on densification and grain growth of alumina at 1200°C–1300°C. *J. Am. Ceram. Soc.*, 87, 692–695 (2004).

128. J. D. French, J. Zhao, M. P. Harmer, H. M. Chan, and G. A. Miller. Creep of duplex microstructures. *J. Am. Ceram. Soc.*, 77, 2857–2865 (1994).

129. R. John, L. P. Zawada, and J. L. Kroupa. Stresses due to temperature gradients in ceramic-matrix-composite aerospace components. *J. Am. Ceram. Soc.*, 82, 161–168 (1999).

130. E. A. V. Carelli, H. Fujita, J. Y. Yang, and F. W. Zok. Effects of thermal aging on the mechanical properties of a porous-matrix ceramic composite. *J. Am. Ceram. Soc.*, 85, 595–602 (2002).

131. G. P. Tandon, D. J. Buchanan, N. J. Pagano, and R. John. Analytical and experimental characterization of thermo-mechanical properties of a damaged woven oxide-oxide composite. *Ceram. Eng. Sci. Proc.*, 22, 687–694 (2001).

132. H. Fugita, G. Jefferson, R. M. McMeeking, and F. W. Zok. Mullite/alumina mixtures for use as porous matrices in oxide fiber composites. *J. Am. Ceram. Soc.*, 87, 261–267 (2004).

133. A. G. Evans and F. W. Zok. The physics and mechanics of fibre-reinforced brittle matrix composites. *J. Mater. Sci.*, 29, 3857–3896 (1994).

134. H. Fujita, C. G. Levi, F. W. Zok, and G. Jefferson. Controlling mechanical properties of porous mullite/alumina mixtures via precursor-derived alumina. *J. Am. Ceram. Soc.*, 88, 367–375 (2005).

135. J. C. McNulty and F. W. Zok. Application of weakest-link fracture statistics to fiber-reinforced ceramic-matrix composites. *J. Am Ceram. Soc.*, 80, 1535–1543 (1997).

136. U. Ramamurty, F. W. Zok, F. A. Leckie, and H. E. Deve. Strength variability in alumina fiber reinforced aluminum matrix composites. *Acta Mater.*, 45, 4603–4613 (1997).

137. N. A. Weil and I. M. Daniel. Analysis of fracture probabilities in nonuniformly stressed brittle materials. *J. Am. Ceram. Soc.*, 47, 268–274 (1964).

138. R. Pipes, R. Blake, J. Gillespie, and L. Carslsson, *Delaware Composites Design Encyclopedia, Vol. 6, Test Methods* (Technomic, Lancaster, PA, 1990).

139. N. K. Naik and R. S. Kumar. Compressive strength of unidirectional composites : evaluation and comparison of prediction models. *Compos. Struct.*, 46, 299–308 (1999).

140. S. Y. Hsu, T. J. Volger, and S. Kyriakides. Compressive strength predictions for fiber composites. *J. Appl. Mech.*, 65, 7–16 (1998).

141. M. K. Cinibulk, K. A. Keller, T. Mah, and T. A. Parthasarathy. Nextel 610 fiber-reinforced alumina-YAG porous matrix composites. *Ceram. Eng. Sci. Proc.*, 23, 629–636 (2002).

142. L. P. Zawada and S. S. Lee. The effect of hold times on the fatigue behavior of an oxide/oxide ceramic composite, in *Thermal and Mechanical Test Methods and Behavior of Continuous-Fiber Ceramic Composites*, edited by M. G. Jenkins, S. T. Goncy, E. Lara-Curzio, N. E. Ashbaugh, and L. P. Zawada (American Society for Testing and Materials, West Conshohocken, PA, 1997).

143. S. G. Steel. Monotonic and Fatigue Loading Behavior of an Oxide/Oxide Ceramic Matrix Composite. Masters Thesis, Air Force Institute of Technology, Wright Patterson Air Force Base, OH (2000).

144. S. Mall and J.-M. Ahn. Frequency effects on fatigue behavior of Nextel 720 (Tm)/alumina at room temperature. *J. Eur. Ceram. Soc.*, 28, 2783–2789 (2008).

145. W. R. Cannon and T. G. Langdon. Review creep of ceramics: part I mechanical characteristics. *J. Mater. Sci.*, 18, 1–50 (1983).

146. S. V. Nair and J. L. Bassani. Macro- and micromechanics of elevated temperature crack growth in ceramic composites, in *High Temperature Mechanical Behavior of Ceramic Composites*, edited by S. V. Nair and K. Jakus (Butterworth-Heinemann, Boston, MA, 1995), pp. 437–470.

147. J. R. Zuiker. A model for the creep response of oxide-oxide ceramic matrix composites, in *Thermal and Mechanical Test Methods and Behavior of Continuous-Fiber Ceramic Composites*, edited by M. G. Jenkins, S. T. Goncy, E. Lara-Curzio, N. E. Ashbaugh, and L. P. Zawada (American Society for Testing and Materials, West Conshohocken, PA, 1997).

148. R. M. McMeeking. Models for the creep of ceramic matrix composite materials, in *High Temperature Mechanical Behavior of Ceramic Composites*, edited by S. V. Nair and K. Jakus (Butterworth-Heinemann, Boston, MA, 1995), pp. 409–436.

149. L. P. Zawada and S. S. Lee. Evaluation of four CMCs for aerospace turbine engine divergent flaps and seals. *Ceram. Eng. Sci. Proc.*, 16, 337–339 (1995).

150. R. John, D. J. Buchanan, and L. P. Zawada. Creep deformation and rupture behavior of a notched oxide/oxide Nextel 720/AS composite. *Ceram. Eng. Sci. Proc.*, 21, 567–574 (2000).

151. M. B. Ruggles-Wrenn and J. C. Braun. Effects of steam environment on creep behavior of Nextel (Tm) 720/alumina ceramic composite at elevated temperature. *Mater. Sci. Eng. A*, 497, 101–110 (2008).

152. M. B. Ruggles-Wrenn, S. Mall, C. A. Eber, and L. B. Harlan. Effects of steam environment on high-temperature mechanical behavior of nextel (Tm) 720/alumina (N720/A) continuous fiber ceramic composite. *Comp. Part A*, 37, 2029–2040 (2006).

153. S. Wannaparhun and S. Seal. Combined spectroscopic and thermodynamic investigation of Nextel-720 fiber/alumina ceramic-matrix composite in air and water vapor at 1100°C. *J. Am. Ceram. Soc.*, 86, 1628–1630 (2003).

154. M. B. Ruggles-Wrenn and N. R. Szymczak. Effects of steam environment on compressive creep behavior of Nextel (Tm) 720/alumina ceramic composite at 1200 degrees C. *Comp. Part A*, 39, 1829–1837 (2008).

155. M. B. Ruggles-Wrenn, G. Hetrick, and S. S. Baek. Effects of frequency and environment on fatigue behavior of an oxide-oxide ceramic composite at 1200 degrees C. *Int. J. Fatigue*, 30, 502–516 (2008).

156. J. M. Mehrman, M. B. Ruggles-Wrenn, and S. S. Baek. Influence of hold times on the elevated-temperature fatigue behavior of an oxide-oxide ceramic composite in air and in steam environment. *Compos. Sci. Technol.*, 67, 1425–1438 (2007).

157. M. B. Ruggles-Wrenn and B. A. Whiting. Cyclic creep and recovery behavior of Nextel (Tm) 720/alumina ceramic composite at 1200 degrees C. *Mater. Sci. Eng. A*, 528, 1848–1856 (2011).

158. S. Mall, A. R. Nye, and G. Jefferson. Tension-tension fatigue behavior of Nextel (Tm) 720/alumina under combustion environment. *Int. J. Appl. Ceram. Tech.*, 9, 159–171 (2012).

159. S. P. Timoshenko and J. N. Goodier, *Theory of Elasticity* (McGraw-Hill, New York, 1970).

160. S. C. Tan, *Stress Concentrations in Laminated Composites* (Technomic, Lancaster, PA, 1994).

161. D. J. Buchanan, V. A. Kramb, R. John, and L. P. Zawada. Effect of small effusion holes on creep rupture behavior of oxide/oxide Nextel(tm)720/AS composite. *Ceram. Eng. Sci. Proc*, 22, 659–666 (2001).

162. R. John, D. J. Buchanan, V. A. Kramb, and L. P. Zawada. Creep rupture behavior of oxide/oxide Nextel720/AS and MI SiC/SiC composite with effusion holes. *Ceram. Eng. Sci. Proc*, 23, 617–628 (2002).

163. J. M. Whitney and R. J. Nuismer. Stress fracture criteria for laminated composites containing stress concentrations. *J. Compos. Mater.*, 8, 253–265 (1974).

164. J. C. McNulty, F. W. Zok, G. M. Genin, and A. G. Evans. Notch-sensitivity of fiber-reinforced ceramic-matrix composites: effects of inelastic straining and volume dependent strength. *J. Am. Ceram. Soc.*, 82, 1217–1228 (1999).

165. L. P. Zawada and K. E. Goecke. Testing methodology for measuring transthickness tensile strength for ceramic matrix composites, in *Mechanical, Thermal and Environmental Testing and Performance of Ceramic Composites and Components*, edited by M. G. Jenkins, E. Lara-Curzio, and S. T. Gonczy (American Society for Testing and Materials, West Conshohocken, PA, 2000), pp. 62–85.

166. M. Kumosa, G. Odegard, D. Armentrout, L. Kumosa, K. Searles, and J. K. Sutter. Comparison of the +/-45 tensile and Iosipescu shear tests for woven fabric composite materials. *J. Compos. Tech. Res.*, 24, 3–16 (2002).

167. G. Odegard, D. Armentrout, K. Searles, L. Kumosa, J. K. Sutter, and M. Kumosa. Failure analysis of +/-45 off-axis woven fabric composite specimens. *J. Compos. Tech. Res.*, 23, 205–224 (2001).

168. S. W. Tsai and E. M. Wu. A general theory of strength for anisotropic materials. *J. Compos. Mater.*, 5, 58–80 (1971).

169. L. J. Hart-Smith. The role of biaxial stresses in discriminating between meaningful and illusory composite failure theories. *Compos. Struct.*, 25, 3–20 (1993).

170. N. P. Bansal and J. I. Eldridge. Effects of interface modification on mechanical behavior of hi-nicalon fiber-reinforced celsian matrix composites. *Ceram. Eng. Sci. Proc.*, 18, 379–389 (1997).

171. R. W. Goettler, S. Sambasivan, and V. P. Dravid. Isotropic complex oxides as fiber coatings for oxide-oxide CFCC. *Ceram. Eng. Sci. Proc.*, 18, 279–286 (1997).

172. M. H. Lewis, A. Tye, E. Butler, and I. Al-Dawery. Development of interfaces in oxide matrix composites. *Key Eng. Mater.*, 164–165, 351–356 (1999).

173. P. E. D. Morgan, D. B. Marshall, and R. M. Housley. High temperature stability of monazite-alumina composites. *Mater. Sci. Eng.*, A195, 215–222 (1995).

174. P. E. D. Morgan and D. B. Marshall. Ceramic composites of monazite and alumina. *J. Am. Ceram. Soc.*, 78, 1553–1563 (1995).

175. K. K. Chawla, H. Liu, J. Janczak-Rusch, and S. Sambasivan. Microstructure and properties of monazite (LaPO4) coated saphikon fiber/alumina matrix composites. *J. Eur. Ceram. Soc.*, 20, 551–559 (2000).

176. R. L. Callender and A. R. Barron. Formation and evaluation of highly uniform aluminate interface coatings for sapphire fiber reinforced ceramic matrix composites (FRCMCs) using carboxylate-alumoxane nanoparticles. *J. Mater. Sci.*, 36, 4977–4987 (2001).
177. D.-H. Kuo, W. M. Kriven, and T. J. Mackin. Control of interfacial properties through fiber coatings: monazite coatings in oxide-oxide composites. *J. Am. Ceram. Soc.*, 80, 2987–2996 (1997).
178. T. A. Parthasarathy, D. B. Marshall, and R. J. Kerans. Analysis of the effect of interfacial roughness on fiber debonding and sliding in brittle matrix composites. *Acta Met.*, 42, 3773–3784 (1994).
179. K. A. Keller, T. Mah, E. E. Boakye, T. A. Parthasarathy, and P. Mogilevsky. Evaluation of dense monazite fiber-coatings in oxide-oxide minicomposites. 27th Annual Cocoa Beach Conference & Exposition on Advanced Ceramics & Composites, Cocoa Beach, FL, January 29 (2003).
180. J. B. Davis, R. S. Hay, D. B. Marshall, P. E. D. Morgan, and A. Sayir. Influence of interfacial roughness on fiber sliding in oxide composites with la-monazite interphases. *J. Am. Ceram. Soc.*, 86 (2003).
181. R. S. Hay. Monazite and scheelite deformation mechanisms. *Ceram. Eng. Sci. Proc.*, 21, 203–218 (2000).
182. K. A. Keller, T. Mah, E. E. Boakye, T. A. Parthasarathy, and P. Mogilevsky. Evaluation of Nextel 610/monazite/alumina composites at high temperature. *105th Annual Meeting of The American Ceramic Society*, Nashville, TN, 2003.
183. M. B. Ruggles-Wrenn, T. Yeleser, G. Fair, and J. B. Davis. Effects of steam environment on creep behavior of Nextel^{TM}610/monazite/alumina composite at 1,100°C. *Appl. Compos. Mater.*, 16, 379–392 (2009).
184. M. B. Ruggles-Wrenn, S. S. Musil, S. Mall, and K. A. Keller. Creep behavior of Nextel (Tm) 610/monazite/alumina composite at elevated temperatures. *Compos. Sci. Technol.*, 66, 2089–2099 (2006).
185. A. Cazzato, M. Colby, D. Daws, J. B. Davis, P. Morgan, J. Porter, S. Butner, and B. Jurf. Monazite interface coatings in polymer and sol-gel derived ceramic matrix composites. *Ceram. Eng. Sci. Proc.*, 18, 269–278 (1997).
186. J. B. Davis, D. B. Marshall, and P. E. D. Morgan. Oxide composites of Al_2O_3 and $LaPO_4$. *J. Eur. Ceram. Soc.*, 19, 2421–2426 (1999).
187. C. Kaya, E. G. Butler, A. Selcuk, A. R. Boccaccini, and M. H. Lewis. Mullite (Nextel 720) fibre-reinforced mullite matrix composites exhibiting favourable thermomechancial properties. *J. Eur. Ceram. Soc.*, 22, 2333–2342 (2002).
188. K. Keller and T. Mah. Development of monazite (LaPO4) coated-fiber reinforced oxide-oxide composites. *High Temp. Ceram. Matrix Compos.*, 5, 149–154 (2005).
189. J. B. Davis, D. B. Marshall, R. M. Housley, and P. E. D. Morgan. Machinable ceramics containing rare-earth phosphates. *J. Am. Ceram. Soc.*, 81, 2169–2175 (1998).
190. J. H. Weaver, J. Yang, C. G. Levi, F. W. Zok, and J. B. Davis. A method for coating fibers in oxide composites. *J. Am. Ceram. Soc.*, 90, 1331–1333 (2007).
191. Y. H. Bao and P. S. Nicholson. AlPO4 coating on alumina/mullite fibers as a weak interface in fiber-reinforced oxide composites. *J. Am. Ceram. Soc.*, 89, 465–470 (2006).
192. R. S. Hay, G. E. Fair, E. E. Boakye, P. Mogilevsky, T. A. Parthasarathy, M. Ahrens, and T. J. Godar. Transmission electron microscopy of rare-earth orthophosphate fiber-matrix interphases that deform by transformation plasticity during fiber push-out. *Ceram. Eng. Sci. Proc*, 32 (2011).
193. E. Boakye, R. S. Hay, M. D. Petry, and T. A. Parthasarathy. Sol-gel synthesis of zircon-carbon precursors and coatings of Nextel 720 fiber tows. *Ceram. Eng. Sci. Proc. A*, 20, 165–172 (1999).
194. M. K. Cinibulk, T. A. Parthasarathy, K. A. Keller, and T. Mah. Porous rare-earth aluminate fiber coatings for oxide-oxide composites. *Ceram. Eng. Sci. Proc.*, 21, 219–228 (2000).
195. J. B. Davis, J. P. A. Lofvander, A. G. Evans, E. Bischoff, and M. L. Emiliani. Fiber coating concepts for brittle matrix composites. *J. Am. Ceram. Soc.*, 76, 1249–1257 (1993).
196. A. G. Evans, F. W. Zok, R. M. McMeeking, and Z. Z. Du. Models of high temperature, environmentally assisted embrittlement in ceramic-matrix composites. *J. Am. Ceram. Soc.*, 79, 2345–2352 (1996).
197. K. A. Keller, T. Mah, C. Cooke, and T. A. Parthasarathy. Fugitive interfacial carbon coatings for oxide/oxide composites. *J. Am. Ceram. Soc.*, 83, 329–336 (2000).
198. E. Lara-Curzio, M. Ferber, and R. A. Lowden. Effect of fiber coating thickness on the interfacial properties of a continuous fiber ceramic matrix composite. *Ceram. Eng. Sci. Proc.*, 15, 989–1003 (1994).
199. J. H. Weaver, J. Yang, and F. W. Zok. Control of interface properties in oxide composites via fugitive coatings. *J. Am. Ceram. Soc.*, 91, 4003–4008 (2008).
200. O. Sudre, A. G. Razzell, L. Molliex, and M. Holmquist. Alumina single-crystal fibre reinforced alumina matrix for combustor tiles. *Ceram. Eng. Sci. Proc.*, 19, 273–280 (1998).
201. B. Saruhan, M. Schmucker, M. Bartsch, H. Schneider, K. Nubian, and G. Wahl. Effect of interphase characteristics on long-term durability of oxide-based fibre-reinforced composites. *Compos. Part A*, 32, 1095–1103 (2001).
202. P. W. M. Peters, B. Daniels, F. Clemens, and W. C. Vogel. Mechanical characterization of mullite-based ceramic matrix composites at test temperatures up to 1200C. *J. Eur. Ceram. Soc.*, 20, 531–535 (2000).

203. L. C. Lev and A. S. Argon. Development of oxide coatings for matching oxide fiber-oxide matrix composites. *Ceram. Eng. Sci. Proc.*, 15, 743–752 (1994).
204. W. Y. Lee, E. Lara-Curzio, and K. L. More. Multilayered oxide interphase concept for ceramic-matrix composites. *J. Am. Ceram. Soc.*, 81, 717–720 (1998).
205. J. A. DiCarlo and M. V. Roode. Ceramic composite development for gas turbine engine hot section components, in *ASME Turbo Expo 2006: Power for Land, Sea, and Air* (Barcelona, Spain, 2006).
206. J. Lane. Oxide-based CMCs for combustion turbines. 31st International Cocoa Beach Conference & Exposition on Advanced Ceramics and Composites, Daytona Beach, FL (2007).
207. A. Szweda, S. Butner, J. Ruffoni, C. Bacalski, J. Lane, J. Morrison, G. A. R. M. Merrill, A. Fahme, N. Miriyala, and D. Leroux. Development and evaluation of hybrid oxide/oxide ceramic matrix composite combustor liners. in *Proceedings of the ASME Turbo Expo 2005*, Vol 1, 315–321.
208. G. B. Merrill and J. A. Morrison. Insulated Ceramic Matrix Composite and Method of Manufacturing. US Patent No. 7,179,524 (2007).
209. K. L. More, L. R. Walker, Y. Wang, E. Lara-Curzio, T. M. Brummett, M. van Roode, J. R. Price, A. Szweda, and G. Merrill. Microstructural and mechanical characterization of a hybrid oxide CMC combustor liner after 25,000-hour engine test, in *Proceedings of the ASME Turbo Expo 2009, Vol. 1*, 255–263
210. M. van Roode. Ceramic gas turbine development: need for a 10 year plan. *J. Eng. Gas Turbine Power*, 132, 011301–011308 (2010).
211. B. Jurf. Fabrication and test of insulated CMC exhaust pipe. Presented at the *ASM International Aeromat Conference*, (Dayton, OH, 2003).
212. J. A. Morrison and K. M. Krauth. Design and analysis of a CMC turbine blade tip seal for a land-based power turbine. *Ceram. Eng. Sci. Proc.*, 19, 249–256 (1998).
213. F. Collier, R. Thomas, C. Burley, C. Nickol, C.-M. Lee, and M. Tong. Environmentally responsible aviation - real solutions for environmental challenges facing aviation. 27th International Congress of the Aeronautical Sciences (2010).
214. J. Heidmann. Improving engine efficiency through core developments. The *AIAA Aero Sciences Meeting* Orlando, FL (2011).
215. Boeing. Continuous Lower Energy, Emissions and Noise (CLEEN) Technologies Development. The *CLEEN Consortium Meeting*, Atlanta, GA (2011).
216. M. Petervary and T. Steyer. Ceramic matrix composites for structural aerospace applications. *4th International Congress on Ceramics*, Chicago, IL (2012).
217. M. L. Hand, B. Chakrabarti, L. L. Lehman, and G. P. Mathur. Exhaust Washed Structure and Associated Composite Structure and Method of Fabrication. US Patent No. 8043690 (2011).
218. C. Epstein. GE's passport 20 engine program is on schedule for 2016 entry into service. *AIN online*, October 22 (2013), Web: October 29, 2013.
219. M. Gerendas, Y. Cadoret, C. Wilhelmi, T. Machry, R. Knoche, T. Behrendt, T. Aumeier, S. Denis, J. Goring, D. Koch, et al. Improvement of oxide/oxide CMC and development of combustor and turbine components in the HiPOC program. *ASME Conference Proceedings*, 2011, 477–490 (2011).
220. T. J. McMahon. Advanced hot gas filter development. *Ceram. Eng. Sci. Proc.*, 21, 47–56 (2000).
221. M. G. Holmquist, T. C. Radsick, O. H. Sudre, and F. F. Lange. Fabrication and testing of all-oxide CFCC tubes. *Comp. Part A*, 34, 163–170 (2003).
222. L. J. Korb, C. A. Morant, R. M. Calland, and C. S. Thatcher. The shuttle orbiter thermal protection system. *Am. Ceram. Soc. Bull.*, 60, 1188 (1981).
223. J. B. Davis, D. B. Marshall, K. S. Oka, R. M. Housley, and P. E. D. Morgan. Ceramic composites for thermal protection systems. *Compos. Part A*, 30, 483–488 (1999).
224. V. Heng, R. A. Dichiara, E. Chu, and D. Zorger. Thermal Insulation Assemblies and Methods for Fabricating the Same. US Patent No. 7887937 (2011).
225. J. A. Riedell. Method and System of Thermal Protection. US Patent No. 6827312 (2004).
226. H. J. Kim, G. Fair, H. Lee, K. Keller, T. A. Parthasarathy, and R. Hay. Processing and transparency of polycrystalline yttrium aluminum garnet (YAG) fibers for optical applications. *Proc. SPIE 7912*, Solid State Lasers XX: Technology and Devices (7912)1T (March 03, 2011).

CHAPTER 9

ULTRAHIGH TEMPERATURE CERAMIC-BASED COMPOSITES

YUTAKA KAGAWA[1,2] AND SHUQI GUO[2]

[1]*Research Center for Advanced Science and Technology, The University of Tokyo, Tokyo, Japan*
[2]*Hybrid Materials Unit, National Institute for Materials Science, Tsukuba, Ibaraki, Japan*

9.1 INTRODUCTION

Zirconium diborides (ZrB_2) and hafnium diborides (HfB_2) belong to the refractory transition metal diborides from the fourth to sixth group of the periodic table. Most of these diborides have melting point above 3000°C, high thermal and electrical conductivities, chemical inertness against molten metals, and good thermal shock resistance, making them potential candidates for a variety of high temperature structural applications [1–3]. However, the densification of ZrB_2 or HfB_2 powders generally requires very high temperature (>2000°C) and external pressure because of strong covalent bond and low self-diffusivity [4–6]. In addition, the use of those single-phase materials for high temperature structural applications is limited by their poor oxidation and ablation resistance, as well as by their poor damage tolerance.

The composite approach has been successfully adopted in order to improve the densification, mechanical properties, and the oxidation and/or ablation resistance of the single-phase ceramics. Recently, the ZrB_2- or HfB_2-based composites developed include three compositions as follows: (i) ZrB_2- or HfB_2-based composites with carbides and disilicide particles, typically, ZrB_2-SiC, HfB_2-SiC, ZrB_2-$MoSi_2$, and HfB_2-$MoSi_2$, (ii) short carbon or silicon carbide fiber-reinforced ZrB_2- or HfB_2-SiC matrix composites, and (iii) continuous carbon fiber or silicon carbide fiber-reinforced ZrB_2- or HfB_2-based matrix composites. Historically, the ZrB_2 or HfB_2-SiC composites have been the dominant ceramic-based composites in ultrahigh temperature ceramics. Later, ZrB_2- or HfB_2-$MoSi_2$, ZrB_2-$ZrSi_2$, and carbon or SiC fiber-reinforced ZrB_2-SiC matrix composites evolved as the common ultrahigh temperature ceramic materials. Although various kinds of ZrB_2- and HfB_2-based composites are available, poor fracture toughness is still an open problem.

The fabrication process and physical and mechanical properties of the ZrB_2- and HfB_2-based composites with SiC, $MoSi_2$, and $ZrSi_2$ particles are described here. The emphasis is directed toward presenting recent advances and providing an evaluation of studies for these ultrahigh temperature ceramic-based composites.

9.2 ULTRAHIGH TEMPERATURE CERAMIC-BASED COMPOSITES WITH PARTICULATES

9.2.1 Fabrication Methods

In studies that were conducted in the 1970s and earlier, densification of ZrB_2- or HfB_2-based composites with SiC was only accomplished by hot pressing (HP). Later, spark plasma sintering (SPS), reactive sintering (RS), and

Ceramic Matrix Composites: Materials, Modeling and Technology, First Edition. Edited by Narottam P. Bansal and Jacques Lamon.

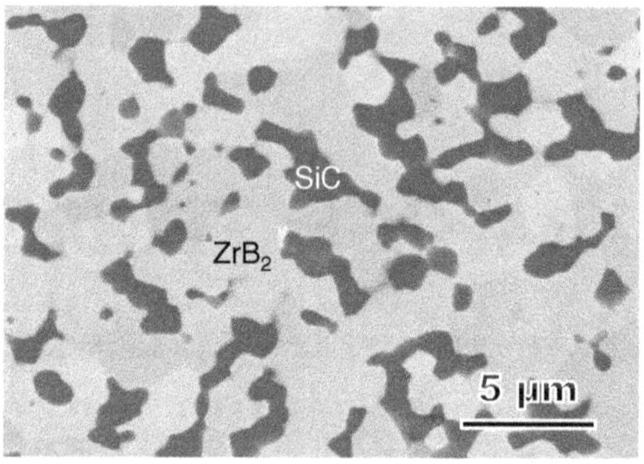

FIGURE 9.1

An example of backscattered electron FE-SEM images of the hot-pressed ZrB_2-20 vol.% SiC composite at 2000°C and 20 MPa for 60 minutes in Ar.

pressureless sintering (PS) processes evolved as the most common densification methods. Recently, the ZrB_2- or HfB_2-based composites with SiC, $MoSi_2$, and $ZrSi_2$ have been fabricated by various methods, including HP, SPS, RS, and PS processes.

9.2.1.1 Hot Pressing Typically, HP of single-phase ZrB_2 or HfB_2 required a temperature of 2100°C or above and moderate pressure (20–30 MPa) [4–6] or lower temperatures (~1800°C) and extremely high pressures (>800 MPa) [7, 8]. For comparison, highly dense ZrB_2- or HfB_2-based composites with SiC may be fabricated at lower temperatures and lower pressures [6, 9–14], as a result of producing an intergranular amorphous phase during HP sintering. The formation of the intergranular amorphous phases during the sintering was traced back to the interactions among the SiC and oxides such as SiO_2, B_2O_3, and ZrO_2 or HfO_2. Figure 9.1 showed a typical SEM micrograph of the microstructure of ZrB_2-based composites with SiC particles [6]. The microstructure of the ZrB_2-based composites with SiC particles consists of the equiaxed ZrB_2 (gray contrast) and SiC (dark contrast) grains. The SiC particles are randomly dispersed in the ZrB_2 matrix.

Results of HP studies on the ZrB_2-based composites with SiC are summarized in Table 9.1 [6, 10, 12], which includes details of the starting powders, HP conditions, mixing methods, and final densities. It has been found that highly dense ZrB_2-based composites with SiC could be fabricated by HP at 1650°C or above. The addition of SiC improved the sinterability, inhibited grain growth, and increased the oxidation and ablation resistance of single-phase ZrB_2 ceramic as well [6, 9–14]. For comparison, HP of single-phase ZrB_2 powder at 2000°C with a pressure of 20 MPa achieved only a relative density of ~91%. The improvement in densification due to addition of SiC is linked with the amount

TABLE 9.1 Starting Powder Size, Sintering Conditions and Final Densities of the Hot-Pressed ZrB_2-Based Composites with SiC

Compositions (vol.%)	Particles size of SiC (μm)	Remarks	Hot press conditions	Relative density (%)	References
ZrB_2		Ball-milled	2000°C/60 min/30 MPa/Ar	90.2	[6]
ZrB_2-5.7 SiC	1.7	Ball-milled	1650°C/120 min/60 MPa/vacuum	81.7	[12]
ZrB_2-11.4 SiC	1.7	Ball-milled	1650°C/120 min/60 MPa/vacuum	91.5	[12]
ZrB_2-22.4 SiC	1.7	Ball-milled	1650°C/120 min/60 MPa/vacuum	97.6	[12]
ZrB_2-22.4 SiC	0.6	Ball-milled	1650°C/120 min/60 MPa/vacuum	98.6	[12]
ZrB_2-22.4 SiC	0.04	Ball-milled	1650°C/120 min/60 MPa/vacuum	99.6	[12]
ZrB_2-5 SiC	0.04	Ball-milled	2000°C/60 min/30 MPa/Ar	95.6	[6]
ZrB_2-10 SiC	0.04	Ball-milled	2000°C/60 min/30 MPa/Ar	97.4	[6]
ZrB_2-20 SiC	0.04	Ball-milled	2000°C/60 min/30 MPa/Ar	97.3	[6]
ZrB_2-10 SiC	0.7	Attrition-milled	1950°C/45 min/30 MPa/Ar	93.2	[10]
ZrB_2-20 SiC	0.7	Attrition-milled	1950°C/45 min/30 MPa/A	99.7	[10]

and size of SiC particles added. Increasing the amount of SiC added improved densification of the ZrB_2-SiC powder mixtures. Early studies in the ZrB_2-SiC composites showed that exceeding 10 vol.% SiC is required for obtaining highly dense composites compacts. In addition, the addition of 22.4 vol.% nanosized SiC, with average particles sizes ranging from ~40 nm to 0.6 μm, sharply reduced the HP temperature necessary to achieve full density to 1650°C (pressed for 120 minutes at 60 MPa) [12]. However, the improvement of densification upon addition of ultrafine SiC is effective only for a uniformly dispersed ZrB_2-SiC system [6]. The agglomeration of the ultrafine SiC particles led to reduced improvement in densification of ZrB_2, even with nanosized SiC particles [6]. Furthermore, the attrition-milled ZrB_2-SiC powder mixture required HP at 1900°C and 32 MPa for 45 minutes to achieve full density [10], whereas the ball-milled ZrB_2-SiC powder mixture required HP at 2000°C and 30 MPa for 60 minutes. The lower HP temperature was attributed to reduction of starting particle size by attrition-milling. Recently, Si_3N_4 [15,16], AlN [17,18], HfN, or ZrN [19,20], have been added to ZrB_2- or HfB_2-SiC powder mixtures producing an intergranular secondary phase and/or reducing oxygen content, both of which assists in the densification of composites. It has been reported that HP of 3 vol.% HfN-HfB_2-SiC [19] or 4.5 vol.% ZrN-ZrB_2-SiC [20] required 1900°C for 30 minutes at 40 MPa. The resulting composites showed a fine and homogenous microstructure with secondary phases such as M(C, N), MO_2 (M = Zr and/or Hf) and BN.

9.2.1.2 Spark Plasma Sintering SPS is one of the most recent advanced processing techniques developed for densifying ceramic materials. Recent studies have shown that SPS enhanced densification and refined microstructure of ZrB_2-based composites can be achieved in very short processing cycles [21–23]. Figure 9.2 shows a schematic of the SPS apparatus. Although SPS is similar to HP, in place of indirect heating, the applied electrical field heats the die and the powder compact. Hence, SPS process has two significant advantages, compared with HP process. One advantage of using SPS is to enhance densification of poorly sinterable ceramics, by simultaneously applying a uniaxial load and a direct or pulsed electric current to a powder compact. Another advantage is that the grain growth of starting materials is restricted, since a considerable shorter sintering time (within minutes) is needed, thereby retaining the fine and homogeneous grains. Figure 9.3 shows typical SEM micrograph of microstructure of ZrB_2-ZrC-SiC composites consolidated by SPS at 1950°C and 50 MPa for 2 minutes in argon atmosphere [22]. The microstructure of the composites consists of the equiaxed ZrB_2 (gray contrast), ZrC (bright contrast), and SiC (dark contrast) grains. It is found that ZrB_2, ZrC, and SiC particles are very fine and homogeneously present in the composite.

Table 9.2 summarizes SPS conditions, final density, and grain size of ZrB_2- or HfB_2-based ceramic composites with SiC and $MoSi_2$ produced by SPS [21, 23–26]. The ZrB_2- or HfB_2-based composites could be sintered to a relative density of >97% by SPS technique within very short sintering time (typical soaking time: ~5 minutes). Compared with HP (Table 9.1), although the temperature used during SPS is not significantly lower and/or is the same for some cases, soaking time and total sintering time were significantly shorter for SPS than for HP. In addition, grain size measurement indicated that the grain growth was inhibited during SPS, as a result of very short soaking time. Furthermore, additions

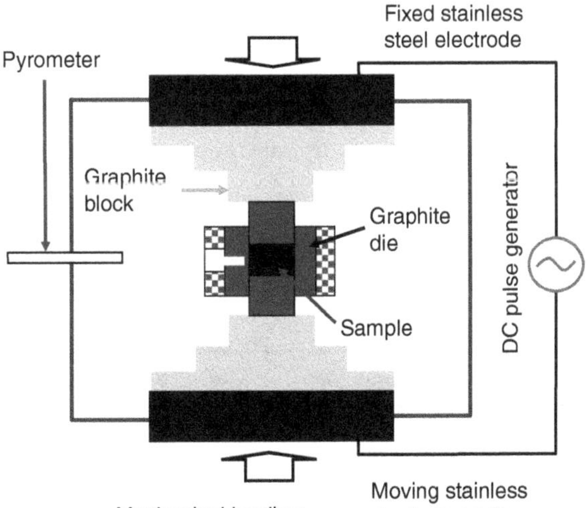

FIGURE 9.2

A schematic of the spark plasma sintering apparatus.

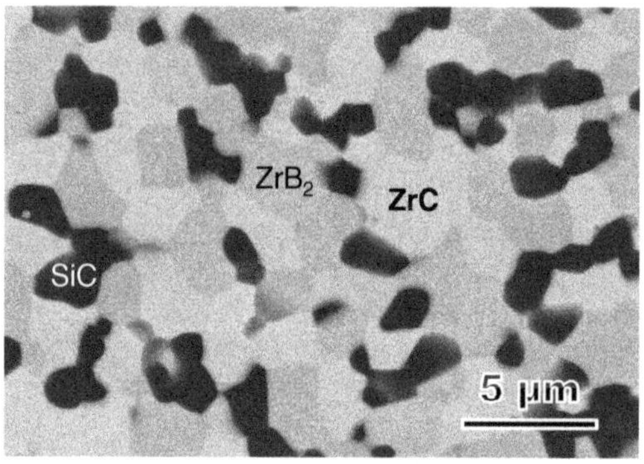

FIGURE 9.3
An example of backscattered electron FE-SEM micrograph of the ZrB_2-ZrC-SiC composites consolidated by SPS at 1950°C and 50 MPa for 2 minutes in Ar.

of AlN, HfN, ZrN, Si_3N_4, etc. could result in a lower sintering temperature and a faster shrinking rate [18, 19, 21], whose effects are similar to those during HP. The enhanced densification resulting from SPS was attributable to mass transfer processes, which are significantly enhanced in the process, effectively promoting densification. The mechanism in SPS technique that enhanced densification—mainly whether or how an electric discharge is involved in accelerating the densification and grain growth—is still the subject of intense debate. However, we suggest that enhancement is most probably due to (i) an efficient heat transfer; (ii) the use of comparatively high pressure; (iii) the presence of an electrical field (use of DC pulses); and (iv) the presence of local spark discharges generated between the powders under high energy electrical pulses.

9.2.1.3 Reactive Sintering RS has been identified as a potential route to produce ZrB_2- or HfB_2-based composites with low impurity levels and high density at a relative lower temperature. According to heating way of green body of powder, the RS can be divided into two main methods: (i) reactive hot pressing (RHP), and (ii) reactive spark plasma sintering (RSPS). In an earlier study, Zhang et al. [27] synthesized the highly dense ZrB_2-based composites with SiC (RD: ~98%) by RHP of Zr, Si, and B_4C powder mixtures at 1900°C and 30 MPa for 60 minutes. Later, Wu et al. [28] successfully consolidated ZrB_2-SiC-ZrC composites (RD: ~97%) by RHP of Zr, Si, and B_4C powders at 1800°C and 20 MPa for 60 minutes in an argon atmosphere. The highly dense ZrB_2-SiC-ZrC composites were synthesized by Qu et al. [29] at a similar HP condition using Zr, Si, and B_4C powder mixtures. Chamberlain et al. [30] fabricated a highly dense single-phase ZrB_2 at 1650°C and 40 MPa for 30 minutes in Ar by RHP of Zr and B powder mixture. In addition, RHP has been used to produce the HfB_2-based composites with SiC and/or ZrC by using Hf, Si, and B_4C precursors as well [31]. Recently, Wu et al. [32] synthesized highly dense ZrB_2–20 vol.% $MoSi_2$ composites by RHP of Zr, Mo, and Si powder mixtures at 1800°C and 20 MPa for 30 minutes. Very recently, authors [33] had fabricated a tough and strong platelet-like ZrB_2-ZrC_x hybrid ceramic bonded by a thin zirconium layer by HP of B_4C and Zr powder mixture at 1900°C using a liquid-phase reaction sintering technique. The microstructure of the as-synthesized hybrid ceramics consists of platelet-like ZrB_2 grains (dark contrast) and equiaxed ZrC_x grains (bright gray contrast) at the grain boundaries [33], as shown in Figure 9.4.

TABLE 9.2 Sintering Conditions, Final Density, and Grain Size of ZrB_2- or HfB_2-Based Composites Produced by SPS Process

Compositions (vol.%)	Spark plasma sintering conditions	Relative density (%)	Grain size (μm)	References
ZrB_2	1900°C/3 min/50 MPa/vacuum	97.6	5.1	[23]
ZrB_2-(10–30) SiC	1900°C/2 min/30 MPa/Ar	99.7	2	[24]
HfB_2-30 SiC	2100°C/2 min/30 MPa/vacuum	100	2	[21]
ZrB_2-15 $MoSi_2$	1750°C/2 min/30 MPa/vacuum	97.7	1.4	[21]
HfB_2-(1–9) $MoSi_2$	1750°C/3 min/100 MPa/vacuum	>97	1.0	[25]
HfB_2-(5–30) SiC-5 $TaSi_2$	1800°C/5 min/30 MPa/vacuum	>98.2		[26]

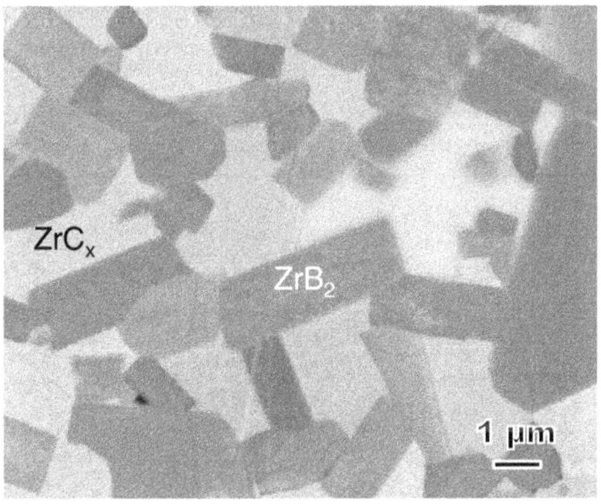

FIGURE 9.4
An example of backscattered electron FE-SEM images of the reactive sintered hybrid ceramic materials, consisting of platelet-like ZrB_2 grains (dark contrast), equiaxed ZrC_x grains (bright gray contrast), and Zr phase (white contrast) at the grain boundaries.

Similarly, the RS of Zr, Si, and B_4C precursors could be conducted by SPS technique. Zhao et al. [34] showed that the RS of Zr, Si, and B_4C precursors could produce the densified ZrB_2-based composites with SiC by SPS at or below 1450°C with a shorter soaking time of 3 minutes. The resulting composites had a finer and more homogeneous microstructure, compared with that of RHP. It is evident that the RS could produce the densified ZrB_2-based composites at a lower temperature by using either HP or SPS, as compared with direct consolidation of commercially available powders. Recently, authors synthesized highly dense single-phase ZrB_2 by using RSPS of ZrH_2-B powder mixture at 1800°C and 50 MPa for 10 minutes in vacuum [35]. For comparison, SPS of commercially available ZrB_2 powder required 1900°C and 3 minutes at 50 MPa for achieving a highly dense compact (RD: >97%) [23]. In addition, HfB_2 ceramic was synthesized by RSPS of a mixture of Hf and B at 1700°C and 95 MPa for 10 minutes in vacuum [36]. RS sintering conditions, precursors, and final density of the ZrB_2- or HfB_2-based composites synthesized by RHP or RSPS technique are summarized in Table 9.3.

9.2.1.4 Pressureless Sintering

Because of the extreme pressures required for densification, PS of ZrB_2- or HfB_2-based composites was considered unlikely or impossible until the late 1980s, when studies of PS actually began to show results. Compared with HP, the development of a PS process would enable almost-net-shaped processing of ceramic parts with complex geometries using standard powder-processing methods, thus reducing processing costs. In PS technique, several additives, typically B_4C, C, WC, $MoSi_2$, and $ZrSi_2$ [37–45], must be used to improve densification of ZrB_2- or HfB_2-based composites. According to the function of additives during densification, the additives used can

TABLE 9.3 Reactive Sintering Conditions, Precursors, and Final Density of ZrB_2- or HfB_2-Based Composites Fabricated by RS Method

Materials	Precursors	Reactive sintering conditions	Relative density (%)	References
ZrB_2	Zr, B	1650°C/30 min/40 MPa/Ar (HP)	>95	[30]
ZrB_2	ZrH_2, B	1800°C/10 min/50 MPa/vacuum (SPS)	97	[35]
HfB_2	Hf, B	1700°C/10 min/95 MPa/vacuum (SPS)	98	[36]
ZrB_2-SiC	Zr, Si, B_4C	1900°C/60 min/30 MPa/Ar (HP)	96.7	[27]
ZrB_2-SiC-ZrC	Zr, Si, B_4C	1800°C/60 min/20 MPa/Ar (HP)	96.8	[28]
ZrB_2-SiC	Zr, Si, B_4C	1450°C/3 min/30 MPa/vacuum (SPS)	98.5	[34]
HfB_2-SiC	Hf, Si, B_4C	1900°C/60 min/50 MPa/vacuum (HP)	100	[31]
ZrB_2-$MoSi_2$	Zr, Mo, Si	1800°C/30 min/30 MPa/vacuum (HP)	99.2	[32]
ZrB_2-SiC-ZrC	Zr, Si, B_4C	1800°C/60 min/30 MPa/Ar (HP)	97	[29]

TABLE 9.4 Pressureless Sintering Conditions, Grain Size, and Final Densities of the Pressureless Sintered ZrB_2- or HfB_2-Based Composites

Compositions	Pressureless sintering conditions	Relative density (%)	Grain size (μm)	References
ZrB_2-(10–40) vol.% $ZrSi_2$	1650°C/60 min/Ar	>95		[37]
ZrB_2-20 vol.% $MoSi_2$	1850°C/30 min/Ar	99.1	2–3	[38]
ZrB_2-4 wt.% B_4C	1850°C/60 min/vacuum	>98	12	[39]
ZrB_2-1.7 wt.% C	1900°C/120 min/Ar	>99	14	[40]
ZrB_2-4 wt.% WC	2050°C/240 min/vacuum	95		[39]
ZrB_2-4 wt.% B_4C-0.5 wt.% C	1800°C/120 min/Ar	>99	<4	[41]
HfB_2-2 wt.% B_4C	2200°C/120 min/vacuum	94.5	22.3	[44]
HfB_2-20 vol.% SiC-(5, 10) wt.% WC	2200°C/120 min/vacuum	>98	6.9, 7.4	[45]
HfB_2-(5–20) $MoSi_2$	1950°C/60 min/Ar	>97	1.5	[42, 43]

be divided into two main groups: (i) liquid phase formers, for example, $MoSi_2$ and $ZrSi_2$ [37, 38, 42, 43], and (ii) solid reactive agents, for example, B_4C, WC [39–41, 44, 45]. During the sintering, the added liquid phase formers reacted with oxides presented on the surface of the starting powders for producing an intergranular amorphous phase, therefore densification. Because formation and continuous action of a liquid phase occurred only at higher contents of added liquid phase formers, the critical addition was required to bring about adequate sintering. Differing from the above mentioned liquid phase formers, reactive agents act as densification aids by reacting with the oxide impurities present on the surface of starting particles, for example, ZrO_2, B_2O_3, which inhibit densification. Table 9.4 summarizes the PS conditions, agents used, grain size, and final density of the resulting ZrB_2- or HfB_2-based composites [37–45]. It is found that PS required a higher sintering temperature for producing highly dense composite compacts, compared to other sintering methods, such as HP, RS, and SPS, resulting in the composites with a larger grain size.

9.2.2 Physical Properties

ZrB_2- or HfB_2-based composites have high strength, high stiffness, and good thermal and electrical conductivities among the transition metal carbides, nitrides, and diborides. High strength and stiffness could be considered for a variety of high temperature, thermomechanical, and structural applications. High thermal conductivity could improve thermal shock resistance by reducing temperature gradients and thermal stress within the materials. In addition, high electrical conductivity allows the fabrication of complex shapes using electrical discharge machining. This paragraph describes the physical properties of these composites, including elastic properties, thermal, and electrical conductivities.

9.2.2.1 Elastic Properties The shear modulus, Young's modulus, and Poisson's ratio is equal to ~220 GPa, ~490 GPa, and ~0.12, respectively, for a fully densified single-phase ZrB_2 ceramic [46]. Similarly, the fully densified single-phase HfB_2 ceramic exhibits almost the same elastic properties, and its shear modulus, Young's modulus, and Poisson's ratio is equal to ~230 GPa, ~500 GPa, and ~0.11 [46], respectively. These shear moduli and Young's moduli are strongly dependent on temperature and they linearly decrease with increasing temperature [46]. For comparison, their Poisson's ratio increases linearly with increasing temperature [46]. In addition, the early studies on the elastic properties showed that the various additives used during densification affect the elastic properties of these ceramic materials. Typically, B_4C alone or a combination of B_4C and C also led to an increase in Young's modulus [41, 47]. In contrast, AlN and Si_3N_4 additions led to a lower Young's modulus [16. 17]. The different changes with the additives are associated with the different grain-boundary phase developed between ZrB_2 grains [16, 17, 41], which results from the interactions of the additives with impurities on the ZrB_2 particle surfaces, because the grain-boundary phase affects Young's modulus of ceramics.

On the other hand, the elastic properties of ZrB_2- and HfB_2-based composites differed with those of single-phase ceramic materials, as a result of the presence of a second phase. The elastic properties of the composites depended on the reinforcement phase added as well as on the amount of additives. Figure 9.5 is the plots of shear modulus and Young's modulus as a function of the volume fraction of SiC in ZrB_2- or ZrB- 40 vol.% $MoSi_2$-based composites with SiC. The shear modulus and Young's modulus of two composites decreased with volume fraction of SiC added, as a result of lower moduli of SiC phase than ZrB_2 phase [48]. Figure 9.6 is the plots of shear modulus and Young's modulus

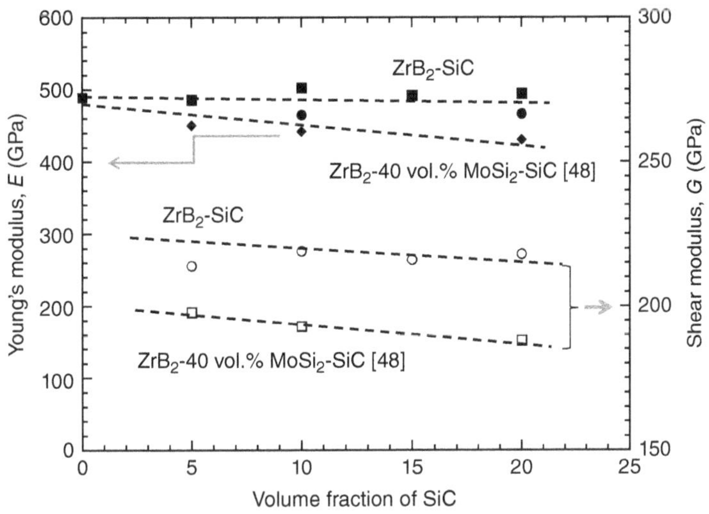

FIGURE 9.5

Plots of shear modulus and Young's modulus as a function of volume fraction of SiC for hot-pressed ZrB_2- and ZrB_2-40 vol.% $MoSi_2$-based composites with SiC.

as a function of the volume fraction of $MoSi_2$ or $ZrSi_2$ for the ZrB-based composites. Similarly, the shear modulus and Young's modulus of both the composites decrease with increase in volume fraction of $MoSi_2$ and $ZrSi_2$ [48, 49]. One exception is the lower shear modulus and Young's modulus for the ZrB_2-10 vol.% $ZrSi_2$ composite than for other composition materials, as a result of the presence of pores (RD: 96.6%) [18]. In contrast, Poisson's ratio is insensitive to additives as well as to porosity, and it almost remains constant for the various ZrB_2-based composites [18, 48, 49].

9.2.2.2 ***Thermal Conductivity*** It is known that the ZrB_2- or HfB_2-based composites are good thermal conductors and they have high thermal conductivity, with the thermal conductivity range between 50 and 150 W/(m·K), depending on the thermal conductivity of the components and the interfacial thermal resistance between the components. Recently, a systematic study in the thermal conductivity of the various ZrB_2-ZrC-SiC composites consolidated by SPS

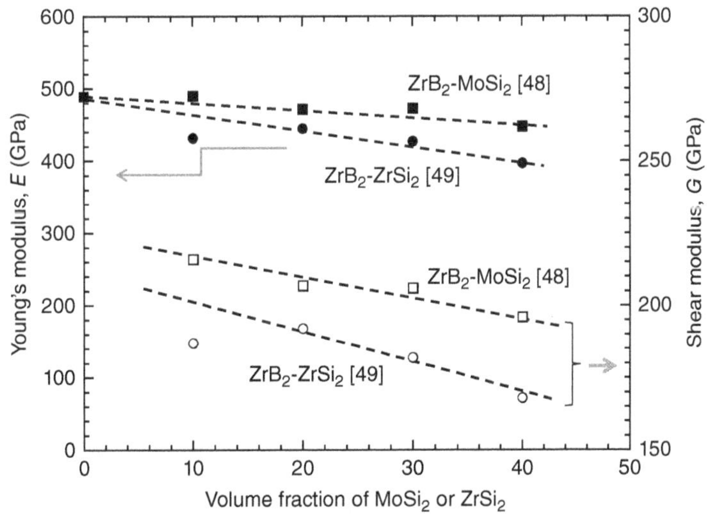

FIGURE 9.6

Plots of shear modulus and Young's modulus as a function of volume fraction of $MoSi_2$ or $ZrSi_2$ for the hot-pressed ZrB_2-based composites with $MoSi_2$ or $ZrSi_2$.

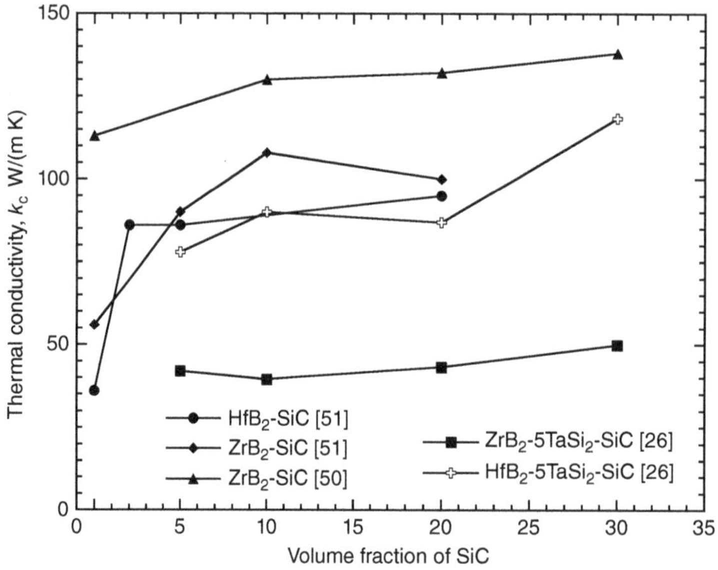

FIGURE 9.7

Plots of thermal conductivity as a function of volume fraction of SiC for the ZrB_2- and HfB_2-based composites with SiC.

showed that the conductivity was in the range of 38–93 W/(m·K), having a strong compositional dependence [22]. The thermal conductivity of the ZrB_2-ZrC-SiC composites increased with increasing ZrB_2 as well as SiC content, whereas the conductivities decreased with increasing ZrC content. Figure 9.7 shows plots of thermal conductivity as a function of volume fraction of SiC for both ZrB_2- or HfB_2-based composites with SiC particles [26,50,51]. In the ZrB_2- or HfB_2-based composites with SiC, it is found that the addition of SiC, whose thermal conductivity is higher than that of ZrB_2 phase, led to an increased thermal conductivity, as a result of decreased resistance for the heat flow through the components and their interfaces. In particular, in the case of the composition with higher concentration of SiC, the higher conductivity phase could form a network-like structure that improved heat capacity and enhanced heat transport as well. Hence, for these composites, although their starting powders and densification route are different, the thermal conductivity increases with increasing volume fraction of SiC. In addition, the addition of SiC with larger size led to a higher thermal conductivity, such as shown in Figure 9.7, the ZrB_2-based composite with ~9 μm SiC starting powder particles [50] appeared to have a higher thermal conductivity than ~1.0 μm SiC particles containing ZrB_2 [51]. However, when several additives with lower thermal conductivity are incorporated into the composites to improve their densification, the resulting composites generally show a lower thermal conductivity. For example, in order to improve densification of ZrB_2- or HfB_2-based composites with SiC particles at lower temperature, 5 vol.% $TaSi_2$ was added to ZrB_2- or HfB_2-SiC composites [26]. Both the resulting composites showed a lower thermal conductivity, compared with $TaSi_2$-free ZrB_2- or HfB_2-based composites with SiC particles [26,50,51]. This decrease in the thermal conductivity probably results from the high thermal transport resistance of the second phase as well as from a higher intergranular thermal resistance.

The compositional dependence of thermal conductivity is also observed in ZrB_2-$ZrSi_2$ and ZrB_2-$MoSi_2$-SiC composite materials, as shown in Figure 9.8 [37, 52]. It is found that the increase in $MoSi_2$ or $ZrSi_2$ led to a decrease in thermal conductivity of the composites. In addition, the thermal conductivity is higher for ZrB_2-$ZrSi_2$ composites than for ZrB_2-$MoSi_2$ composites. However, when an SiC particle with higher thermal conductivity was added to ZrB_2-$MoSi_2$ composite, the thermal conductivity of the composites increased significantly. This improvement was enhanced with increasing SiC amounts, resulting from the enhanced thermal transport in the ZrB_2-$MoSi_2$-SiC materials, but the content of added SiC must exceed 5 vol.%. In addition, for the ZrB_2- or HfB_2-based composites with SiC particles, it was found that the thermal conductivity showed strong temperature dependence, in the range of ~400 to ~1700°C [51,53]. Furthermore, the thermal conductivity of the composites decreased with temperature, regardless of SiC content.

9.2.2.3 Electrical Conductivity The ZrB_2- or HfB_2-based composites are a good electrical conductor, which exhibited metallic-like electrical conductivity. The measurements of electrical resistivity for the ZrB_2- or HfB_2-based

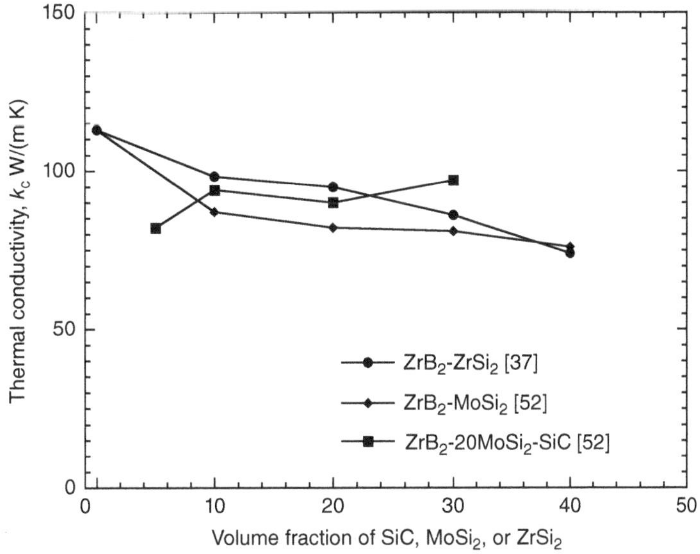

FIGURE 9.8

Plots of thermal conductivity as a function of volume fraction of SiC, $MoSi_2$, or $ZrSi_2$ for the ZrB_2-based composites with SiC, $MoSi_2$, or $ZrSi_2$.

composites showed that the electrical resistivity depended on the composition of the composites [26, 37, 52, 54], as shown in Figure 9.9. Among the ZrB_2-SiC, ZrB_2-$MoSi_2$, and ZrB_2-$ZrSi_2$ composites, it is found that the electrical resistivity of the three compositions has the order as follows: ZrB_2-SiC > ZrB_2-$MoSi_2$ > ZrB_2-$ZrSi_2$. In addition, the electrical resistivity of the ZrB_2-$ZrSi_2$ composites [37] was almost constant when the amount of $ZrSi_2$ was equal to or less than 30 vol.%; however, the resistivity slightly increased at 40 vol.% $ZrSi_2$. For the ZrB_2-$MoSi_2$ composite [52], on the other hand, the electrical resistivity increased as the amount of $MoSi_2$ added increased from 10% to 20 vol.%, subsequently the resistivity almost remains constant with further increase in $MoSi_2$. Furthermore, the additions of $\geq$10 vol.% SiC led to increase in the electrical resistivity of the ZrB_2-$MoSi_2$ composites further [52]. However, an addition

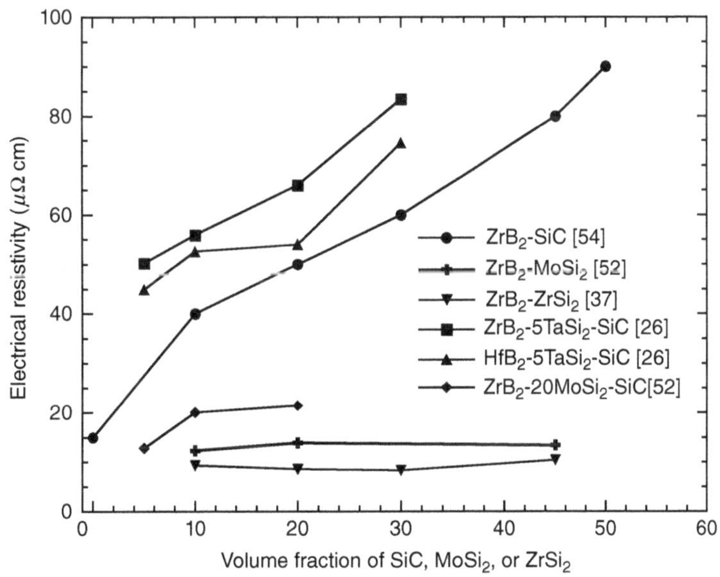

FIGURE 9.9

Plots of electrical resistivity as a function of volume fraction of SiC, $MoSi_2$, or $ZrSi_2$ for the ZrB_2- and HfB_2-based composites with SiC, $MoSi_2$, or $ZrSi_2$.

of 5 vol.% SiC was ineffective for increasing the electrical resistivity. Although the additions of $MoSi_2$ and SiC decreased electrical conductivity of ZrB_2 materials, the electrical resistivity of the ZrB_2-$MoSi_2$-SiC composites are in the range characteristic of conductors [52].

On the other hand, for ZrB_2-SiC composites without additives [54, 55], a systemic study on the effect of the amount of SiC added on the electrical conductivity showed that the electrical resistivity of the composites sharply increased with SiC content over 70 vol.%. This means the presence of an electrical percolation threshold in the ZrB_2-SiC system. Presumably, below the critical amount of SiC, the network-like structure formed by the ZrB_2 phase with high electrical conductivity provides an electrical path with lower resistance, thereby retaining high electrical conductivity. In addition, when ZrC was incorporated in the ZrB_2-SiC system [22], the electrical resistivity of the resulting ZrC-ZrB_2-SiC composites increased with increasing ZrC and/or SiC content, with an electrical resistivity range of 22–10.9 ($\mu\Omega$ cm). Besides, for ZrB_2-SiC composites with 5 vol.% $TaSi_2$ addition [26], the electrical resistivity is higher than that of ZrB_2-SiC composites without additives. Similarly, for HfB_2-SiC composites with 5 vol.% $TaSi_2$ addition [26], the electrical resistivity increased with increasing amount of SiC added. Furthermore, the electrical resistivity is lower for the HfB_2-SiC composites with 5 vol.% $TaSi_2$ than for the ZrB_2-SiC composites with 5 vol.% $TaSi_2$.

9.2.3 Mechanical Properties

9.2.3.1 Fracture Toughness ZrB_2- or HfB_2-based ceramic composites showed intrinsic brittleness, which is similar to single-phase ceramic materials. Table 9.5 summarized fracture toughness measured in several ZrB_2- or HfB_2-based composites with SiC particles. It is found that fracture toughness of both composites was in the range ~3 to ~5 MPa·m$^{1/2}$ [6,21,45,51,56,57]. For the composites with low fracture toughness, intragranular cracking is common with the cracks propagated across grains without being deflected along the grain boundaries. For the composites with high fracture toughness, on the other hand, intergranular cracking is partially present. Early study in ZrB_2-based composites with SiC showed that fracture toughness is dominated by the SiC particle size and distribution in the composites [14]. Increased fracture toughness produced by SiC addition is attributed to the crack deflection that occurs near the SiC particles and/or at ZrB_2/SiC interfaces. The contribution of crack deflection to increasing fracture toughness depends on the total number of crack deflections and the crack deflection angle, that is, crack propagation path. The increase in fracture toughness that accompanies crack deflection is also associated with the elastic and/or frictional bridging mechanism of grains. Elastic bridging is enhanced in larger diameter grains, while the frictional bridging mechanism is in operation only when crack deflection and grain pullout occur, which is prevalent in smaller diameter grains. In addition, fracture toughness of ZrB_2-SiC composites increased with the ratio of ZrB_2 to SiC grain size. This increase in fracture toughness is attributed to two major causes: one cause is large ZrB_2 grains enhanced elastic and/or frictional bridging contribution to increasing toughness. The other is small SiC grains increased the number of crack deflections and pullout grains. Recently, the ZrB_2-ZrC-SiC system reached fracture toughness of ~6 MPa m$^{1/2}$ by optimizing the combination of composition with microstructure [22]. Furthermore, very recently, authors had synthesized the platelet-like ZrB_2-ZrC_x-Zr hybrid ceramic materials [33], whose fracture toughness is in the range ~6.6 to ~11 MPa m$^{1/2}$ and increased by approximately twofold or more, compared to the existing ZrB_2-based composites. The observations of crack propagation path showed that the crack deflection, microcracking and crack arrest within the grains are main toughening mechanism [33]. Figure 9.10 showed an example of observed crack deflection and bridging of platelet-like ZrB_2 grains for ZrB_2-ZrC_x composite.

TABLE 9.5 Fracture Toughness and 4-Point Flexural Strength of ZrB_2- and HfB_2-Based Composites with SiC.

Compositions (vol.%)	Densification method	Fracture toughness (MPa·m$^{1/2}$)	Flexural strength (MPa)	References
ZrB_2-20SiC	HP	3.5	531 ± 10	[6]
ZrB_2-30SiC	HP	5.5	1063 ± 91	[14]
ZrB_2-30ZrC-10SiC	SPS	3.5	723 ± 136	[21]
ZrB_2-20SiC	HP	3.4	460 ± 53	[56]
HfB_2-19SiC	HP	4.3	560 ± 100	[57]
HfB_2-30SiC	SPS	3.9	590 ± 50	[21]
HfB_2-20SiC-5WC	PS	3.4	547 ± 58	[45]
HfB_2-20SiC-10WC	PS	4.8	563 ± 46	[45]

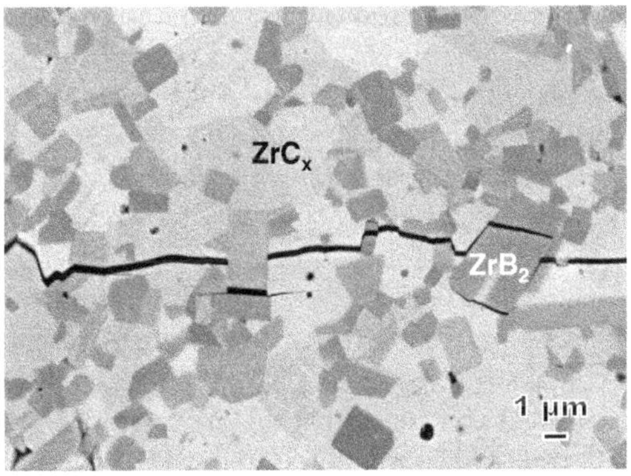

FIGURE 9.10
An example of backscattered electron SEM micrographs of the crack propagation behavior for the reactive sintered hybrid ceramic materials.

Figure 9.11 shows the effect of volume fraction of SiC, or $MoSi_2$ or $ZrSi_2$ on fracture toughness of the ZrB_2- or HfB_2-based composites with SiC, $MoSi_2$, and $ZrSi_2$. Although the ZrB_2-$ZrSi_2$ composite showed slightly improved fracture toughness due to addition of $ZrSi_2$ among the ZrB_2- or HfB_2-based composites, the significant improvement was not observed. In addition, the fracture toughness of these composites showed a very weak dependence on the amount of additive, that is, fracture toughness was not substantially increased with increasing the amount of additive. This indicated that the crack deflection was still the most toughening mechanism for these composites and its contribution to increase fracture toughness of the composites is almost the same for these composites, regardless of kinds of reinforcement phase and the amount the reinforcement phase added. Recently, Wu et al. had synthesized a ZrB_2-20 vol.% $MoSi_2$ composite containing elongated ZrB_2 grains via reactive HP of Zr, Si, B, and Mo mixture at 1800°C and 20 MPa for 30 minutes in argon atmosphere [32]. The as-synthesized ZrB_2-20 vol.% $MoSi_2$ composite shows the improved fracture toughness, with an average toughness value of 4.9 MPa $m^{1/2}$. Obviously, the improvement in fracture toughness is attributed to the presence of the elongated ZrB_2 grains in the composite. Hence, the fracture toughness of the composites could be significantly improved by designing as well as optimizing the microstructure.

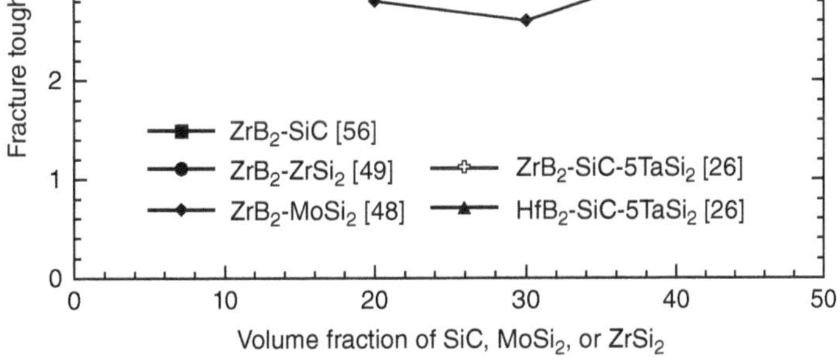

FIGURE 9.11
Plots of fracture toughness as a function of volume fraction of SiC, $MoSi_2$, or $ZrSi_2$ for the ZrB_2- and HfB_2-based composites with SiC, $MoSi_2$, or $ZrSi_2$.

9.2.3.2 Flexural Strength Room-temperature four-point flexural strengths of the ZrB_2- and HfB_2-based composites with SiC particles are also summarized in Table 9.5. It was found that the flexural strengths are almost the same for the ZrB_2- and HfB_2-based composites with SiC, with the strength range from ~450 to ~560 MPa. One exception was the significant high flexural strength for the ZrB_2-30 vol.% SiC composite consolidated by HP of attrition-milled powder at 1850°C and 32 MPa for 45 minutes, whose strength is reached over 1 GPa [14]. The significant increase in strength was attributed to a decrease in grain size due to attrition milling. Similarly, Chamberlain et al. [10] fabricated highly dense ZrB_2-based composites with SiC by HP attrition-milled ZrB_2-SiC powder mixture at 1900°C and 32 MPa for 45 minutes in Ar. The flexural strength of ~1 GPa was reached in the resulting ZrB_2-based composites containing 20 or 30 vol.% SiC. Furthermore, the effects of microstructure and SiC grain size on room-temperature flexural strength were examined in HP 30 vol.% SiC-containing ZrB_2 ceramics by Zhu et al. [13] and Rezaie et al. [14], based on the linear elastic fracture mechanics. Both studies concluded that the maximum SiC grain size in the ZrB_2-SiC composites is the strength-limiting factor, and that the strength is not correlated with average ZrB_2 grain size. Recently, the author's study in ZrB_2-20 vol.% SiC composites showed that the addition of nanosized SiC improved the flexural strength of the ZrB_2-SiC composites before (~610 MPa) and after oxidation (~700 MPa) for 10 hours at 1400°C in air [6]. The increase in strength after oxidation was attributed to the presence of a thin and dense oxide layer on the surface of the post-oxidized samples, which can heal the surface flaws without creating new cracks and defects at the oxidized surface. For comparison, the strength of the ZrB_2-based composite with 20 vol.% micron-sized SiC decreased from ~530 MPa before oxidation to ~500 MPa after oxidation. The loss of strength due to oxidation exposure at high temperature is attributed to the formation of a thicker glassy layer on the surface after oxidation and to the presence of a defect at the interface between the oxide layer and bulk ZrB_2-SiC composites.

Figure 9.12 shows a plot of room-temperature 4-point flexural strength as a function of amount of additives for the HP ZrB_2-based composites with SiC, $MoSi_2$, and $ZrSi_2$. It is found that the flexural strength of the ZrB_2-SiC composites increased with increasing amount of SiC added, with a strength range of ~400 and 500 MPa [56]. Similarly, for the ZrB_2-$ZrSi_2$ composites the flexural strength increases with increase in $ZrSi_2$ up to 30 vol.%, and these strengths are greater than those of the ZrB_2-SiC composites [49]. However, the addition of 40 vol.% $ZrSi_2$ led to a significant degradation of the strength (~30%). For comparison, for the ZrB_2-$MoSi_2$ composites, the flexural strength is in the range of ~750 to 800 MPa [48], showing a weak dependence on the amount of $MoSi_2$ added in the range of 10–40 vol.%. In addition, Bellosi et al. [25] reported that the flexural strengths for the 15 vol.% $MoSi_2$-containing ZrB_2 composites consolidated by SPS and HP are ~640 MPa and 700 MPa, respectively. These studies indicated that although ZrB_2 grain size is an important factor affecting the room-temperature flexural strength of ZrB_2-based composites with carbide and disilicides, it is not the limiting factor.

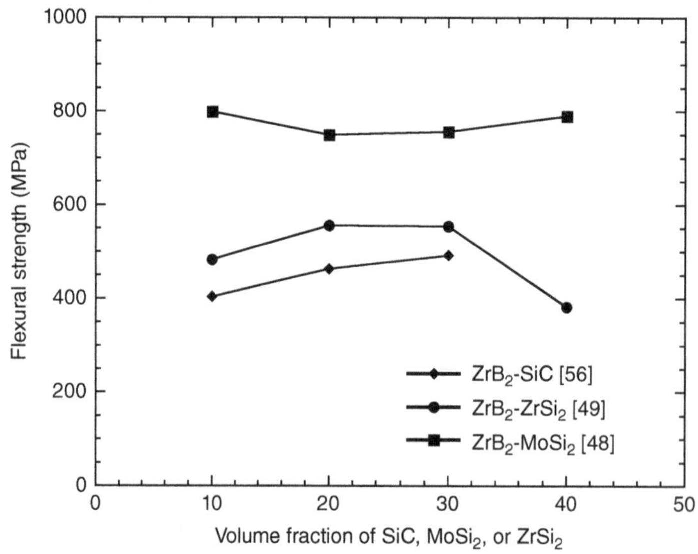

FIGURE 9.12

Plots of room-temperature 4-point flexural strength as a function of volume fraction of SiC, $MoSi_2$, or $ZrSi_2$ for hot-pressed ZrB_2-base composites with SiC, $MoSi_2$, or $ZrSi_2$.

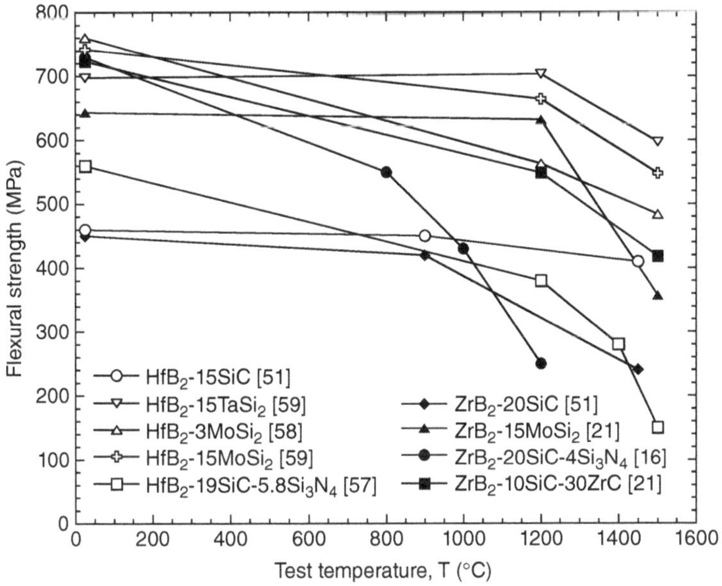

FIGURE 9.13
Plots of 4-point flexural strength as a function of test temperature for hot-pressed ZrB_2- and HfB_2-based composites with various additions.

The high temperature strength of the composites is a more important factor for the ZrB_2- or HfB_2-based composites in high temperature structural applications, compared to room-temperature strength. Early studies in ZrB_2- or HfB_2-based composites [16, 21, 51, 57, 58] showed that the flexural strength of the composites in air decreased with increasing temperature. The decrease in strength of the composites is strongly dependent on the reinforcement phase added to ZrB_2- or HfB_2-matrix (Figure 9.13), as a result of softening of the intergranular amorphous phase. For the ZrB_2-20 vol.% SiC composite [51], the strength decreased sharply at or above 1000°C. The addition of 4 vol.% Si_3N_4 to ZrB_2-SiC composite led to the reduction in strength at lower temperature of 800°C. For the $60ZrB_2$-30ZrC-10SiC (vol.%) composite [21], the strength decreases gradually with increase in temperature. For comparison, the ZrB_2-15 vol.% $MoSi_2$ composite retained the constant strength at temperatures approaching 1200°C [21]; however, the strength degraded rapidly above 1200°C. On the other hand, for the HfB_2-15 vol.% SiC composite [51], the strength almost remained constant from room temperature to 1500°C. However, the addition of 5.8 vol.% Si_3N_4 to HfB_2-SiC composite led to the sharp reduction in strength at or above 1200°C. For comparison, for the HfB_2-3 vol.% $MoSi_2$ composite the strength gradually decreases with increase in temperature [58]. Increasing the amount of $MoSi_2$ led to better high temperature strength [59]. In addition, the addition of $TaSi_2$ to HfB_2 resulted in improved high temperature strength [59], and the improvement was better than the addition of $MoSi_2$. Presumably, the decrease in strength at high temperature is closely linked with the presence of intergranular amorphous phase as well as of impurities. To improve high temperature strength of ZrB_2- or HfB_2-based composites, it is required to increase the refractory index of the intergranular phase and minimize impurities, as well as to promote crystallization of the intergranular amorphous phase.

9.3 ULTRAHIGH TEMPERATURE CERAMIC-BASED COMPOSITES WITH SHORT FIBERS

In order to improve fracture resistance or damage tolerance as well as reduce density of ZrB_2- or HfB_2-based composites, short carbon fiber and silicon carbide fiber are being attempted to apply to the ultrahigh temperature ceramic composites for fabricating discontinuous fiber-reinforced ZrB_2- or HfB_2-based ceramic matrix composites. However, the attempt is not enough for completely understanding the physical and mechanical behavior of the composites. This paragraph describes recent advances obtained in the studies on short fiber-reinforced ZrB_2- or HfB_2-based ceramic matrix composites.

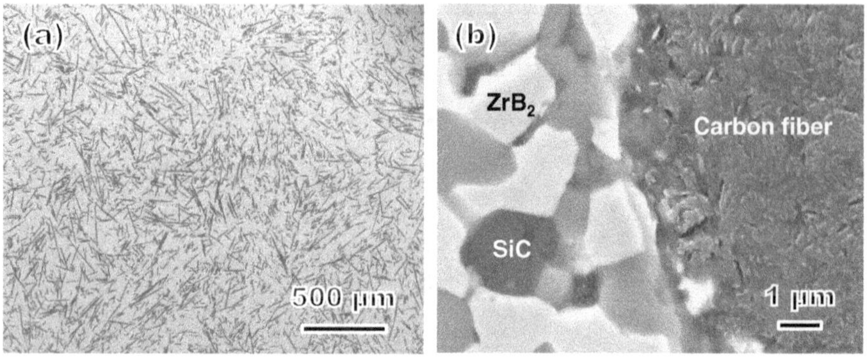

FIGURE 9.14

Examples of typical FE-SEM micrographs of the pitch-based carbon fiber-reinforced ZrB_2-30 vol.% SiC matrix composites ($V_f = 30\%$). (a) Low magnification and (b) high magnification.

9.3.1 Carbon Fiber-Reinforced ZrB_2- or HfB_2-Based Ceramics Matrix Composites

Yang et al. [60] consolidated a highly dense (relative density: 99.3%) PAN-based (T800H, length: 2 mm) carbon fiber-reinforced ZrB_2-20 vol.% SiC matrix composite at 2000°C and 30 MPa for 60 minutes in vacuum by HP. They showed that the fracture toughness of the composite increased from 4.3 MPa $m^{1/2}$ to 6.6 MPa $m^{1/2}$, leading to approximately 54% increase compared with monolithic matrix. For comparison, the room-temperature flexural strength, elastic moduli, and hardness of the composite decreased. Very recently, authors fabricated pitch-based carbon fiber-reinforced HfB_2-20 vol.% SiC or ZrB_2-20 vol.% SiC matrix composites by HP at ≤2000°C and 20 MPa for 60 minutes in Ar [61]. Figure 9.14 showed typical examples of FE-SEM images of the microstructures for the composites. It is found that the short fibers were uniformly dispersed in the matrix without the clumps of fibers (Figure 9.14a). In addition, the defects such as pores and cracks were not observed. Under high magnification (Figure 9.14b), the surface of fiber exhibits an inherent saw-toothed front shape. A noticeable interface reaction layer was not showed between the fiber and the matrix.

The thermal conductivity of the composites is in the range of 54.4–93.8 W/(m K), depended on the fiber volume fraction. Increasing fiber volume fraction led to a significant decrease in the thermal conductivity of the composites, as a result of the increase in interfacial thermal resistance at the fiber/matrix interface [61] (Figure 9.15). In addition, the electrical conductivities of the composites decreased with increase in fiber volume fraction as well. On the other hand,

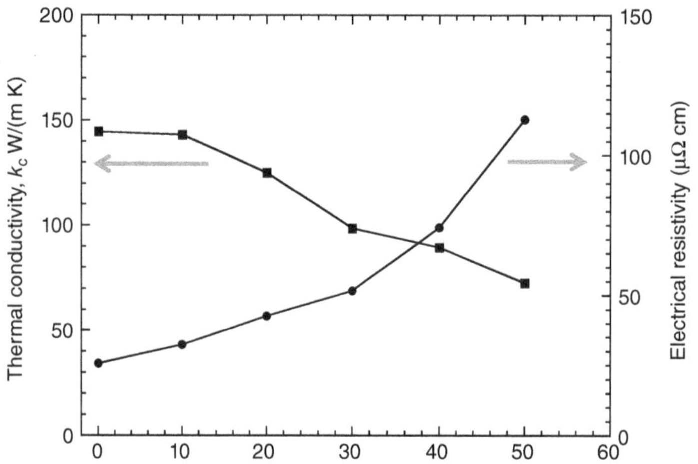

FIGURE 9.15

Plots of thermal conductivity and electrical resistivity as a function of volume fraction of fibers for the pitch-based carbon fiber-reinforced HfB_2-based ceramic matrix composites.

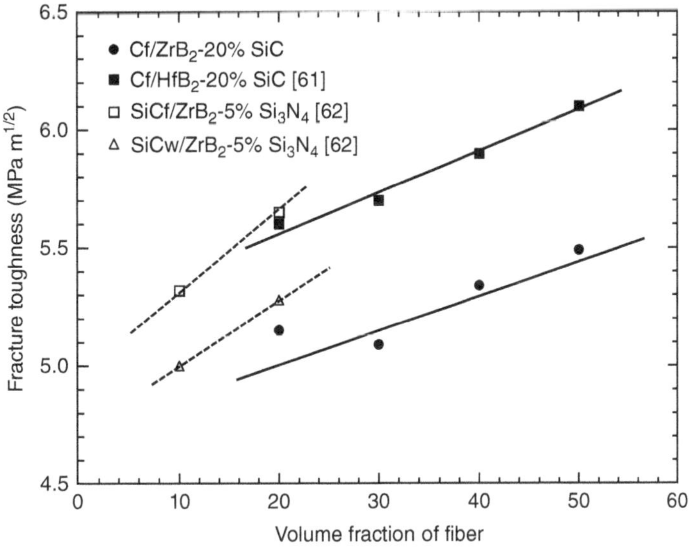

FIGURE 9.16

Plots of fracture toughness as a function of volume fraction of fibers for the pitch-based carbon fiber-reinforced ZrB_2-20 vol.% SiC- and HfB_2-20 vol.% SiC matrix composites and SiC whisker or SiC fiber-reinforced ZrB_2-5 vol.% Si_3N_4 matrix composites.

the fracture toughness of both the composites was significantly improved, compared with the monolithic HfB_2-20 vol.% SiC ceramic ($K_C = 4$ MPa m$^{1/2}$) [61,62], as shown in Figure 9.16. The fracture toughness increased with increasing fiber volume fraction, with the toughness range of 5.7–6.1 MPa m$^{1/2}$ for fiber volume fractions between 20% and 50%. The increase in fracture toughness due to addition of carbon fiber is attributed to the crack deflection and/or bridging. In contrast, the room-temperature flexural strength was lower for the composites than for the monolithic HfB_2-20 vol.% SiC ceramic matrix [61] (Figure 9.17). In addition, the flexural strengths of both the composites were significantly higher at 1600°C than at room temperature. Presumably, the reduction in the strength of the composites due to addition of

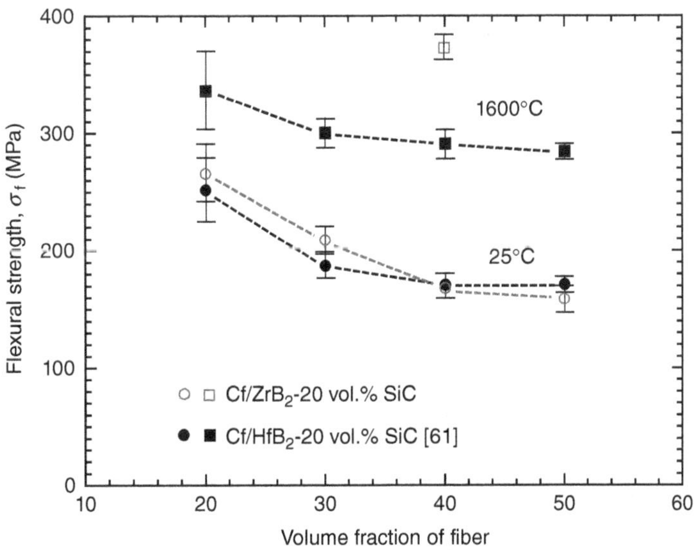

FIGURE 9.17

Plots of 4-point flexural strength at room and high temperatures as a function of volume fraction of carbon fiber for the hot-pressed ZrB_2-20 vol.% SiC- and HfB_2-20 vol.% SiC-matrix composites with short pitch-based carbon fibers.

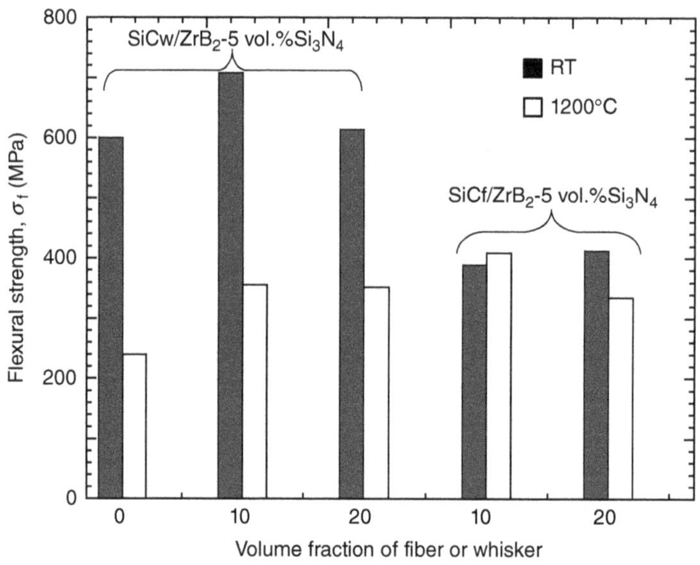

FIGURE 9.18

Plots of 4-point flexural strength at room and high temperatures as a function of volume fraction of carbon fiber for the hot-pressed SiC whisker or SiC fiber-reinforced ZrB_2-5 vol.% Si_3N_4 matrix composites.

fiber was associated with the presence of the thermal residual stress by thermal expansion misfit between the fiber and the matrix. Increasing fiber volume fraction from 20% to 30% led to a significant reduction of the flexural strength. However, further increase in fiber volume fraction led to a slight reduction of the strength. In particular, for the fiber volume fraction greater than 40 vol.%, the flexural strength almost remains constant. Similar dependency of the flexural strengths on fiber volume fraction was also observed for the samples fractured at 1600°C.

9.3.2 Silicon Carbide Fiber-Reinforced ZrB_2-Based Ceramics Matrix Composites

Silvestroni et al. [62] fabricated SiC whisker or SiC fiber (Hi-Nicalon, length: 1 mm) chop-reinforced 5 vol.% Si_3N_4-doped ZrB_2 matrix composites by HP at 1700°C and 50 MPa for 10–15 minutes in low vacuum. Fracture toughness of both composites was increased and the increase is larger for the SiC fiber-reinforced ZrB_2 matrix composite than for the SiC whisker-reinforced ZrB_2 composite (Fig. 9.16). In addition, the fracture toughness of both composites increased with increase in the amount of reinforcement phase; however, the substantial increase was not observed. On the other hand, the room-temperature flexural strength decreased for SiC fiber-reinforced ZrB_2-based ceramic matrix composite [62] (Figure 9.18), while for SiC whisker-reinforced ZrB_2-based ceramic matrix composite the flexural strength increased significantly. At 1200°C, the flexural strength of SiC whisker-reinforced ZrB_2-based ceramic matrix composite significantly decreased, while the strength of SiC fiber-reinforced ZrB_2-based ceramic matrix composite slightly increased ($V_f = 10\%$) or decreased ($V_f = 20\%$).

9.4 SUMMARY REMARKS AND FUTURE OUTLOOK

The studies on single-phase ZrB- or HfB_2- ceramics began in the 1960s. Throughout the research and developments of approximately 50 years, in particular in the last decade, several new synthesis methods and new densification processes have been developed. In addition, new additives are used to improve the sinterability, the mechanical properties, and oxidation or ablation resistance. HP is the most common densification method for ZrB_2- or HfB_2-based composites in historic studies. Recently, RS, PS, and SPS evolved as the most common densification methods. RS could produce the ZrB_2- or HfB_2-based composites at lower temperatures, by *in situ* reaction of precursors during heating or isothermally heating, compared with HP. SPS could densify various compositions of ZrB_2- or HfB_2-based composites at a lower temperature and very short soaking time, compared to the HP, RS, and PS methods. Furthermore, the addition of carbides, nitrides, and disilicides, as well as reducing particle sizes, improved sinterability and lowered the densification temperature. The resulting ZrB_2- or HfB_2-based composites are good thermal conductor materials and good electrical conductors as well.

On the other hand, Young's modulus and shear modulus of the ZrB_2- or HfB_2-based composites depended on composition, whereas Poisson's ratio was insensitive to compositions. Young's modulus and shear modulus of highly dense composites could be predicted by the rule of mixtures. Fracture toughness of ZrB_2- or HfB_2-based composites was in the range of 2–6 MPa $m^{1/2}$. The platelet-like ZrB_2-ZrC_x-Zr hybrid materials showed superior fracture toughness, with toughness value range ~6.6 to ~11 MPa $m^{1/2}$. Room temperature and high temperature flexural strengths of the ZrB_2- or HfB_2-based composites were significantly increased by additions of second phase particles, and the improvement depended on their particle size and distribution as well as on the grain-boundary phase. Compared with ZrB_2-based composites, HfB_2-based composites showed higher room temperature and high temperature strengths. These attempts and efforts made by many researchers have promoted the development in thermomechanical and structural applications of these ultrahigh temperature composites.

However, there are three unsolved problems for these applications as follows: (i) brittle fracture or absence of damage tolerance; (ii) weight; and (iii) poor oxidation or ablation resistance. It is difficult to solve brittle fracture or absence of damage tolerance and reducing weight in monolithic ceramic materials. One of the most promising solutions for preventing catastrophic fracture as well as for improving thermal shock resistance of the monolithic ceramics is to apply continuous ceramics fiber to monolithic ceramics for fabricating fiber/ceramic matrix composites, which could show a noncatastrophic fracture and a high thermal shock resistance, compared to monolithic ceramics. The route for increasing oxidation or ablation resistance is to control inward diffusion of oxygen and outward diffusion of composition elements from bulk material, by means of designing the effective compositions and microstructure of ZrB_2- or HfB_2-based composites.

From previous research on continuous fiber-reinforced ceramic matrix composites, it is quite clear that incorporation of continuous ceramic fibers into these ceramic matrices is the most effective way to increase the fracture resistance and to avoid noncatastrophic fracture behavior. Small diameter SiC fiber is the most promising available fibers for ZrB_2- and HfB_2 matrix composites. Figure 9.19 shows an example of SiC fiber (Hi-Nicalon™)-ZrB_2 matrix composite with BN coating interface [63]. However, the problem is the deformation of fiber during densification of the matrix ceramics. Carbon fiber is another candidate and applied for ZrB_2 matrix composites. To avoid high temperature processing, composites are fabricated by polymer-infiltration pyrolysis (PIP) process. Melt infiltration process is also tried to fabricate ZrB_2 matrix composite. For both the cases, thermal stress in a matrix caused by thermal expansion mismatch is unavoidable problem. The thermal expansion coefficient of SiC fiber is 3.9×10^{-6}°C^{-1} and that of carbon fiber in axial direction is ~0°C^{-1}. The mismatch stress is extremely high and easily exceeds tensile strength level of the matrix. In

FIGURE 9.19

A typical example of polished section of SiC-ZrB_2 matrix composite and fracture surface of the composite.

addition to processing, control of interface between fiber and matrix is also important. The reaction between fiber and matrix during fabrication process and service condition are unavoidable; thus, relatively thick coatings are requested. For example, preliminary study using large diameter SCS-6 SiC fiber-ZrB_2 matrix composite suggests requirement of relatively thick coating to protect fibers and to keep optimum interface mechanical properties. The motivation for future challenge should be the development of nanograined ZrB_2- or HfB_2-based composites and continuous fiber-reinforced matrix composites.

REFERENCES

1. K. Upadhya, J.-M. Yang, and W. P. Hoffmann. Materials for ultrahigh temperature structural applications. *Am. Ceram. Soc. Bull.*, 76, 51–56 (1997).
2. E. Wuchina, E. Opila, M. Opeka, W. Fahrenholtz, and I. Talmy. UHTCs: Ultra-high temperature ceramic materials for extreme environment applications. *Interface*, 16, 30–36 (2007).
3. A. Paul, D. D. Jayaseelan, S. Venugopal, E. Zapata-Solvas, J. Binner, B. Vaidhyanathan, A. Heaton, P. Brown, and W. E. Lee. UHTS composites for hypersonic applications. *Am. Ceram. Soc. Bull.*, 91, 22–29 (2012).
4. G. A. Meeson and A. F. Gorbunow. Activated sintering of zirconium boride. *Inorg. Mater.*, 4, 267–270 (1968).
5. M. Pastor. Metallic borides: preparation of solid bodies, sintering methods and properties of solid bodies, in *Boron and Refractory Borides*, edited by V. I. Matkovich (Springer, New York, 1977), pp. 457–493.
6. S. Q. Guo, J.-M. Yang, H. Tanaka, and Y. Kagawa. Effect of thermal exposure on strength of ZrB_2-based composites with nano-sized SiC particles. *Compos. Sci. Technol.*, 68, 3033–3040 (2008).
7. D. Kalish and E. V. Clougherty. Densification mechanisms in high-pressure hot-pressing of HfB_2. *J. Am. Ceram. Soc.*, 52, 26–30 (1969).
8. D. Kalish, E. V. Clougherty, and K. Kreder. Strength, fracture mode, and thermal stress resistance of HfB_2 and ZrB_2. *J. Am. Ceram. Soc.*, 52, 30–36 (1969).
9. W. C. Tripp, H. H. Davis, and H. C. Graham. Effect of SiC addition on the oxidation of ZrB_2 *Am. Ceram. Soc. Bull.*, 52, 612–616 (1973).
10. A. L. Chamberlain, W. G. Fahrenholtz, G. E. Hilmas, and D. T. Ellerby. High-strength zirconium diboride-based ceramics. *J. Am. Ceram. Soc.*, 87, 1170–1172 (2004).
11. F. Monteverde. Beneficial effects of an ultra-fine α-SiC incorporation on the sinterability and mechanical properties of ZrB_2. *Appl. Phys. A*, 82, 329–337 (2006).
12. S. S. Hwang, A. L. Vasiliev, and N. P. Padture. Improved processing and oxidation-resistance of ZrB_2 ultra-high temperature ceramics containing SiC nanodispersoids. *Mater. Sci. Eng. A*, A464, 216–224 (2007).
13. S. Zhu, W. G. Fahrenholtz, and G. E. Hilmas. Influence of silicon carbide particle size on the microstructure and mechanical properties of zirconium diboride-silicon carbide ceramics. *J. Euro. Ceram. Soc.*, 27, 2077–2083 (2007).
14. A. Rezaie, W. G. Fahrenholtz, and G. E. Hilmas. Effect of hot-pressing time and temperature on the microstructure and mechanical properties of ZrB_2-SiC. *J. Mater. Sci.*, 42, 2735–2744 (2007).
15. F. Monteverde and A. Bellosi. Effect of the addition of silicon nitride on sintering behavior and microstructure of zirconium diboride. *Scripta Mater.*, 46, 223–228 (2002).
16. F. Monteverde, S. Guicciardi, and A. Bellosi, Advances in microstructure and mechanical properties of zirconium diboride based ceramics. *Mater. Sci. Eng. A*, A346, 310–319 (2003).
17. F. Monteverde and A. Bellosi. Beneficial effects of AlN as sintering aid on microstructure and mechanical properties of hot-pressed ZrB_2. *Adv. Eng. Mater.*, 5, 508–512 (2003).
18. S. Q. Guo, Y. Kagawa, T. Nishimura, and H. Tanaka. Elastic properties of spark plasma sintered (SPSed) ZrB_2-ZrC-SiC composites, *Ceram. Int.*, 34, 1811–1817 (2008).
19. F. Monteverde and A. Bellosi. Efficacy of HfN as sintering aid in the manufacture of ultra-high-temperature metal diborides-matrix ceramics. *J. Mater. Res.*, 19, 3576–3585 (2004).
20. F. Monteverde and A. Bellosi. Development and characterization of metal-diboride-based composites toughened with Ultra-Fine SiC particulates. *Solid State Sci.*, 7, 622–630 (2005).
21. A. Bellosi, F. Monteverde, and D. Sciti. Fast densification of ultra-high-temperature ceramics by spark plasma sintering *Int. J. Appl. Ceram. Technol.*, 3 32–40 (2006).
22. S. Q. Guo, Y. Kagawa, T. Nishimura, D. Chung, and J.-M. Yang. Mechanical and physical behavior of spark plasma sintered ZrC-ZrB_2-SiC multiphase composites. *J. Eur. Ceram. Soc.*, 28, 1279–1285 (2008).

23. S. Q. Guo, T. Nishimura, Y. Kagawa, and J.-M. Yang. Spark plasma sintering of zirconium diborides. *J. Am. Ceram. Soc.*, 91, 2848–2855 (2008).
24. M. Ikegami, S. Q. Guo, and Y. Kagawa, Densification behavior and microstructure of spark plasma sintered ZrB_2-based composites with SiC particles. *Ceram. Int.*, 38, 769–774 (2012).
25. D. Sciti, S. Guicciardi, and M. Nygren. Densification and mechanical behavior of HfC and HfB_2 fabricated by spark plasma sintering. *J. Am. Ceram. Soc.*, 91, 1433–1440 (2008).
26. C. F. Hu, Y. Sakka, B. Jang, H. Tanaka, T. Nishimura, S. Q. Guo, and S. Grasso. Microstructure and properties of ZrB_2-SiC and HfB_2-SiC composites fabricated by spark plasma sintering (SPS) using $TaSi_2$ as sintering aid. *J. Ceram. Soc. Japan*, 118, 997–1001 (2010).
27. G.-J. Zhang, Z.-Y. Deng, N. Kondo, J.-F. Yang, and T. Ohji. Reactive hot pressing of ZrB_2-SiC composites. *J. Am. Ceram. Soc.*, 83, 2330–2332 (2000).
28. W.-W. Wu, G.-J. Zhang, Y.-M. Kan, and P.-L. Wang. Reactive hot pressing of ZrB_2-SiC-ZrC ultra high-temperature ceramics at 1800°C. *J. Am. Ceram. Soc.*, 89, 2967–2969 (2006).
29. Q. Qu, X. Zhang, S. Meng, W. Han, C. Hong, and J. Han. Reactive hot pressing and sintering characterization of ZrB2-SiC-ZrC composites, *Mater. Sci., Eng. A*, A491, 117–123 (2008).
30. A. L. Chamberlain, W. G. Fahrenholtz, and G. E. Hilmas. Low-temperature densification of zirconium diboride ceramics by reactive hot pressing. *J. Am. Ceram. Soc.*, 89, 3638–3645 (2006).
31. F. Monteverde Progress in the fabrication of ultra-high-temperature ceramics: in situ synthesis, microstructure and properties of a reactive hot-pressed HfB_2-SiC composite, *Compos. Sci. Technol.*, 65, 1869–1879 (2005).
32. W.-W. Wu, Z. Wang, G.-J. Zhang, Y.-M. Kan, and P.-L. Wang. ZrB_2-$MoSi_2$ composites toughened by elongated ZrB_2 grains via reactive hot pressing. *Scripta Mater.*, 61, 316–319 (2009).
33. S. Q. Guo, Y. Kagawa, and T. Nishimura, Tough hybrid ceramic-based material with high strength, *Scripta Mater.*, 67, 474–477 (2012).
34. Y. Zhao, L.-J. Wang, G.-J. Zhang, W. Jiang, and L.-D. Chen. Preparation and microstructure of a ZrB_2-SiC composite fabricated by the spark plasma sintering-reactive synthesis (SPS-RS) method. *J. Am. Ceram. Soc.*, 90, 4040–4042 (2007).
35. S. Q. Guo, T. Nishimura, and Y. Kagawa. Preparation of zirconium diboride ceramics by reactive spark plasma sintering of zirconium hydride-boron powders. *Scripta Mater.*, 65, 1018–1021 (2011).
36. U. Anselmi-Tamburini, Y. Kodera, M. Gasch, C. Unuvar, Z. A. Munir, M. Ohyanagi, and S. M. Johnson. Synthesis and characterization of dense ultra-high temperature thermal protection materials produced by field activation through spark plasma sintering (SPS): I. hafnium diboride, *J. Mater. Sci.*, 41 3097–3104 (2006).
37. S. Q. Guo, Y. Kagawa, T. Nishimura, and H. Tanaka. Pressureless-sintering and physical properties of ZrB_2-based composites with $ZrSi_2$ additive. *Scripta Mater.*, 58, 579–582 (2008).
38. D. Sciti, S. Guicciardi, A. Bellosi, and G. Pezzotti. Properties of a pressureless-sintered ZrB_2-$MoSi_2$ ceramic composite. *J. Am. Ceram. Soc.*, 89, 2320–2322 (2006).
39. S. C. Zhang, G. E. Hilmas, and W. G. Fahrenholtz. Pressureless densification of zirconium diboride with boron carbide additives. *J. Am. Ceram. Soc.*, 89 1544–1550 (2006).
40. S. Zhu, W. G. Fahrenholtz, G. E. Hilmas, and S. C. Zhang. Pressureless sintering of carbon-coated zirconium diboride powders. *Mater. Sci. Eng. A*, A459, 167–171 (2007).
41. W. G. Fahrenholtz, G. E. Hilmas, and S. C. Zhang, and S. Zhu. Pressureless sintering of zirconium diboride: particle size and additive effects. *J. Am. Ceram. Soc.*, 91, 1398–1404 (2008).
42. L. Silvestroni and D. Sciti. Effects of $MoSi_2$ additions on the properties of Hf- and Zr-B_2 composites produced by pressureless sintering, *Scripta Mater.*, 57, 165–168 (2007).
43. R. Savino, M. D. S. Fumo, L. Silvertroni, and D. Sciti. Arc-jet testing on HfB_2 and HfC-based ultra-high temperature ceramic materials. *J. Euro. Ceram. Soc.*, 28, 1899–1907 (2008).
44. J. Zou, G.-J. Zhang, Y.-M. Kan, and T. Ohji. Pressureless sintering mechanisms and mechanical properties of hafnium diboride ceramics with pre-sintering heat treatment. *Scripta Mater.*, 62, 159–162 (2010).
45. D. W. Ni, J.-X. Liu, and G.-J. Zhang. Pressureless sintering of HfB_2-SiC ceramics doped with WC. *J. Euro. Ceram. Soc.*, 32, 3627–3635 (2012).
46. D. E. Wiley, W. R. Manning, and O. Hunter Jr. Elastic properties of polycrystalline TiB_2, ZrB_2 and HfB_2 from room temperature to 1300K. *J. Less-Common Metals*, 18, 149–157 (1969).
47. S. Zhu, W. G. Fahrenholtz, G. E. Hilmas, and S. C. Zhang. Pressureless sintering of zirconium diboride using boron carbide and carbon additions. *J. Am. Ceram. Soc.*, 90, 3660–3663 (2007).
48. S. Q. Guo, T. Nishimura, T. Mizuguchi, and Y. Kagawa. Mechanical properties of hot-pressed ZrB_2-$MoSi_2$-SiC composites. *J. Eur. Ceram. Soc.*, 28, 1891–1898 (2008).

49. S. Q. Guo, Y. Kagawa, and T. Nishimura. Mechanical behavior of two-step hot-pressed Zrb_2-based composites with $ZrSi_2$, *J. Eur. Ceram. Soc.*, 29, 787–794 (2009).

50. M. Ikegami, K. Matsumura, S. Q. Guo, Y. Kagawa, and J.-M. Yang. Effect of SiC particle dispersion on thermal properties of SiC particle-dispersed ZrB_2 matrix composites. *J. Mater. Sci.*, 45, 5420–5423 (2010).

51. R. Loehman, E. Corral, H. P. Dumm, P. Kotula, and R. Tandon. Ultra high temperature ceramics for hypersonic vehicle applications, Sandia Report SAND 2006-2925 (Sandia National Laboratories, Albuquerque, NM and Livermore, CA, 2006).

52. S. Q. Guo, T. Nishimura, H. Tanaka, and Y. Kagawa. Thermal and electrical properties in hot-pressed ZrB_2-$MoSi_2$-SiC composites. *J. Am. Ceram. Soc.*, 90, 2255–2258 (2007).

53. J. W. Zimmenmann, G. E. Hilmas, W. G. Fahrenholtz, R. B. Dinwiddie, W. D. Porter, and H. Wang. Thermophysical properties of ZrB_2 and ZrB_2-SiC ceramics. *J. Am. Ceram. Soc.*, 91, 1405–1411 (2008).

54. R. Jimbou, K. Takahashi, Y. Matsushita, and T. Kosugi. SiC-ZrB_2 electroconductive ceramic composite. *Adv. Ceram. Mater.*, 1, 341–345 (1986).

55. K. Takahashi and R. Jimbou. Effect of uniformity on the electrical resistivity of SiC-ZrB_2 ceramic composites *J. Am. Ceram. Soc.*, 70, C369–C373 (1987).

56. S. C. Zhang, G. E. Hilmas, and W. G. Fahrenholtz. Mechanical properties of sintered ZrB_2-SiC ceramics. *J. Eur. Ceram. Soc.*, 31, 893–901 (2011).

57. F. Monteverde and A. Bellosi. Microstructure and properties of an HfB_2-SiC composite for ultra high temperature applications. *Adv. Eng. Mater.*, 6, 331–316 (2004).

58. D. Sciti, G. Bonnefont, G. Fantozzi, and L. Silvestroni. Spark plasma sintering of HfB_2 with low additions of silicides of molybdenum and tantalum. *J. Eur. Ceram. Soc.*, 30, 3253–3258 (2010).

59. D. Sciti, L. Silvestroni, G. Celotti, C. Melandri, and S. Guicciardi. Sintering and mechanical properties of ZrB_2-$TaSi_2$ and HfB_2-$TaSi_2$ ceramic composites. *J. Am. Ceram. Soc.*, 91, 3285–3291 (2008).

60. F. Yang, X. Zhang, J. Han, and S. Du. Characterization of hot-pressed short carbon fiber reinforced ZrB_2–SiC ultra–high temperature ceramic composites. *J. Alloys Compd.* 472, 395–399 (2009).

61. S. Q. Guo, K. Naito, and Y. Kagawa. Mechanical and physical behaviors of short pitch-based carbon fiber-reinforced HfB_2–SiC matrix composites. *Ceram. Int.* 39, 1567–1574 (2013).

62. L. Silvestroni, D. Sciti, C. Melandri, and S. Guicciardi. Toughened ZrB_2-based ceramics through SiC whisker or SiC chopped fibers reinforced additions. *J. Euro. Ceram. Soc.*, 30, 2155–2164 (2010).

63. S.H. Lee and Y. Kagawa. Unpublished study.

PART III

ENVIRONMENTAL EFFECTS AND COATINGS

CHAPTER 10

ENVIRONMENTAL EFFECTS ON OXIDE/OXIDE COMPOSITES

MARINA B. RUGGLES-WRENN

Department of Aeronautics & Astronautics, Air Force Institute of Technology, OH, USA

10.1 INTRODUCTION/BACKGROUND

Advances in power generation systems for aircraft engines, land-based turbines, rockets, and, most recently, hypersonic missiles and flight vehicles have raised the demand for structural materials that have superior long-term mechanical properties and retained properties under high temperature, high pressure, and varying environmental factors, such as moisture [1–3]. Typical components include combustors, nozzles, and thermal insulation. Ceramic matrix composites (CMCs), capable of maintaining excellent strength and fracture toughness at high temperatures, are prime candidate materials for such applications. Because these applications require exposure to oxidizing environments, the thermodynamic stability and oxidation resistance of CMCs are vital issues. The need for environmentally stable composites motivated the development of CMCs based on environmentally stable oxide constituents [4–9].

The main advantages of CMCs over monolithic ceramics are their superior toughness, tolerance to the presence of cracks and defects, and non-catastrophic mode of failure. It is widely accepted that in order to avoid brittle fracture behavior in CMCs and improve the damage tolerance, a weak fiber/matrix interface is needed, which serves to deflect matrix cracks and to allow subsequent fiber pullout [10–12]. It has been demonstrated that similar crack-deflecting behavior can also be achieved by means of a finely distributed porosity in the matrix instead of a separate interface between matrix and fibers [13]. This concept has been successfully demonstrated for oxide–oxide composites [4, 7, 9, 14–16]. Resulting oxide/oxide CMCs exhibit damage tolerance combined with inherent oxidation resistance. An extensive review of the mechanisms and mechanical properties of porous-matrix CMCs is given in [1, 17].

High-temperature mechanical properties and behavior of oxide–oxide CMCs based on different fiber types and matrix systems, processing techniques, and sintering conditions were investigated by several research groups [18–25]. Oxide–oxide CMCs have been shown to exhibit good stability in ambient air at elevated temperatures [26–29]. A number of studies evaluated effects of exposure to simulated service environments on microstructural stability and tensile properties of oxide–oxide CMCs [2, 30–34]. Nextel™720/aluminosilicate composite exhibited excellent strength retention and showed no evidence of environmental attack following ground testing on a General Electric F110 turbofan engine [33]. Exposure to moisture or to salt fog had no effect on the high-temperature fatigue performance or on the retained strength values of the N610/aluminosilicate and N720/alumina composites [2]. Long-term exposure (3000 hours) to high temperatures and pressures like those in advanced gas turbine combustor environment had minimal effect on the microstructure and strength of N720/alumina composite [32]. Likewise exposure in a combustor rig caused only a slight reduction in tension strength of the oxide–oxide CMCs [30, 31]. In these studies the test specimens were exposed to degrading environments (long-term exposure) before testing and the test results differed little from the results obtained for as-received (unexposed) specimens. However, if the specimen is subjected to mechanical loading in a degrading environment, the load may change the microstructure thereby accelerating environmental degradation.

Ceramic Matrix Composites: Materials, Modeling and Technology, First Edition. Edited by Narottam P. Bansal and Jacques Lamon.

TABLE 10.1 Average Physical Properties of the Nextel™720/Alumina and Nextel™720/Alumina–Mullite Composites

Property	Nextel™720/alumina	Nextel™720/alumina–mullite
Fiber volume (%)	44	40
Porosity (%)	23	28
Density (g/cm^3)	2.79	2.57

The objective of this chapter is to review recent efforts to assess the effects of steam environment on high-temperature mechanical behavior of the oxide–oxide CMCs. High-temperature water vapor is one of several combustion products and thus a significant component of combustion environment. Therefore in order to evaluate the performance of oxide–oxide CMCs in combustion environment it is important to understand the effects of steam on their mechanical performance and durability. Results discussed in this chapter were obtained in experiments which combined environmental exposure (i.e., air or steam and elevated temperature) with mechanical loading. The details of the experimental setup are given elsewhere [19, 35–37].

10.2 MECHANICAL BEHAVIOR—EFFECTS OF ENVIRONMENT

The two representative materials discussed here are Nextel™720/alumina (N720/A) and Nextel™720/alumina–mullite (N720/AM), oxide–oxide CMCs consisting of a porous oxide matrix reinforced with 0/90 woven Nextel™720 fibers. The N720/AM composite contained approximately 12.5% (by volume) of mullite in the matrix composition. Basic properties of the two composites are summarized in Table 10.1.

10.2.1 Tensile Stress–Strain Behavior

Tensile tests were carried out in laboratory air with the displacement rate of 0.05 mm/s. Tensile properties of the N720/A and N720/AM composites at various temperatures [35–37] are summarized in Table 10.2. The stress–strain curves shown in Figure 10.1 typify the tensile response of the two composites. The tensile properties and stress–strain behaviors of the two composites are similar and change little as the temperature increases from 800°C (900°C in the case of N720/AM) to 1200°C. Both composites behave nearly linearly to failure, exhibiting typical fiber-dominated behavior. Ultimate tensile strength (UTS) and elastic modulus of both materials are fairly independent of temperature in the 900–1200°C range.

TABLE 10.2 Average Tensile Properties of the Nextel™720/Alumina and Nextel™720/Alumina–Mullite Composites (After [35–37]) at Elevated Temperatures in Laboratory Air

Temperature (°C)	Elastic Modulus (GPa)	UTS (MPa)	Failure Strain (%)
Nextel™720/alumina			
900	70.1	190	0.33
1000	73.5	188	0.32
1100	69.6	190	0.33
1200	74.7	189	0.38
Nextel™720/alumina–mullite			
900	74.3	187	0.36
1000	72.1	190	0.36
1100	70.7	186	0.35
1200	74.0	169	0.37

All data are adjusted for $V_f = 0.44$.

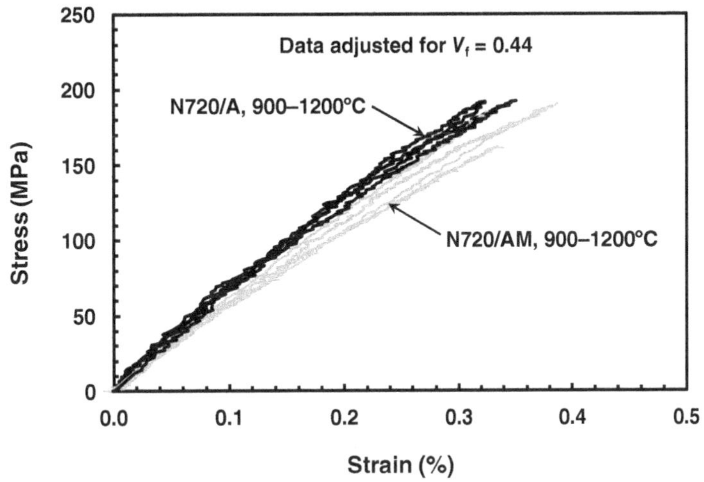

FIGURE 10.1

Tensile stress–strain curves for N720/A and N720/AM ceramic composites obtained in tests conducted at 0.05 mm/s at various temperatures in laboratory air. All data are adjusted for $V_f = 0.44$.

The effect of loading rate on the UTS of N720/A and N720/AM composites was also investigated [35–38]. At 1100°C and 1200°C in steam, the tensile strength and tensile stress–strain behavior of both composites were strongly influenced by the loading rate (Figure 10.2), suggesting the environmentally assisted crack growth as the dominant failure mechanism.

This significant dependence of tensile strength on stress rate (Figure 10.2) is similar to that observed for monolithic ceramics at elevated temperatures. For glass and monolithic ceramics the rate dependence of tensile strength, also termed delayed failure, is known to proceed by environmentally assisted subcritical crack growth [39–41]. The subcritical (slow) crack growth rate can be described by the empirical power law [42–44]:

$$\frac{da}{dt} = A\left[\frac{K_I}{K_{IC}}\right]^N \tag{10.1}$$

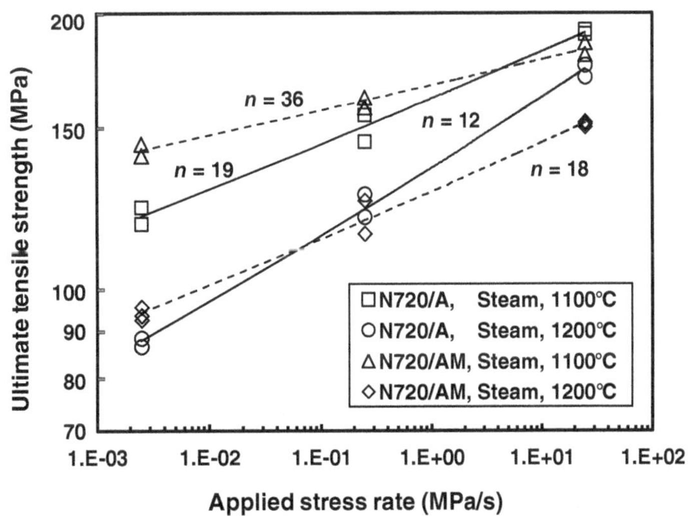

FIGURE 10.2

Ultimate tensile strength as a function of applied stress rate for N720/A and N720/AM ceramic composites at 1100°C and 1200°C in steam. The solid and dashed lines represent the best-fit linear regression based on Equation (10.2). All data are adjusted for $V_f = 0.44$.

where a is the crack size, t is time, K_I is the mode I stress intensity factor, K_{IC} is the critical stress intensity factor (or fracture toughness) under mode I loading, and A and N are the slow crack growth (SCG) parameters. In the case of loading at constant stress rate, the fracture strength σ_f can be derived as a function of applied stress rate $\dot{\sigma}$ as [42–44]

$$\sigma_f = D(\dot{\sigma})^{\frac{1}{N+1}} \tag{10.2}$$

A crack growth parameter D is associated with inert strength σ_i, parameter N, and crack geometry:

$$D = \left[\frac{2(N+1)K_{IC}^2\sigma_i^{N-2}}{AY^2(N-2)}\right]^{\frac{1}{N+1}} \tag{10.3}$$

where Y is a geometry factor related to flaw shape and its orientation with respect to the direction of applied stress. By taking logarithms of both sides, Equation (10.2) is expressed as

$$\log \sigma_f = \frac{1}{N+1}\log\dot{\sigma} + \log D \tag{10.4}$$

where N and D are the SCG parameters, σ_f is the tensile strength, and $\dot{\sigma}$ is the applied stress rate.

Following the procedure in [43, 44], the SCG parameters for N720/A in steam were determined as $N = 19$ and $D = 162.59$ at 1100°C, $N = 12$ and $D = 136.42$ at 1200°C. For the N720/AM composite in steam, the parameters were $N = 36$ and $D = 168.19$ at 1100°C, $N = 18$ and $D = 128.13$ at 1200°C. The crack growth exponent N represents a measure of susceptibility of the material to subcritical crack growth. Typically for brittle materials, the susceptibility is considered high for $N \leq 20$, intermediate for $30 \leq N \leq 50$, and low for $N > 50$. Thus the N720/A composite exhibits significant susceptibility at 1100°C and 1200°C in steam. In the case of N720/AM, the susceptibility to subcritical crack growth is moderate at 1100°C and considerable at 1200°C in steam.

10.2.2 Tensile Creep

The tensile creep behavior of N720/A and N720AM composites with 0/90 fiber orientation was investigated at temperatures in the 1000–1200°C range in laboratory air and in steam [35–37, 45, 46].

In tests performed at 1000°C with creep stresses ranging from 80 to 160 MPa, both composites exhibit primary and secondary creep, but no tertiary creep regime (Figure 10.3). At 1000°C both materials achieve a 100-hour creep run-out in all tests, regardless of test environment. While the presence of steam has no effect on creep lifetimes

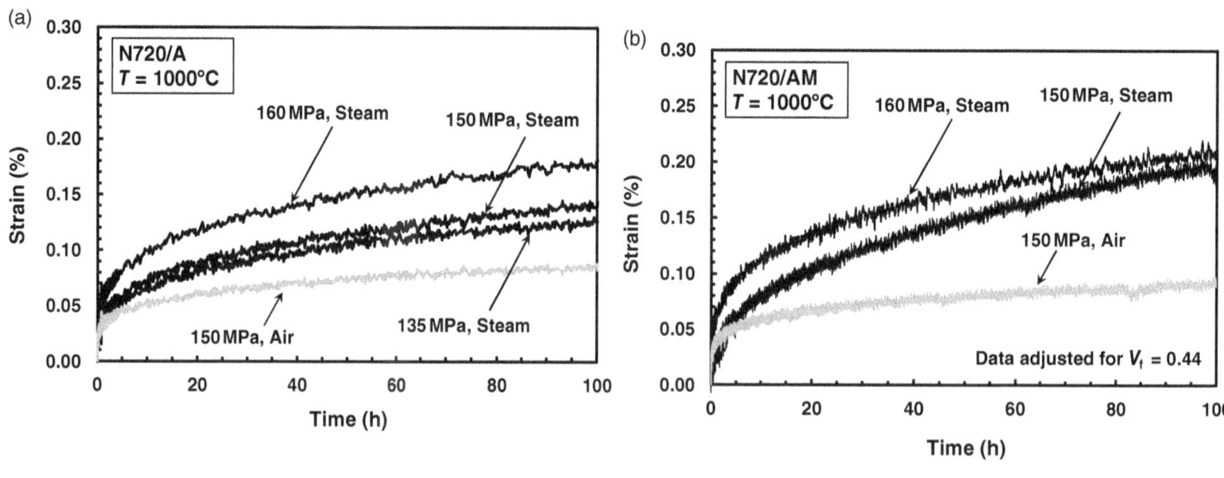

FIGURE 10.3

Creep strain versus time curves for (a) N720/A and (b) N720/AM ceramic composites at 1000°C in laboratory air and in steam. All data are adjusted for $V_f = 0.44$ (after Ruggles-Wrenn et al. [35,37]).

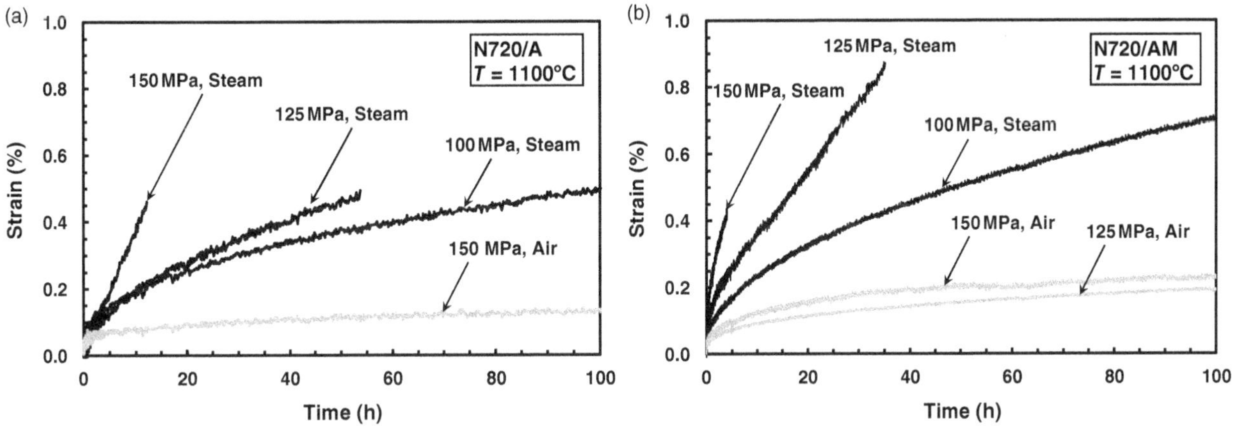

FIGURE 10.4

Creep strain versus time curves for (a) N720/A and (b) N720/AM ceramic composites at 1100°C in laboratory air and in steam. All data are adjusted for $V_f = 0.44$ (after Ruggles-Wrenn et al. [35, 37]).

(up to 100 hours), larger creep strains are accumulated in steam. Still, all creep strains remain below 0.25%, which is considerably less than the failure strains obtained in tension tests. At 1100°C the presence of steam has a noticeable effect on creep strain and creep lifetime of both composites (Figure 10.4). At 1100°C in air, creep run-out stress for both composites was 150 MPa. In contrast, at 1100°C in steam, both materials achieved creep run-out only at 100 MPa. Creep strains produced at 1100°C in steam can be three to four times those obtained in air. Detrimental effects of steam are somewhat more pronounced for N720/AM than for N720/A.

At 1200°C steam drastically degrades the creep performance of both N720/A and N720/AM composites. As expected, creep strains increase with temperature (Figures 10.5 and 10.6). Creep strains accumulated in air at 1000°C and 1100°C were considerably lower than the corresponding failure strains obtained in tension tests. Contrastingly, all creep strains accumulated in air at 1200°C significantly exceed the failure strain obtained in tension tests at 1200°C. Creep run-out was achieved in all tests conducted in air at 1000°C and 1100°C. Conversely at 1200°C in air, creep run-out was achieved only at 80 MPa for N720/A and at 100 MPa for N720/AM. Furthermore, neither composite achieved creep run-out at 1200°C in steam. At 1200°C, creep strain produced at 80 MPa in steam was five times that produced in air for N720/A and four times that obtained in air for N720/AM.

Minimum creep rates of both N720/A and N720/AM composites are profoundly influenced by environment (Figure 10.7). As expected, the creep strain rates increase with increasing applied stress as well as with increasing

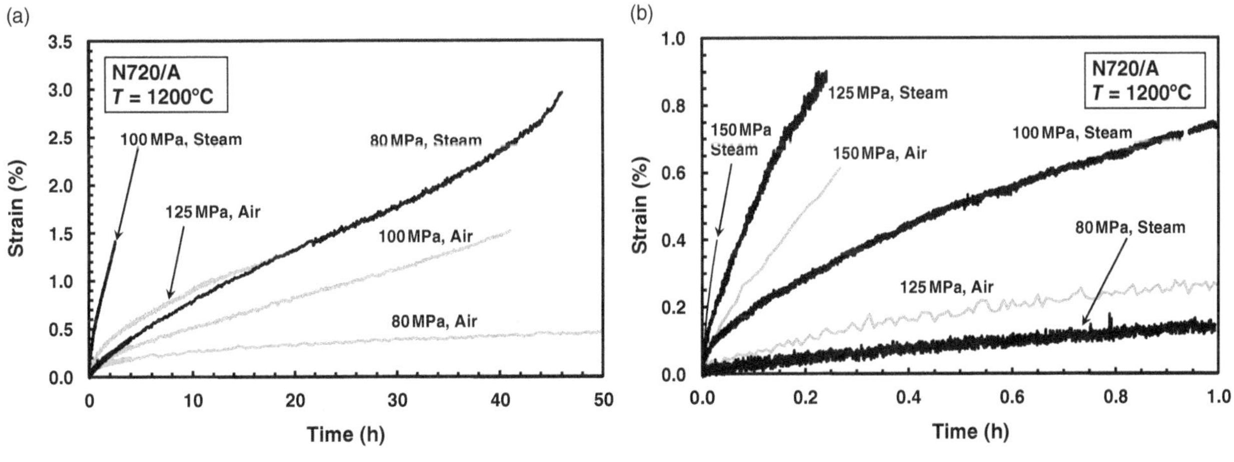

FIGURE 10.5

Creep strain versus time curves for N720/A ceramic composite at 1200°C in air and in steam: (a) at 80 and 100 MPa and (b) at stresses ≥125 MPa. All data are adjusted for $V_f = 0.44$ (after Ruggles-Wrenn et al. [35]).

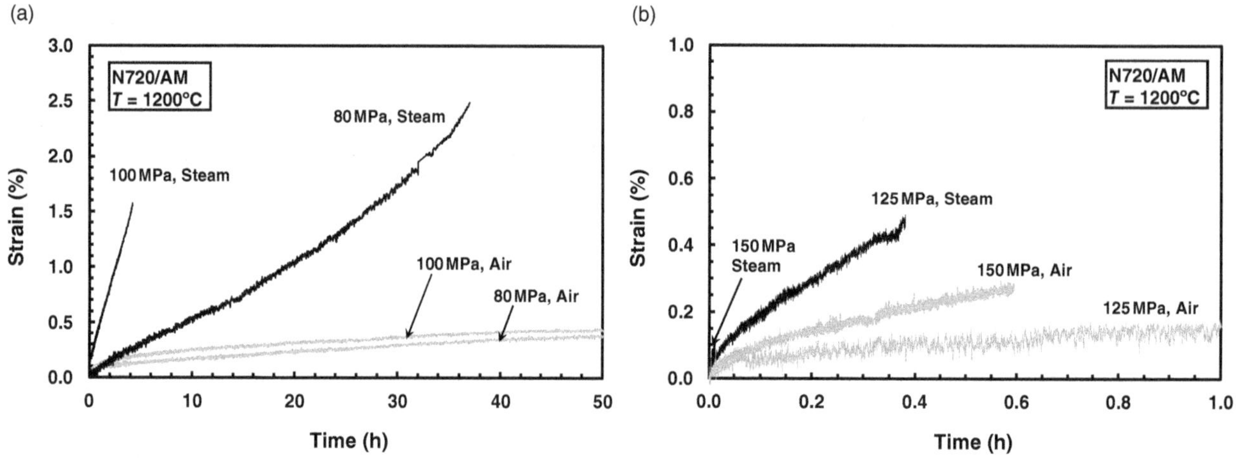

FIGURE 10.6

Creep strain versus time curves for N720/AM ceramic composite at 1200°C in air and in steam: (a) at 80 and 100 MPa and (b) at 125 and 150 MPa. All data are adjusted for $V_f = 0.44$ (after Ruggles-Wrenn et al. [36, 37]).

temperature. The secondary creep rates produced at 1200°C in air can be as high as 10^3 times those at 1000°C and 1100°C. At 1000°C the presence of steam has negligible effect on creep rate of both composites. All creep rates obtained at 1000°C are $\leq 10^{-9}$ per second. As the temperature increases to 1100°C, the creep rates in air remain close to those obtained at 1000°C, but the creep rates in steam accelerate dramatically. At 1100°C, the creep rates obtained in steam are nearly two orders of magnitude higher than the rates obtained in air. It is noteworthy that at 1100°C the N720/AM creep rates are close to those observed for N270/A in a given environment. At 1200°C, creep rates of both composites in steam are approximately one order of magnitude higher than the creep rates obtained in air. Note that at 1200°C in air creep rates of N720/AM are an order of magnitude lower than those of N720/A. At 1200°C in steam, creep rates of N720/AM are close to those of N720/A.

Stress-rupture behavior of N720/A and N720/AM composites is summarized in Figure 10.8. At 1000°C the presence of steam has no effect on creep lifetime of the two composites up to 100 hours. With the creep run-out condition defined as 100 hours, 160 MPa is the run-out stress in air and in steam at 1000°C. At 1100°C steam dramatically reduced creep lifetimes. In air at 1100°C both composites achieved creep run-out at 150 MPa (~80% UTS). In steam, at 1100°C creep run-out was achieved only at 100 MPa (~50% UTS). At 1200°C in air, run-out stress was 80 MPa (~40% UTS) for

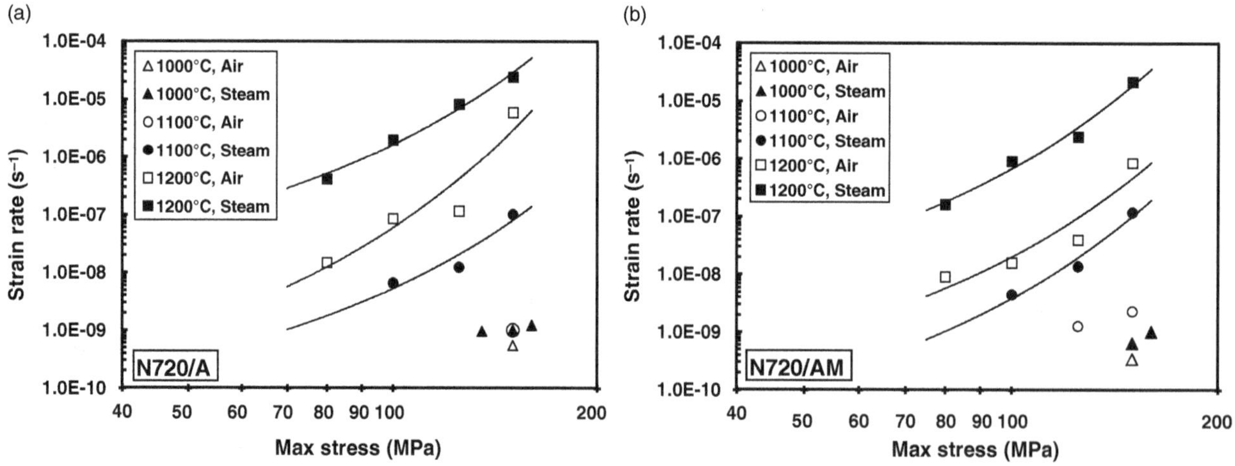

FIGURE 10.7

Secondary creep rate versus applied stress at various temperatures in laboratory air and in steam for (a) N720/A and (b) N720/AM ceramic composites. All data are adjusted for $V_f = 0.44$ (after Ruggles-Wrenn et al. [35–37]).

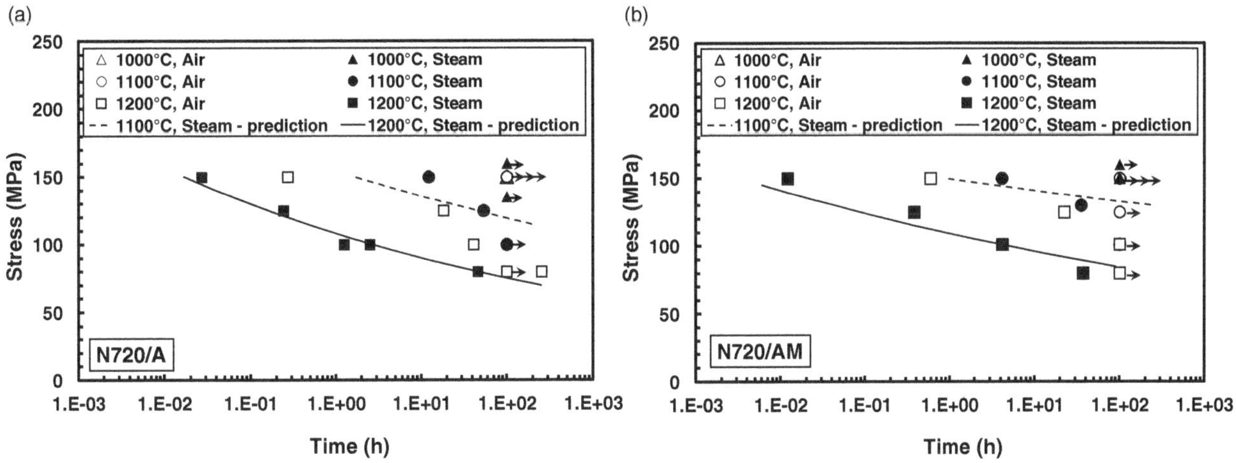

FIGURE 10.8

Creep stress versus time to rupture at various temperatures in laboratory air and in steam for (a) N720/A and (b) N720/AM ceramic composites. All data are adjusted for $V_f = 0.44$ (after Ruggles-Wrenn et al. [35–37]).

N720/A and 100 MPa (~50% UTS) for N720/AM. Neither composite achieved creep run-out at 1200°C in steam. For stresses ≥100 MPa, creep lifetimes of both composites were reduced by at least 90% due to steam.

Recall that N720/A and N720/AM composites exhibited moderate to significant susceptibility to subcritical crack growth at 1100°C and 1200°C in steam. For many glass and ceramic materials that exhibit subcritical (slow) crack growth as the dominant time-dependent failure mechanism, time to failure under constant stress (creep lifetime) can be predicted from constant stress-rate test data by using the linear elastic crack growth model [47,48]. Using the empirical power-law crack-velocity formulation in Equation (10.1) the time to failure under constant stress can be obtained in the form [49]

$$t_f = \left[\frac{2K_{IC}^2\sigma_i^{N-2}}{AY^2(N-2)}\right]\sigma^{-N} \tag{10.5}$$

Combining Equations (10.5) and (10.3), creep lifetime can be expressed in terms of the SCG parameters obtained from constant stress-rate data:

$$t_f = \left[\frac{D^{N+1}}{(N+1)}\right]\sigma^{-N} \tag{10.6}$$

where t_f is the time to failure and σ is the applied creep stress.

The predicted N720/A and N720/AM creep lifetimes in steam are shown in Figure 10.8 as dashed lines (1100°C) and solid lines (1200°C). Good agreement between the predictions and experimental results indicates that the environmentally assisted SCG is indeed the governing failure mechanism for both composites at 1200°C in steam. At 1100°C, the predicted creep lifetimes are in reasonable agreement with the experimental data, suggesting that the delayed failure of both composites at 1100°C in steam is also controlled by the power-law type SCG.

Retained tensile strength of each specimen that achieved creep run-out was evaluated in a tensile test performed at the same temperature as the prior creep test. Prior creep did not degrade the tensile strength of the N720/AM composite. N720/AM specimens that achieved creep run-out in air or steam at all temperatures investigated retained 100% of their tensile strength. Prior creep in air at 1000°C and at 1100°C also had little effect on retained strength of N720/A. However, prior creep in steam decreased the tensile strength of N720/A by ~9% at 1000°C and by nearly 16% at 1100°C. Moreover, prior creep at 1200°C in air reduced the tensile strength of N720/A by ~8%.

Fracture surfaces of N720/A (Figure 10.9) and N720/AM (Figure 10.10) composites produced in creep tests reflect the influence of test temperature and environment. The fracture surfaces produced in air at 1000°C (Figures 10.9a and 10.10a) are dominated by fibrous fracture. Notably, at 1000°C steam appears to have little effect on the fracture surface topography of the two composites (Figures 10.9b and 10.10b). All N720/A and N720/AM fracture surfaces obtained at

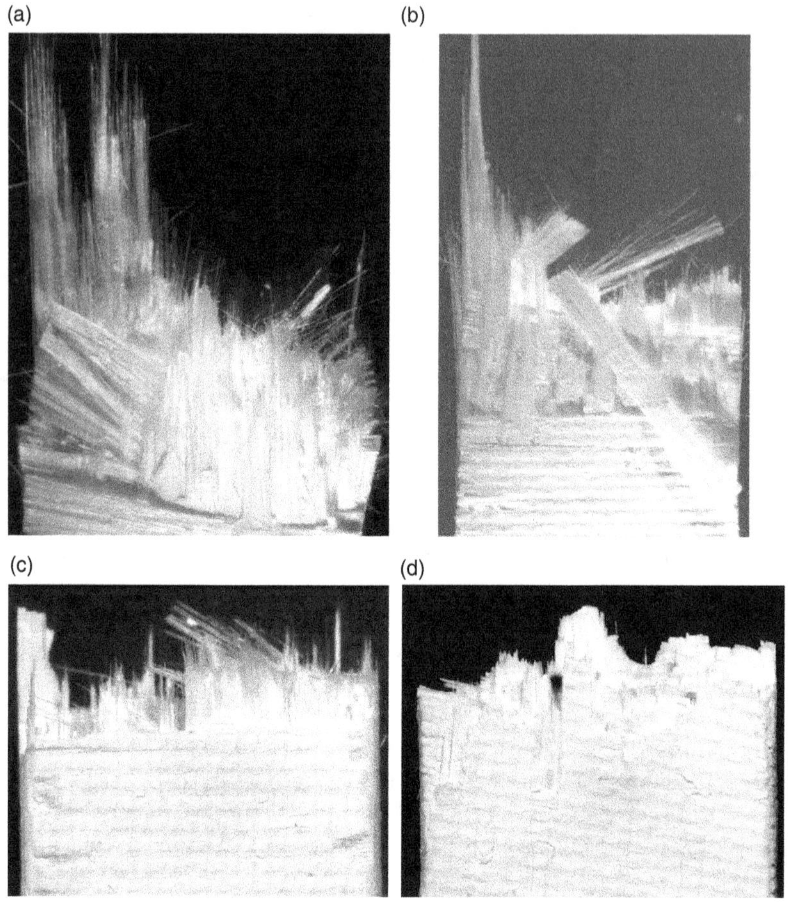

FIGURE 10.9

Fracture surfaces of N720/A ceramic composite obtained in creep tests at 150 MPa: (a) at 1000°C in air, (b) at 1000°C in steam, (c) at 1200°C in air, (d) at 1200°C in steam (after Ruggles-Wrenn et al. [35]).

1000°C still show considerable amounts of fibrous fracture. In contrast, the fracture surfaces obtained at 1200°C in air (Figures 10.9c and 10.10c) and in steam (Figures 10.9d and 10.10d) are dominated by near-planar fracture.

Further insight into the influence of temperature and environment on the fracture surface topography and the microstructure of N720/A and N720/AM specimens tested in creep can be gained by examining SEM micrographs in Figures 10.11 and 10.12. The N720/A and N720/AM fracture surfaces obtained at 1000°C in air (Figures 10.11a and 10.12a) exhibit sizable regions of uncorrelated fiber fracture, where individual fibers are clearly discernible. Only small amounts of matrix material remain bonded to the fiber surfaces (Figures 10.11b and 10.12b). Contrastingly, the fracture surfaces produced at 1200°C in air (Figures 10.11c and 10.12c) are dominated by planar regions of coordinated fiber failure. The amount of matrix remaining bonded to the fibers (Figures 10.11d and 10.12d) is significantly greater than that in the specimens tested in creep at 1000°C in air.

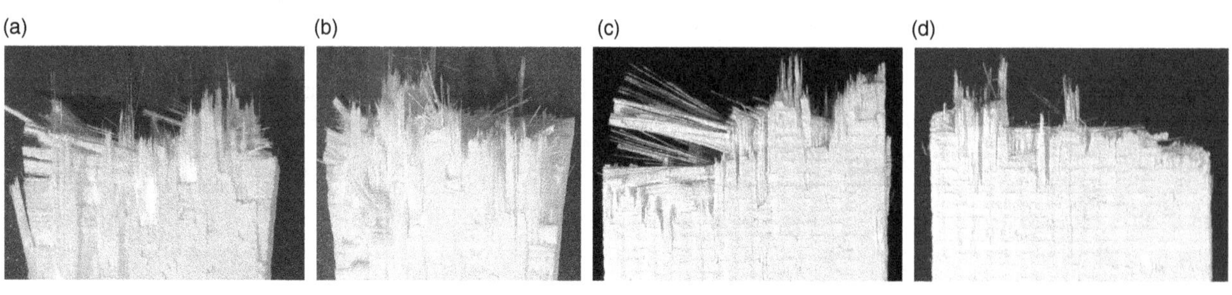

FIGURE 10.10

Fracture surfaces of N720/AM ceramic composite obtained in creep tests at 150 MPa: (a) at 1000°C in air, (b) at 1000°C in steam, (c) at 1200°C in air, (d) at 1200°C in steam (after Ruggles-Wrenn et al. [36, 37]).

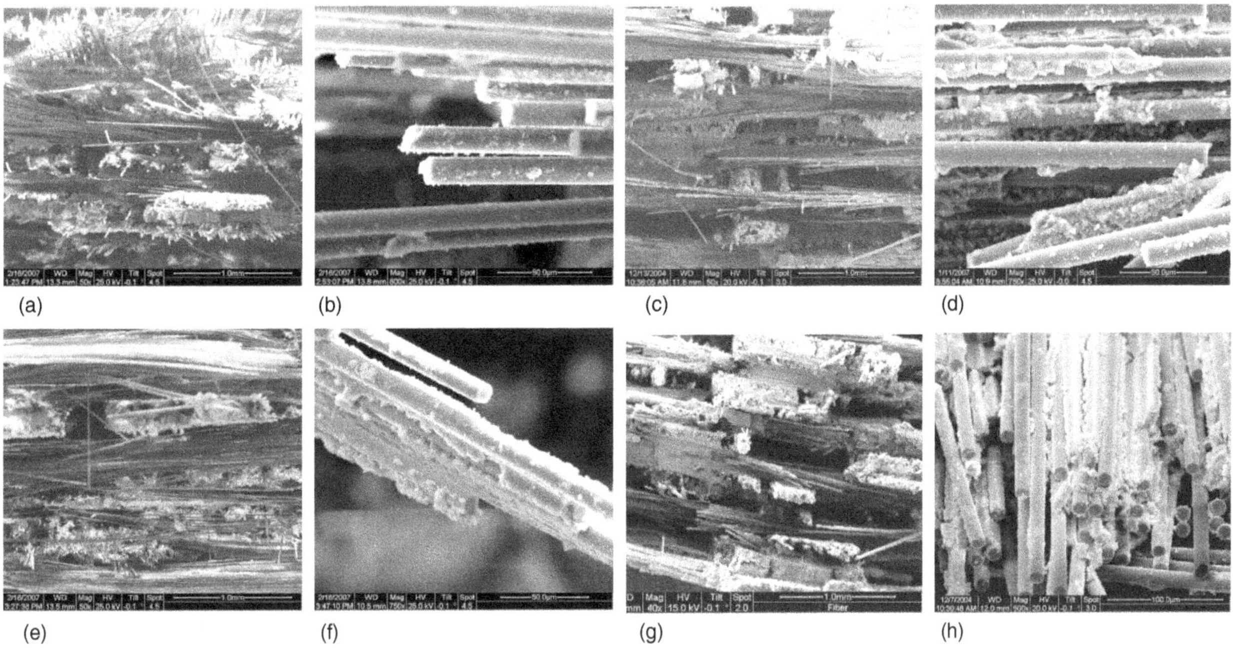

FIGURE 10.11

Fracture surfaces of N720/A ceramic composite obtained in creep tests at 150 MPa: (a) and (b) at 1000°C in air; (c) and (d) at 1200°C in air; (e) and (f) at 1000°C in steam; (g) and (h) at 1200°C in steam (after Ruggles-Wrenn et al. [35]).

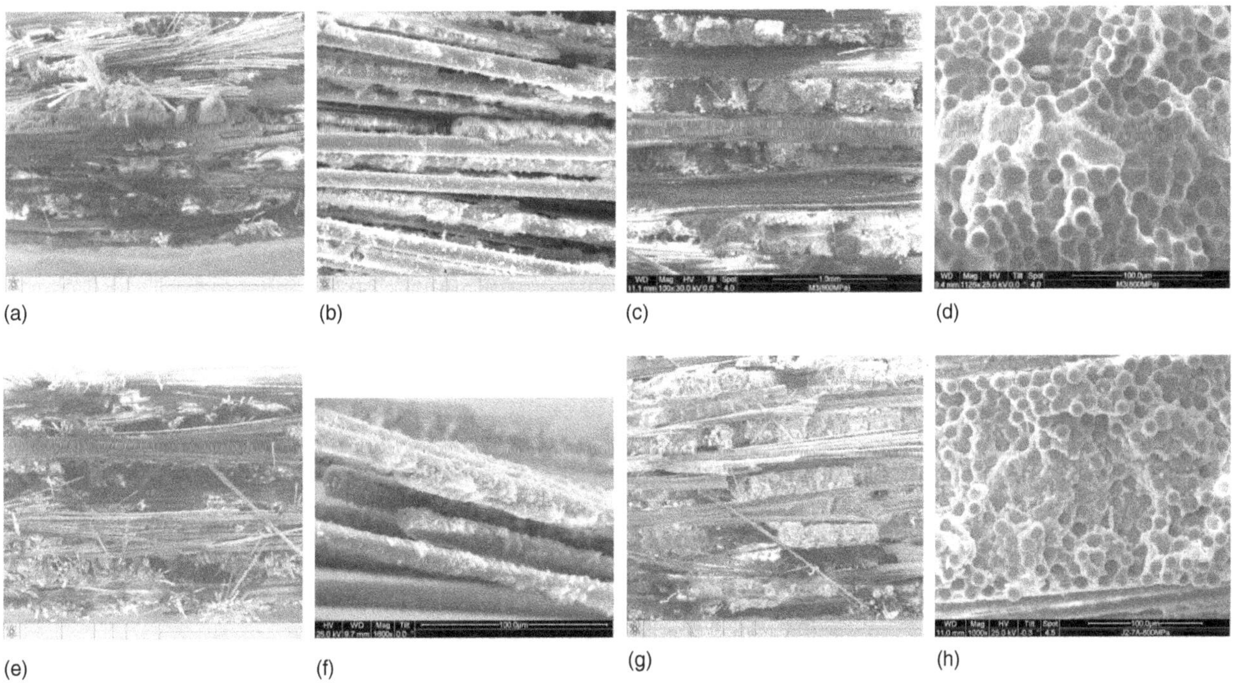

FIGURE 10.12

Fracture surfaces of N720/AM ceramic composite obtained in creep tests at 150 MPa: (a) and (b) at 1000°C in air; (c) and (d) at 1200°C in air; (e) and (f) at 1000°C in steam; (g) and (h) at 1200°C in steam (after Ruggles-Wrenn et al. [36,37]).

Steam environment has a dramatic effect on the N720/A fracture surface appearance. The effect of steam is further amplified at higher temperature. While the N720/A fracture surface produced at 1000°C in steam (Figure 10.11e) still exhibits some areas of uncoordinated fibrous failure, considerable areas of planar fracture are also visible. In contrast, the fracture surface produced at 1200°C in steam (Figure 10.11g) is almost entirely planar. Recall that the N720/A fracture surface obtained at 1200°C in air (Figure 10.11c) still showed some limited areas of fibrous fracture. In the case of the N720/A specimen tested at 1200°C in steam, the amounts of matrix material adhering to the fibers are much greater than those seen in other specimens (Figure 10.11h). Surprisingly, steam environment has little effect on the N720/AM fracture surface appearance. The N720/AM fracture surface produced at 1000°C in steam (Figure 10.12e) is dominated by areas of uncoordinated fibrous failure, as is the N720/AM fracture surface produced at 1000°C in air (Figure 10.12a). The N720/AM fracture surfaces produced at 1200°C in air and in steam (Figures 10.12c, 10.12d, 10.12g, and 10.12h) also have essentially the same topography—both are almost entirely planar.

Notably, as the temperature increases from 1000°C to 1200°C, the extent of correlated fiber failure increases and the creep lifetime decreases. The increase in the spatial correlation in the fiber failure locations is among the main manifestations of the matrix densification [50, 51]. The near-planar fracture surfaces obtained at 1200°C in air and in steam for both composites indicate the loss of matrix porosity and subsequent matrix densification due to additional sintering. As a result, the N720/A and N720/AM composites exhibit decreased damage tolerance and reduced creep lifetimes at 1200°C.

When considering the retained tensile strength of the N720/A and N720/AM specimens that achieved creep run-out at 1000°C, it is instructive to examine their fracture surfaces in Figures 10.11 and 10.12. The fracture surfaces of the N720/AM run-out specimens tested at 1000°C in air and in steam (Figures 10.12a and 10.12e) are fairly brushy. The brushy appearance suggests that matrix changes and, consequently, loss of toughness and degradation of tensile strength were limited. Indeed these specimens retained 100% of their tensile strength. The fracture surface of the N720/A run-out specimen tested at 1000°C in air (Figure 10.11a) is also brushy. This specimen retained nearly 100% of its tensile strength. In contrast, the fracture surface of the N720/A run-out specimen tested at 1000°C in steam (Figure 10.11e) shows fairly extensive areas of coordinated fiber failure and increased fiber–matrix bonding. Apparently 100-hour exposure under load at 1000°C in steam caused changes in the alumina matrix, which led to loss of damage tolerance and resulted in lower retained strength. In the case of oxide–oxide composites, additional matrix sintering is accelerated at higher temperatures [50–52]. It appears that sintering of the alumina matrix in the N720/A composite is also accelerated in the presence of steam.

10.2.3 Tensile Creep and Recovery

It is recognized that the creep strain recovery effect in fiber-reinforced ceramic composites has important implications for life prediction for components subjected to cyclic creep loading. The creep rate and total strain accumulation under cyclic creep loading will be considerably lower than those under sustained creep loading. Therefore life predictions that do not consider strain recovery may significantly underpredict component life. Several studies have investigated the creep-recovery behavior of the SiC-fiber reinforced CMCs at temperatures over 1000°C in air [53–58]. The results revealed that, when compared to sustained creep loading, elevated-temperature cyclic creep loading can decrease the overall creep strain accumulation.

In many potential applications oxide–oxide CMCs will be subject to sustained creep and/or cyclic creep loading histories at elevated temperatures. The cyclic creep and recovery behavior of the N720/A composite with 0/90 fiber orientation was investigated at 1200°C in air and in steam [59]. Cyclic creep and recovery test (schematically shown in Figure 10.13) consisted of loading and unloading at 20 MPa/s between maximum (creep) stress and minimum (recovery) stress (fixed at 2 MPa) with a 1-hour hold time at σ_{max} and at σ_{min}. The creep stress levels σ_{max} of 100 and 125 MPa were selected to exceed the run-out stresses established in sustained creep tests [45]. The run-out was defined as 100 hours (cumulative) at creep stress. Because creep lifetime at 125 MPa in steam was only 0.24 hour [45], a shorter hold time of 0.05 hour was used in the 125 MPa creep-recovery test in steam.

Strain versus time curves produced in cyclic creep-recovery tests at 1200°C in air (Figure 10.14) show appreciable strain recovery at 2 MPa. Careful inspection of the strain–time curves also reveals a significant change in primary creep behavior with the creep-recovery cycles. Initially considerable primary creep is observed. However, after the first few cycles merely a hint of primary creep is seen upon reloading. This significant reduction in the duration of primary creep with cycles is a consequence of the strain recovery and the associated changes in the residual stress state of the composite that occur during unloading [59].

To compare creep strains accumulated during an equivalent length of time in cyclic creep-recovery tests and in sustained creep tests, the cyclic creep strains were plotted versus compressed time (Figure 10.15). In these plots the

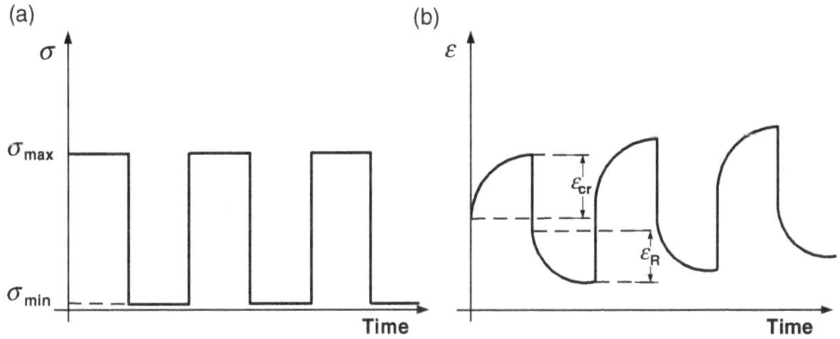

FIGURE 10.13

Schematic of stress input for cyclic creep and recovery experiment (a) and the associated strain response (b), where creep strain, ε_{cr}, and strain, ε_R, recovered during a recovery period are also shown. In all tests the recovery stress σ_{min} was 2 MPa, loading and unloading was performed at 20 MPa/s.

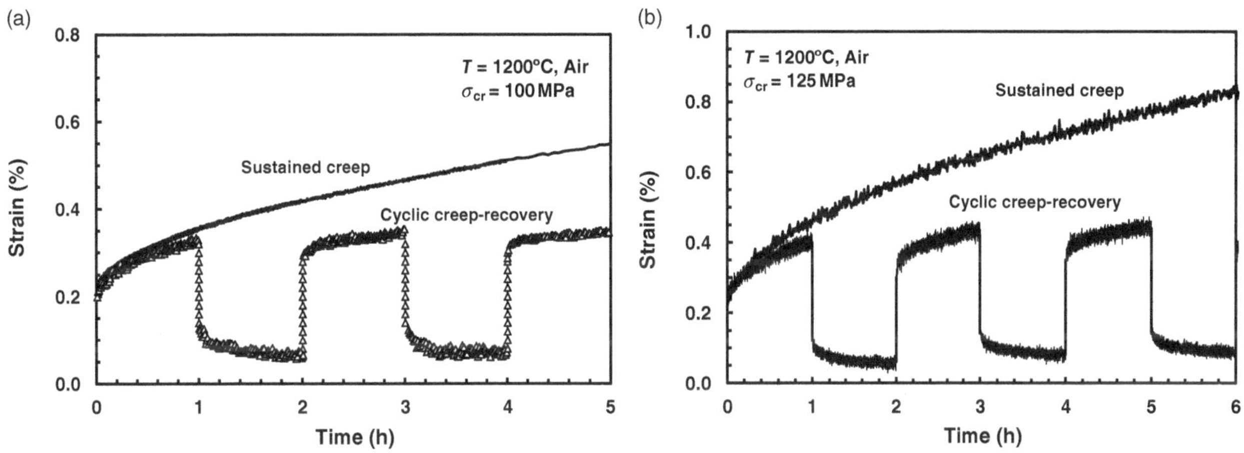

FIGURE 10.14

Cyclic creep recovery and sustained creep of N720/A composite at 1200°C in laboratory air for creep stress of (a) 100 MPa and (b) 125 MPa. Recovery stress is 2 MPa, hold time is 1 hour (after Ruggles-Wrenn et al. [45, 59]).

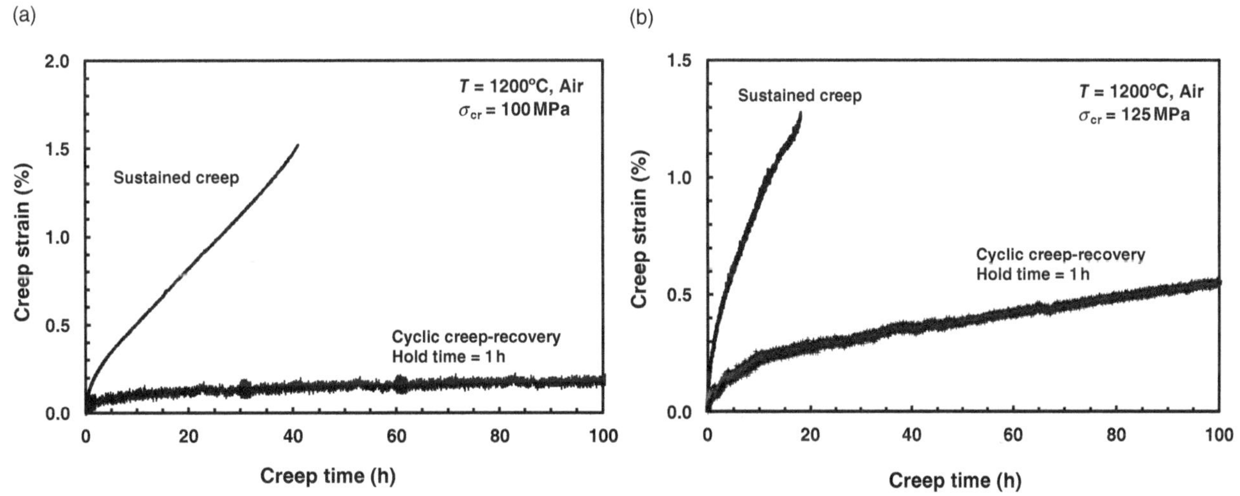

FIGURE 10.15

Comparison of creep strains accumulated during cyclic creep recovery and during sustained creep at 1200°C in air at the creep stress of (a) 100 MPa and (b) 125 MPa. Only creep portions of the cyclic creep-recovery test are shown. The recovery segments were removed and resulting curves were shifted to the left (after Ruggles-Wrenn et al. [45, 59]).

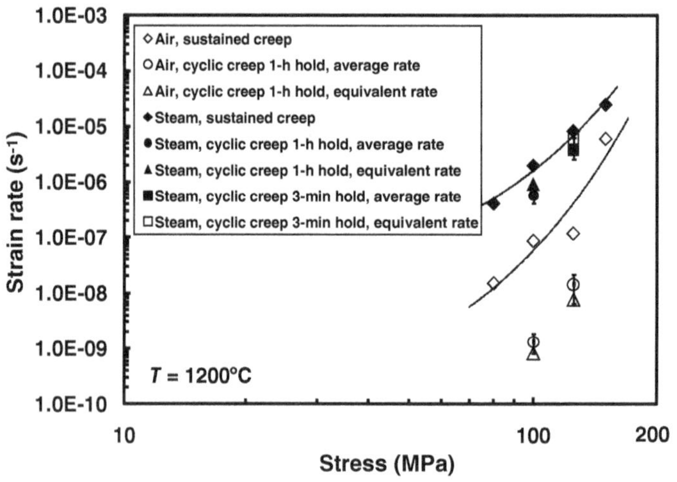

FIGURE 10.16

Minimum creep rate versus creep stress curves obtained in cyclic creep and in sustained creep tests for N720/A ceramic composite at 1200°C in air and in steam (after Ruggles-Wrenn et al. [45, 59]).

cyclic creep-recovery data were compressed by removing loading, unloading, and recovery periods and plotting only strain accumulated during creep portions of each cycle. The resulting strain versus time curves were shifted to the left. At 1200°C in air the introduction of the unloading and recovery at near-zero stress into the creep cycle considerably reduces the amount of accumulated creep strain. Strains accumulated in the cyclic creep-recovery test can be nearly an order of magnitude lower than that accumulated under sustained creep loading during an equivalent period of time. At 1200°C in air, the overall creep rate produced in a cyclic creep-recovery test is significantly lower than that produced in a sustained creep test for a given creep stress.

An average creep rate and an equivalent creep rate were calculated for every cyclic creep-recovery test. An equivalent creep rate was the secondary creep rate determined from the cyclic creep strain versus compressed time curves. An average creep rate was calculated by averaging the secondary creep rate values determined for each individual cycle. For all cyclic creep-recovery tests performed in air the average creep rate is close to the equivalent creep rate (Figure 10.16). Furthermore, creep rates obtained in the cyclic creep-recovery test are one to two orders of magnitude lower than those produced under sustained creep. Strain recovery and the associated changes in the residual stress state of the composite that occur during unloading result in a reduced duration of primary (transient) creep. The reduced duration of primary (transient) creep causes reductions in average and equivalent creep rates as well as the reduction in accumulated creep strain.

At 1200°C in air, the introduction of intermittent periods of unloading and recovery at near-zero stress considerably improves creep lifetime of the N720/A composite (Figure 10.17). Strain recovery is a positive phenomenon and should be taken into account when predicting the life of cyclically loaded components operating in air. Because the accumulated creep strain is frequently an important design consideration, life predictions that do not account for strain recovery may significantly underestimate service life of the component as is evidenced in Figure 10.17.

The steam environment has a profound influence on the cyclic creep-recovery behavior of the N720/A composite at 1200°C. In steam creep-strain recovery is negligible. Examination of the creep strain versus compressed time curves produced at 100 MPa in steam (Figure 10.18) reveals little change in primary creep behavior with cycles. Whereas in air the duration of primary creep was dramatically reduced with cycling, in steam the behavior of primary creep becomes relatively stable after the first cycle. As a consequence, in steam the average creep rate and the equivalent creep rate obtained in the cyclic creep-recovery test are very close to the creep rate produced under sustained creep. Moreover, in steam similar creep strains (Figure 10.18) and similar creep lifetimes (Figure 10.17) are produced in cyclic creep-recovery tests and under sustained creep.

Fracture surfaces obtained in the cyclic creep-recovery tests at 1200°C in air (Figures 10.19a–10.19c) exhibit larger amounts of fiber pullout and longer damage zones than those obtained in sustained creep (Figures 10.19d–10.19f). While the sustained creep fracture surface (Figures 10.19d and 10.19e) is dominated by planar regions of coordinated fiber failure, the regions of uncorrelated fiber fracture are prevalent in the cyclic creep-recovery fracture surface (Figures 10.19a

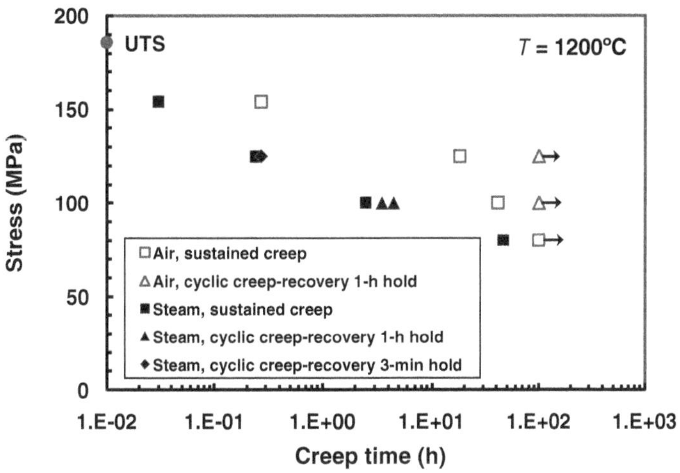

FIGURE 10.17

Creep stress versus time to rupture for N720/A ceramic composite at 1200°C in laboratory air and in steam. For the cyclic creep-recovery tests, the recovery time segments of each cycle are removed and only time at creep stress is shown (after Ruggles-Wrenn et al. [45, 59]).

and 10.19b). Only small amounts of matrix material remain bonded to the fiber surfaces of the specimen subjected to cyclic creep recovery (Figure 10.19c).

Steam environment has a dramatic effect on the fracture surface appearance. The creep-recovery specimen tested in steam (Figures 10.20a–10.20c) produced a considerably less brushy fracture surface than that tested in air. In steam the test history (cyclic creep recovery vs. sustained creep) has little influence on the fracture surface topography. The fracture surface produced in the cyclic creep-recovery test (Figures 10.20a and 10.20b) is similar to that obtained under sustained creep (Figures 10.20d and 10.20e) at 1200°C in steam; both fracture surfaces are nearly planar. Large amounts of matrix material adhere to the fiber surfaces of both specimens (Figures 10.20c and 10.20f). Moreover, amounts of matrix material that remain bonded to the fibers are much greater in the case of the specimens tested in steam (Figures 10.20c and 10.20f) than in the case of those tested in air (Figures 10.19c and 10.19f).

Results of the cyclic creep-recovery tests and sustained creep tests suggest that for the N720/A composite tested at 1200°C in air and in steam, the fracture surface appearance could be correlated with the failure time, with a

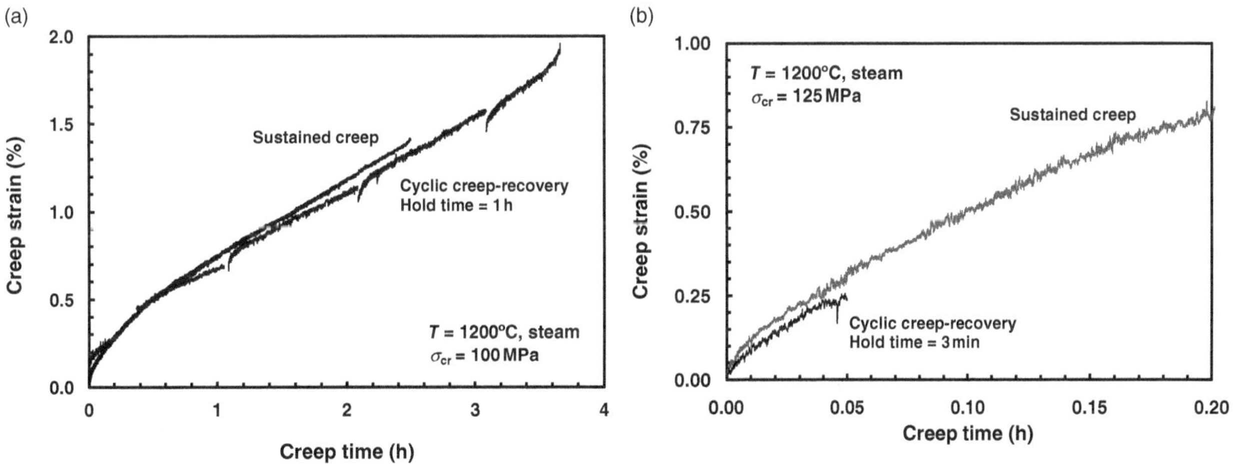

FIGURE 10.18

Comparison of creep strains accumulated during cyclic creep recovery and during sustained creep at 1200°C in steam at the creep stress of (a) 100 MPa and (b) 125 MPa. Only creep portions of the cyclic creep-recovery test are shown. The recovery segments were removed and the resulting curves were shifted to the left (after Ruggles-Wrenn et al. [45, 59]).

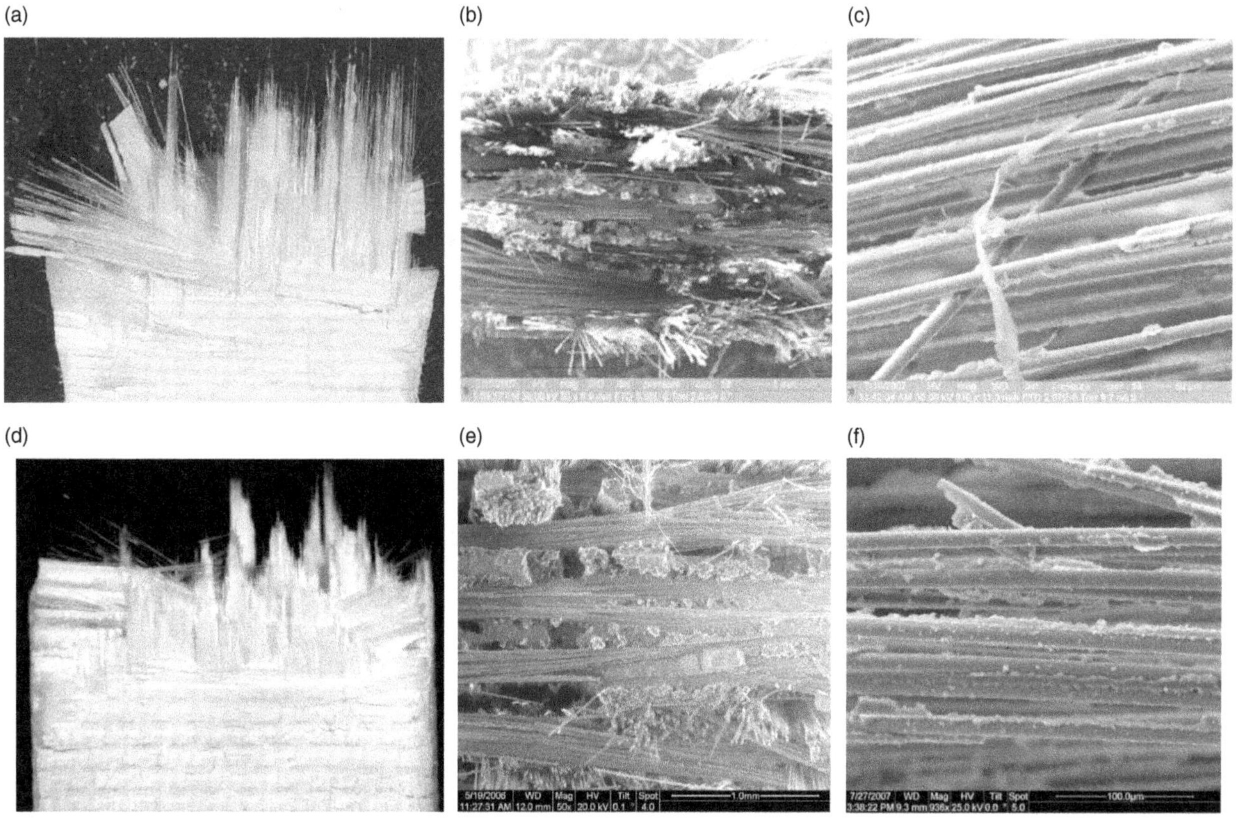

FIGURE 10.19

Fracture surfaces obtained at 1200°C and σ_{max} = 125 MPa in air: (a)–(c) cyclic creep recovery, t_f > 100 hours; (d)–(f) sustained creep, t_f = 18.1 hours (after Ruggles-Wrenn et al. [45, 59]).

predominantly planar fracture surface corresponding to a short lifetime and fibrous fracture indicating longer lifetime. The extent of correlated fiber failure increases and the creep lifetime decreases in the presence of steam regardless of the test history, indicating matrix densification. For an oxide–oxide composite, which relies on matrix porosity for its toughness, the matrix densification is a key mechanism for degradation of mechanical properties and performance [50–52]. Results presented here indicate that for N720/A additional matrix sintering is accelerated in the presence of steam.

The N720/A composite exhibits considerable creep-strain recovery at 1200°C in air, indicating that creep strains can be largely recovered under zero-load conditions, provided that the applied stress is sufficiently low to prevent widespread microstructural damage during initial loading and subsequent creep. The presence of steam at 1200°C accelerates additional sintering of the matrix, causing the loss of matrix porosity and subsequent matrix densification and increased fiber/matrix bonding. As a result, the N720/A composite exhibits decreased damage tolerance and reduced creep lifetimes at 1200°C in steam [45, 59]. In addition, matrix densification and increased fiber/matrix bonding fundamentally change the way the stresses are redistributed between the fibers and matrix during creep portion of the creep-recovery cycle. The redistribution of stresses influences the residual stress state of the fibers and matrix, which develops upon unloading and is largely responsible for the creep-strain recovery [56]. At 1200°C in steam, matrix densification and increased fiber/matrix bonding inhibit the development of the residual stresses that provide a mechanical driving force for the creep-strain recovery process.

10.2.4 Tensile Creep—N720/A and N720/AM Composites with ±45 Fiber Orientation

The woven CMC materials developed for use in aerospace engine components are typically made from 0/90 fiber architectures. Because oxide–oxide CMCs contain an exceptionally weak porous matrix, the composite properties depend almost entirely on the fibers. Hence the material design represents a source of significant anisotropy. Mechanical

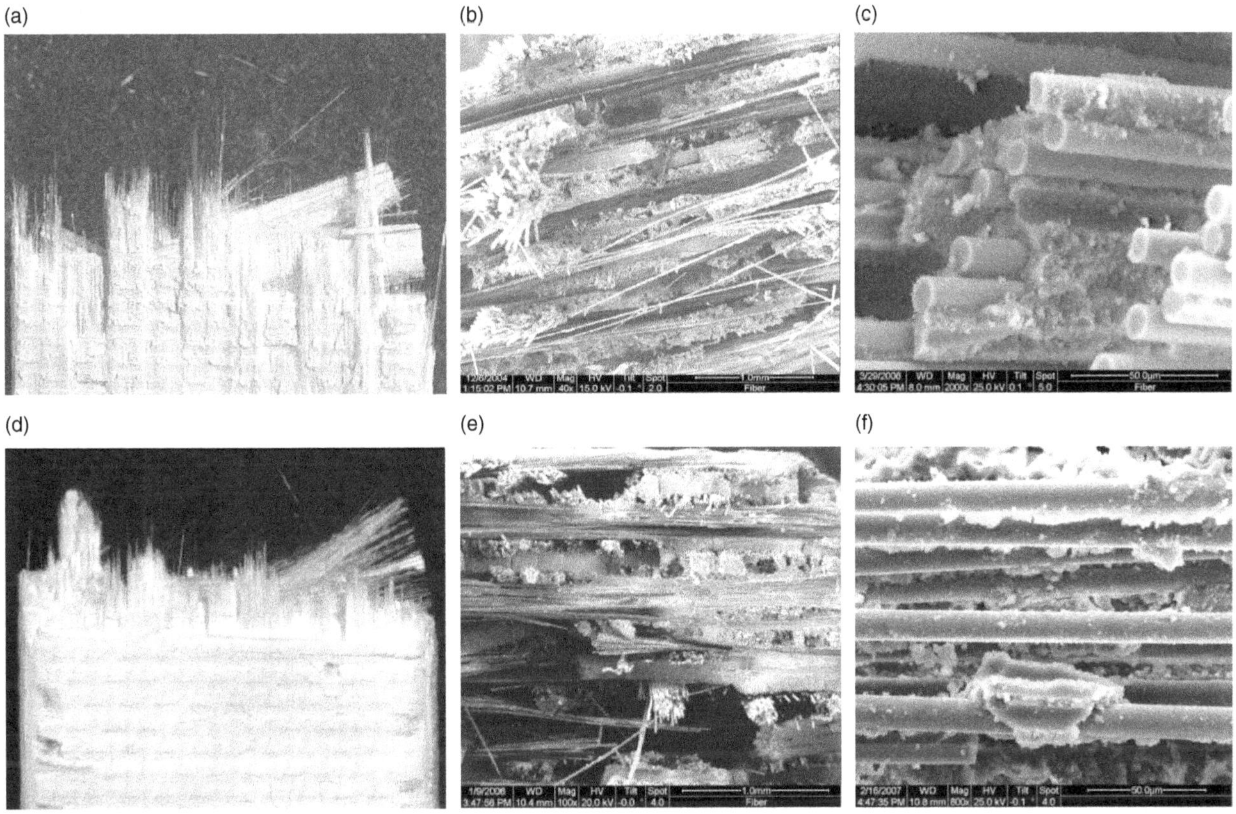

FIGURE 10.20

Fracture surfaces obtained at 1200°C and σ_{max} = 125 MPa in steam: (a)–(c) cyclic creep recovery, t_f = 0.052 hours; (d)–(f) sustained creep, t_f = 0.24 hours (after Ruggles-Wrenn et al. [45, 59]).

property data collected for N720/aluminosilicate composites with 0/90 and the ±45 orientations [22, 60–62] show a substantial level of anisotropy in the elastic modulus, tensile strength, and failure strain. Moreover, the anisotropy of the N720/aluminosilicate and N720/alumina CMCs was amplified dramatically under high-temperature creep in air [22, 33, 61, 63–65]. These findings demonstrate that the effect of anisotropy on the material performance cannot be underestimated.

The combined effects of anisotropy and environment on tensile and creep behaviors of N720/A and N720/AM composites were assessed at 1200°C [65,66]. Tensile stress–strain behavior of the two composites is typified in Figure 10.21. Whereas the stress–strain curves for the 0/90 fiber orientation are nearly linear to failure, the stress–strain behavior for the ±45 orientation becomes nonlinear at fairly low stresses (~15 MPa). As the stress exceeds a threshold value (~50 MPa for N720/A, ~32 MPa for N720/AM), appreciable inelastic strains develop rapidly. Both composites achieve a strain of ~0.3% at the maximum load. After the UTS is reached, appreciable inelastic strains develop at near-constant stress. The elastic moduli (46 GPa for N720/A and 38.6 GPa for N720/AM) and UTS values (55 MPa for N720/A and 37 MPa for N720/AM) obtained for the ±45 orientation are considerably lower than those for the 0/90 specimens.

At 1200°C in air, both composites exhibit primary and secondary creep regimes, but no tertiary creep (Figures 10.22 and 10.23). In contrast, the creep curves obtained in steam show primary, secondary, and tertiary creep. In steam, tertiary creep is observed at 45 MPa (82% UTS) for N720/A and at 32 MPa (86% UTS) for N720/AM. In air, N720/AM achieved a 100-hour creep run-out in all tests, while N720/A achieved run-out only at stress levels <70% UTS. In steam, both composites achieved run-out only at stresses <70% UTS. Moreover, both materials accumulated higher creep strains in steam than in air. This trend is particularly pronounced for N720/AM, where the creep strain produced in steam can be nearly five times that in air.

Secondary creep rates of N720/A and N720/AM composites are profoundly affected by fiber orientation (Figure 10.24). In air, the secondary creep rate of N720/A with ±45 orientation can be nearly 10^6 times that of the 0/90 orientation. Similarly, in air the creep rate of N720/AM composite with ±45 orientation can be nearly 10^3 times that of the

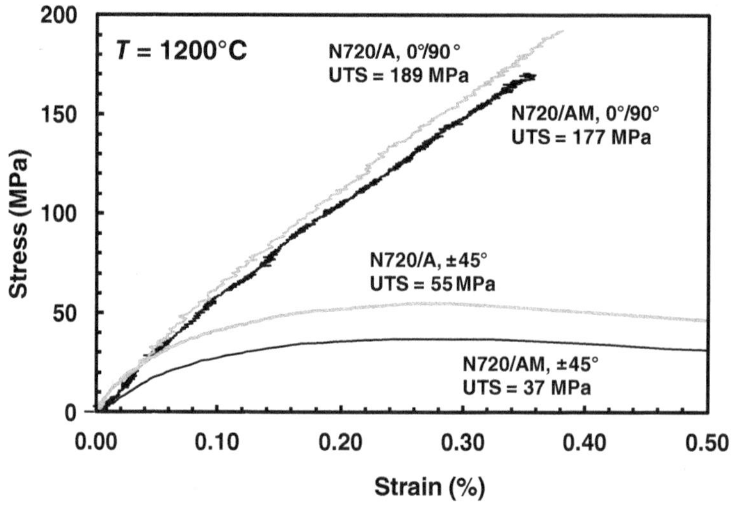

FIGURE 10.21

Tensile stress–strain curves obtained for N720/A and N720/AM ceramic composites with 0°/90° and ±45° fiber orientations. All data are adjusted for $V_f = 0.44$ (after Ruggles-Wrenn et al. [65, 66]).

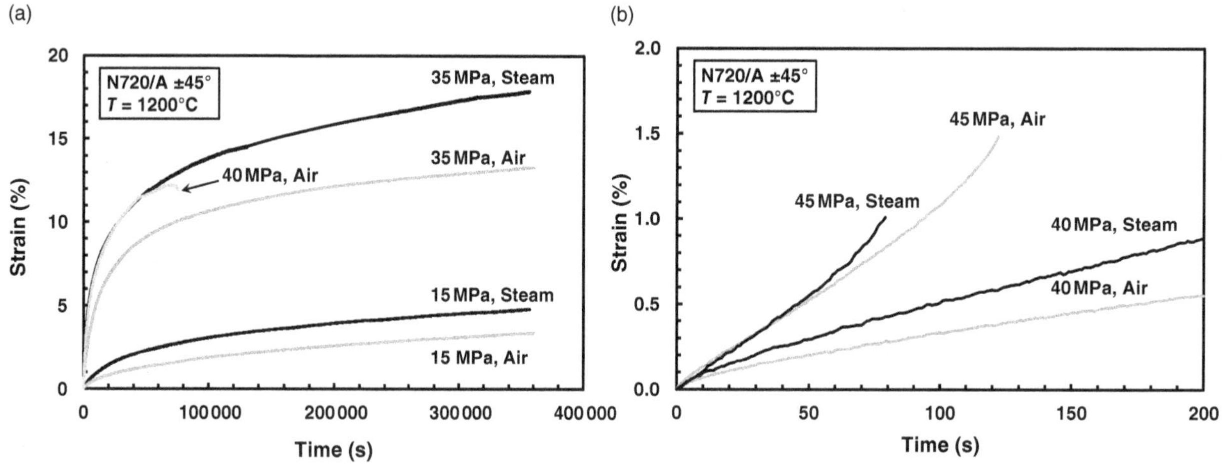

FIGURE 10.22

Creep strain versus time curves obtained for N720/A ceramic composite with ±45° fiber orientation at 1200°C in air and in steam: (a) at 15 and 35 MPa and (b) at 40 and 45 MPa (after Ruggles-Wrenn et al. [65]).

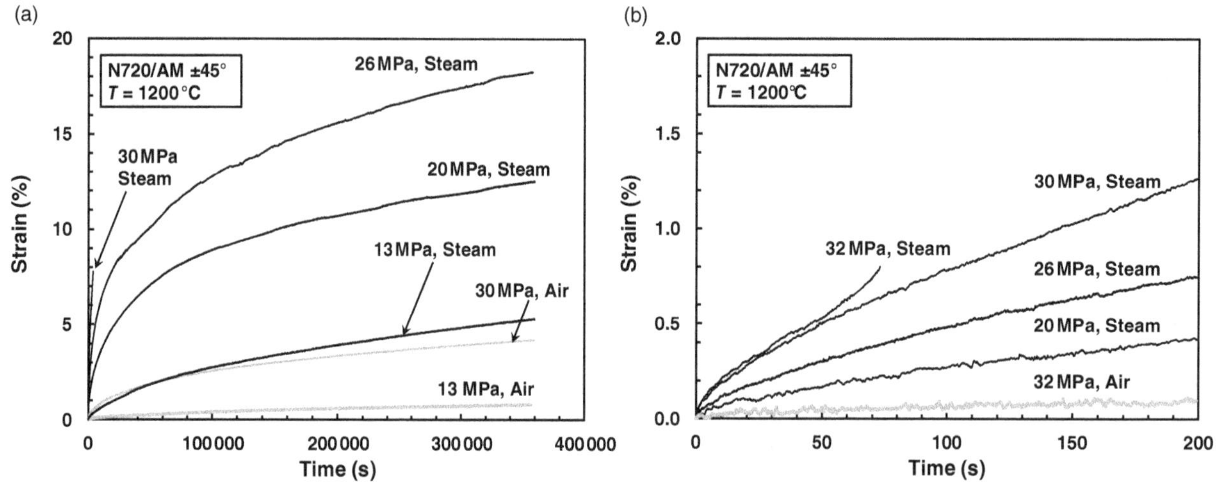

FIGURE 10.23

Creep strain versus time curves obtained for N720/AM ceramic composite with ±45° fiber orientation at 1200°C in air and in steam: (a) at stresses ≤30 MPa and (b) at 32 MPa (after Ruggles-Wrenn et al. [66]).

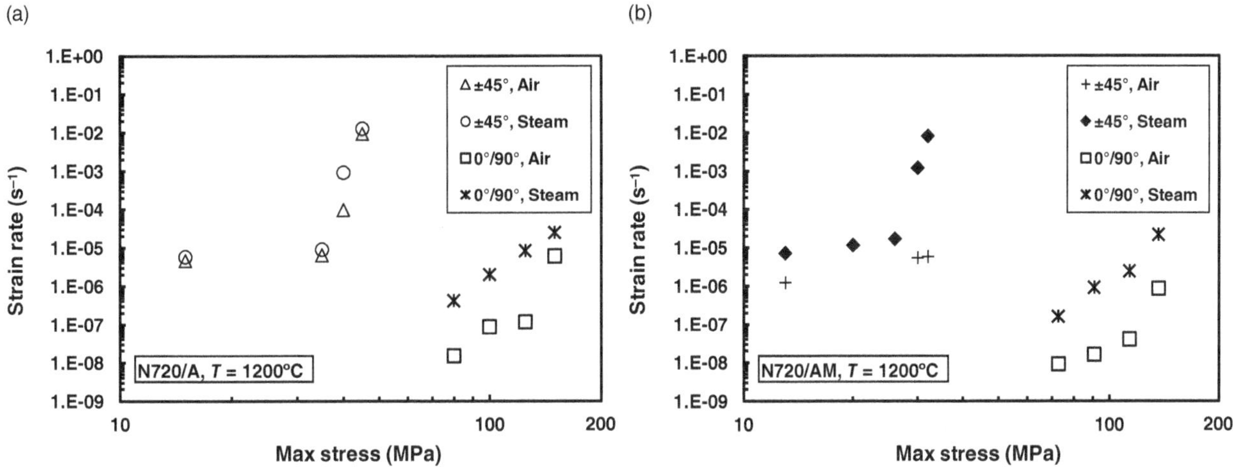

FIGURE 10.24

Minimum creep rate as a function of applied stress for (a) N720/A and (b) N720/AM ceramic composites with 0°/90° and ±45° fiber orientations at 1200°C in air and in steam (after Ruggles-Wrenn et al. [65, 66]).

0/90 orientation. In steam, the ±45 creep rate of N720/A can be as high as 10^5 times the 0/90 rate. Likewise, the ±45 creep rate of N720/AM can be as high as 10^4 times the 0/90 rate. This result is hardly surprising, considering that the creep rupture of the 0/90 orientation is governed by the highly creep-resistant Nextel 720 fibers, while the creep rupture of the ±45 orientation is dominated by an exceptionally weak porous alumina matrix.

The presence of steam has little influence on the ±45 creep rates of N720/A (Figure 10.24a). Conversely, the ±45 creep rates of N720/AM are strongly affected by steam (Figure 10.24b). For creep stresses ≥80%UTS, the creep rates of N720/AM in steam can be three orders of magnitude higher than those obtained in air. The ±45 creep rates of N720/AM are compared to those of N720/A in Figure 10.25, where the applied stresses are normalized by the average UTS at 1200°C. In air the secondary creep rate of N720/A can be as high as 10^3 times the rate of the N720/AM composite. In contrast, creep rates of N720/A were comparable to those of N720/AM in the presence of steam.

Stress-rupture behavior of the two composites with ±45 fiber orientation is summarized in Figure 10.26, where the applied stresses are normalized by the average UTS at 1200°C. In air, the creep run-out stress was ~86% UTS for N720/AM, but only ~64% UTS for the N720/A. Steam environment has little effect on the creep lifetimes (up to 100 hours) of both composites for applied stresses <65% UTS. The N720/AM composite nearly achieved creep run-out

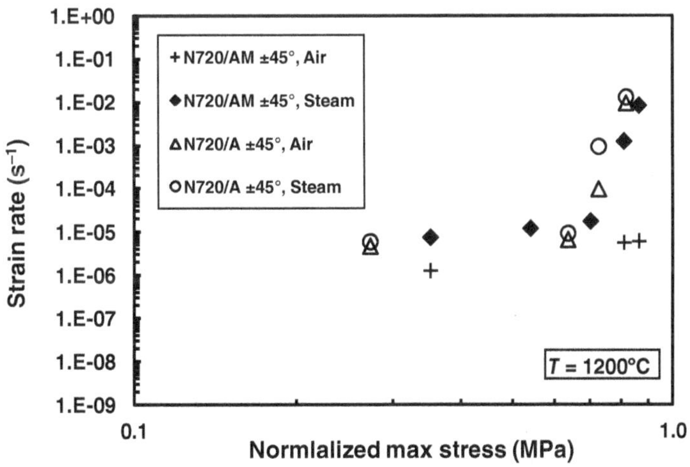

FIGURE 10.25

Minimum creep rate as a function of normalized applied stress for N720/AM and N720/A ceramic composites with ±45° fiber orientation at 1200°C in air and in steam (after Ruggles-Wrenn et al. [65, 66]).

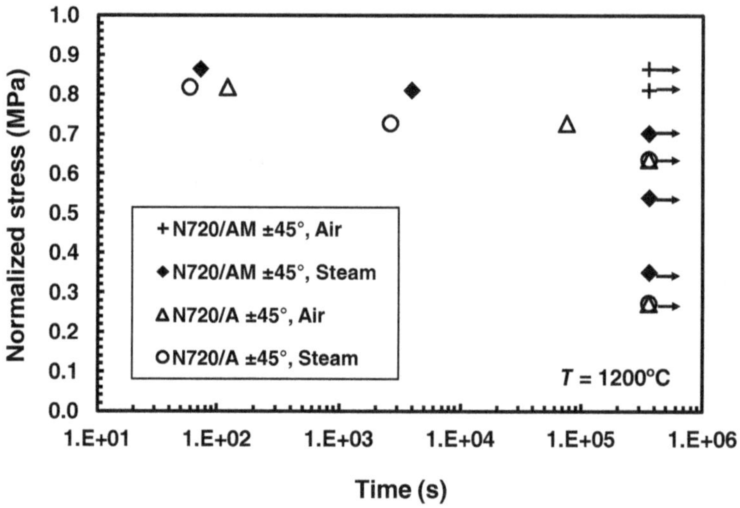

FIGURE 10.26
Normalized maximum stress versus time to rupture for N720/AM and N720/A ceramic composites with ±45° fiber orientation at 1200°C in air and in steam (after Ruggles-Wrenn et al. [65, 66]).

of 100 hours at the creep stress of 70% UTS. Yet at higher creep stresses, creep lifetimes of both composites were reduced by at least 97% in the presence of steam.

Retained tensile strength of all specimens that achieved creep run-out was evaluated at 1200°C. Prior creep appears to have a beneficial effect on tensile strength. Prior creep in air or in steam environment increased the tensile strength of both composites. However, prior creep considerably reduced the materials' capacity for inelastic straining. Since the tensile properties of the composite with the ±45 fiber orientation are largely dominated by the matrix, these results indicate that matrix strengthening (most likely due to additional sintering) takes place during the 100 hours of creep.

The fracture surfaces of the N720/A composite are strongly influenced by the duration of creep test. In creep tests with higher creep stress and, consequently, shorter duration, the N720/A specimens with ±45 fiber orientation fracture along the planes at ~45° to the loading direction (Figures 10.27a and 10.27c). The failure occurs primarily through the matrix, with only minimal fiber fracture. The SEM micrographs (Figures 10.27b and 10.27d) reveal relatively small amounts of matrix particles bonded to the fiber surfaces. Moreover, in creep tests of short duration, steam environment has little effect on the failure mechanism and/or fracture surface appearance.

Contrastingly, the failure of the N720/A specimens subjected to 100 hours of prior creep at a lower stress level (Figures 10.28a and 10.28c) includes extensive fiber fracture. The N720/A specimen tested in air (Figure 10.28a) shows areas of fiber failure, while retaining some of the V-shape fracture typically seen in the porous matrix composites with ±45 fiber orientation. However, specimens subjected to 100 hours of creep in steam fractured along the planes nearly orthogonal to the loading direction (Figure 10.28c). Furthermore, the amount of matrix material remaining bonded to the fibers is greater in the specimens subjected to 100 hours of creep than that in the specimens subjected to creep of shorter duration (Figures 10.28b and 10.28d). This difference is particularly pronounced in specimens tested in steam.

Notably at 1200°C in air, duration of the creep test has little effect on the failure mechanism and/or appearance of the N720/AM fracture surface. All N720/AM specimens subjected to creep in air fractured along the planes at ~45° to the loading direction (as typified in Figure 10.29a). All N720/AM fracture surfaces obtained in air are predominantly fibrous (Figure 10.29b). Only small amounts of matrix particles remain bonded to the fiber surfaces (Figure 10.29c). In the presence of steam, the effects of the creep test duration (i.e., exposure to steam environment under load) are amplified considerably. Specimens subjected to creep of short duration at higher stresses fail primarily through matrix damage and interply delamination with little fiber fracture (Figure 10.30). The V-shaped fracture (Figure 10.30a) and predominantly fibrous fracture surface (Figure 10.30b) are typical. By contrast, the specimen failed in a tensile test after having achieved a 100-hour creep run-out in steam (Figure 10.31a) partially fractured along the plane nearly orthogonal to the loading direction. In this case, the fracture surface is dominated by planar regions of coordinated fiber failure (Figure 10.31b) and large amounts of matrix material remain bonded to the fiber surfaces (Figure 10.31c).

Results of this study suggest that additional sintering of the matrix occurred during 100-hour creep tests, especially those conducted in steam. For N72/A and N720/AM composites, the resultant strengthening of the matrix and the

FIGURE 10.27

Fracture surfaces of N720/A composite with ±45° fiber orientation tested in creep at 45 MPa at 1200°C. (a) and (b) in air; and (c) and (d) in steam. Test duration ≤ 120 seconds.

fiber–matrix interfaces is also manifested in the retained properties. After 10 hours of prior creep, the tensile strength values of both composites increase, and the inelastic straining capabilities decrease. In addition, the matrix strengthening is manifested through the change in the failure mechanism. The failure of the specimens subjected to 100 hours of prior creep in steam is dominated by fiber fracture, while the as-processed materials fail predominantly through matrix damage and interply delamination.

10.2.5 Compression and Compression Creep

Monotonic compression and compressive creep behaviors of the N720/A composite were investigated at 1200°C in air and steam [67]. Compression tests were carried out in air with a constant stress rate magnitude of 25 MPa/s. The average compressive modulus (74 GPa) is close to the tensile modulus. Conversely, the average compressive strength (130 MPa) is considerably less than the tensile strength (189 MPa). Likewise the average compressive failure strain of –0.20% is significantly lower than the 0.38% failure strain obtained in tensile test.

Compressive creep-rupture tests were performed at 1200°C with creep stresses ranging from –40 to –100 MPa (30–77% of compressive strength). At 1200°C in air, the composite exhibits primary and secondary creep, but no tertiary creep (Figure 10.32a). Creep run-out of 100 hours was achieved in all tests performed in air. Recall that at 1200°C in air tensile creep run-out was achieved only at 80 MPa (42% UTS). The tensile creep strains accumulated in air ranged from 0.58% to 1.52% and significantly exceeded the failure strain obtained in tension test. Conversely, all compressive creep strain magnitudes remain ≤0.4% and relatively close to the failure strain obtained in compression test. The presence

FIGURE 10.28

Fracture surfaces of N720/A composite with ±45° fiber orientation tested in creep at 35 MPa at 1200°C. (a) and (b) in air; and (c) and (d) in steam. Test duration > 100 hours.

of steam has a dramatic effect on compressive creep at 1200°C (Figure 10.32b). In steam, compressive creep run-out was not achieved. Compressive creep strains accumulated at –40 and –60 MPa in steam are an order of magnitude higher than those accumulated in air. Nevertheless, compression creep strains are lower than the tensile creep strains accumulated in steam.

Like the tensile creep rates, compression creep rates are strongly affected by steam environment at 1200°C (Figure 10.33). In air, compressive creep rates are nearly an order of magnitude higher than the tensile creep rates. This result is readily understood, once we recall that in tension creep-rupture of the N720/A is governed by creep-rupture of the creep resistant Nextel™720 fibers. Conversely, in compression creep-rupture of the composite is governed by a very weak porous matrix. Presence of steam increased tensile creep rates of N720/A approximately by an order or magnitude. Degrading effect of steam on secondary creep strain rates is more pronounced in compression. In steam, compressive creep rate can be 10^5 times that obtained in air. Moreover, in steam, compressive creep rates can be nearly four orders of magnitude higher than the tensile creep rates.

The presence of steam dramatically reduces creep lifetimes in tension and compression (Figure 10.34). At 1200°C in air, compressive creep lifetime (up to 100 hours) is independent of applied stress; a 100-hour creep run-out is achieved in all tests. In steam, compressive creep run-out is not achieved. Furthermore, compressive creep lifetimes can be reduced by as much as 99% due to steam. Recall that steam had an equally detrimental effect on tensile creep performance of N720/A at 1200°C. Although tensile creep run-out was achieved at 80 MPa (42% UTS) at 1200°C in air, it was not achieved in steam. Tensile creep lifetimes were reduced by as much as 90% due to steam.

Retained compressive strength of all specimens that achieved compression creep run-out was measured at 1200°C. Prior compression creep at 1200°C in air did not degrade compressive strength of N720/A. Retained tensile strength

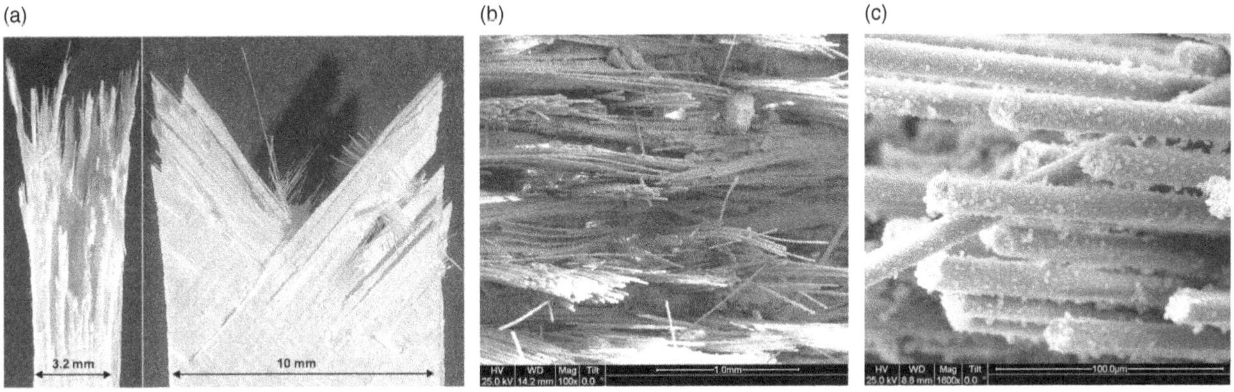

FIGURE 10.29

Fracture surface of N720/AM composite with ±45° fiber orientation tested in creep at 30 MPa at 1200°C in air. Test duration > 100 hours.

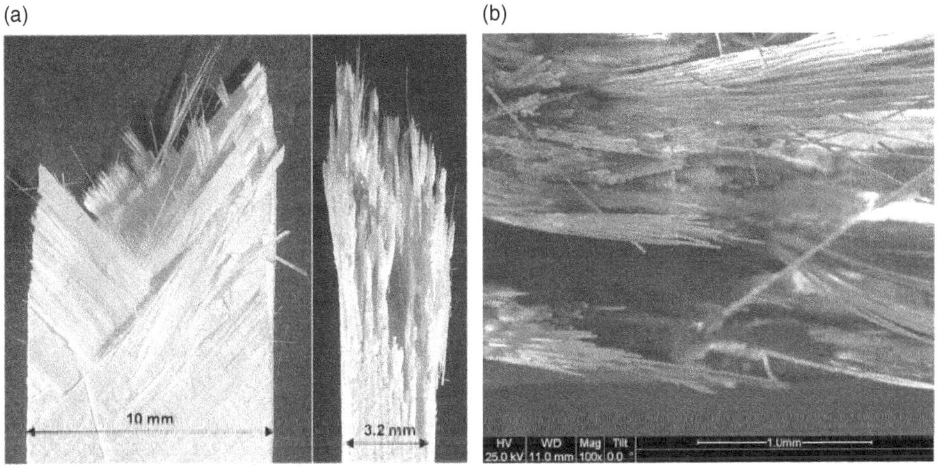

FIGURE 10.30

Fracture surface of N720/AM composite with ±45° fiber orientation tested in creep at 32 MPa at 1200°C in steam. Test duration is 72 seconds.

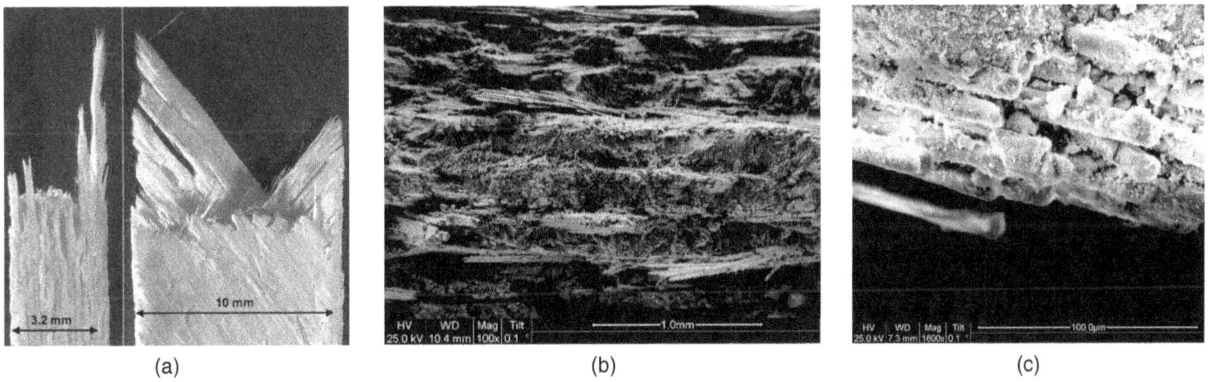

FIGURE 10.31

Fracture surface of N720/AM composite with ±45° fiber orientation tested in creep at 20 MPa at 1200°C in steam. Test duration > 100 hours.

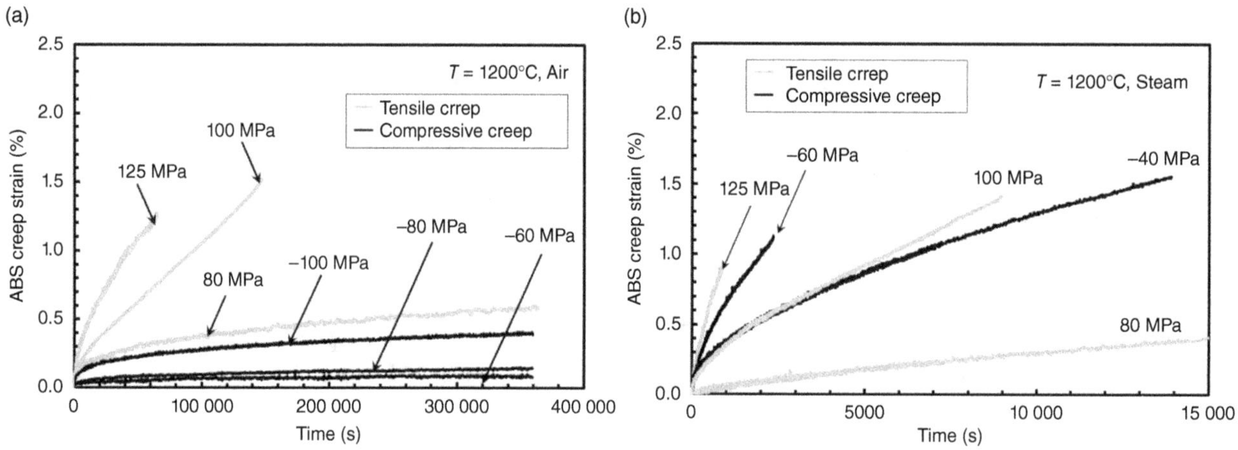

FIGURE 10.32

Compressive and tensile creep strain versus time curves obtained for N720/A ceramic composite at 1200°C: (a) in air and (b) in steam (after Ruggles-Wrenn et al. [67]).

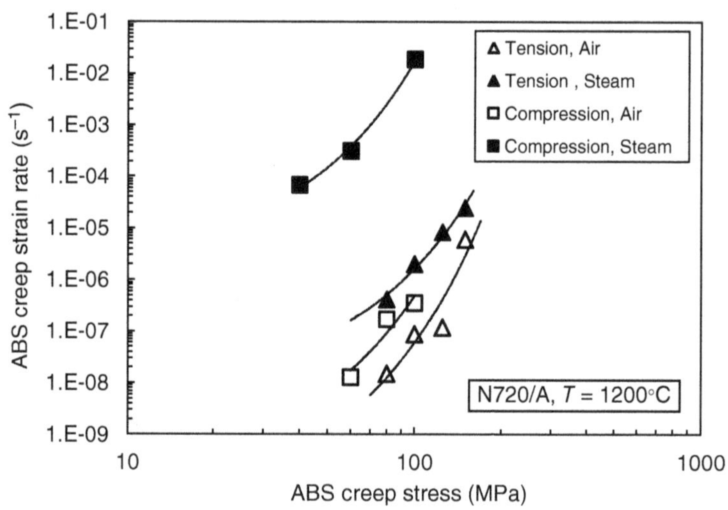

FIGURE 10.33

Minimum creep rate magnitude as a function of applied stress magnitude for N720/A ceramic composite at 1200°C in air and in steam (after Ruggles-Wrenn et al. [67]).

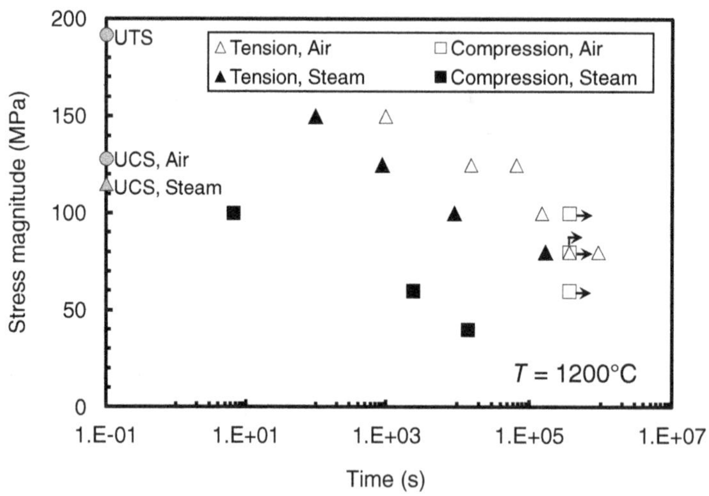

FIGURE 10.34

Creep stress magnitude versus time to rupture for N720/A ceramic composite at 1200°C in air and in steam (after Ruggles-Wrenn et al. [67]).

FIGURE 10.35

Fracture surface of N720/A composite obtained in compression to failure at 1200°C.

of a specimen that achieved a run-out in compression creep test at –80 MPa was also evaluated at 1200°C. Prior compressive creep caused a 30% decrease in tensile strength. Recall that 100 hours of prior tensile creep at 80 MPa in air reduced the tensile strength of N720/A by only 8%.

A typical fracture surface of the N720/A obtained in compression to failure at 25 MPa/s at 1200°C (Figure 10.35) exhibits considerable fibrous fracture and a fairly long damage zone (~46 mm). Similar fracture surfaces are produced in compression tests following 100 hours of creep at 1200°C in air (as typified in Figure 10.36a). Sizable areas of fibrous fracture are still visible, although the damage zone becomes somewhat smaller (~32 mm). Conversely, the fracture surfaces obtained in creep tests in steam (Figure 10.36b) are dominated by coordinated fiber failure and show a much decreased damage zone (approximately 12–16 mm), suggesting that the matrix has densified. Yet the denser matrix did not improve the mechanical performance in compression as may be expected. On the contrary, the most severe degradation of the compressive performance was observed in steam.

10.2.6 Creep in Interlaminar Shear

While CMCs exhibit improved damage tolerance compared with monolithic ceramics, two-dimensional laminated CMCs are more susceptible to failure in the matrix-rich interlaminar regions. The interlaminar failure or delamination may ultimately lead to loss of stiffness and accelerate structural failure [68]. Several studies [69–74] focused on the life-limiting behavior in interlaminar shear of non-oxide CMCS with matrix-dense interlaminar regions, where the failure is controlled by the fiber–matrix interface.

In the case of the porous-matrix oxide–oxide CMCs, interlaminar shear failure is controlled by the weak porous matrix. Creep behavior in interlaminar shear of an oxide–oxide N720/A composite was studied at 1200°C in air and in steam [75]. To investigate the interlaminar shear properties and stress rupture behavior, the double-notch shear (DNS) test specimens (Figure 10.37) were loaded in compression along the specimen axis. The DNS test specimen is

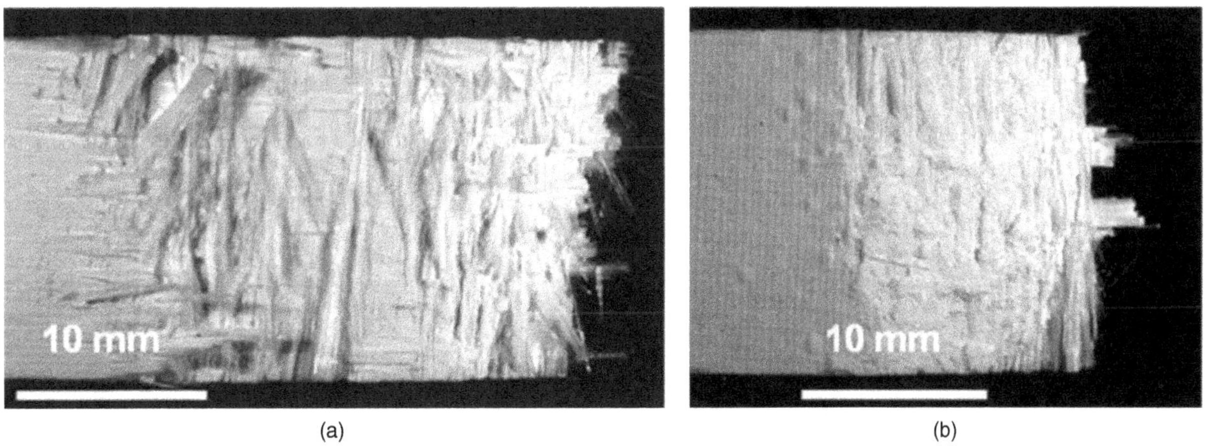

FIGURE 10.36

Fracture surface of N720/A specimens tested in compression creep at –60 MPa at 1200°C: (a) in air, test duration ≥ 100 hours; and (b) in steam, test duration = 0.66 hours.

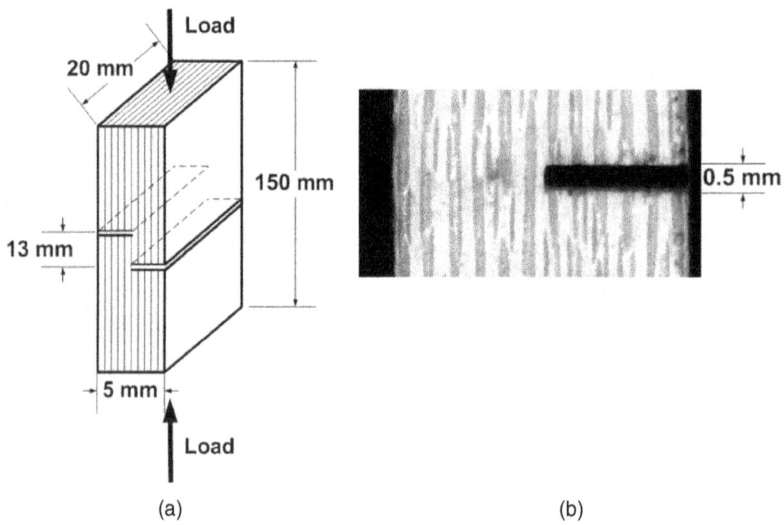

FIGURE 10.37
Double-notch shear specimen: (a) configuration and dimensions and (b) notch details (after Ruggles-Wrenn et al. [75]).

designed to promote shear failure on the plane between the notch tips (Figure 10.38). The 13-mm distance between the notches was chosen specifically to enable the measurement of compressive strain between the notch tips with an MTS high-temperature extensometer of 12.5-mm gage length. The overall specimen length of 150 mm also ensures that the local stress fields at the notch tips are not influenced by the external loading at the specimen ends. The average shear stress, τ, between the notches along the prospective shear plane was calculated as

$$\tau = \frac{P_{\mathrm{a}}}{WL_{\mathrm{n}}} \tag{10.7}$$

where P_a is the applied force, W is the specimen width, and L_n is the distance between the notches. Although the shear stress distribution between the notches is not uniform, the average stresses provided by Equation (10.1) are useful when comparing interlaminar shear strength (ILSS) values obtained in different environments.

The average ILSS of 8.25 MPa was obtained in compression tests performed with the displacement rate of 0.05 mm/s at 1200°C in air. Creep tests in interlaminar shear were performed with creep stresses ranging from 4.0 to 6.5 MPa

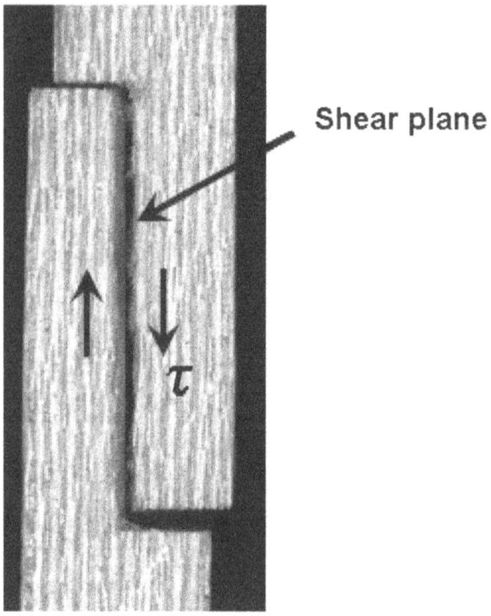

FIGURE 10.38
Micrograph of a typical N720/A specimen after interlaminar shear failure (side view).

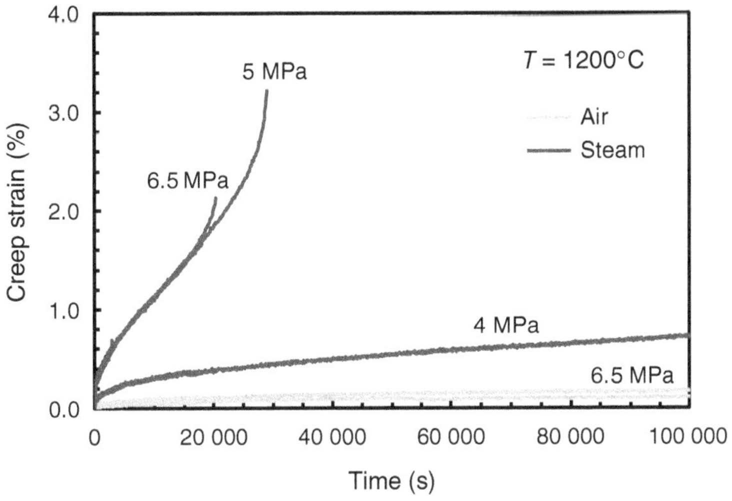

FIGURE 10.39

Creep strain versus time curves for N720/A composite obtained at applied interlaminar shear stresses in the 4–6.5 MPa range at 1200°C in air and in steam (after Ruggles-Wrenn et al. [75]).

(approximately 50–79% ILSS). In air, N720/A exhibits primary and secondary creep (Figure 10.39). In air, creep run-out of 100 hours is achieved at the shear stress of 6.5 MPa (79% ILSS). Creep strains accumulated in air are comparable to those obtained in monotonic tests. In steam, the N720/A composite exhibits primary, secondary, and tertiary creep (Figure 10.39). Creep strains accumulated in steam are at least an order of magnitude higher than the failure strains obtained in monotonic tests. In steam, creep run-out was achieved only at 4 MPa (50% ILSS).

In interlaminar shear secondary creep rate is strongly influenced by the presence of steam at 1200°C (Figure 10.40). In steam, secondary creep rate increases by an order of magnitude when applied stress increases from 4 to 5 MPa. At 6.5 MPa, creep rate in steam is at least two orders of magnitude higher than that in air. In air, creep lifetime in interlaminar shear (up to 100 hours) is independent of applied shear stress up to 78% ILSS; 100-hour creep run-out is achieved in all tests (Figure 10.41). The presence of steam degrades creep performance in interlaminar shear. For applied shear stresses ≥5 MPa, creep lifetimes can be reduced by as much as 94% due to steam. Prior creep at 1200°C in air improved the ILSS by nearly 37%. Conversely, prior creep in steam has decreased the ILSS of N720/A by about 27%.

A typical fracture surface of the DNS specimen tested in compression to failure at 1200°C in air is shown in Figure 10.42. Failure occurred through delamination of the woven 0/90 fiber layers from the matrix-rich regions with minimal fiber fracture. A largely smooth and clean fracture surface indicates that a single fiber layer is associated with delamination (Figures 10.42a and 10.42b). Only relatively small amounts of the matrix material remain bonded to the fibers exposed during delamination (Figure 10.42c). In contrast, failure of the specimen tested in compression following 100 hours of prior creep at 6.5 MPa at 1200°C in air includes considerable fiber fracture (Figures 10.43a and 10.43b). Fracture surface in Figure 10.43c shows rougher areas exposing multiple 0/90 fiber layers as well as limited areas of clean

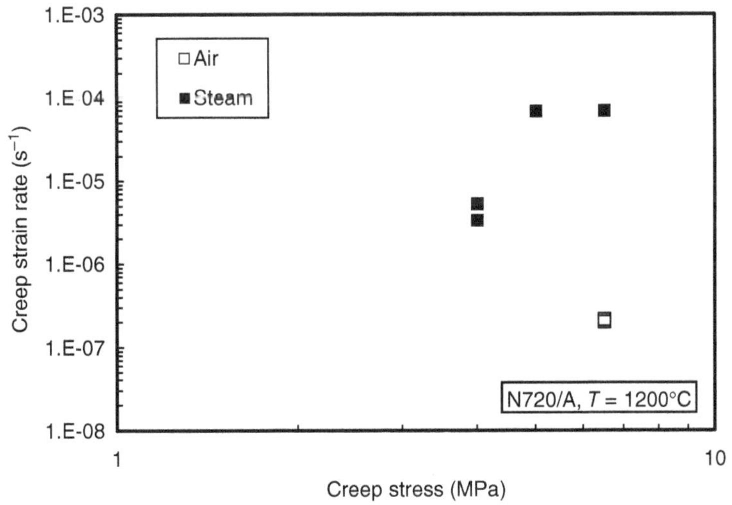

FIGURE 10.40

Minimum creep rate in interlaminar shear as a function of applied shear stress for N720/A ceramic composite at 1200°C in air and in steam (after Ruggles-Wrenn et al. [75]).

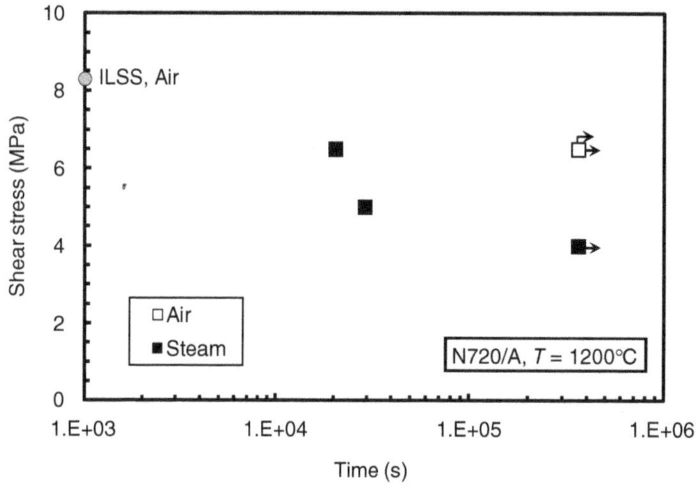

FIGURE 10.41

Interlaminar shear stress versus time to rupture for N720/A ceramic composite at 1200°C in air and in steam (after Ruggles-Wrenn et al. [75]).

(a) (b) (c)

FIGURE 10.42

Fracture surface of the N720/A double-notch shear specimen tested in compression to failure at 1200°C in air. Test duration < 10 seconds.

(a) (b) (c)

FIGURE 10.43

Fracture surface of the N720/A double-notch shear specimen tested in compression to failure following 100 hours at 6.5 MPa at 1200°C in air. Test duration > 100 hours.

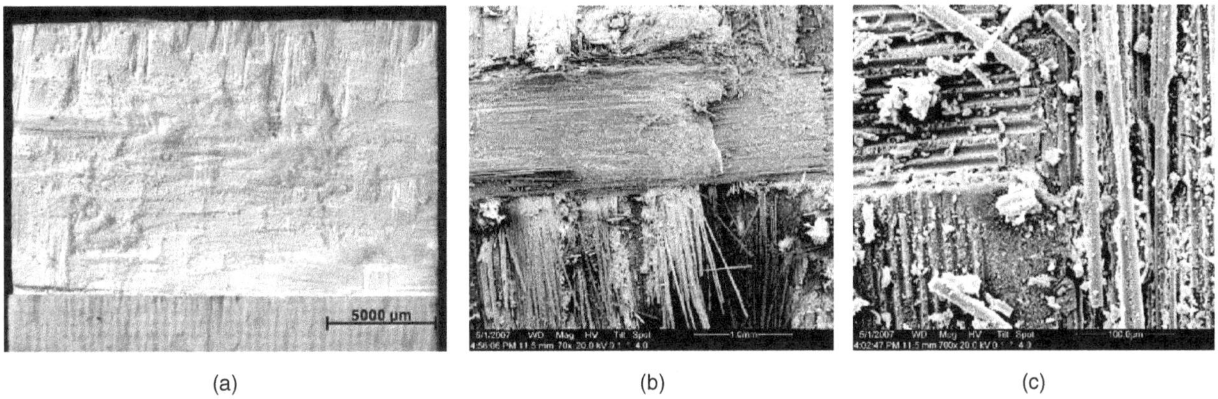

FIGURE 10.44

Fracture surface of the N720/A double-notch shear specimen tested in creep at 6.5 MPa at 1200°C in steam. Test duration ≤ 6 hours.

delamination. It is likely that additional sintering of the matrix, which occurred during 100 hours of creep at 6.5 MPa is behind the change of failure mechanism. The strengthening of the alumina matrix is also manifested in the increased ILSS of the N720/A specimens subjected to prior creep in air.

Fracture surfaces obtained at 1200°C in steam (Figures 10.44 and 10.45) reveal areas of violent failure exposing multiple 0/90 fiber layers and extensive fracture of fiber tows. In steam, failure of DNS specimens is dominated by fiber fracture. The amount of matrix material remaining bonded to the exposed fibers is much greater than that in the specimens tested in air (Figures 10.44c and 10.45c). In tests of long (≥100 hours) duration performed in steam, the fiber tows become bonded together by the matrix material and then fail in coordinated fashion (Figure 10.45c). Additional sintering of the matrix is likely accelerated in the presence of steam. However, whereas prior creep in air causes strengthening of the matrix and increase in ILSS, prior creep in steam results in weakening of the matrix as evidenced by decrease in ILSS. Recent studies [76,77] found that the presence of hydrogen defects in the hydrothermally treated alumina specimens considerably reduced their compressive strength. It is possible that hydrogen defects are introduced into the alumina matrix during creep at 1200°C in steam, which cause weakening of the matrix.

10.2.7 Tension–Tension Fatigue

In many potential applications, porous-matrix oxide/oxide CMCs will be subject to fatigue loading under a wide range of frequencies. The effects of frequency and environment on fatigue behavior of N720/A were studied in cyclic fatigue tests performed at 1200°C in air and in steam with the loading frequencies of 0.1–10 Hz [78].

FIGURE 10.45

Fracture surface of the N720/A double-notch shear specimen tested in compression to failure following 100 hours at 4.0 MPa at 1200°C in steam. Test duration > 100 hours.

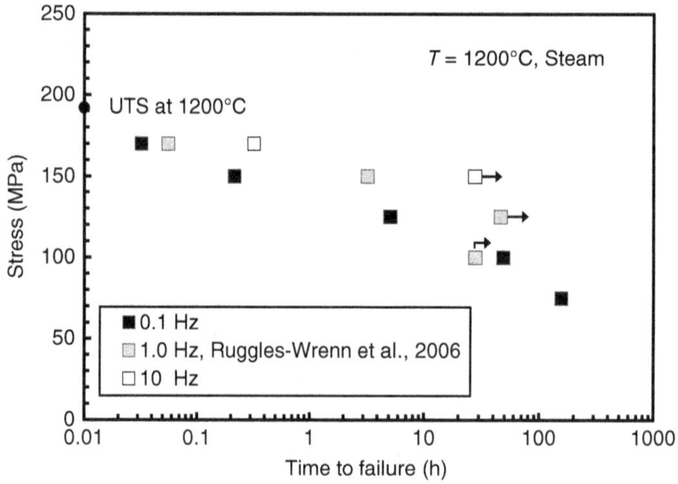

FIGURE 10.46

Fatigue S–N curves for N720/A ceramic composite at 1200°C in steam. Arrow indicates that failure of specimen did not occur when the test was terminated (after Ruggles-Wrenn et al. [46, 78]).

The loading frequency had little effect on fatigue performance at 1200°C in air. Fatigue run-out of 10^5 cycles was achieved in all tests performed at 0.1 and 1.0 Hz. In air, the fatigue limit was 170 MPa (~90%UTS) at 0.1 and 1.0 Hz. In view of the excellent fatigue resistance and high fatigue limit at 0.1 and 1.0 Hz, an equally high fatigue limit could be anticipated at 10 Hz. The presence of steam degraded the fatigue performance at all frequencies (Figure 10.46). Fatigue limits in steam were considerably lower than that in air. Furthermore, in steam the influence of the loading frequency on fatigue life was dramatic. In steam, the fatigue limit was 150 MPa (~79%UTS) at 10 Hz and 125 MPa (66%UTS) at 1.0 Hz. At 0.1 Hz in steam, fatigue run-out was not achieved even at the low fatigue stress of 75 MPa (~39%UTS).

Of importance in cyclic fatigue is the reduction in stiffness (hysteresis modulus determined from the maximum and minimum stress–strain data points during a load cycle), reflecting the damage development during fatigue cycling. Change in modulus is shown in Figure 10.47, where normalized modulus (i.e., modulus normalized by the modulus obtained in the second cycle) is plotted versus fatigue cycles. The first cycle was not used for normalization because of the large permanent strain offset upon unloading. It is noteworthy that although all tests performed in air achieved run-out, a decrease in normalized modulus with cycling was still observed (Figure 10.47a). However, frequency had little effect on the modulus evolution in air. Changes in normalized modulus as well as the influence of loading rate on modulus

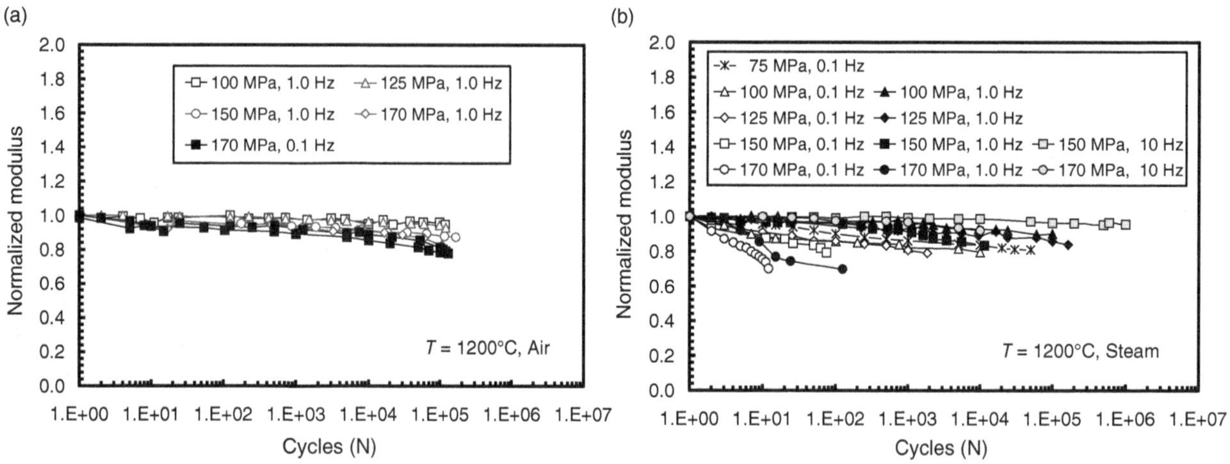

FIGURE 10.47

Normalized modulus versus fatigue cycles for N720/A ceramic composite at 1200°C (a) in laboratory air and (b) in steam (after Ruggles-Wrenn et al. [46, 78]).

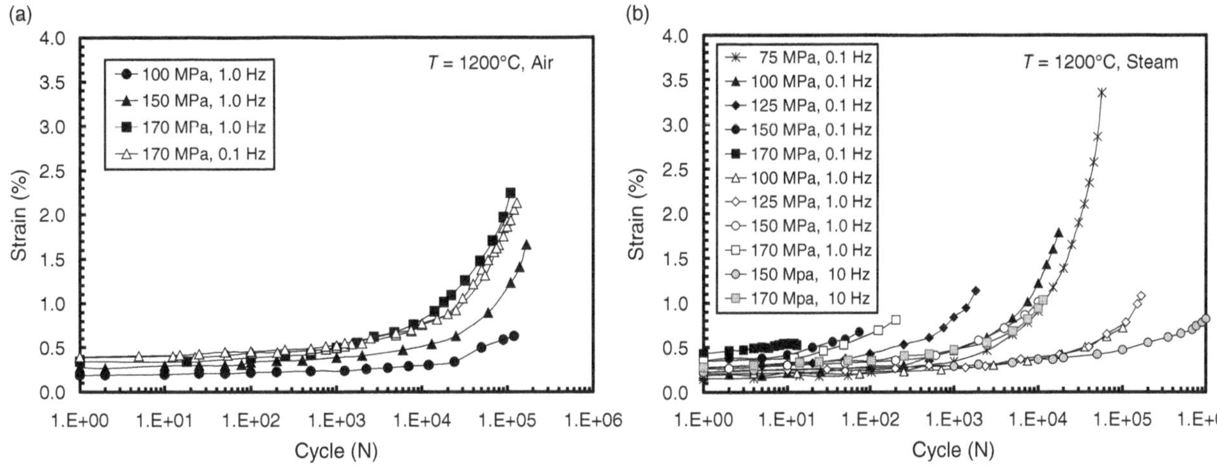

FIGURE 10.48

Maximum strain versus fatigue cycles for N720/A ceramic composite at 1200°C: (a) in laboratory air and (b) in steam (after Ruggles-Wrenn et al. [46, 78]).

evolution become more pronounced in steam (Figure 10.47b). While in air the reduction in normalized modulus was limited to 18%, in steam the normalized modulus loss reached 30%.

Figure 10.48, where maximum cyclic strains are plotted versus fatigue cycles, shows that ratcheting takes place in all tests conducted at 1200°C. In air, the loading frequency does not appear to have a strong effect on strain accumulation rate. Note that strains accumulated in steam are considerably lower than those accumulated in air at the same fatigue stress and loading frequency. Generally, lower strain accumulation with cycling indicates that less damage has occurred, and that it is mostly limited to some additional matrix cracking. However, lower accumulated strains in steam invariably correspond to shorter fatigue lives. In this case lower accumulated strains are more likely due to early bundle failures leading to specimen failure. In steam, the evolution of maximum strain with cycles is strongly influenced by the loading frequency. For a given fatigue stress level, the rate of strain accumulation increases with decreasing frequency. However, higher rate of strain accumulation does not necessarily translate into higher failure strain. The decrease in loading frequency causes a dramatic decrease in fatigue life, thus curtailing strain accumulation.

All specimens that achieved fatigue run-out at 1200°C in air exhibited increased tensile strength (Table 10.3). Yet, considerable stiffness loss (18–33%) was observed. The N720/A composite relies on the porous matrix for crack deflection and damage tolerance. Fatigue cycling promotes progressive matrix cracking, as evidenced by a decrease in stiffness,

TABLE 10.3 Retained Properties of the N720/A Specimens Subjected to Prior Fatigue in Laboratory Air and in Steam at 1200°C. Retained Properties Measured at 1200°C in Laboratory Air (After [46, 78])

Fatigue stress (MPa)	Fatigue environment	Retained strength (MPa)	Strength retention (%)	Retained modulus (GPa)	Modulus retention (%)	Strain at failure (%)
Prior fatigue at 0.1 Hz						
170	Air	194	≥100	51.7	82	0.38
170	Air	196	≥100	55.6	80	0.39
Prior fatigue at 1.0 Hz						
100	Air	194	≥100	56.6	74	0.44
125	Air	199	≥100	54.9	73	0.44
150	Air	199	≥100	43.4	72	0.53
170	Air	192	≥100	40.7	67	0.51
100	Steam	174	90	47.6	84	0.40
125	Steam	168	88	52.0	80	0.43
Prior fatigue at 10 Hz						
150	Steam	184	96	55.4	93	0.48

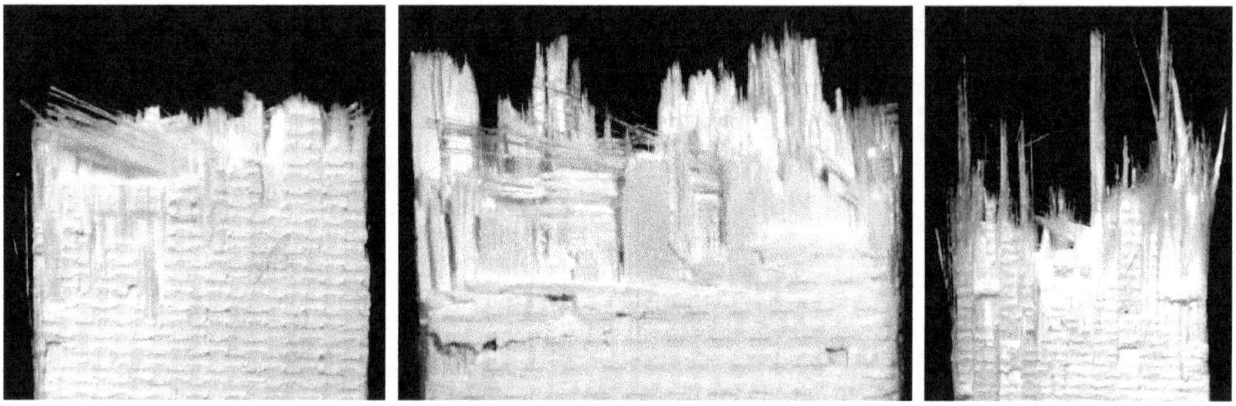

(a) (b) (c)

FIGURE 10.49

Fracture surfaces (optical micrographs) of the N720/A specimens tested in cyclic fatigue with maximum stress of 170 MPa at 1200°C in steam: (a) 0.1 Hz, $t_f = 0.03$ hour; (b) 1.0 Hz, $t_f = 0.06$ hour; and (c) 10 Hz, $t_f = 0.32$ hour. Size of the damage zone and cyclic fatigue lifetime increase with increasing frequency.

and induces the weakening of the fiber–matrix interfaces. By causing additional matrix cracking, prior fatigue in air serves to maintain the matrix porosity at the minimum level needed for crack deflection. As a result, prior fatigue in air effectively improves the damage tolerance and long-term durability of the composite as demonstrated by an improved tensile strength of the pre-fatigued specimens. Conversely, prior fatigue in steam caused reduction in both strength (up to 20%) and stiffness (up to 12%). Strength and stiffness degradation increases with decreasing frequency of prior fatigue in steam.

Fracture surfaces of the specimens tested at 1200°C in steam are strongly influenced by the loading frequency (Figure 10.49). The fracture surface produced at 0.1 Hz is dominated by coordinated failure of the fiber bundles with small isolated areas of fiber pullout. As the frequency increases, the fracture surfaces become progressively brushier. The fiber pullout is extensive and the variation in pull-out length is considerable in the fracture surface produced at 10 Hz. The near-planar fracture surface obtained in the 0.1 Hz test indicates the loss of matrix porosity. As a result, the composite exhibits decreased damage tolerance, brittle fracture behavior, and a short lifetime. Conversely, a fibrous failure surface obtained at 10 Hz demonstrates that mechanical cycling has restored the matrix porosity to the levels required to deflect the matrix cracks and to allow subsequent fiber pullout. The CMC exhibits improved damage tolerance and delayed failure, as evidences by a considerably longer cyclic lifetime obtained at 10 Hz.

The SEM micrographs in Figure 10.50 confirm these conclusions. The fracture surface produced at 0.1 Hz (Figure 10.50a) is dominated by planar regions of coordinated fiber failure, indicative of matrix densification due to additional sintering. The fracture surface obtained at 1.0 Hz (Figure 10.50b) still exhibits large areas of planar fracture; however, areas of uncoordinated brushy failure are also visible. Conversely the fracture surface produced at 10 Hz (Figure 10.50c) is dominated by areas of uncorrelated fiber fracture where individual fibers are clearly discernible. A brushy fracture surface with extensive fiber pullout demonstrates that mechanical cycling at 10 Hz has successfully counteracted the effects of matrix densification by maintaining matrix porosity at the level sufficient to enable crack-deflecting behavior and fiber pullout. For the N720/A composite, the fracture surface appearance may be correlated with time to failure. Predominantly planar fracture surface (Figure 10.50a), indicative of a fast failure process where crack fronts propagate rapidly through both matrix and fibers, corresponds to a shorter lifetime. In contrast, a brushy fracture surface (Figure 10.50c), typically accompanies longer times to failure. The higher levels of fiber pull-out are also consistent with larger failure strains, which are observed in the cases of longer cyclic life.

10.2.8 Tension–Tension Fatigue with Hold Times

Results presented here reveal that creep is significantly more damaging than cyclic fatigue for porous-matrix oxide–oxide CMCs. Hence mechanical performance of oxide–oxide CMCs under combined static-cyclic loading at elevated temperature in oxidizing environments is among the key issues that must be addressed. The effect of hold times at maximum stress on fatigue performance of the N720/A composite was investigated at 1200°C in air and steam [46].

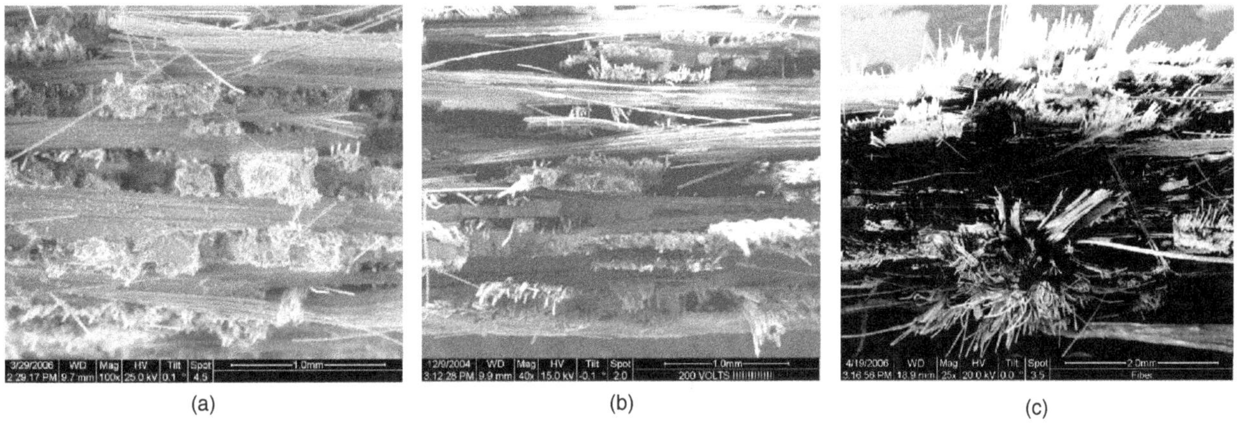

FIGURE 10.50

Fracture surfaces of the N720/A specimens tested in cyclic fatigue with the maximum stress of 170 MPa at 1200°C in steam: (a) 0.1 Hz, (b) 1.0 Hz, and (c) 10 Hz. The spatial correlation in the fiber failure locations within an individual tow increases with decreasing loading frequency.

The maximum stress levels were selected to be at or below the fatigue limit, yet above the creep run-out stress [45]. The ratio *R* (minimum to maximum stress) was 0.05, the loading/unloading frequency was 1 Hz and hold times were 10 and 100 seconds.

Figure 10.51 compares maximum stress versus time to failure data obtained in fatigue with hold times to those produced in creep and in fatigue. At 1200°C in air, introduction of a hold period of either 10-second or 100-second duration into the fatigue cycle has no noticeable effect on time to failure for $\sigma_{max} = 125$ MPa (~66% UTS). All 125 MPa fatigue tests, regardless of the hold time, achieved a 100-hour run-out. However, for $\sigma_{max} = 154$ MPa (~81% UTS), superposition of even a short 10-second hold period noticeably reduces cyclic life. While the specimen tested in fatigue achieved a run-out of 10^5 cycles (~28 hours at 1 Hz), specimen subjected to fatigue with 10-second hold failed after 17.6 hours. The specimen tested in fatigue with 100-second hold survived only 2.73 hours. Finally in creep test, which represents the most damaging loading type, lifetime was further reduced by an order of magnitude to 0.27 hours.

In the presence of steam, superposition of a hold time onto a fatigue cycle dramatically degrades cyclic life of the N720/A composite (Figure 10.51b). In steam, the 100 and 125 MPa fatigue tests achieved run-out. However, introducing a 10-second or a 100-second hold reduced the cyclic lifetimes by one to two orders of magnitude bringing them down to the lifetimes obtained in creep. In steam, run-out was not achieved in fatigue tests with hold time.

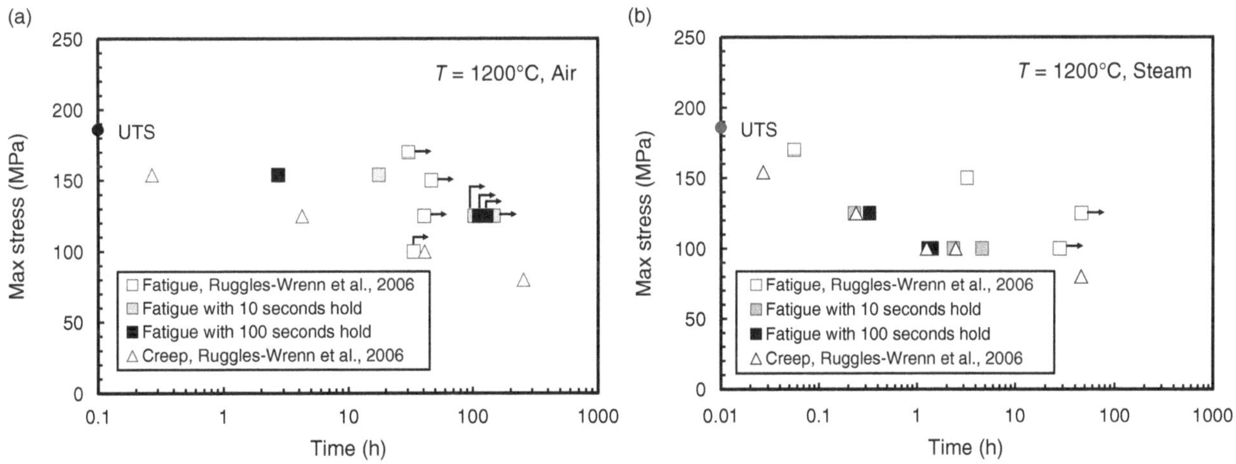

FIGURE 10.51

Maximum stress versus time to failure for N720/A ceramic composite at 1200°C: (a) in air and (b) in steam (after Mehrman et al. [46]).

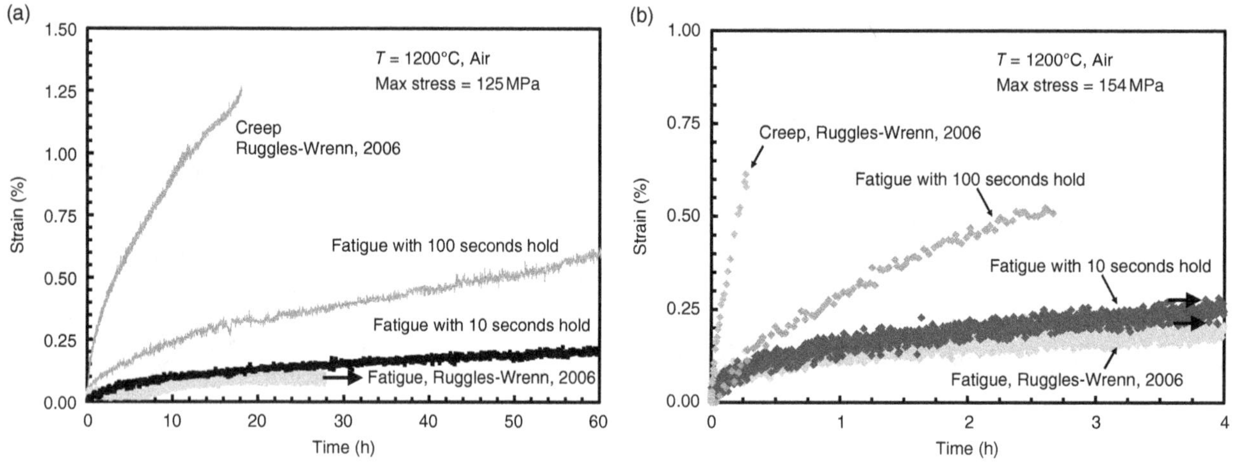

FIGURE 10.52

Maximum strain as a function of time for N720/A ceramic composite at 1200°C in laboratory air: (a) σ_{max} = 125 MPa and (b) σ_{max} = 154 MPa (after Mehrman et al. [46]).

At 1200°C, the N720/A composite accumulated substantial creep strains and exhibited strain ratcheting under cyclic loading. To assess the effect of loading type on the rate of strain accumulation, maximum strains (i.e. strains at maximum stress) were plotted versus time for creep, fatigue, and fatigue with hold times (Figures 10.52 and 10.53). In air, strain accumulations are lowest in fatigue and highest in creep, while strains accumulated in fatigue with hold time fall in the intermediate range (Figure 10.52). In steam, the evolution of maximum strain in fatigue with hold time is akin to that in creep (Figure 10.53). The steady-state strain rate was reached in all tests shown in Figures 10.52 and 10.53. In air, the rate of strain accumulation increases with uninterrupted time spent at maximum stress (Figure 10.54). The presence of steam generally accelerates strain accumulation under both cyclic and creep loading. Strain rates obtained in fatigue tests in steam are at least an order of magnitude higher than those in air. In air, the strain rates increase gradually when hold times of increasing duration are introduced into a fatigue cycle. In steam, strain rates produced in fatigue with hold time of any duration are close to those obtained in creep and significantly higher than those obtained in fatigue.

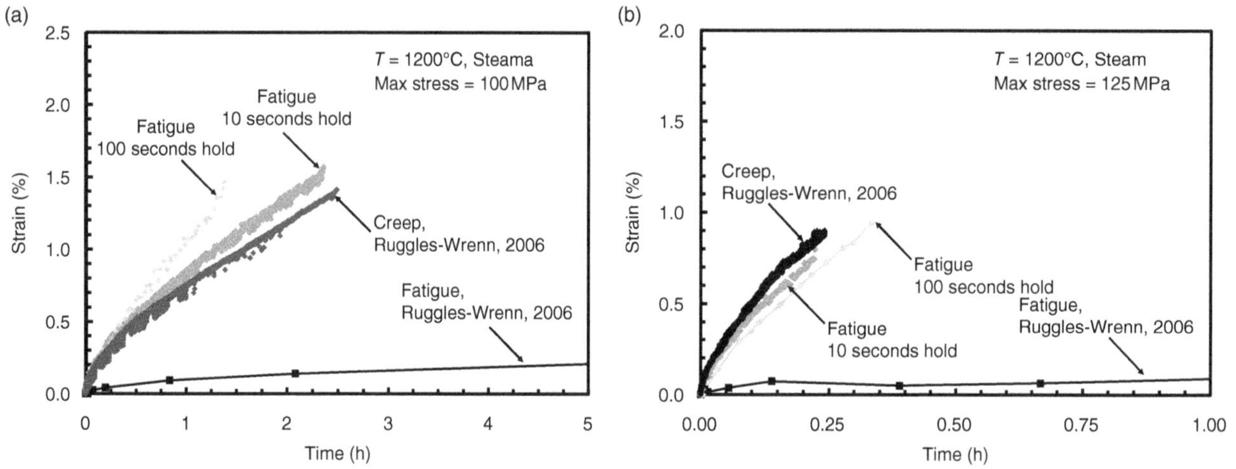

FIGURE 10.53

Maximum strain as a function of time for N720/A ceramic composite at 1200°C in steam: (a) σ_{max} = 100 MPa and (b) σ_{max} = 125 MPa (after Mehrman et al. [46]).

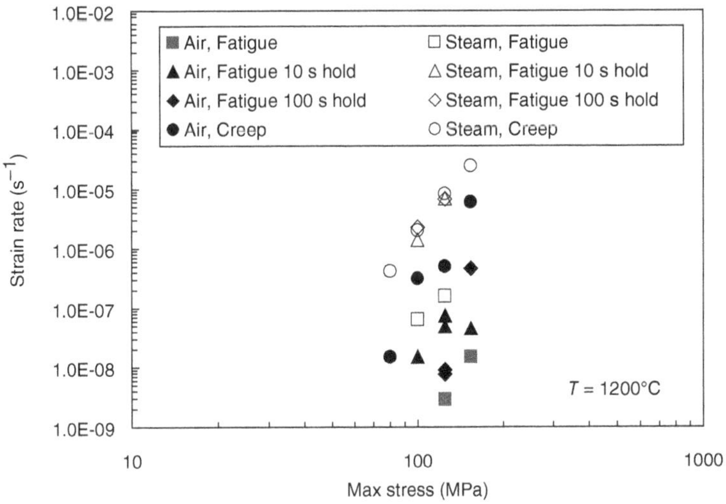

FIGURE 10.54

Strain rate as a function of maximum applied stress for N720/A ceramic composite at 1200°C (after Mehrman et al. [46]).

All N720/A fracture surfaces produced at 1200°C in fatigue or in fatigue with hold times contain brushy regions of fibrous fracture as well as regions of nearly planar failure (Figure 10.55). However, the balance of these two fracture topographies within a given fracture surface is strongly influenced by the test type (Figures 10.56 and 10.57). Extensive fiber pullout is produced in cyclic loading (Figures 10.56a and 10.57a) and predominantly planar fracture, in creep (Figures 10.56d and 10.57d). At 1200°C in air, regions of coordinated failure appear when a hold time is introduced into a fatigue cycle. The planar fracture topography becomes more prevalent with increasing hold time (Figures 10.56b and 10.56c). Conversely, at 1200°C in steam the hold duration has little effect on the fracture topography. In steam, all fracture surfaces produced in fatigue with hold time are dominated by planar fracture (Figures 10.57b and 10.57c). In steam as well as in air, the fracture surface appearance may be correlated with time to failure. Planar fracture surface accompanies short time to failure, while brushy fracture surface is indicative of longer life.

Results demonstrate that at 1200°C in steam the loading frequency has a profound influence on the fatigue life of the N720/A composite. In addition, in steam the superposition of a hold time at maximum stress onto a fatigue cycle drastically reduces cyclic lifetime of N720/A. These findings suggest that a unique time-dependent failure mechanism may be operating in N720/A under both cyclic and static (creep) loading at 1200°C in steam.

It is recognized that environmentally assisted crack growth may be the mechanism behind reduced creep resistance of the N720/A composite at 1200°C in steam. If the cyclic and static fatigue failure mechanisms are indeed identical and dominated by environmentally assisted (slow) crack growth, then the cyclic fatigue lifetime can be predicted from the

(a) (b)

FIGURE 10.55

Fracture surfaces of the N720/A specimens obtained in 125 MPa fatigue tests with 10-second hold conducted at 1200°C: (a) in air ($t_f \geq 100$ hours) and (b) in steam ($t_f = 0.23$ hours).

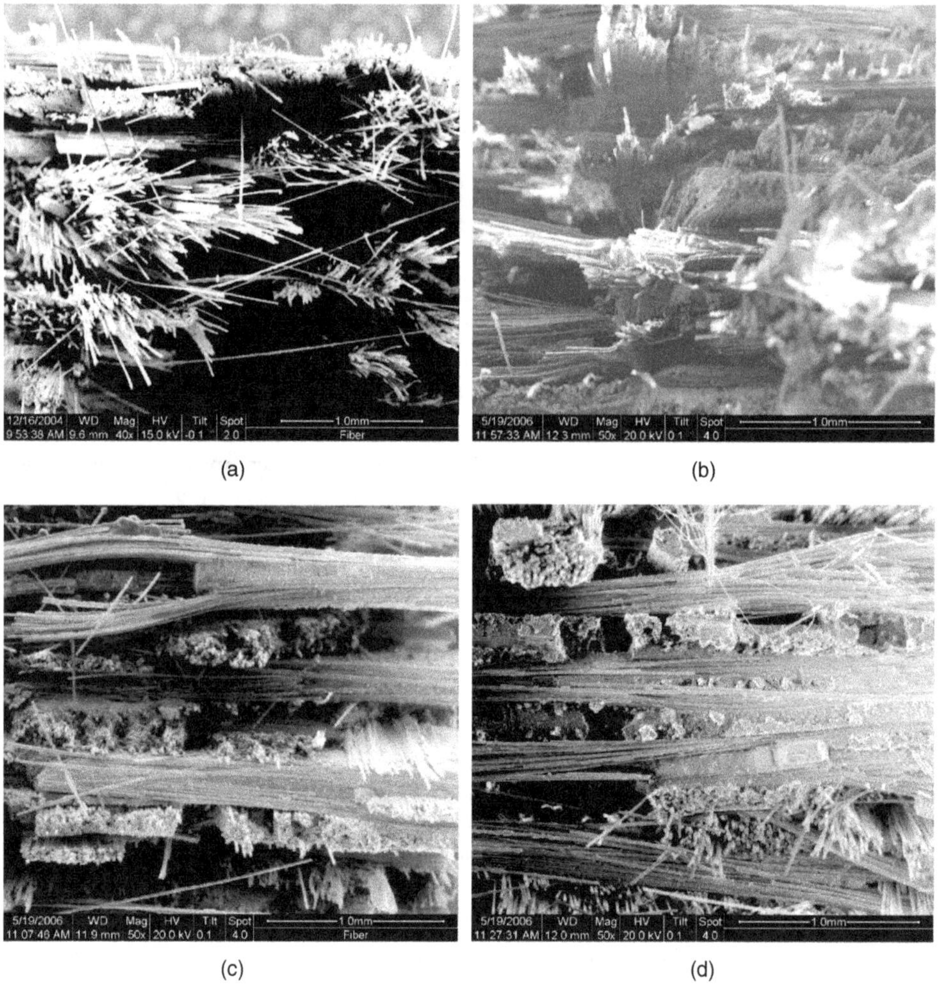

FIGURE 10.56

Fracture surfaces of the N720/A specimens tested at 1200°C in air with $\sigma_{max} = 125$ MPa: (a) fatigue, (b) fatigue with 10-second hold, (c) fatigue with 100-second hold, and (d) creep.

static fatigue data by using a linear elastic crack growth model [42]. The ratio of the time to failure under cyclic fatigue, t_{cf}, to the time to failure under creep, t_{sf}, under the same maximum applied stress σ_{max} can be obtained as [42]

$$r = \frac{t_{cf}}{t_{sf}} = \tau \left\{ \int_0^\tau \left[\frac{\sigma(t)}{\sigma_{max}} \right]^N dt \right\}^{-1} \tag{10.8}$$

For cyclic loading, the applied stress $\sigma(t) = \sigma_{max} f(t)$ is a periodic function of time, τ is the period, and $0 \leq f(t) \leq 1$ in tension. For static loading, the applied stress $\sigma(t) = \sigma_{max}$. For cyclic loading with the ratio R (minimum to maximum stress) and a triangular waveform the periodic function $f(t)$ is expressed as

$$f(t) = \begin{cases} R + (1 - R)\dfrac{2t}{\tau}, & \text{for } 0 \leq t \leq \dfrac{\tau}{2} \\ 1 - (1 - R)\left(\dfrac{2t}{\tau} - 1\right), & \text{for } \dfrac{\tau}{2} \leq t \leq \tau \end{cases} \tag{10.9}$$

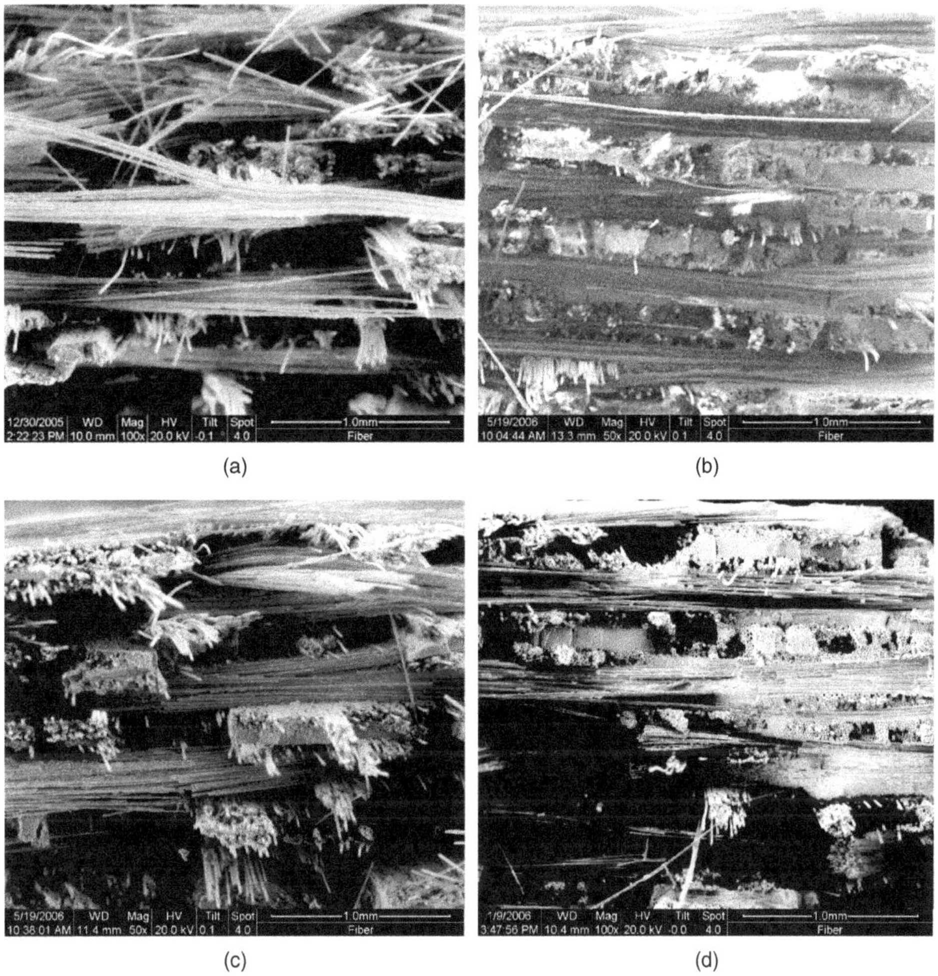

FIGURE 10.57

Fracture surfaces of the N720/A specimens tested at 1200°C in steam with $\sigma_{max} = 125$ MPa: (a) Fatigue, (b) fatigue with 10-second hold, (c) fatigue with 100-second hold, and (d) creep.

For a trapezoidal waveform (i.e., fatigue cycle with hold time at maximum stress), where t_h is the hold time at maximum stress, t_L is the loading/unloading time, and $\tau = (t_h + t_L)$ is the period, the function $f(t)$ is given by

$$f(t) = \begin{cases} R + (1-R)\dfrac{2t}{t_L} & \text{for } 0 \le t \le \dfrac{t_L}{2} \\ 1 & \text{for } \dfrac{t_L}{2} \le t \le t_h + \dfrac{t_L}{2} \\ 1 - \dfrac{2(1-R)}{t_L}\left(t - t_h - \dfrac{t_L}{2}\right) & \text{for } t_h + \dfrac{\tau}{2} \le t \le \tau \end{cases} \tag{10.10}$$

Recall that the SCG parameter $N = 12$ was obtained earlier from the constant stress test data at 1200°C in steam. Using $R = 0.05$ and Equations (10.8)–(10.10), the ratio of cyclic to static lifetimes, r, can be calculated for the triangular waveform as well as for the trapezoidal waveforms with the loading/unloading time $t_L = 1$ second and hold times t_h of 10 seconds and 100 seconds.

The predicted cyclic fatigue lifetimes are compared with the experimental results obtained at 1200°C in steam (Figure 10.58). The predictions are in excellent agreement with the results obtained in fatigue tests at 0.1 Hz (triangular waveform) and with the test data obtained in fatigue tests with hold time (trapezoidal waveform), indicating that there

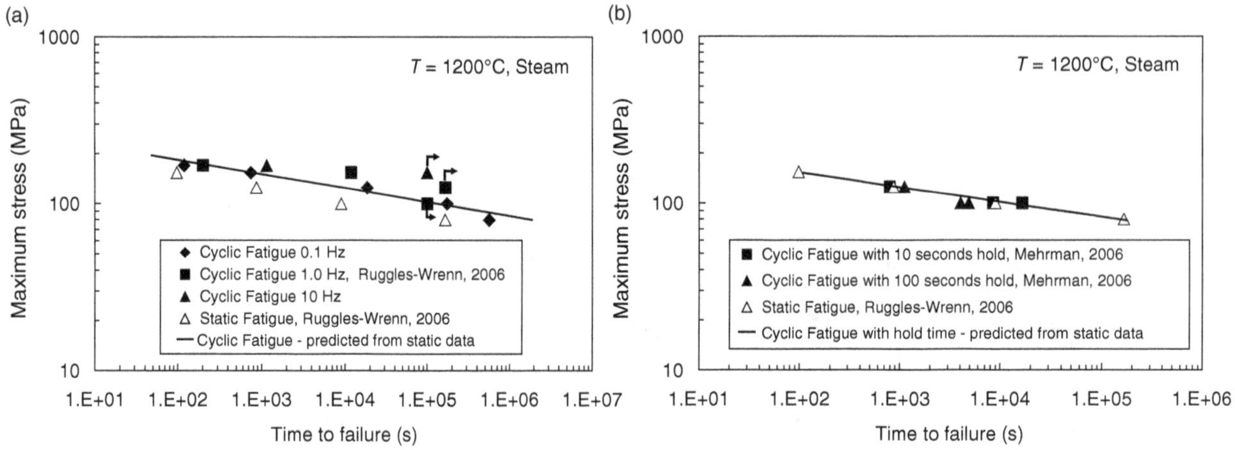

FIGURE 10.58

Comparison between cyclic fatigue lifetimes predicted from static fatigue data and experimental results obtained for N720/A ceramic composite at 1200°C in steam: (a) in cyclic tests with triangular wave form and (b) in cyclic tests with hold time at maximum stress (after Ruggles-Wrenn et al. [46, 78]).

is no apparent cyclic effect on the crack growth rate and that the SCG is the dominant failure mechanism. Conversely, predictions significantly underestimate the cyclic lifetimes measured in fatigue tests conducted with the triangular waveform at higher frequencies. Evidently, cyclic loading with a short duration of maximum stress has a beneficial effect on lifetime compared to static loading.

Because the N720/A composite derives its damage tolerance from a porous matrix the stability of the matrix porosity against densification is critical to the composite's long-term durability. The loss of matrix porosity would inhibit crack deflection, reduce damage tolerance, and accelerate failure. Oxide–oxide CMCs suffer porosity reduction due to additional matrix sintering when exposed to elevated temperatures [50, 51]. It is likely that N720/A specimens tested at 1200°C in steam also undergo additional matrix sintering and subsequent loss of matrix porosity. Under static (creep) loading, the densification of the matrix and the environmentally assisted crack growth work together to accelerate failure and to reduce composite lifetime. In contrast, cyclic loading counteracts matrix densification by causing progressive matrix cracking and weakening of the fiber–matrix interface. The beneficial effects of the mechanical cycling on damage tolerance of the N720/A composite compensate for the negative effects of the environmentally assisted (slow) crack growth, resulting in improved durability and longer lifetimes under cyclic loading.

10.3 CONCLUDING REMARKS AND FUTURE DIRECTIONS

The effects of environment on mechanical behavior and properties of oxide–oxide composites were investigated in experiments that combine environmental exposure (i.e., air or steam and elevated temperature) with mechanical loading. The presence of steam significantly degraded the performance of oxide–oxide CMCs under sustained loading (creep) as well as under low-frequency cyclic loading. The degradation due to steam was minimal at lower temperatures (≤1000°C), but became dramatic as the temperature increased to 1200°C. Fractographic analysis revealed considerable matrix densification during prolonged exposures at elevated temperatures under load. The matrix densification due to additional sintering of the matrix was accelerated in the presence of steam. Matrix densification and subsequent loss of matrix porosity were the key mechanisms responsible for degradation of mechanical properties and performance at elevated temperature in steam. However, in oxide–oxide CMCs the porosity levels in the matrix are generally such that the fibers are also exposed to environmental attack. Thus fiber degradation is a possible source of composite degradation. The effects of steam on the high-temperature performance of oxide fibers is the subject of an ongoing study.

REFERENCES

1. F. W. Zok. Developments in oxide fiber composites. *J. Am. Ceram. Soc.*, 89(11) 3309–3324 (2006).
2. L. P. Zawada, J. Staehler, and S. Steel. Consequence of intermittent exposure to moisture and salt fog on the high-temperature fatigue durability of several ceramic-matrix composites. *J. Am. Ceram. Soc.*, 86(8), 1282–1291 (2003).
3. S. Schmidt, S. Beyer, H. Knabe, H. Immich, R. Meistring, and A. Gessler. Advanced ceramic matrix composite materials for current and future propulsion technology applications. *Acta Astronaut.*, 55, 409–420 (2004).
4. A. Szweda, M. L. Millard, and M. G. Harrison. Fiber-reinforced ceramic-matrix composite member and method for making. *US Patent* 5, 601, 674 (1997).
5. S. M. Sim and R. J. Kerans. Slurry infiltration and 3-D woven composites. *Ceram. Eng. Sci. Proc.*, 13(9–10), 632–641 (1992).
6. E. H. Moore, T. Mah, and K. A. Keller. 3D composite fabrication through matrix slurry pressure infiltration. *Ceram. Eng. Sci. Proc.*, 15(4), 113–120 (1994).
7. F. Lange, W. Tu, and A. Evans. Processing of damage-tolerant, oxidation-resistant ceramic matrix composites by a precursor infiltration and pyrolysis method. *Mater. Sci. Eng. A*, 195, 145–150 (1995).
8. E. Mouchon and P. Colomban. Oxide ceramic matrix/oxide fiber woven fabric composites exhibiting dissipative fracture behavior. *Composites*, 26, 175–182 (1995).
9. W. C. Tu, F. F. Lange, and A. G. Evans. Concept for a damage-tolerant ceramic composite with strong interfaces. *J. Am. Ceram. Soc.*, 79(2), 417–424 (1996).
10. A. G. Evans and F. W. Zok. Review: the physics and mechanics of fiber-reinforced brittle matrix composites, *J. Mater. Sci.*, 29, 3857–3896 (1994).
11. R. J. Kerans and T. A. Parthasarathy. Crack deflection in ceramic composites and fiber coating design criteria. *Comp. A*, 30, 521–524 (1999).
12. R. J. Kerans, R. S. Hay, T. A. Parthasarathy, and M. K. Cinibulk. Interface design for oxidation-resistant ceramic composites. *J. Am. Ceram. Soc.*, 85(11), 2599–2632 (2002).
13. C. G. Levi, J. Y. Yang, B. J. Dalgleish, F. W. Zok, and A. G. Evans. Processing and performance of an all-oxide ceramic composite. *J. Am. Ceram. Soc.*, 81, 2077–2086 (1998).
14. A. G. Hegedus. Ceramic bodies of controlled porosity and process for making same. *US Patent* 5, 0177, 522 (1991).
15. T. Lu. Crack branching in all-oxide ceramic composites. *J. Am. Ceram. Soc.*, 79(1), 266–274 (1996).
16. B. Kanka and H. Schneider. Aluminosilicate fiber/mullite matrix composites with favorable high-temperature properties. *J. Eur. Ceram. Soc.*, 20(5), 619–623 (2000).
17. F. W. Zok and C. G. Levi. Mechanical properties of porous-matrix ceramic composites. *Adv. Eng. Mater.*, 3(1–2) 15–23 (2001).
18. M. B. Ruggles-Wrenn, S. S. Musil, S. Mall, and K. A. Keller. Creep behavior of Nextel™610/monazite/alumina composite at elevated temperatures. *Comp. Sci. Tech.*, 66, 2089–2099 (2006).
19. M. B. Ruggles-Wrenn, T. Yeleser, G. E. Fair, and J. B. Davis. Effects of steam environment on creep behavior of Nextel 610/monazite/alumina composite at 1100°C. *Appl. Comp. Mater.*, 16, 379–392 (2009).
20. P. R. Jackson, M. B. Ruggles-Wrenn, S. S. Baek, and K. A. Keller. Compressive creep behavior of an oxide–oxide ceramic composite with monazite fiber coating at elevated temperatures. *Mater. Sci. Eng. A*, 454–455, 590–601 (2007).
21. L. P. Zawada, R. S. Hay, S. S. Lee, and J. Staehler. Characterization and high-temperature mechanical behavior of an oxide/oxide composite. *J. Am. Ceram. Soc.*, 86(6) 981–990 (2003).
22. D. J. Buchanan, R. John, and L. P. Zawada. Off-axis creep behavior of oxide/oxide Nextel™720/AS-0. *Comp. Sci. Tech.*, 68, 1313–1320 (2008).
23. G. Jefferson, K. A. Keller, R. S. Hay, and R. J. Kerans. Oxide/oxide composites with fiber coatings, in *Ceramic Matrix Composites, Fiber Reinforced Ceramics and Their Applications*, edited by W. Krenkel (Wiley-VCH Verlag, Weinheim, 2008), pp. 187–204.
24. M. Schmucker and P. Mechnich. All-oxide ceramic matrix composites with porous matrices, in *Ceramic Matrix Composites, Fiber Reinforced Ceramics and Their Applications*. edited by W. Krenkel (Wiley-VCH Verlag, Weinheim, 2008), pp. 205–229.
25. S. Hackemann, F. Flucht, and W. Braue. Creep investigations of alumina-based all-oxide ceramic matrix composites. *Comp. A*, 41, 1768–1776 (2010).
26. E. A. V. Carelli, H. Fujita, J. Y. Yang, and F. W. Zok. Effects of thermal aging on the mechanical properties of a porous-matrix ceramic composite. *J. Am. Ceram. Soc.*, 85(3), 595–602 (2002).
27. J. B. Davis, D. B. Marshall, and P. E. D. Morgan. Oxide composites of Al_2O_3 and $LaPO_4$. *J. Eur. Ceram. Soc.*, 19(13–14), 2421–2426 (1999).

28. M. G. Holmquist and F. F. Lange. Processing and properties of a porous matrix composite reinforced with continuous oxide fibers. *J. Am. Ceram. Soc.*, 86(10), 1733–1740 (2003).

29. M.-L. Antti, E. Lara-Curzio, and R. Warren. Thermal degradation of an oxide fibre (Nextel 720)/aluminosilicate composite. *J. Eur. Ceram. Soc.*, 24, 565–578 (2004).

30. T. A. Parthasarathy, L. P. Zawada, R. John, M. K. Cinibulk, R. J. Kerans, and J. Zelina. Evaluation of oxide–oxide composites in a novel combustor wall application. *Int. J. Appl. Ceram. Technol.*, 2(2), 122–131 (2005).

31. M. A. Mattoni, J. Y. Yang, C. G. Levi, and F. W. Zok. Effects of combustor rig exposure on a porous-matrix oxide composite. *Int. J. Appl. Ceram. Technol.*, 2(2), 133–140 (2005).

32. K. L. More, E. Lara-Curzio, P. F. Tortorelli, T. M. Brummet, and A. Szweda. The high-temperature stability of an oxide–oxide composite at high water-vapor pressure [Paper No. GT2005-69065]. Proceedings of ASME Turbo Expo 2005, June 6–9, 2005, Reno-Tahoe, Nevada, USA.

33. J. Staehler and L. P. Zawada. Performance of four ceramic-matrix composite divergent flap inserts following ground testing on an F110 turbofan engine. *J. Am. Ceram. Soc.*, 83(7), 1727–1738 (2000).

34. S. Wannaparhun and S. Seal. A combined spectroscopic and thermodynamic investigation of Nextel-720/Alumina ceramic matrix composite in air and water vapor at 1100°C. *J. Am. Ceram. Soc.*, 86(9), 1628–1630 (2003).

35. M. B. Ruggles-Wrenn and J. C. Braun. Effects of steam environment on creep behavior of Nextel™720 alumina ceramic composite at elevated temperature. *Mater. Sci. Eng. A*, 49, 101–110 (2008).

36. M. B. Ruggles-Wrenn and C. L. Genelin. Creep of Nextel™720/alumina-mullite ceramic composite at 1200°C in air, argon and steam. *Comp. Sci. Tech.*, 69, 663–669 (2009).

37. M. B. Ruggles-Wrenn and T. Kutsal. Effects of steam environment on creep behavior of Nextel™720/alumina-mullite ceramic composite at elevated temperature. *Comp. A*, 41, 1807–1816 (2010).

38. M. B. Ruggles-Wrenn, A.T. Radzicki, S.S. Baek, and K.A. Keller. Effect of loading rate on the monotonic tensile behavior and tensile strength of an oxide-oxide ceramic composite at 1200°C. *Mater. Sci. Eng. A*, 492, 88–94 (2008).

39. S. M. Wiederhorn. Moisture assisted crack growth in ceramics. *Int. J. Fract. Mech.*, 4(2), 171–177 (1968).

40. S. M. Wiederhorn and L. H. Bolz. Stress corrosion and static fatigue of glass. *J. Am. Ceram. Soc.*, 53(10), 543–548 (1970).

41. S. M. Wiederhorn. Subcritical crack growth in ceramics, in *Fracture Mechanics of Ceramics*, Vol. 2, edited by R. C. Bradt, D. P. H. Hasselman, and F. F. Lange (Plenum Press, New York, 1974), pp. 623–646.

42. A. G. Evans. Slow crack growth in brittle materials under dynamic loading conditions. *Int. J. Fract.*, 10(2), 251–259 (1974).

43. ASTM C1368. Standard test method for determination of slow crack growth parameters of advanced ceramics by constant stress-rate flexural testing at ambient temperature. American Society for Testing and Materials, 2001.

44. ASTM C1465. Standard test method for determination of slow crack growth parameters of advanced ceramics by constant stress-rate flexural testing at elevated temperatures. American Society for Testing and Materials, 2000.

45. M. B. Ruggles-Wrenn, P. Koutsoukos, and S. S. Baek. Effects of environment on creep behavior of two oxide/oxide ceramic-matrix composites at 1200°C. *J. Mater. Sci.*, 43, 6734–6746 (2008).

46. J. M. Mehrman, M. B. Ruggles-Wrenn, and S. S. Baek. Influence of hold times on the elevated-temperature fatigue behavior of an oxide-oxide ceramic composite in air and in steam environment. *Comp. Sci. Tech.*, 67, 1425–1438 (2007).

47. S. R. Choi and J. P. Gyekenyesi. Load-rate dependency of ultimate tensile strength in ceramic matrix composites at elevated temperatures. *Int. J. Fatig.*, 27, 503–510 (2005).

48. S. R. Choi, N. P. Bansal, and M. J. Verrilli. Delayed failure of ceramic matrix composites in tension at elevated temperatures. *J. Eur. Ceram. Soc.*, 25, 1629–1636 (2005).

49. J. R. Ritter. Engineering design and fatigue failure of brittle materials, in *Fracture Mechanics of Ceramics*, Vol. 4, edited by R. C. Bradt, D. P. H. Hasselman, and F. F. Lange (Plenum Press, New York, 1978), pp. 661–686.

50. H. Fujita, G. Jefferson, R. M. McMeeking, and F. W. Zok. Mullite/alumina mixtures for use as porous matrices in oxide fiber composites. *J. Am. Ceram. Soc.*, 87(2), 261–267 (2004).

51. H. Fujita, C. G Levi, F. W. Zok, and G. Jefferson. Controlling mechanical properties of porous mullite/alumina mixtures via precursor-derived alumina. *J. Am. Ceram. Soc.*, 88(2), 367–375 (2005).

52. E. A. V. Carelli, H. Fujita, J. Y. Yang, and F. W. Zok. Effects of thermal aging on the mechanical properties of a porous-matrix ceramic composite. *J. Am. Ceram. Soc.*, 85(3), 595–602 (2002).

53. J. W. Holmes, Y. H. Park, and J. W. Jones. Tensile creep and creep recovery behavior of a SiC-fiber-Si_3N_4-matrix composite. *J. Am. Ceram. Soc.*, 76(5), 1281–1293 (1993).

54. X. Wu and J. W. Holmes. Tensile creep and creep-strain recovery behavior of silicon carbide fiber/calcium aluminosilicate matrix ceramic composites. *J. Am. Ceram. Soc.*, 76(10), 2695–2700 (1993).

55. Y. H. Park and J. W. Holmes. Finite element modeling of creep deformation in fiber-reinforced ceramic composites. *J. Mater. Sci.*, 27, 6341–6351 (1992).

56. J. W. Holmes and X. Wu. Elevated temperature creep behavior of continuous fiber-reinforced ceramics, in *High Temperature Mechanical Behavior of Ceramic Composites*, edited by S. Nair and K. Jakus (Butterworth-Hienneman, 1995), pp. 193–259.

57. A. R. Boccaccini, G. West, J. Janczak, M. H. Lewis, and H. Kern. Tensile behavior and cyclic creep of continuous fiber-reinforced glass matrix composites at room and elevated temperatures. *J. Matl. Eng. Perf.*, 6(3), 344–348 (1997).

58. E. Y. Sun, S. R. Nutt, and J. J. Brennan. Flexural creep of coated SiC-fiber-reinforced glass-ceramic composites. *J. Am. Ceram. Soc.*, 78(5), 1233–1239 (1995).

59. M. B. Ruggles-Wrenn and B. A. Whiting. Cyclic creep and recovery behavior of Nextel™720/alumina ceramic composite at 1200°C. *Mater. Sci. Eng. A*, 528, 1848–1856 (2011).

60. L. P. Zawada. Longitudinal and trans-thickness tensile behavior of several oxide/oxide composites. *Ceram. Eng. Sci. Proc.*, 19(3), 327–339 (1998).

61. D. J. Buchanan, R. John, and L. P. Zawada. Creep rupture behavior of ±45 oxide/oxide Nextel720/AS composite. *Ceram. Eng. Sci. Proc.*, 24(4), 451–458 (2003).

62. M. L. Antti and E. Lara-Curzio. Effect of notches, specimen size, and fiber orientation on the monotonic tensile behavior of composites at ambient and elevated temperatures. 25th Annual Conference on Composites, Advanced Ceramics, Materials, and Structures, Cocoa Beach, FL, January 21–27, 2001.

63. E. Lara-Curzio and K. L. More. Stress-rupture behavior of Nextel720/AS fiber reinforced aluminosilicate CFCCs at 1200°C. 24th Annual Conference on Composites, Advanced Ceramics, Materials, and Structures, Cocoa Beach, FL, January 23–28, 2000.

64. L. P. Zawada and S. S. Lee. Evaluation of four CMCs for aerospace turbine engine divergent flaps and seals. *Ceram. Eng. Sci. Proc.*, 16(4), 337–339 (1995).

65. M. B. Ruggles-Wrenn, G. T. Siegert, and S. S. Baek. Creep behavior of Nextel™720/alumina ceramic composite with ±45° fiber orientation at 1200°C. *Comp. Sci. Tech.*, 68, 1588–1595 (2008).

66. M. B. Ruggles-Wrenn and M. Ozer. Creep behavior of Nextel™ 720/alumina-mullite ceramic composite with ±45° fiber orientation at 1200°C. *Mater. Sci. Eng. A*, 527, 5326–5334 (2010).

67. M. B. Ruggles-Wrenn and N. R. Szymczak. Effects of steam environment on compressive creep behavior of Nextel™ 720/Alumina ceramic composite at 1200°C. *Comp. A*, 39, 1829–1837 (2008).

68. N.J. Fang and T. W. Chou. Characterization of interlaminar shear strength of ceramic matrix composites. *J. Am. Ceram. Soc.*, 76(10), 2539–2548 (1993).

69. S. R. Choi and N. P. Bansal. Interlaminar tension/shear properties and stress rupture in shear of various continuous fiber-reinforced ceramic matrix composites. *Ceram. Trans.*, 175, 119–134 (2006).

70. S. R. Choi and N. P. Bansal. Shear strength as a function of test rate for SiC_f/BSAS ceramic matrix composite at elevated temperature. *J. Am. Ceram. Soc.*, 87(10), 1912–1918 (2004).

71. S. R. Choi, N. P. Bansal, A. M. Calomino, and M. J. Verrilli. Shear strength behavior of ceramic matrix composites at elevated temperatures. *Ceram. Trans.*, 165, 131–145 (2005).

72. S. R. Choi, R. W. Kowalik, D. J. Alexander, and N. P. Bansal. Assessments of life limiting behavior in interlaminar shear for Hi-NiC SiC/SiC ceramic matrix composite at elevated temperature, in *Mechanical Properties and Performance of Engineering Ceramics and Composites III*, edited by E. Lara-Curzio, J. Salem, and D. Zhu (Wiley, 2007), pp. 179–189.

73. S. R. Choi and R. W. Kowalik. Interlaminar crack growth resistances of various ceramic matrix composites in mode I and mode II loading. *J. Eng. Gas Turb. Pow.*, 130, 031301-1–031301-7 (2008).

74. S. R. Choi, R. W. Kowalik, D. J. Alexander, and N. P. Bansal. Elevated-temperature stress rupture in interlaminar shear of a Hi-Nic SiC/SiC ceramic matrix composite. *Comp. Sci. Tech.*, 69, 890–897 (2009).

75. M. B. Ruggles-Wrenn and P. D. Laffey. Creep behavior of Nextel™ 720/Alumina ceramic composite at elevated temperature in air and in steam. *Comp. Sci. Tech.*, 68, 2260–2266 (2008).

76. W. P. Tai and T. Watanabe. High-temperature stability of alumina in argon and argon/water-vapor environments. *J. Am. Ceram. Soc.*, 82(1), 245–248 (1999).

77. A. K. Kronenberg, J. Castaing, T. E. Mitchell, and S. H. Kirby. Hydrogen defects in α-Al_2O_3 and water weakening of sapphire and alumina ceramics between 600 and 1000°C – I. infrared characterization of defects. *Acta Mater.*, 48, 1481–1494 (2000).

78. M. B. Ruggles-Wrenn, G. Hetrick, and S. S. Baek. Effects of frequency and environment on fatigue behavior of an oxide-oxide ceramic composite at 1200°C. *Int. J. Fatig.*, 30, 502–516 (2008).

CHAPTER 11

STRESS-ENVIRONMENTAL EFFECTS ON FIBER-REINFORCED SiC-BASED COMPOSITES

GREGORY N. MORSCHER

Department of Mechanical Engineering, The University of Akron, Akron, OH, USA

11.1 INTRODUCTION/BACKGROUND

SiC-based composites are the leading material candidates for extreme temperature–time–oxidative environment applications. They offer greater oxidation protection than carbon/carbon composite for aerospace re-entry applications [1] and higher temperature capability than Ni-based superalloy hot-section jet engine and land-based gas turbine components [1–4]. However, since these composite systems consist of non-oxide materials, they are still subject to oxidation and corrosive degradation in the presence of oxygen [5], water vapor, or corrosive environments such as Na_2SO_4 [6] or CMAS [7]. The degree to which composites are degraded and the mechanisms causing the degradation differ depending on the composition of the composite itself as well as the environment and temperature. If the degradation is too deleterious protective coatings, that is, environmental barrier coatings (EBCs), will be necessary [8].

Therefore, in order to fully grasp the effects of time–stress–temperature–environment degradation of SiC-based CMCs, one must understand (1) the mechanisms that cause degradation corresponding to the particular time–stress–temperature–environment conditions they are exposed to and (2) the composite system of interest itself since particular degradation mechanisms depend on the materials that make up the composite constituents (fiber, interphase, or matrix) and to some extent the fiber architecture of the composite system. If these are well understood, then one can attempt to design composites to minimize environmental degradation and model the time–stress–temperature–environment-dependent mechanisms to develop mechanical performance life models. The outline of this paper will therefore be to discuss the various environment-degradation mechanisms encountered by SiC-based CMCs, the different composite systems and specific mechanisms that apply to them, and finally approaches to model time–stress–temperature–environment degradation toward developing life models.

11.2 MECHANISMS

There are two general types of environment-degradation mechanisms (Figure 11.1): surface recession and interior oxidation of the composite via surface-exposed cracks or other pathways through the material to the environment,

Ceramic Matrix Composites: Materials, Modeling and Technology, First Edition. Edited by Narottam P. Bansal and Jacques Lamon.

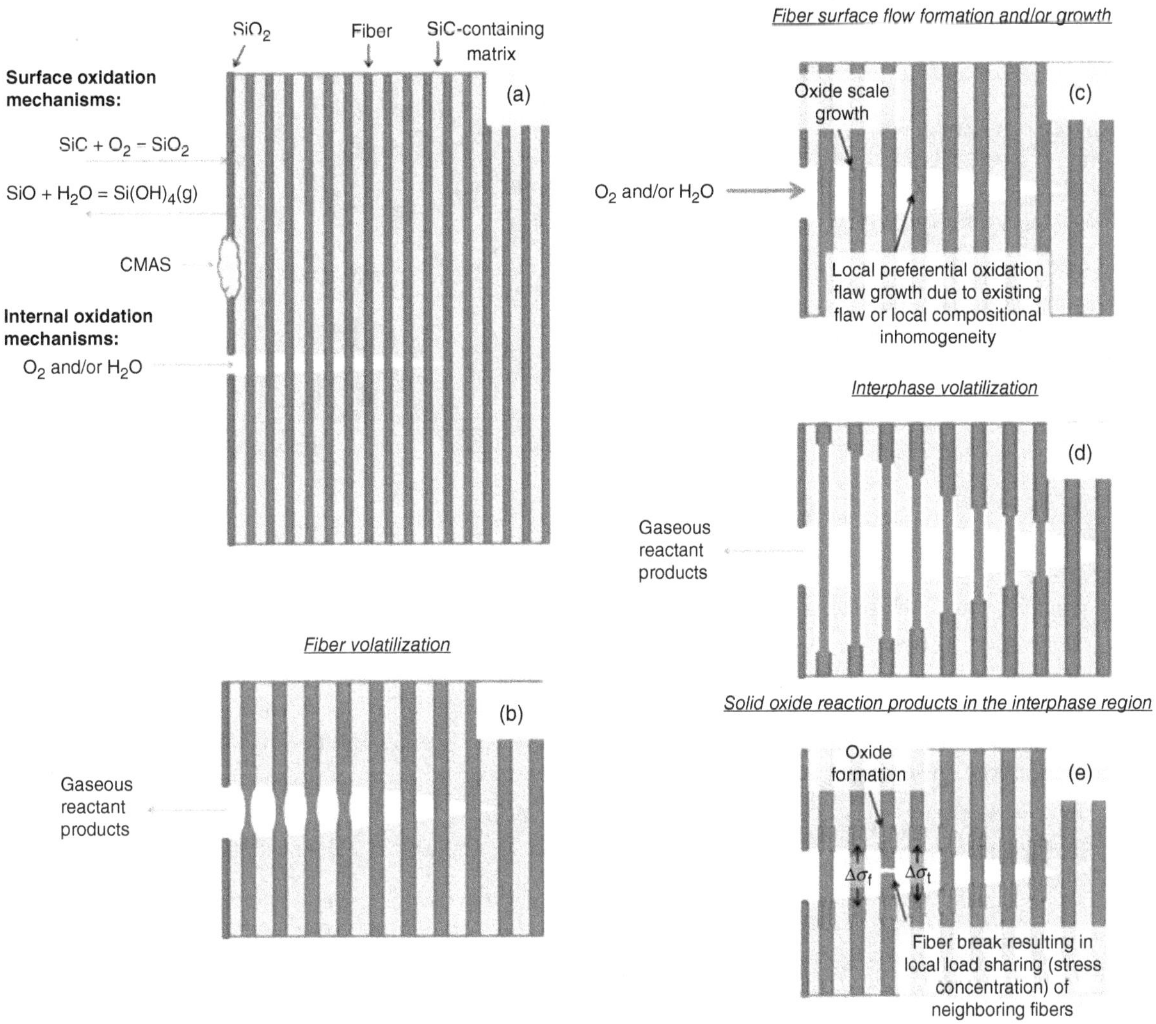

FIGURE 11.1

Schematic representations of (a) general SiC composite oxidation mechanisms and (b–e) specific types of interior oxidation mechanisms.

for example, exposed carbon interfaces. The former is not dependent on stress to any great degree; however, the latter often necessitates matrix cracks and therefore stresses which induce matrix cracking. The effect on mechanical properties can be deleterious for both mechanisms since the former at the very least results in a loss of cross-sectional area of composite and the latter removes and/or replaces the interphase with a strongly bonded oxidation product, altering the critical interfacial properties necessary for load transfer and tough composite behavior often resulting in brittle mechanical properties. It is very possible that both occur at the same time.

The temperature that the composites are exposed to also dictates the nature of the mechanisms due to mechanical and chemical interactions. For the purposes of this paper we define temperature ranges accordingly.

Low temperature: The temperature below which there is no time-dependent degradation. Typically this is because there is no environmental effect.

Intermediate temperature: The temperature range where the effect of time-dependent creep in the fibers and/or the matrix is nonexistent or minimal and mechanical degradation is primarily due to environmental mechanisms. Degradation is dominated in this temperature range by internal oxidation, because surface oxidation is slow, and is typically due to volatile reactions between the interphase and the environment and/or environmentally assisted slow-crack growth or oxidation of the fibers. For SiC-based composites, this temperature can be as

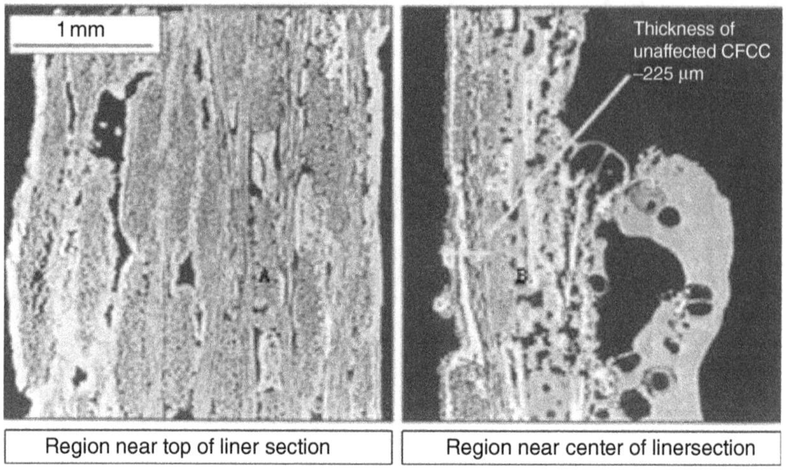

FIGURE 11.2
Micrographs of solar turbine (5018 hours) tested CMC in a region of little oxidation (~2.4-mm-thick composite) and an oxidized region with only ~225 μm of composite remaining. Reproduced with permission from Reference 12. Reproduced with permission from Elsevier Limited, Journal of European Ceramic Society, 2005.

low as ~450°C (oxidation of BN to form boria liquid or $H_xB_yO_z$ volatile gases) or ~500°C for carbon interphases or fibers (volatile CO or CO_2 formation). In some cases this intermediate temperature range mechanical degradation can be even worse than high temperatures and is referred to as a "pest" regime [9].

High temperature: The temperature above which intrinsic time-dependent creep and/or fatigue mechanisms of the reinforcing fibers and matrix become active. At these temperatures mechanical degradation would occur in an inert environment but is enhanced in the presence of oxidizing or corrosive environments. Both surface recession and internal oxidation mechanisms can be active in this temperature range depending on the chemical content of the environment as well as the applied stress state.

11.2.1 Surface Recession

A variety of surface recession mechanisms can occur depending on temperature, environment, and the composite constituents. SiC-based composites are primarily composed of the SiC-containing fibers, BN or C interphases, SiC, Si, and/or Si–O–N–C matrix-containing phases. Surface oxidation of the composite is due to the reaction of the solid phases of the constituents with the oxidizing gas phases (e.g., O_2 and/or H_2O) resulting in solid, liquid, and/or gaseous reaction products. For most SiC-based CMCs, assuming a high content of oxidizing species (O_2 and/or H_2O), the outer surface will be SiC (or Si in melt-infiltrated composites) and the reaction product will be SiO_2 (s), CO (g), H_2 (g), and/or SiO(OH) gasses [10, 11]. For pure SiC, oxidation would exhibit parabolic (protective) oxide scale growth as well as parabolic SiC recession when oxidation is dominated by the solid SiO_2 oxidation product (low H_2O contents and high enough O_2 contents, e.g., typical air environments), linear surface recession when dominated by volatile reaction products or a paralinear behavior when both mechanisms are active. For these composite, the oxidation of the SiC matrix would be the slowest oxidizing constituent of the composite. Therefore, composites with thick SiC surface layers could offer some protection. When the oxidation front reaches the interphase then a number of reactions involving volatile/molten reaction products will occur resulting in drastic material consumption (Figure 11.2) [12]. For long time applications in the presence of high water content atmospheres and/or more corrosive environments (e.g., salt or CMAS) environmental protective coatings will be required [8].

11.2.2 Interior Oxidation and "Oxidation Embrittlement" (Via Cracks or Exposed Volatile Pathways)

When the environment has access to the interior of the composite via diffusion through surface-exposed cracks or for carbon-containing composites (interphases or fibers), internal oxidation of the fibers, interphase, and matrix can result in reduced load-carrying ability. There are a variety of mechanisms which cause reduced mechanical properties depending on the fiber, interphase, and matrix phases. The potential mechanisms include the following.

- *Fiber volatilization* (Figure 11.1b) resulting in reduced fiber cross-sectional area and lower load carrying ability
- *Fiber surface flaw formation and/or growth* (Figure 11.1c) due to local surface oxidation mechanisms resulting in time-dependent reduction in fiber strength. There are a number of different mechanisms described in the

literature: SiO_2 scale formation where the scale itself acts as the strength limiting flaw [13, 14], at intermediate temperatures where SiO_2 scale formation is slow significant time-dependent strength degradation under stress-rupture conditions have been observed for SiC fiber types containing excess carbon related to a slow-crack growth mechanism associated with oxidation of carbon in grain boundaries or bulk of fiber depending on the location of the excess carbon [15], and at higher temperatures slow-crack growth phenomenon associated with creep mechanisms have been postulated [16].

- *Interphase volatilization* (Figure 11.1d) which causes time-dependent lengths of fiber decoupling from the matrix as well as lower interfacial sliding strengths (typically these reactions occur initially at the fiber/interphase interface). This results at least in longer lengths of fibers carrying higher loads resulting in reductions in composite strength due to the statistical nature of fiber failure associated with longer local fiber gage lengths.
- *Solid oxide reaction products in the interphase region* (Figure 11.1e) due to oxidation of the interphase, fiber, and matrix. Locally, the composite loses the toughening mechanisms associated with a weak fiber/matrix debonding and sliding. As a result the locally strongly bonded fibers are subject to local stress concentrations and local load sharing resulting in local fiber failure, that is, unbridged crack growth.

11.3 COMPOSITE SYSTEMS

Ultimately, the specific types of stressed-oxidation mechanisms depend on the available phases of a specific SiC-based composite system. In addition, the structure of the composite, for example, fiber architecture and/or matrix type may have a bearing on the nature of composite oxidation degradation.

11.3.1 Carbon-Fiber-Reinforced SiC Composites (C_f/SiC_m)

C/SiC composites typically consist of carbon fibers reinforcing a SiC matrix. These composites can be fabricated from a number of techniques: chemical vapor deposition, polymer pyrolysis or liquid Si infiltration; however, most oxidation studies have been performed on CVI matrix C/SiC composites. During cooling from the high temperature fabrication temperature, matrix cracking will occur due to the difference in thermal expansion coefficient between the fibers and the matrix [17]. In other words, this results in a precracked composite which is susceptible to interior oxidation even at zero or very low tensile stresses. The primary mechanism for degradation due to oxidation is fiber volatilization due to the oxidation of carbon.

This can occur at temperatures as low as 500°C since passive oxidation of SiC is slow and C oxidation is rapid (Figure 11.3a). In fact, degradation of mechanical properties is often worse [18] (Figure 11.3b) in the intermediate temperature regime (600–1000°C) due to the fact that residual tension in the matrix is relieved (i.e., cracks close) as the temperature approaches the processing temperature [17].

Oxidation tests on C/SiC have also been performed for different steam-containing environments balanced with Argon [19] with similar behavior as in air (weight loss, massive carbon fiber removal, and reduced rupture times at 600°C and 1200°C). Interestingly, the degradation in air at 600°C was worse than in 20% H_2O + 80% Argon environment at 600°C. Although, higher concentration steam environments at 1200°C (80% H_2O) did result in greater strength degradation compared to air under similar temperature/stress conditions.

11.3.1.1 Approaches to Mitigate Stressed-Oxidation Degradation

Since oxidation of the fibers is the degradation mechanism, preventing oxygen from interacting with the fibers is critical. Therefore, methods to prevent oxidation of the fibers include surface coatings and matrix oxygen getters which include crack-sealing reaction products which diminish diffusion of oxygen to the fibers. These can take the form of coatings and matrix gettering phases [18] which in some cases are multilayer in structure [19].

Halbig [18] compared CVI C/SiC composites with a CVI C-B-Si-containing layer (CBS), a matrix "enhanced" with B-containing oxygen getters, and combinations of the two. Stress-rupture tests were performed in air at 800°C and 1454°C at 69 and 172 MPa. For most conditions, improvements (longer lives) were observed for each strategy with decreasing benefit as follows. CBS coating + enhanced matrix > enhanced matrix > CBS coating > as-produced composite (≥ better than). However the CBS coating alone for the 800°C condition at 172 MPa was best compared to the other two strategies.

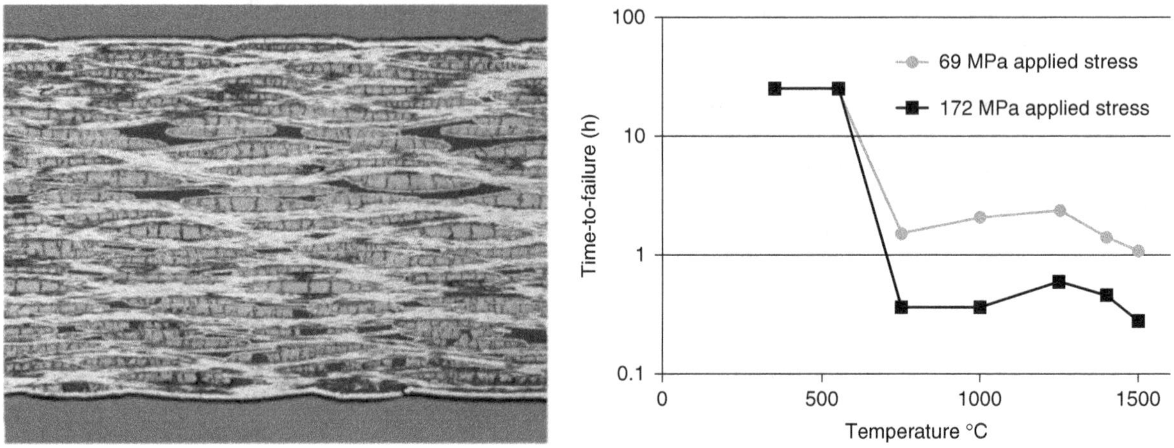

(a) (b)

FIGURE 11.3

(a) Polished cross-section of C/SiC specimen tested in air at 600°C (Reference 19, used with permission). Note the dark regions (oxidized fibers) corresponding to the matrix cracks in the SiC upon cooling of the composite after CVI SiC infiltration. (b) Time-to-failure for stress-rupture of C/SiC composites in air at different temperatures and two different applied stresses [18].

A multilayer approach has been developed [20] for C/SiC composites under the trade name A500 [1] which incorporates multilayers of SiC and a Si–B–C (boron-containing) oxygen getter phase. These composites actually have a net weight gain when oxidized in oxidizing environments compared to standard C/SiC composites which exhibit weight loss at intermediate temperatures (e.g., 700°C). The stress-rupture properties of these composites in Figure 11.4 are quite good as compared to other C/SiC composites, including the composites mentioned above with CBS coatings and enhanced matrices.

11.3.2 SiC-Fiber-Reinforced SiC Matrix Composites with Carbon Interphases ($SiC_f/C_i/SiC_m$)

The interphase region is the critical microstructural feature controlling the stressed oxidation of SiC-fiber-reinforced SiC matrix composites. For carbon interphases a volatile reaction product dominates, for BN interphases, condensed

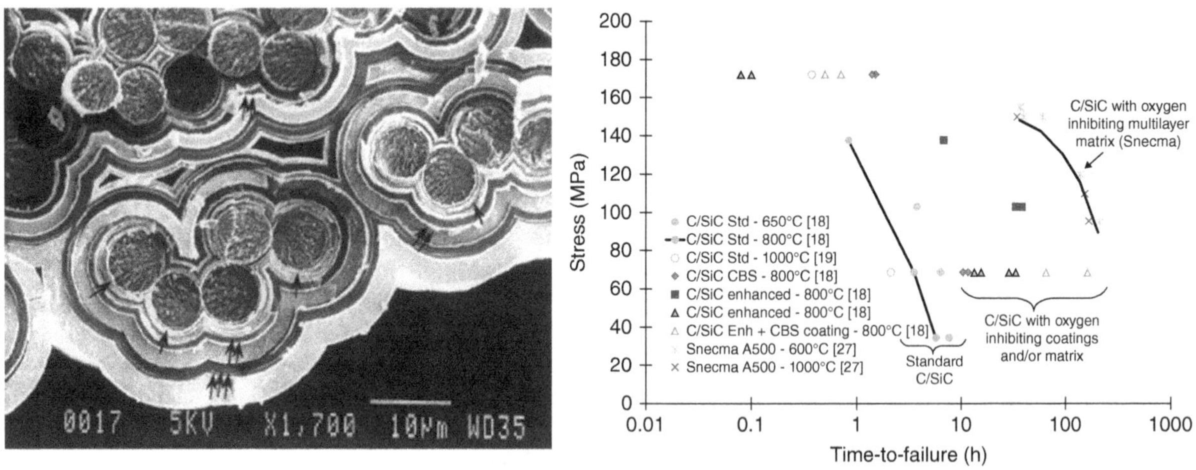

(a) (b)

FIGURE 11.4

(a) Example of inhibiting multilayer matrix. Reproduced with permission from Reference 20 and from Elsevier Limited, Composites Science and Technology, Vol 59, Iss 7, 1999. (b) Stress-rupture of different C/SiC composites tested in air.

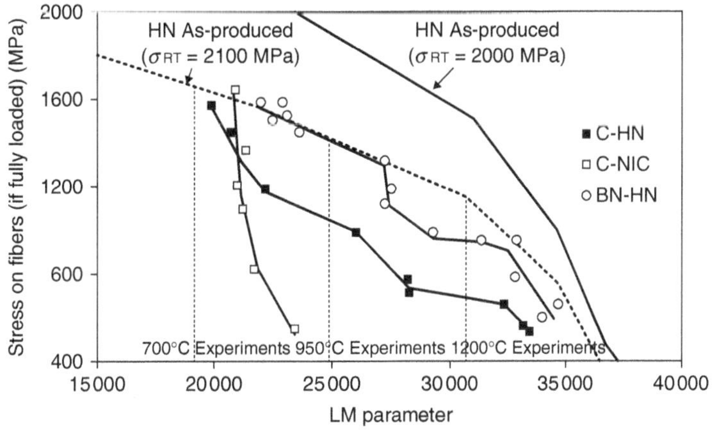

FIGURE 11.5

Plot of stress in fibers if fully loaded, that is, fiber stress in a through-thickness matrix crack, for single tow minicomposites tested under constant stress conditions at elevated temperatures in air from Reference 23. The LM parameter is a temperature compensated time parameter equal to $T(\log tR + C)$, where T is temperature in K, tR is the rupture time in hours, and C is a constant found to be 22 (based on fiber rupture properties [16]). The LM parameter is used for comparative purposes for different fiber/interphase CVI SiC matrix minicomposite rupture tests under different temperature conditions.

phases as well as volatile species can occur. In general, BN interphases are more environmentally durable than carbon interphases in oxidizing environments [21–23] (compare C–HN and BN–HN minicomposite rupture data in Figure 11.5) and are aimed at longer time oxygen-containing environment applications such as gas turbine engines [2] whereas carbon interphase composites are aimed at lower temperature/shorter time applications, for example, exhaust nozzles [1] or hypersonic applications [1, 4], or lower oxygen content environments, for example, nuclear applications [4].

For SiC/C/SiC, two modes of internal oxidation are possible, either from exposed interphases at the surface, for example, cut edges or via matrix cracks which deflect and debond at fiber/interface matrix or within the carbon interphase. The former can be alleviated by a seal coat of SiC if cut edges are exposed due to machining. Therefore, this discussion will focus on interphase oxidation via environment-exposed matrix cracks.

At intermediate temperatures, C interphase oxidation is very rapid and SiC passive oxidation is slow resulting in removal of the interphase (similar to removal of C fibers in C/SiC composites) [24]. However, the fibers are still intact and carry load, but decoupling of fibers from the matrix can be over large distances in relatively short periods of time [21]. With increasing time (and temperature) fiber decoupling from the matrix gives way to fiber bonding to the matrix, at least locally, due to oxidation of fibers and matrix to form SiO_2. This is evident from the change in pull-out lengths at different temperatures for minicomposites tested under different stress/temperature rupture condition in air (Figure 11.6) where after some period of time little or no pull-out occurs due to fibers strongly bonded to the matrix.

11.3.2.1 Approaches to Mitigate Stressed-Oxidation Degradation Since oxidation of the C interphase is the key mechanism for stressed-oxidation degradation of SiC/C/SiC composites, protection strategies similar to that for C/SiC composites, that is, protective coatings and/or oxygen gettering compositions in the matrix, including multilayer coatings, are logical approaches. There have been studies of composites with the enhanced matrix for the Nicalon/C/SiC system; however, the benefits were not that significant if not at all as evidenced by the little difference in Nic/C/SiC stress-rupture in air for CVI SiC matrix composites with standard SiC matrix [21,22, 25] and with enhancement [26,27] (Figure 11.7). In fact the data is almost the same (above the matrix cracking limit where run-out occurs) for all the Nicalon-reinforced CVI SiC matrix composites with carbon interphases tensile tested in air for all temperatures (700–1000°C).

A more significant approach to prohibit stressed oxidation is to use similar multilayer strategies as described earlier for C/SiC composites where multiple layers of SiC and a Si–B–C-containing phase constitute the matrix. Consequently, matrix cracks deviate between the layers, gettering oxygen and sealing cracks from the condensed phase oxidation product (borosilicate glass). One such composite from Snecma, A415, [1, 28] reportedly is capable of greater than 5000 hour stress-rupture at 850°C in air at 150 MPa. This capability actually competes with run-out stresses for some of the best BN interphase composites at intermediate temperatures (see the next section).

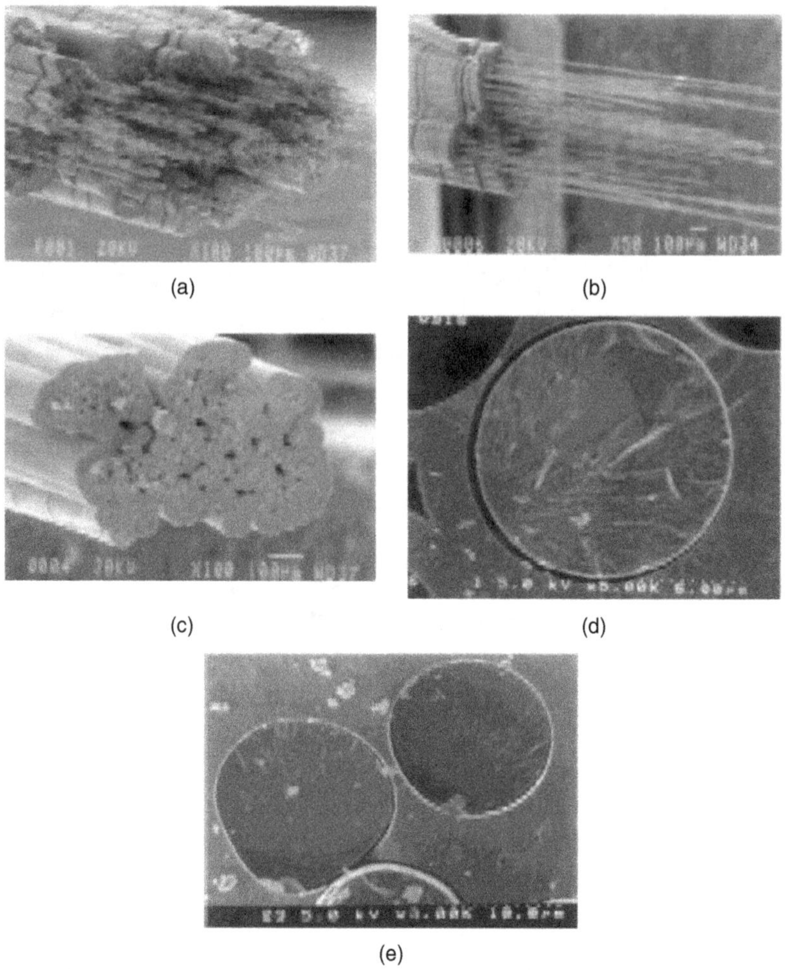

FIGURE 11.6

Minicomposite stress-rupture fracture surfaces of carbon interphase minicomposites for (a) 0.8 hours, 55 kg (b) 2.3 hours, 33 kg, and (c) 115 hours, 22 kg at 700°C. Higher magnification micrographs of Figures (a) and (c) are shown in Figures (d) and (e), respectively. From Reference 21.

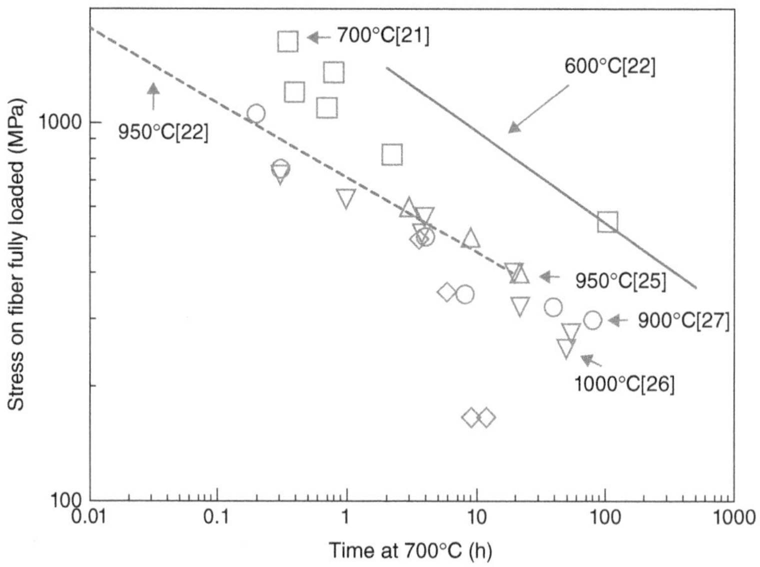

FIGURE 11.7

Stress-rupture data for Nicalon-reinforced CVI SiC matrices with carbon interphases. The data from [26] and [27] are for the enhanced matrix composites.

A similar multilayer concept referred to as an "oxygen inhibited matrix" [29] using multilayers of SiC and B_4C has recently been developed by Hypertherm Inc. in Huntington Beach, CA. Composites (Hi-Nicalon/CVI SiC matrix) with standard C interphases as well as the multilayer matrix were tested in fatigue (0.1 and 1 Hz) at 1200°C in both air and near 100% steam environments. It was found that the multilayer composites were superior in both air and steam environments. Steam environments were more degrading than air environments for comparable loading conditions.

11.3.3 SiC-Fiber-Reinforced SiC Matrix Composites with BN Interphases ($SiC_f/BN_i/SiC_m$)

For BN interphase composites, the primary mechanism for stressed-oxidation degradation is the oxidation of the BN interphase, SiC fibers and matrix within surface-exposed matrix cracks resulting in a solid oxidation reaction product which bonds fibers to neighboring fibers and/or the matrix (Figure 11.8a) [21]. Multiple reactions occur including oxidation of the BN to form boria liquid, oxidation of the SiC fibers and matrix to form SiO_2, fluxing of the boria melt with SiO_2 which can also further enhance SiC oxidation. Initial oxidation within the matrix occurs similar to that depicted in Figure 11.9 where debonding occurs between the BN interphase and the fiber and initial oxidation occurs in the fiber–BN interface and on the surface of the BN interphase. For some cases, a thin C layer exists on the fiber surface which can exacerbate oxidation between the fiber and BN [30]. As the borosilicate melt increases in SiO_2 content, the viscosity of the borosilicate will increase and eventually solidify into a high SiO_2 content glass. In addition, this process is aided by small H_2O gas concentrations in the atmosphere where B reacts with H_2O to form $H_xB_yO_z$-containing volatile species (HBO_2, H_3BO_3, and $H_3B_3O_6$ are all possible depending on temperature range) depleting the B content

(a)

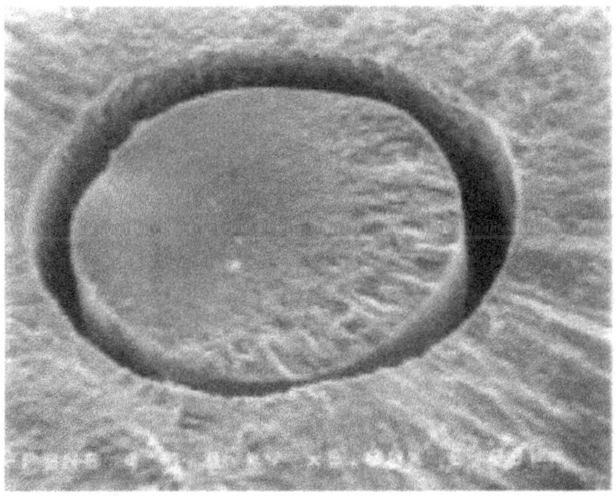

(b)

FIGURE 11.8

FESEM micrographs Hi-Nicalon-reinforced BN interphase minicomposite "that failed at 950°C after 98 hours (a) a portion of fracture surface in the hot-zone and (b) a portion of a fracture surface ~22 mm from the hot-zone after the experiment). Furnace temperature at the location in (b) was ~500°C. From Reference 21.

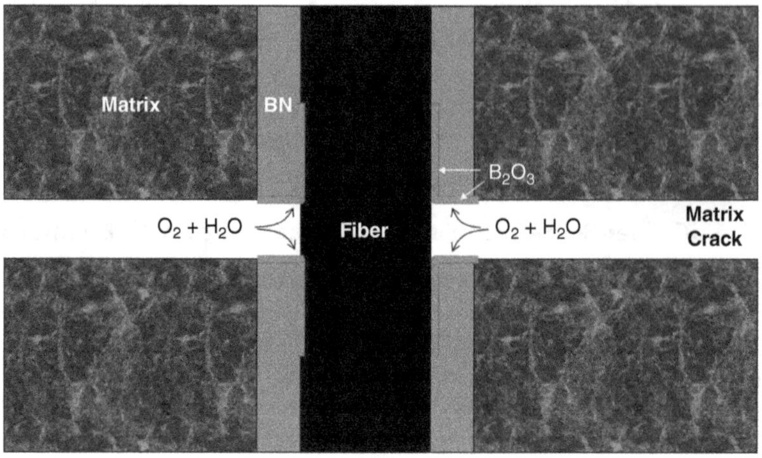

FIGURE 11.9

Schematic of initial oxidation of interphase region for BN interphase composite.

in the borosilicate and hastening the solidification of the melt [31]. If the temperature is low enough and enough H_2O is present in the environment, the volatilization of the boria liquid is so rapid and oxidation of SiC is so slow that BN depletion can occur similar to C volatilization, though not as rapid (see Figure 11.8b).

When a solid oxide glass is the reaction product at the interphase, the result is regions of strongly bonded fibers within matrix cracks which when observed after failure appear as flat regions on a fracture surface (Figure 11.10a) [9, 32]. The extent of these "embrittled" regions within matrix cracks depends on the rate controlling reactions and depletion of oxidizing species within the environment. It is the nearest-neighbor fibers with the thinnest regions of BN

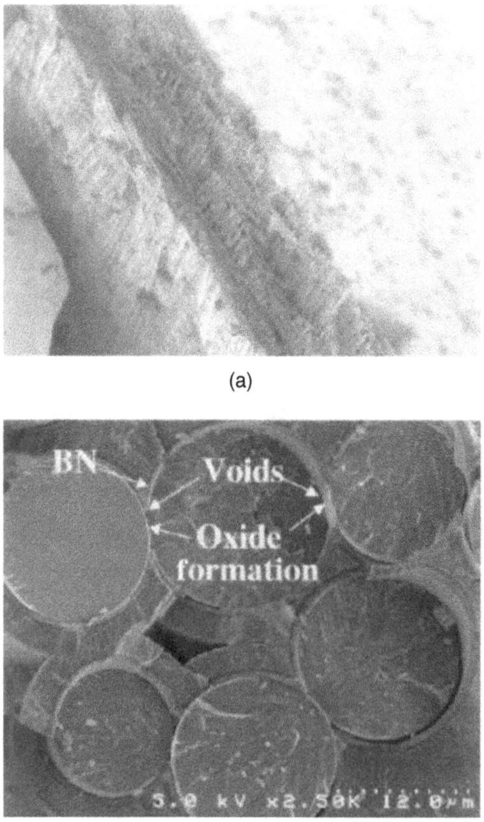

FIGURE 11.10

Fracture surface of regime I specimen tested at 185 MPa. The rupture time was 1.3 hours. (b) Higher magnification region showing embrittled fibers. From Reference 32.

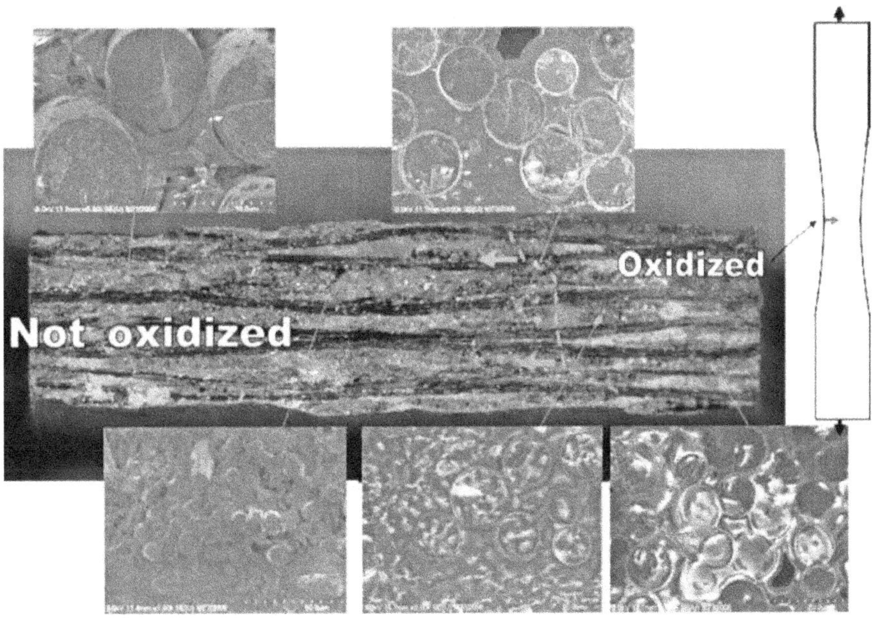

FIGURE 11.11

Example of unbridged crack growth during rupture or fatigue in SiC/BN/SiC (1200°C, air, 179 Mpa, 30 Hz fatigue for 8.1 × 106 cycles, ~75 hours). Note the fibers in the oxidized region of the fracture surface are glazed over with a SiO_2 layer indicating fiber failure occurred at some time prior to ultimate composite failure, that is, unbridged crack growth. From Reference 33.

separating fibers from one another which are the most susceptible to fiber-to-fiber strong bonding [32]. At intermediate temperatures such as 815°C in air, the time to strongly bond fibers which are close to one another, as is typical for woven composites, for a constant stress test was found to take at most 20 minutes for fibers near the surface of the composite (Figure 11.8b).

For SiC/BN/SiC, the mechanism for fiber failure and growth of unbridged matrix cracks is the failure of at least one strongly bonded fiber and the subsequent stress concentration on nearest-neighbor fibers (Figure 11.1e). Due to the statistical nature of fiber failure, local fiber failure could occur due to a slight increase in stress and/or due to fiber weakening from either intrinsic or extrinsic slow-crack growth or creep-induced strength degradation mechanisms. When a local fiber fails and transmits load to the nearest-neighbor fibers due to a lack of global load sharing as a result of fiber-to-fiber or matrix strong bonding, a cascade of strongly bonded fiber failures occur resulting in an unbridged portion of the transverse crack. Note that the fracture mirrors in Figure 11.10b occur at the nearest-neighbor contact points indicating that fiber fracture occurred where the fibers were strongly bonded. For longer time tests and/or higher temperature tests where the fracture surface was removed immediately upon composite rupture, regions of unbridged crack growth are evident by the fact that fiber fracture surfaces are coated with SiO_2 [32,33] (Figure 11.11). If the load transmitted to the region of the cross-section which is not embrittled, that is, intact interphases and global load sharing, is too high for the remaining fibers to carry, catastrophic failure will occur [34].

These processes are all more exacerbated in the presence of steam. Depending on the temperature, H–B–O-containing gases are produced easily resulting in faster oxidation of BN as well as SiC and a loss in the ability of the oxide reaction product to seal cracks from incoming oxidizing species. The degradation in rupture and fatigue properties in steam environments has been demonstrated for a number of different SiC/BN/SiC systems at intermediate and high temperatures [35–40]. The data in the literature for O_2/H_2O environments of SiC/BN/SiC composites does show a "knockdown" in stress–time capability at temperature for fatigue or rupture when higher steam contents are present; however, the run-out condition is still about the same as for air testing since this depends on the matrix cracking stress limit.

11.3.3.1 Approaches to Mitigate Stressed-Oxidation Degradation

Since the interphase is the main oxidizing constituent of the composite, either reducing oxidation of the interphase via strategies such as matrix getters or multilayer matrix as in the Snecma material described above would increase stressed-oxidation capability. A similar concept has been to ensure debonding of the BN interphase region between the matrix and the interphase itself (outside debonding) rather than between the interphase and the fiber (inside debonding) [41]. This has a similar effect to the multilayer concept where the path of oxidation is removed from the fibers themselves enabling them to retain a composite nature.

Another approach has been to improve the oxidation and hydrolytic resistance of the BN interphase itself. Typical BN interphases are infiltrated at relatively low temperatures (<1000°C) [42, 43] resulting in nanocrystalline BN which is less stable to reactions with O_2 and H_2O as compared to long-range order of crystalline or even turbostratic BN [44, 45]. Higher temperature processed BN interphases can offer greater environmental durability properties. Doping BN with a sufficient concentration of Si [34, 46] offers even superior oxidation/hydrolytic resistant properties due to the preferential formation of SiO_2 rather than boria liquid [47]. Either of these approaches requires higher processing temperatures resulting in high deposition rates and consequently either continuous CVD coating of individual tows [44, 47] or CVI coating of pieces of fabric [46], that is, stacking multilayers of fabric as in a typical composite preform would not be possible which is a limitation of these approaches.

Finally a third approach is to increase the distance between neighboring fibers. Fibers that are very near to one another can form strong bonds very rapidly as mentioned above. If the nearest-neighbor spacing is increased, that is, local interphase dimension is greater than 50 or 100 nm, a greater amount of interphase must be oxidized and formed into a glass in order to create strong bonds between the neighboring fibers or between the fibers and the matrix. This can be accomplished in a number of ways. Woven fibers have been "fluffed" so as to increase the dimension of the tow size [34]. Woven Sylramic™ fibers have also been treated in a N_2-containing environment so as to produce a ~100 nm BN layer on the surface of each fiber before BN interphase infiltration (Sylramic-iBN) resulting in greater separation of woven fibers in the preform [48]. Finally, a true tape lay-up approach, for example, GE Hypercomp, naturally spreads fibers out far from one another enabling some of the best matrix cracking as well as stressed-oxidation properties for relatively small fiber volume fractions [49].

Figure 11.12 shows some intermediate temperature stress-rupture results for three of the above mentioned BN interphase improvements with the same five-harness satin architecture melt-infiltrated composite system: fiber spreading, Si-doped BN and outside debonding for Sylramic-type woven fiber melt-infiltrated composites. The outside debonding composites show the best improvement in rupture properties since this prohibits fiber surface oxidation. Note that the run-out condition for these composites is just slightly higher than the matrix cracking stress for Syl-BN or fiber-spread Sylramic composites.

Finally, it should be mentioned, for the SiC-fiber-reinforced SiC matrix system, it would be ideal to have an interphase other than C or BN, that is, a stable oxide which will not react or at least react considerably less in oxygen and water containing environments [14, 50]. One CVD approach [50] using a SiO2/ZrO2/SiO2 coating on the fibers was shown to be a feasible coating without degradation of the fibers during processing. A further modification of that coating did demonstrate some benefit for single tow CVI matrix minicomposite rupture testing at elevated temperatures in air over BN interphase minicomposites [51].

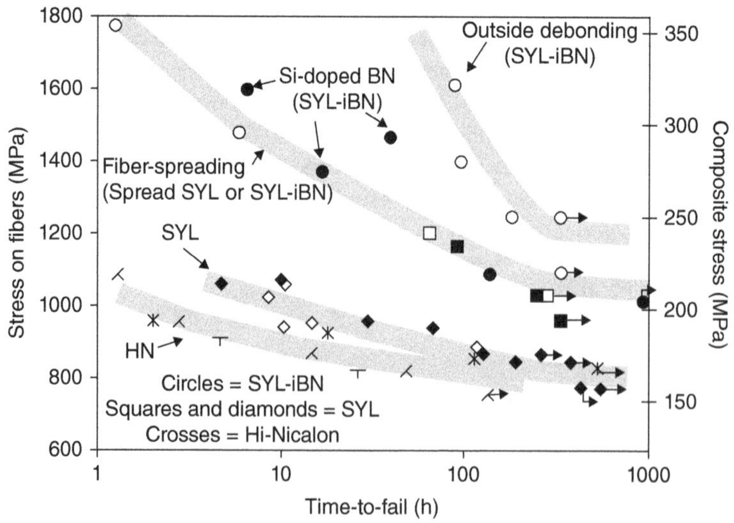

FIGURE 11.12

Stress-rupture properties at 815°C in air of conventional woven BN interphase composites and composites with interphase improvements. From Reference 34.

11.4 MODELING AND DESIGN FOR STRESS-OXIDATION DEGRADATION

Modeling the stress-oxidation degradation of SiC-based composites is quite complex and at this point an area that needs much more study. One must consider which mechanisms (Figure 11.1) are active for a given temperature–chemical environment and how they affect the different constituents of the composite of interest. Even for a given component with a thermal gradient, stress-oxidation mechanisms can change along the length or even through-the-thickness. In general these factors must be well understood to model the behavior:

- Degree of matrix cracking—What is the local crack density, are cracks through-thickness or are they microcracks? This will determine the pathways for oxidizing species.
- What oxidation reactions will occur to what constituents at what rate—As described above, C fiber and interphases will volatilize in oxygen. BN will form a liquid initially as well as volatile reactions and the formation of solid oxide products as SiO_2 concentration increases in the melt. These chemical reactions occur within matrix cracks and down interfaces or oxidized interphases resulting in depletion of oxidizing species and if the pathways for oxidation are very small then boundary effects, that is, diffusion-controlled phenomena or reaction-rate-dependent phenomena. Gas concentrations, pressures, and velocity can have dramatic effects on the kinetics for these reactions.
- How the oxidation of the constituents impacts the mechanics of the matrix crack region—For C/SiC, oxidation of the carbon fibers is the critical constituent. For SiC/SiC, oxidation of the C or BN interphase will most likely occur initially or at least more rapidly than oxidation of the SiC fibers and/or matrix resulting in decoupling of the fiber from the matrix (longer lengths of fibers exposed to higher stresses and wider crack openings). Consequently, as load-bearing constituents either disappear, for example, carbon fibers, or the stress distribution between composites is redistributed due to interphase removal, for example, removal or reduction of the stress carried by the matrix in the bridged matrix crack region for SiC/SiC, time-dependent strain occurs.
- The mechanism for degradation and ultimate failure—One or more mechanisms can ultimately be the cause of the reduction in load-carrying ability or composite embrittlement. The key for modeling is to understand the impact of the above phenomena on the load-carrying ability of the fibers. For C/SiC, loss of fibers increases the stress on remaining fibers. For SiC/C/SiC interphase oxidation results in increased loads on the fibers as well as greater lengths of fibers exposed to the environment and the potential for fiber strength degradation due to oxidation-induced slow-crack growth mechanisms. For longer times, thinner interphase regions or higher oxygen concentrations, strong bonding can occur between neighboring fibers and/or the matrix. For SiC/BN/SiC composites strong bonding will occur between the fibers and or between the fiber and matrix. This will induce a local stress concentration since loads cannot be shared globally if a locally strongly bonded fiber fails. This typically results in large scale unbridged crack growth. For this case it is important to understand the nature of the stress rise that will activate the stress concentration. At a constant load, there is no stress rise when fiber-to-fiber strong bonding occurs [52]. A local stress increase is required for the effect of stress concentration to occur. This could occur from an increase in the external stress or from a local fiber failure. Most likely, local fiber failure will occur from the time-dependent slow-crack growth mechanism for the given fiber in the given environment. However, one difference between the C interphase and BN interphase SiC/SiC composites is the length scale at which the fibers are at maximum stresses within and near the matrix crack. Fiber lengths subjected to the maximum stress in the BN interphase composites are typically much shorter due to the creation of condensed phase oxidation products.

Therefore, based on the above, there are no "general" models to date for stressed-oxidation degradation in SiC-based composites. Modeling attempts are relegated to specific composite systems. A few examples will be discussed here.

11.4.1 Modeling Stressed-Oxidation Degradation in C/SiC Composites—C Fiber Removal

As described above, the main mechanism for stressed-oxidation degradation in C-fiber-reinforced SiC composites is the oxidation (removal) of fibers through pre-existing matrix cracks formed on cooling the composite from processing (although additional cracks can form with applied load) or fiber ends exposed to the environment. Therefore, modeling

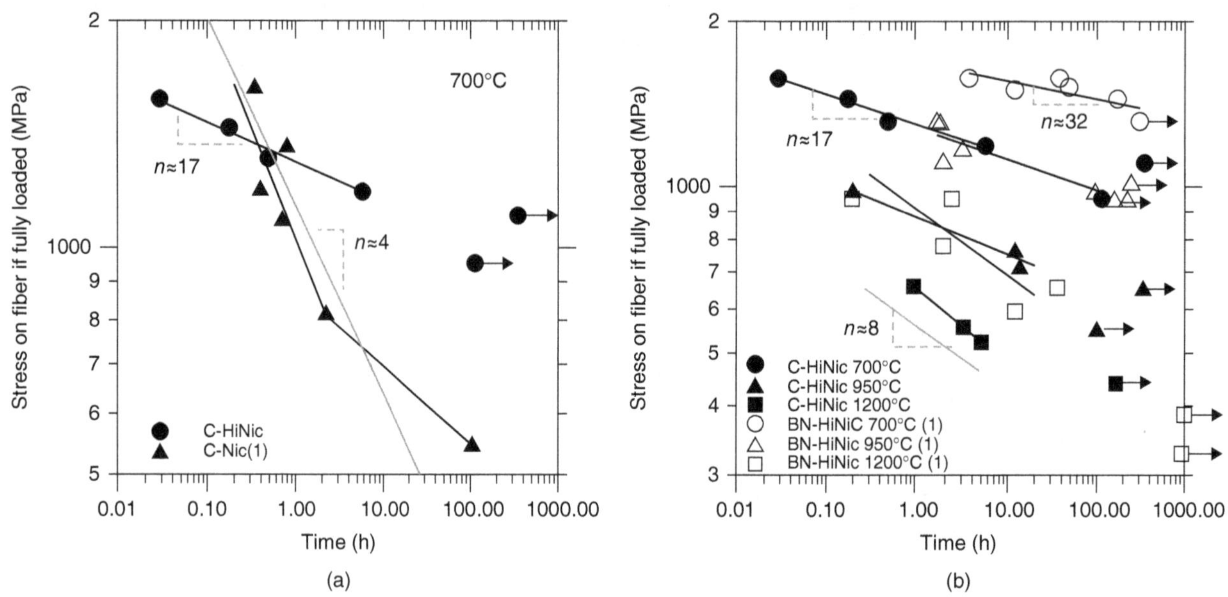

FIGURE 11.13

Stress-rupture of (a) carbon interphase minicomposites with two different fiber types (Hi-Nicalon and Nicalon) in air at 700°C and (b) carbon and BN interphase minicomposites with Hi-Nicalon fibers in air at several temperatures. The C-Nic data is from Reference 21. Slope lines added to figures are from Reference 23.

must incorporate the loss of carbon fibers due to the environment which can be verified by the ability to predict weight loss [53], predict interphase recession [54], which can then be related to the resulting reduction in fiber area. In addition one must consider the oxygen and/or steam content in the environment [55] and the effect that stress has on crack opening and/or enhanced oxidation [56].

11.4.2 Modeling Stressed-Oxidation Degradation in SiC/C/SiC Composites: Interphase Recession and Fiber Flaw Growth

Lara-Curzio [57] modeled composites at intermediate temperature assuming fiber surface oxidation (SiO_2 scale growth) as the cause for fiber degradation after carbon interphase removal. Oxide scale growth is parabolic, that is, oxide thickness $\alpha t^{\frac{1}{2}}$. Assuming that the flaw is the oxide scale, fracture mechanics would necessitate fiber strength proportional to the square root of flaw size, that is, scale thickness, which would lead to rupture strength[1] proportional to $t^{-\frac{1}{4}}$ corresponding to a stress-rupture exponent, n, of 4:

$$\sigma = At^{-1/n} \tag{11.1}$$

Lara-Curzio did measure a $t^{-\frac{1}{4}}$ dependence for rupture strength at 950°C in air for Nicalon-fiber-reinforced CVI SiC composites with carbon interphases [25]. Morscher also measured a similar relationship at 700°C for Nicalon-fiber-reinforced CVI SiC matrix precracked minicomposites with carbon interphases [21] (Figure 11.6). However, Hi-Nicalon fiber precracked minicomposites with carbon interphases showed significantly milder time dependence for strength degradation, of the order of $t^{-1/20}$ (Figure 11.13).

Lamon et al. [15, 58, 59] performed single fiber and fiber tow stress-rupture experiments in oxygen-containing environments between 500°C and 800°C on a number of different SiC fiber types. Considerable fiber degradation occurred for excess C-containing fibers due to a surface flaw growth mechanism where oxidation of the excess C in the bulk of the fiber or at the grain boundaries resulted in a slow-crack growth type of fiber strength degradation instead

[1] There would be an incubation time related to the growth of the scale exceeding the intrinsic flaw size of the fibers themselves. The $t^{1/4}$ relationship would be for time greater than the short incubation time and for stresses above the stress to cause through-thickness matrix cracking [57], unless the specimen was precracked prior to the stress-rupture test.

TABLE 11.1 Intermediate Temperature Fiber Stress-Rupture Properties [15]

Fiber type	Temperature (°C)	n	A (MPa^n s)	Ea (kJ/mol)
Nicalon	600	2.6	2.0×10^{29}	103
Hi-Nicalon	500	8.4	1.05×10^{30}	185
Hi-Nicalon	800	8.3	3.36×10^{26}	185
Hi-Nicalon type S	600	7.2	3.15×10^{26}	177
Hi-Nicalon type S	800	7.2	3.33×10^{24}	177

of surface SiO_2 scale growth. Table 11.1 shows the rupture constants measured for single tow experiments. Note that even at the lower temperature, n is ~2.6, similar to the 700°C single tow Nicalon/C interphase minicomposite data. For Hi-Nicalon, n was found to be ~8.3 (Table 11.1), significantly greater than Nicalon but much less than that observed for single tow minicomposites. Obviously the rates for time-dependent rupture at these intermediate temperatures are strongly dependent on interphase oxidation and the fiber rupture properties. It is interesting that the "n" for HN/C/SiC and HN/BN/SiC minicomposites are very similar for rupture at 700°C, 950°C, and 1200°C (Figure 11.13) even though the C interphase minicomposites have lower rupture strengths. This most likely reflects the longer length of highly stressed fibers in the C interphase system due to volatilization of C.

11.4.3 Modeling Stressed-Oxidation Degradation in SiC Fiber Composites—Unbridged Crack Growth Due to Fiber Weakening and Stress Concentration at Unbridged Crack Tip

Evans et al. [60] model an embrittled region within a matrix crack where fiber weakening occurs at the unbridged matrix crack tip. Initially, weakening of fibers near the surface in a surface-exposed crack occurs due to oxidation of the fibers, fiber failure, and subjection of new pristine fibers to the environment and subsequent weakening. As the effects of narrow crack opening and oxidation reaction product concentration increases within the matrix crack, oxygen gradients develop resulting in the outermost fibers oxidizing and degrading to a greater degree than more interior fibers. This results in outermost fiber failure at the unbridged–bridged crack region and growth of the unbridged crack into the interior of the composite. The two stages resemble stage I (reaction-rate-controlled) and stage II (diffusion-controlled) stress-controlled cracking [61].

This model incorporates a stress concentration on the outermost bridged fibers adjacent to the unbridged crack region. Therefore, oxidation, subsequent fiber degradation (proportional to the square root of oxide scale growth), and the applied stress concentration are modeled in order to determine fiber failure and growth of unbridged cracks. The model does not take into account the presence of strongly bonded fibers.

11.4.4 Modeling Stressed-Oxidation Degradation in SiC/BN/SiC Composites: Unbridged Crack Growth Due to Strongly Bonded Fibers and Local Fiber Failure

Morscher and Cawley [34] modeled the intermediate temperature embrittlement of SiC/BN/SiC composites consisting of two different fiber types according to the strong-bonding mechanism within surface-exposed bridged matrix cracks (Figure 11.1e). Failure occurred when the remaining pristine composite could not carry the load shed from the unbridged crack growth region. Essentially the likelihood that a strongly bonded matrix crack would fail was determined based on the kinetics for fiber-to-fiber oxidation within the depth of the composite as well as the time-dependent degradation of individual fibers due to a slow-crack growth mechanism. The approach was to iterate over time to determine if a strongly bonded fiber would fail. If one strongly bonded fiber failed then it was assumed that all the strongly bonded fibers would fail due to the local stress concentration imposed on neighboring fibers as a result of local load sharing rather than global load sharing, that is, loss of local composite behavior.

The fraction of fibers which embrittled was described as

$$\phi_{\text{emb}} = \phi_{t\,\max} - \phi = \frac{K_\phi}{C^m_{\text{frupture}} C^{2m/n}_{\text{ox}}} \left[x^{2m/n}_{\max} - x^{2m/n} \right] \quad (11.2)$$

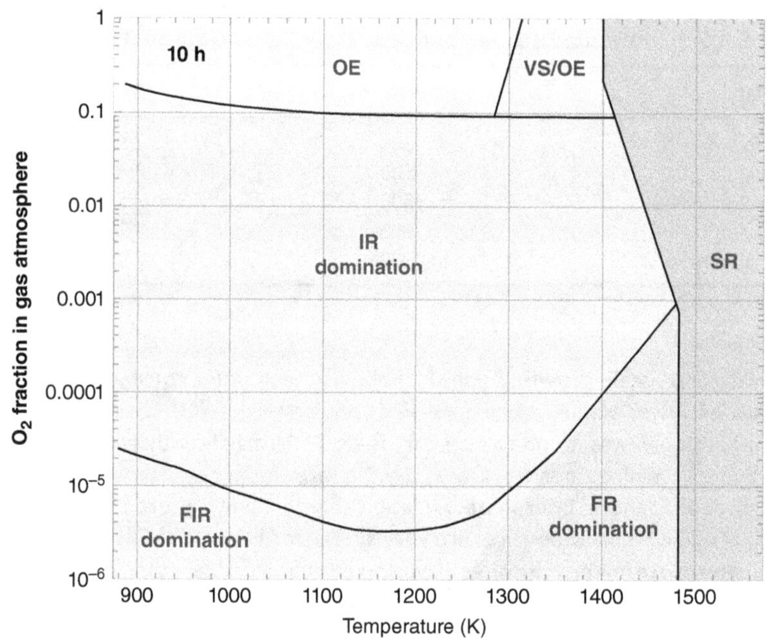

FIGURE 11.14

Map showing regions of dominant crack growth mechanism as a function of oxygen concentration in the environment and temperature for SiC/SiC composites. Used with permission from Reference 62 and 2008, John Wiley &Sons, Journal of the American Ceramic Society.

where ϕ_{max} is the fraction of all fibers that failed in a matrix crack with respect to time and ϕ_t are the fibers that failed in the matrix crack prior to being embrittled (strongly bonded) when global load sharing conditions existed. K_ϕ incorporates the length scale (sliding length and crack opening displacement) for fiber failure probability based on the applied stress state. $C_{frupture}$ is the fiber rupture coefficient (A in equation 11 1). C_{ox} is the coefficient from the parabolic embrittlement depth, $x = C_{ox}t^{1/2}$ which was determined empirically from fracture surfaces of failed rupture specimens. In order to determine when an embrittled fiber would fail, ϕ_{emb} was integrated with respect to depth of oxidation (a function of time) as well as multiplied by the number of cracks. When one fiber fails then all the embrittled fibers fail in the crack, that is, an unbridged crack. Failure would occur when the remaining pristine fibers in the cross-section could not carry the load shed onto them from the embrittled fiber failure.

This approach successfully modeled woven Hi-Nicalon- and Sylramic-fiber-reinforced melt-infiltrated composites with BN interphases at 815°C. For low stress, long time rupture, the run-out limit was shown to correspond to the failure of a single fiber failure (provided at least one matrix crack existed). For higher stress, shorter rupture times, failure was controlled by the kinetics for strong fiber-to-fiber bonding depth. Interestingly, the rate (depth) of fiber-to-fiber oxidation was approximately a factor of two greater for HN composites compared to Sylramic fiber-reinforced composites. This would be expected since HN has a lower modulus than Sylramic fiber and the interfacial shear strength of the sliding interface for the HN composites was lower than that of the Sylramic composites.

11.4.5 Mechanism Map: Henegar and Jones

Jones and Henegar et al. [62, 63] have identified several mechanisms causing elevated temperature time-dependent failure depending on oxygen content for which they have created crack growth mechanism maps (Figure 11.14). The mechanisms considered are FR: fiber relaxation due to creep, FIR: irradiation enhanced creep of fibers and induced creep crack growth, IR: interphase removal due to oxidative volatilization, VS: viscous sliding of oxide in interphase region at higher temperatures, OE: oxidation embrittlement, and SR: fiber stress-rupture. This approach can be useful for simple oxidizing environments and when the kinetics for oxidation reactions is well known. However, new maps would presumably have to be constructed for different composites with different constituents (fibers, interphases, or matrices), different environments, and/or possibly different fiber architectures.

11.4.6 Simple Design Approach: Matrix Cracking Stress

The matrix cracking stress, often determined from a graphical deviation from linearity proportional limit approach [64], has often been considered a method for determining a use-stress for long time composite operation. However, the

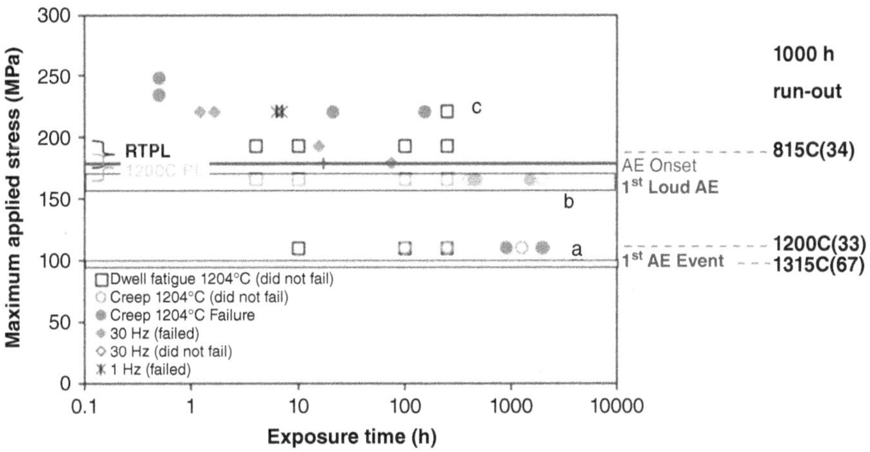

FIGURE 11.15

Relationship between cracking parameters, both graphical and based on AE, with stress-rupture and fatigue data at 1200°C for Sylramic-iBN/MI-SiC woven composites from Reference 33. Also shown is the 1000 hour run-out stress for 815C [34] and 1315C [67] air-rupture tests on the same composite system. From Reference 65).

use of graphical techniques often overestimates the actual stress where stressed-oxidative deleterious matrix crack formation occurs. For woven SiC-fiber-reinforced, BN interphase, melt-infiltrated matrix composites, Morscher [65] has used acoustic emission to determine the onset of microcrack and macrocrack stresses in the composites and shown that these stresses correspond to "run-out" conditions for different use-temperatures depending on the mechanisms of stressed-oxidation degradation. Though this is not a model of actual composite behavior, it does offer a useful approach to determine an approximate upper use-stress for long time conditions.

Microcrack formation for these composites is typically unbridged tunnel cracks or cracks propagating in surface tows perpendicular to the stress direction. The stress at which these cracks form are typically below the long-time stress-rupture run-out condition for these composites because these types of cracks do not intersect load-bearing fibers. They are also relatively low energy acoustic events. Macrocracks do propagate a significant depth into the composite, if not through-thickness, thus bridging load-bearing fibers and enabling oxidizing species to degrade fibers and oxidize interphases. These types of cracks also are higher in energy, in fact, for MI composites, the acoustic energy for these types of cracks are the highest acoustic energy events measured during a tensile test [65].

Figure 11.15 shows a plot of stress versus time for a Sylramic-iBN, BN interphase, melt-infiltrated composite tested for different stress conditions (rupture and fatigue) at 1200°C in air. Also shown is the various AE parameters used to define matrix cracking in these composites: AE onset stress, the stress at which high energy events dominates the AE behavior, first loud AE event, the lowest stress high energy AE event, and first AE event, the lowest stress AE event, typically a low energy event. The former two stresses are indicative of large fiber bridged matrix crack formation whereas the latter stress corresponds to the first microcrack formed, typically associated with a 90° tow tunnel crack. Also shown in the figure is the proportional limit measured at room and elevated temperature which corresponds with AE onset stress. Finally, the run-out stresses for rupture tests performed at 815°C [34], 1200°C [33], and 1315°C [67] are shown on the right-hand side of the figure (however, only 1200°C data is shown in the figure). It is evident that the stress for macrocrack formation (high energy AE events) corresponds to the intermediate temperature stressed-oxidation run-out condition whereas for higher temperature conditions where creep mechanisms are also operative, microcrack formation dictates long time rupture. Therefore, the lowest stress for high energy events is a good stress for intermediate temperature rupture whereas at high temperatures, the lowest stress for AE activity (first AE stress) corresponding to microcrack formation is appropriate. Evidently, microcracks due to creep as well as stressed-oxidation mechanism can lead to crack growth from microcrack formation and ultimate failure. It is important to note that for every fracture surface examined where long time failure occurred, the corner of a rectangular cross-section specimen with a surface 90° tow appeared to be the origin of the initial microcrack that leads to failure. It should also be noted that when rupture tests were performed on the same composite system in vacuum at 1200°C [68], significantly better rupture properties were observed; therefore, confirming the presence of a stressed-oxidation effect at 1200°C. Also, for composite architectures where surface 90's are not present, higher stresses are necessary than that to cause high energy cracks because lower stress interior cracks appear to be stopped by surface tows oriented in loading direction [69].

11.5 CONCLUDING REMARKS AND FUTURE DIRECTIONS

The degradation of mechanical properties due to stressed-oxidation mechanisms in fiber-reinforced SiC matrix composites ultimately limits the use of these materials at high temperatures and especially for long time applications. This is especially true for conditions where the stress exceeds the stress to form matrix cracks, enabling oxidation into the interior of the composite. One must be weary of the nature of the specific type(s) of degradation that a desired composite system could undergo as they are varied depending on the constituent content of the composite as well as the stress–temperature–environment condition it is to be subjected to. Based on the research performed to date, $SiC_f/BN_i/SiC_m$ is more durable than $SiC_f/C_i/SiC_m$ which is more durable than C_f/SiC_m composites. However, there are interphase, matrix, and coating strategies which can further the effectiveness of these composite systems, even C_f/SiC_m, enabling them to survive thousands of hours in extreme environments.

Developing effective life models and health-monitoring approaches are important areas for further study. The range of composite systems and different oxidative/corrosive environments results in a number of different stress-oxidation mechanisms which makes general life models very difficult. Because of this, health-monitoring approaches will need to be developed which can measure the extent of stress-oxidation degradation in SiC-based composites in order to verify life-model predictions and/or provide independent assessment of component viability.

ACKNOWLEDGMENTS

Most of this work has resulted from the author's affiliation with, collaboration with and/or support from NASA, Air Force, and the Navy over the past 23 years for which he is extremely grateful.

REFERENCES

1. F. Christin. Design, fabrication, and application of thermostructural composites (TSC) like C/C, C/SiC, and SiC/SiC composites. *Adv. Eng. Mat.*, 4(12), 903–912 (2002).
2. D. Brewer. HSR/EPM combustor materials development program. *Mater. Sci. Eng. A*, 261, 284–291 (1999).
3. G. Norris. Weight revolution- GE pushes for a breakthrough in lighter materials on back of JSF 136 engine. *Aviat. Week Space Tech.*, 9, 36 (2009).
4. R. Naslain. Design, preparation and properties of non-oxide CMCs for application in engines and nuclear reactors: an overview. *Comp. Sci. Tech.*, 64, 155–170 (2004).
5. J. L. Smialek, R. C. Robinson, E.J. Opila, D.S. Fox, and N.S. Jacobson. SiC and Si_3N_4 recession due to SiO_2 scale volatility under combustion environments. *Adv. Compos. Mater.*, 8(1), 33–45 (1999).
6. N. S. Jacobson, J. L. Smialek, and D. S. Fox. Molten salt corrosion of SiC and Si_3N_4, in *Handbook of Ceramics and Composites, Synthesis and Properties,* edited by N. P. Cheremisinoff (Marcel Dekker, New York, 1990), Vol. 1, pp. 99–136.
7. A. G. Evans, D. R. Clarke, and C. G. Levi, The influence of oxides on the performance of advanced gas turbines. *J. Eur. Ceram. Soc.*, 28(7), 1405–1419 (2008).
8. K. N. Lee, D. S. Fox, and N. P. Bansal. Rare earth silicate environmental barrier coatings for SiC/SiC composites and Si_3N_4 ceramics. *J. Euro. Ceram. Soc.*, 25, 1705–1715 (2005).
9. F. E. Heredia, J. C. McNulty, F.W. Zok, and A.G. Evans, Oxidation embrittlement probe for ceramic-matrix composites. *J. Am. Ceram. Soc.*, 78(8), 2097–2100 (1995).
10. E. J. Opila and R. E. Hann. Paralinear oxidation of CVD SiC in water vapor. *J. Am. Ceram. Soc.*, 80(1), 197–205 (1997).
11. E. J. Opila. Variation of the oxidation rate of silicon carbide with water-vapor pressure. *J. Am. Ceram. Soc.*, 82(3), 625–636 (1999).
12. H. E. Eaton and G.D. Linsey. Accelerated oxidation of SiC CMCs by water vapor and protection via environmental barrier coating approach. *J. Euro Ceram. Soc.*, 22, 2741–2747 (2002).
13. E. Lara-Curzio. Stress ± rupture of Nicalon/SiC continuous ®bers ceramic matrix composites in air at 950°C. *J. Am. Ceram. Soc.*, 80(12), 3268–3272 (1997).
14. E. E. Boakye, P. Mogilevsky, T. A. Parthasarathy, R. S. Hay, J. Welter, and R. J. Kerans. Monazite coatings on SiC fibers I: fiber strength and thermal stability. *J. Am. Ceram. Soc.*, 89(11), 3475–3480 (2006).
15. W. Gauthier, F. Pailler, J. Lamon, and R. Pailler. Oxidation of silicon carbide fibers during static fatigue in air at intermediate temperatures. *J. Am. Ceram. Soc.*, 92(9), 2067–2073 (2009).

16. H. M. Yun and J. A. DiCarlo. Time/temperature dependent tensile strength of SiC and Al2O3-based fibers. in *Ceramic Transactions, Vol. 74, Advances in Ceramic-Matrix Composites III*, edited by N. P. Bansal and J. P. Singh, American Ceramic Society (1996), pp. 17–26.
17. F. Lamoroux, R. Naslain, and J. M. Jouin. Kinetics and mechanisms of oxidation of 2D woven C/SiC composites: II, theoretical approach. *J. Am. Ceram. Soc.*, 77(8), 2058–2068 (1994).
18. M. C. Halbig. Oxidation kinetics and strength degradation of carbon fibers in a cracked ceramic matrix composite. *Ceram. Eng. Sci. Proc.*, 137–144 (2004).
19. M. J. Verrilli, E. J. Opila, A. Calomino, and J. D. Kiser. Effect of environment on the stress-rupture behavior of a carbon-fiber-reinforced silicon carbide ceramic matrix composite. *J. Am. Ceram. Soc.*, 87(8), 1536–1542 (2004).
20. F. Lamouroux, S. Bertrand, R. Pailler, R. Naslain, and M. Cataldi. Oxidation-resistant carbon-fiber-reinforced ceramic-matrix composites. *Comp. Sci. Tech.* 59(7), 1073–1085 (1999).
21. G. N. Morscher. Tensile stress rupture of SiCf/SiCm minicomposites with carbon and boron nitride interphases at elevated temperatures in air. *J. Am. Ceram. Soc.*, 80(8), 2029–2042 (1997).
22. H. T. Lin and P. F. Becher. Effect of coating on lifetime of Nicalon fiber-silicon carbide composites in air. *Mater. Sci. Eng.*, A231, 143–150 (1997).
23. J. Martinez-Fernandez and G. N. Morscher. Room and elevated temperature tensile properties of single tow Hi-Nicalon, carbon interphase, CVI SiC matrix minicomposites. *J. Eur. Ceram. Soc.*, 20, 2627–2636 (2000).
24. L. Filipuzzi, G. Camus, and R. Naslain. Oxidation mechanisms and kinetics of 1D-SiC/SiC composite materials: I, an experimental approach. *J. Am. Ceram. Soc.*, 77(2), 459–466 (1994).
25. E. Lara-Curzio. Stress-rupture of Nicalon/SiC continuous fiber ceramic matrix composites in air at 950°C. *J. Am. Ceram. Soc.*, 80(12), 3268–3272 (1997).
26. P. Lipetzky, N. S. Stoloff, and G. J. Dvorak. Atmospheric effects on high-temperature lifetime of ceramic composites. *Ceram. Eng. Sci. Proc.*, 18(4), 355–362 (1997).
27. T. E. Steyer, F. W. Zok, and D. P. Walls. Stress rupture of an enhanced Nicalon^{TM}/SiC composite at intermediate temperatures. *J. Am. Ceram. Soc.* 81(8), 2140–2146 (1998).
28. E. Bouillon, S. Bertrand, E. Pestourie, and P. Spriet. Ceramic SiC/SiC ceramic matrix composites for gas turbine engines. ASME Turbo Expo, Barcelona, Spain, May 9–11 (2006).
29. M. Ruggles-Wrenn, J. Delapasse, A. L. Chamberlain, J. E. Lane, and T. S. Cook. Fatigue behavior of a Hi-Nicalon^{TM}/SiC-B_4C composite at 1200°C in air and steam. *Mat. Sci. Eng. A*, 534, 1190128 (2012).
30. L. U. J. T. Ogbuji. A pervasive mode of oxidation degradation in a SiC–SiC composite *J. Am. Ceram.*, 81(11), 2777–2784 (1998).
31. N. S. Jacobson, G. N. Morscher, D. R. Bryant, and R. E. Tressler. High-temperature oxidation of boron nitride: II, boron nitride layers in composites. *J. Am. Ceram. Soc.*, 82(6), 1473–1482 (1999).
32. G. N. Morscher, J. Hurst, and D. Brewer. Intermediate-temperature stress rupture of a woven Hi-Nicalon, BN-interphase, SiC-matrix composite in air. *J. Am. Ceram. Soc.*, 83(6), 1441–1449 (2000).
33. G. N. Morscher, G. Ojard, R. Miller, Y. Gowayed, U. Santhosh, J. Ahmed, and R. John. Tensile creep and fatigue of Sylramic-iBN melt-infiltrated SiC matrix composites: retained properties, damage development, and failure mechanisms. *Comp. Sci. Tech.*, 68, 3305–3313 (2008).
34. G. N. Morscher and J. D. Cawley. Intermediate temperature strength degradation in SiC/SiC composites. *J. Eur. Ceram. Soc.*, 22(14–15), 2777–2788 (2002).
35. S. Mall and K. J. LaRochelle. Fatigue and stress-rupture behaviors of SiC/SiC compositor under humid environments at elevated temperature. *Comp. Sci. Tech.*, 66, 2925–2934 (2006).
36. K. J. LaRochelle and G. N. Morscher. Tensile stress rupture behavior of a woven ceramic matrix composite in humid environments at intermediate temperature—Part I. *Appl. Comp. Mater.*, 13, 147–172 (2006).
37. S. Mall and J. L. Ryba. Effects of moisture on tensile stress rupture behavior of a SiC/SiC composite at elevated temperatures. *Comp. Sci. Tech.*, 68, 274–282 (2008).
38. T. T. Kim, S. Mall, L. P. Zawada. Fatigue behavior of Hi-Nicalon type-STM/BN/SiC ceramic matrix composites in a combustion environment. *Int. J. Appl. Ceram. Technol.*, 8(2), 261–272 (2011).
39. M. B. Ruggles-Wren and V. Sharma. Effects of steam environment on fatigue behavior of two SiC/[SiC+Si_3N_4) ceramic composites at 1300°C. *Appl. Compos. Mater.*, 18, 385–396 (2011).
40. M. B. Ruggles-Wren, D. T. Christensen, A. L. Chamberlain, J. E. Lane, and T. S. Cook. Effect of frequency on fatigue behavior of SiC/SiC ceramic matrix composites at 1200°C. *Comp. Sci. Tech.*, 71, 190–196 (2011).
41. G. N. Morscher, H. M. Yun, J. A. DiCarlo, and L. Thomas-Ogbuji. Effect of a boron nitride that debonds between the interphase and the matrix in SiC/SiC composites. *J. Am. Ceram. Soc.*, 87(1), 104–112 (2004).
42. R. Naslain, O. Dugne, A. Guette, J. Sevely, C. R. Brosse, J. P. Rocher, and J. Cotteret. Boron nitride interphase in ceramic-matrix composites. *J. Am. Ceram. Soc.*, 74(10), 2482–2488, (1991).

43. N. Patibandla and K. L. Luthra. Chemical vapor deposition of boron nitride. *J. Electrochem. Soc.*, 139(12), 3558–3565 (1992).

44. A. W. Moore, H. Sayir, S. C. Farmer, and G. N. Morscher. Improved interface coatings for SiC fibers in ceramic composites. *Ceram. Eng. Sci. Proc.*, 16(4), 409–416 (1995).

45. C. G. Cofer and J. Economy. Oxidative and hydrolytic stability of boron nitride—a new approach to improving the oxidation resistance of carbonaceous structures. *Carbon*, 33(4), 3889–3395 (1995).

46. G. N. Morscher, H. Y. Yun, and F. I. Hurwitz. High temperature Si-doped BN interphases for woven SiC/SiC composites. *Ceram. Eng. Sci. Proc.*, 23(3), 295–302 (2002).

47. G. N. Morscher, D. Bryant, and R. E. Tressler. Environmental durability of BN-based interphases (for SiCf/SiCm composites) in H_2O containing atmospheres at intermediate temperatures. *Ceram. Eng. Sci. Proc.*, 18(3), 525–533 (1997).

48. H. M. Yun, J. Z. Gyekenyesi, Y. L. Chen, D. R. Wheeler, and J. A. DiCarlo. Tensile behavior of SiC/SiC composites reinforced by treated Sylramic SiC fibers. *Ceram. Eng. Sci. Proc.*, 22(3), 521–531 (2001).

49. G. S. Corman and K. L. Luthra. Silicon melt infiltrated ceramic composites (HiPerCompTM), in *Handbook of Ceramic Matrix Composites*, Kluwer Academic Publishers (2005) pp. 99–115.

50. W. Y. Lee, E. Lara-Curzio, and K. L. More. Multilayered oxide interphase concept for ceramic-matrix composites. *J. Am. Ceram. Soc.*, 81(3), 717–720 (1998).

51. H. Li, G. N. Morscher, J. Lee, and W. Y. Lee. Tensile and stress-rupture behavior of SiC/SiC minicomposite containing chemically vapor deposited zirconia interphase. *J. Am. Ceram. Soc.*, 87(9), 1726–1733 (2004).

52. W. H. Glime and J. D. Cawley. Stress concentration due to fiber-matrix fusion in ceramic-matrix composites. *J. Am. Ceram. Soc.*, 81(10), 2597–2604 (1998).

53. F. Lamouroux, R. Naslain, and J. M. Jouin. Kinetics and mechanisms of oxidation of 2D woven C/SiC composites: II, theoretical approach. *J. Am. Ceram. Soc.*, 77(8), 2058–2068 (1994).

54. A. J. Eckel, J. D. Cawley, and T. A. Parthasarathy. Oxidation kinetics of a continuous phase in a nonreactive matrix. *J. Am. Ceram. Soc.*, 78(4), 972–980 (1995).

55. E. J. Opila and J. L. Serra. Oxidation of carbon fiber-reinforced silicon carbide matrix composites at reduced oxygen partial pressures. *J. Am. Ceram. Soc.*, 94(7), 2185–2192 (2011).

56. M. C. Halbig, J. D. McGuffin-Cawley, A. J. Eckel, D. N. Brewer. Oxidation kinetics and stress effects for the oxidation of continuous carbon fibers within a microcracked C/SiC ceramic matrix composite. *J. Am. Ceram. Soc.*, 91(2), 519–526 (2008).

57. E. Lara-Curzio. Analysis of oxidation-assisted stress-rupture of continuous fiber-reinforced ceramic matrix composites at intermediate temperatures. *Composites A*, 30 549–550 (1999).

58. P. Forio, F. Lavaire, and J. Lamon. Delayed failure at intermediate temperature (600°–700°C) in air in silicon carbide multifilament tows. *J. Am. Ceram. Soc.*, 87(5), 888–893 (2004).

59. W. Gauthier and J. Lamon. Delayed failure of Hi-Nicalon and Hi-Nicalon S multifilament tows and single filaments at intermediate temperatures (500°–800°C). *J. Am. Ceram. Soc.*, 92(3), 702–709 (2009).

60. A. G. Evans, F. W. Zok, R. M. McMeeking, and Z. Z. Du. Models of high temperature, environmentally assisted embrittlement in ceramic-matrix composites. *J. Am. Ceram. Soc.*, 79(9), 2345–2352 (1996).

61. S. M. Spearing and F. W. Zok. Matrix crack spacing in brittle matrix composites. *Acta Metall. Mater.*, 42, 1933–1945 (1994).

62. R. H. Jones and C. H. Henager. Subcritical crack growth processes in SiC/SiC ceramic matrix composites. *J. Eur. Ceram. Soc.*, 25, 1717–1722 (2005).

63. R. H. Jones, C. H. Henager, C. A. Lewinsohn, and C. F. Windisch Jr. Stress-corrosion cracking of silicon carbide fiber/silicon carbide composites. *J. Am. Ceram. Soc.*, 83(8), 1999–2005 (2000).

64. ASTM C-1275, Standard Test Method for Monotonic Tensile Behavior of Continuous Fiber-Reinforced Advanced Ceramics with Solid Rectangular Cross-Section Test Specimens at Ambient Temperature, ASTM, 100 Barr Harbor Drive, West Conshohocken, PA (2000)

65. G. N. Morscher. Stress, matrix cracking, temperature, environment, and life of SiC/SiC woven composites, in *High Temperature Ceramic Materials and Composites (HTCMC7 Proceedings)*, edited by W. Krenkel and J. Lamon, AVISO Verlagsgesellschaft mbH, Berlin, Germany (2010), pp. 251–260.

66. G. N. Morscher. Modal acoustic emission source determination in silicon carbide matrix composites, in *Review of Progress in Quantitative Nondestructive Evaluation*, edited by D. O. Thompson and D. E. Chimenti, CP 509, (American Institute of Physics, NY, 2000), pp. 383–390.

67. G. N. Morscher. Tensile creep and rupture of 2D-woven SiC/SiC composites for high temperature applications. *J. Eur. Ceram. Soc.*, 30(11), 2209–2221 (2010).

68. G. N. Morscher, R. John, L. Zawada, D. Brewer, G. Ojard, A. Calomino. Creep in vacuum of woven Sylramic-iBN melt-infiltrated composites. *Comp. Sci. Tech.*, 71(1), 52–59 (2011).

69. G. N. Morscher. Tensile creep of melt-infiltrated SiC/SiC composites with unbalanced Sylramic-iBN fiber architectures. *Int. J. Appl. Ceram. Technol.*, 8(2), 239–250 (2011).

CHAPTER 12

ENVIRONMENTAL EFFECTS: ABLATION OF C/C MATERIALS—SURFACE DYNAMICS AND EFFECTIVE REACTIVITY

GERARD L. VIGNOLES, JEAN LACHAUD, AND YVAN ASPA

Laboratoire des Composites Thermostructuraux, Université de Bordeaux, Pessac, France

12.1 INTRODUCTION/BACKGROUND

Ablation, that is, irreversible material loss, is a complex set of phenomena taking place in several very demanding applications. We will discuss here some of the most severe contexts; the baseline materials that will be considered here are carbon/carbon (C/C) composites, which are amongst the most popular for those applications. Interestingly, other families of ablative materials are mostly variations or evolutions of these C/C composites. There are several extreme applications in which the C/C composites are used; in each of them, there are specific issues that we will describe now, but there are also several common points that can be identified.

Atmospheric reentry: In this application, the most extreme temperatures can be encountered up to ~4000 K, and there may be important mechanical solicitations such as shear and compression by the surrounding gases, as well as internal stresses due to the thermal gradients. The leading principle of a thermal protection system (TPS) material is that an appreciable amount of the heat flux received on the exposed side has to be absorbed by endothermic processes like pyrolysis, melting, boiling, chemical reaction, and sublimation, and transfer rate of the remaining heat through the material should be slow enough to guarantee that the maximum viable temperature at the TPS lower wall is not reached by the end of the mission. Simultaneously, the material has to withstand the important stresses to which the hypersonic flow exposes it. So, there is a necessity of keeping as much as possible a solid part at high temperatures—this is achieved through the use of carbon fibers and refractory ceramic components—and of having large enough mechanical resistance, toughness, and thermal shock resistance; this is the case when the thermal expansion coefficient is low and displays a very small mismatch between constituents, or when the material is porous. Last but not least, the material has to be as light as possible. C/C composites fulfill all those requirements at extreme temperatures, while carbon/phenolic composites are more adapted to less severe environments, as summarized in Table 12.1. Indeed, carbon/phenolic composites may be thought of like "parent materials" with respect to C/C composites. The pyrolysis of the phenolic matrix gives a char with a high carbon yield while absorbing heat, and the resulting material behaves as a low-quality C/C composite.

Ceramic Matrix Composites: Materials, Modeling and Technology, First Edition. Edited by Narottam P. Bansal and Jacques Lamon.

TABLE 12.1 Comparison of C/C and C/Phenolic Composites in Ablative Conditions

Property	C/C	Carbon/phenolic
Favorable		
Low heat diffusivity	+	++ (Phenolic resin, pyrolyzed char)
Endothermic processes and counteracting mass flow	++ (sublimation)	++ (Dehydration, pyrolysis, fiber and char sublimation)
Strength	++	–– (Weak matrix)
Toughness	++	– (Weak matrix, pyrolysis cracks)
Thermal shock resistance	++ (low fiber and matrix CTE)	+ (Porous char absorbs dilatation)
Low density	+	++
Detrimental		
Exothermal processes	– (Oxidation by air)	–– (Faster oxidation by air)
Roughness	–	–– (Porous char)

For both materials, no melting or boiling occurs; the vaporization of the extreme surface results either from sublimation or from chemical etching, and it induces surface recession [1]. Surface roughening then appears: this banal but uncontrolled phenomenon has several consequences of importance in this specific case. First, it increases the chemically active surface area of the wall, and second, it promotes the laminar-to-turbulent transition in the surrounding flow [2,3]. Both of these modifications to the physicochemistry lead to an increase in heat transfer, resulting in an acceleration of the surface recession [4]. The TPS thickness design has to take this rather strong effect into account.

Solid/liquid rocket motor nozzles and divergents: The temperatures are somewhat lesser than in the previous case, between 2000 and 3000 K. In this case, sublimation plays a very minor role, but oxidation is rather endothermic, since the oxidative species are usually CO_2 and H_2O [5]. Moreover, as previously, the high velocity flow induces large shear stresses on the walls. Accordingly, the requirements for the thermal protection materials [6] are globally very similar as for severe atmospheric reentry. The ablation of the composite walls has the macroscopic effect of modifying the device geometry, that is, the throat section increases. This may lead to a thrust loss compared to the nominal performances of the propulsion system. Simultaneously, there is a modification of the microscopic structure of the composite characterized by an increase of the surface roughness. Similar to atmospheric reentry, the development of roughness may lead to an increase by a factor of up to 1.7 of the convective heat flux received by the composite [7], which in turn has an impact on the recession velocity. These observations illustrate the fact that ablation is a multiscale and multiphysics process involving strong couplings between recession, surrounding flow, heat, and mass transfer. Accordingly, it is not a surprise that the phenomenon of *overablation* is not fully understood. The problem of overablation is the following: general trends of recession rate versus density are known, but some materials display a surprisingly high rate, neatly escaping from the tendency. This may sometimes be the result of the activation of mechanical erosion, but not in all cases—thus, overablation is poorly understood.

Tokamak armor: C/C composites are good candidate materials, in competition with refractory metals like tungsten, for plasma-facing components [8]. Here, the material requirements are somewhat distinct [9]. First, there is no need for a lightweight material; second, the maximal temperatures are neatly lower (<1500 K); third, the physical phenomena which induce ablation are neither oxidation nor sublimation, but etching by hydrogen (deuterium, tritium) ion fluxes, and sporadically by electron fluxes arising from plasma discharges (also called disruptions) [10, 11]; fourth, there is always an interplay between etching and deposition [12], and the most important issue of this dynamic situation is the resulting retention of deuterium in the walls [13, 14], particularly in the case of carbon-based materials due to the specific carbon–hydrogen chemistry [15]. Finally, the material will have to withstand high neutrons fluxes when producing energy by nuclear fusion. Again, we are facing here a very complex multiphysics process, one of the key issues being the sputtering yield of the material, and again, the acquisition of a given surface roughness has its impact on this issue.

Another application in which C/C ablation plays a prominent role is ionic propulsion of space objects. Ion thrusters feature a grid for ion optics, which is subject to erosion [16]. C/C composites have been validated as one of the most performing classes of materials for this purpose. The sputtering yield of the grid [17, 18] is one of the most important factors in the determination of the thruster lifetime.

Summarizing the problems posed in the listed applications, one can find some central questions which require a detailed investigation, among which there are: (i) the development of roughness and (ii) the effective recession rate

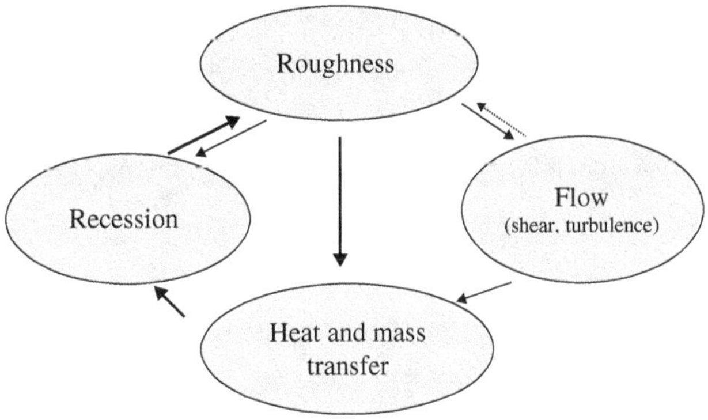

FIGURE 12.1
Multiphysics processes in ablative C/C TPS.

(or ablation resistance) of the material, for a given arrangement of constituents and surface state. Indeed, if general phenomenological tendencies are predictable, the understanding of the interaction between the flow and the material has to be improved. Figure 12.1 is a sketch of the very intricate relationships existing between heat & mass transfer, surface recession, acquisition of surface roughness, and the surrounding heat flow. One of the key points of ablative material design is the understanding of the relation between the global effective recession rate of the composite and the individual recession rates of its constituents. This chapter will focus on these issues. First, we will summarize the experimental data on surface recession and roughness; then, a modeling approach is presented, and its results are discussed.

12.1.1 Materials Description

There are a large variety of ablative C/C composites, depending on their reinforcement and matrix processing history [19]. Reinforcements may be made from ex-PAN, ex-pitch, or ex-cellulose fibers. They can be either woven (fabrics) or nonwoven (felts), needled (punctured) or not, or grouped into straight bundles. The matrix can be produced by impregnation and pyrolysis of phenolic, furannic, or any other high-carbon-yield resin, or of pitch, or by chemical vapor infiltration of hydrocarbons; it is common that the matrix is prepared in several steps. Such a large choice for reinforcement and for matrices gives a very broad panel of materials. Here, we will try to give an overview of the principal features of the surfaces, as they acquire roughness during ablation, with emphasis on a selection of materials solicited in several types of testing.

The materials studied here for atmospheric reentry application are:

1. A 3D C/C composite, made from a 3D ex-PAN carbon fiber preform and a pitch-based carbon matrix (Figure 12.2) [20]. It is a heterogeneous multiscale material. The composite has a "skeleton" of unidirectional bundles made of several thousands of fibers linked together by a pitch-based matrix (mesostructure). The bundles are fit together into a 3D orthogonal pattern repeated by translation on a cubic lattice. This macrostructure

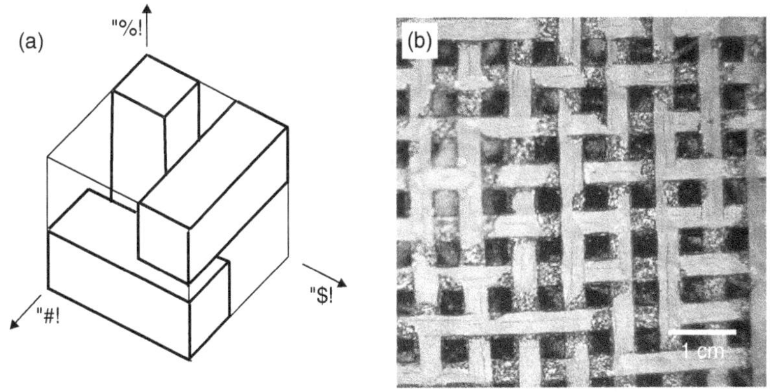

FIGURE 12.2
(a) Scheme of the reinforcements in a 3D composite. (b) View of material 1.

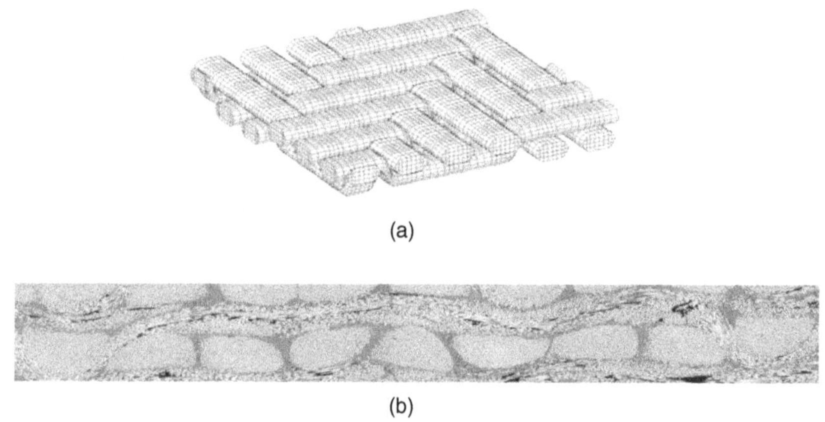

FIGURE 12.3

(a) Scheme of the weaving pattern in a 2D C-phenolic material. (b) Transverse view of the composite.

leads to a network of parallelepipedic macropores (located near each node of the lattice), which are partially filled with a pitch-based carbon matrix. These macropores are named octets. The composite has then been graphitized. This class of C/C is used in tokamak divertors [11], in reentry vehicles as thermal shields [21], and in the throat of rocket nozzles [22, 23].

2. A pyrolyzed C/phenolic resin composite, made of ex-cellulose fibers grouped in yarns, which are woven in satin plies; the plies are stacked together and impregnated with phenolic resin (Figure 12.3) [20]. Under typical ablation conditions, and at the material–environment interface, where the material is ablated, the phenolic resin is not any more present as such, but has already suffered pyrolysis and has been transformed into a highly porous carbon with an approximate density of 700 kg/m^3, that is, only one-third of graphite density.

   For solid rocket motor (SRM) nozzle application, two other kinds of materials have been studied:

3. An equilibrated 4D C/C composite, processed in the same way as the 3D composite (Figure 12.4) [24]. The "4D" composite is reinforced by millimetric bundles of parallel fibers organized following the four inner diagonals of a cube, as explained in 25, 26 and illustrated by the elementary pattern represented in Figure 12.4 where the cylindrical reinforcements are shown in dark, and the interbundle matrix is shown in light. Their typical diameter size is millimetric. The bundles are themselves constituted by a dense packing of parallel fibers embedded in a carbon matrix.

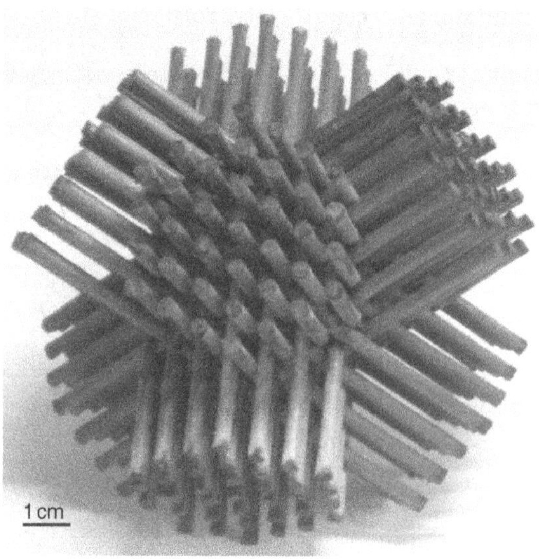

FIGURE 12.4

Example of bundle arrangement in the equilibrated 4D material (material (3)), before matrix filling.

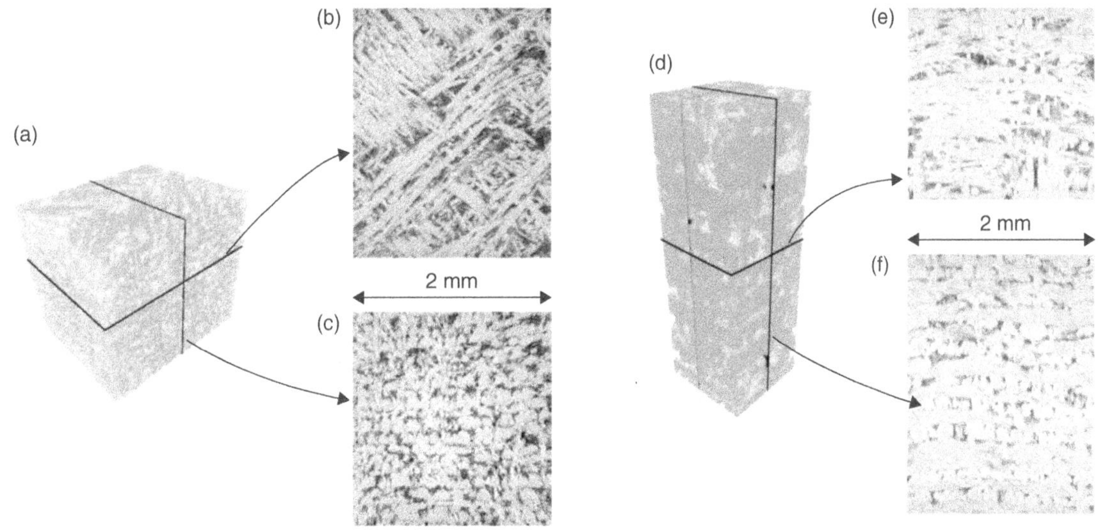

FIGURE 12.5
Tomographic imaging of two "2.5D" C/C composites. (a) 3D rendering, (b) horizontal slice, and (c) vertical slice of material 4A; (d) 3D rendering, (e) horizontal slice, and (f) vertical slice of material 4B. Reproduced with permission from Reference 24, Pergamon (Elsevier), Ceramic Transactions 191.

4. A CVI-processed "2.5D" composite, prepared from preforms made of woven ex-PAN carbon fiber fabrics [27, 24]. The matrix is infiltrated between the fibers by chemical vapor infiltration (CVI), that is, a gas-phase route [28]. During this process, the tortuosity of the preform plays an important role by limiting the deposition in the composite core [29]. As a consequence, the composite material at the end of the process still contains about 10% pore volume. Two different woven architectures were compared (they will be referred to as (4A) and (4B); see Figure 12.5). As the description of such architectures by analytic means is complex and poorly realistic, the structure of the native composite was obtained using x-ray tomography with a voxel dimension of $20 \times 20 \times 20\ \mu m^3$. The scans allow recognizing three phases: the yarns, the surrounding matrix, and the pores. The initial structure of composite A is illustrated in Figure 12.5a. The porosity was determined from the scans to be 10.4% for A and 9.7% for B.

12.1.2 Materials Tests

In the atmospheric reentry application, it is quite difficult to recover samples from real flight experiments; however, ground tests are supposed to be representative of real flight conditions to some extent [30]. These tests are carried out with plasma arc jets, in which a high-velocity air flow is heated through electric arc discharges; heat flux may reach tens of MW/m^2, temperature more than 3800°C, and stagnation pressure several atmospheres [31]. Other classical tests for ablative materials are principally oxyacetylene torch tests, in which temperatures may reach as high as 2900°C, the heat flow rate going to several MW/m^2, while maintaining very oxidative conditions (O_2, H_2O); in contrast, the stagnation point pressure and the shear stresses induced by the flow are modest. So, comparing these conditions to application conditions, it is expected that oxidation is given an exceedingly high role, and that mechanical erosion is minimized. Finally, materials intended for SRM applications are directly tested in *ad hoc* SRM benches, the conditions of which are surely representative of the application, though they are difficult to assess directly. Several comprehensive modeling approaches have been developed in order to elucidate the actual conditions suffered by the nozzle material [32, 23, 33, 34].

Table 12.2 summarizes data reported by several authors [20, 22–24, 35–51] on ablation tests on various pure C/C composites, by plasma torch, rocket nozzle, laser beam, oxyacetylene torch, or plain oxidation tests. We can have a partial overview of the testing conditions, and of the material response in terms of surface recession velocity. A mass oxidation rate (here in mg/s) is often reported, but since the size of the affected surface is not very precisely given (nor the volume of the sample), there is some difficulty in using this quantity elsewhere than in comparative tests. A linear

TABLE 12.2 Summary of Ablation Tests Performed on C/C Composites

Ref	Material	Density (g/cm^3)	Test	Heat flux (MW/m^2)	Temp (K)	Press (MPa)	Time (s)	Linear rate (μm/s)	Mass rate (mg/s)
[35]	Pyrolyzed C/phenolic	1.39	Ar/H_2 plasma arc jet		3300		91	**79.1**	141.3
[35]	C/phenolic	1.4	Ar/H_2 plasma arc jet		3300		69	**95.2**	361.5
[35]	Pyrolyzed C/phenolic + CVI	1.44	Ar/H_2 plasma arc jet		3300		80	**90.3**	120.5
[35]	Pyrolyzed C/phenolic + 2 × furannic alcohol PIP	1.524	Ar/H_2 plasma arc jet		3300		67	**109.4**	231.3
[36]	2D needled felt CVI/PIP	1.8	Plasma arc jet		3300	4.5		**1216–1554**	456–484
[20]	3D HIPIC	2.08	Plasma arc jet	50	4000	5	60	**1000**	
[20]	Pyrolyzed 2D C/phenolic	1.45	Plasma arc jet	~5	3000	2	60	**500**	
[37]	8-satin + punctured		Plasma arc jet						15–27
[38]	Needled fabric + ICVI	1.8	Solid rocket motor		3600–3750	7	8	**70–300**	
[22]	Needled fabric + ICVI	1.8	Solid rocket motor	2–4	3400	5	40	**100–200**	
[23]	Needled fabric + ICVI	1.8	Solid rocket motor	~10	<2800	1.5	120	**100–200**	
[39]	Needled fabric + ICVI		Solid rocket motor	2–4	3000–3500	8		**90**	
[40]	Needled felt ICVI + PIP	1.48	Solid rocket motor plume		<3327	~7	4	**484**	847.2
[41]	2D needled felt + TG-CVI	1.8	Small solid rocket motor		3200	4	3–4	**20–90**	
[41]	2D needled felt + TG-CVI	1.8	Small solid rocket motor		3200	8	3–4	**100–130**	
[24]	Needled fabric + CVI	1.8	Solid rocket motor	8	2800			500	
[24]	4D HIPIC	2.0	Solid rocket motor	8	2800			**500**	
[42]	4D = 3 + 1D pitch IP @ 0.1 MPa	1.5	Liquid rocket motor	45	3300			**630 ± 510**	9060 ± 520
[42]	4D = 3 + 1D pitch IP @ 60 MPa	1.83	Liquid rocket motor	45	3300			**210 ± 80**	1880 ± 310
[42]	4D = 3 + 1D pitch IP @ 90 MPa	1.88	Liquid rocket motor	45	3300			**50 ± 40**	470 ± 40
[43]	3D needled	1.78	Laser beam 100 W 20 mm 2000 mm/min 4 cycles		3200			**30**	
[43]	3D needled	1.78	Laser beam 2000 W 20 mm 2000 mm/min 4 cycles		3200			**130**	
[44]	Needled felt TG-CVI	1.8	Oxyacetylene torch GJB322A-96		3100		30	**6.7**	3.83
[45]	3D	1.8	Oxyacetylene torch		3100		60	**150**	
[46]	4D bundle rods HIPIC (+ CVI) matrix	1.77–1.85	Oxyacetylene torch	3.9	3000–3300		30	**11.3**	3.4
[47]	2D needled felt PIP/TG-CVI	1.72	Oxyacetylene torch	4.2	3300		60	**3**	2.6
[48]	2D woven I-CVI		Oxyacetylene torch		3100		60	**40**	
[49]	2D needled felt PIP/TG-CVI		Oxyacetylene torch		3300		30	**12.3**	5.35
[50]	2D needled felt PIP/TG-CVI	1.8	Oxyacetylene torch		3300		30	**2**	2
[51]	3D HIPIC	2.08	Oxidation dry air (N_2/O_2)	0.05	898	0.1	8×10^4	**45**	
[51]	Pyrolyzed 2D C/phenolic	1.45	Oxidation dry air (N_2/O_2)	0.05	898	0.1		**1411**	

recession rate (here in μm/s) is a preferable quantity. The linear recession rate is commonly measured at the most affected part of the surface, while the mass ablation rate is a more global quantity.

Let us now describe more precisely the testing conditions for the Materials 1–4 presented here.

Materials 1 and 2 have been submitted to plasma jet tests in stagnation point configuration. In this case, the material temperature is high enough (3000 K) to enable both oxidation and sublimation. The plasma jet test of this study reproduces conditions close to atmospheric reentry [1]. The 3D C/C has been tested in stagnation point configuration under a high thermal flux of air plasma (50 MW/m^2) at Mach 4. It leads approximately to a wall temperature of 4000 K and a pressure at the stagnation point of 5 MPa. Under these conditions, the composite is mainly ablated by sublimation, but oxidation also occurs [21]. Mechanical erosion sporadically occurs, but it is shown to be negligible [21]. The recession velocity of the material surface is about 1 mm/s.

The C/R is tested in plane flow configuration, and pyrolyzed *in situ*. It leads to a blowing of pyrolysis gases that tends to cool down the sample surface [52]. The mass loss caused by pyrolysis is around 18% of the initial mass. The flow of pyrolysis gases is then about 22% of the total mass flow lost by ablation. In these conditions, the wall temperature reaches 3000 K. The recession velocity of the material surface is around 0.5 mm/s.

Samples of materials (3) and (4) have been tested in actual SRM ground experiments. The testing conditions are not directly measurable at the materials surface; however, inverse identification coupled to indirect data acquisition (mass ejection rate, outlet temperature, etc.) allows providing rough estimates of the surface temperature (~2500 K) and heat flux (~8 MW/m^2). Here again, the recession velocity of the material surface is of the order of 0.5 mm/s.

An unconventional ablation test has been designed to blend five main requirements for the first steps into the modeling of surface roughness evolution: (i) isothermal conditions on the sample; (ii) laminar flow on the sample; (iii) negligible advection close to the wall (no mechanical erosion); (iv) sample size much larger than the composite unit pattern size; and (v) test leading to typical roughness features obtained by ablation.

The reactor [51] is cylindrical and heated by electrical resistances. Samples of both materials are reduced to a 1 cm^2 square surface incorporated in the reactor. It should be noticed that for this ablation test, the C/R is fully pyrolyzed at 1473 K under vacuum before the experiment. First, this enables the study of its C/C form that is of interest for ablation; also, it prevents outgassing effects [53] that can thus be decoupled from pure ablation. Experiments have been carried out at a controlled temperature of 898 K under dry air at atmospheric pressure [51, 54], and it was checked that the samples were exposed to negligible thermal gradients. The recession velocity was as low as 50 μm/s for the dense C/C composite.

12.1.3 Observation of Roughness Features

The sample surfaces have been observed by binocular magnifier (BM), optical microscope (OM), scanning electron microscopy (SEM), laser profilometry, and x-ray computed microtomography (CMT).

Three scales of surface roughness have been identified: macroscale, mesoscale, and microscale. They are strongly linked to the material's architecture. Indeed, at each scale of the composite architecture, a typical surface roughness develops. There are thus epimicrostructural, epimesostructural, and epimacrostructural features, which will be described and discussed in the following, material by material. An interesting point is the comparison of the surface of the same material ablated under distinct conditions.

Material 1: 3D C/C tested under plasma jet and under oxidation.

The material samples, tested in the plasma jet and in the oxidation furnace show comparable surface states, depicted in Figures 12.6 and 12.7. Basically, ablation reveals the architecture of the composite. Epimacrostructural surface roughness seems to result from the difference of reactivity between parallel bundles, perpendicular bundles, and interbundle matrix (cited by increasing order of reactivity). The recession of the most reactive phase is the fastest until a steady situation settles. As represented in Figure 12.6a., octets are more reactive, hence they recess faster, creating crenels. The section of emerging bundles (parallel or perpendicular) is slightly undulated. Indeed, if edges of initially square sections of bundles emerge, they are smoothed out to a wavy form. On the OM visualization of a transverse cut (Figure 12.6b), this phenomenon is obvious. From left to right: the tip of a perpendicular bundle is concave; next, the parallel bundle section is convex; then appears another concave tip of a perpendicular bundle. Between the bundles, a deep crack results from the preferential ablation of the interbundle matrix. It is difficult to exonerate thermomechanical erosion to explain these cracks. Under the parallel bundle, a pitch-based matrix octet features the edge of a residual macropore. This kind

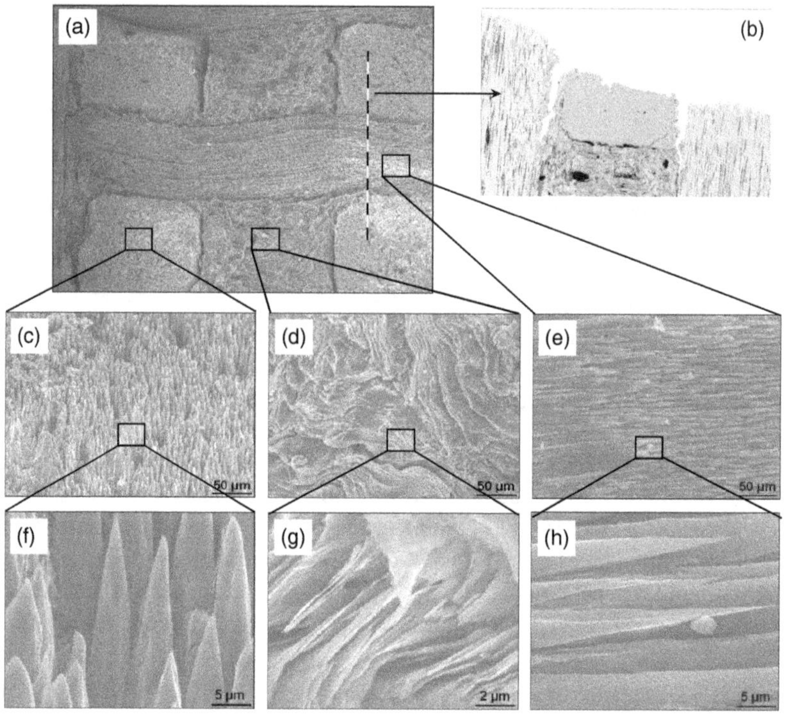

FIGURE 12.6
Observation of material 1 after a plasma arc jet test: (a) large-scale SEM view of the surface; (b) optical micrograph of a transverse slice; (c, d, e) SEM mesoscopic details of a vertical bundle, a matrix "octet," and a transverse bundle, respectively; (f, g, h) SEM microscopic details of micrographs (c, d, e) showing the ablated fiber and matrix morphologies [59]. Images b,c,f Reproduced with permission from Reference 60, Amer. Ceram. Soc.

of macropores lies in the middle of all the octets and is roughly spherical. Mechanical erosion may sporadically occur through the detachment of a matrix octet or of some fibers emerging from parallel bundles.

Looking closer at the bundle scale, we can see that, due to an important recession of some weak phase, the fibers, which are more resistant, are partially stripped, become thinner, and acquire a needle shape. The matrix seems less resistant, but some parts of it are still emerging, at a lower height than the perpendicular fibers (Figures 12.6c and 12.6f). In the literature, many micrographs show similar roughness features on carbon-based composites during ablation by oxidation [55–57] or both by oxidation and sublimation [35, 58, 42]. Parallel yarn ablation (Figures 12.6e and 12.6h) differs from the perpendicular one by the fact that the fibers are stripped layer by layer, often on many levels; the fibers have a sharper tip angle. The ex-pitch matrix of the interbundle "octets" (Figure 12.6d and 12.6g) appears as a stacking of wavy carbon foils separated by gaps, as reported earlier [42]. Indeed, ablation reveals the layered structure of the carbon matrix, leading to deep and narrow notches.

In the oxidation reactor, mechanical erosion cannot occur; there, ablation is only driven by gasification. The most reactive phases progressively uncover the least reactive phases, which increase their exposed surface area [60]. In these conditions, as shown in Figure 12.7a, the perpendicular bundles look like truncated pyramids; their tips display a plateau. The height of the truncated pyramids is approximately equal to the square section edge size. The presence of this plateau may appear surprising; it is analyzed in the modeling section. The mesoscale and microscale morphologies are similar to the previous case, but with the following differences: the denuded fibers (Figure 12.7b) are sharper, and one can see matrix foils surrounding some of them like sleeves. From this, we can infer that there is a weak interphase between the fibers and the intrabundle matrix. This part of the material has a profound effect on its behavior [61].

The emerging perpendicular bundles of material 1 after plasma arc jet or oxidation tests have been investigated by x-ray CMT [63]. Figure 12.8 shows CMT slices of the material in both cases. We can easily see that the features are extremely similar: the only difference is the depth of the affected zone around the fibers and the fiber tip angles. It has been possible to quantify the morphological parameters on materials 1 and 2 from the CMT data, as reported in Table 12.3.

The interphase between fibers and matrix has been investigated in closer detail, by Transmission Electron Microscopy (TEM) [51, 64]. In particular, in Figure 12.9, an "interphase carbon" is neatly visible, with a smaller lateral extension of the graphitic domains, even though the material has been prepared without any intention of creating a separate interphase; moreover, pores and cracks are concentrated either directly between the matrix and the fibers or in the matrix close to the fibers.

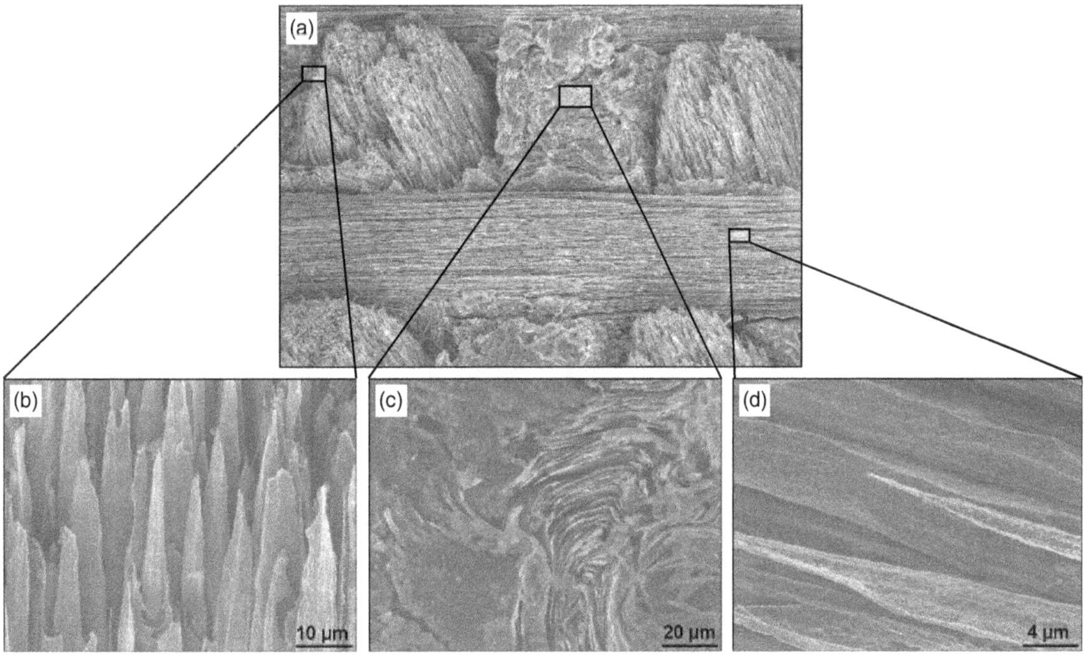

FIGURE 12.7
SEM observation of material 1 after a plain oxidation test: (a) large-scale view of the surface; (b, c, d) meso/microscopic details of a vertical bundle, a matrix "octet," and a transverse bundle, respectively. Reproduced with permission from Reference 62, Pergamon (Elsevier).

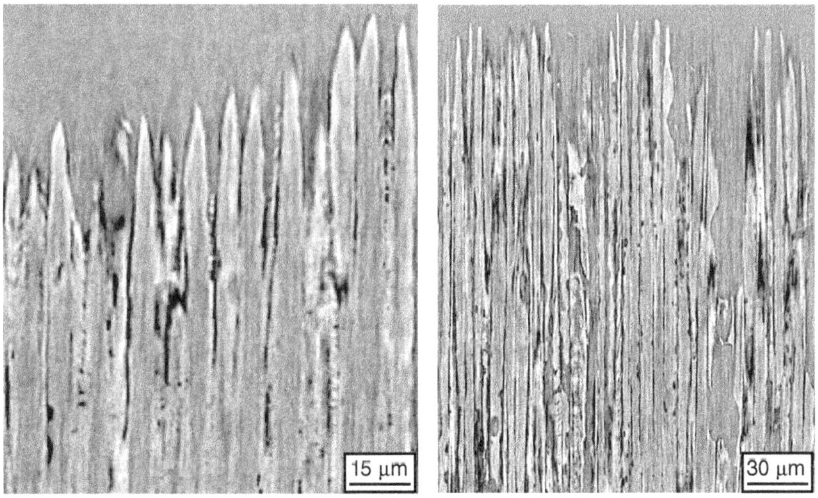

FIGURE 12.8
X-ray CMT slices of material 1 tested (a) in a plasma arc jet facility and (b) in a plain oxidation reactor [59].

TABLE 12.3 Morphological Parameters on Material 1 for Two Different Testing Conditions

Test conditions	Denuded height	Tip angle
Plasma jet	37 ± 2 μm	20.5° ± 2°
823 K dry air oxidation	111 ± 6 μm	7° ± 1°

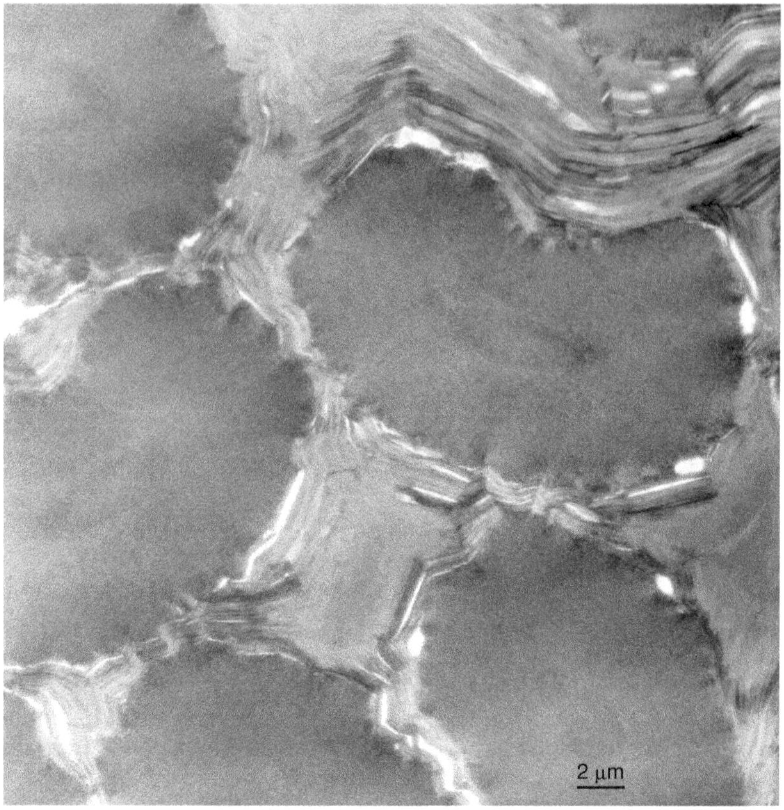

FIGURE 12.9
Bright field TEM observation of the interphase lying between fibers and intrabundle matrix.

Material 2: 2D C/R tested in plasma jet and in oxidation furnace.

The C/R sample orientations were different in the two tests. For the oxidized sample, the tissue ply and the weaving inside the ply have been placed perpendicular to the surface, while there was a 30° downstream tilt in the plasma jet test to withstand the shear stress of the high-speed flow with minimal mechanical erosion. The yarn weaving inside the tissue is inclined by 45° with the surface of the sample. The photograph (Figure 12.10a) shows a surface characteristic of this material. The surface features look like saw teeth spaced out by one-ply thickness. This morphology is caused by the preferential ablation of the thin layer of interply matrix. Nevertheless, in some places, the interyarn matrix (inside the plies) appears protected by the yarns (Figures 12.10d and 12.10g). Ablation uncovers the yarns, which emerge slanted with respect to the surface (see micrograph, Figure 12.10b). The smoothing, induced by advection and/or diffusion limitation, creates a wavy surface state. The amplitude of the waves is close to the interply matrix thickness, while their wavelength is equal to yarns section. Some chain yarns sporadically emerge perpendicularly to the wavy surface (Figures 12.10b, 12.10e, and 12.10h). The smaller-scale micrographs show that, similarly to material 1, fibers were denuded and sharpened; however, here, their morphology is more irregular, probably due to a less homogeneous texture of their constituting carbon (ex-cellulose instead of ex-PAN). The matrix has been converted into a low-density char (around 0.7 g/cm^3), which is ablated much faster than the fibers. We can see on Figure 12.10g its porous nature.

Figure 12.11 shows the morphology of the same material, after a plain oxidation test. At large scale (Figure 12.10a), the surface roughness of the sample is quite different from Figure 12.10. Perpendicular yarns, which seem less reactive than parallel yarns, emerge by about 400 μm from the surface level of parallel yarns. Interply and interyarn matrices have been deeply ablated and are hardly visible at the surface. Fibers Figures (12.11b–12.11g) are extremely sharp and thin, with an even more irregular surface with respect to Figure 12.10. Some of them are broken, probably because of sample manipulation between the test and the observation.

Material 3: 4D C/C tested in SRM facility.

Figure 12.12 is a multiscale montage of the material surface morphology. Figure 12.12b shows a bundle surrounded by matrix—the roughness is not important at that scale. The interbundle matrix (Figure 12.12d) displays characteristic features of stacked foils, as for material 1. The surface of the bundle (Figure 12.12c) has an important, evenly distributed

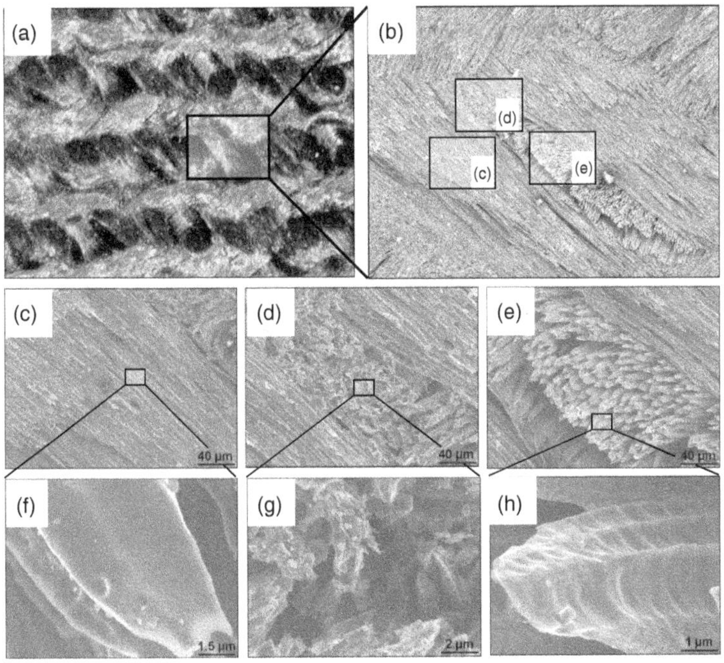

FIGURE 12.10

Observation of a pyrolyzed C/phenolic composite (material 2) after plasma arc jet testing. (a) Overall view (optical micrograph); (b) intermediate zoom; (c, d, e) mesoscale features of yarns and interyarn char; (f, g, h) microscale details of fibers and char [59].

roughness. Figure 12.12e shows that the recession has been faster in a thin zone (0.5–1 μm) around the fibers than for the fibers themselves or the matrix. This reveals again the presence of a more reactive interphase located between the fibers and the intrabundle matrix, which have more comparable reactivities. Moreover, the microscopic observation of the surface does not reveal any influence of the surrounding flow: bundles emerging in distinct directions all display the same features. Finally, it is noted that the material porosity is low—around 3%.

Material 4: 2.5D C/C tested in SRM facility.

Figure 12.13 shows the surface morphology at several scales. At the largest scale (Figure 12.13a), one notes that the matrix surrounding the bundles is only visible in depth. Some of the existing pores (~10% vol.) have been opened. The

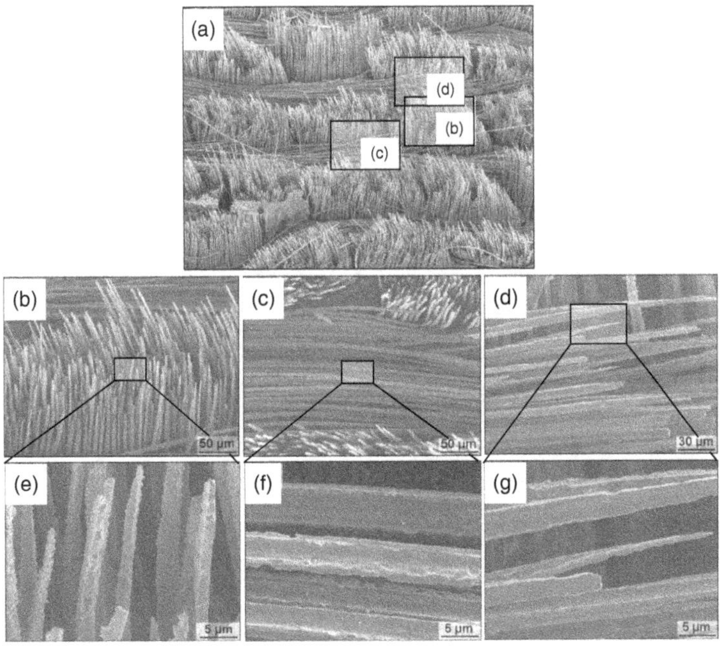

FIGURE 12.11

Observation of a pyrolyzed C/phenolic composite (material 2) after plain oxidation testing. (a) Overall view; (b, c, d) mesoscale features of perpendicular and parallel yarns; (e, f, g) microscale details of fibers [59].

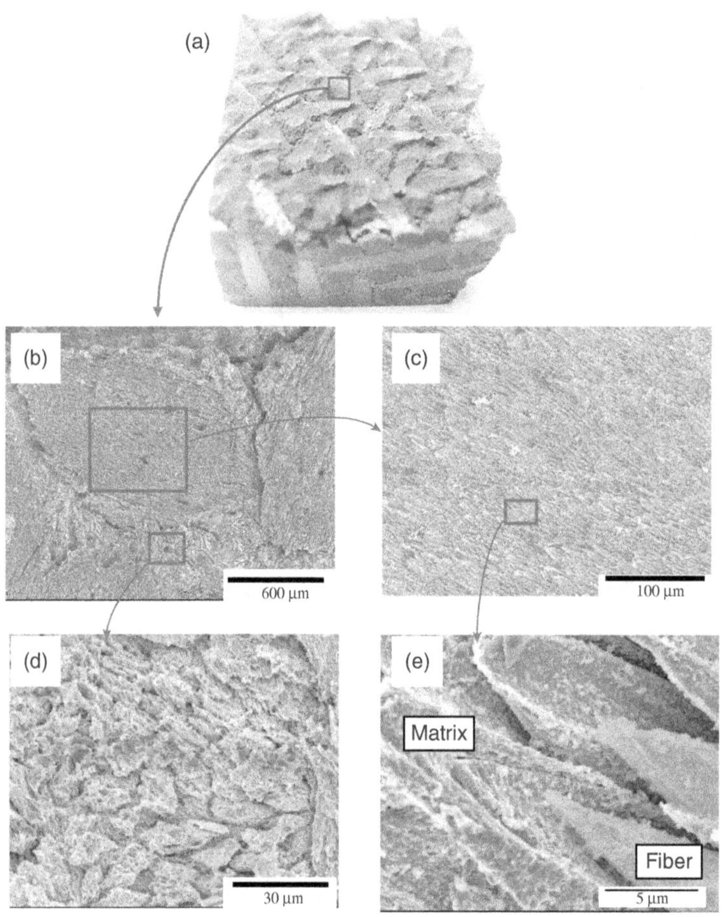

FIGURE 12.12

Observation of a 4D C/C composite (material 3) after SRM testing. (a) Overall view; (b) mesoscale features of bundles and interbundle matrix; (c, d) close-ups on bundle and matrix; (e) microscale view of the fibers and intrabundle matrix [65].

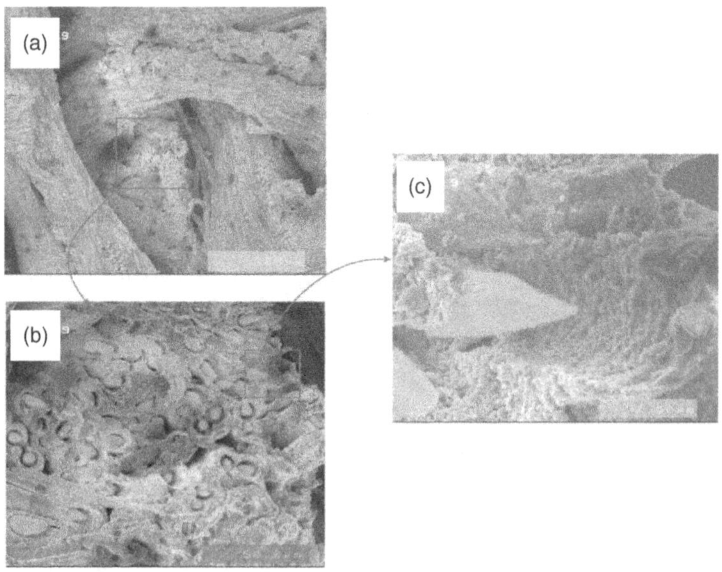

FIGURE 12.13

SEM observation of a 2.5D C/C composite (material 4) after ablation in an SRM. (a) View of the parallel and perpendicular bundles and matrix; (b) close-up on an emerging bundle; (c) microscopic view of a single fiber [65].

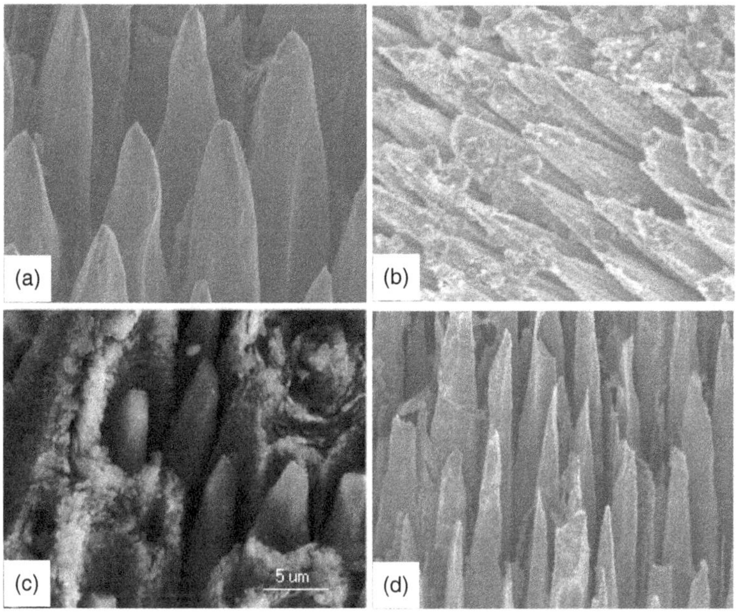

FIGURE 12.14
Comparison of "needle clusters" obtained on 3D or 4D C/C composites in distinct conditions. (a) Plasma arc jet; (b) solid rocket motor [69]; (c) tokamak reactor, reproduced with permission from Reference 8, Elsevier BV; (d) oxidation reactor, reproduced with permission from Reference 62, Pergamon (Elsevier).

close-ups on an emerging bundle (Figures 12.13b and 12.13c) display a morphology that is closely related to materials 1 and 3. Again, it is seen that recession is faster in an interfacial zone between matrix and fibers, which have a quite equal reactivity. This had already been demonstrated in a previous study on the same class of materials [66–68]. There is no influence of the flow direction on the morphology.

As a conclusion, at all scales, ablation phenomenon uncovers the material structure. Indeed, the constituents have distinct ablation resistances, which provoke a difference in the recession rate. So, the surface acquires rapidly some characteristic features.

At macroscopic scale, it is possible to distinguish between fiber bundles, parallel and perpendicular, on the one hand, and matrix blocks on the other hand. In the case of material 2, the matrix is very weak, so it recedes much more than the bundles; in the other cases, there often appears an intermediate phase around the bundles and matrix blocks, with a higher recession rate.

At the microscale, the surface of bundles perpendicular to the average material surface look like "needle clusters"; on the other hand, tangent bundles look like "needle layers." Similarly, the ex-pitch matrix acquires a characteristic appearance of stacked layers separated by small voids. Looking closer, we can see that fiber tips are faceted (Figures 12.6f, 12.7b, 12.12e, and 12.13c). Some materials also show holes on the top of the fibers. Of particular interest is the evidence of interphases lying between fibers and matrix, since it appears that they are weaker than any other component of the material.

Some of these features are rather ubiquitous, as in particular the "needle clusters." Figure 12.14 is a comparison of the ablation morphologies obtained in four totally distinct sets of conditions. This indicates that in a dynamical sense, the underlying phenomena are common—mostly, the competition between phases with distinct ablation resistances. What makes them differ (precise shape, angles, depths, etc.) is linked to the specific conditions in which the tests have been carried out.

12.2 MATERIALS OBSERVATION: RECESSION RATE

Tests in rocket nozzles have been extensively studied in terms of kinetics with respect to H_2O and/or CO_2 attack [22, 23, 70]. Table 12.4 summarizes the rate laws reported by several authors, cast into the single formalism of the following relation:

$$V_a(\mathrm{m/s}) = (A/\rho_s)\, p_{ox}^{a} \exp(-E_a/\mathcal{R}T_s) \tag{12.1}$$

TABLE 12.4 Summary of Oxidation Rate Law Parameters (see Equation 12.1) Reported by Several Authors in Various Conditions

Ref.	Gas	A (kg/(m^2 s atm^a))	a	E_a (kcal/mol)
[22] Upper bound	H_2O, CO_2	4.25–4.48	1	20.07
[22] Lower bound	H_2O, CO_2	2.12–2.23	1	18.52
[23]	H_2O	18.3	1	7.0
[70]	H_2O	$4.8 \cdot 10^5$	$^1/_2$	68.8
[70]	CO_2	$9.0 \cdot 10^5$	$^1/_2$	68.1
[71]	H_2O, O_2	$2.43 \cdot 10^3$	1	41.9
[72]	CO_2	158	1	40
[73] Upper bound	O_2	150	0.6	42.6
[73] Lower bound	O_2	0.2	0.6	42.6

where A is a pre-exponential factor, ρ_s is the material density (kg/m^3), p_{ox} is the oxidative species pressure in atm, a is the reaction order, and E_a is the activation energy. Since V_a is dependent from the quantity of gaseous reactant, it is chosen to compare $V_a/p_{ox}{}^a$, which is in some way a measure of the intrinsic reactivity of the material. In Figure 12.15, the experimental data are compared to the kinetic rate laws associated to the oxidation of various kinds of carbons in several conditions: by O_2/H_2O mixtures [71], by CO_2 [72], and by dioxygen [73].

Recession rates are usually reported from the experiments by dividing the recession length (or mass) at the end of the test by the test duration. Such quantities are not easy to interpret, since it has been reported many times that they vary with time. For instance, in a simple oxidation test [20], material I has a large transient period during which the ablation rate increases neatly to more than twice its initial value (Figure 12.16). One of the principal reasons for this comes from the fact that the initially smooth surface is acquiring its roughness features during this transient period. Indeed, the most reactive components react first, and this leads to the progressive uncovering of the most resistant components. The contribution of the latter to the overall mass loss increases continuously, while the contribution of the former is roughly constant. One then reaches a steady situation in which the most reactive phase is buried so deep inside the material that diffusion ends up preventing its consumption rate to be higher than for the less reactive phase when the regime is not reaction limited. In the reaction-limited regime, the steady state is reached only when the reactive surface area of the less reactive phase (S_{lr}) has sufficiently increased compared to the reactive surface area of the more reactive phase (S_{mr}) to reach the equality $k_{lr}\ S_{lr} = k_{mr}\ S_{mr}$.

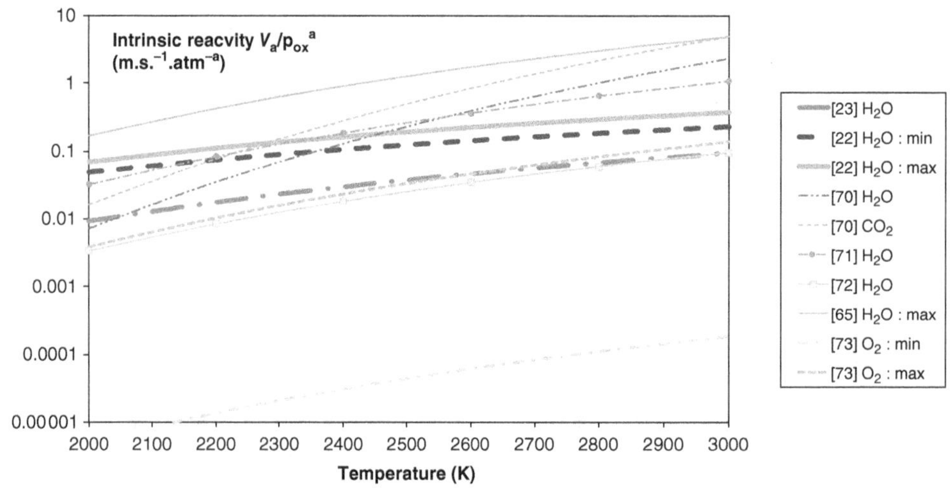

FIGURE 12.15

Arrhenius plot of the intrinsic reactivities ($V_a/p_{ox}{}^a$, see Equation 12.1) for several reported oxidation laws.

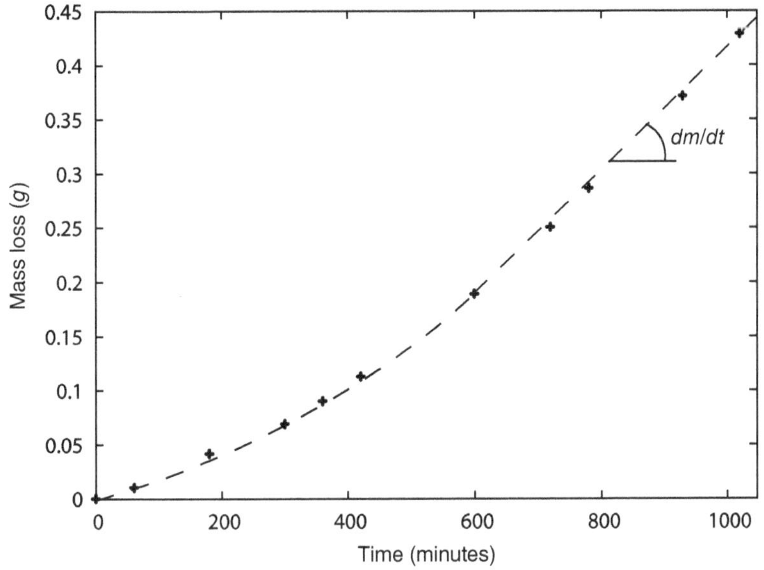

FIGURE 12.16

Mass loss kinetics of material 1 in a plain oxidation test (dry air, $T = 898$ K, $p = 1$ atm). The dotted line is a guide to the eye.

Another very important point, related to the preceding one, is the phenomenon of "overablation." An old "rule of thumb" relating the ablation rate (or the ablation resistance) to the material density had been set up by engineers. Although a general tendency has always been observed (the denser, the more resistant), some spurious data points are present, that is, materials with largely higher ablation rates than expected from the tendency.

A more precise illustration has been given with specific oxidation tests [51]. It has been managed to obtain separate samples of the fibers and matrix for materials 1 and 2; oxidation tests have been carefully performed in order to extract the intrinsic reactivities of the components, that is, the ablation molar rate (in mol/m^2.s) divided by the oxygen concentration (in mol/m^3), as well as for the whole composites. The results are plotted as reaction rates versus density in Figure 12.17. The very striking fact of this graph is that the effective intrinsic reactivity of the composite is higher than the fiber and the matrix. This is the confirmation that there is a weak interphase present in the material.

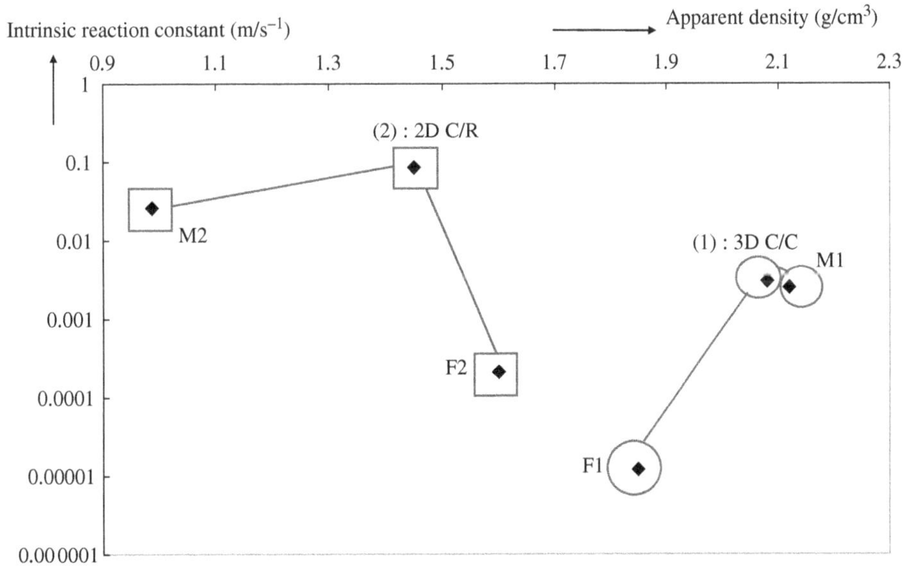

FIGURE 12.17

Intrinsic reactivities versus apparent densities for materials 1 (3D C/C) and 2 (2D C/R) and their constituting fibers and matrices (F1, M1 and F2, M2, respectively).

Summarizing, the kinetic data provide very useful information on the material behavior, in complement to the observations discussed in the previous part. The principal conclusions are

i. The morphology is the result of the competition between phases with distinct ablation resistances.
ii. The acquisition of the morphology is related to a transient period in kinetics.
iii. There is usually a weak interphase between the components, which has a strong impact on the effective material behavior.
iv. The morphological indicators vary with the ablation conditions.

These elements will be quantified in the next sections, devoted to modeling.

12.2.1 Theory

All the preceding observations motivate the setup of a modeling approach that may allow to better understand and to quantify the observed effects, both in terms of morphology and of effective recession rate.

12.2.1.1 Model Bases and Strategy A minimal model suited to the representation of receding ablative surfaces should feature the following elements [20, 62, 74, 75]:

i. a surface recession equation, where the recession velocity results from the impinging flux normal to the surface;
ii. the mass flux is related to the heterogeneous reaction rate (or the sublimation rate), which depend on the local value of the gas species concentrations and of the temperature, and on the local reactivity of the solid;
iii. conservation equations for the quantities influencing the ablation rate, that is, species concentration and/or temperature;
iv. some parts of the material surface may differ from other ones by their intrinsic reactivity, that is, their heterogeneous mass transfer constant.

From these principles a variety of models can be derived, depending on the relative importance of mass or heat transfer; moreover, an averaging procedure may be designed in order to obtain an effective reaction rate from the average behavior of the surface at a given scale. Models may be coupled in order to account for a multiscale description—for example, considering the materials described earlier, a two-scale approach can suffice. A first, "micro-scale" model, is focused on the fiber bundles or tows and allows the description of their effective behavior on a ~50 μm scale; then, the results of this model are inserted in a second, "macro-scale" model, defined at a ~500 μm scale, featuring the bundles and the interbundle matrix. Figures 12.18 and 12.19 give examples for materials 1 and 3 described in the section 12.1.1.

If the microscale models can safely be designed in an isothermal approximation, it is not so safe to do so at the macroscale. However, this remains possible in the special case of small heat fluxes [62]. Another interesting point is that the microscale model may be simplified by converting the gas transport problem into a simple diffusion equation restricted to the interior of some boundary layer, the thickness of which is dictated exteriorly by the gas tangential velocity. The problem domain may be sketched as indicated in Figure 12.20.

Let us now recall the main equations governing the ablation problem [72]. It is convenient to describe the surface by an implicit equation:

$$S(x, y, z, t) = 0 \tag{12.2}$$

When the surface does not contain any overhang, it is possible to write down the implicit function S as

$$S(x, y, z, t) = z - h(x, y, t) \tag{12.3}$$

where h is the height of the surface. The propagation equation for the surface is a Hamilton–Jacobi equation which reads

$$\frac{\partial S}{\partial t} + \mathbf{v} \bullet \nabla S = 0 \tag{12.4}$$

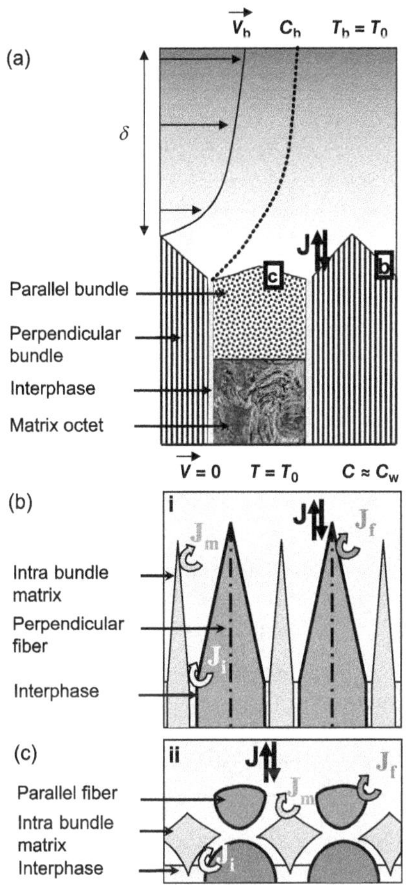

FIGURE 12.18

Multiscale modeling strategy for material 1—3D C/C composite. Reproduced with permission from Reference 62, Pergamon (Elsevier).

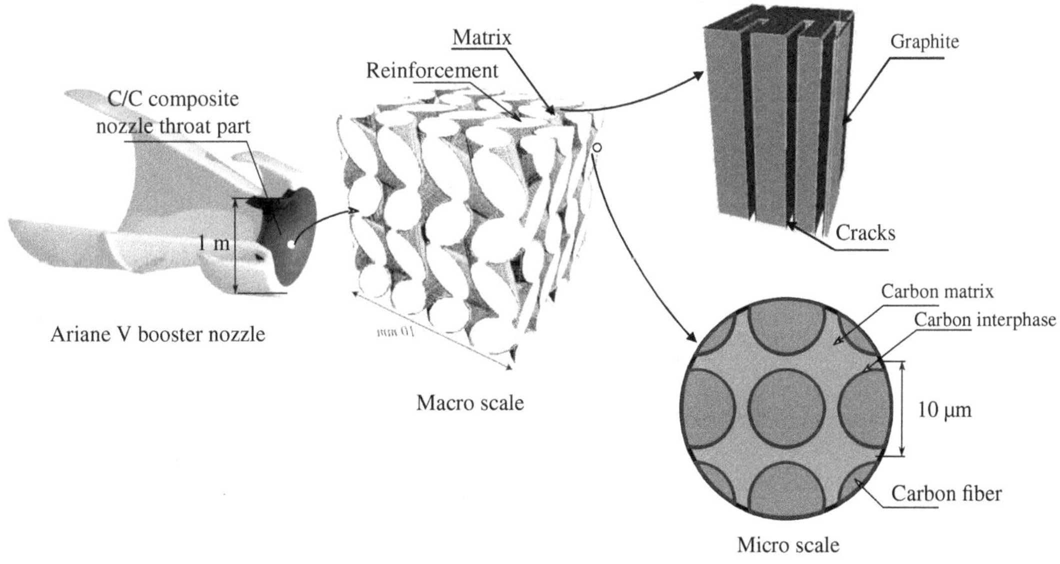

FIGURE 12.19

Multiscale modeling strategy for material (3)—4D C/C composite.

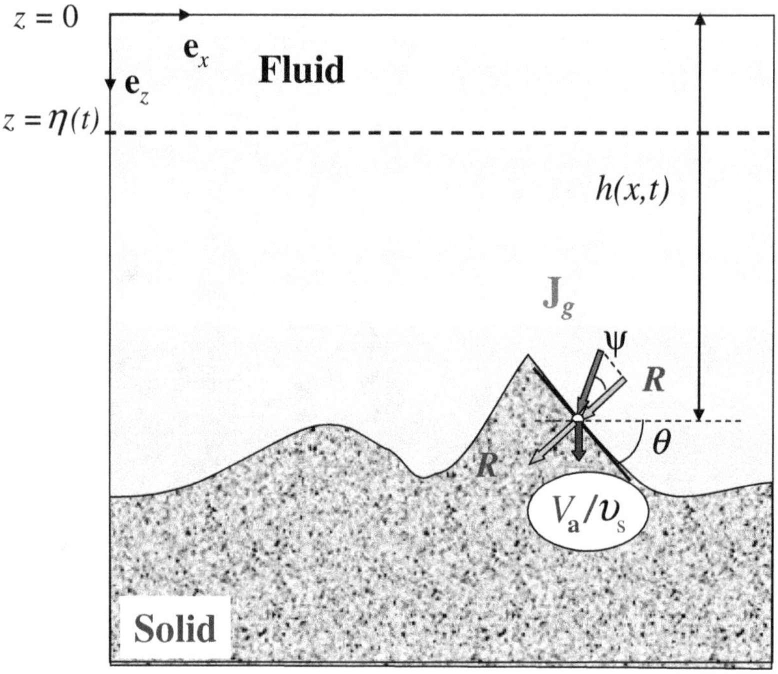

FIGURE 12.20
Typical resolution domain for a simple diffusion-heterogeneous transfer problem.

The Hamiltonian $H = \mathbf{v}\bullet\nabla S$ contains the actual normal surface recession velocity, **v**. Note that the apparent ablation velocity V_a defined in Equation (12.1) is a vertical projection of **v**:

$$V_a = \mathbf{v} \bullet \mathbf{e}_z \quad (12.5)$$

The normal surface recession velocity depends on the ablation rate R, expressed in mol/m^2/s:

$$\mathbf{v} = \upsilon_s R\, \mathbf{n} \quad (12.6)$$

where $\upsilon = M_s/\rho_s$ is the molar volume of the solid. When ablation is linked to oxidation, this rate depends a priori on surface temperature and of local concentration of reactant gas:

$$R = k_0 \exp(-E_a/\mathcal{R}T_s) \bullet C(z = h)^a \quad (12.7)$$

Comparison of Equations (12.1) and (12.7) gives a relation between A and k_0:

$$k_0 = A(\mathcal{R}T_s)^a/M_s \quad (12.8)$$

In the case of sublimation, **v** depends on the local partial pressure of sublimed species, compared to the equilibrium value, for example, through a Hertz–Knudsen–Langmuir relationship:

$$\mathbf{v} = \upsilon_s\, \alpha_0\, \exp(-E_a/\mathcal{R}T_s)\,(2\pi\mathcal{R}TM_g)^{-1/2}[p(z = h) - p^{eq}(T(z = h))]\,\mathbf{n} \quad (12.9)$$

Equation (12.9) is very similar to Equations (12.6) and (12.7); the problems of sublimation and oxidation are dynamically equivalent when: (i) one replaces the concentration C by the undersaturation p-p^{eq} and (ii) one defines an equivalent reactivity $k'_0 = \alpha_0(2\pi\,\mathcal{R}TM)^{-1/2}$.

Note that the normal to the surface is obtained from the gradient of the implicit function $S(x,y,z,t)$:

$$\mathbf{n} = \nabla S/||\nabla S|| \quad (12.10)$$

When the surface height is explicit, one has

$$\mathbf{n} = \frac{1}{\sqrt{1 + (\partial h/\partial x)^2 + (\partial h/\partial y)^2}} \bullet \begin{pmatrix} -1 \\ \partial h/\partial x \\ \partial h/\partial y \end{pmatrix} \tag{12.11}$$

In Equations (12.6) and (12.7) or (12.9), it is seen that the surface temperature and the gas concentration have to be determined. In this goal, the model has to now account for the transport of gas and/or heat through balance equations for gas species and/or enthalpy. In the absence of homogeneous reactions, the gas species mass balance reads:

$$\frac{\partial C}{\partial t} + \nabla \bullet (\mathbf{J}_g) = 0 \tag{12.12}$$

where the gas mole flux is given by the sum of convective and diffusive contributions:

$$\mathbf{J}_g = \mathbf{v}_g C - D_g \nabla C \tag{12.13}$$

Here, $\mathbf{v}_g$ is the local gas velocity, and D_g the species diffusion coefficient. The consumption or production by the surface is given by

$$\mathbf{J}_g \mathbf{n} = -R \text{ for } S = 0, \text{ that is, for } z = h \text{ if the surface is explicit.} \tag{12.14}$$

Gas-phase transport should feature diffusion (possibly multicomponent), and convection. This last element requires knowledge of the gas phase velocity field $\mathbf{v}_g$, in possibly turbulent flow conditions, which may be extremely difficult to obtain in a realistic fashion.

Similar to the gas concentration, the local surface temperature is evaluated by a heat balance equation featuring transport in the solid gas phases:

$$\rho_g c_{pg} \left(\frac{\partial T}{\partial t} + \mathbf{v}_g \bullet \nabla T \right) + \nabla \bullet (-\lambda_g \nabla T) = 0 \tag{12.15}$$

$$\rho_s c_{ps} \frac{\partial T}{\partial t} + \nabla \bullet (-\lambda_s \nabla T) = 0 \tag{12.16}$$

where c_{pi} and λ_i are the heat capacities and conductivities of the gas (i = g) and solid (i = s). The interfacial heat consumption by ablation reads

$$[-\lambda \nabla T.\mathbf{n}]_{g->s} = L_r R \tag{12.17}$$

where L_r is the molar heat of ablation (either by sublimation or by chemical reaction).

A very simple model may be built using a set of restrictive assumptions.

- Isothermal conditions. This is suited to two kinds of situations: (1) isothermal oxidation tests; (2) microscale simulations where it appears that temperature differences across a characteristic roughness feature length are small enough to be neglected.
- Gas transport is restricted to the pure diffusion of a single species between the surface and a gas source (or sink) located at a large-enough distance above the surface; at this place, it is considered that convection ensures the constancy of gas concentration. The relation between source concentration and source-to-surface distance may be obtained through a boundary-layer analysis. The criterion for the choice of the source-to-surface distance is based on a perturbation analysis argument, which shows that it can be as small as a few times the transverse characteristic length of the roughness pattern.
- Surface gas transfer is first order.

The basic equations are

$$\begin{cases} \frac{\partial S}{\partial t} + \mathbf{v} \bullet \nabla S = 0 \\ \mathbf{v} = -\upsilon_s R\mathbf{n} \\ R = kC \quad \text{at} \quad S(x, y, z, t) = 0 \\ \nabla \cdot (-D\nabla C) = 0 \quad \text{at} \quad S(x, y, z, t) \geq 0 \\ -D\nabla C \cdot \mathbf{n} = -R \quad \text{at} \quad S(x, y, z, t) = 0 \end{cases} \tag{12.18}$$

Even though this model looks extremely restrictive, it has the merit of being easily tractable and of capturing the essentials of the bulk transport/interfacial transfer competition. Moreover, it will appear that most of the roughness set-up dynamics is governed by the material itself, which will be correctly described by the model; on the other hand, it is absolutely not claimed to describe accurately the gas flow. We will see later that such an oversimplified model has the merit of giving analytical solutions, which are of interest to understand the influence of the parameters on the material response.

On the other hand, numerical approaches have been built for more complex situations, as described in the following section.

12.2.1.2 Numerical Aspects Two distinct numerical approaches have been developed, with some common points, and many distinct features.

These codes contain three parts: one devoted to image handling, another one to the gas transfer, and the last one is a driver. The "driver" part is essentially the same whatever the details of the other parts are, as illustrated in Figure 12.21: it handles the progressive modification of the surface and saves periodically material images and recession data.

In the image manipulation part, the essential point lies in the representation of the surface, and on the technique used to make it recede. We have chosen here volume-of-fluid (VOF) techniques [77], in which a local phase indicator gives the relative amount of fluid phase ε_f. One can think of this function as a transformation of the S function described sooner, in which ε_f is 1 for S above a given threshold and 0 if S is below another given threshold. The interface is given as a piecewise discrete approximation of the isosurface of ε_f (or of S, as would be the case in a level-set method [78]).

The first model [75, 79] makes use of a planar interface construction (PLIC) discretization of the interface boundary [80]. In the PLIC description, the interface is modeled by local planar surfaces, whose positions and slopes are obtained from the phase indicator variable, as illustrated in Figure 12.22. We adapted this technique to our problem with the idea to keep the possibility of sharp edges, as it is a characteristic feature of ablated surfaces. The mass transfer equation was discretized using a finite volume (FV) scheme.

The second numerical model [60, 81] is based on a Monte Carlo random walk (MC/RW) algorithm for gas diffusion, with a suited sticking probability law for surface reaction, and a simplified marching-cube (SMC) discretization of the interface [82, 83]. Again, this scheme is naturally able to handle sharp surface cusps without numerical difficulties (see figure 12.23 for an illustration).

Both solvers yield the same result: a steady morphology eventually appears, and it is in very good agreement with the analytical results, when available.

In order to complete this first approach, another model [65, 84] featuring momentum transfer equations and advection, has been treated numerically in 2D, with nonevolving surface geometries, using commercial finite element software.

One of the outputs of these models is, for a given consumed gas flux $|J|$ (mol/(m^2/s)) due to recession, the value of the local gas concentration C (mol/m^3) at the same position. By comparison, one obtains an effective reactivity k^{eff} (m/s):

$$k^{\text{eff}}(z) = |J|/C(z) \tag{12.19}$$

This quantity is fundamental for the validation of the approaches with respect to actual data, and to perform micro-to-macro changes of scale.

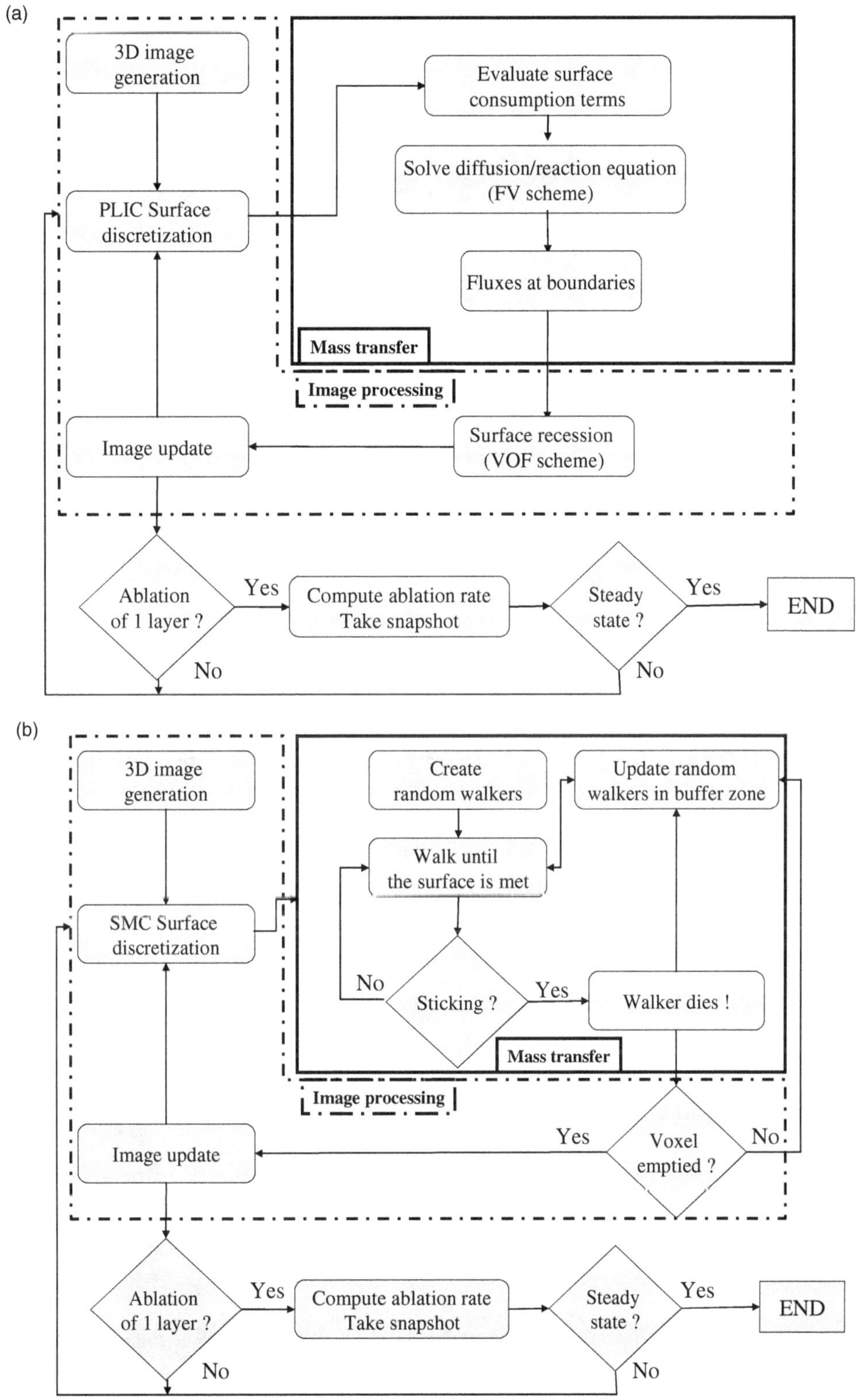

FIGURE 12.21

Algorithms for ablation solvers. (a) FV/VOF-PLIC [69]; (b) MCRW/SMC [62].

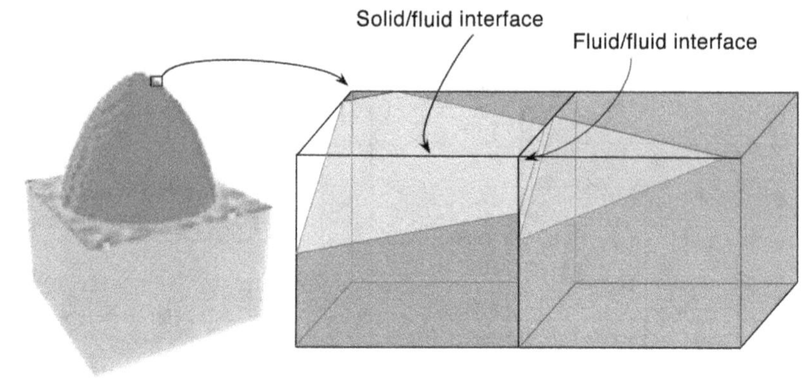

FIGURE 12.22
Scheme of a VOF/PLIC discretization [24]. Reproduced with permission from Reference 24, Pergamon (Elsevier), Ceramic Transactions 191.

12.2.2 Results and Discussions

12.2.2.1 Analytical Results: Steady Morphologies A very characteristic situation, which may serve as a benchmark, is the "needle forest" situation: unidirectional bundles of fibers bonded by a weak matrix are oriented perpendicularly to the average surface; in an ablation experiment, the fibers are progressively denuded from the matrix and they acquired an ogive-like tip shape.

The peak-to-valley roughness is the fiber denudation height h_f^s and is given by the following formula [62]:

$$h_{\rm f}^s = \frac{D}{k_{\rm m}}\left(\sqrt{1 + 2Sh_{\rm f}\sqrt{A^2-1} - Sh_{\rm f}^2} - 1\right) \tag{12.20}$$

where D is the gas diffusion coefficient (m^2/s), k_m and k_f are, respectively, the matrix (or weak phase), and fiber first-order reaction constants (m/s). The number $A = \frac{k_m \upsilon_m}{k_f \upsilon_f}$ is the matrix-to-fiber contrast ratio ($\upsilon_i = M_i/\rho_i$ are molar volumes) and $Sh_f = k_m R_f/D$ is a Sherwood number, characteristic of matrix reaction/diffusion competition, based on the fiber radius, R_f. These two dimensionless numbers play a central role in the approach: indeed, the morphology depends only on them, plus the fiber volume fraction, as illustrated in Figure 12.24, which reports the numerically (or analytically) computed evolution of the tip shape for various values of the Sherwood number, fiber volume fraction, and reactivity

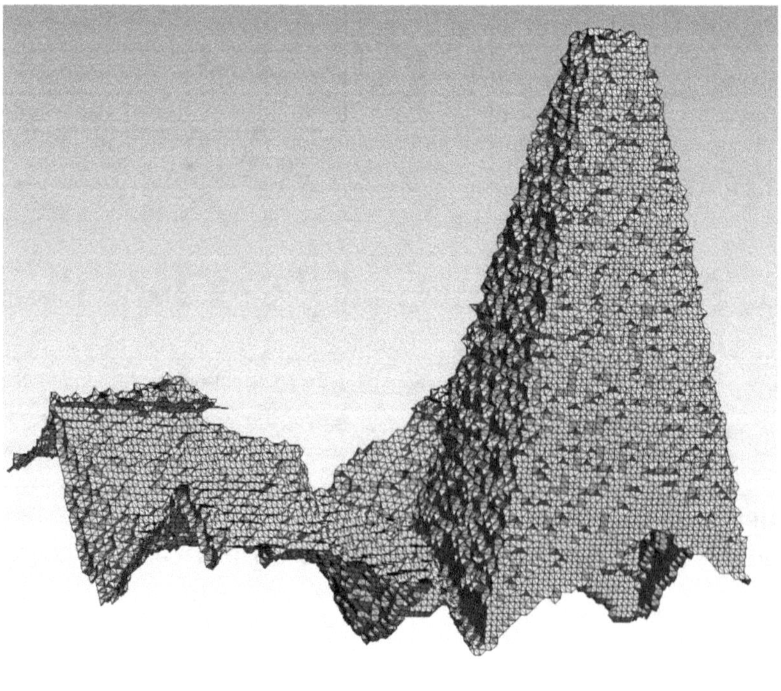

FIGURE 12.23
Rendering of a simplified marching cubes interface discretization.

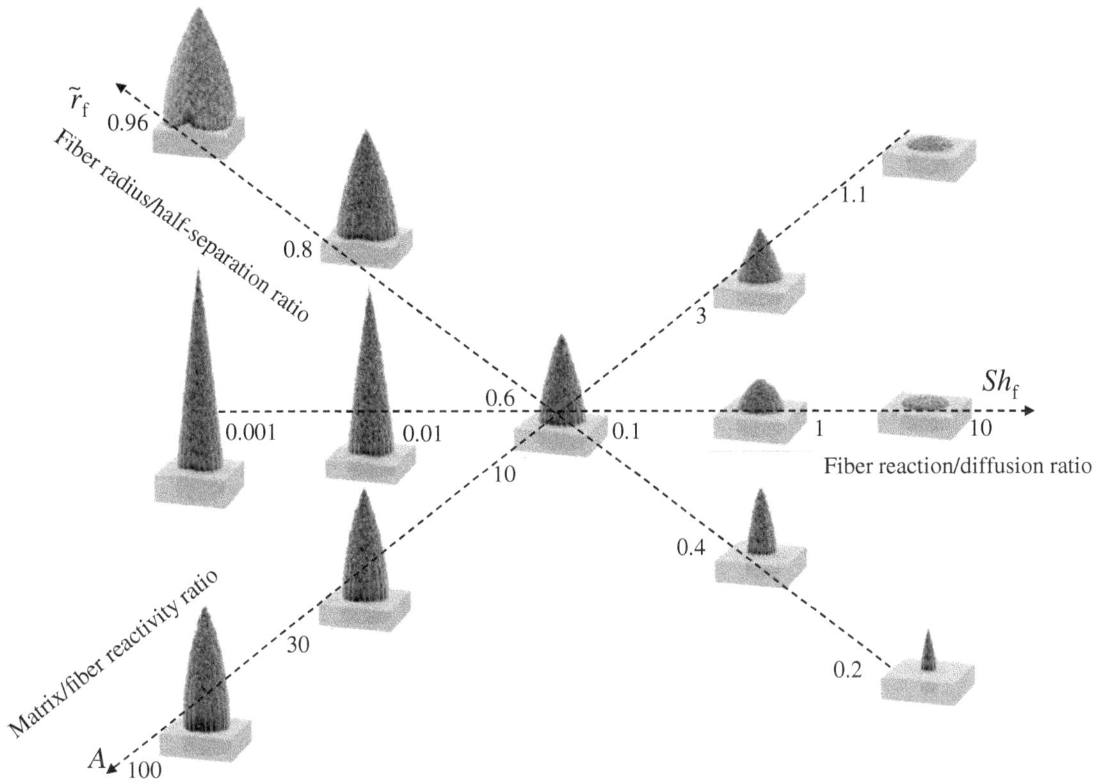

FIGURE 12.24

Parametric influence on the roughness morphology in the analytical case of a single fiber in a matrix. Reproduced with permission from Reference 20, Pergamon (Elsevier)

contrast. The roughness tends to disappear when Sh_f becomes high, because bulk gas diffusion is the only limiting factor and the effect of reactivity contrast vanishes. Conversely, when heterogeneous reaction becomes limiting, the sharpest features are obtained. The curvature radius of the ogival shape being equal to $La = D/k_m$, the fiber tip becomes conical for low values of Sh_f; the value of the relative height h_f/r_f tends toward the contrast factor A. Finally, it can be pointed out that the relative tip height is not a function of the fiber volume fraction ϕ_f (or relative radius $\tilde{r}_f = \frac{r_f}{a/2} = \sqrt{\frac{\phi_f}{\pi}}$, a being the unit cell size).

Another important consequence of the analytical model results is that it is possible to estimate the dimensionless parameters from the study of an actual morphology, in a sort of "inverse approach." This has been extensively used for a physical validation of the model, as described in the following parts.

12.2.2.2 Macroscopic Roughness and Mass Loss (Model Validation) Model validation has been obtained on material I in completely controlled oxidation conditions [62]. First, the separate reactivities of the whole composite and of the lone fibers have been measured separately [51]. For this purpose, a tubular reactor has been used, in which the air flow and oxygen composition have been modeled by CFD analysis [54] in order to ensure a correct interpretation of the data in terms of actual intrinsic reactivities. Then, the analysis of the morphological parameters allowed inferring the reactivity of the weakest phase, which is of utter importance in the present case because of a weakest-link law. First, the fiber-scale parameters have been determined: the very sharp conical shape with $h_f/r_f = 32$ ensures that one lies locally in the chemically controlled regime, that is, $Sh_f \ll 1$; accordingly, the contrast factor A is also equal to 32, which gives the weakest-phase reactivity as being 32 times that of the fiber's. The experimental determination of the fiber reactivity gives $k_f = 1.2\times10^{-5}$ m/s; so we infer that for the interphase $k_i = 3.8\times10^{-4}$ m/s. From these values, the composite effective reactivity of a bundle has been determined in steady state: in the given conditions, since the weakest-link rule holds, its value is very close to the weakest-phase reactivity. It has been checked that the transient time at microscale (a few seconds) is very short with respect to the macroscale transient time (close to 1 day). At a larger scale, the

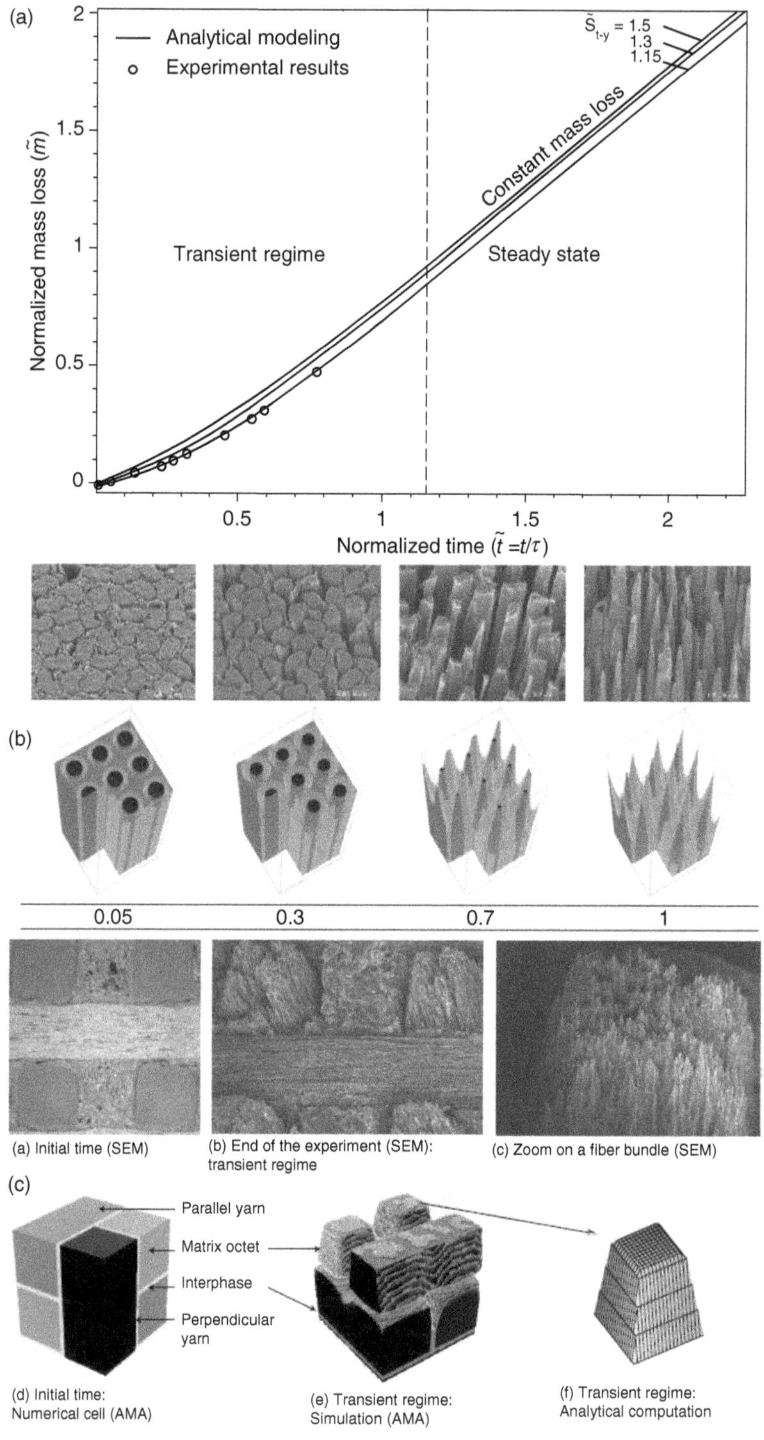

FIGURE 12.25

Validation of the model on material 1 in mild oxidation conditions: (a) mass loss curve, (b) morphology at fiber scale, and (c) bundle scale.

image analysis of Figure 12.7 gave the following results: the bundle-scale Sherwood number Sh_b is also very low, and the contrast factor between the bundles and the weakest phase (interphase around the bundles) is $A = 8$. This gives weakest-phase and average reactivities equal to 3.1×10^{-3} m/s. The final result is within 3% of the experimental value (3.0×10^{-3} m/s).

With the obtained data, it has been possible to reproduce simultaneously and quantitatively the mass loss curve and the ablated surface morphology, as depicted in Figure 12.25. Though absolutely not evident from Figure 12.16, the model predicts that the steady state for the morphology evolution is not yet reached at the last data point ($t = 960$ minutes),

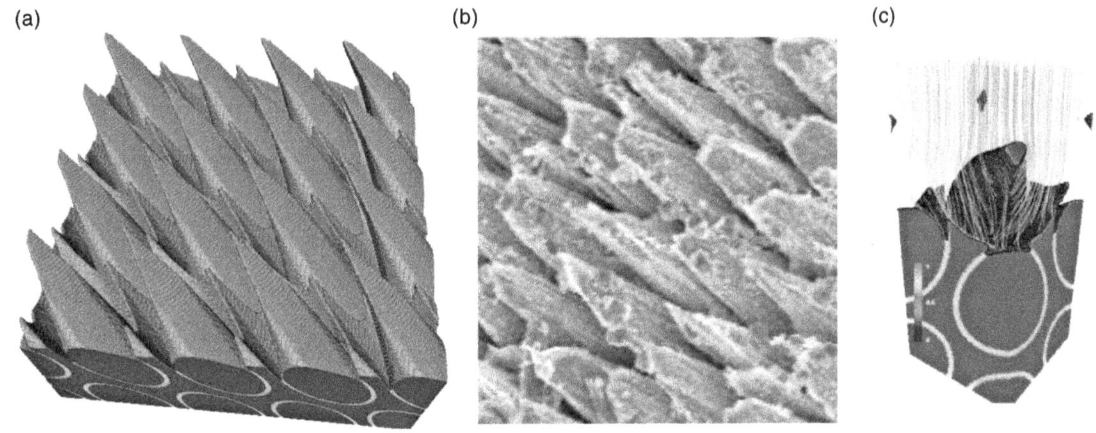

FIGURE 12.26

(a) Modeled and (b) actual morphology of an ablated, inclined fiber bundle of material 3. (c) Rendering of the diffusion flux lines.

as confirmed by direct observation of the sample after removal from the reactor. On the other hand, the steady-state morphology had been reached much sooner for the fiber-scale features. This illustrates the potential of the method in the extraction of material parameters from simple experimental data.

The same methodology has been applied in the case of material 3; in that case, the physicochemical conditions were much harsher [24]. However, at fiber scale, the assumption of small temperature gradient holds. The analytical model helped in getting a first hint of the fiber/interphase reactivity ratio A; then, a parameter variation study of the numerical model allowed finalizing the identification—a value of $A = 8.5$ was optimal for the reproduction of the tilted fiber morphology, as illustrated in Figure 12.26. Then, the effective reactivity of the interbundle matrix has been determined by direct simulations, taking into account the porous character of the matrix, as seen in Figure 12.27, and the reactivity of the ex-pitch carbon, which was previously known. The overall value of the material reactivity has been obtained from the actual recession rate by an analysis of the turbulent mass transfer rate [65]. With all these elements in hand, it

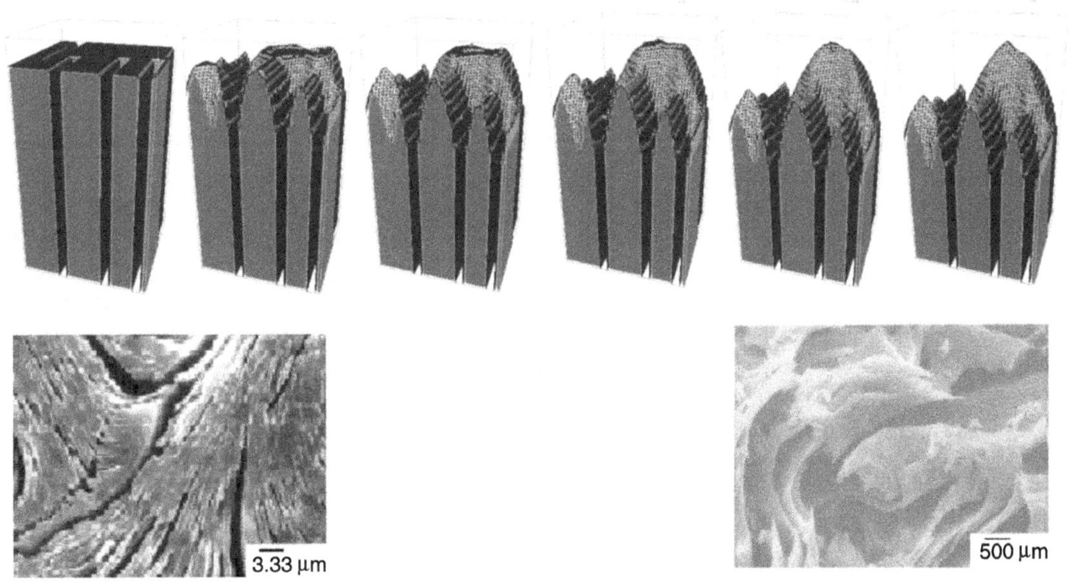

FIGURE 12.27

Top: simulation of the ablation of an ex-pitch carbon matrix. Bottom: micrographs of the matrix before (left) and after (right) an ablation test. Reproduced with permission from Reference 35, Pergamon (Elsevier).

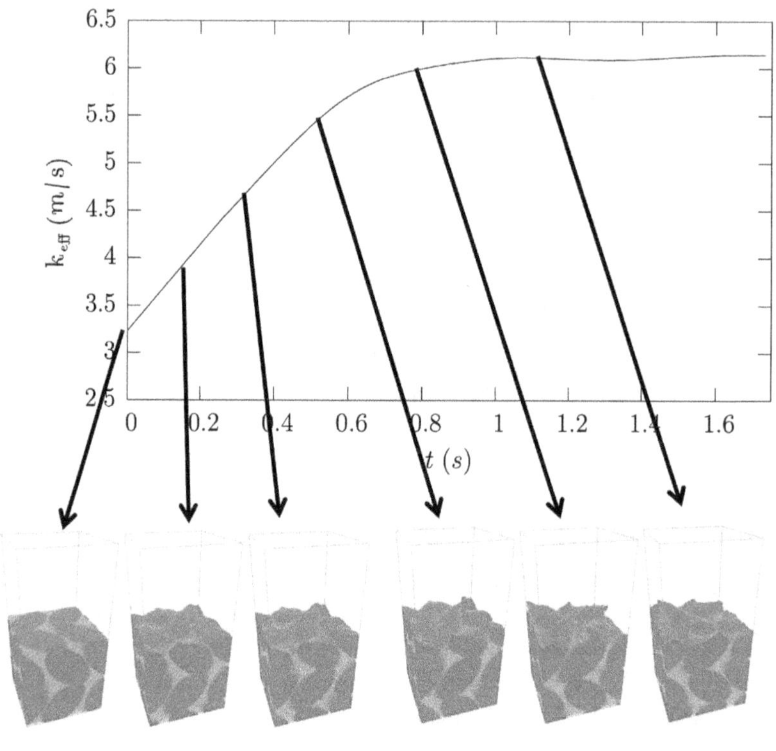

FIGURE 12.28
Morphology and effective reactivity evolution of material 3 with time.

has been possible to reproduce perfectly the observed morphology of the ablated material 3, as seen in Figure 12.28. In addition, the effective reactivity is obtained as a function of the transient time: we can see again that the material reactivity increases until it reaches its steady-state morphology. Taking this phenomenon into account might help reduce the current mismatch between detailed rocket nozzle modeling and actual ablation results (see Figure 12.18 of Reference 33).

12.2.2.3 Roughness Features: Numerical Results Using numerical solvers is extremely interesting in approaching actual morphologies. In addition to the preceding ones, a large variety of forms have been successfully reproduced, as shown below.

Material 1 has been tested in a variety of conditions. In addition to the conditions described in the first part, two other tests have been carried out. The first one was performed at Von Kármán Institute (VKI) in a plasma arc jet fed with air [85], with an approximate temperature of ~2300 K; the second one was carried out with an arc image furnace under inert atmosphere, the temperature reaching as high as 3000 K [86]. The resulting morphologies are clearly different from the previously observed ones. In the case of very high flux plasma jet ablation (Figure 12.6), the fibers are less pointed than in a mild oxidation case; this comes from the lower value of the Sherwood number. The analytical study has shown that if this number decreases, then the tip morphology becomes more ogival than conical. Figures 12.29a and 12.29b illustrate the quality of the agreement, obtained directly with the analytical model and the inverse identification of the ablation rate constants. In the case of air plasma arc jet ablation (Figure 12.29c), the fibers emerge again; however, their internal structure gets revealed. Indeed, it is well known [56] that the T300 ex-PAN fiber has a "core" and a "skin," separated by a "flesh," that is, a less resistant intermediate phase. This has been well reproduced by the numerical simulation, as illustrated in Figure 12.29d. The second test rests only on sublimation, as opposed to active oxidation. Surprisingly, in this case, the fibers have a larger recession rate than the matrix (Figure 12.29e); adjusting the fiber/matrix reactivity ratio A in the numerical simulations yields a very similar result (Figure 12.29f).

Materials 4A and 4B have been numerically ablated in two distinct conditions. At low temperature (1850 K), the Sherwood number is fairly low, whereas it is much higher at 3000 K. It is found that material 4A is 24% more reactive than material 4B at 1850 K, while there is less than 1% difference at 3000 K. From the images displayed in Figure 12.30, this behavior can be understood in the following way: at low T, material B has a rougher, more tortuous surface, thus delaying more efficiently the diffusion of gases to the lower parts of the sample; at high T, there is no more gas penetration, and diffusion—being more limiting—tends to wipe out the material differences.

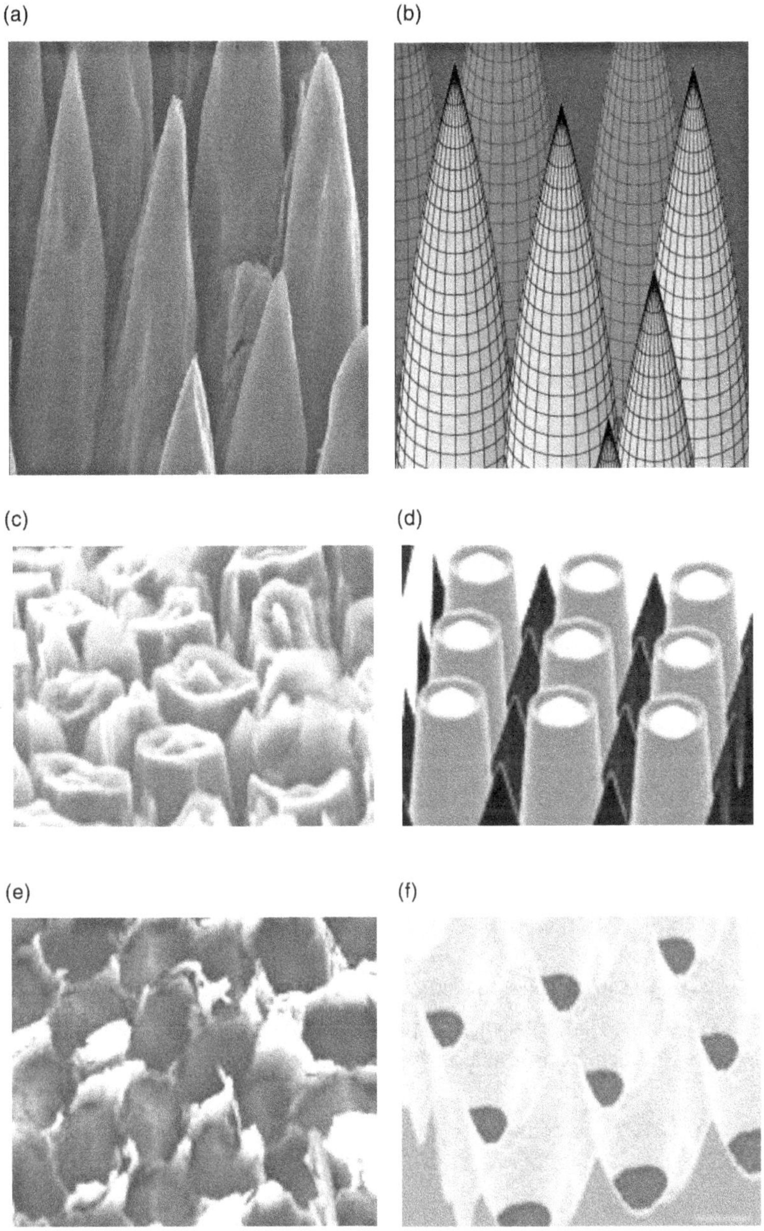

FIGURE 12.29
Material 1 tested in three distinct ablation conditions. (a, c, e) SEM micrographs; (b) 3D analytical and (d, f) numerical simulations. Conditions were (a, b) air-fed high flux plasma jet facility, $T \sim 4000$ K, $J_{th} \sim 50$ MW/m^2; (c, d) air-fed plasma jet (plasmatron), $T \sim 2000$K; (e, f) arc image furnace, Ar atmosphere, $T \sim 3000$ K.

Figure 12.31 displays the case of the oxidation of the space shuttle reentry protection tiles, due to the presence of pinholes in the outer layer, which can be considered inert with respect to oxygen etching. These composites protect a 2D C/C composite that is considered uniform at macroscale. When cracks occur in the protective layer the atmospheric oxygen can diffuse toward the C/C composite. This oxygen causes a mass loss by gasification, as described by Jacobson [87] who reports experiments of TPS oxidation resistance. The oxidation of the C/C composite layer causes the formation of a cavity, as exposed in Figure 12.31a. Its cross-section is compared to the morphology obtained by numerical modeling using equivalent conditions in Figure 12.31b. It can be shown that the morphology of the cavity (here close to a hemisphere) is once more deeply influenced by the ratio of the diffusion velocity to the reaction rate. That is why this computation allows obtaining the reactivity of the C/C composite from the experimental morphology and flow parameters by inverse analysis [79]. Figure 12.31c reveals that the mass transfer resistance is rather high at the level of the pinhole, which acts as a point source of oxygen; this is why the pinhole is hemispherical.

The case of polycrystalline graphite may be treated by the numerical models, adding as an extra feature a dependence of the reactivity to orientation, which illustrates the anisotropy of graphite. Indeed, graphene reactivity at the edges is

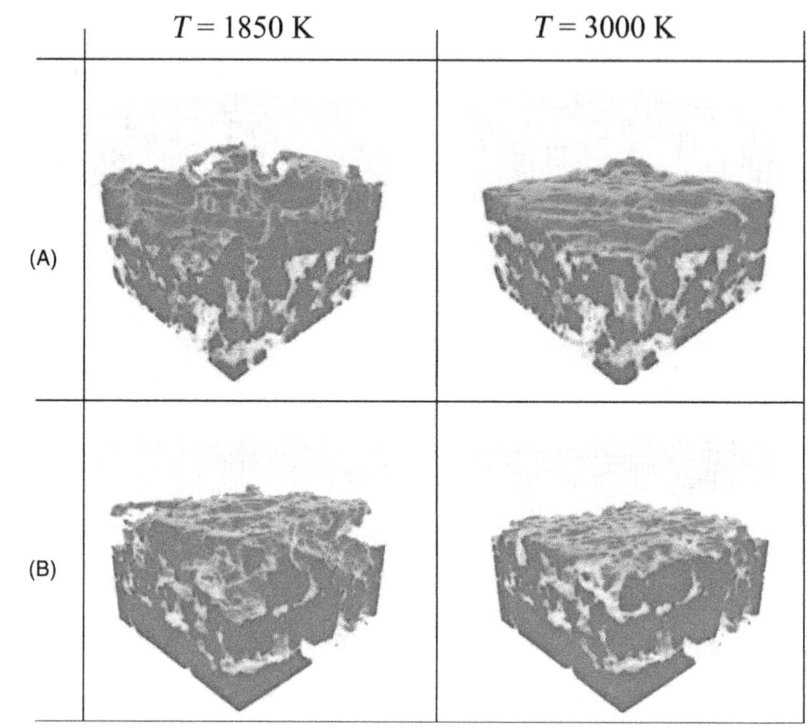

FIGURE 12.30
Numerical ablation of samples (4A) and (4B) at two temperatures. At 1850 K the Sh_f parameter is much lower than at 3000 K.

known to be much larger than in the perpendicular direction. The following law for the orientation-dependent reactivity has been chosen:

$$k_{\text{eff}}(\theta) = \frac{k_{//}|\sin\theta| + k_{\perp}|\cos\theta|}{|\sin\theta| + |\cos\theta|} \tag{12.21}$$

The numerical simulation performed with the MC/RW–SMC algorithm yields typical morphologies as illustrated in Figure 12.32b, a comparison with Figure 12.32a shows that a very good qualitative agreement may be obtained. In this case, the grain-related peak-to-valley roughness is roughly one-third of the grain size [88].

12.2.2.4 Effective Reactivity: Composite Laws In addition to the successful results on morphology and mass loss reproduction, there is also another interesting output of the study, which is the prevision of effective ablation

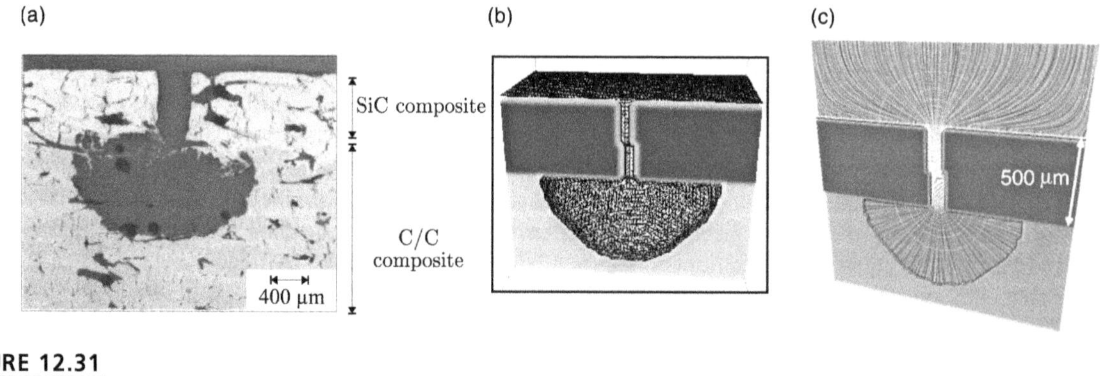

FIGURE 12.31
Oxidation of a protected C/C behind a pinhole. (a) Actual morphology, obtained at T = 1400°C with 0.092 bar air, during 1 hour, reproduced with permission from Reference 87, Pergamon(Elsevier); (b) simulated morphology; (c) visualization of the diffusion flux lines.

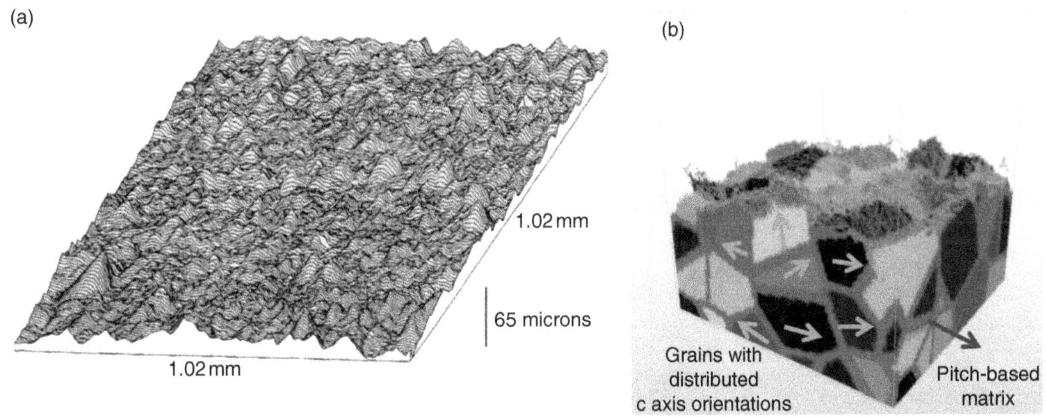

FIGURE 12.32
Polycrystalline graphite roughness after ablation. (a) Profilometric observation, reproduced with permission from Reference 74, Pergamon (Elsevier); (b) numerical simulation with an orientation-dependent grain reactivity [88].

resistances (or reactivities) from the constituent individual properties, under the form of "composite laws." Initial (flat) surfaces, steady-state surfaces, and transient surfaces have been investigated [89], either analytically or numerically.

Results have been obtained for the initial reactivity of the composite, when its surface still is flat: in that case, the values range between the arithmetic average when reaction becomes limiting and the harmonic average when diffusion is limiting, as shown in Figure 12.33. Those two limits are easy to understand. First, in the case of reaction limitation, there is no impact of the bulk transport on the effective transfer; so, all constituents act in a purely parallel fashion. The equivalent conductivity is the sum of the constituents' ones. On the other hand, in the case of diffusion limitation, the constituents are strongly in competition between themselves, since bulk transfer is slow: they are now acting as if they were in series, which explains the harmonic average behavior. From Figure 12.33 it is seen that the role of gas flow is small with respect to bulk diffusion [65]; an increase by four orders of magnitude of the Reynolds number $Re = R_f U/\nu$ (U = fluid velocity, ν = kinematic viscosity) has the same effect as an increase by less than one order of magnitude of the Sherwood number Sh_f. However, the inspected values of Re did not enter the domain of turbulence, in which many other effects should appear, like the enhancement of the effective diffusion coefficient, etc.

For rough surfaces, the effective reactivity defined at the highest point of the surface can be estimated in a rather simple way. It does not depend on the volume fractions of the constituents, under the assumptions of steady morphology and vertical flux approximation. When the contrast is low or the diffusion is limiting, then the equivalent reactivity tends to the harmonic average; on the other hand, when the reaction is limiting and the contrast is strong, then weakest-link rule works. This is illustrated in Figure 12.34.

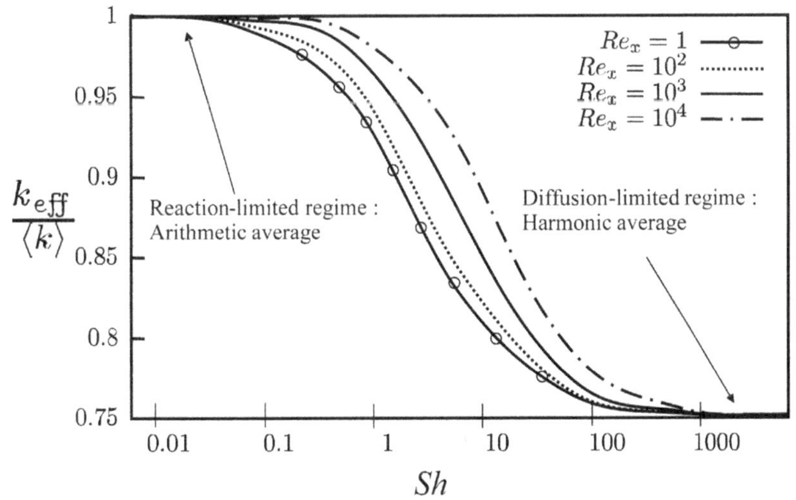

FIGURE 12.33
Effective reactivity of a composite with a flat surface, scaled by the arithmetic average. $A = 3$; fiber volume fraction $\phi_f = 0.5$. Reproduced with permission from Reference 89.

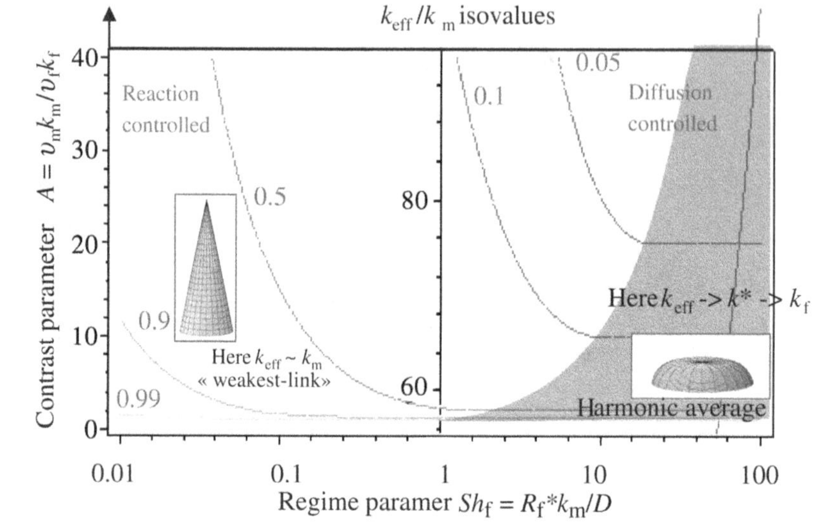

FIGURE 12.34

Evolution of the effective reactivity, as compared to the harmonic average (left) and weakest-phase reactivity (right) in the (Sh_f, A) parameter plane.

Choosing the surface defined as the arithmetic average of all heights (i.e., an "average surface") yields a more complicated expression also involving the fiber volume fraction [65], but the asymptotic results for steady morphologies are quite similar to the highest position reactivity; in the diffusion-controlled regime, the effective reactivity tends to the harmonic average (and the steady surface is flat), while in the reaction-controlled regime, the effective reactivity tends to the maximal value, that is, to the weakest-link rule.

The transient regime makes the link between the flat surface situation and the steady morphology situation; in the case of reaction control, this brings out an increase in the reactivity, as demonstrated sooner on materials 1 and 3; in the converse case, the reactivity may decrease with time. The transient time is given in the analytical model by

$$\tilde{t}_{tot} = \tau \frac{\sqrt{A^2 - 1}}{A - 1} \qquad (12.22)$$

where $\tau = R_f / (C_0 \upsilon_s k_f)$ is the characteristic time magnitude.

Note that $\tilde{t}_{tot}/\tau$ is close to unity when the contrast is important, but increases without bounds when the contrast number A goes to 1. Figure 12.35 shows that, as the Sherwood number increases, this transient time first increases,

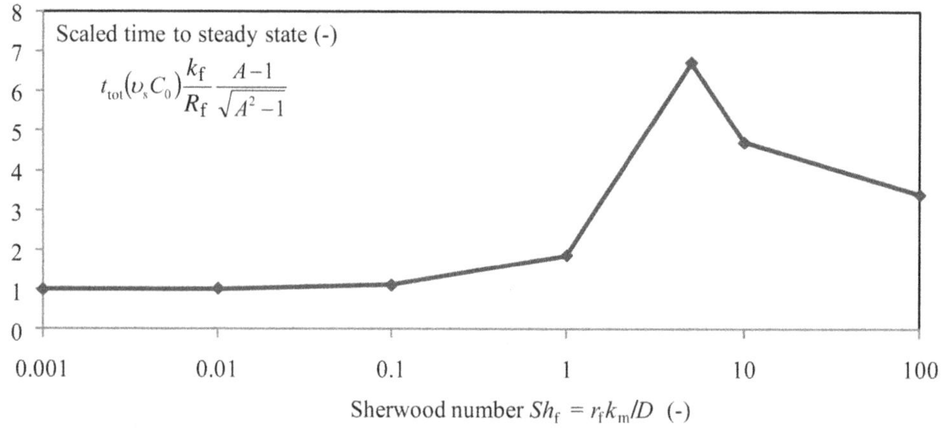

FIGURE 12.35

Scaled time to steady state as a function of the fiber Sherwood number. Reproduced with permission from Reference 89.

TABLE 12.5 Average Reactivity for a Composite (at the Highest Point of Surface) According to Conditions

	Initial (flat)	Steady state
Reaction-limited ($Sh_f \ll 1$)	Arithmetic average	Weakest link
Diffusion-limited ($Sh_f \gg 1$)	Harmonic average	

and then decreases when diffusion becomes the limiting factor. The extreme value does not exceed 10 times the value given in Equation (12.22).

Though many simplifying hypotheses have been used, some useful conclusions can be drawn on the effective reactivities (or resistances). For instance, the weakest-link rule [90] has been confirmed, but only when the heterogeneous transfer is limiting; moreover, it is shown that it does not hold either in the initial and transient periods. Table 12.5 collects the results for the initial and final (steady) morphologies.

Other points of interest are

- **i.** When the reaction is limiting, the strongest (less reactive) phase brings a protection to the weakest phase at the beginning of ablation; this protective effect diminishes until the equilibrium morphology is attained.
- **ii.** In the converse case, one always has a harmonic average; however, no roughness is developed and the true limiting factor is not the material reactivity but the gas transfer to the wall.
- **iii.** The influence of the surrounding fluid velocity (Reynolds number) seems to be much less important than that of the diffusion/reaction ratio.
- **iv.** The transient time is a complex function of the contrast, regime, and fiber volume fraction; however, it can be noted that its order of magnitude τ is mostly a function of the fiber reactivity and of the contrast number A, which diverges when A diminishes to 1, that is, when the contrast fades away.

12.3 CONCLUDING REMARKS AND FUTURE DIRECTIONS

In this chapter we have presented a description and a modeling approach of C/C composite ablation, based on a material point of view. The emphasis has been set on the relation between the material structure and its behavior in the specific conditions of ablation, principally gasification either by oxidation or by sublimation.

A comprehensive analytical and numerical effort has allowed explaining many observations and experimental results. The main conclusions are:

- Roughness develops itself at the material surface because of a reactivity contrast between the constituents (fibers, interphases, matrices), until a steady state is attained.
- The competition between diffusion and reaction plays an important role in modifying the roughness features and the effective behavior.
- The effective reactivity of the material varies during the transient period.
- The effective reactivity may follow a weakest-link rule when diffusion is not limiting, and only when the steady state has been reached. So, for long ablation times, it is more important to strengthen the weakest phase than the strong phase. The presence of the weakest-link rule is the key to understanding the phenomenon of overablation.
- On the other hand, for short ablation times, the strongest phase has a protective role.

These conclusions may serve as guidelines for the conception of optimized materials; as a matter of fact, it appears impossible to have a "polyvalent material" since the optimization direction may change drastically depending on the physicochemical conditions considered. Disposing of numerical tools is thus a precious help.

The current trend in thermal protection materials is to incorporate ultra-high temperature refractory materials; even if this subject has not been developed here, some of the rules derived above could be of interest.

Further work is oriented toward the incorporation in the models of more physicochemistry, such as nonlinear reaction rates, multicomponent gas diffusion, thermal gradients, more effects of the fluid velocity, like turbulence, and mechanical erosion. Improving the relation with aerothermochemical codes is an important issue.

Another exciting perspective is the fundamental study of the reactivity of each carbon phase, and of its relation to the material organization at all scales, from the micrometer or more down to the nanometer. This point is a key issue in building comprehensive models for the reactivity and behavior of C/C composites.

ACKNOWLEDGMENTS

This book chapter summarizes many research efforts which have been funded by Commissariat à l'Energie Atomique (CEA) through a PhD grant to J. L. and Snecma Propulsion Solide (Safran group, nowadays Heraklès) for a PhD grant to Y. A.

The authors wish to thank P. Weisbecker for the TEM image of Figure 12.9.

REFERENCES

1. J. Couzi, J. de Winne, and B. Leroy. Improvements in ablation predictions for reentry vehicle nosetip, in *ESA Conferences Proceedings*, Vol. SP-426, *Proceedings of the 3rd European Symposium on Aerothermodynamics for Space Vehicles*, edited by R. A. Harris. (European Space Agency, The Netherlands, 1998). pp. 493–499.
2. M. D. Jackson. *Roughness Induced Transition on Blunt Axisymmetric Bodies*, Vol. XV of Interim Report SAMSOTR7486 of Passive Nosetip Technology (PANT) Program, (Air Force Space and Missile Systems Organization, Los Angeles, CA, 1974).
3. D. C. Reda. *Correlation of Nosetip Boundary Layer Transition Data Measured in Ballistic Range Experiments*, Sandia Report SAND 790649 (Sandia National Laboratories, Livermore, CA, 1979).
4. R. G. Batt and H. H. Legner. A review of roughness-induced nosetip transition. *AIAAJ.*, 21(1), 7–22 (1983).
5. K. Kuo and S. Keswani. A comprehensive theoretical model for carbon-carbon composite nozzle recession. *Combust. Sci. Technol.*, 42(3), 177–192 (1986).
6. J.-J. Choury. Carbon–carbon materials for nozzles of solid propellant rocket motors, in *Proceedings of AIAA/SAE 12th Propulsion Conference*, Palo Alto, CA, AIAA Papers 76-609 (1976).
7. V. Borie, J. Brulard, and G. Lengellé. An aerothermochemical analysis of carbon–carbon nozzle regression in solid-propellant rocket motors. *J. Propul. Power*, 5(6), 665–673 (1989).
8. S. Pestchanyi, V. Safronov, and I. Landman. Estimation of carbon fibre composites as ITER divertor armour. *J. Nucl. Mater.*, 329–333 part A, in *Proceedings of the 11th International Conference on Fusion Reactor Materials (ICFRM-11), Kyoto, Japan, 2003*, edited by A. Kohyama, T. Muroga, A. Kimura, K. Abe, F. W. Wiffen, and R. E. Stoller (Elsevier, The Netherlands, 2003), pp. 697–701.
9. D. M. Duffy. Fusion power materials. *Adv. Mater. Processes*, 168(7), 15–18 (2010).
10. H. Pacher, G. Vieider, U. Mszanowski, Y. Igitkhanov, G. Janeschitz, J. Schlosser, and L. Plöchl. Lifetime of Be-, CFC- and W-armoured ITER divertor plates. *J. Nucl. Mater.*, 233–237 part A, in *Proceedings of the 7th International Conference on Fusion Reactor. Materials (ICFRM-7), Obninsk, Russia, 1995*, edited by V. M. Chernov, W. M. Gauster, D. S. Gelles, A. B. Hull, R. L. Klueh, and H. Stamm (Elsevier, The Netherlands, 1996), pp. 701–707.
11. T. Burtseva, A. Hassanein, I. Ovchinnikov, and V. Titiov. Study of brittle destruction and erosion mechanisms of carbon-based materials during plasma instabilities, *J. Nucl. Mater.*, 290–293, in *Proceedings of the 14th International Conference on Plasma-Surface Interactions in Controlled Fusion Devices, Rosenheim, Germany, 2000*, edited by J. Roth, K. Ertl, Ch. Linsmeier (Elsevier, The Netherlands, 2001), pp. 1059–1063.
12. L. Alves, E. Alves, N. Barradas, R. Mateus, P. Carvalho, J. Coad, A. Widdowson, J. Likonen, and S. Koivuranta. Erosion and re-deposition processes in JET tiles studied with ion beams. in *Nucl. Instr. & Meth. in Phys. Res. B*, Vol. 268 (11–12), *Proceedings of the 19th International Conference on Ion Beam Analysis, Cambridge, UK, 2009*, edited by R. Webb, M. Bailey, C. Jeynes and G. Grime. (Elsevier, The Netherlands, 2010), pp. 1991–1996.
13. N. Bernier, C. Brosset, F. Bocquet, E. Tsitrone, W. Saikaly, H. Khodja, V. Alimov and J. Gunn. Localization by TEM and EELS of deuterium trapping sites in CFC exposed to plasma irradiation in Tore Supra. *J. Nucl. Mater.*, 385(3), 601–605 (2009).
14. E. Tsitrone, B. Pégourié, Y. Marandet, J. Artaud, C. Brosset, J. Bucalossi, Y. Corre, T. Dittmar, E. Gauthier, P. Languille, et al. Multi machine scaling of fuel retention in 4 carbon dominated tokamaks. *J. Nucl. Mater.*, 415 (1, suppl.), in *Proceedings of the 19th*

International Conference on Plasma-Surface Interactions in Controlled Fusion, edited by A. Leonard, S. Allen, R. Doerner, D. Hill, T. Rognlien, and P. Stangeby (Elsevier, The Netherlands, 2011), pp. S735–S739.

15. P. B. Wright, J. W. Davis, R. G. Macaulay-Newcombe, C. G. Hamilton, and A. A. Haasz. Chemical erosion of DIII-D divertor tile specimens. *J. Nucl. Mater.*, 313–316, in *Proceedings of the Conference on Plasma-Surface Interactions in Controlled Fusion Devices, Gifu, Japan, 2002*, edited by N. Ohyabu, N. Noda, K. Morita, S. Takamura, and T. Tanabe (Elsevier, The Netherlands, 2003), pp. 158–162.
16. R.E. Wirz, J. R. Anderson, and I. Katz. Time-dependent erosion of ion optics. *J. Propul. Power*, 27(1), 211–217 (2011).
17. I. Funaki, H. Kuninaka, K. Toki, Y. Shimizu, and K. Nishiyama. Verification tests of carbon–carbon composite grids for microwave discharge ion thruster. *J. Propul. Power*, 18(1) 169–175 (2002).
18. R. P. Doerner, D. G. Whyte, and D. M. Goebel. Sputtering yield measurements during low energy xenon plasma bombardment. *J. Appl. Phys.*, 33(9), 5816–5823 (2003).
19. G. Savage. *Carbon/Carbon Composites* (Chapman & Hall, London, 1993).
20. G. L. Vignoles, J. Lachaud, Y. Aspa, and J.-M. Goyhénèche. Ablation of carbon-based materials: multiscale roughness modelling. *Compos. Sci. Technol.*, 69(9), 1470–1477 (2009).
21. G. Duffa. Ablation (Monograph), in French. (CEA-CESTA, Bordeaux, 1996).
22. V. Borie, Y. Maisonneuve, D. Lambert, and G. Lengellé. *Ablation des matériaux de tuyères de propulseurs à propergol solide* (in French), Technical Report ONERA 1990–2013 (ONERA, Châtillon, France, 1990).
23. Y. Maisonneuve. Ablation of solid-fuel booster nozzle materials. *Aerosp. Sci. Technol.*, 1(4), 277–289 (1997).
24. G. L. Vignoles, Y. Aspa, and M. Quintard. Modelling of carbon-carbon composite ablation in rocket nozzles. *Compos. Sci. Technol.*, 70(9), 1303–1311 (2010).
25. X. Aubard, C. Cluzel, L. Guitard, and P. Ladevèze. Modelling of the mechanical behaviour of 4D carbon/carbon composite materials. *Compos. Sci. Technol.*, 58(5), 701–708 (1998).
26. X. Aubard, C. Cluzel, L. Guitard, and P. Ladevèze. Damage modeling at two scales for 4D carbon/carbon composites, *Comput. Struct.*, 78(1–3), 83–91 (2000).
27. O. Coindreau and G. L. Vignoles. Assessment of geometrical and transport properties of a fibrous C/C composite preform as digitized by X-ray computed micro-tomography. Part I: image acquisition and geometrical properties. *J. Mater. Res.*, 20(9), 2328–2339 (2005).
28. G. L. Vignoles. Modelling of the CVI processes, in *Advances in Science and Technology*, Vol. 50, *Advanced Inorganic Fibrous Composites for Structural Applications-V, part of CIMTEC 11, Acireale, Italy, 2006*, edited by P. Vincenzini and M. Singh (Trans Tech Publications, Zürich, Switzerland, 2006), pp. 97–106.
29. I. Golecki. Rapid vapor-phase densification of refractory composites. *Mater. Sci. Eng., R.*, R20(2), 37–124 (1997).
30. M. R. Wool. *Summary of Experimental and Analytical Results*, Volume X of Interim Report SAM-SOTR7486 of Passive Nosetip Technology (PANT) Program (Air Force Space and Missile Systems Organization, Los Angeles, CA, 1975).
31. E. Venkatapathy, B. Laub, G. Hartman, J. Arnold, M. Wright, and G. A. Allen Jr. Thermal protection system development, testing, and qualification for atmospheric probes and sample return missions: examples for Saturn, Titan and Stardust-type sample return. *Adv. Space Res.*, 44(1), 138–150 (2009).
32. S. T. Keswani and K. K. Kuo. Validation of an aerothermochemical model for graphite nozzle recession and heat-transfer process. *Combust. Sci. Technol.*, 47(3–4), 177–192 (1986).
33. P. Thakre and V. Yang. Chemical Erosion of carbon–carbon/graphite nozzles in solid-propellant rocket motors. *J. Propul. Power*, 24(4), 822–833 (2008).
34. R. Acharya and K. K. Kuo. Effect of reaction kinetic schemes on graphite rocket nozzle erosion rates. *Int. J. Energetic Mater. Chem. Propul.*, 9(1), 71–90 (2010).
35. D. Cho and B.-I. Yoon. Microstructural interpretation of the effect of various matrices on the ablation properties of carbon-fiber-reinforced composites. *Compos. Sci. Technol.*, 61(2), 271–280 (2001).
36. J. Yin, X. Xiong, H.-B. Zhang, and B.-Y. Huang. Microstructure and ablation performances of dual-matrix carbon/carbon composites. *Carbon*, 44(9), 1690–1694 (2006).
37. G.-H. Bai, S.-H. Meng, B.-M. Zhang, and Y. Liu. Prediction on carbon/carbon composites ablative performance by artificial neutral net (*SiC*). *J. Mater. Sci. Technol.*, 24(6), 945–952 (2009).
38. R. L. Geisler. The relationship between solid-propellant formulation variables and nozzle recession rates, in *Proceedings of JANNAF Rocket Nozzle Technology Subcommittee Meeting* (Lancaster, CA, 1978).
39. B. Evans, K. K. Kuo, P. J. Ferrara, J. D. Moore, P. Kutzler, and E. Boyd. Nozzle throat erosion characterization study using a solid-propellant rocket motor simulator, in *Proceedings of the 43rd AIAA/ASME/SAE/ASEE Joint Propulsion Conference and Exhibit* (Cincinatti, OH, 2007), *AIAA Papers* 2007-5776.

40. B. Chen, L.-T. Zhang, L.-F. Cheng, and X.-G. Luan. Erosion resistance of needled carbon/carbon composites exposed to solid rocket motor plumes. *Carbon*, 47(6), 1474–1479 (2009).
41. K.-Z. Li, X.-T. Shen, H.-J. Li, S.-Y. Zhang, T. Feng and L.-L. Zhang. Ablation of the carbon/carbon composite nozzle-throats in a small solid rocket motor. *Carbon*, 49(4), 1208–1215 (2011).
42. Y. J. Lee and H.-J. Joo. Investigation on ablation behavior of CFRC composites prepared at different pressures. *Composites Part A*, 35(11), 1285–1290 (2004).
43. Q.-M. Liu, L.-T. Zhang, F.-R. Jiang, J. Liu, L.-F. Cheng, H. Li, and Y.-G. Wang. Laser ablation behaviors of SiC–ZrC coated carbon/carbon composites. *Surf. Coat. Technol.*, 205(17–18), 4299–4303 (2011).
44. S.-P. Li, K.-Z. Li, H.-J. Li, Y.-L. Li, and Q.-L. Yuan. Effect of HfC on the ablative and mechanical properties of C/C composites. *Mater. Sci. Eng., A*, 517(1–2), 61–67 (2009).
45. W. Sun, X. Xiong, B.-Y. Huang, G.-D. Li, H.-B. Zhang, Z.-K. Chen, and X.-L. Zheng. ZrC ablation protective coating for carbon/carbon composites. *Carbon*, 47(14), 3368–3371 (2009).
46. S. Farhan, K.-Z. Li, L.-J. Guo, Q.-M. Gao, and F.-T. Lan. Effect of density and fibre orientation on the ablation behaviour of carbon-carbon composites. *New Carbon Mater.*, 25(3), 161–167 (2010).
47. X.-T. Shen, K.-Z. Li, H.-J. Li, H.-Y. Du, W.-F. Cao, and F.-T. Lan. Microstructure and ablation properties of zirconium carbide doped carbon/carbon composites. *Carbon*, 48(1), 344–351 (2010).
48. Z.-K. Chen, X. Xiong, G.-D. Li, W. Sun, and Y. Long. Texture structure and ablation behavior of TaC coating on carbon/carbon composites. *Appl. Surf. Sci.*, 257(2), 656–661 (2010).
49. H.-H. Sun, H.-J. Li, X.-T. Shen, G.-X. Cao, X.-F. Qiang, X.-B. Ren, and Q.-G. Fu. Microstructure and ablation behavior of C/C composites doped with ZrB_2. *J. Inorg. Mater.*, 26(6), 669–672 (2011).
50. X.-T. Shen, K.-Z. Li, H.-J. Li, Q.-G. Fu, S.-P. Li, and F. Deng. The effect of zirconium carbide on ablation of carbon/carbon composites under an oxyacetylene flame. *Corros. Sci.*, 53(1), 105–112 (2011).
51. J. Lachaud, N. Bertrand, G. L. Vignoles, G. Bourget, F. Rebillat, and P. Weisbecker. A theoretical/experimental approach to the intrinsic oxidation reactivities of C/C composites and of their components. *Carbon*, 45(14), 2768–2776 (2007).
52. M. Dauchier and J.-C. Cavalier. Matériaux composites phénoliques ablatifs. *Techniques de l'Ingénieur*, AM 5325, 1–9 (2002). (in French)
53. K. A. Trick and T. E. Saliba. Mechanisms of the pyrolysis of phenolic resin in a carbon/phenolic composite. *Carbon*, 33(11), 1509–1515 (1995).
54. N. Bertrand, J. Lachaud, G. Bourget, F. Rebillat, and G. L. Vignoles. Identification of intrinsic carbon fiber oxidation kinetics from experimental data and CFD modeling, in *Proceedings of the 12th European Conference on Composite Materials, Biarritz, France, September 2006*, edited by J. Lamon and A. Torres-Marques (AMAC, Paris, 2006).
55. D. Cho, J.-Y. Lee, and B. I. Yoon. Microscopic observations of the ablation behaviours of carbon fibre/phenolic composites. *J. Mater. Sci. Lett.*, 12(24), 1894–1896 (1993).
56. E. Duvivier. Cinétique d'oxydation d'un composite carbone/carbone et influence sur le comportement mécanique (in French). Ph.D. Thesis, No. 1692. (University Bordeaux, 1997).
57. E. Duvivier, C. Robin-Brosse, and R. Naslain. Oxidation of 3D-C/C composites in presence (or not) of catalytic impurities: kinetics and effect on mechanical properties, in *Ceramic Engineering and Science Proceedings, Vol. 27(2), Mechanical Properties and Performance of Engineering Ceramics II, Cocoa Beach, 2006*, edited by R. Tandon, A. Wereszczak, and E. Lara-Curzio (John Wiley & Sons, New York, 2006), pp. 225–231
58. J.-C. Han, X.-D. He, and S.-Y. Du. Oxidation and ablation of 3D carbon–carbon composite at up to 3000°C. *Carbon*, 33(4), 473–478 (1995).
59. J. Lachaud. Modélisation physico-chimique de l'ablation de matériaux composites en carbone. Ph. D. Thesis No. 3291. (University Bordeaux, 2006).
60. J. Lachaud, G. L. Vignoles, J.-M. Goyhénèche, and J.-F. Epherre. Ablation in C/C composites: microscopic observations and 3D numerical simulation of surface roughness evolution, in *Ceramic Transactions, Vol. 191, 6th Pacific Rim Conference on Ceramic and Glass Technology, Hawaii, 2005*), edited by M. Affatigato (The American Ceramic Society, Westerville, OH, 2006), pp. 149–160
61. J. Lachaud, Y. Aspa, G. Vignoles, and G. Bourget. Experimental characterization and 3D modelling of carbon/carbon composites oxidation: role of the interphase, in *Proceedings of the 12th European Conference on Composite Materials, Biarritz, France, September 2006*, edited by J. Lamon and A. Torres-Marques (AMAC, Paris, 2006).
62. J. Lachaud, Y. Aspa, and G. L. Vignoles. Analytical modeling of the steady state ablation of a 3D C/C composite. *Int. J. Heat Mass Trans.*, 51(9–10), 2614–2627 (2008).
63. G. L. Vignoles, C. Mulat, C. Germain, O. Coindreau, and J. Lachaud. Benefits of X-ray CMT for the modelling of C/C composites. *Adv. Eng. Mater.*, 13(3), 178–185 (2011).

64. P. Weisbecker, M. Alrivie, J. Lachaud, and G. L. Vignoles. Microstructural oxidation behaviour of C/C composites: TEM characterisation and modelling, in *Proceedings of Euromat 2007 Congress, Nürnberg, Germany, 2007*, edited by F. Delannay, P. D. Portella, T.W. Clyne, E. Zschech, K. J. Kurzydlowski, S. van der Zwaag, P. P. Schepp (DGM, Germany, 2007).
65. Y. Aspa. Modélisation de l'ablation des composites C/C dans les tuyères (in French). Ph.D. Thesis No.2431. (INPT Toulouse, 2006).
66. J. Lahaye, F. Louÿs, P. Ehrburger. The reactivity of carbon–carbon composites. *Carbon*, 28(1), 137–141 (1990).
67. S. Labruquère, X. Bourrat, R. Pailler, and R. Naslain. Structure and oxidation of C/C composites: role of the interface. *Carbon*, 39(7), 971–984 (2001).
68. R. Naslain, R. Pailler, X. Bourrat, S. Labruquère, and E. Duvivier. The role of interfaces in the oxidation of 3D-carbon/carbon composites, *Ceram. Eng. Sci. Proc.*, 23(3), in *26th Annual Conference on Composites, Advanced Ceramics, Materials, and Structures: A*, edited by H. T. Lin and M. Singh (John Wiley & Sons, New York, 2002), pp. 287–294.
69. Y. Aspa, M. Quintard, J. Lachaud, and G. Vignoles. Identification of microscale ablative properties of C/C composites using inverse simulation, in *A Collection of Technical Papers. 9th AIAA/ASME Joint Thermophysics and Heat Transfer Conference Proceedings, Vol. 1, San Francisco, CA, 2006. AIAA Papers* 2006-2907 (AIAA, 2006), pp. 52–64
70. H. K. Chelliah, A. Makino, I. Kato, N. Araki, and C. K. Law. Modeling of graphite oxidation in a stagnation-point flowfield using detailed homogeneous and semiglobal heterogeneous mechanisms with comparisons to experiments. *Combust. Flame*, 104(4), 469–480 (1996).
71. P. A. Libby and T. R. Blake. Burning carbon particles in the presence of water vapor. *Combust. Flame*, 41, 123–147 (1981).
72. E. S. Golovina. The gasification of carbon by carbon dioxide at high temperatures and pressures. *Carbon*, 18(3), 197–201 (1980).
73. I. W. Smith. The intrinsic reactivity of carbons to oxygen. *Fuel*, 57(7), 409–414 (1978).
74. G. Duffa, G. L. Vignoles, J.-M. Goyhénèche, and Y. Aspa. Ablation of carbon-based materials: investigation of roughness set-up from heterogeneous reactions. *Int. J. Heat Mass Trans.*, 48(16), 3387–3401 (2005).
75. Y. Aspa, G. L. Vignoles, M. Quintard, F. Plazanet, and C. Descamps. Ablation of carbon/carbon composite: direct numerical simulation and effective behavior, in *Ceramic Engineering and Science Proceedings, Vol. 26 (2), Mechanical Properties and Performance of Engineering Ceramics and Composites*, edited by E. Lara-Curzio (John Wiley & Sons, New York, 2005), pp. 99–106.
76. G. L. Vignoles, J. Lachaud, and Y. Aspa. Roughness evolution in ablation of carbon-based materials: multi-scale modelling and material analysis, in *ESA Conference Proceedings, Vol. SP-631, Proceedings of the 5th European Symposium on Thermal Protection Systems and Hot Structures*, edited by K. Fletcher (ESA Publications, Noordwijk, The Netherlands, 2006).
77. D. Gueyffier, J. Li, A. Nadim, R. Scardovelli, and S. Zaleski. Volume-of-fluid interface tracking with smoothed surface stress methods for three-dimensional flows. *J. Comput. Phys.*, 152(2), 423–456 (1999).
78. S. Osher and J. A. Sethian. Fronts propagating with curvature-dependent speed: algorithms based on Hamilton–Jacobi formulations. *J. Comput. Phys.*, 79(1), 12–49 (1988).
79. Y. Aspa, J. Lachaud, and G. L. Vignoles. Simulation of C/C composites ablation using a VOF method with moving reactive interface. *Proceedings of the 12th European Conference on Composite Materials, Biarritz, France, September, 2006*, edited by J. Lamon and A. Torres-Marques (AMAC, Paris, 2006).
80. S. W. J. Welch and J. Wilson. A volume of fluid based method for fluid flows with phase change. *J. Comput. Phys.*, 160(2), 662–682 (2000).
81. J. Lachaud, Y. Aspa, and G. L. Vignoles. A Brownian motion simulation technique to simulate gasification and its application to ablation. *Comput. Mater. Sci.*, 44(4), 1034–1041 (2009).
82. G. L. Vignoles. Modelling binary, Knudsen and transition regime diffusion inside complex porous media. *J. de Phys IV France, Vol. 5(C5), Proceedings of the 10th European Chemical Vapor Deposition Conference, Jesolo Lido, Italy, 1995*, edited by G. A. Battiston (EDP Sciences, Les Ulis, France, 1995), pp. 159–166.
83. M. Donias, G. L. Vignoles, C. Mulat, C. Germain, and J.-F. Delesse. Simplified marching cubes: an efficient discretization scheme for simulations of deposition/ablation in complex media. *Comput. Mater. Sci.*, 50(3), 893–902 (2011).
84. Y. Aspa, M. Quintard, J. Lachaud, and G. L. Vignoles. Effective surface approach for the design of ablative composites, in *Proceedings of the 1st International ARA days, Arcachon, France, 2006*, edited by E. Arquis (ARA, Bordeaux, 2006).
85. B. Vancrayenest and D. G. Fletcher. Emission spectroscopic survey of graphite ablation in the VKI plasmatron, in *A Technical Collection of Papers, 9th AIAA/ASME Joint Thermophysics and Heat Transfer Conference, Vol. 1, San Francisco, CA, 2006. AIAA Papers* 2006-2907 (AIAA, 2006), pp. 1–13
86. J. Lachaud, G. L. Vignoles, V. Ducamp, J.-F. Epherre, J.-M. Goyheneche, G. Duffa, and B. Vancrayenest. A new experimental/theoretical approach to measure and extrapolate the ablative properties of TPS materials, in *Proceedings of IPPW5: 5th ESA International Planetary Probe Workshop, Bordeaux, 2007*, edited by J.-M. Bouilly (ARA, Bordeaux, 2007).
87. N. S. Jacobson, T. A. Leonhardt, D. M. Curry, R. A. Rapp. Oxidative attack of carbon/carbon substrates through coating pinholes. *Carbon*, 37(3), 411–419 (1999).

88. J. Lachaud, Y. Aspa, G. Vignoles, and J.-M. Goyhénèche. 3D modeling of thermochemical ablation in carbon-based materials: effect of anisotropy on surface roughness onset, in *ESA Publications, Vol. SP-616, Proceedings of the 10th International Symposium on Materials in a Space Environment*, edited by M. Dinguirard and J. Kleiman (ESA, Noordwijk, The Netherlands, 2006).

89. G. L. Vignoles, J. Lachaud, Y. Aspa, and M. Quintard. Effective surface recession laws for the physico-chemical ablation of C/C composite materials, in *Ceramics Engineering and Science Proceedings, Vol. 31: Mechanical Properties and Performance of Engineering Ceramics and Composites V, Daytona Beach, FL, 2010*, edited by D. Singh, J. Salem, S. Mathur, and T. Ohji (John Wiley & Sons, New York, 2010), pp. 351–360.

90. W. H. Glime and J. D. Cawley. Oxidation of carbon fibers and films in ceramic matrix composites: a weak link process. *Carbon*, 33(8), 1053–1060 (1995).

CHAPTER 13

RADIATION EFFECTS

YUTAI KATOH

Materials Science and Technology Division, Oak Ridge National Laboratory, Oak Ridge, TN, USA

13.1 INTRODUCTION

Energetic radiation may impose a variety of effects on materials ranging from subtle and transient ones such as radioluminescence to severe and permanent effects including plastic deformation, strength loss, and thermal conductivity degradation. Understanding the effects of radiation is therefore essential when certain material is considered, designed, or qualified for application in harsh radiation environments. Effects of radiation on ceramic matrix composites are in general complex not only because these materials consist of multiple constituents that behave in different manners under radiation but also because of the complexity of physical processes of radiation effects in ceramics as multielement compounds. This chapter attempts to provide an overview of radiation effects on ceramic matrix composites, starting from the radiation damage processes in an atomistic scale, microstructural evolutions, and the resulting property modifications in ceramics. The discussion is then continued to examine the modifications in dimensions and properties for ceramic matrix composites as the consequences of radiation effects on the constituents and interactions of individual radiation effects through the composite architectures.

The materials of main interest here are SiC fiber-reinforced SiC matrix (SiC/SiC) composites and carbon fiber-reinforced carbon matrix (C/C) composites. Among various ceramic matrix composites, SiC/SiC composites are of greatest technological importance in terms of radiation effects, since they are considered for use as the key-enabling materials for innovative safety features in the existing nuclear reactors [1], partial to entire core constructions of various advanced fission reactor concepts [2–4], and certain thermostructural functions in future fusion energy systems [5, 6]. These high expectations for the SiC/SiC composites are based on the existing data that demonstrate the exceptional radiation resistance of SiC and its composites of certain grades [7, 8]. The C/C composites are also attractive materials particularly for high temperature nuclear applications because of the superior very high temperature performances and the moderate radiation resistance [9]. Therefore, the material systems of main focus for discussion in this chapter will be SiC/SiC and C/C composites. Other ceramic systems may occasionally be mentioned for the purpose of discussion on fundamental aspects of radiation effects.

13.2 THEORY OF RADIATION DAMAGE

13.2.1 What Is Radiation Damage?

Particles, quantum, or electromagnetic radiation can cause a variety of effects in solid materials depending on the kind, specie, and energy of the radiant. While some of the radiation effects such as electronic excitation and ionization are only transitional in most cases, other radiation effects including atomic displacement and nuclear transmutation often cause permanent and significant modifications to dimensions and properties of materials. When certain radiation

Ceramic Matrix Composites: Materials, Modeling and Technology, First Edition. Edited by Narottam P. Bansal and Jacques Lamon.

results in permanent and detrimental effects in materials, it is referred to as radiation damage. For ceramic matrix composites that are used primarily for structural purposes, radiation damage of particular importance is related to the atomic displacements introduced by high energy neutron irradiation in composites for applications in nuclear reactors, fusion energy systems, and nuclear propulsion systems. The typical consequences of atomic displacement damage in ceramic matrix composites include dimensional instability, change in mechanical properties (often but not necessarily detrimental), and decrease in thermal conductivity. Nuclear transmutation becomes a concern for SiC in case the neutron spectrum extends to very high energy beyond ~10 MeV such as in fusion reactors or spallation targets. B present in certain sintered SiC burns very quickly with much lower energy neutrons often causing substantial detrimental effects. However, the following discussion will be mostly limited to radiation effects related with the atomic displacement damage.

13.2.2 Radiation Defect Production

When a particle of atomic mass M_1 and kinetic energy E_1 collides elastically with a target atom of atomic mass M_2, the kinetic energy transferred to the target atom E_2 is given by

$$E_2 = \frac{4M_1M_2}{(M_1+M_2)^2}E_1 \cdot \sin^2\frac{\theta}{2} \quad (13.1)$$

using the scattering angle θ. When E_2 exceeds the displacement threshold energy E_d, the target atom is displaced from the lattice position. E_d is a function of element of target atom, material, and the crystallographic orientation; however, the typical E_d in common ceramics falls in a range 20–60 eV as listed in Table 13.1 [10]. In case E_2 is only slightly greater than E_d, which is a case for most electron irradiation-induced atomic displacement, a self-interstitial atom (SIA) is created leaving a vacant lattice site (vacancy). This constitutes the simplest form of displacement damage called Frenkel pair. The neighboring vacancy and interstitial mostly annihilate through correlated mutual recombination when thermally activated migration is possible for at least one of the point defects. A very small fraction of defects that survive the mutual recombination process, possess the ability to significantly alter properties of the target materials.

In a nuclear environment, fast neutrons with kinetic energy greater than 0.1 MeV cause the bulk of the radiation damage. Equation (13.1) indicates that these fast neutrons are capable of transferring kinetic energy on order of tens of keV to the target atoms. In such case that $E_2 \gg E_d$, the primary knock-on atom (PKA) produces a large number of high energy secondary knock-ons, creating three-dimensional collision cascades. The number of atomic displacement events caused by single PKA of the initial kinetic energy E_p is defined by modified Kinchin–Pease model [11] as

$$N_d = \frac{\kappa E_p}{2E_d} \quad (13.2)$$

where factor κ is typically taken to be 0.8 as later proposed in Norgett–Robinson–Torrens (NRT) model [12]. Numerous computational simulation studies based on molecular dynamics have demonstrated that the number of individual atomic displacement events, as predicted by Equation (13.2), is significantly greater than the number of Frenkel defects that

TABLE 13.1 Threshold Energy for Atomic Displacement Recommended in Reference 10 for Various Materials

Material	Element	E_d (eV)
Graphite	C	~30
Diamond	C	~40
SiC	Si	~40
	C	~20
Al_2O_3	Al	~20
	O	~50
MgO	Mg	~55
	O	~55

survive the collision cascade. However, NRT model is widely accepted today as a standard tool to define damage levels for materials irradiated by different particles or with different energy spectra. A radiation damage unit of displacement-per-atom (dpa) is most commonly used based on NRT model, roughly indicating in average how many times each atom in the target material was displaced from its equilibrium lattice site by energetic collisions.

Collision cascade has important implications including (1) spatial separation of SIAs and vacancies preventing correlated recombination, (2) direct production of clusters of point defects that are far less mobile than single defects, and (3) creation of a small zone that has experienced very high density collision events (e.g., cascade mixing and cascade-induced local amorphization). Therefore, radiation that produces high energy PKAs (such as fast neutrons in fission and fusion systems and heavy ions in nuclear fuel) often imposes characteristic effects on materials that differ from the other radiations that result in PKA energies only slightly above the displacement threshold (e.g., electrons in a transmission electron microscope).

13.2.3 Defect Migration and Evolutions

Irradiation of energetic particles leaves isolated Frenkel defects and possibly small defect clusters after the collision cascade process is over. In compounds like most ceramics, antisite defects may also be produced. In general, isolated single defects are relatively easy to migrate through thermal activation. Examples of activation threshold energy for point defect migration in ceramics are shown in Table 13.2 [10, 13, 14]. Diffusivity of defects is expressed below using the diffusivity pre-exponential (D_0) that is determined by the vibrational frequency and the number of possible jump sites for the particular defect along with an Arrhenius term that takes into account the migration energy (E_m), and Boltzmann constant (k).

$$D = D_0 \exp\left(-\frac{E_m}{kT}\right) \tag{13.3}$$

Thermal diffusion of a mobile defect results in defect annihilation by mutual recombination, defect absorption at existing extended defects (such as defect clusters, dislocations, twin boundaries, grain boundaries, etc.; these extended defects thus accommodate mutual recombination), creation of a new defect cluster or antisite, or conversion into a point defect of other element through reaction with an antisite. The defect clusters attract the migrating defects through the strain field interactions and often grow in size as radiation continues. For example, SIA clusters in certain fcc systems often grow as $\frac{a}{3}$<111> extrinsic dislocation loops and eventually forming $\frac{a}{2}$<011> dislocation network through dislocation reactions [15]. On the other hand, vacancy clusters may form intrinsic dislocation loops, stacking fault tetrahedra, and three-dimensional voids. Relative stability of different cluster configurations may alter depending on the cluster size and temperature and differs widely among various materials and crystal structures.

TABLE 13.2 Examples of Migration Activation Energy Estimated or Calculated for Point Defects in Ceramics

Material	Point defect	E_m (eV)	Reference
Graphite	V	1.0 (a-axis), 4.7 (c-axis)	13, (simulation)
	I	<0.03 (a-axis), 3.9 (c-axis)	
SiC (3C)	V_{Si}	>2.4	14, (simulation)
	V_C	>3.5	
	I_{Si}	3.5	
	I_C	>0.5	
Al_2O_3	V_{Al}	1.8–2.1	10
	V_O	1.8–2.0	
	I_{Al}, I_O	0.2–0.8	
MgO	V_{Mg}	2.0–2.3	10
	V_O	2.0–2.5	
	I_{Mg}, I_O	0.2–1.5	

13.2.4 Resultant Macroscopic Effects

Accumulation of radiation-produced defects, in isolated, clustered, or extended forms, alters macroscopic properties of materials. Representative effects of radiation on ceramic materials include volumetric swelling, anisotropic deformation in non-cubic systems, irradiation creep, thermal conductivity degradation, and various changes in mechanical properties. At low temperatures where thermal diffusion is limited for the various defects, amorphization may be observed. Radiation enhanced diffusion results in phase separation in systems that are otherwise kinetically stable at low temperatures while accommodating radiation-induced solute segregation that modifies the local alloy composition eventually leading to phase instability. In addition to these direct effects of radiation, polycrystalline and multiphase materials may experience the secondary effects such as grain boundary cracking in AlN [16] and reaction-bonded SiC [17] due to anisotropic deformation of crystal grains and differential swelling between two phases, respectively.

Radiation effects on composite materials occur as results of the radiation effects on the constituents. For example, the macroscopic physical properties such as the composite dimensions and thermal conductivity may evolve under radiation following the applicable constitutive laws as far as the mechanical integrity is retained. The radiation effects on mechanical properties are often not simple because they are determined by a complex interplay among the mechanical property changes in fibers, matrices, and interface or interphase, differential deformation, anisotropy effects, and mechanical constraints specific to the material systems and architectures. In this sense, understanding the constitutive mechanisms dictating the property of interest is the key to understand the radiation effects on composite materials.

13.3 RADIATION EFFECTS ON CERAMICS

13.3.1 Silicon Carbide

Silicon carbide (SiC), perceived as among the most radiation-resistant engineering materials, is the ceramic material of greatest technological importance for high radiation environment applications [1]. Studies on radiation effects on SiC started in late-mid-twentieth century as the coating material for nuclear fuels [18, 19]. It is important to note that materials that are nominally referred to as SiC may include various SiC-based ceramics in addition to true SiC. Among them, SiC in high purity, stoichiometric, and fully crystalized forms (often simply "SiC" hereafter) are known to be most and exceptionally stable in high radiation environments. Among true SiC, cubic (β-phase or 3C-) SiC is expectedly most stable due to the crystallographic isotropy; however, there has been no reported convincing experimental evidences to indicate inferior radiation stability for hexagonal polytypes. Vast majority of the experimental data on radiation effects are for chemically vapor-deposited (CVD) polycrystalline cubic SiC. The data show that this material does not deteriorate, except the thermal conductivity decrease, to as much damage level as data exist in a temperature range between the amorphization threshold ($\sim$150°C) and the progressive void swelling regime ($> \sim$1500°C, more explanation to follow) [7].

It is a general trend that the radiation resistance deteriorates as the quality of material deviates from the above-mentioned three attributes for the "true SiC." For example, non-pure SiC sintered with B is known to degrade as nuclear reaction ^{10}B (n, α) ^{7}Li burns B producing He and Li. Radiation stability of pressure-less liquid phase-sintered (LPS) SiC will be limited by the rather poor radiation resistance of the oxide ceramic matrix. The reaction-bonded (RB) SiC is a good example of non-stoichiometry resulting in radiation instability. As mentioned in the previous section, the differential swelling between grains of SiC and unreacted metallic Si severely spoils strength of RB-SiC [17]. Moreover, amorphous or nano-crystalline SiC, commonly derived from polymeric precursors, undergoes crystallization in radiation environment accompanying volume contraction, extensive microcracking, and possible phase decomposition.

13.3.2 Radiation Damage Evolution in SiC

When SiC is irradiated with energetic particles, primary defects produced are SIAs of both elements (I_{Si} and I_C) and vacancies in both sublattices (V_{Si} and V_C). Damage production in C sublattice is often more aggressive due to the smaller E_d for C atoms in SiC. As shown in Table 13.2, migration energy for I_C is significantly smaller than point defects of other types, making it readily mobile at most temperatures of technological interest, whereas the higher migration energies make vacancies and I_{Si} far less mobile (or practically immobile). When migrating I_C reacts with V_{Si}, they likely make into an antisite C_{Si}, although it is not impossible that a metastable I_C–V_{Si} pair is produced. When SIAs encounter with each other (or when the strain fields overlap, to be more precise), they will likely make a di-interstitial, or the smallest SIA cluster, that is energetically preferred over two isolated SIAs. If a migrating SIA merges an existing SIA cluster, then the

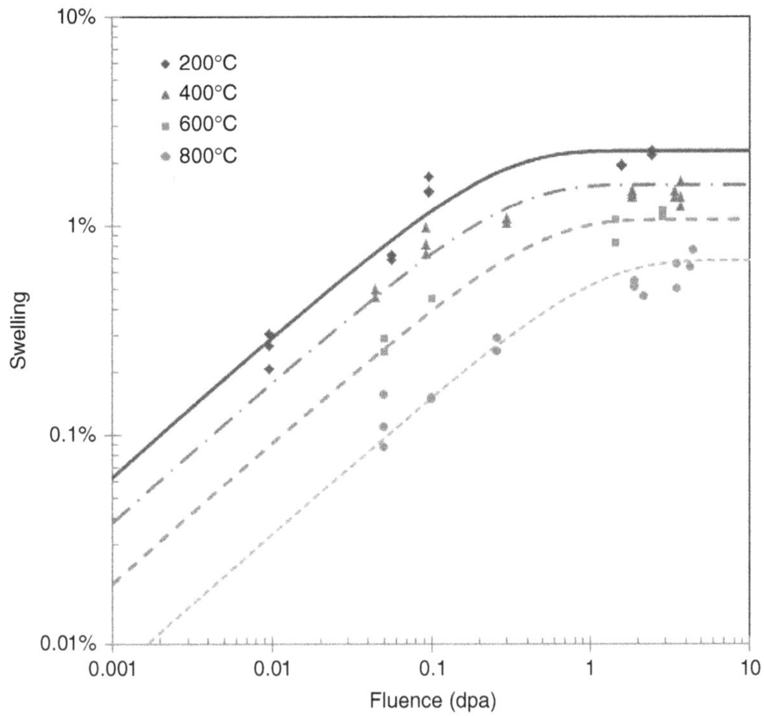

FIGURE 13.1

Fluence-dependent evaluation of swelling for SiC at various temperatures. Experimental data points are replotted from Ref [7].

SIA cluster will grow in size. The SIA clusters are considered far less immobile than single isolated SIAs with the mobility further decreases as the size increases. Binding energy for SIA clusters is generally high enough to practically prohibit the thermal dissociation.

In this way, during the early stages of irradiation, defects with immobile (or very limited mobility; vacancies, antisites, and SIA clusters) accumulate rapidly. Once the immobile defects are established to a very high density, migrating defects no longer produce new clusters but react with existing defects. When the vacancy concentration becomes high enough to make nearly all of newly displaced SIAs and the growing SIA clusters react with vacancies, a state of dynamic equilibrium in radiation damage process is achieved. In this case, the macroscopic radiation effects will saturate. Therefore, it is radiation-induced defect microstructure itself that provides just the correct amount of defect sinks that prohibits further progressive development of radiation damage, making SiC an exceptionally radiation-resistant material. The defect accumulation in SiC is clearly represented by swelling, because the lattice point separation into vacancies and SIAs appears as the volume expansion. The evolution of swelling for SiC is shown in Figure 13.1 as a function of damage level, depicting the gradual accumulation of defects until apparent saturation is reached. Results shown in Figure 13.1 are consistent with the classic rate theory equation where at higher temperatures, the characteristic diffusion length of the defects is extended and a smaller number of sinks are necessary to achieve the dynamic equilibrium between defect generation and annihilation.

During irradiation at high enough temperatures (>~800°C) where vacancies are significantly mobile, such saturations in radiation effects may never be achieved. Now the migrating vacancies produce vacancy clusters that may eventually develop into large vacancy loops or three-dimensional voids. Because SIAs in general are surrounded by more extensive strain fields, SIA clusters and dislocations tend to capture more migrating SIAs than migrating vacancies. This mechanism, historically called the dislocation–interstitial bias, promotes the creation and growth of voids as vacancy clusters [20]. Unlike the transient swelling due to point defects and SIA clusters, the volume expansion due to vacancy cluster growth (often referred to as void swelling) does not saturate but exhibits unstable increase. Rapid and unstable void swelling has been observed in SiC at temperatures exceeding ~1400°C [21]. The void swelling also accompanies unstable growth of SIA clusters and possible extensive dislocation evolutions [22].

13.3.3 Radiation Effects in SiC

In addition to swelling, accumulation of point defects results in various property changes in SiC. The transient swelling accompanies decrease in Young's modulus by a few percent at irradiation temperature ~600°C and above and up to

~10% at lower temperatures as the amorphization threshold temperature approaches [7]. The strength and fracture toughness tend to increase with irradiation probably because the defect clusters blunt stress concentration at crack tips and/or deflect cracks increasing fracture surface [23], whereas the excessive void swelling at very high temperatures likely degrades strength [24]. Existing data suggest that the irradiation-induced or assisted creep of SiC is minimal [25], though more experimental data are needed to determine the creep deformation of SiC at high fluence levels.

While most other radiation effects on SiC appear relatively minor, the radiation-induced decrease in thermal conductivity is significant. It is not surprising that the accumulation of point defects and clusters, in particular vacancies, very effectively reduces thermal conductivity that relies on the phonon transport in ceramics. It is partly because of the very high thermal conductivity for SiC in the nonirradiated condition that the radiation effect appears so dramatic, lowering nonirradiated room temperature thermal conductivity up to ~400 W/(m K) for CVD SiC to <20 W/(m K) after irradiation at $T < 800°C$. It is particularly interesting to note that the radiation defect thermal resistivity ($1/K_{rd}$), defined as

$$\frac{1}{K_{rd}} = \frac{1}{K_{irradiated}} - \frac{1}{K_{nonirradiated}} \tag{13.4}$$

is linearly correlated with the magnitude of transient swelling [7], indicating that the thermal conductivity decrease is proportional to the total excess volume from all radiation defects. More comprehensive descriptions of radiation effect phenomena in SiC are found in References 7 and 8.

13.3.4 Carbon Materials

Radiation effects in carbon materials are briefly described here because graphite (or carbon with similar properties) is extensively used as the interphase materials in the SiC/SiC composites as well as the C/C composite is an important ceramic composite for nuclear applications.

Graphite is a carbon material made of hexagonal basal planes (or graphenes) loosely bonded together via Van del Waals interaction. Engineering graphite is a complex composite made of highly defective graphitic constituents in various forms with exceptions of single crystals and highly oriented polycrystalline graphite (HOPG). In a simple model of radiation damage for graphite crystallite, carbon SIAs that survived recombination form new basal planes leaving vacancies in the original basal plane [26]. This atom transfer from the original basal plane to the new one causes shrinkage along the *a*-axis and expansion along the *c*-axis, resulting in anisotropic deformation of crystallites. Because the pre-existing defects (cracks and pores) along basal planes in the engineering graphite accommodate the *c*-axis expansion, polycrystalline graphite initially densifies under radiation. Once those defects are closed, graphite starts to swell until eventually losing shape and structural integrity shortly after exceeding the original volume [26]. In case of Grade IG-110 nuclear graphite irradiated at 600°C, the contraction maximum occurs at ~15 dpa and regression to the original volume is reached at ~30 dpa [27]. During the contraction process, strength, elastic constants, and electrical resistivity increases, whereas the thermal conductivity monotonically decreases.

Pyrolytic or glassy carbon-like graphite, used as the interphase material in nuclear-grade SiC/SiC composites, is anticipated to behave in similar manner to engineering graphite because they are also highly defective graphite-based materials with specific mass significantly lower than HOPG, although no high fluence irradiation data are available for the bulk properties of these materials. Radiation effects in carbon fibers will be discussed later.

13.4 RADIATION EFFECTS IN CERAMIC MATRIX COMPOSITES

13.4.1 SiC/SiC Composites

The materials discussed here are continuous SiC fiber-reinforced, SiC matrix (SiC/SiC) composites. Like monolithic SiC, radiation stability of SiC/SiC composites largely varies depending on quality of their constituents. While having the radiation-resistant CVD SiC as the matrix, early SiC/SiC composites with the chemically vapor-infiltrated (CVI) matrices failed to meet the expectation of reasonable radiation stability [28]. The observed significant strength degradation for these composites was attributed to the interfacial debond and fiber strength decrease due to the radiation-induced volume shrinkage and degradation of the ceramic-grade Si–C–O fibers (Figure 13.2) [29]. The electron beam-cured Generation II SiC fibers with reduced oxygen suffered the same problem though to a lesser extent [30]. Later, the

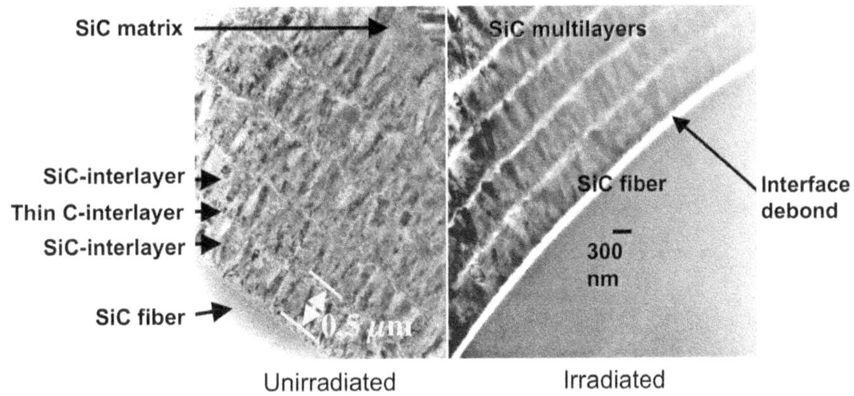

FIGURE 13.2

TEM images from fiber–matrix interface regions in unirradiated (left) and irradiated to ~1.1 dpa at 385°C (right) conditions, showing irradiation-induced interfacial debond. Material is ceramic-grade Nicalon™-reinforced CVI SiC matrix composite with PyC/SiC multilayer interphase. (Micrographs used in Reference [29] are used after rearrangement.)

near-stoichiometric Hi-Nicalon™ Type S fiber composite finally demonstrated the significant improvement in radiation resistance, proving the theory that stoichiometry is the key feature for radiation performance in SiC materials [30, 31]. Another commercial near-stoichiometric SiC fiber, Tyranno™-SA3, demonstrated similar resistance to radiation environments [32]. Today these near-stoichiometric Generation III SiC fibers are considered reasonably radiation resistant.

For the reasons discussed in Section 13.3, quality of the matrices determined by the raw materials and the processing route is among key factors dictating the composite behavior in radiation environments. Literature data indicate that the matrix derived by CVI or a specific liquid phase-sintering called nano-infiltration and transient eutectic-phase (NITE) process offers excellent radiation resistance [33]. On the other hand, matrices densified through polymer impregnation and pyrolysis (PIP) and melt-infiltration (MI) are considered inadequate for radiation services due to the known radiation-instability issues [7].

The SiC/SiC composites with radiation-resistant fibers and matrix are indeed stable in high radiation environments. As an example is shown in Figure 13.3 for the Hi-Nicalon™ Type S fiber CVI SiC matrix composite, the crystalline

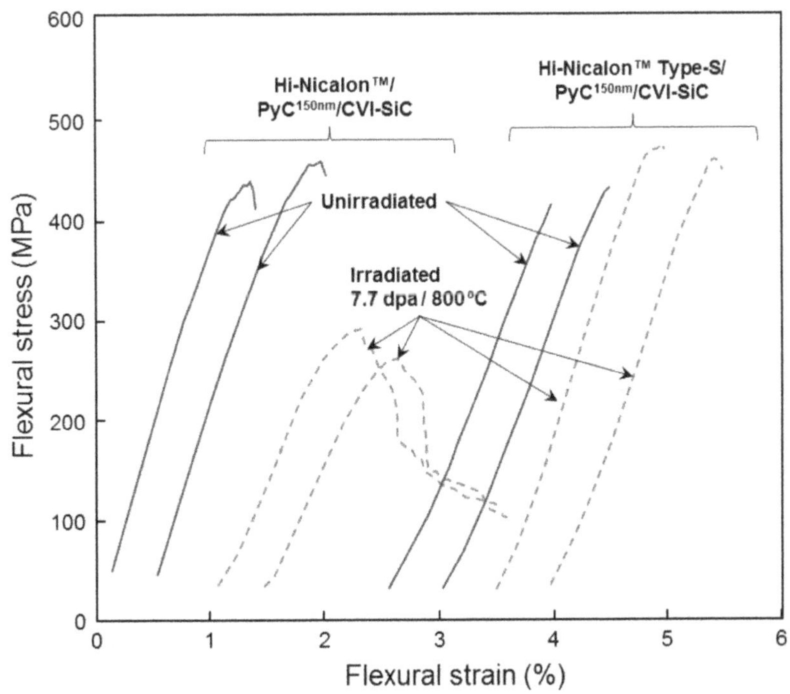

FIGURE 13.3

Effects of neutron irradiation on strength of CVI SiC matrix composites with non-stoichiometric or near-stoichiometric SiC fibers. Replotted from Reference 30.

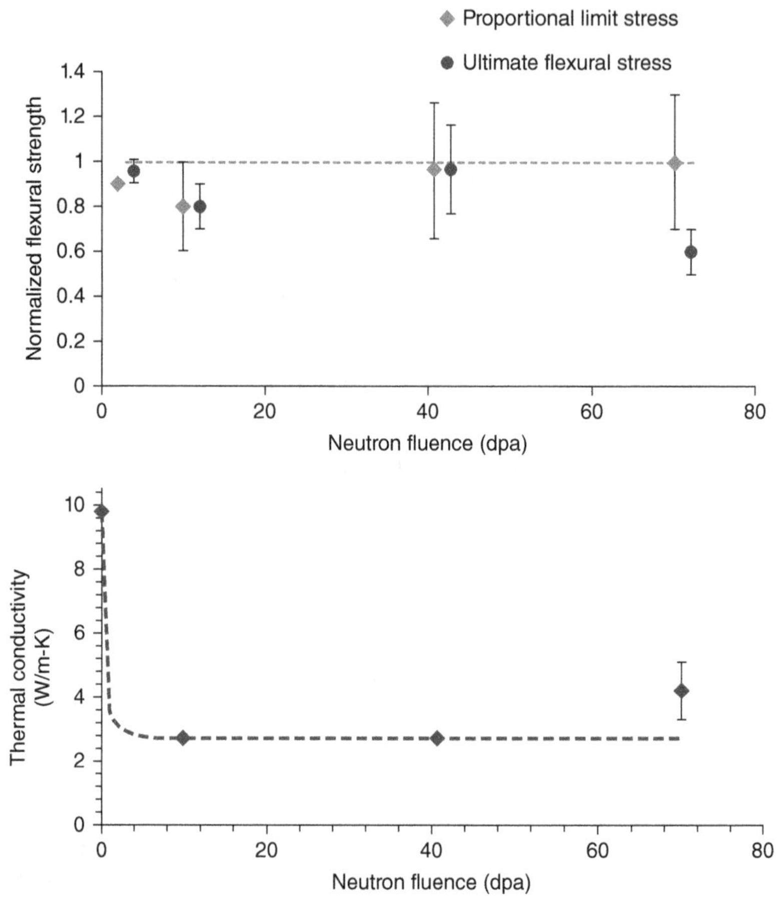

FIGURE 13.4

Fluence dependence of strength and thermal conductivity for Hi-Nicalon^{TM} Type S, CVI SiC/SiC composite neutron irradiated to very high fluences at ~800°C [34].

SiC matrix composites with near-stoichiometric SiC fiber reinforcement do not exhibit significant changes in tensile strength, proportional limit stress, and tensile modulus after neutron irradiation to significant fluence levels at elevated temperatures, whereas the CVI composite with earlier generation SiC fibers indeed degrade [30]. The data shown in Figure 13.4 indicate the retention of mechanical properties for the Hi-Nicalon^{TM} Type S CVI SiC/SiC up to ~40 dpa at 800°C, followed by modest strength degradation at ~70 dpa [34]. It also shows the complete lack of the progressive change in thermal conductivity beyond the saturation in the initial transient. The reason for the strength degradation observed at a very high exposure level is yet to be clarified.

Dimensional stability of SiC/SiC composites during neutron irradiation is compared with other nuclear materials and ceramics in Figure 13.5 [9,35–37]. Dimensional stability is an extremely important attribute for core materials for nuclear reactors. For example, the spacing between fuel rods determines the neutronic reactivity of a fission nuclear reactor core. The location and alignment of the fuel assembly have to be appropriately maintained for the reaction control system and the emergency shutdown systems to properly operate. In spite of the initial swelling transient, the lack of unstable fluence-dependent swelling appears a notable advantage for SiC/SiC composites over other material systems.

Thermal conductivity of SiC/SiC composites significantly decreases following the trend of irradiation effects on thermal conductivity of monolithic SiC. Figure 13.6 shows, for example, thermal conductivity for near-stoichiometric SiC fiber, CVI SiC matrix composites measured before and after neutron irradiation [38]. Note that the through-thickness thermal conductivity for two-dimensional composites is significantly lower than that of monolithic material due primarily to the presence of interlaminar pores. Low thermal conductivity of PyC interlayers (2–5 W/(m K)) is also contributing to further lowering the through-thickness thermal conductivity when high thermal conductivity fiber like Tyranno^{TM}-SA3 (~64 W/(m K) at room temperature) is used. A close examination of data has revealed that the radiation defect thermal resistivity values defined by Equation (13.3) are significantly different among CVD SiC, Tyranno^{TM}-SA3, and Hi-Nicalon^{TM} Type S fibers [38]. However, once the radiation defect thermal-resistivity values are known for the individual constituents for the irradiation condition of interest, thermal conductivity of the composite materials in any architecture can be predicted with an appropriate finite element model. Alternatively, a simple parallel-serial approach,

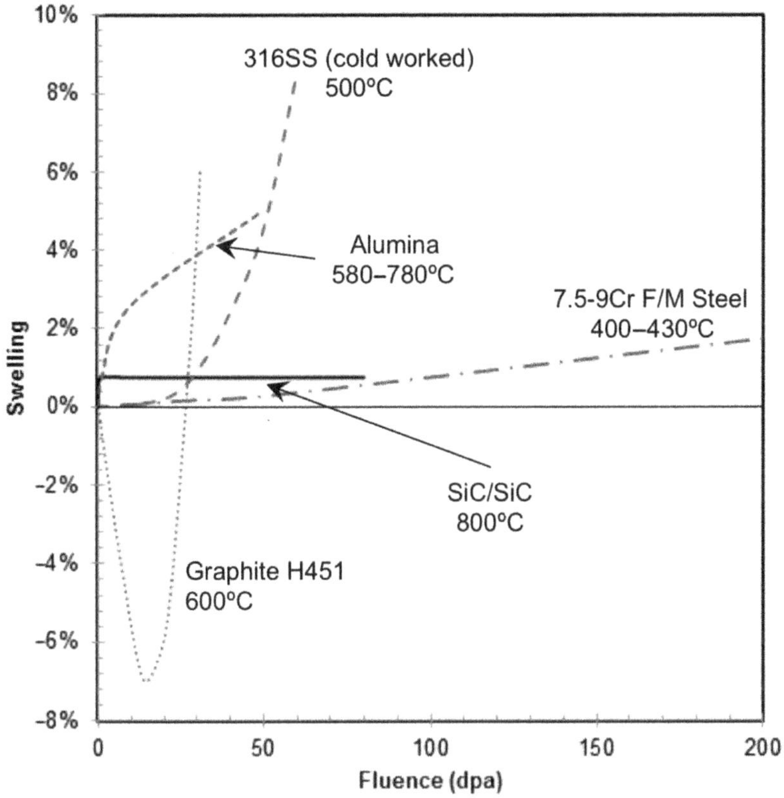

FIGURE 13.5

Neutron-induced swelling of SiC/SiC composite as compared with other materials. Original data are found in References [9, 35–37].

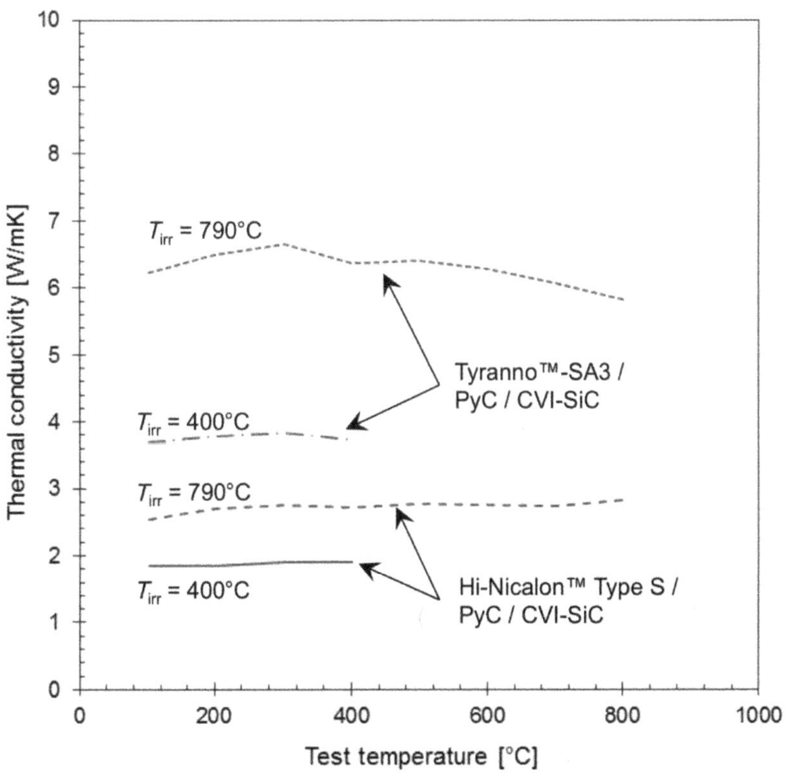

FIGURE 13.6

Effect of neutron irradiation on through-thickness thermal conductivity of 2D SiC/SiC composites. Materials are 2D CVI SiC matrix composites with Hi-Nicalon™ Type S or Tyranno™-SA3 fibers. Replotted from Reference 38.

proposed by Ishikawa, et al. for elastic modulus estimation, gives reasonable approximation of thermal conductivity for orthotropic architectures [39, 40].

13.4.2 Interphase in SiC/SiC Composites

For the debond/sliding interlayer material in the radiation-resistant SiC/SiC composites, pyrolytic carbon (PyC) has almost exclusively used. The PyC/SiC multilayer interphase has also been used; however, such multilayer interphase is considered as a single PyC interlayer between the fiber and matrix from the standpoint of fast fracture properties. BN, another interphase material commonly used in the SiC/SiC composites, is not considered very attractive mainly because of the large cross-section for neutron capture of B and the undesirable transmutation products He and Li. However, neutron irradiation data demonstrating inferior performance for the SiC/SiC composites with BN interphase and otherwise radiation-resistant constituents are not available. The poor radiation resistance of the SiC/SiC composite with B-sintered SiC fibers is reported [41].

The PyC interphase as deposited onto the fiber surface may take various forms but most commonly that of turbostratic carbon. As described above, PyC is anticipated to volumetrically shrink in the beginning of irradiation followed by swelling, and the elastic constants and strength increase during shrinkage and decrease during swelling. Figure 13.7 compares the high resolution lattice images within a PyC interlayer in an SiC/SiC composite before and after irradiation to ~10 dpa at 600°C. The basal planes in the irradiated PyC are clearly more fragmented with the basal plane spacing significantly increased. Continued irradiation largely beyond this fluence level will amorphize the PyC. It is interesting to note that the very significant irradiation-induced microstructural modification and the property changes that must have accompanied do not seem to impose a strong impact on the properties of the composites, as the near-stoichiometric SiC/SiC composites easily survives neutron radiation to 10 dpa at this temperature. One possible explanation is that the changes in PyC properties by a factor of magnitude do not significantly affect the composite properties and fracture behavior when the strength and modulus for SiC constituents are greater by an order or more. The theories for strength of ceramic matrix composites by Evans [42] and Curtin [43] in fact predict relatively weak influences of the interfacial mechanical properties on the composite properties.

The radiation effects on the fiber–matrix interfacial sliding properties have also been examined experimentally [44–46]. In general, a fair number of reports from low fluence neutron irradiation experiments suggest decrease in interfacial sliding stress after irradiation based on single filament push-out, fracture surface examination, or tensile hysteresis analysis. Nozawa et al. recently reported the initial decrease followed by increase at higher fluence levels in

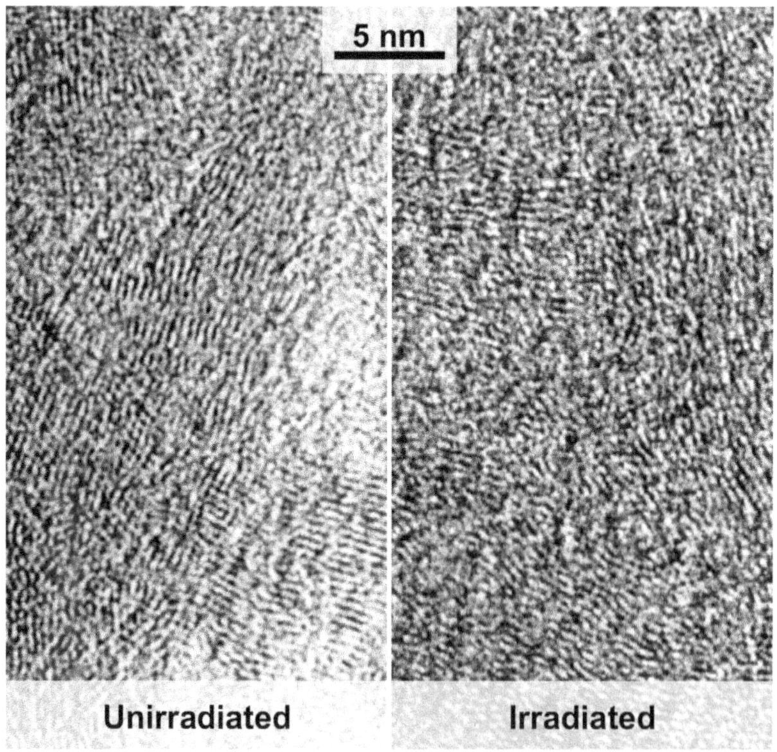

FIGURE 13.7

High resolution transmission electron images of pyrocarbon interphase in as-deposited (left) and irradiated (right, 10 dpa at 600°C) conditions.

both shear debond stress and sliding stress in neutron-irradiated Hi-Nicalon™ Type S, PyC interphase, CVI SiC matrix composites based on the single filament push-out experiment [44]. This report implies the change in fiber clamping stress due to volume change of PyC interphase as the possible governing mechanism.

13.4.3 Carbon Fiber Composites

Carbon-based materials are useful for radiation environment applications, in spite that their radiation-resistance is incomparable with SiC of good quality. Engineering graphite survives high radiation nuclear environment up to tens of dpa at or beyond the upper operating temperature limit for any code-qualified heat-resistant metallic alloys [47]. While the engineering graphite may be used under moderate compressive mechanical loading, continuous carbon fiber-reinforced carbon matrix (C/C) composite is the only form of carbon materials that can withstand substantial non-compressive stresses with adequate reliability as structural materials. C/C composites are particularly designer-friendly because of the lack of properties degradation at elevated temperatures (except thermal conductivity) and the lack of thermal creep/stress rupture at the operating temperatures. Moreover, C/C composites are often useful for their light weight and the small and tailorable thermal expansivity.

Most C/C composites are considered to belong to a category of non-rigid matrix composites, with elastic modulus of the matrix being by order of magnitude smaller than that for the fibers. In fact, typical Young's moduli for the carbon matrix and the carbon fibers are ~10 and >200 GPa, respectively. Many properties of the C/C composites are thus determined primarily by the fiber properties and the reinforcement architecture. Therefore, radiation effects on C/C composites are governed mainly by the radiation effects on carbon fibers.

The graphitic microstructures in carbon fibers are significantly oriented with basal planes more or less aligned in parallel with the fiber axis. Examples of the (somewhat) aligned radial and hourglass structures in typical pitch-based carbon fibers and the core-sheath structure in PAN (polyacrylonitrile)-based carbon fibers are schematically illustrated in Figure 13.8 [48]. The higher elastic modulus carbon fibers have more aligned graphitic microstructure for an obvious

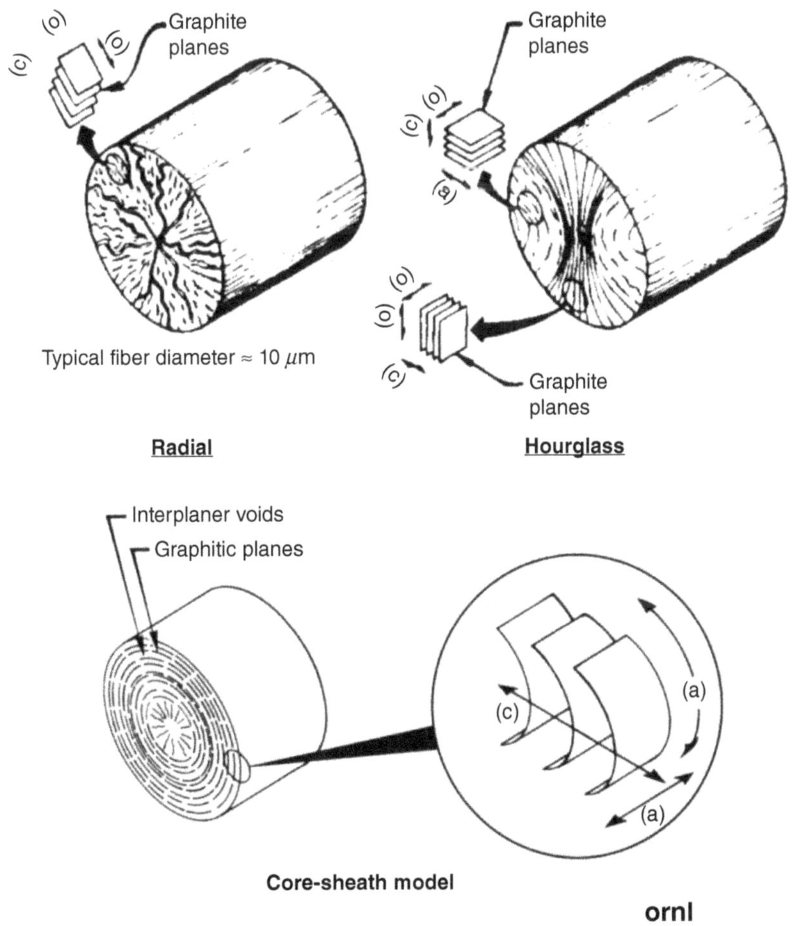

FIGURE 13.8
Schematic illustrations of aligned graphitic microstructures in typical pitch and PAN-based carbon fibers. Reproduced with permission from Dr. Tim Burchell, Physica Scripta 1996.

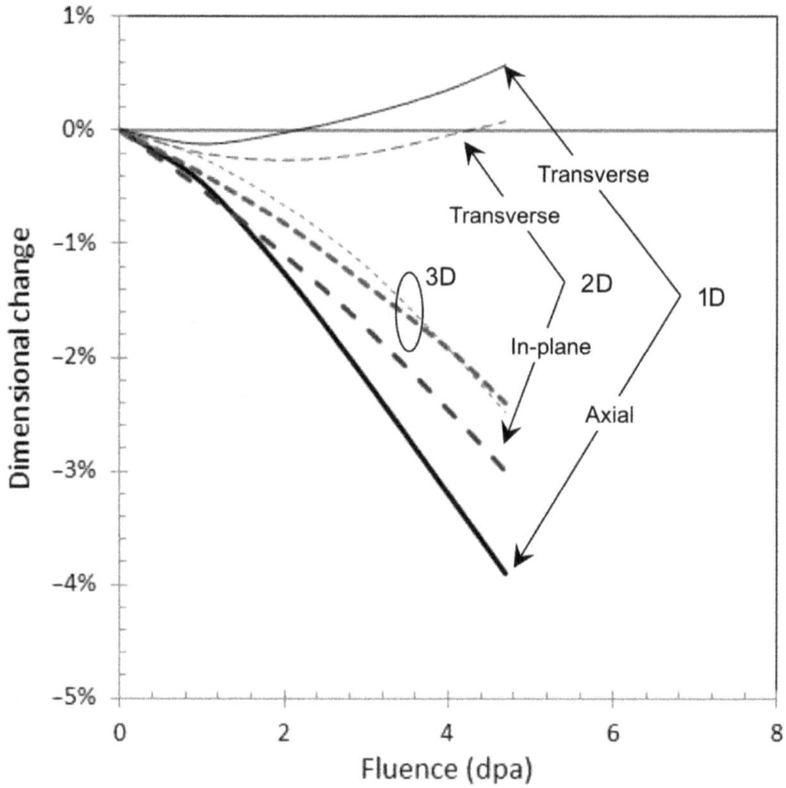

FIGURE 13.9
Dimensional evolutions of uni-directional (1D), two-dimensional (2D), and three-dimensional balanced-weave (3D) C/C composites during neutron irradiation at 600°C. Replotted from Reference 48.

reason. Because the atomic displacement damage creates new interstitial basal planes while shrinking the existing ones, the aligned microstructures make carbon fibers under energetic radiation increase in diameter and decrease in axial length. The degree, rate, and temperature dependence of these dimensional changes will vary depending on the carbon fiber grade.

This characteristic dimensional evolution for carbon fibers is reflected in dimensional changes for C/C composites. The radiation-induced dimensional instability is indeed among the primary design-limiting factors for nuclear applications of C/C composites. Examples of radiation-induced dimensional evolutions for C/C composites in a few different architectures are shown in Figure 13.9 [48]. The uni-directional composite is considered to undergo deformation almost exactly following the fiber deformation. Two-dimensional C/C composites expand in thickness as the fiber diameter enlarges and shorten in in-plane orientations during irradiation. However, the extent of in-plane contraction is determined in a complex manner by the fibers' axial shrinkage and the diametral expansion mechanically constraining each other. The dimensional changes in three-dimensional C/C composites are obviously more complex. Snead et al. report the significant dependence on the absolute dimension of the relative dimensional changes in an irradiated balanced three-dimensional weave composite due to the reduced constraint effect at near surface [9,49,50].

As the carbon fibers increase strength and elastic modulus during neutron irradiation until detrimental radiation effects start to impose toward the end of useful life, so do the C/C composites, as shown in Figure 13.10 [50,51]. While the adequate strength in relation to the original strength is retained up to moderate fluence levels, it has been reported that the ductility of C/C composites deteriorate at significantly lower fluences [52]. Although the exact reason for the radiation-induced ductility degradation of C/C composites has not clearly been demonstrated, the possible explanations include the significant *in situ* tensile loading to the fibers due to the axial shrinkage of fibers themselves and enhanced chemical bonding between the fibers and matrix.

In radiation environments, C/C composites also exhibit thermal conductivity decrease [48]. The thermal conductivity decrease is caused not only by the radiation-induced fragmentation of basal planes in the fiber and matrix but also by the extensive matrix microcracking due to the anisotropic fiber deformation. An example of experimentally recorded effect of neutron irradiation on thermal conductivity of C/C composite is shown in Figure 13.11. Unlike SiC/SiC composites that exhibit early saturation in thermal conductivity decrease, C/C composites' thermal conductivity continue decreasing after the initial drop, although the initial drop is often less severe than for SiC/SiC composites.

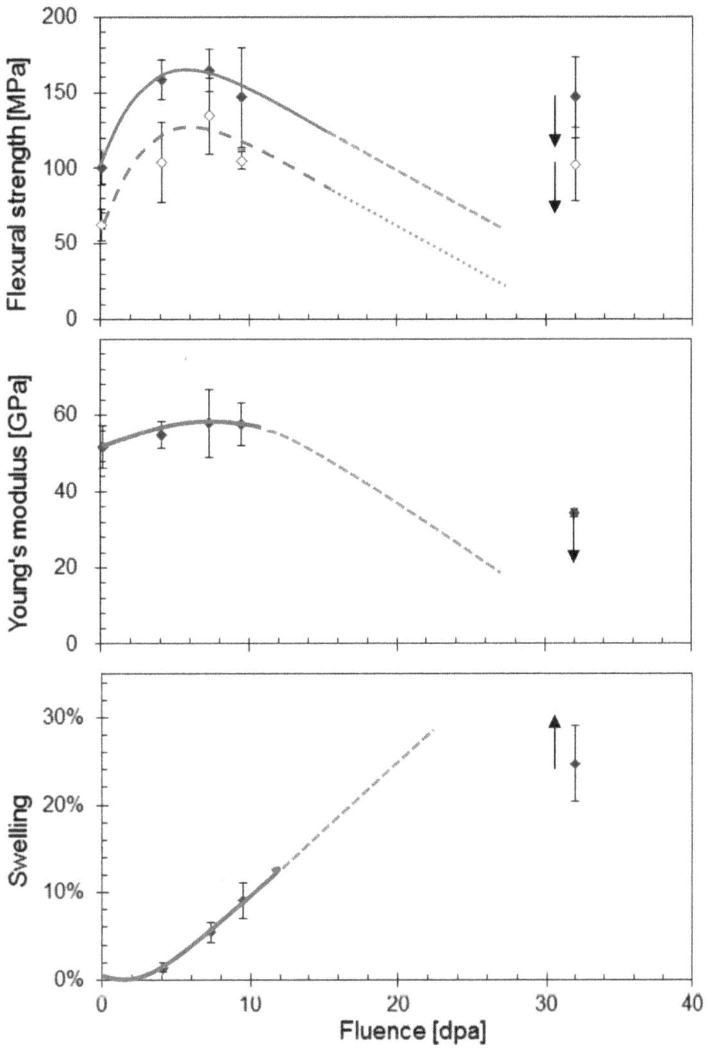

FIGURE 13.10

Neutron fluence-dependent evolutions of volumetric swelling, strength and elastic modulus of FMI-222 pitch-fiber, three-dimensional balanced-weave C/C composite. Irradiation temperature was ~800°C. (Replotted from Reference 50).

13.5 CONCLUDING REMARKS AND FUTURE DIRECTIONS

Energy security associated with geopolitics, international relations, economy, and environment protection are among the most critical issues in modern and future world. Ceramic matrix composites, in particular SiC/SiC composites, are very important materials that may enable the innovative safety features of nuclear fission energy systems and future fusion energy modules that may significantly influence the energy strategy of many nations and may possibly resolve the energy issue. Properly addressing the long-term environmental degradation, including the radiation effects, is of critical importance for the future of these materials in the field of nuclear energy.

Although SiC/SiC composites are considered as the materials of outstanding radiation resistance, they can be further developed toward achievement of extended service life in harsh radiation environments. While radiation-induced mechanical degradation has not been reported for high purity CVD SiC, radiation and environmental resistance of the current generation near-stoichiometric SiC fibers is limited probably by the presence of excess carbon and undesired impurities. The carbon-based interphase is the weakest link for the environmental degradation in most applications

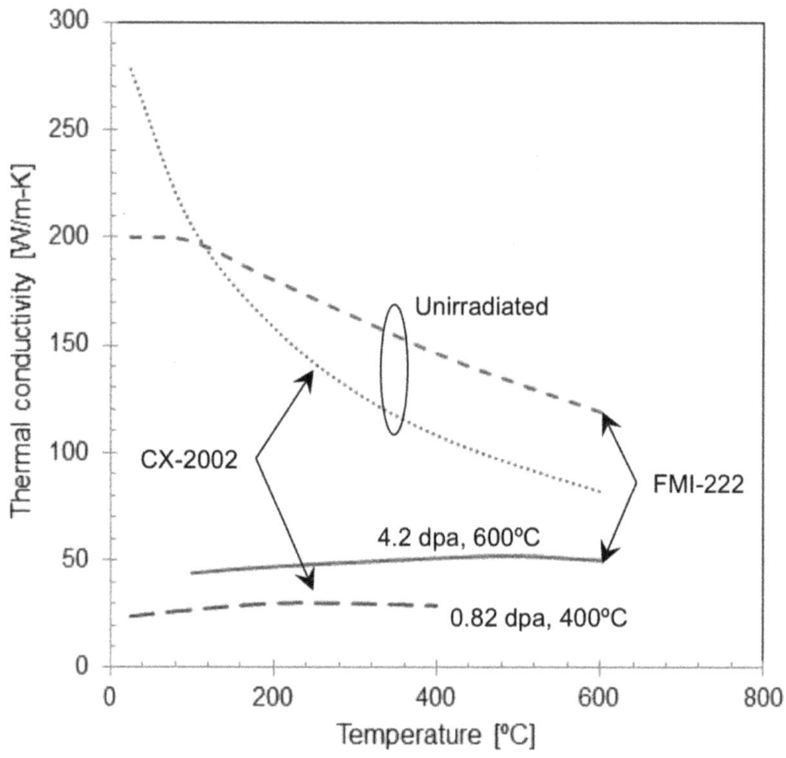

FIGURE 13.11
Thermal conductivity change in C/C composite after neutron irradiation. Materials are FMI-222 pitch fiber three-dimensional balanced-weave composite (data taken from Reference [48]) and CX-2002U chopped PAN fiber composite from Reference [53]).

and the radiation degradation at high doses. Development of SiC fibers and interphases with enhanced radiation and environmental stability will be the breakthrough for SiC/SiC composites for these applications.

Finally, SiC is currently recognized to be uniquely radiation-tolerant, partly because not many ceramic systems have been studied for radiation effects. There will be opportunities of search for other ceramic systems and syntheses of novel ceramics that may offer radiation resistance comparable with or superior to SiC. Once such ceramic systems are identified, it is a role of ceramic composite scientists to make them into structural materials.

ACKNOWLEDGMENT

Research sponsored by the US Department of Energy under contract DE-C05-00OR22725 with UT-Battelle, LLC, at Oak Ridge National Laboratory.

REFERENCES

1. Y. Katoh et al. Radiation effects in SiC for nuclear structural applications. *Curr. Opin. Solid State Mater. Sci.*, 16(3), 143–152 (2012).
2. L. L. Snead et al. Ceramic composites for near term reactor application, in *The 4th International Topical Meeting on High Temperature Reactor Technology (HTR-2008)* (ASME, Washington, DC, 2008).
3. L. Chaffron and J.-L. Se'ran. *Innovative SiCf/SiC Composite Materials for Fast Reactor Applications, Vol. 102* (Transactions of the American Nuclear Society, 2010), pp. 905.
4. C. W. Forsberg, P. F. Peterson, and R. A. Kochendarfer, Design options for the advanced high-temperature reactor. International Conference on Advances in Nuclear Power Plants, ICAPP 2008, June 8–12, 2008, pp. 733–749, Anaheim.
5. S. J. Zinkle and A. Kohyama. Advanced materials for fusion energy. 19th IEEE/IPSS Symposium on Fusion Engineering, January 21–25, 2002, Atlantic City.

6. Y. Katoh et al. Current status and critical issues for development of SiC composites for fusion applications. *J. Nucl. Mater.*, 367, 659–671 (2007).
7. L. L. Snead et al. Handbook on SiC properties for fuel performance modeling. *J. Nucl. Mater.*, 371, 329–377 (2007).
8. L. L. Snead, Y. Katoh, and T. Nozawa. Radiation effects in SiC and SiC-SiC, in *Comprehensive Nuclear Materials*, edited by R. J. M. Konings (Elsevier, Amsterdam, 2012), pp. 215–240.
9. L. L. Snead, T. D. Burchell, and Y. Katoh. Swelling of nuclear graphite and high quality carbon fiber composite under very high irradiation temperature. *J. Nucl. Mater.*, 381, 55–61 (2008).
10. S. J. Zinkle and C. Kinoshita. Defect production in ceramics. *J. Nucl. Mater.*, 251(11), 200–217 (1997).
11. G. H. Kinchin and R. S. Pease. The displacement of atoms in solids by radiation. *Rep. Prog. Phys.*, 18, 1–51 (1955).
12. M. J. Norgett, M. T. Robinson, and I. M. Torrens. A proposed method of calculating displacement dose rates. *Nucl. Eng. Des.*, 33(1), 50–54 (1975).
13. C. H. Xu, C. L. Fu, and D. F. Pedraza. Simulations of point-defect properties in graphite by a tight-binding-force model. *Phys. Rev. B*, 48(18), 13273–13279 (1993).
14. M. Bockstedte, A. Mattausch, and O. Pankratov. Ab initio study of the migration of intrinsic defects in 3C-SiC. *Phys. Rev. B*, 68, 205201 (2003).
15. F. A. Garner and D. S. Gelles. Irradiation creep mechanism: an experimental perspective. *J. Nucl. Mater.*, 159, 286–309 (1988).
16. T. Yano et al. Neutron irradiation effects on isotope tailored aluminum nitride ceramics by a fast reactor up to 2 x l026 n/m2. *J. Nucl. Mater.*, 329–333, 1471–1475 (2004).
17. R. B. Matthews. Irradiation damage in reaction-bonded silicon carbide. *J. Nucl. Mater.*, 51, 203–208 (1974).
18. R. J. Price. Properties of silicon carbide for nuclear fuel particle coatings. *Nucl. Technol.*, 35, 320–336 (1977).
19. R. J. Price and J. C. Bokros, Mechanical properties of neutron-irradiated pyrolytic carbons. *J. Nucl. Energy*, 21, 158–174 (1967).
20. P. T. Heald. The preferential trapping of interstitials at dislocations. *Philos. Magazine*, 31(3), 551–558 (1975).
21. L. L. Snead, Y. Katoh, and S. Connery. Swelling of SiC at intermediate and high irradiation temperatures. *J. Nucl. Mater.*, 367, 677–684 (2007).
22. S. Kondo et al. High temperature ion-irradiation effects on microstructural evolution in b-SiC. *Fus. Sci. Technol.*, 44, 181–185 (2003).
23. Y. Katoh and L. L. Snead. Mechanical properties of cubic silicon carbide after neutron irradiation at elevated temperatures. *J. ASTM Internat.*, 2(8), 12377-1-13 (2005).
24. S. Kondo, Y. Katoh, and L. L. Snead. Concentric ring on ring test for unirradiated and irradiated miniature SiC specimens. *J. Nucl. Mater.*, 417(1–3), 406–410 (2011).
25. Y. Katoh et al. Observation and possible mechanism of irradiation induced creep in ceramics. *J. Nucl. Mater.*, 434, 141–151 (2013).
26. B. T. Kelly. The behavior of graphite under neutron irradiation. *J. Vac. Sci. Technol.*, A4(3), 1171–1178 (1986).
27. S. Ishiyama et al. The effect of high fluence neutron irradiation on the properties of a fine-grained isotropic nuclear graphite. *J. Nucl. Mater.*, 230, 1–7 (1996).
28. G. W. Hollenberg et al. The effect of irradiation on the stability and properties of monolithic silicon carbide and SiCf/SiC composites up to 25 dpa. *J. Nucl. Mater.*, 219, 70–86 (1995).
29. L. L. Snead and E. Lara-Curzio. Interphase integrity of neutron irradiated SiC composites. *Mater. Res. Soc. Symp. Proc.*, 540, 273–278 (1999).
30. Y. Katoh et al. Neutron tolerance of advanced SiC-Fiber/CVI-SiC composites. 19th IAEA Fusion Energy Conference, October 14–19, 2002, Lyon.
31. L. L. Snead et al. Evaluation of neutron irradiated near stoichiometric silicon carbide fiber composites. *J. Nucl. Mater.*, 283, 551–555 (2000).
32. K. Ozawa et al. Mechanical properties and microstructure of advanced SiC/SiC composites after neutron irradiation. *J. Nucl. Mater.*, 367, 713–718 (2007).
33. A. Kohyama. Present status of NITE-SiC/SiC for advanced nuclear energy systems. 31st International Conference on Advanced Ceramics and Composites, January 21–26, 2007, Daytona Beach.
34. Y. Katoh et al. *Properties of Hi-Nicalon Type-S CVI SiC Composite Irradiated to 70 dpa at Elevated Temperatures* (Oak Ridge National Laboratory, 2012).
35. Y. Miwa et al. Swelling of F82H irradiated at 673 K up to 51 dpa in HFIR. *J. Nucl. Mater.*, 283–287, 334–338 (2000).
36. F. A. Garner. Irradiation performance of cladding and structural steels in liquid metal reactors, in *Materials Science and Technology, A Comprehensive Treatment, Nuclear Materials, Part 1*, edited by R. W. Cahn, P. Haasen, and E. J. Kramer (VCH Verlagsgesellschaft mbH, Weinheim, 1994), pp. 419–543.

37. T. Yano et al. Neutron irradiation damage in aluminum oxide and nitride ceramics up to a fluence of 4.2×1026 n/m2. *J. Nucl. Mater.*, 283–287, 947–951 (2000).
38. Y. Katoh et al. Thermophysical and mechanical properties of near-stoichiometric fiber CVI SiC/SiC composites after neutron irradiation at elevated temperatures. *J. Nucl. Mater.*, 403(1–3), 48–61 (2010).
39. T. Ishikawa et al. Experimental stress/strain behavior of SiC-matrix composites reinforced with Si-Ti-C-O fibers and estimation of matrix elastic modulus. *Comp. Sci. Technol.*, 58, 51–63 (1998).
40. Y. Katoh et al. Property tailorability for advanced CVI silicon carbide composites for fusion. *Fus. Eng. Des.*, 81, 937–944 (2006).
41. G. A. Newsome et al. Evaluation of neutron irradiated silicon carbide and silicon carbide composites. *J. Nucl. Mater.*, 371, 76–89 (2007).
42. A. G. Evans and D. B. Marshall. The mechanical behavior of ceramic matrix composites. *Acta Metall.*, 37(10), 2567–2583 (1989).
43. W. A. Curtin. Theory of mechanical properties of ceramic-matrix composites. *J. Am. Ceramic Soc.*, 74(11), 2837–2845 (1991).
44. T. Nozawa et al. Strength and microstructures of the fiber/matrix interface of neutron-irradiated SiC/SiC composites. 15th International Conference on Fusion Reactor Materials, October 16–21, 2011, Charleston.
45. T. Nozawa, Y. Katoh, and L. L. Snead. The effect of neutron irradiation on interfacial shear properties of multilayered interphase SiC/SiC composite. *J. Nucl. Mater.*, 367, 685–691 (2007).
46. T. Nozawa, Y. Katoh, and L. L. Snead. The effect of neutron irradiation on the fiber/matrix interphase of silicon carbide composites. *J. Nucl. Mater.*, 384(3), 195–211 (2009).
47. L. L. Snead et al. Candidate structural materials for in-core VHTR application. *Trans. Am. Nucl. Soc.*, 98, 1019–1020 (2008).
48. T. D. Burchell. Radiation damage in carbon-carbon composites: structure and property effects. *Phys. Scripta*, T64, 17–25 (1996).
49. L. L. Snead, T. D. Burchell, and A. L. Qualls. Strength of neutron-irradiated high-quality 3D carbon fiber composites. *J. Nucl. Mater.*, 321, 165–169 (2003).
50. L. L. Snead, Y. Katoh, and K. Ozawa. Stability of 3-D carbon fiber composite to high neutron fluence. *J. Nucl. Mater.*, 417(1–3), 629–632 (2011).
51. S. Sato et al. Thermal shock resistances and fracture toughnesses of graphites and C/C-composites as plasma-facing first wall components for fusion reactor devices. *Fus. Eng. Des.*, 13(2), 159–176 (1990).
52. Y. Katoh and L. L. Snead. *Post Irradiation Evaluation of SiC/SiC and C/C Composites After 10 DPA Exposure* (Oak Ridge National Laboratory, Oak Ridge, 2006).
53. T. Maruyama and M. Harayama. Neutron irradiation effect on the thermal conductivity and dimensional change of graphite materials. *J. Nucl. Mater.*, 195, 44–50 (1992).

CHAPTER 14

FOREIGN OBJECT DAMAGE IN CERAMIC MATRIX COMPOSITES

SUNG R. CHOI

Naval Air Systems Command, Patuxent River, MD, USA

14.1 INTRODUCTION/BACKGROUND

The aeroengines, particularly advanced aeroengines, are aimed to achieve improved power and specific fuel efficiency and lower emission. These goals can be achieved through higher turbine-inlet temperature, reducing cooling air, and using much lighter propulsion materials. In a material's standpoint, ceramic matrix composites (CMCs) are the only enabling materials to meet those goals. Because of this strategic standing of the materials, a large amount of works/investments has been made in CMCs for advanced aeroengine applications via government, academia, and industry alike. However, there are still persistent and unsolved issues of CMCs in advanced engine applications particularly when harsh engine operational conditions/missions are considered. The issues are related to reliability/durability, environmental stability, and manufacturing/affordability, all affecting the transition of CMCs into aeroengine hot-section components.

One of those issues that hinders widespread CMC acceptance into aeroengines is associated with the degradation of CMCs with impact damage by small objects such as sand, metallic, or other particles ingested into or formed within aeroengines. The degradation is attributed to the inherent brittle nature of CMCs, as compared with a ductile nature of metals. Since CMCs are brittle and since some sections of CMC components such as airfoils are very thin in their configurations, the impact generates a varying degree of damage ranging from localized surface damage/cracking to complete penetration, depending on the severity of the impact event. Therefore, FOD can result in premature component failure, significantly limiting the use of CMCs in aeroengine applications. From perspectives of durability, reliability, and affordability, it is very important to understand and assess appropriately the effects of FOD in CMCs, with which pertinent component and material designs can be assured preventing from undesirable, premature failure.

Extensive work on impact damage has been done in monolithic ceramic materials for the last several decades [1–24], including armor ceramics recently [e.g., 25–27]. The same is true for polymer matrix composites [e.g., 28–33]. However, this is not the case for CMCs since the materials are still relatively new and have various practical issues particularly in reliability, manufacturability, and affordability (cost). Notwithstanding, a span of FOD work in CMCs [34–61] exists and is still being conducted to better understand their FOD phenomena via assessment and simulation efforts. This paper, primarily compiled from a series of previous FOD work [41–60], presents a summary of FOD in gas-turbine grade SiC/SiCs and oxides/oxides CMCs, focusing on materials' impact responses with regard to target support, velocity, impact morphology, residual strength, temperature, preloading, and environments (CMAS and Salt). Although important and under active progress, FOD modeling will not be described in this paper because of the scope of the paper. Comparison in FOD between gas-turbine grade silicon nitrides and current CMCs will also be briefly discussed.

Ceramic Matrix Composites: Materials, Modeling and Technology, First Edition. Edited by Narottam P. Bansal and Jacques Lamon.

14.2 EXPERIMENTAL TECHNIQUES

14.2.1 FOD Test Rig

Typically there are two representing FOD test methods used in CMCs: one is to utilize ballistic impact gun [36–38,41–61] and the other is to use drop weight/hammer impact tester [34,35,40]. In both cases, pertinent projectiles or impactors are used to achieve their respective test conditions. In some cases, a quasi-static indentation is also used using spherical or any shaped indenters to simulate low velocity impact conditions [39,40]. For aeroengine applications, ballistic impact gun has been exclusively used because of its efficacy to achieve a wide range of impact conditions reaching up to or over Mach 1. Consequently, this paper will describe impact phenomena of CMCs subject to ballistic impact.

A schematic of a ballistic impact gun is shown in Figure 14.1 [42,43]. A hardened (HRC $\geq$ 60) chrome steel (or any type of materials) spherical projectiles with a diameter of 1.59 mm is inserted into a 300- to 500-mm-long gun barrel of an inner diameter of 1.59 mm. A 2000-psi helium gas cylinder and relief valves are used to pressurize the reservoir (pressure vessel) to a specific level, depending on a prescribed impact velocity to be used. Upon reaching a specific level of pressure, a solenoid or any quick-acting valve is instantaneously opened accelerating a steel-ball projectile through the gun barrel to impact a target specimen. The target specimen is supported in various support modes depending on requirements, which will be described in the later sections. Exit velocity of each projectile leaving the barrel is determined using two pairs of laser transmitter and receiver, incorporated with two holes at the end of the gun barrel. The exit velocity is also referred to as a *nominal* impact velocity. It is also feasible to determine the impact velocity right before a projectile hits the target. The range of impact velocity employed in the work was from 100 to 400 m/s. For conservative estimations, normal incidence impact angle has been widely used as was the case in this program.

14.2.2 Materials: CMC and Projectile Materials

Basic properties of the target materials of CMCs used in the program are summarized in Table 14.1 and have also been described elsewhere [42, 43, 45, 59]. The state-of-the-art, melt-infiltrated (MI) SiC/SiC CMC used Sylramic™ fibers, woven into 2D 5 harness-satin cloth and converted to Sylramic™ iBN fibers. The Sylramic™ iBN cloth was 0/90°, 8 ply-stacked and chemically vapor infiltrated with a thin BN-based interface coating followed by SiC matrix over-coating. Remaining matrix porosity was filled with SiC particulates and then with molten silicon at 1400°C [62]. The second SiC/SiC CMC used was a commercial, prepreg MI SiC/SiC, manufactured by GE Ceramic Composite Products. The composite was fabricated in a proprietary process with Nicalon™ Type S fibers with a fiber volume fraction of about 0.30. The fiber interface coating was presumably in BN. The CMC was 0/90° cross-plied with a total of 8 plies.

The oxide/oxide CMCs used Nextel™ 720 oxide fibers, woven into 2D 8 harness-satin cloth. The cloth was cut into a proper size, slurry-infiltrated with aluminosilicate or alumina, and 0/90°, 12 ply-stacked followed by consolidation and

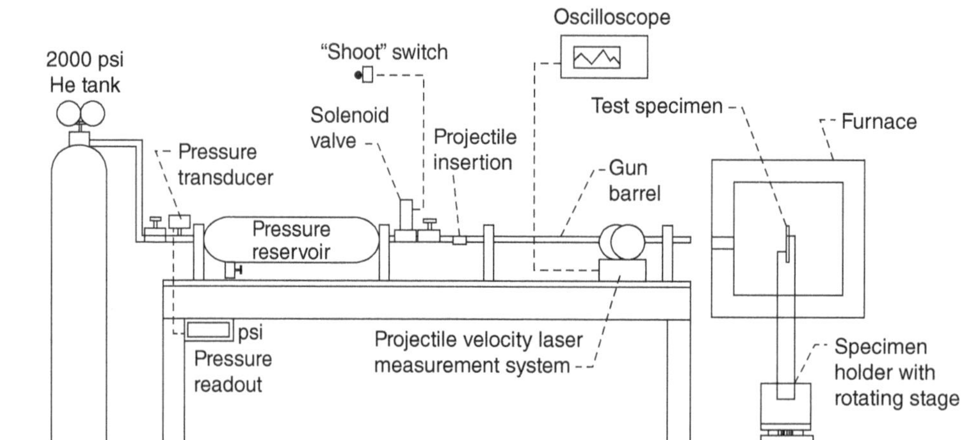

FIGURE 14.1
Ballistic impact gun used in the program [42, 43]. From J. Am. Ceram. Soc.

TABLE 14.1 Basic Physical and Mechanical Properties of Target CMCs and Projectiles Used in the Program [42,43,45,59]

Materials		Architecture	Fiber/matrix	Fiber volume fraction	Bulk density (g/cm^3)	Elastic modulus (GPa)	Flexure strength (MPa)	Manufacturer
CMC Targets	MI SiC/SiC	2D woven; 0/90°; 5 HS; 8 plies; $t = 2.2$ mm	Sylramic™ SiC/SiC	0.34	2.36	197	578 ± 56	GE PSC (Newark, DE)
	MI SiC/SiC	2-D Prepreg 0/90°; 8 plies; $t = 2.2$ mm	Hi-Nic Type S/SiC	0.30	3.0	300	550 ± 43	GE CCP
	Oxide/oxide	2D woven; 0/90°; 8 HS; 12 plies; $t = 3.0$ mm	Nextel™ 720/ aluminosilicate	0.45	2.55	67	141 ± 4	GE
	Oxide/oxide	2D woven; 0/90°; 8 HS; 12 plies; $t = 3.0$ mm	Nextel™ 720/alumina	2.74	0.45	81	140–145	ATK/COIC
Projectiles	Chrome steel (SAE 52100), hardened	–	–	–	7.78	200	–	–

sintering. No interface fiber coating was employed in the oxides/oxides. The porosity of both oxides/oxides was around 20%. Note that the elastic moduli of the oxides/oxides were significantly lower as 70–80 GPa than those (200–300 GPa) of the SiC/SiC counterparts, primarily attributed to the substantial level of porosity present in the oxides/oxides CMCs. Typical microstructures of the Sylramic MI SiC/SiC and the N720/aluminosilicate CMCs are shown in Figure 14.2.

Different CMCs in architectures and materials have been also used by other investigators, which include 2D woven mullite-fiber/mullite-matrix [36], 2D woven carbon-fiber/SiC [35], 2D cross-plied carbon-fiber/glass-LAS (lithium aluminosilicate)-matrix [34], 2D woven SiC/SiCs [38], 3D woven SiC/SiCs [37,39,40].

Typical projectiles were of chrome steel (SAE 52100) hardened to a Vickers hardness of 8.2 GPa [22,23], typically used as rolling elements of ball bearings in Grade 5. Any appropriate projectile materials, ranging from tungsten carbide to chrome steels to brass to silicon nitride to glass to graphite, can be used to determine their specific FOD/impact responses to any target materials.

14.2.3 Configuration and Support of CMC Targets

CMC target specimens were cut into rectangular plates of 8–12 mm in width and 45–50 mm in length or into tensile test specimens of 10 mm width by 150 mm length, depending on the test matrices and materials. The target thickness was

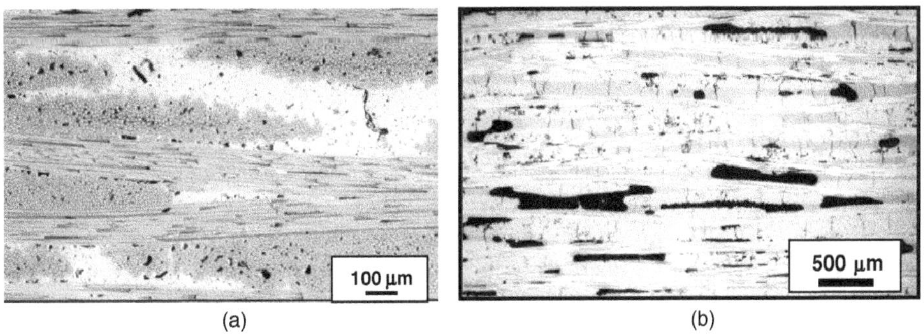

FIGURE 14.2

Typical microstructures of (a) MI Sylramic/SiC [42] and (b) oxide/oxide (N720/aluminosilicate) [46]. CMCs presented in this work.

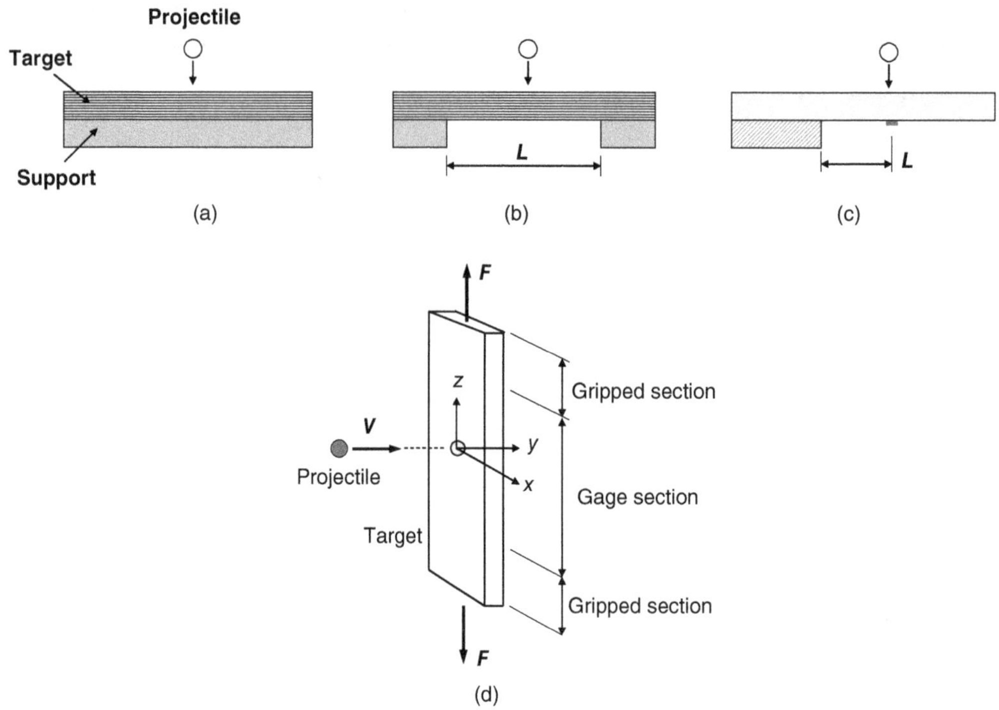

FIGURE 14.3

Different types of target supports used: (a) full support; (b) partial support; (c) cantilever support; (d) free-standing support under tensile loading. The span in (b) and (c) is $L = 20$ mm.

used as as-furnished with about 2.2 mm for MI SiC/SiCs or 3.0 mm for oxides/oxides. The use of the small rectangular plate specimens was primarily due to constraints of materials related to cost. Ballistic impact is stochastic in nature particularly to brittle materials, so increased number of specimens per condition has to be used for improved data reproducibility.

In FOD testing, usually two different types of target supports—full and partial supports—were employed to take accounts of target support effects, as shown in Figure 14.3. For the partial support, the span L was kept to be $L =$ 20 mm. The targets were supported in a rigid steel block. Two additional target supports were a cantilever support with $L = 20$ mm [59, 60] and a free-standing support with the target under tensile loading during impact [53]. The use of those target supports was to simulate the structural/operational configurations of gas-turbine airfoil components such as blades, vanes, and shrouds, etc. Square target plates with their sides/peripheries being rigidly fixed have been also used in impact testing for SiC/SiCs by other researchers [37–40].

Usually three target specimens were used for a given impact condition of velocity and target support. Three to four different impact velocities were typical of many experiments together with use of normal incidence angle.

14.2.4 Impact Damage Assessments

It has been observed that residual strength estimation for impacted CMC targets provided a very convenient and readily quantitative way to assess the degree of overall impact damage associated [42, 43]. Other approaches to impact damage assessments include optical or scanning electron microscopy, 3D topology, radiology, computed tomography, and/or pulsed thermography [47]. Recently, an electrical resistivity (ER) method has been introduced experimentally and analytically to provide the nature and degree of impact damage where damage is related to the level of electrical resistivity in electrically conducting CMCs such as SiC/SiCs [57, 58]. This ER technique, if further developed, may provide a way to *in situ* monitor the initiation and accumulation of impact damage (or any defects) of structural CMC components.

It is also recommended that projectiles after impact events be retrieved to get further information regarding how impact energy is associated with the impact event with regard to energy dissipation and its deformation or fracture.

Postimpact residual strength testing is typically conducted in four-point flexure with 20-mm inner and 40-mm outer spans [42,43] or in tension [53] using an appropriate test frame at a test speed of 0.1–0.5 mm/min. In testing, impact damage is placed at the center of the tensile side of a flexure target specimen. As-received strength of the composites is also determined for a baseline reference.

In situ visualization of impact event can be achieved via appropriate high speed imaging or strain visualization to provide more detailed information on impact dynamics such as restitution of projectiles or strain history during impact, etc.

14.3 PHENOMENA OF FOREIGN OBJECT DAMAGE IN CMCs

14.3.1 Residual Strength

The results of postimpact (or residual) strength testing after impact for the four different CMCs in Table 14.1 are presented in Figure 14.4, where postimpact strength was plotted as a function of impact velocity for two different target supports, full and partial supports [42,43,59,60]. Included in the figure are the respective as-received ("As-R") strengths of the four CMCs. All the specimens impacted did fracture from impact sites. Postimpact strength decreased with increasing impact velocity, regardless of the type of materials, attributed to increased impact damage with increasing

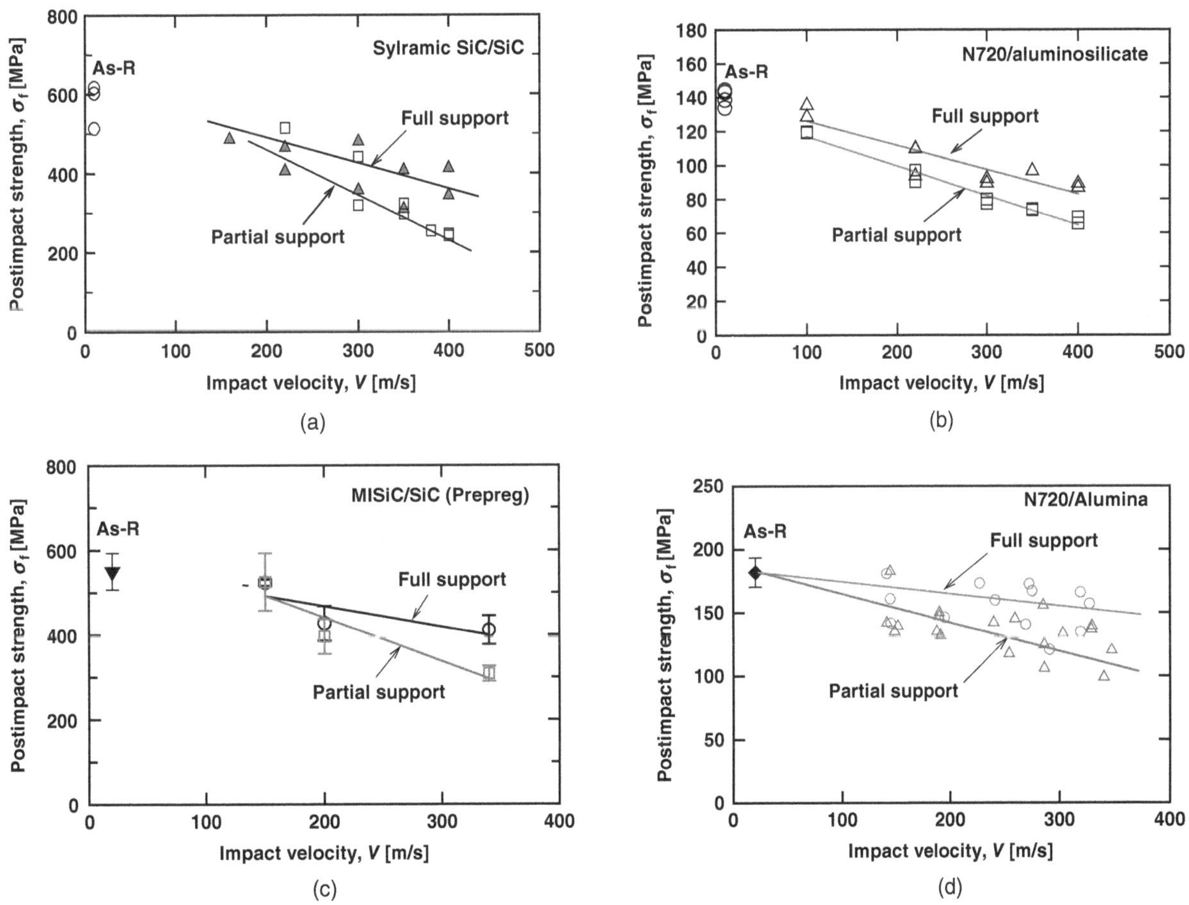

FIGURE 14.4

Summary of postimpact strength as a function of impact velocity in full and partial supports, impacted by 1.59-mm-diameter steel ball projectiles at ambient temperature in four different CMCs: (a) MI Sylramic SiC/SiC [42]; (b) N720/aluminosilicate [43]; (c) prepreg MI SiC/SiC [59]; (d) N720/alumina [60].

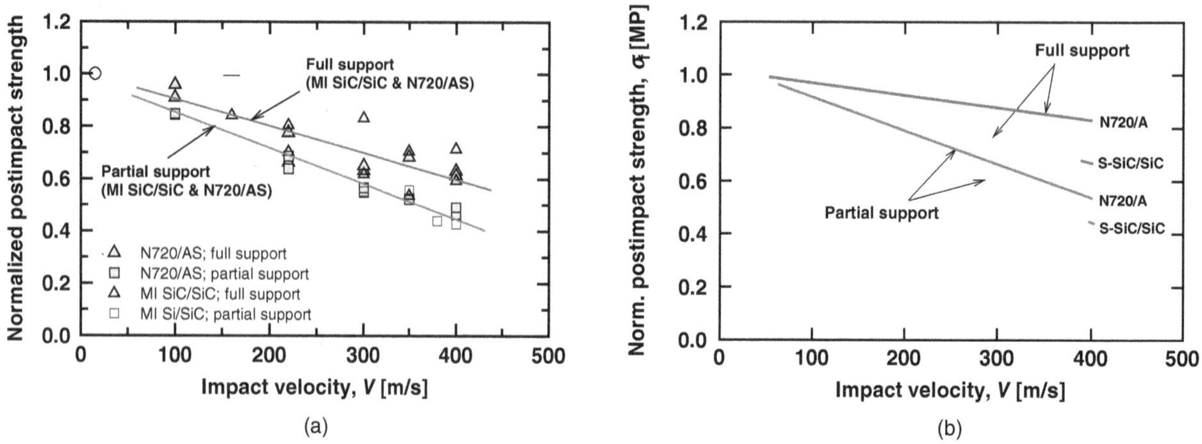

FIGURE 14.5

Normalized postimpact strength as function of impact velocity in full and partial supports, reconstructed from the data in Figure 14.4: (a) MI Sylramic SiC/SiC and N720/aluminosilicate; (b) prepreg MI SiC/SiC and N720/alumina. The normalized impact strength is the impact strength (data in Figure 14.4) normalized with respect to as-received strength of a target CMC.

impact velocity. Invariably, for a given impact velocity, the strength degradation was greater in partial support than in full support, indicating that greater damage occurred in partial support and might be caused primarily by the presence of tensile stresses on the backside of target specimens.

Figure 14.5 compares normalized strength, that is, postimpact strength normalized with respect to as-received strength, as a function of impact velocity for the four different CMCs presented in Figure 14.4. The normalized scheme can provide a quick and easy comparison of the degree of impact damage among different materials. As seen from the figure, for a given type of target support, the degree of strength degradation was almost identical in either MI Sylramic SiC/SiC or N720/aluminosilicate. The strength degradation at 400 m/s, for example, was about 40% and 60%, respectively, in full and partial supports. The strength degradation for prepreg MI SiC/SiC and N720/alumina was around 20% and 40% in full support, and was about 50% and 60% in partial support. This indicates that N720/alumina exhibited the least degradation, while the other three CMCs showed an almost similar degradation. One thing to note is that postimpact strength tests in flexure may not be accurately representing the overall impact damage since the calculated stresses were based on the elementary beam theory and since the volume subjected to tensile stresses may not be sufficient. Although tensile tests are favored, a major concern still arises due to a need of substantially increased material cost, particularly when one considers impact response in brittle solids to be significantly stochastic (which requires a reasonably large number of target specimens).

Figure 14.6 shows an example of the load–displacement curves determined for N720/aluminosilicate from postimpact strength testing in full and partial supports [43]. The displacements were determined using an LVDT (linear variable differential transformer) by placing it on the tensile side of a flexure target specimen. As impact velocity increased, both stiffness and strength reduced. For a given impact velocity, the decrease in stiffness was much greater in partial support than in full support, due to substantially increased damage accompanying additional backside cracking and delamination in partial support. This backside damage will be described in Section 14.3.2. The shape of load–displacement curves is consistent to the outcome of residual strength as to how one aspect (damage) affects the others (strength and stiffness).

14.3.2 Impact Damage Morphology

14.3.2.1 Projectiles The hardened chrome steel ball projectiles differed significantly in their response to ballistic impact, depending on the impact velocity and type of CMC materials. The steel ball projectiles impacting the MI SiC/SiCs were severely deformed (flattened) at impact velocities $\geq$350 m/s. By contrast, the projectiles impacting the oxides/oxides exhibited no visual significance in damage or plastic deformation even at the highest impact velocity of 400 m/s, as seen in Figure 14.7 [41,46]. Of course, the degree of projectile deformation in the SiC/SiCs depended on impact velocity: the higher impact velocity yields the greater projectile deformation, and *vice versa*. In the case of the

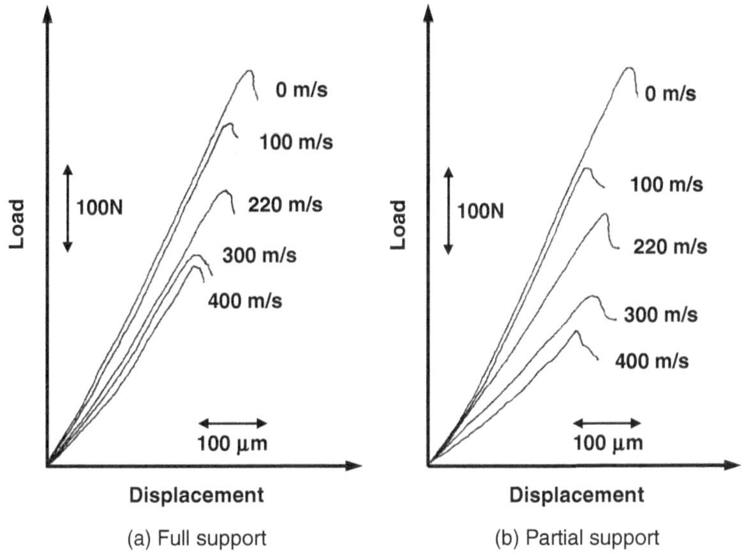

FIGURE 14.6
Examples of load versus displacement curves of in flexure testing for N720/aluminosilicate impacted by 1.59-mm-diameter steel ball projectiles at different impact velocities: (a) full support and (b) partial support [43].

oxides/oxides, projectiles were embedded into target specimens at higher impact velocities ≥350 m/s. This is attributed to the materials' *open* and "*soft*" structures. A similar case can be found in ductile metals (Ni-based superalloys) impacted by sharp brittle particles [63].

The difference in projectile impact response was due to difference in elastic modulus of the target materials, associated with the varying degree of Hertzian contact stresses occurring at the impact sites. The two SiC/SiCs exhibited higher elastic modulus ($E = 200$–300 GPa) than the oxides/oxides ($E = 70$–80 GPa) and therefore, the former was harder than the latter, thus creating greater contact stresses and consequently greater impact damage. The converse is also true in which a target material is fixed while projectile material is varied in property (i.e., hardness). In a previous study, four different projectile materials of silicon nitride, hardened steel, annealed steel, and brass were used to determine their damage effects on gas-turbine grade silicon nitride targets. The hardness of projectile materials was found to be a key material parameter affecting most the degree of impact damage of the targets [22, 23].

14.3.2.2 Front Impact Damage The front impact damages generated in target specimens, in general, were in the forms of indents, craters (spallation), fiber-tows/matrix breakage, and/or layer-to-layer breakage. Of course, the severity of impact damage increased with increasing impact velocity, also depending on the type of CMCs. For a given impact velocity, the dependency of front impact damage on type of target supports was hardly distinguishable. Typical examples showing such damage features are shown in Figure 14.8, where the damage was generated at an impact velocity of 340 m/s in full support for prepreg MI SiC/SiC [59] and N720/alumina [49]. Well-developed crater and layer/tow damage can be seen in N720/alumina. By contrast, the SiC/SiC showed less-developed damage feature, compared to the oxide/oxide. However, when observed in detail, the SiC/SiC still showed fiber/matrix and layer-to-layer breakages observed from a 3D topological image, see Figure 14.9.

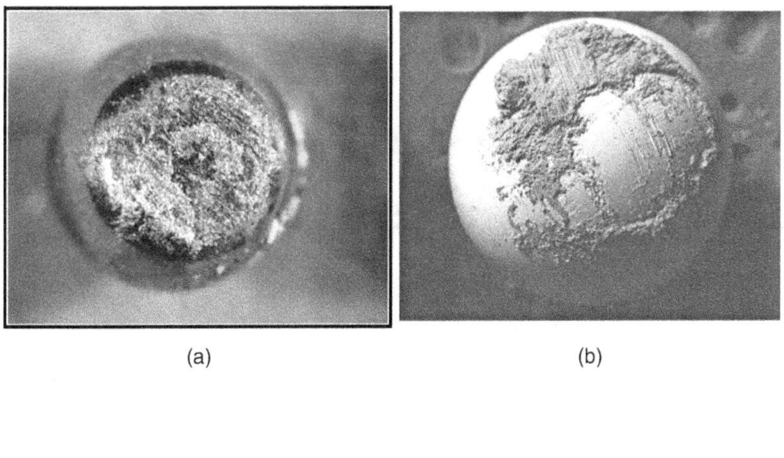

FIGURE 14.7
Steel ball projectiles (1.6-mm diameter) impacted at 350–400 m/s under full support onto: (a) MI Sylramic SiC/SiC [41] and (b) N720/aluminosilicate oxide/oxide [46]. Please note in (b) the occurrence of some material (fibers and matrix) transfer to the projectile from the target material.

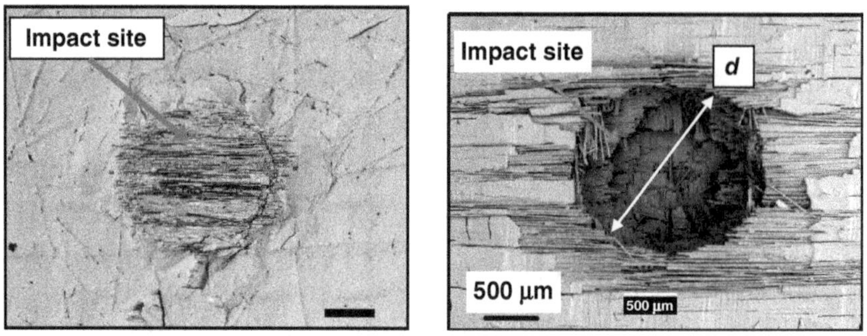

(a) MI SiC/SiC (b) N720/alumina

FIGURE 14.8

Typical examples showing overall front impact damage generated at 340 m/s under full support in (a) prepreg MI SiC/SiC [59] and (b) N720/alumina [49] impacted by 1.59-mm-diameter steel ball projectiles. The damage size *d* is defined in (b). Bar = 500 μm.

Since the impact damage of the oxides/oxides was reasonably symmetric and hemispherical in geometry (because of the materials' unique porous and soft microstructure), an attempt was made to determine the average damage diameter (*d*) and its depth (*z*). The result is presented in Figure 14.10, where all impact data from various impact velocities are included. Also included is an idealized geometrical relationship between the damage diameter *d* and the damage depth *z*

$$d^2 = 4Dz - 4z^2 \tag{14.1}$$

where *D* is diameter of ball projectiles (*D* = 1.59 mm). A general trend is evident such that the impact damage in diameter and depth increased with impact velocity. However, the impact damage in geometry (*d* and *z*) does not follow the idealized relationship of Equation (14.1), indicative of a violent and random nature of impact events occurring in the material. By contrast, the geometrical relations between *d* and *z* in the case of static indentation showed an excellent agreement with the idealized relationship [49].

Complete penetration by projectiles did not occur at 340–400 m/s in either the SiC/SiCs or the oxides/oxides CMCs. The slurry-cast MI SiC/SiCs (with Tyranno SA3 fibers/8-plies/5HS-satin and a fiber-volume fraction of 0.4), however, did show a complete penetration at an impact velocity of 340 m/s by 1.59-mm-diameter steel ball projectiles [57]. An example showing such a complete penetration is seen in Figure 14.11. Also included is a CT image of an MI Sylramic SiC/SiC coupon showing complete piercing by steel ball projectiles at 450 m/s [64].

Although the frontal impact damage is certainly one of important damage features in CMCs, it alone does not contribute to overall degree of degradation in composite strength. Overall impact damage in CMCs is rather dependent

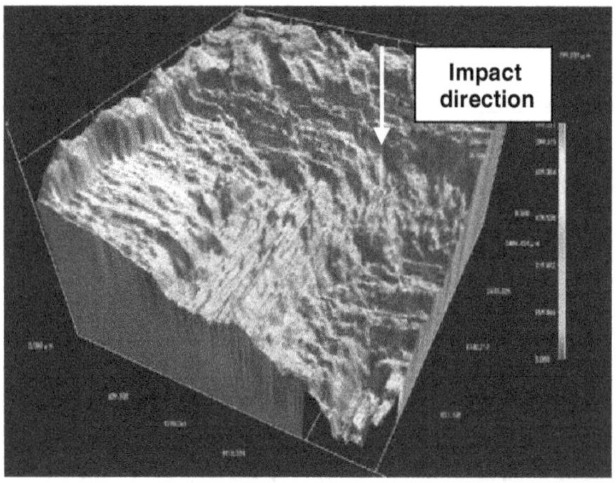

FIGURE 14.9

Typical example of a 3D topological image (one quarter) from an impact site generated at 340 m/s in prepreg MI SiC/SiC impacted by 1.59-mm-diameter steel ball projectiles (in cantilever support) [59]. The image was taken via optical microscopy using Model KH-7700, HiRox (Japan).

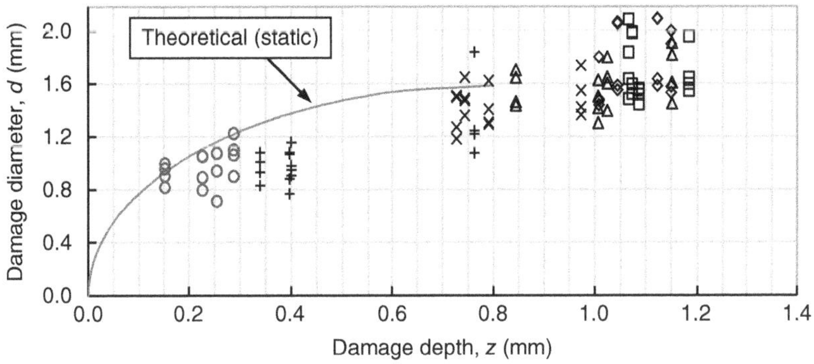

FIGURE 14.10

Relationship between impact damage diameter (*d*) and depth (*z*) in N720/alumina impacted by 1.59-mm-diameter steel ball projectiles from 100 to 340 m/s [49]. Included is a theoretical (idealized geometrical) relation of Equation (14.1).

on other features such as backside damage, internal cracking, and delamination, etc, which will be discussed in the next sections.

14.3.2.3 Backside Impact Damage As mentioned before, the frontal damage in CMCs was independent of the type of target supports for a given impact velocity. This was not the case for backside impact damage. Figures 14.12 and 14.13 show summaries of backside impact damage with respect to impact velocity in both N720/aluminosilicate [46] and prepreg MI SiC/SiC [59] impacted by 1.59-mm-diameter steel ball projectiles in full and partial supports. For N720/aluminosilicate, lower impact velocities $\leq$220 m/s generated indistinguishable backside damage between full and partial supports. However, a remarkable difference started at impact velocity $\geq$300 m/s: No damage was observed in full support; whereas significant damage was seen in partial support, primarily generated by the presence of backside tensile stresses upon impact. This appreciable damage was related to breakage and removal of fibers/matrix, fiber-tows fracture, and fiber-tows bending/protrusion toward the impact direction, as seen from the figures.

The prepreg MI SiC/SiC exhibited a similar but a little different response. The backside damage started at a low impact velocity of 150 m/s in a form of ring cracking in either full or partial support. At $\geq$200 m/s, well-developed backside cracking was formed in full support, while significant backside damage with multiple plies' spallation and scabbing took place in partial support. In terms of the degree of impact damage particularly at higher impact velocities $\geq$200 m/s, the backside damage was always greater in partial support than in full support in many structural CMCs [37–43,59,60]. The backside ring cracking, in fact, was a result of the formation and growth of a cone crack initiated from the front impact site. The backside damage in partial support was about five times greater than front impact damage, attributed to the formation of cone cracks as well as to target loading configuration and to ballistic dynamic effect (scabbing).

14.3.2.4 Cross Sections of Impact Damage Cross-sectional views of impact damage, although not always feasible and available in preparation, can provide very valuable information on the shape and size of craters, damage beneath impact sites, formation of cone-cracks, interlaminar delaminations, penetration, and backside damage including scabbing, etc. It can be also observed how overall damage progressed with increasing impact velocity. Examples are shown in Figures 14.14 and 14.15, respectively, for N720/aluminosilicate [46] and prepreg MI SiC/SiC [59], where the cross-sectional views of impact damage is depicted with respect to impact velocity in full and partial supports.

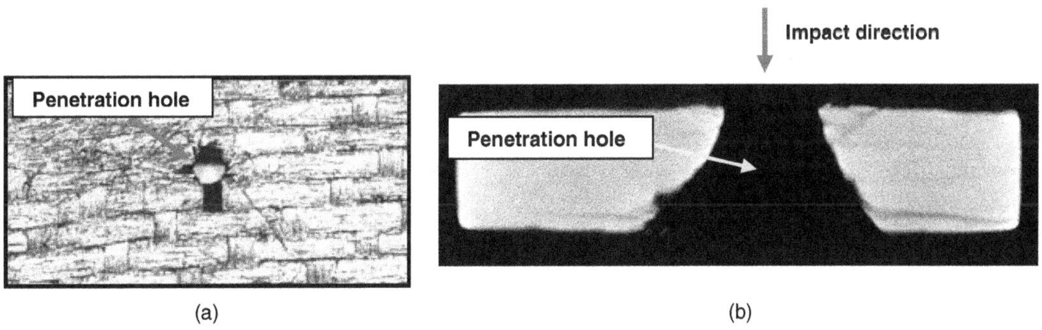

FIGURE 14.11

Examples of complete penetration by 1.59-mm-diameter steel ball projectiles: (a) a top view of MI Tyranno SA SiC/SiC [57] impacted at 340 m/s; (b) a CT image of MI Sylramic SiC/SiC impacted at 450 m/s [64].

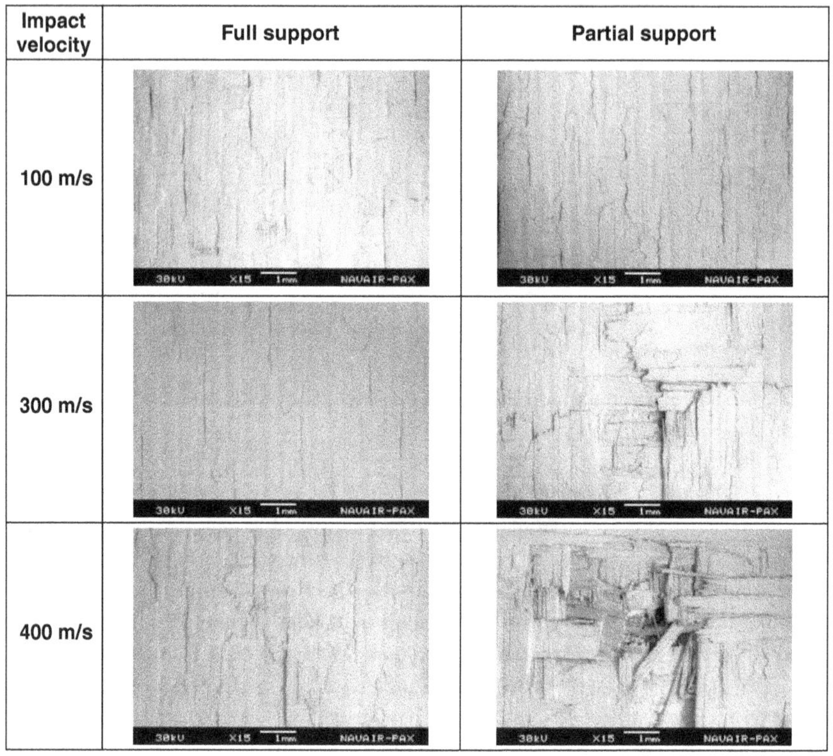

FIGURE 14.12
Summary of backside impact damage with respect to impact velocity in N720/aluminosilicate oxide/oxide impacted by 1.59-mm-diameter steel ball projectiles in full and partial supports [46].

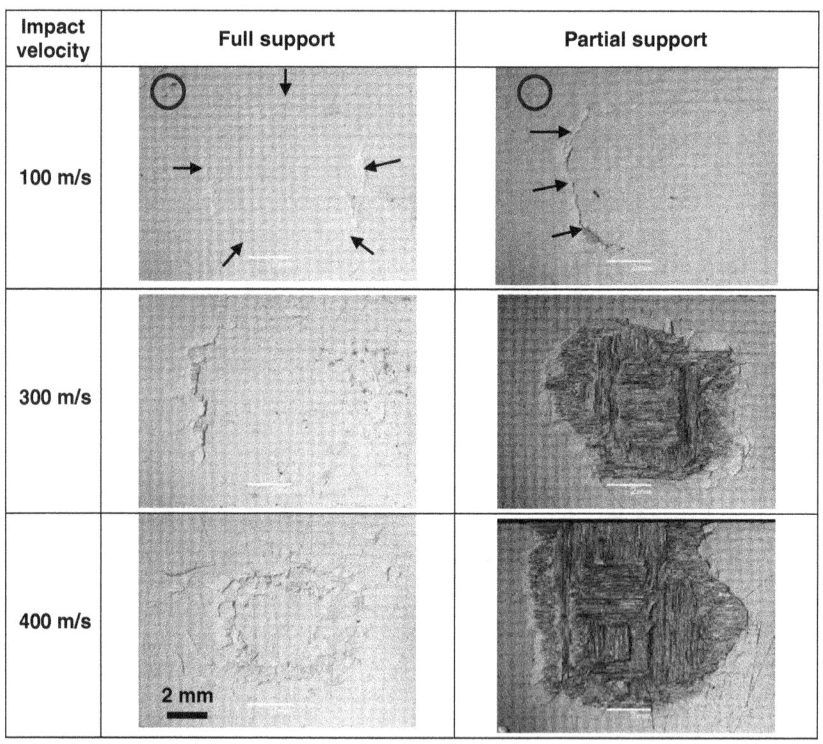

FIGURE 14.13
Backside impact damage with respect to impact velocity in full and partial supports for a prepreg MI SiC/SiC CMC impacted by 1.59-mm-diameter chrome steel ball projectiles [59]. The open circle in the figure represents the size (1.59 mm) of projectiles utilized. Backside damage in a form of a ring crack is marked as arrows. All the images are under the same magnification. Bar = 2 mm.

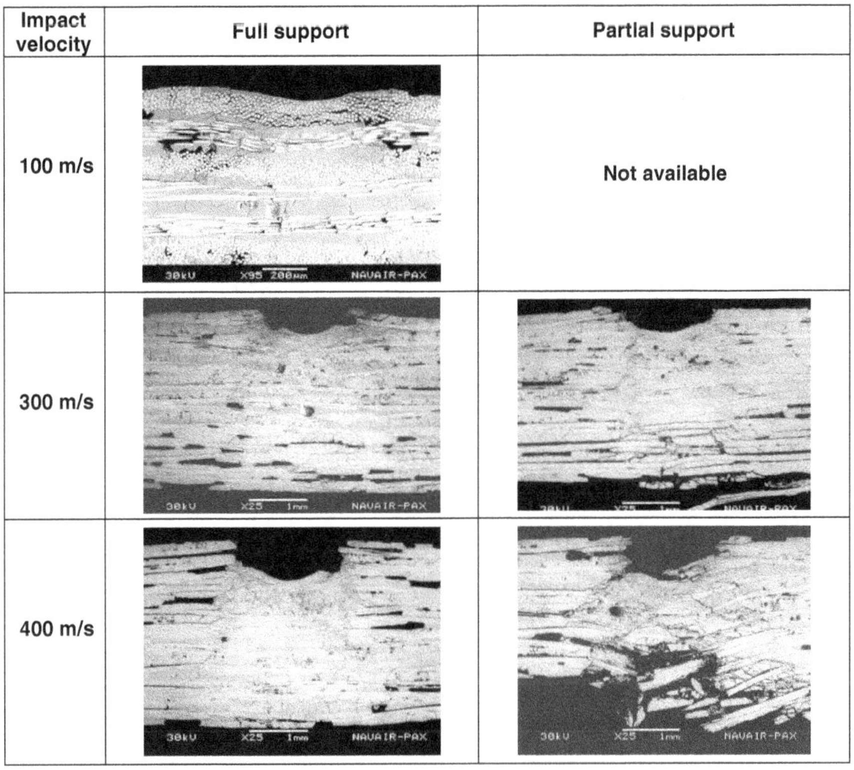

FIGURE 14.14

Cross-sectional views of impact sites with respect to impact velocity in N720/aluminosilicate oxide/oxide composite impacted by 1.59-mm-diameter steel ball projectiles in full and partial specimen supports [46]. Note features of compaction and cone crack configurations. Bar = 1 mm (except "200 μm" for the 100 m/s in full support).

For N720/aluminosilcate in full support, the impact site at 100 m/s was barely developed only with a slight contour of deformation, as shown in Figure 14.14. With increasing impact velocities ≥300 m/s, the impact damage increased in size and degree. Of a special importance was the material beneath the impact that showed compaction or densification upon impact. This compaction also can be seen from the figure where the area beneath the impact no longer showed the horizontal pores that were inherently present in the material prior to impact. This compaction can be understood from the fact that the CMC was *soft* and *open* structure, as aforementioned. No backside damage was observed even at the highest impact velocity of 400 m/s.

The N720/aluminosilcate also showed a similar aspect of compaction in partial support. However, with increasing impact velocities ≥300 m/s, the backside damage became much more predominant in partial support than in full support, significantly differentiating its degree of impact damage. The damage was related to breakage and removal of fibers/matrix, fiber-tows delamination/fracture, and fiber-tows protrusion toward the impact direction, as described in Section 14.3.2.2. Although the N720/aluminosilicate CMC targets at 400 m/s was on the verge of penetration by the projectile, they still survived the impact. This was in contrast with monolithic silicon nitrides that exhibited catastrophic fracture upon impact at much lower velocities of 180–260 m/s [20]. Also note that the overall damage configuration toward the impact direction was cone shaped. The impact response/phenomenon of N720/alumina, although not presented here, was very similar to those of N720/aluminosilicate because of their similarity in microstructure [51].

Figure 14.15 shows the cross sections of prepreg MI SiC/SiC in association with different impact velocities and target supports [59]. Impact craters were not well developed, compared to those in N720/aluminosilicate. No compaction of material underneath the impact site was observed. Progressive damage evolution with increasing impact velocity was evident in either full or partial support. The degree of damage, as seen in the N720/aluminosilicate CMC (Figure 14.14), was significantly greater in partial support than in full support. The material was typified of a peculiar feature of cone cracking, regardless of the type of target supports, with their shape being better defined with increasing impact velocity. The well-defined cone cracking was presumably due to the dense and hard nature of the material (Elastic modulus of 300 GPa, see Table 14.1), as seen from monolithic ceramic materials [5–7, 17, 21]. The backside damage of

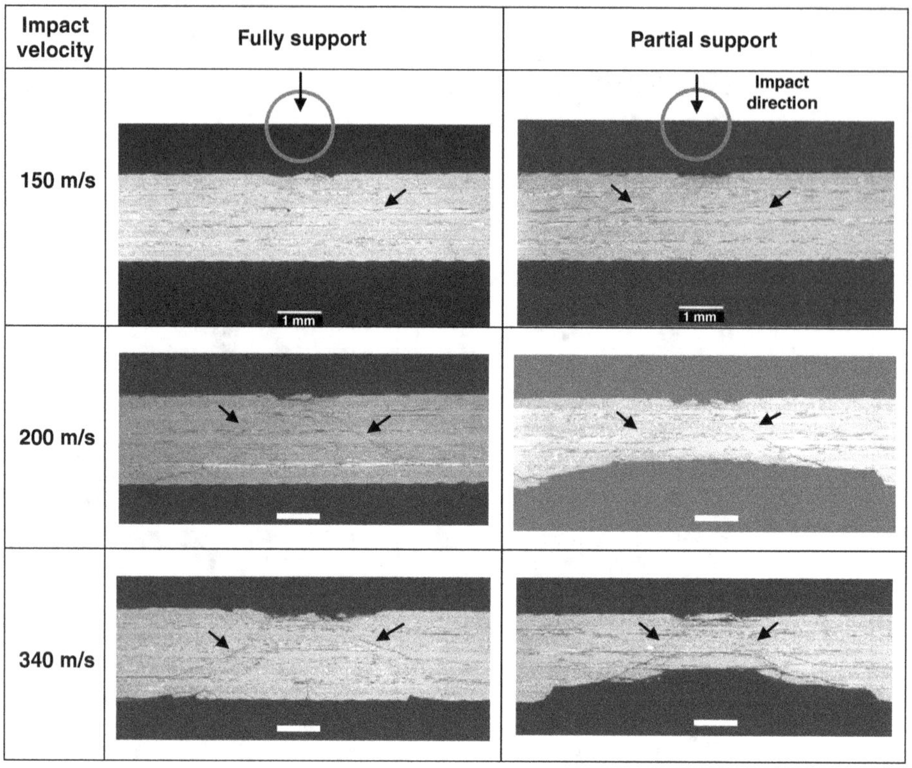

FIGURE 14.15

Cross-sectional views of impact sites with respect to impact velocity in prepreg MI SiC/SiC impacted by 1.59-mm-diameter steel ball projectiles in full and partial supports [59]. The arrows indicate cone cracks. The circle represents the size of ball projectiles (1.59 mm). Bar = 1 mm.

prepreg MI SiC/SiC in partial support was characterized mainly by material loss (scabbing or spallation) and interlaminar cracking/delamination.

14.3.2.5 Cone Cracking As seen from the cross-sectional views of both the oxide/oxide and the MI SiC/SiC CMCs in Figures 14.14 and 14.15, one of the important features of impact damage was cone cracking starting from the front impact site and propagating in a 3D fashion (conically) toward the backside of targets. This would make damage more substantial at the backside than at the front impact site, particularly when combined with backside tensile stresses and stress–wave interactions (producing scabbing), as seen from the prepreg MI SiC/SiC in partial support at ≥240 m/s (Figure 14.15). The formation of cone cracking has been observed in many brittle solids subject to ballistic impact by spherical projectiles, including alumina, zirconia, silicon nitrides, silicon carbide, and glasses [e.g., 1–3, 5–7, 17, 21, 67]. In the same way, cone cracking occurred in those materials subject to static spherical indentation [e.g., 10, 23, 65, 66]. Other CMC material systems also exhibited the occurrence of cone cracking in 2D or 3D woven SiC/SiCs either in ballistic impact or in quasi-static indentation [37–40]. Examples of such cone cracking occurring in CMCs and silicon nitride are shown in Figure 14.16 . Note in the figure (Figure 14.16g) that silicon nitride (AS800) revealed a contour of a well-developed cone crack from its fracture surface, leaving a cone cavity therein [21]. Note a *volcanic-island*-like cone was separated from the cavity, as also observed by Hara et al. [17].

A large amount of studies has been done for the last several decades to characterize and analyze the cone-cracking behavior of monolithic ceramics and glasses in static spherical indentation as well as in dynamic spherical impact [1, 5–7, 10, 17, 21, 23, 65–67]. The analysis of cone cracking has been based on the classical analysis of Hertzian contact stresses between two contacting bodies by Timoshenko [68], and has extended with some simplifying assumptions to the cases of quasi-static impact in "isotropic" brittle solids. Although the fundamental basics would remain similar, the analysis of cone cracking in CMCs would impose greater difficulties, due to materials' inhomogeneous and complex nature in architectures and constituents in addition to complicated dynamic effects. Appropriate experimental and analytical approaches needs to be made to explore the cone-cracking phenomenon in CMCs.

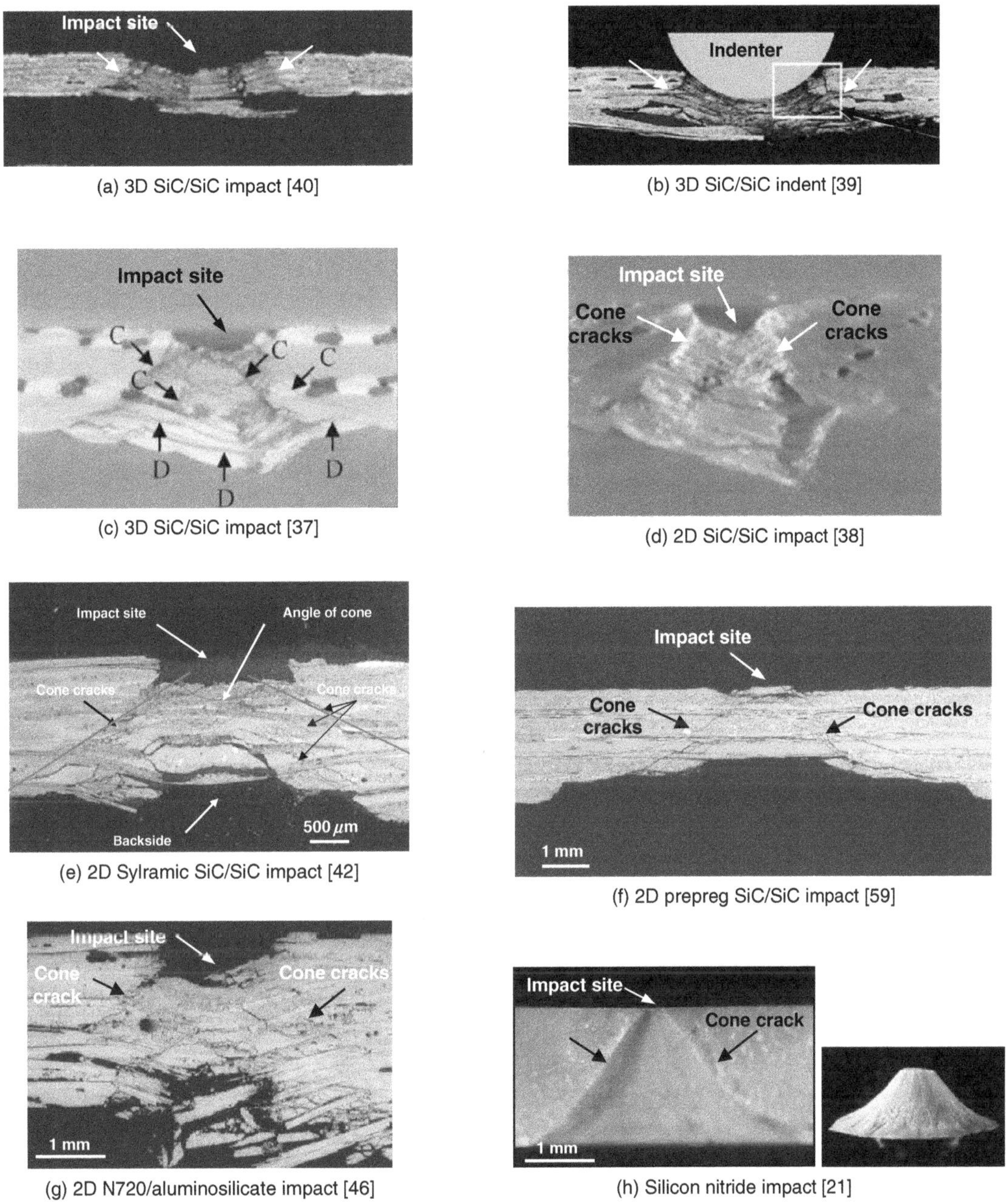

FIGURE 14.16

Examples of cone cracking in various CMCs and a silicon nitride impacted or indented by spherical projectiles or indenter [21, 37–40, 42, 46, 59]. (a) 3D SiC/SiC impact [40]; (b) 3D SiC/SiC indent [39]; (c) 3D SiC/SiC impact [37]; (d) 2D SiC/SiC impact [38]; (e) 2D Sylramic SiC/SiC impact [42]; (f) 2-D prepreg SiC/SiC impact [59]; (g) 2-D N720/aluminosilicate impact [46]; (h) Silicon nitride impact [21].

14.3.3 Prediction of Impact Force

One may wonder as to how much impact force was involved in impact with different CMCs. By assuming that the impact event was quasi-static, a first-order approximation of impact force can be made using the energy balance principle in conjunction with the contact yield pressure analysis [22, 44]. The energy balance in impact may be written as follows:

$$U_k = U_{el} + U_{pl} + U_L + U_{k,bc} + U_{k,tp} \tag{14.2}$$

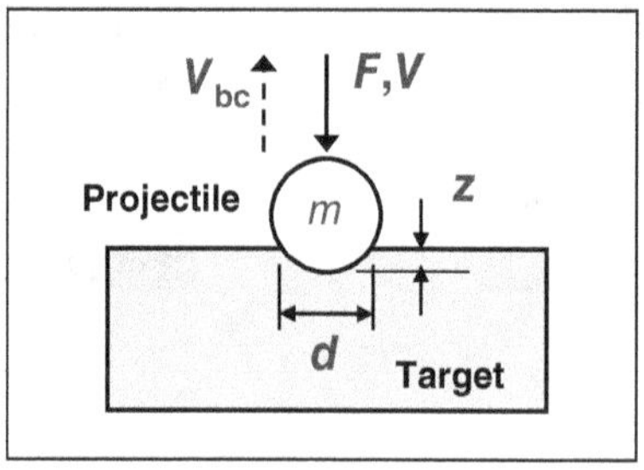

FIGURE 14.17
A schematic diagram showing a spherical projectile impacting onto a target specimen under an idealized quasi-static impact condition.

where U_k is the kinetic energy of a projectile, U_{el} is the elastic deformation energy of projectile and target, U_{pl} is the plastic deformation energy of projectile and target, U_L is the energy loss associated with friction and cracking such as spallation and delamination, $U_{k,bc}$ is the bouncing-back kinetic energy of a projectile, and $U_{k,tp}$ is the kinetic energy of target in contact with projectile right after impact but before the projectile's bouncing-back. Assume that the impacting kinetic energy is mainly consumed to deform the target both plastically and elastically with little energy loss, since the projectile is much more rigid than the soft targets. The related energies may be expressed as follows:

$$\begin{aligned} U_k &= mV^2/2 \\ U_{pl} &= \int Fdz \\ U_{k,bc} &= mV_{bc}^2/2 \\ U_{el} &= \alpha F^2L^3/(2EI) \end{aligned} \tag{14.3}$$

where m is the mass of a projectile, V is the impact velocity, F is the time-independent impact force, V_{bc} is the bouncing-back velocity of a projectile, L is the span length of a target support defined in Figure 14.3, E is the elastic modulus of the target, I is the second moment of inertia of the target, and α is a constant depending on the type of target supports ($\alpha = 1/3$, $1/48$, 0, respectively, in cantilever, partial, and full supports). The U_{el} is simply an elastic energy associated with target deflection due to F. The impact depth z is defined in Figure 14.17 and is geometrically related to the impact impression diameter d and the projectile diameter D assuming no violent impact to occur[1]

$$d = (4Dz - 4z^2)^{1/2} \tag{14.4}$$

A relationship between impact velocity and bouncing-back velocity of a projectile is defined as

$$e = -V_{bc}/V \tag{14.5}$$

where e is a coefficient of restitution. Considering the impact event as a quasi-static indentation, the impact force may be expressed as

$$F = (\pi d^2/4)\, p_y = [\pi(Dz - z^2)]\, p_y \tag{14.6}$$

[1] It was observed from a previous study [44] on the oxide/oxide material, as also seen from Figure 14.10, that the impact damage was greater and more violent than the static indentation counterpart. The geometrical relationship of Equation 14.4 used in analysis was thus idealized for analysis purpose.

where p_y is the contact yield pressure which can be estimated through static indentation onto the target material by the ball projectiles [22,44]. A linear relationship between contact impression area ($A = \pi d^2/4$) and indent force (P) indicates an average "contact yield pressure" (P/A) remains constant, so that

$$p_y = \frac{dP}{dA} \approx \frac{\Delta P}{\Delta A} \tag{14.7}$$

Using Equations (14.2)–(14.7), one can obtain the following equation:

$$m(1-e^2)V^2/2 = \pi p_y(Dz^2/2 - z^3/3) + \alpha \pi^2 p_y^2 (Dz - z^2)^2 L^3/(2EI) \tag{14.8}$$

This quartic equation of z in Equation (14.8) can be solved numerically or analytically as a function of impact velocity (V) for the given values of m, e, p_y, D, α, L, E, and I. Once z is solved, the impact force F can be calculated using Equation (14.6).

The values of p_y were found to be p_y = 3800 and 1270 MPa, respectively, for MI Sylramic SiC/SiC and N720/aluminosilicate CMCs [44]. Using the known values of D and m for steel ball projectiles (D = 1.59 mm) and $\alpha = 0$ (e.g., in full support for an example in this section), impact force can be obtained and resulting prediction is shown in Figure 14.18 for e = 0, 0.3, and 0.5. F increases monotonically or linearly with increasing V. For a given V, the impact force is greater in the SiC/SiC than the oxide/oxide since the former is much harder than the latter. For a given impact velocity, the difference in impact force between e = 0.0 and 0.5 is somewhat insignificant.

Since proper impact-force measurements (e.g., via strain gaging) were not available due to target's geometrical limitation as well as the violent nature of impact (causing strain gages to be popped off upon impact), the prediction of impact force made here was not verified directly. Instead, the data on front impact damage size (d) available in the oxide/oxide were used for an indirect verification. This result is presented in Figure 14.19, where the predicted d (solid lines) is compared with the experimental data for the oxide/oxide. The prediction for the oxide/oxide is in good agreement with the experimental data. This indicates that despite many simplifying assumptions and uncertainties associated with the violent and random nature of impact events, the prediction of impact force would be reasonable at least as a first-order approximation. But no verification can be made for the SiC/SiC (because of no available data), although it is expected that the prediction would be still in a reasonable level as a first-order simulation. In-depth efforts on FOD modeling as well as on advanced experimental techniques are of necessity to explore systematically the dynamic behavior and history of impact events.

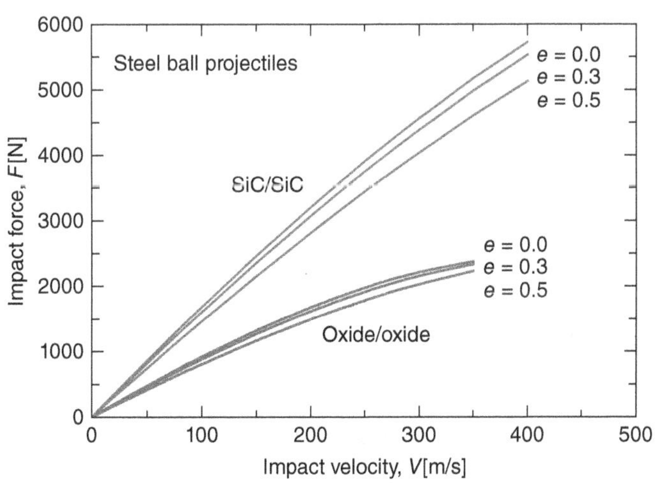

FIGURE 14.18

Predicted impact force as a function of impact velocity in full support with e = 0–0.5 for the MI Sylramic SiC/SiC and the N720/aluminosilicate CMCs impacted by 1.59-mm-diameter steel ball projectiles [44].

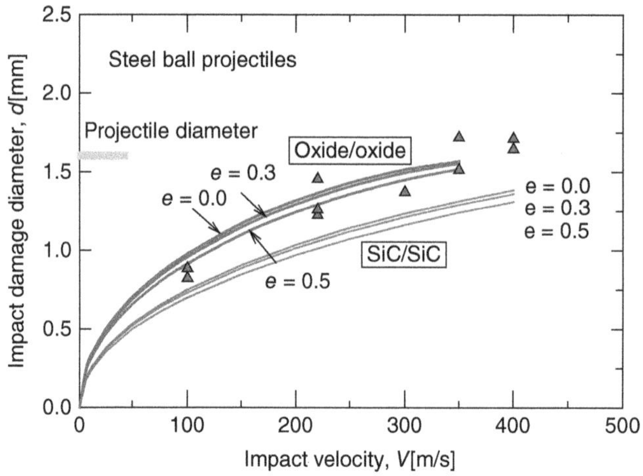

FIGURE 14.19

Comparison in front impact damage size (*d*) between the prediction (solid lines) and the experimental data (the triangles) obtained in full support for N720/aluminosilicate impacted by 1.59-mm-diameter steel ball projectiles [44].

14.3.4 Other Effects in FOD

Other effects associated with FOD will be briefly discussed in this section. The effects include temperature, preloading, environments (CMAS, salt), and cantilever support.

14.3.4.1 Temperature Effect The postimpact strength of MI Sylramic SiC/SiC impacted at 1316°C (2400°F) in air is shown in Figure 14.20 [45]. The strength was determined at ambient temperature for target specimens that were impact at 1316°C by 1.59-mm diameter steel ball projectiles in full and partial supports. As compared with the ambient-temperature counterpart (see Figure 14.4a), the difference in strength between 25°C and 1316°C was insignificant in full support. By contrast, the strength difference was significant in partial support. Strength degradation was about 25% when impact at 1316°C, indicating that high temperature impact generated more damage particularly in partial support than ambient-temperature impact. Invariably, strength degradation at 1316°C was greater in partial support than in full support.

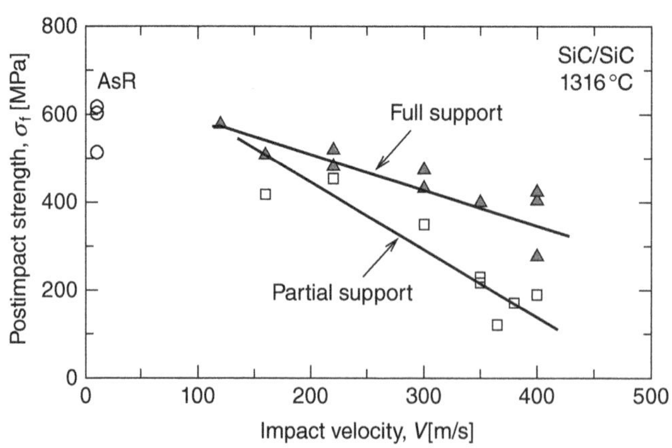

FIGURE 14.20

Postimpact strength as a function of impact velocity for MI Sylramic SiC/SiC CMC impacted at 1316°C in air by 1.59-mm-diameter steel ball projectiles in full and partial supports. "As-R"=As-received strength determined at ambient temperature with no impact [45].

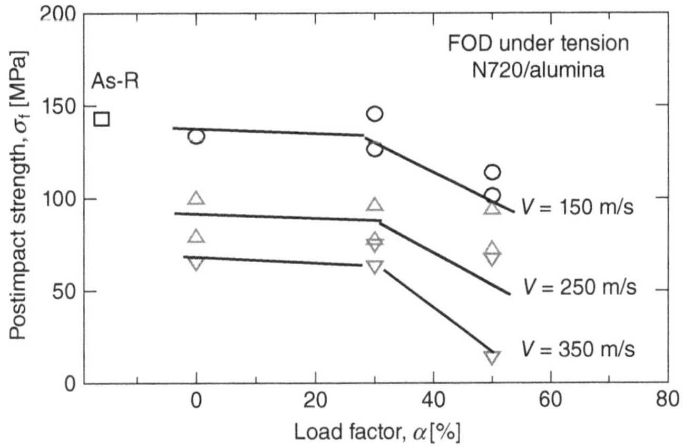

FIGURE 14.21

Postimpact tensile strength as a function of load factor in N720/alumina oxide/oxide CMC under different levels of impact velocities, impacted by 1.59 steel ball projectiles [53]. As-R: as-received tensile strength.

14.3.4.2 Preloading Effects Tensile targets of the N720/alumina oxide/oxide CMC were ballistic-impacted under tensile preloading with three different levels of preloads [53]. A similar work has been done with silicon nitride tensile specimens at high temperatures [69]. The test was performed to simulate FOD behavior of rotating aeroengine components such as blades where tensile stresses are generated due to centrifugal force. The target tensile specimens (10–12-mm wide) were tensile preloaded via two wedge grips, as illustrated in Figure 14.3d. Each target specimen was impacted at its center at a normal incidence angle by 1.6-mm-diameter steel ball projectiles. Applied tensile preloading (or stress) during ballistic impact was normalized with respect to the composite strength to get a load factor (α) defined as

$$\alpha = \sigma_{FOD}/\sigma_f \tag{14.9}$$

where σ_{FOD} is the applied stress during impact and σ_f is the as-received tensile composite strength (140 MPa). Three levels of load factor, $\alpha = 0$, 30%, and 50%, were used.

The results are shown in Figure 14.21, where postimpact tensile strength was plotted as a function of load factor for different levels of impact velocities. Despite some inherent data scatter, it is evident that postimpact strength decreased with increasing load factor, particularly at higher load factors of $\alpha > 30\%$, independent of impact velocity. At higher preloading of 50%, impact damage was oriented more to the direction perpendicular to tensile preloading, as seen from impact morphologies.

14.3.4.3 Environmental Effects: Combined CMAS and Salt A small amount of the dry, pulverized sand (aka CMAS, calcium magnesium aluminosilicate) of about 10 mg was placed in an impact site at the central region of each of impacted CMC target specimens. Also a mixture (about 10 mg) of CMAS and salt (sodium sulfate, Na_2SO_4) in a 50 to 50 weight percent was placed in another set of impacted CMCs target specimens. Those target specimens, first impacted at 340 m/s by 1.59-mm-diameter steel ball projectile in partial support and then covered with CMAS or CMAS/salt, were then exposed to an isothermal condition of 1200°C for 10 hours in air. The severity of combined FOD and CMAS attacks was quantified by conducting residual strength testing for the FOD-and-CMAS/salt-treated specimens to determine their strength degradation. The salt has been considered as prevailing corrosion species formed from NaCl and sulfur oxides from fuel combustion particularly in naval gas turbine engines [70].

The results of residual strength determinations are summarized in Figure 14.22 [56], where strength was normalized with respect to as-received strength of each of CMC materials. Somewhat systematic strength degradation was observed to occur from FOD to FOD-and-CMAS to FOD-and-CMAS/salt, resulting in the worst strength degradation occurring in FOD-and-CMAS/salt condition, regardless of the type of CMCs used. As seen from the figure, the FOD itself induced about 34–47% strength degradation, the combined FOD and CMAS degraded by 29–55%, and the combined FOD and CMAS/salt reduced by 43–81%, in which the degree of strength degradation was dependent on the type of the CMCs used. Therefore, FOD generated the greatest damage, FOD and CMAS/salt did intermediate damage, and FOD and CMAS generated the least damage. Unlike other materials, the Sylramic SiC/SiC exhibited some strength

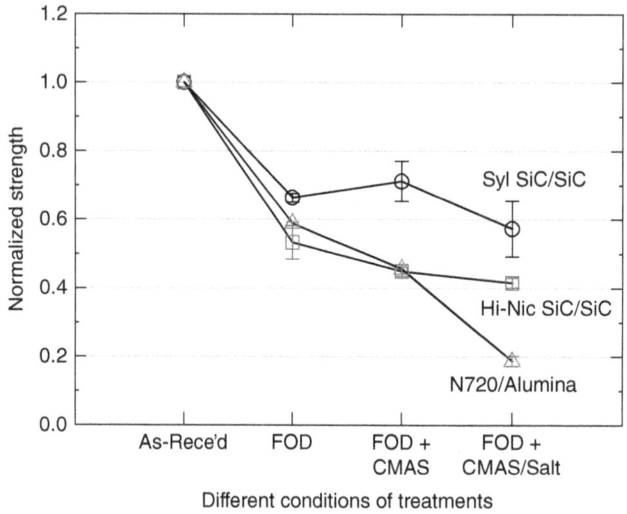

FIGURE 14.22

Normalized residual strength of FOD, FOD-and-CMAS, and FOD-and-CMAS/salt-treated test specimens of MI Sylramic SiC/SiC, MI Hi-Nic SiC/SiC, and N720/alumina oxide/oxide CMCs [56]. The strength was normalized with respect to the respective as-received strength of each of CMCs. Error bars indicate the normalized ± 1.0 standard deviation.

increase in FOD-and-CMAS treatments. This could be attributed to *dominant* crack healing effect by melt CMAS glass. However, the exact cause is yet to be determined. Of particular interest is the N720/alumina oxide/oxide that showed considerable strength degradation (81%) in CMAS/salt exposure. Their fracture surfaces showed a complete absence of fiber pullout and a mode of brittle fracture as well. Significant reactions of related species with fibers causing the composite degradation were noted.

14.3.4.4 Effects of Cantilever Support Figure 14.23 represents the results of impact testing for both prepreg MI SiC/SiC and N720/alumina CMCs conducted in an additional target support, cantilever support (Figure 14.3c). The target specimens with a length of 45–50 mm were cantilever-supported with $L = 20$ mm and impacted by 1.59-mm-diameter steel ball projectiles [59, 60]. The figure compares with the data obtained in full and partial supports. Despite some inherent data scatters, the overall trend is obvious such that regardless of the type of CMCs, the greatest strength degradation occurred in cantilever support, the intermediate degradation occurred in partial support, and the least strength degradation occurred in full support.

An important observation acquired from this work is that the overall impact damage or strength degradation was greater in cantilever support than in partial support. Unlike in partial support, there should be no static backside tensile stresses in cantilever support at its impact point because of zero moment-arm therein. This should have created much less damage in cantilever than partial support. However, the outcome was quite different, opposite, and contradictory to our basic understanding of mechanics. This implies that the overall impact behavior in cantilever support would have been governed more by dynamic aspects (stress wave interactions, etc.) than by static alone. Further study should be sought to validate using appropriate analytical or experimental dynamics approaches.

14.4 FOD RESPONSE OF ENVIRONMENTAL BARRIER COATINGS

It has been well known that current SiC-based CMC material systems alone may not be used in aeroengine hot-section components without appropriate protective coatings due to recession by water vapor. The coatings, termed environmental barrier coatings (EBCs), are mostly ceramics so that their brittleness is naturally prone to cracking or spallation when subjected to impact by foreign objects. A previous study of BSAS (barium strontium aluminosilicate)/mullite-based EBCs on an MI Sylramic SiC/SiC substrate showed their significant susceptibility to FOD, when impacted at either 25°C or 1316°C by 1.59-mm-diameter steel ball projectiles [71]. The thickness of the EBCs used was about 325 μm excluding Si bond coat. Figure 14.24 shows an impact response of the EBCs with respect to spallation size. The spallation size,

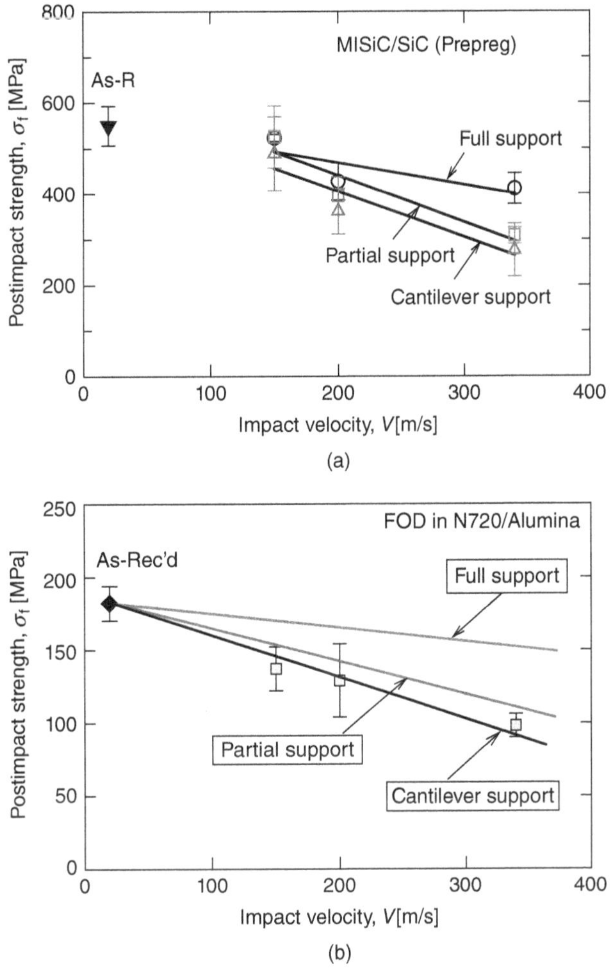

FIGURE 14.23

Postimpact strength as a function of impact velocity in cantilever support determined in (a) prepreg MI SiC/SiC [59] and (b) N720/alumina [60] CMCs, impacted by 1.59-mm-diameter steel ball projectiles. The data on full and partial supports [59, 60] are included for comparison. "As-rec'd" or "As-R" indicates "as-received" flexure strength.

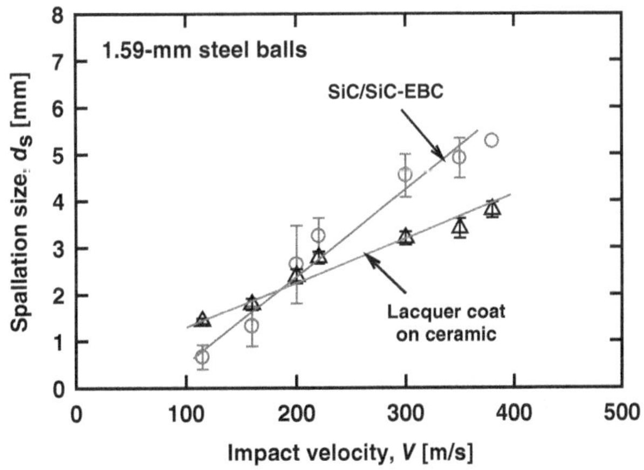

FIGURE 14.24

Spallation size as a function of impact velocity in a BSAS/mullite-based EBCs on an MI Sylramic SiC/SiC substrate, impacted at 25°C by 1.59-mm-diameter steel ball projectiles [64, 71]. Data from a polymeric lacquer coating on a silicon nitride substrate impacted by the same projectiles is also included for comparison [64].

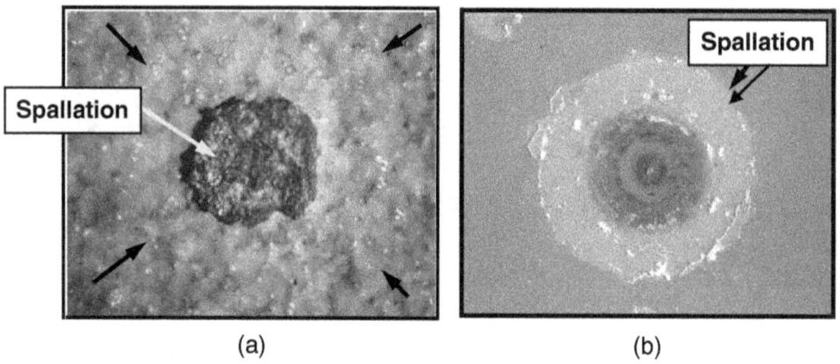

FIGURE 14.25

Examples showing impact damage as well as spallations in (a) BSAS/mullite-based EBCs and (b) lacquer coating on a silicon nitride substrate, impacted at 25°C by 1.59-mm-diameter steel ball projectiles [64]. The arrows in (a) indicate a contour of subsurface delamination of the EBC coatings.

averaged from a total of five data points per velocity, increases almost linearly with increasing impact velocity. Also included in the figure is data obtained from lacquer coating (~150-μm thick) on a silicon nitride substrate impacted by the same projectiles [64]. A similar trend can be found between the brittle EBCs and the somewhat "ductile" polymeric coating. However, the overall dependency of damage on impact velocity was greater in the EBCs than in the lacquer coating. The use of the lacquer coating was to make an attempt to model its impact behavior by facilitating clear visible response to spallation.

Figure 14.25 presents typical examples of impact damage of the two coatings, showing well-developed, circular spallation taking place in both of the coatings. Particularly, a contour or locus (indicated with arrows) of delamination, greater in size than spallation, is seen from the EBCs, but their exact size could not be determined by means of optical microscopy. By contrast, the pulsed thermography or CT technique has demonstrated to be very useful to assess quantitatively and accurately the degree of subsurface damage [47,71]. In conclusions, current EBCs either of BSAS/mullite [71] or of rare earth silicates [72] have exhibited insufficient durability to projectile/particle impact and therefore need to be improved in FOD resistance to be used in aeroengine functional/structural subcomponents.

14.5 COMPARISON OF CMCs AND SILICON NITRIDES

Figure 14.26 compares FOD response between two CMCs and two gas-turbine grade silicon nitrides. Specifically, the figure shows normalized postimpact strength ($\bar{\sigma}_f$) as a function of impact kinetic energy (U_k) for MI Sylramic SiC/SiC

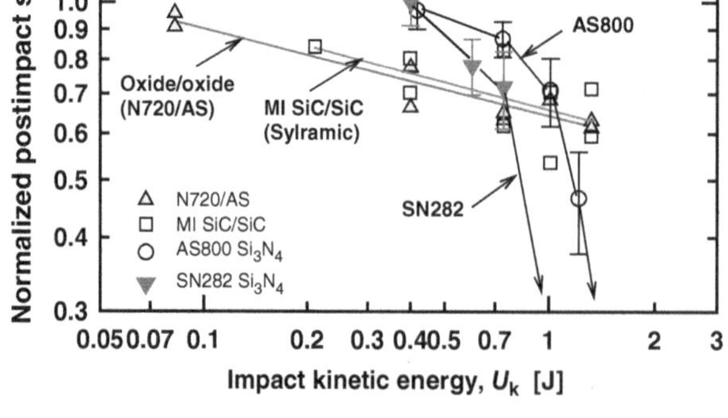

FIGURE 14.26

Comparison of normalized postimpact strength as a function of impact kinetic energy (U_k) for MI Sylramic SiC/SiC CMC [42], N720/aluminosilicate [43], and AS800 and SN282 silicon nitrides [20], impacted by 1.59-mm-diameter steel projectiles in full support.

[42], N720/aluminosilicate oxide/oxide [43], and AS800 and SN282 silicon nitrides [20], all impacted in full support by 1.59-mm-diameter steel ball projectiles. A monotonic degradation in strength with increasing impact energy was typified for the two CMCs, yielding a linear relationship between log ($\bar{\sigma}_f$) and log (U_k). By contrast, monolithic AS800 and SN282 silicon nitrides exhibited greater strength degradation, approaching their respective "critical" impact energies of about 1.3 J (with $V = 400$ m/s) and 0.8 J (at $V = 310$ m/s), in which flexure test specimens fractured completely upon impact. However, at lower kinetic energy, $U_k \leq 0.4$ J (corresponding to $V = 220$ m/s), the two silicon nitrides showed a negligible strength degradation, while the two CMCs yielded about 20–30% degradation. Although not included here, two other CMCs, prepreg MI SiC/SiC and N720/alumina CMCs, revealed the same trend as the MI Sylramic SiC/SiC and N720/aluminosilicate CMCs (as also can be seen by comparison from the data in Figure 14.5). It can be summarized that the gas-turbine grade CMCs exhibited greater FOD resistance over the silicon nitrides particularly at greater impact velocities or kinetic impact energies.

14.6 CONSIDERATION FACTORS OF FOD IN CMCs

Designing aeroengine CMC components to withstand FOD is a highly complex and daunting task. Considerations of many affecting factors are needed and may be listed as follows:

- effect of primary impact variables such as impact velocity and incidence angle;
- effect of the material, geometry, and size of projectiles;
- effect of target material in architectures, material constituents (fiber, matrix, interfacial coating), and processing routes;
- effect of the type, size, and material of target support;
- effect of temperature, environment, vibration, and thermomechanical loading; and
- effect of protective coatings such as impact barrier coatings and/or thermal/environmental barrier coatings (T/EBCs).

Figure 14.27 illustrates those affecting factors in FOD in CMC material systems and/or CMC airfoils.

Impact by foreign objects in CMCs is random (stochastic) and violent and thus so complex in nature from both characterization and analytical points of view. Appropriate FOD modeling should be able to serve in a feedback loop to characterization techniques, one complementing and improving the other. Because of the scope of this paper, FOD modeling was not described in this paper. However, it should be noted that considerable efforts on FOD modeling for various CMCs, particularly for SiC/SiCs, are undergoing, including the quantification of sensitivity of materials parameters to impact damage [73–79]. Undoubtedly, this helps and reinforces our understanding of complex nature of FOD

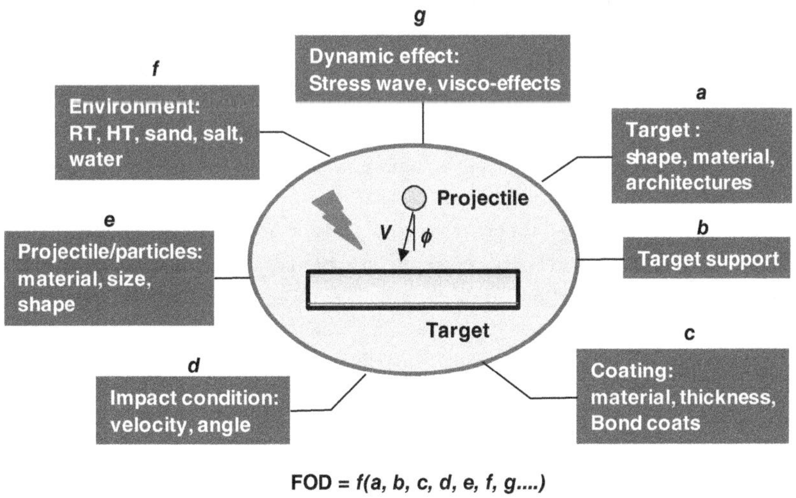

FIGURE 14.27
A schematic showing various affecting factors in foreign object damage (FOD) in ceramic matrix composites (CMCs) and CMC airfoils.

and consequently can serve as valuable tools to design and fabricate more robust FOD-resistant CMC materials or components for future advanced aeroengines.

14.7 CONCLUDING REMARKS

The responses of foreign object damage (FOD) on various ceramic matrix composites (CMCs) were presented as a function of variables such as material, support, temperature, preload, and environments (CMAS and salt). The current gas-turbine grade CMC material systems exhibited substantial vulnerability to FOD at an impact velocity of around Mach 1, although they exhibit much more enhanced FOD resistance than gas-turbine grade silicon nitrides. Foreign object impact is random and violent, and the damage is further complicated by the complex nature of materials in architecture and constituents. More systematic and multidisciplinary approaches to FOD issue on CMCs can help advance our understanding of FOD so that transition of CMC material systems can be accelerated to future advanced aeroengine components.

ACKNOWLEDGMENTS

The author acknowledges the continued support of the Office of Naval Research (ONR) by Dr. David Shifler. The author is indebted to C. Faucett, D. Alexander, J. Bilko, W. Lightel, and R. Kowalik from NAVAIR and R. Pawlik from NASA Glenn Research Center for their valuable inputs and contributions during the courses of this work.

REFERENCES

1. S. M. Wiederhorn and B. R. Lawn. Strength degradation of glass resulting from impact with spheres. *J. Am. Ceram. Soc.*, 60(9–10), 451–458 (1977).
2. S. M. Wiederhorn and B. R. Lawn. Strength degradation of glass impact with sharp particles: I, annealed surfaces. *J. Am. Ceram. Soc.*, 62(1–2), 66–70 (1979).
3. K. Breder, G. de Portu, J. E. Ritter, and D. D. Fabbriche. Erosion damage and strength degradation of zirconia-toughened alumina. *J. Am. Ceram. Soc.*, 71(9), 770–775 (1988).
4. J. E. Ritter, S. R. Choi, K. Jakus, P. J. Whalen, and R. G. Rateick. Effect of microstructure on the erosion and impact damage of sintered silicon nitride. *J. Mater. Sci.*, 26, 5543–5546 (1991).
5. Y. Akimune, Y. Katano, and K. Matoba. Spherical-impact damage and strength degradation in silicon nitrides for automobile turbocharger rotors. *J. Am. Ceram. Soc.*, 72(8), 1422–1428 (1989).
6. D. A. Shockey, D. J. Rowcliff, K. C. Dao, and L. Seaman. Particle impact damage in silicon nitride. *J. Am. Ceram. Soc.*, 73(6), 1613–1619 (1990).
7. C. G. Knight, M. V. Swain, and M. M. Chaudhri. Impact of small steel spheres on glass surfaces. *J. Mater. Sci.*, 12, 1573–1586 (1977).
8. A. M. Rajendran and J. L. Kroupa. Impact design model for ceramic materials. *J. Appl. Phys.*, 66(8), 3560–3565 (1989).
9. L. N. Taylor, E. P. Chen, and J. S. Kuszmaul. Microcrack-induced damage accumulation in brittle rock under dynamic loading. *Comp. Meth. Appl. Mech. Eng.*, 55, 301–320 (1986).
10. R. Mouginot and D. Maugis. Fracture indentation beneath flat and spherical punches. *J. Mater. Sci.*, 20, 4354–4376 (1985).
11. A. G. Evans and T. R. Wilshaw. Dynamic solid particle damage in brittle materials: an appraisal. *J. Mater. Sci.*, 12, 97–116 (1977).
12. B. M. Liaw, A. S. Kobayashi, and A. G. Emery. Theoretical model of impact damage in structural ceramics. *J. Am. Ceram. Soc.*, 67, 544–548 (1984).
13. M. van Roode, et al. "Ceramic gas turbine materials impact evaluation. ASME Paper No. GT2002-30505 (2002).
14. D. W. Richerson and K. M. Johansen. "Ceramic gas turbine engine demonstration program. Final Report, DARPA/Navy Contract N00024-76-C-5352, Garrett Report 21-4410 (1982).
15. G. L. Boyd and D. M. Kreiner. AGT101/ATTAP ceramic technology development. Proceeding of the twenty-fifth automotive technology development contractors' coordination meeting, October 26–29, Dearborn, MI, 1987. p. 101.

16. M. van Roode, et al. Ceramic stationary gas turbine development—fourth annual summary. ASME Paper No. 97-GT-317 (1997).
17. Y. Hara, K. Matsubara, K. Mizuno, T. Shimamori, and H. Yoshida. Development and evaluation of silicon nitride components for ceramic gas turbine. ASME Paper 98-GT-498 (1998).
18. H. Yoshida, M. M. Chaudhri, and Y. Hoshi. Quasistatic indentation and spherical particle impact studies of turbine-grade silicon nitrides. *Philos. Mag. A*, 82(10), 2031–2040 (2002).
19. Peralta and H. Yoshida. *Ceramic Gas Turbine Component Development and Characterization*, edited by M. van Roode, M. K. Ferber, D. W. Richerson, Vol. 2, (ASME, New York, NY, 2003), pp. 665–692.
20. S. R. Choi, J. M. Pereira, L. A. Janosik, and R. T. Bhatt. Foreign object damage in flexure bars of two gas-turbine grade silicon nitrides. *Mater. Sci. Eng. A*, 379, 411–419 (2004).
21. S. R. Choi, J. M. Pereira, L. A. Janosik, and R. T. Bhatt. Foreign object damage in disks of gas-turbine-grade silicon nitrides by steel ball projectiles at ambient temperature. *J. Mater. Sci.*, 39, 6173–6182 (2004).
22. S. R. Choi. Foreign object damage behavior in a silicon nitride ceramic by spherical projectiles of steels and brass. *Mat. Sci. Eng. A.*, 497, 160–167 (2008).
23. S. R. Choi, Z. Racz, R. T. Bhatt, and D. N. Brewer. Foreign object damage in a gas-turbine grade silicon nitride by spherical projectiles of various materials. NASA TM-2006-214330, National Aeronautics & Space Administration, Glenn Research Center, Cleveland, OH (2006).
24. S. R. Choi and Z. Racz. Effects of target size on foreign object damage in gas-turbine grade silicon nitrides by steel ball projectiles. *J. Eng. Gas Turbines Power*, 134, 051301-1-8, Trans. ASME (2012).
25. P. Lundberg, R. Renström, and B. Lundberg. Impact of conical tungsten projectiles on flat silicon carbide targets: transition from interface defeat to penetration. *Int. J. Impact Eng.*, 32, 1842–1856 (2006)
26. J. C. LaSalvia and J. W. McCauley. Inelastic deformation mechanics and damage in structural ceramics subjected to high-velocity impact. *Int. J. Appl. Ceram Tech.*, 7(5), 595–605 (2010).
27. V. S. Deshpande and A. G. Evans. Inelastic deformation and energy dissipation in ceramics: a mechanism-based constitutive model. *J. Mech. Phys. Solids*, 56, 3077–3100 (2008).
28. R. L. Sierakowski. Impact damage-tolerant composite structural design, in *Impact Behavior of Fiber-Reinforced Composite Materials and Structures*, edited by S. R. Reid and G. Zhou (Woodhead Publishing Limited, Cambridge, England, 2000), pp. 106–132.
29. ASTM D 7136. Test method for measuring the damage resistance of a fiber-reinforced polymer matrix composite to a drop-weight impact event, in *Annual Book of ASTM Standards* (ASTM, West Conshohocken, PA) (2012).
30. S. Abrate. Impact on laminated composites: recent advances. *Appl. Mech. Rev.*, 47(11), 517–544 (1994).
31. W. J. Cantwell and J. Morton. The impact resistance of composite materials—a review. *Composites*, 22(5), 347–362 (1991)
32. S. Abrate. Impact on laminated composite materials. *Appl. Mech. Rev.*, 44(4), 155–190 (1991).
33. E. Sevkat, B. Liew, F. Delale, and B. Raju. Drop-weight impact of plain-woven hybrid glass-graphite/toughened epoxy composites. *Comp. Part A*, 40, 1090–1110 (2009).
34. D. C. Phillips, N. Park, and R. J. Lee. The impact behavior of high performance ceramic matrix composites. *Compos. Sci. Technol.*, 37, 249–265 (1990).
35. Y. Leijiang, F. Ziyang, and C. Qiyou. Low velocity impact damage evaluation of 2D C/SiC composite material. *Adv. Mater. Res.*, 78–82, 11835–11838 (2009).
36. A. R. Boccaccini, S. Atiq, D. N. Boccaccini, I. Dlouhy, and C. Kaya. Fracture behavior of mullite fiber reinforced mullite matrix composites under quasi-static and ballistic impact loading. *Compos. Sci. Technol.*, 65, 325–333 (2005).
37. K. Ogi, T. Okabe, M. Takahashi, S. Yashiro, and A. Yoshimura. Experimental characterization of high-speed impact damage behavior in a three-dimensionally woven SiC/SiC composite. *Compos. Part A*, 41(4), 489–498 (2010).
38. S. Yashiro, K. Ogi, and M. Oshita. High-velocity impact damage behavior of plain-woven SiC/SiC composites after thermal loading. *Compos. Part B*, 43, 1353–1362 (2012).
39. V. Herb, G. Gouégnat, and E. Martin. Damage assessment of thin SiC/SiC composite plates subjected to quasi-static indentation loading. *Compos. Part A*, 41(11), 1677–1685 (2010).
40. V. Herb, E. Martin, and G. Couégnat. Damage analysis of thin 3D-Woven SiC/SiC composite under low velocity impact loading. *Compos. Part A*, 43, 247–253 (2012).
41. S. R. Choi and D. J. Alexander. Foreign object damage by steel ball projectiles in a SiC/SiC ceramic matrix composite at ambient and elevated temperatures. NAWCADPAX/TR-2008/3, Naval Air Warfare Center Aircraft Division, Patuxent River, MD (2008).
42. S. R. Choi. Foreign object damage phenomenon by steel ball projectiles in a SiC/SiC ceramic matrix composite at ambient and elevated temperatures. *J. Am. Ceram. Soc.*, 91(9), 2963–2968 (2008).
43. S. R. Choi, D. J. Alexander, and R. W. Kowalik, Foreign object damage in an Oxide/Oxide composite at ambient temperature. J. Eng. Gas Turbines Power, Trans. ASME, Vol. 131, 021301 (2009).

44. S. R. Choi, D. J. Alexander, and D. C. Faucett. Comparison in foreign object damage between SiC/SiC and oxide/oxide ceramic matrix composites. *Ceram. Eng. Sci. Proc.*, 30(2), 177–188 (2009).
45. S. R. Choi, D. C. Faucett, and D. J. Alexander. Foreign object damage in an N720/alumina oxide/oxide ceramic matrix composite. *Ceram. Eng. Sci. Proc.*, 31(2), 221–232 (2010).
46. S. R. Choi and D. J. Alexander. Characterization of foreign object damage in an oxide/oxide composite at ambient temperature. *Ceram. Eng. Sci. Proc.*, 29(2), 167–178 (2008).
47. L. M. Cosgriff, R. Bhatt, S. R. Choi, and D. S. Fox. Thermographic characterization of impact damage in SiC/SiC Composite materials, *Proc. SPIE*, 5767, 363–372 in *Nondestructive Evaluation & Health Monitoring of Aerospace Materials, Composites, and Civil Structure IV* (2005).
48. S. R. Choi, D. J. Alexander, and D. C. Faucett. Foreign object damage vs. static indentation damage in an oxide/oxide ceramic matrix composite, in *Processing & Properties of Advanced Ceramics & Composites. Ceram. Trans.*, 203, 171–180 (2009).
49. D. C. Faucett, D. J. Alexander, and S. R. Choi. Static-contact and foreign-object damages in an oxide/oxide (N720/Alumina) ceramic matrix composite: comparison with N720/aluminosilicate, in *Processing & Properties of Advanced Ceramics & Composites II. Ceram. Trans.*, 220, 243–255 (2010).
50. S. R. Choi. Foreign object damage in ceramic matrix composite and monolithic silicon nitrides, in *Processing & Properties of Advanced Ceramics & Composites II. Ceram. Trans.*, 220, 221–242 (2010).
51. S. R. Choi, D. C. Faucett, and D. J. Alexander. Foreign object damage in an N720/alumina oxide/oxide ceramic matrix composite. *Ceram. Eng. Sci. Proc.*, 31(2), 221–232 (2010).
52. D. C. Faucett, D. J. Alexander, and S. R. Choi. Static contact damage in an N720/alumina oxide ceramic matrix composite with reference to foreign object damage. *Ceram. Eng. Sci. Proc.*, 31(2), 233–244 (2010).
53. D. C. Faucett and S. R. Choi. Foreign object damage in an N720/alumina oxide/oxide ceramic matrix composite under tensile loading. *Ceram. Trans.*, 225, 99–107, edited by N. P. Narottam, J. P. Singh, J. Lamon, and S. R. Choi (2011).
54. S. R. Choi and D. J. Alexander. Characterization of foreign object damage in an oxide/oxide composite at ambient temperature. *Ceram. Eng. Sci. Proc.*, 29(2), 167–178 (2008).
55. R. T. Bhatt, S. R. Choi, L. M. Cosgriff, D. S. Fox, and K. N. Lee. Impact resistance of uncoated SiC/SiC composites. *Mater. Sci. Eng. A*, 476, 20–28 (2008).
56. S. R. Choi and D. F. Faucett. Combined effects of CMAS and FOD in ceramic matrix composites. ASME Paper GT2012-70049. Proceedings of ASME Turbo Expo 2012, June 11–15, Copenhagen, Denmark, 2012.
57. G. N. Morscher, C. Baker, S. Ramasamy, S. Choi, and C. Faucett. Use of electrical resistivity and acoustic emission to assess impact damage states in two SiC-based CMCs. ASME Paper GT2012-69167. Proceedings of ASME Turbo Expo 2012, June 11–15, Copenhagen, Denmark, 2012.
58. G. N. Morscher, C. Baker, A. Gyekenyesi, C. Faucett, and S. Choi. Damage detection and tensile performance of various SiC/SiC composites impacted at high speed projectiles. ASME Paper GT2013-95638. Proceedings of ASME Turbo Expo 2013, June 3–7, San Antonio, TX, 2013.
59. D. C. Faucett, J. Wright, M. Ayre, and S. R. Choi. Effects of the mode of target supports on foreign object damage in an MI SiC/SiC ceramic matrix composite, in *Processing & Properties of Advanced Ceramics & Composites IV. Ceram. Trans.*, 234, 231–243 (2012), edited by J. P. Singh, N. P. Bansal, T. Goto, J. Lamon, S. R. Choi, M. M. Mahmoud, and G. Link.
60. D. C. Faucett, J. Wright, M. Ayre, and S. R. Choi. Effects of target supports on foreign object damage in an oxide/oxide CMC. *Ceram. Eng. Sci. Proc.*, 33(2) (2012).
61. M. Verrilli, S. Foster, and J. Subit. Impact behavior of a SiC/SiC composite at an elevated temperature. Proceedings of the 7th International Conference on High Temperature Ceramic Matrix Composites (HTCMC-7), September 20–22, Bayreuth, Germany, 2010. 519–530, edited by W. Krenkel and J. Lamon.
62. D. Brewer. "HSR/EPM" combustor materials development program. *Mat. Sci. Eng.*, A262, 284–291 (1999).
63. S. R. Choi. Unpublished work on erosion and impact damage in Ni-based superalloys (1989).
64. S. R. Choi. Unpublished work on FOD in CMCs and coatings (2004).
65. T. R. Wilshaw. The Hertzian fracture test. *J. Phys. D. Appl. Phys.*, 4, 1567–1581 (1971).
66. B. Lawn and R. Wilshaw. Review: indentation fracture: principles and applications. *J. Mater. Sci.*, 10, 1049–1081 (1975).
67. M. M. Chaudhri and C. R. Kurkjian. Impact of small steel spheres on the surfaces of "normal" and "anomalous" glasses. *J. Am. Ceram. Soc.*, 69(5), 404–410 (1986).
68. S. P. Timoshenko and J. N. Goodier. *Theory of Elasticity*, Chapter 12, (McGraw-Hill Kogakusa, Ltd, Tokyo, 1970).
69. H. Yoshida, M. M. Chaudhri, T. Fukudome, and S. Tsuruzono. Impact fracture behavior of turbine-grade silicon nitride ceramic under tensile stress at elevated temperatures. ASME Paper GT2005-68182. Proceedings of ASME Turbo Expo 2005, June 6–9, Reno, NV, 2005.
70. J. Tschinkel. Formation of sodium sulfate in gas turbine combustors. *Corrosion*, 28(5), 161–169 (1972).

71. R. T Bhatt, S. R. Choi, L. M. Cosgriff, D. S. Fox, and K. M. Lee. Impact resistance of environmental barrier coated SiC/SiC composites. *Mater. Sci. Eng.*, A476, 8–19 (2008).

72. D. Hass. Impact and erosion resistance of novel thermal/environmental barrier coating systems. Ceramic Matrix Composites Symposium, Materials Science & Technology (MS&T) 2011 Conference & Exhibition, October 16–20, Columbus, OH, 2011.

73. F. Abdi, H.-M. Yun, C. Godines, and G. Morscher. Modeling foreign object damage to CVI MI SiC/SiC and oxide/oxide ceramic composite components in gas-turbine engines at ambient and elevated temperatures. ASME Paper GT2011-46851. Proceedings of ASME Turbo Expo 2012, June 11–15, Copenhagen, Denmark, 2012.

74. Y. Xue, F. Abdi, G. Morscher, and S. Choi. Non destructive ceramic matrix composite impact modeling validation. ASME Paper GT2013-94728. Proceedings of ASME Turbo Expo 2013, June 3–7, San Antonio, TX, 2013.

75. F. Abdi, W. Troha, G. Morscher, and S. Choi. ES2012 ballistic impact and parameters sensitivity of ceramic aerospace materials at room temperature. Proceedings of ASCE, Earth & Space 2012 Conference, April 15–18, Pasadena, CA, 2012.

76. F. Abdi, G. Morscher, W. Troha, and S. Choi. Multi-scale analysis of foreign object damage (FOD) in ceramic matrix composites (CMCs) using and adaptive meshing and damage evolution technique. 36th Annual Conference on Composites, Materials, and Structures, January 23–26, Cape Canaveral, FL (Restricted Sessions), 2012.

77. C. Iwano, E. Klock-McCook, and B. Sullivan. Innovative approaches to developing foreign object damage resistant ceramic matrix composites. 36th Annual Conference on Composites, Materials, and Structures, January 23–26, Cape Canaveral, FL (Restricted Sessions), 2012.

78. R. S. Fertig and D. Robbins. Efficient multiscale modeling for predicting effect of foreign object damage in SiC/SiCs. 36th Annual Conference on Composites, Materials, and Structures, January 23–26, Cape Canaveral, FL (Restricted Sessions), 2012.

79. I. Miller and E. Madenci. Foreign object damage prediction in ceramic matrix composites. 36th Annual Conference on Composites, Materials, and Structures, January 24–27, Cape Canaveral, FL (Restricted Sessions), 2011.

CHAPTER 15

ENVIRONMENTAL BARRIER COATINGS FOR SiC_F/SiC

KANG N. LEE

Rolls Royce Corporation, Indianapolis, IN, USA

15.1 INTRODUCTION

Economical and environmental concerns, that is, higher efficiency, higher thrust, and reduced emissions, are the main driving force behind the ever increasing demand for higher gas turbine engine inlet temperatures. Technology improvements in cooling, materials, and coatings are required to achieve higher inlet temperatures [1]. Advances in the development of airfoil cooling designs have been achieved by combining high convective cooling efficiencies with film cooling. As the temperature capability of Ni-based superalloys approaches their intrinsic limit, further improvements in their temperature capability have become increasingly difficult. SiC/SiC ceramic matrix composites (CMCs) are the most promising material system that has the temperature and structural capability to meet the needs of next generation gas turbine engines [2].

Original Equipment Manufacturers (OEMs) have plans to gradually implement SiC/SiC CMCs in hot section components, such as blade tracks, combustor liners, vanes, and blades. One major barrier to the implementation of CMC hot section components is the lack of environmental durability in combustion environments [3–6]. Water vapor, a combustion reaction product, reacts with protective silica scale that forms on CMC, forming gaseous reaction products, such as $Si(OH)_4$. In high pressure, high gas velocity combustion environments, this reaction results in rapid recession of CMCs. CMCs also suffer from severe hot corrosion in molten salt [7–9].

Initial coating work for Si-based ceramics (SiC and Si_3N_4) in the late 1970s to mid 1980s was to provide protection from molten salt corrosion and contact stress damages as summarized in Reference 10. Mullite coating for SiC with superior crack resistance compared to previous mullite coatings was developed in the early 1990s at NASA [11]. After the discovery of SiC recession in high pressure, high velocity combustion environments a new class of coatings, now known as environmental barrier coatings (EBCs), was developed in the late 1990s in NASA's HSCT-EPM (high speed civil transport-enabling propulsion materials) Program to protect CMCs from water vapor in gas turbine engines [12, 13]. The first generation EBCs consist of three layers: silicon bond coat, mullite-based intermediate coat, and barium–strontium–aluminosilicate ($1-x$BaO·xSrO·Al_2O_3·$2SiO_2$, $0 \leq x \leq 1$) top coat. CMC combustor liners coated with the first generation EBCs were retrofitted in a Solar Turbines' industrial gas turbine engine and successfully completed a ~14 000-hour field test in the late 1990s [14–17]. Following the successful field test of the first generation EBCs, second generation EBCs based on rare earth silicates having higher temperature capabilities than the first generation EBCs were developed in the early 2000s in NASA's UEET (ultraefficient engine technology) Program [18]. Next generation EBCs under development are based on the first and second generation EBCs.

EBC is a prime reliant coating and therefore developing a reliable EBC lifing method is critical for successful implementation of CMC components. Considering the similarity between EBC and TBC (thermal barrier coating) lifing

Ceramic Matrix Composites: Materials, Modeling and Technology, First Edition. Edited by Narottam P. Bansal and Jacques Lamon.

approaches similar to those developed for TBC seem to be logical options. Multiple sources responsible for EBC degradation, such as water vapor, Calcium-magnesium aluminosilicate (CMAS), stresses, foreign objects, etc., are present simultaneously in gas turbines and therefore strong interactions between various degradation modes are expected. EBC lifing will require thorough understanding of individual degradation mode as well as interactions among them. A model that incorporates all these factors needs to be developed to successfully project the life of EBC-coated CMCs in gas turbines. Generation of life validation data in relevant environments is as important. Simulating engine environments relevant for EBCs requires specialized test rigs, such as steam cycling rig, laser rig, and burner rig. Since there is no one rig that is capable of simulating all key sources critical for EBC degradation, combinations of rigs are utilized to validate EBCs.

15.2 BACKGROUND

SiC/SiC CMCs are the most promising material system that has the temperature and structural capability to meet the needs of next generation gas turbines [2].

Figure 15.1 shows turbine components that can be benefited by CMCs. Potential benefits include reduced SFC (specific fuel consumption), reduced cooling air, reduced emissions, and reduced noise. One major barrier to implementing CMCs in gas turbines is surface recession due to evaporation of silica scale by water vapor in combustion environments [3–6].

15.2.1 Silica Volatility

SiC oxidizes according to the following reaction [7]:

$$SiC + 1.5O_2(g) = SiO_2 + CO(g) \tag{15.1}$$

SiC/SiC CMCs have excellent high temperature oxidation resistance in dry air because SiO_2 is one of the slowest-growing oxide scales in dry air, growing even slower than Al_2O_3 at $T > \sim 1000°C$. In an environment containing water vapor and oxygen, SiC oxidizes according to reaction (15.1) and the following reaction [3]:

$$SiC + 3H_2O(g) = SiO_2 + 3H_2(g) + CO(g) \tag{15.2}$$

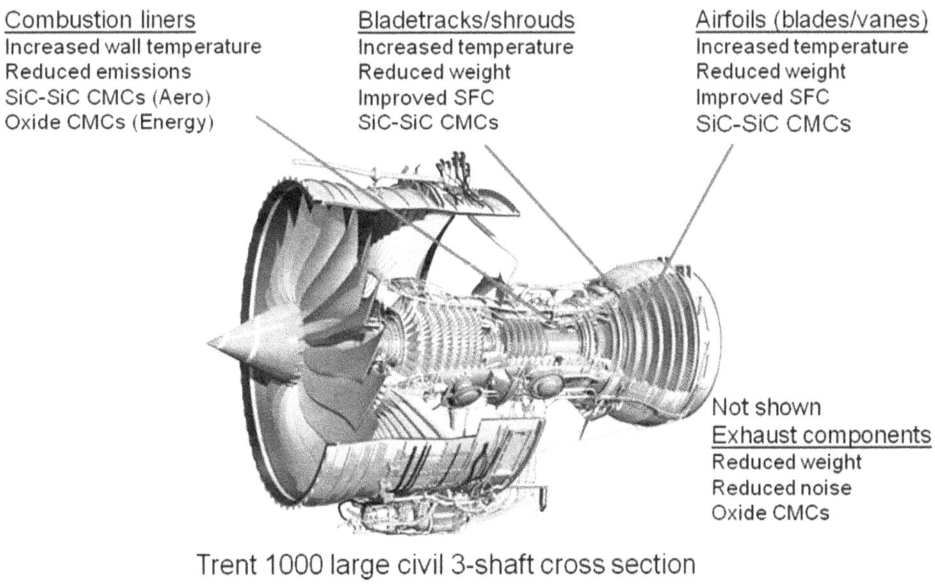

FIGURE 15.1
Candidate CMC turbine components.

Jacobson calculated that that the products in combustion contain about 10% water vapor independent of the fuel-to-air ratio [19]. It was also shown that, in a high velocity, high water vapor pressure environment, simultaneous oxidation of SiC to form SiO_2 and volatilization of SiO_2 to form $Si(OH)_4$ according to reaction (15.3) occur [3]:

$$SiO_2 + 2H_2O(g) = Si(OH)_4(g) \tag{15.3}$$

Opila et al. developed a silica volatility model based on laminar boundary layer controlled flux equation in NASA's HSCT-EPM (high speed civil transport-enabling propulsion materials) Program, which shows the following dependency of silica volatility on gas velocity and pressure [3]:

$$\text{Volatility} \propto v^{1/2} \times p(H_2O)^2/(P_{total})^2 \tag{15.4}$$

Here, v = velocity, $P(H_2O)$ = partial pressure of water vapor, and P_{total} = total pressure. Robinson et al. generated a large amount of SiC volatility data in a high pressure burner rig and performed linear regression to fit the data in the laminar boundary layer flux equation. The results are Equations (15.5) and (15.6), for rich and lean burn, respectively [20]:

$$\text{Volatility (rich)} = 82.5 \exp(-159(\text{kJ/mol})/RT) P^{1.74} v^{0.69} \tag{15.5}$$

$$\text{Volatility (lean)} = 2.04 \exp(-108(\text{kJ/mol})/RT) P^{1.50} v^{0.50} \tag{15.6}$$

Here, the units are volatility in mg/(cm^2 h), T in Kelvin, P in atm, and v in m/s. SiC recession rate (μm/h) can be obtained by multiplying the volatility in Equations (15.5) and (15.6) by 3.1.

Figure 15.2 shows the projected 100 hours recession rate of CMC as a function of gas velocity in a lean burn combustion environment at P_{total} = 7 atm and T = 1400°C using Equation (15.6) [21]. Several mils of CMC recession at 100 hours, as projected in Figure 15.2, will rapidly consume CMC components. Projected recession rates will be even higher, according to Equation (15.4), as the temperature and pressure increase in next generation gas turbines. Solar Turbines' SiC/SiC combustor liners in an industrial gas turbine in the world's first CMC component field test in the late 1990s lost as much as 80% of the CMC liner thickness after 5000 hours with the maximum liner temperature at ~1250°C [14–17]. These results prove that CMCs are not viable in the gas turbine hot section without mitigating the recession.

15.2.2 Key Requirements for EBC

An external barrier coating, now known as EBC, is needed to protect CMCs from water vapor–induced recession. Key requirements for a successful EBC include [12]: (i) environmental stability; (ii) adherence; (iii) chemical compatibility; and (iv) low stresses. Figure 15.3 schematically shows the key EBC requirements. It is also desirable to have a low thermal conductivity for cooled components to save cooling air.

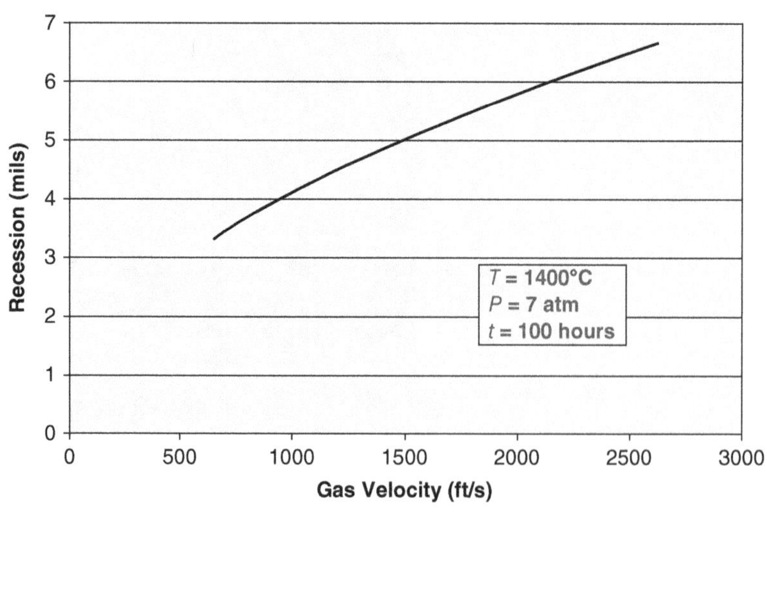

FIGURE 15.2
Projected 100 hours recession rate of CMC as a function of gas velocity in a lean burn gas turbine at P_{total} = 7 atm and T = 1400°C using Equation (15.6).

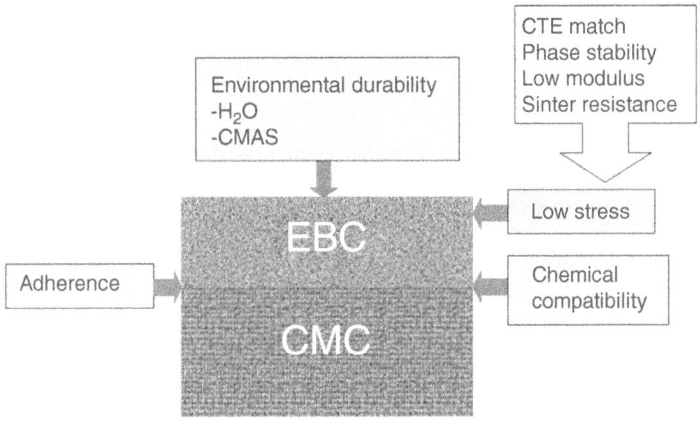

FIGURE 15.3

A schematic of key EBC requirements. Reproduced with permission from Reference 12, Elsevier S.A. Surface Coatings and Tech, Vol 133–134.

Thermal Expansion: For a long life under thermal cycling environments, the coefficient of thermal expansion (CTE) should be the first criterion to be considered for selecting coating candidates.

Figure 15.4 compares the thermal expansion of refractory oxides and silicates with the thermal expansion of SiC and Si_3N_4 [10]. Note that traditional high temperature refractory oxides, such as alumina, zirconia, and hafnia, all have significantly higher thermal expansion than SiC and Si_3N_4. In contrast refractory silicates, such as mullite, BSAS, and rare earth silicates, have fairly close CTE match with SiC and Si_3N_4. Coatings research for CMCs, therefore, has focused on these high temperature silicate materials.

Water Vapor Stability: The next key criterion for selecting coating candidates is the volatility in water vapor.

Figure 15.5 shows calculated volatility versus temperature for SiC, mullite, and BSAS at $T = 1500°C$, $p(H_2O) = 0.5$ atm, $P_{total} = 1$ atm, and $v = 4.4$ cm/s, using silica volatility model and thermodynamics and kinetics database [21]. When thermodynamics and kinetics data are not available, first principle calculations were used to estimate the data. The calculated SiC volatility in Figure 15.5 was validated by comparing it against the experimental volatility in Reference 20. Note the disparity in volatility between SiC, mullite, and BSAS, with BSAS showing the lowest volatility.

Figures 15.6 and 15.7, respectively, show experimentally measured volatility of selected rare earth disilicates and rare earth monosilicates at $T = 1500°C$, $p(H_2O) = 0.5$ atm and $P_{total} = 1$ atm [18]. The experimental volatility of BSAS is also plotted in Figure 15.6 for a comparison. The volatility ranking based on Figures 15.5, 15.6, and 15.7 is SiC > mullite > BSAS, rare earth disilicate > rare earth monosilicate.

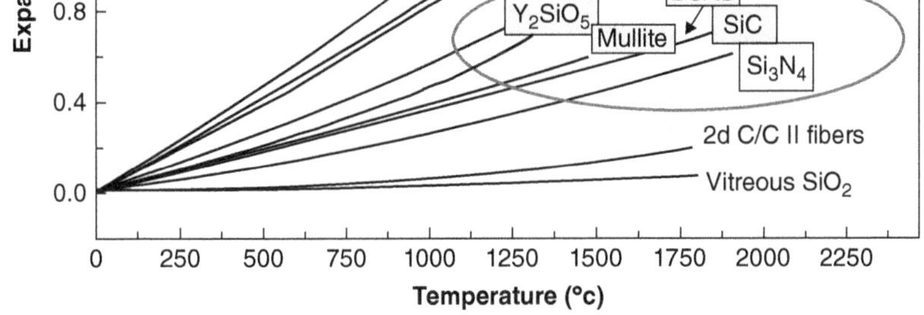

FIGURE 15.4

Comparison of the thermal expansion of refractory oxides and silicates with the thermal expansion of SiC and Si_3N_4 [10]. Permission from ASME Press, Chapter 27: Coatings for Engineering Ceramics, 2003.

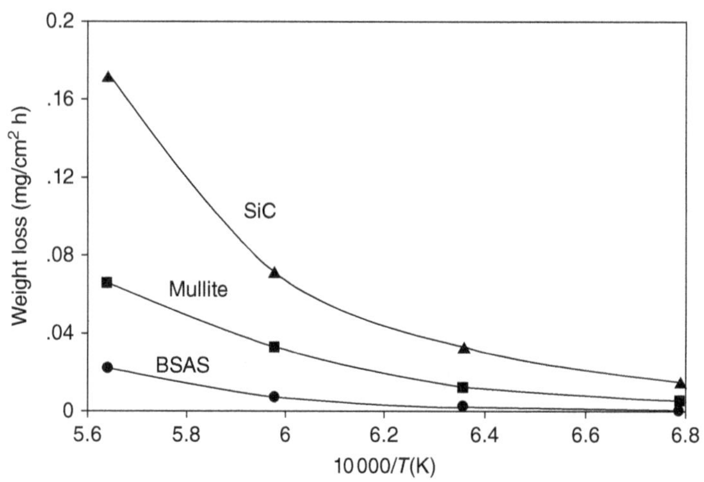

FIGURE 15.5

Calculated volatility versus temperature for SiC, mullite, and BSAS at $T = 1500°C$, $p(H_2O) = 0.5$ atm, $P_{total} = 1$ atm, and $v = 4.4$ cm/s, using the silica volatility model (Equation 15.4) and thermodynamic and kinetics database.

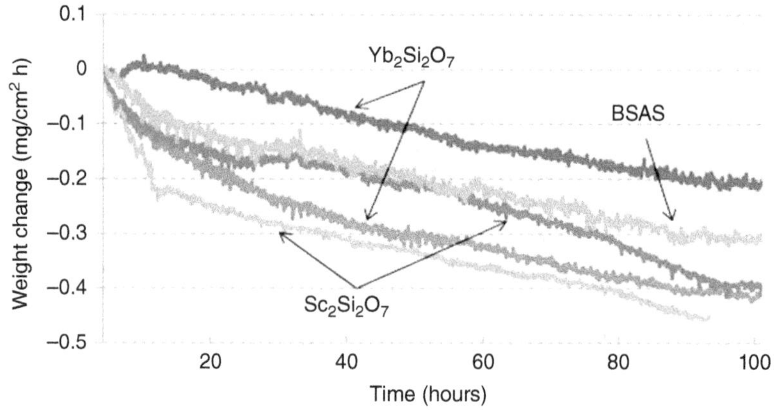

FIGURE 15.6

Experimentally measured volatility of selected rare earth disilicates at $T = 1500°C$, $p(H_2O) = 0.5$ atm, and $P_{total} = 1$ atm. Reproduced with permission from Reference 18, Elsevier, Journal of European Ceramic Society, Vol. 25, Iss 10, 2005.

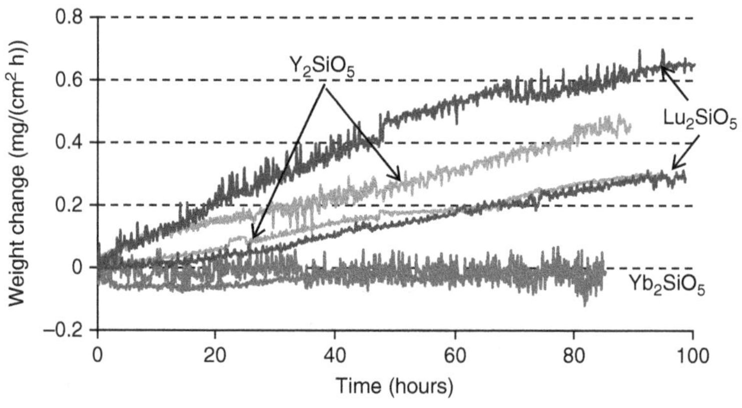

FIGURE 15.7

Experimentally measured volatility of selected rare earth monosilicates at $T = 1500°C$, $p(H_2O) = 0.5$ atm, and $P_{total} = 1$ atm. Reproduced with permission from Reference 18, Elsevier, Journal of European Ceramic Society, Vol 25, Iss 10, 2005.

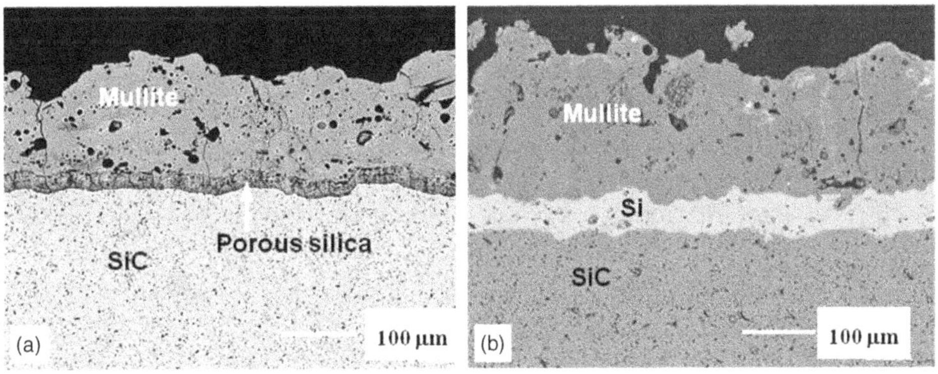

FIGURE 15.8

Cross section of (a) mullite-coated SiC and (b) silicon/mullite-coated SiC after 200 hours at $T = 1300°C$ with 2 hour cycles and 300 hours at $T = 1380°C$ with 1 hour cycles, respectively, at $p(H_2O) = 0.9$ atm, $P_{total} = 1$ atm, and $v = 2.2$ cm/s.

Adherence: Refractory oxide and silicate coatings need a bond coat because they do not bond very well on SiC or Si_3N_4. Silica scale grows fast at the oxide or silicate coating and SiC interface in steam environments via Equation (15.2). Stresses due to thermally grown low CTE silica scale can lead to cracking in the silica scale under thermal cycling [22, 23]. The coating eventually spalls off when these cracks grow horizontally and coalesce. Currently silicon is the most effective bond coat. It provides excellent oxidation resistance and thus much enhanced thermal cycling life, which is attributed to the combination of strong chemical bonding and close CTE match with CMC [12, 13].

Figures 15.8a and 15.8b compare the cross section of (a) mullite-coated SiC and (b) silicon/mullite-coated SiC after 200 hours at $T = 1300°C$ and 300 hours at $T = 1380°C$, respectively, at $p(H_2O) = 0.9$ atm, $P_{total} = 1$ atm, and $v = 2.2$ cm/s [24]. The cycling frequency was 2 hours (Figure 15.8a) and 1 hour (Figure 15.8b) at temperature and 12 minutes at room temperature. A thick (~25 μm) porous silica scale formed at the mullite/SiC interface (Figure 15.8a), whereas silica scale was too thin to be visible with the silicon bond coat (Figure 15.8b), demonstrating the effectiveness of Si bond coat.

Chemical Compatibility: As oxides and silicates have a high tendency to react with each other at high temperatures, the chemical compatibility at interfaces is a key factor for a long life.

Figure 15.9 shows BSAS-coated CMC after 100 hours at $T = 1300°C$, $p(H_2O) = 0.9$ atm, $P_{total} = 1$ atm, and $v = 2.2$ cm/s [13]. Each cycle was 2 hours at 1300°C and 12 minutes at room temperature. A thick glassy reaction zone developed at the silica scale/BSAS interface with multiple pores in the reaction zone. The glassy reaction zone is attributed to a low melting (~1300°C) eutectic in the $BaO–Al_2O_3–SiO_2$ ternary system [13, 25], while pores are attributed to bubbling of gaseous species through the low viscosity eutectic glass [13]. In long exposures, pores continue

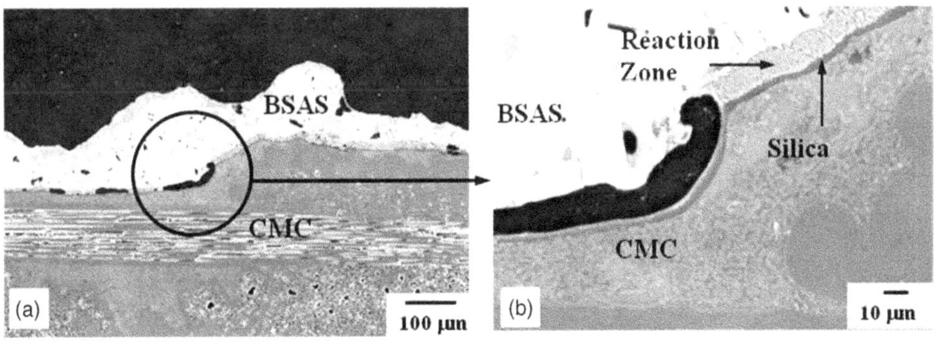

FIGURE 15.9

Cross section of BSAS-coated CMC after 100 hours with 2 hour cycles at $T = 1300°C$, $p(H_2O) = 0.9$ atm, $P_{total} = 1$ atm, and $v = 2.2$ cm/s. Reproduced with permission from Reference 13, Journal of the American Ceramic Society, Vol 86, Iss 8, 2003.

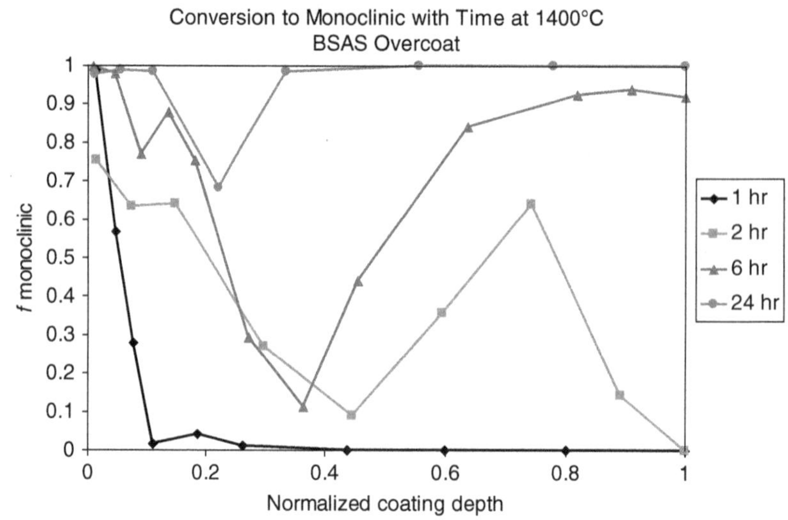

FIGURE 15.10

Fraction of monoclinic celsian ($f_{monoclinic}$) across the BSAS thickness after 1, 2, 6, and 24-hour heat treatment at 1400°C. Reproduced with permission from Reference 13, Journal of the American Ceramic Society, Vol 86, Iss 8, 2003.

to grow and coalesce, potentially leading to complete coating spallation. The low viscosity reaction zone can also lead to EBC spallation under shear stresses.

Phase Stability: Phase stability is another factor that will affect EBC stresses and thus life. BSAS has two polymorphs: celsian (monoclinic) and hexacelsian (hexagonal) [26–29]. Plasma-sprayed BSAS was mostly amorphous as sprayed but readily converted to hexacelsian after a thermal exposure [13, 30]. For example, hexacelsian was the dominant phase after 130 minutes at 1100°C and after less than 10 minutes at 1200°C. In contrast, the conversion from hexacelsian to celsian was much more sluggish.

Figure 15.10 shows the fraction of monoclinic celsian ($f_{monoclinic}$) across the BSAS thickness after 1, 2, 6, and 24 hours heat treatment at 1400°C [13]. The celsian phase preferentially nucleated at the outer surface and later near the BSAS/bond coat interface, and the conversion slowly progressed inward from both surfaces. High energy transmission x-ray diffraction is a useful tool to determine stresses in EBCs [31–34].

Figure 15.11 shows the CTE and stress in various phases present in Si/mullite + 20wt% SAS/BSAS-coated CMC after 20 hours annealing at 1300°C [34]. BSAS becomes SAS when all BaO is replaced with SrO. Note that phases with

Phase	CTE ($\times 10^{-6}$/C)
BSAS (Hexacelsian)	8.37
BSAS (Celsian)	4.28
Mullite	5.50
SAS	4.28
Silicon	4.44
SiC	5.06

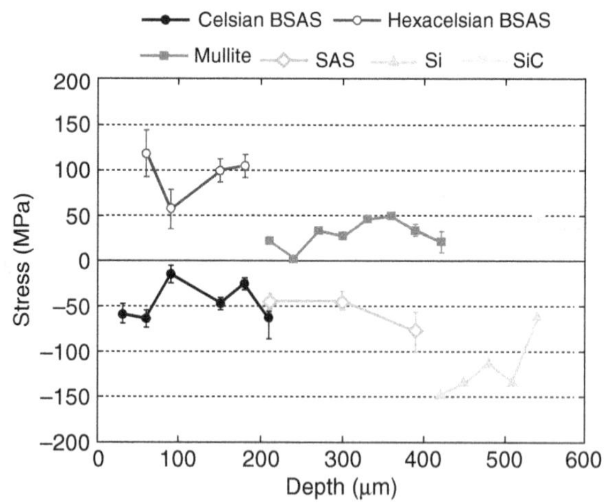

FIGURE 15.11

CTE and stress in various phases present in Si/(mullite + 20 wt% SAS/BSAS-coated CMC after 20 hours annealing at 1300°C determined by high energy transmission x-ray diffraction. Reproduced with permission from Reference 34, Journal of the American Ceramic Society, Vol 86, Iss 8, 2003.

the CTE higher than SiC are in tensile and phases with the CTE lower than SiC are in compression. This indicates that stresses are relaxed at high temperatures and that stresses build up according to the CTE mismatch as the coated CMC is cooled. This is also consistent with the observation that as-sprayed BSAS develops vertical cracks that close up after thermal exposure as compressive stresses develop due to hexacelsian to celsian phase transformation. Phase stability of mullite is also critical to the EBC durability as will be discussed later in the Section 15.3.1.

No single material satisfies all key requirements and therefore current EBCs evolved into multilayer coatings [12, 13, 18]. An EBC, at the minimum, comprises a bond coat and a top coat. Bond coat facilitates adherence and top coat provides protection from water vapor and thermal insulation. Additional layers may be incorporated between the bond coat and the top coat and/or above the top coat to facilitate various functions as needed, such as chemical compatibility, CTE transition, compliance, water vapor stability, higher temperature capability, environmental durability, and low thermal conductivity.

15.3 EVOLUTION OF EBCs

15.3.1 Mullite Coating

Early coatings research for SiC and Si_3N_4 was conducted in the late 1970s to mid 1980s at Garrett Turbine Engine Company, Oak Ridge National Laboratories, Solar Turbines, and GTE laboratories to protect these ceramics from contact stress damages and molten salt corrosion [10]. Mullite has attracted the most interest because of its good CTE match and chemical compatibility with SiC and Si_3N_4 ceramics. A major breakthrough occurred in the early 1990s at NASA. Lee et al. [11] identified the crystallization of amorphous mullite and the accompanying shrinkage under thermal cycling in conventionally plasma-sprayed mullite as the root cause for poor thermal cycling durability of previous plasma-sprayed mullite coatings. The NASA group subsequently developed a modified plasma-spray process that enabled the deposition of crystalline mullite coating that exhibited dramatically improved thermal cycling durability.

Figures 15.12a and 15.12b compare two mullite coatings processed by conventional plasma spraying and modified plasma spraying, respectively, after 48 hours with 2 hour cycles at 1000°C in air [11]. Mullite by conventional plasma spraying was severely cracked and delaminated (Figure 15.12a), while mullite by modified plasma spraying showed excellent adherence with only a few vertical cracks (Figure 15.12b). The new mullite coating remained intact after 1200 hours with 2 hour cycles at 1300°C in air as well as after 50 hours in a 6 atm pressure hot corrosion rig containing Na_2SO_4 [23,35]. CVD mullite coating was developed at Boston University [36–39] in the mid 1990s, which also showed excellent durability in air thermal cycling and hot corrosion test [40–42]; however, it was never followed up for scaling up.

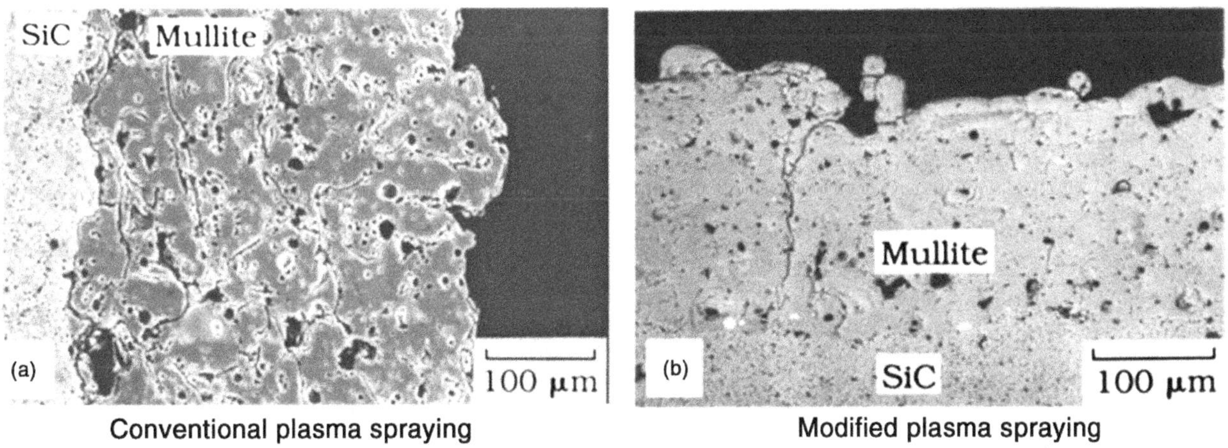

FIGURE 15.12

Comparison of two mullite coatings processed by (a) conventional plasma spraying and (b) modified plasma spraying, after 48 hours at 1000°C with 2 hour cycles in air. Permission from Reference 11, Journal of the American Ceramic Society, 78 [3] 705-10 (1995).

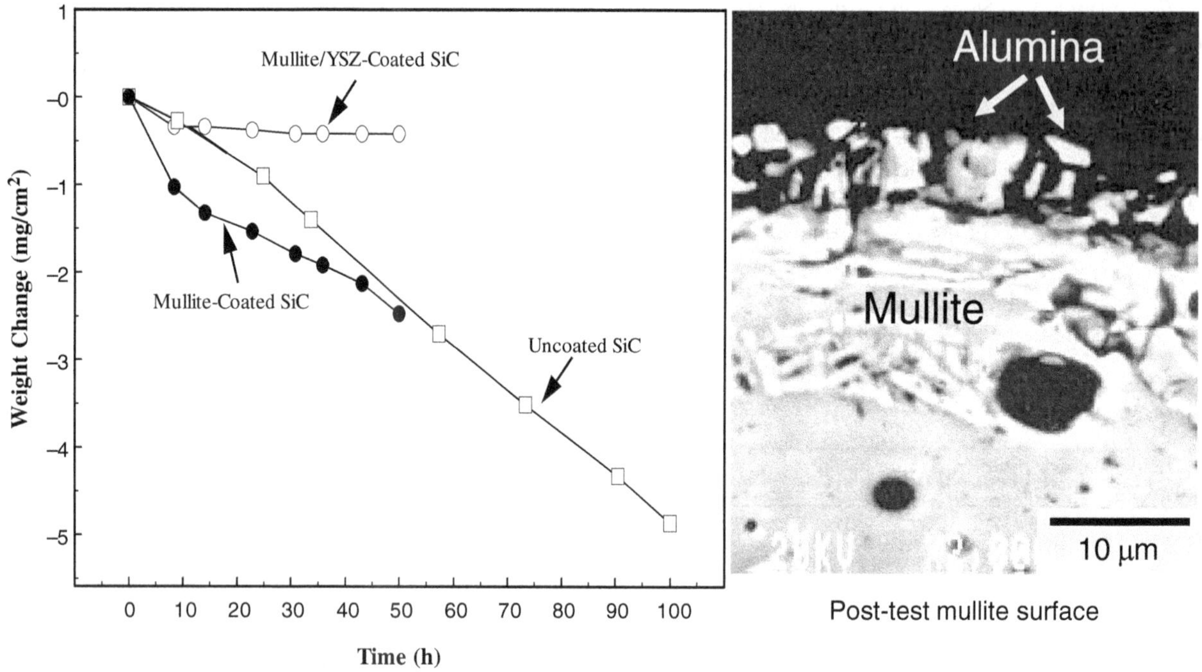

FIGURE 15.13

Recession of SiC, mullite, and 7YSZ in a high pressure burner rig test ($T = 1230°C$, $P_{total} = 6$ atm, $p(H_2O) = 0.6$ atm, $v_{gas} = 24$ m/s) [23]. Permission from ASME Press, Key Durability Issues with Mullite-Based Environmental Barrier Coatings for Si-Based Ceramics, Journal of Engineering for Gas Turbines and Power, Volume 122, 2000.

15.3.2 First Generation Environmental Coatings

After the discovery of water vapor–induced SiC recession in the HSCT-EPM Program, a NASA-GE-PW coating team was established in the Program to develop a coating that can protect CMCs from high velocity, high pressure combustion environments. As was discussed in the Section 15.2.2 mullite does not have adequate water vapor stability for a long life in gas turbines. The relatively high silica activity in mullite (0.3–0.4) causes selective volatilization of silica and recession of mullite in high velocity combustion environments.

Mullite recessed fairly fast in a 50-hour high pressure burner rig test at $T = 1230°C$, $P_{total} = 6$ atm, $p(H_2O) =$ 0.6 atm, and $v_{gas} = 24$ m/s, leaving a 6- to 7-μm-thick porous alumina surface layer due to selective evaporation of silica (Figure 15.13) [23]. The porous alumina surface layer readily crumbled and spalled off. Note that 7YSZ did not lose any weight, indicating its superior recession resistance.

Extensive search for EBC candidates in the HSCT-EPM Program identified barium–strontium–aluminosilicate (BSAS: $1\text{-}xBaO{\cdot}xSrO{\cdot}Al_2O_3{\cdot}2SiO_2$, $0 \leq x \leq 1$) as the most promising candidate. It has a few key attributes required for a successful EBC, that is, low silica activity (~0.1), good CTE match with SiC (Figure 15.4), and a low modulus (~100 GPa for dense BSAS). The low silica activity translates to high stability in water vapor (Figure 15.5), while the low CTE and low modulus translate to low stresses in thermal cycling. BSAS, however, forms detrimental glassy reaction zone in contact with CMC as discussed in Section 15.2.2. Using the new mullite coating as a bond coat eliminated the chemical incompatibility issue with BSAS; however, mullite bond coat does not have a long durability in steam cycling as shown in Figure 15.8a. Two key breakthroughs had to occur to transform the coating based on mullite and BSAS into a viable EBC. The first breakthrough was the development of mullite + BSAS mixture bond coat [12, 13]. Adding about 20 wt.% BSAS in the mullite bond coat significantly improved the thermal cycling durability.

Figure 15.14 is a cross section of Si/mullite + 20 wt.% BSAS/BSAS-coated CMC after 100 hours with 1 hour cycles at $T = 1300°C$, $p(H_2O) = 0.9$ atm, $P_{total} = 1$ atm, and $v = 2.2$ cm/s [13]. There was no visible oxidation or degradation. The improved thermal cycling durability is attributed to the reduced strain in the mixture bond coat due to the low modulus BSAS second phase.

Figure 15.15 is a cross section of Si/mullite + 20 wt.% BSAS/BSAS-coated CMC after 1000 hours with 1 hour cycles at $T = 1316°C$, $p(H_2O) = 0.9$ atm, $P_{total} = 1$ atm, and $v = 2.2$ cm/s [18]. The EBC showed no visible interfacial glass

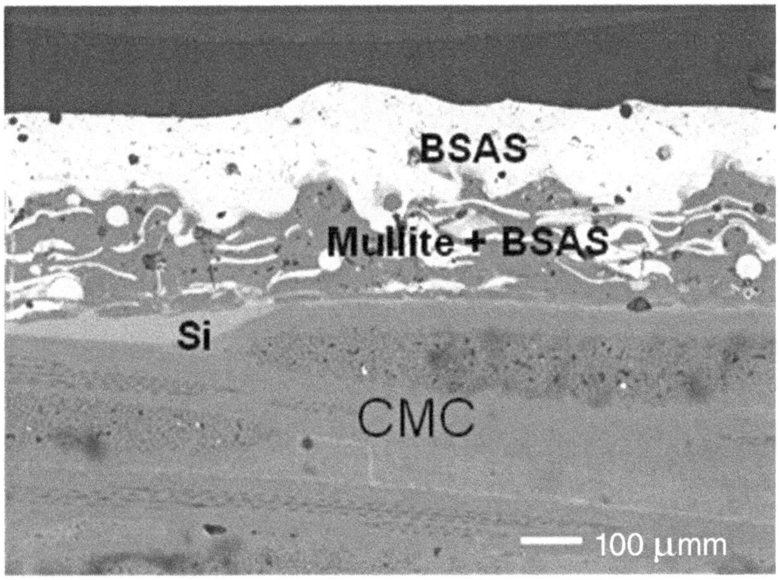

FIGURE 15.14

Cross section of Si/mullite + 20 wt% BSAS/BSAS-coated CMC after 100 hours with 1 hour cycles at $T = 1300°C$, $p(H_2O) = 0.9$ atm, $P_{total} = 1$ atm, and $v = 2.2$ cm/s. Reproduced with permission from Reference 13, Journal of the American Ceramic Society, Vol 86, Iss 8, 2003.

formation or thick silica scale. Pockets of glassy phases developed within the BSAS top coat, which are likely beneficial for compliance but detrimental to water vapor stability. As the temperature increases the glass formation reaction kicks off, eventually leading to EBC spallation.

Figure 15.16 shows a cross section of Si/mullite + 20 wt.% BSAS/BSAS-coated CMC and after 300 hours with 1 hour cycles at $T = 1400°C$, $p(H_2O) = 0.9$ atm, $P_{total} = 1$ atm, and $v = 2.2$ cm/s [13], and the phase diagram showing BAS-silica eutectic [13] and Figure 15.17 shows a cross section of mullite + 20 wt.% BSAS/BSAS-coated SiC after 20 hours at 1440°C in air [13]. At 1400°C EBC remained adherent, but a section of EBC transformed to a glass (Figure 15.16), while at 1440°C EBC spalled off and was completely surrounded by glass phases. The composition of the glass on both sides of the spalled coating in Figure 15.17 was similar to the BSAS–silica eutectic composition, indicating that the glass is a product of the BSAS–SiO_2 eutectic reaction. The glass formation due to BSAS–SiO_2 reaction suggests that EBCs based on mullite and BSAS should not be used above the eutectic temperature (~1300°C) for long exposures. The second breakthrough was the identification of silicon as a bond coat (see Figure 15.8b) [12, 13, 24]. The introduction of silicon bond coat and mullite + BSAS mixture intermediate layer improved the thermal cycling life of EBCs based on mullite and BSAS by at least an order of magnitude.

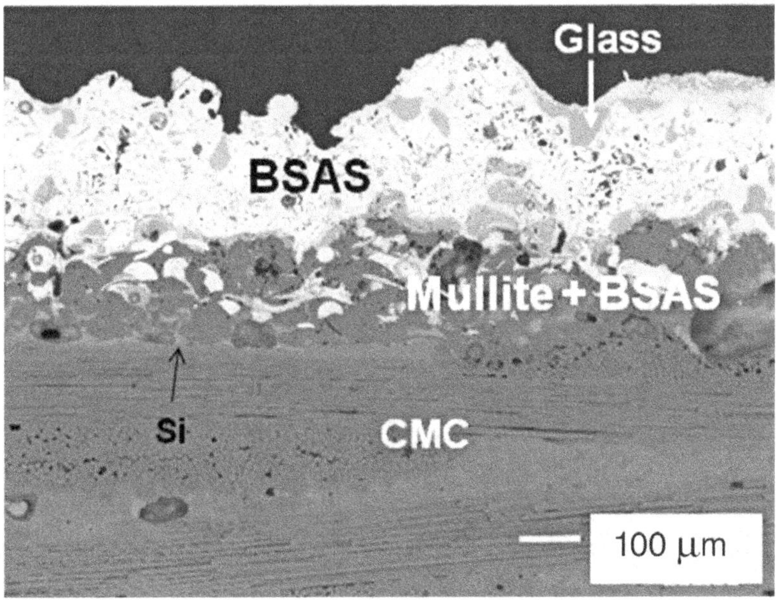

FIGURE 15.15

Cross section of Si/mullite + 20 wt.% BSAS/BSAS-coated CMC after 1000 hours with 1 hour cycles at $T = 1316°C$, $p(H_2O) = 0.9$ atm, $P_{total} = 1$ atm, and $v = 2.2$ cm/s. Reproduced with permission from Reference 18, Elsevier, Journal of European Ceramic Society, Vol 25, Iss 10, 2005

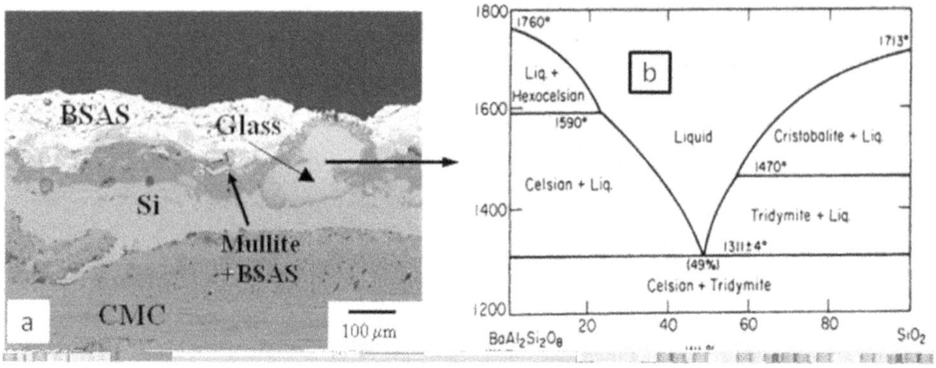

FIGURE 15.16

Cross section of Si/mullite + 20 wt.% BSAS/BSAS-coated CMC after 300 hours with 1 hour cycles at $T = 1400°C$, $p(H_2O) = 0.9$ atm, $P_{total} = 1$ atm, and $v = 2.2$ cm/s and the phase diagram showing BSAS–Silica eutectic. Reproduced with permission from Reference 13, Journal of the American Ceramic Society, Vol 86, Iss 8, 2003.

Another logical EBC candidate from water vapor stability point of view was yttria-stabilized zirconia (7YSZ: ZrO_2 + 7 wt.% Y_2O_3). 7YSZ has been successfully used as a TBC for superalloy components in gas turbine engines [43]. Key disadvantage of YSZ as an EBC are high CTE (see Figure 15.4) and sintering at $T > 1200°C$. YSZ is not adherent in thermal cycling without a bond coat because of the large CTE mismatch with CMC. Figure 15.13 shows mullite/YSZ-coated SiC after a 50 hours high pressure burner rig test at $T = 1230°C$, $P_{total} = 6$ atm, $p(H_2O) = 0.6$ atm, and $v_{gas} = 24$ m/s [23]. Note the absence of weight loss indicating excellent stability of YSZ in water vapor. Large in-plane tensile stresses due to CTE mismatch and sintering cause vertical cracks in YSZ in thermal cycling. Vertical cracks provide easy paths for water vapor, resulting in rapid oxidation of CMC at the interface. Vertical cracks can propagate horizontally, especially in thermal cycling under temperature gradients, leading to EBC spallation. Modifications to reduce strains due to CTE mismatch and sintering are necessary to make zirconia or hafnia-based TBCs a viable EBC.

Si/mullite + BSAS/BSAS EBC was applied on SiC/SiC CMC combustor liners in Solar Turbines' Centaur 50S gas turbine engines under the DOE Ceramic Stationary Gas Turbines (CSGT) Program [14–17]. One engine used by Texaco in Bakersfield, CA, successfully completed a ~14 000-hour field test ($T_{average}$ = ~1200°C; T_{max} = ~1250°C). The higher operating temperature due to CMC liners resulted in emissions consistently below 15 ppmv nitrogen oxides (NO_x) and

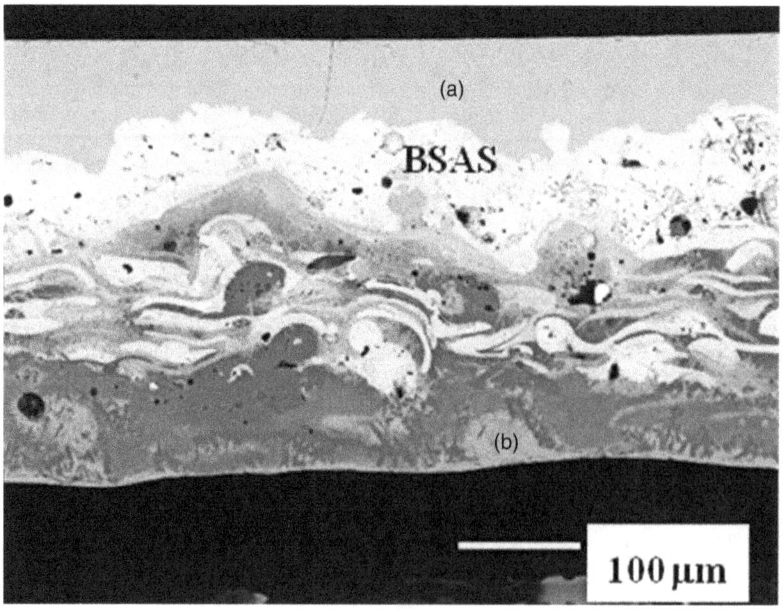

FIGURE 15.17

Cross section of mullite + 20 wt.% BSAS/BSAS-coated SiC after 20 hours at 1440°C in air. (a) glass on EBC surface; (b) glass at EBC bottom. Reproduced with permission from Reference 13, Journal of the American Ceramic Society, Vol 86, Iss 8, 2003.

below 10 ppmv carbon monoxide (CO) throughout, roughly reducing the NO_x and CO loads on the environment by factors of about 2 and 5, respectively.

15.3.3 Second Generation Environmental Coatings

In 1999, research was initiated at NASA under the UEET (ultraefficient engine technology) Program, the follow-on program to the HSCT-EPM Program, to develop EBCs with 1482°C (2700°F) surface temperature and 1316°C (2400°F) EBC/CMC temperature capabilities. It was obvious based on the durability data discussed in the Section 15.3.2 that EBCs based on mullite and BSAS were not going to meet the goal. After another round of extensive search and screening tests, rare earth silicates were selected as prime candidates [12, 18]. Key attributes include high stability in water vapor, high melting points, and, for some selected rare earth silicates, close CTE match with CMC. There are two types of rare earth silicates: monosilicate (RE_2SiO_5) and disilicate ($RE_2Si_2O_7$), where RE = rare earth elements. Advantages of monosilicates are superior water vapor stability (see Figure 15.7) and high melting point (~1950°C for Yb_2SiO_5 vs. ~1850°C for $Yb_2Si_2O_7$) [25]. One advantage of disilicates is better CTE match with CMC. Volatility data (Figures 15.6 and 15.7) indicate that rare earth monosilicates are significantly less volatile than BSAS in water vapor, while the volatility of rare earth disilicates ($RE_2Si_2O_7$) is similar to that of BSAS.

Rare earth disilicates ($RE_2Si_2O_7$) were applied on Si_3N_4 ceramics by a slurry process followed by sintering [44, 45]. Short-term exposures at relatively low temperatures showed promising results. RE monosilicates, similar to YSZ, were not adherent in thermal cycling without a bond coat because of the CTE mismatch with CMC [18]. RE monosilicate-coated CMC suffered rapid oxidation, forming a thick scale at the coating/CMC interface. Premature EBC spallation occurred along the interface due to CTE mismatch stresses coupled with the low toughness of silica scale.

Figure 15.18 shows a cross section of Si/mullite/Yb_2SiO_5-coated CMC after 1000 hours with 1 hour cycles at $T = 1380°C$, $p(H_2O) = 0.9$ atm, $P_{total} = 1$ atm, and $v = 2.2$ cm/s [46]. The EBC maintained excellent adherence and crack resistance. There was no glassy phase formation in contrast to the Si/mullite + 20 wt.% BSAS/BSAS EBC that formed multiple pockets of glassy phases after 1000 hours at 1300°C.

Figure 15.19 is a cross section of Si/mullite/Yb_2SiO_5-coated CMC after a 100-hour high pressure burner rig test at $T = 1400°C$, $P_{total} = 6$ atm, $p(H_2O) = 0.6$ atm, and $v_{gas} = 24$ m/s [46, 47]. The EBC still looked pristine after the exposure. A CMC vane coated with the Si/mullite/Yb_2SiO_5 EBC was successfully tested in a high pressure burner rig (1400°C, 100 hours, 6 atm, $v_{gas} = 24$ m/s) at NASA [47]. The combination of long durability in steam cycling (Figure 15.18) and successful validation in a simulated engine environment (Figure 15.19) demonstrated the viability of EBCs based on rare earth silicates in gas turbine applications.

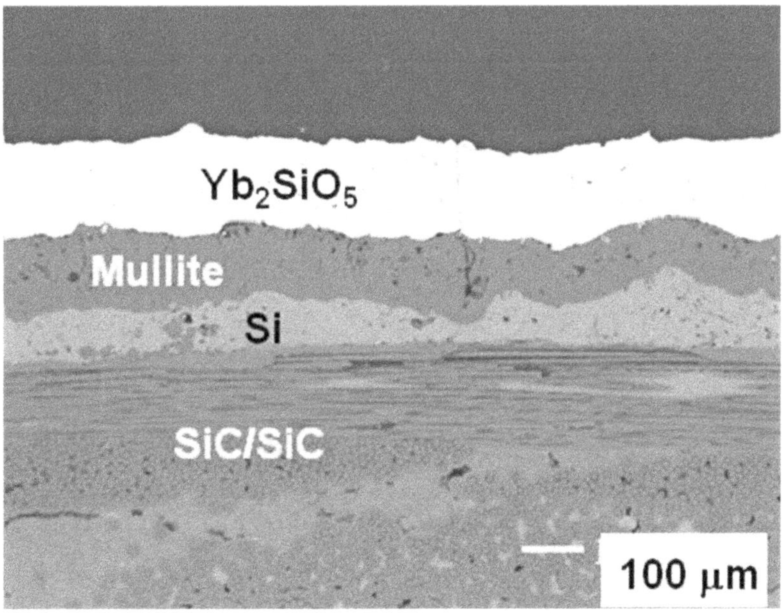

FIGURE 15.18
Cross section of Si/mullite/Yb_2SiO_5-coated CMC after 1000 hours with 1 hour cycles at $T = 1380°C$, $p(H_2O) = 0.9$ atm, $P_{total} = 1$ atm, and $v = 2.2$ cm/s.

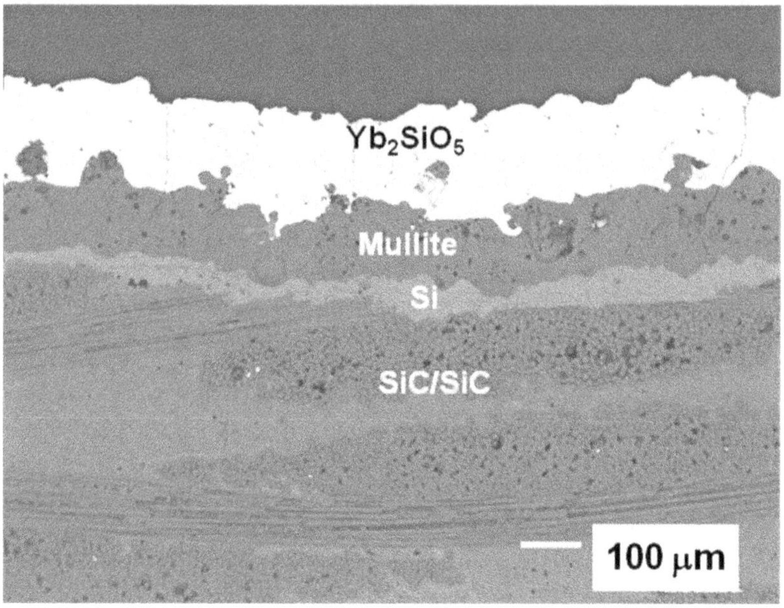

FIGURE 15.19

Cross section of Si/mullite/Yb_2SiO_5-coated CMC after a 100 hours high pressure burner rig test at $T = 1400°C$, $P_{total} = 6$ atm, $p(H_2O) = 0.6$ atm, and $v_{gas} = 24$ m/s.

15.3.4 Next Generation Environmental Coatings

As the implementation of EBC-coated CMC components is around the corner, OEMs and government laboratories continue improving and validating next generation EBCs in relevant environments. Patent searches indicate that the majority of current efforts is on optimizing EBC technologies developed in the first and second generation EBCs, developing cost-effective and robust processing methods, and developing validation test methods that are relevant to gas turbines. There is also a continuous push for higher temperature capabilities, up to 1650°C (3000°F) top coat and 1482°C (2700°F) bond coat. Several engine tests have been conducted with next generation EBCs but test results are not available in the public domain due to the competition-sensitive nature of the technology.

15.4 PROCESSING, TESTING, AND LIFING

15.4.1 Processing

APS (air plasma spraying) and EB-PVD (electron beam physical vapor deposition) are most widely used methods to fabricate coatings for turbine hot-section components [48].

Figures 15.20a and 15.20b show as-deposited APS and EB-PVD 7YSZ, respectively. In plasma-spraying process, powders of coating material are melted in plasma flame and are propelled to the substrate by the flame. Upon impingement

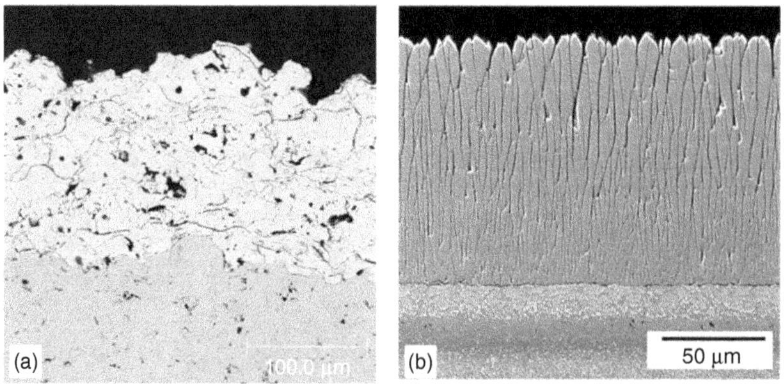

FIGURE 15.20

As-deposited 7YSZ: (a) APS; (b) EB-PVD.

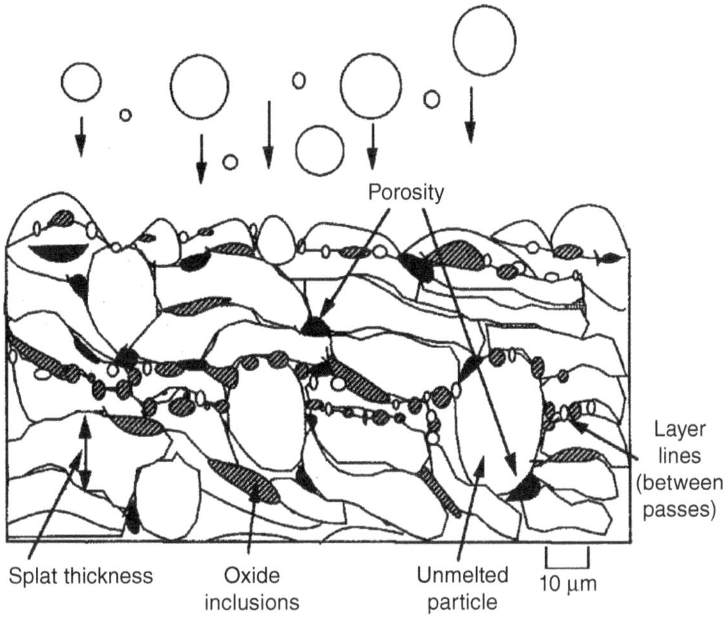

FIGURE 15.21
Thermal spray coating microstructure showing common features. Permission from Reference 49, ASM International, 2004.

on the substrate, the molten droplets are quenched and solidified. A coating is produced by the successive impingement of the droplets, referred to as splats, on the substrate. In EB-PVD process, coatings are produced by condensation of vapor on the substrate. A focused electron beam is used to evaporate the coating material. Multiple beams are employed to produce coatings containing components with differing vapor pressures.

Figure 15.21 schematically shows common features in plasma-sprayed coatings [49]. Two prominent crack morphologies found within as-plasma-sprayed coatings are (1) elongated crack-like separations between flattened splats that melted during the spray deposition. These are oriented nominally parallel to the interface. They form because of thermal contraction as the splats cool; (2) large, more equiaxed, voids contiguous with equiaxed particles. These are particles that did not fully melt during the deposition. The separations between splats are efficient in lowering thermal conductivity, but this is at the expense of surface finish and erosion resistance.

EB-PVD coating has a columnar microstructure (see Figure 15.20b), which imparts excellent strain tolerance and thus longer cyclic life than APS coating. Other advantages of EB-PVD coating include aerodynamically favorable smooth surface finish and good erosion resistance. But the columnar structure with open porosity parallel to the direction of heat conduction results in a higher thermal conductivity compared to APS coating.

APS is currently the most successful and widely used process to fabricate EBCs [10]. With EB-PVD processing of silicate EBCs, the high vapor pressure of silica compared to the other elements makes it difficult to control the coating composition [10]. Other coating processes include DVD (directed vapor deposition) [50, 51], an enhanced EB-PVD process at DVTI (Charlottesville, VA), PS-PVD (plasma-sprayed physical vapor deposition) [52, 53], an enhanced plasma-spraying process that has PVD capability at Sulzer (Wohlen, Switzerland) and NASA (Cleveland, OH), chemical vapor deposition (CVD) [36–39] and slurry-based processes [54–58]. APS and EB-PVD are line-of-sight processes that can leave bald spots on complex-shaped components, whereas DVD, PS-PVD, CVD, and slurry-based process have non-line-of-sight process capability. Non-line-of-sight processes are desirable for CMCs because CMC components cannot afford bald spots due to water vapor recession.

15.4.2 Testing

There are three key components essential for validating EBCs: water vapor, gas velocity, and temperature gradient. Table 15.1 lists various rigs currently being utilized for EBC testing and their capabilities. Steam cycling rig is a low cost, low maintenance rig that is ideal for screening test of candidate EBC systems and validating long-term oxidation of EBC-coated CMC in steam environments [5, 6, 13, 18]. Water vapor is the most important component for the oxidation of Si bond coat and CMC. Gas velocity has no effect on oxidation because the Si bond coat and CMC are shielded from the gas flow by EBC. High velocity–high steam rig has been proven to be useful for long-term EBC recession test

TABLE 15.1 EBC Test Rigs

Rig	Typical test condition	Capability
High steam cycling rig	$P(H_2O)$ = up to 1 atm, v = a few cm/s, P_{total} = 1 atm	Screening test, long-term cyclic steam oxidation test
High velocity–high steam burner rig	$p(H_2O)$ = 0.27 atm, v = 100 m/s, P_{total} = 1 atm	Recession test
High heat flux laser rig	$P(H_2O)$ = very low (ambient air), v = zero, P_{total} = 1 atm	Temperature gradient cycling, time-dependent thermal conductivity
High pressure burner rig	$p(H_2O)$ = 0.6 atm, v = 24 m/s, P_{total} = 6 atm	Recession test

[59–61]. High heat flux laser rig is ideal for temperature gradient cycling test and time-dependent thermal conductivity measurement [62–64]. High pressure burner rig is useful for recession test [4, 20]. Since there is no one rig that is capable of simulating all three components simultaneously, combinations of these rigs are utilized to validate EBCs.

15.4.3 Lifing

In TBCs, it is generally agreed that a TBC fails when the TBC strain energy exceeds the TBC/substrate interfacial toughness [65]. The same principle may apply to EBCs. One problem with EBC lifing is that, unlike TBCs in which 7YSZ is an industry standard and has been in service for more than two decades, EBCs are still evolving and therefore there is no industry standard yet. An industry standard or a model system, which is relevant to various versions of EBCs, needs to be established on which the modeling effort can be focused.

Sources for EBC degradation can be categorized into three groups: chemical, mechanical, and physical. Table 15.2 lists typical modes of degradation in each category and ultimate failure modes. It should be noted that multiple sources for EBC degradation are present simultaneously in gas turbines and therefore strong interactions among various degradation modes are expected.

Oxidation: Oxidation plays a critical role in TBC failure [65]. TGO (thermally grown oxide) resulting from oxidation is a key contributor to TBC strain energy that eventually leads to TBC spallation. Similarly in EBCs, a thick silica TGO has been shown to lead to EBC delamination along the TGO [18, 66], indicating that silica TGO is likely to play a critical role in EBC life.

Recession: BSAS recession calculations using the silica volatility model projects ~70 μm recession after 1000 hours at $T = 1400°C$, $P_{total} = 6$ atm and $v = 24$ m/s, which is unacceptably too high for gas turbine components [67]. Actual gas turbines operate at significantly higher gas velocities and higher total pressures, raising the projected recession much higher. Si/mullite + BSAS/BSAS and Si/mullite + BSAS/SAS EBCs in Solar Turbines field tests suffered significant recession in some areas, losing most of the BSAS and SAS layer, after ~14 000- and ~8400-hour tests, respectively [17].

Table 15.3 compares experimentally measured Y and Yb disilicate volatility [68]. Volatilities by three research groups were converted to volatilities at $T = 1500°C$, $v_{gas} = 13$ cm/s, $p(H_2O) = 0.3$ atm, $P_{total} = 1$ atm using the silica volatility model (see Equation 15.6). Note that the volatilities measured in furnace tests (Lee [18] and Maier [68]: low gas velocity) are within a factor of two, whereas the volatilities measured in a burner rig (Klemm [59]: high gas

TABLE 15.2 Sources for EBC Degradation

Category	Source	Degradation mode	Failure mode
Chemical	H_2O, CMAS, chemical incompatibility	Oxidation, volatilization, chemical reactions, delamination	Recession, spallation
Mechanical	Thermal stress, external stress, foreign objects	Cracking, erosion, delamination	Recession, spallation
Physical	Phase transformation, sintering	Cracking, delamination	Spallation

TABLE 15.3 Volatility of Y and Yb Disilicate

EBC	$Yb_2Si_2O_7$			$Y_2Si_2O_7$		SiO_2
Group	Maier[a]	Lee[b]	Klemm[c]	Maier[a]	Klemm[c]	Maier[a]
Volatility ($\times 10^{-3}$ mg/(cm^2 h))	2.6	1.5	0.1	1.9	0.1	14.4

[a]Steam furnace (alumina tube), $T = 1500°C$, v_{gas}=13 cm/s, $p(H_2O) = 0.3$ atm, $P_{total} = 1$ atm [68].
[b]Steam TGA furnace (quartz tube), $T = 1500°C$, $v_{gas} = 4.4$ cm/s, $p(H_2O) = 0.5$ atm, $P_{total} = 1$ atm [18].
[c]Burner rig, $T = 1450°C$, $v_{gas} = 100$ m/s, $p(H_2O) = 0.27$ atm, $P_{total} = 1$ atm [59].

velocity) are significantly lower, by about an order of magnitude, than the volatilities measured in steam furnaces. RE silicates are expected to volatilize in water vapor according to reactions (15.7) and (15.8):

$$RE_2Si_2O_7 + 2H_2O(g) \rightarrow RE_2SiO_5 + Si(OH)_4(g) \tag{15.7}$$

$$RE_2SiO_5 + 2H_2O(g) \rightarrow RE_2O_3 + Si(OH)_4(g) \tag{15.8}$$

Figures 15.22a and 15.22b show the surface and cross section of $Yb_2Si_2O_7$ after 224 hours high velocity–high steam burner rig test at $T = 1450°C$, $v = 100$ m/s, $p(H_2O) = 0.27$ atm, and $P_{total} = 1$ atm [59]. Note the formation of a porous Yb_2SiO_5 surface layer, which is consistent with Equation (15.7), that is, RE disilicate transforms to RE monosilicate as silica is selectively vaporized by water vapor. Unlike mullite in which the alumina surface layer after selective volatilization of silica readily crumbled and spalled off [23], the RE monosilicate surface layer remained adherent even under the impact of high velocity flame. This explains the lower volatility of RE disilicates in high velocity–high steam burner rig than in low velocity steam furnaces. The recession of RE disilicates was too slow in the low velocity furnace test to form a substantial amount of RE monosilicates [18, 68]. The high velocity burner rig results suggest that the volatility of RE disilicates in gas turbines can be as low as the volatility of RE monosilcate if the RE monosilicate surface layer stays adherent. The volatility of Y and Yb monosilicates could not be measured even in the high velocity–high steam burner rig test, let alone in steam furnaces [18], because no weight changes were detected even after several hundred hours of burner rig exposures [59].

CMAS: CMAS results from ingestion of siliceous minerals (dust, sand, volcanic ashes, runway debris) in the air [69–72]. CMAS deposit is universally present in engine components. CMAS composition varies depending on the quality of air. A model CMAS composition (33Ca-9Mg-13Al-45Si in mol% cation) was established for laboratory CMAS tests based on the average melts detected on turboshaft shrouds operating in a desert environment with the melting point of ~1235°C [70]. CMAS is benign as long as it stays in the solid state. Molten CMAS, however, caused severe chemical and mechanical degradation to TBCs, leading to TBC delamination and spallation [69–72]. The degradation is so rapid and devastating that CMAS melting point is the upper-use temperature for current TBCs. EBCs are expected to encounter molten CMAS as EBCs are most likely to operate at temperatures above the CMAS melting point because higher temperature capabilities are the key advantage of EBC-coated CMCs over TBC-coated metallic components.

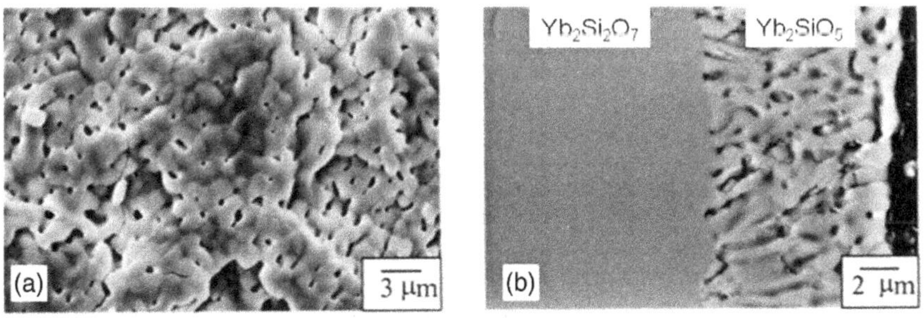

FIGURE 15.22
$Yb_2Si_2O_7$ after 224 hours high velocity burner rig test at $T = 1450°C$, $v = 100$ m/s, $p(H_2O) = 0.27$ atm, and $P_{total} =$ 1 atm. (a) surface; (b) cross section. Permission from Reference 59, The 28th International Conference and Exposition on Advanced Ceramics and Composites, Cocoa Beach, FL, January 25–30, 2004.

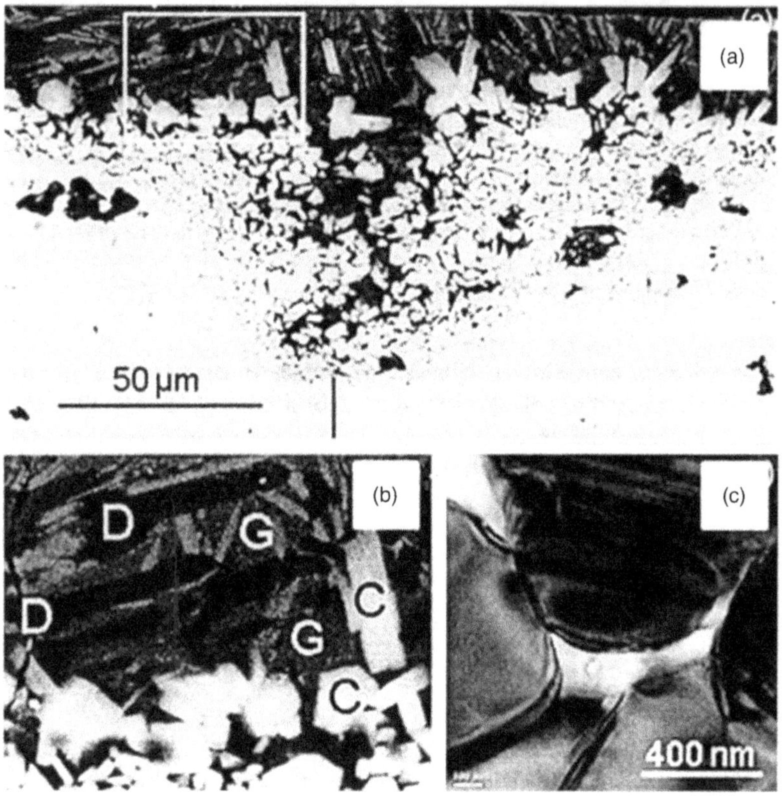

FIGURE 15.23
Hot pressed BSAS loaded with 10 mg/cm^2 of model CMAS and exposed to 1300°C for 4 hours. (a) BSAS penetrated by CMAS; (b) BSE image of identified phases. C: celsian; D: diopside; G: CMAS; (c) Sr-rich pocket. Reproduced with permission from Reference 73, Elsevier S.A. Surface Coatings and Tech, Vol 133–134.

Figure 15.23 shows hot pressed BSAS loaded with 10 mg/cm^2 model CMAS on the surface and exposed to 1300°C for 4 hours [73]. BSAS dissolved in the CMAS melt and re-precipitated as a modified celsian phase along with secondary crystalline phases of anorthite ($CaAl_2Si_2O_8$) and diopside ($CaMgSi_2O_6$). The attack was further aggravated by preferential penetration along the BSAS grain boundaries, probably associated with the presence of a Sr-rich second phase. CMAS can penetrate a hot pressed BSAS with a typical thickness in a few hours. CMAS attack on silica TGO and Si bond coat can lead to spallation of EBC and water vapor stability of BSAS is likely to be compromised due to chemical reactions with CMAS.

Figures 15.24b–15.24d show plasma-sprayed Si/mullite/BSAS-coated CMC loaded with 35 mg/cm^2 model CMAS on the surface and exposed to 1300°C after 4, 24, and 48 hours, respectively (Figure 15.25a is cross section with no CMAS and Figure 15.25e shows stress evolution in the EBC vs. CMAS exposure time) [34]. CMAS penetration in plasma-sprayed BSAS was slower than in hot pressed BSAS. Stress evolution in EBC was measured using a high energy transmission x-ray diffraction. CMAS increased residual stress (compressive) in EBC (Figure 15.25e), which is likely detrimental to EBC durability. CMAS accelerated hexacelsian to celsian phase transformation, which, on the other hand, is likely to be beneficial due to reduced CTE mismatch stresses.

Figures 15.25a–15.25c, and 15.25e show hot pressed Y_2SiO_5 loaded with 12–13 mg/cm^2 model CMAS on the surface and exposed to 1300°C for 1, 4, 24, and 100 hours, respectively [74]. Y_2SiO_5 dissolved into molten CMAS and reprecipitated as apatite ($(Ca_2Y_8(SiO_4)_6O_2)$). The reaction is sustained by the transport of CMAS along the apatite boundaries, decorated by continuous nanoscale amorphous films. The growth proceeds with continuous re-nucleation of new crystals at the interface. CMAS penetration was slower than in hot pressed BSAS, but faster than in RE zirconates, a low thermal conductivity TBC system. Y_2SiO_5 was more effective than $Y_2Si_2O_7$ in slowing down CMAS penetration (~100 μm penetration after 100 hours).

These results indicate that BSAS and rare earth silicate EBCs are more resistant to CMAS than 7YSZ; however, both are projected to be fully penetrated by CMAS and as a result suffer damages and premature spallation. CMAS mitigation of EBCs, therefore, is a critical durability issue that must be addressed to fully utilize the temperature benefits of EBC-coated CMC components.

Foreign Object Impact: Foreign objects injected with the air are known to cause impact damages on turbine components.

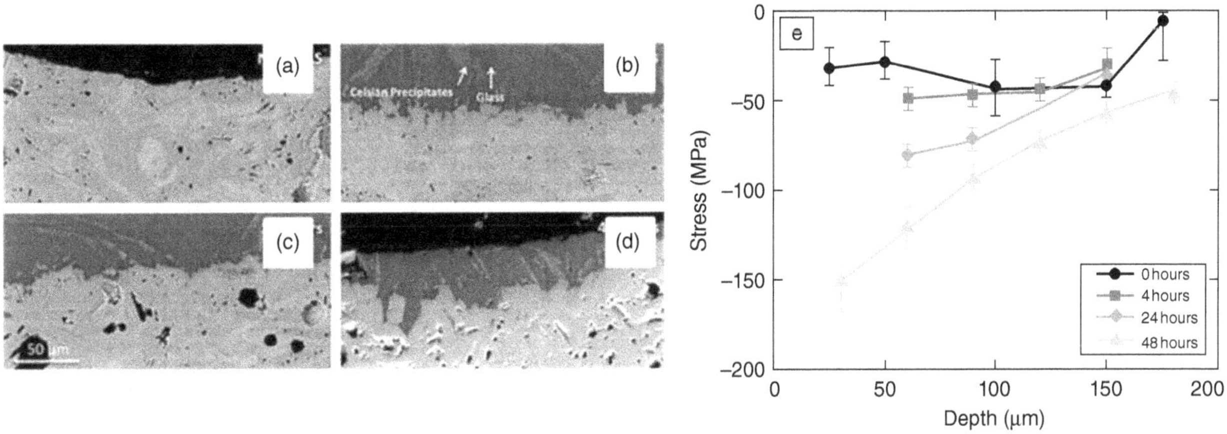

FIGURE 15.24

Plasma-sprayed Si/mullite/BSAS-coated CMC loaded with 35 mg/cm^2 of model CMAS and exposed to 1300°C. (a) no CMAS; (b) 4 hours; (c) 24 hours; (d) 48 hours; (e) Stress evolution in the EBC versus CMAS exposure time. Reproduced with permission from Reference 34, Journal of the American Ceramic Society, Vol 86, Iss 8, 2003.

Figure 15.26 is a cross section of 525-μm-thick Si/mullite + 20 wt.% BSAS/BSAS-coated CMC impact tested at room temperature at 290 m/s and then oxidized for 500 hours at $T = 1316°C$, $p(H_2O) = 0.9$ atm, $P_{total} = 1$ atm, and $v = 2.2$ cm/s [75,76]. It shows EBC spallation and CMC oxidation below the impact site (area A) and EBC delamimation and CMC damages on the backside (area B). Impact damages can be devastating and thus needs be incorporated in EBC lifing.

Other Degradation Sources: BSAS–silica chemical reaction produces low melting (~1300°C) glasses that lead to EBC degradation and delamination as discussed in the Section 15.2.2 Phase instability in BSAS and its impact on EBC stresses were discussed in the Section 15.2.2. EBC can be eroded by particles, leading to thinning and eventual loss of EBC. Thermal and external stresses coupled with other sources for EBC degradation are likely to further aggravate EBC degradation and thus accelerate EBC failure.

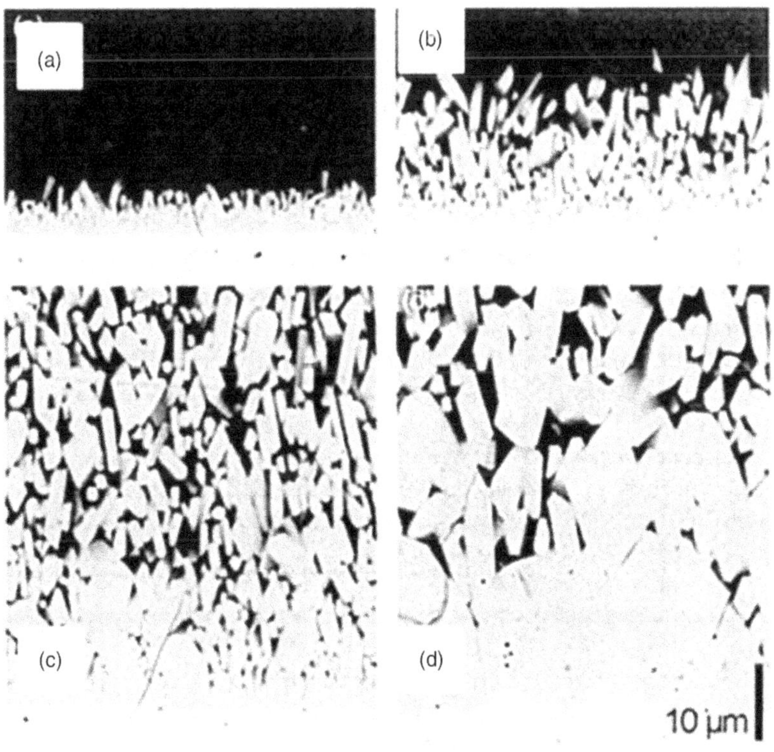

FIGURE 15.25

Hot pressed Y_2SiO_5 loaded with 12–13 mg/cm^2 of model CMAS and exposed to 1300°C. (a) 1 hour; (b) 4 hours; (c) 24 hours; (d) 100 hours. Reproduced with permission from Reference 74, Journal of the American Ceramic Society, Vol 86, Iss 8, 2003.

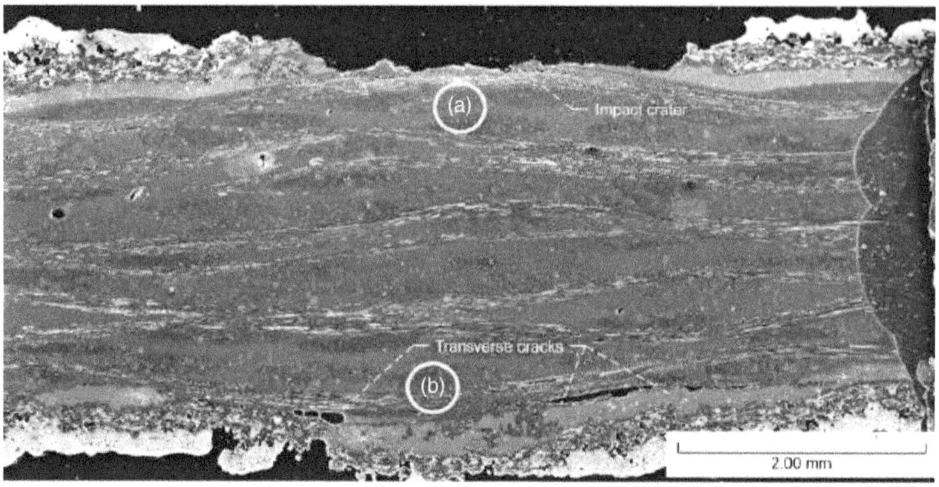

FIGURE 15.26

Cross section of 525-mm-thick EBC-coated CMC impact tested at room temperature at 290 m/s and then oxidized for 500 hours at $T = 1316°C$, $p(H_2O) = 0.9$ atm, $P_{total} = 1$ atm, and $v = 2.2$ cm/s. Reproduced with permission from Reference 76, Elsevier S.A. Surface Coatings and Tech, Vol 133–134.

Strong interactions among various degradation modes are expected. EBC lifing will require thorough understanding of individual degradation mode as well as interactions among them. A model that incorporates all these factors needs to be developed to successfully project the life of EBC-coated CMCs in gas turbines. Nondestructive technologies to monitor EBC health will be a useful tool to complement the EBC life model.

15.5 CONCLUDING REMARKS AND FUTURE DIRECTIONS

EBCs are an enabling technology for SiC/SiC CMCs. Within two decades after the development of a fully crystalline mullite coating, the first viable coating for SiC, in the early 1990s at NASA, intense collaborations between the government and OEMs led to the successful development of two generations of EBCs and a number of successful engine tests. With OEMs preparing for the implementation of EBC-coated CMC components in the near future, there are still a number of challenges remaining, including developing a reliable lifing method to accurately project the life of EBCs, generating an EBC validation database in conditions relevant to gas turbine engines, and developing a non-line-of-sight process.

Current EBC R&D is confined mostly within gas turbine OEMs due to the competition-sensitive nature of EBC technologies. Understanding EBC degradation mechanisms and developing mechanism-based EBC lifing, however, require collaborative EBC research between industry, academia, and government laboratories. Another barrier to EBC lifing is the absence of a standard EBC. It is, therefore, imperative that an industry standard or a relevant model EBC system is established and a broad collaboration between government, academia, and industry is initiated to tackle EBC lifing and other challenges.

EBCs are an exciting new field and a successful implementation of EBC-coated CMC components in gas turbines will open new R&D opportunities to further improve the reliability and robustness of EBC and to develop advanced EBC process technologies and lifing methods.

REFERENCES

1. F. O. Soechting. Thermal barrier coating workshop. NASA Conference Publication 3312 (NASA Lewis Research Center, Cleveland, OH, 1995), pp. 1–15.

2. D. Anson and D. W. Richerson. The benefits and challenges of the use of ceramics in gas turbines. *Progress in Ceramic Gas Turbine Development*, vol. 2, edited by M. van Roode, M. Ferber, and D. W. Richerson (ASME Press, New York, NY, 2003), pp. 1–10.
3. E. J. Opila and R. Hann. Paralinear oxidation of CVD SiC in water vapor. *J. Am. Ceram. Soc.*, 80(1), 197–205 (1997).
4. J. L. Smialek, R. C. Robinson, E. J. Opila, D. S. Fox, and N. S. Jacobson. SiC and Si_3N_4 scale volatility under combustor conditions. *Adv. Composite Mater.*, 8(1), 33–45 (1999).
5. K. L. More, P. F. Tortorelli, M. K. Ferber, L. R. Walker, J. R. Kiser, N. Miriyala, W. D. Brentnall, and J. R. Price. ASME Paper 99-GT-292. ASME Turbo Expo 1999, Indianapolis, IN, June 7–10, 1999.
6. M. K. Ferber, H. T. Lin, V. Parthasarathy, and W. D. Brentnall. ASME Paper 99-GT-265. ASME Turbo Expo 1999, Indianapolis, IN, June 7–10, 1999.
7. N. S. Jacobson. Corrosion of silicon-based ceramics in combustion environments. *J. Am. Ceram. Soc.*, 76(1), 3–28 (1993).
8. N. S. Jacobson, J. L. Smialek, and D. S. Fox, *Handbook of Ceramics and Composites* edited by N. S. Cheremisinoff (Marcel Dekker, New York, NY, 1990) 1, pp. 99–135.
9. J. I. Federer. Corrosion of SiC ceramics by Na_2SO_4. *Adv. Ceram. Mater.*, 76(1), 56–61 (1988).
10. K. N. Lee, H. Fritze, and Y. Ogura. Coatings for engineering ceramics. *Progress in Ceramic Gas Turbine Development*, vol. 2. edited by M. van Roode, M. Ferber, and D. W. Richerson (ASME Press, New York, NY, 2003) pp. 641–664.
11. K. N. Lee, R. A. Miller, and N.S. Jacobson. New generation of plasma-sprayed mullite coatings on silicon-carbide. *J. Am. Ceram. Soc.*, 78(3), 705–710 (1995).
12. K. N. Lee. Current status of environmental barrier coatings for Si-based ceramics. *Surf. Coat. Technol.*, 133–134, 1–7 (2000).
13. K. N. Lee, D. S. Fox, J. I. Eldridge, D. Zhu, R. C. Robinson, N. P. Bansal, and R. A. Miller. Upper temperature limit of environmental barrier coatings based on mullite and BSAS. *J. Am. Ceram. Soc.*, 86(8), 1299–1306 (2003).
14. H. E. Eaton, G. D. Linsey, K. L. More, J. B. Kimmel, J. R. Price, and N. Miriyala. ASME Paper 2000-GT-0631. International Gas Turbine and Aeroengine Congress and Exposition, Munich, Germany, May 8–11, 2000.
15. H. E. Eaton, G. D. Linsey, E. Y. Sun, K. L. More, J. B. Kimmel, and J. R. Price. ASME Paper 2001-GT-0513. ASME Turbo Expo 2001, New Orleans, LA, June 4–7, 2001.
16. K. L. More, P. F. Tortorelli, L. R. Walker, J. B. Kimmel, N. Miriyala, and J. R. Price. ASME Paper 2002-GT-30630. ASME Turbo Expo 2002, Amsterdam, The Netherlands, June 3–6, 2002.
17. UTRC Final Report. ONR N00014-03-0477, October 2006.
18. K. N. Lee, D. S. Fox, and N. P. Bansal. Rare earth silicate environmental barrier coatings for SiC/SiC composites and Si_3N_4 ceramics. *J. Euro. Ceram. Soc.*, 25, 1705–1715 (2005).
19. N. S. Jacobson. High-temperature durability considerations for HSCT combustor. NASA TP3162, January 1992.
20. R. C. Robinson and J. L. Smialek. SiC recession caused by SiO2 Scale volatility under combustion conditions: I, experimental results and empirical model. *J. Am. Ceram. Soc.*, 82(7), 1817 1825 (1999).
21. K. N. Lee and F. Zhang. Recession of environmental barrier coatings for ceramic matrix composites. Thermal Barrier Coatings III, Irsee, Germany, August 7–12, 2011.
22. K. N. Lee, E. J. Opila, and R. A. Miller. Oxidation and pore formation at the Mullite/SiC interface in air and water vapor, in *Fundamental Aspects of High Temperature Corrosion*, edited by D. A. Shores, R. A. Rapp, and P. Y. Hou (The Electrochemical Society, Pennington, NJ, 1997) 124–138.
23. K. N. Lee. Key durability issues with mullite-based environmental barrier coatings for Si-based ceramics. *J. Eng. Gas Turbines Power*, 122(10), 632–636 (2000).
24. K. N. Lee. Evolution of environmental barrier coatings for Si-Based ceramics. The 28th International Conference and Exposition on Advanced Ceramics and Composites, Cocoa Beach, FL, January 25–30, 2004.
25. Phase Equilibria Diagrams, CD-ROM Database Version 2.1 (American Ceramic Society, Westerville, OH, 1998).
26. M. J. Hyatt and N. P. Bansal. Crystal growth kinetics in $BaOAl_2O_32SiO_2$ and $SrOAl_2O_32SiO_2$ glasses. *J. Mater. Sci.*, 31, 172–184 (1996).
27. D. Bahat. Kinetic study on the Hexacelsian–Celsian phase transformation. *J. Mater. Sci.*, 5, 805–810 (1970).
28. N. P. Bansal, M. J. Hyatt, and C. H. Drummond. Crystallization and properties of Sr-Ba aluminosilicate glass-ceramic matrices. *Ceram. Eng. Proc.*, 12(7–8), 1222–1234 (1991).
29. C. H. Drummond, W. E. Lee, N. P. Bansal, and M. J. Hyatt. Crystallization of barium aluminosilicate glass. *Ceram. Eng. Proc.*, 10(9–10), 1485–1502 (1989).
30. K. N. Lee, J. I. Eldridge, and R. C, Robinson. Stress evolution and its effects on the durability of environmental barrier coatings for SiC ceramics, *J. Am. Ceram. Soc.*, 88(12), 3483–3488 (2005).
31. B. J. Harder, J. Almer, C. M. Weyant, K. N. Lee, and K. T. Faber. Residual stress analysis of multilayer environmental barrier coatings. *J. Am. Ceram. Soc.*, 92(2), 452–459 (2009).

32. B. J. Harder, J. Almer, K. N. Lee, K. T. Faber. In situ stress analysis of multilayer environmental barrier coatings. *Powder Diffr.*, 24(2), 94–98 (2009).
33. K. T. Faber, C. Weyant, B. Harder, J. Almer and K. N. Lee. Internal stresses and phase stability in multiphase environmental barrier coatings. *Int. J. Mat. Res.*, 98, 1188–1195 (2007).
34. B. Harder, J. Ramirez-Rico, J. Almer, K. Lee, and K. Faber. Chemical and mechanical consequences of environmental barrier coating exposure to calcium–magnesium–aluminosilicate. *J. Am. Ceram. Soc.*, 94(SI), SI178–SI185 (2011).
35. N. S. Jacobson, K. N. Lee, and T. Y. Yoshio. Corrosion of mullite by molten salts. *J. Am. Ceram. Soc.* 79(8), 2161–2167 (1996).
36. V. K. Sarin. Systematic development of customized CVD coatings. *Surf. Coat. Technol.*, 73, 23–33 (1995).
37. R. P. Mulpuri and V. K. Sarin. Synthesis of mullite coating by chemical vapor deposition. *J. Mater. Res.*, 11(6), 1315–1324 (1996).
38. M. L. Auger and V. K. Sarin. The development of CVD mullite coatings for high temperature corrosive applications. *Surf. Coat. Technol.*, 94-95, 46-52 (1997).
39. S. N. Basu, P. Hou, and V. K. Sarin. Formation of mullite coatings on silicon-based ceramics by chemical vapor deposition. *Int. J. Refract. Met. Hard Mater.*, 16, 343-352 (1998).
40. W. Y. Lee, K. L. More, and Y. W. Bae. Characterization of Si3N4 coated with chemically-vapor-deposited mullite after Na_2SO_4-Induced corrosion. *J. Am. Ceram. Soc.*, 79(9), 2489–2492 (1996).
41. J. A. Haynes, K. M. Cooly, D. P. Stinton, R. A. Lowden, and W. Y. Lee. Corrosion-resistant CVD mullite coatings for Si_3N_4. *Ceram. Eng. Sci. Proc.*, 20(4), 355–362 (1999).
42. J. A. Haynes, M. J. Lance, K. M. Cooly, M. K. Ferber, R. A. Lowden, and D. P. Stinton. Protective capacity of CVD mullite coatings in a high-temperature, high-pressure, Air-H_2O environment. *J. Am. Ceram. Soc.*, 83(3), 657–659 (2000).
43. R. A. Miller. Thermal barrier coating workshop. NASA Conference Publication 3312 (NASA Lewis Research Center, Cleveland, OH 1995), pp. 17–34.
44. T. Ohji. Environmental barrier coating on silicon nitride: challenges and critical issues. The 28th International Conference and Exposition on Advanced Ceramics and Composites, Cocoa Beach, FL, January 25–30, 2004.
45. S. Ueno, D. D. Jayaseelan, N. Kondo, T. Ohji, and S. Kanzaki. Recession mechanism of $Lu_2Si_2O_7$ EBC layer in water vapor environment. The 28th International Conference and Exposition on Advanced Ceramics and Composites, Cocoa Beach, FL, January 25–30, 2004.
46. K. N. Lee. Current status of environmental barrier coatings for SiC/SiC composites and Si3N4 ceramics. 107th Annual Meeting and Exposition of the American Ceramic Society, Baltimore, MD, April 11–13, 2005.
47. K. N. Lee. Overview of EBC research in the NASA UEET program. Third Environmental Barrier Coatings Workshop, Nashville, TN, November 17–18, 2004.
48. Y. Tamarin. *Protective Coatings for Turbine Blades* (ASM International, Materials Park, OH, 2002).
49. *Handbook of Thermal Spray Technology*, edited by J. R. Davis (ASM International, Metals Park, OH, 2004).
50. D. Hass. Coatings for advanced engineering systems. Materials Science & Technology 2009, Houston, TX, October 17–21, 2010.
51. H. N. G. Wadley and J. F. Groves, inventors. Directed vapor deposition of electron beam evaporant. U.S. Patent 5, 534, 314 (July 9, 1996).
52. K. von Niessen and M. Gindrat. Plasma spray-PVD: a new thermal spray process to deposit out of the vapor phase. *J. Therm Spray Technol.*, 20(4), 736–743 (2011).
53. B. J. Harder and D. Zhu. Physical vapor deposition of thermal and environmental protection systems. ITSC 2012, Houston, TX, May 21–24, 2012.
54. B. Armstrong. Slurry deposition of environmental protection systems. Fourth Environmental Barrier Coatings Workshop, Nashville, TN, November 16–17, 2005.
55. S. Ramasamy, S. N. Tewari, K. N. Lee, R. T. Bhatt, and D. S. Fox. Environmental durability of slurry based Mullite–Gadolinium silicate on silicon carbide. *J. Eur. Ceram. Soc.*, 31, 1123–1130 (2011).
56. S. Ramasamy, S. N.Tewari, K. N. Lee, R. T. Bhatt, and D. S. Fox. Mullite–Gadolinium silicate environmental barrier coatings for melt infiltrated SiC/SiC composites. *Surf. Coat. Technol.*, 205, 3578–3581 (2011).
57. S. Ramasamy, S. N. Tewari, K. N. Lee, R. T. Bhatt, and D. S. Fox. Slurry based multilayer environmental barrier coatings for silicon carbide and silicon nitride ceramics—I. Processing. *Surf. Coat. Technol.*, 205, 258–265 (2010).
58. S.Ramasamy, S. N.Tewari, K. N. Lee, R. T. Bhatt, and D. S. Fox. Slurry based multilayer environmental barrier coatings for silicon carbide and silicon nitride ceramics—II. Oxidation resistance. *Surf. Coat. Technol.*, 205, 266–270 (2010).
59. H. Klemm, M. Fritsch, and B. Schenk. Corrosion of ceramic materials in hot gas environment. The 28th International Conference and Exposition on Advanced Ceramics and Composites, Cocoa Beach, FL, January 25–30, 2004.
60. M. Fritsch and H. Klemm. The Water Vapor Hot Gas Corrosion Behavior of Al_2O_3-Y_2O_3 Materials, Y_2SiO_5 and $Y_3Al_5O_{12}$-Coated Alumina in a Combustion Environment. The 30th International Conference and Exposition on Advanced Ceramics and Composites, Cocoa Beach, FL, January 25–30, 2006.

61. H. Klemm. Silicon nitride for high-temperature applications. *J. Am. Ceram. Soc.*, 93(6) 1501–1522 (2010).
62. S. R. Choi, J. I. Eldridge, K. N. Lee, R. A. Miller, and D. Zhu. Surface cracking and interface reaction associated delamination failure of thermal and environmental barrier coatings, *Ceramic Engineering and Science Proceedings*, 24(3), 469–475 (2003), (The American Ceramic Society, Westerville, OH).
63. D. Zhu, K. N. Lee, and R. A. Miller. Thermal gradient cyclic behavior of a thermal/environmental barrier coating system on SiC/SiC ceramic matrix composites[Paper GT2002-30632; NASA/TM-2002-211593]. Proceedings of the ASME Turbo Expo 2002, Amsterdam, The Netherlands, June 3–6, 2002.
64. D. Zhu, K. N. Lee, and R. A. Miller. Sintering and cyclic failure mechanisms of thermal and environmental barrier coating systems under high thermal gradient test conditions, *Ceramic Engineering and Science Proceedings*, 23(4), 505–516 (2002). (The American Ceramic Society, Westerville, OH).
65. A. G. Evans, D. R. Mumm, J. W. Hutchinson, G. H. Meier, and F. S. Pettit. Mechanism controlling the durability of thermal barrier coatings. *Prog. Mater. Sci.*, 46, 505–553 (2001).
66. T. Bathia. EBC development at UTRC. Fourth Environmental Barrier Coatings Workshop, Nashville, TN, November 16–17, 2005.
67. K. Lee, J. I. Eldridge, D. S. Fox, C. Robinson, and N. P. Bansal. Effects of environment, chemistry, and stress on the durability of environmental barrier coatings. Gordon Research Conference, Colby College, Waterville, ME, August 4–9, 2002.
68. N. Maier, K. G. Nickel, and G. Rixecker. Water vapor corrosion of rare earth disilicate ($RE_2Si_2O_7$) in flowing hot atmosphere containing $Al(OH)_3$. EUROMAT, Prague, Czech Republic, September 5–8, 2005.
69. C. Mercer, S. Kramer, S. Faulhaber, M. G. Evans, and R. Darolia. A delamination mechanism for thermal barrier coatings subject to calcium–magnesium–alumino–silicate (CMAS) penetration. *Acta Mater.*, 53, 1029–1039 (2005).
70. S. Kramer, J. Y. Yang, C. A. Johnson, and C. G. Levi. Thermomechanical interaction of thermal barrier coatings with molten $CaO\text{-}MgO\text{-}Al_2O_3\text{-}SiO_2$ (CMAS) Deposits. *J. Am. Ceram. Soc.*, 89, 3167 (2006).
71. A. G. Evans and J. W. Hutchinson. The mechanics of coating delamination in thermal gradients. *Surf. Coat. Technol.*, 201, 7905–7916 (2007).
72. S. Kramer, S. Faulhaber, M. Chambers, D. R. Clarke, C. G. Levi, J. W. Hutchinson, and A. G. Evans. Mechanism of cracking and delamination within thick thermal barrier systems in aero-engines subject to Calcium-Magnesium-Alumino-Silicate (CMAS) penetration. *Mater. Sci. Eng. A*, 490, 26–35 (2008).
73. K. Grant, S. Kramer, J. Lofvander, and C Levi. CMAS degradation of environmental barrier coatings. *Surf, Coat. Tech.*, 202, 653–657 (2007).
74. K. Grant, S. Kramer, G. Seward, and C Levi. Calcium–magnesium Alumino-Silicate interaction with yttrium monosilicate environmental barrier coatings. *J. Am. Ceram. Soc.*, 93(10), 3504–3511 (2010).
75. R. T. Bhatt, S.R. Choi, L.M. Cosgriff, D.S. Fox, and K.N. Lee. Impact resistance of uncoated MI SiC/SiC composites. *Mater. Sci. Eng. A*, 476, 20–28 (2008).
76. R. T. Bhatt, S.R. Choi, L.M. Cosgriff, D. S. Fox, and K. N. Lee. Impact resistance of EBC coated MI SiC/SiC composites. *Mater. Sci. Eng. A*, 476, 8–19 (2008).

CHAPTER 16

OXIDATION PROTECTIVE COATINGS FOR ULTRAHIGH TEMPERATURE COMPOSITES

QIANGANG FU AND YIGUANG WANG

National Key Laboratory for Thermostructure Composite Materials, Northwestern Polytechnical University, Xi'an, Shaanxi, China

16.1 INTRODUCTION

Ultrahigh temperature composites (mainly C/C and C/SiC) are one of the few materials that can retain their mechanical properties above 2273 K. They possess a series of excellent attributes such as low density, high specific strength, high specific modulus, low coefficient of thermal expansion (CTE), excellent thermal shock resistance, ablation resistance, and erosion resistance for gases containing solid particles. Especially, their strength increases as temperature rises, and this unique property gives their application in aeronautics and aerospace fields as thermal structural parts show incomparable advantage than other materials [1–3]. However, the initial oxidation temperature of carbon fibers is only 643 K, and they will be oxidized rapidly above 773 K [4], and the oxidation rate increases sharply as the temperature rises. Without oxidation protective measures, prolonged use of ultrahigh temperature composites in the high temperature oxidizing environments will cause catastrophic consequences [4, 5]. With the rise of requirements for ultrahigh temperature materials' performance all over the world, the anti-oxidation of ultrahigh temperature composites is of vital importance.

There are two anti-oxidation methods for C/C composites [6]. First is an internal matrix modification technology based on the inhibition of oxidation reaction by the material itself, namely, modifying the carbon fiber or carbon matrix during the preparation process of C/C composites, which can make these materials have higher anti-oxidant capacity. Second is an external anti-oxidation coating technology based on the prevention of contact diffusion of oxygen containing gases, namely, in order to achieve the anti-oxidation purpose, preparing anti-oxidation coatings on the surface of C/C composites, using high temperature coatings to isolate oxygen and C/C composite substrates. The coating technique is also suitable for the protection of C/SiC composites. The anti-oxidation temperature and protection time of the matrix modification technology for C/C composites are limited [7]. Acknowledgedly, high temperature and long life anti-oxidation of C/C and C/SiC composites must rely on coating technology.

In this chapter, basic requirements of oxidation protective coating for C/C and C/SiC composites are provided. Methods for the preparation of oxidation-resistant coatings are introduced. The research status of several oxidation-resistant coatings, including glass, metal, ceramic, and composite coatings are reviewed. The present problems are discussed and the potential development direction on the investigation of oxidation-resistant coatings for C/C and C/SiC composites is also proposed.

Ceramic Matrix Composites: Materials, Modeling and Technology, First Edition. Edited by Narottam P. Bansal and Jacques Lamon.

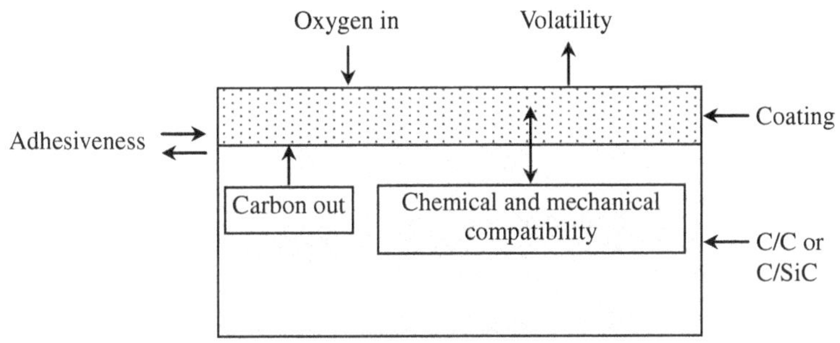

FIGURE 16.1
Schematic of the considerations of the anti-oxidation coating.

16.2 BASIC REQUIREMENTS OF ANTI-OXIDATION COATING FOR C/C AND C/SIC COMPOSITES

Compared with matrix modification technology, anti-oxidation coating technology develops faster because of its higher anti-oxidant capacity. The basic principle of this method is to use coatings to block the contact between oxygen and C/C (or C/SiC) substrates in order to achieve the anti-oxidation purpose at high temperature. To make C/C and C/SiC composites work reliably over a long period of time under high temperature oxidizing atmosphere and thermal shock condition from room temperature to high temperature, several considerations in the design of anti-oxidation coating are shown in Figure 16.1 [8,9].

Therefore, the protective coating must meet the following basic requirements.

16.2.1 Inhibiting Ability of Oxygen Diffusion

Adopting a common anti-oxidant index, effective working time is 100 hours, and the maximum of oxidative mass loss percentage allowed for C/C or C/SiC composites is 1%. According to calculation [1], the maximum oxygen diffusion permeation rate for coatings is about 3×10^{-10} g/(cm s). The oxygen permeation rates of ThO_2, Y_2O_3, HfO_2, and ZrO_2 are high at temperatures above 1273 K [10], indicating they cannot inhibit the diffusion of oxygen effectively. While Al_2O_3 and SiO_2 have a relatively lower rate, they can meet the requirements of inhibiting oxygen diffusion [11]. Based on the rough calculation, dense SiO_2 coating with a thickness of 10 μm can still inhibit the attack of oxygen from SiO_2 coating to the surface of C/C or C/SiC composites after 100 hours at 1873 K. Due to the poor adhesion property between SiO_2 and C/C or C/SiC composites, it is difficult to use SiO_2 as the coating for C/C or C/SiC directly. Therefore, it is usual to use the Si-based ceramic, which would be oxidized to form SiO_2 to obtain the coatings with low oxygen permeation rate.

16.2.2 Good Match of Thermal Expansion with C/C and C/SiC Composites

The mismatch of thermal expansion between coating and C/C or C/SiC composites may cause the coating cracking or debonding. The CTE of coating is generally higher than those of C/C and C/SiC composites [12]. There are two methods to relieve the mismatch in the design of practical anti-oxidation coatings: first, adopting gradient coatings, namely, making the coating materials diffuse to the surface layer of C/C or C/SiC composites, thereby creating certain concentration gradient in the coatings; second, introducing B or B-containing inhibitor to make them form borates that can seal the microcracks in the coatings at high temperature and prevent oxygen from diffusion to the C/C or C/SiC substrate.

16.2.3 Low Volatility during Service Process

In order to avoid the degradation of coating at high temperature and prevent the erosion of coating in high speed flow, the volatility of coating should be low. Refractory carbides such as ZrC, HfC still have a good thermal stability even if the temperature is higher than 2273 K, so do the oxides such as ZrO_2, HfO_2, ThO_2, and Y_2O_3. However, Al_2O_3 with a low oxygen penetration rate can only be used at lower temperatures [13] (<2073 K).

16.2.4 Compatible Stability with C/C or C/SiC Composites

The major concern of keeping the chemical compatibility between coatings and C/C or C/SiC composites at high temperature is to prevent the carbon thermal reduction reaction between coating and carbon substrate. C can generate

CO by reacting with oxide ceramics at high temperature, which will lead to the degradation of the oxide coatings because of the reduction by C. For example, C can generate CO and SiO by reacting with SiO_2 at 1723 K and the vapor pressure of CO is higher than 0.1 MPa at 1773 K, which will cause the outward diffusion of C after the reaction [14]. Therefore, it is not wise to adopt oxide ceramics as the internal coatings of C/C or C/SiC composites.

16.2.5 Good Interfacial Bonding with C/C or C/SiC Composites

In order to avoid the debonding of coatings under thermal shock or at high temperatures, coatings should form a good chemical bonding with C/C or C/SiC substrates.

16.3 PREPARATION METHODS OF ANTI-OXIDATION COATINGS

Currently, there are several methods for preparing anti-oxidation coatings for C/C or C/SiC composites, including pack cementation, chemical vapor deposition, liquid phase reaction, plasma spraying, and slurry method.

16.3.1 Pack Cementation

Pack cementation is the most widely applied method for preparation of the Si-based ceramic coatings for C/C composites. The basic process and forming mechanism of pack cementation is to embed C/C composites among several solid mixed powders, which will be then heated at a certain temperature so that coatings can be formed on the specimen surface by reacting with mixed powder. Compared with other methods, its advantages are as follows. (a) The process is simple since a dense coating can be prepared by a single process. (b) The dimensional change of the substrates before and after pack cementation is negligible. (c) A certain composition gradient layer can be formed at the interface of coating and substrate, resulting in a good interface bonding. This method has the following detects. (a) The process can easily cause fiber damage by chemical reaction, which affects the mechanical properties of C/C substrates; (b) Due to gravity and other factors, the uniformity of coating by pack cementation is difficult to control.

Pack cementation is mainly used to prepare Si-based ceramic coatings for C/C composites, such as SiC, $MoSi_2$, and WSi_2. This method is not very suitable for preparing coatings for C/SiC composites because of the difficulty of synthesis reaction between SiC and coating materials.

16.3.2 Chemical Vapor Deposition

Chemical vapor deposition (CVD) is one of the important methods to prepare anti-oxidation coatings for C/C and C/SiC composites. Its basic principle is to form coatings on the surface of C/C or C/SiC substrates at a certain temperature by a series of physical chemistry processes such as decomposition, synthesis, diffusion, adsorption, and surface spreading. The materials prepared by CVD have high density and purity. It is easy for CVD process to design the microstructure, morphology, and composition as well as control the deposition rate [15]. Therefore, CVD technology has aroused much attention.

The major advantage of CVD is that coatings with a variety of elements and compounds can be deposited at a lower temperature so that substrates can avoid the defects or mechanical property damage caused by high temperature treatment. In addition, the chemical composition and structure of the coatings prepared by CVD are controllable. However, the CVD process needs a vacuum or protective atmosphere, which demands a higher airtightness for the device. The interface bonding between CVD coating and C/C or C/SiC is not very strong.

Currently, the coatings on the C/C or C/SiC composites surface deposited by CVD mainly include SiC, Si_3N_4, HfC, ZrC, TiC, and $MoSi_2$; among them the use of carbides such as SiC and HfC is most extensive.

16.3.3 Liquid Phase Reaction

The liquid phase reaction is a method of forming coatings by impregnating C/C or C/SiC composites into the organometallic compounds, alkoxy metal, metallic salt solution or colloids; then after drying or chemical reaction they decompose or react to form coatings [16]. The conversion yield from liquid phrase to coating is low when adopting this method to prepare coatings; thus, it often requires a reduplicative impregnation. Debrunner [17], Wilson [18], and

McKee [19] all described the details of the anti-oxidation coatings in their patents, including material selection, temperature, and and process control. This method not only is applicable to special ceramic coating materials, such as SiC, $MoSi_2$, and WSi_2, but also can be used for the preparation of glass coatings such as B_2O_3 and SiO_2.

This method has low temperature of preparation, low damage to the strength of substrate, and can obtain the coatings with a strong self-healing ability. Its disadvantages are as follows. The change in the coating composition is strongly affected by the mutual diffusion process between the impregnated liquid components and substrates. A large amount of protective gas is required. In addition, a low wetting angle between the liquid materials and substrate is required in order to obtain a good interface bonding [20].

16.3.4 Plasma Spraying

Plasma can generate a very high temperature that can heat the coating materials to a molten state. The powder is heated and accelerated by plasma stream at high temperature, then is sprayed onto the surface of the substrates to form coatings. The uniformity of the coating by plasma spraying is good and the thickness of the coating can be controlled. In addition, the coatings with almost all kinds of high melting point and wear resistance can be obtained. But the porosity of coatings is high and the interfacial bonding between coating and substrate is weak. Therefore, it is easy to peel or crack under thermal cycles. In order to improve the interfacial bonding strength between coatings and C/C or C/SiC composites, it is general to prepare internal SiC coating on C/C or C/SiC surface by pack permeation or CVD before adopting plasma-spraying method. The outer coating materials of spraying include oxides, silicates, ultrahigh temperature ceramics, and high temperature alloys. In recent years, the supersonic plasma-spraying technique was developed, which can improve the density and the cohesion strength of the oxidation protective coatings greatly.

16.3.5 Sol-Gel Method

The basic process of sol-gel method is to prepare a uniform solution with metallic, organic, or inorganic salt, after being hydrolyzed and polycondensated and other chemical reactions at a low temperature to make the sol transform into gel, then heat the gel to form glass, ceramics, and other coatings. The equipment for this method is simple and the heat treatment temperature is low. In addition, it can coat the whole surface of materials and the uniformity of coating is good. However, the coating thickness prepared by this method is thinner, so it requires multiple dip-coatings, and the coatings crack easily under dry process.

16.3.6 Supercritical Fluid Technique

The so-called supercritical fluid is a gas–liquid coexistent special material when the temperature and pressure of a chemical system exceed a certain critical condition. This liquid has strong dissolvability, low viscosity, and high diffusion degree and almost no effect of surface tension. Therefore, it has a strong delivery capacity and can permeate submicron pores. Berneburg [21] invented this method for the preparation of coating for C/C composites. Its basic principle is to use supercritical fluid as a solvent and carrier, help the coating precursor impregnate C/C, and fill the defects of the internal pores of C/C after thermolysis. Then the anti-oxidation coating is formed on the external after surface heat treatment under an inert atmosphere. The coatings obtained by this method are SiC and B_4C coating. The supercritical fluid not only generates the anti-oxidation coating on the external, but also fills the internal defects and pores, thereby improving the anti-oxidation ability from the inside to the outside. The disadvantage is its high requirement for the equipments, such as a high pressure and temperature condition.

16.3.7 Slurry Method

The process for slurry coating is as follows. First prepare the powder made of coating materials that meet certain requirements; then mix it with solvent to make it into slurry, adding some suitable dispersants and adhesives, and stirring sufficiently. The coatings will be formed by brushing the surface of substrates or impregnating the substrates into slurry, and then being heat treated under an inert atmosphere after being dried at a certain temperature. Slurry coating is a simple and low cost process. But the coatings by slurry have some disadvantages, such as weak bonding between the coating and substrate, poor thermal shock resistance, and microporous structure.

16.3.8 *In Situ* Reaction

In order to prevent the oxidation of C/C or C/SiC composites, the coatings are generally prepared to be under the protection of an inert atmosphere. A new method of preparing a silicate coating under an oxidizing atmosphere was invented, that is, *in situ* reaction. Using Si and Y_2O_3 as original powder, the molten state SiO_2 which is formed by the oxidation of Si at high temperature will react with Y_2O_3 to synthesize yttrium silicate coating for C/C or C/SiC composites. On the one hand, it solves the problem of the weak internal bonding for silicate coatings. On the other hand, it fills the pores of coatings effectively during the expansion process, thus creating high density yttrium silicate coatings [22]. This method can also be applied in the preparation of silicate coating such as mullites and zirconium silicates.

Besides the above methods, there are several coating preparation methods, such as hydrothermal electro-deposition [23] and vapor infiltration laser melting technique [24]. Generally, the above methods are used in combination to obtain the desired coatings in practical applications.

16.4 OXIDATION-RESISTANT COATING SYSTEMS

Up to now, the developed oxidation-resistant coatings mainly include glass, metal, ceramic, and composite coatings. The research status of the as-mentioned four coatings will be discussed in the following sections.

16.4.1 Glass Coatings

In the earlier research, study on anti-oxidation technology of C/C and C/SiC composites was mainly focused on glass coating. It works by using the liquidity and wettability of glass at high temperature to protect C/C or C/SiC substrate. Glass coatings mainly include borate, phosphate, silicate, and composite glass coating. Phosphate glass not only has good wettability with C/C or C/SiC substrate but also can fill surface defects and reduce the oxidation active point in the C/C or C/SiC substrate. Besides, phosphate is cheap and easy to brush. Furthermore, it is also suitable for the operating temperature range of C/C and C/SiC brake disk, which makes it widely used to protect the non-friction surface of carbon brake disk against oxidation. The phosphate glass coating developed by Fu et al. [25] possesses excellent oxidation protective at 973–1473 K. After oxidation in air at 973 K for 66 hours, the mass loss of the coated C/C is only 1.11%, and after oxidation at 1473 K in air for 5 minutes, the mass loss is less than 0.8%. Furthermore, after thermal cycles between 1173 K and room temperature in air for 100 times, the mass loss is 1.6%. The above properties show that this kind of phosphate glass coating could meet the requirements of carbon brake disk against oxidation. The glass coating including B_2O_3, P_2O_5, ZnO, Li_2O, Na_2O, and CuO also have good oxidation protective ability at 1073 K [26,27].

Borosilicate glass has good liquidity at high temperature, and can seal the cracks produced from thermal cycles effectively. Besides, it can obviously prevent oxygen from diffusing to the C/C or C/SiC substrate due to its low oxygen permeability. As a result, borosilicate glass can be used as a good sealing material for C/C composites. The stability of the glass at high temperatures could be improved by introducing ceramic particles into the glass phase. The interface compatibility between borosilicate glass coating and C/C or C/SiC substrate could be improved by adding a SiC transition layer. Smeacetto et al. [28] prepared Y_2O_3 + borosilicate glass/$MoSi_2$ + borosilicate glass coating on β-SiC coated C/C composites, which can efficiently protect C/C composites against oxidation at 1573 K for 150 hours. The protective temperature range of the borosilicate glass coating could be adjusted by controlling the composition and synthetic process of the raw materials. For example, three kinds of borosilicate glass coatings by slurry with different compositions and synthetic process have good oxidation-resistant property at 1173, 1573, and 1773 K, respectively [29–31].

16.4.2 Metal Coatings

Many metals, such as Ir, Hf, Cr, have good oxidation-resistant property due to their high melting points and low oxygen permeabilities, especially Ir (2683 K). Worrel et al. [32] reported two coatings including Ir-Al-Si and Ir-Al. The results showed the mass losses of the above two coatings were only 7.29 and 5.26 mg/cm^2 after oxidation at 1823 and 1873 K for 280 and 200 hours, respectively. However, the coatings containing Ir have not been used widely due to their high cost.

Cr is also a common alloying element used for oxidation-resistant coating, since its oxidation product, Cr_2O_3, can react with SiO_2 to form stable glassy materials. Cr-Al-Si [33, 34] and Si-Mo-Cr [35, 36] were prepared on the surface of SiC-coated C/C composites. It was found that the above two coatings can efficiently protect C/C composites against oxidation at 1773 K for 200 and 342 hours, respectively. Furthermore, the mass of Cr-Al-Si coated sample basically remained unchanged after oxidation for 200 hours. Si-Mo coating [37] prepared by Fang et al. could protect C/C substrate against oxidation for 115 hours at 1643 K with no mass loss. Moreover, a Si-Mo coating [38] was also prepared on the surface of C/SiC gradient inner layer by Zhang et al. It was found the as-mentioned coating had excellent oxidation protective ability at 1773 K and good thermal shock resistance. In the work of Terentieva et al. [39], a Mo-Si-Ti coating was achieved. The testing results showed it not only had good resistance to erosion and thermal shock, but also could protect C/C composites against oxidation at 2048 K for 2 hours with no substantial changes.

16.4.3 Ceramic Coatings

The study about ceramic coatings has been carried out most profoundly and acquired richest results. Ceramic coatings, such as SiC, Si_3N_4, and $MoSi_2$, have outstanding anti-oxidation properties due to the self-sealing performance of SiO_2 formed from oxidation. However, the CTE of ceramic is obviously larger than that of C/C and C/SiC composites. For example, the CTE of $MoSi_2$ and SiC are 8.0×10^{-6} K^{-1} and 4.3×10^{-6} K^{-1}, respectively. As a result, the mismatch of CTE would lead to some cracks in the coating, through which oxygen can attack C/C or C/SiC substrate. In view of the above question, coatings, such as multiphase inlay coating, whisker- or nanowire-toughened coating, nanoparticles-toughened coating, and gradient composite coating, have been widely researched.

16.4.3.1 Multiphase Inlay Ceramic Coatings By introducing ceramic particles with high CTE into the ceramic coating with low CTE, multiphase inlay ceramic coating was constructed. As a result, a large number of phase interfaces could be formed, which can efficiently reduce the thermal expansion mismatch between C/C or C/SiC substrate and coating. Nowadays, various multiphase coatings have been achieved, mostly by pack cementation. The present concern is on the silicide ceramic coatings with a multiphase inlay structure. The silicides were distributed in the SiC ceramic in the coating, forming some silicide-SiC phase interface, which can play a positive effect on inducing the transition of cracks, and releasing and redistributing the thermal stress in the coating, avoiding the formation of penetration cracks, and resolving the mismatch of thermal expansion between silicides and C/C composites. SiC-$MoSi_2$ [40] coating was proved to have an excellent oxidation protective ability. The testing results showed it could efficiently protect C/C composites against oxidation for 400 hours at 1773 K and 200 hours at 1873 K, respectively. The reported Si-$MoSi_2$ coating [41] had a mass loss of 0.36% after oxidation at 1673 K for 1000 hours, and $MoSi_2$/SiC coating [42] could efficiently protect C/C composites against oxidation for 52 hours at 1773 K. Using the pack cementation, the SiC-$CrSi_2$ [43] (shown in Figure 16.2), $MoSi_2$-SiC-Si [44] and SiC-$MoSi_2$/WSi_2 [45] coatings were achieved. The testing results showed SiC-$CrSi_2$ coating could prolong the service time of single SiC-coated C/C composites about 10 times. $MoSi_2$-SiC-Si [28] and SiC-$MoSi_2$/WSi_2 [46] could efficiently protect C/C composites against oxidation at 1773 K for 200 and

FIGURE 16.2

SEM image of the surface of SiC-$CrSi_2$ coating.

FIGURE 16.3
SEM image of the SiC nanowires on C/C composites.

315 hours, respectively. In the work of Huang et al. [47], mullite-Al_2O_3-SiC coating was achieved by a two-step pack cementation. It was found that the above coating could efficiently protect C/C composites against oxidation at 1773 K for 130 hours and at 1873 K for 80 hours, respectively.

16.4.3.2 Whisker-Toughened Coatings Thermal expansion mismatch would result in the crazing and debonding of the coating. To improve the toughness of ceramic coatings and reduce their cracking trend, whisker-toughed ceramic coating was put forward to protect C/C composites against oxidation [48–50]. Preparation process of these coatings was as follows. First, using Si, C, and SiC whisker as raw materials, distilled water was used as dispersant and adding amount of binder, the porous whisker layer was achieved on the surface of C/C composites by slurry coating. Second, using Si and silicide as the raw materials, by the introduction of oxide as the penetration enhancer, protective outer layer was prepared on the whisker layer by pack cementation. According to the above method, SiC whisker-toughened SiC [48], $MoSi_2$-SiC-Si [49], and SiC-$CrSi_2$ [50] coatings were achieved. The testing results of the mentioned three coatings were as follows. As the content of SiCw is 10 wt.%, the SiC coating had the best oxidation resistance, and the mass loss of the coated sample was only 0.63% after oxidation at 1773 K for 310 hours. As for $MoSi_2$-SiC-Si coating, the mass loss of the coated sample with 10 wt.% SiCw was reduced from 2.31% to 0.33% after oxidation at 1773 K for 200 hours. It was also found that oxidation protective ability of SiC-$CrSi_2$ was greatly improved between 1473 K and 1773 K with the introduction of SiCw.

In order to better exhibit the effect of whisker, high purity of SiCw [51] and SiC nanowires [52] was achieved (shown in Figure 16.3). It was expected to have better effect if used as the toughening phase in ceramic coating.

16.4.3.3 Nanowire-Toughened Coatings SiC nanowire network exhibits great potential as nanofiller in composites. The nanostructures with special morphology are possible by controlling the synthesis process, such as bamboo-shaped, necklace-shaped, and ribbon-shaped, which are the promising reinforcements. SiC nanowire network with special morphology could be obtained by CVD, and they have been attempted to be used as the reinforcement for oxidation-resistant coatings for C/C composites [53–57]. Their incorporation makes the ceramic's microhardness, elastic modulus, and fracture toughness increase obviously. SiC nanowires in the coating suppressed the cracking of the coating via various toughening mechanisms including nanowire pullout, microcrack bridging by nanowire, and microcrack deflection, which is advantageous to the oxidation, ablation, and thermal shock resistance of the ceramic coatings. Most interestingly, novel toughening mechanisms are observed that show the SiC nanowires anchoring themselves in the ceramic to effectively transfer load from the matrix to the nanowires by nanoscale mechanical interlocking between the nodes and the surrounding matrix. Moreover, the incorporated nanowires by this method also play a positive role in enhancing the interface bonding between ceramic coating and C/C composites due to the excellent anchoring of the nanowires at their interface.

16.4.3.4 Nanoparticle-Toughened Coatings Nanoparticle-toughened coating is of great interest since the toughness of the coating is expected to improve by grain refining, indicating the number and size of crack can be reduced. Moreover, the mechanical properties of the ceramic coating would be improved without affecting its compactness. For example, SiC nanoparticle-toughened SiC coating [58] and ZrO_2 nanoparticle-toughened SiC coating [59] were achieved. The testing results showed that the number and size of the crack in the above two coatings were greatly reduced. Besides, the oxidation-resistant property was also improved. SiC nanoparticle-toughened SiC coating could efficiently protect C/C composites against oxidation at 1773 K for 215 hours. As for ZrO_2 nanoparticle-toughened SiC coating, the mass loss was reduced from 4.52% to 1.05% after oxidation at 1873 K for 70 hours.

16.4.3.5 Gradient Coatings Thermal expansion mismatch between coating and substrate can also be eliminated by designing a gradient structure. Various gradient coatings have been prepared. In the work of Kowbel [60], Zr-BN gradient coating was achieved by Chemical Vapor Infiltration (CVI). It was found that the coating could effectively protect C/C composites from oxidation in air above 1773 K. In the work of Zhu et al. [61], a $(SiC/Si_3N_4)/C$ gradient coating was prepared by the diffusion reaction method, which had good oxidation resistance at 1823 K.

The CTE of ZrO_2 is eight times higher than that of C/C or C/SiC composites, and it has high oxygen permeability, which is considered difficult to be applied as the oxidation-resistant coating for C/C or C/SiC. In the work of Huang et al. [62], the gradient ZrO_2-SiO_2 coating structure was designed and was prepared on SiC-coated C/C composites using $ZrOCl_2{\cdot}8H_2O$ and tetraethyl orthosilicate as the precursors with their gradient-changing concentration ratios, successfully applying ZrO_2 coating on C/C composites by a sol-gel method. The gradient transition of thermal expansion was realized by the gradient design of phase composition in the coating, effectively releasing thermal stress and resolving the problem of cracking and debonding in the monolayer ZrO_2 coating. The design of gradient coating structure provided a route for the refractory materials with high CTE to be applied as the oxidation-resistant coatings for C/C composites.

The coatings prepared by conventional plasma spraying were thought difficult as the long-term oxidation-resistant coatings for C/C and C/SiC composites due to their porous structure and weak interface bonding. This method has been applied to prepare the gradient yttrium silicate coating [63] as the oxygen barrier layer, reported by Huang et al. The gradient distribution of composition could effectively relax the mismatch of CTE, improving the bonding strength between the inner and the outer layers. The three-layer SiC/yttrium silicate/glass coating could protect C/C composites against oxidation at 1773 K in air for 164 hours. The novel plasma-spraying process for preparing gradient yttrium silicate resolves some key problems of the coating prepared by conventional plasma spraying, including weak bonding strength and porous structure, and realizes the preparation of long-term anti-oxidation coatings for C/C composites.

To better reduce the tendency of cracking, the idea of preparing a carbon layer between SiC coating and C/C matrix was put forward [64, 65]. Preparation processes were as follows: In the first step, using phenolic resin, graphite as the raw materials, the carbon layer was prepared on the surface of C/C composites. As a result, the surface structure of the C/C composite became consistent, which could ensure the uniformity for the subsequent coating. In the second step, with the method of molten silicon impregnation reaction, the carbon layer was transformed into C-SiC gradient coating at a high temperature. The testing result showed the service life of the coated sample was increased nearly three times at 1773 K.

16.4.3.6 Alternative Multilayer Coatings Alternative multilayer coatings have been attracting much interest in recent years for applications in ultrahigh temperature coatings [66,67]. Ultramet company developed HfC-SiC alternative multilayer coatings by using CVD. According to the report by Ultramet company itself, the layered coatings could withstand temperatures greater than 2093°C for shorter periods and a temperature of 1799°C for extended periods (hours) [66]. Such coatings demonstrated zero ablation at heat flux levels up to 363 W/cm^2 and ablation rate of about 5 $\mu m/s$ at 595 W/cm^2 [66].

Alternative multilayer coating ZrC-SiC was fabricated on C/SiC composites by Houbu Li et al. The coatings showed better ablation resistance than the SiC monolayer and ZrB_2-SiC mixed layer under the oxyacetylene flame [67]. Liu et al. [68] studied the oxidation behavior of alternative multilayer ZrC-SiC coatings in the temperature range of 1800–2000°C based on mini-composites. With the layered ZrC-SiC coatings, the composites showed much better oxidation resistance than those with only SiC layer. The retaining strength of layered ZrC-SiC coated C/SiC mini-composites after oxidation at 1800°C is about twice than that of SiC coated C/SiC mini-composites. The layered ZrC-SiC coated C/SiC mini-composites still had strength even after oxidation for 5 minutes at 2000°C, while SiC-coated C/SiC mini-composites almost disappeared. Further studies [68] indicated that the formed ZrO_2 effectively blocked the activation oxidation of SiC below. Meanwhile, the formed ZrO_2 could also act as a silica buffer to retard the evaporation of silica at ultrahigh temperatures.

16.5 COMPOSITE COATING

Glass coatings possess the ability of self-sealing, but liquidity and volatility at high temperature limit their applications. Metal and ceramic coatings are easy to crack, which is difficult to heal. If these coatings are combined together, it should be expected to achieve better oxidation-resistant ability.

Nowadays, various composite coatings have been achieved. The simplest composite coating is a double-layer coating. SiC is usually used as a transition layer or bonding layer to alleviate thermal stress, and refractory oxide, high temperature glass, or alloy is chosen for the outer layer. Morimoto et al. [69] put forward the double coating using rare earth element refractory oxide as the outer layer and SiC-C as the transition layer. The testing results showed its anti-oxidation temperature could reach to 1923 K. In the work of Aparicio et al. [70], SiC/yttrium silicate coating was studied. It was found the coating could effectively protect the C/C composites from oxidation at 1873 K for 53 hours with a mass loss of 3%.

Multilayer coating could combine the function of different oxidation coatings and play their respective advantages. The composite coating with a four-layer structure was put forward by Savage [1]. From inside to outside, it includes transition layer, barrier layer, seal layer, and anti-ablation layer. Transition layer is to solve the CTE mismatch between the coating and C/C or C/SiC substrate. Barrier layer is to provide barriers for the diffusion of oxygen and prevent material against oxidation. Seal layer is to provide glass state materials and seal the crack produced in the barrier layer at high temperature. Anti-ablation layer is to reduce the scour loss in high speed airflow, the evaporation loss at high temperature, and corrosion loss in harsh atmosphere. The as-mentioned composite coating with the four-layer structure is considered to be suitable for application above 2073 K. Based on this idea, Hai-Ming et al., Ruscher et al., and Wang et al. prepared TiC/SiC/ZrO_2-$MoSi_2$ [71], Si/SiC/mullite [72], and LaB_6-Si/PCS/SiO_2 coatings [73]. The oxidation-resistant temperature of the above three coatings stayed at about 1573 K, meaning that their protection effect was not ideal as expected. Si/SiC/mullite and Ir-C/Ir/$SrZrO_3$(Al_2O_3) coatings [74] were achieved by Yamauchi et al. It was found Si/SiC/mullite possessed good oxidation-resistant ability at 1873 K. As for Ir-C/Ir/$SrZrO_3$(Al_2O_3) coating, the results showed it could work above 2273 K, but could only hold on for 17 minutes. In the work of Sekigawa et al. [75], TiC/Ir/Y_2O_3 and Ir-C/Ir/Y_2O_3 coatings were achieved. The mass loss of TiC/Ir/Y_2O_3 coating was 6.6% after oxidation in air at 2213 K for 30 minutes. As for Ir-C/Ir/Y_2O_3 coating, the mass losses were 3.1% at 2073 K and 6.0% at 2213 K after oxidation in air for 30 minutes, respectively. HfC/Ir/HfO_2 coating was achieved by Sugahara et al. [76]. Due to the well bonding of HfC and matrix and the existence of Ir, it was proved to be suitable for application above 2173 K. But the report was not detailed. In the work of Hsu [77], SiC/(Si-$ZrSi_2$)/$ZrSi_2$ coating was prepared by slurry spraying and heat treatment. It was found that the coating could protect C/C composites against oxidation at 1923 K for 120 hours.

16.6 SUMMARY

Research on oxidation protective coating is always a hot issue and a difficulty for applications of C/C and C/SiC composites at high temperatures. In recent years, many coatings have been developed, such as multiphase inlay ceramic coating, whisker-toughening ceramics, and so on. The testing results show they all have good oxidation protective ability at temperatures of 1773–1873 K, and parts of them have excellent thermal shock resistance.

However, there are still many problems that need to be solved. Emphasis on the next step research is mainly focused on the following aspects. (i) CTE mismatch of coating and C/C or C/SiC substrate needs to be solved by a suitable way. Currently, cracking is the most common failure form for the coatings. Though the tendency of cracking can be relieved to some extent by forming gradient coating or other coatings, thermal shock resistance and self-healing of the coating still need to be further improved. (ii) High temperature stability of the coating needs to be further increased. The existing coatings mainly depends on the self-sealing performance of SiO_2 formed from oxidation to achieve long life service at high temperature, but the viscosity of SiO_2 above 1923 K reduces while the tendency of volatilization is enhanced, indicating that it is not suitable for service at high temperatures for a long time. By introducing some metal oxide in SiO_2, the viscosity and stability of SiO_2 at high temperatures can be reinforced, which is expected to extend protection life of the coating. (iii) Oxidation protective coatings effective in the whole service temperature range need to be further developed. The range of service temperature about the as-developed coatings is narrow. Introducing boride or designing a gradient structure in the coating might be a possible way. (iv) Erosion resistance at high temperatures needs to be improved. Up to now, the reported coatings cannot solve the problem of erosion and oxidation in the dynamic environment, that is to say, they still have a big gap for actual applications. Erosion resistance can be improved by the following two methods: one is to introduce anti-ablation components in the coating; the other is to improve the bonding strength of the coating. (v) Research on the coatings for large-size C/C or C/SiC parts need to be further developed. The reported coatings primarily concentrate on small samples. If they were applied to large complex parts surface, it would still need to study their stability, uniformity, and integrity during the preparation process. This problem could be better solved by the following three aspects: designing the reasonable clamping mold and choosing the right placed ways to improve the uniformity of coating; optimizing the design of the coating structure, preconditioning the edges and corners or using the multiple coating to ensure the integrity of the coating; and controlling the preparation process of the coating strictly to ensure stability on the surface of large and complex components.

REFERENCES

1. G. Savage. *Carbon–Carbon Composites* (London: Chapman & Hall, London, 1993), pp. 198–209.
2. H. J. Li. Carbon/carbon composites. *New Carbon Mater.*, 16(2), 79–80 (2001).
3. H. Z. Fu. Challenge and development trends of future aero engine materials. *J. Aeronaut. Mater.*, 18(4), 52–61 (1998).
4. J. D. Buckley. Carbon/carbon, an overview. *Am. Ceram. Bull.*, 67(2), 364–368 (1988).
5. N. S. Jacobson and D. M. Curry. Oxidation microstructure studies of reinforced carbon/carbon. *Carbon*, 44(7), 1142–1150 (2006).
6. J. F. Huang, H. J. Li, X. B. Xiong, X. R. Zeng, K. Z. Li, Y. W. Fu, and M. Huang. Progress on the oxidation protective coating of carbon/carbon composites. *New Carbon Mater.*, 20(4), 373–379 (2005).
7. X. T. Shen, K. Z. Li, H. J. Li, F. T. Lan, and T. Feng. Effect of ZrO_2 from the oxidation of ZrC on ablation of ZrC modified carbon/carbon composites. *J. Inorg. Mater.*, 24(5), 943–947 (2009).
8. H. M. Guo, W. B. Shu, S. R. Qiao, and G. W. Meng. Preparation and properties of the multilayer oxidation protection coating for carbon–carbon composites. *Areosp. Mater. Technol.*, 28(5), 31–40 (1998)
9. K. Jian, H. F. Hu, and C. H. Chen. High temperature oxidation resistance coatings for carbon–carbon composites. *Mater. Protect.*, 36(1), 22–24 (2003).
10. P. Ehrburger. Inhibition of the oxidation of carbon/carbon composite by boron oxide. *Carbon*, 24(4), 495–499 (1986).
11. F. J. Buchan and J. A. Little. Particulate-containing glass sealants for carbon–carbon composites. *Carbon*, 33(4), 491–497 (1995).
12. R. W. Lynch, B. Morosin. Thermal expansion, compressibility, and polymorphism in hafnium and zirconium titanates. *J. Am. Ceram. Soc.*, 55(8), 409–413 (1972)
13. J. R. Strife. Ceramic coatings for carbon–carbon composites. *Ceram. Bull.*, 67(2), 369–374 (1988).
14. T. Jacques. Kinetics and mechanisms of oxidation of 2D woven C/SiC composites: I, experimental approach. *Am. Ceram. Soc.*, 77(8), 2049–2057 (1994)
15. Z. H. Xu and H. J. Li. Orthogonal analysis of CVD SiC coating processing. *Ordnance Material Science and Engineering*, 23(5), 35–40 (2000).
16. L. F. Cheng and L. T. Zhang. Liquid reaction-formation method: new preparation technique of oxidation protective coating for carbon/carbon composites. *High Technol. Lett.*, 2(9), 4–7 (1992).
17. R. E. Debruuner and C. Patricia, inventors. Method of protection carbonaceous material from oxidation at high temperature. U.S. Patent 3, 713, 882 (January 30, 1973).
18. F. W. Wilson, inventor. Oxidation retardant for graphite. U.S. Patent 4, 439, 491 (May 27, 1974).
19. D. W. McKee. Borate treatment of carbon fibers and carbon/carbon composites for improved oxidation resistance. *Carbon*, 24(6), 737–741 (1986)
20. L. F. Cheng and L. T. Zhang. Investigation of liquid spreading out in the preparation of oxidation protection coating for C/C by liquid reaction-formation method. *Acta Aeronautica et Astronautica Sinica*, 17(4), 508–510 (1996).
21. L. P. Berneburg and J. Val Krukonis, inventors. Processing of carbon–carbon composites using supercritical fluid technology. U.S. Patent 5,035,921 (July 30, 1991).
22. P. R. Skowronski and D. Kramer, inventors. Coating for carbon–carbon composites and method for producing same. U.S. Patent 5,955,197 (September 21, 1999).
23. J. F. Huang, H. J. Li, and X. R. Zeng. Bioactive hydroxypatite coating on carbon/carbon composite prepared by sonoelectrodeposition and alkaline-heat treatment. *Rare Metal Mater. Eng.*, 34(9), 1489–1492 (2005).

24. J. Santon. Laser coating extends applicability of carbon–carbon composites. *Mater. Perform.*, 39(7) 47-49 (2000).
25. Q. G. Fu, H. J. Li, J. F. Huang, X. H. Shi, B. Shi and K. Z. Li. Oxidation resistance of phosphate coating for carbon/carbon composite materials. *Mater. Protect.*, 38(3), 52–54 (2005).
26. R. Delaval, G. Palavit, J. Rey, M. Laxague, and J. Thebault, inventors. Method for protecting a porous carbon-containing material from oxidation, and material obtained thereby. U.S. Patent 5,714,244 (February 3, 1998).
27. J. Thebault, M. Laxague, J. Rey, R. Delaval, and G. Palavit, inventors. Method for applying an anti-oxidative coating on brake disks of a carbon-containing composite material. U.S. Patent 5,686,144 (November 11, 1997).
28. S. Federico, F. Monica, and S. Milena. Multilayer coating with self-sealing properties for carbon–carbon composites. *Carbon*, 41(11), 2105–2111 (2003).
29. Q. G. Fu, H. J. Li, K. Z. Li, X. H. Shi, and M. Huang. A SiC/glass oxidation protective coating for carbon/carbon composites for application at 1173K. *Carbon*, 45(4), 892–894 (2007).
30. Q. G. Fu, H. J. Li, X. H. Shi, K. Z. Li, C. Wang, and M. Huang. Double-layer oxidation protective SiC/glass coatings for carbon/carbon composites. *Surf. Coat. Tech.*, 200(11), 3473–3477 (2006).
31. Q. G. Fu, H. J. Li, X. H. Shi, K. Z. Li, J. Wei, and M. Huang. Oxidation protective glass coating for Sic coated carbon/carbon composites for application at 1773K. *Mater. Lett.*, 60(3), 431–434 (2006).
32. W. L. Worrell and K. N. Lee. High temperature alloys. U.S. Patent 6,127,047 (October 3, 2000).
33. M. Huang, K. Z. Li, H. J. Li, Q. G. Li, and G. D. Sun. A Cr-Al-Si oxidation resistant coating for carbon/carbon composites by slurry dipping. *Carbon*, 45(5), 1124–1126 (2007).
34. M. Huang, K. Z. Li, H. J. Li, Q. G. Li, and G. D. Sun. Double-layer oxidation protective SiC/Cr-Al-Si coating for carbon–carbon composites. *Surf. Coat. Tech.*, 201(18), 7842–7846 (2007).
35. Y. L. Zhang, H. J. Li, Q. G. Fu, X. Y. Yao, K. Z. Li, and G. S. Jiao. An oxidation protective Si-Mo-Cr coating for C/SiC coated carbon/carbon composites. *Carbon*, 46(1), 179–182 (2008).
36. Y. L. Zhang, H. J. Li, X. Y. Yao, Q. G. Fu, and K. Z. Li. Mechanical properties of C/SiC/Si-Mo-Cr multilayer coated carbon/carbon composites. *J. Inorg. Mater.*, 23(4), 725–728 (2008).
37. H. T. Fang, J. C. Zhu, Z. D. Yin, J. H. Jeon, and Y. D. Hahn. A Si-Mo fused slurry coating for oxidation protection of carbon–carbon composites.
J. Mater. Sci. Lett., 20(2), 175–177 (2001).
38. Y. L. Zhang, H. J. Li, Q. G. Fu, K. Z. Li, D. S. Hou, and J. Fei. A Si-Mo oxidation protective coating for C/SiC coated carbon/carbon composites. *Carbon*, 45(5), 1130–1133 (2007).
39. V. S. Terentieva, O. P. Bogachkova, and E. V. Goriatcheva. Method for protecting products made of a refractory material against oxidation, and resulting produced products. U.S. Patent 5,677,060 (October 14, 1997).
40. X. R. Zeng, H. J. Li, and Z. Yang. Oxidation protective mechanism of $MoSi_2$-SiC multi-phase ceramic coatings for carbon/carbon composites. *J. Chinese Ceram. Soc.*, 27(1), 8–15 (1999).
41. J. Zhao, Q. G. Guo, J. L. Shi, G. T. Zhai, and L. Liu. SiC/Si-$MoSi_2$ oxidation protective coatings for carbon materials. *Surf. Coat. Tech.*, 201(3–4), 1861–1865 (2006).
42. P. Ran, M. Z. Yi, and J. X. Jiang. Preparation and structure of a $MoSi_2$/SiC high temperature anti-oxidative gradient coating for carbon/carbon composites. *New Carbon Mater.*, 21(03), 231–236 (2006).
43. Q. G. Fu, H. J. Li, X. H. Shi, X. L. Liao, K. Z. Li, and M. Huang. Microstructure and anti-oxidation property of $CrSi_2$-SiC coating for carbon/carbon composites. *Appl. Surf. Sci.*, 252(10), 3475–3480 (2006).
44. H. J. Li, H. Xue, Y. J. Wang, Q. G. Fu, and D. J. Yao. A $MoSi_2$-SiC-Si oxidation protective coating for carbon/carbon composites. *Surf. Coat. Tech.*, 201(24), 9444–9447 (2007).
45. K. Z. Li, D. S. Hou, H. J. Li, Q. G. Fu, and G. S. Jiao. Si-W-Mo coating for SiC coated carbon/carbon composites against oxidation. *Surf. Coat. Tech.*, 201(24), 9598–9602 (2007).
46. D. S. Hou, K. Z. Li, H. J. Li, Q. G. Fu, and Y. L. Zhang. Investigation of the multi-coating for high temperature and long life oxidation protection of C/C composites. *J. Inorg. Mater.*, 23(2), 213–218 (2008).
47. J. F. Huang, X. R. Zeng, H. J. Li, X. B. Xiong, and M. Huang. Mullite-Al_2O_3-SiC oxidation protective coating for carbon/carbon composites. *Carbon*, 41(14), 2825–2829 (2003).
48. H. J. Li, Q. G. Fu, X. H. Shi, K. Z. Li, and Z. B. Hu. SiC whisker-toughened SiC oxidation protective coating for carbon/carbon composites. *Carbon*, 44(3), 602–605 (2006).
49. Q. G. Fu, H. J. Li, K. Z. Li, X. H. Shi, Z. B. Hu, and M. Huang. SiC whisker-toughened $MoSi_2$-SiC-Si coating to protect carbon/carbon composites against oxidation. *Carbon*, 44(9), 1866–1869 (2006).
50. Q. G. Fu, H. J. Li, X. H. Shi, K. Z. Li, W. Zhang, and M. Huang. A SiC whisker-toughened SiC-$CrSi_2$ oxidation protective coating for carbon/carbon composites. *Appl. Surf. Sci.*, 253(8), 3757–3760 (2007).

51. Q. G. Fu, H. J. Li, X. H. Shi, K. Z. Li, Z. B. Hu, and J. Wei. Microstructure and growth mechanism of SiC whiskers on carbon/carbon composltes prepared by CVD. *Mater. Lett.*, 59(19–20), 2593–2597 (2005).
52. Q. G. Fu, H. J. Li, X. H. Shi, J. Wei, and Z. B. Hu. Synthesis of silicon carbide nanowires by CVD without using a metallic catalyst. *Mater. Chem. Phy.*, 100(1), 108–111 (2006).
53. Y. H. Chu, Q. G. Fu, C. W. Cao, H. J. Li, K. Z. Li, and Q. Lei. SiC nanowire-toughened SiC-$MoSi_2$-$CrSi_2$ oxidation protective coating for carbon/carbon composites. *Surf. Coat. Tech.*, 205(2), 413–418 (2010).
54. Y. H. Chu, Q. G. Fu, H. J. Li, K. Z. Li, X. Zou, and C. G. Gu. Influence of SiC nanowires on the properties of SiC coating for C/C composites between room temperature and 1500K. *Corros. Sci.*, 53 (9), 3048–3053 (2011).
55. Y. H. Chu, H. J. Li, Q. G. Fu, X. H. Shi, L. H. Qi, and B. B. Wei. Oxidation-protective and mechanical properties of SiC nanowire-toughened Si-Mo-Cr composite coating for C/C composites. *Corros. Sci.*, 58, 315–320 (2012).
56. Y. H. Chu, H. J. Li, Q. G. Fu, H. P. Wang, X. H. Hou, X. Zou, and G. N. Shang. Oxidation protection of C/C composites with a multilayer coating of SiC and Si+SiC+SiC nanowires. *Carbon*, 50(3), 1280–1288 (2012).
57. Y. H. Chu, Q. G. Fu, H. J. Li, X. H. Shi, K. Z. Li, X. Wen, and G. N. Shang. Effect of SiC nanowires on the mechanical and oxidation protective ability of SiC coating for C/C composites. *J. Am. Ceram. Soc.*, 95(2), 739–745 (2012).
58. H. J. Li, Y. L. Zhang, Q. G. Fu, K. Z. Li, J. Wei, and D. S. Hou. Oxidation behavior of SiC nanoparticle-SiC oxidation protective coating for carbon/carbon composites at 1773K. *Carbon*, 45(13), 2704–2707 (2007).
59. Y. L. Zhang, H. J. Li, Q. G. Fu, and K. J. Li. High performance SiC oxidation protective coating with ZrO_2 particle dispersion for carbon/carbon composites. *Adv. Eng. Mater.*, 10(10), 986–989 (2008).
60. W. Kowbel. Graded-codeposited ZrC-BN coating for the thermal protection of carbon–carbon composites. Proceedings of the Third International Symposium, Las Vegas, CA, November 27–30, 1988. pp. l5–27.
61. Y. C. Zhu, S. Ohtani, Y. Sato, and N. Iwamoto. Formation of a functionally gradient(Si_3N_4+SiC)/C layer for the oxidation protection of carbon–carbon composites. *Carbon*, 1999, 37(9): l417–1723.
62. J. F. Huang, X. R. Zeng, H. J. Li, X. B. Xiong, and G. L. Sun. ZrO_2-SiO_2 gradient multilayer oxidation protective coating for SiC coated carbon/carbon composites. *Surf. Coat. Tech.*, 190(2–3), 255–259 (2005).
63. J. F. Huang, H. J. Li, X. R. Zeng, and K. Z. Li. Yttrium silicate oxidation protective coating for SiC coated carbon/carbon composites. *Ceram. Inter.*, 32(4), 417–421 (2006).
64. Y. L. Zhang, H. J. Li, Q. G. Fu, K. Z. Li, J. Wei, and P. Y. Wang. A C/SiC gradient oxidation protective coating for carbon/carbon composites. *Surf. Coat. Tech.*, 201(6), 3491–3495 (2006).
65. Y. L. Zhang, H. J. Li, Q. G. Fu, K. Z. Li, and H. B. Ouyang. Preparation and performance of pre-coated carbon layer for carbon/carbon composites. *J. Inorg. Mater.*, 24(1), 125–128 (2009).
66. Ultramet Ceramic Protective Coating http://www.ultramet.com/ceramic_protective_coatings.html (2004).
67. H. B. Li, L. T. Zhang, L. F. Cheng, and Y. G. Wang. Ablation resistance of different coating structures for C/ZrB_2-SiC composites under oxyacetylene torch flame. *Int. J. Appl. Ceram. Tech.*, 6(2), 145–150 (2009).
68. Q. M. Liu, L. T. Zhang, J. Liu, L. F. Cheng, and Y. G. Wang. The oxidation behavior of SiC-ZrC-SiC coated C/SiC minicomposites at ultrahigh temperatures. *J. Am. Ceram. Soc*, 93(12), 3990–3992 (2010).
69. T. Morimoto, Y. Ogura, M. Kondo, and T. Ueda. Multilayer coating for carbon–carbon composites. *Carbon*, 33(4), 351–357 (1995).
70. A. Mario, and A. Duran. Yttrium silicate coatings for oxidation protection of carbon-silicon carbide composites. *J. Am. Ceram. Soc.*, 83, 1351–1355 (2000).
71. H. M. Guo, W. B. Shu, S. R. Qiao, S. H. Bai, and G. M. Meng. Preparation and properties of the multilayer oxidation protection coating for carbon–carbon composites. *Aerosp. Mater. Technol.*, 28(5): 37–40 (1998).
72. C. H. Ruscher, H. Fritze, G. Borchardt, T. Witke, and B. Schultrich. Mullite coatings on SiC and C/C-Si-SiC substrates characterized by infrared spectroscopy. *Am. Ceram. Soc.*, 80(12), 3225–3228 (1997).
73. R. Wang, H. Sano, Y. Uchiyama, and K. Kobayashi. Oxidation behaviours of carbon/carbon composite with multi-coatings of LaB_6-Si/polycarbosilane/SiO_2. *J. Mater. Sci.*, 31(23), 6163–6169 (1996).
74. Y. Hiroshi, K. Katsuaki, and K. Katsuhiro. Study of anti-oxidation coating system for advanced C/C composites. The 8th Symposium on High Performance Materials for Severe Environments, Tokyo, Japan, September 17–18, 1997. pp. 283–294.
75. T. Sekigawa, F. Takeda, and M. Taguchital. High temperature oxidation protection coating for C/C composites. The 8th Symposium on High Performance Materials for Severe Environments. Tokyo, Japan, September 17–18, 1997. pp. 307–315.
76. N. Sugahara, T. Kamiyama, and O. M. Yamamoto. Stabilized HfO_2/Ir/HfC Oxidation Resistant Coating System for C/C Composites. The 8th Symposium on High Performance Materials for Severe Environments. Tokyo, Japan, September 17–18, 1997. pp. 397–408.
77. S. E. Hsu, H. D. Wu, T. M. Wu, S. T. Chou, K. L. Wang, and C. I. Chen. Oxidation protection for 3D carbon/carbon composites. *Acta Astronaut.*, 35(1), 34–41 (1995).

PART IV MODELING

CHAPTER 17

DAMAGE AND LIFETIME MODELING FOR STRUCTURE COMPUTATIONS

PIERRE LADEVÈZE, EMMANUEL BARANGER, MARTIN GENET, AND CHRISTOPHE CLUZEL

LMT-Cachan (ENS Cachan / CNRS UMR 8535 / PRES UniverSud Paris), Cachan CEDEX, France

17.1 INTRODUCTION

Ceramic matrix composites (SiC fibers, PyC interphase, and mutistacked Si-B-C matrix in this paper) combine good mechanical properties with the capability of sustaining high temperatures in oxidizing environments. The use of a multi-layer self-healing matrix improves the lifetime of ceramic matrix composites (CMCs) considerably [1]. This technology along with the use of SiC fibers paves the way for applications with very long lifetimes (several thousands of hours).

Structural applications of CMCs need to answer to the main and critical question that is at the center of this chapter: how to predict, that is, to compute, the lifetime of engineering structures. The answer is a multiphysic macroscopic model able to reproduce couplings between the mechanical and environmental fields. For such materials, confidence in the simulation is related directly to one's understanding of each degradation mechanism and to the introduction of such mechanisms in the model; lifetimes being too large for a purely experimental validation over long lifetimes (tens of thousands of hours).

The first part of the chapter deals with the modeling of the different microcracking mechanisms. The key phenomenon is the deflection of the matrix cracks in the fiber matrix interphase [2–4]: this mechanical fuse protects the fibers from the cracks and enables the development of multiple cracks in the matrix, giving the composite some level of ductility, even though all of its constituents are brittle. The matrix cracking scenario has already been established for woven CMCs [5]. Since the matrix between the yarns is stiffer, but more brittle, and has larger pores than the yarns themselves, it begins to crack first. For a well-densified composite, the corresponding cracks are oriented by the main loading direction and are orthogonal to the applied stress direction, which is a main modeling difficulty. Then, when this crack network nearly reaches saturation, new cracks mainly appear in the matrix inside the yarns. The corresponding cracks are oriented by the fibers: they are orthogonal to the fibers in the longitudinal yarns, and parallel to the fibers in transverse yarns if the main load direction corresponds to the longitudinal yarns. In summary, there are three types of cracks in the damaged material: (i) inter-yarn cracks, (ii) intra-yarn longitudinal cracks, and (iii) intra-yarn transverse cracks.

Among composites, CMCs are the most difficult situation for damage modeling. Rather complex closure effects occur for the damage mechanism associated with microcracks when direction is orthogonal to the loading direction and, therefore, is not known *a priori*. Here, we are developing our answer with a general anisotropic damage theory including microcracks closure effects [6,7]. In order to differentiate between the crack opening and closing effects, which are very pronounced in CMCs, the strain energy is divided into three parts associated with damage due to pure traction,

Ceramic Matrix Composites: Materials, Modeling and Technology, First Edition. Edited by Narottam P. Bansal and Jacques Lamon.

damage due to traction and compression, and initial stiffness for compression stresses. First applications to CMCs and C/C were proposed in Reference 8. A major evolution of the model, suitable for composites with self-healing matrix was proposed in References 7,9, and 10. In this approach, the behavior is initially anisotropic and the kinematics of damage is not defined *a priori*; it is the damage evolution laws that define which components of the compliance tensor are affected. Each of the different mechanical degradation mechanisms—inter-yarn matrix cracking, intra-yarn matrix cracking, and wear of the fiber/matrix interface—is represented by a damage variable. Then, an evolution law is associated with each macroscopic damage variable, that is, with each mechanism. Finally, the global damage results from the accumulation of the various degradation mechanisms. A first identification for a CMC family has been done in Reference 11.

The second part of the chapter is devoted to the modeling of couplings with the environment characterized by temperature, oxygen pressure, and humidity. Self-healing mechanism, thanks to multilayer matrix is also taken into account. The idea, here, is to develop micro-multiphysic models that provide information for the computational macromodel. In fact the last version is very suitable for that [11, 12]. The various experimental results obtained by SAFRAN Snecma Propulsion Solide or by Reference 13 were used to identify the mechanical model. In this description of material cracking, the evolution laws and crack opening indicators are derived from assumptions about the physical cracking mechanisms. For any complex static or fatigue loading [14], mechanical macro-modeling enables the calculation of the opening indicator value for the cracks perpendicular to the fiber yarns. This indicator constitutes the link from the mechanical model to the oxidation model. The link from the oxidation model to the mechanical model relies on the fiber's strength evolution as well as a PyC damaging. Indeed, oxygen diffusion toward the fibers, the PyC oxidation, and the creation of an oxide plug that heals the yarn are directly related to the opening of the inter-yarn cracks [15]. SAFRAN Snecma Propulsion Solide proposes a range of self-healing, multilayer CMCs, one of which is CERASEPA410, whose model is presented below. In an oxidizing environment, certain matrix layers become oxidized and form a glass that fills the cracks and plays the role of a plug that slows the diffusion of oxygen [2]. A model of the crack opening, matrix healing, and fiber strength degradation mechanisms as functions of oxygen concentration proposed in Reference 12 are described here. The methodology leading to a macro-multiphysic model giving the lifetime is put forward.

The last part of the chapter deals with computational aspects. Such a damage model allows to predict the macro-crack's initiation and their propagation. Classical numerical difficulties are displayed and answers are given. To end, the lifetime model is applied to several structural tests [16]. The first test concerns a dog-bone specimen subjected to tension. The combined action of the mechanical and thermal fields is analyzed in both hot- and cold-grip experiments (i.e., homogeneous and heterogeneous temperature fields), and elements of validation are presented. The fact that the model is capable of predicting the location of the specimen's fracture zone is found to be very useful for the design of lifetime experiments on structures. The second structural test concerns a plate with a circular opening subjected to tension (open-hole test). This test enables us to assess the effectiveness of the strategy in the case of multiaxial stress fields with high stress gradients. Then, having established a certain degree of confidence in the strategy, we begin to outline a damage tolerance analysis through the prediction of a specimen's residual properties and lifetime after impact. This answers one of the main industrial concerns regarding the development of new composite structures [17, 18]. Of course, this model is also used to *a posteriori* analyze the lifetime for a known state in terms of stresses or strains and temperature. An illustration is also given here.

17.2 DAMAGE MODELING BASED ON AN ANISOTROPIC DAMAGE THEORY INCLUDING CLOSURE EFFECTS

The CMCs' damage models given here are based on an anisotropic and unilateral damage theory proposed in References 6,7, and 9.

17.2.1 General Notations

First-order tensors, or vectors (e.g., displacement), will be denoted by $\underline{u}$, $\underline{v}$, etc. Second order tensors, that is, linear operators over vectors, or matrices (e.g., strains and stresses), will be denoted by $\underline{\underline{\epsilon}}$, $\underline{\underline{\sigma}}$, etc. Fourth order tensors, that is, linear operators over matrices (e.g., Hooke's operator, compliance operator) will be denoted by $\underline{\underline{\underline{\underline{K}}}}$, $\underline{\underline{\underline{\underline{C}}}}$, etc.

The Eulerian transpose will be denoted by ${}^t\underline{u}$, ${}^t\underline{v}$, etc., such that the scalar product of $\underline{u}$ and $\underline{v}$ is ${}^t\underline{u}\underline{v}$. If there is no ambiguity, this scalar product can also be written $\underline{u} \cdot \underline{v}$.

Work and power bring into play the trace operator:

$$\mathrm{Tr}(\underline{\underline{\sigma}}\,\underline{\underline{\epsilon}}) = \sum_{i,j=1,2,3} \sigma_{ij}\epsilon_{ij}$$

where σ_{ij} and ϵ_{ij} denote the components of the tensors $\underline{\underline{\sigma}}$ and $\underline{\underline{\epsilon}}$ in an orthogonal basis.

For working with anisotropic operators, it is suitable to use the "engineering notation," which is introduced here for two-dimensional (2D) problems. The stress and strain are represented by

$$\underline{\hat{\epsilon}} = \begin{bmatrix} \epsilon_{11} \\ \epsilon_{22} \\ \sqrt{2}\epsilon_{12} \end{bmatrix} \qquad \underline{\hat{\sigma}} = \begin{bmatrix} \sigma_{11} \\ \sigma_{22} \\ \sqrt{2}\sigma_{12} \end{bmatrix}$$

so that $\mathrm{Tr}(\underline{\underline{\sigma}}\,\underline{\underline{\epsilon}}) = {}^t\underline{\hat{\sigma}}\,\underline{\hat{\epsilon}}$, and the Hooke operator by

$$\underline{\underline{\hat{K}}} = \begin{bmatrix} K_{1111} & K_{1122} & K_{1112} \\ & K_{2222} & K_{2212} \\ \text{sym} & & K_{1212} \end{bmatrix}$$

17.2.2 Thermodynamic Framework

17.2.2.1 Energy Potential and State Laws After choosing a damage kinematics, the construction of a mathematical model does not cause any particular difficulty as long as one uses the classical framework of thermodynamics, which can be found in Reference 19. The scale can be macro or meso or even micro. Small displacement assumption is used here.

Let us consider a damage kinematics defined by the sets of damage variables $\underline{d}$ and $\underline{\delta}$. $\underline{\underline{\epsilon}}$, $\underline{\underline{\epsilon_e}}$, and $\underline{\underline{\epsilon_i}}$ designate the total, elastic, and inelastic strains, such that

$$\underline{\underline{\epsilon}} = \underline{\underline{\epsilon_e}} + \underline{\underline{\epsilon_i}} \tag{17.1}$$

At time t at point $\underline{M}$ of the domain Ω, the state of the material is assumed to be correctly defined by the values at time t and point $\underline{M}$ of $\underline{\underline{\epsilon}}$, $\underline{\underline{\epsilon_i}}$, $\underline{d}$, $\underline{\delta}$, $\underline{X}$, where $\underline{X}$ designates additional internal variables such as hardening variables.

The free energy $\rho\Psi$ has the following functional dependence:

$$\rho\Psi(\underline{\underline{\epsilon_e}}, \underline{d}, \underline{\delta}, \underline{X}) \tag{17.2}$$

To satisfy thermodynamic conditions, it must be continuously differentiable and convex. The damage forces conjugates of $\underline{d}$ and $\underline{\delta}$ are, for fixed $\underline{\underline{\epsilon_e}}$ and $\underline{X}$:

$$\begin{cases} \underline{Y_d} = -\rho\dfrac{\partial\Psi}{\partial\underline{d}} \\ \underline{Y_\delta} = -\rho\dfrac{\partial\Psi}{\partial\underline{\delta}} \end{cases} \tag{17.3}$$

In many cases, one also has

$$\begin{cases} \underline{Y_d} = -\rho\dfrac{\partial e}{\partial\underline{d}} \\ \underline{Y_\delta} = -\rho\dfrac{\partial e}{\partial\underline{\delta}} \end{cases} \tag{17.4}$$

where e is the strain energy density. One also has

$$\begin{cases} \underline{Y_X} = \rho \dfrac{\partial \Psi}{\partial \underline{X}} \\ \underline{\underline{\sigma}} = \rho \dfrac{\partial \Psi}{\partial \underline{\underline{\epsilon_e}}} \end{cases} \tag{17.5}$$

17.2.2.2 Damage Evolution Laws The damage forces govern the evolution of damage and, subsequently, rupture. The corresponding laws are

$$\begin{cases} \underline{d}(t) = \underline{A_d}(\underline{Y_d}(\tau), \underline{Y_\delta}(\tau); \tau \leq t) \\ \underline{\delta}(t) = \underline{A_\delta}(\underline{Y_d}(\tau), \underline{Y_\delta}(\tau); \tau \leq t) \end{cases} \tag{17.6}$$

that state that the values at time t of the damage variables depend on the past time history of the damage forces. $\underline{A_d}$ and $\underline{A_\delta}$ characterize the material and have to satisfy the following thermodynamic condition:

$$\underline{Y_d} \cdot \underline{\dot{d}} + \underline{Y_\delta} \cdot \underline{\dot{\delta}} \geq 0 \tag{17.7}$$

In practice, for composites subject to quasi-static loading, the values of $\underline{d}$ and $\underline{\delta}$ at time t depend on the maximum over $[0, t]$ of some combination of the elementary forces. For fatigue loading, one has

$$\begin{cases} \underline{d} = \underline{d_s} + \underline{d_f} \\ \underline{\delta} = \underline{\delta_s} + \underline{\delta_f} \end{cases} \tag{17.8}$$

where $(\underline{d_s}, \underline{\delta_s})$ designate the quasi-static part. The fatigue part $(\underline{d_f}, \underline{\delta_f})$ follows classical fatigue damage models.

Remark It is possible to introduce a variable transformation for the damage variables $\underline{d}$ and $\underline{\delta}$. Thus, the conjugate forces will also be modified. Such a variable transformation does not affect the prediction of the material's behavior.

Remark Serious difficulties appear at the structural level concerning localization phenomena and final fracture. Their cause is now rather well understood: the classical damage models present serious shortcomings. For example, they do not contain classical fracture mechanics, which works quite well in many cases. However, let us note that such models provide the correct response before the initiation point for a structure subject to complex loading. One solution is given by the concept of localization limiter [20–23]. These are regularization procedures; the additional terms are built from a nonlocal approach or from a second-gradient approach. Viscoplastic regularization can also be introduced.

Our solution for composites depends on the scale chosen. The best approach, whenever possible, is to use what we call a "damage mesomodel." It is a semi-discrete modeling approach in which the damage state is locally uniform within meso-constituents [24]. A more general approach is to introduce a damage model with a delay effect combined with a dynamic analysis. Then, a characteristic length—actually a characteristic volume with a strong physical significance—is introduced. Calculations with such mesomodels are given in References 25–27. We will come back to that problem in Section 17.4.

17.2.2.3 Inelastic Strains: Coupling Damage with Plasticity (or Viscoplasticity) Microdefects, that is, damage itself, lead to sliding with friction and, thus, to inelastic strains. More generally, it is possible to get coupling phenomena between damage and what we classically call plasticity or viscoplasticity. A method that seems to work quite well for composites is applying plasticity or viscoplasticity type models to quantities that are called effective (effective stress $\underline{\underline{\tilde{\sigma}}}$ and effective inelastic strain $\underline{\underline{\tilde{\epsilon}_i}}$) [28], which have to satisfy

$$\mathrm{Tr}(\underline{\underline{\sigma}}\, \underline{\underline{\dot{\epsilon}_i}}) = \mathrm{Tr}\left(\underline{\underline{\tilde{\sigma}}}\, \underline{\underline{\dot{\tilde{\epsilon}}_i}}\right) \tag{17.9}$$

All other internal variables are, in fact, effective quantities. Then, the dissipation condition becomes

$$\mathrm{Tr}\left(\underline{\underline{\tilde{\sigma}}}\,\dot{\underline{\underline{\tilde{\epsilon}_i}}}\right) + \underline{Y_d} \cdot \underline{\dot{d}} + \underline{Y_\delta} \cdot \underline{\dot{\delta}} \geq 0 \tag{17.10}$$

Here, we do not follow the strain equivalence assumption advocated in Reference 29, which does not work, in general, for composites. A choice that works for many composite materials is

$$\begin{cases} \underline{\underline{\tilde{\sigma}}} = \underline{\underline{K_0}}\,\underline{\underline{K}}^{-1}\underline{\underline{\sigma}} \\ \dot{\underline{\underline{\tilde{\epsilon}_i}}} = \underline{\underline{K}}^{-1}\underline{\underline{K_0}}\dot{\underline{\underline{\epsilon_i}}} \end{cases} \tag{17.11}$$

where $\underline{\underline{K}}$ designates the eventually damaged Hooke's tensor, and $\underline{\underline{K_0}}$ the initial Hooke's tensor.

In order to predict the state of the material and its time dependence, one uses the following state evolution law:

$$\frac{d}{dt}\begin{bmatrix} \underline{\underline{\tilde{\epsilon}_i}} \\ -\underline{X} \end{bmatrix} = \underline{\underline{B}}\begin{bmatrix} \underline{\underline{\tilde{\sigma}}} \\ \underline{Y_X} \end{bmatrix} \tag{17.12}$$

where $\underline{\underline{B}}$ is an operator that depends on the material. It must be positive in order to satisfy the second law of thermodynamics. For noneffective quantities, this type of model is classical for plasticity and viscoplasticity.

17.2.3 A First CMC Damage Model

A first characteristic of ceramic composites is that very different responses can be observed for compression and traction loading due to the opening and the closing of microcracks. Another characteristic concerns the microcracking mechanisms: microcracks are not only parallel or orthogonal to the fibers, they can be orthogonal to the loading direction. The family of CMCs studied and identified here are woven SiC/SiC composites with self-healing matrix made by Snecma Propulsion Solide [18].

17.2.3.1 *Energy Potential* Modeling closure effects for isotropic materials or for microcracks having a fixed direction is not very difficult [6,30,31]. Similar models have been developed, in particular in References 32–34. The difficulty is to develop models that lead to continuous relations between stresses and strains in the unilateral case and that are also associated with a free energy. Another much more serious difficulty is to model microcracking mechanisms orthogonal to the maximum positive traction stress direction for anisotropic materials. The first step in that direction, introduced in Reference 6 and developed in Reference 25, is to express this unilateral character in terms of energy rather than by distinguishing between "traction" and "compression" behavior using stresses and strains. The latter, at any rate, would not make sense from a general point of view. We split the elastic energy of the undamaged material into three parts: a "traction" energy, a "compression" energy, and an energy identical for traction and compression behavior:

$$e = e_t + e_c + e_{tc} \tag{17.13}$$

Moreover, we introduce three sets of damage variables, $(\underline{d_t}, \underline{\delta_t})$, $(\underline{d_c}, \underline{\delta_c})$ and $(\underline{d_{tc}}, \underline{\delta_{tc}})$, which now represent the damage state of the material. We obtain

$$e(\underline{\underline{\epsilon_e}}, \underline{d_t}, \underline{\delta_t}, \underline{d_c}, \underline{\delta_c}, \underline{d_{tc}}, \underline{\delta_{tc}}) = e_t(\underline{\underline{\epsilon_e}}, \underline{d_t}, \underline{\delta_t}) + e_c(\underline{\underline{\epsilon_e}}, \underline{d_c}, \underline{\delta_c}) + e_{tc}(\underline{\underline{\epsilon_e}}, \underline{d_{tc}}, \underline{\delta_{tc}}) \tag{17.14}$$

For a large number of materials, and in particular for CMCs, damage does not increase under compression loading, therefore $\underline{d_c} = \underline{\delta_c} = 0$. And for many materials, one also has $\underline{d_{tc}} = \underline{\delta_{tc}} = 0$.

Following the damage geometrical description introduced in Reference 6, the simplest anisotropic damage model is defined by

$$\begin{cases} e_t = \dfrac{1}{2E_0}\text{Tr}\big(\underline{\underline{H}}\langle\underline{\underline{\sigma}}\rangle_+^H\underline{\underline{H}}\langle\underline{\underline{\sigma}}\rangle_+^H\big) \\ e_c = \dfrac{1}{2E_0}\text{Tr}(\langle\underline{\underline{\sigma}}\rangle_-\langle\underline{\underline{\sigma}}\rangle_-) \\ e_{tc} = -\dfrac{\nu_0}{2E_0}(\text{Tr}(\underline{\underline{\sigma}})^2 - \text{Tr}(\underline{\underline{\sigma}}^2)) \end{cases} \tag{17.15}$$

where E_0 and ν_0 are the initial Young's modulus and Poisson coefficient, and $\underline{\underline{H}}$ a positive second symmetric operator directly linked to $\underline{d_t}$ and δ_t. Let us introduce the spectral decomposition of $\underline{\underline{\sigma}}$ relative to $\underline{\underline{H}}$, where eigenvectors and eigenvalues are denoted $\underline{T_i}$ and λ_i:

$$\begin{cases} \underline{\underline{\sigma}}\underline{T_i} = \lambda_i \underline{\underline{H}}^{-1}\underline{T_i} \quad \forall i \\ {}^t\underline{T_i}\underline{\underline{H}}^{-1}\underline{T_j} = \delta_{ij} \quad \forall i,j \end{cases} \tag{17.16}$$

Definition Taking the positive part of the eigenvalues, we get the positive part of stress:

$$\langle\underline{\underline{\sigma}}\rangle_+^H = \sum_i \langle\lambda_i\rangle_+ (\underline{\underline{H}}^{-1}\underline{T_i})^t(\underline{\underline{H}}^{-1}\underline{T_i}) \tag{17.17}$$

This positive part depends on the stress but also on the damage state. (Note that if the stress is positive in the classical sense, one has $\langle\underline{\underline{\sigma}}\rangle_+^H = \langle\underline{\underline{\sigma}}\rangle_+ = \underline{\underline{\sigma}}$.) $\langle\underline{\underline{\sigma}}\rangle_-$ designates the classical negative part of $\underline{\underline{\sigma}}$.

Remark An important property of this model is that the energy is positive and can be differentiated [6]:

$$\delta e_t = \frac{1}{E_0}\text{Tr}\big(\underline{\underline{H}}\langle\underline{\underline{\sigma}}\rangle_+^H\underline{\underline{H}}\delta\underline{\underline{\sigma}}\big) + \frac{1}{E_0}\text{Tr}\big(\langle\underline{\underline{\sigma}}\rangle_+^H\underline{\underline{H}}\langle\underline{\underline{\sigma}}\rangle_+^H\delta\underline{\underline{H}}\big) \tag{17.18}$$

Remark The previous model has been used for CMCs [25,35], as well as for other materials such as concrete [36,37] and metal alloys [38].

Remark Other attempts to model closure effects have been made for anisotropic materials [39–41]. It seems, in practice, that the thermodynamics damage force is defined under the assumption that microcracking mechanisms have fixed directions; a critical analysis is given in Reference 42. However, the recent model derived in Reference 43 is consistent; it is based on the homogenization of distributed directional cracks, which has common points with the directional approach introduced in References 7, 24, and 44 and extended in Reference 45.

17.2.4 An Advanced CMC Damage Model

17.2.4.1 Energy Potential The framework of this advanced version has been introduced in Reference 7. Its interest is to facilitate and also to better take into account the damage mechanisms, which at the microscale are rather simple to describe. The first identification of the mechanical behavior has been done in References 9 and 10. The main idea is that the damage kinematics is not specified *a priori*; then the positive part of the stress is defined for any compliance operator. The damage kinematics is, in fact, specified through the choice of the damage evolution laws.

The starting point is the energy splitting as previously:

$$\begin{cases} e(\underline{\sigma}, \underline{\underline{C}}, \underline{\underline{Z}}) = \frac{1}{2}\mathrm{Tr}\big(\underline{\underline{C}}\langle\underline{\sigma}\rangle_+^{C}\langle\underline{\sigma}\rangle_+^{C}\big) + \frac{1}{2}\mathrm{Tr}\big(\underline{\underline{C_0}}\langle\underline{\sigma}\rangle_-^{C_0}\langle\underline{\sigma}\rangle_-^{C_0}\big) + \frac{1}{2}\mathrm{Tr}(\underline{\underline{Z}}\,\underline{\sigma}\,\underline{\sigma}) \\ \underline{\underline{C}}(t=0) = \underline{\underline{C_0}} \\ \underline{\underline{Z}}(t=0) = \underline{\underline{Z_0}} = \underline{\underline{K_0}}^{-1} - \underline{\underline{C_0}} \end{cases} \tag{17.19}$$

where $\underline{\underline{C}}$ and $\underline{\underline{Z}}$ are two compliance operators that can vary. Several options exist for the choice of $\underline{\underline{C_0}}$ and then $\underline{\underline{Z_0}}$. In References 11 and 16 we used $\underline{\underline{C_0}} = \underline{\underline{K_0}}^{-1}$ so that $\underline{\underline{Z_0}} = 0$. However, for an orthotropic material it is also possible to use (example in 2D with $\underline{n_1}$ and $\underline{n_2}$ the orthotropic axes):

$$\underline{\underline{C_0}} = \begin{bmatrix} 1/E_1 & 0 & 0 \\ & 1/E_2 & 0 \\ \text{sym} & & 1/2G_{12} \end{bmatrix}_{(\underline{n_1},\underline{n_2})} \quad \textit{and} \quad \underline{\underline{Z_0}} = \begin{bmatrix} 0 & -\nu_{12}/E_1 & 0 \\ & 0 & 0 \\ \text{sym} & & 0 \end{bmatrix}_{(\underline{n_1},\underline{n_2})} \tag{17.20}$$

$\underline{\underline{C}}$ has the classical restrictions:

$$\begin{cases} \mathrm{Tr}(\underline{\underline{C}}\,\underline{\sigma_i}\,\underline{\sigma_j}) = \mathrm{Tr}(\underline{\underline{C}}\,\underline{\sigma_j}\,\underline{\sigma_i}) \quad \forall \underline{\sigma_i}, \underline{\sigma_j} \\ \mathrm{Tr}(\underline{\underline{C}}\,\underline{\sigma}\,\underline{\sigma}) \geq 0 \quad \forall \underline{\sigma} \end{cases} \tag{17.21}$$

but other than that no assumption is made about the compliance; one considers the most general damage kinematics by the internal variable $\underline{\underline{C}}$. The first point is to consider that $\underline{\underline{C}}$, rather than a conventional damage variable, is the internal variable defining damage. A second point, again according to the ideas developed previously, is to build the square root of $\underline{\underline{C}}$.

Definition One then introduces the symmetric positive operator $\underline{\underline{H}}$ such that $\underline{\underline{C}} = \underline{\underline{H}}^2$. $\underline{\underline{H}}$ is easy to define by introducing the classical spectral decomposition of $\underline{\underline{C}}$:

$$\begin{cases} \underline{\underline{C}}\underline{\sigma_i} = \lambda_i \underline{\sigma_i} \quad \forall i \\ \mathrm{Tr}(\underline{\sigma_i}\underline{\sigma_j}) = \delta_{ij} \quad \forall i, j \end{cases} \tag{17.22}$$

Using the engineering representation, one gets

$$\begin{cases} \underline{\underline{\widehat{C}}}\,\underline{\widehat{\sigma}_i} = \lambda_i \underline{\widehat{\sigma}_i} \quad \forall i \\ {}^t\underline{\widehat{\sigma}_i}\,\underline{\widehat{\sigma}_j} = \delta_{ij} \quad \forall i, j \end{cases} \tag{17.23}$$

Thus, one gets

$$\underline{\underline{\widehat{H}}} = \sum_i \sqrt{\lambda_i}\,\underline{\widehat{\sigma}_i}\,{}^t\underline{\widehat{\sigma}_i} \tag{17.24}$$

Definition At this stage, we introduce the positive and negative parts of the stress:

$$\begin{cases} \langle\underline{\sigma}\rangle_+^C = \underline{\underline{H}}^{-1}\langle\underline{\underline{H}}\underline{\sigma}\rangle_+ \\ \langle\underline{\sigma}\rangle_-^{C_0} = \underline{\underline{H_0}}^{-1}\langle\underline{\underline{H_0}}\underline{\sigma}\rangle_- \end{cases} \tag{17.25}$$

where $\langle\rangle_{+/-}$ designated the classical positive/negative part of a symmetric operator.

Theorem The traction energy is positive and can be differentiated [6, 7]:

$$\delta e_t = \frac{1}{2}\mathrm{Tr}(\underline{\underline{C}}\langle\underline{\sigma}\rangle_+^C\delta\underline{\sigma}) + \frac{1}{2}\mathrm{Tr}(\delta\underline{\underline{C}}\langle\underline{\sigma}\rangle_+^C\langle\underline{\sigma}\rangle_+^C) \tag{17.26}$$

Remark With $\underline{\underline{H}}^2 = \underline{\underline{C}}$, the traction energy could be written:

$$e_t = \frac{1}{2}\mathrm{Tr}(\underline{\underline{C}}\langle\underline{\sigma}\rangle_+^C\langle\underline{\sigma}\rangle_+^C) = \frac{1}{2}\mathrm{Tr}(\langle\underline{\underline{H}}\underline{\sigma}\rangle_+\langle\underline{\underline{H}}\underline{\sigma}\rangle_+) \tag{17.27}$$

Theorem Another important result is that if $\underline{\underline{C}}$ is positive, the traction energy is a convex function. Let us introduce $\underline{A} = \underline{\underline{H}}\underline{\sigma}$. Thus, the traction energy writes $\mathrm{Tr}(\langle\underline{A}\rangle_+\langle\underline{A}\rangle_+)/2$. The proof comes from the remark that the function $e_t(\underline{A})$ is the Legendre–Fenchel transform of the function

$$\begin{aligned} l : \mathbb{E} &\rightarrow \mathbb{R} \\ \underline{\epsilon} &\mapsto \frac{1}{2}\mathrm{Tr}(\langle\underline{\epsilon}\rangle_+\langle\underline{\epsilon}\rangle_+) + \Psi_K(\underline{\epsilon}) \end{aligned} \tag{17.28}$$

and then a convex function. Ψ_K is the indicator function of the convex set $K = \{\underline{\epsilon}/\inf_{\|\underline{V}\|=1} {}^t\underline{V}\underline{\epsilon}\underline{V} \geq 0\}$, that is:

$$\Psi_K(\underline{\epsilon}) = \begin{cases} 0 \text{ if } \underline{\epsilon} \in K \\ +\infty \text{ otherwise} \end{cases} \tag{17.29}$$

We have

$$l^*(\underline{A}) = \sup_{\underline{\epsilon}\in\mathbb{E}}(\mathrm{Tr}(\underline{A}\underline{\epsilon}) - \frac{1}{2}\mathrm{Tr}(\langle\underline{\epsilon}\rangle_+\langle\underline{\epsilon}\rangle_+) - \Psi_K(\underline{\epsilon})) \tag{17.30}$$

$$= \sup_{\underline{\epsilon}\in\mathbb{E},\underline{\epsilon}\in K}\left(\mathrm{Tr}(\langle\underline{A}\rangle_+\underline{\epsilon}) + \underbrace{\mathrm{Tr}(\langle\underline{A}\rangle_-\underline{\epsilon})}_{\leq 0} - \frac{1}{2}\mathrm{Tr}(\langle\underline{\epsilon}\rangle_+\langle\underline{\epsilon}\rangle_+)\right) \tag{17.31}$$

$$= \frac{1}{2}\mathrm{Tr}(\langle\underline{A}\rangle_+\langle\underline{A}\rangle_+) \tag{17.32}$$

It follows that the admissible domain $\boldsymbol{\omega}$ for the damage variable $\omega = (\underline{\underline{C}}, \underline{\underline{Z}})$ is such that the energy $e(\underline{\sigma}, \omega)$ is a convex function of $\underline{\sigma}$. Moreover, for $\omega \in \boldsymbol{\omega}$, the energy is positive or equal to zero. The proof is trivial:

$$\forall\underline{\sigma} \in \mathbb{E} \quad e\left(\frac{\underline{\sigma}+0}{2}, \omega\right) = \frac{1}{4}e(\underline{\sigma}, \omega) \leq \frac{1}{2}e(\underline{\sigma}, \omega) + 0 \tag{17.33}$$

$$\Rightarrow \forall\underline{\sigma} \in \mathbb{E} \quad 0 \leq \frac{1}{4}e(\underline{\sigma}, \omega) \tag{17.34}$$

It is also easy to see that $\boldsymbol{\omega}$ is a convex set. For ω at the interior of $\boldsymbol{\omega}$, $e(., \omega)$ is strictly convex. Moreover, from Equations (17.33) and (17.34) we get that $e(., \omega)$ is positive for $\underline{\underline{\sigma}} \neq 0$.

At the boundary $\partial\, \boldsymbol{\omega}$, $e(., \omega)$ is no more strictly convex and such ω states are limit damage states for the material. They are associated to the initiation of a macrocrack. A characteristic of these limit states is

$$\exists \underline{\underline{\sigma}} \neq 0 / e(\underline{\underline{\sigma}}, \omega) = 0 \tag{17.35}$$

Remark Let us define the energy in terms of the elastic strain $\underline{\underline{\epsilon_e}}$. For $\omega \in \boldsymbol{\omega}$:

$$g(\underline{\underline{\epsilon_e}}, \omega) = \sup_{\underline{\underline{\sigma}} \in \mathbb{E}} (\mathrm{Tr}(\underline{\underline{\sigma}}\,\underline{\underline{\epsilon_e}}) - e(\underline{\underline{\sigma}}, \omega)) \tag{17.36}$$

All the properties of the energy $e(., \omega)$ are transfered to the energy $g(., \omega)$ for $\omega \in \boldsymbol{\omega}$.

Definition A symmetric and positive compliance operator $\underline{\underline{\underline{C}}}$ is totally positive if $\underline{\underline{\underline{C}}}\ \underline{\underline{\sigma}}$ is a positive operator for any positive $\underline{\underline{\sigma}}$.

Definition A damaged compliance operator $\underline{\underline{\underline{C}}}$ is non-transverse relative to the direction $\underline{n}$ if ${}^t\underline{t}(\underline{\underline{\underline{C}}} - \underline{\underline{\underline{C_0}}})(\underline{n}\,{}^t\underline{n})\underline{t} = 0 \quad \forall \underline{t} / {}^t\underline{t}\underline{n} = 0$.

17.2.4.2 Damage Kinematics Damage evolution laws define the damage kinematics. Here, we consider situations in which it is possible to get information about the damage evolution law (e.g., Mode-I or Mode-II microcracking). In such situations, which are encountered very often, the approach described here leads to rather efficient damage models, of which the previous one related to a $\underline{H}$-damage kinematics is a particular case. For the sake of simplicity, 2D problems and quasi-static loadings are considered.

Writing damage evolution laws in terms of stress or strain is mathematically equivalent. However, to the right instability behavior, it is easier to see them in terms of strains. That is why we have introduced some changes regarding our classical CMC model.

17.2.4.2.1 Damage evolution laws describing microcracking mechanisms orthogonal to the loading direction (mechanism M0) Let us consider that the dissipation due to damage is

$$\dot{D} = \mathrm{Tr}(\underline{\underline{\sigma}}\dot{\underline{\underline{\epsilon_e}}}) - \dot{e}(\underline{\underline{\sigma}}, \underline{\underline{\underline{C}}}, \underline{\underline{\underline{Z}}}) = \frac{1}{2}\mathrm{Tr}\big(\underline{\underline{\underline{\dot{C}}}}\langle\underline{\underline{\sigma}}\rangle_+^C\langle\underline{\underline{\sigma}}\rangle_+^C\big) + \frac{1}{2}\mathrm{Tr}(\underline{\underline{\underline{\dot{Z}}}}\underline{\underline{\sigma}}\underline{\underline{\sigma}}) \tag{17.37}$$

The second thermodynamic principle prescribes

$$\dot{D} \geq 0 \tag{17.38}$$

Many choices are possible for the damage forces that control the damage evolution. Here the strain is privileged and thus we take

$$\begin{cases} \underline{\underline{\widehat{Y'}}} = \big(\underline{\underline{\underline{\widehat{C}}}}\langle\underline{\underline{\widehat{\sigma}}}\rangle_+^C\big)^t\big(\underline{\underline{\underline{\widehat{C}}}}\langle\underline{\underline{\widehat{\sigma}}}\rangle_+^C\big) \\ \underline{\underline{\widehat{Y''}}} = \big(\underline{\underline{I_2}}\underline{\underline{\underline{\widehat{C}}}}\langle\underline{\underline{\widehat{\sigma}}}\rangle_+^C\big)^t\big(\underline{\underline{I_2}}\underline{\underline{\underline{\widehat{C}}}}\langle\underline{\underline{\widehat{\sigma}}}\rangle_+^C\big) \end{cases} \tag{17.39}$$

where $\underline{\underline{I_2}}$ is the $\pi/2$ rotation operator.

Let us assume that the damage evolution depends on only one scalar effective damage force $\bar{z}$. Here, we introduce a mode-I microcracking by taking

$$\begin{cases} \bar{z}(t) = \sup_{\tau \leq t} z(t) \\ z(t) = (a\text{Tr}(\underline{\underline{\widehat{Y'}}})^{n+1} + (1-a)\text{Tr}(\underline{\underline{\widehat{Y'}}}^{n+1}))^{1/n+1} \end{cases} \tag{17.40}$$

where a and n are two coefficients that depend on the material. The corresponding damage evolution laws are

$$\begin{cases} \underline{\underline{\widehat{C}}}^{-1}\dot{\underline{\underline{\widehat{C}}}}\underline{\underline{\widehat{C}}}^{-1} = \dot{\alpha}\dfrac{a\text{Tr}(\underline{\underline{\widehat{Y'}}})^{n}\underline{\underline{1}} + (1-a)\underline{\underline{\widehat{Y'}}}^{n}}{\bar{z}^{n}} \\ \underline{\underline{\widehat{C}}}^{-1}\dot{\underline{\underline{\widehat{Z}}}}\underline{\underline{\widehat{C}}}^{-1} = \dot{\alpha}\dfrac{b\underline{\underline{\widehat{Y''}}}^{n}}{\bar{z}^{n}} \end{cases} \tag{17.41}$$

where b is a coupling coefficient that depends on the material. The function $\bar{z} \mapsto \alpha(\bar{z})$ is a material function. If $a = 0$ and n is large, $\underline{\underline{\widehat{Y'}}}^{n}$ and $\underline{\underline{\widehat{Y''}}}^{n}$ depend on the maximum positive strain and its direction. Then, the damage kinematics described by the evolution laws simulates a microcracking mechanism orthogonal to the maximum positive deformation direction. On the other side, if $a = 1$ and $b = 0$, one gets an isotropic damage.

The dissipation is equal to

$$\dot{D} = \frac{\dot{\alpha}}{\bar{z}^{n}}(a\text{Tr}(\underline{\underline{\widehat{Y'}}})^{n+1} + (1-a)\text{Tr}(\underline{\underline{\widehat{Y'}}}^{n+1}) + b\text{Tr}((\underline{\underline{\widehat{C}}}\underline{\widehat{\sigma}})\underline{\underline{\widehat{Y''}}}^{n\,t}(\underline{\underline{\widehat{C}}}\underline{\widehat{\sigma}}))) \tag{17.42}$$

It is positive as long as $0 \leq a \leq 1$ and $b \geq 0$. Let us also note that the second is null if $\langle\underline{\underline{\sigma}}\rangle_{+}^{C} = \underline{\underline{\sigma}}$.

Remark Let us study the case where $\underline{n_1}$ and $\underline{n_2}$ denote the orthotropic axes of the material, and where the material has been damaged with a stress $\sigma\underline{n_1}{}^t\underline{n_1}$. It follows that the strain energy for any stress $\underline{\underline{\sigma}}$ has the following expression:

$$e = \frac{1}{2}\left(\frac{\sigma_{11}^{+\,2}}{E_1^0}(1+h) + \frac{\sigma_{11}^{-\,2}}{E_1^0}\right) + \frac{1}{2}\left(\frac{\sigma_{22}^2}{E_2^0} - \frac{2\nu_{12}}{E_1^0}\sigma_{11}\sigma_{22}\right) + \frac{1}{2}\frac{\sigma_{12}^2}{G_{12}^0}(1+\gamma) \tag{17.43}$$

where h and γ depend on the damage state. This expression characterizes a damaged material whose microcracks are parallel to $\underline{n_2}$.

Remark The compliance $\underline{\underline{C}}$ is totally positive if $\underline{\underline{C_0}}$ is totally positive. Under the hypothesis of the previous Remark, $\underline{\underline{C}}$ is non-transverse relative to the direction $\underline{n_2}$.

17.2.4.2.2 Damage evolution laws describing microcracking mechanisms associated with a fiber direction (mechanism F1) The damage mechanisms involved here are related to the yarn. Here, we describe a first model for which the damage evolution is controlled by only one effective damage force. More complex behaviors will be introduced in Section 17.3, devoted to the impact of environmental conditions.

Let $\underline{n_1}$ be the fiber direction and $\underline{n_2}$ the orthogonal direction. We define the projectors:

$$\begin{cases} \underline{\underline{P_1}} = \underline{n_1}{}^t\underline{n_1},\ \underline{\underline{P_2}} = \underline{n_2}{}^t\underline{n_2},\ \underline{\underline{P_{12}}} = \left(\underline{n_1}{}^t\underline{n_2}\right)_{\text{sym}} \\ \underline{\underline{\widehat{Q_1}}} = \widehat{\underline{P_1}}{}^t\widehat{\underline{P_1}},\ \underline{\underline{\widehat{Q_2}}} = \widehat{\underline{P_2}}{}^t\widehat{\underline{P_2}},\ \underline{\underline{\widehat{Q_{12}}}} = \widehat{\underline{P_{12}}}{}^t\widehat{\underline{P_{12}}} \end{cases} \tag{17.44}$$

The damage mechanisms associated with $\underline{n_1}$ are defined by the effective scalar damage force:

$$\begin{cases} \bar{z}_1(t) = \sup\limits_{\tau \le t} z_1(t) \\ z_1(t) = ((\mathrm{Tr}(\widehat{\underline{\underline{Q_1}}}\,\widehat{\underline{\underline{Y'}}}) + c\mathrm{Tr}(\widehat{\underline{\underline{Q_2}}}\,\widehat{\underline{\underline{Y'}}}))^2 + e\mathrm{Tr}(\widehat{\underline{\underline{Q_{12}}}}\,\widehat{\underline{\underline{Y'}}})^2)^{1/2} \end{cases} \tag{17.45}$$

where c and e are two material constants. The corresponding damage evolution laws are

$$\begin{cases} \widehat{\underline{\underline{C}}}^{-1}\dot{\widehat{\underline{\underline{C}}}}\widehat{\underline{\underline{C}}}^{-1} = \dot{\alpha}_1 \dfrac{\mathrm{Tr}(\widehat{\underline{\underline{Q_1}}}\,\widehat{\underline{\underline{Y'}}}) + c\mathrm{Tr}(\widehat{\underline{\underline{Q_2}}}\,\widehat{\underline{\underline{Y'}}})}{\bar{z}_1}(\widehat{\underline{\underline{Q_1}}} + c\widehat{\underline{\underline{Q_2}}}) \\ \widehat{\underline{\underline{C}}}^{-1}\dot{\widehat{\underline{\underline{Z}}}}\widehat{\underline{\underline{C}}}^{-1} = \dot{\alpha}_1 \dfrac{e\mathrm{Tr}(\widehat{\underline{\underline{Q_{12}}}}\,\widehat{\underline{\underline{Y'}}})}{\bar{z}_1}\widehat{\underline{\underline{Q_{12}}}} \end{cases} \tag{17.46}$$

where the function $\bar{z}_1 \mapsto \alpha_1(\bar{z}_1)$ is a material function.

The dissipation for these mechanisms is

$$\dot{D} = \frac{\dot{\alpha}_1}{2\bar{z}_1}((\mathrm{Tr}(\widehat{\underline{\underline{Q_1}}}\,\widehat{\underline{\underline{Y'}}}) + c\mathrm{Tr}(\widehat{\underline{\underline{Q_2}}}\,\widehat{\underline{\underline{Y'}}}))^2 + e(\mathrm{Tr}(\widehat{\underline{\underline{Q_{12}}}}\,\widehat{\underline{\underline{Y'}}}))^2) \tag{17.47}$$

It is positive as long as $c \geq 0$ and $e \geq 0$.

17.2.4.3 *Residual Strains*

The total strain is divided into elastic and inelastic parts:

$$\dot{\underline{\underline{\epsilon}}} = \dot{\underline{\underline{\epsilon_e}}} + \dot{\underline{\underline{\epsilon_i}}} \tag{17.48}$$

where the inelastic part is due essentially to microcracking mechanisms, in fact sliding with friction. The evolution of the inelastic part is modeled thanks to two uncoupled plasticity models, one for each tow direction. For instance, regarding the longitudinal tow, the plasticity model is an associated one with isotropic hardening. For unilateral damage, the definition of inelastic strain rate is not clear. Then, we prefer to operate in a different way. For instance, for the direction $\underline{n_1}$, we introduce a stress criteria that can be differentiated:

$$g(\underline{\underline{\sigma}}) = \left(\left(\frac{(\widehat{\underline{\underline{H_0}}})^2_{11}}{(\widehat{\underline{\underline{H}}})^2_{11}}((\langle\underline{\underline{H}}\underline{\underline{\sigma}}\rangle_+)^2_{11} + (\langle\underline{\underline{H_0}}\underline{\underline{\sigma}}\rangle_-)^2_{11}\right) + d\left(\frac{(\widehat{\underline{\underline{H_0}}})^2_{11}}{(\widehat{\underline{\underline{H}}})^2_{11}}((\langle\underline{\underline{H}}\underline{\underline{\sigma}}\rangle_+)^2_{12} + (\langle\underline{\underline{H_0}}\underline{\underline{\sigma}}\rangle_-)^2_{12}\right)\right)^{1/2} \tag{17.49}$$

Then, one has

$$\begin{aligned} \dot{\underline{\underline{\epsilon_i}}} = \dot{p}\frac{\partial g}{\partial \underline{\underline{\sigma}}} = \frac{\dot{p}}{R(p)} \Bigg(& \left(\frac{(\widehat{\underline{\underline{H_0}}})^2_{11}}{(\widehat{\underline{\underline{H}}})^2_{11}}((\langle\underline{\underline{H}}\underline{\underline{\sigma}}\rangle_+)_{11}\underline{\underline{\underline{\underline{HP_1}}}} + ((\langle\underline{\underline{H_0}}\underline{\underline{\sigma}}\rangle_-)_{11}\underline{\underline{\underline{\underline{H_0P_1}}}}\right) \\ & + d\left(\frac{(\widehat{\underline{\underline{H_0}}})^2_{11}}{(\widehat{\underline{\underline{H}}})^2_{11}}((\langle\underline{\underline{H}}\underline{\underline{\sigma}}\rangle_+)_{12}\underline{\underline{\underline{\underline{HP_{12}}}}} + ((\langle\underline{\underline{H_0}}\underline{\underline{\sigma}}\rangle_-)_{12}\underline{\underline{\underline{\underline{H_0P_{12}}}}}\right)\Bigg) \end{aligned} \tag{17.50}$$

with

$$\begin{cases} \dot{p} \geq 0 \\ g(\underline{\underline{\sigma}}) - R(p) \leq 0 \\ \dot{p}(g(\underline{\underline{\sigma}}) - R(p)) = 0 \end{cases} \tag{17.51}$$

The model depends on the material constant f and the material hardening function $R : p \mapsto R(p)$ with

$$\dot{p} = \frac{\mathrm{Tr}(\underline{\underline{\sigma}}\,\underline{\underline{\dot{\epsilon}_i}})}{R(p)} \tag{17.52}$$

Remark Inelastic behaviors such as viscoplasticity have been introduced in Reference 46. However, the previous plastic model can be easily extended to viscoplasticity.

17.2.4.4 The Final Model Residual strains are associated to a plasticity model with isotropic hardening. The internal variables are

- inelastic strain rate: $\underline{\underline{\dot{\epsilon}_i}} = \underline{\underline{\dot{\epsilon}_i}}^{\underline{n_1}} + \underline{\underline{\dot{\epsilon}_i}}^{\underline{n_2}}$;
- cumulated inelastic strain: $p^{\underline{n_1}}, p^{\underline{n_2}}$;
- inelastic threshold: $R^{\underline{n_1}}, R^{\underline{n_2}}$.

For an equilibrated material, the complete plasticity model depends on the hardening function $R : p \mapsto R(p)$ and the constant f.

The damage behavior is described by the damage variables $\underline{\underline{C}}$ and $\underline{\underline{Z}}$ such that

$$\begin{cases} \underline{\underline{C}} = \underline{\underline{C}}^{M0} + \underline{\underline{C}}^{F1} + \underline{\underline{C}}^{F2} \\ \underline{\underline{Z}} = \underline{\underline{Z}}^{M0} + \underline{\underline{Z}}^{F1} + \underline{\underline{Z}}^{F2} \end{cases} \tag{17.53}$$

Each damage mechanism has its own variables. For an equilibrated material, the complete damage model for quasi-static loadings depends on two functions h and k and four material constants a, b, c, and e. These material characteristics do not depend upon environmental conditions.

For fatigue behavior, one uses what we have introduced previously:

$$\begin{cases} \underline{\underline{C}} = \underline{\underline{C}}^{\text{quasi-static}} + \underline{\underline{C}}^{\text{fatigue}} \\ \underline{\underline{Z}} = \underline{\underline{Z}}^{\text{quasi-static}} + \underline{\underline{Z}}^{\text{fatigue}} \end{cases} \tag{17.54}$$

Damage laws will be seen in the Section 17.2.4.5 for static loadings and for fatigue 17.3.1.5 loadings.

Two rupture criteria (σ^f_{n1} and σ^f_{n2}) are added to describe the brittle failure of the two yarn directions.

$$\mathrm{Tr}(\underline{\underline{P_1}}\,\underline{\underline{\sigma}}) < \sigma^f_{n1} \tag{17.55}$$

$$\mathrm{Tr}(\underline{\underline{P_2}}\,\underline{\underline{\sigma}}) < \sigma^f_{n2} \tag{17.56}$$

The two associated strengths depend upon environmental conditions (temperature, oxygen concentration, humidity, etc.), and that will be seen in Section 17.3.3.

17.2.4.5 Identification at Room Temperature for Quasi-Static Loading The material studied is woven SiC-SiC; a model using the $\underline{\underline{H}}$-damage kinematics was developed in References 25 and 35. According to the microscopic description derived in Reference 47, one can say that the first damage mechanism to appear is formed by microcracks starting from microvoids. The microcracks are orthogonal to the maximum positive stress direction. Toward the saturation of this mechanism and for a longitudinal stress, a new damage mechanism occurs in the transverse fiber bundle. Near the end, damage appears and increases in the longitudinal fiber bundle. The enhanced model, like the previous one, is able to simulate such phenomena. However, only the enhanced model is able to take into account strong initial anisotropy. Here, it has been identified for a particular composite [9]; one must add the damage contributions of the different damage mechanisms. The identified material functions at room temperature h and k are given in Figures 17.1 and 17.2. (For details on the calibration of the material parameters a, b, c, etc., see References 14 and 48.) Comparisons between tests and simulations are given in Figure 17.3.

17.2.5 Some tools for the Implementation of the Anisotropic Framework

Computation of the positive part of a symmetric operator $\underline{\underline{\sigma}}$ Let us introduce the associated eigenvalue problem:

$$\begin{cases} \underline{\underline{\sigma}}\,\underline{T_i} = \lambda_i \underline{T_i} & \forall i \\ \underline{T_i} \cdot \underline{T_j} = \delta_{ij} & \forall i,j \end{cases} \tag{17.57}$$

The positive part of $\underline{\underline{\sigma}}$ is

$$\langle \underline{\underline{\sigma}} \rangle_+ = \sum_i \langle \lambda_i \rangle_+ \underline{T_i}^t \underline{T_i} \tag{17.58}$$

where

$$\langle \lambda_i \rangle_+ = \begin{cases} \lambda_i & \text{if } \lambda_i \geq 0 \\ 0 & \text{otherwise} \end{cases} \tag{17.59}$$

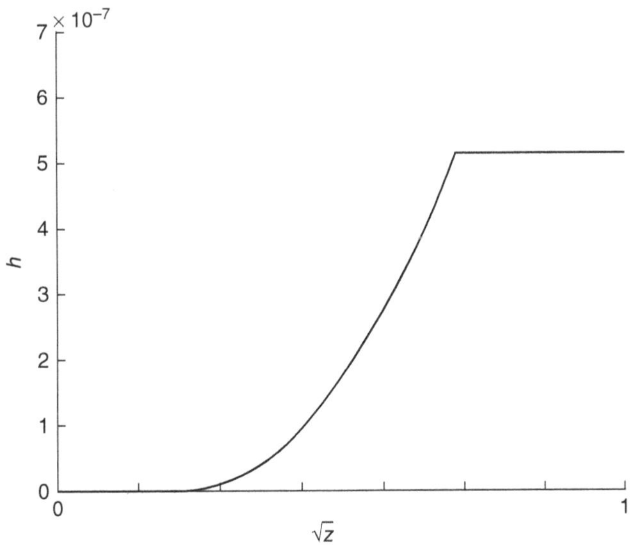

FIGURE 17.1

Identification of the material function h for the matrix damage mechanism.

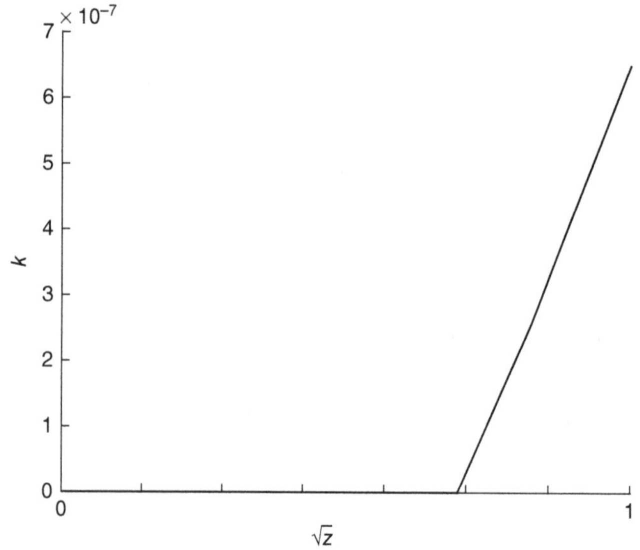

FIGURE 17.2

Identification of the material function k for the fiber bundle damage mechanism.

Computation of the square root of a symmetric operator $\underline{\underline{C}}$ Let us introduce the associated eigenvalue problem:

$$\begin{cases} \underline{\underline{C}}\underline{\sigma_i} = \lambda_i \underline{\sigma_i} & \forall i \\ \hat{\underline{\sigma_i}} \cdot \hat{\underline{\sigma_j}} = \delta_{ij} & \forall i,j \end{cases} \tag{17.60}$$

The square root of $\underline{\underline{C}}$ is

$$\hat{\underline{\underline{H}}} = \sqrt{\hat{\underline{\underline{C}}}} = \sum_i \sqrt{\lambda_i} \hat{\underline{\sigma_i}}^t \hat{\underline{\sigma_i}} \tag{17.61}$$

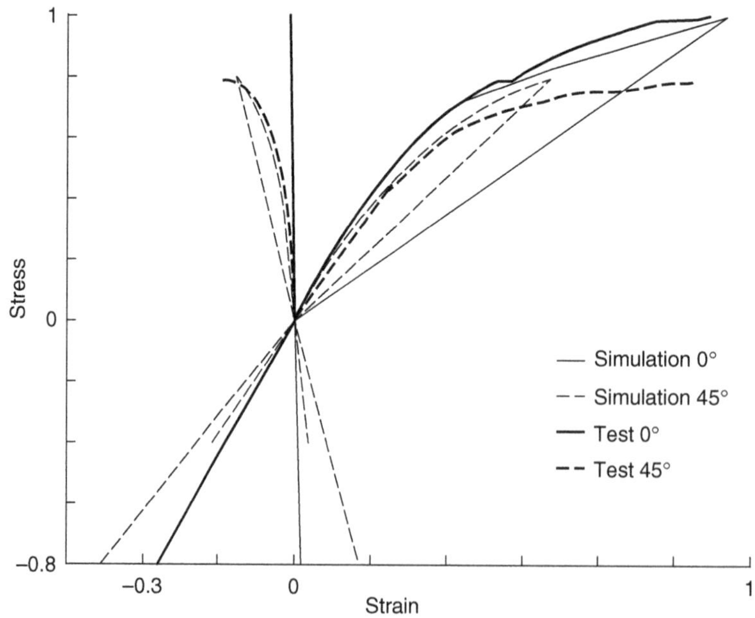

FIGURE 17.3

Stress–strain response: tests and simulations. Reproduced with permission from Elsevier, Composites Science and Technology, V59, Iss7, 1999.

One also has

$$\underline{\underline{H}}\,\underline{\sigma} = \sum_i \sqrt{\lambda_i}\mathrm{Tr}(\underline{\sigma_i}\underline{\sigma})\underline{\sigma_i} \tag{17.62}$$

Inversion of the law $\underline{\epsilon_e} = \underline{\underline{C}}\langle\underline{\sigma}\rangle_+^C + \underline{\underline{C_0}}\langle\underline{\sigma}\rangle_-^{C_0} + \underline{\underline{Z}}\underline{\sigma}$ Let us introduce the energy function:

$$e(\underline{\sigma}) = \frac{1}{2}\mathrm{Tr}\left(\underline{\underline{C}}\langle\underline{\sigma}\rangle_+^C\langle\underline{\sigma}\rangle_+^C\right) + \frac{1}{2}\mathrm{Tr}\left(\underline{\underline{C_0}}\langle\underline{\sigma}\rangle_-^{C_0}\langle\underline{\sigma}\rangle_-^{C_0}\right) + \frac{1}{2}\mathrm{Tr}(\underline{\underline{Z}}\underline{\sigma}\underline{\sigma}) \tag{17.63}$$

It is strictly convex as long as the damage state $\omega = (\underline{\underline{C}}, \underline{\underline{Z}})$ belongs to the interior of $\boldsymbol{\omega}$. It follows that the minimization problem for $\omega \in \boldsymbol{\omega}^0$:

$$\text{find } \underline{\sigma} \in \mathbb{E} \text{ minimizing } \begin{cases} \mathbb{E} \rightarrow \mathbb{R} \\ \underline{\sigma}' \mapsto e(\underline{\sigma}') - \mathrm{Tr}(\underline{\sigma}'\underline{\epsilon_e}) \end{cases} \tag{17.64}$$

admits a unique solution for any $\underline{\epsilon_e} \in \mathbb{E}$. Moreover, it is easy to see that the solution $\underline{\sigma}$ satisfies $\underline{\epsilon_e} = \underline{\underline{C}}\langle\underline{\sigma}\rangle_+^C + \underline{\underline{C_0}}\langle\underline{\sigma}\rangle_-^{C_0} + \underline{\underline{Z}}\underline{\sigma}$.

In practice, to get an approximated solution $\underline{\sigma}_h$, we propose an iterative technique to minimize e. This is a classical quasi-Newton method where the conditioner is

$$\underline{\underline{\widehat{R}}} = \frac{1}{2}(\underline{\underline{\widehat{C}}} + \underline{\underline{\widehat{C_0}}}) + \underline{\underline{\widehat{Z}}} \tag{17.65}$$

The algorithm is

Initialization: $\underline{\sigma}_0$

Iteration m: $\underline{\underline{\widehat{R}}}(\underline{\widehat{\sigma}}'_m - \underline{\widehat{\sigma}}_{m-1}) + \underline{\underline{\widehat{C}}}\langle\underline{\widehat{\sigma}}_{m-1}\rangle_+^C + \underline{\underline{\widehat{C_0}}}\langle\underline{\widehat{\sigma}}_{m-1}\rangle_-^{C_0} + \underline{\underline{\widehat{Z}}}\underline{\widehat{\sigma}}_{m-1} - \underline{\widehat{\epsilon_e}} = 0$

$$\underline{\widehat{\sigma}}_m = \lambda\underline{\widehat{\sigma}}'_m \text{ with } \lambda = \begin{cases} \mathrm{Tr}\left(\underline{\widehat{\sigma}}'_m\underline{\widehat{\epsilon}_e}\right)/2e\left(\underline{\sigma}'_m\right) \text{ if } \mathrm{Tr}\left(\underline{\widehat{\sigma}}'_m\underline{\widehat{\epsilon}_e}\right) > 0 \\ \mathrm{Tr}\left(\underline{\widehat{\sigma}}'_m\underline{\widehat{\epsilon}_e}\right)/2e\left(-\underline{\sigma}'_m\right) \text{ if } \mathrm{Tr}\left(\underline{\widehat{\sigma}}'_m\underline{\widehat{\epsilon}_e}\right) < 0 \end{cases}$$

Stopping criterion: $\dfrac{{}^t(\underline{\underline{\widehat{R}}}(\underline{\widehat{\sigma}}_m - \underline{\widehat{\sigma}}_{m-1}))(\underline{\widehat{\sigma}}_m - \underline{\widehat{\sigma}}_{m-1})}{{}^t(\underline{\underline{\widehat{R}}}\underline{\widehat{\sigma}}_m)(\underline{\underline{\widehat{R}}}\underline{\widehat{\sigma}}_m)} \leq \epsilon$

17.3 MULTISCALE MODELING OF THE OXIDATION/DAMAGE COUPLING AND THE SELF-HEALING EFFECTS

As introduced before, many degradation and healing mechanisms can occur in the studied material. Handling the competition between these different mechanisms purely experimentally would lead to a very expensive campaign. In fact, the available information is spread over different scales and different physics. In order to have a complete overview of the mechanisms and their competitions, a model able to handle all the competitions is necessary. The framework of the proposed model [11] is depicted in Figure 17.4. It is a macroscopic model enriched with micro-descriptions of the competitions of groups of mechanisms. It relies on the global description of how the material works:

1. different cracks appear in the yarns and between the yarns;
2. oxygen diffuses quickly through the transverse intra-yarn crack network;

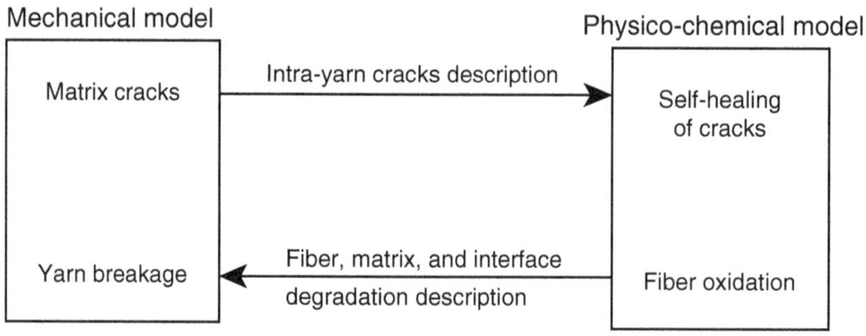

FIGURE 17.4
Modeling framework.

3. the high concentration of oxygen attacks the fibers;
4. matrix layers react with oxygen and form a glass;
5. oxygen diffuses slowly through the glass;
6. the low concentration of oxygen attacks slowly the fiber;
7. the boron carbide layer oxidation decreases the glass plug efficiency;
8. the oxidation under stress of some fibers reduces the threshold of brittle yarns;
9. mechanical loading may lead to yarn breakage and leads to the failure of the composite.

The macroscopic mechanical model presented previously is the first brick of the model. It allows the description of matrix cracks and to separate the influence of each crack network. To this damage modeling is added a brittle failure stress criterion to model the fiber breakage. The other bricks will be described in the present section:

- use of a microcomposite model for the transverse intra-yarn cracks description (summarized as a crack opening indicator);
- use of a diffusion/reaction micromodel for the self-healing description;
- use of a cumulative oxygen concentration indicator for the fiber degradation description.

When available and reliable, the data identified at the microscale are used. When the microscopic data are not available directly due to the complexity of the experimental device or due to a lack of information like the statistical repartition of defects, the complementary microscopic data are identified by an inverse problem from experimental macroscopic tests. In order to increase the robustness of the predictions, the micromodels are used to give the form of the evolution laws and complete the missing macroscopic informations.

It is important to note that the model developed hereafter is used independently in the longitudinal and transverse directions.

17.3.1 Modeling of Fatigue and of the Transverse Intra-yarn Crack Opening

During the mechanical loading, different crack networks appear. For a given macroscopic loading, if the porosity is low, the main cracks first appear rather orthogonal to the loading direction. This first network is noted (M0) in Equation (17.53). It corresponds to cracks starting from macropore. In the second network, the crack directions are mainly dependent on the fiber directions. It is noted (Fi) in Equation (17.53). Among this second network, most of the matrix cracks in a given yarn, are orthogonal to the fiber direction. They deviate at the fiber/matrix interphase. This interphase made of Pyrocarbon layers (PyC) brings some ductility to the material, but those transverse cracks allow the oxygen to reach the fibers.The self-healing mechanism is based on the oxidation of boron carbide matrix layers, which leads to the formation of a higher volume of boron oxide used to fill the crack. The efficiency of the healing plug involves the volume of the cracks; therefore, an accurate description of the crack opening is a major key of such a modeling.

Two kinds of information are available on fatigue as well as crack opening of CMC yarns. The first kind is experimental data, almost inexistent for crack openings and for coupled static/fatigue loadings, some data are available on fatigue [13, 49, 50]. The other kind relies on models describing the behavior of a mini- or microcomposite at the fiber scale [3, 4, 51–53]. Our aim being to predict the behavior and lifetime at the macro-scale, the relevant experimental

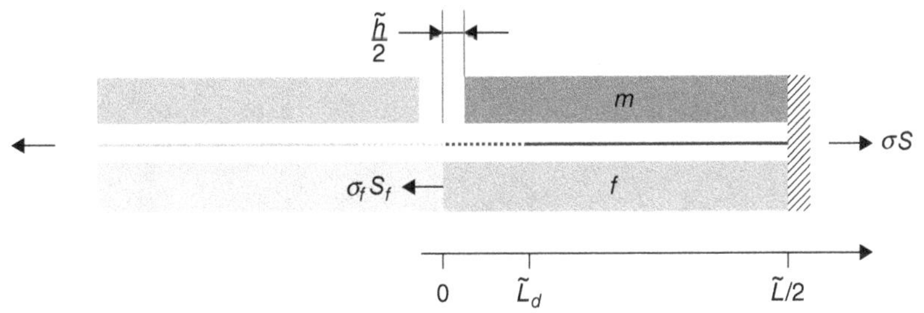

FIGURE 17.5
Mini composite modeling for a null consumption of PyC.

information relies at this scale. The models are used here to couple the available parsimonious information to get the form of the different evolution laws and indicators. These mechanical models are of great interest to handle properly the competition between different mechanisms, that is, transverse cracking, crack deviation at the PyC interphase, and wear of the interphase. The aim of such a detailed modeling is to define a relation between macroscopic damage variables and the transverse crack opening values for a given loading.

17.3.1.1 Micromodel Description and Identification Let us introduce the modeling of a microcomposite (a single fiber embedded in a matrix layer) as sketched in Figure 17.5. In this first model, the PyC consumption in oxidizing atmosphere is supposed to be neglectible. It is in agreement with models and experimental results in the healing loading range. The matrix and the fiber is denoted by m and f subscript letters. $\tilde{L}$ is the crack spacing, which is a parameter in this model. Slipping occurs on a length $\tilde{L}_d$. σ represents a homogenized stress, u a displacement, and $\tilde{h}$ is the crack opening. v_f and v_m denote the fiber and matrix volume ratios in a yarn.

The fiber equilibrium local equation reads

$$\pi\frac{d_f^2}{4}\frac{d\sigma_f}{dx} + \pi d_f \tau = 0 \ \forall x \in [0, \tilde{L}_d] \tag{17.66}$$

where $\tau(x)$ represents the shear stress in the interface between fiber and matrix at position x in the decohesion zone and d_f the fiber diameter.

With small displacement assumption, the displacement of point O is given by

$$-u_f(x=0) = \frac{\sigma}{E}\left(\frac{\tilde{L}}{2} - \tilde{L}_d\right) + \int_0^{\tilde{L}_d} \frac{\sigma_f}{E_f}dx \tag{17.67}$$

$$= \frac{\sigma}{E}\left(\frac{\tilde{L}}{2} - \tilde{L}_d\right) + \frac{\sigma\tilde{L}_d}{E_f v_f} - \frac{4}{E_f d_f}\int_0^{\tilde{L}_d}\int_0^x \tau(\alpha)d\alpha dx \tag{17.68}$$

where

$$E = E_m v_m + E_f v_f \tag{17.69}$$

The strain is defined from the displacement:

$$\tilde{\varepsilon} = -\frac{u_f(0)}{\tilde{L}/2} \tag{17.70}$$

$$= \frac{1}{E}\sigma + \left(\frac{1}{E_f v_f} - \frac{1}{E}\right)\frac{\tilde{L}_d}{\tilde{L}/2}\sigma - \frac{4}{E_f d_f \tilde{L}/2}\int_0^{\tilde{L}_d}\int_0^x \tau(y)dydx \tag{17.71}$$

This stress/strain law can be rewritten in the same form as the macroscopic one:

$$\tilde{\varepsilon} = \left(C^0\sigma + \Delta\tilde{C}_{stat+fat}\sigma + \tilde{\varepsilon}^p_{stat+fat}\right) \tag{17.72}$$

The *stat* and *fat* are relative to the static and fatigue damages in the yarns. The macroscopic maximal stress of the yarn is defined by

$$\bar{\sigma}(t) = \sup_{\tau \leq t} \sigma(t) \tag{17.73}$$

The shear stress distribution at the interface when the microcomposite is unloaded is noted τ^p. If the anelastic term $\tilde{\varepsilon}^p_{stat+fat}$ is defined as the residual strain after unloading, the identification gives

$$C^0 = \frac{1}{E} \tag{17.74}$$

$$\Delta\tilde{C}_{stat+fat} = \frac{E_m v_m C^0 \tilde{L}_d}{E_f v_f \tilde{L}/2} - \frac{8}{E_f d_f \tilde{L}\bar{\sigma}} \int_0^{\tilde{L}_d} \int_0^x \left(\tau(\alpha) - \tau(\alpha)^p\right) d\alpha dx \tag{17.75}$$

$$\tilde{\varepsilon}^p_{stat+fat} = -\frac{8}{E_f d_f \tilde{L}} \int_0^{\tilde{L}_d} \int_0^x \tau(\alpha)^p d\alpha dx \tag{17.76}$$

Like in Reference 51, the fatigue mechanism is attributed to the wear of fiber/matrix interface. To split damage parameters of Equation (17.75), the fatigue part corresponds to the variation of the shear stress, and the static part corresponds to creation of a new crack, or to the evolution of the length of the decohesion zone: $\tilde{L}_d$. To create a new crack, it is necessary to increase the loading. Then, the probability of transverse matrix crack creation is defined using the law:

$$P = 1 - \exp\left(-2\frac{S_m}{V_0}\left(\frac{\tilde{L}}{2} - \tilde{L}_d\right)\left(\frac{\sigma_m}{\sigma_0}\right)^n\right) \tag{17.77}$$

The presented model is identified in order to represent the stress/strain behavior of a minicomposite (Figure 17.6). It has to be noted that for static loading without fatigue, the shear stress is a constant noted τ_0 in the modeling.

17.3.1.2 Fatigue Modeling

Fatigue damage evolution law For a fatigue loading, the slip conditions at the PyC interface between fiber and matrix along $\tilde{L}_d$ decrease the shear stress. In this modeling, the wear of interface is not uniform, but it is assumed to be driven by the cumulative slip on each point of the interface. This driving parameter L_{cs} is defined by

$$L_{cs}(x) = \int_{time} |u_f(x) - u_m(x)| dt \tag{17.78}$$

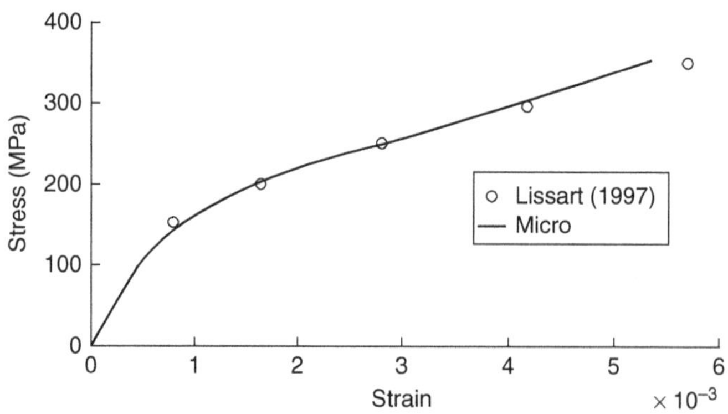

FIGURE 17.6
Behavior of a minicomposite, comparison of the micromodel and experimental data.

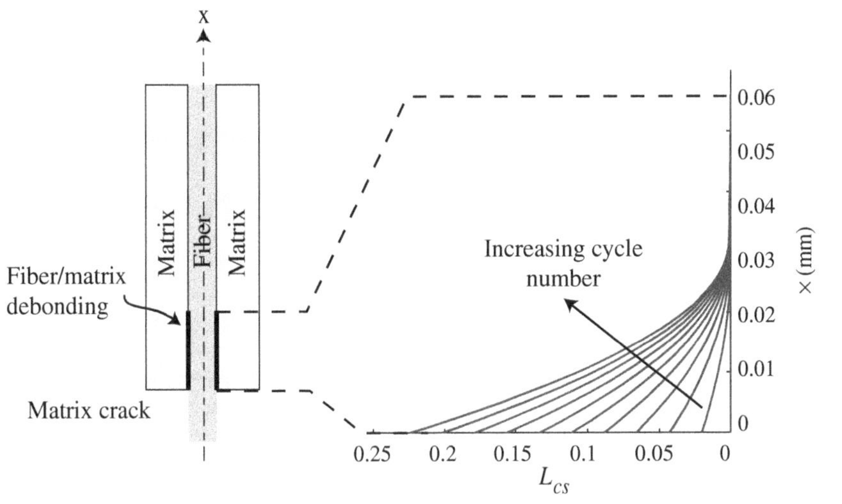

FIGURE 17.7
Influence of the cyclic loading on the cumulative slip length.

An example of evolution of L_{cs} in case of a fatigue loading is given in Figure 17.7.

The wear effect is a progressive decrease of the interface shear stress τ for initial value τ_0 to the final value τ_{min} such as

$$|\tau| = \tau_0 \frac{\exp(-\gamma_1 L_{cs}) + \beta_1}{1 + \beta_1} \tag{17.79}$$

$\tilde{L}_d$ is defined so that the stress in the fibers goes from $\frac{1}{v_f}\sigma$ to $\frac{E_f}{E}\sigma$ by stress transfer (τ) on the matrix through the interface. In order to link the micromodeling to the macroscopic damage modeling, we define a mean shear stress by

$$\tilde{L}_d \tau_{mean} = \int_0^{\tilde{L}_d} \tau(y)dy = \left(\frac{1}{v_f} - \frac{E_f}{E}\right)\frac{d_f}{4}\sigma = \frac{E_m v_m}{E v_f}\frac{d_f}{4}\sigma = c\,\sigma \tag{17.80}$$

Let us introduce the following functional:

$$T(\tau) = \int_0^1 \int_0^\eta \tau(\mu \tilde{L}_d) d\mu d\eta \tag{17.81}$$

With those functional, the damage of Equation (17.75) reads

$$\Delta\tilde{C}_{stat+fat} = \frac{E_m v_m}{E_f v_f} C^0 \frac{\tilde{L}_d}{\tilde{L}/2} - \frac{8}{E_f d_f \tilde{L}\sigma} \tilde{L}_d^2 T(\tau - \tau^p) \tag{17.82}$$

$$\Delta\tilde{C}_{stat+fat} = \left(\frac{E_m v_m}{E_f v_f} C^0 \frac{c}{\tau_{mean}} - \frac{4}{E_f d_f} \frac{c^2}{\tau_{mean}^2} T(\tau - \tau^p)\right)\frac{\bar{\sigma}}{\tilde{L}/2} \tag{17.83}$$

The function α is defined by

$$\Delta\tilde{C}_{stat+fat} = \alpha \frac{\bar{\sigma}}{\tilde{L}} \tag{17.84}$$

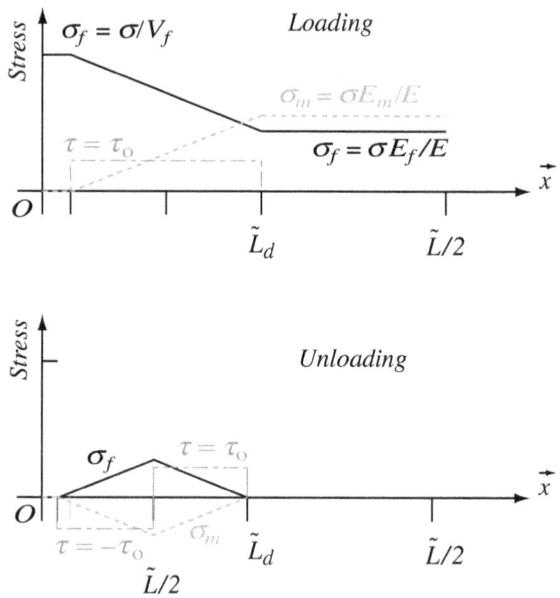

FIGURE 17.8
Behavior of a minicomposite: normal stress in fiber and matrix and shear stress in PyC interface for initial static loading and unloading.

Taking the time derivative to get the damage evolution law leads to

$$\Delta\dot{\tilde{C}}_{stat+fat} = \left(\frac{\dot{\bar{\sigma}}}{\tilde{L}} - \frac{\bar{\sigma}\dot{\tilde{L}}}{\tilde{L}^2}\right)\alpha + \frac{\bar{\sigma}}{\tilde{L}}\dot{\alpha} \tag{17.85}$$

1. The case of a pure static loading corresponds to the creation of a new crack. The stress distribution in case of loading and unloading is described (Figure 17.8).

   During the loading, $\tau_{mean} = \tau_0$ and $T(\tau - \tau^p) = \tau_0/4$ and then $\dot{\alpha} = 0$, which leads to

$$\Delta\dot{\tilde{C}}^0_{stat} = \left(\frac{\dot{\bar{\sigma}}}{\tilde{L}} - \frac{\bar{\sigma}\dot{\tilde{L}}}{\tilde{L}^2}\right)\alpha_0 \tag{17.86}$$

$$\Delta\tilde{C}^0_{stat} = \frac{\bar{\sigma}}{\tilde{L}}\alpha_0 \tag{17.87}$$

   For this static case, the static damage increment is a function of the stress increment that leads to increase the decohesion zone $\tilde{L}_d$ and a function of the creation of new crack that involves $\tilde{L}$.

2. In case of fatigue after a static loading, $\dot{\bar{\sigma}} = 0$ and $\dot{\tilde{L}} = 0$, this leads to

$$\Delta\dot{\tilde{C}}^0_{fat} = \Delta\tilde{C}^0_{stat}\frac{\dot{\alpha}}{\alpha_0} \tag{17.88}$$

$$\Delta\tilde{C}^0_{fat} = \Delta\tilde{C}^0_{stat}\frac{\alpha - \alpha_0}{\alpha_0} \tag{17.89}$$

In the general case

$$\Delta\dot{\tilde{C}}_{stat+fat} = \Delta\dot{\tilde{C}}^0_{stat}\left(1 + \frac{\Delta\tilde{C}^0_{fat}}{\Delta\tilde{C}^0_{stat}}\right) + \Delta\tilde{C}^0_{stat}\frac{\dot{\alpha}}{\alpha_0} \tag{17.90}$$

In the micromodel, the effect of wear is modeled by Equation (17.79). The function $\frac{\alpha_0}{\alpha}$ is a function of the wear of fiber/matrix interface and its evolution follows the same kind of function. Introducing ϕ such as, for N fatigue cycles

$$\phi(N) = \frac{\alpha_0}{\alpha} = \frac{\exp(-\gamma_2 N) + \beta_2}{1 + \beta_2} \tag{17.91}$$

The macroscopic damage evolution law reads

$$\frac{d\Delta\tilde{C}_{fat}}{dN} = -\Delta\tilde{C}^0_{stat} \frac{\phi'(N)}{\phi(N)^2} \tag{17.92}$$

Introducing the $\phi(N)$ function to describe the evolution of the interface wear leads to represent the evolution of α by a value of number of cycles N. The case of a static loading after N_1 cycles of fatigue leads to modify the τ_{mean} value and thus the α value. An equivalent number of cycles has to be recalculated by inversion of Equation (17.89).

Fatigue inelastic strain evolution law The inelastic part of the strain reads

$$\tilde{\varepsilon}^p_{stat+fat} = -\frac{8}{E_f d_f \tilde{L}} \int_0^{\tilde{L}_d} \int_0^x \tau(\alpha)^p \, d\alpha dx \tag{17.93}$$

$$= -\frac{8\tilde{L}_d^2}{E_f d_f \tilde{L}} T(\tau^p) \tag{17.94}$$

$$= -\frac{8c^2}{E_f d_f} \frac{\bar{\sigma}^2}{\tilde{L}} \frac{T(\tau^p)}{\tau^2_{mean}} \tag{17.95}$$

$$= \frac{\bar{\sigma}^2}{\tilde{L}} \beta \tag{17.96}$$

This leads to the following evolution law:

$$\dot{\tilde{\varepsilon}}^p_{stat+fat} = \left(\frac{2\bar{\sigma}\dot{\bar{\sigma}}}{\tilde{L}} - \frac{\bar{\sigma}^2 \dot{\tilde{L}}}{\tilde{L}^2} \right) \beta + \frac{\bar{\sigma}^2}{\tilde{L}} \dot{\beta} \tag{17.97}$$

1. In case of a pure static loading without previous fatigue:

$$\dot{\tilde{\varepsilon}}^{p0}_{stat} = \left(\frac{2\bar{\sigma}\dot{\bar{\sigma}}}{\tilde{L}} - \frac{\bar{\sigma}^2 \dot{\tilde{L}}}{\tilde{L}^2} \right) \beta_0 \tag{17.98}$$

$$\tilde{\varepsilon}^{p0}_{stat} = \frac{\bar{\sigma}^2}{\tilde{L}} \beta_0 \tag{17.99}$$

2. In case of a fatigue loading after a static loading:

$$\dot{\tilde{\varepsilon}}^p_{fat} = \tilde{\varepsilon}^{p0}_{stat} \frac{\dot{\beta}}{\beta_0} \tag{17.100}$$

$$\tilde{\varepsilon}^p_{fat} = \tilde{\varepsilon}^{p0}_{stat} \frac{\beta - \beta_0}{\beta_0} \tag{17.101}$$

In the general case

$$\dot{\tilde{\varepsilon}}^p_{stat+fat} = \dot{\tilde{\varepsilon}}^{p0}_{stat} \left(1 + \frac{\tilde{\varepsilon}^p_{fat}}{\tilde{\varepsilon}^{p0}_{stat}} \right) + \tilde{\varepsilon}^{p0}_{stat} \frac{\dot{\beta}}{\beta_0} \tag{17.102}$$

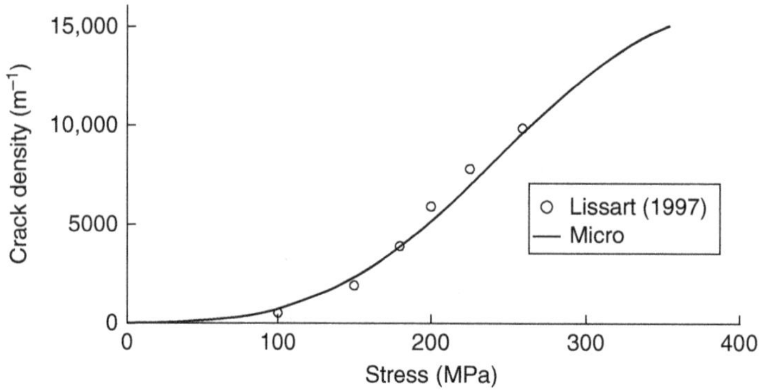

FIGURE 17.9
Crack density of a minicomposite, comparison of the micromodel and experimental data.

Introducing Φ such as, for N fatigue cycles

$$\Phi(N) = \frac{\beta_0}{\beta} = \frac{\exp(-\gamma_3 N) + \beta_3}{1 + \beta_3} \tag{17.103}$$

The α and β functions are not the same, but as the source of fatigue is the same wear of interface, we assume that $\gamma_2 = \gamma_3$. The β_2 and β_3 parameters affect the level of minimum value of the fatigue functions; thus the two functions can be approximated by a single one and a proportionality factor φ:

$$\left(\frac{1}{\Phi(N)}\right)' = \varphi \left(\frac{1}{\phi(N)}\right)' \tag{17.104}$$

The fatigue anelastic evolution law reads

$$\frac{d\bar{\varepsilon}^p_{fat}}{dN} = -\bar{\varepsilon}^{p0}_{stat} \frac{\Phi'(N)}{\Phi(N)^2} \tag{17.105}$$

Introducing the $\Phi(N)$ function to describe the evolution of the interface wear leads to represent the evolution of β by a value of number of cycles N. The case of a static loading after N_1 cycles of fatigue leads to modify the τ_{mean} value and thus the β value. An equivalent number of cycles has to be recalculated by inversion of Equation (17.101).

17.3.1.3 *Crack Density Indicator*

The crack density is defined by

$$\tilde{d} = \frac{1}{\tilde{L}} \tag{17.106}$$

Numerical simulations of the model are in good agreement with the available experimental results (see Figure 17.9).

The crack density does not have a simple form in the general case but in our special case, this law can be approximated as linear (Figure 17.10).

17.3.1.4 *Crack Opening Indicator*

Using the sketch (Figure 17.5), the crack opening is described as differential displacement between fiber and matrix. It reads

$$\frac{\tilde{h}}{2} = uf(0) - u_m(0) = \int_0^{\tilde{L}_d} \frac{\sigma_f}{E_f} dx - \int_0^{\tilde{L}_d} \frac{\sigma_m}{E_f} dx \tag{17.107}$$

$$= \int_0^{\tilde{L}_d} \frac{\sigma_f}{E_f} dx \left(1 + \frac{E_f v_f}{E_m v_m}\right) - \tilde{L}_d \frac{\sigma}{E_m v_m} \tag{17.108}$$

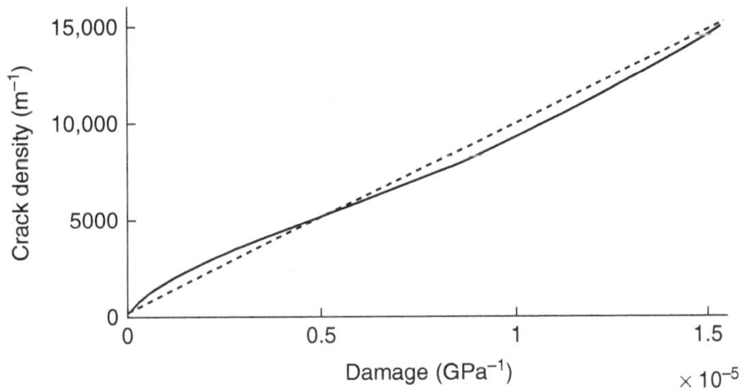

FIGURE 17.10

Crack density of a minicomposite versus damage without fatigue.

Note that

$$\int_0^{\tilde{L}_d} \frac{\sigma_f}{E_f} dx = \left(\Delta\tilde{C}_{stat+fat}\sigma + \tilde{\varepsilon}^p_{stat+fat}\right)\tilde{L}/2 + C^0\sigma\tilde{L}_d \tag{17.109}$$

Thus:

$$\frac{\tilde{h}}{2} = \left(\left(\Delta\tilde{C}_{stat+fat}\sigma + \tilde{\varepsilon}^p_{stat+fat}\right)\tilde{L}/2 + C^0\sigma\tilde{L}_d\right)\right)\left(1 + \frac{E_f v_f}{E_m v_m}\right) - \tilde{L}_d \frac{\sigma}{E_m v_m} \tag{17.110}$$

$$\frac{\tilde{h}}{2} = \tilde{L}/2\left(\Delta\tilde{C}_{stat+fat}\sigma + \tilde{\varepsilon}^p_{stat+fat}\right)\left(1 + \frac{E_f v_f}{E_m v_m}\right) \tilde{L}_d\sigma\left(-\frac{1}{E_m v_m} + C^0\left(1 + \frac{E_f v_f}{E_m v_m}\right)\right) \tag{17.111}$$

$$\frac{\tilde{h}}{2} = \tilde{L}/2\left(\Delta\tilde{C}_{stat+fat}\sigma + \tilde{\varepsilon}^p_{stat+fat}\right)\left(1 + \frac{E_f v_f}{E_m v_m}\right) \tag{17.112}$$

17.3.1.5 *Macroscopic Identification* From that study at the microscale, only the form of the laws is kept as the scale transition and statistics mean definition is not obvious between the studied scale and the macroscale.

Using the Voigt notations, the link between each mechanism and yarn macroscopic damage functions in the fiber longitudinal direction is associated to

1. creation of new crack and increase of decohesion length $\tilde{L}_d$ represented by $\Delta\dot{C}^0_{stat\ 11} = \dot{C}$. Its evolution law is defined in Chapter 17.2.4.2.
2. residual crack opening after unloading is represented by $\varepsilon^p_{stat\ 1}$. It's evolution law is defined in Chapter 17.2.4.3.
3. fatigue loading wears the PyC interface and decrease the interface shear stress. This shear is not uniform at the microscpic scale but its effects are taken into account at the macroscopic scale by the following fatigue damage evolution and inelastic strain laws:

$$\frac{d\Delta C_{fat\ 11}}{dN} = \Delta C_{stat\ 11}\left(-\frac{\phi'(N)}{\phi(N)^2}\right)\frac{\Delta\sigma_1}{\sigma_1^{max}}\begin{bmatrix} f & 0 & 0 \\ 0 & 0 & 0 \\ 0 & 0 & 0 \end{bmatrix} \tag{17.113}$$

$$\frac{d\varepsilon^p_{fat\ 11}}{dN} = \varepsilon^p_{stat\ 1}\left(-\frac{\phi'(N)}{\phi(N)^2}\right)\frac{\Delta\sigma_1}{\sigma_1^{max}}\begin{bmatrix} h \\ 0 \\ 0 \end{bmatrix} \tag{17.114}$$

where $\Delta\sigma_1$ is the longitudinal stress variation in the yarn and σ_1^{max} the maximum value of this stress in a loading cycle. And the fatigue function defined by

$$\phi(N) = \frac{\exp(-\gamma_0 N) + \beta_0}{1 + \beta_0} \tag{17.115}$$

4. decrease of interface shear stress increases the length of fiber/matrix slipping $\tilde{L}$ by

$$\dot{\Delta C}_{stat\ 11} = \dot{\Delta C}^0_{stat\ 11}\left(\frac{\Delta C_{fat\ 11}}{\Delta C^0_{stat\ 11}} + 1\right) \tag{17.116}$$

5. when the fatigue loading has different phases with increasing stress levels, all cracks created do not wear out simultaneously. For that case, an additive fatigue damage law is defined based on Miner's law. When there is a variation of fatigue loading amplitude, an equivalent cycle number (in the sense of α) is computed from

$$\frac{\alpha}{\alpha_0} = \frac{\Delta C_{fat\ 11}}{f\Delta C_{stat\ 11}}\frac{\sigma_1^{max}}{\Delta\sigma_1} + 1 \tag{17.117}$$

$$\phi(N_{eq\ 1}) = \frac{\alpha_0}{\alpha} \tag{17.118}$$

The damage macromodel allows to describe the state of degradation of each strand including a fairly accurate description of the longitudinal cracks. It is used to describe the opening of cracks, necessary for modeling the healing process. The macrocrack opening indicator reads

$$\tilde{h}_1 = \tilde{L}_1 K\left(\Delta C_{stat+fat\ 11}\ \sigma_1 + \varepsilon^p_{stat+fat\ 1}\right) \tag{17.119}$$

where the macrocrack density indicator reads

$$\tilde{d}_1 = \frac{1}{\tilde{L}_1} = a\Delta C_{stat\ 11} \tag{17.120}$$

Similar laws are written in the direction 2. The introduced parameters are identified using fatigue tests at the macroscale and a unique value (K over a parameter) of lifetime for the macrocrack opening indicator.

In order to validate the macroscopic representation, different cyclic loading paths (Figure 17.11) have been simulated using the micromodel. The comparison between the results of the micro- and macromodels are shown in Figure 17.12, which shows a good agreement.

17.3.2 Modeling of the Healing Process

The main effect of the presence of transverse cracks in the strands is to allow oxygen diffusion to fibers. At high temperature, silicon carbide fibers are sensitive to the presence of oxygen as well as the PyC interphase. The principle of self-healing is thus to limit oxygen concentration around the fibers by the introduction of a plug. The plug is made of glass resulting from the oxidation of matrix layers of boron carbide. When the volume of boron oxide is sufficient to fill the cracks in the yarns, the oxygen flow is limited by diffusion through the plug. The oxygen concentration around the fiber becomes low, which allows a significant increase in service life. Details of this principle and some morphology of self-healing C/SiC materials can be found in References 1,54 (see Figure 17.13). On that understanding of the healing mechanism, the simplified model presented in Figure 17.14 is deduced. It represents a group of several fibers protected by a self-healing matrix barrier on a perimeter noted P_e. Assuming a large radius of yarns in comparison to the plug length, the perimeter can be eliminated from the following equations. The material morphology and reaction kinetics are identified from References 55–57.

The diffusion model is solved using a wireframe linear discretization presented in Figure 17.15.

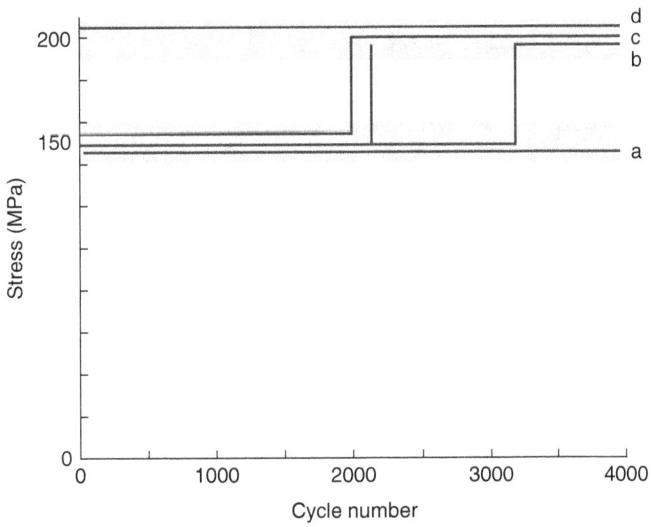

FIGURE 17.11

Cyclic loading scenarios.

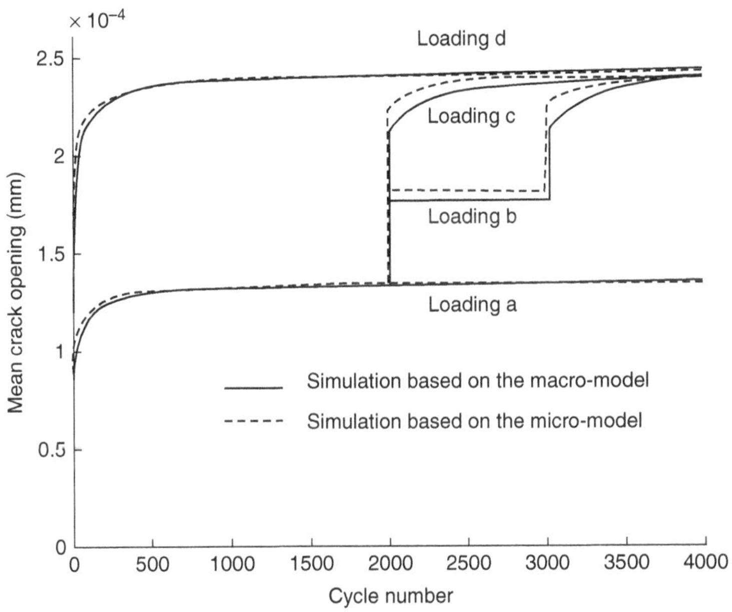

FIGURE 17.12

Comparison between micro- and macromodels for various cyclic loadings.

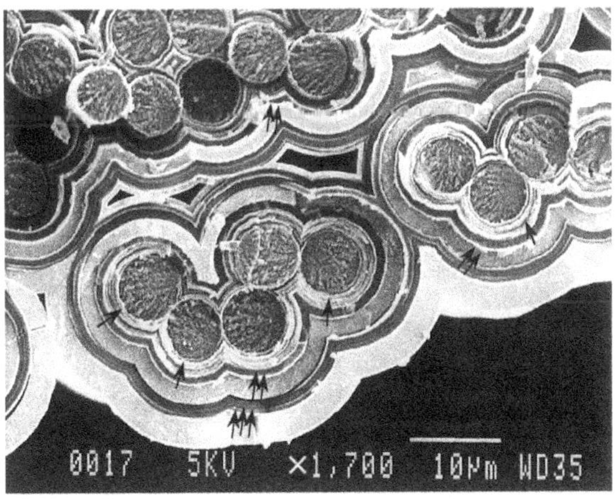

FIGURE 17.13

Morphology of self-healing matrix inside a yarn from Reference 1.

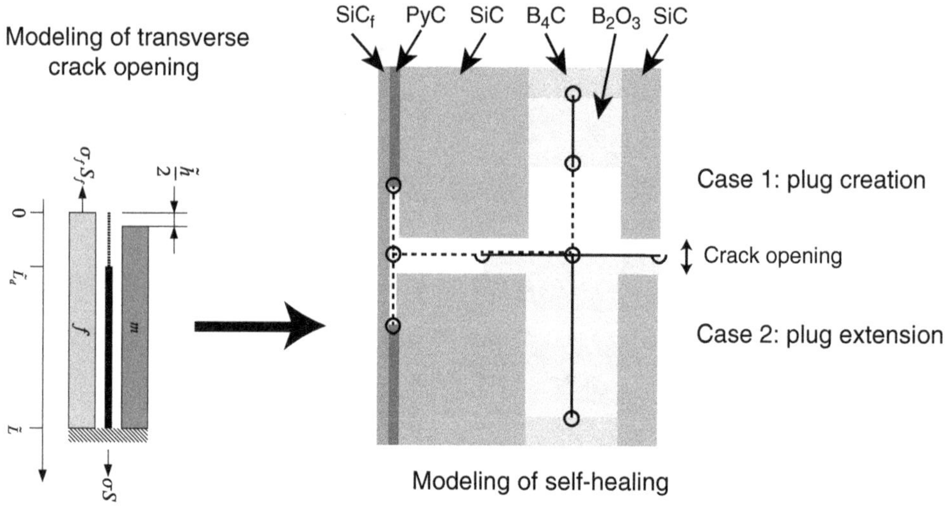

FIGURE 17.14

Sketch of the multilayered matrix and healing scenarios.

17.3.2.1 Healing Scenario Description Healing is obtained by the oxidation of boron carbide matrix layers. In that case, the diffusion of oxygen through the oxide plug controls the plug creation and extension as the reaction kinetics are too fast. When the plug is active, the diffusion model allows to estimate the amount of oxygen around the fibers. To compute the diffusion across the plug with a complex geometry that changes during the simulation, a simple solution is to make a wireframe discretization of the plug. The relationship between wireframe and 3D models of the plug is based on the description of the flow of oxide. Thus, several volumes have to be defined.

The total volume of oxide in the plug is

$$V_{B_2O_3} = P_e(e_m e_{B_4C} + (L_{fi} + L_{fe})h) \tag{17.121}$$

where $e_m \in [0, h/2 + h_m]$ is the thickness of the matrix plug, e_{B_4C} the thickness of the healing matrix layer, h the crack opening, and e_{SiC} the total thickness of the SiC matrix layers, $L_{fi} \in [0, L_{fi0}]$ and $L_{fe} \in [0, L_{fe0}]$ the interior and exterior limit lengths of the oxide plug. h_m is the length of oxidized B_4C matrix. $S = h$. *Perimeter* is the crack surface perpendicular to the flow of glass (Perimeter being the perimeter of the healing layers around a group of fibers).

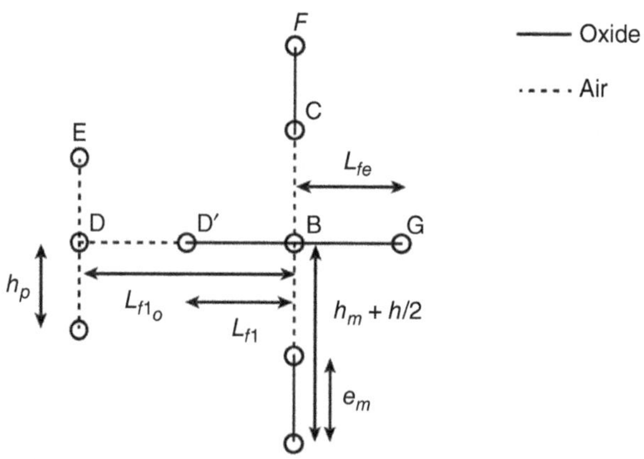

FIGURE 17.15

Wireframe discretization of the model crack.

Let us introduce some volumes to define the considered healing scenario:

$$V_0 = (h + 2h_m)P_e e_{B_4C} \tag{17.122}$$
$$V_1 = V_0 + 2L_{fi0}P_e h \tag{17.123}$$
$$V_2 = V_0 + (L_{fi0} + L_{fe0})P_e h \tag{17.124}$$

In order to calculate the flow of oxygen through the plug, a wireframe discretization is defined. But two cases are distinguished: the plug creation (case 1 of Figure 17.14) and the extension of the plug in the crack (case 2 of Figure 17.14).

1. $V_{B_2O_3} \leq V_0$—closure of the crack by the oxide: G and D′ are in B.

$$e_m = \frac{1}{2}\frac{V_{B_2O_3}}{P_e e_{B_4C}} \; ; \; L_{fi} = L_{fe} = 0 \tag{17.125}$$

2. $V_{B_2O_3} > V_0$—extension of the plug within the crack: C is in B.

$$e_m = \frac{h}{2} + h_m \tag{17.126}$$

a. $V_0 < V_{B_2O_3} \leq V_1$—symmetric plug extension.

$$L_{fi} = L_{fe} = \frac{V_{B_2O_3} - V_0}{2P_e h} \tag{17.127}$$

b. $V_1 < V_{B_2O_3} < V_2$—interior plug truncation.

$$L_{fi} = L_{fi0}; \; L_{fe} = \frac{V_{B_2O_3} - V_0}{P_e h} - L_{fi0} \tag{17.128}$$

c. $V_2 = V_{B_2O_3}$—interior and exterior plug truncation.

$$L_{fi} = L_{fi0} \; ; \; L_{fe} = L_{fe0} \tag{17.129}$$

The plug truncation corresponds to the impossibility of the plug to grow as it reaches the fibers (interior truncation), or when it flows in the macro-porosity (exterior truncation).

Concerning the oxidative environment, the matrix cracks appear first in the composite. They are noted "M0" in Equation (17.53). They start from macropores and they cross the composite thickness up to the seal coat. These cracks are widely opened and for a temperature less than 700°C, the self-healing mechanism in the seal coat is not efficient. The composite being quite porous, the oxygen concentration in the pores is supposed to be equal to the concentration at the surface of the composite. While the plug in the seal coat is not efficient (cf. Chapter 17.3.4.2), the oxygen concentration at point G is related to the oxygen partial pressure by

$$C_G = \frac{P_{O_2}}{RT} \tag{17.130}$$

The other boundary conditions are related to the chemical reactions at the PyC interface and B_4C matrix layer. For the moment other healing matrix layers are not considered and the fibers are not supposed to react consequently with oxygen.

$$\Phi_{DE} = C_E P_e e_{pyc} k_{PyC} \left(= \frac{1}{2}\Phi_{BD}\right) \tag{17.131}$$
$$C_F = 0 \tag{17.132}$$

where Φ_{DE} is the flux of oxygen through wire DE, in (mol/s). The flux Φ_{BD} is used for robustness when $h_p = 0$.

The reaction constant k_{PyC} is dependent on the temperature:

$$k_{PyC} = k^0_{PyC} \exp\left(-\frac{E_{PyC}}{RT}\right) RT \tag{17.133}$$

This law is identified from thermo-gravimetric analyses [58] depending on the oxygen partial pressure, this justify the second RT term appearing here.

The evolution of the diffusion coefficient of oxygen in B_2O_3 with temperature is given by References 59–63:

$$D_{ox} = k^0_p \exp\left(-\frac{E_a}{RT}\right) RT \tag{17.134}$$

The parameter identification is obtained by oxidation tests of B_4C pellets described in Reference 56.

The diffusion of oxygen in the air in a Knudsen and Fickian regime is given by Reference 64:

$$D_{gf} = \frac{C_M T^{1.75}}{P_{O_2}} \text{ (Fickian regime)} \tag{17.135}$$

$$D_{gk} = \frac{2\,h}{3}\left(\frac{8RT}{\pi M_o}\right)^{\frac{1}{2}} \text{ (Knudsen regime)} \tag{17.136}$$

$$D_g = \frac{1}{\frac{1}{D_{gf}} + \frac{1}{D_{gk}}} \tag{17.137}$$

where C_M is a parameter function of air and oxygen molar mass, M_o the molar mass of oxygen and h the crack opening. The diffusion coefficient in the air must be defined because it allows a robust modeling of oxygen diffusion, particularly to the layer of PyroCarbon, but this is not a sensitive parameter of the model. Indeed, the diffusion rate in the B_2O_3 plug is very slow compared to diffusion rate in the air.

The description of the oxidation reaction of boron carbide is used to describe the flow of boron oxide formed as a function of oxygen flow. The main chemical balance equations [56] are

$$B_4C + 4O_2 \rightarrow 2B_2O_3 + CO_2 \tag{17.138}$$

$$B_4C + \frac{7}{2}O_2 \rightarrow 2B_2O_3 + CO \tag{17.139}$$

$$B_4C + 3O_2 \rightarrow 2B_2O_3 + C \tag{17.140}$$

$$C + O_2 \rightarrow CO_2 \tag{17.141}$$

C and CO are supposed to be recombined because O is in excess and so only the first equation is used.

The evolution laws for matrix consumption (V_{B_4C}), oxide creation ($V_{B_2O_3}$) and interface consumption (V_{PyC}) are function of oxygen flux. It reads

$$\dot{V}_{B_4C} = 2P_e e_{B4C}\dot{h}_m = \frac{2}{4}V_{m.m}\Phi_{CF} \tag{17.142}$$

$$\dot{V}_{PyC} = \dot{h}_p P_e e_{PyC} = V_{m.pyc}\frac{\Phi_{BD}}{2} \tag{17.143}$$

where: $V_{m.m}$ and $V_{m.pyc}$ are, respectively, the molar volume of B_4C matrix and PyC interphase in (m^3/mol). The volume of boron oxide used to create the plug is noted $V_{B_2O_3}$. It is both a function of the oxidation reaction of boron carbide and of oxide removal. The functions describing the evacuation of the oxide depend on the state of healing. The first case corresponds to the closure of the crack with a thickness of plug defined by Equation (17.125). The second case corresponds to the extension of the plug into the crack with sizes defined by equations from (17.126) to (17.129).

17.3.2.2 *Case 1 Description* The evolution of the plug volume is

$$\dot{V}_{B_2O_3} = 2V_{m.ox}\frac{\dot{V}_{B_4C}}{V_{m.m}} - Q_v P_e e_{B4C} \tag{17.144}$$

where $Q_v P_e h$ is the $B_2 0_3$ oxide volatilization flux. This volatilization is activated by water partial pressure and its evolution on a free surface (in m/s) is given by Reference 65:

$$Q_v = k_v \exp\left(-\frac{E_v}{RT}\right) P^n_{H_2O} \tag{17.145}$$

The different flux expressions on the branches of the circuit are

$$\Phi_{ED} = D_g P_e e_{pyc}\frac{C_E - C_D}{h_p} \tag{17.146}$$

$$\Phi_{BD} = D_g P_e h\frac{C_B - C_D}{L_{fi0}} \tag{17.147}$$

$$\Phi_{BC} = D_g P_e e_{B4C}\frac{C_B - C_C}{h_m + \frac{h}{2} - e_m} \tag{17.148}$$

$$\Phi_{CF} = D_{ox} P_e e_{B4C}\frac{C_C - C_F}{e_m} \tag{17.149}$$

The flux balance equations read

$$2\Phi_{ED} + \Phi_{BD} = 0 \tag{17.150}$$

$$\Phi_{BC} - \Phi_{CF} = 0 \tag{17.151}$$

17.3.2.3 *Case 2 Description* The evolution of the plug volume is

$$\dot{V}_{B_2O_3} = 2V_{m.ox}\frac{\dot{V}_{B_4C}}{V_{m.m}} - Q_v P_e h - Q_{ej}(e_m - h_m)P_e \tag{17.152}$$

with $Q_{ej} = Q^1_{ej}\Delta h + Q^2_{ej}$ being the flux of ejected oxide during a cyclic loading and caused by static flow of oxide for high temperature. Both Q^1_{ej} and Q^2_{ej} are parameters related to the viscosity of the oxide and to the wetability properties of the matrix by the oxide. But the modeling of the oxide by a fluid at microscale has so far not succeeded. Identification is performed in Reference 66 for one temperature by differential measurement of cyclic fatigue tests and creep tests. Associated to that law, the mean crack opening indicator over a loading cycle is used, allowing to treat complex fatigue cases. The different flux expressions on the branches of the circuit are

$$\Phi_{ED} = D_g P_e e_{pyc}\frac{C_E - C_D}{h_p} \tag{17.153}$$

$$\Phi_{DD'} = D_g P_e h\frac{C_D - C_{D'}}{L_{fi0} - L_{fi}} \tag{17.154}$$

$$\Phi_{BD'} = D_{ox} P_e h\frac{C_B - C_{D'}}{L_{fi}} \tag{17.155}$$

$$\Phi_{BF} = D_{ox} P_e e_{B4C}\frac{C_B - C_F}{h_m + \frac{h}{2}} \tag{17.156}$$

The flux balance equations read

$$2\Phi_{ED} + \Phi_{BD} = 0 \tag{17.157}$$

$$\Phi_{BD'} + 2\Phi_{BF} + \Phi_{BG} = 0 \tag{17.158}$$

$$\Phi_{BD'} - \Phi_{D'D} = 0 \tag{17.159}$$

17.3.2.4 Case Transition, Model Continuity, Integration Scheme, and Identification Initially, $V_{B_2O_3} = 0$, $V_{B_4C} = 0$, and $V_{PyC} = 0$. The problem to solve is partially nonlinear and time dependent. The balance equations are solved analytically. The evolution laws are solved using a Runge–Kutta 45 algorithm and an adaptive time step. The continuity between the cases 1 and 2 is very important to ensure the convergency of the nonlinear solver. The model is driven by the volume of oxide for that purpose.

17.3.2.5 Case of Cyclic Loading : Evacuation of Boron Oxide When the plug heals the crack, if there is a variation of the stress related to mechanical fatigue loading, boron oxide present in the crack can migrate to a macropore. It is very complex to model at the microscopic level, the boron oxide flux leaving and returning to a macropore during a variation of crack opening. An average modeling of this mechanism is realized in Reference 66 for very high temperatures. The evacuated glass flow during a cycle is defined proportional to the change of opening by the equation

$$V_{ejection} = C_{ejec.}.(h_{max} - h_{min}) \tag{17.160}$$

where $C_{eject.}$ is a material coefficient.

17.3.3 Modeling of the SubCritical Cracking of Fibers

Once the surrounding matrix has cracked, the fibers are no longer protected from the aggressive environment: air (oxygen, water, etc.) can penetrate the crack network and damage the fibers through oxidation or corrosion [67, 68]. There have been attempts to introduce an environmental protection of the fibers within the composite (using, for example, self-healing multilayered ceramic matrices or fiber treatments, referred to as chemical fuses), but even then the environmental protection is not complete and the fibers remain one of the key factors in the material's lifetime [11, 69]. The study of a fiber's lifetime under given environmental conditions is denoted static fatigue analysis because most works use the Paris law as in classical fatigue analysis, or subcritical cracking. Subcritical propagation, also known as slow crack growth, has been widely studied in the literature, especially for ceramics [67, 69–74]. These works describe the crack's propagation using a Paris-like law relating the crack's velocity to the crack's stress intensity factor calculated under the assumption of elastic behavior. This law, which depends on the temperature, agrees fairly well with experimental results [71, 72], but cannot be applied directly to varying environmental conditions such as oxygen or water concentration around the fibers. Moreover, it was found [67, 70, 72, 75] that as the temperature increases two different propagation stages take place. In the first stage, called the reaction-controlled stage, the crack's tip is damaged directly. The second stage, called the diffusion-controlled stage, is different in the sense that oxygen and water, prior to damaging the crack's tip, must first diffuse through an oxide layer formed on the surface of the fiber. This second stage only appears at high temperatures (above 750°C). Corresponding models can be found in Reference 70 for the fibers and in Reference 75 for the multifilament tow.

The subcritical cracking mechanism depends of the applied stress, the oxygen that has reached the fiber and is activated by temperature. Reference 69 proposes a simple model for constant loadings (applied stress, temperature, oxygen concentration). It links the yarn strength to the lifetime:

$$t\sigma_{yarn}^{n} = A_0 \exp\left(\frac{E_a}{RT}\right) \tag{17.161}$$

The lifetime is also inversely proportional to the oxygen partial pressure. This law is then simply rewritten introducing the cumulative oxygen concentration:

$$(C\,t)\,\sigma_{yarn}^{n} = B_0 \exp\left(\frac{E_a}{RT}\right) \tag{17.162}$$

From that equation, a failure criteria is deduced:

$$\sigma < \sigma^f_{yarn} \tag{17.163}$$

$$\Theta(t,T)\,\sigma^f_{yarn} = B_0 \tag{17.164}$$

The strength of the yarn (σ^f_{yarn}) evolution is piloted by the cumulative oxygen concentration seen by the fibers (concentration at point D) and weighted by the temperature T:

$$\Theta(t,T) = \int_0^t C_D(\tau)\exp\left(-\frac{E_a}{RT(\tau)}\right)d\tau \tag{17.165}$$

This allows to follow the evolution of the strength for variable loadings of the fibers (temperature, healing, crack opening). All the information is condensed in Θ. It can be written as an evolution law and integrated simultaneously as the diffusion/reaction model:

$$\frac{d\sigma^f_{yarn}}{dt} = -\frac{1}{n}B_0^{-1}C_D\exp\left(-\frac{E_a}{RT}\right)\left(\sigma^f_{yarn}\right)^{n+1} \tag{17.166}$$

$$\sigma^f_{yarn}(t=0) = \sigma^{f,static}_{yarn} \tag{17.167}$$

$\sigma^{f,static}_{yarn}$ is the strength of the yarn in neutral environment.

To pass to the composite failure criterion, a constant scale factor is introduced, k:

$$\sigma^f_{composite} = k\,\sigma^f_{yarn} \tag{17.168}$$

$$\frac{d\sigma^f_{composite}}{dt} = -\frac{k^{-n}}{n}B_0^{-1}C_D\exp\left(-\frac{E_a}{RT}\right)\left(\sigma^f_{composite}\right)^{n+1} \tag{17.169}$$

In practice, transverse cracks completely cross the yarns so that k is taken equal to the fiber volume ratio in the considered direction (either longitudinal or transverse).

The proposed evolution law is in agreement with the available experiments on yarns from Reference 69 at different temperatures and oxygen partial pressures. Nevertheless, one should note that in the available experiments the imposed stress is equal to the strength, that is, $\sigma = \sigma^f_{composite}$. As said before, the strength of the fiber must depend of the applied load as in References 76,77. A variant of the previous evolution law can be postulated to take into account the influence of the applied stress and still respect the available experimental data; the strength is split into two contributions:

$$\frac{d\sigma^f_{composite}}{dt} = -\frac{k^{-n}}{n}B_0^{-1}C_D\exp\left(-\frac{E_a}{RT}\right)\left(\sigma^f_{composite}\right)^{n+1-m}\sigma^m \tag{17.170}$$

In order to identify the exponent m, interrupted static fatigue tests can be used to get the residual strength of the fibers. Without available data, $m = 0$ is assumed.

17.3.4 Simulation of Degradation Mechanisms and Prediction of Lifetime

The first use of this modeling is to describe the competition of different degradation mechanisms leading to rupture of the composite under complex static or fatigue mechanical loadings and under various environmental conditions. To give a good confidence in the predictive simulations, some of the tests of static fatigue on short times are used for additional identification and another part with a different temperature range is used as validation. The associated identification and validation has been developed in Reference 48 for CERASEP A410.

17.3.4.1 Identification Concerning the fatigue modeling, cyclic fatigue tests have been conducted on coupons in Reference 13 and are directly used for identification of the macromodeling of fatigue.

Concerning the reaction/diffusion model, the experimental data from References 65 and 78 are used for the reaction kinetics, diffusion, and volatilization constants. The morphology is directly taken from micrographs, the sequential layers deposition being controlled and leading to a clear morphology definition. Concerning the crack opening indicator, only the form is kept and one parameter needs to be determined at the macroscale: (K) of Equation (17.119). For that, a static fatigue test is used. This test gives the lifetime of a composite for a constant 200MPa longitudinal loading at 500°C in air [49].

The ejection law due to fatigue described by Equation (17.160) can be identified [79] comparing two experiments: static fatigue lifetimes and cyclic fatigue lifetimes at the same mean stress. Both tests have been performed but not with the same mean crack opening; this does not allow us to identify this law properly. The identification is performed at three scales, micro, meso, and macroscopic.

- Microparameters: multi-layered matrix morphology, chemistry, subcritical crack growth of fibers/yarns;
- Mesoparameters: scale transition (k);
- Macroparameters: mechanical behavior for static and fatigue loadings based on static tests shown in Figure 17.3. The crack opening magnification (K) uses the mean lifetime value for a test under axial stress of 200 MPa, with a temperature of 500°C, 20 kPa of oxygen partial pressure and 1.8 kPa water vapor partial pressure. The corresponding mean lifetime of 250 hours can be seen via the set of tests in the middle of Figure 17.17.

17.3.4.2 *Validation and Mechanisms Analysis* Further identification of the lifetime modeling for static fatigue loading is realized on coupons at 200 MPa and for a temperature of 500°C. The influence of stress and temperature is present in the whole model via the damage mechanisms (and corresponding crack networks), via the oxidizing of the matrix, via the oxide volatilization, and via the subcritical cracking of the fibers. It is thus interesting to compare the model response with different cases of loading (applied stress and temperature levels) for which experimental results are available (mainly in tension). From an experimental point of view, the test leading to the highest lifetime value (five months and a half or 4000 hours) corresponds to an applied stress of 100 MPa. The experimental lifetimes are quite short, these experiments are complex to control beyond few months while the expected lifetimes for industrial applications is about 5 years ($4.4\ 10^4$ hours). For the sake of simplicity, the first comparison relates to the influence of the applied stress. The experiment/simulation comparison is presented in Figure 17.16 and shows the influence of the tension loading level for a constant temperature of 500°C and oxygen and water vapor partial pressures of 200 kPa and 1.8 kPa.

On the range of the experimental results, the model is in good agreement considering the mean slope as well as the highest variation for low stress levels. On the other hand, for applied stresses close to the static limit, the model predicts a sharp decline of the strength before the initial strength level.

It is interesting to note (in Figure 17.17) that this decline also appears on some experiments. This decline seems to be quite sensitive to the stress variation. The analysis of the mechanisms via the model shows, for high load levels, that the cracks are too widely opened to be healed correctly. To analyze this mechanism for a high load level, the

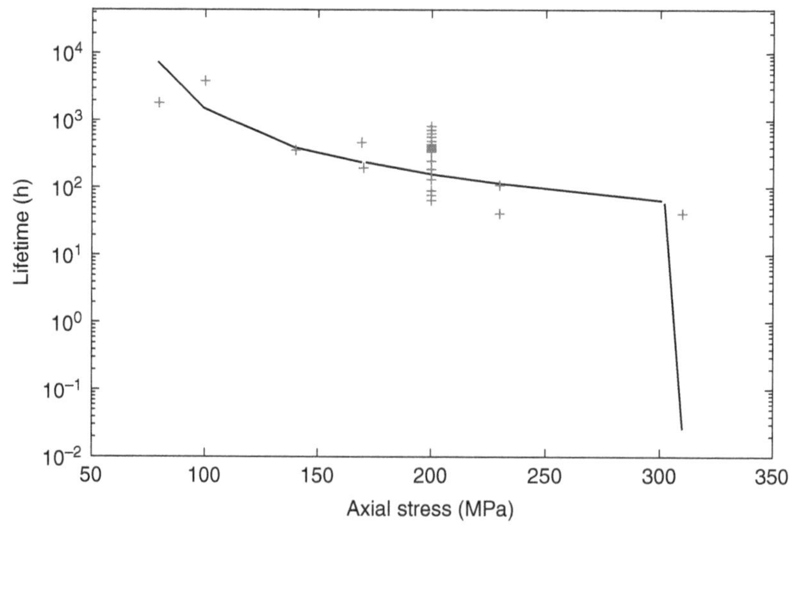

FIGURE 17.16
Lifetime versus applied stress for a temperature of 500°C, 20 kPa oxygen partial pressure and 1.8 kPa water vapor partial pressure.

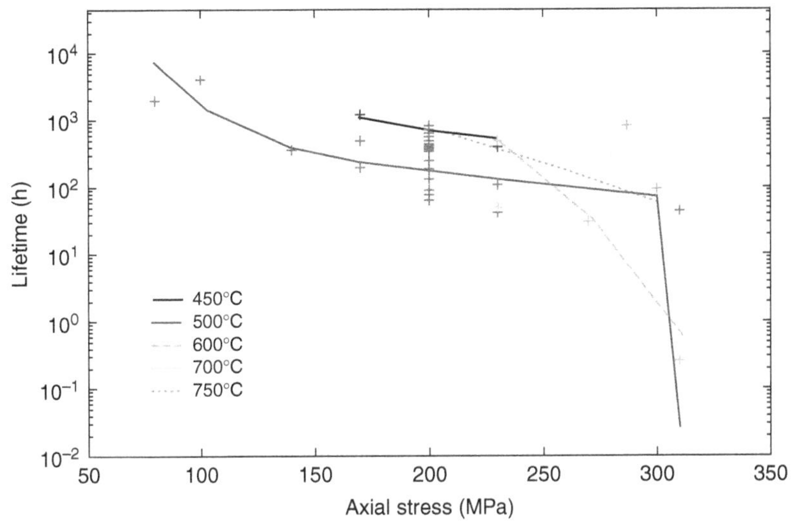

FIGURE 17.17

Lifetime versus applied stress for different temperatures.

simulations help to understand the protecting role of the oxide plug on the subcritical cracking of fibers. In Figure 17.18, several simulations have been conducted at 500°C under 20 kPa of oxygen partial pressure and 1.8 kPa water vapor partial pressure. The applied stress is here no more linked to the crack opening, this crack opening is imposed to three different values: low opening (hx1), wide opening (hx2), infinite opening (i.e., without possible healing). In Figure 17.18, the evolution of the failure stress threshold (the strength) of the longitudinal yarns is presented as a function of the exposure time. For example, for a 200 MPa loading, the lifetime corresponds to the intersection between the strength curve and the applied stress horizontal line. For the three prescribed crack openings, it corresponds to 160 hours (h×1), 40 hours (h×2), and 25 hours (without healing). The plateaux on the curves (h×1) and (h×2) correspond to the creation of a plug; as the crack opening increases, the oxide plug creation duration increases. The boron carbide consumption limits its capacity to regenerate, and beyond a certain point, the oxide plug looses its efficiency. The oxygen then reaches quite easily the fibers and the strength declines sharply (in log scale, nevertheless). Widely opened cracks lead to a fiber-driven lifetime while closed cracks lead to healing-driven lifetime. All curves merge for very long lifetimes. This plateau of plug efficiency is almost constant and for high stress levels, a small variation of crack opening (and thus of stress level) can switch a long lifetime to a short one.

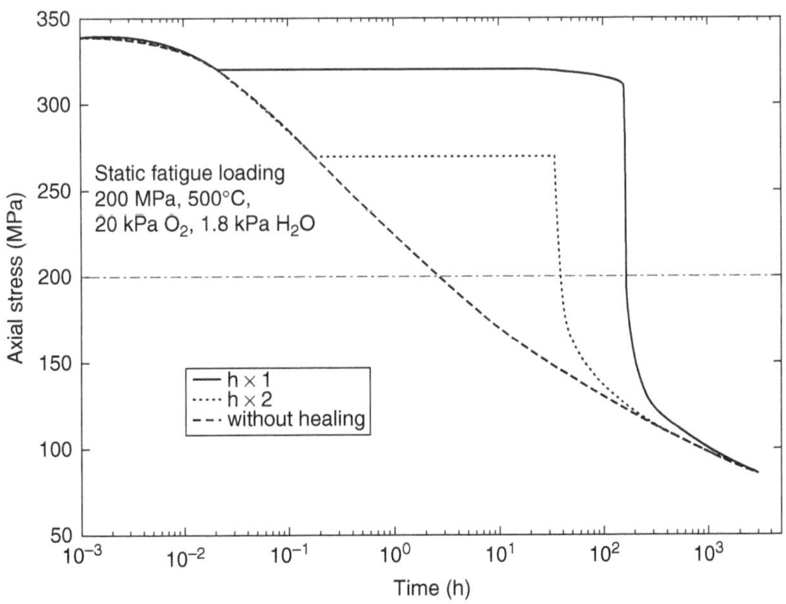

FIGURE 17.18

Strength evolution with times for different crack openings.

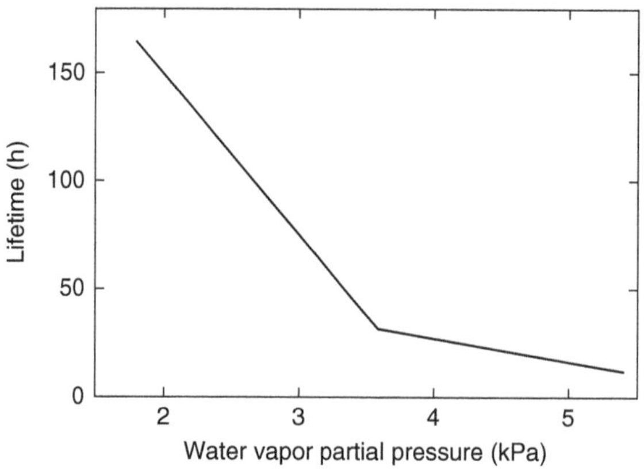

FIGURE 17.19

Lifetime evolution for different partial pressures of water vapor.

Although macrotests are not available to determine the influence of oxygen and water vapor partial pressures, simulations allow the predictive quantification of their influence. For example, for an applied stress of 200 MPa and 500°C, Figure 17.19 shows the evolution of the lifetime versus the applied water vapor partial pressure. The negative influence of the water vapor is exhibited. In Figure 17.20, the influence of oxygen partial pressure is shown. It presents a non-monotonous evolution due to the competition of the healing efficiency and due to the fiber degradation rate.

In Figure 17.21, the evolution of the lifetime versus temperature can be observed. It is non-monotonic. For an applied stress of 150 MPa, a maximum lifetime can be found around 480°C, which is in agreement with experimental results on coupons [16].

For higher temperatures, the modeling with a plug localized at the heart of the material in the yarns is not sufficient. For high temperatures (around 700°C), a plug can form in the seal coat. Indeed, with thick layers of matrix in the seal coat, the cracks are too widely opened to allow the formation of a plug in the seal coat for temperatures between 450°C and 600°C. At 700°C the Boron oxide is sufficiently abundant to form a protection at the surface of the composite besides the protection at the heart of the composite. To understand this additional mechanism, the crack spacing has been measured directly on morphologies on composite coupons after loading as well as the matrix layers thicknesses in the seal coat. Simulations have been conducted in the presence of these two potential healing zones in a serial framework as shown in Figure 17.22.

Figure 17.23 shows the evolution of the oxygen concentration behind the plug in the seal coat (noted: seal-coat healing), which constitutes the entry for the modeling of the inner plug but also the oxygen concentration near the fibers (noted: yarn healing, Yarn + seal-coat healing). The figure shows that the first active plug occurs at the heart of

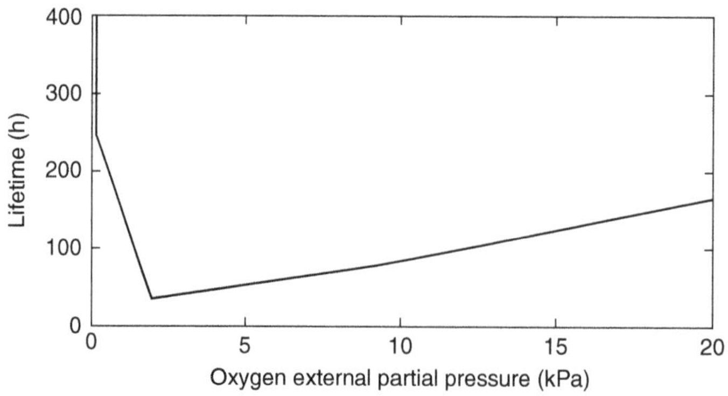

FIGURE 17.20

Lifetime evolution for different partial pressures of oxygen.

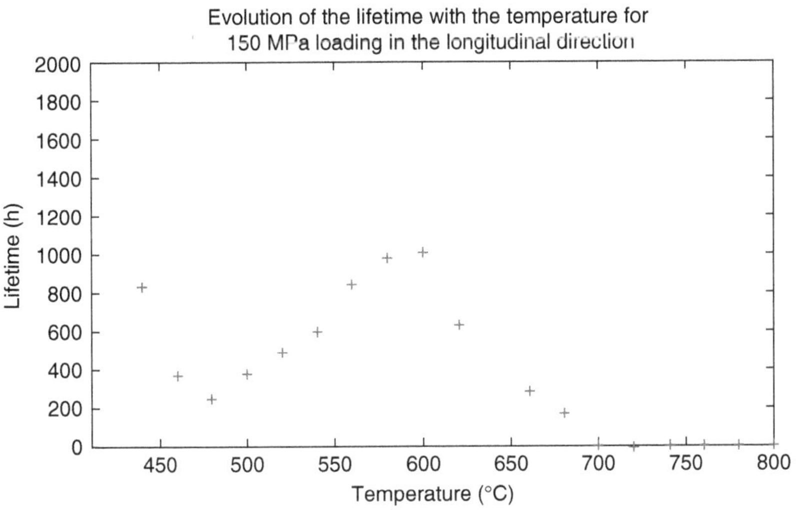

FIGURE 17.21

Lifetime versus temperature.

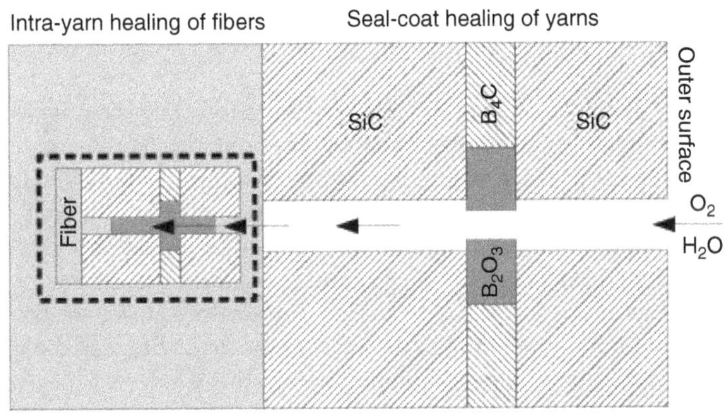

FIGURE 17.22

Sketch of the double plug healing mechanism.

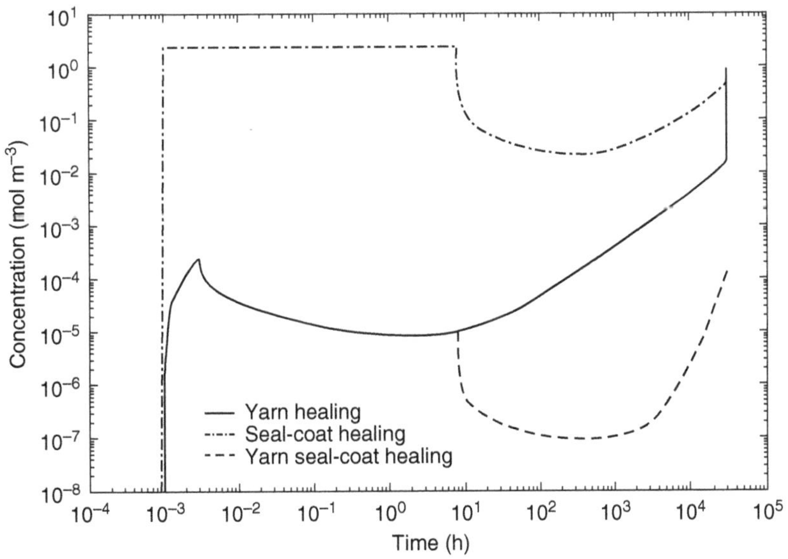

FIGURE 17.23

Influence of the seal coat on oxygen concentration at 700°C.

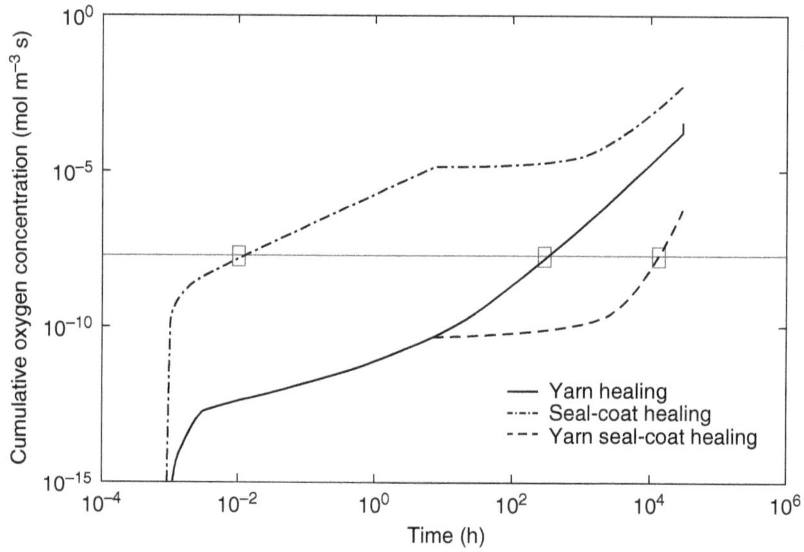

FIGURE 17.24
Influence of the seal coat on oxygen cumulative concentration at 700°C.

the yarns. When the second plug becomes active, it creates a second protecting layer, which limits the yarn degradation in particular by limiting the water vapor content at the heart of the composite.

The simulation of the degradation of the strength of the yarns is obtained from a variable describing the cumulative oxygen concentration. Figure 17.24 shows the evolution of this variable. By fixing a limit of cumulative concentration corresponding to the failure at 150 MPa, this figure shows that the seal-coat plug does not allow an efficient protection by itself. Nevertheless, a combination with the inner plug allows to increase the lifetime of almost two decades.

The competition between the different mechanisms, fiber degradation, plug creation, and evacuation, is obtained from simulation. Figure 17.25 shows the strength evolution curves can cross because the different mechanisms are not affected in the same manner by the temperature. Without modeling these different mechanisms, it is very complex to interpret the experimental values. Indeed, the presented curves present the strength evolution and not only the final failure. The failure corresponds to the intersection of the strength curve and the applied stress history curve. For static fatigue tests, the loading is constant, but the model is able to handle more complex variable loadings.

From simulations, response surfaces can be built as shown in Figure 17.26 that presents the evolution of the lifetime versus applied stress and temperature for constant loadings. It allows to visualize the most pertinent loading ranges. Such a model based on the description of the micromechanisms allows: first to have a better reading and understanding of experimental values, and second, to optimize the material for large lifetimes applications.

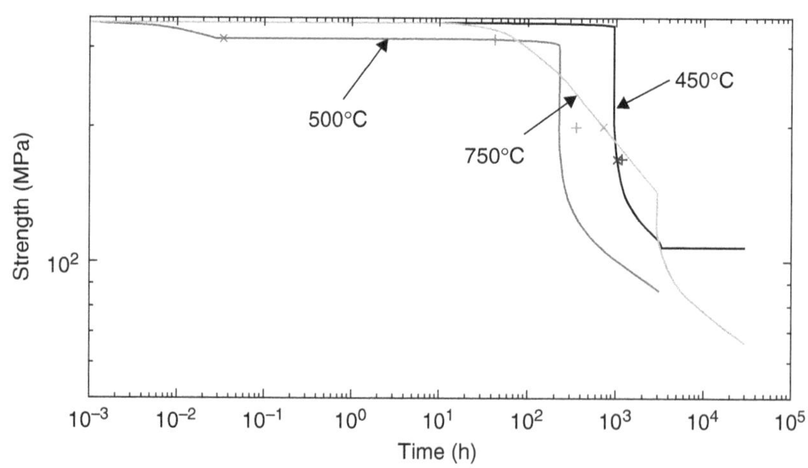

FIGURE 17.25
Strength evolution with times for different temperatures.

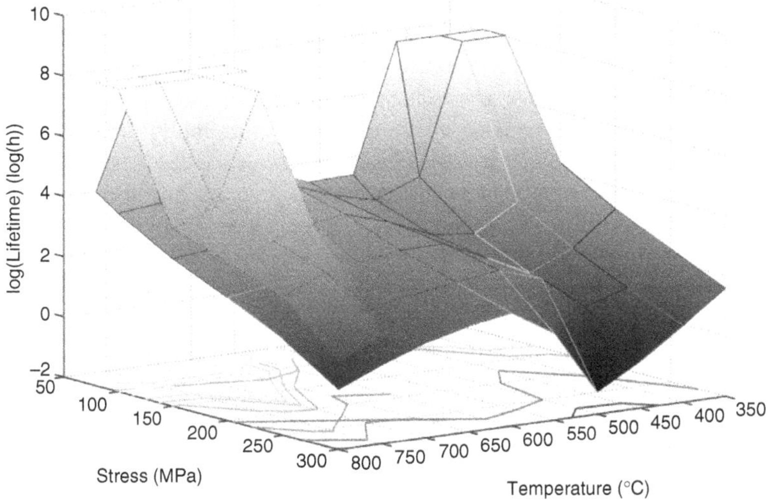

FIGURE 17.26
Influence of the seal coat on lifetime maps.

17.4 PREDICTION CAPABILITIES

This section describes the implementation in a finite element code (in our case: ABAQUS/Standard) of the anisotropic and unilateral damage framework introduced in Section 17.2, with several illustrations of its suitability to predict the failure of industrial structures. Two different failure modes are considered here: failure in static fatigue, where the structure's lifetime depends on a coupled action of mechanics and environment; and static failure, where the environment plays a minor role, but where localized mechanical phenomena must be taken into account in order to satisfactorily predict the structure's behavior until final fracture. More details can be found in References 16, 80.

First, the numerical resolution of the model's local equations is described. It allows to compute the stress and damage fields over any structure and over time. And coupled to the environmental modeling introduced in Section 17.3, it allows to predict the lifetime of any structure in static fatigue.

Secondly, extension of the theory to the case of localized failure is described, together with associated numerical techniques. We illustrate the possibility to predict both the initiation and propagation of macroscopic cracks leading to the structure's failure.

17.4.1 Calculation of the Local Behavior

Classically, in order to compute mechanical fields over space and time in the framework of continuum mechanics, one usually makes the following assumptions: (i) deformation is small (< 1%), such that one can use a measure of the deformation that is linear with regard to displacement; and (ii) inertia effects are small, so that acceleration terms can be removed from the balance equation, making the problem quasi-static. This assumptions lead to the following problem:

$$\begin{cases} \text{find } \underline{u}(x,t),\ \underline{\underline{\epsilon}}(x,t),\ \underline{\underline{\sigma}}(x,t)\ \&\ \underline{\underline{\underline{\underline{K}}}}(x,t) \text{ over } \Omega\times[0;T] \text{ such that} \\ \underline{\text{div}_x}(\underline{\underline{\sigma}}(x,t)) + \underline{f^g}(x,t) = 0 \quad \text{in } \Omega\times[0;T] \\ \underline{\underline{\sigma}}(x,t) = \underline{\underline{\underline{\underline{K}}}}(x,t)\,\underline{\underline{\epsilon}}(x,t) \quad \text{in } \Omega\times[0;T] \\ \underline{\underline{\epsilon}}(x,t) = (\underline{\underline{\text{grad}_x}}(\underline{u}(x,t)))_{\text{sym}} \quad \text{in } \Omega\times[0;T] \\ \underline{u}(x,t) = \underline{u^g} \quad \text{on } \partial\Omega^u\times[0;T] \\ \underline{\underline{\sigma}}(x,t)\,\underline{n} = \underline{F^g} \quad \text{on } \partial\Omega^F\times[0;T] \end{cases} \tag{17.171}$$

where Ω denotes the material domain and $[0;T]$ the time interval, $\underline{f^g}$ the applied volume load, $\underline{u^g}$ the imposed displacement on the domain boundary $\partial\Omega^u$, $\underline{F^g}$ the applied surface load on the domain boundary $\partial\Omega^F$ (one has

$\partial\Omega^u \cup \partial\Omega^F = \partial\Omega$ and $\partial\Omega^u \cap \partial\Omega^F = 0$), $\underline{n}$ the normal to the outward normal to the domain boundary, and $\underline{\underline{K}}$ the eventually nonlinear relation between strains and stresses. See for instance, Reference 19 for more details.

Since for most real-world cases problem (17.171) does not have an analytical, or even closed form, solution (because of the complex material behavior as it is the case here, or simply because of the geometry), one has to use numerical methods to compute approximated solutions. If cutting-edge methods seem to tend toward so-called time–space approximations where the whole space and time domain is handled all along the computation in order to optimally capture the key features of the solution [81,82], more traditional and widespread methods use the following approach. First, the time interval is discretized into several time increments (not necessarily of same duration), and one now looks for the solution at a finite number of time steps, from the initial one to the final one, assuming linear (or quadratic, etc.) evolution of the solution over the time increments. Second, in the same manner, for each time step the spatial domain is discretized into several finite elements (triangles, quadrangles, tetrahedrons, hexahedrons; one usually uses the same discretization for all time steps) over which the solution is assumed to be linear (or quadratic, etc.) and can be expressed from the solution at the element's nodes, so that one now looks for the solution at a finite number of points. This discretization of the equation generates a discrete problem from the initial continuous problem (17.171).

The space resolution for a given time step is usually based on Newton's method, which consists of iterating between two different problems: (i) resolution of global (i.e., structure-wide) equilibrium of the structure with external loading, the variables characterizing the material's behavior being fixed; and (ii) resolution of the local (i.e., point-to-point) material behavior, the displacement field being fixed. Most finite element packages available implement internally the time incrementation as well as the resolution of the structure's global equilibrium, as these are general mechanical equations. Thus, the only part that remains to implement is generally the local resolution of the material's behavior, which is strongly user- and application-dependent. We now describe such implementation for the anisotropic and unilateral damage model introduced in Section 17.2.

17.4.1.1 Local Loop The local loop is run for each integration point and at each global iteration of each load increment. The input is $\underline{\underline{\epsilon}}^{l,i}$, the total strain tensor at load increment l and global iteration i, and the output consists of the damage tensors $\underline{\underline{C}}^{l,i}$ and $\underline{\underline{Z}}^{l,i}$, the corresponding damage "hardening" variable $\alpha^{l,i}$, and the stress tensor $\underline{\underline{\sigma}}^{l,i}$. Starting here, in order to simplify the notation, the subscripts l, i will be omitted. The local problem writes

$$\begin{cases} \text{with } \underline{\underline{\epsilon_e}} \ \& \ \underline{\underline{\underline{\underline{C_0}}}} \text{ known, find } \underline{\underline{\sigma}}, \ \underline{\underline{\underline{\underline{C}}}} \ \& \ \underline{\underline{\underline{\underline{Z}}}} \text{ such that} \\ \underline{\underline{\epsilon_e}} = \underline{\underline{\underline{\underline{C}}}}\langle\underline{\underline{\sigma}}\rangle_+^{C} + \underline{\underline{\underline{\underline{C_0}}}}\langle\underline{\underline{\sigma}}\rangle_-^{C_0} + \underline{\underline{\underline{\underline{Z}}}}\underline{\underline{\sigma}} \end{cases} \tag{17.172}$$

Fixed-point solver We propose the following fixed-point algorithm.

$$\begin{cases} \text{Initialization: } j = 0\,;\ \alpha^j = \alpha^{l-1}\,;\ \underline{\underline{\underline{\underline{C}}}}^j = \underline{\underline{\underline{\underline{C}}}}^{l-1}\,;\ \underline{\underline{\underline{\underline{Z}}}}^j = \underline{\underline{\underline{\underline{Z}}}}^{l-1} \\ \text{Loop:} \\ \quad \left| \begin{array}{l} \text{Stress: } \underline{\underline{\sigma}}^j / \underline{\underline{\epsilon_e}} = \underline{\underline{\underline{\underline{C}}}}^j\langle\underline{\underline{\sigma}}^j\rangle_+^{C} + \underline{\underline{\underline{\underline{C_0}}}}\langle\underline{\underline{\sigma}}^j\rangle_-^{C_0} + \underline{\underline{\underline{\underline{Z}}}}^j\underline{\underline{\sigma}}^j \\ \text{Residual: } R^j = \alpha(\underline{\underline{\sigma}}^j) - \alpha^j \\ \text{Exit test: } \dfrac{|R^j|}{|\alpha^j - \alpha^{l-1}|} < \text{tolerance} \Rightarrow \text{exit} \\ \text{Damage: } \begin{cases} \alpha^{j+1} = \alpha^j + R^j \\ \underline{\underline{\underline{\underline{C}}}}^{j+1} = \underline{\underline{\underline{\underline{C}}}}(\underline{\underline{\underline{\underline{C}}}}^{l-1}, \alpha^{j+1} - \alpha^{l-1}) \\ \underline{\underline{\underline{\underline{Z}}}}^{j+1} = \underline{\underline{\underline{\underline{Z}}}}(\underline{\underline{\underline{\underline{Z}}}}^{l-1}, \alpha^{j+1} - \alpha^{l-1}) \end{cases} \end{array} \right. \\ \text{End loop: } j = j + 1 \end{cases} \tag{17.173}$$

where α and $\underline{\underline{\underline{\underline{C}}}}$ & $\underline{\underline{\underline{\underline{Z}}}}$ are the material function and damage evolution laws defined in Equations (17.41) or (17.46). (In order to simplify the notation we presented the case of a single damage mechanism, but the extension to multiple mechanisms is straightforward.)

Solver acceleration The fixed-point method can be viewed as a Newton–Raphson method with a unit search direction and, therefore, can oscillate greatly. This is particularly true in our case because of the presence of unilateral conditions. Therefore, we propose to use Aitken's relaxation methods in order to improve the algorithm's convergence. This can be viewed as a search direction optimization based on previous iterations. The only modification to algorithm (17.173) concerns the damage increase line, which becomes

$$\alpha^{j+1} = \alpha^j + s^j R^j \qquad \text{with} \quad s^j = \begin{cases} 1 & \text{if } j = 0 \\ -s^{j-1} \dfrac{R^{j-1}}{R^j - R^{j-1}} & \text{if } j > 0 \end{cases} \tag{17.174}$$

17.4.1.2 *Behavior Loop* In the previous discussion of the local loop, we did not described how we calculate $\underline{\sigma}$ when all the internal variables are fixed (second line of algorithm (17.173)). While this step is straightforward for most existing models, it is not for ours. Indeed, because of the partitioning of $\underline{\sigma}$ into positive and negative parts, the state law is nonlinear even with fixed operators, so the problem can be formulated as follows:

$$\begin{cases} \text{with } \underline{\epsilon_e}, \underline{\underline{C}}, \underline{\underline{C_0}} \ \& \ \underline{\underline{Z}} \text{ known, find } \underline{\sigma} \text{ such that} \\ \underline{\epsilon_e} = \underline{\underline{C}} \langle \underline{\sigma} \rangle_+^C + \underline{\underline{C_0}} \langle \underline{\sigma} \rangle_-^{C_0} + \underline{\underline{Z}} \underline{\sigma} \end{cases} \tag{17.175}$$

Newton–Raphson solver This nonlinear problem is solved using a Newton–Raphson algorithm.

$$\begin{cases} \text{Initialization: } k = 0 \, ; \, \underline{\sigma}^k = \underline{\sigma}^{l-1} \\ \text{Loop:} \\ \quad \text{Residual: } \underline{R}^k = \underline{\epsilon_e} - \underline{\underline{C}} \langle \underline{\sigma}^k \rangle_+^C - \underline{\underline{C_0}} \langle \underline{\sigma}^k \rangle_-^{C_0} - \underline{\underline{Z}} \underline{\sigma}^k \\ \quad \text{Exit test: } \|\underline{R}^k\| < \text{tolerance} \Rightarrow \text{exit} \\ \quad \text{Stress: } \underline{\sigma}^{k+1} = \underline{\sigma}^k + \underline{\underline{D}}^k \underline{R}^k \\ \text{End loop: } k = k + 1 \end{cases} \tag{17.176}$$

for which several search directions $\underline{\underline{D}}^k$ can be used. The actual tangent direction is not an option because one cannot derive the state law with respect to $\underline{\sigma}$ in the general case where $\langle \underline{\sigma} \rangle_+^C \neq 0$ and $\langle \underline{\sigma} \rangle_-^{C_0} \neq 0$. One can choose, for example, the initial operator $\underline{\underline{K_0}} = \underline{\underline{C_0}}^{-1}$, which is not a good direction, especially when the damage is significant, but which is fast because the calculation of the operator is very inexpensive. Another option is to use the quasi-secant operator, defined as

$$\underline{\underline{D}}^k = \begin{cases} (\underline{\underline{C}}^{l-1} + \underline{\underline{Z}}^{l-1})^{-1} & \text{if } \text{Tr}(\underline{\sigma}^k) > 0 \\ (\underline{\underline{C_0}} + \underline{\underline{Z}}^{l-1})^{-1} & \text{if } \text{Tr}(\underline{\sigma}^k) < 0 \end{cases} \tag{17.177}$$

which is a much better search direction, but also a more expensive one because these operators are usually not saved and must be recalculated at each iteration.

Solver acceleration In practice, the initial operator converges very slowly, or even does not converge at all if the damage is significant. The secant operator defined in Equation (17.177) generally converges very poorly, too. Therefore, we propose once again to use Aitken's relaxation. The only modification to algorithm (17.176) is the last line, which becomes

$$\underline{\underline{\sigma}}^{k+1} = \underline{\underline{\sigma}}^{k} + s^k \underline{\underline{\underline{\underline{D}}}}^k \underline{\underline{R}}^k \tag{17.178}$$
$$\text{with} \quad s^k = \begin{cases} 1 & \text{if } k = 0 \\ -s^{k-1} \dfrac{\mathrm{Tr}(\underline{\underline{R}}^{k-1}(\underline{\underline{R}}^k - \underline{\underline{R}}^{k-1}))}{\mathrm{Tr}((\underline{\underline{R}}^k - \underline{\underline{R}}^{k-1})(\underline{\underline{R}}^k - \underline{\underline{R}}^{k-1}))} & \text{if } k > 0 \end{cases}$$

17.4.2 Lifetime Predictions and Application to Damage Tolerance Analysis

An additional hypothesis is made in order to predict a structure's final failure in static fatigue: we assume that the mechanical fields do not vary significantly after the application of the mechanical load, and therefore consider them constant in time. Consequently, the mechanical damage model presented in Section 17.2 is used to compute the mechanical response of the structure to the mechanical load, and the environmental damage model presented in Section 17.3 is used as a post-treatment of the mechanical data to compute the environmental degradation of the material with time. Final failure, and then lifetime, are predicted thanks to a simple criterion defined now.

17.4.2.1 Failure Criterion It is well known that local failure criteria cannot predict the final failure of materials properly [83]. Consequently, we use a nonlocal failure criterion, based on Reference 84, and recently applied to polymer matrix composites [85] as well as ceramic matrix composites [86].

The criterion is very simple both in concept and in practice (see Figure 17.27). It consists, for each tow direction, of averaging the stress field over an area Ω defined by a length l_0 considered to be an intrinsic parameter of the material. (In practice, we use the size of a weave pattern, that is, 2 mm.) For instance, for the longitudinal tow, the failure criterion is

$$\exists \underline{x} / \tilde{\sigma}_1(\underline{x}) \geq \sigma_1^R(\underline{x}) \quad \Longrightarrow \quad \text{failure} \tag{17.179}$$
$$\text{with} \quad \tilde{\sigma}_1(\underline{x}) = \frac{1}{|\Omega|} \int_\Omega \sigma_{11}^+(\underline{x}')\, d\underline{x}' \quad \forall \underline{x}$$

The formulation for the transversal tow is similar.

This failure criterion can be rewritten differently based on the normalized residual strength (or strength reserve factor) field [87]. This quantity provides the best illustration of the relation between the stress field (imposed by the

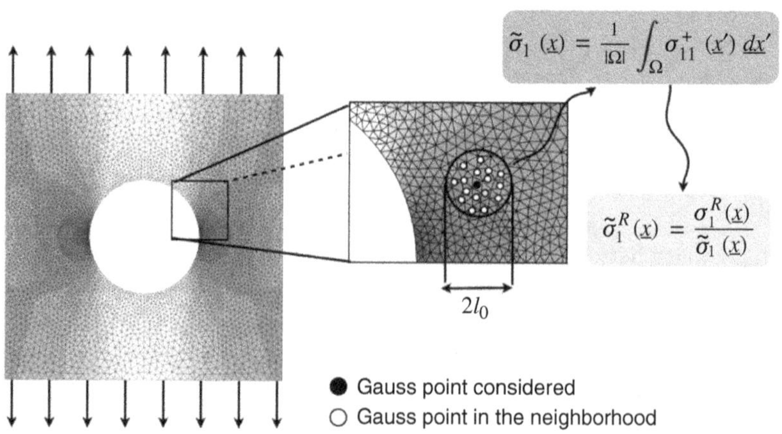

FIGURE 17.27
Nonlocal failure criterion for a structure with a high stress gradient.

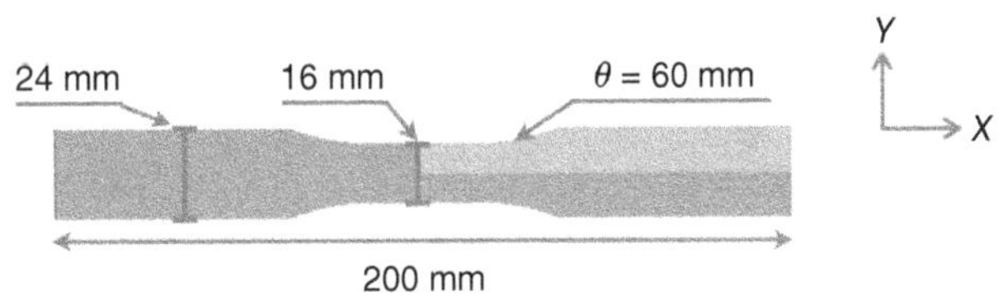

FIGURE 17.28

The geometry of the dog-bone specimen subjected to tension in the x-direction.

boundary conditions) and the strength field (which evolves over time because of environmental damage) during the failure process, and is defined by

$$\tilde{\sigma}_1^R(\underline{x}) = \frac{\sigma_1^R(\underline{x})}{\tilde{\sigma}_1(\underline{x})} \quad \forall \underline{x} \tag{17.180}$$

Then, the failure criterion becomes

$$\exists \underline{x} / \tilde{\sigma}_1^R(\underline{x}) \leq 1 \quad \Longrightarrow \quad \text{failure} \tag{17.181}$$

17.4.2.2 A First Illustration The first example is a dog-bone specimen made of SAFRAN–*Snecma Propulsion Solide*'s CERASEP A410 self-healing CMC (Figure 17.28). The FE mesh was generated using GMSH [88]. The specimen was subjected to a static fatigue loading at high temperature: all the loading conditions, that is, the applied stress and temperature and the other environmental conditions (20 kPa partial oxygen pressure; 1.8 kPa partial water pressure) remained constant over time.

We used an identification of the model for this material provided by References 11, 14, and 15. While some parameters (e.g., the reaction/diffusion coefficients) can be found in the literature, most were identified by fitting the mechanical response of macroscopic specimens (e.g., initial behavior, damage evolution laws) or by analyzing the morphology of the material on the fiber's scale (e.g., the characteristic dimensions of the matrix layers). Still, a few parameters (e.g., the anisotropy of the cracks) remained unidentified because of the lack of experimental data and had to be chosen arbitrarily [14, 15].

We will present first the damage induced by the mechanical loading (i.e., the stiffness reduction), followed by the damage induced by the environment (i.e., the strength reduction).

Damage fields predictions As mentioned previously, the key property of the model is that it distinguishes among the contributions associated with each crack network, which, thus, can be analyzed separately. For instance, for a 50 MPa applied load, inter-yarn damage alone is in effect, whereas for a 250 MPa applied load (Figure 17.29a), both inter-yarn matrix damage and intra-yarn matrix damage are in effect (Figures 17.29b and 17.29c): intra-yarn damage takes over after inter-yarn damage has reached saturation.

The mechanical part also provides an intra-yarn transverse crack opening indicator field, which is the key input to the physicochemical model of strength reduction through oxidation (see Section 17.3). Now, let us illustrate the capability of our strategy to simulate the complex competition between mechanical loading and environmental loading through the analysis of two related test cases: a hot-grip experiment (i.e., with a homogeneous temperature field) and a cold-grip experiment (i.e., with a temperature gradient over the specimen).

Lifetime predictions, case of homogeneous temperature field The use of hot-grip testing machines greatly simplifies the study of the lifetime of self-healing CMCs because these machines allow the analysis of the influence of the stress and temperature fields on the lifetime to be fully uncoupled. Therefore, let us first consider the use of our model to analyze hot-grip experiments.

Figure 17.30 shows the strength degradation in the specimen after 278 hours under 200 MPa loading at two uniform temperatures (500° and 700°). In this case of a homogeneous temperature field, the result of interaction with the damage field (which is nearly homogeneous too) is that the strength decreases mainly in the middle of the specimen, where the stress is higher. For instance, from the initial value of 340 MPa, the strength reduced to about 120 MPa at

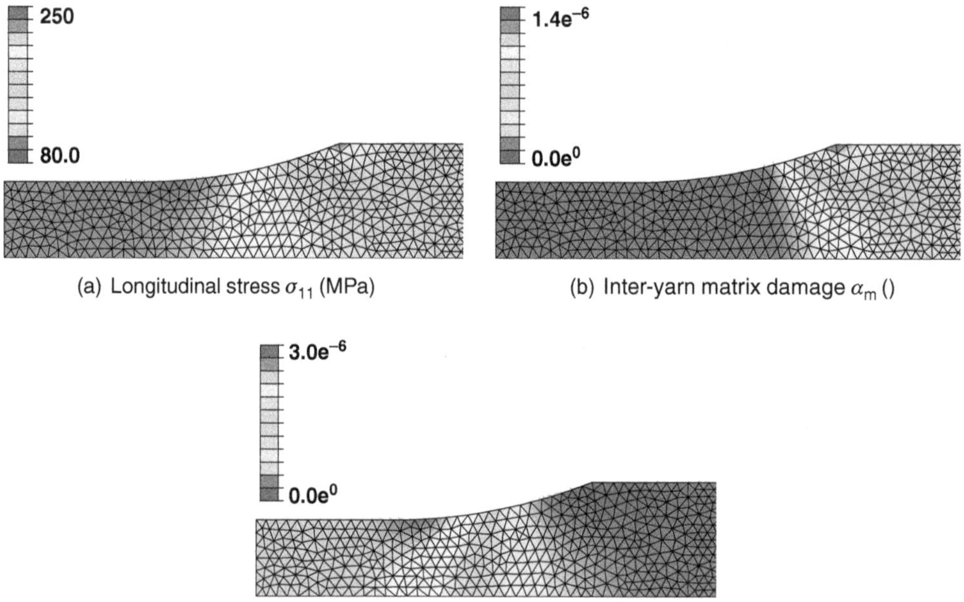

(a) Longitudinal stress σ_{11} (MPa) (b) Inter-yarn matrix damage α_m () (c) Inter-yarn matrix longitudinal damage α_{F_1} ()

FIGURE 17.29

The dog-bone specimen subjected to a 250 MPa traction: (a) longitudinal stress, (b) inter-yarn matrix damage, (c) intra-yarn matrix damage.

500° (Figure 17.30a), and to only about 210 MPa at 700° (Figure 17.30b). Figure 17.31 shows the same type of result, this time for the strength reserve factor field. This shows clearly that at 500° the specimen breaks in the zone of interest whereas at 700° it does not break at all.

This shows that the model is capable of illustrating and quantifying classical types of behavior: (i) an increase in the applied stress decreases the lifetime of the specimen drastically (see Figure 17.32, which indicates a very good match with experimental data); (ii) a modification in the applied temperature enables one to define the domain of applicability of the healing process ([11, 15]). Let us note that the wide scatter in the lifetime data of Figure 17.32 was evidenced only in a recent study [89] and will be analyzed within our modeling framework in a forthcoming paper.

Lifetime predictions, case of temperature gradient Hot-grip testing machines are uncommon and most experiments are carried out using cold-grip machines. This makes the analysis complicated because neither the stress field nor the temperature field, both of which have a significant impact on the lifetime, is uniform [89, 93]. This led us to apply our model to the analysis of cold-grip experiments in order to illustrate how it can help improve their design and understanding. The temperature field considered was obtained by References 91 and 92 and is shown in Figure 17.33.

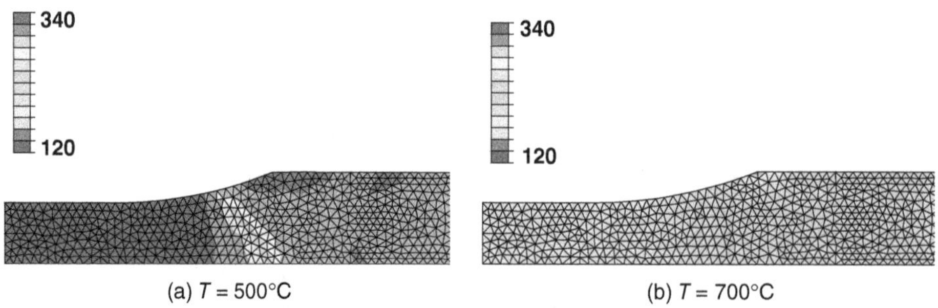

(a) $T = 500°C$ (b) $T = 700°C$

FIGURE 17.30

Residual strength after mechanical (200 MPa) and environmental (278 hours, without temperature gradient) loading.

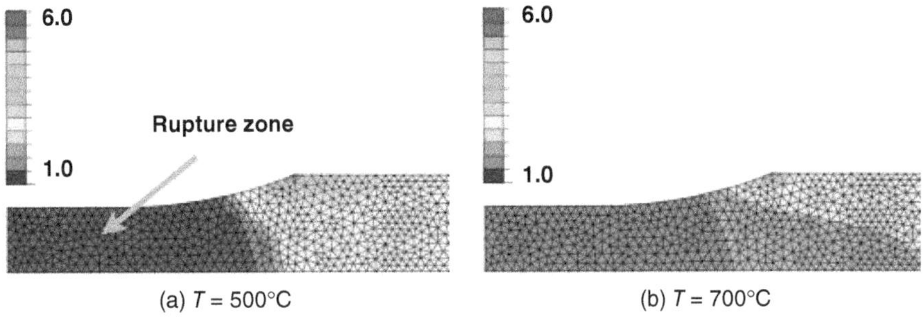

FIGURE 17.31

Strength reserve factor after mechanical (200 MPa) and environmental (278 hours, without temperature gradient) loading.

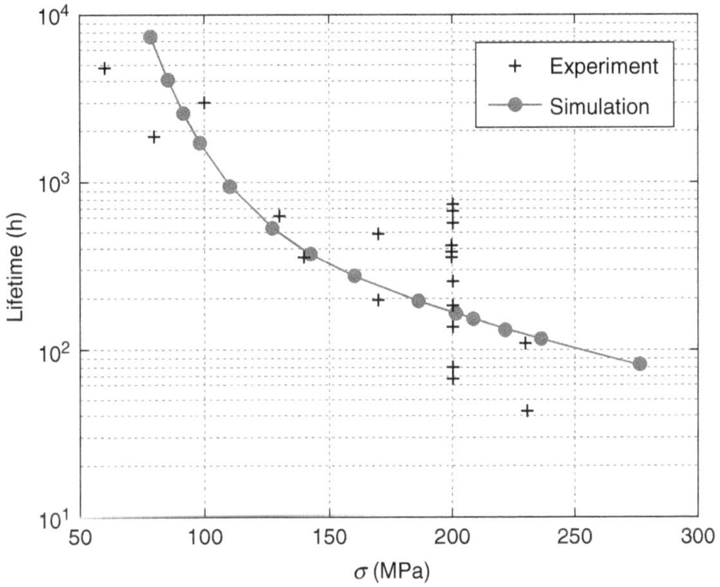

FIGURE 17.32

Comparison of the lifetime predictions between model and experiment: stress dependence at 500° (experimental data from References 89–92).

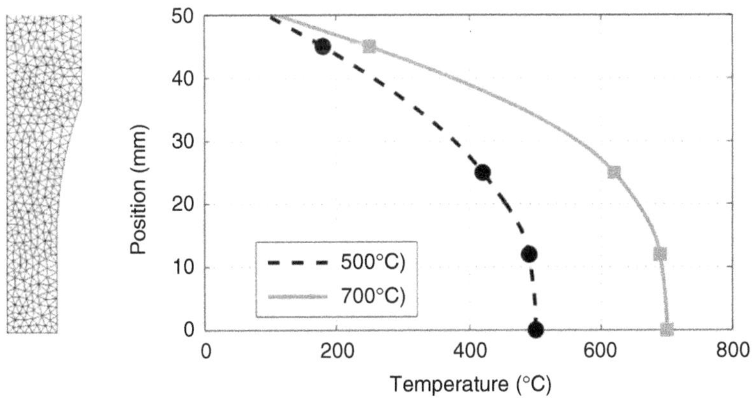

FIGURE 17.33

The temperature gradient in the dog-bone specimen in a cold-grip testing machine at two temperatures [89].

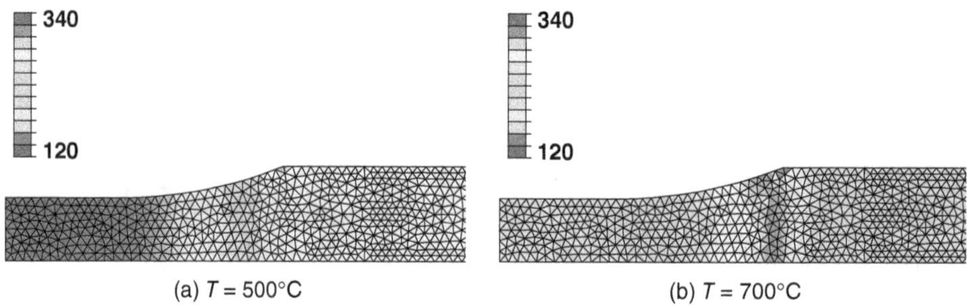

FIGURE 17.34
With a temperature gradient: the residual strength at two temperatures (200 MPa loading for 278 hours).

Figures 17.34 and 17.35 show the residual strength and the strength reserve factor in the specimen after 278 hours under 200 MPa loading at two nonuniform temperatures (500° max and 700° max). For a 500° prescribed temperature in the center of the specimen, the predicted strengths (Figures 17.34a and 17.35a) are very close to those with a uniform temperature (Figures 17.30a and 17.31a). Indeed, in this temperature range the healing process has almost no influence. Conversely, for a 700° prescribed temperature in the center of the specimen, the predicted strengths (Figures 17.34b and 17.35b) are markedly different from those with a uniform temperature (Figures 17.30b and 17.31b): the strength degradation patterns can be very complex. Indeed, due to the fact that the effectiveness of the oxide tip varies significantly, and especially non-monotonically, as a function of the temperature [11], the strength degradation can be much greater even though the mechanical loading is lesser. It was even possible to predict the fracture zone in the grips (Figure 17.35b), a phenomenon that was observed experimentally and that led to major experimental problems [89, 93]. Thus, our model could be used as an experiment design tool in order to ensure that fracture would take place in the zone of interest of the specimen. Besides, this case illustrates the need to consider the whole structure for the lifetime study, if only because the fracture zone cannot always be determined *a priori*.

17.4.2.3 Application to Damage Tolerance Analysis In this section, we illustrate the application of our strategy to the damage tolerance analysis of structures, which is one of the main concerns associated with industrial developments [18, 94]. The objective is to characterize the influence of damaging impacts on the residual lifetime of structures.

According to experimental observations, classical quasi-static indentations generate cone-shaped damaged zones, which can be approximated by cylindrical holes [94]. Let us consider the structural example of an open-hole specimen made of the same material and subjected to the same type of loading as the dog-bone specimen of the previous section. Thus, this test case involves multiaxial stress fields and high stress gradients, especially near the hole. Figure 17.36 shows the geometry of the open-hole plate (with the FE mesh generated once again using GMSH [88]) subjected to tension loading in the x-direction.

Damage fields predictions The model predicts that with an applied stress equal to 25 MPa, inter-yarn matrix cracks initiate around the hole, whereas with an applied stress equal to 55 MPa, intra-yarn matrix cracks initiate, also around

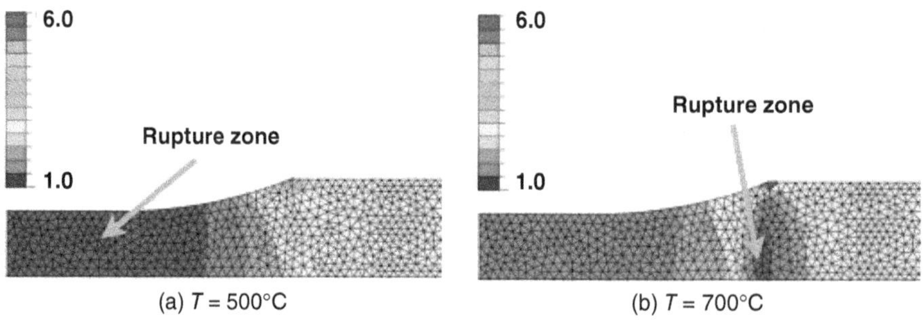

FIGURE 17.35
With a temperature gradient: the strength reserve factor at two temperatures (200 MPa loading for 278 hours).

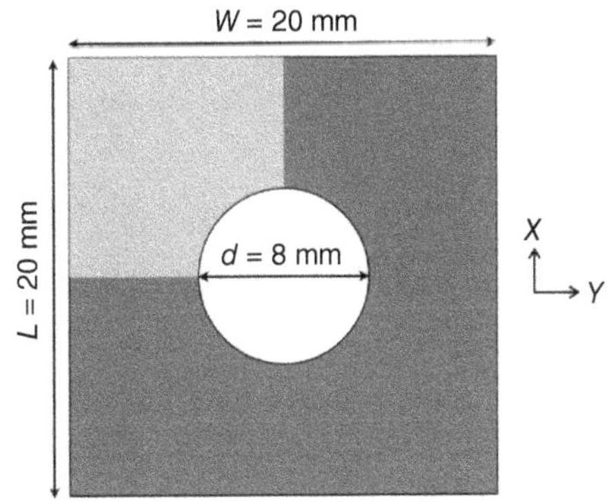

FIGURE 17.36

The geometry of the open-hole specimen subjected to tension loading in the x-direction.

the hole. Moreover, with an applied stress equal to 250 MPa (Figure 17.37), the inter-yarn damage becomes saturated within the structure and intra-yarn matrix damage appears near the hole (Figures 17.37b and 17.37c). These predictions, which are consistent with the available experimental data, show that there is no unstable propagation of the damage leading to purely mechanical fracture of the specimen. Therefore, it is now necessary to address the damage tolerance problem associated with delayed fracture, which conditions the structure's lifetime.

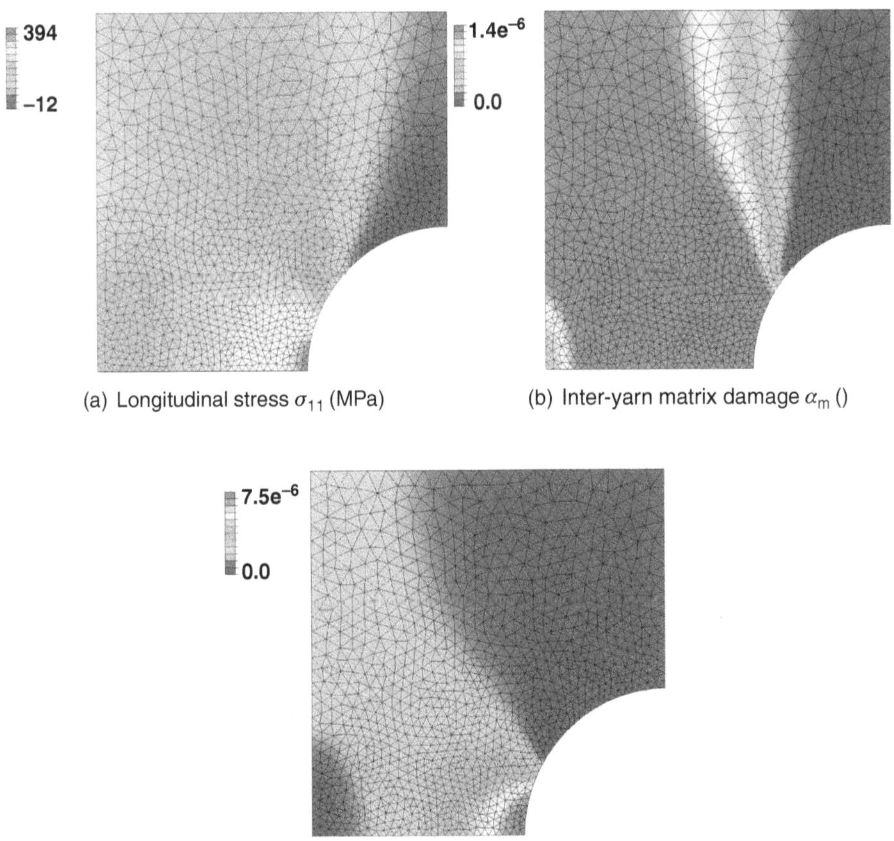

(a) Longitudinal stress σ_{11} (MPa)

(b) Inter-yarn matrix damage α_m ()

(c) Inter-yarn matrix longitudinal damage α_{F_1} ()

FIGURE 17.37

The open-hole specimen under tension loading: (a) longitudinal stress, (b) inter-yarn matrix damage, (c) intra-yarn matrix damage.

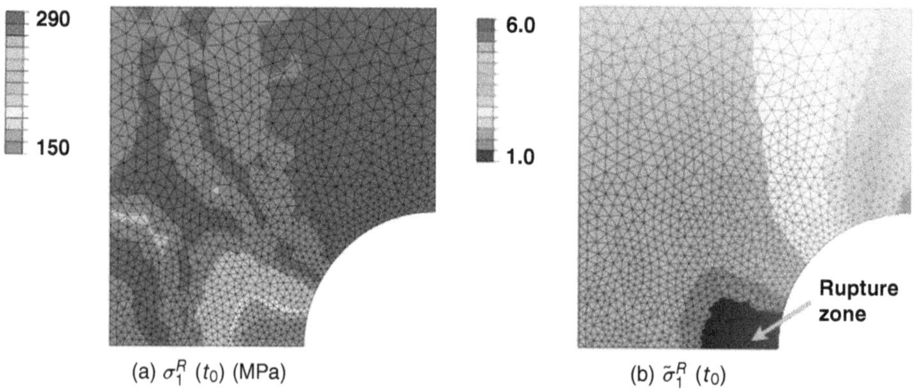

FIGURE 17.38

The open-hole specimen subjected to tension loading: (a) residual strength, (b) strength reserve factor (200 MPa loading at 500° for $t_0 = 110$ hours).

Lifetime predictions and damage tolerance analysis Figure 17.38 shows typical distributions of the indicator. In the uniform temperature case, the model predicts that at any temperature the strength degradation takes place mainly around the hole, where final fracture initiates.

In order to study the effect of the hole on the structure's lifetime, similar simulations were performed on plates with different hole diameters. The influence of the hole was measured by comparing the lifetimes of open-hole and dog-bone specimens under the same effective stress (i.e., the same stress in the midplane of the plates). The results (Figure 17.39) indicate that the influence of the hole is greater at low macroscopic stress levels. Moreover, it appears that even under identical effective stresses large defects have more of an impact on the lifetime than small defects. Indeed, the damage zone is larger, which induces greater chemical degradation. It is important to note that this phenomenon, which is specific to lifetime analysis and does not exist in the case of strength analysis, is inherently part of our strategy, which illustrates its suitability for damage tolerance analysis in an industrial framework.

Also, in Figure 17.39, one should note that the ratio of the open-hole stress (for a given hole) over the dog-bone stress (for a given lifetime) is almost constant. This constant is a stress concentration factor, not in the classical sense, but dedicated to delayed fracture. This stress concentration factor could be used by engineers as a very simple design tool.

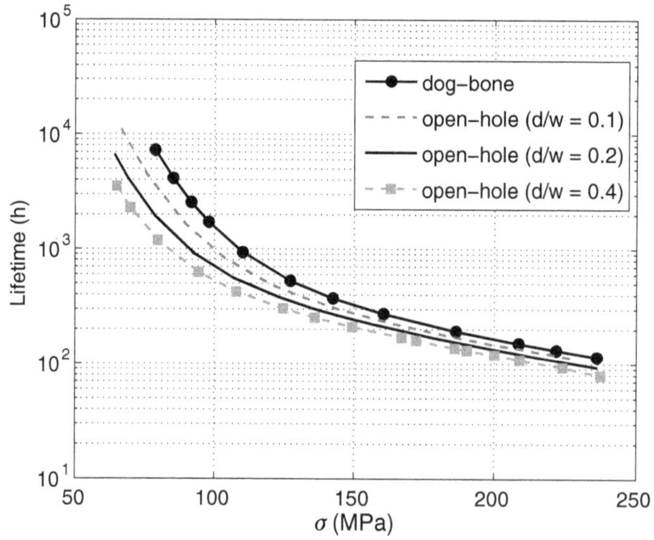

FIGURE 17.39

Damage tolerance analysis: the lifetime as a function of the effective stress at 500° for several hole diameters.

17.4.3 Control of the Damage Localization and Static Failure Predictions

We now demonstrate the ability of the mechanical damage model presented in Section 17.2 to predict the initiation and propagation of the macroscopic crack that will induce the structure's final failure. However, it is well known that the calculation of softening damage laws such as the ones of Equations (17.41) and (17.46) over a domain goes through a critical point where a loss of ellipticity, that is, a loss of uniqueness of the solution [20, 21], occurs and leads to pathological mesh dependencies [27, 95]. Several remedies exist for damage models, including nonlocal formulations [20, 95] and delay-effect approaches [27, 95]. The latter was chosen for our work because it is a local method, and then is easy to implement in classical FE codes.

17.4.3.1 The Delay-Effect Method The delay-effect method for controlling the localization of damage consists of replacing the evolution of the scalar damage variable α (defined in Equation (17.41) for matrix microcracking and in Equation (17.46) for tows microcracking, and calibrated in Section 17.2.4.5) by an evolution with a bounded rate:

$$\dot{\alpha} = \frac{1}{\tau_c}(1 - e^{-\langle \alpha^{\text{static}} - \alpha \rangle_+}) \tag{17.182}$$

where α^{static} is the static function calibrated in Section 17.2.4.5 and τ_c a fictitious time parameter. Equation (17.182) has a strong physical meaning in dynamic [27] (cracks can-not appear and propagate instantaneously), but is used here in quasi-static as a pure numerical method. Thus, the fictitious time parameter is not related to any physical process; it must simply be chosen in order to avoid uncontrolled localization. The numerical resolution of such a simple nonlinear, but scalar, equation is straightforward and will not be discussed here.

The behavior law is modified with the introduction of this viscous-like parameter. Figure 17.40 shows the stress–strain curves obtained with different values of τ_c.

17.4.3.2 Illustrative Examples

CT specimen The effectiveness of the delay-effect method in controlling the localization of damage was evaluated using a simple CT-like test, that is, a pre-cracked specimen subjected to tension (see Figure 17.41). The response and the damage fields for different meshes (generated with GMSH [88]) are shown in Figure 17.42: one can clearly observe that the solution is objective, that is, mesh independent, in terms of both the mechanical response and the damage fields.

Plate with an open hole Now let us consider a first structural example that consists of a plate with an open hole (see Figure 17.43). The model was the same as for the CT specimen, except for $\tau_c = 110^{-2}$s. Figure 17.44 shows the response of the model using different meshes (again generated with GMSH [88]). Once again, one can clearly observe that the method enables an objective prediction of the final failure of the structure.

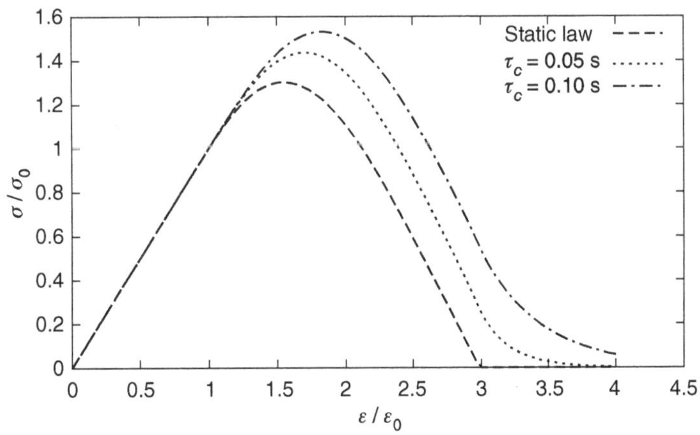

FIGURE 17.40

The response of the model under unidirectional tension loading using different characteristic times for the delay-effect law.

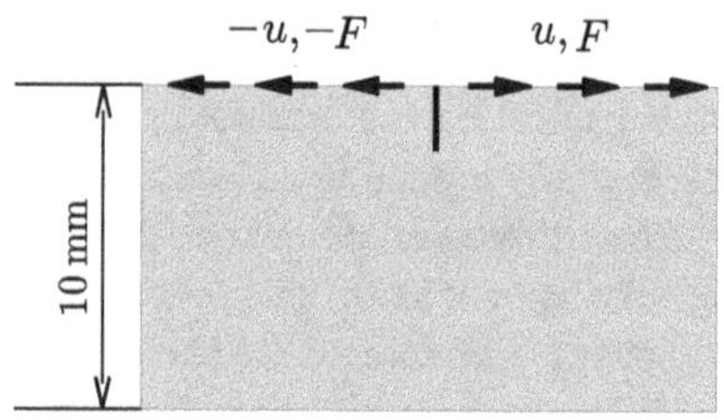

FIGURE 17.41

The problem used for the evaluation of localization control based on delay-effect damage ($\dot{\epsilon}_1^g$ denotes the applied strain).

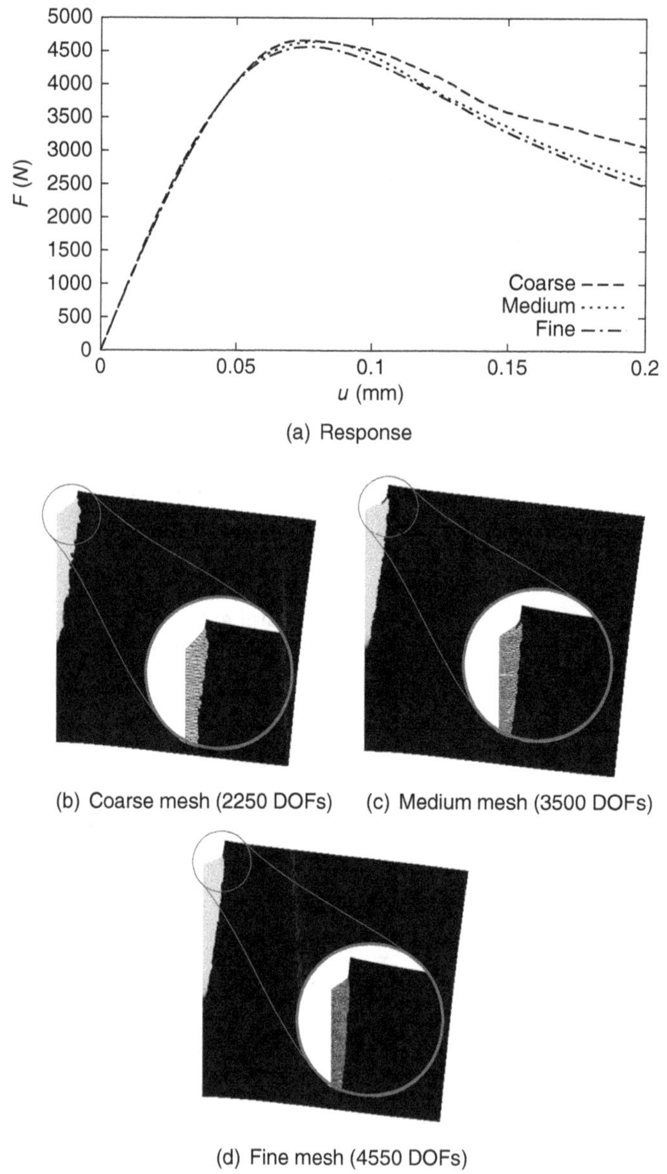

FIGURE 17.42

The response and the damage field for the CT specimen at $t = 0.75$ s (dark grey: no damage; light gray: full damage): objectivity of the mesh and prediction of the crack's propagation ($\dot{\epsilon}_1^g = 0.2\%/\text{s}$), DOFs: Degrees of freedom.

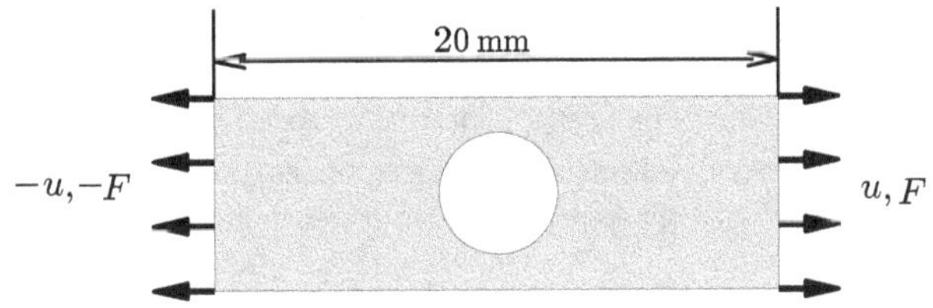

FIGURE 17.43

Structural case: plate with an open hole (ϵ_1^g denotes the applied strain).

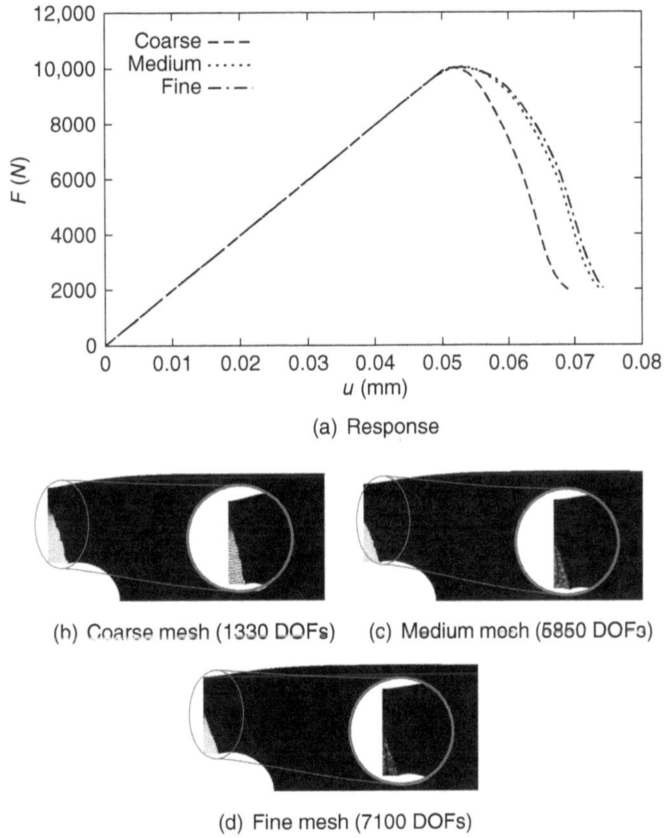

FIGURE 17.44

The response and the damage field for the open hole specimen at $t = 1$ s (dark grey: undamaged; light gray: full damage): objectivity of the mesh and prediction of the final failure ($\dot{\epsilon}_1^g = 0.1\%/\mathrm{s}$), DOFs: Degrees of freedom.

REFERENCES

1. F. Lamouroux, S. Bertrand, R. Pailler, R. Naslain, and M. Cataldi. Oxidation-resistant carbon-fiber-reinforced ceramic-matrix composites. *Comp. Sci. Technol.*, 59(7), 1073–1085 (1999).
2. P. Forio and J. Lamon. Fatigue behavior at high temperatures in air of a 2 D SiC/Si-B-C composite with a self-healing multilayered matrix. *Ceram. Trans. (USA)*, 128, 127–141 (2001).
3. J. Aveston and A. Kelly. Theory of multiple fracture of fibrous composites. *J. Mater. Sci.*, 8(3), 352–362 (1973).
4. A. Evans, J. Domergue, and E. Vagaggini. Methodology for relating the tensile constitutive behavior of ceramic-matrix composites to constituent properties. *J. Am. Ceram. Soc.*, 77(6), 1425–1435 (1994).

5. D. Rouby and N. Louet. The frictional interface: a tribological approach of thermal misfit, surface roughness and sliding velocity effects. *Compos. Part A: Appl. Sci. Manufact.*, 33(10), 1453–1459 (2002).
6. P. Ladevèze. On an anisotropic damage Theory [in French]. Internal Report 34 (LMT-Cachan, March 1983).
7. P. Ladevèze. An anisotropic damage theory with unilateral effects: applications to laminate and three- and four-dimensional composites, in *Continuum Damage Mechanics of Materials and Structures*, edited by O. Allix and F. Hild (Elsevier, 2002), pp. 205–233.
8. A. Gasser, P. Ladeveze, and M. Poss. Damage mechanisms of a woven SiC/SiC composite: modelling and identification. *Compos. Sci. Technol.*, 56(7), 779–784 (1996).
9. P. Ladevèze and S. Letombe. Recent advances on an anisotropic damage theory including unilateral effects. Proceedings of the International Bimestre on Damage Mechanics. Symposium on Continuous Damage and Fracture, Cachan, France, October 23–27, 2000.
10. P. Ladevèze, S. Letombe, and C. Cluzel. A CMC damage model based on micro and macromechanics for high-temperature and complex loading. Proceedings of the 4th International Conference on High Temperature Ceramic Matrix Composites [HTCMC4], Munich, Germany, October 1–3, 2001.
11. C. Cluzel, E. Baranger, P. Ladevèze, and A. Mouret. Mechanical behaviour and lifetime modelling of self-healing ceramic-matrix composites subjected to thermomechanical loading in air. *Compos. Part A: Appl. Sci. Manufact.*, 40, 976–984, (2009).
12. S. Letombe, C. Cluzel, and P. Ladevèze. A macroscopic model coupling oxidation and damage for CMCs. Proceedings of JNC13 Conference, Strasbourg, France, March 12–14, 2003, pp. 713–722.
13. O. Penas, P. Reynaud, D. Rouby, and G. Fantozzi. Self-healing SiCf/SiC composite behaviour under high-temperature cyclic fatigue in air, in *High Temperature Ceramic Matrix Composites*, edited byW. Krenkel, R. Naslain, and H. Schneider. (Wiley-VCH, 2001), pp. 480–485.
14. E. Baranger, C. Cluzel, P. Ladevèze, and A. Mouret. Prediction of the lifetime of self-healing ceramic matrix composites: I—macroscopic modeling of cracking. Proceedings of the 15th National Conference on Composite Materials (JNC15) [in French], Marseille, France, June 6–8, 2007.
15. C. Cluzel, E. Baranger, P. Ladevèze, and A. Mouret. Prediction of the lifetime of self-healing ceramic matrix composites: II—analysis of cracking and oxydation mechanisms. Proceedings of the 15th National Conference on Composite Materials (JNC15) [in French], pp. 885–894, 2007.
16. M. Genet, L. Marcin, E. Baranger, C. Cluzel, P. Ladevèze, and A. Mouret. Computational prediction of the lifetime of self-healing CMC structures. *Compos. Part A: Appl. Sci. Manufact.*, 43, 294–303 (2012).
17. B. Cox and Q. Yang. In quest of virtual tests for structural composites. *Science*, 314(5802), 1102–1107 (2006).
18. M. Bourgeon. Thermostructural materials in aerospace industry: applications and standardization. Proceedings of the 7th International Conference on High Temperature Ceramic Matrix Composites [HTCMC7], Bayreuth, Germany, September 20–22, 2010.
19. J. Lemaître, J.-L. Chaboche, R. Desmorat, and A. Benallal. *Solid Materials Mechanics*, 3rd edition [in French]. (Dunod, 2009).
20. Z. Bazant and G. Pijaudier-Cabot. Nonlocal continuum damage, localization instability and convergence. *J. Appl. Mech.* [Transactions of the ASME], 55(2), 287–93 (1988).
21. A. Needleman. Material rate dependance and mesh sensitivity in localization problems. *Comput. Meth. Appl. Mech. Eng.*, 67, 69–85 (1988).
22. D. Lasry and T. Belytschko. Localization limiters in transient problems. *Int. J. Solids Struct.*, 24, 581–597 (1988).
23. L. Sluys and R. De Borst. Wave propagation and localization in a rate-dependent cracked medium—model formulation and one-dimensional examples. *Int. J. Solids Struct.*, 29, 2945–2958 (1992).
24. P. Ladevèze. On a damage mechanics approach, in *Mechanics and Mechanisms of Damage in Composites and Multi-Materials*, edited by D. Baptiste (MEP, SaintÉtienne, France, 1989), pp. 119–141.
25. P. Ladevèze. Modeling and Simulation of the Mechanical Behavior of CMCs. *High Temp. Ceram. Matrix Compos.*, 47, 53–63 (1995).
26. O. Allix and J.-F. Deü. Delayed-damage modelling for fracture prediction of laminated composites under dynamic loading. *Eng. Trans.*, 45(1), 29–46 (1997).
27. P. Ladevèze, O. Allix, J.-F. Deü, and D. Lévêque. A mesomodel for localisation and damage computation in laminates. *Comput. Meth. Appl. Mech. Eng.*, 183, 105–122 (2000).
28. P. Ladevèze. Inelastic Strains and Damage, in *Damage Mechanics of Composite Materials*, edited by R. Talreja (Elsevier, Amsterdam, The Netherlands, 1994), pp. 117–138.
29. J. Lemaître. How to use damage mechanics. *Nucl Eng. Des.*, 80, 233–245, (1984).
30. P. Ladevèze and J. Lemaître. Damage effective stress in quasi unilateral conditions. IUTAM Congress, Lyngby, Denmark, August 19–25, 1984.

31. P. Ladevèze. On the damage mechanics of composites [in French]. Comptes rendus des Cinquièmes Journées nationales sur les composites, Paris, France, September 9–11, 1986, edited by C. Bathias and D. Menkès. pp. 667–683.
32. M. Ortiz. A constitutive theory for the inelastic behavior of concrete. *Mech. Mater.*, 4, 67–93, (1985).
33. J. Simo and J. Ju. Strain-based and stress-based continuum damage models. 1. Formulation. *Int. J. Solids Struct.*, 23(7), 821–840 (1987).
34. J. Simo and J. Ju. Strain-based and stress-based continuum damage models. 2. Computational Aspects. *Int. J. Solids Struct.*, 23(7), 841–869 (1987).
35. P. Ladevèze, A. Gasser, and O. Allix. Damage mechanisms modelling for ceramic composites. *J. Eng. Mater. Technol.*, 116, 331–336 (1994).
36. J. Dubé, G. Pijaudier-Cabot, and C. Laborderie. Rate dependent damage model for concrete in dynamics. *J. Eng. Mech.*, 122(10), 939–947 (1996).
37. R. Desmorat, F. Gatuingt, and F. Ragueneau. Nonlocal anisotropic damage model and related computational aspects for quasi-brittle materials. *Eng. Fract. Mech.*, 74, 1539–1560 (2007).
38. J. Lemaître, R. Desmorat, and M. Sauzay. Anisotropic damage law of evolution. *Eur. J. Mech.*, 19(2), 187–208 (2000).
39. J.-L. Chaboche. Damage induced anisotropy: on the difficulties associated with the active/passive unilateral condition. *Int. J. Damage Mech.*, 1, 148–171 (1992).
40. J.-L. Chaboche. Development of continuum damage mechanics for elastic solids sustaining anisotropic and unilateral damage. *Int. J. Damage Mech.*, 2, 311–329 (1993).
41. D. Halm and A. Dragon. A model of anisotropic damage by mesocrack growth; unilateral effect. *Int. J. Damage Mech.*, 5, 384–402 (1996).
42. F. Cormery and H. Welemane. A critical review of some damage models with unilateral effect. *Mech. Res. Commun.*, 29, 391–395 (2002).
43. F. Cormery and H. Welemane. A stress-based macroscopic approach for microcracks unilateral effect. *Comput. Mater. Sci.*, 47, 727–738, (2010).
44. P. Ladevèze. On an anisotropic damage theory. in *CNRS International Colloquium on Failure Criteria of Structured Media*, Villard de Lans, France, June 21–24, 1983, edited by J. Boehler (Balkema, Rotterdam, 1993), pp. 335–365.
45. Q.-C. C. He and A. Curnier. A more fundamental approach to damaged elastic stress-strain relations. *Int. J. Solids Struct.*, 32, 1433–1457 (1995).
46. C. Rospars, J.-L. Chermant, and P. Ladevèze. On a first creep model for a 2D SiCf-SiC composite. *Mater. Sci. Eng.: A*, 250, 264–269 (1998).
47. L. Guillaumat and J. Lamon. Multi-fissuration de composites SiC/SiC. *Revue des composites et des matériaux avancés*, 3, 159–171 (1993).
48. E. Baranger, C. Cluzel, P. Ladevèze, and A. Mouret. Identification and validation of a multi-physic macro model for the lifetime prediction of self-healing ceramic matrix composites. Proceedings of the 13th European Conference on Composite Materials [ECCM 13], Stockholm, Sweden, June 2–5, 2008.
49. M. Moevus, M. Mécanismes d'endommagement, émission acoustique et durées de vie en fatigue statique du composite SiC/[Si-BC] aux températures intermédiaires (< 800 C). PhD thesis. INSA-Lyon (2007).
50. H. Mei and L. Cheng. Stress-dependence and time-dependence of the post-fatigue tensile behavior of carbon fiber reinforced SiC matrix composites. *Compos. Sci. Technol.*, 71, 1404–1409 (2011).
51. D. Rouby and P. Reynaud. Fatigue behaviour related to interface modification during load cycling in ceramic-matrix fibre composites. *Compos. Sci. Technol.*, 48(1–4), 109–118 (1993).
52. P. Reynaud, D. Rouby, and G. Fantozzi. Effects of temperature and of oxidation on the interfacial shear stress between fibres and matrix in ceramic-matrix composites. *Acta Mater.*, 46(7), 2461–2469 (1998).
53. N. Lissart and J. Lamon. Damage and failure in ceramic matrix minicomposites: experimental study and model. *Acta mater.*, 45(3), 1025–1044 (1997).
54. F. Rebillat, X. Martin, E. Garitte, and A. Guette. Overview on the self-sealing process in the SiCf/[SI, C, B]m composites under wet atmossphere at high temperature, in *Design, Development, and Applications of Engineering Ceramics and Composites*, edited by D. Singh, D. Zhu, and Y. Zhou. Vol. 215: Ceramic Transactions, edited by M. Singh (John Wiley & Sons, 2010), pp. 151–166.
55. X. Martin, F. Rebillat, and A. Guette. Oxidation behavior of a multilayered (Si-BC) ceramic in a complex atmosphere N2/O2/H2O. *Proc. 4th Int. Symp. High Temp. Corros. Mater. Chem.*, 339 (2003).
56. X. Martin, Oxydation/corrosion de matériaux composites (SiCf/SiBCm) à matrice auto-cicatrisante. PhD thesis. Université Bordeaux I (2003).
57. E. Garitte, Etude de l'oxydation/corrosion des composites céramiques. PhD thesis. Université Bordeaux I (2007).

58. L. Filipuzzi and R. Naslain. Oxidation mechanisms and kinetics of 1D-SiC/C/SiC composite materials: II. Modeling. *J. Am. Ceram. Soc.*, 77(2), 467–480 (1994).
59. F. Norton. Permeation of gaseous oxygen through vitreous silica. *Lett. Nat.*, 191, 701 (1961). Doi:10.1038/191701a0.
60. B. Deal and A. Grove. General relationship for the thermal oxidation of silicon. *J. Appl. Phys.*, 36(12), 3770–3778 (1965).
61. J. Schlichting and S. Neumann. GeO_2/SiO_2-glasses from gels to increase the oxidation resistance of porous silicon containing ceramics. *J. Non-Cryst. Solids*, 48(1), 185–194 (1982).
62. J. Schlichting. Oxygen transport through glass layers formed by a gel process. *J. Non-Cryst. Solids*, 63(1–2), 173–181 (1984).
63. J. Viricelle, P. Goursat, and D. Bahloul-Hourlier. Oxidation behaviour of a multi-layered ceramic-matrix composite (SiC)f/C/(SiBC)m. *Compos. Sci. Technol.*, 61(4), 607–614 (2001).
64. R. Reid, J. Prausnitz, and B. Poling. *The Properties of Gases and Liquids* (McGraw Hill Book Co., New York, 1987).
65. F. Rebillat, X. Martin, and A. Guette. Kinetic oxidation laws of boron carbide in dry and wet environments. Proceedings of HTCMC. 5, 321–326 (2004).
66. S. Letombe, Modélisation du couplage oxydation/endommagement des Composites à Matrice Céramique Autocicatrisante. PhD thesis. ENS-Cachan (2005).
67. A. G. Evans, F. W. Zok, R. M. McMeeking, and Z. Z. Du. Models of high-temperature, environmentally assisted embrittlement in ceramic-matrix composites. *J. Am. Ceram. Soc.*, 79, 2345–2352 (1996).
68. P. Forio, F. Lavaire, and J. Lamon. Delayed failure at intermediate temperatures (600 degrees–700 degrees C) in air in silicon carbide multifilament tows. *J. Am. Ceram. Soc.*, 87, 888–893 (2004).
69. W. Gauthier and J. Lamon. Delayed failure of Hi-Nicalon and Hi-Nicalon S multifilament tows and single filaments at intermediate temperatures (500 degrees–800 degrees C). *J. Am. Ceram. Soc.*, 92, 702–709. (2009).
70. B. R. Lawn. Diffusion-controlled subcritical crack growth in the presence of a dilute gas environment. *Mater. Sci. Eng.*, 13, 277–283 (1974).
71. S. M. Wiederhorn, E. R. Fuller, and T. Thomson. Micromechanics of crack growth in ceramics and glasses in corrosive environments. *Met. Sci.*, 14, 450–458 (1980).
72. J. B. Wachtman. *Mechanical Properties of Ceramics*. (John Wiley & Sons, Inc., 1996).
73. F. Hild, D. Marquis, O. Kadouch, and J. P. Lambelin. Analysis of the failure of ceramics due to subcritical crack growth. *J. Eng. Mater. Technol.*, 118, 343–348 (1996).
74. P. Ladevèze and M. Genet. A new approach to the subcritical cracking of ceramic fibers. *Compos. Sci. Technol.*, 70, 1575–1583, (2010).
75. A. Laforêt and J. Lamon. Static fatigue of multifilament tows at high temperatures above 900°C. Proceedings of the 13th European Conference on Composite Materials [ECCM 13], Stockholm, Sweden, June 2–5, 2008.
76. P. Ladevèze and M. Genet. A new approach to the subcritical cracking of ceramic fibers. *Compos. Sci. Technol.*, 70(11), 1575–1583 (2010).
77. W. Gauthier, F. Pailler, J. Lamon, and R. Pailler. Oxidation of silicon carbide fibers during static fatigue in air at intermediate temperatures. *J. Am. Ceram. Soc.*, 92(9), 2067–2073 (2009).
78. F. Rebillat, E. Garitte, and A. Guette. Quantification of higher SiC fiber oxidation rates in presence of B2O3 under air, in *Design, Development, and Applications of Engineering Ceramics and Composites*, edited by D. Singh, D. Zhu, and Y. Zhou. Vol. 215: Ceramic Transactions, edited by M. Singh (John Wiley & Sons, 2010), pp. 135–149.
79. S. Letombe. Modelling of the damage/oxydation coupling in ceramic matrix composites [in French]. Phd thesis. ENS-Cachan, France (2005).
80. M. Genet, L. Marcin, and P. Ladevèze. On structural computations until fracture based on an anisotropic and unilateral damage theory. *Int. J. Damage Mech.*, 23, 483–506 (2014). Doi:10.1177/1056789513500295.
81. P. Ladevèze. *Nonlinear Computational Structural Mechanics: New Approaches and Non-Incremental Methods of Calculation*. Mechanical Engineering Series (Springer, 1999).
82. P. Ladevèze, J.-C. Passieux, and D. Néron. The LATIN multiscale computational method and the proper generalized decomposition. *Comput. Meth. Appl. Mech. Eng.*, 199(21–22), 1287–1296 (2010).
83. R. J. Whitney and J. M. Nuismer. Stress fracture criteria for laminated composite laminates containing stress concentrations. *J. Compos. Mater.*, 8, 253–265 (1974).
84. J. M. Nuismer and R. J. Whitney. Uniaxial failure of composite failure laminates containing stress concentrations, in *Fracture Mechanics of Composites* (ASTM STP 593, 1975), pp. 117–142.
85. S. Miot, C. Hochard, and N. Lahellec. A non-local criterion for modelling unbalanced woven ply laminates with stress concentrations. *Compos. Struct.*, 92, 1574–1580 (2010).

86. S. Flores, A. G. Evans, F. W. Zok, M. Genet, B. N. Cox, D. B. Marshall, O. Sudre, and Q. D. Yang. Treating matrix nonlinearity in the binary model formulation for 3D ceramic composite structures. *Compos. Part A: Appl. Sci. Manufact.*, 41, 222–229 (2010).
87. Department of Defence. *Composite Materials Handbook*, vol. 5. (Ceramic Matrix Composites, Department of Defence, USA, 2002).
88. C. Geuzaine and J.-F. Remacle. Gmsh: a three-dimensional finite element mesh generator with built-in pre- and post-processing facilities. *Int. J. Numer. Meth. Eng.*, 79, 1309–1331 (2009).
89. O. Loseille and J. Lamon. Prediction of lifetime in static fatigue at high temperatures for ceramic matrix composites. *Adv. Mater. Res.*, 112, 129–140, (2010).
90. O. Penas, Study of SiC/SiBC multi-layered matrix composites in cyclic fatigue at high temperature under air [in French]. Phd thesis. INSA-Lyon, France (2002).
91. M. Moevus, D. Rouby, N. Godin, M. R'Mili, P. Reynaud, G. Fantozzi, and G. Farizy. Analysis of damage mechanisms and associated acoustic emission in two SiC/[Si-B-C] composites exhibiting different tensile behaviours. Part I: damage patterns and acoustic emission activity. *Compos. Sci. Technol.*, 68, 1250–1257 (2008).
92. M. Moevus, N. Godin, M. R'Mili, D. Rouby, P. Reynaud, G. Fantozzi, and G. Farizy. Analysis of damage mechanisms and associated acoustic emission in two SiCf/[Si-B-C] composites exhibiting different tensile behaviours. Part II: unsupervised acoustic emission data clustering. *Compos. Sci. Technol.*, 68, 1258–1265 (2008).
93. S. Momon, M. Moevus, N. Godin, M. R'Mili, P. Reynaud, G. Fantozzi, and G. Fayolle. Acoustic emission and lifetime prediction during static fatigue tests on ceramic-matrix-composite at high temperature under air. *Compos. Part A: Appl. Sci. Manufact.*, 41, 913–918 (2010).
94. V. Herb, G. Couégnat, and E. Martin. Damage assessment of thin SiC/SiC composite plates subjected to quasi-static indentation loading. *Compos. Part A: Appl. Sci. Manufact.*, 41, 1677–1685 (2010).
95. R. Desmorat, M. Chambart, F. Gatuingt, and D. Guilbaud. Delay-active damage versus non-local enhancement for anisotropic damage dynamics computations with alternated loading. *Eng. Fract. Mech.*, 77, 2294–2315 (2010).

CHAPTER 18

APPROACH TO MICROSTRUCTURE–BEHAVIOR RELATIONSHIPS FOR CERAMIC MATRIX COMPOSITES REINFORCED BY CONTINUOUS FIBERS

JACQUES LAMON

CNRS/LMT ENS Cachan, Cachan, France

18.1 INTRODUCTION

The microstructure of a material can influence strongly physical properties such as strength, toughness, ductility, and high temperature behavior, which govern the application of these materials in industrial practice. Because microstructures are complex ensembles of materials, few researchers have attempted to correlate microstructure with these properties [1]. Instead, average material data are used to produce approximate models for material behavior. In some cases, mean-field models work well, but we cannot expect them to be predictive for cases when properties exhibit statistical distributions, or depend on microstructural features or environment.

Ceramic matrix composites (CMCs) reinforced with continuous fibers exhibit several features that differentiate them from homogeneous unreinforced materials. Their microstructure consists of three basic constituents: fibers, matrix, and fiber/matrix interfaces or interphases. Several entities at micro- and mesoscopic length scales can be defined depending on fiber arrangement: single filament-interface-matrix assemblies (represented by microcomposite test specimens), tow-reinforced matrix in woven composites (represented by minicomposites), and unidirectional plies in laminated composites.

CMCs are highly heterogeneous as they are made of constituents with distinct shapes and properties. They also contain various flaws. Being damage tolerant, they can include cracks in the matrix, in the interfaces, and in the fibers. They are versatile, which implies that some properties can be tailored by selecting appropriate constituents, which is made possible by the broad spectrum of potential materials. The stress–strain response to load is dependent on constituent properties: it is linear or nonlinear depending on contrast in elastic properties and fracture characteristics of constituents. The constituents are ceramic materials. As such, they exhibit statistical distributions of strength data, as a result of inherent sensitivity to microstructural flaws with random distribution. The flaw populations govern matrix cracking and fiber failures. Thus, description of microstructure–behavior relations using a bottom-up multiscale approach

Ceramic Matrix Composites: Materials, Modeling and Technology, First Edition. Edited by Narottam P. Bansal and Jacques Lamon.

to mechanical behavior is essential to understand the phenomena at low length scale that influence the behavior at macroscopic scale. It can be the basis of an approach to composite and component design.

The chapter describes the microstructure–behavior relations for woven composites. It focuses on matrix damage by multiple cracking, on ultimate fracture, on delayed fracture at high temperatures, on nonlinear stress–strain behavior, on the influence of constituent properties, and on stochastic features induced by flaw populations. Models of damage and ultimate failure are based on micromechanics and fracture probabilities. Woven SiC-based CMCs are promising materials and they are used in systems running at high temperatures. They provide excellent examples to discuss the microstructure–behavior relationships in CMCs. One-dimensional (1D) composites suffer several limitations as a result of high anisotropy. The stress–strain behavior of 1D composite has been discussed earlier in the literature [2].

18.2 COMPOSITE MECHANICAL BEHAVIOR

The following fiber and matrix characteristics deeply influence the mechanical behavior of CMCs.

- Elastic properties: fiber and matrix elastic moduli range from low (≈ a few GPa) to very large values (≈400 GPa for SiC-based fibers or matrices, and ≈700 GPa for carbon fibers). This situation is quite different from that prevailing in organic matrix composites for which the matrix elastic modulus is always ≈5 GPa, whatever the matrix is.
- Fiber/matrix bond: fibers can be more or less strongly bonded to the matrix.
- Resistance to fracture, and statistical distributions of strength data.
- Volume fractions of fiber and matrix: strengths of ceramic materials are sensitive to scale effects.

18.2.1 Tensile Stress–Strain Behavior of Composites Reinforced by Continuous Fibers

CMCs exhibit an elastic damageable behavior (Figure 18.1). This means that the response of damaged composite is elastic, as indicated by the initial linear portions of curves on reloading. The stress–strain curve depends strongly on the respective deformations of the fibers oriented in the loading direction and of the matrix. The basic principles can be illustrated using the mixtures law, which is appropriate for unidirectional composites:

$$E_c = E_m V_m + E_f V_f \tag{18.1}$$

where E_c, E_m, and E_f are elastic moduli of composite, matrix, and fibers, respectively, and V_m and V_f are volume fractions of matrix and fibers, respectively.

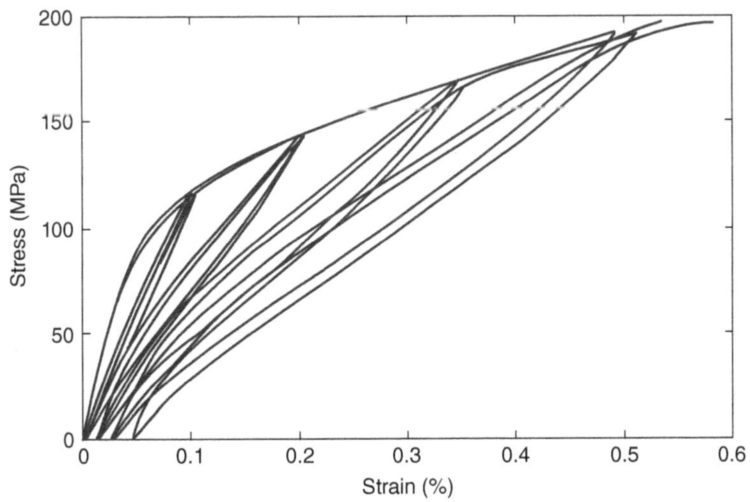

FIGURE 18.1

Typical tensile stress–strain behaviors measured on 2D SiC/SiC reinforced with Hi-Nicalon S fibers. © 2010 The American Ceramic Society, International Journal of Applied Ceramic Technology.

A nonlinear stress–strain curve reflects *damage-sensitive* behavior. It is obtained when matrix stiffness contributes significantly to composite Young's modulus. The matrix carries a significant part of the load. Then, load sharing decreases if matrix is damaged. As E_m degrades as a result of transverse cracking in the matrix, E_c decreases.

A linear stress–strain curve reflects *damage-insensitive* behavior. It is obtained when $E_m \ll E_f$. As shown by the resulting Equation (18.2), composite Young's modulus is commensurate with fiber stiffness:

$$E_C \sim E_f V_f \quad (18.2)$$

The stresses operating on the matrix are low so that cracking is limited or nil. In SiC/SiC and C/SiC composites, the matrix is stiffer than the fibers: E_m (~410 GPa) > E_f (200–400 GPa), $V_m \sim V_f$. The matrix contributes significantly to composite modulus. It carries a substantial amount of the load. The tensile behavior is damage sensitive. In C/C composites, the fibers are much stiffer than the matrix: E_f (200–400 GPa) > E_m (~30 GPa), $V_m \sim V_f$. The matrix is subject to low stresses. The fibers carry most part of the load. The tensile behavior is damage insensitive.

Load-carrying capacity of the matrix is thus a key property that can be adjusted during matrix damage through interface characteristics. This issue is discussed in a subsequent issue and in Chapter 3.

In those damage-sensitive unidirectional composites under on-axis tensile loads, the basic damage phenomenon involves multiple microcracks or cracks that form in the matrix, perpendicular to the fiber direction, and that are arrested by the fibers at fiber/matrix interfaces. In the composites reinforced with fabrics of fiber bundles, matrix damage is influenced by multilength scale structure [3,4]. The microstructure of 2D CVI SiC/SiC is heterogeneous, as a result of the presence of woven infiltrated tows, large pores (referred to as macropores) located between the plies or at yarn intersections within the plies and a uniform layer of matrix over the fiber preform (referred to as the intertow matrix) (Figure 18.2). Much smaller pores are also present within the tows.

In 2D SiC/SiC and 2D C/SiC, damage consists of transverse cracks located, in a first step, between the longitudinal tows and deflected by the tows; in a second step, within the longitudinal tows and deflected by the filaments. In 2D C/SiC, the first step of damage initiates during cooling down from the processing temperature as a result of large carbon fiber coefficient of thermal expansion when compared to SiC matrix.

Damage modes in CVI SiC matrix composites have been identified using *in situ* microscopy during tensile tests and analysis of acoustic emission signals. The steps in matrix cracking (Figure 18.3) depend on applied deformations and microstructure. In 2D CVI SiC/SiC, the following sequence has been observed [3,4].

Step 1: Cracks initiate at macropores where stress concentrations exist (deformations between 0.025% and 0.12%).

Step 2: Cracks form in the transverse yarns and in the interply matrix (deformations between 0.12% and 0.2%).

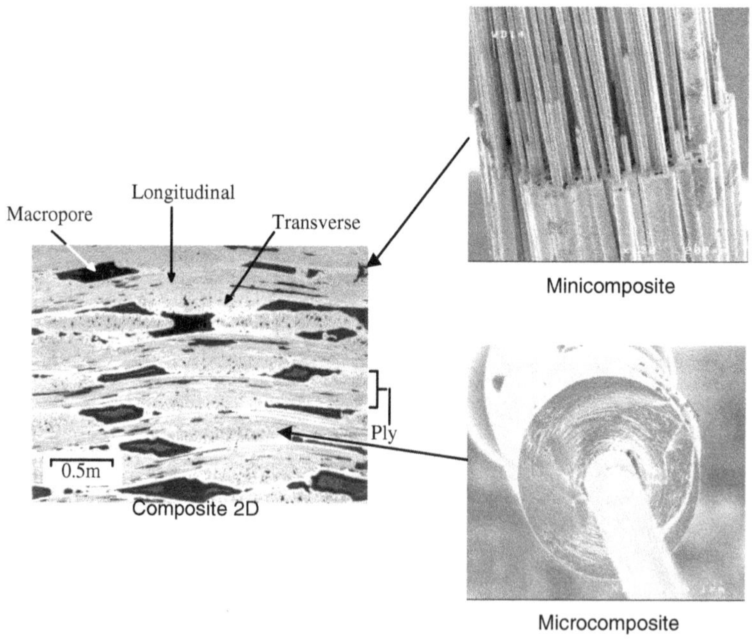

FIGURE 18.2

Microstructure of a 2D woven SiC/SiC CVI composite. Also shown are minicomposite and microcomposite test specimens that correspond to lower and intermediate length scales.

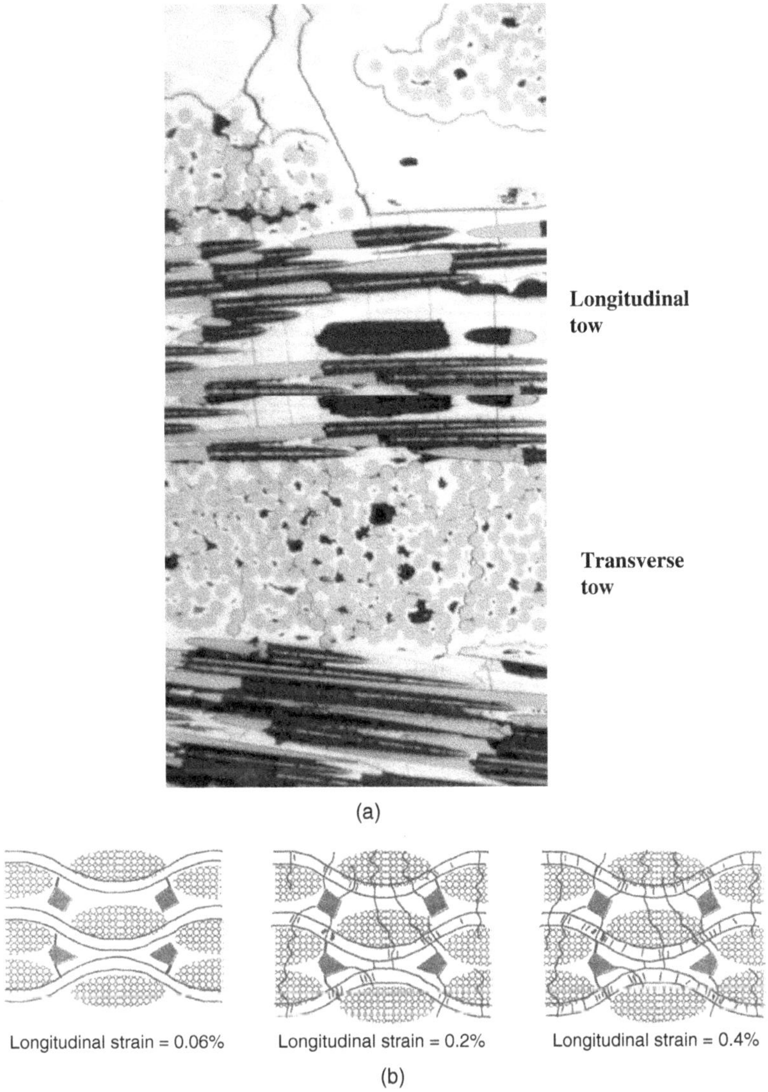

FIGURE 18.3
Successive transverse matrix cracks observed in 2D woven SiC/SiC composite under longitudinal uniaxial traction in the interply matrix, in the transverse tows, and in the longitudinal tows (a) and sequence of multiple cracking under increasing uniaxial tensile load (b). From Krenkel Book.

Step 3: Transverse microcracks initiate in the longitudinal tows (deformations larger than 0.2%). These microcracks are confined within the longitudinal tows. They do not propagate in the rest of material. The matrix in the longitudinal tows experiences a fragmentation process and the crack spacing decreases as the load increases.

In dense 2D composites without macropores, the first cracks initiate in the transverse tows. The directions of principal stresses are dictated by fiber orientations. Under axial tension, all the matrix cracks are perpendicular to the loading direction. Under off-axis tension, those matrix cracks located in the tows are perpendicular to the fiber direction, whereas those located between the tows are perpendicular to the load direction. Only on-axis loading conditions are discussed in this chapter.

The typical stress–strain curves for 2D CVI SiC/SiC composites shown in Figure 18.4 summarize trends in the mechanical behavior. This composite behaves linearly to a strain of about 0.03%, and then it exhibits a curved domain as a result of matrix cracking. Saturation of matrix damage is indicated by a point of inflection (Figure 18.4). Then the ultimate portion of the curve reflects the deformation of fibers and individual fiber breaks. The major modulus loss (70% in 2D SiC/SiC) is caused by the first families of cracks located between the longitudinal tows (deformations <0.2%). By contrast, the microcracks within the longitudinal tows are responsible for a 10% loss only (Figure 18.5). The first substantial modulus drop reflects important changes in load sharing: the load becomes carried essentially by the matrix-coated longitudinal tows (tow reloading). During microcracking in the longitudinal tows, load sharing is affected further,

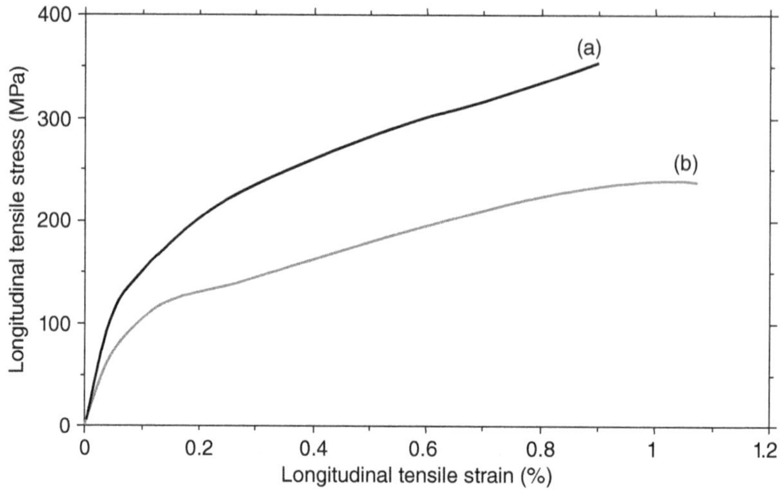

FIGURE 18.4
Typical tensile stress–strain behaviors measured on 2D SiC/SiC composites possessing PyC-based interphases and fabricated from untreated or treated Nicalon (ceramic grade) fibers: (a) strong fiber/coating interfaces and (b) weak fiber/coating. Reproduced with permission from Kluwer Academic Publishers, Handbook of Ceramics and Glasses, pp. 55–76.

and the load becomes carried essentially by the filaments (fiber reloading). The elastic modulus reaches a minimum given by the following equation:

$$E_{\text{min}} = {}^{1}\!/\!{}_{2}\, E_{\text{f}} V_{\text{f}} \qquad (18.3)$$

V_{f} is the volume fraction of fibers oriented in the loading direction.

Equation (18.3) reflects that the mechanical behavior is fully controlled by the fiber tows oriented in the direction of loading (Figure 18.4).

The damage phenomenon exhibits the following features that are of importance for modeling: it is a top-down (from large to small volumes of material), multilength scale, and sequential process. The fibers in the loading direction are loaded progressively, toward a tow-controlled behavior.

18.2.2 Ultimate Failure

Matrix damage and ultimate failure are successive phenomena. Ultimate failure generally occurs after saturation of matrix cracking, when the load is carried totally by the longitudinal tows. In that case, the strains-to-failure of composite and tows are very close: >0.8% for most SiC-based tows (Nicalon and Hi-Nicalon at least [5]) and the 2D SiC/SiC composite

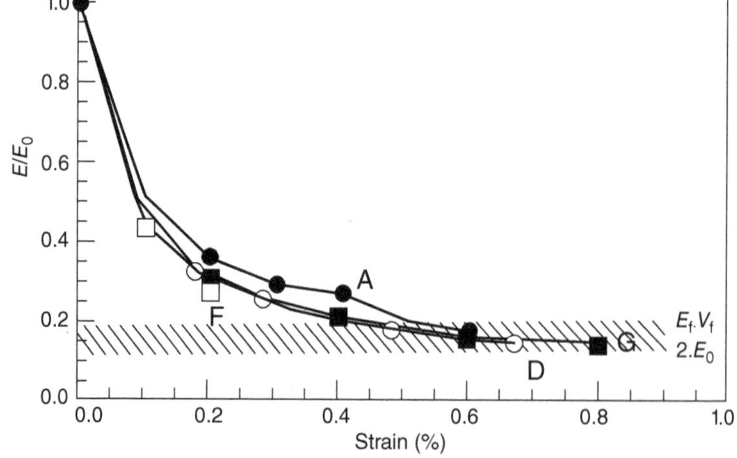

FIGURE 18.5
Relative elastic modulus versus applied strain during tensile tests on various 2D woven SiC/SiC composites reinforced with treated fibers and containing different interphases:
(A) Nicalon/$(PyC_{20}/SiC_{50})_{10}$/SiC, (D) Nicalon/PyC_{100}/SiC, (F) Hi-Nicalon/PyC_{100}/SiC, (G) Hi-Nicalon/$(PyC_{20}/SiC_{50})_{10}$/SiC. Reproduced with permission from Kluwer Academic Publishers, Handbook of Ceramics and Glasses, pp. 55–76.

of Figure 18.4. This shows that fiber weaving has a negligible influence. The analysis of acoustic emission signals indicates that fracture of filaments starts at strains >0.6%. Lower composite strain-to-failures indicate premature failure as a result of filament interactions (like contacts or too strong interfaces).

The *ultimate failure of a tow* of parallel fibers involves two steps, as discussed in Chapter 1 of this book:

- a first step of stable failure, and
- a second step of unstable failure.

During the first step, the fibers fail individually under increasing load. In the absence of fiber interactions, the load is carried by the surviving fibers only (equal load sharing). Fiber interactions cause tow weakening leading to premature failures. The ultimate failure of a tow (second step) occurs when the surviving fibers cannot tolerate the load increment resulting from a fiber failure. This particular filament that initiates instability is designated the critical fiber. It fails after the failure of 12–17% of the filaments present in a tow, for Nicalon and H-Nicalon fibers and more generally for those fibers with Weibull modulus $5 < m < 8$ [5, 6]. The statistical distributions of SiC Nicalon or Hi-Nicalon filament failure data show that the lowest strain-to-failure is about 0.5%.

The *ultimate failure of a longitudinal tow coated with matrix* results from the same two-step mechanism and involves global load sharing when a fiber fails. In the presence of multiple cracks across the matrix and associated interface cracks, the load-carrying capacity of the matrix is tremendously reduced or annihilated. The matrix-coated tows behave like dry tows under the stress-state including shear lags induced by matrix cracks. The ultimate failure of a matrix-coated tow occurs when a critical number of fibers have failed. This mechanism operates in the tows within textile CVI SiC/SiC composites. The ultimate failure of composite is caused by the failure of a critical number of tows (≥ 1), depending on the stress state: ~ 1 under uniaxial tension, >1 in bending.

As a result of the process of fiber flaw population truncation during stable fracture, the magnitude and the scatter of strengths show a significant decrease for tows and composites (Figure 18.6). At both length scales, ultimate failure coincides with the lowest extreme of the strength distribution of filaments in tow or minicomposite and of tows in composite. The gap between tow and minicomposite strengths results from the cross-sectional area involved in strength determination: the number of fibers carrying the load for tows, the total cross-sectional area for minicomposite and composite.

The successive truncations lead to a reduced ultimate population of flaws [7]. As a consequence, composite strength data exhibit a limited scatter and an insignificant dependence on stressed volume and loading conditions (Figures 18.7 and 18.8). Thus, the flexural strength is 1.15 times as large as the tensile strength [5, 8].

The Weibull model does not describe the volume dependence of strength data [7, 9], since the weakest link concept is violated due to the presence of cumulative damage process. However, the Weibull modulus (m) that can be extracted

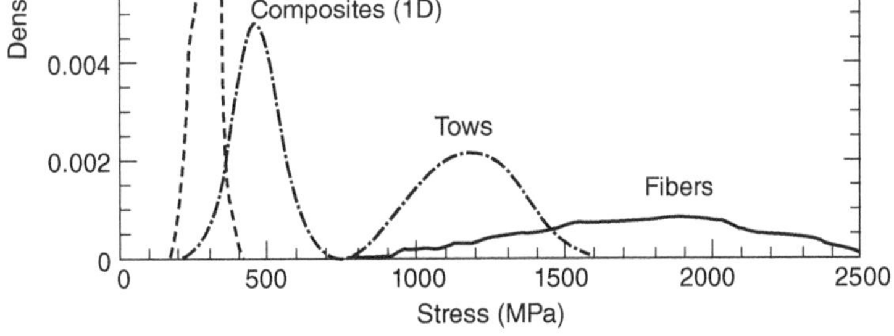

FIGURE 18.6

Strength density functions for SiC fibers (NLM 202), SiC fiber tows, SiC/SiC (1D) minicomposites, and 2D SiC/SiC composites. Reproduced with permission from Kluwer Academic Publishers, Handbook of Ceramics and Glasses, pp. 55–76.

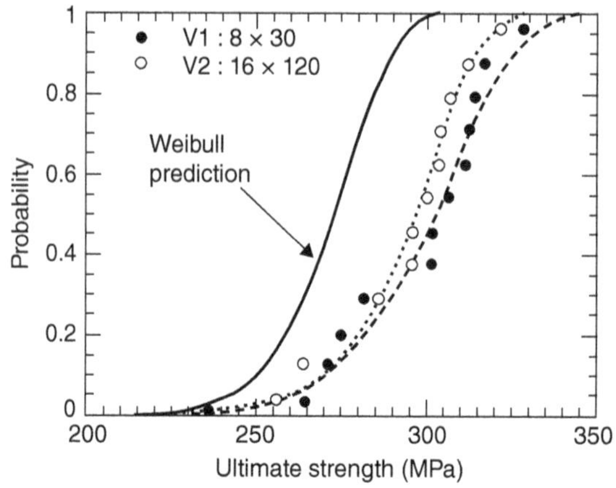

FIGURE 18.7

Scale effects in 2D woven SiC/SiC composites. Influence of specimen dimensions on ultimate failure in tension: (•) 8 × 30 mm^2, (∘) 160 × 120 mm^2 [7]. Also shown is the strength distribution for larger predicted using the Weibull model. Reproduced with permission from Elsevier, CSTE, Vol 62, Iss 3, 2002.

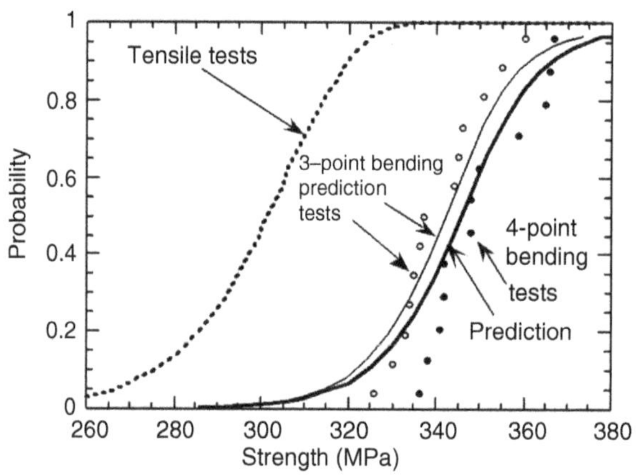

FIGURE 18.8

Strength distributions for 2D woven SiC/SiC composites tested under various loading conditions: tension, 3-point bending, and 4-point bending. The lines show predictions using macroscopic model detailed in Reference 9. Reproduced with permission from Elsevier, CSTE, Vol 62, Iss 3, 2002.

from the statistical distribution of strength data can be regarded as an indicator of scatter. *m* values in the range of 20–29 have been obtained for 2D woven SiC/SiC composite. Such values reflect a small scatter.

The stochastic nature of the stress state, as a result of the presence of flaws, heterogeneities, and cracks, is also a feature that needs to be taken into account at macroscopic scale. Details on the model that predicted satisfactorily bending strengths from tensile data (Figure 18.7) can be found in Reference 9.

18.3 CONSTITUENT PROPERTIES AND LENGTH SCALES

Constitutive properties are the topic of several other chapters. This section focuses on the characteristics of those constituent properties that influence damage and failure of composites, that is, resistance to fracture, populations of fracture-inducing flaws, and interface cracks. Data on fibers are addressed in the chapter on fibers (chapter 1). Besides properties of constituents, lowest and intermediate length scales are of interest for the investigation and modeling of the microstructure/properties relations.

18.3.1 Length Scales: Micro and Minicomposites

Figure 18.2 shows the microcomposite and minicomposite test specimens that can be manufactured and then tested under various conditions of tension (high temperatures, fatigue, and creep). Microcomposite specimens represent the fundamental entity. They are appropriate for the determination of interface characteristics [10], matrix strengths, and *in situ* fiber properties. Minicomposite specimens are reinforced by single tows. They are less fragile than microcomposite specimens, which is an advantage for testing. They are useful to reproduce the behavior of longitudinal tows. They are also used for the determination of matrix, interface, and fiber properties [6, 11, 12]. This chapter refers to results on minicomposites in several sections.

18.3.2 Fracture Strength of Matrix

As usual with brittle ceramics, strength data exhibit a significant scatter, since the fracture-inducing flaws have a random distribution. An important consequence is that the fracture stress is not an intrinsic characteristic. It is instead a statistical variable, which depends on several factors including the test method, the size, and the number of test specimens [13]. Therefore, a universal reference value of fracture strength cannot be recommended.

It is widely accepted that the Weibull model satisfactorily describes the statistical distribution of failure strengths under uniaxial tensile stresses:

$$P = 1 - \exp\left\{-\int (\sigma/\sigma_0)^m \, dV/V_0\right\} \tag{18.4}$$

where P is the probability of failure, σ is the stress, σ_0 is the scale factor, m is the Weibull modulus, V is the volume of specimen, and V_0 is a reference volume (1 m^3 is generally used). m reflects the scatter in data and σ_0 is related to the mean value of strength.

The strength of matrix for various geometries, sizes, and stress states can be determined using Equation (18.4) or Equation (18.7) given in Section 18.2.3. However, m, σ_0, and V_0 must be available. It is important to note that the estimate of σ_0 depends on V_0 [13]. It is substantially different if $V_0 = 1\ m^3$ or $1\ mm^3$. This dependence is ignored in most of the papers. When V_0 is not given, σ_0 is meaningless. The strength cannot be determined safely. The values of Weibull modulus of CVD SiC at room temperature reported in References 6 and 14] span a large range, from 2 to 12. The following values were measured using tensile tests on CVI SiC/SiC minicomposite specimen: $m = 6.1$, $\sigma_0 = 10.5$ MPa ($V_0 = 1\ m^3$) [6, 15].

Figure 18.9 shows a logarithmic plot of fragment strength–size relationships for a SiC CVD matrix [16]. Fragment sizes were measured by scanning electron microscopy (SEM) during tensile tests on SiC/SiC minicomposite specimens. As expected for a ceramic, the strengths obtained for fragments of the same size display a significant scatter, which reflects the statistical distribution of SiC strength data. Matrix failure statistical parameters can be extracted from these

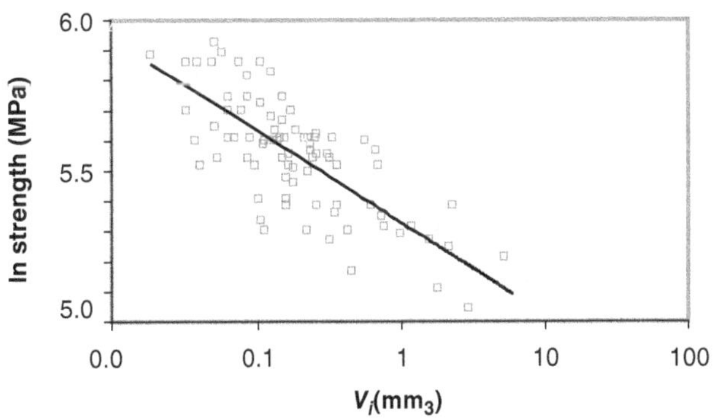

FIGURE 18.9

Plot of fragment strengths (logarithmic values) versus volumes during fragmentation of the SiC matrix of carbon-reinforced minicomposites. The solid line represents the best fit curve [16].

distributions using a large amount of data. Then, the strengths increase as fragments are smaller, according to the scale effect equation derived from the Weibull equation (18.4) for a uniform stress state:

$$\sigma_{Mi}^{m} V_i = \text{constant} \tag{18.5}$$

Best fit was obtained for $m = 7.5$:

$$\ln \sigma_{Mi} = -\frac{1}{7.5} \ln V_i + \text{constant} \tag{18.6}$$

This result demonstrates that the fracture of fragments obeys satisfactorily the Weibull model, which is based on the extreme value theory. This implies that the failure of a fragment is dictated by the most severe among the flaws present within the fragment, and that the flaw density function is independent of fragment size. This latter hypothesis is implicit in most statistical models. Note that the trend is observed down to the smallest fragment lengths obtained at saturation: $L_s = 60$ μm. The above-mentioned m estimate is in the range of Weibull modulus extracted from the statistical distributions of strengths at onset of matrix cracking: 5.1 (±1.0)–6.2 (±1.9) was determined on two sets of SiC/SiC minicomposite specimens [16].

18.3.3 Flaw Strength Distributions

There are various approaches to fracture statistics for brittle materials and ceramics [13, 17–24]. The fundamental ones that recognize the flaws as physical entities are based on flaw severity distributions [13, 20–24]. Flaw severity is defined either as *elemental strength* (i.e., the local stress that causes extension of flaw), or *flaw size*. The corresponding equation of failure probability is appropriate to predict failure in various loading cases including multiaxial stress states [11, 20–24]:

$$P = 1 - \exp\left(- \int_V dV \int_0^X g(X) \mathrm{d}X \right) \tag{18.7}$$

where $g(X)$ is the flaw density function, X measures flaw severity, V is the stressed volume, $X = S$ the elemental strength, and $X = a$ the flaw size.

Various formulae for density function can be introduced. The power law provides a satisfactory approximation of the normal distribution [25].

In Figure 18.10a, the flaw density functions represent a composite with the most severe flaws located in the matrix. This composite may be equivalent to a woven one with matrix including the intertow matrix and the transverse tows (this equivalence is discussed in a subsequent section). By contrast, Figure 18.10b defines a composite with populations of comparable severity. Taking into account the matrix and fiber elastic moduli, which determine the magnitude of stresses, the respective positions of flaw strength distributions reflect the order of damage sequences under uniaxial tension. When $E_m \geq E_f$, cracks initiate first in the matrix for the composite of Figure 18.10a; they initiate in both the matrix and fibers for that of Figure 18.10b. When $E_m < E_f$, low stresses operate on the matrix, and cracks will appear first in the fibers of composite of Figure 18.10b that will fragment. In the composite of Figure 18.10a, the matrix and fibers may crack simultaneously. It is worth pointing out that the trend depends on stress state and size of constituent stressed volumes. It can be refined by computing failure probabilities using Equation (18.7).

The respective position of flaw strength density functions for the matrix and fibers of SiC/SiC composites (Figures 18.6 and 18.11) agrees with the order of damage sequences described above for tensile loading conditions (paragraph 18.2). They indicate that the cracks initiate first from the matrix flaws.

18.3.4 Damage Tolerance and Influence of Macroscopic Flaws

Flaws bigger than the microstructure characteristic size as defined by the local volume of constituents, warrant also some consideration since they may exert a different influence. These macroscopic flaws include the previously mentioned preexisting macropores located at tow intersections, and extrinsic flaws like a notch, a hole, or foreign object damage (FOD).

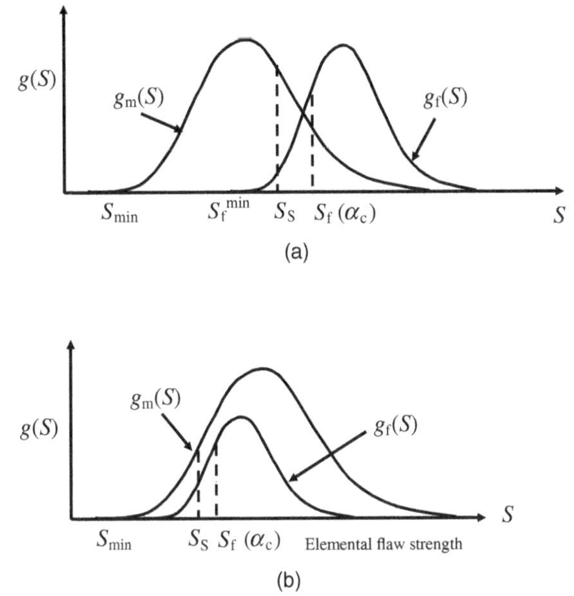

FIGURE 18.10

Schematic diagrams showing examples of flaw strength density functions for matrix and fibers: (a) more severe flaws in the matrix, (b) comparable severity of flaws in the matrix and in fibers.

The macropores initiate the first cracks in the intertow CVI SiC matrix, but they do not cause fracture of composite when the crack arrest mechanisms operate. Although their location is determined by the weave pattern, they exhibit a statistical distribution in severity due to variation in size and shape. Figure 18.11 shows an example of distribution of macropore strengths referred to as first family.

A process zone of matrix micro cracks is generated at the tip of a notch or a hole, which induces stress relaxation. As indicated in a subsequent section, matrix crack density depends on debond length. Cracking from a notch in a fracture mechanics specimen like a CT (compact tension) test specimen, is governed by the failures of the fiber bundles located within the process zone [26], and the fiber matrix interfaces that control the stresses on fibers. The propagation of the crack is stable, as it requires load increases. CMCs like SiC/SiC and C/C are generally insensitive to notch and FOD [27].

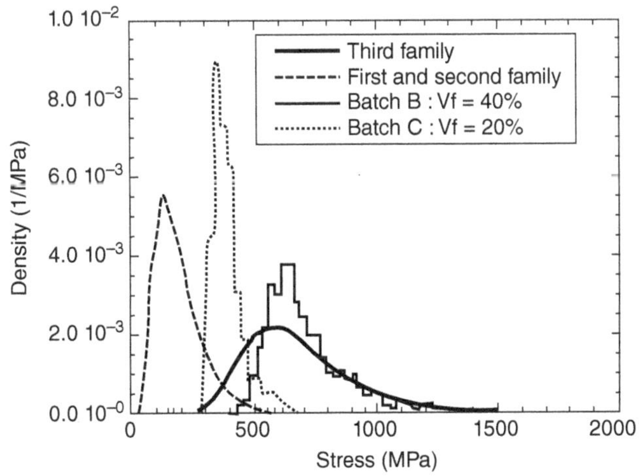

FIGURE 18.11

Flaw strength density functions for the SiC matrix of SiC/SiC minicomposite test specimens with different volumes of matrix, and for the intertow and the intratow matrices in 2D woven SiC/SiC composite.

Damage tolerance and resistance to crack propagation depend strongly on fibers, first through interfaces and second through fiber strength. Matrix cracking is an alternative mechanism of energy dissipation.

18.3.5 Interface Strength and Influence of Interface Cracks on the Mechanical Behavior

Crack arrest results from debonding at fiber/matrix interface. As discussed in chapter 3, debonding is caused by opening tensile stresses induced either by a load perpendicular to fiber axis (transverse tows; see section 18.5.1) or the triaxiality of stress state ahead of a transverse crack in the matrix (longitudinal tows). In the transverse tows, the interface crack propagates in the matrix. Interface strength determines the onset of cracking in dense matrix. In the longitudinal tows, the interface crack propagates in the interface. This crack influences the stress state, and consequently further damage of the matrix and lifetime (Figure 18.12). It is difficult to determine experimentally the tensile strength of interfaces. Debond length in the direction of loaded fibers provides a strength index. It is equivalent to the interfacial shear stress in the presence of residual contact between the interface surfaces. The stress across the debonded interface is assumed to be a constant value. This approximation permits the development of micromechanically based models of composite damage and failure.

The fiber–matrix interfacial domain is critical for composite behavior because it controls crack arrest and further load transfers from the matrix to the fiber and vice versa (depending on matrix and fiber elastic moduli). Most authors recommend weak interfaces to enhance crack arrest through deflection. Toughening is expected from crack bridging and fiber pullout. Weak interfaces are detrimental to composite strength since most of the load is carried by the fibers as a result of the presence of long interface cracks. A high strength requires that the matrix carries a part of the load. This is obtained with short interface cracks and/or significant sliding friction. These latter requirements are met with rather strong interfaces.

Efforts have been directed toward optimization of interfaces in SiC/SiC composites, using interphases and strengthened fiber/coating bond. Fiber/matrix interfaces in CVI SiC/SiC composites consist of a thin layer (less than 1 μm thick) of one or several materials deposited on the fiber (interphase). Strengthened interfaces have been obtained using treated fibers [26, 28]. The concept of strengthened interfaces has been established on CVI SiC/SiC composites with PyC and multilayered $(PyC/SiC)_n$ fiber coatings. Less interesting results have been achieved with BN interphases [29]. In CVI SiC/SiC composites with PyC-based fiber coatings, the interfacial shear stresses range between 10–40 MPa for the weak interfaces, whereas they are larger than 100–300 MPa for the strong interfaces [6, 10, 29–31].

The size of interface crack and associated sliding friction influence matrix damage. In the presence of weak bond of fibers with matrix or interphase, single long interface cracks are created during matrix cracking (adhesive failure type).

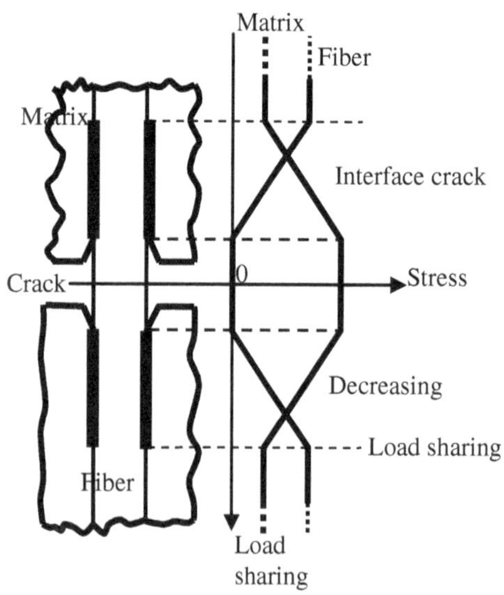

FIGURE 18.12

Shear lag in the axial stresses in the fibers and matrix caused by matrix crack deflection in fiber/matrix interface, with constant shear stress across the debonded fiber–matrix interface.

As a consequence, the volume of matrix that may experience further cracking is reduced, whereas the volume of fibers that carry the load is increased. Matrix crack density is small: the crack spacing distance at saturation as well as the pull-out length tends to be long (>100 μm). Energy dissipation by matrix cracking and composite strength are limited. The tensile stress–strain curve exhibits a narrow curved domain limited by a stress at matrix saturation that is distinct from ultimate strength (Figure 18.4).

In the presence of strengthened fiber/coating bonds, the matrix cracks are deflected within the coating (cohesive failure type), into short and branched multiple cracks. Short interphase cracks as well as improved load transfers allow further cracking of the matrix via a scale effect leading to a higher density of matrix cracks (which are slightly opened). Sliding friction within the coating as well as multiple cracking of the matrix increases energy absorption, leading to toughening. Short interphase cracks and improved load transfers reduce the load carried by the fibers, leading to strengthening. The tensile stress–strain curve exhibits a wide curved domain and the stress at matrix cracking saturation is close to composite strength (Figure 18.4).

The interphase is ineffective when fiber surface is too rough, although deflection of matrix cracks occurs. Because of strong fiber/matrix interactions in the interface cracks, premature fracture of composite occurs under small strains close to the strain at proportional limit. This phenomenon is observed on CVI SiC/SiC reinforced with Tyranno SA3 fibers [32].

18.4 MODELING OF STRESS–STRAIN BEHAVIOR

18.4.1 Stochastic Model of Matrix Fragmentation in 1D Composite and Minicomposite

Modeling of fragmentation is critical to the determination of composite nonlinear stress–strain behavior and ultimate failure, the development of optimized composites, and the design of structural components. Although a lot of papers have been produced on modeling of fragmentation of brittle fibers or ceramic matrices, this issue is of interest not only because the future for structural materials is in systems made of small elements, but also because experimental series of fragmentation stresses have been measured recently using appropriate techniques and test specimens, so that models can be evaluated.

In stress-based approaches, the stress on fragmenting constituent is compared to the strength of elements (chain-of-segment model [15]), or fragments (fragment dichotomy model [15]). These models are based on extreme value theory. They can be integrated into 2D or 3D architectures. The chain-of-element model Is based on Bayesian statistics and flaw strength density function. It has been developed recently [13]. It can be extended to 2D- and 3D-reinforced composite. The models are discussed in this section 18.4.1 in the case of 1D composite.

18.4.1.1 *The Chain-of-Segment Model* The chain-of-segment model has been widely used in the Monte Carlo simulation of fragmentation of fibers in a polymer or metal matrix. The distribution of element strengths is constructed from the weakest link failure of independent fibers. Fibers are assumed to follow the Poisson–Weibull model for strength, where the average number of flaws per unit length with stresses less than or equal to stress σ is of the power form [33]:

$$N(\sigma) = L_o^{-1}\left(\frac{\sigma}{\sigma_o}\right)^m \tag{18.8}$$

where σ_o is the scale parameter relative to the test length L_o, and m is a positive constant.

As a consequence, fiber elements as well as a single fiber tension tested at arbitrary gauge length follow a similar Weibull distribution (extreme value statistics):

$$P(\sigma, L) = 1 - \exp\left[-\frac{L}{L_o}\left(\frac{\sigma}{\sigma_o}\right)^m\right] \tag{18.9}$$

where L designates the length of either the element or the fiber. In Monte Carlo simulations, a fiber length L is subdivided into N_e segments, each of length $L/N_e \ll L$. In some cases, N_e was selected arbitrarily as large as 2000 [34] or 10^4 [35].

In other cases, segment length was the final fragment size at saturation, or was defined as twice the slip length at characteristic strength [33, 36].

Each segment is randomly assigned strength according to the Weibull cumulative probability distribution derived from (18.8) and (18.9):

$$P\left(\sigma, \frac{L}{N_e}\right) = 1 - \exp\left[-\frac{L}{N_e L_o}\left(\frac{\sigma}{\sigma_o}\right)^m\right] \tag{18.10}$$

A uniform stress σ_f is gradually applied along the entire fiber length, and a break is formed at a segment when the stress on the segment equals the assigned strength of that segment. Stress on segment may be determined using various models for shear stress or the slip zone is considered to be an exclusion zone. Comparison with available experimental fragmentation results is very sensitive to the statistical parameters introduced in Equation (18.10) [35, 36]. Agreement was not obtained with those parameters derived from the failure of independent fibers with gauge length L [36, 37]. Drawbacks associated to fiber strength distribution dependence on length and statistical fiber strength characteristics have been addressed in Reference 38. Other experimental results showed that predictions are satisfactory only for low stresses quite far from the saturation of fragmentation [34, 36].

The process of fragmentation is not addressed correctly: an in-parallel arrangement is implicitly considered instead of an in-series one, weakest link statistics are used instead of Bayesian statistics, and the underlying flaw strength distribution refers to the weakest flaws instead of the whole population of flaws present in the material.

18.4.1.2 The Fragment Dichotomy Model [6, 15, 16] The fragment dichotomy model considers fragments (instead of identical segments) of decreasing size as fragmentation proceeds. It reproduces the process of fragment generation. Each fragment results from the failure of a parent fragment. Failure is caused by the most severe flaw in the parent fragment. The fracture-inducing flaw corresponds to the low extreme of the flaw strength distribution pertaining to the parent fragment. This model predicted satisfactorily the distributions of fragment strengths for the SiC matrix in SiC or C fiber-reinforced minicomposites, as well as the tensile stress–strain behavior of minicomposites [6, 15, 16, 39].

The fragment dichotomy model of fragmentation is based on extreme value theory and power form of fragment flaw strength distribution. It has been discussed in References 6, 15, and 16. Only the main features are outlined here. It is demonstrated that Equation (18.7) reduces to the following equation, when the low strength extreme of $g(S)$ is considered only, and when it is described using a power law [13, 23, 24] with constants m and λ_0:

$$P = 1 - \exp\left[-\frac{V}{V_0}K\left(\frac{\sigma_{\text{ref}}}{\lambda_0}\right)^m\right] \tag{18.11}$$

where λ_0 is a scale factor and m is a shape parameter. K is obtained by integrating the stress state over the volume V. K depends on the probabilistic model that is considered. σ_{ref} is a reference stress (peak stress) in V. V_0 is the reference volume ($V_0 = 1$ m^3 when international units are used).

Equation (18.11) applies to the brittle failure of fragments. The flaw density function for the fragment is $g(S)$. In an elemental strength based approach, $g(S)$ is independent of fragment size when the spatial distribution of flaws within the material is homogeneous. However, an important prerequisite is that the fragments contain a sufficient amount of flaws, so that failure obeys fracture statistic laws. The most severe flaw in a fragment is characterized by the smallest elemental strength S. The strength of a fragment is thus derived from the low extreme of $g(S)$, using Equation (18.11), which accounts for size effects. According to Equation (18.11), the fragment strength reduces to

$$\sigma^i_{\text{ref}} = \lambda_0\left[\frac{V_0}{V_i K}Ln\left(\frac{1}{1-P_i}\right)^m\right] \tag{18.12}$$

where V_i is the volume of ith fragment, and P_i is the failure probability for this fragment.

In minicomposites with a uniform cross section S_m under a tensile load along fiber axis, matrix crack is perpendicular to fiber axis, and it propagates catastrophically leading to two new fragments. The stress state is affected by fiber/matrix debonding. Determination of the series of stresses σ^i_{ref} requires the successive fragment volumes V_i and their failure probabilities P_i. When the stress state operating on fragments is introduced into K, the strength of a fragment of length $2l_i$ (or l_i at the ends of specimen) is

$$\sigma_{Mi}(l_i, P_i) = \sigma_{\mathrm{M}}^{\mathrm{R}}(P)\left(\frac{V_{\mathrm{OM}}}{V_i}\right)^{1/m}\left(\frac{l_{ei}}{l_i}\right)^{-1/m} \tag{18.13}$$

where V_{OM} is the initial volume of matrix (in the absence of cracks), V_i is the volume of fragment ($V_i = 2l_iS_m$), $\sigma_{\mathrm{M}}^{\mathrm{R}}$ is the initial reference matrix strength at probability P, and l_{ei} is the equivalent fragment length [15, 16] with

$$l_{ei} = l_i\left(1 - \frac{l_{di}}{l_i}\frac{m+\beta}{m+1}\right) \quad \text{when } l_i > l_{di} \tag{18.14}$$

$$l_{\mathrm{ei}} = \frac{\left(l_i - \beta l_{di}\right)^{m+1}}{(m+1)l_{di}^m(1-\beta)^m} \quad \text{when } l_i \leq l_{di} \tag{18.15}$$

$$\beta = \frac{l_{\mathrm{o}}}{l_d}$$

where l_0 is the length of the zone where there is no contact between the interface surfaces as a result of Poisson effect.

V_i, l_i, and l_{ei} are derived from probability of location of a fracture-inducing flaw in a fragment:

$$P(x) = \frac{x - l_d}{2(l_i - l_d)} = X \tag{18.16}$$

where x is the distance to fragment edge, and l_{d} is the debond length.

Simpler analytical equations can be derived from general equations 18.13–18.16, for: either $P_i = 0.5$, or $X = 0.25$ (cracks in the middle of fragments), or $l_i > l_{di}$. For the condition $P_i = 0.5$ and $X = 0.25$, satisfactory estimates of stresses and strains were obtained in Reference 16.

For the condition $P_i = 0.5$, $X = 0.25$ and $l_i > l_{di}$, Equation (18.13) reduces to, since $V_{\mathrm{OM}}/V_i = k + 1$ at step i:

$$\bar{\sigma}_k = \bar{\sigma}_{\mathrm{M}}^{\mathrm{R}}L^{1/m_{\mathrm{m}}}\left(\frac{L}{k+1} - 2l_d\frac{m+\beta}{m+1}\right)^{-1/m_{\mathrm{m}}} \tag{18.17}$$

$\bar{\sigma}_k$ and $\bar{\sigma}_{\mathrm{M}}^{\mathrm{R}}$ are the strengths for $P_i = 0.5$. L is the gauge length, k is the number of cracks, and β accounts for Poisson's effect in the interface crack ($0 \leq \beta \leq 1$). Note that the statistical parameters correspond to the brittle failure of initial volume of material. They can be derived from the strengths of a set of specimens. Matrix cracking saturation is obtained when the term between brackets is negative. The amount of cracks at saturation is thus

$$n_{\mathrm{s}} = \frac{L}{2l_{\mathrm{d}}}\frac{m+1}{m+\beta} - 1 \tag{18.18}$$

$$n_{\mathrm{s}} \approx \frac{L}{2l_{\mathrm{d}}}$$

The $\bar{\sigma}_k$ versus V_i relationship computed using Equation (18.17) has been satisfactorily compared to experimental results in Reference 6.

18.4.1.3 The Bayesian Model The Bayesian model of fragmentation is the exact solution when fragmentation is assimilated to progressive multiple failure of a chain of elements. Failure of a link in a chain that can experience successive failures involves conditional probabilities since failure of a link depends on the response of the other links. Probability of the kth failure is derived from the following basic equation of conditional probabilities for the kth failure:

$$P(\bar{k} - 1 \cap k) = \bar{P}_{k-1}.P(k/\bar{k} - 1) = P\bar{k} - P\bar{k} - 1 \tag{18.19}$$

The subscript $\bar{k}-1$ designates the event of non-failure for the $(k-1)$th link and k the event of failure for link k. $\bar{P}_{k-1}$ is the probability of no failure of link $k-1$, and P_k is the probability of kth failure.

From Equation (18.18), it comes

$$P(k/\bar{k}-1) = 1 - \frac{1-P_k}{1-P_{k-1}} \tag{18.20}$$

Note that Equation (18.20) represents the operation of truncating the flaw strength distribution at probability P_{k-1}, to obtain the distribution of unbroken links after the $(k-1)$th failure. This is the pertinent distribution for the surviving links that will experience failure under increased load. P_{k-1} and P_k pertain to the non-truncated flaw strength distribution. $P(k/\bar{k}-1)$ corresponds to the truncated distribution after the $(k-1)$th failure. For simplicity, $P(k/\bar{k}-1)$ will be referred to as P^k.

It was assumed for simplicity that

$$P^1 = 1 - \exp\left[-\frac{V}{V_o}\left(\frac{S}{S_o}\right)^p\right] \tag{18.21}$$

for the first failure, that is, for the complete strength distribution when the chain contains all the links. The power form distribution was selected in a first step, although it may not be satisfactory because it is an extreme value theory based equation. More accurate strength distributions can be used. The statistical parameters are designated p and S_o because they characterize a wider strength distribution when compared to the low strength extreme (Equation 18.4).

The following equation for the stress at formation of the kth crack was established in Reference 15:

$$S_k = S_o k^{1/p}\left(\frac{V_o}{V}\right)^{1/P}[-\ln(1-P_\alpha)]^{1/P} \tag{18.22}$$

P_α may be taken to be 1%, 0.1%, or less, depending on the flaw population size. P_α may be also selected randomly, or k dependent, but it must remain smaller than a certain limit. To some extent, this limit depends on the number of flaws in the distribution. It is assumed that the number of remaining flaws is much larger than 1 so that statistics may apply.

Good predictions of fragmentation strengths (Figure 18.13) were obtained with the fragment dichotomy model and the Bayesian model [15]. By contrast, the fragmentation stresses were either underestimated (low stresses) or overestimated (high stresses) by the Monte Carlo simulation method. The upper and lower bounds show that predictions exhibit a significant uncertainty. Although it presents flaws in its foundation, the Monte Carlo simulation method

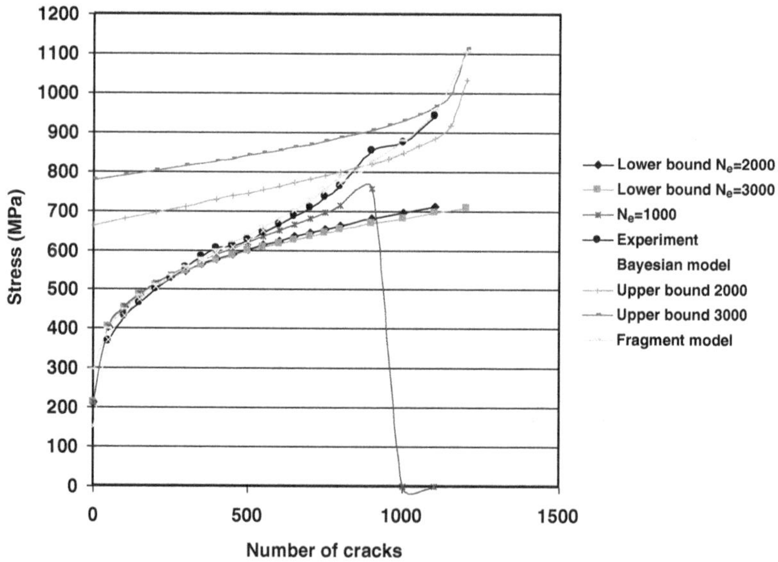

FIGURE 18.13

Strengths versus crack numbers for the SiC matrix in SiC/SiC, a minicomposite test specimen tested in tension. Comparison of experimental results with predictions using the chain-of-segment model, the Bayesian model, and the fragment dichotomy model [15]. Reproduced with permission from Elsevier, Composites Science and Technology, V 69, Iss 10, 2009.

is generally used for the treatment of multiple failures in a chain of links. It may provide a rough approximation of the low fragmentation stresses when the number of elements is similar to the number of fragments at saturation. However, this requirement cannot be met for prediction purposes, since the number of fragments is one of the outputs of analysis. Using larger amounts of elements leads to high variability in predictions, overestimation of most of fragmentation strengths, and misestimation of fragmentation stresses (Figure 18.13).

In Reference 13 the Bayesian model was extended to nonuniform and multiaxial stress states and applied to 2D-reinforced composites by using the multiaxial elemental strength.

18.4.2 Ultimate Failure in 1D Composite and Minicomposite

Ultimate failure is derived from the survival probabilities of fiber volume elements delineated by matrix fragments [6]. Global load sharing is a realistic assumption, as discussed in a section 18.2.2 on ultimate failure. Equivalent fiber length (L_{equi}) is defined as the length subjected uniformly to the peak stress at the failure probability for the actual stress state induced by cracks in the matrix and in the interface. Overlapping of interface cracks is taken into account ($l_i < l_{di}$). Failure probability of a fiber within minicomposite is given by the following equation:

$$P_{\mathrm{f}} = 1 - \exp\left[-\left(\frac{\sigma_{\mathrm{f_{max}}}}{\sigma_{of}}\right)^{m_{\mathrm{f}}} A_{\mathrm{f}} L_{\mathrm{equi}}\right] \tag{18.23}$$

with

$$\sigma_{\mathrm{f_{max}}} = \frac{F}{A_{\mathrm{f}}^{\mathrm{t}}(1-\alpha)} \tag{18.24}$$

F is the applied force, A_f is the cross sectional area of a single filament, A_f^t is the tow cross sectional area, and α is the fraction of individual fiber breaks (the probability of the Nth filament). Ultimate failure of minicomposite results from instability in the evolution of individual fiber failures. It is characterized by $\alpha = \alpha_c$ or $\delta F/\delta\alpha = 0$. Solving Equation (18.23) requires a numerical analysis to determine the spatial distribution of matrix cracks and the associated stress state [6]. Simple closed-form expressions can be obtained when there is no overlapping of interface cracks. These equations allow detailed understanding of fiber fracture.

For $l_i \geq l_{di}$, Equation (18.23) reduces to, in the presence of n cracks in the matrix:

$$P_{\mathrm{f}} = 1 - \exp\left[-\left(\frac{\sigma_{\mathrm{f}}}{\sigma_{of}}\right)^{m_{\mathrm{f}}} V_{\mathrm{F}}\left(1 + \frac{2n\,\bar{l}_{di}}{L}A\right)\right] \tag{18.25}$$

with $\bar{l}_{di}$ the average debond length, V_F the fiber volume, load-sharing parameter $a = \frac{E_m V_m}{E_f V_m}$, $\sigma_f = \frac{\sigma}{(1+a)V_f}$, and σ is the stress applied to minicomposite.

$$A = \left(\frac{u}{l_{\mathrm{d}}} + \beta\right)(1+a)^{m_{\mathrm{f}}} + \left[\frac{1-\beta}{m+1}\left(\frac{(1+a)^{m_{\mathrm{f}}+1} - 1}{a}\right) - 1\right] \tag{18.26}$$

When failure occurs after saturation of matrix cracking (when $l_i = l_{di}$), Equation (18.26) leads to the following equation of strength:

$$p_{\mathrm{f}} = 1 - \exp\left[-\left(\frac{\sigma}{\sigma_{0\mathrm{f}}}\right)^{m_{\mathrm{f}}}\left(\frac{1}{V_{\mathrm{f}}(1+a)}\right)^{m_{\mathrm{f}}}(1+A)\,V_{\mathrm{F}}\right]^{\mathrm{f}} \tag{18.27}$$

It was shown in Reference 16 that Equation (18.27) overestimates failure stresses by 12%. This slight overestimation results from the condition $l_i \geq l_{di}$, which excludes overlapping of debonding. As a consequence, load sharing is more effective, and, thus, higher loads can be carried by the specimen. Since overestimation is moderate, Equations (18.25) and (18.27) can be used to anticipate trends in the behavior.

18.4.3 Toward a Probability-Based General Model of Composite Behavior

The failure probability equations allow the occurrence of fragmentation in the fiber or in the matrix to be anticipated whatever the matrix and the fibers. Qualitative general trends under tensile monotonous loading are discussed in section 18.4.4 for significant contrast between fiber and matrix Young's moduli. Other material combinations for intermediate values can be investigated by computing P_M and P_f using the above equations. However, it is worth mentioning that σ_o and m may influence the trend. Each situation depends on various constituent characteristics and stress state.

Probabilities of formation of the first crack in fiber and in matrix are derived from Equations (18.4) or (18.11) (for the Weibull model):

$$P_f = 1 - \exp\left[-V_F\left(\frac{\sigma_f}{\sigma_{of}}\right)^{m_f}\right] \tag{18.28}$$

$$P_M = 1 - \exp\left[-V_M\left(\frac{\sigma_M}{\sigma_{o_M}}\right)^{m_M}\right] \tag{18.29}$$

When the load sharing parameter is small (like in carbon matrix composites: $E_m < E_f$), stresses in the matrix are initially small values. When $\sigma_M \ll \sigma_{o_M}$, $P_M = 0$. $P_M = 0$ and $P_f > 0$ indicate clearly that cracking will initiate in fibers.

When the load sharing parameter is initially significantly large like in the SiC matrix CMCs ($E_m > E_f$), stresses in the fiber are small values. When $\sigma_f \ll \sigma_{of}$, $P_f = 0$. $P_f = 0$ and $P_M > 0$ indicate that cracking initiates in the matrix.

Then, P_M (first case) or P_f (second case) increases with stresses during cracking in the other constituent. The second case is of interest for the SiC matrix composites reinforced with SiC or C fibers. According to Equations (18.23) and (18.25), P_f increases during matrix cracking with L_{equi}, σ_f, and n. The highest values are obtained after saturation, as L_{equi} approaches L and load sharing declines to 0. At this stage, failure probability reduces to Equation (18.28). Equation (18.28) is the failure probability of a fiber subjected uniformly to the tensile stress σ_f that is still increasing under monotonous loading. This is the critical condition. This indicates that fibers should not fail prior to saturation of matrix cracking unless extra loads or localized phenomena make fiber failure probability to increase. This result agrees with various experimental results such as the coincidence of strains-to-failure of tows and composites and the analysis of acoustic events [3, 6], that have been mentioned in section 18.2.2.

This analysis can be applied to the first case ($E_m < E_f$) and also to brittle polymer matrices. The appropriate expressions for P_M can be derived from Equations (18.23), (18.25), and (18.27).

18.4.4 The Stress–Strain Behavior of 1D Composites and Minicomposites

The tensile stress–strain behavior of composites is dictated by fiber deformations and failure. Deformations are determined from the stress state in fibers induced by the number of fragments:

$$\varepsilon = \frac{1}{LE_f}\int_L \sigma_f(x)\mathrm{d}x \tag{18.30}$$

with L = fiber length.

Figure 18.14 exemplifies the satisfactory predictions of stress–strain behavior that have been obtained on various minicomposites, using the fragment dichotomy model [6, 15, 16, 39, 40]. The characteristics of constituents required for computations (Table 18.1) include intrinsic properties (the thermoelastic constants) and fracture parameters (debonded interface characteristics and statistical parameters). They were measured independently. Thermally induced stresses arise during cooling down from the processing temperature in the C/SiC minicomposites shown here. The resulting multiple cracking is responsible for the initial nonlinear part of the stress–strain curve.

The model can be used for composite design as well as for the investigation of microstructure/behavior relations. As shown in Figure 18.15, the influence of constituent properties and size can be predicted. The following particular trend warrants attention. It was found that the statistical parameters of the matrix exert a significant influence on the stress–strain curve. It can be noted that low matrix Weibull modulus enhances nonlinear stress–strain behavior, unlike high moduli, which suggests that a stochastic matrix with a wide population of microstructural flaws should be preferred to a deterministic matrix. In other words, a dense matrix fractures under constant load, whereas a heterogeneous matrix with flaws of various sizes fractures under increasing load, which favors high toughness (through energy dissipation

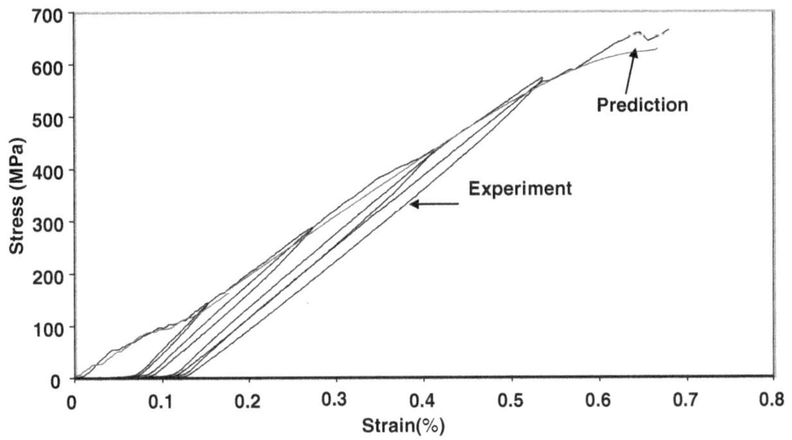

FIGURE 18.14
Comparison of stress–strain curve predicted using the fragment dichotomy model for $P_i = 0.5$ and experimental result (C/SiC minicomposite) [39]. Reproduced with permission from Elsevier, Composites Science and Technology, V 70, Iss 5, 2011.

caused by progressive multiple cracking) and high strength (as a result of associated load sharing as the load increases). Predictions also show that the stress–strain behavior is damage insensitive when matrix elastic modulus is much smaller than the fiber one.

18.4.5 Alternate Approach

There have been a certain number of papers on micromechanical models in CMCs. An approach to calculating 1D composite stress–strain curve as a function of matrix cracking was discussed in Reference 2. It was motivated by the need for a detailed understanding of tensile failure in brittle matrix composites. It is based on Weibull forms for the number of flaws.

The number of matrix flaws in a length L and area A that can propagate at applied stress σ is

$$N(\sigma, L, A) = \frac{LA}{L_0 A_0}\left(\frac{\sigma - (\sigma^* - \sigma_{\text{th}})}{\sigma_0 - \sigma^*}\right)^{\rho} \tag{18.31}$$

where $\sigma_0 - \sigma^*$ is the typical cracking stress in length L_0, σ_{th} is a thermal residual stress, σ^* is the minimum cracking stress found by Aveston et al. [2], and ρ is the Weibull modulus for the flaw strength distribution around σ_0.

For the reinforcing fibers, the two-parameter Weibull form was used for the number of flaws that can fail in length L.

$$N(\sigma, L) = \frac{L}{L_0}\left(\frac{\sigma}{\sigma_0}\right)^{\rho} \tag{18.32}$$

TABLE 18.1 Characteristics of the C/SiC Minicomposites Required for Predictions

S_m (mm^2)	0.31
Gauge length (mm)	75
V_m	0.5
τ (MPa)	20
l_d (μm)	24
E_m (GPa)	400
E_f (GPa)	230
α_L fiber	1.1×10^{-6}
α_L matrix	4.5×10^{-6}
m_f	3.6
m_m	4.9
σ_{of} (MPa)	2
σ_{om} (MPa)	3.3
V_0 (m^3)	1

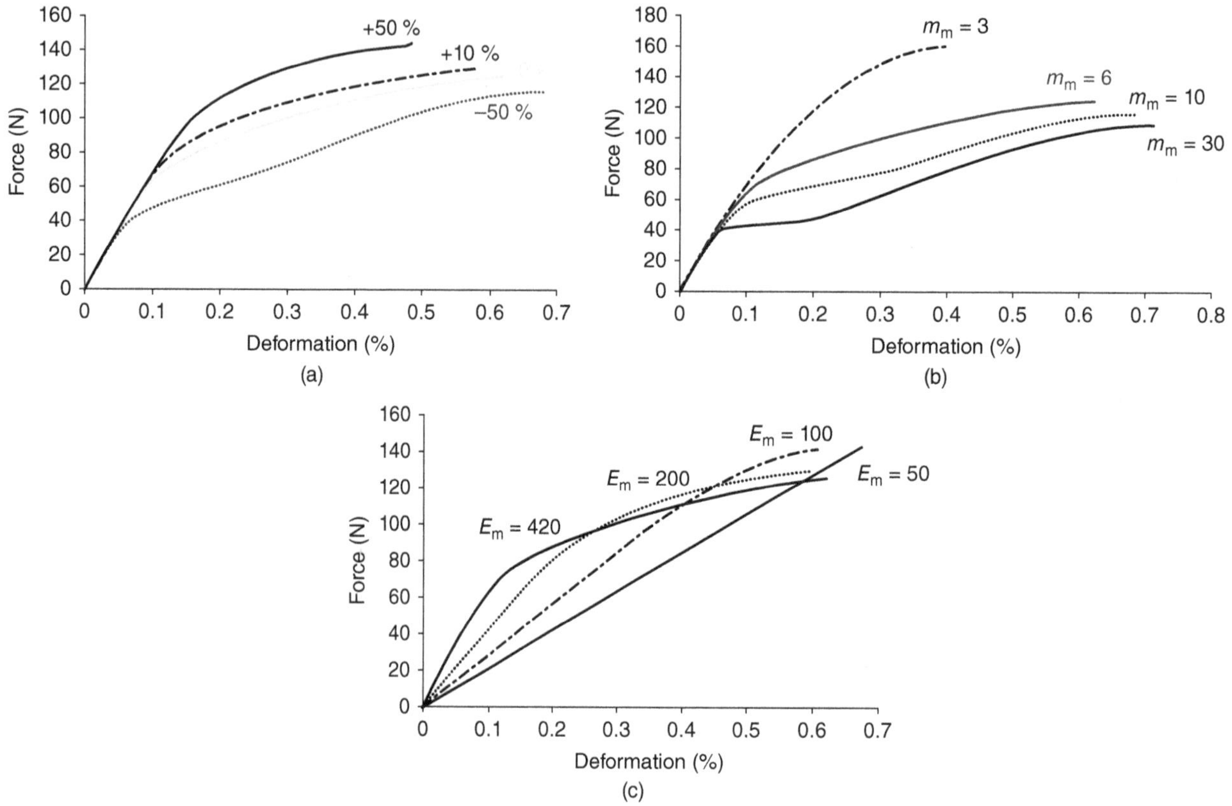

FIGURE 18.15

Predictions of the influence of matrix properties on the stress–strain behavior of minicomposites using the model based on the fragment dichotomy model: (a) influence of matrix strength, (b) influence of matrix statistical parameter m, (c) influence of matrix Young's modulus.

Curtin considered that, in a composite, there must be some characteristic matrix strength that is related to a characteristic volume of composite [2]. This characteristic volume is associated with slip around the matrix cracks. The above-mentioned number of flaws was rewritten as a function of associated critical matrix cracking stress σ_R and critical slip length δ_R. A main outcome of analyses in the literature is that at matrix crack saturation, there is a characteristic matrix crack spacing that is related to the characteristic slip length:

$$\delta_R = \frac{r\sigma_R}{\tau} \tag{18.33}$$

where r is the fiber radius.

Using average crack spacing, the stress profile on fiber is integrated to yield composite strain. The stress–strain equation is a function of various stresses (thermal stress, minimum stress, debond stress, critical matrix, and cracking stress) and effective statistical parameters. All these parameters are not intrinsic by essence, and they show variation. Estimating these parameters is an issue.

As for the matrix strength, it is considered that there is a characteristic length δ_C and an associated characteristic strength σ_C that are relevant for the fibers in the composite.

$$\delta_c = \frac{r\sigma_c}{\tau} = \left[\frac{\sigma_0^m \tau L_0}{r}\right]^{\frac{1}{m+1}} \tag{18.34}$$

where σ_0 is the fiber characteristic strength at a tested gauge length L_0 and m is the Weibull modulus.

The ultimate strength is directly proportional to the characteristic strength:

$$\sigma_u = V_f \sigma_c \left(\frac{2}{m+2}\right)^{\frac{1}{m+1}} \left(\frac{m+1}{m+2}\right) \tag{18.35}$$

Equation (18.35) is based also on global load sharing when the fibers begin to fail, and on the assumption that the fibers broken within a sliding length carry some load. It was found that it underestimates significantly the ultimate strength of SiC/SiC minicomposites. Furthermore, significant underestimations of composite strengths were also reported in Reference 41. Curtin indicated that this equation has no rigorous applicability to the composite performance but it is often utilized nonetheless.

18.5 VIRTUAL TESTING: COMPUTATIONAL APPROACH FOR WOVEN COMPOSITES

18.5.1 Multiple Cracking in Transverse Tows

Cracking in the transverse tows is of primary importance for the composite mechanical behavior, since it affects the limit of elasticity, it determines the subsequent loads on longitudinal tows and load sharing. Multiple cracking in transverse tows is made possible by the presence of: (1) crack initiating defects such as fibers, voids, and fiber/matrix bonds inducing statistical distribution of failures strengths and (2) neighboring longitudinal tows that load the fragments (Figure 18.16). The determination of minimum cracking stresses cannot neglect the presence of the defect populations present in the transverse tows. In Reference 42 the behavior of transverse tows was investigated using virtual tensile experiments on a transverse tow bonded to two longitudinal tows loaded in uniaxial tension. Virtual testing is a powerful alternative to microscopy on specimens under load. It consists of object-oriented finite elements analysis [1], using a mesh that reproduces the features of microstructure, that are the presence of various phases, flaws, heterogeneities, and so on. The finite element mesh was constructed using image analysis of micrographs of transverse tow sections (Figure 18.16). The current properties of SiC fibers and SiC matrix are available from various tensile tests on fibers, microcomposites, and minicomposites (Table 18.1). Cracks across tow were introduced at locations where stresses exceed matrix or interface strength. This transverse crack pattern is usually observed on SiC matrix composites. An important outcome is that there is a competition between the three families of defects (fibers, pores, and interfaces), depending on stress concentrations induced by fibers and pores in the matrix, and resistance of matrix versus interfaces. The pores are more deleterious than fibers. Their influence is outdone by weak interfaces. In the absence of pores and in the presence of strong interfaces, the fibers can act as inclusions that initiate matrix cracks.

The presence of flaw populations affects the stresses that display a statistical Gaussian distribution (Figure 18.17a). In the presence of cracks, the stress distribution becomes multimodal and it is shifted toward low values (Figure 18.17b). Circular domains of stress relaxation centered on crack diameter were identified (Figure 18.18). When the load is increased, the relaxation domains act as exclusion zones for further cracking. As a consequence, the amount of possible

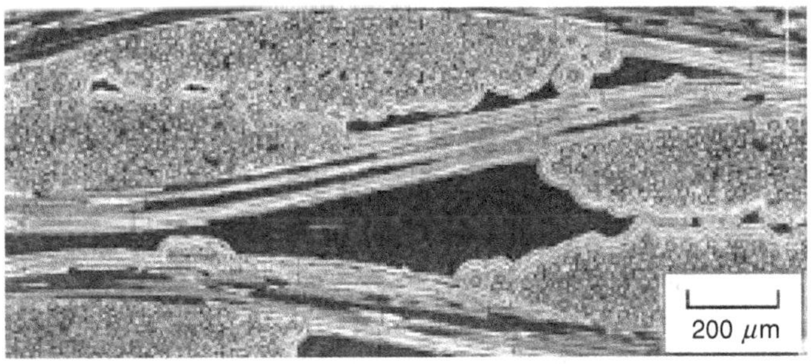

FIGURE 18.16

Micrograph of a 2D woven SiC/SiC showing microstructure of transverse tows and neighboring longitudinal tows. Reproduced with permission from Elsevier, Mechanics Research Communications, Vol 38, Iss 8, 2011.

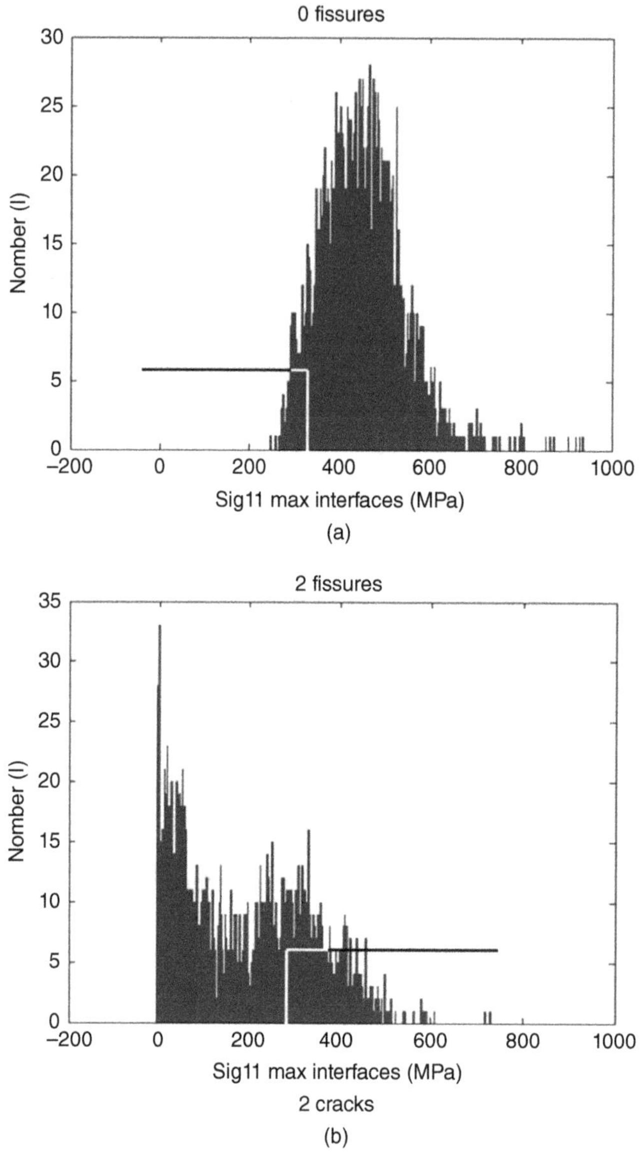

FIGURE 18.17

Statistical stress state in transverse tows induced by the fibers that act as stress concentrators: (a) before cracking, (b) in the presence of 2 cracks.

cracks is limited by tangency of relaxation domains. Thus, the number of transverse cracks at saturation is dictated by tow thickness, according to the equation

$$N_{\max} = E\,(2L/h + 1) \tag{18.36}$$

$E(\cdot)$ stands for integer part, h is the thickness, and L is the length of interface with longitudinal tow.

$N_{\max} = 5$ or 6 was obtained using Equation (18.36) for the tow geometry that was duplicated from real 2D woven SiC/SiC. It is close to virtual testing predictions. It is slightly larger than results of observations. This overestimation is reasonable as debond cracks were not introduced. In the presence of intertow cracks, wider exclusion zones are expected. Equation (18.36) is an upper bound. It can be a guide to tow transverse aspect ratio. It suggests that thin and wide tows enhance multiple cracking. Furthermore, rather strong fiber/matrix interfaces should be sought accordingly.

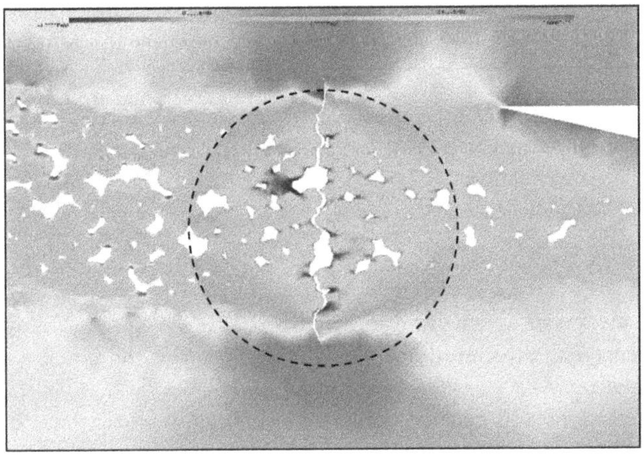

FIGURE 18.18
Effect of a crack on stress field in a transverse tow under tensile load: stress relaxation domain.

This would eliminate a population of flaws weaker than pores. Resistance to onset of matrix cracking and the limit of elasticity would be improved.

The resistance to cracking of transverse tows was found to be much smaller (0.02–0.06%) than that of the CVI SiC matrix (0.1%). These strain data highlight the influence of fracture-inducing flaws and interfaces. Furthermore, for modeling purposes, 2D woven composite can be assimilated to a 1D composite made of an equivalent heterogeneous matrix including transverse tows and reinforced with SiC-coated tows. The model developed for minicomposite can be applied to this equivalent composite when undulations are not pronounced, like in 2D woven SiC/SiC.

18.5.2 Matrix Damage in 2D Woven Composites

Matrix damage can be simulated as a process of multiple failures in volume elements in the matrix and in the tows for a 1D composite, or in the longitudinal and the transverse tows in 2D composite. This approach suits a finite element mesh, assuming that the elements contain a sufficient amount of flaws. The process involves crack initiation from flaws and extension element by element. The criterion for cracking is based on failure probability. The 2D structure with different tow directions, interply matrix, and macropores can be reproduced (Figure 18.19).

In References 43 and 44, finite element analysis of failure probabilities has been developed for the computation of matrix damage evolution in 2D woven SiC/SiC and C/C composites. Stresses in the matrix were computed using a commercial computer code. Failure probabilities are determined using a post processor that includes the multiaxial elemental strength model [13, 23, 24] for multiaxial fracture. This model is based on Equations (18.7) or (18.10) of failure probability. It involves multiaxial elemental strength and the appropriate corresponding flaw density function. The flaw strength parameters for the SiC matrix at various length scales are summarized in Table 18.2.

The deformations were increased stepwise. At each step, stresses and failure probabilities were computed. Cracks were introduced in critical elements using node splitting technique. A critical element is characterized by the failure

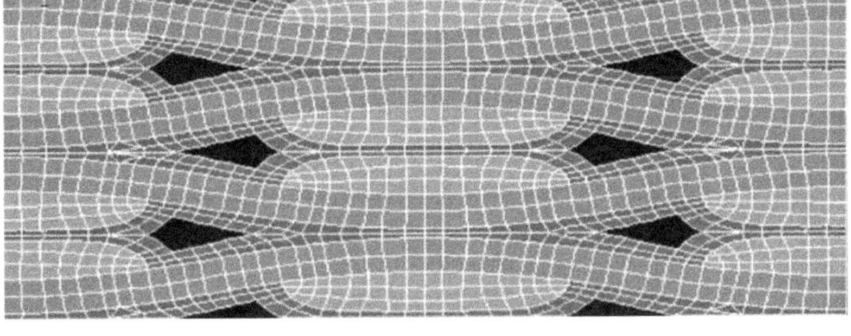

FIGURE 18.19
2D finite element mesh for 2D SiC/SiC reproducing transverse tows, longitudinal tows, intertow matrix, and macropores.

TABLE 18.2 Flaw Strength Parameters for the Matrix

Matrix	Shape parameter	Scale factor
Interply matrix	4.9	1.6
In transverse tows	4.9	0.6
In longitudinal tows	6.2	7.3

probability of 1. The load is increased as long as there is no critical element. When a critical element is obtained, successive computations of failure at constant deformation are done and the crack is increased element by element while failure probability at crack tip is 1.

Computations of crack locations and damage evolution under uniaxial tension compared quite well with *in situ* observations on 2D SiC/SiC and C/C composites (Table 18.3). An example of crack pattern obtained under bending conditions is shown on Figure 18.20. The stress–strain curves derived from average stresses at cell boundaries compared satisfactorily with experimental results. This approach can be applied to 3D meshes (Figure 18.21).

18.6 PREDICTIONS OF RUPTURE TIME

Lifetime is governed by the failure of fibers. It is driven first by the damage induced by loading, which determines the stress state and allows ingress of aggressive species. The sequence of mechanisms that control lifetime may be summarized as follows:

- First, matrix cracking and associated fiber debonding depending on the magnitude of applied load.
- Second, environmentally driven phenomena that cause either weakening or overloading of fibers that carry the loads.
- Time to failure corresponds to the kinetics of decrease of the load-carrying capacity of fibers.

At high temperatures in air, the following elementary phenomena affect fiber load–carrying capacity, depending on temperature.

- The oxidation of interphases increases the stresses operating on fibers; this phenomenon is observed with carbon interphases from 450°C.
- Creep of the matrix at temperatures >1200°C causes increase of stresses on fibers.
- Oxidation of fibers may cause fiber weakening. This phenomenon is observed on carbon fibers. Kinetics laws for carbon consumption are available. The strength degradation can be related to carbon mass loss rate [45].
- Creep of fibers is also responsible for weakening at temperatures >1200°C. Much work has been devoted to creep of fibers. But the rate in strength degradation has not been modeled.

TABLE 18.3 Comparison of Observed and Predicted Damage Modes in the Matrix of 2D woven SiC/SiC Composites Under Tensile Load

	In situ observations		Computations	
Damage mode	Deformation (%)	Occurrence	Deformation (%)	Occurrence
Cracking from macropores	0.035	Yes	0.04	Yes
Matrix cracking in transverse tows	0.085	Yes	0.11	Yes
Matrix cracking in longitudinal tows	0.185	Yes	0.2	Yes

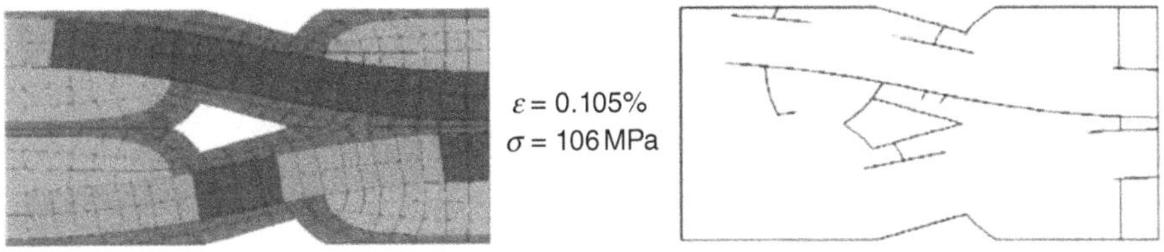

FIGURE 18.20
Predicted crack pattern for a 2D woven SiC/SiC cell under flexural load.

- Slow crack growth in SiC fibers at temperatures <1000°C. This phenomenon is addressed in the chapter 1 on fibers.
- It is worth noting that oxidation can slow down the phenomenon of weakening when a layer of oxide protects the fibers against degradation by environment. Oxide is produced in the matrix cracks of self-healing matrices or at the surface of SiC fibers (at temperatures depending on fiber composition).

Additional structural phenomena take place in composite specimens.

- Overloading of fibers when a fiber fails.
- Catastrophic failure of isolated fibers that cannot withstand overloading or of groups of touching fibers. As a consequence, weak as well as strong fibers can be eliminated during this step.

Finally, various stochastic structural features need to be taken into account.

- The stresses on fibers depend on both the elementary force f operating on each fiber, and the section of fiber. Fiber diameters have been shown to follow a Gaussian distribution [46], so that fiber sections S_i within a tow are also described using a Gaussian distribution $P_G\ (S_i)$.
- Fiber strengths follow a Weibull distribution denoted $P_W(\sigma_f)$.
- The elementary forces f depend on microstructural factors that control load sharing such as the amount of matrix bonded to the fiber (the distance to neighboring fibers), fiber arrangement, density of matrix cracks, and the size of interface cracks. So, the elementary forces follow a statistical distribution.
- When a fiber breaks from slow crack growth, the force it was carrying is shared by the surviving fibers. For the same microstructural features as mentioned above, random load sharing prevails. So, the resulting force

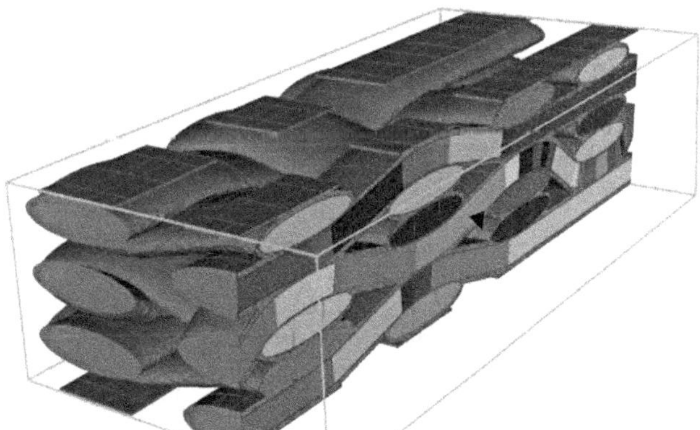

FIGURE 18.21
Example of 3D mesh for woven SiC/SiC composite.

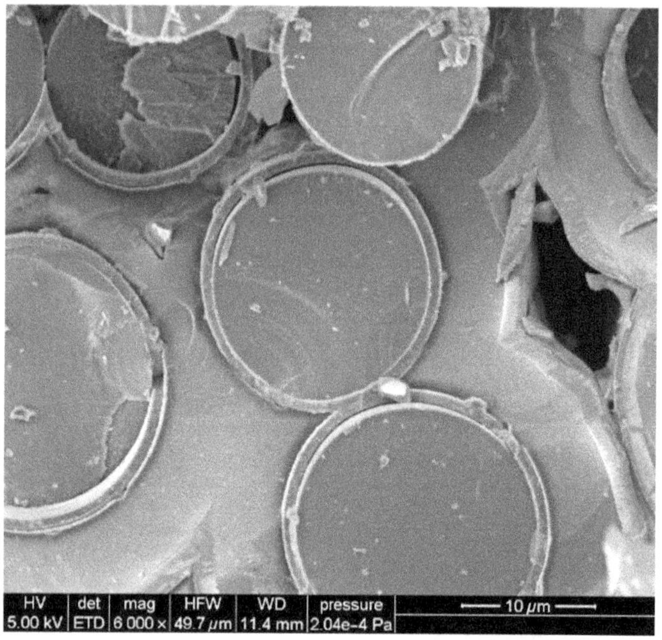

FIGURE 18.22

Evidence of subcritical crack growth at 700°C in SiC Hi-Nicalon filaments in a SiC matrix.

increments Δf on surviving fibers can be assumed to follow a statistical distribution, with $\Sigma\Delta f = f$, where f is the force that was carried by the fiber that broke.

- When a group of fibers fails $\Sigma\Delta f = \Sigma f$, so that a larger force increment operates on the surviving fibers. It depends on the size of fiber cluster. The sizes of fiber clusters exhibit a statistical distribution.

In a multiscale approach to failure, modeling of matrix fragmentation and associated stress state on fibers is the first important step. Then, modeling of stress state changes and fiber strength degradation as a function of time is the second one. The ultimate failure occurs when the stresses reach the resistances of critical fibers.

Attempts to predict lifetime in the frame of a multiscale approach have been made on 1D C/C composites [45] and 2D woven SiC/SiC [47]. The approach to calculating the lifetime of 2D woven SiC/SiC composites at high temperature (<1000°C) under a constant load was based on the rupture time of tows subjected to a uniform stress state and it integrated several statistical features. At these temperatures, SiC filaments experience slow crack growth (Figure 18.22). The model reproduces the above-mentioned failure modes. Elementary forces on fibers f, force increments Δf, were assumed to follow normal distributions ($P_N(f)$, $P_N(\Delta f)$). Fiber strength was described statistically using a Weibull distribution denoted $P_W(\sigma_f)$.

The rupture time of a longitudinal tow is given by the sum of the rupture times due to slow crack growth under load f (for the fiber which failed first) and then under successive overloads ($f + \Delta f$). It was assumed that the force on composite was shared equally by all the longitudinal tows (Equation 18.37).

$$F = \frac{F_{\text{comp}}}{\lambda} \tag{18.37}$$

λ was taken to be the number of longitudinal tows within the specimen (150).

Successive filaments were selected randomly using three parameters (section S_i, elementary force f, and strength σ_f), which were determined using random numbers X, Y, and U (<1; Equations 18.38–18.41).

$$X = P_G(S_i) \tag{18.38}$$

$$Y = P_N(f) \tag{18.39}$$

$$U = P_W(\sigma_f) \tag{18.40}$$

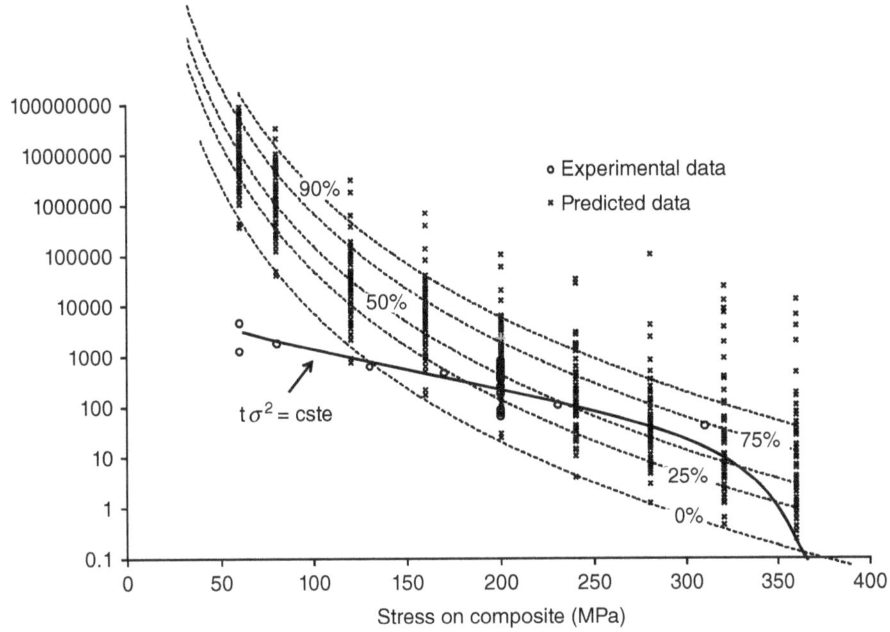

FIGURE 18.23

Stress–rupture time data that were predicted using the model, for 2D woven SiC/SiC composite specimens under constant load at 500°C in air. From Prediction of lifetime in static fatigue at high temperatures for ceramic matrix composites. O. Loseille, J. Lamon. Journal of Advanced Materials Research, Vol. 112, pp. 129–140, 2010.

Then, the overload was also selected using the same method (Equation 18.41).

$$Z = P_N(f + \Delta f) \tag{18.41}$$

The predicted tow lifetimes exhibit a significant scatter as a result of variation in fiber characteristics S_i and σ_f and loads f and Δf (Figure 18.23). They compare with those measured on 2D woven specimens for stresses >150 MPa, when the loading conditions of tows in composite are close to the uniform ones assumed here. According to the fracture mode of 2D woven ceramic composite specimens in tension, which is controlled by the weakest tow, the lifetime of composite should be caused by the fracture of the tow that fails first. Results indicate that in some specimens, the weakest tow did not fail first. This is a logical consequence of the above-mentioned statistical features. Under low stresses, when the stress state on fibers in composite is not uniform, the load is not equally shared by the tows. Thus, nonuniform stress state on filaments must be introduced in the analysis.

18.7 CONCLUSIONS

Damage modes in woven composites are hierarchically organized, sequential, and stochastic as a result of microstructure and the respective properties of constituents. Multiple cracking initiates in the matrix from macroscopic and then microscopic flaws. As it proceeds, the load is progressively transferred to the tows.

The woven structure involves various length scales. Associated scale effects inherent to brittle ceramics delineate steps in the load transfer. Damage in the matrix and then in the tows is responsible for truncation of flaw strength distributions leading to a reduced distribution of critical filaments that governs ultimate failure. As a consequence, the 2D woven SiC/SiC composites exhibit a limited scatter in strength, and a limited sensitivity to scale effects. These remarkable features can be affected by processing defects, such as local strong bonds at interfaces.

The damage modes under tensile load govern the resistance to fatigue at high temperature in aggressive environment like air, when the reinforcing fibers become exposed to overloading and weakening by oxidation or slow crack growth.

Probabilistic models allow the determination of damage kinetics in longitudinal and transverse tows, and in 2D woven composites taking into account flaw strength distribution characteristics and constituent properties. They allow also predictions of the stress–strain behavior, and of the influence of elastic and fracture characteristics of fibers and matrix, and resistance of interfaces.

Composites are more complex than most homogeneous materials. However, they offer a remarkable versatility in microstructure. Microstructure–behavior relations allow one to capitalize on this outstanding advantage.

REFERENCES

1. S. A. Langer, E. R. Fuller, and W. C. Carter. OOF: an image-based finite-element analysis of material microstructures. *Comput. Sci. Eng.*, 3, 1–9 (2001).
2. W. A. Curtin. Stress–strain behavior of brittle matrix composites, in *Comprehensive Composites Materials*, vol. 4, edited by. A. Kelly and C. Zweben (Elsevier Science Ltd., 2000), pp. 47–76.
3. J. Lamon. A micromechanics-based approach to the mechanical behavior of brittle-matrix composites. *Compos. Sci. Technol.*, 61, 2259–2272 (2001).
4. L. Guillaumat and J. Lamon. Multifissuration de composites SiC/SiC, in *Microstructure, Comportements Thermomécaniques et Modélisation des Composites Céramique–Céramique à Fibres*, edited by J. L. Chermant and G. Fantozzi (Revue des Composites et des Matériaux Avancés, 3, 1993), pp. 159–171.
5. V. Calard and J. Lamon. Failure of fibres bundles. *Compos. Sci. Technol.*, 64, 701–710 (2004).
6. N. Lissart and J. Lamon. Damage and failure in ceramic matrix minicomposites: experimental study and model. *Acta Mater.*, 45, 1025 (1997)
7. V. Calard and J. Lamon. A probabilistic–statistical approach to the ultimate failure of ceramic-matrix composites—part I: experimental investigation of 2D woven Sic/Sic composites. *Compos. Sci. Technol.*, 62, 385–393 (2002).
8. J. C. McNulty and F. W. Zok. Application of weakest-link fracture statistics to fiber-reinforced ceramic-matrix composites. *J. Am. Ceram. Soc.*, 80, 1535–1543 (1997).
9. V. Calard and J. Lamon. A probabilistic approach to the ultimate failure of ceramic-matrix composites—part II: macroscopic model. *Compos. Sci. Technol.*, 62(3), 395–399 (2002).
10. J. Lamon, F. Rebillat, and A. G. Evans. Assessment of a microcomposite test procedure for evaluating constituent properties of ceramic matrix composites. *J. Am. Ceram. Soc.*, 78(2), 401–405 (1995).
11. R. Naslain, J. Lamon, R. Pailler, X. Bourrat, A. Guette, and F. Langlais. Micro/minicomposites: a useful approach to the design and development of non-oxide CMCs. *Compos. Part A Appl. Sci. Manuf.*, 30(4), 537–547 (1999).
12. G. N. Morscher and J. Martinez-Frenandez. Fiber effects on minicomposite mechanical properties for several silicon carbide fiber-chemically vapor-infiltrated silicon carbide matrix systems. *J. Am. Ceram. Soc.*, 82(1), 145–155 (1999).
13. J. Lamon. *Mécanique de la rupture fragile et de l'endommagement: Approches Statistiques-Probabilistes* [Mechanics of Brittle Fracture and Damage: Statistical Probabilistic Approaches] (Hermès-Lavoisier, Paris, France, 2007).
14. L. L. Snead, T. Nozawa, Y. Katoh, T.-S. Byun, S. Kondo, and D. A. Petti. Handbook of SiC properties for fuel performance modeling. *J. Nucl. Mater.*, 371, 329–377 (2007).
15. J. Lamon. Stochastic models of fragmentation of brittle fibers or matrix in composites. *Compos. Sci. Technol.*, 70, 743–751 (2010).
16. J. Lamon. Stochastic approach to multiple cracking in composite systems based on the extreme value theory. *Compos. Sci. Technol.*, 69, 1607–1614 (2009).
17. W. Weibull. A statistical distribution of wide applicability. *J. Appl. Mech.*, 18, 293 (1951).
18. A. Freudenthal. *Statistical approach to brittle fracture*, vol. II, edited by H. Liebowitz (Academic Press, New York, 1968), Chapter 6, pp. 591–619.
19. E. Gumbel. *Statistics of Extremes* (Columbia University Press, New York, 1968).
20. A. De, S. Jayatilaka, and K. Trustrum. Statistical approach to brittle fracture. *J. Mat. Sc.* 10, 1426–1430 (1977).
21. A. S. Argon and F. A. McClintock. *Mechanical Behavior of Materials*. (Addison-Wesley, Reading, MA, 1966).
22. S. B. Batdorf and J. G. Crose. Statistical theory for the fracture of brittle structures subjected to nonuniform polyaxial stresses. *J. Appl. Mech.* [Transactions of the ASME], 41(2), 459–464 (1974).
23. J. Lamon and A.G. Evans. The statistical analysis of bending strengths for brittle solids: a multiaxial fracture problem. *J. Am. Ceram. Soc.*, 66, 177–182 (1983).
24. J. Lamon. Statistical approaches to failure for ceramic reliability assessment. *J. Am. Ceram. Soc.*, 71, 106–112 (1988).

25. M. R'Mili, N. Godin, and J. Lamon. Flaw strength distributions and statistical parameters for ceramic fibers: the normal distribution. *Phys. Rev. E Stat. Nonlin. Soft. Matter Phys.*, 85(5), 1106–1112 (2012).
26. C. Droillard and J. Lamon. Fracture toughness of 2D woven SiC/SiC CVI composites with multilayered interphases. *J. Am. Ceram. Soc.*, 79, 849–858 (1996).
27. M. Picard, E. Maillet, P. Reynaud, N. Godin, M. R'Mili, G. Fantozzi, and J. Lamon. Damage sensitivity and acoustic emission of SiC/SiC composites during tensile test and static fatigue at intermediate temperature after impact damage. *Ceramic Engineering and Science Proceedings*, John Wiley & Sons, 33(2), 3–14, 2012.
28. R. Naslain. Fiber-matrix interphases and interfaces in ceramic matrix composites processed by CVI. *Compos. Interfaces*, 1, 253–258 (1993).
29. F. Rebillat, J. Lamon, and A. Guette. The concept of a strong interface applied to SiC/SiC composites with a BN interphase. *Acta Mater.*, 48, 4609–4618 (2000).
30. F. Rebillat, J. Lamon, R. Naslain, E. Lara-Curzio, M. K. Ferber, and T. Besmann. Interfacial bond strength in SiC/C/SiC composite materials as studied by single-fiber push-out tests. *J. Am. Ceram. Soc.*, 81, 965 (1998).
31. F. Rebillat, J.Lamon, R. Naslain, E.Lara-Curzio, M. K. Ferber, and T. Besmann. Properties of multilayered interphases in SiC/SiC chemical-vapor-infiltrated composites with "weak" and "strong" interfaces. *J. Am. Ceram. Soc.*, 81, 2315–2326 (1998).
32. C. Sauder, A. Brusson, and J. Lamon. Influence of interface characteristics on the mechanical properties of Hi-Nicalon type-S or Tyranno-SA3 fiber-reinforced SiC/SiC minicomposites. *Int. J. Appl. Ceram. Tech.*, 7(3), 291–303 (2010).
33. C.-Y. Hui, S. L. Phoenix, M. Ibnabdeljalil, and R. L. Smith. An exact closed form solution for fragmentation of Weibull fibers in single filament composite with applications to fiber-reinforced ceramics. *J. Mech. Phys. Solids*, 43(10), 1551–1585 (1995).
34. C. Baxevanakis, D. Jeulin, and D. Valentin. Fracture statistics of single-fiber composite specimens. *Compos. Sci. Tech.*, 48, 47–56 (1993).
35. H. D. Wagner. Stochastic concepts in the study of size effects in the mechanical strength of highly oriented polymeric materials. *J. Polymer Sci. B Polymer Phys.*, 27, 115–149 (1989).
36. T. Okabe, N. Takeda, J. Komotori, M. Shimizu, and W.A. Curtin. A new fracture mechanics model for multiple matrix cracks of SiC reinforced brittle-matrix composites. *Acta Mater.*, 47(17), 4299–4309 (1999).
37. G. Rousset, J. Lamon, and E. Martin. In situ fiber strength determination in metal matrix composites. *Compos. Sci. Tech.*, 69(15–16), 2580–2586 (2009).
38. V. Glushko, V. Kovalenko, S. Mileiko, and V. Tvardovsky. Evaluation of fiber strength characteristics using fiber fragmentation test, in *Composites: Fracture Mechanics and Technology*, edited by S. Mileiko and V. Tvarkovsky (Russian Composite Society, Chernogolovska, Russia, 1992), pp. 66–76.
39. F. Pailler and J. Lamon. Micromechanics-based model of fatigue/oxidation for ceramic matrix composites. *Compos. Sci. Tech.*, 65, 369–374 (2005).
40. S. Bertrand, R. Pailler, P. Forio, and J. Lamon. Hi-nicalon/SiC minicomposites with (PyC/SiC)n nanoscale-multilayered interphases. *J. Am. Ceram. Soc.*, 82(9), 2465–2473 (1999).
41. W. A. Curtin. Ultimate strengths of fibre-reinforced ceramics and metals. *Composites*, 24(2), 98–102 (1993).
42. P. Pineau, G. Couegnat, and J. Lamon. Virtual testing applied to transverse multiple cracking of tows in woven ceramic composites. *Mech. Res. Comm.* 38(8), 579–585 (2011).
43. L. Guillaumat and J. Lamon. Probabilistic–statistical simulation of the nonlinear mechanical behavior of a woven SiC/SiC composite. *Compos. Sci. Tech.*, 56, 803–808 (1996).
44. J. Lamon, B. Thommeret and C. Percevault. Probabilistic-statistical approach to the matrix damage and stress–strain behavior of 2-D woven SiC/SiC ceramic matrix composites (CMCs). *J. Eur. Ceram. Soc.*, 18, 1797–1808 (1998).
45. G-A. Lebrun, F. Lamouroux, and J. Lamon. Lifetime predictions for C/SiC composites at high temperature in air. *Ceramic Transactions*, [The American Ceramic Society], 103, 331–342 (2000).
46. N. Lissart and J. Lamon., Statistical analysis of failure of SiC fibers in the presence of bimodal flaw populations. *J. Mat. Sci.*, 32, 6107–6117 (1997).
47. J. Lamon. CVI SiC/SiC composites, in *Handbook of Ceramics and Glasses*, edited by N. P. Bansal (Kluwer Academic Publishers, New York, 2005), Chapter 3, pp. 55–76.

PART V JOINING

CHAPTER 19

INTEGRATION AND JOINING OF CERAMIC MATRIX COMPOSITES

MONICA FERRARIS AND VALENTINA CASALEGNO

Department of Applied Science and Technology-DISAT, Politecnico di Torino, Torino, Italy

19.1 INTRODUCTION/BACKGROUND

Joining: the more you know it, the less you use it.

For sure, the best solution from an engineering point of view would be to avoid joining at all: joining is a discontinuity, that is, an intrinsic source of failure.

However, it must be clear that if we want to use CMC, sooner or later we have to join them to themselves or to other materials. Thinking monolithic with CMC is unfeasible. Welding them is impossible. Providing an efficient and reliable joining technology for CMC is still an issue and a challenge for scientists and designers.

On the other hand, a reliable, efficient joining and integration of CMC can be a powerful tool in enabling a speedy drive toward innovation and efficiency of several key industrial products.

A wide number of CMC with exceptional properties such as high thermomechanical, thermochemical resistance, and low density compared to traditional materials are now available on the market, but their use is limited to very few applications, mainly because of problems related to join them in a reliable way.

Just to cite few possibilities, SiC/SiC, C/SiC, C/C, or oxide/oxide composites could be used right now to replace existing automotive or aerospace components; this will contribute to the increasing demand for novel components with improved properties such as light weight for reducing fuel consumption and CO_2 emission.

Integration of CMC to other engineering materials should be easier, more rapid, should permit inspection, should repair the damaged parts, and should possibly allow for disassembly at the end of life. As soon as these issues are solved, CMC can be extensively used to produce several "hybrid" structures, made of two or more different types of material, each contributing with unique properties and complementing the properties of the other.

Excluding welding for obvious reasons, there are some possible techniques for joining CMC to themselves (*joining*, in this chapter) and to dissimilar materials (*integration*, in this chapter). To list them in order of priority, there are techniques based on solutions typical of *mechanics* (fasteners, rivets, bolts screws, etc.), others using *adhesives*, then there is *brazing* and finally some more "*exotic*" joining materials and methods, used in restricted application areas or at a research stage. Some of them will be reviewed in this chapter, with particular reference to the most recent updates.

An important point to be considered when CMC have to be integrated to metallic structures is the difference in coefficient of thermal expansion (CTE). In order to accommodate the CTE mismatch, a number of approaches have

Ceramic Matrix Composites: Materials, Modeling and Technology, First Edition. Edited by Narottam P. Bansal and Jacques Lamon.

been proposed such as suitable design to minimize stress concentration, metallic interlayers, graded joints, or mechanical structuring of the surface to be joined.

In general, the formation of a strong and thermodynamical stable bond between CMC and the joining material or CMC and the metal to be joined is an advantage for the overall joint strength: design of interfaces and prediction of their thermodynamic stability is an important tool for CMC joining and integration.

In any case, the formation of successful joints depends on the intimate contact between the work pieces, the conversion of these contacting surfaces into an atomically bonded interface, and the ability of this interface to accommodate thermomechanical residual stresses generated during joining operations.

It is worth noting that several recent reviews have been published on some topics covered by this chapter; readers may refer directly to them and their references [1–4].

19.2 MECHANICAL JOINING AND INTEGRATION OF CMC

It is the most obvious and conservative approach to join materials; when feasible for joining and integration of CMC, it should be preferred. However, some possible CMC damage due to drilling and issues related to induced stress concentration must be carefully considered.

Some *metallic screw/nut and bolt/rivet systems* are available to join CMC, but the working temperature of the best metallic systems (superalloys) is much lower than the expected one requested to CMC.

A few CMC or ceramic screw/nut and bolt/rivet systems have been proposed [5–7].

CMC components can be integrated to the noncomposite structure with mechanical metal or composite threaded fasteners, CMC "T" joints, Miller fasteners [8, 9], and round braided CMC fasteners. In Reference 10 some SiC/SiC fasteners were used to join SiC/SiC components to metallic ones in an aircraft combustion environment.

An articulating universal joint with a central coupling tube, orthogonal shear pins, and C-Clips for mechanical pin retention have been designed and manufactured to join SiC/SiC for control rods in nuclear plants (Figure 19.1) [11]. The more highly stressed regions of the joint are the shear pins connecting the segments and a triaxially braided architecture for them is to be preferred.

Nanopowder Infiltration and Transient Eutectic-phase (NITE) SiC/SiC screws have been obtained without damaging the composite structure, due to its high density and toughness (Figure 19.2) [12–14].

An interesting improvement of *mechanical joining for CMC* is the *coupling of it with modified adhesives or other joining materials* (brazing alloys or glasses/glass-ceramics) to give reliable and hermetic joints. Actually, the use of adhesives coupled with screws is a way currently used to join metallic or polymeric parts; its adaptation to CMC is a challenging but promising concept [15].

As an example of this concept, a carbon fiber-reinforced commercial adhesive (Graphibond 551RN produced by Aremco Products, Valley Cottage, NY, USA) reinforced by nano- and micro-SiC powders was used to join SiC/SiC composites mechanically machined to have interlocking surfaces in Reference 16. Flexural strength higher than 90 MPa for the joined SiC/SiC was obtained at the working conditions (1100°C, 1 hour, Ar).

FIGURE 19.1

Assembled nuclear grade SiC/SiC control rod sheath joint segment [11]. Reproduced with permission from Elsevier BV, Journal of Nuclear Materials.

FIGURE 19.2
Screw type mechanical joint of 10-mm inner diameter (12-mm outer diameter) NITE SiC/SiC tube. Reproduced with permission from Annales de Chimie Science des materiaux, 30 [6] 659–671 (2005).

Even if a mechanical joining of CMC by CMC screws/bolts has a potential, mechanical joint has some drawbacks to be carefully considered in the case of integration of CMC with metals: first of all CTE mismatch, then stress concentration make this way of joining often unfeasible.

On the other hand, bolted or mechanically locked joints are interesting for ease of assembly/disassembly, in comparison with other joining techniques.

19.3 ADHESIVE JOINING OF CMC

Due to the intrinsic low temperature stability of adhesives compared to the high temperature application of CMC, there are just a few recent examples of adhesive joining of CMC [17–19].

However, adhesives can be of interest to join CMC whenever low temperature applications are requested, for example, when CMC are joined to polymers.

Adhesives can be also modified by adding ceramic particles, fibers, or carbon nanotubes to increase their temperature resistance [16, 20–24] or directly used as precursors of ceramic-based joining materials [25].

Adhesive technology offers now commercial products claiming to withstand working temperatures of 1650°C and mechanical strengths of several tens of MPa, after suitable curing [26–29], thus in the range of most CMC; if validated, these materials may open the way to a more extensive use of adhesives for integration and joining of CMC.

19.4 BRAZING OF CMC

Advances in brazing have been recently reviewed in Reference 3, in particular in Chapters 7 and 13 (referring to C/C) and Chapter 11 (integration of CMC).

Brazing is the most commonly used joining and integration method for CMC; brazing alloys can be considered a sort of metallic "*high temperature glue.*" Copper, nickel, and silver are the most frequently used base metals for brazing alloys; aluminum and gold are also used for specific purposes. For each base metal, many different alloy compositions are available. The additional metals combined with the base metal determine the brazing alloy compatibility for use in joining specific CMC. The melting range for a brazing alloy is defined by the minimum temperature at which the alloy will start to melt (solidus) and the temperature at which the alloy is 100% liquid (liquidus). For most purposes, the actual brazing temperature is 30–110°C above the liquidus. Some alloys (eutectics) have a very narrow melting range while other alloys have a comparatively wide range. High temperature brazing alloys are based on gold, nickel, and copper and are often used for joining CMC to cobalt- or nickel-based superalloys [30].

The critical issues in brazing CMC are wettability of the braze on the substrates, the chemical interactions between the braze and the CMC, and the possible reaction products at the interface.

The brazing alloy is usually placed as a foil or a paste between the two substrates to be joined and heated above its liquidus temperature. The melting point is lower than those of the CMC and/or the metals to be joined. Preparation of the surfaces to be joined and time/temperature process are usually given by the supplier and are important parameters to be possibly optimized and adapted. Several brazing alloys are available on the market and many others are continuously proposed and tested at a laboratory level. Brazing alloys are mainly divided in two categories: *passive and active*, the latter able to react and give interfacial products able to increase wettability.

The main issues to be taken in account with brazing are (i) *wettability*, (ii) *reactivity*, (iii) *thermodynamic stability* of the brazing alloy toward CMC, and (iv) the intrinsic lower *melting temperature* of the brazing alloy compared to the CMC or the metal to be joined. Several books and reviews on these issues are available [31–35]; a brief summary will be given below.

The theory of wettability is based on the contact angle between the molten brazing alloy and the substrate to be joined, the working atmosphere, and pressure. In particular for CMC, the surface qualities (roughness, porosity) also have a role.

Together with wettability, the reactivity and thermodynamic stability of the interfaces in working conditions are fundamentals to guarantee the suitability of the proposed brazing alloy. Wettability of a brazing alloy can be controlled by introduction of reactive elements, such as Ti (active brazing), which are able to promote the formation of compounds on the CMC surface (e.g., titanium carbides). The surface modification of CMC to improve wettability toward the brazing alloys can be also obtained by direct reaction of the CMC surface (e.g., formation of metal carbides), by metallization of the CMC surface or the brazing foil itself.

The thermodynamic stability of the brazed interface(s) can be controlled by deposition of "barrier" coatings, able to stop (or slow) diffusion between brazing alloy and joined materials.

It is, of course, impossible to change the intrinsic lower melting temperature of the brazing alloy compared to the one of CMC or metals to be joined, which constitutes the main drawback for this method.

High temperature brazing alloys up to 1230°C are commercially available [36].

A useful tool to select the most suitable and available brazing alloys for CMC is given in References 30 and 37. In order to deal with the CTE mismatch between CMC and brazing alloys, several options have been proposed: different layers with a gradually changing CTE from CMC to the brazing alloy, compliant metallic layers (e.g., Cu, see also Section 19.9.1), stiff layers (e.g., W, Mo, see also Section 19.9.1), flexible metallic layers, composite brazing alloys obtained by adding short fibers or particles to the brazing alloy matrix [38, 39].

An extensive research on brazing of CMC for their joining and integration has been done by M. Singh and his group at NASA [40–46].

Their results show that CMC can be successfully joined to themselves and integrated with metallic structures by using judiciously selected off-the-shelf (usually active) brazing alloys to yield metallurgical sound joints. Uniform and continuous joints were obtained irrespective of differences in the CMC fiber orientation.

19.5 LIQUID SILICON INFILTRATION

The Liquid Silicon Infiltration (LSI), initially developed to produce CMC [47], has been successfully used to join carbon-based CMC (Figure 19.3) [48]. It can be considered as a sort of brazing, where molten silicon or silicon-based alloys act as a braze. The LSI process allows the joining of CMC parts during manufacturing, that is, C/C-SiC joints can be produced *in-situ*. Porous C/C components are prepared and fixed together and molten silicon flows between the surfaces and react with the carbon material to convert it to silicon carbide and bond the surfaces together. The LSI process is still one of the most effective to join and integrate CMC, in particular for integration of CMC brake disks in brake systems such as the joining of the automotive C/SiC friction rotor with the metallic part [49].

19.6 ARCJOINT

Partially based on silicon infiltration, **ArcJoinT** is an affordable, robust joining technology for ceramics and CMC developed at NASA [50, 51] and consists of applying a carbonaceous mixture, curing it at 110–120°C, then applying a silicon

FIGURE 19.3

CT images of C/C-SiC joined using the LSI-route: cut-out of the material with the joint visible in the middle of the short fibre material. Courtesy of Prof. Krenkel, Ceramic Materials Engineering, Department of Ceramic Materials, University of Bayreuth, Bayreuth, Germany.

or silicon-based alloy paste (or tape) which is heated at 1250–1425°C. A wide variety of silicon carbide-based ceramics and composites, in different shapes and sizes, have been joined using this technique. There is no change in the mechanical strength of joints in silicon carbide matrix composites up to 1200°C in air. This technology is suitable for the joining of large and complex shaped ceramic and composite components, and with certain modifications, can be applied to repair of ceramic components damaged in service.

19.7 "EXOTIC" TECHNIQUES FOR INTEGRATION AND JOINING OF CMC

The joining methods reviewed in this paragraph regard more "exotic" techniques for integration and joining of CMC. Some of them are at a research laboratory level, others have a limited industrial application.

19.7.1 Transient-Liquid-Phase Bonding

Transient-Liquid-Phase Bonding (TLPB) is a process characterized by the formation and controlled disappearance of a liquid film at the joining temperature. The technology is currently being adapted for CMC using "interlayers" based either on glasses or pure metals or alloys (such as Ge and Ge–Si for joining SiC/SiC composites) [52–54]. The chemical reactions and diffusion processes taking place during TLPB give a joint region which remelts at temperature well above the temperature used for the joining process, ideal to produce joints for high temperature applications. The understanding of the liquid wettability at the interface and the nature of reaction products formed is fundamental for TLPB and also to develop new joining approaches. TLPB process has a number of key advantages: joining can take place at relatively lower temperatures than those used to join the final component; the liquid film can infiltrate the CMC surface; the liquid also reduces or eliminate the use of pressure during joining. Even if there is a limited number of references about application of TLPB to CMC integration and joining, this is considered a technology to be explored in particular for high temperature applications (Figure 19.4) [55, 56].

19.7.2 Nanopowder Infiltration and Transient Eutectic Phase

NITE is a method based on a transient oxide based liquid phase. Nanopowder Infiltration and Transient Eutectic-phase (NITE) [14, 57] is a joining technology which uses a mixture of beta-SiC nanopowder below 30 nm and mixtures based on SiO_2 and/or Al_2O_3 and/or Y_2O_3 powders as joining material for SiC/SiC. There is the formation during joining of a transient amorphous liquid phase based on the oxides above. The resulting joint has the same composition of

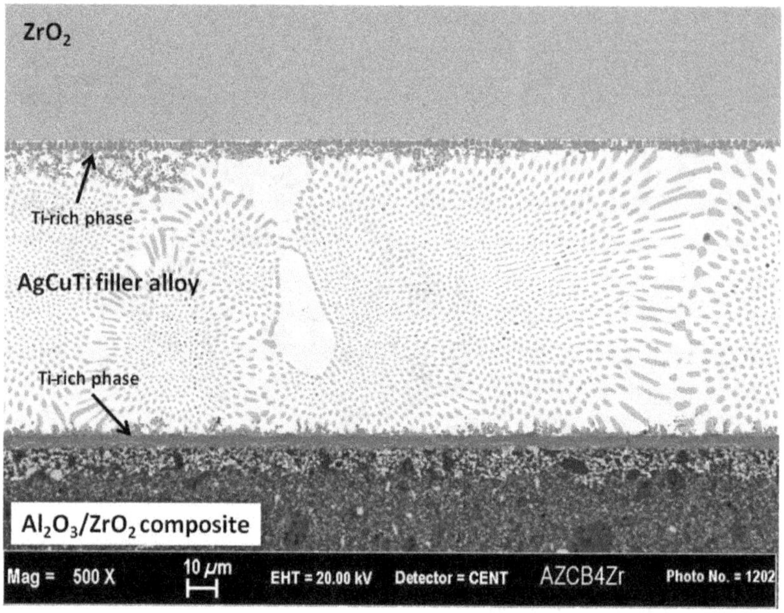

FIGURE 19.4

Alumina/zirconia composite (proportion in weight: 80–20%; produced by CNR-ISTEC-Italy) brazed to yttria stabilized zirconia. Filler layer: Ag27Cu3Ti; $T = 950°C$; holding time: 30 s; cooling rate: 4°C/min; working atmosphere: pure Ar. Ti segregation at metal/ceramic interfaces (dark strips in the figure) promoted wetting and adhesion. Courtesy of Prof. Passerone, Dr. Valenza, National Research Council-Inst. for Energetics and Interphases CNR-IENI-Italy.

the SiC matrix in SiC/SiC made by NITE technology, thus offering comparable mechanical strength for joined and not joined samples. The only drawback is that this joining process requires a pressure of 10–20 MPa at temperature ranging between 1700°C and 1900°C in a controlled environment.

Initially used to sinter ceramics and composites, the following techniques have recently found interesting applications as *pressureless, localized heating* joining and integration methods for CMC. They can work with and without a joining material which can be a glass, a metal, or a metallic alloy.

19.7.3 Spark Plasma Sintering

Spark Plasma Sintering (SPS) realizes a local heating by so-called sparks due to the pulsed current flow in the near gap between the two surfaces to be joined without bulk heating [58–60]. Promising results have been recently obtained (Figure 19.5) with SPS joining of C/SiC, both with a Ti foil and without it [61].

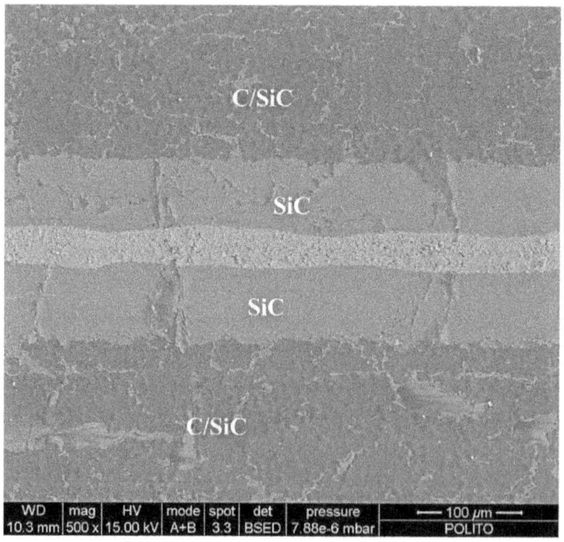

FIGURE 19.5

SPS joining of C/SiC by Ti foil. Courtesy of Nanoforce Ltd. UK, unpublished results.

19.7.4 Micro-Wave-Assisted Joining

Micro-wave-assisted joining gave promising results [62–65] and is still of interest because of its peculiar characteristic of being steered by optics as a means of localized heating. This is important for *in situ* joining and repairing of CMC components [66]. A few examples have been reported on alumina–zirconia composites joined by hybrid heating using microwave radiation (2.45 GHz, 700 W), with sodium silicate glass powder as an interlayer [67].

19.7.5 Laser-Assisted Joining

Laser-assisted joining has been enabled by the progress made in laser-beam technology in the last few years. Moreover, laser-assisted joining technology has no special requirements regarding the applied pressure and the process atmosphere such as vacuum or inert gas, an advantage in terms of process automation.

Relevant progress has been made in the optimization of the laser parameters suitable for oxide [68] and nonoxide joining materials [59] or to directly join ceramics [69]. One of the very few and preliminary applications of this method to SiC/SiC composites is given in Reference 59.

19.7.6 Glass and Glass-Ceramic as Joining Materials for CMC

Glasses and glass-ceramics have been proposed for a long time [70–74] as joining materials for CMC because of their unique properties such as tailorable characteristic temperatures, CTE, and wettability; their thermomechanical stability at working temperature can be high enough to be of interest for applications of CMC and they are not affected by oxidation; due to their tailorable viscosity versus temperature they allow pressureless joining processes and hermeticity and can be self-sealant [75–77].

Particles and fibers can be added to glass-ceramics to obtain a composite joint, suitable for joining and integration of CMC. The refractoriness of a glassy joining material could be enhanced by a controlled crystallization, the so-called *glass-ceramic method*, to obtain joining materials mechanically reliable up to the softening point of the residual amorphous phase, which can be minimized by suitable thermal treatments. Moreover, glasses and glass-ceramics can be used in machined CMC surfaces to improve the mechanical strength and hermeticity of a joint [78].

19.7.7 Solid State Displacement Reactions

Solid state displacements reactions in the system Ti–Si–C have been proposed as joining material and technology in particular for SiC- and C-based CMC [79]. Sound joints for SiC/SiC composites (Figure 19.6) were obtained using tape

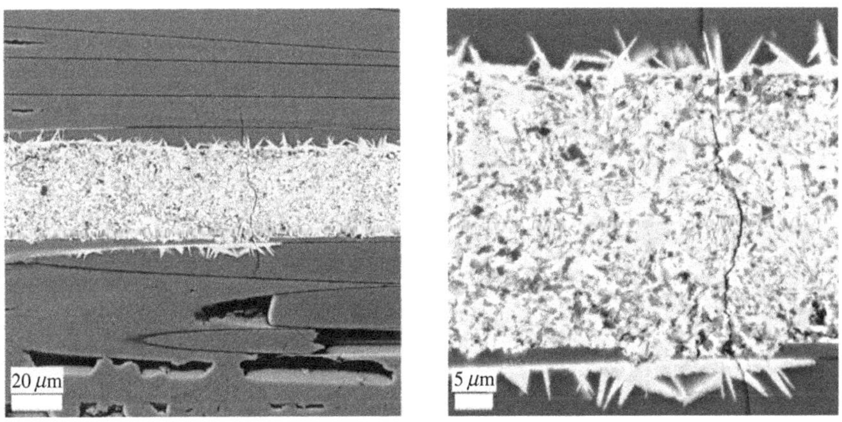

(a) (b)

FIGURE 19.6

Dense joint between SiC-composite blocks processed at 1723 K and 30 MPa for 2 hours in Argon. Joint thickness is about 40 μm and consists of SiC particles (dark regions in joint) and Ti–Si–C matrix (lighter regions) obtained by displacement reaction (a). Note the transverse crack in (b) that also demonstrates crack deflection. C. H. Henager, R. J. Kurtz, Low-activation joining of SiC/SiC composites for fusion applications, J. Nucl. Mat. 417 375–378 (2011).

cast ribbons of Si/TiC powders which are able to give a solid state displacement reaction between 1200°C and 1400°C using 30 MPa applied pressure; the reported shear strength (50 MPa) exceeds the shear strength of SiC/SiC composite materials [80].

19.7.8 Preceramic-Polymer Joints

Several papers report on the use of preceramic polymers as precursors for joining CMC for high temperature applications, discussing the effects of precursor yield, temperature, and heating rate on the maximum stress due to shrinkage associated with the conversion of preceramic precursors to bonded structures [81,82].

Generally, a mixture of polymer precursors acting as a source of carbon and aluminum, boron and/or silicon is applied to the two surfaces (in a tape, paste, or slurry), then heated in an inert atmosphere in a furnace (or air, using a propane torch). The joint forms through pyrolysis of the polymer precursor material, which subsequently reacts with the silicon to form SiC, in the presence of aluminum and boron or other sintering aids. The fact that the joint can be produced using a simple gas torch could have a major impact on the repair or on-site production of joints. So far, various grades of SiC/SiC and C/SiC composites have been joined using this technique.

Conditions where the joint material viscosity and the shrinkage rates are high must be avoided. Fillers and prepyrolysis of the precursor can be used to decrease the stresses that occur during processing of joints. To avoid the formation of cracks in joints derived from preceramic polymer precursors, it is critical that the evolution of gaseous species is completed before the formation of closed porosity in the joint [83].

19.8 BACK TO BASIC: JOINTS FOR CMC LIKE IN WOOD-BASED PRODUCTS

A suitable design of the surfaces to be joined (Figure 19.7) permits in principle to achieve comparable mechanical strength for joined and not-joined composites [78, 84].

This concept mimics what is currently done with wood-based components [85], which are machined and glued together. No fasteners, bolts, or rivets are needed. A mechanical interlocking is used to keep the two parts together: examples include dovetails, tongues and grooves, T-slots, and wedges.

The possibility of doing it with CMC joined to themselves or to other materials open new possibilities.

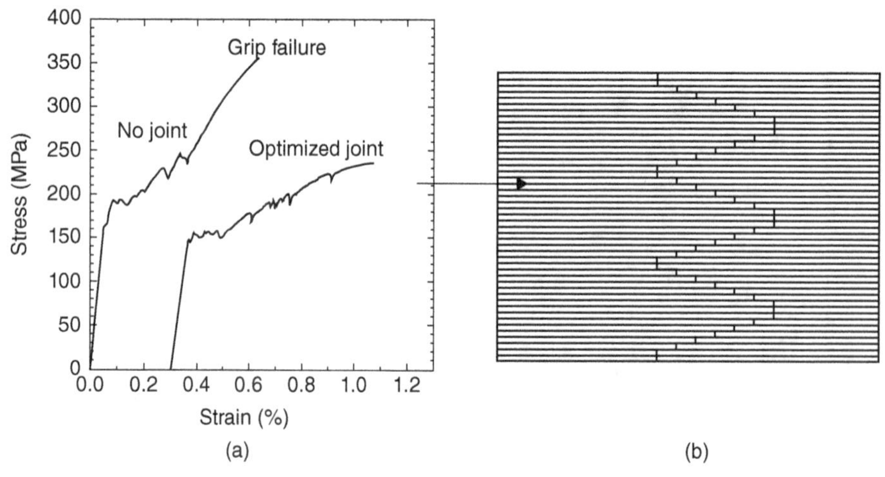

FIGURE 19.7

(a) Comparable mechanical strength for joined ("Optimized Joint" curve) and not-joined composites ("No joint" curve) can be achieved by suitable design of the composite surfaces to be joined (b). Reproduced with permission by Blackwell Publishing, Journal of American Ceramic Society, V81, Iss 12, ISSN:1551-2916, 1998.

FIGURE 19.8
Example of mechanical machining and joining of SiC/SiC (a), laser machining (b), and joining of laser machined C/C (c).

Laser or mechanical structuring of surfaces to be joined and selective matrix removal from composites may allow to obtain "brush-like" surfaces to be joined.

A combination of suitable joining material, mechanical machining or laser structuring (Figure 19.8), surface modification by chemical reactions, and selective matrix removal of the composite matrix can be developed and tailored to specific application.

A new "brush-like" joining technology can be developed to improve the mechanical strength of CMC joined to metals and to polymers, as sketched in Figure 19.9; the composite matrix can be selectively removed by chemical or thermal etching from a thin surface layer in order to infiltrate the joining material and increase the mechanical strength of the whole joined structure.

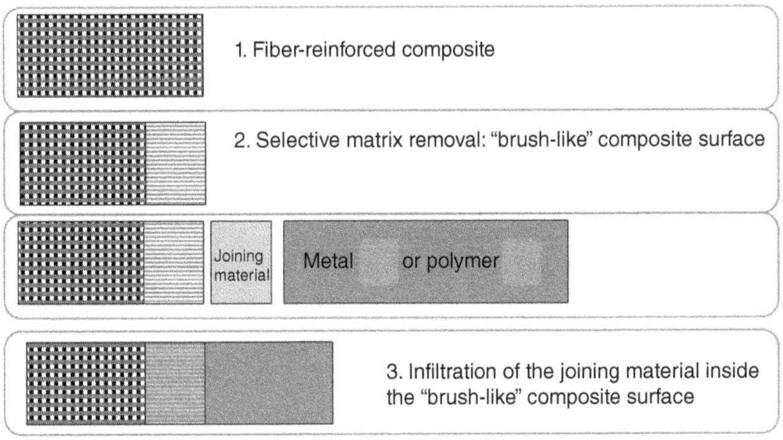

FIGURE 19.9
Brush-like joining process for composite-to-metal or composite-to-polymer surfaces.

19.9 SPECIAL ISSUES

19.9.1 Joining of CMC for Nuclear Applications

In the field of nuclear energy production (both fusion and fission), requirements for joints are extremely severe; the joining material must be compatible with a neutron environment and the joining technique must comply with the fusion nuclear reactor design. It means that CMC components several meters long and few millimeters thick must be joined in a reliable and feasible way, possibly without using any pressure [86, 87].

Among CMC, C/C and SiC/SiC have been proposed for applications in different nuclear environments, namely fusion and fission.

C/C was originally considered as plasma-facing material in one important component of ITER, the divertor. Even if C/C is currently being replaced by W in the final ITER design, one decade long activity developed interesting methods to join C/C to copper and its alloys [88]. The main point was to make C/C wettable by molten copper; one way was the surface modification of C/C by formation of metal carbides. Figure 19.10 shows the cross section of a Cr-modified C/C surface, then brazed to Cu for ITER divertor [89–91]; another option was the use of active brazing alloys able to form metal carbides during brazing; other techniques involved the use of laser structuring of C/C followed by one of the two above mentioned approaches in order to improve the mechanical strength of the joined component.

Knowledge developed for ITER will be hopefully transferred to other applications involving integration of C/C to copper-based alloys.

SiC/SiC have been initially proposed as structural materials for DEMO [92], the fusion reactor beyond ITER. This ambitious plan requires SiC/SiC and a suitable joining technology still not available in the market. Recently, the use of SiC/SiC is being considered for particular components for the next generation fission reactors. Also in this case, research is still needed to assess a reliable joining technology.

The very few available solutions for SiC/SiC joining in the field of nuclear applications are reported in References 80, 83, 93–95 and have been recently reviewed in Reference 11. Currently, the joining processes based on NITE, glass-ceramic, solid-state displacement reactions (TiC + Si), preceramic polymers, and transient liquid phase metal joining are considered sufficiently mature and attractive to proceed to neutron irradiation testing.

Very promising results are being obtained in joining of SiC/SiC for nuclear applications by using a proprietary technology based on infiltration of nanosized SiC and formation of a transient eutectic phase (NITE) [14].

NITE joining is believed to be the most promising candidate, since it is based on a joining zone having the same composition of NITE–SiC/SiC composites matrix; a mixture of beta-SiC nanopowder below 30 nm and SiO2 + Al2O3 + Y2O3 powders is used as joining material in NITE joint process, which involves a transient amorphous liquid phase based on the oxides above. Moreover, this ceramic has been demonstrated to be irradiation stable for those conditions

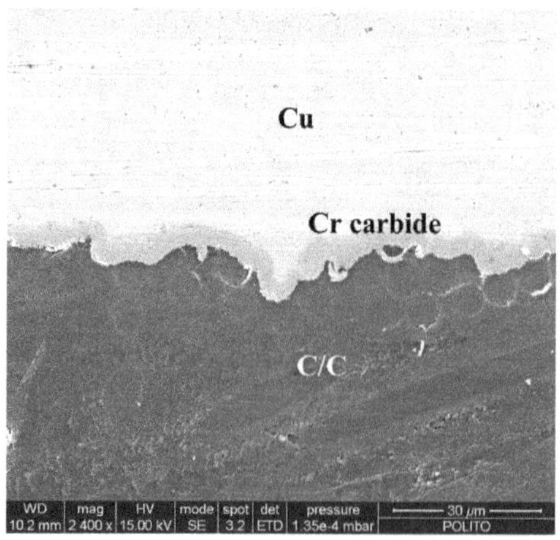

FIGURE 19.10

Scanning electron microscopy of the cross section of Cr-modified C/C-Cu-sample brazed by Cu-based alloy for ITER divertor.

studied. The joining process for NITE requires a 10–20 MPa pressure at temperature ranging between 1700°C and 1900°C in a controlled environment.

Glasses and glass-ceramics have been proposed as joining materials for SiC/SiC because of their unique properties such as tailorable characteristic temperatures, CTE, and wettability on SiC/SiC substrate; glasses and glass-ceramics can be designed to be low activation materials (LAM), transmutation free, and stable under neutron irradiation; their thermomechanical stability at working temperature can be higher than 1000°C and they are not affected by oxidation; due to their tailorable viscosity versus temperature they allow pressureless joining processes and hermeticity and can be self-sealant. The refractoriness of a glassy joining material could be enhanced by a controlled crystallization, the so-called glass-ceramic method, to obtain joining materials mechanically reliable up to the softening point of the residual amorphous phase, which can be minimized by suitable thermal treatments.

However, the use of glass-ceramics as joining materials in a neutron environment is still debatable. A few references are available on neutron irradiation of glass-ceramics: two competing processes such as swelling and densification of amorphous and crystalline phases present in a glass-ceramic are known. It is also known and predictable that differential swelling/shrinking of two phases can lead to reduction of strength and ultimately to separation of the phases and total destruction of the material. Hopefully, a designed balance between swelling and shrinkage could lead to a glass-ceramic joining material to be used by a pressureless joining method. This last feature can be of importance when meters-long components have to be joined.

A few glass-ceramics are currently under evaluation as pressureless joining materials for nuclear environment [78, 94].

The use of glass-ceramics as joining materials in a neutron environment has been recently demonstrated up to 820°C, 5 dpa [96] (comparison of bending and shear strength for neutron irradiated and nonirradiated glass-ceramic-joined SiC and SiC/SiC indicates that the mechanical strength is unaffected by irradiation at the used irradiation conditions.

Mostly used for nonnuclear applications, but of potential interested in this field is **BraSiC**™: several patents [97–99] and a few conference papers describe this family of brazing alloys based on metal silicides. Brazing is obtained by a pressureless process ranging between 1200°C and 1800°C. The main concern from a nuclear standpoint is the residual silicon in the brazing alloy which has been shown, in monolithic SiC, to reduce strength due to anisotropic swelling of the differing phases [100].

In conclusion, further activity is needed in order to test the joined materials near to relevant work conditions in fusion plants.

19.9.2 Joining of CMC for Ultrastable Structures

There are a few available joining materials with a low CTE, coefficient of moisture expansion (CME), and suitable mechanical strength, in order to guarantee the dimensional stability and mechanical reliability of joined C/C sandwich panels for ultrastable space instrumentation. Such instruments will require ultra-high stability, large structures (some square meters), and a near to zero CTE is necessary for both in-orbit and ground-and-orbit operations. Carbon/carbon composites have been selected as key materials for these instruments as their properties satisfy the application requirements. Some brazing alloys, polymeric precursors of ceramic and glass-ceramics have been tested [23, 101]: TiCuNi brazing alloy and pure silicon were successfully used, but the brazing process performed at very high temperature (1450°C) under controlled atmosphere was an issue from a practical point of view. The joining process performed by using a commercial carbon fiber adhesive modified by adding low CTE fillers such as fumed silica and a compound glass-ceramic powder (CTE $= -1.2 \times 10 - 6°C^{-1}$) was found to be the most promising solution for this application.

19.10 MECHANICAL TESTS ON JOINED CMC

19.10.1 Shear Strength Tests

A recognized problem within the scientific and technology community is the lack of standardization in mechanical testing of joined CMC. This leads to two extremely negative issues: first, an impossible comparison of potentially good joining solutions; second, project designers do not consider joining and integration issues of CMC into their projects because they do not have reliable mechanical strength data, shear in particular.

The asymmetrical four-point bending test [102] is currently the only one available standard test for the evaluation of the shear strength of joined ceramic matrix composites. The compression double-notch shear test is used to evaluate the shear strength of continuous fiber-reinforced ceramic matrix composites [103, 104] and seems appropriate for the determination of the shear strength of joined ceramic matrix composites at high temperature [105], but notches must be carefully machined into the joined section. If not, the obtained values are not representative of the joining strength but of the interlaminar shear strength of the CMC itself [106]. Lap tests in compression are useful for a comparison among differently joined CMC, but what they measure is just an "apparent" shear strength, including several other loading modes depending on the sample size.

In practice, a homogeneous shear stress state is never obtained within the joint in most of the test currently used, so that evaluating the shear strength of the same joining material with different test methods could lead to different results.

In Reference 106 the results of an experimental investigation on carbon/carbon composites (C/C) bonded joints tested in shear at room temperature, under seven different configurations, have been discussed: several difficulties appeared in using the asymmetrical four-point bending test. First, the small size of the specimen (4 mm × 3 mm × 36 mm) gives problems to obtain precisely machined specimen and a correct alignment of the two joined parts. Due to the small value of the inner span length (4 mm), the positioning of the loading pins is difficult and leads to scattered measured values. Second, following strictly the ASTM standard requires each specimen to be made by cutting the specimen from a larger (60-mm thick) joined block. This can be very expensive, due to the large amount of wasted material, cutting cost, and number of defective samples obtained during machining and cutting. It can be also simply impossible to perform when ceramic matrix composites of this size are not available. This test requires to measure both the flexural strengths of the base material (CMC) and the joined material (joined CMC). If the flexural strength of the joined materials is lower than 25% of the base material one, then it is possible to use asymmetrical four-point bending test. If it is between 25% and 50%, V- or U-shaped notches must be machined in the joint region before four-point asymmetrical bending testing, leading to other possible waste of material due to induced defects during notches machining. If the flexural strength of the joined material is higher than 50% of the base material one, the asymmetrical four-point bending test cannot be used; it means that in case of a successful joining material, it may be impossible to obtain its shear strength with this ASTM test. In any case, at least 10–15 base material specimens and 20–30 joined ones are necessary to obtain one reliable shear strength value.

The torsion test method is proposed in some ASTM standards but none of them is directly applicable to joined CMC [107]. These two types of tests have been adapted to joined CMC, first by preparing joined cylinders then joined hourglass-shaped samples, as in Reference 106. A pure but nonuniform shear stress distribution has been obtained in the joined area, with an unexpected large scatter in the results is probably due to the maximum stress occurring on the external surface of the joined sample, where surface defects are more likely to promote the propagation of fracture. Further experimental and modeling activity is requested to validate the use of torsion tests on joined CMC.

19.10.2 Nondestructive Tests

Several nondestructive testing methods, such as lock-in thermography, ultrasonic inspection, microtomography, and microradiography, are used to assess the manufacturing quality of joints between CMC themselves (e.g., C/C) and integrated with metals (e.g., Cu alloys) [108]. Results revealed that ultrasonic inspection is critical in the case of CMC integrated with metals, because of a significant difference in the acoustic impedance and sensitivity to irregular shaped joined surfaces; therefore, high ultrasonic echo exists even if the joint is good; microtomography and microradiography offer useful qualitative information on the joints, since CMC are significantly less X-ray sensitive than metals. Lock-in thermography offers information on thermal continuity at interface and can predict the component behavior under critical heat flux event, since it gives a global information about the soundness of the heat path, but not necessarily on the chemical continuity at the interface; the advantage of lock-in thermography is that it is not sensitive to irregular shaped joined surfaces and it can be also used for machined, joined, and integrated CMC.

Generally speaking, nondestructive test on joined and integrated CMC results should be at least initially validated by mechanical tests.

19.11 CONCLUDING REMARKS AND FUTURE DIRECTIONS

It would be extremely beneficial to have *industrials designers, material scientists, and engineers to work together* to find the best novel, reliable, and affordable approach and technologies for joining and integrating CMC to metals or polymers in order to ensure the integrity of the planned component throughout its design life, production, and in-service performance as well as end-of-life issues.

Since the strength of a joint depends on the characteristics of the interfaces between the connected materials, a consistent part of the future work should deal with *interface engineering*, with the purpose of establishing the desired conditions between bonding agent and materials to be joined.

Thermodynamic prediction of high temperature interaction and interface formation in hybrid systems, and advanced wettability studies supported by advanced computer simulations should be extensively used to select joining materials.

International cooperation should be devoted to prenormative *standardization* research of mechanical tests (shear, in particular) of joined CMC, in cooperation with international standard and measurements agencies.

Finally, *dismantling* of joined and integrated CMC components should be addressed, in order to obtain reusable surfaces and for a complete end-of-life analysis of joined and integrated CMC. At the moment, dismantling/disassembling research is focused on structural adhesives.

ACKNOWLEDGMENTS

Colleagues at Politecnico di Torino, Italy (S. Rizzo, M. Salvo, F. Smeacetto, A. Ventrella) are sincerely acknowledged for their valuable help in writing this chapter.

REFERENCES

1. W. Krenkel. *Ceramic Matrix Composites: Fiber Reinforced Ceramics and Their Applications* (John Wiley & Sons, Inc., New York, 2008).
2. K. Mergia. Joining of Cf/C and Cf/SiC composites to metals, in *Nanocomposites with Unique Properties and Applications in Medicine and Industry*, edited by Dr. John Cuppoletti (InTech, 2011), Ch. 12, p. 239. http://www.intechopen.com/books/nanocomposites-with-unique-properties-and-applications-in-medicine-and-industry/joining-of-cf-c-and-cf-sic-composites-to-metals, July 2014.
3. D. P. Sekulic. *Advances in Brazing Science, Technology and Applications* (Woodhead Publishing Limited, Sawston, Cambridge, 2013).
4. M. Singh, T. Ohji, R. Asthana, and S. Mathur. *Ceramic Integration and Joining Technologies: From Macro to Nanoscale* (John Wiley & Sons, Inc., New York, 2011).
5. M. Ortelt, F. Rühle, H. Hald, and H. Weihs. Dynamic qualification of a new CMC fastener. in *High Temperature Ceramic Matrix Composites*, edited by W. Krenkel, R. Naslain, and H. Schneider (Wiley-VCH, Weinheim, 2001), pp. 760–766.
6. A. R. Frost and J. D. Riehl. Ceramic matrix composites capped bolt attachment. US patent 7967562 (2011).
7. D. C. Jarmon. Internal pocket fastener system for ceramic matrix composite heat exchanger. US patent 20120168121 (2012).
8. R. J. Miller, M. E. Palusis, and D. C. Jarmon. Composite fastener for use in high temperature environments. US patent 6045310 (2000).
9. H. Böhrk and U. Beyermann. Secure tightening of a CMC fastener for the heat shield of re-entry vehicles. *Comp. Struct.*, 92(1), 107–112 (2010).
10. M. J. Verrilli, D. Brewer, and H. R. Gray. *Characterization of Ceramic Matrix Composite Fasteners Exposed in a Combustion Liner Rig Test. Presented at IGTI 2002* (ASME Turbo Expos Amsterdam, The Netherlands, 2002).
11. L. L. Snead, T. Nozawa, M. Ferraris, Y. Katoh, R. Shinavski, and M. Sawan. Silicon carbide composites as fusion power reactor structural materials. *J. Nucl. Mat.*, 417(1), 330–339 (2011).
12. T. Hinoki and A. Kohyama. Current status of SiC/SiC composites for nuclear applications. *Ann. Chim. Sci. Mat.*, 30(6), 659–671 (2005).
13. T. Nozawa, T. Hinoki, A. Hasegawa, A. Kohyama, Y. Katoh, L. L. Snead, C. H. HenagerJr., and J. B. J. Hegeman. Recent advances and issues in development of silicon carbide composites for fusion applications. *J. Nucl. Mat.*, 386–388, 622–627 (2009).
14. Y. Katoh, A. Kohyama, T. Nozawa, and M. Sato. SiC/SiC composites through transient eutectic-phase route for fusion applications. *J. Nucl. Mat.*, 329—333 (1–3 PART A), 587–591 (2004).
15. J. Kweon, J. W.Jung, T. H. Kim, J. H.Choi, and D. H. Kim. Failure of carbon composite-to-aluminum joints with combined mechanical fastening and adhesive bonding. *Comp. Struct.*, 75(1–4), 192–198 (2006).
16. M. Salvo, S. Rizzo, V. Casalegno, K. Handrick, and M. Ferraris. Shear and bending strength of SiC/SiC joined by a modified commercial adhesive. *Int. J. Appl. Ceram. Technol.*, 9(4), 778–785 (2012).

17. J. Zhang, R. Luo, M. Jiang, Q. Xiang, and J. Li. The preparation and performance of a novel room-temperature-cured heat-resistant adhesive for ceramic bonding. *Mat. Sci. Eng. A*, 528, 2952–2959 (2011).
18. Xiaozhou Wang, Jun Wang, and Hao Wang. Synthesis of a novel preceramic polymer (V-PMS) and its performance in heat-resistant organic adhesives for joining SiC ceramic. *J. Eur. Ceram. Soc.*, 32(12), 3415–3422 (2012)
19. E. A. S. Marques, D. N. M. Magalhes, and L. F. M. da Silva. Experimental study of silicone-epoxy dual adhesive joints for high temperature aerospace applications. *Mat.-wiss. u.Werkstofftech.*, 42(5) (2011).
20. Y. Zhang, R. Luo, J. Zhang, and Q. Xiang. The reinforcing mechanism of carbon fiber in composite adhesive for bonding carbon/carbon composites. *J. Mater. Proc. Tech.*, 211, 167–173 (2011).
21. J. Zhang, R. Luo, and C. Yang. A multi-wall carbon nanotube-reinforced high-temperature resistant adhesive for bonding carbon/carbon composites. *Carbon*, 50(13), 4922–4925 (2012).
22. V. K. Srivastava. Effect of carbon nanotubes on the strength of adhesive lap joints of C/C and C/C–SiC ceramic fibre composites. *Int. J. Adhes. Adhes.*, 31(6), 486–489 (2011).
23. M. Salvo, V. Casalegno, Y. Vitupier, L. Cornillon, L. Pambaguian, and M. Ferraris. Study of joining of carbon/carbon composites for ultra-stable structures. *J. Eur. Ceram. Soc.*, 30, 1751–1759 (2010).
24. J. Li, R. Luo, Y. Bi, Q. Xiang, C. Lin, Y. Zhang, and N. An. The preparation and performance of short carbon fiber reinforced adhesive for bonding carbon/carbon composites. *Carbon*, 46(14), 1957–1965 (2008).
25. C. H. Henager, Y. Shin, Y. Blum, L. A. Giannuzzi, B. W. Kempshall, and S. M. Schwarz. Coating and joining for SiC and SiC-composites for nuclear energy systems. *J. Nucl. Mat.*, 367–370, 1139–1143 (2007).
26. http://www.eurobond-adhesives.co.uk/ultra-high-temperature-adhesives.html (accessed October 15, 2012).
27. www.cotronics.com (accessed October 15, 2012).
28. www.masterbond.com (accessed October 15, 2012).
29. www.loctite.co.uk (accessed October 15, 2012).
30. http://www.inductionatmospheres.com/brazing/alloy-suppliers.html (accessed October 15, 2012).
31. A. Passerone and M. L. Muolo. Metal-ceramic interfaces: wetting and joining processes. *Int. J. Mater. Prod. Tec.*, 20(5–6), 420–439 (2004).
32. A. Passerone. Interfacial phenomena in metal-ceramic systems. *Mater. Chem. Phys.*, 15(3–4), 263–279 (1986).
33. N. Eustathopoulos, D. Chatain, and L. Coudurier. Wetting and interfacial chemistry in liquid metal-ceramic systems. *Mater. Sci. Eng. A*, 135(C), 83–88 (1991).
34. F. Moret and N. Eustathopoulos. Ceramic to metal direct brazing. *J. Phys-Paris*, 3(7), 1043–1042 (1993).
35. B. Drevet and N. Eustathopoulos. Wetting of ceramics by molten silicon and silicon alloys: a review. *J. Mater. Sci.*, 47, 8247–8260 (2012).
36. www.jm-metaljoining.com (accessed October 15, 2012).
37. http://www.silvaloy.com/brazing_table.php (accessed October 15, 2012).
38. G. Blugan, J. Kuebler, V. Bissig, and J. Janczak-Rusch. Brazing of silicon nitride ceramic composite to steel using SiC-particle-reinforced active brazing alloy ceram. *Ceram. Int.*, 33, 1033–1039 (2007).
39. G. Lin, J. Huang, and H. Zhang. Joints of carbon fiber-reinforced SiC composites to Ti-alloy brazed by Ag–Cu–Ti short carbon fibers. *J. Mater. Process. Tech.*, 189(1–3), 256–261 (2007).
40. M. Singh and R. Asthana. Brazing of advanced ceramic composites: issues and challenges. *Ceram. Trans.*, 198, 9–14 (2007).
41. M. Singh, T. Matsunaga, H.-T. Lin, R. Asthana, and T. Ishikawa. Microstructure and mechanical properties of joints in sintered SiC fiber-bonded ceramics brazed with Ag-Cu-Ti alloy. *Mater. Sci. Eng. A*, 557, 69–76 (2012).
42. M. Singh, R. Asthana, and T. Shpargel. Brazing of carbon–carbon composites to Cu-Clad molybdenum for thermal management applications. *Mater. Sci. Eng. A*, 452–453, 699–704 (2007).
43. M. Singh. Design, fabrication and characterization of high temperature joints in ceramic composites. *Key Eng. Mat.*, 415, 164–165 (1999).
44. M. Singh, R. Asthana, and T. Shpargel. Brazing of ceramic-matrix composites to Ti and hastealloy using Ni-base metallic glass interlayers. *Mat. Sci. Eng. A*, 498, 19–30 (2008).
45. R. Asthana and M. Singh. Joining of ZrB2-based ultra-high-temperature ceramic composites using Pd-based braze alloys. *Scrip. Mater.*, 61(3), 257–260 (2009).
46. R. Asthana and M. Singh. Joining of partially sintered alumina to alumina, titanium, hastealloy and C–SiC composite using Ag–Cu brazes. *J. Eur. Ceram.*, 28(3), 617–631 (2008).
47. W. Krenkel and P. Schanz. Fiber ceramic structures based on liquid impregnation technique. *Acta Astronaut.*, 28(C), 59–169 (1992).
48. W. Krenkel, T. Henke, and N. Mason. In-situ joined CMC components. *Key Eng. Mat.*, 127–131, 313–320 (1997).

49. R. Renz, G. Seifert, and W. Krenkel. Integration of CMC brake disks in automotive brake systems. *Int. J. Appl. Ceram. Tec.*, 9(4), 712–724 (2012).
50. M. Singh, J. D. Kiser, and S. C. Farmer. Joining of SiC-based ceramics by reaction forming approach. *Ceram. Eng. Sci. Proc.*, 18(3), 161 (1997).
51. M. Singh. A reaction forming method for joining of silicon carbide-based ceramics. *Scrip. Mater.*, 37(8), 1151–1154 (1997).
52. G. O. CookIII and C. D. Sorensen. Overview of transient liquid phase and partial transient liquid phase bonding. *J. Mater. Sci.*, 46(16), 5305–5323 (2011).
53. K. Kakegawa and A. M. Glaeser. Transient FGM joining of silicon carbide ceramics: a feasibility study. *Compos. Part B-Eng.*, 28(1–2), 85–91 (1997).
54. A. M. Glaeser. The use of transient FGM interlayers for joining advanced ceramics. *Compos. Part B-Eng.*, 28(1–2), 71–84 (1997).
55. G. W. Liu, F. Valenza, M. L. Muolo, and A. Passerone. SiC/SiC and SiC/Kovar joining by Ni-Si and Mo interlayers. *J. Mater. Sci.*, 45(16), 4299–4307 (2010).
56. N. Saito, H. Ikeda, Y. Yamaoka, A. M. Glaeser, and K. Nakashima. Wettability and transient liquid phase bonding of hafnium diboride composite with Ni-Nb alloys. *J. Mater. Sci.*, 47(24), 8454–8463 (2012).
57. K. Shimoda, T. Hinoki, and A. Kohyama. Effect of additive content on transient liquid phase sintering in SiC nanopowder infiltrated SiCf/SiC. *Compos. Sci. Technol.*, 71(5), 609–615 (2011).
58. W. Chen and Y. Miyamoto. Adhesion properties of tungsten and SiC ceramic-bonded carbon. *Int. J. Refract. Met. H.*, 35, 279–283 (2012).
59. E. Jacques, L. Maillé, Y. LePetitcorps, C. Lorrette, and C. Sauder. Preliminar results on joining of thin SiC/SiC composites by silicides compounds and local heating. *Ceram. Eng. Sci. Proc.*, 32(9), 73–83 (2011).
60. W. R. Pinc, M. Di Prima, L. S. Walker, Z. N. Wing, and E. L. Corral. Spark plasma joining of ZrB_2-SiC composites using zirconium-boron reactive filler. *J. Am. Ceram. Soc.*, 94(11), 3825–3832 (2011).
61. S. Grasso, Nanoforce Ltd. UK, private communication (2012).
62. S. Aravindan and R. Krishnamurthy. Microwave joining of Al2O3-ZrO2 composites. *Ceram. Eng. Sci. Proc.*, 20(4), 71–78 (1999).
63. S. Das, A. K. Mukhopadhyay, S. Datta, and D. Basu. Prospects of microwave processing: an overview. *B. Mater. Sci.*, 32(1), 1–13 (2009).
64. J. P. Singh, K. Y. Lee, E. D. Case, and D. Reinhard. Chapter 64: Microwave joining and repair of ceramics and ceramic composites, in *Proceedings of the 21st Annual Conference on Composites, Advanced Ceramics, Materials, and Structures—B*, Ceramic Engineering and Science Proceedings, Volume 18, Issue 4 (John Wiley & Sons, 2008).
65. I. Ahmad, R. Silberglitt, Y. L. Tian, and J. D. Katz. Microwave joining of sic ceramics and composites, in *Microwaves: Theory and Application in Materials Processing IV*, Ceramic Transactions 80, edited by D. E. Clark, W. H. Sutton, and D. A. Lewis (The American Ceramic Society, Westerville, OH, 1997), pp. 455–450.
66. http://www.fm-technologies.com/joining_info.html. (accessed October 15, 2012).
67. S. Aravindan and R. Krishnamurthy. Joining of ceramic composites by microwave heating. *Mater. Lett.*, 38(4), 245–249 (1999).
68. W. Lippmann, J. Knorr, R. Wolf, R. Rasper, H. Exner, A. M. Reinecke, M. Nieher, and R. Schreiber. Laser joining of silicon carbide—a new technology for ultra-high temperature resistant joints. *Nucl. Eng. Des.*, 231(2), 151–161 (2004).
69. W. Lippmann, M. Herrmann, C. Hille, and A. Hurtado. Laser joining of ceramics—a contribution to high temperature range application of ceramic components, in *Proceedings of the International Conference on Nuclear Engineering*, ICONE 1 (Springer, 2008), pp. 827–834.
70. M. Ferraris, C. Badini, M. Montorsi, P. Appendino, and H. W. Scholz. Joining of SiCf/SiC composites for thermonuclear fusion reactors. *J. Nucl. Mater.*, 212—215 (PART B), 1613–1616 (1994).
71. M. Salvo, M. Ferraris, and P. Lemoine. Joining of CMC's for thermonuclear fusion applications. *J. Nucl. Mat.*, 233–237(PART II), 949–953 (1996).
72. C. Isola, M. Salvo, M. Ferraris, and M. A. Montorsi. Joining of surface modified carbon/carbon composites using a barium-aluminum-boro-silicate glass. *J. Eur. Ceram. Soc.*, 18(8), 1017–1024 (1998).
73. M. Ferraris, M. Salvo, C. Isola, M. Appendino Montorsi, and A. Kohyama. Glass-ceramic joining and coating of SiC/SiC for fusion applications. *J. Nucl. Mat.*, 258–263(PART 2 B), 1546–1550 (1998).
74. Y. Katoh, M. Kotani, A. Kohyama, M. Montorsi, M. Salvo, and M. Ferraris, Microstructure and mechanical properties of low-activation glass-ceramic joining and coating for SiC/SiC composites. *J. Nucl. Mater.*, 283–287(PART II), 1262–1266 (2000).
75. H. Peng, Q. G. Fu, B. L. Jia, and H. J. Li. Thermal shock resistance of SiC/MAS-LAS multilayer joint of carbon/carbon composites to LAS glass-ceramic. *Mater. Sci. Eng. A-Struct.*, 556, 607–611 (2012).

76. K.-Z. Li, J. Wang, X. B. Ren, H. J. Li, W. Li, and Z. Q. Li. The preparation and mechanical properties of carbon-carbon/lithium-aluminum-silicate composite joints. *Mater. Design.*, 44, 346–353 (2013).
77. L. Esposito and A. Bellosi. Joining ZrB2-SiC composites using glass interlayers. *J. Mater. Sci.*, 40(17), 4445–4453 (2005).
78. M. Salvo, V. Casalegno, A. Ciampichetti, F. Smeacetto, M. Zucchetti, and M. Ferraris. Joining of machined SiC/SiC composites for thermonuclear fusion reactors. *J. Nucl. Mat.*, 375, 410–415 (2008).
79. R. Radhakrishnan, C. H Henager Jr., J. L. Brimhall, and S. B. Bhaduri. Synthesis of Ti_3Si_2/SiC and $TiSi_2$/SiC composites using displacement reactions in the Ti-Si-C system. *Scrip. Mater.*, 34(12), 1809–1814 (1996).
80. C. H. Henager, Y. Shin, Y. Blum, L. A. Giannuzzi, B. W. Kempshall, and S. M. Schwarz. Coating and joining for SiC and SiC-composites for nuclear energy systems. *J. Nucl. Mater.*, 367–370, 1139–1143 (2007).
81. C. A. Lewinsohn, R. H. Jones, P. Colombo, and B. Riccardi. Silicon carbide-based materials for joining silicon carbide composites for fusion energy applications. *J. Nucl. Mater.*, 307–311, 1232–1236 (2002).
82. P. Colombo, B. Riccardi, A. Donato, and G. Scarinci. Joining of SiC/SiCf ceramic matrix composites for fusion reactor blanket applications. *J. Nucl. Mater.*, 278, 127–135 (2000).
83. C. A. Lewinsohn, P. Colombo, and I. Reimanis. Stresses occurring during joining of ceramics using preceramic polymers. *J. Am. Ceram. Soc.*, 84(10), 2240–2244 (2001).
84. M. K. Brun. Formation of tough composite joints. *J. Am. Ceram. Soc.*, 81(12), 3307–3312 (1998).
85. H. Tokgoz, A. Ozcifci, M. Atar, and B. Uysal, Shear and bending strength of some end to end grained joints prepared from scotch pine, *Turk. J. Agric. For.*, 23, 621–625 (1999).
86. M. Ferraris, M. Salvo, and V. Casalegno. Joining and integration issues of ceramic matrix composites for the nuclear industry, in *Ceramic Integration and Joining Technologies: From Macro to Nanoscale*, edited by M. Singh, T. Ohji, R. Asthana, and S. Mathur (John Wiley & Sons Inc., 2011), pp. 39–89.
87. M. Ferraris, M. Salvo, V. Casalegno, S. Rizzo, and A. Ventrella. Joining and integration issues of ceramic matrix composites for nuclear applications, in *Ceramic Transactions 220*, edited by N. P. Bansal, J. P. Singh, J. Lamon, S. R. Choi, M. M. Mahmoud (John Wiley & Sons Inc., 2010), pp. 173–186.
88. L. L. Snead and M. Ferraris. Carbon as a fusion plasma-facing material, in *Comprehensive Nuclear Materials* (Elsevier, Konings R.J.M, Amsterdam, 2012), pp. 583–620.
89. P. Appendino, M. Ferraris, V. Casalegno, M. Salvo, M. Merola, and M. Grattarola. Proposal for a new technique to join CFC composites to copper. *J. Nucl. Mater.*, 348, 102–107 (2006).
90. V. Casalegno, M. Salvo, and M. Ferraris. Surface modification of carbon/carbon composites to improve their wettability by copper. *Carbon*, 50(6), 2296–2306 (2012).
91. G. Pintsuk, V. Casalegno, M. Ferraris, T. Koppitz, and M. Salvo. Thermal fatigue characterization of CFC divertor modules using a one step brazing process. *J. Nucl. Mat.*, 426(1–3), 78–84 (2012).
92. L. Giancarli, H. Golfier, S. Nishio, R. Raffray, C. Wong, and R. Yamada. Progress in blanket designs using SiCf/SiC composites. *Fus. Eng. Des.*, 61–62, 307–318 (2002).
93. C. H. Henager Jr., Y. Shin, Y. Blum, L. A. Giannuzzi, B. W. Kempshall, and S. M. Schwarz. Coatings and joining for SiC and SiC-composites for nuclear energy systems. *J. Nucl. Mater.*, 367–370, 1139–1143 (2007).
94. M. Ferraris, M. Salvo, V. Casalegno, S. Han, Y. Katoh, H. C. Jung, T. Hinoki, and A. Kohyama. Joining of SiC-based materials for nuclear energy applications. *J. Nucl. Mater.*, 417(1–3), 379–382 (2011).
95. T. Hinoki and A. Kohyama. Current status of SiC/SiC composites for nuclear applications. *Ann. Chim. Sci. Mater.*, 30(6), 659–671 (2005).
96. M. Ferraris, V. Casalegno, S. Rizzo, M. Salvo, T. O. Van Staveren, and J. Matejicek. Effects of neutron irradiation on glass ceramics as pressure-less joining materials for SiC based components for nuclear applications. *J. Nucl. Mater.*, 429, 166–172 (2012).
97. A. Gasse. Method for assembling parts made of materials based on SiC by non-reactive refractory brazing, brazing composition and joint and assembly obtained by said method. US patent 7,318,547 B2 (2008).
98. V. Chaumat and J. F. Henne. Method of joining parts made of SiC-based materials by non-reactive brazing with addition of a reinforcement, brazing compositions and joint and assembly that are obtained by such a method. WO2011 113760 (2011).
99. V. Chaumat, J. F. Henne, and N. Miloud-Ali. Method of joining parts made of SiC-based materials by non-reactive brazing, brazing compositions and joint and assembly that are obtained by this method. WO2011 113758 (2011).
100. J. C. Corelli, J. Hoole, J. Lazzaro, and C. W. Lee. Mechanical, thermal, and microstructural properties of neutron-irradiated SiC. *J. Am. Ceram. Soc.*, 66, 529 (1983).
101. M. Lutz, L. Cornillon, Y. Vitupier, and L. Pambaguian. Evaluation of ultrastable carbon/carbon sandwich structures joined with ceramic cement, in *61st International Astronautical Congress* (Prague, CZ, 2010), pp. 4829—4832.

102. ASTM C1469–00. Standard test method for shear strength of joints of advanced ceramics at ambient temperature, 2005.

103. E. Lara-Curzio and M. K. Ferber. Shear strength of continuous fiber reinforced ceramic composites, in *Thermal and Mechanical Test Methods and Behavior of Continuous Fiber Ceramic Composites*, edited by M. G. Jenkins, S. T. Gonczy, E. Lara-Curzio, N. E. Ashbaugh, and L. P. Zawada (West Conshohocken, PA: American Society for Testing and Materials, 1997), pp. 31–49.

104. ASTM STP 1309. W. Conshohocken, PA: American Society for Testing and Materials; 1997.

105. M. Singh and E. Lara-Curzio. Design, fabrication and testing of ceramic joints for high temperature SiC/SiC composites. NASA/CR-2000-209809, 2000.

106. M. Ferraris, A. Ventrella, M. Salvo, M. Avalle, F. Pavia, and E. Martin. Comparison of shear strength tests on av119 epoxy-joined carbon/carbon composites. *Compos. Part B-Eng.*, 41(2), 182–191 (2010).

107. F1362-97. Standard test method for shear strength and shear modulus of aerospace glazing interlayer materials, 2003.

108. V. Casalegno, M. Salvo, M. Ferraris, F. Smeacetto, M. Merola, and M. Bettuzzi. Non-destructive characterization of carbon fiber composite/Cu joints for nuclear fusion applications. *Fusion Eng. Des.*, 83, 702–712 (2008).

PART VI NONDESTRUCTIVE EVALUATION

CHAPTER 20

USE OF ACOUSTIC EMISSION FOR CERAMIC MATRIX COMPOSITES

GREGORY N. MORSCHER[1] AND NATHALIE GODIN[2]

[1] *Department of Mechanical Engineering, The University of Akron, Akron, OH, USA*
[2] *Institute of Applied Science, Mateis Laboratory, Batiment Blaise Pascal, Villeurbanne Cedex, France*

20.1 INTRODUCTION/BACKGROUND

Acoustic emission (AE) is a technique that acquires transient sound waves produced by mechanical events typically associated with the creation of local damage or failure. It has been used to detect the formation of mechanical damage in ceramic matrix composites (CMCs) for over 20 years [1]. CMCs have multiple damage mechanisms including transverse matrix cracks, longitudinal cracks, fiber breakage, and/or various sliding mechanisms, whether fibers or interlaminar in nature, which all emit sounds when they occur, resulting in large quantities of AE during typical tensile tests which has made it difficult to quantify source mechanisms. However, more recently with the improvement in computer technology, software development and sorting techniques, location, and damage source identification, quantification has improved dramatically, enabling AE to become a powerful technique for monitoring damage and assessing the health of a specimen or structure.

In most CMCs, sound travel is quite good and can be detected tens of centimeters from the source. Early on, it was shown that AE was detected at very low stresses corresponding to matrix microcracking during a tensile test which was well below any discernible nonlinearity in the stress–strain curve or discernible from microscopic observation [1–4], which highlights one of AE's most useful attributes: the ability to detect low stress damage. As computer technology has increased in acquisition speed and storage capability, the characteristics of the actual waveforms can be better measured, resulting in better identification of different damage sources and how they relate to mechanical behavior, failure, and life.

AE is based on similar principles as ultrasonic or acousto-ultrasonic methods [5]; however, the fundamental difference between AE and ultrasonic testing is the absence of external wave excitation since the emissions originate from the fracture events within the materials. The ultrasonic methods, for example, are able to detect the presence or even shape of a defect using an artificially generated source (sound) signal and a receiver, whereas AE detects the elastic waves radiated by a growing fracture. The AE technique identifies defects only while they develop under stress or other degradations factors. So, AE consists of the recording and analysis of elastic waves that are generated during material damage, and, if done properly, can give real-time information on initiation and evolution of damage.

Ceramic Matrix Composites: Materials, Modeling and Technology, First Edition. Edited by Narottam P. Bansal and Jacques Lamon.

20.2 AE PRINCIPLES AND PRACTICE

A characteristic of most CMCs is that they produce a large amount of AE activity over the entire volume of the specimen. It is critical that the AE used for analysis is from the volume of interest, for example, the gage section of a tensile bar. It has been shown for typical AE tensile tests on dogbone specimens that over 70% of the AE activity recorded can come from the region outside the 25 mm gage section, much of that coming from the grip or near-grip region [6]. Therefore it is essential to locate AE and only use the AE which comes from the volume of interest. AE is not a "black box" technique; however, understanding several principles of sound creation and transmission and applying that understanding enables effective analysis of AE activity as it pertains to location and identification of damage events.

When a mechanical, AE source event occurs, for example, a fracture or sliding, sound is created as part of the energy released due to a pressure acting on some volume. AE is the sound that is able to be measured from the event on one or more sensors. The sound wave received from an event on a given sensor is often referred to as a "hit". For composites, there are many event or damage sources due to the multiple fracture and sliding mechanisms associated with the damage in the composites, for example, matrix cracking, interlaminar debonding and/or sliding, fiber sliding, fiber breakage, and so on. AE occurs at a source but also must travel some distance through the composite to the sensors. Therefore, the composite is both what is being damaged as well as a waveguide of the acoustic waves produced from the damage. Therefore, AE is dependent on three factors.

- The source of the emission (i.e., damage event) will dictate the modes of sound waves created as well as frequency content.
- The physical properties, E/ρ (elastic modulus/density) of the material the sound must travel through will also determine which modes and frequencies can transmit easily or poorly as well as the speed of the sound waves. In addition, porosity which also affects E/ρ may also increase the damping of acoustic signals.
- The geometry of the structure that the sound must travel through also has a strong effect on sound travel. Most CMC structures are thin-plate structures, that is, one dimension (thickness) is much smaller than the wavelength [7]. For thin plates, only two waveform modes are possible: the extensional (longitudinal) wave and the flexural wave. The former is analogous to pressure (dilatational) waves in infinite and semi-infinite materials. There are several key features to these two waveforms which will be discussed below. In addition, geometry and other structural features such as holes, notches, corners, joints, and so on, also affect reflections and transmission of sound.

The "traditional" approach to AE was to process data as damped sine curves [8]. Typically, resonant frequency transducers were used (or preamps with frequency filters), which force the AE waveform into a damped sine wave. Figure 20.1a shows a typical narrow frequency waveform. Traditional AE (TAE) would then process the damped sine

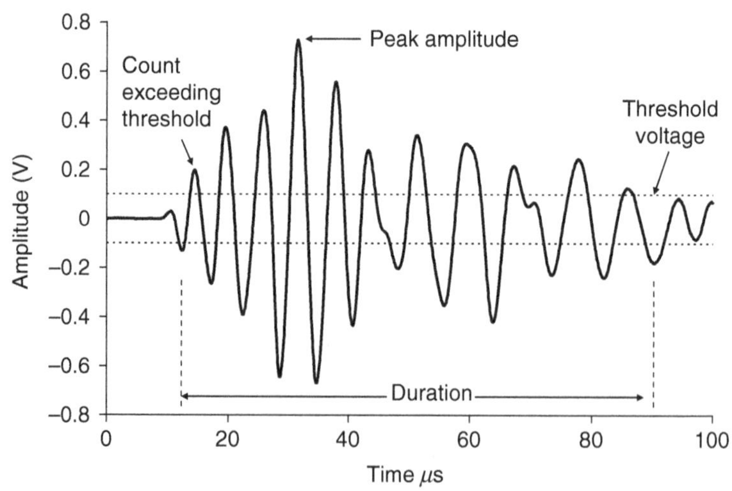

FIGURE 20.1

Typical resonant frequency or narrow frequency range filtered waveform with typical AE parameters.

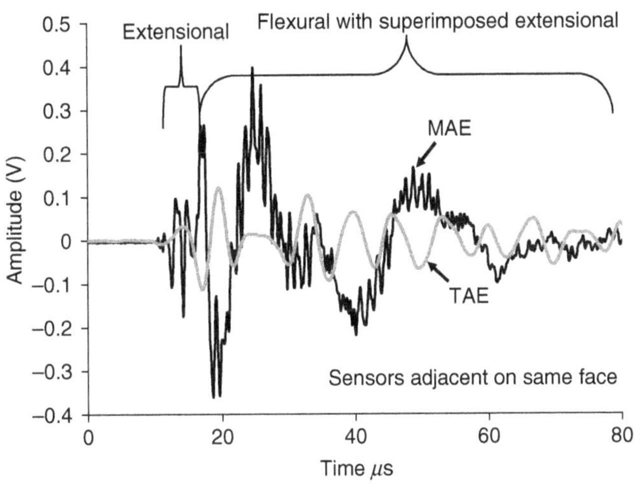

FIGURE 20.2

The same waveform from a woven SiC/SiC CMC fracture event captured on wide-band (50–2000 kHz) transducers (MAE) and narrower-band (100–800 kHz) transducers with a typical 100–300 kHz frequency filter in the preamplifier (TAE). The transducers were adjacent to one another, the center of each transducer separated by 7 mm, so that the waveform hit the MAE sensor first.

wave into specific parameters, for example, peak amplitude, duration of peaks above a threshold, rise time from first peak to amplitude, energy, and so on. This enabled AE data to be stored as text files without saving the waveforms digitally, which was necessary with older computers due to data processing and memory storage limitations. Today, however, those limitations do not exist and modern AE equipment enables the full waveform and the whole frequency spectrum to be captured (unless resonant frequency transducers or frequency filters are used).

The digital capture of full frequency spectrum waveforms is often referred to as "modal acoustic emission" (MAE) [9–11] because the different frequency content of the waveform typically corresponds to different modes of wave propagation. Figure 20.2 shows a comparison of the exact same AE waveform as processed for a TAE and MAE system. Note that the content and shape of the waveform are considerably different and that the MAE waveform contains considerably more information. See Section 20.3. For typical CMC specimens where the thickness is less than a few millimeters, the specimen can be considered a thin plate, that is, one dimension is much smaller than the wavelength of the traveling wave (typical wavelengths for the frequencies that are transmitted in CMCs are on the order of 1–2 cm). Characteristics of the two thin-plate waveforms, extensional and flexural, which are useful for analysis and understanding of AE are as follows:

- Extensional waves are always higher frequencies; are always faster in speeds of sound, and are essentially nondispersive (speed of sound is not or only mildly dependent on frequency).
- Flexural waves always have lower frequencies, always slower than extensional waves, and are highly dispersive.

Figure 20.3 shows a dispersion curve for a CVI SiC matrix composite [12], which shows the frequency dependence for speed of sound for the different waveforms. Also shown on the figure are typical frequency ranges for extensional and flexural waveforms.

To illustrate the nature of sound travel in CMCs, Figure 20.4 shows a typical tensile test setup with three sensors (sensor 3 in the gage section) of a tensile bar. An event occurred in or near the bottom grip and the sound travelled up the specimen where it hit sensor 2, then sensor 3, and then sensor 1. Note that the lower amplitude, higher frequency region at the beginning of the waveform recorded on each sensor increases in the time domain as the sensors are farther away from the source event. This is due to the fact that the extensional wave speed is faster and more extensional wave content "pulls away" from the flexural wave content the farther the distance travelled (as in a marathon race where the faster runners pull away farther from the pack the further into the race). Also note that the first peak occurs later in time the further from the source. If the first peak can be discerned, for example, from a threshold-crossing technique

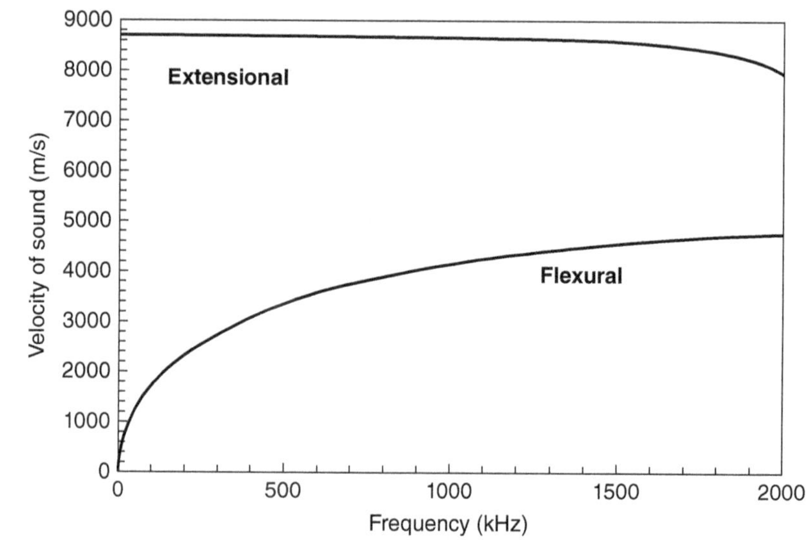

FIGURE 20.3

Dispersion curve for woven Hi-Nicalon reinforced CVI SiC matrix composite calculated assuming isotropic plate theory where $E = 200$ GPa and $n = 0.15$. From Reference 12.

or a peak signal-to-noise ratio technique, and if the distance between sensors is known, the speed of sound can then easily be determined from the distance divided by the difference in times of the first peaks.

Ideally then, the protocol for analyzing AE data is to first sort out the AE that only comes from the desired location (e.g., gage section) then to analyze the AE events from the region of interest for general waveform features such as energy or amplitude or more detailed features such as frequency content, which can then be related to damage sources. There appear to be two predominant approaches to AE acquisition and analysis as applied to CMCs. In the following, these current two most reported approaches in the literature will be highlighted. The first is an "event-based" approach to AE acquisition and analysis (Section 20.3). The second is a "signal-" or "hit"-based approach where pattern recognition techniques are used to characterize and classify sources (Section 20.4).

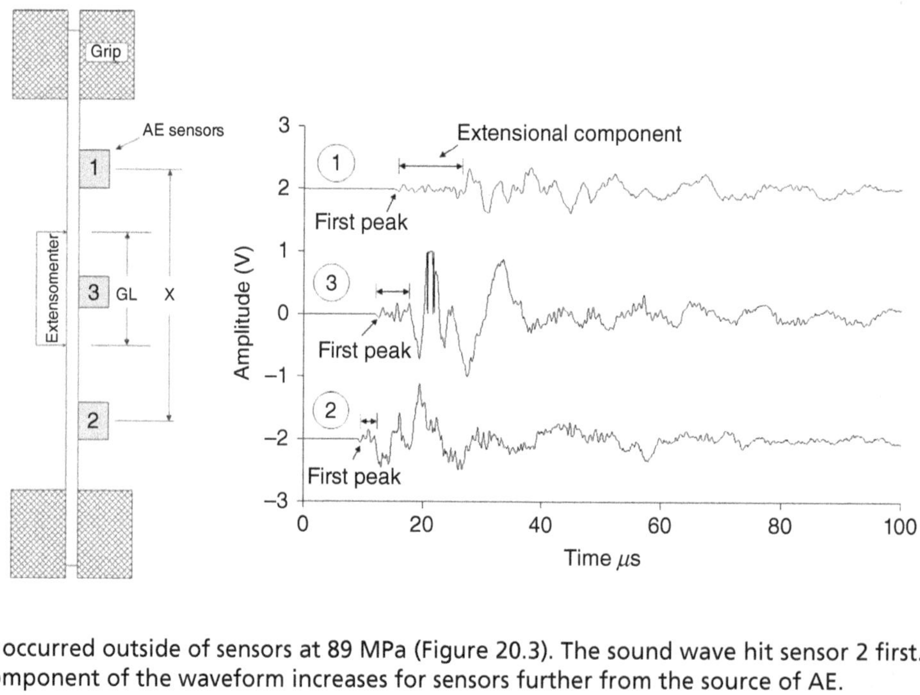

FIGURE 20.4

AE event that occurred outside of sensors at 89 MPa (Figure 20.3). The sound wave hit sensor 2 first. Note how the extensional component of the waveform increases for sensors further from the source of AE.

20.3 EVENT-BASED AE MONITORING OF CMCs

A sound wave produced by a damage source transmits sound in all directions which is received on however many sensors one uses in the experimental setup. For Figures 20.4 and 20.5, three sensors are used for that tensile test and for every event, three different signals (hits or waveforms) are received from that event on the three different sensors which are in different geometric locations on the specimen. The key emphasis and attribute of event-based AE is the ability to locate source events based on speeds of sound and location of sensors, which enables sorting out of AE activity in the desired region of a specimen, and then to verify estimated locations from saved waveforms if needed. Waveform features can then be analyzed of the entire event or of a waveform of a particular sensor of the event depending on if one is interested in the channel that is closest to or perhaps farthest away from the source. In some cases, one may be interested in the nature of sound travel from a sensor to another from a given source (e.g., attenuation of the waveform or particular frequency ranges).

An important experimental detail for event-based AE data collection is that the channels receiving the sound waves from the sensors are synchronized (i.e., when one channel is triggered, all channels record and save the waveform data simultaneously at exactly the same time). In this way, the time domains for each sensor are in sync with or relative to one another. Most AE systems, with the exception of the Digital Wave (Centennial, Colorado) by default have channels which trigger independent of one another. This is fine for large sensor separation distances, for example, pressure vessels. However, for high E/ρ materials such as CMCs (5000 to over 10 000 m/s) and sensor separation on the order of less than 10 cm, differences in times of arrival are very small between the sensors (μs). In order for sensors to acquire the next set of data after saving a given waveform or computing the AE parametrics, there is a lockout time where the computer has to rearm the channel for collecting the next set of data from the sensor. A consequence of independently triggered channels is that the first (closest to) sensor that triggers from a given source will be displaced in time from the next sensor(s) to trigger. If sensors act independently, and event acquisition rates approach a hundred events per second (not uncommon for a CMC tensile test), a condition can easily occur where one sensor's locked out period of time overlaps with another sensor's data capture time, rendering that data useless, since the location of the source cannot be determined (at least two sensors are required for linear location and three sensors for planar location). If sensors are synchronized, this cannot occur; however, it is of course possible to lose AE data during channel rearming.

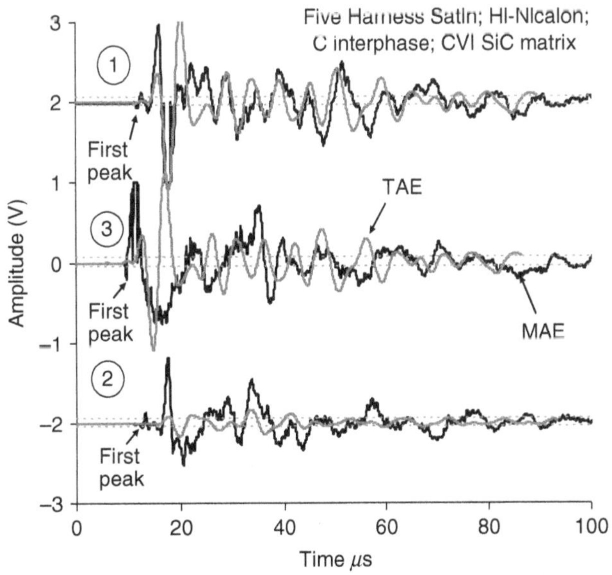

FIGURE 20.5

A typical event that occurred very near the center of the gage section. Wide-band frequency sensors were used; however, a splitter was used so that the wave went through a line driver to a typical waveform-analysis device and the signal went through a preamplifier with a 100–300 kHz filter to a traditional AE device. The dashed lines correspond to a typical threshold voltage (50 mV).

If one desires to determine exact location of AE events, it is important to be able to discern the first peak of the extensional wave. Flexural wave speeds of sound can vary by a factor of two over the frequency ranges of typical waveforms which mean that location estimates can be very inaccurate. Typical location analysis are based on the time (arrival time) at which the waveform crosses a threshold voltage. For linear location, for example, along a tensile bar using two sensors, the location can be determined from

$$\text{location} = x/2(\Delta t/\Delta t_x) \tag{20.1}$$

where location = 0 is the center of the gage section, x is the separation distance of the outer two sensors (Figure 20.4), Δt is the difference in arrival times for the two sensors that receive the signal from the given event, and Δt_x is the difference in arrival times for the two sensors for an event that occurs outside the two sensors, that is, x/C_e, where C_e is the velocity of the extensional wave. Care must be taken to setting the threshold voltage for location analysis. If the threshold voltage setting is set too high, the threshold crossing will occur at a time that corresponds to the flexural wave and not the extensional wave; however, if the threshold voltage setting is too low, background noise interference can also give erroneous arrival time values.

Other approaches to isolate AE for a given region of the specimen include the use of guard sensors. For example, in the sensor setup in Figures 20.4 and 20.5, three sensors are used on a tensile bar with the middle sensor in the center of the specimen. If the sensor separation in sensor 3 is a distance equal to "GL" (Figure 20.4) from sensor 1 and sensor 2, then if an AE event hits sensor 3 first (Figure 20.5), that event would occur in the gage section whereas events which hit sensors 1 or 2 first (e.g., Figure 20.4) would occur outside the gage. AE data can then easily be sorted for analysis according to the AE event data that triggered sensor 3 as opposed to event data that triggered sensors 1 or 2.

The key to event-based AE is that the waveforms from each sensor pertaining to the individual event are saved simultaneously. This affords easy verification of whether or not waveforms are actually occurring, where the software sorting routine estimates them to be.

20.3.1 Event-Based AE of CMC Stress–Strain Behavior

An example of AE monitoring during a room temperature tensile test is shown in Figure 20.6. Figure 20.6a shows the stress–strain curve for an unload/reload hysteresis test to failure as well as the individual event locations within the gage section (extensometer knife edges) with different energy ranges and the cumulative AE energy of each event over the entire test (these data are from Reference 6, though analyzed in greater detail than presented there). Note that significant AE activity occurs for strains above 0.035% up to about 0.12% (60–130 MPa), which corresponds to the range where transverse matrix cracking occurs. Note that the proportion of cumulative AE energy over that strain/stress range accounts for 90% of the total energy measured, whereas it only corresponds to a little more than 50% of the total number of events that occurred in the gage. For CMCs with high modulus, dense matrices, such as SiC fiber-reinforced CVI or MI SiC matrices, matrix cracking corresponds to the loudest events. In fact, it has been experimentally verified that cumulative AE energy is essentially directly related to matrix crack density (Figure 20.7) [13]. Consequently, if the final matrix crack density is known, the normalized cumulative AE energy versus stress relationship of Figure 20.6c for a given MI or CVI SiC/SiC CMC[1] can be multiplied by final matrix crack density to establish a stress-dependent matrix crack distribution relationship, which can then be used to model stress–strain behavior [13–15].

An example of a very accurate location analysis performed on a 3D-orthogonal architecture SiC fiber-reinforced SiC MI matrix composite is shown in Figure 20.8 [16]. For this study, the first peak of the extensional wave was determined for each event so that location accuracy was on the order of 0.25 mm in the axial direction. The 3D-orthogonal architecture had a repeatable structure every 1.5 mm. For this specific composite system, the Z fiber tows were very large and were the source of initial micro-tunnel cracks (low energy) followed by larger energy macrocracks in the same region. With increasing stress, the other 0/90 structure of the architecture is cracked. Event-based AE was able to locate these different regions of cracking.

Further analysis can be applied such as waveform frequency content to determine source identification which can then be related to damage and failure mechanisms. Figure 20.9 shows several different event waveforms and fast fourier transform (FFT) spectrums. Figure 20.9a is typical of large matrix cracks which extend to the surface of a composite

[1] Note that this does not apply to C fiber reinforced SiC or polymer infiltrated SiC/SiC composites. For those systems, the matrix is already cracked or damaged to some extent from processing and any further damage would have to be added to the existing damage.

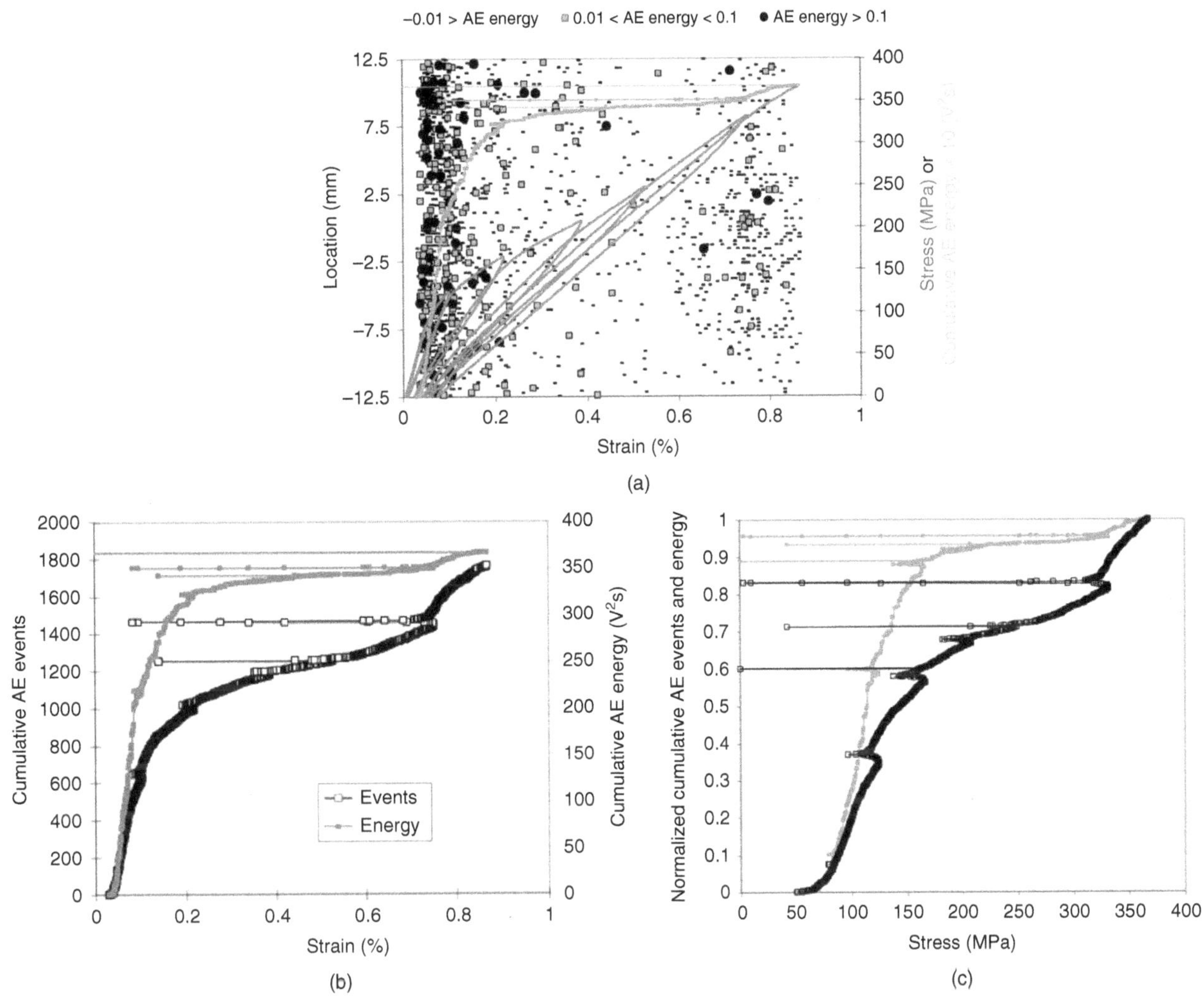

FIGURE 20.6

AE behavior during a room temperature tensile unload/reload hysteresis test of a typical woven Hi-Nicalon-reinforced CVI SiC matrix composite. From Reference 12. (a) Stress–strain behavior, location of events during the tests in the gage for different event energy ranges, as well as the cumulative energy of AE events as a function of strain. (b) Comparison of gage events and cumulative energy of those events versus strain. (c) Comparison of gage events and cumulative energy of those events normalized by maximum value (b). Note that the strain or stress reversals of cumulative curves correspond to events which occurred during the unload/reload loop.

exciting a strong flexural mode in the waveform. Figure 20.9b corresponds to an internal transverse matrix microcrack which excites a stronger extensional component of the waveform. Events which occur during the unload/reload region of a tensile hysteresis test (Figure 20.6) at higher strains often produce the damped sound waves similar to Figure 20.9d, which corresponds to longitudinal (interlaminar) type cracks. This type of data has been used to separate fiber breaks from matrix cracks for HN/CVI composites by analyzing the frequency content of the extensional wave portion of the waveform [6, 12].

One final experiment detail for CMCs is the fact that during the test, the speed of sound is decreasing due to matrix cracking and other damage mechanisms, that is, Δt_x (Equation 20.1) increases during the test. Figure 20.10 shows the reduction in speed of sound and the square root of the instantaneous unloading modulus versus strain for an HN/CVI SiC composite [12, 17]. The speed of sound was measured from the events that occurred outside of the outer sensors, that is, in or near the grips. The data are normalized by undamaged speed of sound of the extensional wave (C_e) and undamaged elastic modulus. There is a direct relationship between the two. What is significant is that C_e decreases by

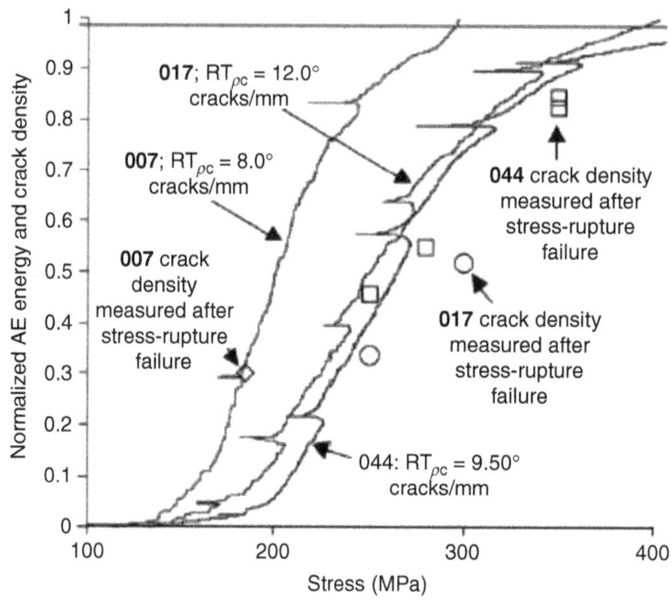

FIGURE 20.7

Crack density measured in the load-bearing minicomposites after stress rupture at 815°C in air compared with the AE activity at room temperature for specimens from the same panel for two different panels.

about 50% during the test. Therefore, it is essential to know the stress (damage)-dependent C_e in order to get accurate location of events. Additionally, this setup allows for an *in situ* measure of the elastic modulus during the test as well as damping behavior of sound waves which may relate to damage states as well [17].

Finally, for woven SiC fiber-reinforced MI matrix composites, long-time life conditions appear to be dependent on the formation of matrix cracks where oxygen or other deleterious species can enter through matrix cracks when

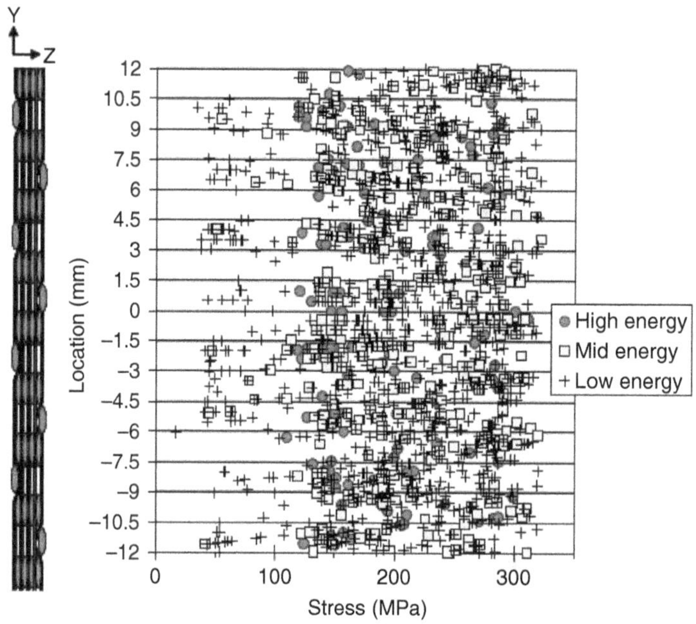

FIGURE 20.8

(a) Location of AE events along the length of an ZMI specimen oriented in the *Y*-direction.

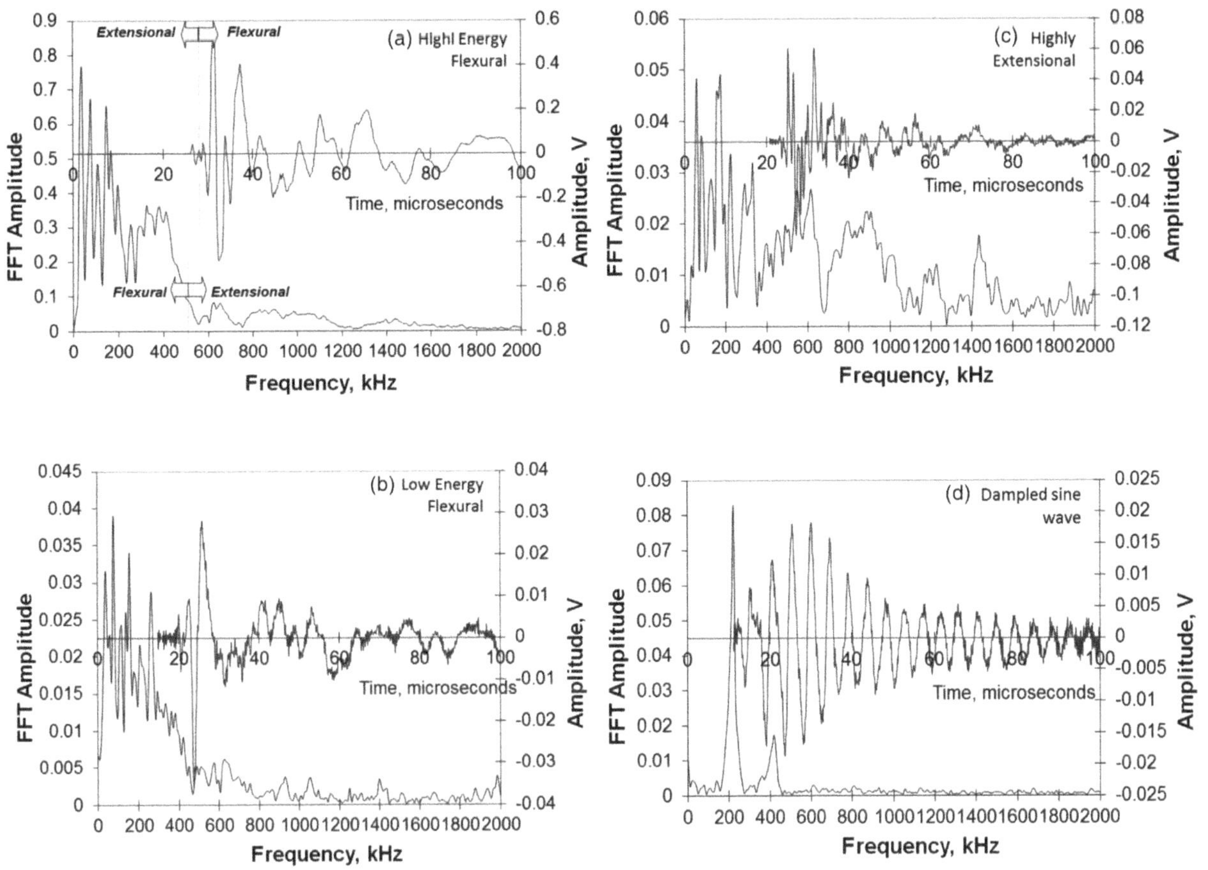

FIGURE 20.9

Waveforms and FFT spectrum for different events which occurred during a tensile test. From Reference 6.

the composite is under stress at temperature and react with the fibers, interphase, and/or matrix to either degrade fiber strengths and/or strongly bond fibers to the matrix. It has been shown [18] that the lowest stress AE events for the woven Sylramic-iBN, BN interphase, and MI matrix composites correspond to long-time creep run-out conditions above 1200°C and the lowest stress high energy AE event stress corresponds to run-out conditions at intermediate temperatures (815°C). The former corresponds to the formation of small microcracks which grow under creep conditions, whereas the latter corresponds to large fiber-bridged matrix cracks which enable oxygen ingress into the interior of the composite, resulting in oxidation embrittlement mechanisms.

20.3.2 Event-Based AE of CMC-Elevated Temperature Stress-Rupture

AE has been performed at intermediate and high temperatures during stress-rupture type tests [19,20]. For these tests, two AE sensors were directly attached to the end of the tensile bar within the water-cooled grips (i.e., only extensional waves can be detected). Figure 20.11 shows the results of AE activity versus time for two different stress rupture tests. For the higher temperature rupture test (960°C—Figures 20.11a and 20.11b), failure actually occurred outside the hot zone, that is, pest temperature. Location analysis showed the cumulation of AE activity in this location which eventually resulted in failure. The same progressive damage accumulation was observed for a lower temperature test (815°C—Figure 20.11c) except in the hot zone. At these temperatures, failure is associated with the growth of unbridged cracks due to fiber failure as a result of local load sharing, resulting from oxidation-induced fiber-to-fiber and fiber-to-matrix strong bonding.

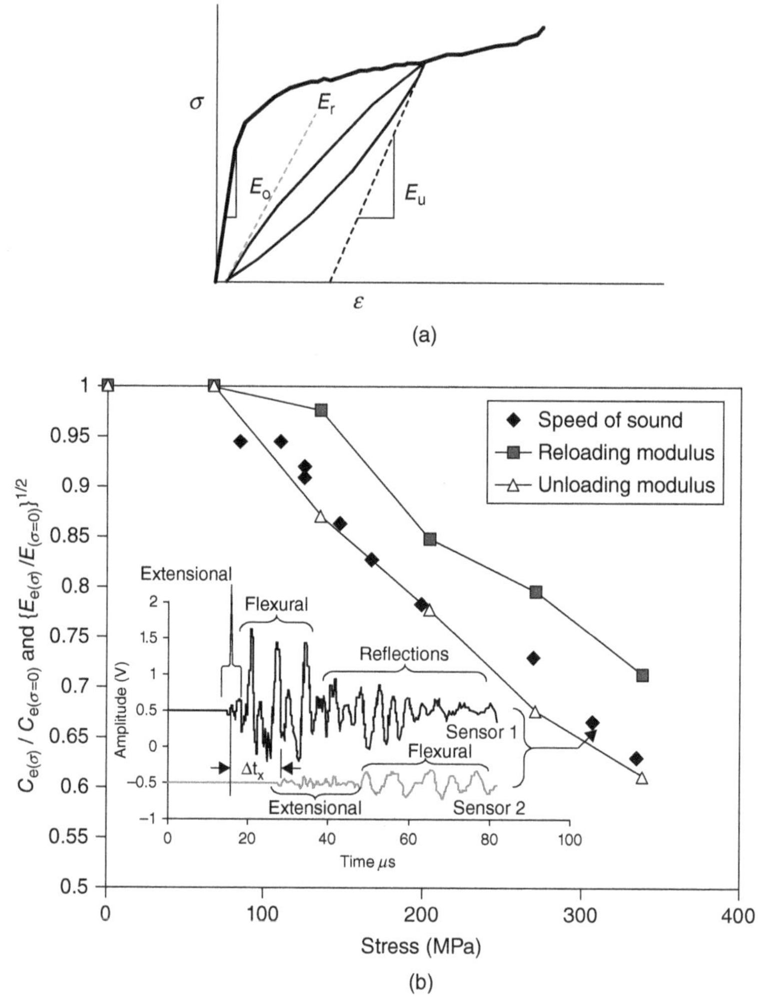

FIGURE 20.10

(a) Schematic of initial modulus, E_0, unload modulus E_u, and reload modulus after unloading, E_r. E_r is typically higher due to crack closure effects. (b) Normalized velocity of sound and elastic moduli for unloading and reloading portions of hysteresis loops for a HN/C/SiC matrix composite. From Reference 17. In the inset of (b) are typical waveforms for an event which occurred outside of the sensors used to measure time.

20.3.3 Event-Based AE of CMC C-Coupon Testing

Perhaps the most underutilized use of AE is with other types of tests such as interlaminar shear and tensile tests where first cracking is very difficult to obtain from nonlinearity in deflection curves or other methods such as digital image correlation (DIC). AE is ideal for acquiring this type of information. An example of this was for C-coupon tests on SiC/SiC composites aimed at determining the interlaminar tensile strength of several CMC variations [21]. Figure 20.12 shows a C-coupon test setup, AE data, and a micrograph of radial interlaminar cracks and tensile hoop cracks. AE was very successful at determining the first crack which was undetectable from nonlinearity, which grew from the center of the specimen around the radius followed by subsequent radial cracks. Also, at higher loads transverse cracks occurred which were higher in energy.

20.4 AE SIGNAL ANALYSIS USING PATTERN RECOGNITION TECHNIQUES

CMC composites are attractive candidates for many high temperature structural applications, because of their excellent creep resistance, high temperature strength and light weight. Future engine applications in civil aircrafts are foreseen for such composites, which require very long lifetimes under in-service conditions. Even though both fibers and matrix are made of brittle ceramics (mostly carbon and silicon carbide), CMCs can reach a high strain-to-failure as a result of energy dissipation through multiple matrix cracking and deflection of cracks at fiber/matrix interfaces. Matrix cracking is the first main damage mechanism [13, 14, 22, 23]. In static fatigue at high temperature under low stresses, composite

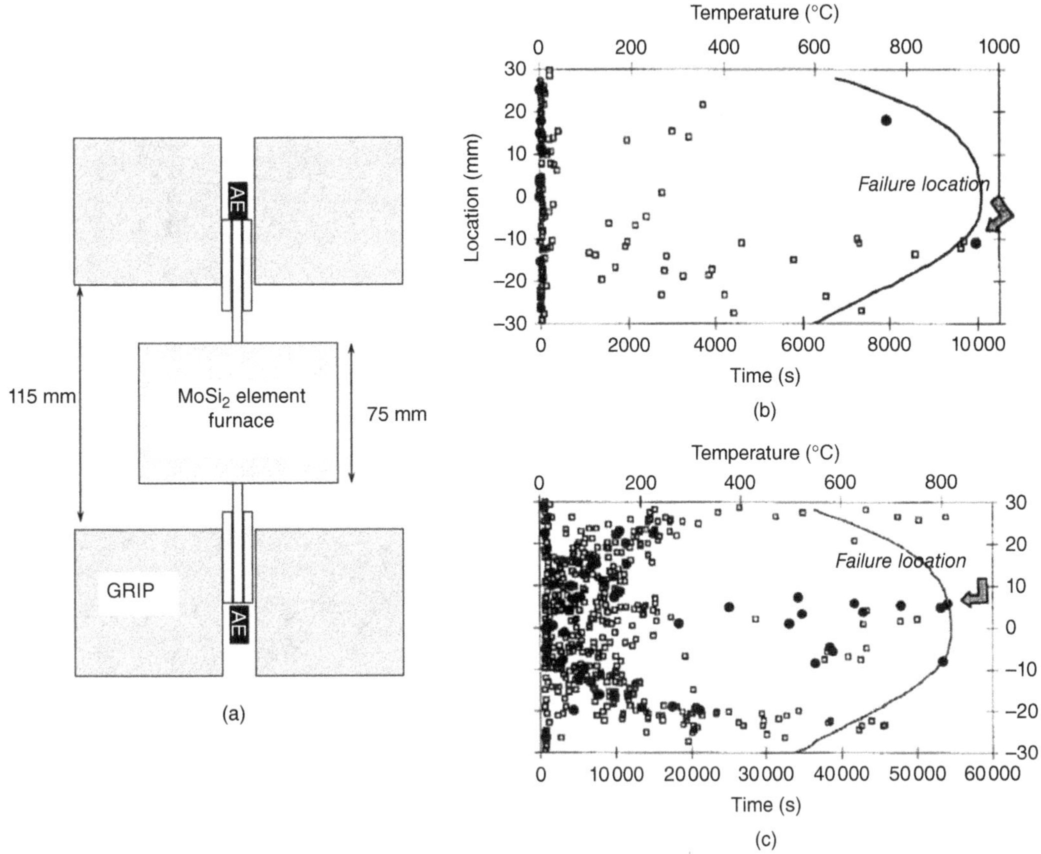

FIGURE 20.11

Load and AE event location versus time for (a) an applied stress of 163 MPa at 960°C hot zone temperature and (b) an applied stress of 148 Mpa at 815°C hot zone temperature. The AE event location data are separated open squares (second highest order of magnitude energy events) and filled circles (highest order of magnitude energy events). The lowest energy AE data are not shown. Also plotted is location versus furnace temperature profile (line). From Reference 20.

ultimate failure involves oxidation-activated slow crack growth in fibers and associated stress redistributions when fibers fail [24–29]. So, damage monitoring of composite materials is a key point for the control of durability and reliability in service. Beyond characterization of damage, the current challenge is to predict structures lifetime. To achieve this goal, a quantification of damage is needed but so is an identification of the various damage mechanisms, for which AE is a suitable technique. In the case of composite materials, various mechanisms act as AE sources including matrix cracking, fiber/matrix interface debonding, and fiber failure (Figure 20.13). During long-term tests, lifetime is often controlled by the ageing of fibers.

AE signals can reasonably be considered as a representation of their sources and then the acoustic signature of damage mechanisms can be determined. Common analysis of AE is based on diagrams of cumulative hits or counts or histograms of amplitude. Nevertheless, in the case of composite materials, this approach is sometimes insufficient in order to identify the various damage mechanisms. It can be improved by grouping signals of similar shapes into clusters using classifier parameters [30]. Many works [31–33] have shown that AE technique and multivariable classification techniques are the basis of pattern recognition tool.

In this type of study, the following assumptions are done:

1. Sources with different energy release (in terms of intensity and speed) generate AE signals with different waveforms.
2. AE signal energy is related to the energy released at the source. It depends in particular on the source size and amount of elastic energy stored. This parameter is thus very important for sources discrimination.

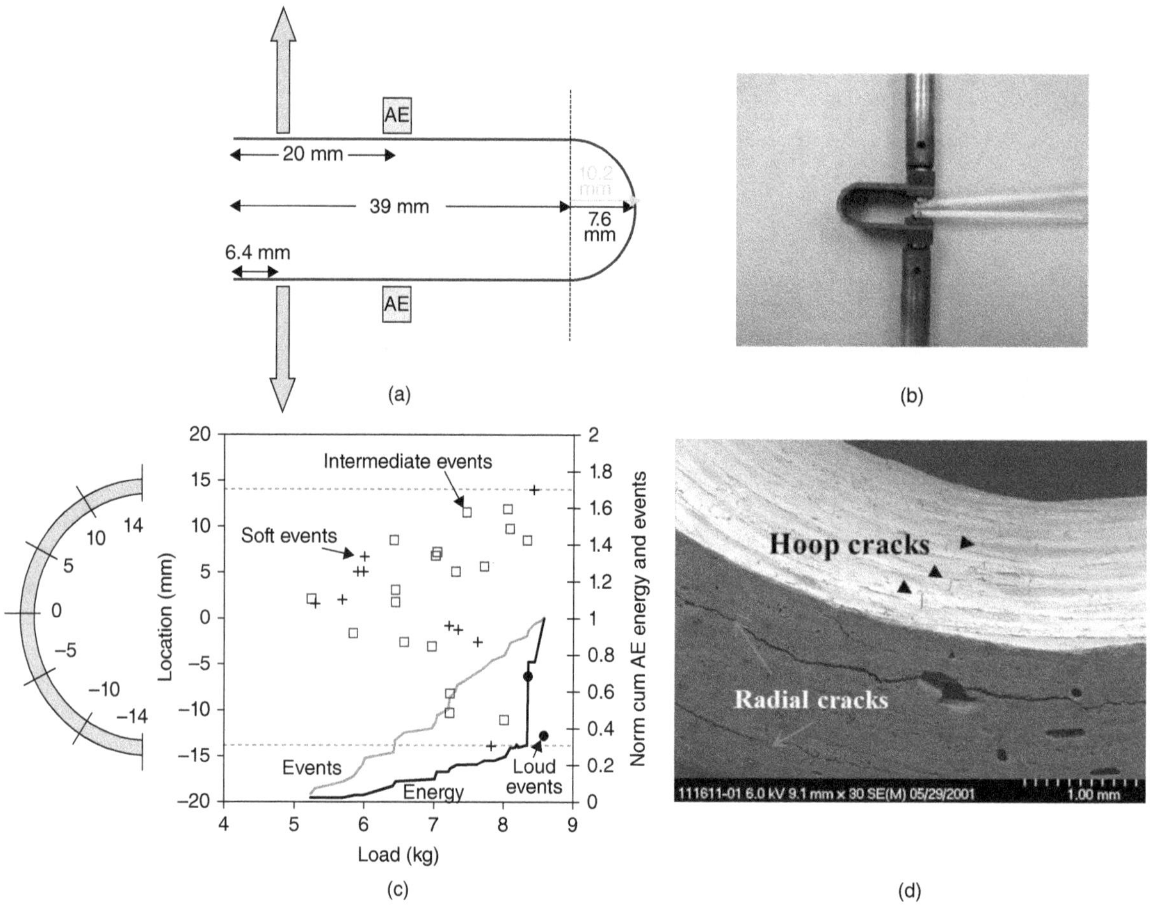

FIGURE 20.12

(a) Schematic representation of C-coupon test with sensor placement. (b) Optical photograph of actual specimen. (c) AE data acquired versus load showing location of AE events. Note the first event corresponds to the center of the C-coupon followed by crack jumps up around the radius followed by subsequent radial cracks. There are also transverse hoop cracks which occur on the tensile surface at higher loads (loud events). From Reference 21.

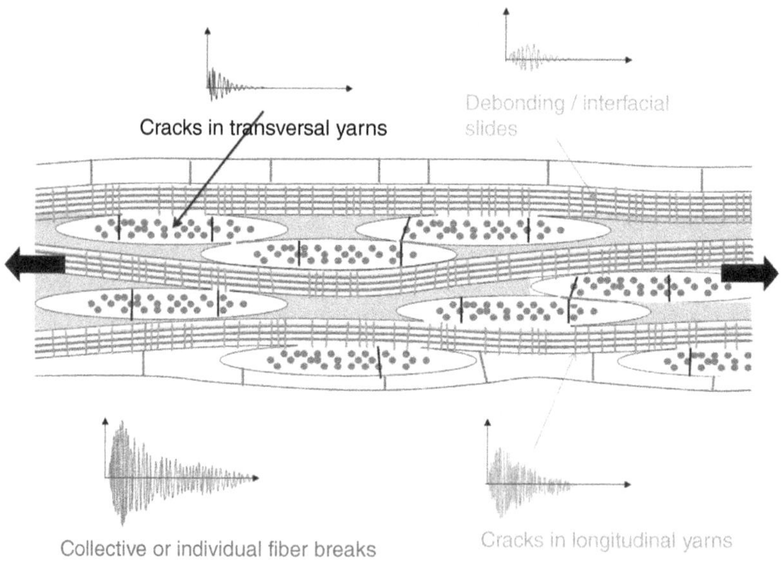

FIGURE 20.13

Schematic diagram illustrating the damage mechanisms involved in CMC composites.

3. Signals are affected by propagation but they remain images of sources. The recorded signals depend jointly on the events they have their origin in, material properties of the propagation medium, and sensors features.

Each recorded signal depends on the source mechanism features, but also on material properties and sensor characteristics. In fact, AE signals are subjected to many effects caused by variations in the material along the ray path (heterogeneities, anisotropy) and by the recording system (coupling, sensor characteristic, etc.). So, the AE signal is the convolution of the source function, the material's Green's function, the transfer function of the sensor, and coupling and transfer function of the recording system. Accordingly, there is no universal signature for a given damage source. However, under defined setup conditions, similarities exist between AE signals originating from similar damage.

20.4.1 Unsupervised Clustering Methodology Applied to AE Signals

Clustering refers to unsupervised classification. If no labeled data is available, an unsupervised methodology is required to perform an exploratory pattern analysis. "Unsupervised pattern recognition" is used to describe the complete methodology consisting of procedures for descriptors selection, cluster analysis, and cluster validity. AE signals with similar characteristics are segmented into clusters (which reveal the natural structure of data) based on similarity measures. A major issue in the use of AE technique is to associate every AE signature to a specific damage mechanism.

AE signals are described by a number d of features, or descriptors (amplitude, duration, rise time, counts, etc.). A typical pattern clustering involves the following steps (Figure 20.14):

- Representation of AE signals by relevant descriptors (feature selection/extraction procedure). The descriptors selection is an important step [34]. For the unsupervised pattern recognition, the selected descriptors should be relevant and limited in number.

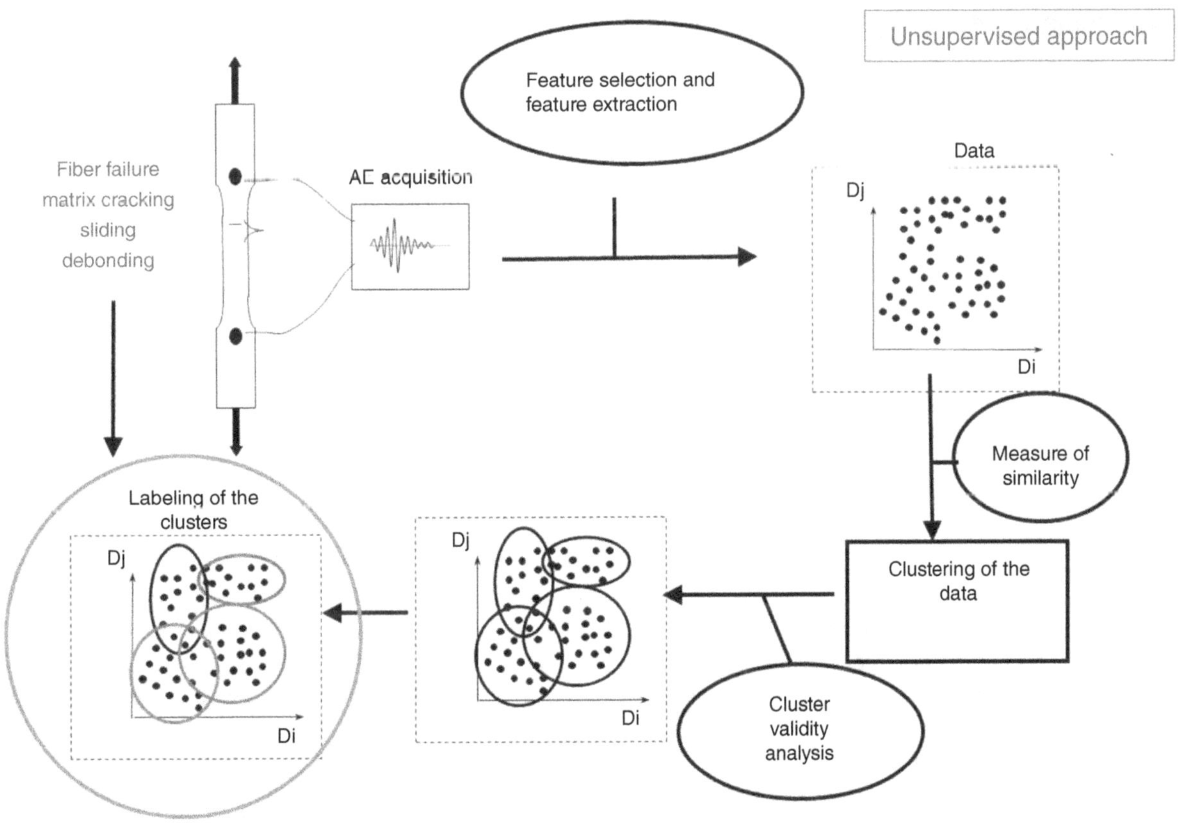

FIGURE 20.14

Flow chart representation of the unsupervised pattern recognition method for the analysis of AE data.

- Definition of a similarity measure appropriate to the data domain (Euclidean distance, Mahalanobis distance, etc.).
- Clustering or grouping of similar pattern signals (k-means, Fuzzy k-means, etc.).
- Cluster validity analysis (Davies–Boudin index, Silhouette value, Jaccard coefficient, Rand statistic, etc.). For instance, questions like "how many clusters exist in the dataset?" call for clustering results validation.
- Cluster labeling (identification of the AE sources).

After a successful classification according to criteria such as Davies–Bouldin index, the resulting clusters should be correlated with the material's damage mechanisms. After clustering, the maximum possible number of clusters may be equal to the number of mechanisms if all the acoustic signatures are distinguishable. But some mechanisms may produce similar AE signals that would not be separated by the clustering algorithm. In this case, the number of clusters would be lower than the number of mechanisms.

The most challenging point in unsupervised pattern recognition technique is to identify damage mechanisms that these clusters correspond to. This is due to the fact that the classification process does not lead to a unique solution and there does not exist an indisputable criterion to determine which classification result is more appropriate and representative of the damage mechanisms. The link between clusters and damage mechanisms is specific to each material and must be developed for each study based on knowledge of the material. AE-based identification of damage modes is not straightforward for two reasons. First, the AE signals have complex shapes that must be characterized by multiple pertinent descriptors. Second, the acoustic signatures of damage modes are not known a priori and they are scattered.

Through the identification of damage mechanisms by AE using unsupervised method, it is sometimes difficult to validate the classification and labeling of AE data. Indeed, most of the time, different source mechanisms are activated simultaneously and in unknown proportions. In most work on AE, the assignment of a signal to a damage mechanism is very difficult to validate, mostly impossible without a modeling work. Microstructural observations of damage are not sufficient to perform this validation.

20.4.2 Supervised Pattern Recognition Method

A supervised classification technique can also be used to analyze AE signals recorded during fatigue of CMC composites. This method requires a database of signals that have been labeled: the training set. Classification is the process of identifying the source of an event by comparing its features to those of previous events whose source is known. For example, k-nearest neighbor rule classifies an AE signal as follows: (i) find the k-closest signals in the training set, (ii) predict the class by majority vote, that is, choose the class that is most common among those k neighbors.

20.5 HIGH TEMPERATURE TESTING AND AE MONITORING

Moevus [35,36] and Momon [37] analyzed C_f/[Si-B-C] and SiC_f/[Si-B-C] composites with self-healing matrix under static fatigue under air at temperatures of 700°C, 1000°C, and 1200°C for the C_f/[Si-B-C] composite and at 450°C and 500°C for the SiC_f/[Si-B-C] composite. The composites were manufactured by SAFRAN Snecma Propulsion Solide (France). In order to protect the interface against oxidation, matrix contains phases which produce sealants at high temperatures preventing oxygen from reaching the interphase [38]. Both composite materials have a multilayered [Si–B–C] matrix and PyC interface layer. They are reinforced with different fibers: Hi-Nicalon fibers for the first one and carbon fibers (ex-PAN-HR) for the second one. The external surface was protected by a seal coat. For the tests at 450°C or 500°C under air on the pneumatic machine, two sensors were positioned directly on the specimen. Medium viscosity vacuum grease was used as a coupling agent.

For the tests above 700°C, two wave guides made of heat-resistant steel were put on the specimen. Each wave guide was attached to the corresponding sensor using vacuum grease as a coupling agent. For AE sources identification, AE wave velocity was determined before the test using a pencil lead break procedure: several sources were generated on the specimen at various locations between sensors. The difference in time of arrival at both sensors was calculated using the first peak of each signal. Elastic modulus decrease during the test was monitored using repeated unloading/reloading cycles performed every 6 or 12 hours and wave velocity was corrected [13, 17]. At the end of the test, the velocity was found to be equal to 6480 m/s, instead of 10 000 m/s in the undamaged state.

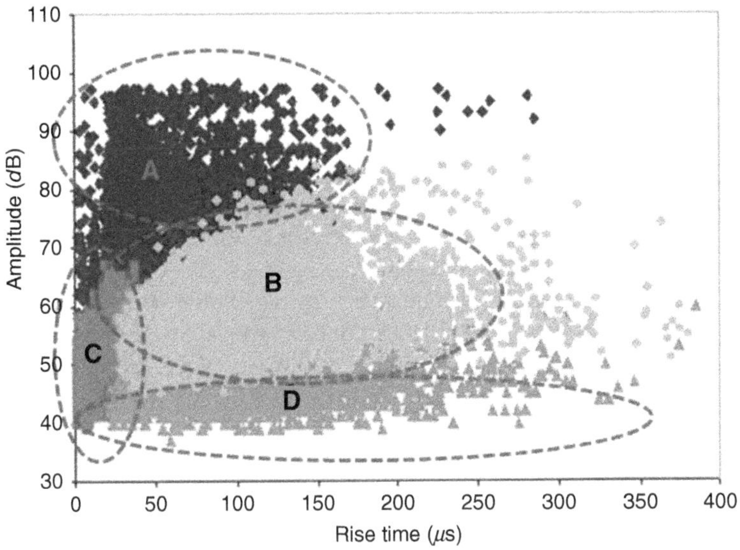

FIGURE 20.15

Clustering solution plotted in amplitude versus rise time for a static fatigue test carried out at 450°C.

20.5.1 Identification of Damage Mechanism on SiC_f/[Si-B-C] Composite at Intermediate Temperatures (450–750°C)

For the SiC_f/[Si-B-C] composite, four clusters are identified (denoted A, B, C, and D on Figure 20.15). Their activity can be determined during the first loading and during static fatigue (Figure 20.16). In such CMCs, matrix cracking can be classified into several kinds of cracking. First, cracks initiate in the external sealcoat and in the macropores inside the composite and propagate through the inter-yarn matrix. Then cracks propagate inside the transverse yarns through fiber–matrix interfaces. Multiple matrix cracking finally occurs inside the axial yarns. These cracks are deflected by the fiber–matrix interphase layer, leading to fiber debonding and overloading. Some fiber breaks are expected under high stresses. They rapidly lead to unstable fracture of entire yarns and of the composite. The sources of AE are the different matrix cracks, interfacial debonding, individual fiber fractures, and yarn fractures.

Cluster A appears for the most part at the beginning of test, during initial loading (Figure 20.16), so it should correspond to some matrix cracking. This cluster contains signals having the highest energy level, which are certainly the biggest observed cracks. This mechanism is expected to saturate, but in reality some activity is recorded in the failure

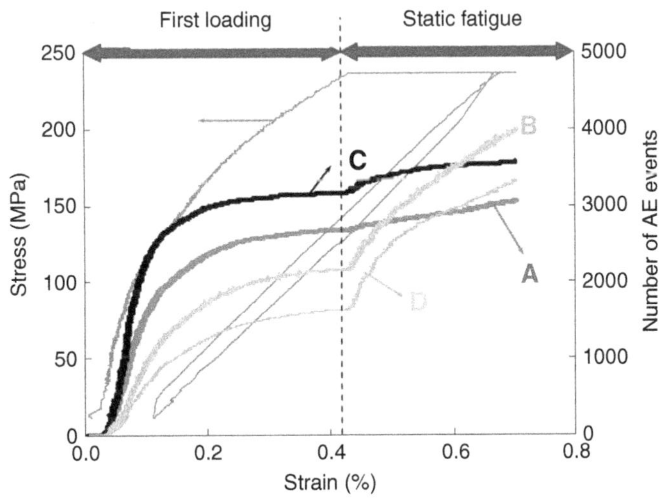

FIGURE 20.16

Activities of clusters during a static fatigue test.

zone just before failure (Figure 20.16). This suggests that Cluster A also contains fracture signals associated with the composite's final fracture. These signals are attributed to some yarns fracture or collective fiber breaks (individual fiber failures will rather be associated to cluster B of lower energy). As a conclusion, cluster A contains signals from two damage mechanisms, which are chronologically well separated.

Cluster B is also active from the beginning of initial loading; it is thus associated with another type of matrix cracking in transverse yarns. According to Figure 20.16, cluster B is also linked with the composite's final fracture, since its activity locally increases in the fracture zone just before failure. These last signals can thus be associated with fiber breaks. It seems difficult for the clustering algorithm to isolate the fiber fractures from the rest of AE data, certainly because the number of such fractures is very small in comparison with the global number of cracks.

Cluster C contains signals with relatively short duration, short rise time, and low amplitude when compared to the others. Its activity during initial loading (Figure 20.16) suggests that it corresponds to the last type of matrix cracks in the transverse yarns. These are the shortest cracks. Moreover, a saturation of the activity of this cluster is observed.

Cluster D is the last one to be activated and it becomes more active as strain increases (Figure 20.16). D-type signals seem to be a consequence of the existing damage within the material and are more likely attributed to fiber–matrix interfacial debonding. The D-type signals have a longer rise time when compared to other signals. Therefore, the shapes of these signals are well differentiated from the others. As for C-type signals, some D-type signals were probably not recorded because of their low amplitude level.

The different types of matrix cracking are successfully distinguished by AE analysis. Nevertheless, the signals associated with fiber failures, which correspond to the most severe mechanism, are not separated, probably because of the small number of related signals [34].

20.5.2 Identification of Damage Mechanism on C_f/[Si-B-C] Composite at High Temperature (700–1200°C)

After an unsupervised analysis, a supervised classification [37] technique was also used to analyze AE signals recorded during fatigue of C_f/[Si-B-C] composites. The training set was created by merging data from AE collected during several static fatigue tests (1200°C; 150 MPa) and cyclic fatigue tests (700°C; 0/130 MPa). The unsupervised classification gave reproducible clustering solutions of AE signals recorded during tests in different classes (A, A′, B, C, C′, and D), from their acoustic signature, without requiring preliminary knowledge of the source mechanisms. The knowledge gained from studying mechanical behavior, microscopic observations, and acoustic activity of the different classes then allowed a labeling of the recorded signals: Class A: collective fiber breaks, Class A′: individual fiber breaks, Class B: matrix cracking, Class C: fiber/matrix debonding, Class C′: yarn/yarn decohesion, and Class D: sliding at fiber/matrix interfaces and closing of matrix cracks after unloading. In order to establish the training set of labeled signals for the supervised analysis, the same amount of signals of each class (A, A′, B, C, C′, and D) was used. This training set was created to improve the classification and use six classes of signals. It included all the damage mechanisms that may be involved in this composite and allowed a more accurate classification. The method consists in comparing each detected signal to those of the library and to assign it to the class that is the most represented among its K-nearest neighbors. This method is interesting because it allows monitoring the activation of different source mechanisms. Using the supervised classification technique with the six class library, it was observed that in addition to classes A, B, C, and D, in all the tests a small amount of type A′ and type C′ signals was identified. This was not necessarily the case with the unsupervised classification technique. This effect may be explained by the fact that at 700°C, some individual fiber breaks and debonding between yarns in static fatigue may be involved. But these mechanisms were certainly in too small quantities to be represented by specific classes in the unsupervised classification. The supervised method is very interesting because it is easier to use and allows to label signals recorded in small number, whereas the unsupervised method was unable to identify them.

20.6 ACOUSTIC EMISSION AND LIFETIME PREDICTION DURING STATIC FATIGUE TESTS

In CMCs, global rupture is not controlled by a single event, as occurs in a homogeneous material, but rather by a succession of events in general, corresponding to diffuse damage. In this case, failure can be considered as a cooperative phenomenon. Many researchers investigated the elastic energy release during the failure process of materials. Smith

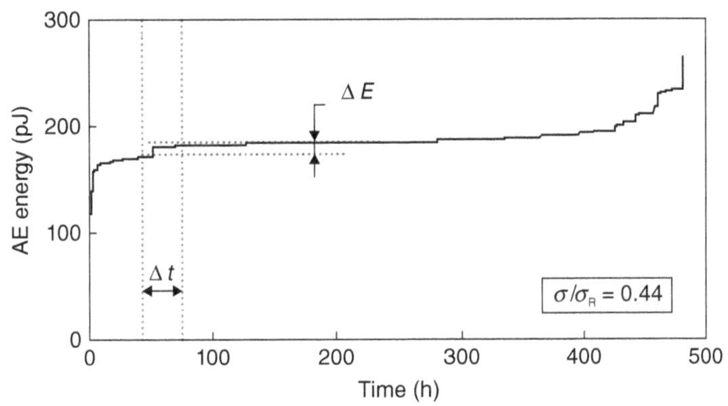

FIGURE 20.17

Cumulative energy releases as a function of time during a static fatigue test.

and Phoenix [39], Curtin [40], and Newman and Phoenix [41] studied the critical point hypothesis using fiber model. The results indicated that prior to material failure, energy release would accelerate in the form of a power law. Many authors (Johansen and Sornette [42], Guarino et al. [43], Nechad et al. [44], and Deschanel et al. [45]) have since observed characteristic power laws that describe the failure process of heterogeneous materials and more particularly an acceleration of energy release prior to failure using AE monitoring during load controlled test. The energy of the recorded AE events represents a part of the elastic energy released by CMC specimens. It is interesting to quantify this energy in order to assess the amount of damage of the composites. Thus, it is possible to investigate the evolution of elastic energy released by analyzing the energy of AE events (Figure 20.17).

20.6.1 Detection of Energy Release Acceleration

A criterion denoted R_{AE} was defined [46, 47],

$$R_{AE}(t) = \frac{\Delta E(t)}{E_{\text{loading}} \Delta t} \tag{20.2}$$

where E_{loading} is the cumulative AE energy recorded during initial loading up to the nominal test load, $\Delta E(t)$ is the cumulative energy AE recorded during the interval (Figure 20.17) $[t\ ;\ t + \Delta t]$. This coefficient (Figure 20.18) is observed

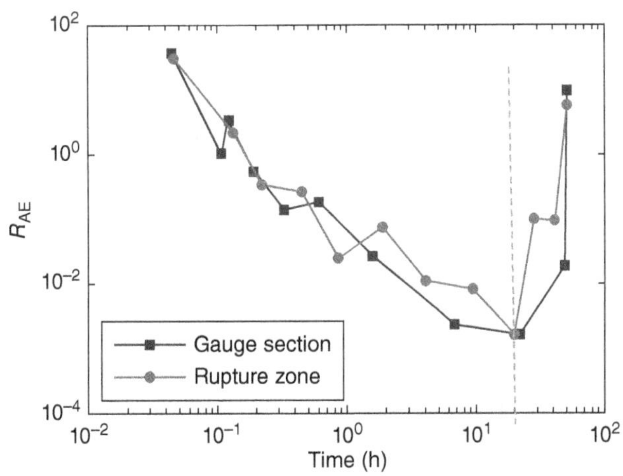

FIGURE 20.18

Comparison between R_{AE} calculated with signals recorded in all the gauge section and signals recorded only in the rupture zone for a specimen tested at 650°C with $\sigma/\sigma_R = 0.6$.

to decrease first, then to reach a minimum value for $t = t_m$, and finally to increase up to failure at time t_R. This ratio appears to be quite reproducible: $t_m/t_R = 0.57 \pm 0.06$. So the R_{AE} ratio can be used as a criterion for predicting the remaining lifetime of a specimen under static loading.

20.6.2 Application of the Benioff Law to Assess Lifetime of the CMCs

In order to evaluate a composite's lifetime, the Benioff law [48,49] can be applied to the AE data recorded during static fatigue tests at high temperatures. This law is based on a description by a power law of the cumulative energy (Ω) released by the composite and is characterized by four parameters (Ω_r, γ, φ, and t_r).

$$\Omega(t) = \sum_{i=1}^{N(t)} \sqrt{E_i} = \Omega_r + B\,(t_r - t)^{1-\gamma} \tag{20.3}$$

where E_i is the energy of the *i*th AE signal detected and *N(t)* is the number of AE signals located along the gauge length until time *t*. Ω_r is the value of $\Omega(t)$ when $t = t_r$, t_r being the critical time or failure time. The coefficient B is defined by the relation (3), where γ and φ are constants:

$$B = -\frac{\varphi}{1-\gamma} \tag{20.4}$$

Momon [46] showed on C/SiC and SiC/SiC composites that data beyond the minimum of R_{AE} were in better agreement with the Benioff law than those before the minimum. The Benioff law parameters were then estimated using linear regression technique in log–log plot on data collected beyond the minimum of R_{AE}.

Here is the procedure used to get an estimate of the time-to-failure (t_r):

1. Determination of Benioff law parameters (Ω_r, γ, φ, and t_r) from AE data recorded until rupture, using the $\log(\Omega_r - \Omega) = f[\log(t_r - t)]$ plot.
2. Estimation of t_r using a partial AE dataset, corresponding to data recorded between t_m (minimum of R_{AE}) and $t = t_m + t$. The linear fit is applied to the truncated data in order to deduce t_r as one of the four variables of this fit (Ω_r, γ, φ, and t_r) by the least-squares method.

For the sample tested at 500°C with $\sigma/\sigma_r = 0.44$, at *t* equal to 63% of the lifetime, the estimated value of t_r was only 5% lower than the experimental value. It was less than 1% lower after 85% of the lifetime.

The method for real-time analysis of energy release allows the identification of a characteristic time at 55% of the measured rupture time. This characteristic time reflects a local critical behavior described by the Benioff law and indicates a second damage phase when subcritical crack growth in fibers is predominant. However, this procedure requires preliminary tests until rupture to determine γ and φ for the studied material. The identification of the law parameters is crucial. Nevertheless, this evaluation is done for given testing conditions and sample geometry. Size effects are not taken into account and should be investigated. Moreover, the relevance of this approach is closely related to the quality of the source's energy measurement [47].

20.7 CONCLUDING REMARKS AND FUTURE DIRECTIONS

AE monitoring of CMCs is a very useful and sometimes quantitative technique for assessing damage in CMCs as well as the potential for inferring life expectancy from the accumulated damage. It is expected that with familiarity of the technique and the need for more complex testing, such as subcomponents, AE will be utilized even more due to the ability of the technique to locate and sense damage at low applied stresses and in regions of a composite which are difficult to instrument. However, AE is not a "black box" technique. In order to utilize AE successfully, one must understand the nature of sound travel through the system (composite) being examined, the limitations of the AE equipment being used (e.g., synchronized versus independent channels), and the sources and significance of damage and noise that pertain to the composite being tested (e.g., an already microcracked C/SiC composite compared to a SiC/SiC composite with little or no preexisting cracks).

REFERENCES

1. R. Y. Kim and N. J. Pagaon. Crack initiation in unidirectional brittle matrix composites. *J. Am. Ceram. Soc.*, 74(5), 1082–1090 (1991).
2. Y-J. Luo, S-C. Chang, and I. M. Daniel. Acoustic emission study of failure mechanisms in ceramic matrix composites under longitudinal tensile loading. *J. Compos. Mater.*, 29, 1946–1961 (1995).
3. M. G. Jenkins, J. P. Piccola, and E. Lara-Curzio. Onset of cumulative damage and the effects of test parameters on the tensile behavior of continuous fiber-reinforced ceramic composites (CFCC), in *Fracture Mechanics of Ceramics*, Vol. 12, edited by R. C. Bradt, D. P. H. Hasselman, D. Munz, M. Sakai, and V. Y. Shevchenko (Plenum Publishing, 1996), pp. 267–282.
4. M Surgeon, E. Vanswijgenhoven, M. Wevers, and O. Van Der Biest. Acoustic emission during tensile testing of SiC-fibre-reinforced BMAS glass-ceramic composites. *Compos. A: Appl. Sci. Manuf.*, 28A, 473–480 (1997).
5. Y. Bar-Cohen and A. K. Mal. Ultrasonic inspection, in *Non-Destructive Evaluation and Quality Control (Metals Handbook)*, Vol. 17 (ASM International, Metals Park,OH, 1989), pp. 231–240.
6. G. N. Morscher. Modal acoustic emission source determination in silicon carbide matrix composites, in *Review of Progress in Quantitative Nondestructive Evaluation*, edited by D. O. Thompson and D. E. Chimenti, CP 509 (American Institute of Physics, 2000), pp. 383–390.
7. K. F. Graff. *Wave Motion in Elastic Solids* (Dover, Inc., New York, 1975).
8. A. A. Pollock. *Non-Destructive Evaluation and Quality Control (Metals Handbook)*, Vol. 17 (ASM International, Metals Park, OH, 1989), pp. 278–294.
9. M. R. Gorman. Plate wave acoustic emission. *J. Acoust. Soc. Am.* 90(1), 358–364 (1991).
10. M. R. Gorman. New technology for wave based acoustic emission and acousto-ultrasonics, in *Wave Propagation and Emerging Technologies*, AMD-Vol. 188 (ASME, 1994), pp. 47–59.
11. W. H. Prosser, K. E. Jackson, S. Kellas, B. T. Smith, J. McKeon, and A. Friedman. Advanced, waveform based acoustic emission detection of matrix cracking in composites. *Mater. Eval.*, 53(9), 1052–1058 (1995).
12. G. N. Morscher. Modal acoustic emission of damage accumulation in a woven SiC/SiC composite. *Compos. Sci. Technol.*, 59, 687–697 (1999).
13. G. N. Morscher. Stress-dependent matrix cracking in 2D woven SiC-fiber reinforced melt-infiltrated SiC matrix composites. *Compos. Sci. Technol.*, 64, 1311–1319 (2004).
14. G. N. Morscher, M. Singh, J.D. Kiser, M. Freedman, and R. Bhatt. Modeling stress-dependent matrix cracking and stress-strain behavior in 2D woven SiC fiber reinforced CVI SiC composites. *Compos. Sci. Technol.*, 67, 1009–1017 (2007).
15. G. N. Morscher and V. V. Pujar. Design guidelines for in-plane mechanical properties of SiC fiber-reinforced melt-infiltrated SiC composites. *Int. J. Appl. Ceram. Technol.*, 6(2), 151–163 (2009).
16. G. N. Morscher, H. M. Yun, and J. A. DiCarlo. Matrix cracking in 3D orthogonal melt-infiltrated SiC/SiC composites with various Z-fiber types. *J. Am. Ceram. Soc.*, 88(1), 146–153 (2005).
17. G. N. Morscher and A. L. Gyekenyesi. The velocity and attenuation of acoustic emission waves in SiC/SiC composites loaded in tension. *Compos. Sci. Technol.*, 62, 1171–1180 (2002).
18. G. N. Morscher. Stress, matrix cracking, temperature, environment, and life of SiC/SiC woven composites, in *High Temperature Ceramic Materials and Composites (HTCMC7 Proceedings)*, edited by W. Krenkel and J. Lamon (AVISO Verlagsgesellschaft mbH, Berlin, Germany, 2010), pp. 251–260.
19. G. N. Morscher. Modal acoustic emission of damage accumulation in woven SiC/SiC at elevated temperatures, in *Review of Progress in Quantitative Nondestructive Evaluation*, Vol. 18A, edited by D. O. Thompson and D. E. Chimenti (Kluwer Academic/Plenum Publishers, 1999), pp. 419–426.
20. G. N. Morscher, J. Hurst, and D. Brewer. Intermediate-temperature stress rupture of a woven Hi-Nicalon, BN-Interphase, SiC-matrix composite in air. *J. Am. Ceram. Soc.*, 83(6), 1441–1449 (2000).
21. G. N. Morscher, F. I. Hurwitz, and A. M. Calomino. C-coupon studies of SiC/SiC composites. Part I: acoustic emission monitoring. *Ceram. Eng. Sci. Proc.*, 23(3), 379–386 (2002).
22. J. Lamon. A micromechanics-based approach to the mechanical behavior of brittle matrix composites. *Compos. Sci. Technol.*, 61, 2259–2272 (2001).
23. J. Lamon. *Handbook of Ceramics Composites*, edited by N. P. Bansal (Springer, 2005).

24. O. Loseille and J. Lamon. Prediction of lifetime in static fatigue at high temperatures for ceramic matrix composites. *J. Adv. Mater. Res.*, 112, 129–140 (2010).
25. P. Forio, F. Lavaire, and J. Lamon. Delayed failure at intermediate temperatures (600–700°C) in air in silicon carbide multifilament tows. *J. Am. Ceram. Soc.*, 87(5), 888–893 (2004).
26. W. Gauthier and J. Lamon. Delayed failure of Hi-Nicalon and Hi-Nicalon S multifilament tows and single filaments at intermediate temperatures. *J. Am. Ceram. Soc.*, 92(3), 702–709 (2009).
27. P. Reynaud, D. Rouby, and G. Fantozzi. Effects of temperature and of oxidation on the interfacial smear stress between fibres and matrix in ceramic-matrix composites. *Acta Mater.*, 46/7, 2461–2469 (1998).
28. G. Farizy, J. L. Chermant, J. Vicens, and J. C. Sangleboeuf. Understanding of the behaviour and the influence of oxidation during creep of SiCf-SiBC composites in air. *Adv. Eng. Mater.*, 7/6, 529–534 (2005).
29. A. Dalmaz, P. Reynaud, D. Rouby, G. Fantozzi, and F. F. Abbe. Mechanical behaviour and damage evolution during cyclic fatigue at high temperature of a 2,5D Cf/SiC composite. *Compos. Sci. Technol.*, 58, 693–699 (1998).
30. A. K. Jain, M. N. Murty, and P. J. Flynn. Data clustering: a review. *ACM Comput. Surv.*, 31(3), 264–323 (1999).
31. V. Kostosopoulos, T. H. Loutas, A. Kontsos, G. Sotiriadis, and Y. Z. Pappas. On the identification of the failure mechanisms in oxide/oxide composites using acoustic emission. *NDT & E Internat.*, 36, 571–580 (2003).
32. Y. Z. Pappas, A. Kontsos, T. H. Loutas, and V. Kostosopoulos. Failure mechanisms analysis of 2D carbon/carbon using acoustic emission monitoring. *NDT & E Internat.*, 31, 571–580 (1998).
33. T. H. Loutas and V. Kostopoulos. Health monitoring of carbon/carbon, woven reinforced composites. Damage assessment by using advanced signal processing techniques. Part I: Acoustic emission monitoring and damage mechanisms evolution. *Compos. Sci. Technol.*, 69(2), 265–272 (2009).
34. A. Sibil, N. Godin, M. R'Mili, E. Maillet, and G. Fantozzi. Optimization of acoustic emission data clustering by a genetic algorithm method. *J. Nondestr. Eval.*, 31(2), 169–180 (2012).
35. M. Moevus, D. Rouby, N. Godin, M. R'Mili, P. Reynaud, G. Fantozzi, and G. Fayolle. Analyse of damage mechanisms and associated acoustic emission in two SiC/[Si-B-C] composites exhibiting different tensile curves. Part I: Damage patterns and acoustic emission activity. *Compos. Sci. Tech.*, 68(6), 1250–1257 (2008).
36. M. Moevus, N. Godin, D. Rouby, M. R'Mili, P. Reynaud, G. Fantozzi, and G. Farizy. Analyse of damage mechanisms and associated acoustic emission in two SiC/[Si-B-C] composites exhibiting different tensile curves. Part II: Unsupervised acoustic emission data clustering. *Compos. Sci. Technol.*, 68(6), 1258–1265 (2008).
37. S. Momon, N. Godin, P. Reynaud, M. R'Mili, and G. Fantozzi. Unsupervised and supervised classification of AE data collected during fatigue test on CMC at high temperature. *Compos. A: Appl. Sci. Manuf.*, 43, 254–260 (2012).
38. L. Quemard, F. Rebillat, A. Guette, H. Tawil, and C. Louchet-Pouillerie. Self-healing mechanisms of a SiC fiber reinforced multi-layered ceramic matrix composite in high pressure steam environments. *J. Eur. Ceram. Soc.*, 27, 2085–2094 (2007).
39. R. L. Smith and S. L. Phoenix. Asymptotic distributions for the failure of fibrous materials under series-parallel structure and equal load-sharing. *J. Appl. Mech.*, 48, 75–82 (1981).
40. W. A. Curtin. Theory of mechanical properties of ceramic-matrix composites. *J. Am. Ceram. Soc.*, 74, 2837–2845 (1991).
41. W. I. Newman and S. L. Phoenix. Time dependent fiber-bundles with local load sharing. *Phys. Rev. E*, 63(2), (2001).
42. A. Johansen and D. Sornette. Critical ruptures. *Eur. Phys. J. B*, 18, 163–181 (2000).
43. A. Guarino, A. Garcimartin, and S. Ciliberto. An experimental test of the critical behaviour of fracture precursors. *Eur. Phys. J. B*, 6, 13–24 (1998).
44. H. Nechad, A. Helmstetter, R. El Guerjouma, and D. Sornette. Andrade and critical time-to-failure laws in fiber-matrix composites: experiments and model. *J. Mech. Phys. Solids*, 53(5), 1099–1127 (2005).
45. S. Deschanel, L. Vanel, G. Vigier, N. Godin, and S. Ciliberto. Experimental study of crackling noise: conditions on power law scaling correlated to fracture precursors. *J. Stat. Mech.*, P01018 (2009).
46. S. Momon, M. Moevus, N. Godin, M. R'Mili, P. Reynaud, G. Fantozzi, and G. Fayolle. Acoustic emission and lifetime prediction during static fatigue tests on ceramic-matrix-composite at high temperature under air. *Compos. A: Appl. Sci. Manuf.*, 41, 913–918 (2010).
47. E. Maillet, N. Godin, M. R'Mili, P. Reynaud, J. Lamon, and G. Fantozzi. Analysis of acoustic emission energy release during static fatigue tests at intermediate temperatures on ceramic matrix composites: towards rupture time prediction. *Compos. Sci. Technol.*, 72, 1001–1007 (2012).
48. C. G. Bufe and D. J. Varnes. Predictive modelling of the seismic cycle of the greater San Francisco Bay region. *J. Geophys. Res.* 98, 9871–9883 (1993).
49. Y. Kawada and H. Nagahama. Cumulative Benioff strain-release, modified Omori's law and transient behaviour of rocks. *Tectonophysics*, 424, 157–166 (2006).

PART VII APPLICATIONS

CHAPTER 21

CMC APPLICATIONS TO GAS TURBINES

PATRICK SPRIET

Herakles, Safran Group (Retired), Aeronautics & Hi-Temp Composites B.U., Le Haillan, France

21.1 INTRODUCTION

This section is related to CMC applications to aero engines as well as land-based gas turbines.

Ceramic matrix composites (CMCs) are able to withstand extensive high temperature thermal cycling with little degradation or physical distortion. Therefore, they are considered as serious material candidates to replace superalloys for high temperature applications, including the core of gas turbine engines. Their introduction in the core of gas turbine engines aims at operating the system at higher temperatures with reduced cooling air, thus enabling superior fuel efficiency and reduced harmful emissions.

Over the past 20–30 years, different companies and R&D organizations, mainly in the United States, in Europe, and in Japan have placed interest on those materials, developing various carbide and oxide technologies, targeting among others gas turbine applications.

In the field of aero-engines, R&D activities have involved, at first, military applications targeting the augmentor sections of large military engines and covering the exhaust and the afterburner components. In most of those cases, the main drivers for introducing those new materials are to provide component durability with less maintenance, along with reduced weight.

Using the feedback resulting from those demonstrations, and thanks to new and more performing materials concepts, the technology development was later on transitioned to the core engine, including HP and LP turbines as well as combustion chambers of both military and commercial engines.

R&D efforts in the field of land-based gas turbines have been mostly developed over the past 20 years, stemming from the work performed on military engines.

A global engineering approach, combining design analysis and modelization, testing of subelements and prototype component for product development, along with manufacturing process development is a way to achieve successful developments.

However, various constraints linked to confidentiality and defense restrictions (i.e., ITAR regulations) tend to make it harder to get a good overview of the applications, especially in the defense domain.

Ceramic Matrix Composites: Materials, Modeling and Technology, First Edition. Edited by Narottam P. Bansal and Jacques Lamon.

21.2 CMC DEVELOPMENTS FOR MILITARY ENGINES

21.2.1 Demonstration and Developments for Exhaust Section

21.2.1.1 Snecma Jet Engine Applications [1,2]

21.2.1.1.1 Safran/Herakles background The first development of high temperature composite materials started at SEP in the seventies, with C/C composite materials targeting performance and reliability improvement of solid rocket motors (weight reduction and improved erosion behavior of the nozzle throat). Later on, the Sepcarb® technology was very successfully transitioned to aircraft brake application, leading to significant mass-production ramp-up using CVI manufacturing process.

The CMC technology was developed by SEP in the 1980s, particularly C/SiC materials, in order to meet thermal protection system applications for future space planes like the Hermes shuttle, as well as long durations of service in aircraft jet engines.

21.2.1.1.2 Snecma M53 engine [2] The first development step of the technology for jet engine applications was focused on subcomponent evaluations for flame holder, exhaust cone, flaps using 2D C/SiC and SiC/SiC composites for feasibility purpose. The next step targeted components rig-testing and then engine ground testing using the Snecma M53-2 engine as test bed.

The convergent–convergent nozzle includes a set of inner flaps exposed to the primary exhaust flow and a set of outer flaps exposed to mixed primary–secondary flow—2D SiC/SiC using Nicalon C.G. fiber from NGS Advanced Fibers (formerly a Division of Nippon Carbon Company), embedded in a SiC–CVI matrix, with appropriate finishing treatment. The component designed for a maximum temperature of 850°C achieved 60% weight saving compared to cobalt-base alloy.

Ground tests were performed on master and slave flaps for the inner and outer convergent sections, for a total of 350 hours without visible change of the components. This phase was then followed by in-flight testing on a Mirage 2000 fighter (Figure 21.1). The test program covered several tens of hours of flight without any incident, and was highlighted by a first worldwide demonstration performed at the 1989 Paris Air Show.

21.2.1.1.3 Snecma M88 engine

Outer nozzle application The C/SiC technology was then successfully transitioned to the outer nozzle of the Snecma M88 engine powering the Dassault Rafale fighter (Figure 21.2).

FIGURE 21.1

CMC components on Snecma M53-2 Engine.

FIGURE 21.2
Snecma M88-2 engine exhaust Sepcarbinox® outer flap for M88 engine.

The 2D C/SiC material was engineered with an improved carbon multilayer reinforcement and an enhanced finishing treatment to improve its behavior in oxidative atmosphere.

The baseline material Sepcarbinox®A262 is characterized by a nonbrittle behavior with failure strength of about 250 MPa at room temperature, and life time duration, for temperatures below 700°C, matching the application requirements.

The component design achieved a 30% weight saving over Inconel alloy and the technology was qualified for mass production in 1996.

Since then, this material has been baselined for all M88 engines in serial production and several thousands of components have been delivered so far. The in-service behavior of the Sepcarbinox outer flap has been very positive, meeting the life objective of the component and showing very good durability and very good tolerance to external damage mechanisms.

Inner nozzle experience The M53-2 engine demonstration of the 2D SiC/SiC material outlined a positive behavior of the inner flaps. However, the performance capability of the first generation of SiC/SiC material (Cerasep®A300) was not sufficient to sustain the high thermal and mechanical loads met in the M88 component counterpart, which were significantly exceeding the elastic limit of the material stress–strain curve. As a result, early material damage, leading to insufficient durability was observed during the initial ground testing of the components.

Those conclusions initiated the development of a new generation of SiC/SiC material using a self-sealing matrix concept which was proposed for engine evaluation by SEP in the late 1990s.

21.2.1.2 US Military Engines All US military jet fighters fly with aerospace turbine engines designed with a convergent–divergent exhaust nozzle. Extensive evaluation of CMC exhaust nozzle hardware has been performed by the USAF and the US Navy, respectively on F100 and F110 legacy engines. The publication [3] presented at the HT-CMC5 conference in 2005 at Seattle provides a good overview of the experience gained through those evaluations.

21.2.1.2.1 US Air Force experience [3] CMCs have been evaluated for both the General Electric F110 engine and the Pratt & Whitney F100 engine that power the F-15 Eagle as well as the F-16 Fighting Falcon.

GE F110 Engine [3, 4] Evaluation of CMCs for the nozzle of the F110 engine occurred in the1990s [4].

Four different CMC systems were first evaluated on a ground test engine, with two of the CMCs (SiC/C and SiC/SiNC) showing no degradation after testing to 10% of life. Additional ground testing to 70% of life demonstrated the durability of the CMC flaps and their suitability for the variable exhaust nozzle (VEN).

A field service evaluation (FSE) at Hill AFB, Utah was then performed on an operational F-16. The flaps generally exhibit no visual sign of degradation for the first 100-200 engine flight hours (EFH).

After a few hundred flight hours, degradation in the SiC/C CMC was observed at the aft end of the flaps. That may be related to excessive acoustic excitation due to the design of the CMC hardware and metallic frame.

Degradation in the SiC/SiNC material was significantly different than what was observed on the SiC/C CMC, and consisted of delaminations, with no significant wear observed in any of the flaps. Delaminations seem to be amplified by acoustic loads and thermal gradients, and the damage increased in areas where the CMC flaps came in contact with warped adjacent metallic seals.

Several of the F110 flaps were removed and cut up into test specimens for measuring the remaining tensile strength. Both the ground-tested and flight-tested material showed significant strength degradation, with nonsymmetric distribution. That was attributed to temperature effects in relation with hot streaks in the exhaust nozzle.

Pratt & Whitney F100 engine [3, 5] CMCs have also been ground tested on Pratt & Whitney F100 engines as exhaust nozzle divergent seals in the early 2000s. Three different CMC systems have been ground tested on both F100-PW-229 and 220 engines through robust accelerated mission tests (AMT) aimed at subjecting the hardware to the most aggressive conditions seen in actual use. The three CMCs were S200® (SiC/SiNC) from ATK/COI-C and Cerasep® A410 and Sepcarbinox® A500 seals from Safran/Herakles.

The A410 and A500 seals from Herakles utilized a unique multilayer woven fiber architecture and a sequential self-sealing matrix (carbide type) with no external specific coating. The A410 material used Hi-Nicalon fiber from NGS Advanced Fibers as reinforcement and the A500 material used a carbon fiber. Most of the seals were installed in hot-streak locations to evaluate their behavior in the worst conditions.

After various ground endurance testing phases carried out at the West Palm P&W testing facility and AEDC testing facility of the US Air Force, the A500 seals have accumulated 11 161 tactical air cycles (TACs) corresponding to approximately 5000 engine flight hours (EFH)—including 176 hours in augmented conditions. This endurance testing went well beyond the full-life objective of 4300 TACs and significantly exceeded the life of the current metal hardware.

The seals experienced no delamination and no wear and appeared visually to be in excellent condition after ground testing (Figure 21.3). Mechanical posttest analysis was performed for comparison with the as-processed material database. This residual strength testing showed no loss for the A410 material and only 6% for the A500 material. The S200 seals experienced delaminations in only the surface ply (30% of the surface ply delaminated) and 15% of the surface ply was missing

The A500 seals have been installed on operational F15 and F16 fighter aircraft for FSE (Figure 21.4). The flight experience accumulated exceeded 1000 hours and showed no visible damage on the hardware and confirmed the high performance potential of the material. Some loose metal aft fasteners have been observed on F15 aircraft, and are probably linked to metallic attachment nonconformities and design issues.

The SepcarbinoxA500 seals have performed outstandingly on F15 and F16 and are considered by USAF/AFRL as one of the most successful flight demonstration of CMC hardware to date.

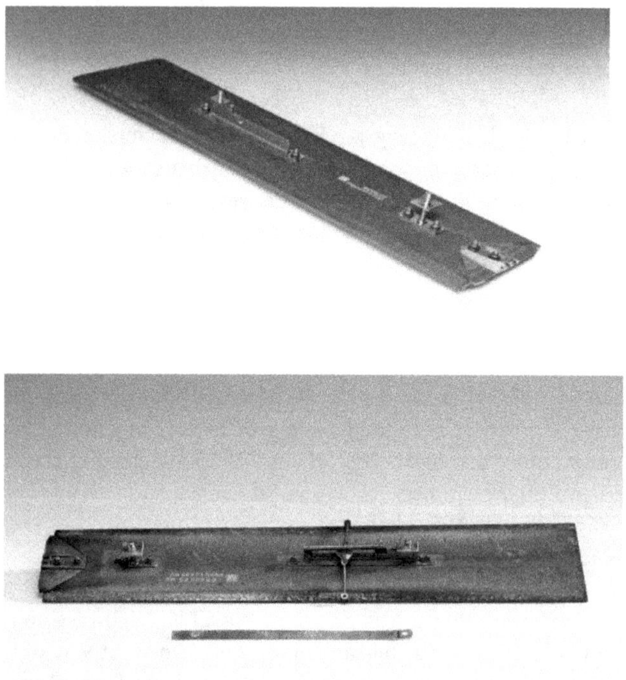

FIGURE 21.3
Sepcarbinox® A500 divergent seal for F100-PW-229 engine (as processed and after endurance testing).

FIGURE 21.4

Exhaust of F100-PW-229 powering F15 and F16 fighters with Sepcarbinox® A500 divergent seals.

Pratt & Whitney PYBBN vectoring nozzle Flaps using A500 material have also been designed for flight evaluation on the PYBBN thrust vectoring nozzle fitted on F15 experimental aircraft tested out of NASA Dryden.

Three parts have been evaluated during a 3-year test period, cumulating almost 200 hours of flight without any material damage (Figure 21.5).

21.2.1.2.2 US Navy CMC application [6] CMC flaps and seals have been qualified in the divergent section of the F414-GE-400 engine powering the F/A18-E/F Super Hornet because of the high temperature environment of the engine.

The material originally qualified and baselined for the secondary seals in the augmentor VEN section of the F414-400 engine consists of a SiC fiber-reinforced carbon matrix (SiC/C) material (Figure 21.6).

FIGURE 21.5

NASA F15 #837 aircraft with PYBBN nozzle and CMC divergent flap.

FIGURE 21.6

F414-GE-400 with SiC/C exhaust nozzle.

A second material source has been qualified and introduced by General Electric into the F414 program: an oxide–oxide configuration developed by GE was introduced into production in 2011 on the F414.

21.2.1.3 CMC Mixer Nozzle for Helicopter Demonstration Evaluation of CMC technology in the exhaust section of a turbine engine for military helicopter is underway at Herakles, as outlined by the mixer primary nozzle designed and manufactured for flight evaluation on a military platform (Figure 21.7).

The objectives of the demonstration are

- weight reduction;
- extension of life in service;
- compatibility of the component with increased thermal environments; and
- flight endurance evaluation.

21.2.2 Demonstration of CMC Afterburner Components

Flame holder components are located in the augmentor module at the turbine exhaust and are designed to provide additional thrust when needed by the mission profile of jet fighter aircraft. Those components are submitted to high temperature environment and to high thermal gradients and aerodynamic loads of the primary flow of the engine.

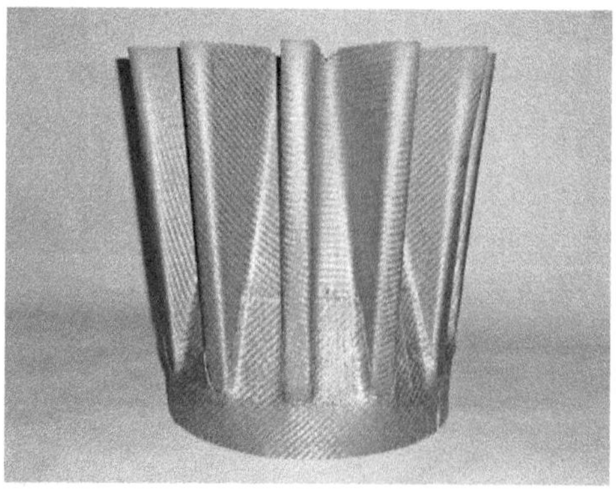

FIGURE 21.7

CMC mixer primary nozzle for military helicopter (Herakles).

FIGURE 21.8
CMC flame holders submitted to thermal fatigue and partial mechanical rig test (Herakles).

The introduction of CMC material is a way to initiate weight saving and to increase life in service of the component.

CMC technology has certainly been evaluated on various military engines; however, limited public information has been made available in that domain.

21.2.2.1 Safran/Herakles Experience [1] A demonstration program has been recently achieved by Safran, using the Snecma M88 engine as a test bed.

The preliminary design phase of the component highlighted that the thermal gradient, typical of this component, was generating significant thermal stresses. Therefore, the Sepcarbinox A500 material, less sensitive to thermal gradient, thanks to a lower Young modulus, was preferred for the detailed design phase of the component.

Subelement feasibility and rig test (Figure 21.8) The preliminary engineering of the components was achieved on subelements for manufacturing feasibility purpose and design validation.

The flame holder subelements were then submitted to thermal cycling tests and mechanical testing in order to validate the material behavior.

Thermal fatigue testing, performed on a burner rig cumulated the equivalent of 2000 hours (mixed engine cycle), reaching 1180°C max. temperature. After the thermal fatigue program, material analysis performed on subelements, confirmed the absence of material damage and the good efficiency of the self-sealing matrix.

Mechanical testing took place on a multiaxial mechanical bench. Subelements were submitted to tangential and axial loads representative of the application. Afterwards, one component was submitted to LCF testing and no damage was detected after the representative cycling program.

Components manufacturing and testing A 30% weight gain over the reference metallic component was finally achieved.

The manufactured prototype components were inserted on a Snecma M88 engine for performance and then ground endurance AMT (Figure 21.9). The objective of 1000 hours mixed mission cycles was achieved with no damage of the A500 material.

Posttest morphological and mechanical test analysis confirmed the absence of material damage as a result of the test and confirmed that the material had the potential to sustain extended life in service.

21.2.3 Turbine Component Demonstrations on Military Engines [7,8]

Over the past 20 years, various R&T programs have been implemented in the United States, in Japan, and in Europe on CMC technologies for the improvement of turbine engine technology.

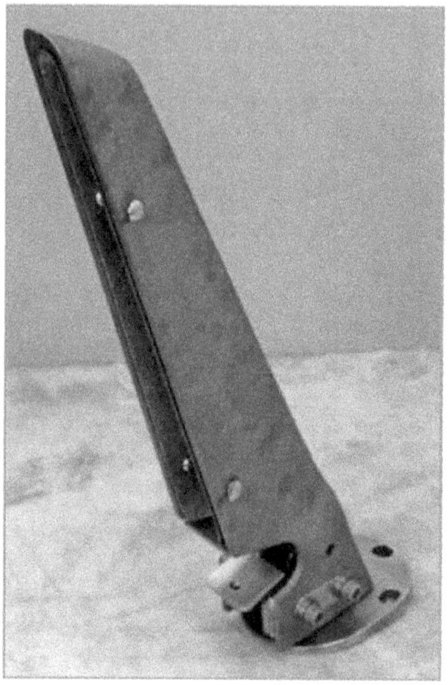

FIGURE 21.9

Sepcarbinox® A500 flame holder for Snecma M88-2 engine (Herakles).

In the United States, collaborative programs such as IHPTET, VAATE, UEET, etc., managed by US DoD, NASA and the Industry, involved CMC components development and evaluation in the core engine in order to drastically increase engine performance and efficiency.

As part of those program achievements, typical research activity leading to rig and engine testing include:

- SiC/SiC composite technology for combustor liner capable of 1350°C, allowing cooling flow reduction.
- CMC LP turbine vane using 3D fiber architecture for increased strength and durability.
- CMC hollow HP turbine vane targeting 50% weight reduction and 20% cooling reduction than a typical nickel-based superalloy vane.
- HP turbine shrouds using CMC materials have also been evaluated.

GE evaluations [9] In 2009, the GE/Rolls-Royce Fighter Engine Team announced R&T activities on CMCs for combustor and turbine areas of the F136 engine of the Joint Strike Fighter (JSF), quoting undergoing CMC evaluation on the F136 engine for the third stage low-pressure turbine vanes made by General Electric (GE Aviation website).

In November 2010, GE has tested a CMC turbine blade in a modified F414 engine, representing the first GE application of CMC material in a rotating engine part.

The 4-hour test taking place in an operational environment provided a technology readiness level of six, meaning that the hardware is eligible to be incorporated in GE's next generation of commercial and military engines.

The F414 test was funded by the US Navy's energy task force, targeting turbine blade upgradation for future applications of the engine.

21.3 CMC R&D FOR COMMERCIAL ENGINES

Developments of future commercial airplanes are challenged by lower fuel burn and noise reduction which call for more efficient engines.

Typical performance levels of future engine generation for single-aisle aircraft target fuel consumption reduction of the engine (15%), combined with ICAO noise reduction, and reduction of nitrous oxide emission to 60% of the current level.

Such improvements can be achieved by increasing engine thermal efficiency (increased temperatures and pressure cycles) and dilution ratio.

However, that leads to engines with larger diameters and higher weights. Technologies providing weight savings are thus of primary importance. CMCs deliver weight savings in the range of 30–40% compared to similar metallic designs and are, therefore, considered in the design of the next generation of turbofan engine in the turbine and exhaust sections.

21.3.1 CMC R&D for Combustion Chamber

Incorporation of CMC liners in the combustion chambers of future engine designs aims at reducing cooling airflow requirements, thus leading to reduced fuel burn and harmful emissions levels (primarily NOx).

Prototype evaluations of CMC technologies have already been performed at lab-scale levels on flame tubes, as part of the US and European R&T cooperative programs.

GE Aviation demonstration [10] A full-scale demonstration of CMC combustor liners has been performed several years ago by General Electric as part of the NASA Ultra Efficient Engine Technology (UEET) Program on the CFM56 engine, with the objective of increasing the technology readiness level.

An SiC/SiC system using the GE HyperComp slurry-cast MI material initially developed under the NASA EPM program has reached the engine-test stage.

Safran demonstration [11] Similar initiative has been performed by Safran, using also the CFM 56 engine as a test bed.

The full scale engine demonstration was performed by Snecma in charge of the overall combustor design and engine integration, in cooperation with Herakles in charge of the CMC components detailed design and manufacturing (Figure 21.10).

The program aimed at developing a CMC design expertise for combustor liners and cowls, as well as demonstrating the manufacturing feasibility of full-scale liners for engine testing.

The main steps of the demonstration included:

- components design (liners with effusion cooling);
- material down selection and characterization (Cerasep® A415 using Hi-Nicalon fiber and the Herakles self-sealing matrix);
- subelements development for feasibility and flame testing; and
- manufacturing of prototype components.

Snecma took care of the assembly and instrumentation of the combustion chamber prior to rig testing.

This multiyear demonstration program was then completed with a full scale rig test of the CMC combustion chamber (Figure 21.11).

FIGURE 21.10
Subelement (flame tube) for effusion cooling evaluation and Cerasep® A415 liner for CFM56 combustion chamber (Herakles).

FIGURE 21.11
Cerasep® A415 liners tested on a CFM56 Combustion Chamber (Safran).

A 35% reduction of combustor cooling air was achieved, resulting with equivalent reduction of NOx formation.

The project has been a major milestone in terms of CMC design and integration in a complex environment and resulted with significant TRL enhancement. However, additional rig testing would be required prior to engine evaluation, and should include adjustment of the air flow split, in order to enable a temperature profile compatible with HP turbine life.

21.3.2 R&D on CMC Turbine Components

Due to competitive constraints and high stakes involved in this domain, limited information has been made public by the industry so far.

Safran experience with low pressure turbine blades for CFM56-5B type engines [12] Following on from the initial feasibility studies conducted in 2008, Herakles has designed and manufactured CMC turbine blades (Cerasep® A40C material) for the LP turbine wheels of a CFM56-5B engine (Figure 21.12).

The subsequent tests of CMC blades performed by Snecma, mid of 2010, in an engine were a world first.

The tests demonstrated the viability of these materials for the moving parts of aeronautical engines and initiated the development of the CMC technology for LP turbine of future commercial engines.

General Electric next-generation engines [9] The preliminary evaluation of CMC components performed on military engines (F136 with static components and F414 with rotating LPT part, cf. Section 21.2.3) have made the technology eligible to be incorporated in GE's next generation of commercial engines.

CMC materials are likely to prove attractive on the next generation of wide-body engine, with key potential benefit expected on weight savings. For example, GE has estimated that incorporating CMC turbine blades on a GE90-sized engine could reduce the overall weight by about 455kg which represents about 6% of the dry weight of the GE90-115 engine.

FIGURE 21.12
Cerasep® A40C LPT blades for CFM 56 engine test (Safran).

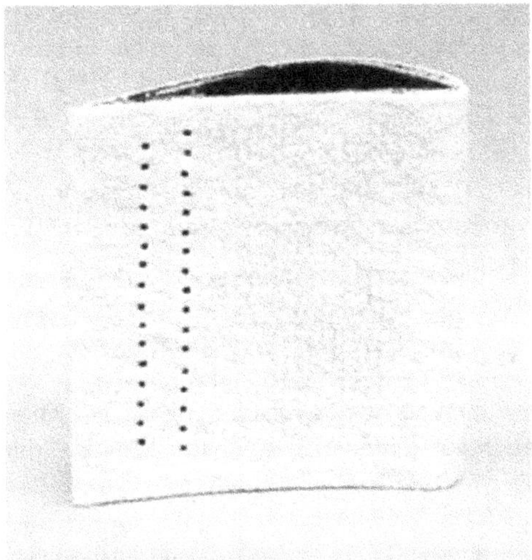

FIGURE 21.13

As-fabricated SiC/SiC vane with an EBC (NASA).

NASA UEET program [10] The NASA UEET program has sponsored efforts to develop CMC turbine technology, particularly for vane components.

The activities [13, 14] involved R&D at NASA GRC on CMC vane subelement fabrication and testing. Research work dedicated to developing innovative fiber architecture led to the manufacturing of SiC/SiC subelement vanes (SiC CVI and MI slurry cast process) with a NASA-developed EBC (Figure 21.13).

Testing was performed in turbine environment using the NASA high pressure burner rig (HPBR) (Figure 21.14). The test program consisted of 50 hours of steady-state operation and 102 thermal cycles (2 minutes) with gas temperatures from 900–1050°C to 940–1440°C. The EBC surface temperature reached ~1300°C. Inadvertent exposure to standing water and deposition of molten alloy prior to combustion exposure led to damage in the vanes resulting from material spallation. However, the overall test results showed the potential of SiC/SiC composites for airfoil applications.

FIGURE 21.14

Burner-rig test of a CMC turbine vane assembly (NASA).

Similar R&D has been implemented by UTRC [15] which has developed an internally cooled CMC vane including the evaluation of several design options to reduce the thermal and mechanical stresses.

The targeted surface temperatures were 1482°C on the EBC and 1315°C at the CMC/EBC interface. The selected CMC materials used the Sylramic-iBN slurry-cast MI SiC/SiC systems from GE and Goodrich with advanced four-layer (Si/Mullite/BSAS/Mullite) EBC configuration developed by UTRC.

CMC airfoils were fabricated and vane assembly that could be incorporated in the Pratt & Whitney 27 MW FT8 engine, were satisfactorily tested in an atmospheric burner-rig test up to 1315°C during 6 hours and then to 100 thermal cycles (2-minute duration each) between 1315°C and 482°C.

21.3.3 Development Activities in the Exhaust Section

The trends of the ongoing engine developments are to run the engines at higher temperatures and to look for weight savings on the engines and the integration components. The exhaust sections which are traditionally made from metallic materials are affected by the following.

- State-of-the-art titanium alloys are reaching their temperature allowable.
- Nickel-based superalloys are alternate solutions, but they are adding weight to the components.

CMCs may be viable and improved alternatives, bringing better thermal margins, as a result of their higher temperature resistance, and component weight savings, as a result of their lower specific gravity. Life durability and acoustic behavior might be also criteria in favor of CMC technology for exhaust components.

R&D activities are ongoing in the United States and in France targeting exhaust section components.

21.3.3.1 FAA CLEEN Program in the United States [16] The Continuous Lower Energy, Emissions, and Noise (CLEEN) program is the FAA's principal NextGen environmental effort to develop and demonstrate new technologies, procedures, and sustainable alternative jet fuels. One of those new technologies evaluated is a noise reducing engine nozzle (Figure 21.15).

Under the CLEEN program, Boeing is teaming with partners ATK/COI-C, Albany Engineered Composites (AEC), to develop and demonstrate an acoustic CMC (oxide) exhaust nozzle on a Rolls Royce Trent 1000 engine. The CMC technology will reduce weight and increase the temperature capability of future engine nozzles, reducing aircraft fuel burn by 1%.

The exhaust assembly includes a CMC nozzle and a CMC centerbody.

As a support to the final design, various studies have been performed on subcomponents manufacturing, thermal testing on cylinder, and dynamic test on centerbody subcomponent.

The prototype component supplied by AEC and ATK/COIC is an oxide CMC (2D Nextel 610 fiber/aluminosilicate matrix). The design of the component performed by Boeing provides a 15% weight saving over titanium.

The plan is to conduct ground tests on a Rolls-Royce Trent 1000 engine by the end of 2012, followed by a flight test late 2013 on the B787 "ecoDemonstrator" aircraft.

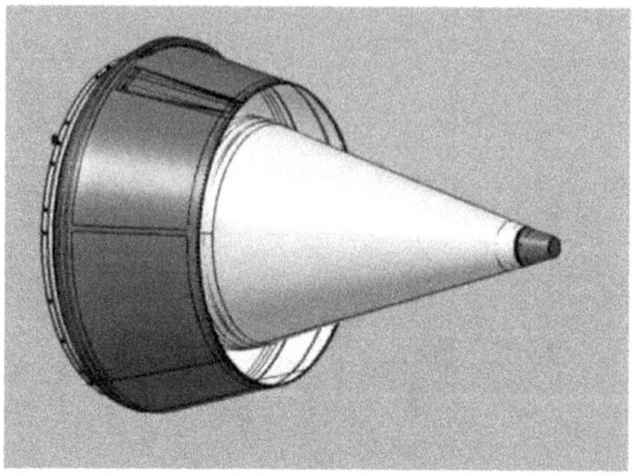

FIGURE 21.15

CMC exhaust system components for CLEEN demonstration (Boeing).

21.3.3.2 Safran R&D on CMC Exhaust Components

CMC mixer demonstration for CFM56 engine A first technology demonstration dedicated to the evaluation of a CMC mixer for a CFM56-5C engine, powering the A340-200/300 jets, has been performed by Snecma, Aircelle, and Herakles, sponsored by a French DGAC initiative.

The main functional features of the mixer component are:

- to contribute to the overall engine exhaust performance, including its weight reduction;
- a direct assembly to the aft flange of the engine; and
- complex shape and large-size component enabling the efficient mixing of the hot primary gas flow with the fan cold air flow.

The component design, jointly performed by Snecma and Herakles, showed a 30% weight saving over Inconel. Additional double digit weight saving could be achieved by optimizing the overall design, especially at the attachment interface with the turbine flange.

It provides also extended temperature capability over Ti alloys for enhanced performance engines.

As part of the engineering phase, subelements components have been developed for validation of the manufacturing process due to the complex shape of the component; they were also dedicated to dynamic, thermal, and mechanical tests aimed at validating the design singularities.

The prototype mixer, manufactured with Cerasep® A40C, was ground tested in 2007 for performance evaluation and then submitted to a ground endurance phase covering 700 engine cycles and including 70 take-off hours, with no material damage identified by the posttest analysis (Figure 21.16).

CMC centerbody demonstration [17] Safran has proceeded in February 2012 to the first flight test of CMC centerbody on an Airbus A320 test flight aircraft equipped with a CMC exhaust centerbody designed and produced by Safran/Herakles (Figure 21.17).

That is a second important milestone in the CMC exhaust development program aimed at reducing the carbon footprint of commercial aviation.

The component is using an improved Cerasep® A40C material for enhanced life duration. The design phase indicates a weight saving of 9 kg (centerbody only) over the metallic counterpart, while it reaches 40 kg for the overall exhaust (nozzle and centerbody) without acoustic reduction system.

In order to match the weight of the baseline metallic component, which was a requirement for the flight test, the CMC exhaust included an acoustic reduction subsystem providing an acoustic reduction of 0.5 dB.

Prior to the flight program, the prototype component sustained rig testing for acoustic evaluation on a dedicated GE aviation test rig, then engine ground testing for performance measurement, and risk mitigation. In parallel, extensive material testing has been performed on test specimens to evaluate the material behavior in environment (thermal cycling, vibration, lightning, etc.).

The prototype component cumulates 71 hours of test, including 18 flight hours on an Airbus A320 test aircraft.

FIGURE 21.16
CFM56-5C Cerasep® A40C mixer demonstration (Safran)®.

FIGURE 21.17
Flight demonstration of a Cerasep® A40C centerbody.

A certification modification has been requested and is expected by the end of 2012, in order to prepare the next step consisting of an in-service endurance evaluation on an Air France A320 jet, due to start in 2013 for 2 years operation. That phase should deliver 5000 hours endurance and extend significantly the field service behavior experience for the technology.

21.3.4 CMC Program Demonstrations on Industrial Gas Turbine

SiC- and oxide-based CMCs have been extensively evaluated since the 1990s for the hot section components of industrial gas turbines with the objective of improving fuel efficiency and reducing exhaust emissions of NOx and CO.

Most of the demonstration activities have been conducted under US government-sponsored programs and performed by industrial consortiums led by Solar Turbines, Inc., and General Electric, and ample literature has been published through the ASME Turbo Expo conferences.

Solar Turbines field evaluations [18, 19] Solar Turbines, Inc., has been evaluating CMC combustor liners in test rigs and Solar Centaur 50S engines since 1992, targeting the development of advanced low emissions combustors. The field testing of CMC combustor liners, which started in 1997, at first with SiC/SiC materials and later on with Ox/Ox components, has been a way to gain in-service long-duration test experience.

The combustor inner and outer liners have been designed under the CSGT program to be interchangeable with the Centaur 50S Hastelloy X combustor liners.

After subscale combustor liner tests which demonstrated the potential of the technology to produce low NOx and CO emissions, full-scale combustors were then tested in ambient pressure rig and then in engine environment with temperatures reaching 1200°C, in order to qualify the engine for longer term field testing.

The first field tests performed on SiC/SiC liners protected by a SiC seal-coat at a Texaco site confirmed the recession of the seal-coat on the gas path surface after a few thousands of hours exposure and confirmed the predictions issued by ORNL after lab-furnace testing.

The oxidation/corrosion effect observed on the surface of the component exposed to the hot-gas path in the combustor environment is due to the water vapor formed in the combustion process which then reacts with the oxide species (SiO_2) formed on the CMC surface and result by surface recession caused by the oxide volatilization ($Si(OH)_4$).

To overcome such recession effects and extend life-in-service of SiC/SiC combustor liners, specific environmental barrier coatings (EBC) have been developed from a combination of water-resistant oxide compositions (mainly BSAS and mullite). Such coatings are required on the hot-gas path CMC surface for service time exceeding 5000 hours at temperatures above 1000°C.

Engine field tests of SiC/SiC liners with EBC manufactured by Goodrich and GE (ex Dupont Lanxide Composites) have been operated on Solar Centaur 50S gas turbines at two commercial sites (ChevronTexaco Bakersfield site and Malden Mills textile facility). Over the 1997–2007 period, 15 144 and 13 937 hours have been logged on SiC/SiC combustor liners with EBCs at each location.

The various evaluations performed show:

- in the absence of EBC, an improved durability of the melt-infiltrated SiC/SiC compared to SiC/SiC fabricated by SiC CVI and
- ceramic oxide-based EBCs improve SiC/SiC liner life by a factor of 2–3 and provide to both kinds of CMCs an equivalent life.

The concept of hybrid oxide CMC has been developed by a Siemens Power Generation led team in a NIST-sponsored program.

The technology has also been evaluated by Solar Turbines, Inc., on combustor liners with Centaur 50S engine.

The hybrid oxide/oxide CMC concept combines an oxide substrate (e.g., alumina or aluminosilicate matrix with Nextel fiber from COI Ceramics, Inc.) with a Friable Graded Insulation (FGI) patented by Siemens.

The FGI is a thick (3–5 mm) coating, consisting of hollow microspheres, which provides a refractory thermal barrier layer keeping the CMC temperature below 1000°C.

In the 2006–2007 timeframe, a hybrid oxide CMC combustor outer liner has survived for more than 25 000 hours in a Solar Turbines Centaur 50S engine field test, setting a new record for CMC combustor liners.

GE CMC shrouds for gas turbines [10] Like combustor liners, CMC turbine shrouds have been identified as offering opportunity for fuel savings, reduction in emissions, and cost reductions in gas turbine applications.

The higher temperature capability of CMCs allows reduction of cooling air which results with improvement of fuel efficiency and higher output.

R&D in this domain has been performed in Japan (IHI Ltd for the ESPR program) and in the United States with General Electric Power Systems.

After several risk reduction testing at lab scale, then rig and small engine testing, GE PSC has proceeded in 2002–2003 with the successful evaluation of MI SiC/SiC turbine shrouds on its 7FA advanced gas turbine (~170 MW in combined cycle).

In a rainbow test, 9 out of 96 first stage inner shrouds were exposed to more than 5000 hours of firing with 15 start/stop cycles and temperatures up to ~1250°C. The MI SiC/SiC shrouds were fabricated with the GE Hi-Nicalon HyperComb Prepreg and slurry-cast systems protected by a BSAS-type EBC. Localized EBC spoliation was observed, however, no shroud failures or performance anomalies were reported.

21.4 SUMMARY AND INSERTION ISSUES

A wide variety of CMC evaluations have been implemented in the hot section of gas turbines over the past 20 years, including complex-shaped components and lately rotating components.

Beside the testing experience gained on the components, especially in the field of land-based gas turbines for long-term endurance, those R&D initiatives have contributed to significant improvement of the design and lifing methodologies, to the emergence of more performing material solutions, as well as innovative fiber-reinforcement architectures. Manufacturing feasibility issues have also been addressed as part of those projects.

Outstanding successes have been accomplished with the latest material technologies, especially on static components. The application to rotating components is a major challenge which will bring a major technology breakthrough for future jet engines. It requires continuous material improvements to match the high strength requirements, as well as design adjustments to act on the most critical areas of the component.

So far, mass production of CMC components has been limited to a few components, mainly for military applications. As a consequence of those niche market applications, some challenges remain, especially in term of cost-effective solutions and database generation.

Because of the current limited production, the relatively high acquisition cost of hot section CMC components must be balanced by life cycle cost benefits to bring a positive business case to new applications. In this domain production volume will be required to cut the cost down as it has been highlighted in the past with the significant production development of carbon/carbon material for aircraft brakes.

Generation of substantial database is another challenge to bring confidence to the end-user community and to achieve certification process for civil applications.

REFERENCES

1. E. P. Bouillon, S. Bertarnd, G. Habarou, and P. Spriet. Ceramic matrix composite for aircraft engine applications. 2nd EUCASS Conference, Brussels, Belgium, July 2007.
2. P. Spriet and A. Lasalmonie. New trends for the manufacturing in the aeronautic industry. Hegan Conference, San Sebastian, Spain, May 24–25, 2000.
3. L. P. Zawada, G. Y. Richardson, G. Doppes, and P. C. Spriet. CMCs for aerospace turbine engine exhaust nozzles. Proceedings of the 5th International Conference on High Temperature Ceramic Matrix Composites, HT-CMC5, conference, Seattle, WA, September 12–16, 2004.
4. J. M. Staehler and L. P. Zawada. Performance of four ceramic-matrix composite divergent flap inserts following ground testing on an F110 turbofan engine. *J. Am. Ceram. Soc.*, 83(7), 1727–1738 (2000).
5. L. Zawada, E. Bouillon, G. Ojard, G. Habarou, C. Louchet, D. Feindel, P. Spriet, C. Logan, T. Arnold, K. Rogers, et al. Manufacturing and flight test experience of ceramic matrix composites seals and flaps for the F100 gas turbine engine. ASME Paper No. GT2006-90448 (2006).
6. Super Hornet Boeing's multi-mission strike fighter. *Air International*, July, 57–61 (2012).
7. AFRL Brochure: IHPTET and VAATE Turbine Engine Technology.
8. GE primes CMC for turbine blades. *Flight International*, November 17 (2010).
9. Turning up the heat on CMCs. *Flight International*, November 23–29 (2010).
10. J. A. DiCarlo and M. van Roode. CMC developments for gas turbine engine hot section components. ASME Paper No. GT2006-90151 (2006).
11. P. Spriet. Aerospace & industrial applications of C/C, C/SiC & SiC/SiC composites at SPS. MS&T'11 Conference & Exhibition, Columbus, OH, October 17, 2011.
12. Safran/Herakles. *Civil Aeronautics*, http://www.herakles.com/aeronautique/aeronautique-civile/?lang=en (accessed June 26, 2014).
13. A. Calomino and M. Verrilli. CMC vane subelement fabrication. ASME Paper No. GT2004-53974 (2004).
14. M. Verrilli, A. Calomino, R. C. Robinson, and D. J. Thomas. CMC vane subelement testing in a gas turbine environment. ASME Paper No. GT2004-53970 (2004).
15. V. Vedula, J. Shi, D. Jarmon, S. Ochs, L. Oni, T. Lawton, K. Green, L. Prill, J. Schaff, G. Linsey, et al. CMC turbine vanes for gas turbine engines. ASME Paper No. GT2005-68229 (2005).
16. M. Petervary and T. Steyer, The Boeing Company. Ceramic matrix composites for structural aerospace applications. 4th International Congress on Ceramics, Chicago, IL, July 17, 2012, and FAA/CLEEN Internet releases.
17. Les thermostructuraux sur avions civils. *Air & Cosmos*, April 27 (2012).
18. M. van Roode, J. Price, J. Kimmel, N. Miriyala, D. Leroux, A. Fahme, and K. Smith. CMC combustor liners: a summary of field evaluation. ASME Paper No. GT2005-68420 (2005).
19. M. van Roode and A. K. Bhattacharya. Durability of oxide/oxide CMCs in gas turbine combustors. ASME Paper No. GT2012-668974 (2012).

CHAPTER 22

CERAMIC MATRIX COMPOSITES: NUCLEAR APPLICATIONS

CÉDRIC SAUDER

CEA, DEN, DMN, SRMA, F-91191 Gif Sur Yvette, France

22.1 INTRODUCTION

The previous chapter brought to light specific uses for ceramic matric composites (CMCs) in aerospace and industry. It was also shown that these materials have strong potential for development in these applications, although there are still several obstacles to be cleared for their larger-scale use. In the nuclear industry, while some constraints are shared with aerospace and conventional industry (such as operating temperature, resistance under oxidizing and weakly oxidizing conditions, and high temperature mechanical strength), there are additional constraints (such as the effects of irradiation and containment of radioactive substances), which make their use in nuclear power plants more complex and only conceivable in the long or very long term (>20 years). CMCs are mainly under consideration for use as a cladding material (providing fuel containment and heat exchange) or as a structural material, for both fission and fusion reactors [1–10]. They would therefore be directly exposed to a neutron flux (whose intensity depends on the type of reactor), which could significantly modify their dimensional, thermomechanical, or thermal properties. Furthermore, the very high level of safety required for the construction of a nuclear reactor involves long and costly qualification phases. This explains why, until the 2000s, studies were exclusively led by R&D programs for experimental fusion reactors [1–7]. However, over the past 10 years or so, various applications have been developed for fission reactors [7–11], in particular thanks to the appearance of "third-generation," high purity SiC fibers [12, 13].

The use of composite materials as cladding or structural materials in the nuclear industry is based on requirements that are common to fusion and fission reactors. Some of the common, indispensable properties expected of CMCs for this type of application are given below:

- High thermomechanical strength over a wide range of temperatures.
- High thermal conductivity (including under irradiation).
- Low activation under irradiation (very strict purity requirements).
- High melting/decomposition point (ideally >2000°C).
- Low swelling under irradiation.
- Low impact of irradiation on thermomechanical properties.

In the case of fusion reactors, CMCs are under consideration as plasma-facing components (PFCs), both as complete structures and components thereof, and would therefore be exposed to both a high heat flux and a high energy

Ceramic Matrix Composites: Materials, Modeling and Technology, First Edition. Edited by Narottam P. Bansal and Jacques Lamon.

(14.1 MeV) neutron flux. They must therefore meet very-specific extreme constraints for these applications, which are described in the first part of this chapter.

For fission reactors, research is mainly focused on increasing the high temperature strength of materials for better safety on facilities (while higher temperatures can also lead to higher power-generation efficiencies) [14, 15], and for the design of fourth-generation reactors [9–11, 16]. A more comprehensive description of the various possible applications for CMCs in fission reactors is given later in this chapter, and provides a glimpse of applications that are probably more easily accessible than cladding in the medium term.

Chapter 13 showed that the CMCs which have the greatest potential for nuclear applications are long-fiber carbon composites (C), carbon fiber-reinforced carbon matrix composites (C/C) and, in particular, long-fiber silicon carbide composites (SiC) and silicon carbide fiber-reinforced silicon carbide matrix composites (SiC/SiC). C/C composites were the first to reach industrial maturity such that their use in the nuclear industry could be considered. However, their behavior under irradiation (see Chapter 13) means that they are limited to uses where radiation doses do not exceed 10 displacements per atom (dpa), that is, a fluence of approximately 1×10^{26} neutrons per m^2 [17–19]. Despite this, C/C composites have been used and tested in experimental fusion reactors and, thanks to this experience feedback, many optimizations have been implemented in the manufacturing processes. The third part of this chapter describes these manufacturing processes and the properties of the C/C materials obtained.

Finally, SiC/SiC composites are the most promising for use in nuclear reactors. In particular, they provide better resistance under neutron flux, even under high doses (>40 dpa) [20]. They also have better high temperature resistance under weakly oxidizing and oxidizing conditions than C/C composites. However, to be compatible with such applications, new concepts need to be developed and the most suitable manufacturing processes chosen. This is the subject of the last part of this chapter.

22.2 CMC FUSION APPLICATIONS

22.2.1 Introduction

The long-term goal for nuclear fusion reactors is to maintain a continuous, or semi-continuous, fusion reaction and to extract the energy to produce heat and/or electricity. It is generally accepted that these reactors will use the Deuterium–Tritium (D–T) fusion reaction:

$$D + T \rightarrow He\,(\sim 3.5\ \text{MeV}) + n\,(\sim 14.1\ \text{MeV}) \tag{22.1}$$

As helium is produced as charged particles, they are confined by the magnetic field and thus heat the plasma. The neutrons produced leave the plasma carrying their energy out.

As deuterium is abundant in seawater (33 g/m^3), fuel for the fusion reaction is considered inexhaustible. Furthermore, unlike fission reactions, this reaction does not generate radioactive products. However, for this fusion reaction to occur, the fuel (D + T) must be maintained at an extremely high temperature, of the order of 10^8 K, and at a certain pressure (10^{-5} Pa), in the form of a plasma confined in a reaction chamber. A magnetic field is used to prevent any contact between the plasma and the walls of the containment vessel, which makes it possible to maintain the reaction over significant periods. Several research reactors use this principle and are devoted, in particular, to improving understanding of plasma physics: Tore Supra and ITER in France, Wendelstein 7-X in Germany, Joint European Torus (JET) in the United Kingdom, Tokamak Fusion Test Reactor (TFTR) in the United States, HT-7 in China, and JT60U and JT60SU in Japan.

Currently, the concept that has been most studied is plasma confinement using a toroidal magnetic field, called the "tokamak" system (Russian acronym for toroidal magnetic chamber). In particular, this is the concept used for the International Thermonuclear Experimental Reactor (ITER) project (Figure 22.1) [21]. Its tokamak will be the largest ever built, with a volume of 840 m^3 and will operate at 150 million degrees. It should produce 500 MW of power from an input power of 50 MW. The planned pulse width (burn cycle) is approximately 400 s. The aim of ITER is to demonstrate the scientific and technological possibility of power generation from nuclear fusion. If it succeeds, it should open the way to the design of a Fusion Power Reactor (FPR). ITER itself is not designed for power generation and has no system for converting the thermal power produced by the plasma into electricity. Furthermore, it has no system for producing and

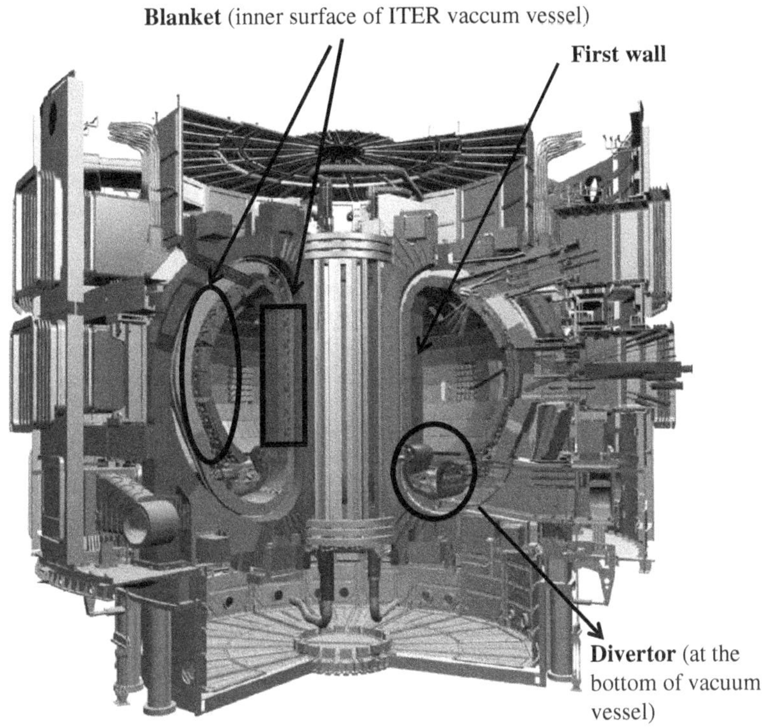

FIGURE 22.1
Schematic illustration of ITER with localization of plasma-facing components [21].

extracting tritium from the "blanket" to continuously supply fuel for the reaction. These systems must exist on future FPR, which represents a major technological challenge, in particular with regard to structural materials.

The development of components and new materials that are able to meet the requirements of research and power reactors is a key issue. This is especially true for PFCs. Such components are directly subject to irradiation from neutrons and helium from the plasma. While the helium atoms only penetrate a few microns thickness, the high energy (14.1 MeV) neutrons produced can interact with these components over a thickness of approximately 1 m. These PFCs, in particular, the first wall and the structures behind it, will therefore be subject to an intense neutron flux of several tens of dpa per Full-Power-Year (FPY) and a very high heat flux of between 2 and 6 MW/m^2 in a power reactor [22]. Finally, it should be specified that, although the heat flux is relatively uniform around the circumference of the vessel during normal operation of such a fusion reactor, the plasma breakdown phase can be violent (plasma disruption). This can lead to plasma instabilities which instantly release its energy onto the PFCs, leading to a significant rise in heat flux compared with normal operating conditions. This "accident-condition" heat flux specifies the boundary condition for these components. In addition to high thermal loads, they are therefore also susceptible to significant erosion during operation and during plasma disruptions. This would lead to impurities being introduced into the plasma, leading to its cooling by loss of energy as largely discussed elsewhere [6]. It is clear that, ideally, PFCs must be made of materials based on light elements (i.e., with low atomic number), which also have a low tendency to erode and migrate into the plasma. Clearly, the best compromise between the material's atomic number and its erosion behavior must therefore be determined.

It is mainly for the manufacture of these components (PFCs), where the temperature and irradiation conditions are the most intense, that CMCs (C/C and SiC/SiC) have been, and are currently, under consideration for use in experimental reactors and future power reactors. Worldwide, there are three fusion reactor concepts currently being studied whose operation is based on the use of SiC/SiC CMCs: (i) TAURO led by the European Union, (ii) DREAM (DRastically EAsy Maintenance) led by Japan, and (iii) ARIES led by the United States [9].

PFCs essentially divide into three distinct components as shown in Figure 22.1: (i) the divertor, (ii) the first wall, and (iii) the blanket, located immediately behind the first wall, for which specific operating constraints impose particularly stringent properties for CMCs. Among the CMCs under consideration (C/C and SiC/SiC), SiC/SiC composites probably offer the greatest potential for power reactors, given their behavior under irradiation (see Chapter 13) and the low activation of SiC under neutron irradiation [1, 5, 23]. This is because the use of low activation materials (LAMs), in

FIGURE 22.2
Illustration of ITER divertor [25].

particular for short-term activity, is perceived as a determining constraint to ensure the safety of the facility (easier maintenance) and low environmental impact (simpler waste management) [24].

22.2.2 The Divertor

22.2.2.1 Description The divertor is located on the floor of the vacuum vessel (Figure 22.1). It is a device with three main functions. First, it removes charged particles no longer participating in the fusion reaction (such as helium), and plasma impurities, from the plasma. Without this function, the plasma would cool and fusion reactions would no longer be possible. Second, it extracts heat (particles that hit it transfer their energy to it). Finally, it can control the plasma density via its pumping capacity. As an example, the ITER divertor is shown in Figure 22.2 [25]. It comprises 54 modules (or cassette body) that can be removed by remote handling. Each module (Figure 22.3) contains three PFCs [26].

These components are also called "targets": the inner vertical target (Figure 22.3a), the outer vertical target (Figure 22.3b), and the dome (Figure 22.3c). The targets are located at the intersection of magnetic field lines, where the high energy particles from the plasma strike the components. The kinetic energy of these particles is transformed into heat. The heat flux received by these targets is extremely intense and requires active water cooling. Therefore, whatever material is used for the targets, this material must be bonded to the body of the module to ensure good heat transfer into the coolant. It is precisely for the manufacture of these targets that CMCs are planned for use on ITER [27, 28] and on power fusion reactor concepts such as DREAM [2]. CMC targets have also been tested in other experimental

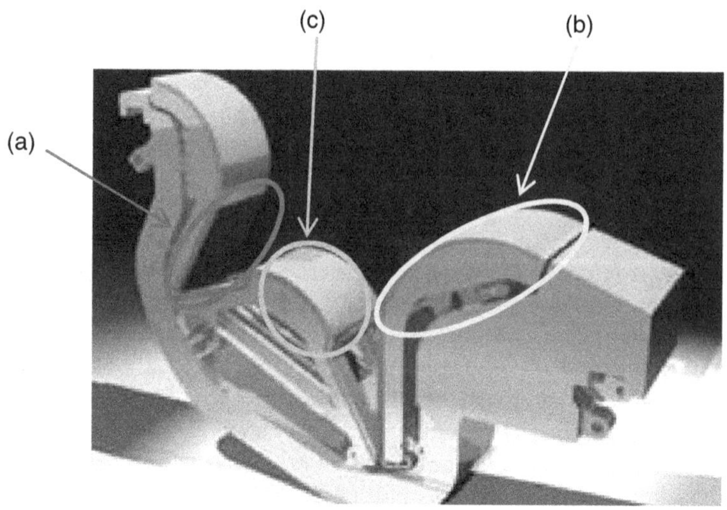

FIGURE 22.3
Illustration of a cassette body of ITER divertor [26] with: (a) inner vertical target, (b) outer vertical target, and (c) dome.

TABLE 22.1 Main Operating Conditions of the ITER Inner Vertical Target [27, 28]

	Heat flux (MW/m^2)	Number of cycles	Duration	Temperature (K)
Normal operation	5–10	>3000	<450 s	Up to 1773
Off-normal operation	20	>300	10 s	Up to 3273
Disruptions	10–30	–	0.1–3 ms	–

reactors such as Tore Supra [4, 6]. The many constraints associated with this application impose particular specifications for CMCs, which are described in section 22.2.2.2.

22.2.2.2 CMC Specifications for Divertor Structural Component Applications The minimum specifications required for divertor component applications, or other fusion applications, directly depend on the technical solutions adopted and therefore on reactor design. To simplify this analysis, only constraints specific to the ITER experimental reactor are given, while the constraints associated with power-reactor concepts may be slightly different [2]. Table 22.1 gives the main thermal constraints associated with the application "inner vertical targets of the ITER divertor." As stated in Section 22.2.2.1, these targets must be bonded to a copper-based alloy (CuCrZr) to remove the heat produced [6, 27, 28]. Figure 22.4 shows the two designs (flat tile and monobloc) under consideration for the shape of the CMCs [29], although the monobloc configuration is considered to be the reference design for ITER [30, 31].

The main constraint is therefore the removal of a heat flux of several MW/m^2 over significant periods of time (up to 450 s). For comparison, the heat flux on the Space Shuttle's nose cone was approximately 1 MW/m^2. Heat fluxes of this magnitude mean that the CMC material must have excellent resistance to thermal shock. These materials must therefore have very high thermal conductivities and low coefficients of thermal expansion to limit as far as possible the mechanical stresses generated within the material. For example, on ITER, C/C material should have an average thermal conductivity ≥300 W/(m K) at room temperature (RT) and ≥150 W/(m K) at 1000°C for the direction of highest thermal conductivity [32]. This is therefore a significant technological challenge, especially as the material will be subject to irradiation by high energy particles which can lead to significant erosion of the material. The CMC must therefore possess significant erosion resistance to limit plasma pollution. An additional constraint is to select a material that has low retention of hydrogen (deuterium and tritium). This is because the safety authorities impose a maximum tritium inventory. For example, on ITER, the authorized limit is 700 g of tritium present in the vacuum vessel. Finally, the radiation doses received by the material will be relatively small on ITER (~0.2 dpa) but would be much larger on a power reactor, which is the long-term objective [2]. Ideally, therefore, a material is required whose characteristics (thermal conductivity, erosion resistance, and thermomechanical properties) change little under irradiation.

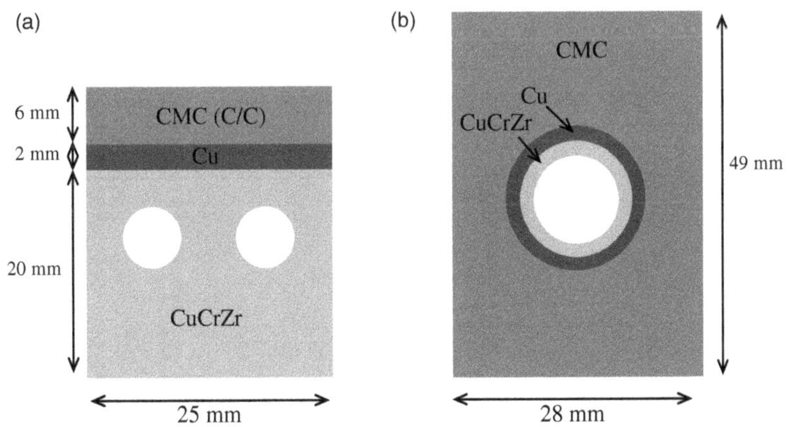

FIGURE 22.4
Geometry of PFCs: (a) flat tile of tore Supra, (b) monoblock of ITER.

22.2.3 First Wall and Blanket

22.2.3.1 Description The first wall is the first non-gaseous, plasma-facing material. This material is exposed to very intense heat and neutron fluxes. The blanket, which is located directly behind the first wall, will perform several roles in fusion power reactors. Its first role will be to recover the energy from the fusion reaction and (via a coolant) convert it into electricity. It will also need to continuously produce and transport tritium (to maintain the fusion reaction) using the neutron flux produced by the fusion reaction. This function is made possible by the reaction in Equation (22.2), where the neutron is provided by (22.1).

$$\mathrm{Li} + n \rightarrow^{4} \mathrm{He} + \mathrm{T} \tag{22.2}$$

Finally, the blanket must protect other components (such as the magnets) and the environment from the neutrons produced by the fusion reaction (1). Furthermore, the higher the maximum operating temperature of the blanket, the higher the power-generation efficiency and therefore the lower the cost of electricity production.

The doses received by these materials will be of the order of 20–30 dpa per year, which clearly excludes C/C composites (see Chapter 13). The use of SiC/SiC composites as structural materials for the blanket can be viewed as a high risk, long-term challenge, but one that would bring significant benefits if successful. Indeed, the main advantage of SiC (and therefore of SiC/SiC composites) with respect to the other candidates (Reduced Activation Ferritic Martensitic steels or Vanadium alloys) is associated with its low activation [5, 23, 24], along with its operating capacities and its mechanical strength at high temperatures. Furthermore, in Chapter 13, it was seen that SiC has good behavior even under very high doses of radiation. For these reasons, SiC/SiC composites are under consideration for the manufacture of the blanket in the three fusion power-reactor concepts (TAURO, DREAM, and ARIES) described elsewhere [2].

Table 22.2 lists the strengths and weaknesses of the three main candidates considered for fusion-reactor PFCs: C/C composites, SiC/SiC composites, and tungsten. Although it has many good qualities, Beryllium (Be) has not be used as a candidate in this comparison because it is not considered refractory ($T_f \ll 2000°C$).

22.2.3.2 Constraints Associated with CMC for First Wall and Blanket Applications In the case of the first wall, the main stresses to which the material is subjected are the heat flux (of several MW/m^2) and the high energy (14 MeV) neutron flux. The blanket and first wall may be difficult to separate in the event of the use of CMC, which explains why only the combination first wall + blanket is covered here. As an example, and to give an idea of the difficulties inherent in the manufacture of this blanket, Figure 22.5 shows a module corresponding to the blanket design for TAURO [2]. Currently, no industrial manufacturer can produce such large systems in CMC. Their manufacture is one of the major challenges for the future.

As the blankets have a "structural" role, they are also subject to many constraints. Table 22.3 summarizes operating conditions and design requirements on SiC/SiC composites for the three blanket concepts: TAURO, DREAM, and ARIES [9, 23, 33].

In addition to the manufacturing constraints regarding the shape of the planned structures, the CMC material will be subject to a surface heat flux and a neutron flux through a significant thickness. It must therefore have high thermal conductivity in operation (under neutron flux) and high resistance to irradiation for doses much greater than 100 dpa. For the TAURO design, compatibility with Pb17Li coolant up to approximately 700°C must also be guaranteed.

TABLE 22.2 Strengths and Weaknesses of the Three Main Candidates Considered for Fusion-Reactor PFCs

Critical PFCs properties for fusion reactors	Requirement on PFCs	C	SiC	W
Plasma compatibility	Low atomic number	+	+	–
Plasma life	Low erosion rate	–	–	+
High operating temperature	High melting temperature (T_f)	+	+	+
Cooling down capability	High thermal conductivity (λ)	+	–	+
H (D + T) chemical compatibility	Low retention/permeation rate of H (D + T)	–	+	+
Compatibility with neutron irradiation	Stability of properties for high dose (>100 dpa)	–	+	–
Integration—joining	Large systems processing capability	–	–	+
Waste management	Low activation and toxicity	–	+	–

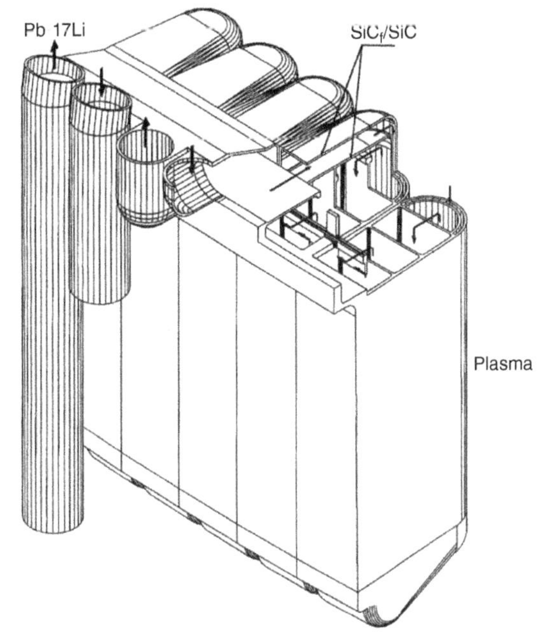

FIGURE 22.5
Outboard module of the TAURO Blanket design [2]. Reproduced with permission by Elsevier BV, Fusion Engineering and Design, Vol 55, Iss 1, 2001.

TABLE 22.3 Operating Conditions and Design Requirements on SiC/SiC Composites in TAURO, DREAM, and ARIES Blanket Concepts [9, 23, 33]

	TAURO	DREAM	ARIES
Surface heat load (MW/m^2)	0.5	0.5	0.5
Neutron wall load (MW/m^2)	2	5	3.3
Coolant	Pb17Li (2 MPa)	He (10 MPa)	He (10 MPa)
Inlet/outlet temperature (K)	528/948	973/1173	623/923
Breeding material	Pb17Li (liquid)	Li_2O (pebbles)	Li_2ZrO_3 (pebbles)
Neutron multiplier	Pb17Li	Be	Be
First-wall protection	SiC CVD	SiC CVD	SiC CVD
Density (g/cm^3)	2.5	2.5	2.5
Porosity (%)	10	10	10
Fiber content (%)	40		40
Thickness (mm)	6–10		6–10
Thermal conductivity in plane and through thickness (W/(m K))	~15	~15	~12–15
Electrical conductivity $(\Omega\ m)^{-1}$	5–500		5–500
Thermal expansion ($\times 10^{-6}\ K^{-1}$)	~4	3.3	~4
Mechanical properties			
Allowable stress (MPa)	300	200	~200
Young's modulus (GPa)	200	200	~360
Interlaminar shear stress (MPa)	44		44
Poisson's ratio	0.18	0.18	0.16
Maximum allowable temperature (K)	1273	1373	1273

Compatibility experiments (1000 hours at 800°C) on monolithic SiC specimens lead to the conclusion that no evidence of reaction or wetting between SiC and Pb-17Li was observed [34] (while compatibility with He coolant has already been validated for the other concepts). Finally, as the specifications regarding the mechanical properties of CMC are strongly affected by the thermal conductivity, each new version of the material should be taken into account by updating the mechanical characteristics required.

22.3 CMC FISSION APPLICATIONS

22.3.1 Introduction

Unlike fusion reactors, fission reactors are already widely used for power generation worldwide. Light water reactors (LWRs), boiling water reactors (BWRs), and pressurized water reactors (PWRs) represent most (~90%) of the installed capacity worldwide. Currently, although the fission industry is not under question, the Fukushima Daiichi accident has impacted attitudes and relaunched the debate on the safety of current facilities of all types and of planned future designs. This accident, while due to an exceptional natural catastrophe (which had not been adequately taken into account in the design of the safety systems), has highlighted the weak points of the cladding and structural materials used (mainly zirconium alloys) under severe accident conditions (loss of cooling systems or of control of reactivity). Under such conditions, these materials do not ensure prevention of (partial or total) core melt. A real technological leap that remains to be made is to develop intrinsically safe reactors, that is, reactors with no risk of core melt following a severe accident (such as the one that occurred on the Fukushima Daiichi nuclear power plant). One way to ensure better safety would be to develop materials (such as CMCs) that are more refractory than those currently used, and also less reactive under oxidizing conditions (air + water) during accident situations. While the search for new core materials for this type of reactor was started several years ago [35–37], it remains valid even though these reactors probably do not represent the future of fission reactors. These reactors have the drawback of only using one tiny (<1%) part (^{235}U) of natural uranium (which is mostly ^{238}U) and therefore do not represent a long-term solution for power generation (beyond about a hundred years), given the available uranium resources in the Earth's crust.

It was partly for these reasons (improving safety and better management of natural resources) along with others (such as limiting the quantity of long-lived radioactive waste produced and providing access to hydrogen production) that a wave of projects was launched by the US Department of Energy (DOE) under the title Generation IV Nuclear Energy Systems Initiative in 2001. This Initiative has succeeded in bringing together the nuclear research bodies of 13 countries (the United States, France, Japan, the Republic of South Africa, the Republic of Korea, the United Kingdom, Argentina, Brazil, Canada, and Switzerland in 2002; Euratom in 2003; and the People's Republic of China and the Russian Federation in 2006) into an informal cooperative structure called the Generation IV International Forum (GIF). Figure 22.6 shows the changes between various generations of nuclear power systems and the place that Generation IV Systems are to take in the future [16].

Following two years of work by a hundred international experts, six systems were chosen as the most promising for meeting the four main objectives: (i) sustainability, (ii) safety and reliability, (iii) economics, and (iv) proliferation resistance and physical protection. These six systems are

- Gas-Cooled Fast Reactor (GFR)
- Lead-Cooled Fast Reactor (LFR)
- Molten Salt Reactor (MSR)
- Sodium Fast Reactor (SFR)
- Supercritical-Water Reactor (SCWR)
- Very-High Temperature Reactor (VHTR)

Table 22.4 gives the characteristics of each of these systems [16].

Among these systems, SFRs have been the most studied, and a high power reactor (Superphénix) was even built and operated in France from 1986 to 1998. Other than SFRs, the two most studied systems are VHTRs and GFRs. One of the key points for the deployment of VHTRs and GFRs is the development of materials that provide high temperature

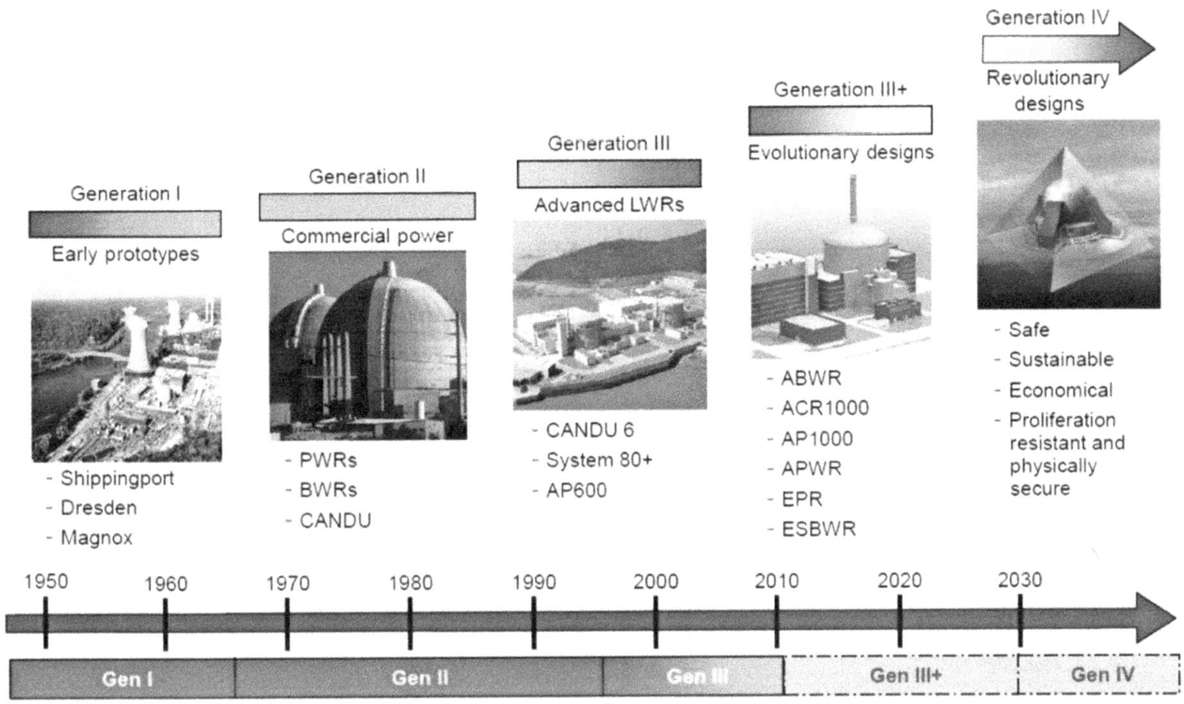

FIGURE 22.6
Evolution of past and future generations of fission nuclear energy [16].

resistance under neutron flux. It is mainly in this context that CMCs (C/C and, especially, SiC/SiC composites) have been the subject of research, whose final goal is to validate their use as structural or cladding materials.

Various applications under consideration for CMCs in the various types of reactor are described in sections 22.3.2 and 22.3.3, with a more precise description of the proposed concepts and the characteristics that CMCs will need to be used in LWRs or Generation IV reactors. A distinction will be made between structural materials (such as for guide tubes, control rods, diagrids, and fuel assemblies) and in-core cladding materials (for fuel cladding).

TABLE 22.4 Overview of the Generation IV Systems

System	Neutron spectrum	Fuel cycle	Size (MWe)	Applications	R&D needed
Very-high temperature reactor (VHTR)	Thermal	Open	250	Electricity, hydrogen production, process heat	Fuels, materials, H_2 production
Supercritical-water reactor (SCWR)	Thermal, fast	Open, closed	1500	Electricity	Materials, thermal-hydraulics
Gas-cooled fast reactor (GFR)	Fast	Closed	200–1200	Electricity, hydrogen production, actinide management	Fuels, materials, thermal-hydraulics
Lead-cooled fast reactor	Fast	Closed	50–150 300–600 1200	Electricity, hydrogen production	Fuels, materials
Sodium-cooled fast reactor	Fast	Closed	300–1500	Electricity, actinide management	Advanced recycle options, fuels
Molten salt reactor	Epithermal	Closed	1000	Electricity, hydrogen production, actinide management	Fuel treatment, materials, reliability

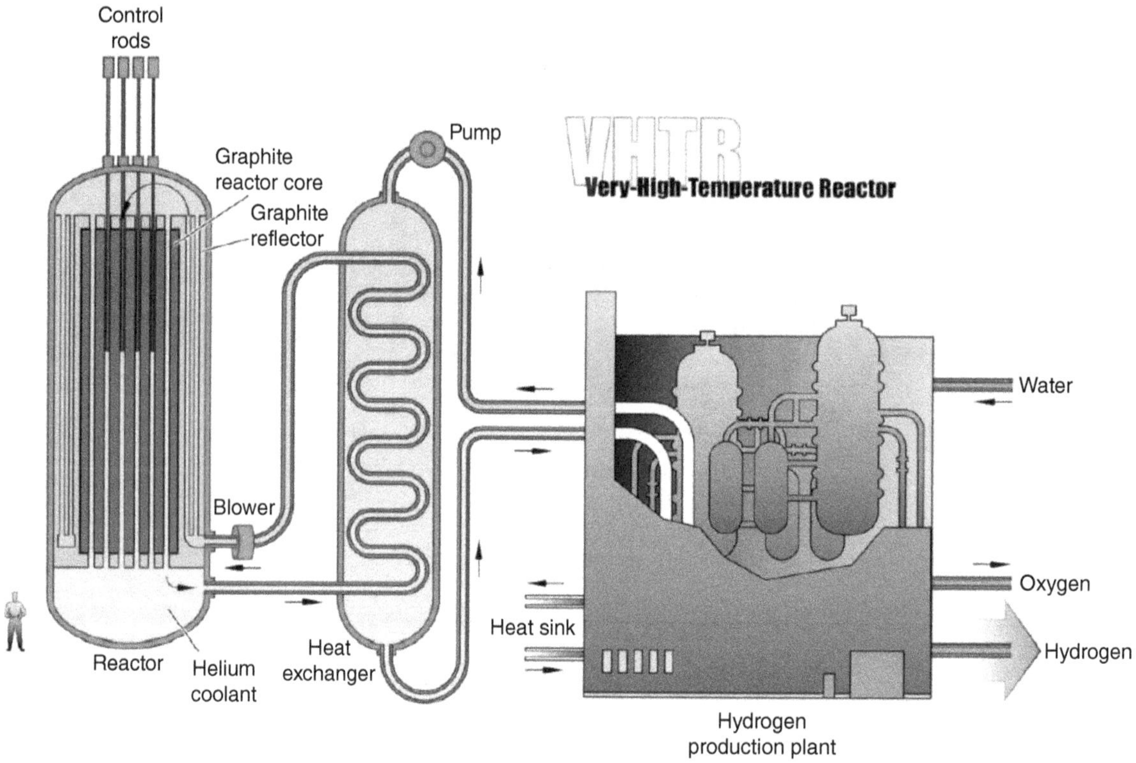

FIGURE 22.7
Schematic illustration of a very-high temperature reactor (VHTR) [16].

22.3.2 CMCs for Structural Materials in VHTRs and SFRs

22.3.2.1 VHTR Applications VHTR is a thermal-spectrum reactor using a graphite moderator and gaseous helium coolant (Figure 22.7). The fuel is in the form of millimeter-sized particles (TRISO particles) [38], fabricated into cylindrical "compacts" which are inserted into graphite-block fuel elements. This type of reactor has the advantage of being intrinsically safe [39] because, according to the designers, it has a negative temperature coefficient that leads the reactor to becoming spontaneously subcritical in the event of accidental situation. Furthermore, the helium coolant would have a core-outlet temperature of about 1000–1100°C. This would provide high temperature heat for various industrial uses such as seawater desalination or hydrogen production via high temperature electrolysis (HTE) or thermochemistry.

The design of this type of reactor is directly dependent on the development of materials with high temperature resistance such as CMCs. Figure 22.8 [40] gives a description of the structural components of the Gas Turbine-Modular Helium Reactor (GT-MTR), which is based on the same concept as that of VHTR, but at lower operating temperatures (850°C instead of 1100°C). The various possible applications of CMCs in this type of reactor are shown in Figure 22.8.

The application for which CMCs have been studied in most detail is for control rods. The associated operating conditions are given in Table 22.5. Note that the accident conditions used in Table 22.5 correspond to hot-gas-duct failure.

The temperatures and data given are overall values, which provide an approximation to the values to which the hottest parts of the reactor will be subject. Furthermore, control rods are subject to temperature gradients (in particular, axial gradients from 350–400°C in the upper part to ~1100°C in the lower part). The radial gradients should be small, especially if the coolant can flow inside the control rods (configuration to be specified).

These control rods are long (~10 m) and made of several sections, each approximately 500-mm long, connected by flexible joints. This is because the rods must not be too rigid, so that they can be inserted into the reactor core under all possible operating conditions, including earthquake. Each section comprises two coaxial tubes in which the absorber (B_4C, HfC, or other) is enclosed. However, as the exact design of the control rods is not yet fixed, there is uncertainty over the dimensions of each section (diameter, length, and thickness). The same applies to the system for connecting sections together which has yet to be specified (such as whether it is a screw, pivot, or swivel-type joint).

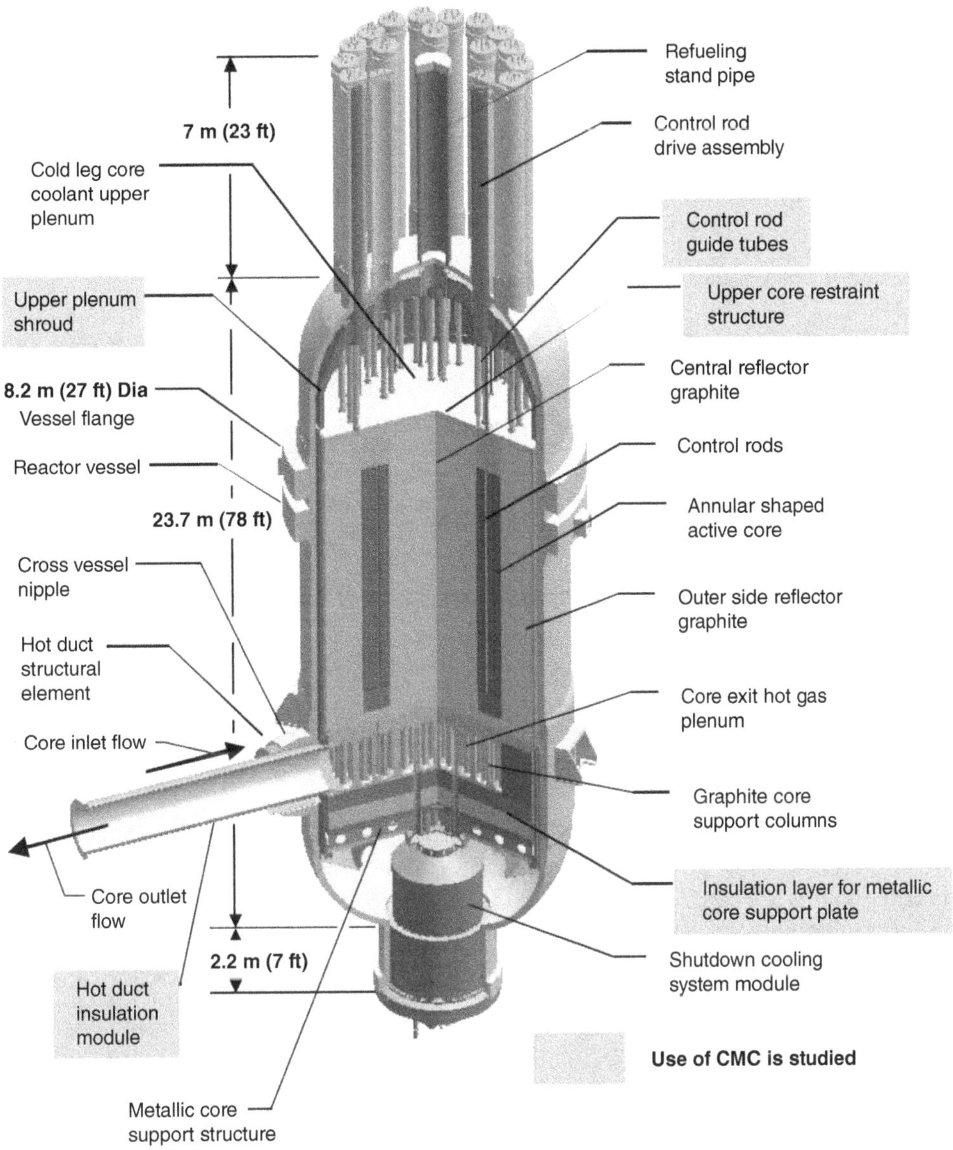

FIGURE 22.8

Details of core structures of GT-MHR reactor with CMC possible applications [40].

TABLE 22.5 Operating Conditions of Control Rods Use in VHTR

	Control rods for start/stop	Control rods for reaction control
Nominal operating conditions (Under helium)	400°C	400–1100°C
Accidental situation (without neutron flux, in air)	$1200 < T < 1600$°C during 150 h	
Maximum fluence ($E > 0.18$ meV)	≈0	Up to 6×10^{20} n/cm²/y
Irradiation damage	≈0	Up to 40 dpa for reactor life
Environment in nominal situation	He 40–70 bars, $V = 35$ m/s Impurities: CO (<500–1570 μbar), H_2O (<10 μbar), H_2O/H_2 (>0.01), CH_4/H_2O (≪100), N_2 (<5 μbar)	

TABLE 22.6 Thermal, Mechanical, and Dimensional Specifications for CMC (C/C) Control Rods Structural Material for VHTR

	Properties at 25°C (without irradiation)	Properties at $T = 400$–1100°C under neutron irradiation
Thermal conductivity (W/(m K))	30–150	>30
Mechanical strength σ_R (MPa)	>80	
Toughness (MPa(m$^{1/2}$))	>3	
Thermal shock resistance $\lambda\sigma_R/(E\alpha)$ (W/m)	>65,000	
Swelling (%)	–	As low as possible As an indication: $\Delta L/L < 2\%$
Reactivity in air (mg/g/ h)	–	As low as possible (As an indication "burn-off rate" of graphite = 0.2)
Wear resistance	≥ the one of graphite	

The mechanical stresses on the control rods are small, and the main risks associated with this application are:

- Mechanical failure during a sudden vertical drop or under seismic loads.
- Thermomechanical failure due to thermal gradients.
- Deterioration under irradiation: changes in control-rod dimensions (leading to a risk of control-rod jamming), changes in the thermal and mechanical properties (leading to a risk of premature failure), and swelling of the absorber (leading to a risk of failure or jamming).
- Corrosion of the composite under oxidizing conditions (leading to loss of component integrity and risk of failure).

These various failure modes, and the requirement for thermal compatibility with the graphite block into which the control rods are inserted in the low position, have led to the specifications given in Table 22.6.

Finally, one criterion that is directly associated with the nuclear application concerns the very low rate of impurities that must be present in the control-rod materials. This limits the formation of radioactive elements that cause problems in waste management and also limits the presence of oxidation catalysts.

Both C/C and SiC/SiC composites are under consideration as serious candidates for the structural components of VHTR control rods [9, 11–19]. However, SiC/SiC composites are preferred for control rods due to their better dimensional stability under irradiation [19]. This is the context in which the International Nuclear Energy Research Initiative (I-NERI) was launched in 2005. This was originally an American multi-laboratory program for the development of tubular SiC/SiC composites, involving the Oak Ridge National Laboratory (ORNL), the Idaho National Laboratory (INL), and the Pacific Northwest National Laboratory (PNNL). The CEA (the French Alternative and Atomic Energies Commission) and LCTS (the French Thermostructural Composites Laboratory) joined this program in 2006.

22.3.2.2 SFR Applications A recent patent [41] has highlighted the potential benefits of manufacturing a hexagonal-tube fuel-assembly body (containing a bundle of fuel pins) from a SiC/SiC-type CMC for an SFR. Currently, the hexagonal-tube body is made of an austenitic, ferritic, or ferritic-martensitic steel alloy. This body has two main roles. First, it provides the fuel assembly with structural integrity and provides rigidity to hold it fast when the foot is mounted in the diagrid. It also guides the upward flow of the coolant (sodium) required for cooling the fuel pins and ensures that the hydraulic diameters are maintained in each section of the assembly. This maintenance of hydraulic diameters is an important condition for ensuring operating stability. Radial deformation of the tube must therefore be limited.

The patent proposes the production of a composite assembly body (Figure 22.9) with end components connected via a metal structure fitted with openings and a central cylinder in CMC material whose negative influence on neutronic reactions is reduced. This material could thus have a positive influence (no creep, low swelling).

As shown in Figure 22.9, components 4, 6 (the ends of the hexagonal body), and 8 (the bar connecting sleeves 4 and 6) would be made of metal alloys. The whole of the central part (component 10) would therefore be made of a hexagonal CMC sleeve (possibly in sections) with a cross-flats dimension of approximately 200 mm and a thickness of 3–4 mm. The constraints for this CMC sleeve would be those associated with the operation of SFRs. The CMC would have to be chemically compatible with sodium up to 700°C (or ideally up to its boiling temperature: $T_{b\,Na} = 883$°C),

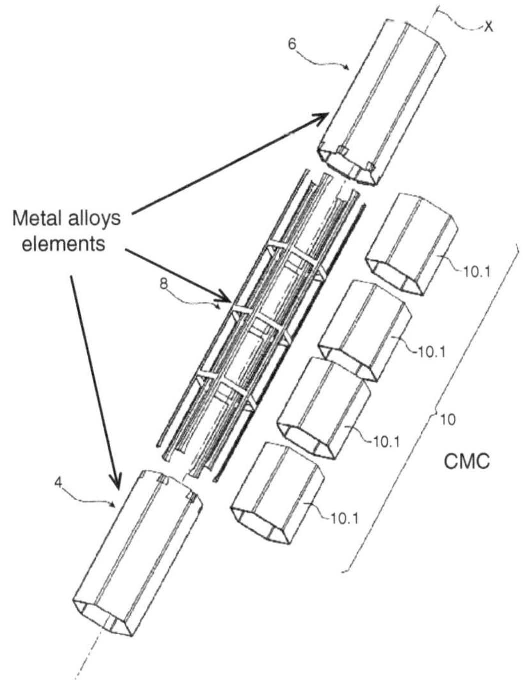

FIGURE 22.9

Hexagonal composite (metal alloy/CMC) assembly body for SFR [41].

have low swelling in the 400–700°C temperature range, be resistant to an intense neutron flux (~100 dpa or more by end-of-service life), and be thermomechanically and thermochemically compatible with the alloys used for the assembly body. As research into this application started very recently, there are no more precise specifications.

22.3.3 CMCs for Pin Cladding or Structural Materials in GFRs and LWRs

22.3.3.1 GFR Applications A schematic diagram of a GFR is shown in Figure 22.10.

The operating temperatures and the coolant are the same as on VHTRs, but GFRs are fast-spectrum reactors that optimize natural resources and minimize waste. Therefore, the structural materials, while subject to the same operating temperatures as in VHTRs, would also be subject to intense neutron radiation during their service life (probably over 100 dpa). The main technological sticking point for this type of reactor is the development of the fuel element (fuel and fuel cladding). This constitutes a major challenge that must be overcome before their deployment could be considered. Here again, CMCs (SiC/SiC only) are considered as a (or perhaps the only) reference candidate for the fuel cladding, due to their refractivity, low neutron absorption, resistance under oxidizing conditions, and behavior under neutron irradiation (see Chapter 13). Two fuel concepts were initially explored: (i) the macrostructural plate concept [10, 42] and (ii) the classic fuel pin concept [10, 43, 44].

Currently, mainly for reasons of manufacturability, the fuel-pin concept is the reference for researchers even though questions remain as to the maximum length allowed (due to handling constraints) for CMC pins [10, 44]. The main characteristics of a GFR are given in Table 22.7, along with the associated consequences for the required properties of the CMC cladding material.

The mechanical properties required for cladding are highly dependent on the thickness and thermal conductivity of the material used (heat exchanger function) [43, 44]. This is because the mechanical loading on the cladding (excluding the undesirable pellet-cladding interaction) is mainly due to the thermal gradients between inner and outer faces, which are directly associated with the thermal conductivity and thickness of the cladding. As the operating conditions of a GFR are particularly severe (power density of 100 MW/m^3), the cladding application requires implementation of CMCs in the form of a thin-walled ($\leq$1 mm) tube [43, 44].

Of course, if the major challenge of manufacturing a fuel element was resolved, it would also be possible to use CMCs as structural materials (such as for guide tubes, fuel assembly structures, and the diagrid) in GFRs. This is because the loads associated with these structures are generally lower than those associated with the cladding material.

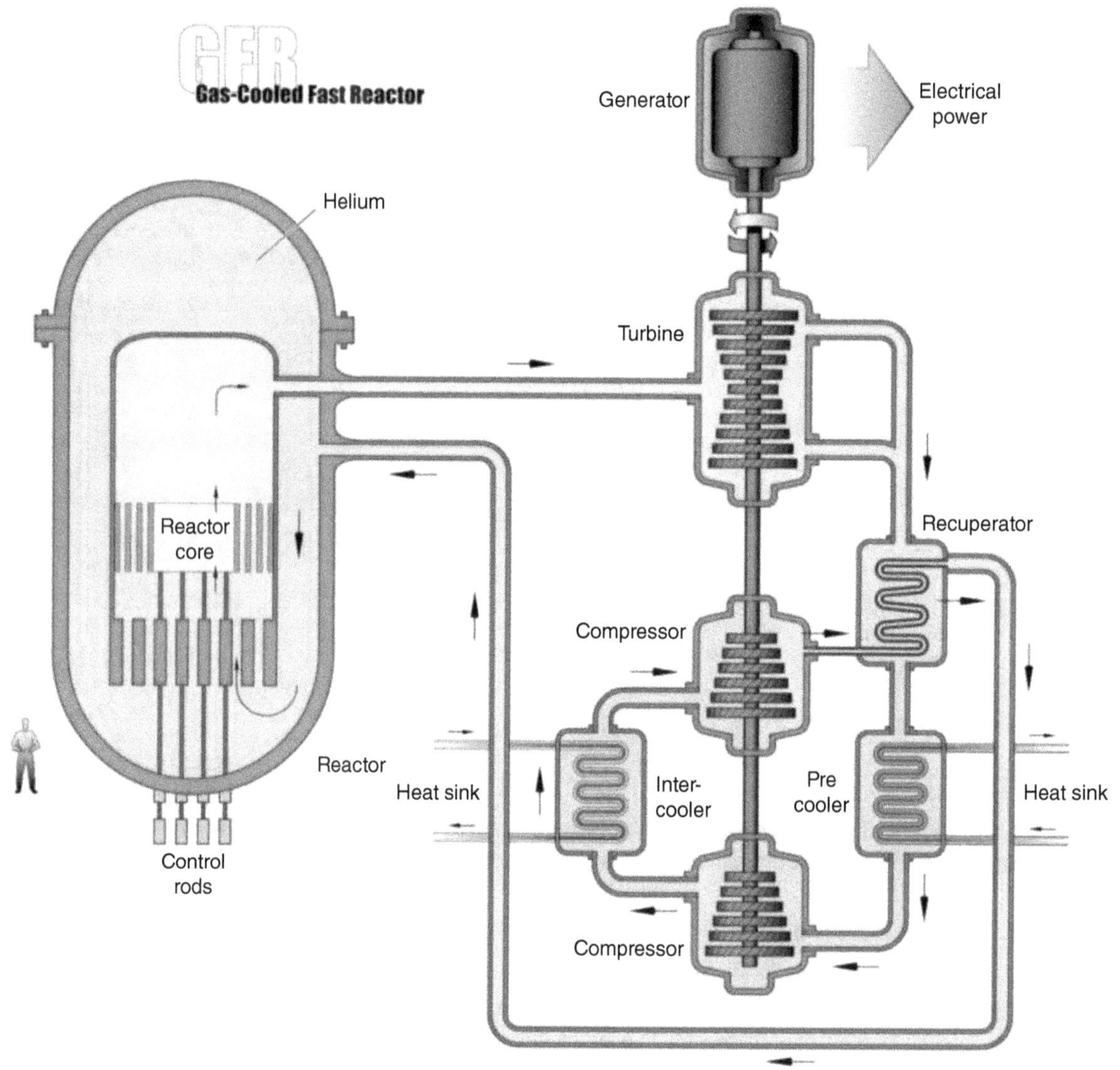

FIGURE 22.10
Schematic representation of GFR [16].

22.3.3.2 BWR or PWR Applications As specified in Section 22.3.1, the accident on the Fukushima Daiichi nuclear power plant in Japan has clearly highlighted the need for in-core materials with better performance at high temperature and under oxidizing conditions (presence of steam) than those currently used. This accident has renewed the interest of the nuclear industry in SiC-based (and thus SiC/SiC-based) cladding, which is more resistant at high temperatures under oxidizing conditions, to replace zirconium alloys [45–48]. The main goal would be to limit the heat released and the hydrogen generated during high temperature oxidation of these alloys in steam/air [45]. This would reduce core damage following a loss of coolant accident and maintain high temperature fuel integrity for longer periods. Although there has been a recent upsurge in interest, CMC were already under consideration for this type of application before the Fukushima accident [35–37]. The constraints on the SiC/SiC material associated with this application are similar to the constraints given in Table 22.7 for GFRs, even though the coolant and neutron spectrum (thermal spectrum) differ, which leads to some differences as shown in Table 22.8.

The main differences observed concern the operating temperature (which is lower) and the coolant (which is water). Although radiation damage would be lower than in a GFR, numerous constraints remain. The main problem (other than leaktightness) is associated with the operating temperature of these reactors, which is much lower than for

TABLE 22.7 Operating Conditions and Consecutive Requirement on CMC Cladding for GFRs

	Value for GFR (mainly estimations)	Consecutive requirement on CMC cladding material
Manutention/transport	Dynamic solicitation on pin limited to 5 g	Length of pin limited to 1.5–2 m for 1 mm wall thickness
Coolant	He (see Table 5 for impurities)	Leaktightness relative to He and fission product Resistance to corrosion in low oxidizing environment Good thermal conductivity (at least 10 W/(m K))
Pressure (and velocity) of coolant	40–70 bar (50–100 m/s)	good resistance to erosion
Maximum fluence ($E > 0.18$ meV)	~100 dpa SiC	Low swelling and conservation of good thermal conductivity and mechanical properties Low neutron absorption
Nominal operating temperature of coolant	400/900°C	Good thermo-mechanical behavior and low swelling over this range of temperature
Abnormal operating temperature of coolant (drop of coolant pressure)	Cladding temperature up to 1600°C	Leaktightness relative to He and fission product No severe degradation of thermo-mechanical properties
Accidental situation (loss of coolant)	Cladding temperature up to 2000°C in oxidizing environment (air, water?)	Keep mechanical integrity (cladding geometry = no fusion or severe degradation of cladding) to allow cooling of fuel Leaktightness not require
Fuel	(U,Pu)C	Thermochemical compatibility

a GFR. In this temperature range, the thermal conductivity of SiC is highly degraded for low doses (<1 dpa) [20, 49] and swelling is greater [50]. Finally, even though SiC is highly resistant under high temperature oxidizing conditions [45,46], it should be checked that the results still hold for CMCs, depending on their manufacturing procedure.

However, studies performed over several years mean that innovative concepts can be proposed for this type of application and these will be described in last part of this chapter.

TABLE 22.8 Operating Conditions and Consecutive Requirements on CMC cladding for LWRs

	Value for LWR (mainly estimations)	Consecutive requirement on CMC cladding material
Manutention/transport	Dynamic solicitation on pin limited to 5 g	Length of pin limited to 1.5–2 m for 1 mm wall thickness
Coolant	H_2O	Leaktightness relative to H_2O and fission product Resistance to corrosion in water Good thermal conductivity
Pressure of coolant	155 bar (PWR), 70–80 bar (BWR)	Good resistance to erosion
Maximum fluence ($E > 0.18$ meV)	<10 dpa SiC	Low induced swelling Good residual thermal conductivity (at least 10 W/(m K) ?) and mechanical properties Low thermal neutron absorption
Nominal operating temperature of coolant	280–330°C (PWR) ~300°C (BWR)	Good thermomechanical behavior and low swelling over this range of temperature
Severe accidental situation (loss of coolant)	Cladding temperature >1000°C in oxidizing environment (water or water + air)	Keep mechanical integrity (cladding geometry = no fusion or severe degradation of cladding) to allow cooling of fuel Leaktightness require?
Fuel	$(U,Pu)O_2$	Thermochemical compatibility

22.4 PROCESSING OF C/C COMPOSITES FOR NUCLEAR APPLICATIONS

22.4.1 Introduction

The behavior of C/C composites under irradiation, especially anisotropic swelling [17–19], limits their use to nuclear applications with low doses of radiation (less than a few dpa maximum) [6] as discussed in Sections 22.2.2 and 22.3.2.1. This part will discuss the proposed concepts, manufacturing processes, and properties associated with the composites (including thermomechanical characteristics) that have been developed specifically as part of nuclear programs (fusion and fission) by the main suppliers (SGL, Herakles, Dunlop, and Toyo Tanso).

C/C composites developed as part of the ITER project and those currently used in prototypes of experimental fusion reactors (Tore Supra and JET) will be discussed in detail (see Section 22.4.2) due to their large database. For this application, and for reference, the main characteristic sought for these materials is the highest possible thermal conductivity (see Section 22.2.2.2), given that the planned doses are mostly <1 dpa [27].

As stated in Section 22.3.2.1, the other applications (significantly less studied and documented) concern gas-cooled VHTRs. C/C composites in these reactors are, or rather were, primarily used for control rods and guide tubes to control the reactor. The Japan Atomic Energy Research Institute (JAERI), for example, has developed a design for a control rod for the High Temperature Test Reactor (HTTR) with a C/C composite cladding in place of a Ni–Fe–Cr alloy (A800H) [51]. The planned temperature range is 400–1200°C for a maximum fluence of 10^{21} n/cm^2 for this reactor. The approach for selecting C/C composites for the JAERI is given in Section 22.4.3 as an example of fission application of C/C composites.

22.4.2 C/C Composites for Fusion Applications

22.4.2.1 Manufacturing Process

Choice of Fibers/Tows The criterion of excellent thermal conductivity requires the use of PAN or, even better, anisotropic pitch carbon fibers. Thermal conductivity (just like Young's modulus) of carbon fibers is directly related to texture and microstructural organization (see Chapter 2). The more organized the microstructure of the fibers (low d_{002} and high L_a and L_c) and the more anisotropic their texture (low Z), the higher their heat conductivity (and Young's modulus) [52]. PAN (Polyacrylonitrile) fibers can be divided into two categories: high modulus (HM) and high resistance (HR). The latter (with a Young's modulus between 200 and 300 GPa) are used more frequently throughout the world due to their lower cost and higher rupture elongation (2%). In contrast, HM fibers (Young's modulus >400 GPa) are more expensive and have a rupture elongation that is generally <1%, making them more difficult to use with traditional textile methods (weaving, braiding, etc.). For their part, anisotropic pitch fibers have extremely variable microstructures and textures (see Chapter 2) and may have particularly high moduli (up to 950 GPa). This category provides the fibers with highest thermal conductivity. HM PAN and anisotropic pitch fibers are graphitizable, that is, their structural organization (and thus their thermal conductivity) can be improved with heat treatments at very high temperature.

For the application in question, both types of fibers are often used in combination so that composite thermomechanical properties (e.g., thermal conductivity) can be adjusted in the various directions of the structure.

Creating a Reinforcing Architecture This operation is used to make significant changes in the properties of the composite, as shown in section 4.2. Since the main concern for the composite is thermal conductivity, manufacturers have opted to retain textures that can easily be made dense (to limit ultimate porosity) and adapted to HM carbon fibers. As a result, 2D weaving and assemblies were most often used with an additional needling operation by some firms (Dunlop and Herakles, especially) for increased resistance to delamination and better thermal conductivity over the thickness.

Fiber–Matrix Bonding The microstructure on the surface of the fibers controls fiber/matrix adherence, which plays a primordial role in the mechanical behavior of the composite. The interfacial shear strength τ (fiber/matrix coupling intensity) can be controlled by creating deposits on the fibers [53]; using controlled oxidation treatments of fibers, which changes their surface state [53,54]; or eliminating the use of a sizing agent on the fibers [55]. These treatments can, for example, compensate for deterioration in fiber adherence caused by subsequent steps of the process, especially during graphitization treatments at very high temperature (>2200°C), which may cause a significant decrease in fiber/matrix

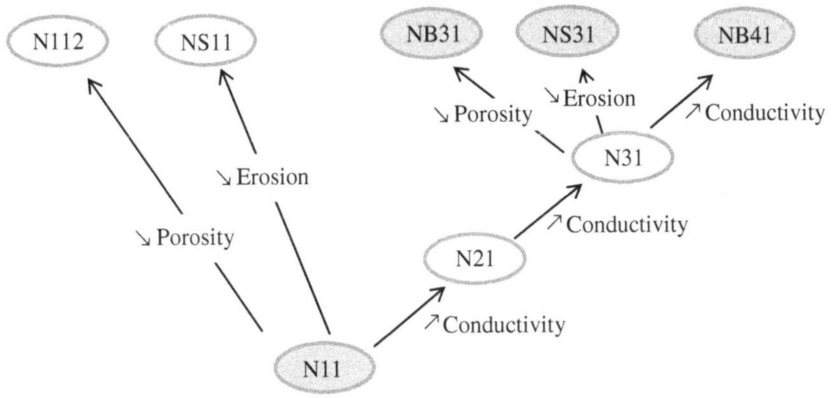

FIGURE 22.11

Evolution of Herakles C/C composites development for fusion application.

coupling [54]. Since firms consider this type of treatment an industrial secret, it is usually never mentioned in public even if they use it.

Densification Process The need for high thermal conductivity and high purity levels in nuclear applications requires the use of chemical vapor infiltration (CVI) and/or pitch impregnation processes (see Chapter 5).

The CVI process may result in a very pure and graphitizable carbon matrix [56], which, after thermal graphitization treatment (≥2200°C), will have excellent thermal conductivity. Due to industrial advances, it is possible to manufacture a product of any shape and size. It nevertheless results in significant residual porosity (in general, between 10% and 20%) and does not produce uniform densification over large thicknesses.

In order to make very thick products more dense and limit residual porosity in C/C composite material (resulting in loss of conductivity), the current industrial solution is to use the CVI process in combination with liquid phase impregnation (LPI) or pitch impregnation [6, 57, 58]. This latter operation, judiciously placed in the manufacturing process, significantly reduces porosity by impregnating the composite with a graphitizable carbon phase (only for pitch impregnation). Pitch impregnation nevertheless presents numerous disadvantages (see Chapter 5), such as difficulty in controlling microstructure, ballooning during carbonization phase, maintaining pitch quality, and so on. In addition, and in contrast with CVI, the carbon matrix using pitch impregnation is not deposited around the fibers like onion skins, which may make the pitch carbon matrix slightly adherent in the composite [59]. Even if the search for significant density may produce composites with high thermal conductivity, it may also make manufacturing final components (e.g., divertor cassette for ITER) more difficult. During manufacture, the C/C composite must be welded to a copper alloy (see Section 22.2.2.2). During this step, the spread of copper in the C/C composite can be stopped and make welding more difficult [59]. Some researchers even recommend limiting C/C composite density to 1.9 g/cm^3 for this type of application [59]. Since assembly has been discussed earlier in this book (see Chapter 22), it will not be considered here.

For a divertor-type application, the strong attraction of carbon and thus of C/C composites to hydrogen may result in making hydrocarbon species volatile with high erosion, as well as high retention of tritium due to its large specific surface area. The two phenomena are highly limiting for this application, as shown in Section 22.2.2.2. To resolve this problem, most industrial firms have proposed a final siliconization treatment (liquid silicon infiltration) of the material. The last step, however, results in increased manufacturing costs and difficulties (including reproducing previous productions), so that materials using this final treatment are no longer being considered [27].

Herakles was one of the leaders in developing C/C composites for the fusion application. Work began by developing the N11 material, which is used on the first wall of the Tore Supra reactor. Subsequent development to respond to specific requirements for fusion applications is presented in Figure 22.11. Table 22.9 gives specifications for the most studied commercial grades of C/C composites developed by various firms for fusion applications. Finally, analyses [32] have shown that the total quantity of impurities in these materials may be <100 ppm.

22.4.2.2 Thermomechanical Properties of C/C Composites Developed for Fusion Specifications for materials (mainly listed in Table 22.9) developed for fusion programs are more or less complete, and a compilation of thermomechanical properties of some of these C/C composites has been summarized in Table 22.10 [6, 59, 60].

Mechanical properties of C/C composites are highly dependent on the direction of stress but are particularly stable for temperatures up to 2500°C [61–63]. Up to at least 1500°C, these characteristics are thus always higher than those

TABLE 22.9 Characteristics of Main Commercial C/C Composites Developed for Fusion

Name	Producer	Texture/fiber	Densification process	Density (g/cm^3)	Porosity (%)
N11	Herakles	3D ex-PAN (*x,y*) and needled fibers (*z*, ex-PAN)	CVI + HTT + CVI	1.75	15
N21	Herakles	3D ex-PAN (*x,y*) and needled fibers (*z*, ex-PAN)	CVI + HTT+ Pitch impregnation + HTT	2	1-4
NS11	Herakles	3D ex-PAN (*x,y*) and needled fibers (*z*, ex-PAN)	CVI + HTT + Siliconization	2–2.2	1–4
N31	Herakles	3D ex-pitch (*x*), ex-PAN (*y*) and needled fibers (*z*, ex-PAN)	CVI + HTT + CVI	1.75	14
NB31	Herakles	3D ex-pitch (*x*), ex-PAN (*y*) and needled fibers (*z*, ex-PAN)	CVI + HTT + CVI + Pitch impregnation + THT	>1.9	8
NS31	Herakles	3D ex-pitch (*x*), ex-PAN (*y*) and needled fibers (*z*, ex-PAN)	CVI + HTT + Siliconization	>1.98	5
NB41	Herakles	3D ex-pitch (*x*), ex-PAN (*y*) and needled fibers (*z*, ex-PAN)	CVI + HTT + CVI + Pitch impregnation + HTT	1.9	5
CX270G	TOYO TANSO	2D woven ex-PAN fibers	LPI + Pitch impregnation + HTT 2800°C	1.69	
CX-2002U	TOYO TANSO	3D ex-PAN fibers	Pitch impregnation	1.65–1.7	
Dunlop 3D	Dunlop	3D ex-pitch P25 (*x*), ex-PAN (*y*) and needled fibers (*z*, ex-PAN)	CVI + HTT	>1.80	
DMS780	Dunlop	2D woven ex-PAN fibers	CVI + HTT		
DMS814	Dunlop	3D ex-pitch P120 (*x*), ex-PAN (*y*), and needled fibers (*z*, ex-PAN)	CVI + HTT		
FMI-222	Fiber Materials Inc.	3D Balance weave ex-pitch fibers (P120)	Pitch impregnation	1.96	
Sigrabond	SGL	3D woven	CVI		

HTT, high temperature treatment; CVI, chemical vapor infiltration; LPI, liquid phase impregnation.

TABLE 22.10 Thermophysical Properties of Some C/C Composites of Interest for Fusion Application

Properties		N11	N31	NB31	NB41	DMS780	DMS814	Dunlop 3D
Tensile stress (MPa) at 20°C	*X/Y*	40	110/50	110/20	150/20			176/20
	Z	27	7	5	9			10
Tensile modulus (GPa) in plane (*X,Y*) at 20°C		25		110	150			
α (10^{-6}/°C) at 1000°C	*X/Y*	1.8		<1	<1	<1		
	Z	2.85		2.85		3.8		
λ_{XX}	20°C	240	300	260	427	256	282	320
	1000°C	80	150	120	186	100		140
λ_{YY}	20°C	240	115	100	111	256	126	110
λ_{YY}	1000°C	80	60	48	55	100		
λ_{ZZ}	20°C	150	90	85	80	48	62	60
	1000°C	55	45	40	42	25		25

measured at ambient temperature. Thermal conductivity is highly dependent on the type of fibers used in both directions of the weaving plane. It is also clearly lower in the transversal direction, even for needled materials. Finally, the method of heat conduction (using phonons) in these materials results in a deterioration of thermal conduction that increases with temperature. This deterioration stabilizes above 1000°C (see Chapter 11).

Resistance to damage from thermal shock (major constraint of the fusion application) is currently evaluated using basic thermomechanical characteristics with the following equation:

$$R_{\mathrm{th}} = \frac{\sigma_{\mathrm{r}}\lambda}{E\alpha} \tag{22.3}$$

where σ_r designates the mechanical resistance, λ the thermal conductivity, E the Young's modulus, and α the coefficient of thermal expansion. Thermal shock resistance evolution is mainly related to thermal conductivity evolution [62], so that composites have significantly better resistance in the direction of pitch fibers.

22.4.3 C/C Composites for Fission Reactor Applications

Very few published studies that include results have examined this type of application. In fact, nearly all work on C/C composites for nuclear applications is on fusion programs; however, one grade of C/C composite was selected by JAERI as part of the preliminary design for HTTR control rods [64], commercial grade CX270G (Table 22.9) produced by Toyo Tanso. Reinforcement consists of a 2D weaving of (HR) PAN fibers, which is densified using liquid impregnation (LPI + pitch impregnation). Various production samples have demonstrated that tensile strength improves significantly if the sizing polymer is removed before densification with a liquid. Breaking strain increases from 0.15% to 0.7% and breaking stress increases from 150 to 300 MPa. No explanation has been given for this behavior. The results also indicate that optimum tensile strength is achieved by using 3K and 6K tows. The fiber content was adjusted to 50% to ensure optimum mechanical properties. A thermal graphitizing treatment of 2800°C is performed after a final pitch infiltration to obtain the best stability during radiation, even if mechanical properties are inferior compared with graphitizing heat treatments at lower temperature.

In a study, Ishiyama shows [55] that resistance during axial compression of tube sections is significantly less than the compression resistance value obtained from standard tests performed on flat specimens. Diametrical compression tests on rings likewise produce tensile strength values that are twice as small as typical tensile strength values provided by the supplier. In the latter test, shearing forces supplement tension/compression constraints, which can produce parasitic effects and make test interpretation more difficult [65]. For all cases, the results demonstrate well that transposing standard tests on flat specimens to the case of tubes cannot be direct, given the different reinforcing architectures that can exist between flat and tubular specimens. To our knowledge, there is also no European or international standard (ISO) for performing mechanical tests on tubes or rings that apply to CMC.

22.5 PROCESSING OF SiC/SiC COMPOSITES FOR NUCLEAR APPLICATIONS

22.5.1 Introduction

It is widely admitted [5, 8, 23, 24] that SiC/SiC composites are the most capable of meeting the needs of nuclear power reactors due to their low activation and excellent behavior under irradiation conditions, even with significant doses (see Chapter 13). This conclusion is only valid when using the latest generation of Hi-Nicalon Type-S (HNLS) and Tyranno SA3 (TSA3) SiC fibers, which are the only ones of which the stability under irradiation has been more or less verified below 1000°C [9, 20, 66].

The three leaders in research on SiC/SiC composites for nuclear applications are the United States, Japan, and the European Union (France, in particular). Research in the United States focuses mainly on the application for VHTR (control rods essentially). There are fewer restrictions for this application than applications for fission reactor cladding or for fusion reactor PFC (mainly blanket application). For these, the composites must be at least leaktight, highly conductive, easy to weld/join, and must not interact with the fuel. These are very difficult, even impossible, critical requirements to meet with the SiC/SiC composites being manufactured at this time. Despite all this, studies regarding the application of cladding for LWRs have been led, resulting in the recommendation of two designs to be described later in this document. In Japan, research focuses mainly on the fusion application and the manufacturing of composites using

Tyranno SA3 fiber and an original densification process called NITE (Nano powder Infiltration and Transient Eutectic-phase). For the last decade, French research has focused on use for GFR reactor cladding, which represents a significant challenge for the viability of this type of reactor. An original cladding design meeting numerous requirements for this application that other usual composites do not meet was recently developed.

Finally, regardless of the envisaged application, one of the critical points for the use of these materials is the development of standard normalized tests adapted for nuclear applications (for tubular structures in particular) and the development of behavior models. In the literature, the mechanical characterizations are mainly performed using bending tests, which are easier to implement but are not adapted to the rigorous mechanical characterization of CMCs. During bending tests, the field of stress in the material is not uniform; results are thus highly dispersed and difficult to interpret/compare. Finally, traction-compression and interlaminar shear stress tests are required for the development of CMC constitutive models aimed at structural design [67–69]. Results obtained during tensile tests will thus be favored in the rest of this chapter, since they provide easier comparison and more rigorous analysis of the influence of manufacturing parameters.

22.5.2 Selection of SiC Fibers

Only two SiC fibers are compatible with nuclear applications—Hi-Nicalon S (HNLS) fibers produced by NGS Advanced Fibers and Tyranno SA3 (TSA3) fibers produced by Ube Industries. These fibers are prepared using two distinct processes leading to similar thermomechanical characteristics as indicated in Table 22.11 [70].

Following the manufacturing process (final high temperature sintering), the Tyranno SA3 fiber has better thermal conductivity (as a result of more advanced grain growth) and better thermal stability in particular as presented in Figure 22.12 [71]. Finally, HNLS fibers are more resistant to creep than TSA3 fibers below 1500°C even if creep remains negligible below 1200°C for these fibers [70].

This makes TSA3 fiber more interesting from the point of view of intrinsic properties for nuclear applications (where thermal conductivity and strength at high temperatures are required). Nonetheless, it should be mentioned that to this day no nD ($n \geq 2$) composite material with TSA3 reinforcement exhibited a strain to failure of >0.3%, whereas similar materials with HNLS reinforcement frequently exhibit strain to failure equivalent to that of the tow (approximately 0.6–0.7%). This point is particularly critical for nuclear applications, as this means that the fiber/matrix coupling (strong) in CMCs with TSA3 reinforcement leads to a low ductility material regardless of the process of densification.

The basic chemical composition of the two fibers has recently been analyzed. Elements C, O, N, and H were dosed using the IGA (Instrumental Gas Analysis), and the other trace elements were dosed using GDMS (Glow Discharge Mass Spectroscopy). Table 22.12 lists all results obtained by the two techniques. Measurement uncertainty is estimated

TABLE 22.11 Properties of Hi-Nicalon S and Tyranno SA3 SiC Fibers [70]

Suppliers	NGS Advanced Fibers, Japan	Ube Industries Ltd, Japan
Type of fiber	Hi-Nicalon S	Tyranno SA3
Diameter (μm)	13	7 or 10
Density (g/cm^3)	3.0	3.1
Tensile strength (GPa)	2.5	2.8
Tensile modulus (GPa)	375	380
λ (W/(m K))	18	65
Grains size (nm)	20	60–70
XRD	10–50	50–400
Chemical composition (wt/at.%)		
Si	68.4/48.1	69.1/49 (edge) 66.1/45.6 (core)
C	31.3/51.5	30.5/50.7 (edge) 33.5/54.1 (core)
O	0.3/0.3	0.1/0.1
Al	–	0.3/0.2
C/Si (at. %)	1.07	1.03 (edge) 1.19 (core)

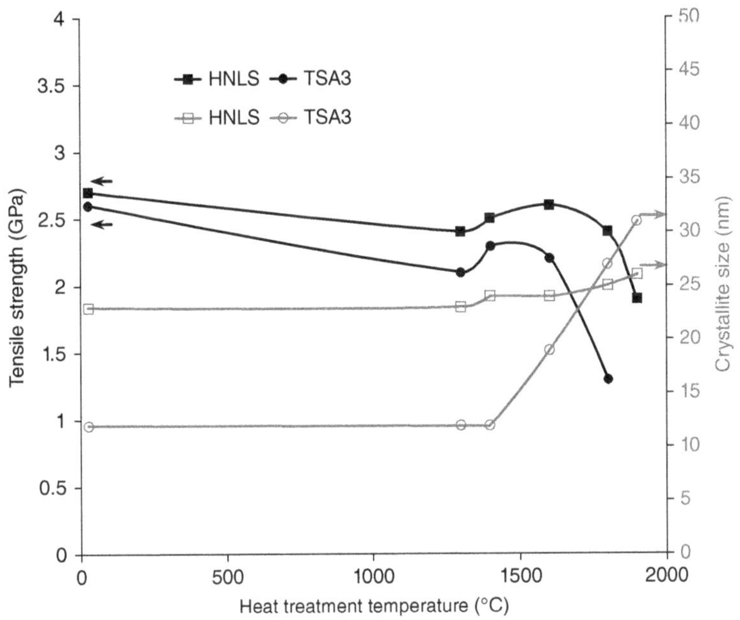

FIGURE 22.12
Average room temperature tensile strength of fibers and apparent crystallite size of β-SiC under heat treatment at elevated temperatures in Ar for 1 h [71].

at 20% with regards to the value obtained for components present in trace form. The total accuracy of the carbon and silicon element measurement is estimated between 0.5% and 1% in absolute value.

These analyses show that there is more excess carbon in TSA3 fibers, as discussed elsewhere [72]. Oxygen is present in greater quantity in HNLS fibers. This is mainly attributed to the manufacturing process, which is different for the two types of fiber. The HNLS fiber manufacturing process uses a step involving electron beam cross-linking and a final cross-linking thermal treatment at a lower temperature. This process makes the fiber particularly sensitive to any trace of oxygen in the environment. In addition, the maximum temperature at the end of the treatment does not provide for total elimination of the oxygen in its core. On the contrary, TSA3 fiber is created using a step involving air cross-linking, which leads to a high oxygen content in the fiber before its final, extremely high temperature treatment (~1800°C). This last treatment eliminates oxygen from the fiber more completely in the form of mainly SiO(g) and CO(g) (or SiO_2(g)). The presence of aluminum in significant quantities in TSA3 fibers is also explained by the manufacturing process, since aluminum serves as a sintering agent during the processing of the fiber (see sections 1.2 and 7.4.1). The most surprising result is that of boron, of which the concentration (~3000 ppm) is high for TSA3 fibers, well above the results obtained

TABLE 22.12 Chemical Composition of Hi-Nicalon S and Tyranno SA3 Fibers Samples Determined by Interstitial Gas Analysis (Elements C, N, O, H) and Gas Discharge Mass Spectrometry (Other Elements)

Chemical elements	HNLS	TSA3	Chemical elements (ppm)	HNLS	TSA3
C (at. %)	50.0	53.0	K	0.12	0.77
Si (at. %)	49.1	46.4	Ca	1.9	16
C/Si	1.02	1.14	Ti	0.05	46
N (ppm)	170	630	Cr	7.7	0.54
O (ppm)	8500	770	Fe	50	1.1
H (ppm)	300	460	Co	0.04	<0.02
B (ppm)	2.0	3000	Ni	0.41	0.19
Na (ppm)	0.31	2.2	Cu	0.08	0.03
Mg (ppm)	0.17	0.99	As	0.53	0.35
Al (ppm)	2.8	1250	Zr	0.05	0.12
P (ppm)	1.7	0.43	Ta	<1	<1
S (ppm)	57	2.8	W	0.23	0.02
Cl (ppm)	44	10	Nb, Mo, Sn, Hf, Pt	<0.01	<0.01

(1000–1400 ppm) in previous analyses [66]. The presence of boron (generated by the polymerization catalyst used) in the material can have a prohibitory effect for the targeted application (nuclear), as the isotope ^{10}B (20% of the natural boron) leads to the production of helium (~600 ppm considering the measurements taken) during neutron irradiation as described in Equation (22.4):

$$^{1}_{0}n + ^{10}_{5}\mathrm{B} = ^{7}_{3}\mathrm{Li} + ^{4}_{2}\mathrm{He} \quad (22.4)$$

This helium production could considerably alter the mechanical properties of the fibers (swelling, embrittlement) and thus those of the final material when exposed to radiation [73]. The presence of boron in non-negligible quantities could therefore call into question the use of TSA3 fibers in a nuclear environment. Nevertheless, Katoh [66] demonstrates that under neutron irradiation, the mechanical behavior of TSA3 fibers and composites with TSA3 reinforcement is unaltered. However, a basic elemental analysis of irradiated fibers was not performed, and the boron concentration estimated using the elemental analysis of the irradiated composite remains lower (two times lower) in the reference state (~1400 ppm). To confirm the GDMS results, NRA (Nuclear Reaction Analysis) analyses, using the $^{11}B(p,\alpha)^{8}Be$ nuclear reaction described in the literature [74], were therefore performed to verify the boron content in the TSA3 fibers. The results revealed a concentration of 2400 ± 400 ppm. This corresponds to the measurements taken by GDMS if the uncertainty of this technique is taken into account (±20% of the measured value, for 3000 ± 600 ppm). This result clearly shows the significance of the measurement of boron concentration in the TSA3 fibers, which may vary greatly from batch to batch, especially for an application where the presence of this element can be very detrimental. There are, however, no results providing for the definition of the acceptable limit concentration for boron in TSA3 fibers to prevent deterioration of their characteristics when exposed to neutron radiation. This study would undoubtedly be relevant and would make it possible to impose maximum boron content for the fiber on the manufacturer for nuclear applications. Finally, the high level of purity of these two reinforcements must be emphasized (if the presence of sintering elements B and Al in the TSA3 fiber are considered irrelevant). These fibers may thus be considered materials with low activation.

Lastly, the surface of these fibers is mainly made up of a carbon phase (concentration >80%) of sp^2 hybridization in the first nanometers and not of a SiC phase [72].

22.5.3 Selection of Architecture

Unlike C/C composites, the thermal conductivity of the SiC/SiC composites is not essentially linked to the direction of fiber tows in the composite architecture, since the SiC fibers have an isotropic microstructure. The major obstacle to the thermal conductivity of these materials is therefore the porous network of the material. The maximum final porosity of the material must thus be limited to obtain high thermal conductivity. The creation of textures using *n*D ($n > 2$) weaving of these rigid fibers is easy, provided that the tows are already insulated (by a polymer covering) to protect them during use. Having described the architecture in Chapter 4, we will also present the main effect of the nature of filament winding or braiding (2D or 3D) type textures on the characteristics of the composites. In fact, these two techniques are the most appropriate for the manufacturing of parts with an axis of rotation, often necessary for nuclear applications (fuel cladding, control rods, channel boxes for BWR, etc.). These techniques do not necessarily require the prior insulation of HNLS and TSA3 tows if adapted machines are used.

These techniques have been used for the manufacturing of SiC/SiC composite tubes (densified by CVI) with small internal diameters (~8 mm) using HNLS fibers. The surface and cross-section aspects of the tubes obtained are represented in Figure 22.13. Only the filament winding technique preserves the circular shape of the tubes' internal surface. This technique and 2D braiding result in a high fiber volume fraction ($V_f \sim 50\%$), with filament winding resulting in closed structure with very small porosities that could be detrimental to the uniform densification of thick tubes (>1–2 mm). The 3D braiding texture was the cause of the greatest deviation from circularity observed on the internal section, leading to lower fiber volume fractions ($V_f < 40\%$) and higher porosity.

22.5.4 Choice of Interphase

The nature of the nuclear application simplifies the choice of interphase material. In fact, the behavior of boron when exposed to neutron radiation makes the use of BN interphases impossible. Furthermore, operating feedback involving the irradiation of composites with pyrocarbon interphases is positive, even at high doses [20]. No degradation of the mechanical behavior was observed under irradiation conditions for $e_{PyC} \leq 150$ nm interphase composites [7, 20, 75–77]. However, it seems that a maximum thickness must not be exceeded for this type of application. Some authors report

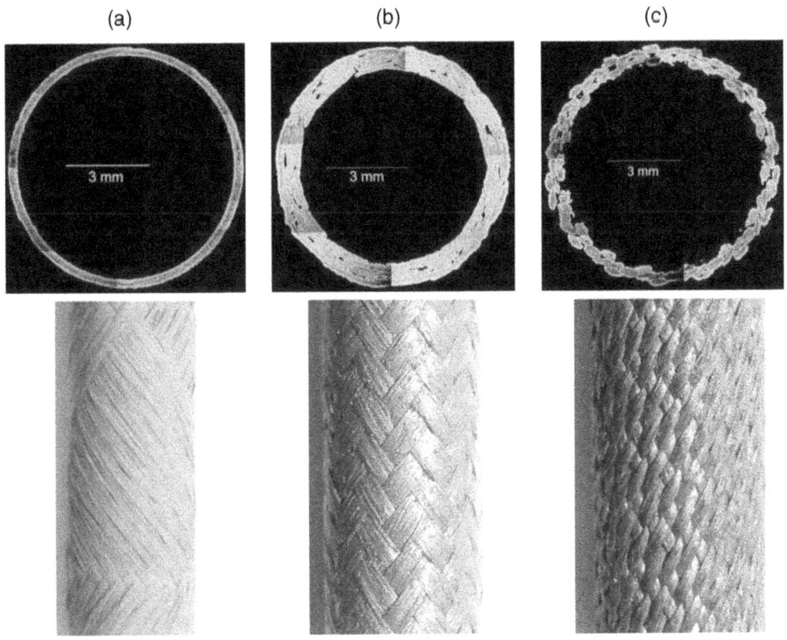

FIGURE 22.13 Surface and cross-section aspects of CVI tubes of various textures: (a) filament winding, (b) 2D braiding, and (c) 3D braiding.

degradation of mechanical behavior after irradiation (weakening of the fiber–matrix connection) for larger interphase thickness ($e_{PyC} \geq 300$ nm) [75, 77,78]. It has also been demonstrated that the thickness of the PyC interphase influences the intensity of the fiber–matrix coupling. The thicker the interphase, the weaker the interface shear stress [7, 79,80], even if this phenomenon is not as important as the one observed for composites reinforced by SiC fibers of earlier generation (Nicalon and Hi-Nicalon fibers). Thus, and for nuclear applications, PyC thickness (monolayer or PyC/SiC multilayers) should be reduced to its minimum so that no degradation of mechanical behavior of composite occurs during irradiation [78]. Finally, the interest of multilayer interphases $(PyC/SiC)_n$ is still in question. Some studies [81, 82] do not provide support for their interest, since the crack deviations are mainly observed on the fiber–interphase interface and not within the multiple layers. The only advantage of multilayer interphases may be to enhance the rupture time governed by the interphase recession in an oxidative environment [83, 84].

In conclusion, it is surprising that no studies have been performed (or published) to determine the influence of the pyrocarbon microstructure or the chemistry of the surface of fibers on the mechanical behavior of the composites, both before and after irradiation. It is highly probable that the values of the strains to failure, varying from 0.1% to 0.6%, obtained for the SiC/SiC composites with HNLS fibers [7,8, 66, 77, 82] are linked to these parameters, but this is never discussed in literature.

22.5.5 Densification Processes and Associated SiC/SiC Composite Properties

22.5.5.1 CVI Process and Associated Composites Properties The CVI process is essential for the creation of the pyrocarbon interphase. In fact, there are not yet any alternative solutions for the uniform deposit of a thin pyrocarbon interphase. The CVD process is used to achieve the SiC phases that are the most stable under irradiation conditions [49]. This is, of course, only valid for high purity SiC deposits without a large amount of free silicon or free carbon (see chapter 13). Since the CVI process used for the densification of composites is similar, it is natural that it is the most appropriate for nuclear applications. This process is thus, for the scientific community, the reference SiC/SiC composite densification process for nuclear applications. Its main drawback is linked to the residual porosity ratio of materials densified in this way. This is often between 10% and 25% [9, 80, 81], which negatively impacts thermal conductivity and most often does not permit the achievement of leaktight materials. Nonetheless, recent manufacturing at CEA has repeatedly achieved porosity ratios lower than 10% on thin tubes (~1 mm) of which the texture was created with HNLS fiber using filament winding and/or 2D weaving (bi- or tri-axial). Manufacturing of leaktight tubes in the linear elastic range was even demonstrated. Despite this, as soon as the mechanical stress exceeds the linear yield strength, the composite immediately loses its hermeticity (Figure 22.14).

Finally, this process provides materials in high purity. Table 22.13 presents the results of the analyses performed (by the same technique as that used in Section 22.5.3) on CVI composite reinforced with HNLS manufactured at CEA,

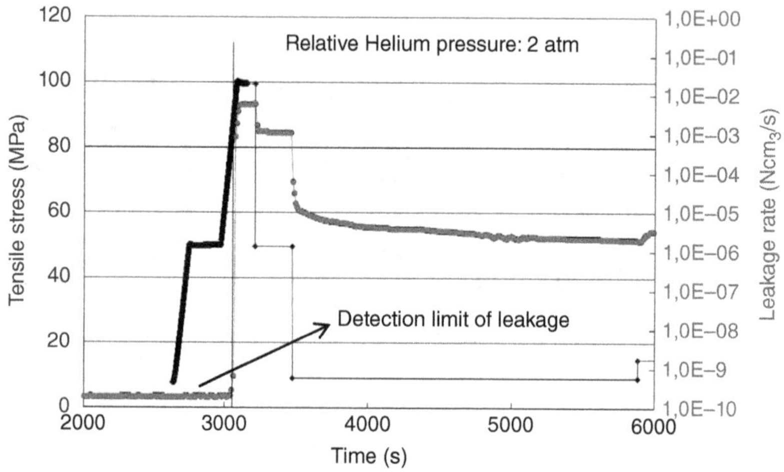

FIGURE 22.14
Evolution of leakage rate during tensile test of 2D braiding HNLS CVI composite.

TABLE 22.13 Chemical Composition of Monolithic β-SiC CVD and Hi-Nicalon S Reinforced Composites Samples Determined by Interstitial Gas Analysis (Elements C, N, O, H) and Gas Discharge Mass Spectrometry (Other Elements)

Chemical elements	HNLS CVI composite (CEA)	HNLS CVI composite [66]	SiC–CVI (CEA)(for $V_f = V_m = 50\%$)	SiC CVD R&H
C (at. %)	51.7	–	50.4	50.7
Si (at. %)	48.3	–	49.6	49.3
C/Si	1.07	–	1.02	1.03
N (ppm)	89	13 120	5.7	<7
O (ppm)	3500	11 480	100	43
S (ppm)	15.6	906	3.7	4.4
H (ppm)	320	–	400	260
B (ppm)	0.22	21.5	<0.02	2.54
Na (ppm)	<0.04	24.6	<0.09	<0.04
Mg (ppm)	<0.04	12.2	<0.09	<0.04
Al (ppm)	0.52	20.6	<0.007	<0.007
P (ppm)	0.22	7.7	0.006	0.058
Cl (ppm)	2.86	95.1	0.005	0.12
K (ppm)	0.056	1	<0.005	<0.03
Ca (ppm)	1.5	14.8	0.25	0.15
Ti (ppm)	<0.04	0.04	<0.04	<0.04
Cr (ppm)	<0.1	0.1	<0.1	<0.1
Fe (ppm)	5.36	13.8	7.53	<0.02
Co (ppm)	<0.01	–	<0.01	<0.01
Ni (ppm)	0.34	0.8	0.58	0.03
Cu (ppm)	<0.01	0.01	<0.01	<0.01
As (ppm)	<0.01	0.4	<0.02	<0.01
Zr (ppm)	0.02	1.7	<0.003	<0.002
Nb (ppm)	<0.01	–	<0.02	<0.01
Mo (ppm)	<0.01	–	<0.02	<0.01
Sn (ppm)	<0.008	–	<0.015	<0.008
Hf (ppm)	<0.005	–	<0.01	<0.005
Ta (ppm)	<1	–	<2	<1
W (ppm)	<0.001	0.032	<0.001	0.001
Pt (ppm)	<0.005	–	<0.009	<0.005

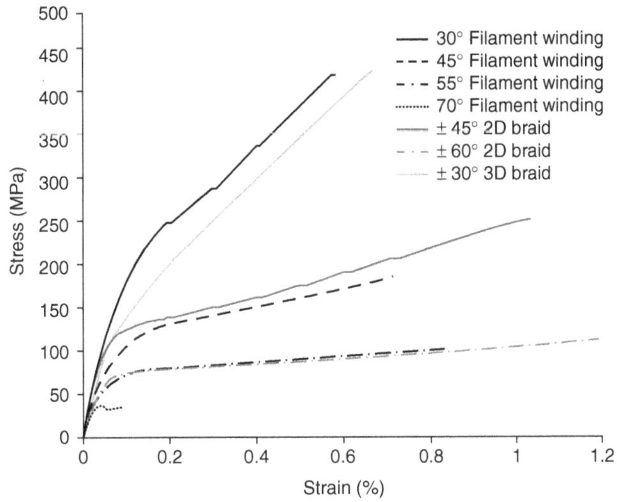

FIGURE 22.15

Influence of filament winding or braiding angle on tensile stress–strain behaviors of HNLS tubular CVI composites.

other HNLS CVI composite [66] and Rohm & Haas CVD SiC. The composition of the CVI matrix for CEA materials is estimated using the results obtained for the composite and fibers (see Section 22.5.2), taking into account the volume fractions of the fibers and the equivalent matrix $V_f \sim V_m \sim 50\%$. The results show that the composites prepared at CEA have a very high level of purity, comparable with Rohm & Haas CVD SiC, and are thus compatible with nuclear applications [21,22, 38]. Furthermore, the major impurities present in the composite come mainly from the HNLS fiber reinforcement. This shows that the SiC matrix prepared using the I-CVI process used at CEA is especially pure.

A recent unpublished study (led by CEA) has shown the influence of the filament winding or braiding angle on the tensile mechanical behavior of tubular composites with HNLS reinforcement manufactured using the CVI process (Figure 22.15). These tubes have an internal diameter $\phi_i = 8$ mm and thickness $e \sim 1$ mm (characteristic dimensions of GFR cladding). As expected, the winding or braiding angle influences mechanical behavior: the more parallel the tows are to the axis of the stress, the greater the mechanical strength and the lower the strain to failure.

Tubes using 2D braiding have greater mechanical strength (break stress and especially breaking strain) for equivalent angles and for similar volume fractions (~45%) compared with materials obtained using filament winding. The use of 3D braiding presents, like 2D braiding, greater mechanical properties (especially breaking strain) for identical angles, despite the reduced fiber volume fraction (35% compared to ~45%).

A summary of the properties of 2D- [66] and 2.5D-woven CVI composites based on HNLS and TSA3 reinforcement is presented in Table 22.14. This table also lists additional characteristics.

A recent (unpublished) study was led at CEA on 2.5D composites (Table 22.14) with HNLS reinforcement treated at different temperatures (between 1500 and 2100°C) in an argon atmosphere for 2 hours. Residual mechanical tensile behavior at room temperature was examined. The relative change in failure characteristics obtained with regard to untreated materials ($\varepsilon_R/\varepsilon_0$ and σ_R/σ_0) as a function of the thermal treatment temperature is presented in Figure 22.16. Mechanical properties remain significant even after treatment at 1700°C. A clear deterioration in behavior is observed at 1900°C; this is obviously linked to the strength degradation of the HNLS fiber at high temperatures (Figure 22.12).

The results of high temperature mechanical tests on this type of material are extremely rare. Nonetheless, changes in the mechanical bending behavior up to 1600°C for tubes (2D weaving at 55°) and plates (satin woven 0/90°) densified by CVI with HNLS reinforcement (interphase thickness of $e_{PyC} = 150$ nm) were studied [85]. The main characteristics are summarized in Table 22.15.

Two strain rates were used for these tests so as to examine time-dependent phenomena. The authors observe that the maximum temperature for the use of these materials made of HNLS fiber is approximately 1300°C due to the high fiber thermal creep at this temperature and the loss of mechanical strength of fibers at higher temperatures. Nevertheless, other results conclude that no significant thermal creep is observed below 1300°C for HNLS

TABLE 22.14 Typical Characteristics and Properties of 2D and 2.5D Weaving CVI-Composites Reinforced with HNLS and TSA3 Fibers

	HNLS CVI-composites		TSA3 CVI-composites	
Property Interphase	2D weaving [66] Multilayered interphase 5x (PyC_{20nm}/SiC_{100nm})	2.5D weaving Monolayer (e_{PyC} = 150 nm)	2D weaving [66] Multilayered interphase 5x (PyC_{20nm}/SiC_{100nm})	2.5D weaving Monolayer (e_{PyC} = 150 nm)
Density (g/cm^3)	2.74	2.85	2.65	2.85
Porosity (%)	11%	~9%	~12%	~9%
Fiber volume fraction V_f (%)	~31%	~49%	~33%	~49%
Thermal diffusivity (mm^2/s)	5.1	11.2	13.6	17.8
Tensile modulus E (GPa)	230	300	218	320
Ultimate tensile stress σ_R (MPa)	281	330	266	240
Strain at load maximum ε_R (%)	0.28	0.63	0.22	0.25

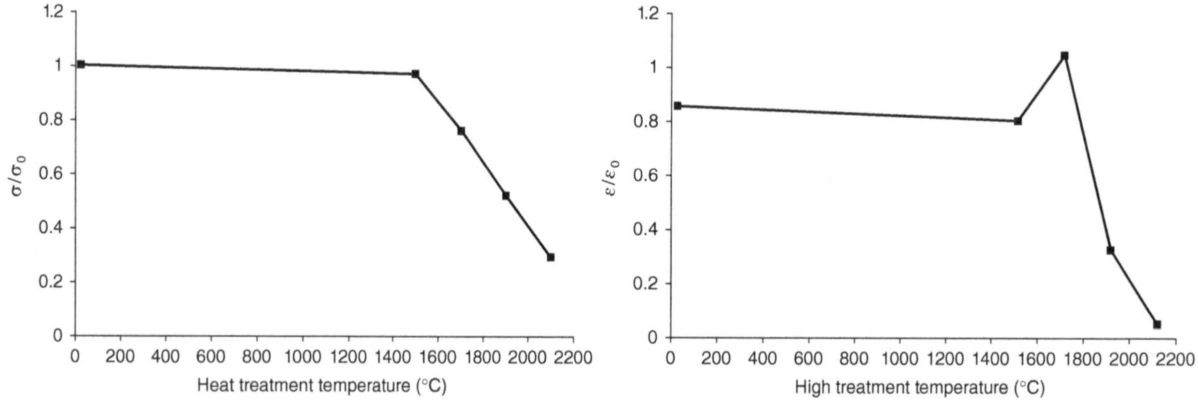

FIGURE 22.16

Relative evolutions of tensile strength and strain to failure with heat treatment temperature of 2.5D HNLS CVI composites.

TABLE 22.15 Evolution of Flexural Mechanical Properties of 2D CVI-Composites with Testing Temperature [85]

Composite material	Temperature (K)	Strength (MPa)	Elastic modulus (GPa)	Proportional limit stress (MPa)	Strain to failure (%)
0/90 5-harness satin weave	298	740	256	115	0.8
	1273	703	223	120	0.8
	1573	613	192	110	0.7
	1673	519	202	100	0.6
	1873	295	182	100	0.4
± 55° braid	298	120	155	65	0.2
	1273	123	178	60	0.2
	1573	114	225	45	0.15
	1673	113	246	40	0.2
	1873	104	–	–	–

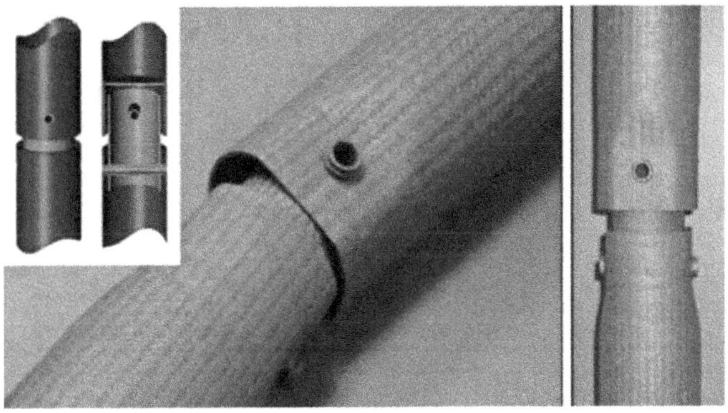

FIGURE 22.17

Prototype SiC/SiC control rods with articulating joints. Reproduced with permission by Pergomon, Current Opinion in Solid State & materials Science, Vol 16, Iss 3, 2012

fibers [70]. At this time, published works are not sufficient to give precisely the maximum temperature use for HNLS CVI composites.

The results of mechanical tensile tests (Table 22.14) available for CVI composites reinforced by TSA3 fibers and manufactured with a PyC interphase thickness compatible with nuclear applications (e_{PyC} < 300 nm) indicate that they can be likened to monolithic ceramics with a very low ductility (ε_R < 0.3%) [7, 66, 80]. This is a problem for major nuclear applications where damage tolerance is required. The exact reason for this behavior is not clearly identified, even if the interfacial shear stress in these materials is very high [72, 80, 82], which is attributed by some to the high roughness of TSA3 fibers [82]. In any case, this behavior is not a result of deterioration of the TSA3 reinforcement during the manufacturing process. In fact, significant tensile strain to failure (>0.5%) was obtained with 1D composites (mini-composites) with TSA3 reinforcement and PyC interphase (e_{PyC} = 150 nm) prepared using the CVI process [82]. Nevertheless, the observation of this type of behavior (high ductility) is unusual, even for mini-composites, and the mechanical tensile behavior is systematically fragile (ε_R < 0.3%) with very strong interface connection characteristics when the texture is more complex (>1D) as shown in Table 22.14. One validated method for diminishing the fiber/matrix coupling in the composites reinforced by SiC fiber of earlier generation (Nicalon and Hi-Nicalon) is to increase the thickness of the interphase so as to absorb and compensate for thermal residual stress and/or roughness of the fibers [80]. Given this, and for interphase thicknesses varying from 50 to 500 nm (thus well below the limit tolerated for nuclear applications), no decrease in the intensity of the fiber/matrix connection sufficient to obtain composites with TSA3 reinforcement with significant tensile strain to failure (ε_R >0.3%) has been observed [7, 80]. Studies must be led to find solutions for the achievement of mechanical behaviors similar to those achieved with HNLS reinforcement.

In the case of the application of control rods for VHTRs, articulated control rod segments have been developed for 10 years in the United States. Figure 22.17 [19] provides a concrete example from the manufacturing of control rods (with articulation) densified by the CVI process using HNLS fiber for this type of application. This is the only concrete and documented case for this type of application. Most of the published studies only cover manufacturing processes or characterizations of model objects (plates or tubes).

22.5.5.2 The NITE Process and Properties of Associated Composites The NITE process is based on liquid-phase sintering [86]. The main goal of this process is to decrease residual porosity to a maximum in order to obtain hermetic composites with high thermal conductivity. The LPS (Liquid Phase Sintering) process has been improved, thanks to progress made in the development of fibers that are more stable at high temperatures (TSA3 fiber) and the wide availability of SiC nanopowders. A slurry of β-SiC nanopowders and additives (Al_2O_3, Y_2O_3, and SiO_2) is infiltrated into the texture (1D or 2D) made of SiC fibers and dried to manufacture prepreg sheets. After drying, these prepreg sheets are heated (~1800°C) and isostatically compressed (15–20 MPa, in general) to make SiC/SiC composites. This type of manufacturing is only possible using TSA3 fiber; thermal stability of HNLS fibers is not sufficient to withstand the sintering conditions (essentially temperature) of the process (Figure 22.12). The NITE composite matrix is made up of polycrystalline

TABLE 22.16 Room Temperature Properties of NITE-SiC Composites [87]

Property	TSA3 NITE-composites	
Reinforcement	UD	2D Cross plied
Fiber content (%)	53	51
Density (g/cm^3)	3.11	3.06
Porosity (%)	0.6	3.8
Proportional limit stress (MPa)	358	148
Tensile strength (MPa)	408	167
Straint to failure (%)	0.13	0.08
Young's modulus (GPa)	358	288
Thermal conductivity (W/(m K)	32	

SiC with a non-negligible quantity of oxides (<20%). Table 22.16 lists the characteristics generally obtained for NITE composites [87].

The thermal conductivity is high (~30 W/(m K)) so as the linear yield strength (>100 MPa). This last characteristic is promoted as very advantageous for nuclear applications [87]. This argument should be qualified, as the breaking strain is very low (<0.2%), as was observed for composites with TSA3 reinforcement densified using the CVI process (Table 22.14). These materials thus exhibit a rather fragile behavior with a low damage tolerance characteristic of a strong interfacial shear stress (τ). Thicker pyrocarbon interphase ($e_{Pyc} \geq 500$ nm) allows reducing τ but strain to failure never rise above 0.3% [88, 89]. Anyway, as interphase thickness should be reduced below 300 nm (see Section 22.5.4) for nuclear applications this route is not an issue. The strain to failure does not change, even after heat treatment at high temperature [90], confirming the existence of a strong fiber/matrix bonding. These results also indicate that the NITE composites are not degraded (conservation of residual mechanical strength) after heat treatment at temperatures up to 1800°C [90, 91] due to the good thermal stability of TSA3 fibers (Figure 22.12). While a clear improvement in the porosity ratio (and thus thermal conductivity) is observed as a result of this process, several issues remain. The best composites ever obtained using this process have a maximum breaking strain of 0.3%, and materials produced on a greater scale most often have breaking strains of below 0.1%. This is thus a composite that can be compared to a monolithic ceramic. Furthermore, a thick interphase (generally $e_{PyC} = 500$ nm) must be used to protect the fiber during the manufacturing process (during the sintering step in particular), and the use of such interphases for nuclear applications is widely questioned (see Section 22.5.4). The effect of additives on the behavior of these composites under irradiation conditions has been studied little. Only a few studies have been dedicated to this issue concerning the NITE composite matrix [92, 93], and they do not provide assurance of good behavior during exposure to high doses.

Another similar densification process (SITE process) has been recently proposed by a Slovenia team for nuclear applications [94, 95]. This process is supposed to prepare SiC/SiC composite with a SiC matrix without sintering additives, using a PIP (Polymer Infiltration and Pyrolysis) process. The porosity is lowered by filling the voids in the SiC-fabric with SiC powder (by electrophoretic deposition (EPD) principle) prior to polymer infiltration. As this process is new and weakly documented (especially mechanical characterization results), it will not be described in this chapter.

It seems therefore that the main interest of NITE composites is providing better thermal conductivity (due to lower porosity) than that of the CVI composites [96]. They also provide hermetic composites in the absence of mechanical stress [97, 98]. As in the case of composites densified using CVI (see Section 22.5.5.1), this hermeticity is not conserved once the material is mechanically stressed beyond its linear yield strength (as soon as the first matrix crack appears). For major nuclear applications (fuel or structure cladding for fission and fusion in particular), safety authorities will certainly not accept the use of a material for which tightness is not guaranteed for a large strain range and thus in the matrix cracking range. For example, and for the studies led in France by CEA, the GFR cladding material must satisfy the criterion of a minimum breaking strain of 0.5%, regardless of the direction of the mechanical stress. Similarly, the tightness must be guaranteed until the failure of the cladding and therefore of the envisaged composite. These particularly strict requirements imposed on CMC materials used for nuclear applications are impossible to satisfy using the CVI or NITE processes, whether alone or combined. New SiC/SiC material designs for these demanding applications must therefore be developed. This has been the subject of several studies led by different research teams over recent years. These studies will be presented in section 22.5.

22.5.6 Designs for Hermetic Composites Adapted for Nuclear Applications

22.5.6.1 Introduction The ultimate goal for nuclear applications is thus to develop a composite material with a high strength and strain to failure and that remains hermetic even under high mechanical stress. This requirement is, to our knowledge, incompatible with the use of CMCs. In fact, their strain tolerance is the consequence of the controlled multicrack matrix phenomenon, which results in high permeability. As observed in Figure 22.14, the SiC/SiC composite loses its gas tightness upon the appearance of the first crack for very low strains ($\ll$0.1%), regardless of the densification process used.

Several teams have worked on this issue for fission reactor fuel cladding. American teams have mainly focused on cladding for LWRs, whereas French teams have mainly focused on cladding for GFRs. The designs developed and the characteristics associated with the manufactured objects are also presented.

22.5.6.2 US Designs for LWR Pin Cladding As explained in Section 22.3.3.2, the use of CMC composites as a cladding or structure material in LWRs may be useful for increasing the safety of this type of nuclear installation. The first proposal of a cladding design using CMC was patented in the United States in 1993 [35]. This patent is based on the use of composite cladding made of zirconia (ZrO_2) or alumina (Al_2O_3) fiber composite and a zirconia or alumina matrix. While interesting, this design does not satisfy several critical criteria for this type of application. Such composites do not have significant tightness (especially after cracking of the matrix), and the behavior under irradiation conditions of these phases (ZrO_2 and Al_2O_3) is not compatible with nuclear applications (significant swelling). Moreover, the thermal conductivity was very low, resulting in increased secondary thermal stresses in the cladding and an increase in the temperature of the fuel. Finally, these materials do not have very high resistance to creep at high temperatures.

A second, more recent design for the use of SiC/SiC composites was patented in 2006 [36]. A so-called "triplex" cladding is made up of an internal tube made of monolithic β-SiC, a central layer of SiC/SiC composite, and an external CVD SiC coating. The internal monolithic SiC is supposed to be compressed by the layer of SiC/SiC composite, which should increase its tightness range under internal pressuring stress (effective stress in operation). This design was used several times for manufacturing and characterization, and the effects of irradiation have also been analyzed for low doses [37, 99]. The results of several of the characteristics obtained for cladding of this type are presented in Table 22.17 [37, 99].

This solution, fully made up of ceramic phases, has some problems. The first is related to the tightness range of this type of ceramic cladding. The tightness of the cladding relies on the fragile internal monolithic SiC tube. Manufacturing of long cladding is performed, according to the designers, by assembling small sections using brazing, the behavior of which under irradiation conditions is unknown. Even without taking into account the behavior of brazing under irradiation conditions, the manufacturing of long cladding in large quantities seems unrealistic, and the fragility of these structures would be very high. Moreover, as partially discussed elsewhere [19], in operating conditions ($T \sim 300°C$ and under neutron flux), the thermal conductivity over the thickness of the cladding will not exceed 5 W/(m K), resulting in a thermal gradient of >100 K over the thickness. This thermal gradient will thus cause non-negligible stresses in the cladding. Swelling will also be significant, and the combination of these phenomena (among others) will lead to elastic deformations in the cladding beyond the yield strength of the composite and/or monolithic SiC tube. This

TABLE 22.17 Compilation of Non-irradiated and Irradiated Triplex Cladding Pin Properties [37, 99]

Property		Triplex SiC composites
External diameter (mm)		10–10.1
Internal diameter (mm)		8.3–8.4
Hoop stress at failure (MPa) at 25°C	Non-irr	~230
	300°C ~ 1 dpa	~103
Thermal diffusivity (mm^2/s) at 25°C	Non-irr	~25
	300°C ~ 1 dpa	~3
Thermal diffusivity (mm^2/s) at 300°C	Non-irr	~13
	300°C ~ 1 dpa	~2

is even more critical if the internal tube/composite connection is deteriorated. The monolithic tube is then supporting the entire mechanical load, resulting in a rapid loss of tightness under significant mechanical stress. Calculations performed at CEA on this type of cladding confirm that the integrity of the cladding would be largely jeopardized in certain operating conditions. The studies (Table 22.17) led on short cladding (35–50 mm) in LWR conditions were not very conclusive, as the breaking stress drops nearly 50% after irradiation for radiation doses of 1 dpa in LWR conditions [99].

A more simple design called the SiC CMC Metal Liner Hybrid Design, consisting in the manufacturing of mixed metal/composite cladding, was recently proposed by the INL [100]. Although it is currently little documented, it uses mixed cladding made up of an internal Zircaloy tube and a SiC/SiC composite coating. The design seems more robust than the previous design due to the tightness provided by Zircaloy, but several uncertainties remain. For example, no values for the thickness of the Zircaloy and composite layers are indicated. It is clear that a minimum thickness for the Zircaloy layer is necessary to ensure its strength in operating conditions, with respect to both the coolant and the fuel. If it is too thick, the thickness of the layer of composite will be limited, thus limiting the advantage of the design (mechanical reinforcement at high temperatures especially in accidental situations). Furthermore, in the event of a serious accident, if the SiC and/or Zircaloy cladding is not very thick, this may be detrimental. In these conditions, the chemical interactions between Zircaloy and SiC would also be very significant. Other issues undoubtedly exist, but without more thorough documentation on this design, and without any characterization of the corresponding objects, it is impossible to perform a more serious analysis at the moment.

22.5.6.3 The CEA Sandwich Design Since 2002, CEA (the French Alternative and Atomic Energies Commission) has been running a large R&D program for the design of GFR fuel cladding (complementary to related studies on both reactor and systems, on the one hand, and materials, on the other hand). Studies have been led on the design of GFR cores, taking into account SiC/SiC structure elements and cladding [44]. This work has shed light on the problem related to pellet/cladding interaction. In fact, the carbide fuel that would be used in a GFR is relatively cold (compare to its melting temperature) and cannot accommodate the pellet clad mechanical interaction because of the lack of carbide fuel plasticity at operating temperatures, unlike an oxide fuel. It must therefore be ensured that there is no pellet/cladding contact by introducing a controlled gas (He) seal between the fuel and the composite. The thickness of this seal cannot exceed a defined value due to the significant increase in fuel temperature and thus the release of fission gases that would result, in particular. This compensates for the rather low burnup (5 at.%). Furthermore, the thermochemical compatibility of the fuel and SiC is limited and results in the formation of metal phases with low melting points. An unpublished study led at CEA, performed with $UC_{1.04}$ and SiC/SiC heated to 2000°C, showed that half of the cladding thickness could be melted within 15 minutes. Finally, the He seal does not ensure that the fuel pellets will remain centered in the cladding. Given the low thermal conductivity of the composite cladding under irradiation conditions ($\lambda \sim 10$ W/(m K) considered for the calculations at the moment), off-centering produces additional thermal stresses on the cladding (136 MPa total) with regard to those (76 MPa) observed in normal situations (pellet centered in the cladding).

To solve this problem, CEA is proposing an innovative design [101] using a porous carbon fiber seal ($V_f = 20$–40%)—for example, a 2D braided layer, instead of a He seal. This solution increases the pellet-cladding play without thermal penalties, favoring heat exchange. It allows the fuel to expand without stressing the cladding (experimental tests showed that a 10 MPa compressive stress is sufficient to achieve near-complete densification of the initial structure). It minimizes fuel pellet off-centering problems. Finally, it prevents chemical interaction between the fuel and the SiC/SiC cladding, by obtaining a source of free carbon that prevents the decarburization of the fuel (phenomenon recognized during tests performed at 2000°C for 15 minutes that did not show degradation of the fuel pellet with a buffer seal). The advantage of this type of seal has also been demonstrated by core calculations recently performed [44], showing that the maximum burnup can be twice the one with He seal (from 5 at.% to approximately 10 at.%), thus compatible with the GIF goal [16].

Nonetheless, the problem of the tightness of damaged SiC/SiC composites was a key technological issue that needed to be solved in order to use them as a GFR cladding material. CEA thus proposed a new innovative design called the sandwich design [102] illustrated in Figure 22.18.

This design proposes to integrate a thin ductile refractory metal liner between two layers of SiC/SiC composites. The liner confined between two thicker, more rigid composite layers is subject to little stress during mechanical loading of the cladding and uniquely ensures the tightness of the cladding. Moreover, its high ductility should ensure that its integrity is maintained, even under significant mechanical stress.

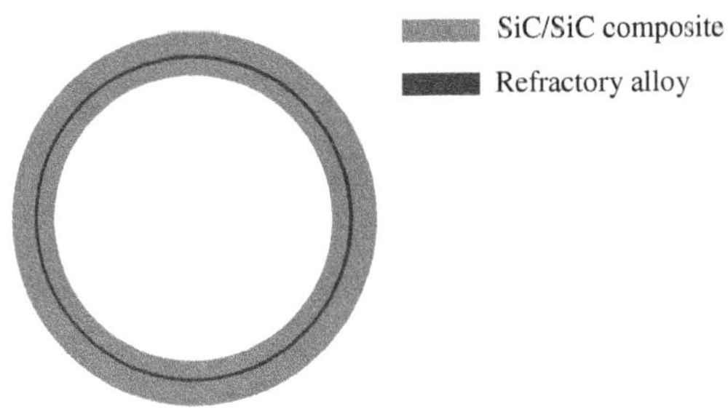

FIGURE 22.18

Sandwich concept design for GFR cladding.

The choice of the refractory liner was guided by its thermochemical compatibility with SiC. For this reason, a review of the kinetics of interactions between refractory metals and SiC was performed [103–106]. The results are presented in Figure 22.19.

It is clear that ductile metals with the best compatibility with SiC at high temperatures are tantalum (Ta) and niobium (Nb). Since these refractory metals have high neutron absorption, their thickness must be limited to a maximum (targeted thickness between 50 and 100 μm). Recent studies have made it possible to perfect a robust process for the manufacturing of such objects for lengths of ~150–200 mm and for the two metallic Ta and Nb liners. It is important to specify that the lengths produced are only limited by the size of the CVI installation used for their manufacturing (research installation). No step of the process currently seems to limit the manufacturing of cladding longer than 1 m.

A tube with an internal diameter of ~8 mm, manufactured by this process, is presented in Figure 22.20. The internal SiC/SiC tube has a thickness of ~0.3 mm, the metal (Nb) liner a thickness of 0.07 mm, and the external SiC/SiC tube a thickness of ~0.6 mm.

In order to verify the relevance of the proposed design, helium permeation testing during mechanical tensile stress was performed on "sandwich" tubes at ambient temperature in CEA. The test consists in performing tensile tests coupled with a measurement of leak detection linked to the inside of the tube. The exterior of the tube is confined and

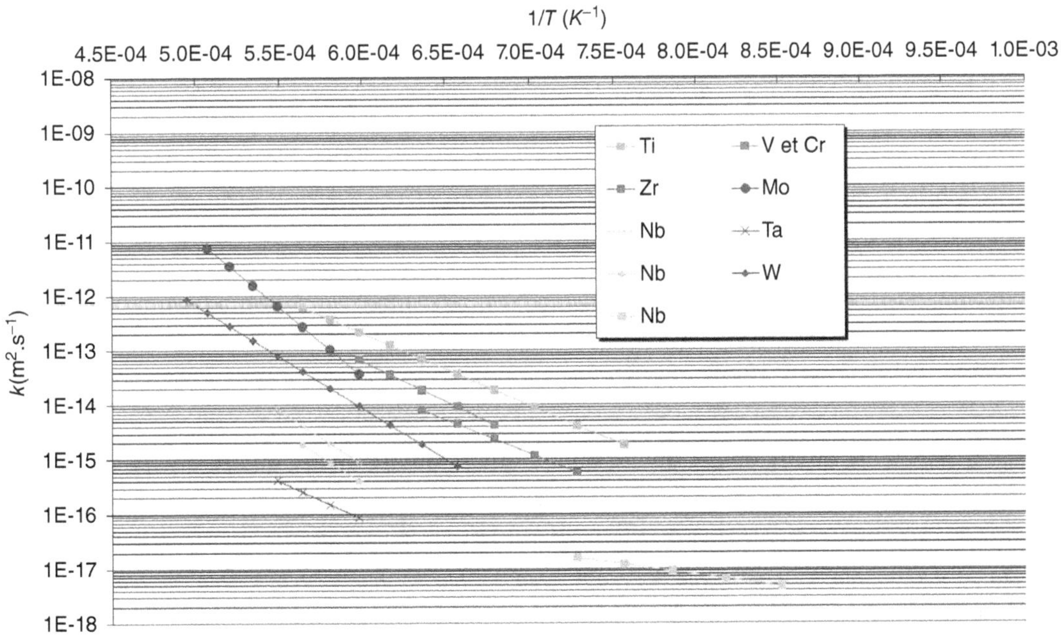

FIGURE 22.19

Evolution of chemical diffusivities between SiC and refractory metals with temperature [103–106].

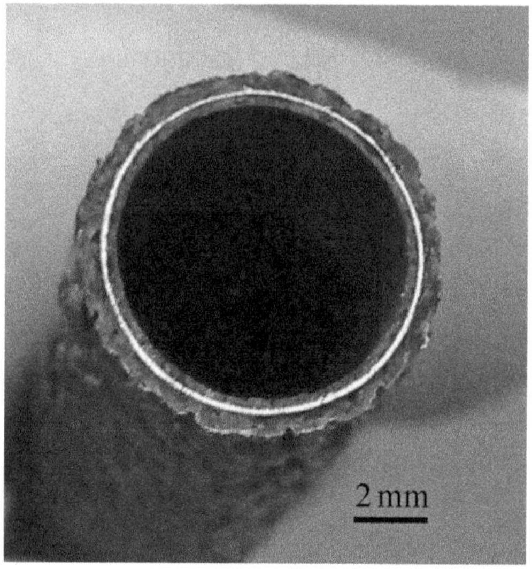

FIGURE 22.20

Picture of a niobium liner sandwich SiC/SiC tube densified by CVI process.

subject to a helium pressure of 2 bars. The detection limit of the system is ~3×10^{-11} Ncm^3/s. The result is presented in Figure 22.21. For the entire mechanical stressing phase (in stages) of the tube until its failure, it is not possible to measure a signal greater than background noise. In the conditions of the test, this means that the permeation coefficient of the tube remains below 10^{-17} m^2/s (detection limit taking into account the total thickness of the tube) until its failure.

Studies are still underway to obtain a robust cladding closure process. Finally, 3D dimensional checks recently performed on a dozen tubes have demonstrated the high manufacturing quality. The internal and external dimensions are obtained with a tolerance better than ±0.02 mm. The external and internal cylindricities are always lower than 0.05 mm. Concentricity is always lower than 0.05 mm. Straightness on the exterior and interior is always lower than 0.04 mm. Finally, average roughness R_a is <5 μm, and the maximum roughness R_z < 30 μm.

This innovative design is a significant advancement in the development of leaktight SiC/SiC composites for high temperature applications in nuclear cores. Nonetheless, there are still a lot of unknowns, including the behavior of the liner under irradiation conditions (excessive fragilization for example) with very high doses (~100 dpa). Studies are also being performed to check the thermochemical compatibility of the envisaged liners with the SiC/SiC composite at high temperatures (~1000°C) and over long periods of time (>1 year) even if estimations based on the data presented in Figure 22.19 are encouraging. Moreover, the behavior of this type of cladding and metal liner in an impure helium

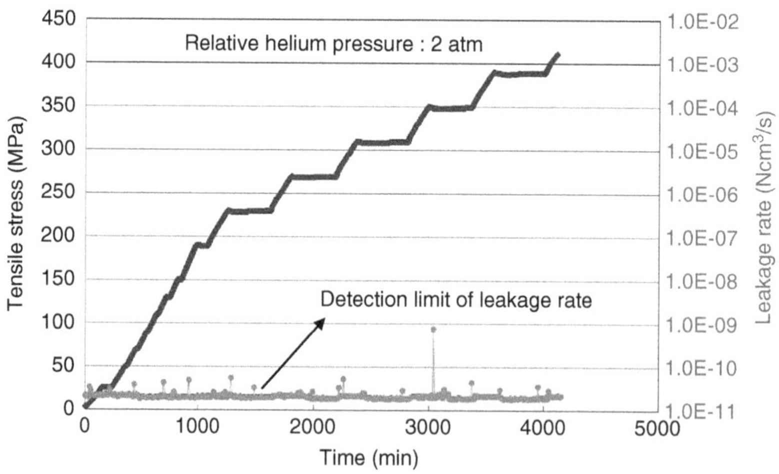

FIGURE 22.21

Evolution of leakage rate during tensile test of a niobium liner "sandwich" CVI composite.

atmosphere must be studied. A European project is proposed to address the behavior of this type of cladding in a high temperature He loop in GFR conditions (excluding irradiation).

Finally, this design may also be of interest for LWRs even if the type of metal liner must be chosen to withstand a highly oxidizing environment.

22.6 CONCLUSIONS AND PERSPECTIVES

Two types of CMCs have been and are still envisaged for nuclear applications: C/C and SiC/SiC. C/C have the advantage of unarguable industrial maturity and are already widely used for aerospace and industrial applications. Nonetheless, the significant requirements associated with nuclear power reactors do not allow for their use for applications involving radiation doses of greater than a few dpa. They may, for example, be used in VHTRs for applications (hot gas duct, core supports, and restraints) involving low doses. Furthermore, they are used in nuclear fusion research reactors and are still planned for use as divertor targets for ITER. Since these research reactors are not intended to produce energy, the maximum radiation doses to which the materials are subject in terms of plasma remain low ($\ll$1 dpa). This will not be the case in the future if a fusion power reactor is developed. The radiation doses to which the materials of the first walls will be subject will be such that C/Cs will no longer be able to be used.

Thanks to progress in the development of high purity HNLS and TSA3 (so-called third-generation fibers) SiC fiber at the end of the 1990s, SiC/SiC composites are currently the only CMCs of interest for nuclear applications in which the materials are subject to significant radiation doses (greater than a few dpa). The main handicap of these materials is their manufacturing cost, which highly limits their development. This is mainly linked to the low availability and cost of third-generation SiC fibers (>10 000 dollars/kg). The recent creation (2012) of the company NGS (N for Nippon Carbon, G for General Electric, and S for Snecma Propulsion Solide, today Herakles), formerly Nippon Carbon, for the production of Hi-Nicalon S fiber is, however, excellent news. This ensures long-term supply in significant quantities and shows the interest of two major manufacturers in the field of CMCs in terms of SiC/SiC composites in general.

While there are many publications discussing the development of materials with third-generation SiC reinforcement, little progress has been made (the same results are often repeated), and the mechanical characterizations presented are most often the result of bending tests. This does not provide for an easy comparison of results from different sources. Nonetheless, some conclusions can be drawn based on the studies that have been published or are in progress.

Of the two fiber reinforcements, HNLS reinforcement is the only currently providing for composites with high ductility (ε_R > 0.5%). While TSA3 reinforcement has better conductivity and better strength at very high temperatures, the low ductility behavior of the composites resulting from its use is hardly compatible with nuclear applications for the moment. There is still significant work to be done before the mechanisms governing this fragility will be understood, and no solution is currently proposed.

The pyrocarbon interphase seems to be the only one that satisfies the requirements of the application. Its thickness must be <300 nm to conserve good mechanical properties under irradiation conditions. The best operating feedback concerns single-layer interphases having a thickness of 150 nm and multilayered PyC/SiC interphase.

The CVI process is the reference process for densification, since it provides for a high purity SiC matrix with low activation and is perfectly stable under irradiation conditions. The NITE process can only be used with TSA3 fiber reinforcement, and its behavior under high doses of neutron irradiation must be analyzed in greater depth.

Finally, the greatest progress made over the past years has been driven by fission reactor applications. SiC/SiC control rod demonstrators have thus been manufactured in the United States. Original fuel cladding designs have also been proposed for LWRs and GFRs. Of these designs, the "sandwich" design for GFR cladding is the most promising, since it ensures maintenance of tightness in the cladding until its failure. The refractory metal liner used to ensure this tightness is also used to create an airtight closure without using brazing as done elsewhere (see Chapter 22). Characterizations have also been performed for short objects (<200 mm), showing good mechanical characteristics and excellent tolerances on the internal and external sides and geometries.

Of course, several characterization campaigns are still necessary to validate the CMC designs proposed for each application. These very long-term studies require significant human and financial means. The manufacturers in the sector are thus hesitant to finance such long-term research. This explains why most of research led on these materials is supported by state entities (national, European Community, or international programs). It would undoubtedly be useful to establish international collaborations and pool means allocated for research on fusion reactors and fourth generation fission reactors in order to make more rapid progress. The operating conditions of these reactors are often very similar, and their implementation is largely linked to the development of new materials satisfying similar, particularly demanding specifications. SiC/SiC for nuclear applications therefore represent a most difficult challenge, yet are highly attractive, due to their low activation under irradiation and the operating temperatures they make it possible to achieve.

ACKNOWLEDGMENT

The author would like to thank J.-L. Séran, M. Le Flem, L. Chaffron, D. Nunes, M. Zabiego, E. Buet, L. Briottet, B. Ricetti, G. Ioupias, D. Bossu, C. Lorrette, I. Hugon, L. Veiller, D. Bossu, L. Gelebart, F. Bernachy-Barbe, C. Chateau, C. Cossu, E. Suzon, A. Michaux, and G. Vaux.

REFERENCES

1. S. Nishio, S. Ueda, I. Aoki, R. Kurihara, T. Kuroda, H. Miura, T. Kunugi, Y. Seki, T. Nagashima, M. Ohta, et al. Improved tokamak concept focusing on easy maintenance. *Fusion Eng. Des.*, 41, 357–364 (1998).
2. A. R. Raffray, R. Jones, G. Aiello, M. Billone, L. Giancarli, H. Golfier, A. Hasegawa, Y. Katoh, A. Kohyama, S. Nishio, et al. Design and material issues for high performance SiC_f/SiC-based fusion power cores. *Fusion Eng. Des.*, 55, 55–95 (2001).
3. K. Ehrlich, E. E. Bloom, and T. Kondo. International strategy for fusion materials development. *J. Nucl. Mater.*, 283–287, 79–88 (2000).
4. J. Schlosser, P. Chappuis, M. Chatelier, J. J. Cordier, P. Deschamps, L. Garampon, D. Guilhem, M. Lipa, and R. Mitteau. In-service experience feedback of the Tore-Supra actively cooled inner first wall. *Fusion Eng. Des.*, 27, 203–209 (1995).
5. T. Muroga, M. Gasparotto, and S. J. Zinkle. Overview of materials research for fusion reactors. *Fusion Eng. Des.*, 61–62, 13–25 (2002).
6. L. L. Snead and M. Ferraris. Carbon as a fusion plasma-facing material, in *Comprehensive Nuclear Materials*, Vol. 4, edited by R. J. M. Konings (Elsevier Ltd., 2012), pp. 583–618.
7. Y. Katoh, L. L. Snead, C. H. Henager Jr., A. Hasegawa, A. Kohyama, B. Riccardi, and H. Hegeman. Current status and critical issues for development of SiC composites for fusion applications. *J. Nucl. Mater.*, 367, 659–671 (2007).
8. Y. Katoh, L. L. Snead, T. Nozawa, N. B. Morley, and W.E. Windes. Advanced radiation-resistant ceramic composites. *Adv. Sci. Technol.*, 45, 1915–1924 (2006).
9. L. L. Snead, T. Nozawa, M. Ferraris, Y. Katoh, R. Shinavski, and M. Sawan. Silicon carbide composites as fusion power reactor structural materials. *J. Nucl. Mater.*, 417, 330–339 (2011).
10. D. Petti, D. Crawford, and N. Chauvin. Fuels for advanced nuclear energy systems. *MRS Bull.*, 34, 1–6 (2009).
11. J.-P. Bonal, A. Kohyama, J. V. D. Laan, and L. L. Snead. Graphite, ceramics, and ceramic composites for high-temperature nuclear power systems. *MRS Bull.*, 34, 28–34 (2009).
12. A. Kohyama, Y. Katoh, L. L. Snead, and R. H. Jones. Development of SiC/SiC composite for fusion application. Proceedings of the 18th Conference on IAEA Fusion Energy 2000 18th conference proceedings, Sorrento, Italy, October 4–10, 2000.
13. A. Hasegawa, A. Kohyama, R. H. Jones, L. L. Snead, B. Riccardi, and P. Fenici. Critical issues and current status of SiC/SiC composites for fusion. *J. Nucl. Mater.*, 283, 128–137 (2000).
14. M. S. Kazimi, J. Dobisesky, D. Carpenter, J. Richards, E. E. Pilat, and E. Shwageraus. *Feasibility and Economic Benefits of PWR Cores with Silicon Carbide Cladding*, MIT-ANP-PR-134 (Massachusetts Institute of Technology, 2011).
15. S. Ray, E. Lahoda, and F. Franceschini. Assessment of different materials for meeting the requirement of the future fuel designs. Reactor Fuel Performance Meeting, Paper A0115, Manchester, UK, September 2–6, 2012.
16. A Technology Roadmap for Generation IV Nuclear Energy Systems. http://www.gen-4.org/Technology/roadmap.htm, Issued by the U.S. DOE Nuclear Energy Research Advisory Committee and the Generation IV International Forum, Report USDOE-GIF-002-00 (2002).
17. L. L. Snead, T. D. Burchell, and A. L. Qualls. Strength of neutron-irradiated high-quality 3D carbon fiber composite. *J. Nucl. Mater.*, 321(2–3), 165–169 (2003).
18. L. L. Snead, T. D. Burchell, and Y. Katoh. Swelling of nuclear graphite and high quality carbon fiber composite under very high irradiation temperature. *J. Nucl. Mater.*, 381(1-2), 55–61 (2008).
19. Y. Katoh, L. L. Snead, I. Szlufarska, and W. J. Weber. Radiation effects in SiC for nuclear structural applications. *Curr. Opin. Solid State Mater. Sci.*, 16(3), 143–152 (2012).

20. Y. Katoh, T. Nozawa, L. L. Snead, K. Ozawa, and H. Tanigawa. Stability of SiC and its composites at high neutron fluence. *J. Nucl. Mater.*, 417(1-3), 400–405 (2011).
21. http://www.iter.org/fr/mach
22. M. E. Sawan, N. M. Ghoniem, L. Snead, and Y. Katoh. Damage production and accumulation in SiC structures in inertial and magnetic fusion systems. *J. Nucl. Mater.*, 417(1–3), 445–450 (2011).
23. P. Fenici, A. J. Frias Rebelo, R. H. Jones, A. Kohyama, and L. L. Snead. Current status of SiC/SiC composites R & D. *J. Nucl. Mater.*, 258, 215–263 (1998).
24. P. Rocco and M. Zucchetti. Criteria for defining low activation materials in fusion-reactor applications. *Fusion Eng. Des.*, 15(3), 235–244 (1992).
25. http://www.iter.org/fr/mach/divertor
26. http://www.iter.org/fr/mach/divertor
27. M. Ferraris, M. Salvo, and V. Caselegno. Joining and integration issues of ceramic matrix composite for nuclear industry, in *Ceramic Integration and Joining Technology*, edited by M. Singh, T. Ohji, E. Asthana, and S. Mathur (John Wiley & Sons, Inc., Hoboken, 2011), pp. 39–89.
28. M. Merola, M. Akiba, V. Barabash, and I. Mazul. Overview on fabrication and joining of plasma facing and high heat flux materials for ITER. *J. Nucl. Mater.*, 307–311, 1524–1532 (2002).
29. G. Chevet, E. Martin, J. Boscary, G. Camus, V. Herb, J. Schlosser, F. Escourbiac, and M. Missirlian. Characterization and damaging law of CFC for high heat flux actively cooled plasma facing components. *J. Nucl. Mater.*, 417(1-3), 643–646 (2011).
30. M. Merola and G. Vieider. On the use of flat tile armour in high heat flux components. *J. Nucl. Mater.*, 258, 672–676 (1998).
31. R. Tivey, M. Akiba, D. Driemeyer, I. Mazul, M. Merola, and M. Ulrickson. ITER R&D: vacuum vessel and in-vessel components: divertor cassette. *Fusion Eng. Des.*, 55(2–3) 219–229 (2001).
32. G. Pintsuk, J. Compan, J. Linke, P. Majerus, A. Peacock, D. Pitzer, and M. Rödig. Mechanical and thermo-physical characterization of the carbon fibre composite NB31. *Phys. Scr.*, T128, 66-71 (2007).
33. L. Giancarli, H. Golfier, S. Nishio, R. Raffray, C. Wong, and R. Yamada. Progress in blanket designs using SiCf/SiC composites. *Fusion Eng. Des.*, 61–62, 307–318 (2002).
34. B. A. Pint, K. L. More, M. H. Meyer, and J. R. DiStefano. Recent progress addressing compatibility issues relevant to fusion environments. *Fusion Sci. Technol.*, 47(4), 851–855 (2005).
35. H. Feinroth. Water cooled nuclear reactor and fuel elements therefor. US Patent 5,182,077 (1993).
36. H. Feinroth and B. R. Hao. Multi-layered ceramic tube for fuel containment barrier and other applications in nuclear and fossil power plants. US Patent 0039,524 (2006).
37. D. Carpenter, G. E. Kohse, and M. S. Kazimi. *An Assessment of Silicon Carbide as a Cladding Material for Light Water Reactors*, MIT-ANP-TR-132 (Massachusetts Institute of Technology, 2010).
38. L. L. Snead, T. Nozawa, Y. Katoh, T.-S. Byun, S. Kondo, and D. A. Petti. Handbook of SiC properties for fuel performance modeling. *J. Nucl. Mater.*, 371(1–3) 329–377 (2007).
39. U. Cleve, K. Knizia, and K. Kugeler. The technology of high-temperature-reactors. *Proceedings of ICAPP 2011*, Nice, France, May 2011. Paper 11076.
40. G. Baccaglini, S. Ball, T. Burchell, B. Corwin, T. Fewell, M. LaBar, P. MacDonald, P. Rittenhouse, E. Shaber, F. Southworth, et al. Very High Temperature Reactor (VHTR): Survey of materials research and development needs to support early deployment. *GEN IV Nuclear Energy Systems*, INEEL/EXT-03-00141 (2003).
41. A. Ravenet. Corps d'assemblage de combustible nucléaire et un assemblage de combustible nucléaire comportant un tel corps. French Patent 2,951,312 (2011).
42. A. Ravenet. Element combustible de type plaque macrostructurée. French Patent 2,889,765 (2007).
43. J. K. Lee and M. Naganuma. Design factor using a SiC/SiC composites for core component of gas cooled fast reactor. 2: Thermal stress, in *Mechanical Properties and Performance of Engineering Ceramics and Composites III*, Vol. 27, edited by E. Lara-Curzio, J. Salem, and D. Zhu (John Wiley & Sons, Inc., Hoboken, 2008), pp. 55–63.
44. J.-J. X. Ingremeau. Methodologie d'optimisation d'un Coeur de réacteurs à neutrons rapides, application à l'identification de solutions (combustible, cœur, système) permettant des performances accrues. Thèse. Université de Paris-Sud 11, 2011.
45. K. A. Terrani, G. W. Chinthaka Silva, J. R. Keiser, M. P. Brady, T. Cheng, B. A. Pint, and L. L. Snead. High temperature oxidation of silicon carbide and advanced iron-based alloys in steam-hydrogen environments. Reactor Fuel Performance Meeting, Manchester, UK, September 2–6, 2012.
46. T. Cheng, J. R. Keiser, M. P. Brady, K. A. Terrani, and B. A. Pint. Oxidation of fuel cladding candidate materials in steam environments at high temperature and pressure. *J. Nucl. Mater.*, 427(1–3), 396–400 (2012).
47. J. Dobisesky, E. E. Pilat, and M. S. Kazimi. *Reactor Physics Considerations for Implementing Silicon Carbide Cladding into PWR Environment*, MIT-ANP-TR-136 (Massachusetts Institute of Technology, 2011).

48. L. Hallstadius, S. Johnson, and E. Lahoda. Cladding for high performance fuel. *Progr. Nucl. Technol.*, 57, 71–76 (2012).
49. L.L. Snead, Y. Katoh, and T. Nozawa. Radiation effects in SiC and SiC-SiC, in *Comprehensive Nuclear Materials*, Vol. 4, edited by R.J. M. Konings (Elsevier Ltd., 2012), pp. 215–239.
50. L.L. Snead, Y. Katoh, and S. Connery. Swelling of SiC at intermediate and high irradiation temperatures. *J. Nucl. Mater.*, 367, 677–684 (2007).
51. M. Eto, S. Ishiyama, and H. Ugachi. Development of C/C composite control rod for HTT-preparation of elements and their fracture tests. *JAERI Res.*, 96–043 (1996).
52. J. G. Lavin, D. R. Boyington, J. Lahijani, B. Nysten, and J. P. Issi. The correlation of thermal-conductivity with electrical-resistivity in mesophase pitch based carbon-fiber. *Carbon*, 31(6), 1001–1002 (1993)
53. J. P. Singh, D. Singh, and M. Suturia. Role of fibre and interface. *Compos. A Appl. Sci. Manuf.*, 30, 445–450 (1999).
54. D. D. L. Chung. *Carbon Fiber Composites* (Butterworth-Heinemann, Newton, MA, 1994).
55. S. Ishiyama, K. Fukaya, M. Eto, Strength test of C/C composite high performance control rod for FTGR. *Journal of the Atomic Energy Society of Japan*, 41(10), 1092–1099, 1999.
56. C. Sauder, J. Lamon, and R. Pailler. The tensile properties of carbon matrices at temperatures up to 2200°C. *Carbon*, 43(10), 2054–2065 (2005).
57. I. Berdoyes and J. Thebault. Thermostructural composites materials: from space to advanced fission applications. Proceedings of European Congress on Advanced Materials and Processes EUROMAT, Prague, Czech Republic, September 5–8, 2005. pp. 11–17.
58. J. D. Buckley and D. D. Edie. *Carbon-Carbon Materials and Composites*, edited by J. D. Buckley and D. D. Edie (Noyes Publications, Park Ridge, NJ, 1993).
59. A. T. Peacock, M. Merola, M. A. Pick, and R. Tivey. Status of CFC development in Europe for ITER. *Phys. Scr.*, T128, 23–28 (2007).
60. M. Richou, M. Missirlian, B. Riccardi, P. Gavilla, C. Desgranges, N. Vignal, V. Cantone, and S. Constans. Assessment of CFC grades under thermal fatigue for the ITER inner vertical target. *Phys. Scr.*, T145, 1–5 (2011).
61. T. Windhorst and G. Blount. Carbon-carbon composites: a summary of recent developments and applications. *Mater. Des.*, 18(1), 11–15 (1997).
62. J. Deng, W. Liu, and H. Du. Thermal shock resistance of C/C composites. *Carbon*, 37(11), 1868–1869 (1999).
63. H. Lemaire. Influence de l'interface fibre/matrice sur le comportement mécanique en température des composites C/C. Doctoral Thesis. Université de Bordeaux I, 2001.
64. M. Eto, S. Ishiyama, K. Fukuya, T. Saito, M. Ishihara, and S. Hanawa. Development of C/C composites control rod for HTTR, concept, specification and mechanical test of materials. *JAERI Res.*, 98–2003 (1998).
65. M. Roedig, W. Baur, and B. Woschek. Burst tests to measure the tangential strength of CFC tubes. Proceedings of a Specialists Meeting, International Atomic Energy Agency, Tokai-mura, Japan, September 9–12, 1991.
66. Y. Katoh, L. L. Snead, T. Nozawa, S. Kondo, and J. T. Busby. Thermophysical and mechanical properties of near-stoichiometric fiber CVI SiC/SiC composites after neutron irradiation at elevated temperatures. *J. Nucl. Mater.*, 403(1–3), 48–61 (2010).
67. R. El Bouazzaoui, S. Baste, and G. Camus. Development of damage in a 2D woven C/SiC composite under mechanical loading. 2. Ultrasonic characterization. *Compos. Sci. Technol.*, 56(12), 1373–1382 (1996).
68. G. Camus, L. Guillaumat, and S. Baste. Development of damage in a 2D woven C/SiC composite under mechanical loading. 1. Mechanical characterization. *Compos. Sci. Technol.*, 56(12), 1363–1372 (1996).
69. F. Abbe, V. Herb, G. Camus, and E. Martin. Characterization and modeling of the nonlinear mechanical behavior of SEPCARB (R) NB41 carbon/carbon composite. *Fusion Sci. Technol.*, 56(2), 1009–1012 (2009).
70. C. Sauder and J. Lamon. Tensile creep behavior of SiC-based fibers with a low oxygen content. *J. Am. Ceram. Soc.*, 90(4), 1146–1156 (2007).
71. J. J. Sha, T. Nozawa, J. S. Park, Y. Katoh, and A. Kohyama. Effect of heat treatment on the tensile strength and creep resistance of advanced SiC fibers. *J. Nucl. Mater.*, 329, 592–596 (2004).
72. E. Buet, C. Sauder, S. Poissonnet, P. Brender, R. Gadiou, and C. Vix-Gutherel. Influence of chemical and physical properties of the last generation of silicon carbide fibres on the mechanical behaviour of SiC/SiC composite. *J. Eur. Ceram. Soc.*, 32(3), 547–557 (2012).
73. G. Newsome. The effect of neutron irradiation on silicon carbide fibers. *Proceedings of the 21st Annual Conference on Composites, Advanced Ceramics, Materials, and Structures—A: Ceram. Eng. Sci. Proc.*, 18(3), 579–583 (1997).
74. M. Suzuki, M. Tomita, and A. Murakoshi. Application of NRA to evaluation of boron implants in Si for shallow junctions. *Nucl. Instr. Meth. Phys. Res. B*, 190, 552–555 (2002).
75. Y. Katoh, T. Nozawa, L. L. Snead, and T. Hinoki. Effect of neutron irradiation on tensile properties of unidirectional silicon carbide composites. *J. Nucl. Mater.*, 367–370, 774–779 (2007).

76. G. Newsome, L. L. Snead, T. Hinoki, Y. Katoh, and D. Peters. Evaluation of neutron irradiated silicon carbide and silicon carbide composites. *J. Nucl. Mater.*, 371(1–3) 76–89 (2007).
77. T. Nozawa, Y. Katoh, and L. L. Snead. The effect of neutron irradiation on the fiber/matrix interphase of silicon carbide composites. *J. Nucl. Mater.*, 384(3), 195–211 (2009).
78. Y. Katoh, A. Kohyama, T. Hinoki, and L. L. Snead. Progress in SiC-based ceramic composites for fusion applications. *Fusion Sci. Technol.*, 44(1), 155–162 (2003).
79. Y. Katoh, L. L. Snead, T. Nozawa, T. Hinoki, and A. Kohyama. Mechanical properties of chemically vapor-infiltrated silicon carbide structural composites with thin carbon interphases for fusion and advanced fission applications. *Mater. Trans.*, 46(3), 527–535 (2005).
80. T. Hinoki, L. L. Snead, T. Tagushi, N. Igawa, T. Nozawa, Y. Katoh, and A. Kohyama. Optimization and characterization of chemical vapor infiltrated SiC/SiC composites, in *Advanced SiC/SiC Ceramic Composites: Developments and Applications in Energy Systems*, Vol. 144, edited by A. Kohyama, M. Sing, H.-T. Lin, and Y. Katoh (John Wiley & Sons, Inc., Hoboken, 2012), pp. 207–214.
81. T. Nozawa, K. Hironaka, Y. Katoh, A. Kohyama, T. Tagushi, S. Jitsukawa, and L. L. Snead. Tensile strength of chemical vapor infiltrated advanced SiC fiber composites at elevated temperatures, in *Advanced SiC/SiC Ceramic Composites: Developments and Applications in Energy Systems*, Vol. 144, edited by A. Kohyama, M. Sing, H.-T. Lin, and Y. Katoh (John Wiley & Sons, Inc., Hoboken, 2012), pp. 215–219.
82. C. Sauder, A. Brusson, and J. Lamon. Influence of interface characteristics on the mechanical properties of Hi-Nicalon type-S or Tyranno-SA3 fiber-reinforced SiC/SiC minicomposites. *Int. J. Appl. Ceram. Technol.*, 7(3), 291–303 (2010).
83. Y. Kim, J. Lee, M. Kim, and J. Park. Effect of multilayer coating on mechanical properties of Nicalon-fibre-reinforced silicon carbide composites. *J. Mater. Sci.*, 31(2), 335–338 (1996).
84. R. Naslain, R. Pailler, and J. Lamon. Single- and multilayered interphases in SiC/SiC composites exposed to severe environmental conditions: an overview. *Int. J. Appl. Ceram. Technol.*, 7(3), 263–275 (2010).
85. C. Henager Jr. Fracture strength and time dependent properties of SiC/SiC Type S fiber compoites. Fusion Materials Semiannual Progress Report for Period Ending, Vol. 41, DOE/ER-0313/41 (2006), pp. 33–50.
86. J.-S. Park, A. kohyama, T. Hinoki, K. Shimoda, and Y.-H. Park. Efforts on large scale production of NITE-SiC/SiC composites. *J. Nucl. Mater.*, 367–370, 719–724 (2007).
87. A. Kohyama. CMC for nuclear applications, in *Ceramic Matrix Composites: Fiber Reinforced Ceramics and Their Applications*, edited by W. Krenkel (Wiley-VCH Verlag GmbH, Weinheim, Germany, 2008), pp. 353–384.
88. K. Shimoda, A. Kohyama, and T. Hinoki. High mechanical performance SiC/SiC composites by NITE process with tailoring of appropriate fabrication temperature to fiber volume fraction. *Compos. Sci. Technol.*, 69(10), 1623–1628 (2009).
89. K. Shimoda, J.-S. Park, T. Hinoki, and A. Kohyama. Influence of pyrolytic carbon interface thickness on microstructure and mechanical properties of SiC/SiC composites by NITE process. *J. Nucl. Mater.*, 68(1) 98–105 (2008).
90. K. Shimoda, J.-S. Park, T. Hinoki, and A. Kohyama. Microstructural optimization of high-temperature SiC/SiC composites by NITE process. *J. Nucl. Mater.*, 386–388, 634–638 (2009).
91. W. Yang, H. Araki, A. Kohyama, Q. Yang, and T. Noda. Effects of heat treatment on the microstructure and flexural properties of CVI-Tyranno-SA/SiC composite. *Ceram. Int.*, 33(2), 141–146 (2007).
92. T. Koyanagi, S. Kondo, and T. Hinoki. The influence of sintering additives on the irradiation resistance of NITE SiC. *J. Nucl. Mater.*, 417(1–3), 435–439 (2011).
93. T. Nozawa, T. Hinoki, A. Hasegawa, A. Kohyama, Y. Katoh, L. L. Snead, C. H. Henager Jr, and J. B. J. Hegeman. Recent advances and issues in development of silicon carbide composites for fusion applications. *J. Nucl. Mater.*, 386–388, 622–627 (2009).
94. S. Novak and A. Iveković. Preparation of SiCf/SiC composites by the slip infiltration and transient eutectoid (SITE) process. *J. Nucl. Mater.*, 399(2–3), 167–174 (2010).
95. S. Novak, G. Dražić, K. König, and A. Iveković. Fabrication of SiCf/SiC composites by SITE-P process. *J. Nucl. Mater.*, 427(1–3), 110–115 (2012).
96. K. Shimoda, T. Hinoki, Y. Katoh, and A. Kohyama. Development of the tailored SiC/SiC composites by the combined fabrication process of ICVI and NITE methods. *J. Nucl. Mater.*, 384(2), 103–108 (2009).
97. T. Hino, Y. Hirohata, Y. Yamauchi, M. Hashiba, A. Kohyama, Y. Katoh, Y. Lee, T.Jinushi, M. Akiba, K. Nakamura, et al. Plasma material interaction studies on low activation materials used for plasma facing or blanket component. *J. Nucl. Mater.*, 329–333, 673–677 (2004).
98. T. Hino, E. Hayashishita, Y. Yamauchi, M. Hashiba, Y. Hirohata, and A. Kohyama. Helium gas permeability of SiC/SiC composite used for in-vessel components of nuclear fusion reactor. *Fusion Eng. Des.*, 73(1), 51–56 (2005).
99. J. D. Stempien, D. Carpenter, G. E. Kohse, and M. S. Kazimi. *Behavior of Triplex Silicon Carbide Fuel Cladding Designs Tested Under Simulated PWR Conditions*, MIT-ANP-TR-135 (Massachusetts Institute of Technology, 2011).
100. G. Griffith. *LWRS Advanced LWR nuclear Fuels*, 2010 American Nuclear Society Winter Meeting and Nuclear Technology Expo, Las Vegas, NV, November, 2010.

101. P. David, A. Ravenet, D. Rochais, and M. Zabiego. Joint d'interface solide à porosité ouverte pour crayon de combustible nucléaire et pour barre de commande nucléaire. French Patent 2,961,623 (2011).

102. C. Sauder, E. Buet, and S. Poissonnet. Processing and characterization of SiC/SiC Pin cladding reinforced with Hi-Nicalon S and SA3 SiC fibers. Materials Challenges in Alternative & Renewable Energy (MCARE), Annual Meeting, Clearwater Beach, February 26–March 3, 2012.

103. M. Naka, T. Fukai, J. Osaka, and J. C. Schuster. Reaction phases and diffusion paths in SiC/metals systems, in *Surfaces, Interfaces, and the Science of Ceramic Joining*, edited by K. S. Weil, I. E. Reimanis, and C. A. Lewinsohn, *Proceedings of the 106th Annual Meeting of the American Ceramic Society* (Indianapolis, Indiana, 2004), pp. 19–27.

104. M. Naka and J. C. Feng. Phase reaction and diffusion path of SiC/Nb system, in *Advanced Materials '93, III/B : Composites, Grain Boundaries and nanophase Materials*, Vol. 16, edited by T. Masumoto, A. Inoue, A. Makiwshima, L. Arnberg, M. Doyama, M. I. Baskes, R. Nieminen, K. Kuribayashie, H. Hashimoto, M. Z. E. Saghir, A. Sawaoka, R. S. Sokolowski, H. U. Walter, and T. Kishi (Elsevier Science, Amsterdam, Netherlands, 1994), pp. 1143–1146.

105. A. Joshi, H. S. Hu, L. Jesion, J. J. Stephens, and J. Wadsworth. High temperature interactions of refractory metal matrices with selected ceramic reinforcements. *Metall. Trans. A*, 21, 2882–2837, 1990.

106. K. H. Lee, J. K. Yoon, J. K. Lee, J. M. Doh, K. T. Hong, and W. Y. Yoon. Growth kinetics of W5Si3 layer in WSi2/W system. *Surf. Coat. Technol.*, 187, 146–153, 2004.

CHAPTER 23

CERAMIC MATRIX COMPOSITES FOR FRICTION APPLICATIONS

WALTER KRENKEL[1] AND JACQUES GEORGES THÉBAULT[2]

[1] Ceramic Materials Engineering, University of Bayreuth, Bayreuth, Germany
[2] Thermostructural composites expert, SAFRAN group (Retired), Bordeaux, France

23.1 INTRODUCTION

In a braking system mechanical loads are applied to absorb kinetic energy and generate heat. This kind of application requires materials maintaining high mechanical characteristics at high temperature, that is, high thermomechanical properties. Thermostructural composite materials are manufactured from light and refractory components like carbon and silicon carbide, associated to get composites in which a refractory matrix is reinforced by high performance fibers. In this chapter it is illustrated, how these composites have proven their interest as constituents in systems for high performance friction-based applications.

Two main classes of CMC materials are used for friction applications.

- Carbon/carbon composites, mainly used for aircraft braking systems and racing cars [1].
- Carbon/ceramic composites, more suitable for terrestrial applications (e.g., passenger cars, elevators) [2].

23.2 CARBON/CARBON FOR FRICTION APPLICATIONS

23.2.1 Aircraft Braking Particularities

The main application for C/C composites in terms of volume, mass, and market are aircraft brakes for commercial and military applications. As for every flying object the first aspect is to be as light as possible. A lighter aircraft has a longer range and fuel consumption is decreased. Aircraft braking systems must fulfill different requirements coming from standards (e.g., FAA, TSO; (US) MIL 5013; (F) Air 8507) and from aircraft manufacturer's specifications [3, 4]. These documents are related to the wheel and the brake which forms a unique single system linked with the landing gear [4, 5]. To begin, these specifications can be described in global characteristics like energy level to be absorbed, stopping distance, etc., and are later translated into technical items such as number and disk size, weight, etc. For example, the energies to be absorbed for different applications are listed in Table 23.1. Globally, two main great topics must be optimized by the wheel and brake designer, namely safety and economics.

Ceramic Matrix Composites: Materials, Modeling and Technology, First Edition. Edited by Narottam P. Bansal and Jacques Lamon.

TABLE 23.1 Braking Energies

(a) Airbus-per brake [4]

Braking sequence	Taxi	Service	Normal	Overload	RTO
Speed (m/s)	5–10	47	67	78	92
Energy (MJ)	1	14	34	47	86

(b) [8]

Application		Concorde	Airbus	Falcon 900	F1	Tank
Energy (kJ/kg)	Normal	850	400–1000	1000	500	450–800
	Emergency	2200	2400	1800	1000	1000

23.2.1.1 Safety Aspects In terms of brake design and materials definition, furthermore to normal braking efficiency, safety implies in particular satisfying "RTO" (rejected takeoff) conditions [3]. This test implies for the brakes to be able to stop an aircraft safely, fully loaded and just at the takeoff speed, in a rather short distance which simulates the rest of the available runway remaining in front of the plane at that "no-go" instant. This stopping distance matters in the airplane manufacturer's specifications for brakes.

During such an emergency stop, all of the kinetic energy of the plane-maximum in that configuration (Table 23.1a), which implies also that the brake unit is considered to be at the maximum worn state [3]-is transferred into the brake unit. Taking into account that the braking duration is short, no cooling can be efficient and all the thermal energy generated will be stored into the brake which acts as a heat sink. Consequently, extremely high temperatures are reached and more than 2000°C have been experimented locally [1]. So the "RTO" implies very high temperature considerations for the design of the whole brake and the choice of the thermostructural materials. Safety is the major improvement given by introducing C/C brakes [1].

23.2.1.2 Economical Aspects Decreasing costs is for sure a hard driving force for progress. For an aircraft the global cost can be described as "CPL" (cost per landing). This cost value takes into account the price of the braking unit and the price for maintenance and refurbishment divided by the number of landings the braking unit is able to sustain before overall level operation. So, this driving force implies two directions for progress, one is to decrease the manufacturing cost of the materials used for disks and the other is to decrease the wear per landing which is the sum of three different sets of braking: "taxi out" when the aircraft leaves the park to takeoff, then "landing," and finally "taxi in" when after landing the aircraft goes to dock at the airport. Under these conditions the brake unit and the disk materials must be optimized for low wear at relatively low temperature (e.g., 650°C) [6, 7].

So, at the end it appears that in its definition an aircraft brake system must fulfill two antagonistic conditions of operation simultaneously: low wear at medium temperature for a large number of landing cycles (>2500) [1, 5] and high efficiency at very high temperature once (RTO). For safety and efficiency reasons the introduction of carbon/carbon for disk brakes is "one of the most significant breakthrough in the history of brake technology" [5].

23.2.2 History and Materials Choice

Nowadays, all the aircraft braking systems use disk brakes. In small aircraft, disks and brake pads are still applied, but more and more disk-on-disk configurations are used. The multidisk configuration (Figure 23.1) [7] consists of a set of disks linked to the wheel drum, and for that named rotors (Figure 23.2), alternatively spaced with disks linked to the wheel axle (or torque tube), named stators. When braking, these disks are axially pressed together between a pressure plate (hydraulic or now electric) and a torque plate at the other end of the assembly.

Initially the rotating disks were made from metals or alloys. Iron (Figure 23.3) was selected for its low cost and relatively high temperature capability, but is heavy. Copper has good thermal properties but is limited in service temperature. Beryllium is light and has good characteristics but is expensive and presents health hazards in use and machining.

These rotor disks were initially associated with stators made with a circular assembly of pads. These pads were manufactured from different mineral refractory powders or fibers (asbestos) [9] linked by an organic polymer (phenol resin). Later on, for thermal stability reasons, polycrystalline graphite multipad stators were also used [3, 4, 10]. This

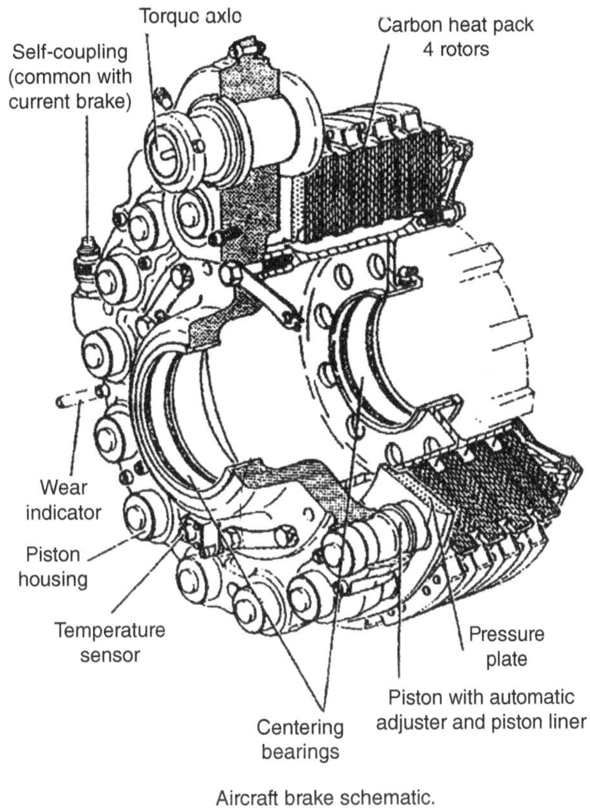

FIGURE 23.1

Multidisk brake configuration of an aircraft. Courtesy of Herakles company.

graphite on iron configuration is interesting to remind because it shows that an all carbon material pad, as polycrystalline graphite, can be used as a braking constituent [11].This is rather odd because graphite was generally known and used as a solid lubricant or as a low-friction coefficient and wear material for electric motor brushes, for example. But a new carbon behavior was experienced especially during the Second World War when military aircraft began to fly at stratospheric altitude and encountered great problems with their electric motor brushes: these graphite brushes were rapidly worn under these conditions and the problem was named as "dusting."

FIGURE 23.2

Rotors and stators. Courtesy of Herakles company.

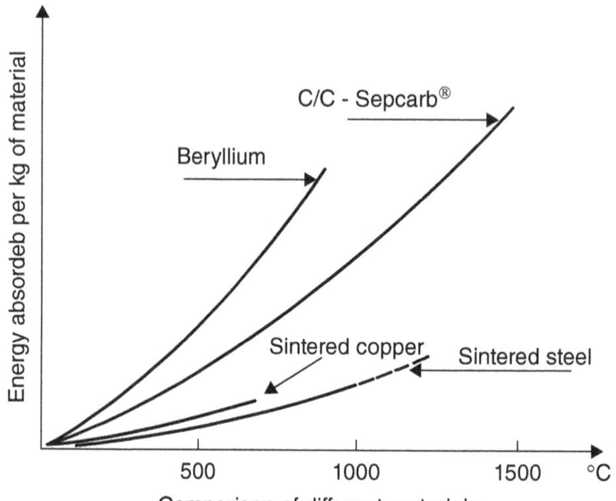

FIGURE 23.3

Energy absorbed per kilogram of different disk materials [8]. Courtesy of Herakles company.

This phenomenon has been studied and understood [9, 12, 13] and the results can be synthesized as: graphite (or carbon) can be used and behaves as a lubricant only in the presence of some gases able to be physicochemically "adsorbed" on the new worn surfaces, and the most efficient ordinary gas is water vapor [14]. At high altitude the partial pressure of water is so low that the behavior of carbon drastically changes and from solid lubricant goes to abrasive wear. On the other hand, it means that to make brakes, a high coefficient of friction (CoF) can also be obtained with graphite or carbon materials if the temperature can quickly reach a sufficient level to "desorb" (essentially) the "adsorbed" gases (water, oxygen) near the rubbing surfaces [6] (Figure 23.4).

The last step was in the 1960s with the supersonic aircraft program "Concorde." For such an airplane weight savings were once more crucial. But, related to its delta wing geometry, the takeoff and landing speeds were also particularly high so that the energy to be absorbed by the brakes increased significantly (Table 23.1b). It appears quickly that the weight of the landing system and especially the brakes were a major problem. At that time DUNLOP (UK, now MEGGIT) got the Concorde brake and wheel contract and worked to find the best definition [15]. As said before, iron was too heavy, beryllium too expensive and dangerous (but was initially used on C5A Galaxy and on the US shuttle) [4, 5], and copper was limited in temperature.

Taking into account the density, the melting point temperature, and the specific heat capacity (Table 23.2) of different elements, carbon appears to be a major candidate and, based on its thermal characteristics, the best choice to manufacture these high performance brakes.

Then the problem lies in the level of the mechanical properties required to manufacture structural components such as disks [5]. The order of magnitude of the mechanical strength required to manufacture disks was estimated to be around 80 MPa (12 000 Psi) [16]. At that time the mechanical properties of available carbon materials like polycrystalline graphite were insufficient to manufacture reliable brake disks.

Fortunately, a few years before, DUNLOP had bought a license to manufacture carbon/carbon composite materials and had been looking for using these materials in brakes since 1971 [7]. These thermostructural composites (discovered in 1958 [3]) offer a performance level sufficient to fulfill the requirements to manufacture brake disks [17] (Table 23.3).

An additional interest is that the mechanical properties increase with temperature [19]. In the end, since this material was convenient to manufacture rotors, stators were also manufactured with the same material [5]. The tribological pair of C/C rotors rubbing against C/C stators was then used to manufacture the Concorde brakes. This new brake system was first fly tested on a VC 10 in 1973 and then used as a standard fitment on Concorde 1 year later, with a weight saving over 600 kg [7, 17].

It was the first time C/C brakes were used on a commercial airplane. Previously, C/C rotors and stators were first prototyped and tested in 1968 on a Beech "Debonair" plane and in 1972 a US military fighter (F 15) has been fitted with C/C brakes using C/C rotors and C/C pads and then C/C stators [1, 5].

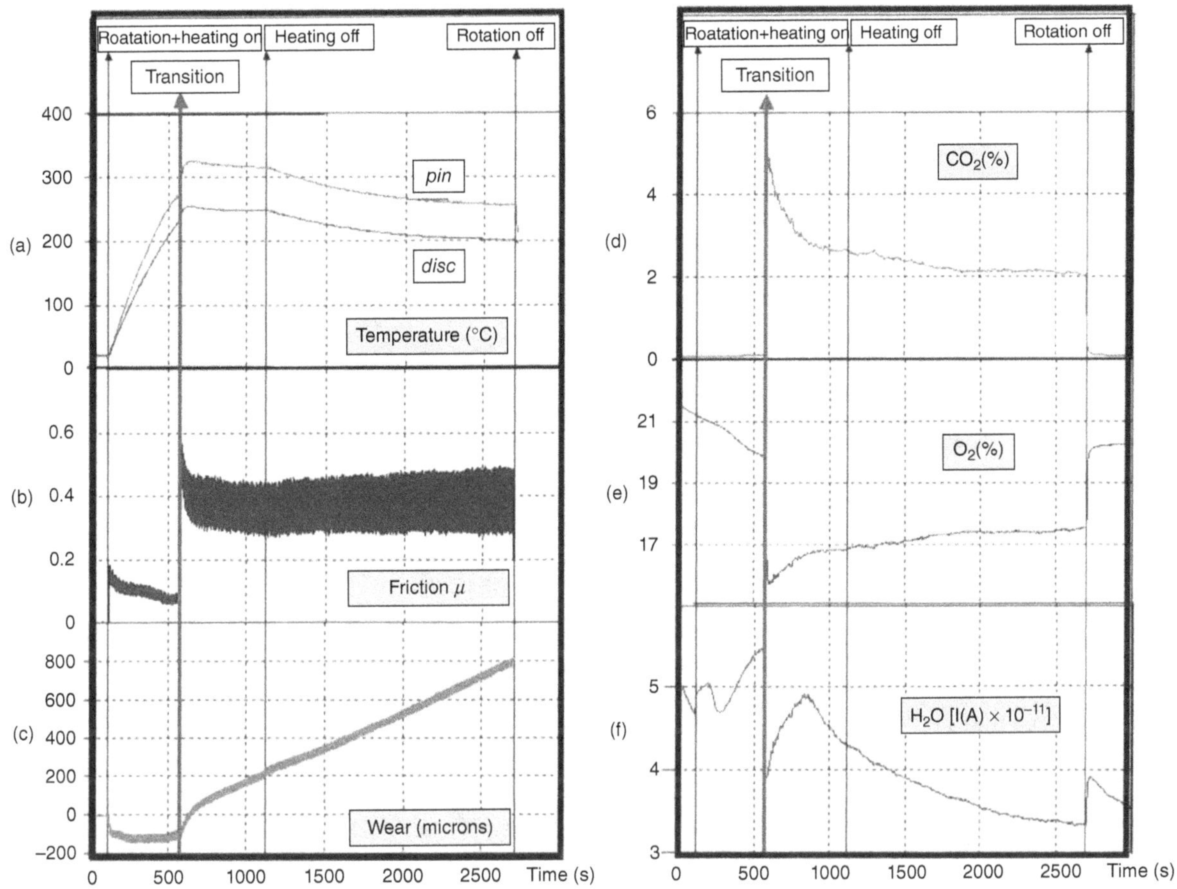

FIGURE 23.4

Coefficient of friction transition and tribochemical correlations between *T*, *μ*, wear, and *in situ* gas analysis. Reproduced with permission from Reference 19, Elsevier, "Wear."

23.2.3 C/C Tribological Behavior and Constituents

23.2.3.1 Tribological Behavior Generally, a tribological couple implies two different materials and often a "hard" component against a "soft" one. So, the carbon/carbon disk brake system represents a nonusual tribological configuration in which the same material is rubbing against itself. Another particularity of carbon (or graphite) materials is that, in any case, these materials are all-carbon made, but with multi carbonaceous phases, and fiber reinforced composites even more!

TABLE 23.2 Typical Thermomechanical Properties of Materials Used for Disk Brakes

	C/C	Steel	Copper	Beryllium
Specific heat (J/g·K)	(20°C) 0.8 (600°C) 1.6	0.55	0.40	1.82
Tensile strength (MPa)	60–125	410	240	550
Density	1.6–1.8	7.9	9.0	1.85
Thermal conductivity (J/m·s·K)	100 (XY) 30 (axial)	80	385	201
Melting point (°C)	>3000	1450	1083	1278

TABLE 23.3 Mechanical Properties of C/C Used for Brakes [20, 25]

Tensile strength			
xy direction	60–125 MPa		
z (axial)	5 MPa		
Young's modulus			
xy directions	40 GPa		
Flexural strength	140–170 MPa		
Flexural modulus	50 GPa		
Density	1.7–1.8 g/cm^3		
Heat capacity (J/g·K)	**24°C**	**511°C**	**1100°C**
	0.7	1.6	2.0
Thermal conductivity (W/cm·K)	**23°C**	**300°C**	**500°C**
Max.	0.125	0.188	0.226
Min.	0.046	0.066	0.075
Coefficient of thermal expansion	**10^{-6}/°C**	**100°C**	**800°C**
	xy direction	0.3	1.14
	z (axial)	5	10

The tribological behavior of carbon materials has been studied for a long time (e.g., [6, 18]) and recent works validate some previous insight. To summarize, it clearly appears [12, 19] that the tribological behavior of carbon on carbon presents generally two main domains of tribological behavior: a domain of low CoF ($\mu = 0.1$) associated with a low wear behavior and a high μ domain ($\mu = 0.4$) with a higher but controllable wear. The first domain is more related to sliding and the second to braking applications. The pathway from the first to the second domain is described as a true transition because it is sharp, short, and reproducible [20] (Figure 23.4).

During the transition from low to high μ, a high wear regime is observed. From Figures 23.4 and 23.5 it clearly appears (and confirms and validates previous works and assumptions) that the transition takes place at the moment when carbon oxidation starts [6, 9, 12]. It is generally allowed that the carbon oxidation starts at about 350–425°C and the temperature we have to take into account is that of the rubbing surfaces, which is at every time difficult to measure directly. So in fact, the C/C tribocontact behaves like a triboreactor in the presence of the reactive gases which are here (on earth) oxygen and water vapor.

On the basis of mass spectrometry analysis (Figures 23.4 and 23.5) and starting from room temperature, we notice that at first water vapor, from physically adsorbed water, is desorbed at above 100°C and then water vapor is still expelled at higher temperature and increases with temperature till the transition is reached. At that moment the amount of oxygenated groups (O or OH) is still sufficient to keep the CoF low. This "water" excess comes probably from the

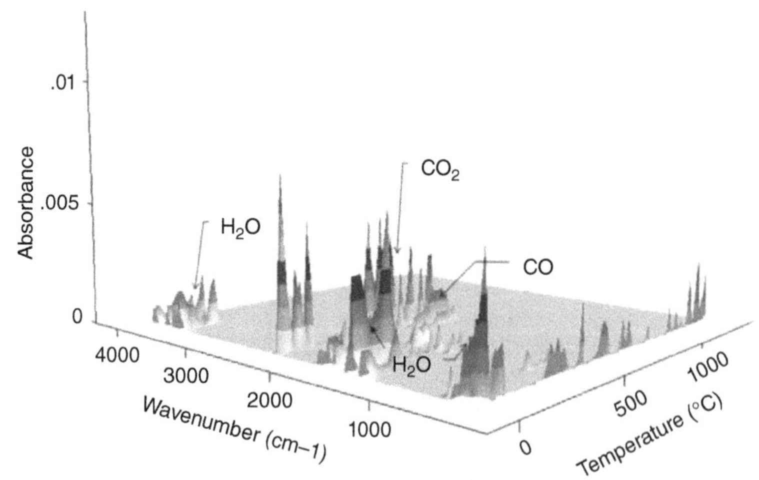

FIGURE 23.5

Gas analysis during TGA of the friction layer sample. Reproduced with permission from Reference 12, Elsevier, "Carbon."

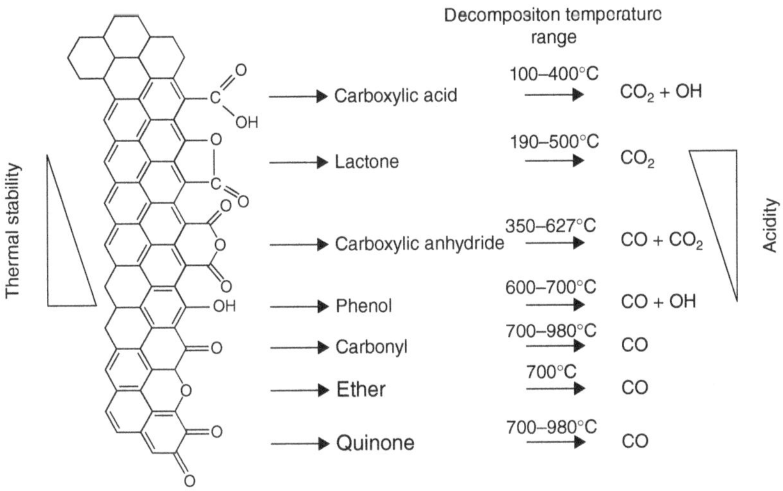

FIGURE 23.6
Functional groups and their thermal stability. Reproduced with permission from Reference 21, Elsevier, "Carbon."

decomposition of hydroxyl-oxygenated carbon groups preexisting at the surface of the material. These groups are well known and their decomposition temperature is globally described in Figure 23.6 [21].

Arriving at the transition point (high temperature) we can observe a fast modification of all parameters. The most significant modifications are related to the oxidation of the carbon simultaneously by oxygen and water which leads to CO_2 and CO release (CO and H_2 release has also been identified but not quantified in Reference 19). The evolution of these gases then follows the amount of locally available gases. This amount is also linked to the geometry of the system through the renewal of these reactants by diffusion, gas flow, etc.

It is said [9] that, at that atomic level, these conditions imply that dandling carbon bonds are formed on each of the rubbing surfaces, and that the face–to-face reaction of these bonds is responsible of the high friction coefficient. In fact, the behavior of carbon disks in real-operation is much more complicated, taking into account that the parts are manufactured with different carbons: fiber and matrix, and different carbons for the matrix (see Chapter 5). Each of these carbons have their own mechanical, thermal, and oxidative characteristics, but globally the same behavior with different set off [12, 22, 23].

23.2.3.2 Materials A brake is a system, so the definition of the C/C material and the brake design are interdependent. That means that, at the end, a brake is a compromise and that different C/C materials can be used by different brake manufacturers with different brake designs [5].

Today (2013) the C/C carbon disk brake market can be estimated as 2000 metric tons or about 250 000 disks a year. This is a very large quantity for a high tech material. The major (western) producers are: GOODRICH (UTC Aerospace-USA, materials: Duracarb, Supercarb); MESSIER BUGATTI DOWTY (SAFRAN-EU, materials: Sepcarb, Supercarb); MEGGIT (ex ABSC (US) + DUNLOP ABS-UK, materials: DMS); and HONEYWELL (USA, materials: Carbenix 2000–4000). Other producers are known from Russia and Asia (China, Korea) but information on materials and markets are not available.

Structural C/Cs are fiber-based composites, so the first constituent is the C fiber reinforcement. At the beginning, for Concorde, the only available carbon fiber was the ex-Rayon one [17], and only on 2D weaved form. As the aim of these fibers is to give high mechanical properties and toughness [24], the fiber volume fraction must be sufficiently high (20–30%) and fiber architectures like felts (5–10%) were not well suited [25]. So these brake disks were 2D (in plane) composite materials. The matrix selected was pyrocarbon obtained by chemical vapor infiltration (CVI). This kind of process is time consuming ($\approx$1000 hours for CVI [8]) and expensive (£550/kg in 1974 [17]). At that time this price was acceptable for such a plane but was too high for general purpose on subsonic aircraft. As the fuel price increases, weight savings were more and more economically driven and justified.

Since that time a lot of work has been done to decrease the costs [26] and improve the mechanical and tribological characteristics. Different evolutions were done such as, for the reinforcement, from ex-rayon C fibers to ex-PAN C fibers [24], ex-pitch C fibers, 3D fibrous preforms, short-fiber mats, ex-pitch carbon foams [27]. For the matrix, the use of organic resins (phenol) was implemented to get a hardened fibrous preform [26]. In particular, low-cost 3D-needled

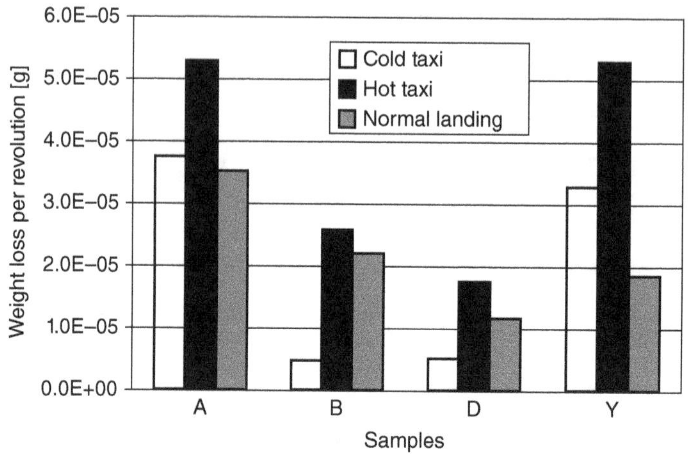

FIGURE 23.7

Wear rates for C/C under all stop conditions [26]. Reproduced with permission from Reference 25, Elsevier, "Carbon."

preforms of ex-PAN C fibers prepared by an automated method were developed in EU and then also used by US companies and represent about 75% of the market [8, 26, 28].

Today, different disk materials are produced and used on aircraft brakes, for example [19, 29]: 3D-needled preforms densified by pyrocarbon, as these preforms are free standing, and also short ex-pitch C fibers, resin hardened, then carbonized, and finally densified by pyrocarbon, pitch densification with different processes, etc.

Related to the disk geometry and the symmetry of this application these parts do not require a full isotropic material and, on the contrary, anisotropy can be useful. For example, the axial properties (through the thickness) of the disk can be different of those in plane (radial and circumferential) and optimized to manage the interlaminar shear strength (ILSS) and the thermal properties (Table 23.2). These axial properties can be modified through the amount of fibers in this direction (disk thickness) (with 3D preforms) which induce the mechanical link between the layers (ILSS) and the conductivity along the fibers and the deposited matrix. These properties can then be optimized by a high temperature thermal treatment (HTT or "graphitization" up to 2800°C, [3]) which has simultaneously an influence on the thermal conductivity and on the hardness, and consequently on the wear behavior of the disk.

At that place also a balance must be found in the thermal conductivity. If the HTT is high, the thermal conductivity (of the graphitizable carbon phase) is high and the heat sink behavior is favored. This is more convenient for high energy landings and it avoids overheating of the brake (design problem). On the other hand, if the composite has a lower thermal conductivity, the temperature of the rubbing surfaces increases quickly, the high μ regime is quickly reached, and the efficiency of the brake is increased at low energy (but local oxidation is favored) [7].

This behavior can also partly explain why the major disk wear does not occur during the landing braking [25, 30] (Figure 23.7). When landing, the disks are cold and high energy (Table 23.1a) enters quickly into the disks. Heating is fast and the high μ domain is quickly reached and then the tribological characteristics are stable and at the end wear is low. During that braking there are quite no exchanges with the atmosphere (low duration and all disks pressed together and in contact) and there is almost no oxidative contribution to wear at that moment.

On the other hand, after landing, the disks are hot and the pressure on the disks is released. So, during this "taxi in" an oxidative interaction with the atmosphere can take place and increase wear. At the same time, as the heat diffuses to the core of the disks, the surface temperature of the disks decreases. During the stop at the airport the brake cools down but during that time oxidizes and adsorbs water from atmospheric humidity. Then, later on, on "taxi out" when the aircraft leaves the dock, disk surfaces are oxidized, water is present, μ is low, and for the low speed and low energy stops before takeoff (Table 23.1a), the brake needs a higher pressure at low temperature and then a higher wear is observed. This thermochemical behavior is described for the whole C/C composite but must be separately applied to each of the carbon constituents of the material (fiber and matrices). In the end the recommended way to use steel or C/C brakes are different [30].

23.2.3.3 Antioxidation Protection Another important item for carbon brakes is the antioxidation protection (AOP) system. During service landings the mass temperature of the whole disks can reach temperatures above 1000°C.

At these temperatures carbon oxidation by air is rather fast and can lower the life duration of the brake significantly. So, an AOP is necessary to protect the disks, especially the inner and outer lateral faces where the notches are machined (Figure 23.2). For such a protection a tight layer is efficient and glass-based sealants are described [26].

But a specific problem arises with aircraft brakes; in winter, runways need to be deiced and to do that airports use deicers, such as potassium acetate solutions. Aircrafts are landing and running in this salt solution which splashes the C/C disks. The problem is that, in general, these salts are catalysts for the carbon oxidation and unfortunately potassium is one of the most efficient catalysts [31]. The same problem is encountered with the sodium catalyst from sea water (airports near the sea, Navy aircraft). This catalyzed oxidation is fast since the low temperatures [32], and specific anticatalytic AOP must be used.

This kind of AOP is known and belongs generally to the borates or phosphates families [31, 33]. For an efficient protection at higher temperatures, a tight coating is required and is generally based on silicate or glass glaze. At the end, an AOP for brakes can be a complex system to fulfill both the high temperature and the anticatalytic protection.

A final problem is that this AOP must stay outside of the rubbing surfaces, during manufacture and in service, as it is known and experienced that on entering into the brake (triboreactor) it will modify the chemical surface reactions previously described, and lowers or destabilizes the friction coefficient μ and wear.

23.2.4 Other C/C Tribological Applications

A well-known tribological application for C/C composites is brake disks for racing cars and especially for Formula One. For this kind of application the requirements are rather close to the aircraft ones at lower energy level (Table 23.1a).

A low weight is of general interest but much more as it is an unsprung mass and interacts directly with the road holding of the car. A low weight on a rotating part (wheels) implies a lower moment of inertia and impacts also the deceleration and acceleration of the racing car. As racing car brakes stay hot during the race, the stability of the friction coefficient μ is good and the braking efficiency is very high with very short braking distances. A major concern for using C/C for racing cars brakes is safety. With the high thermomechanical loads and stresses applied to the disks during the race, the metallic disks (e.g., spheroidal graphite cast iron) can be radially cracked and then broken in service. As C/C composite materials are not sensitive to thermal cycling and crack propagation, their use increases the pilot's safety.

In a single race lap, disks are heated and "glow" (>800°C) many times (Figure 23.8) and an efficient cooling is necessary. That means also, that an efficient high temperature AOP is required for these applications. If the disk-on-disk configuration has been studied, Formula One racing cars still use pads on disk configuration (Figure 23.9). In this case the C/C manufactured for the pads can have a slightly different composition from that of the disk [34]. Carbon/carbon brakes were first tested in Formula One in 1977 and won for the first time in 1984 and were then world champion. Today all the Formula One teams use C/C brakes [1, 8]. Later on they were also used for endurance races, and the MAZDA team won "Le Mans 24 hours" race in 1991 with C/C discs made by Carbone Industries (now MBD) [8].

For racing cars C/C composites are also used for clutches with the same interest of weight and inertia gain on high RPM engines (18 000 RPM for Formula One) [1]. Another interest is to lower the size and especially the diameter of

FIGURE 23.8

Racing car with "red glowing" brakes. Courtesy of Herakles company.

FIGURE 23.9

Racing car brakes: ventilated disks and pads. Courtesy of Herakles company.

the clutch system to lower the engine in the chassis and increase the road holding of the car. For that kind of application (in air) high mechanical properties, reliability, and static CoF at moderate temperature are important. The design is, like for aircraft brakes, a thin cranked multidisk configuration (Figure 23.10). The C/C material is manufactured with high performance C fibers, probably without thermal treatment to quickly heat the rubbing surfaces and so have a high μ at start. Another C/C clutch application is said to be on the US fighter F35, between the engine and the fans for vertical takeoff.

Other kinds of brakes have been studied, for example, to stop helicopter rotors, stop turbo-engine propellers (ATR aircraft), braking high speed trains, and motorcycles. Another tribological use of C/C composites is that of shock absorbers or clutches rubbing in motor oil. A very good behavior of C/C has been experienced in this kind of medium, especially a stable μ in static and dynamic use and a low wear. For example a C/C made with a single layer of densified C fabric replaces paper in wet clutches, traction limited slip differentials [35], and automatic gearboxes [36]. Friction in

FIGURE 23.10

Racing car clutch. Courtesy of Messier Bugatti Dowty.

oil is also used for high energy shock absorbers for heavy vehicles like tanks [8]. Tribological behaviors in other liquid medium are not well described in the literature. One can think that use in water (or sea water) could be interesting. One example cited is pyrocarbon (LTI) coated heart valves, initially manufactured many years ago (wear resistance and biocompatibility in blood, with or without SiC additions).

23.2.5 C/C Limitations and Trends

Carbon/carbon limitations for tribological applications lie, as for other applications, in costs for manufacturing and use. Although aircraft disk brakes have shown that when manufactured in large quantities, the cost can be lowered [17, 27] and being acceptable especially if other advantages can be brought forward (safety, inertia, etc.) [1].

Decreasing C/C costs is a strong basic trend for all the manufacturers since the beginning [25]. The actualized price of the first Concorde brakes gives about 3500 $/kg and this price has been cut by five in 1980 [17]. This cost is compatible with data from Reference 27 (440–2900 $/kg in 1996).That means, that C/C industrial manufacturers have done a huge effort to decrease the costs. Now we can consider that these products are mature and are on the plateau phase in their product life cycle. Simultaneously, the number of landings before being worn, has been improved from 700 (for steel disks) to 2000–2500 [5] and is expected to reach 3000 landings (with refurbishment).

So, in the end for an aircraft company, the cost per landing has been dramatically decreased with a large increase in safety for the passengers simultaneously. With these technical–economical improvements these high-tech C/C brakes are now competitive with steel brakes even for smaller aircraft (B 737, A 320) [30] taking also into account the interest of weight savings (500–900 kg/aircraft) and corresponding fuel savings and CO_2 decreases.

These materials are also technologically on a plateau phase. For the material by itself, improvements can come from a better and better control and management at the nanolevel. On the other hand, the future lies, perhaps, in an improved new brake concept and design associated with the "electric brake" development. From a technical point of view, oxidation is also a limitation (above 350°C for long duration), but is inherent to the carbon itself and is not specific to tribological applications. Another tribological limitation for C/C is their low static CoF μ. This low μ is mainly due, at low temperature, to the presence of water physicochemically "adsorbed" (linked). For aircraft applications, this is a problem only when the plane is in parking and the brakes are used as park brakes. This problem has been identified till the beginning, and for CONCORDE it had required a higher pressure for hydraulics [17].

Another example of that limitation is given by the US "V 22 OSPREY" brakes. The service specifications for this tilt-rotor plane imply that it must stay on aircraft carrier vessels, and at sea the deck of these vessels would be inclined (20° slope) [37]. In this case a high static μ was required to avoid an uncontrolled motion of the plane on the deck. The technical choice was to change the composite material and use a C/C whose C surface has been transformed to SiC. By this siliconizing process the disk surface was transformed into SiC on a thickness of 1–2 mm [37]. The static CoF was significantly increased by the SiC layer ($\mu > 0.3$) and sufficient for this specification. As this plane is a VSTOL (vertical or short takeoff and landing) the landing speed was not high, the SiC wear was low, and the compromise acceptable.

Another limitation is the transition from low to high μ when temperature increases (Figure 23.4) [20]. That means, in the case of use for a passenger car—on which brakes are generally cold—the braking efficiency will be low at the beginning and requires a high pressure for efficiency and then, when temperature increases, the transition to high μ occurs but not at the same time for all the wheels. This unbalanced braking efficiency can be dangerous for road holding and is detrimental for C/C brakes use on ordinary cars and trucks.

These kind of limitations show an example and reasons for an évolution from C/C materials to CMC for some tribological applications. For example, the use of refractory additives (SiO_2, SiC, B_4C, etc.) to C/C has been experienced long time ago [11] to avoid these problems. By the end, this évolution from carbon/carbon to carbon/ceramic composites has opened other tribological applications.

23.3 CARBON/CERAMIC FOR FRICTION APPLICATIONS

As described, for friction applications in passenger cars and other terrestrial brake systems an additional interior protection of C/C is necessary to improve the CoF and wear stability. Replacing most of the carbon matrix by silicon carbide is the most promising approach to increase the oxidation stability and to decrease the porosity of C/C. Silicon carbide is particularly suitable as a matrix material due to its superior temperature and thermal shock stability as well as its high creep and wear resistance. Such SiC matrices can be produced either by gas phases (CVI process) or by liquid phases

(PIP and LSI processes). For cost-sensitive applications like brake disks and pads the reactive melt infiltration of silicon into porous C/C composites (LSI process) is the most economical process. Today, nearly all of the C/SiC (also called C/C-SiC) materials which are used in friction applications are fabricated by the LSI process.

Research on the infiltration of carbon/carbon with molten silicon and its subsequent chemical reaction to silicon carbide started in the 1970s and showed that the avoidance of the chemical reaction of the carbon fibers with silicon is the main problem of this process [38–40]. The infiltration of conventional C/C showed a high tendency of fiber degradation and results in a more or less monolithic SiSiC ceramic with a very low fracture toughness. Only the use of high modulus carbon fibers with their lower chemical reactivity or the precoating of the fibers with SiC led to C/SiC materials with acceptable damage tolerance behavior.

The German Aerospace Center (DLR) in Stuttgart (Germany) succeeded in 1988 to manufacture suitable carbon/carbon preforms without any fiber coatings and based on low cost HT fibers, which result in C/SiC composites of sufficiently high damage tolerance [41]. New carbon/carbon preforms were developed which consist of dense, crack-free segments of fiber bundles, and a translaminar microcrack pattern. These C/C composites were specially adapted to the reactive melt infiltration of silicon (so-called siliconizing) and restricted the carbon fiber degradation to a minimum. The whole LSI process for the manufacture of C/SiC materials for friction applications, which was originally developed for thermal protection systems of spacecraft, is schematically shown in Figure 23.11.

The temperatures involved are at least beyond the melting point of silicon (1415°C) and vacuum conditions support the fast infiltration of Si due to the high capillary forces of the cracks in the C/C preform. The following reaction to SiC is based on a diffusion process of silicon into the carbon matrix which generally results in a composite material of three phases: carbon (fibers and residual matrix), silicon carbide (matrix), and silicon (nonreacted part in the matrix). Figure 23.12 shows typical microstructures of C/SiC composites with a 2D (fabric) fiber reinforcement and a short-fiber reinforcement, respectively.

The prime chemical reaction of the LSI process is the solid–liquid reaction of molten silicon with the amorphous carbon of the matrix. Due to the fact that the molar volume of the resulting silicon carbide is approximately 30% smaller than the sum of the molar volumes of the two reaction partners Si and C, the composition and design of the microstructure of LSI–C/SiC materials is always a compromise between a low porosity and a low amount of free silicon. The process allows either a dense matrix containing residual silicon (less than 5%) or a Si-free matrix with a certain amount of porosity (1–5%). Matrices showing no pores and no elemental silicon are not directly achievable via the LSI

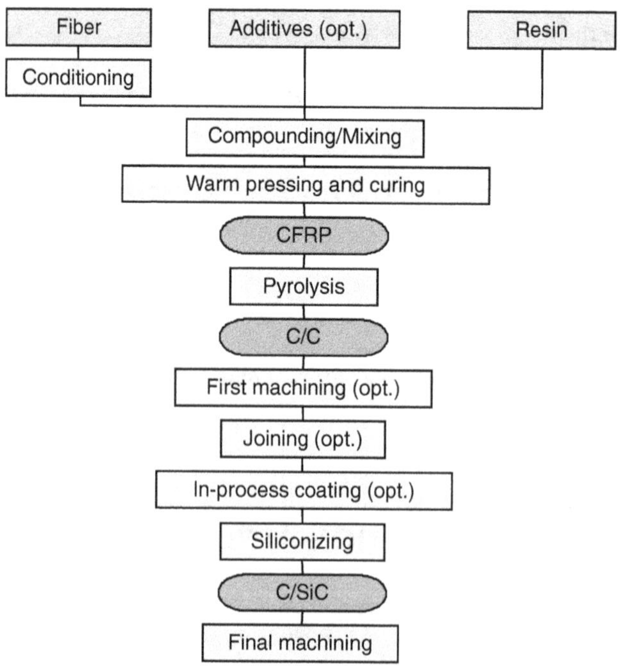

FIGURE 23.11

Manufacturing scheme of the modified LSI process for friction applications.

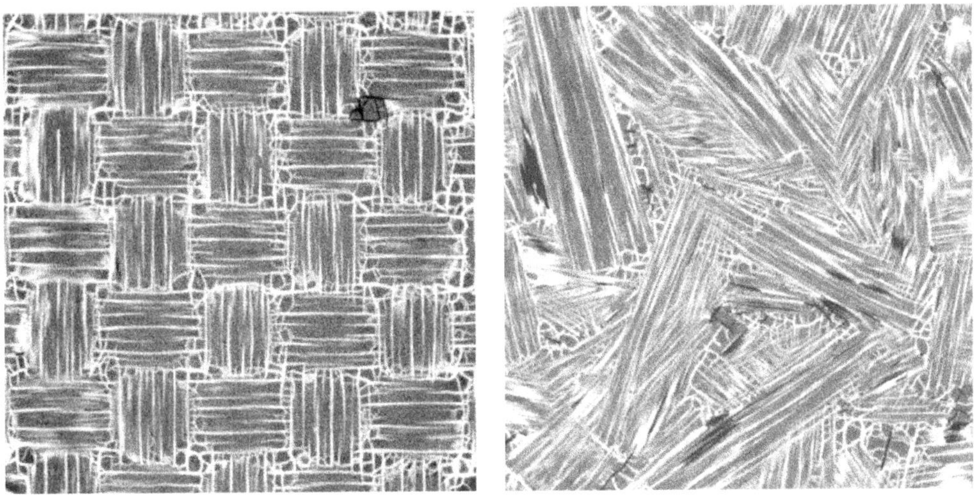

FIGURE 23.12
Microstructures of C/SiC composites with a 2D fabric reinforcement (left) and a short-fiber reinforcement (right) (From Fraunhofer HTL).

process. These inherent existing particularities of the LSI process have to be taken into account when optimizing the microstructures and properties of C/SiC for friction applications.

23.3.1 History and Milestones of the C/SiC Development

The development of C/SiC composites for their use in friction systems started in the early 1990s and represents a successful spin-off from space technologies to terrestrial applications. First projects dealt with the prototype manufacture of brake disks in dimensions of up to 640 mm in diameter for high speed trains to take advantage of the low density and high thermal stability of the new material [42]. The technological challenge at that time lied in the upscaling of the LSI process from thin aerospace structures (thickness 3–6 mm), reinforced with 2D fabrics, to thick (up to 60 mm) and voluminous disks without any delaminations.

The development of first C/SiC brake pads and brake disks for automotive applications started in 1994 at DLR [43]. In cooperation with the sports car manufacturer Porsche in Germany, different prototypes of novel carbon/ceramic brake disks were developed between 1994 and 1996. Although first tribological tests showed their superior properties in comparison to carbon/carbon and grey cast iron in general, the LSI process and the resulting carbon/ceramic material had to be improved basically. The following four technological challenges represent the milestones in the development of carbon/ceramic brakes.

1. Modification of the composition and microstructure of C/SiC to increase the transverse thermal conductivity in order to increase and stabilize the CoF.
2. Reduction of the wear rates by the development of a process-integrated surface coating (goal: lifetime brakes).
3. Use of short-fiber reinforcements instead of 2D fabric reinforcements for a near-net-shape and cost-efficient manufacture of brake disks.
4. Development of an *in situ* joining method for the manufacture of internally ventilated brake disks (goal: disk design of high complexity).

In principal, monolithic SiC is a very good friction material showing extreme low wear rates and high coefficients of friction. These properties predestine this nonoxide ceramic material for friction applications, but many attempts in the 1970s to use it in brake disks failed because of its high brittleness, resulting in a catastrophic behavior after thermal shocks or mechanical impacts. Within the matrix of C/C–SiC materials one is able to take advantage of the good frictional properties of both main constituents: C/C leads to a high damage tolerance and silicon carbide offers high and stable friction coefficients. Low energy disk-on-disk tests show that the friction coefficient of C/C–SiC can be

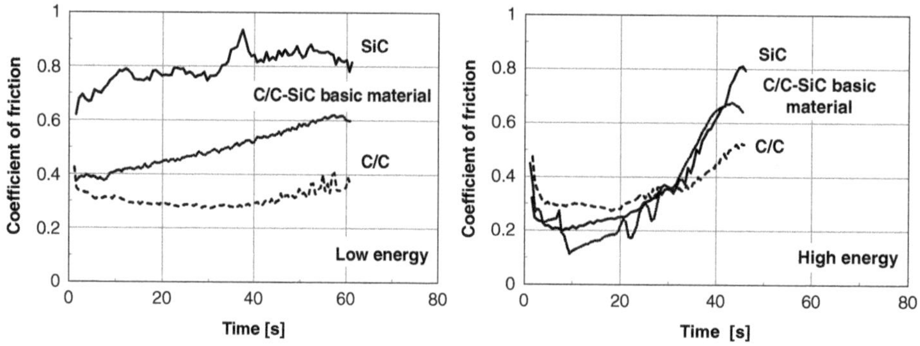

FIGURE 23.13

Coefficient of friction of 2D C/C-SiC and its two main constituents C/C and SiC at low energy ($v = 6$ m/s, $p = 0.1$ MPa, left) and at high energy brakings ($v = 16$ m/s, $p = 0.35$ MPa, right).

regarded as the superposition of its two main constituents, carbon/carbon and silicon carbide (Figure 23.13). At higher energies, the differences between C/C and SiC are much less pronounced and the CoF becomes unstable [44, 59].

The main reason for this CoF instability was the low transverse (through thickness) thermal conductivity (about 9 W/m K) of the originally used 2D C/SiC material which had to be improved considerably with respect to this property. Higher transverse thermal conductivities were achieved by modifications of the microstructure, for example, by the increase of the ceramic fraction (Si, SiC) or the addition of highly conductive raw materials (e.g., graphite fibers). Higher ceramic contents can be achieved by reducing the fiber content resulting in a higher silicon uptake, a higher degree of SiC formation and consequently in a higher density of the C/SiC material [45]. Generally, the lower the fiber volume content and the higher the density, the higher the transverse thermal conductivity of the C/SiC material. However, the higher densities coincide with the decrease in strength and toughness of the material. Therefore, the carbon/ceramic composite must be designed as a compromise between sufficiently high thermal and mechanical properties.

Modified C/SiC materials show fiber volume contents of 30–50%, porosities below 5%, and transverse thermal conductivities of about 40 W/m K at room temperature. As for most other materials, the thermal conductivity of C/SiC decreases with higher temperatures [46]. The high transverse thermal conductivity ensures high heat fluxes from the outer friction surface to the center of the brake disk and avoids an overheating of the surface. Lower temperatures result in a higher stability and a higher level of the CoF (Figure 23.14), whereas the wear rates are only slightly influenced by the higher thermal conductivities.

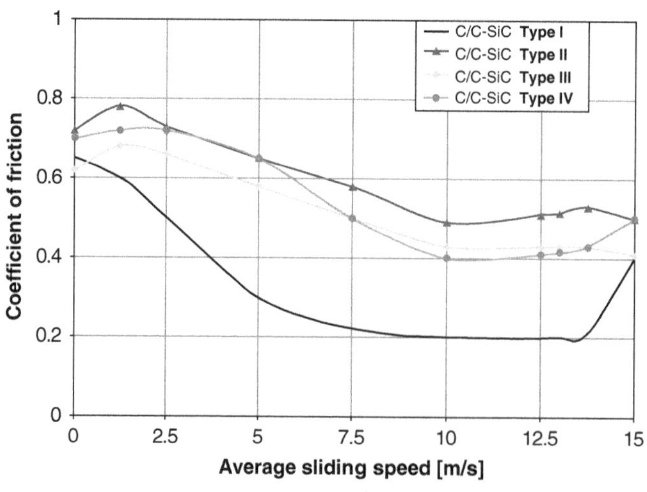

FIGURE 23.14

Coefficient of friction of C/SiC materials with different transverse thermal conductivities (Type I: low conductivity (2D reinforcement), Type II–IV: high conductivity (graphite fibers, transverse fiber orientation, high SiC content) [45].

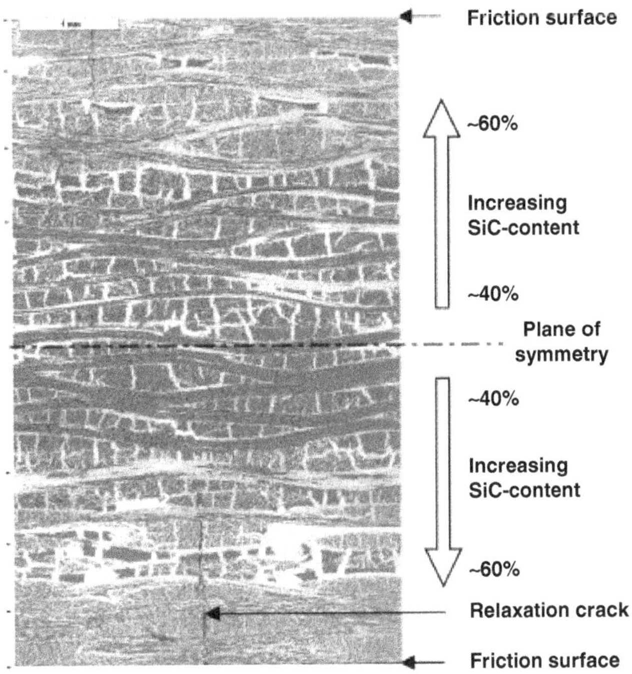

FIGURE 23.15

Cross section of a graded C/SiC composite (white: silicon and silicon carbide; grey: carbon fibers and carbon matrix) [48].

The wear rates of these modified C/SiC materials were already considerably lower in comparison to carbon/carbon but still too high for commercial applications. To justify the higher costs of carbon/ceramic brakes in comparison to grey cast iron, extremely low wear rates are required for passenger cars, allowing a lifetime up to 300 000 km for the brake disks. To increase the wear stability, a higher amount of ceramic in the composition of the friction surfaces was considered to be necessary.

Two different approaches were pursued to fulfill the conflicting requirements on a high material's ductility in general and high hardness and good wear resistance on the surfaces. For brakes with a limited lifetime (e.g., sports cars), gradient C/SiC materials with a gradual increase of SiC from the center to the surfaces were developed (Figure 23.15). For brakes requiring a longer lifetime, for example, for sedans, a method to manufacture thin SiSiC coatings on the surfaces was developed which ensures their deposition and formation during the siliconizing of the C/C preforms. Hence, the manufacture of such SiSiC friction layers could be completely integrated in the LSI production process [46, 47]. This additional friction layer results in a more oxidation resistant as well as wear resistant friction material compared to carbon/carbon [52].

The coefficient of thermal expansion (CTE) of the ceramic-rich surface is considerably higher than the CTE of the C/SiC substrate. Depending on this CTE mismatch, a more or less pronounced microcrack pattern occurs in the SiSiC coating as a result of tensile stresses during the cooling-down processing (Figure 23.16). The formation of this microcrack pattern in the friction surface depends on the fiber architecture of the substrate material. The highest amount of cracks can be observed for 2D C/SiC whose shrinkage during cooling is close to zero due to the thermal in-plane stability of the carbon fibers. Short-fiber reinforcements with their considerably higher CTE can result in nearly crack-free surfaces.

Both approaches, graded and coated carbon/ceramic disks, were successfully developed for automotive applications. For series brakes so-called wear indicators are inserted in the friction surfaces in order to visualize the end of the disk's life time [49].

The first C/SiC brake disks were fabricated by cutting out rings from bidirectionally reinforced rectangular plates (Figure 23.17). It was obvious that the use of technologies, originally developed and applied for aerospace products, like autoclave curing and usage of woven fabrics as reinforcements were not suitable for a series production of friction materials. Therefore, a method to manufacture brake disks and pads in a near-net-shape process, using short carbon fibers instead of fabrics and pressing techniques instead of autoclave curing, was developed [50].

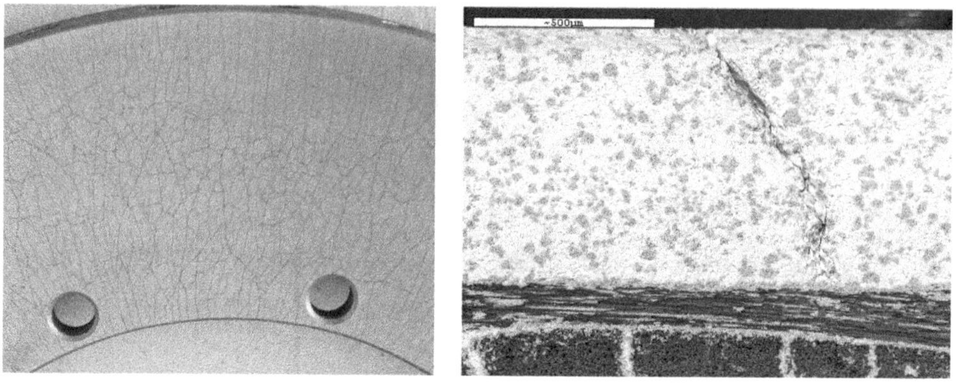

FIGURE 23.16

SiSiC coatings on C/SiC brake disks for improved wear stability. Left: top view on the microcracked friction surface; right: cross section of the coating with a relaxation crack [46].

Compounds of chopped fiber rovings and powdery phenolic resins were used as suitable new raw materials. Mixed and filled in a mold with a geometry similar to the desired shape of the disk a CFRP green body was consolidated and cured in a warm press. During the following pyrolysis and siliconizing steps the shrinkage rates of about 1% in the in-plane direction and about 5% in the press direction are very low and necessitate a minimum of final machining. Due to the more isotropic fiber orientation and the higher amount of ceramic, C/SiC composites with short-fiber reinforcements show a higher transverse thermal conductivity and a lower strength level in comparison to fabric reinforcements. In principle, C/SiC composites reinforced with fibers of about 40 mm in length show higher strength values, whereas short fibers with less than 10 mm are easier to handle and increase the thermal conductivity perpendicular to the friction surface.

Generally, to avoid overheating of the brake surface and to extend the lifetime of the brake system, brake disks of high performance cars are ventilated and the cooling channels often show a very complex shape. Therefore, different approaches for the design and manufacture of internally ventilated C/SiC brake disk had to be developed. Except in the case, where lost cores are used to form the cooling channels in the CFRP stage, a joining method for a force- and form-locking assembly of single C/C parts prior to the siliconizing step is necessary to design the brake disk in a modular design. Figure 23.18 depicts schematically the principles of the manufacture of internally ventilated brake disks. An *in situ* joining method was developed based on the use of a carbonaceous paste which reacts simultaneously to the composite's matrix formation to a heat-resistant SiC joint [51]. By this joining method, preshaped and pyrolized C/C parts can be joined easily and safely to C/SiC components of very high complexity. One is able to optimize and manufacture the

FIGURE 23.17

First generation of carbon/ceramic brake disks with 2D fiber reinforcement (From DLR).

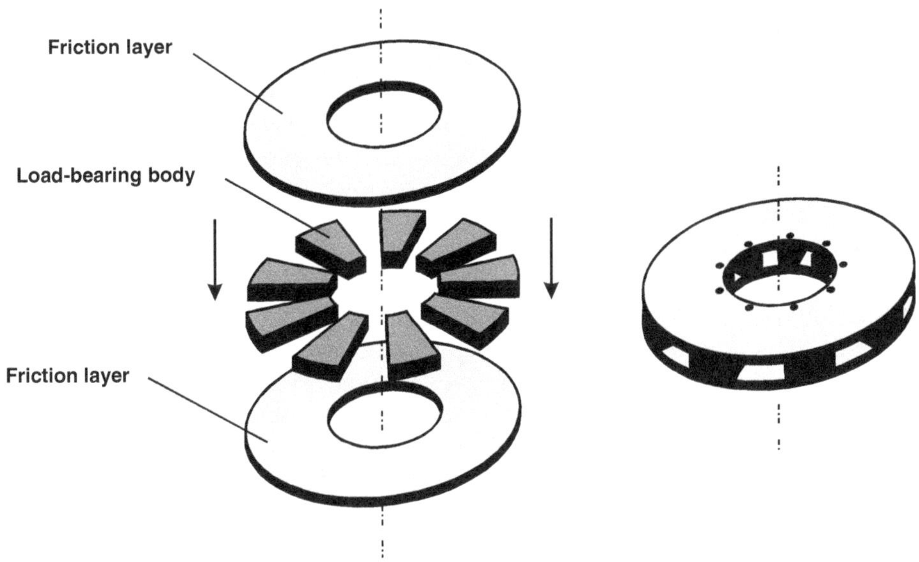

FIGURE 23.18
Principle of the modular manufacture of C/SiC brake disks with internal ventilation [47].

friction parts and the load-bearing body separately with respect to their functions (low wear, high CoF and high thermal conductivity, high mechanical strength, respectively), prior to their joining.

All described improvements in the material's properties and the brake disk's design and manufacture were patented by DLR and result in first prototypes of ventilated carbon/ceramic brake disks in 1996 (Figure 23.19). Dynamometer tests showed in principal the applicability of C/SiC materials for friction applications and the feasibility of the LSI process for a cost-efficient production of brake disks and pads.

FIGURE 23.19
First prototype of a ventilated carbon/ceramic brake disk for automotive applications (From DLR).

23.3.2 C/SiC Brake Disks for Passenger Cars

After these technological breakthroughs, the industrial brake disk development started in 1997 in a joint cooperation between Porsche and SGL Carbon (Germany), at that time in competition with the C/SiC brake development at Daimler in Germany. In 2000, a limited edition of passenger cars was firstly equipped with C/SiC brake disks by Daimler. In 2001 SGL started the series production of carbon/ceramic brake disks for Porsche 911 GT2. Historically, Porsche was the first car manufacturer worldwide using C/SiC brake disks in series cars. A joint venture of Daimler and Brembo (Italy) was established in 2004. In 2009, the two competitive manufacturing routes of Brembo and SGL were merged in a joint venture to the company Brembo SGL Carbon Ceramic Brakes. This company is currently the only commercial producer of C/SiC brake disks for sedans.

In general, C/SiC materials combine the good thermomechanical properties of carbon/carbon with the high wear stability and constant frictional properties of silicon carbide. The damage tolerance behavior of C/SiC lies in the same order of magnitude as for grey cast iron. Table 23.4 shows comparatively some material data of the different brake materials C/SiC, grey cast iron (GG 20), aluminum reinforced with SiC particles (Al-MMC) and carbon/carbon. The low density of about 2.4 g/cm^3 of C/SiC offers a big lightweight potential and reduces the unsprung mass significantly compared to cast iron. However, the volume-specific heat capacity of C/SiC is lower compared to grey cast iron. Typically, this leads to a larger diameter for carbon/ceramic brake disks of about 1 inch and reduces the weight savings from theoretically 65% to approximately 50% compared to grey cast-iron brakes.

Due to the lower thermal conductivity transverse to the friction surface, the temperatures of C/SiC brake disks are generally higher than in conventional metallic brakes. As a result, more attention has to be focused on the heat protection of surrounding parts in a vehicle in order to avoid an overheating of the brake fluid for example. The maximum temperature of C/SiC is lower in comparison to C/C and limited to about 1350°C because of the residual free silicon.

Disk-on-disk tests demonstrate the high wear stability of C/SiC materials. Using disks of identical materials, pairs of carbon/ceramic materials show considerably lower wear rates in comparison to carbon/carbon [59, 60]. Figure 23.20 depicts exemplarily the mass loss of both types of material after a deceleration test with a starting velocity of 16 m/s and a contact pressure of 0.7 MPa.

So far, only few studies are published about the frictional behavior of carbon/ceramic disks in combination with organic or sintered metallic bounded pads in currently used vehicles. Since numerous organic pads are on the market, general conclusions on the tribological properties are difficult. In addition to a variety of company-specific test programs a so-called AK-master test is commonly used to examine the friction properties like CoF level and stability by varying pressure, velocity, and temperature. Typically, dry friction values of carbon/ceramic brake disk in combination with specific organic pads are in the range of 0.35–0.45. The high stability against fading of C/SiC (decrease of friction due to thermal effects) is depicted exemplarily in Figure 23.21 [48]. Although the surface temperature of the C/SiC brake disk increases over the number of brakings from room temperature to about 650°C, the CoF remains considerably stable between 0.45 and 0.35 during repeated deceleration tests.

TABLE 23.4 RT Properties of Typical Brake Materials [53–58]

		Short-fiber C/SiC (Sigrasic SGL)	GG-20	Al-MMC (SiC-Particles)	C/C
Density	kg/dm^3	2.3–2.45	7.25	2.7	1.7–1.8
Mass-specific heat capacity	J/kg K	800	500	820–886	700
Volume-specific heat capacity	J/dm^3 K	1800	3600	2350	1200
CTE (in-plane)	10^{-6} 1/K	1 (RT) 2 (300°C)	9 (RT) 12 (300°C)	14–21	0.3
Thermal conductivity (transverse)	W/m K	40	54	160–185	13
Tensile strength (in-plane)	MPa	20–40	150–250	310–370	70–100
Young's modulus (in-plane)	GPa	30	90–110	86–125	40
Bending strength	MPa	50–80	150–250		
Strain (in-plane)	%	0.3	0.3–0.8	0.4–1.2	
Thermal shock resistance	W/m	>27 000	<14 000		
Maximum temperature	°C	1350	700	400	>1350

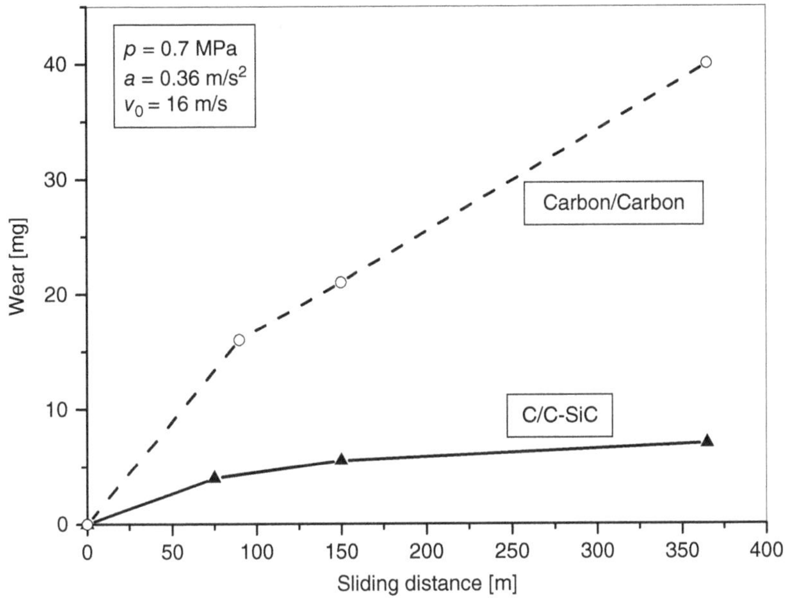

FIGURE 23.20 Wear rates versus sliding distance (disk-on-disk with identical materials).

The wider operation temperature range of C/SiC brake disks due to their lower heat conductivity and heat capacity in comparison to metals require more thermally stable brake pads, acting as the friction counterpart to the disk. As a consequence, the composition of the pads should contain a lower amount of organic constituents as phenolics, rubber, and other organics limit the thermal stability.

For passenger cars, not only high and stable CoFs and low wear rates are common requirements, but also the NVH (noise, vibration, and harshness) performance is of high interest. The currently used "low-met" materials represent a compromise between comfort and frictional properties. The tribology of C/SiC brake disks is not fully understood and topic of international research [61–72]. The ceramic system is much more complicated as typically no friction interphase, the so-called third body, occurs which plays a major role in the formation of a wear-resistant surface on the pad. A special preconditioning of the disk and pad is necessary for example by a first "running-in" of the virgin system. Otherwise the lack of a third body can result in extreme high pad and disk wear rates, low NVH performance, or even a destruction of the disk and pad.

Generally, carbon/ceramic brakes are characterized by a high and widely constant CoF, resulting in a spontaneous response behavior and lower pedal forces. The high thermal stability ensures a constant dosing even after a large number of high-level brake stops. The low CTE reduces the tendency of hot juddering and heat-related distortions significantly

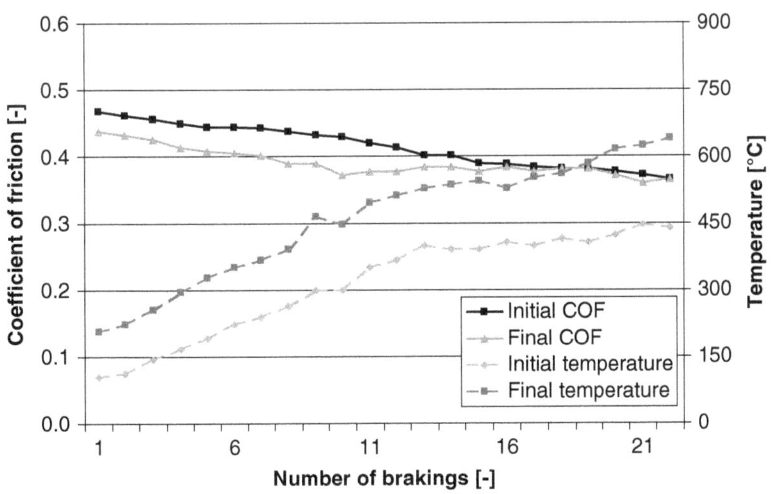

FIGURE 23.21 Fading tests of C/SiC brake disks (repeated brakings on the hot friction surface by vehicle's deceleration from 134 to 44 km/h).

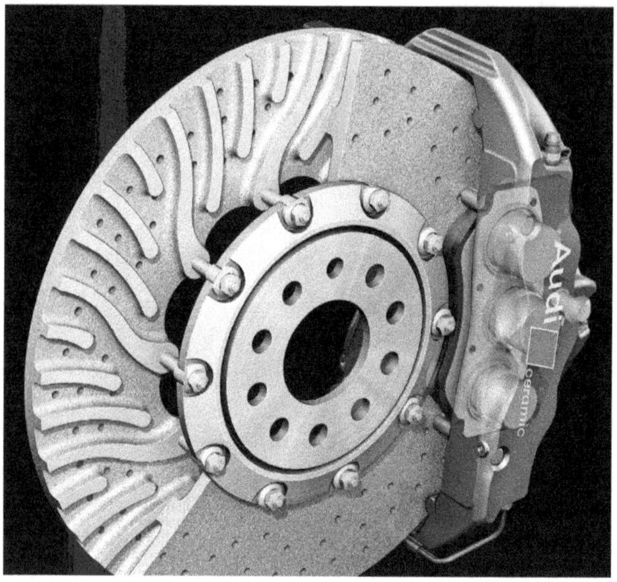

FIGURE 23.22
Ventilated carbon/ceramic brake disk for high performance cars (From Audi AG).

compared to grey cast iron. The high wear resistance prevents from local washouts and minimizes the risk of cold juddering. The corrosion stability of the ceramic-rich friction surfaces guarantees a good braking and noise behavior even on winter road conditions. In comparison to carbon/carbon, the stability of the CoF, in particular under wet and cold conditions, is superior and represents the main advantage of C/SiC besides the extremely high wear resistance.

Carbon/ceramic brake disks are meanwhile established in many sports cars, high performance luxury sedans, and SUVs (Figure 23.22). Starting in 2001 with a small production volume of some thousand disks for Porsche cars exclusively, the number increased to approximately 60 000 in 2006 [73]. More than 100 000 brake disks are currently produced every year for different car producers like Porsche, Audi, BMW, and Ferrari with a growing tendency. To increase the production volume and to broaden the use of carbon/ceramic brake disks to the midsize car segment, the still high manufacturing costs have to be reduced significantly. To achieve this goal, highly automated manufacture processes must be developed. This includes continuous processes for the high temperature steps of pyrolysis and siliconizing. Also, the CFRP green body manufacturing, currently dominated by manual labor, should be improved for example by an adapted design in combination with injection technologies [49].

23.3.3 C/SiC Materials for Industrial Applications

Besides automotive applications there is also a big interest in high performance brake systems in special fields of engineering. In many mechanical engineering and conveying systems, conventional brake systems with metallic disks and organic friction linings have reached their thermal limits. Mostly designed as emergency brake systems, the wear stability in these cases is of minor interest, but high power input to the friction surface requires extreme mechanical and frictional stabilities of the brake materials [74, 75]. Two applications are described in the following where C/SiC composites are successfully used in commercial applications since several years.

23.3.3.1 Brake Disks for Emergency Brake Systems Emergency stop brakes are used exemplarily in electric drives for lifts, cranes, and conveying systems. In case of power failure or an emergency case electromagnetic spring applied brakes stop machines or keep loads in position. The brake system often comprises of a rotating and two stationary disks (rotor and stators) similar to a heat pack of aircraft brakes. Due to this compact construction almost the whole kinetic energy has to be stored for a short period in the brake disks and conventional organic linings are thermally overloaded. The high temperatures result in unstable and low coefficients of friction as depicted in Figure 23.23. Using C/SiC materials for both brake disks, the CoF remains high and stable with tolerable wear rates even under severe conditions.

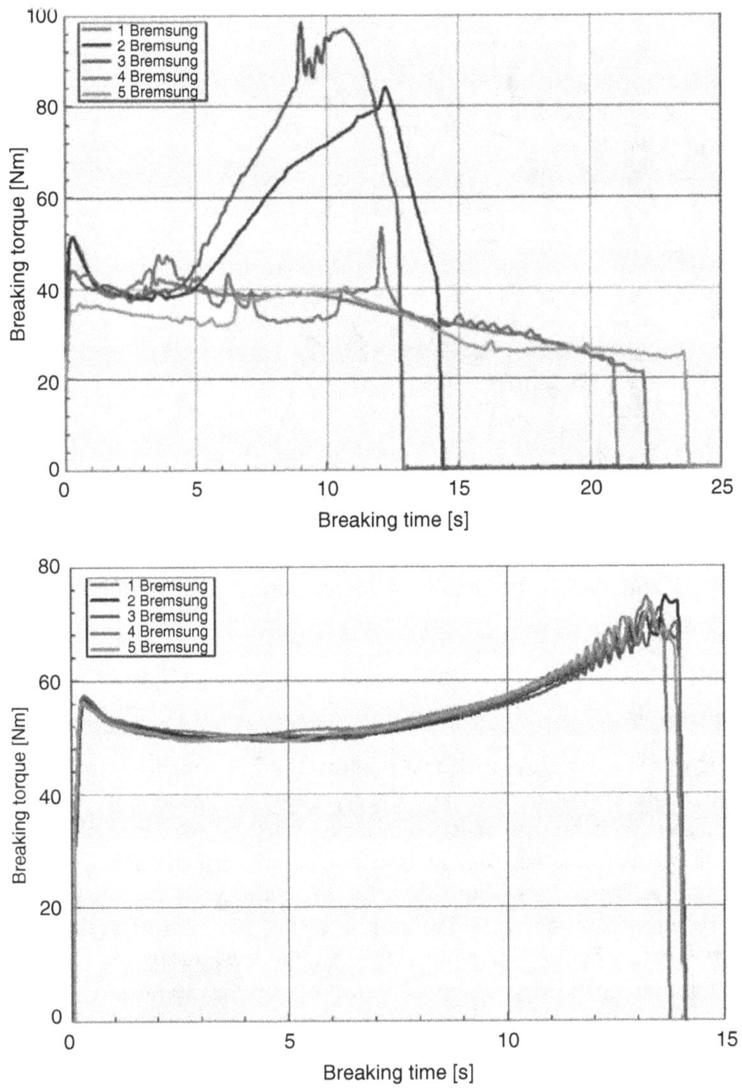

FIGURE 23.23

Braking torque versus braking time of an organic/metallic system (above) in comparison with a C/SiC rotor–C/SiC stator system (below) during a high energy stop braking (From Mayr Antriebstechnik).

Figure 23.24 shows the rotor and two stators of such a brake system manufactured of C/SiC materials with a high heat capacity and high transverse thermal conductivity. In this brake system, the thermomechanical loads are much higher than in automotive brakes, for example, with the result that the disks must be reinforced with 2D fabrics instead of short fibers. This reinforcement prevents a near-net-manufacture of the disks and results in a considerable amount of waste and high costs which are about one order of magnitude higher than for conventional friction materials. Therefore, the market is still small but the outstanding frictional characteristics of C/SiC materials widen the applicability of electromagnetic emergency brakes enormously.

23.3.3.2 Brake Pads for High-Rise Lifts Another application for the commercial use of carbon/ceramic friction materials are brake systems of high performance elevators. The drive units of these lifts are designed for buildings of several hundred meters in height and high velocities of the cabin of up to 10 m/s. If the speed exceeds this value, the elevator is automatically brought to a controlled emergency stop. Thereby, a high and constant CoF even under static conditions is necessary to keep the load in position. Also, a high thermomechanical stability because of the abrupt impact of the pad against the metallic guide bar and resulting temperatures of up to 1200°C are major requirements. At such high temperatures, conventional organic or sintered metallic pad materials are far beyond their thermal stability. Graded C/SiC materials comprising of abrasive-resistant SiC-rich friction surfaces and fiber-rich cores (Figure 23.15) show almost no wear, high coefficients of friction, and sufficiently high mechanical properties.

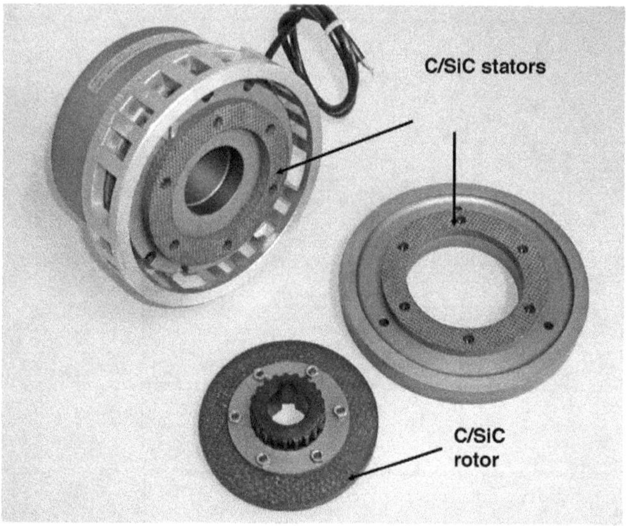

FIGURE 23.24

Emergency brake system comprising of a rotor and two stators, made of 2D-reinforced C/SiC materials (From Mayr Antriebstechnik).

23.3.4 Future Challenges for C/SiC Materials

Carbon/ceramic disks and pads are increasingly used in different frictional applications. Their main advantages lie in the low density, the high thermal shock resistance, the low wear rates, and the high level as well as stability of the CoF. C/SiC materials are used when no further alternatives exist, that is, metals, polymers, or even monolithic ceramics have reached their limits. However, they will remain materials only for niche applications if the problem of the high manufacture costs cannot be solved in the future. Predictions of drastic cost reductions with the implementation of new technologies (e.g., one-piece manufacture, heavy tow fibers, continuous processing) and with increasing production numbers were too optimistic [53]. The costs for the end user are almost on the same level during the last decade. To spread the technology to the midsize car or common elevator segment, the manufacture has to be automated and the raw materials, in particular the carbon fibers, must be available at much lower costs.

The introduction of carbon/ceramic disks in passenger cars in 2001 created also a market for a new type of brake pad. The wider operation range and the high temperatures between the disk and the brake pads required new pad materials with a reduced content of organic materials and higher mechanical strength levels in order to fulfill the future NVH characteristics. To take profit of the low density of the disk material, future pad materials must be reduced in size and weight and should contain no organic ingredients. The reinforcement with chopped fibers must be improved in order to replace costly fabric reinforcements and to use the potential of a near-net-shape manufacture.

The industrial manufacture of C/SiC materials for brake disks (2001) and pads (2003) has a comparatively young history. As a consequence, the designers and also the users are still lacking practical experiences with this new ceramic material. There is a necessity for compiling and standardizing rules and guidelines, design methods, and concepts for an efficient component testing and validation. Also, there is a need for modeling the tribological behavior in the interphase between C/SiC and the different counterparts steel (e.g., for elevators), inorganic pad material (e.g., for passenger cars), or C/SiC (e.g., for emergency brakes). Moreover, a closed simulation process chain is necessary to predict the lifetime and the reliability of the carbon/ceramic brake system with special consideration of water and salt and other environmental influences.

23.4 CONCLUSIONS

Today, tribological applications are the major production, use and market for thermostructural composites. The Western production is estimated to be 2000 metric tons of C/C and 500 tons of C/SiC a year. There is a large competition between manufacturers to improve the products and decrease the costs. For aircraft braking applications the market and materials seem to be on a mature plateau phase but incremental evolutions can reasonably be expected.

C/SiC materials are still in an early phase of market penetration and have been successfully developed for different frictional applications. For example, almost wear-free brake disks for passenger cars and pads of high CoF stability under dynamic and static conditions for elevators have proven the feasibility of this type of material for serial brake systems.

For both materials, carbon/carbon and carbon/ceramic, the manufactured quantities are constantly growing but still burdened with very high material costs compared to conventional metal/organic materials. Future developments comprise faster, automated processes for the manufacture and final machining as well as simpler geometries and novel designs of the still metal-like pads and disks.

These thermostructural composites are at the beginning of a wider industrial use. Extensive interdisciplinary research (material science, tribology, process technology, etc.) is still necessary in order to take full benefit of this new class of material. Besides frictional applications, these composites offer excellent perspectives of development for both, academia as well as industry.

ACKNOWLEDGMENTS

Thanks to Messier Bugatti Dowty and Herakles (SAFRAN group) for providing information and documents as well as Mrs. Petra Jelitschek (University of Bayreuth) for preparing the manuscript.

REFERENCES

1. D. L. Schmidt, K. E. Davidson, and L. S. Theibert. Unique applications of carbon/carbon composite materials, part 1, *SAMPE J.*, 25, 3 (1999).
2. W. Krenkel and R. Renz. CMCs for frictional applications, in *Ceramic Matrix Composites*, edited by Walter Krenkel (Wiley-VCH, Weinheim, Germany, 2008).
3. D. J. Holt. Aircraft braking systems. *Aerosp. Eng.*, June, 7–11 (1993).
4. J. Guichard. Freinage à très haute performances. *Techniques de l'ingenieur, Traite de genie mécanique, ref.* B5580 (August 10, 1996).
5. J. P. Ruppe. Today and the future of aircraft wheel and brake development. *Can. Aeronaut. Space J.*, 26, 3 (1980).
6. K. Lafdi and K. M. Teo. Friction of carbons. International Carbon Conference, Providence, RI, July 11–16, 2004.
7. Concorde brakes by DUNLOP are C/C, *Aircraft Eng.*, 48, 9 (1976).
8. J. L. Cullerier. The carbon/carbon story: from rocket propulsion to high performance brakes. GEC Alsthom Technical Review, No. 8 (1992).
9. J. K. Lancaster. Carbons and graphite in tribology. Proceedings of the 4th International Conference on Carbon and Graphite, London, September 23–27, 1974.
10. J. F. Hamann and P. Girod. Le freinage aéronautique carbone/carbone: de l'éprouvette à l'avion. *Nouvelle revue d'aéronautique et d'astronautique*, No. 2 (1996).
11. R. C. Bill. Friction and wear of carbon-graphite materials for high-energy brakes. AVRADCOM technical report, 27 (78904) STAR category 27, 268–279 (1978).
12. K. Peszynska, K. B. Anderson, T. Szymanski, M. Krkoska, and P. Filip. Thermal analysis of bulk C/C composite friction products derived from it during simulated aircraft braking, *Carbon*, 45 (2007).
13. K. Lafdi and K. M. Teo. Friction of carbons, Part 1 and 2, International Carbon Conference, Providence, RI, July 11–16, 2004.
14. R. H. Savage. Carbon brush contact films, *G.E. Rev.*, 48, 13–20 (1945).
15. D. Marshall. C/C composites. *AC Eng.*, 231(6) (1991).
16. R. J. Diefendorf. Personal communication, 36th International Conference on Advanced Ceramics and Composites, Daytona Beach, FL, USA, January 22–27, 2012.
17. I. L. Stimson and R. Fisher. Design and engineering of carbon brakes. *Philos. Trans. R. Soc. London, Ser. A*, 294 (1980).
18. G. M. Savage. Carbon/carbon composites. *Chem. Ind.*, July, 525–529 (1992).
19. M. Gouider, Y. Berthier, P. Jacquemard, B. Rousseau, S. Bonnamy, and H. Esrade-Szwarckopf. Mass spectrometry during C/C composite friction: carbon oxidation associated with high friction coefficient and high wear rate. *Wear*, 256 (2004).

20. D. Gibson and C. Taccini. C/C friction materials for dry and wet brake and clutch applications. *SAE Trans.*, 98(5) (1989).
21. H. Muckenhuber and H. Grothe. The heterogeneous reaction between soot and NO_2 at elevated temperature. *Carbon*, 44, 546–559 (2006).
22. J. G. Pruett. Wear mode hypothesis for C/C composite brakes. International Carbon Conference, Pennstate University, PA, July 18–23, 1997.
23. J. G. Pruett. Modeling of friction performance in C/C brake materials. International Carbon Conference, Pennstate University, PA, July 18–23, 1997.
24. F. P. Kirhart. Carbon aircraft brakes-a description. 12th Conference on Carbon, Pittsburg, 1975
25. C. Byrne and Z. Wang. Influence of thermal properties on friction performance of carbon composites. *Carbon*, 39 (2001).
26. D. L. Schmidt, K. E. Davidson, and L. S. Theibert. Evolution of carbon/carbon composites. *SAMPE J.*, 32(4) (1996).
27. F. Dillon, C. A. Parker, D. E Hayes, and N. Murdie. Friction and mechanical properties of pitch-densified carbon foams, International Carbon Conference, Berlin, Germany, July 9–13, 2000.
28. Aviation Week and Space Technology, vol. 141, no. 17 (October 24, 1994).
29. S. Yu, G. Liu, X. Li, B. Pu, and X. Xiong. Contrast and analysis of properties of C/C braking discs from different producers. *AMCS*, 20(3) (2003).
30. Boeing. NN: Taxi braking recommendations for carbon and steel brakes [Company document] (March 2009).
31. C. A. Mims. Catalytic gasification of carbon: fundamentals and mechanism, in *Fundamental Issues in Control of Carbon Gasification Reactivity*, edited by J. Lahaye and P. Ehrburger, NATO ASI Series E: Applied Sciences, Vol. 192, Advanced Research Workshop, Cadarache, France (Kluwer Academic Publishers, Netherlands; NATO, Cadarache, 1990).
32. T. B. Walker, L. A. Booker, and R. Donaldson. Catalytic oxidation of C/C composite friction materials, International Carbon Conference, Pennstate University, PA, July 18–23, 1997.
33. T. B. Walker and L. A. Brooker. Inhibition of catalytic oxidation in C/C composite friction materials, International Carbon Conference, Berlin, Germany, July 9–13, 2000.
34. T. B. Walker and D. L. Johnson. Method for producing C/C brake material with improved initial friction coefficient. US patent 0271876 A1 (2005).
35. S. King. Eaton offers range of C/C products. ACM Monthly, July 320 (1998).
36. Textron. Carbon friction material for automotive transmission system, *Tech. Ceram. Internat.*, December (1995).
37. R. W. Froberg and B. A. Grider. High friction C/C aircraft brakes, 40th International SAMPE Symposium, May 8–11, 1995.
38. C. C. Evans, A. C. Parmee, and R. W. Rainbow.Silicon treatment of carbon fiber-carbon composites, Proceedings of the 4th International Conference on Carbon and Graphite, London, September 23–27, 1974.
39. E. Gelhorn, U. Gruber, and H. Leis. Silicon-carbide coated carbon–carbon-an advanced material for rocket systems. AGARD Paper CP-449, Bath, UK, 1988.
40. R. Gadow. Die Silizierung von Kohlenstoff. Doctoral Thesis, PPN: 013801430 (University of Karlsruhe, Germany, 1986).
41. W. Krenkel and H. Hald. Liquid infiltrated C/SiC-an alternative material for hot space structures, in *Spacecraft Structures and Mechanical Testing*, ESA SP-289 (European Space Agency Publication Division, Paris, France, 1989).
42. H. Pfeiffer, W. Krenkel, B. Heidenreich, and L. Schlenk. Bremsscheiben aus keramischen verbundwerkstoffen für schienenfahrzeuge, in *Werkstoffe für die Verkehrstechnik*, edited by U. Koch (DGM Informationsgesellschaft, Frankfurt, Germany, 1997).
43. W. Krenkel. CMC materials for high performance brakes, Proceedings of the 27th International Symposium on Automotive Technology and Automation (ISATA), Conference on Hybrid and Alternative Fuel Vehicles and Supercars, Aachen, Germany, October 31–November 04, 1994.
44. B. Heidenreich and W. Krenkel. Development of C/C-SiC materials for frictions applications, in *Ceramics-Processing, Reliability, Tribology and Wear*, EUROMAT 99, vol. 12, edited by G. Müller (Wiley-VCH, Weinheim, Germany, 1999).
45. W. Krenkel, B. Heidenreich, and R. Renz. C/C-SiC composites for advanced friction systems. *Adv. Eng. Mater.*, 4(7) (2002).
46. W. Krenkel, R. Renz, and B. Heidenreich. Lightweight and wear resistant CMC brakes, in *Ceramic Materials and Components for Engines*, edited by J. G. Heinrich and F. Aldinger (Wiley-VCH, Weinheim, Germany, 2001).
47. W. Krenkel. Design of ceramic brake pads and disks. *Ceram. Eng. Sci. Proc.*, 23(3) (2002).
48. W. Krenkel. Carbon fiber reinforced CMC for high-performance structures. *Int. J. Appl. Ceram. Technol.*, 1(2) (2004).
49. R. Renz, G. Seifert, and W. Krenkel. Integration of CMC brake disks in automotive brake systems. *Int. J. Appl. Ceram. Technol.*, 9(4) (2012).
50. B. Heidenreich, R. Renz, and W. Krenkel. Short fibre reinforced CMC materials for high performance brakes, in *High Temperature Ceramic Matrix Composites*, edited by W. Krenkel, R. Naslain, and H. Schneider (Wiley-VHC, Weinheim, Germany, 2001).
51. W. Krenkel, T. Henke, and N. Mason. In situ joined CMC components. *Key Eng. Mater., Proceedings of the International Conference on Ceramic and Metal Matrix Composites CMMC 96, Part I* (Trans Tech Publications Ltd., Zürich, Switzerland, 1997).

52. R. Raether, J. Meinhardt, and A. Kienzle. Oxidation behaviour of carbon short fibre reinforced c/sic composites. *J. Eur. Ceram. Soc.*, 27(2–3) (Elsevier, Oxford, England, 2007).
53. D. Neudeck and A. Wüllner. Bremsen mit nichtmetallischen bremsscheiben, in *Bremsenhandbuch*, edited by B. Breuer and K. H. Bill (Vieweg & Teubner/GWV Fachverlage, Wiesbaden, Germany, 2006).
54. Deutsches Institutfür Normung. Gießereiwesen-gusseisen mit lamellengraphit. DIN EN 1561 (Beuth Verlag, Berlin, Germany, 1997).
55. Duralcan *Composites for Gravity Castings* (Data Sheet Duralcan, San Diego, CA, USA, 1992).
56. Lanxide, ASM Al-342, Lanxide, 91-X-1060-30P-T6 (aluminum alloy-matrix composite). *Alloy Digest* (ASM International, Materials Park, OH, USA, Al-342, 2002).
57. P. Morgan. *Carbon Fibers and Their Composites*, 1st Edition (CRC Press, Boca Raton, FL, USA, 2005).
58. SGL Group. SIGRASIC 6010 GNJ-Faserverstärkte Keramik für Bremsscheiben von SGL Carbon (Data Sheet SGL Technik, Meitingen, Germany, 1999).
59. J. Föhl and J. Wiedemeyer. Forschungsberichte der. *DKG*, 15, 1 (2000).
60. J. Föhl and J. Wiedemeyer. Qualifikation von faserverbundkeramik für friktionsanwendungen [Kap. 8.3.1.3], in *DKG-Handbuch*, edited by J. Kriegesmann (Deutscher Wirtschaftsdienst, Köln, Germany, 2001).
61. N. Langhof, R. Voigt, H. Mucha, and W. Krenkel. The effect of residual silicon in CMC brake pads on friction and wear, Proceedings of the 6th European Conference on Braking, Lille, France, 2010.
62. N. Langhof, H. Mucha, and W. Krenkel. The effect of additives on CMC friction pads at elevated temperatures, in *High Temperature Ceramic Materials and Composites*, edited by W. Krenkel and J. Lamon (AVISO Verlagsgesellschaft mbH, Berlin, Germany, 2010).
63. D. Richter, C. Hane, M. Rabenstein, H. Mucha, W. Krenkel, F. Yang, D. Kunz, H.-S. Park, M. Willert-Porada, J. Rosenlöcher, et al. Fibre ceramic based friction linings. *Proceedings Braking 2009* (Woodhouse Publishing Ltd., Cambridge, England, 2009).
64. N. Langhof, B. Alber-Laukant, C. Greuel, M. Hilpert, A. Kozub, R. Voigt, and W. Krenkel. Fiber reinforced ceramic frictions pads for automotive applications-the dependence of friction and wear on speed and braking pressure. *Proceedings of Eurobrake, Dresden, Germany,* 2012.
65. Y. Xu, Y. Zhang, L. Cheng, L. Zhang, J. Lou, and J. Zhang. Preparation and friction behaviour of carbon fibre reinforced silicon carbide matrix composites. *Ceram. Int.*, 33 (2007).
66. H. Zhou, S. Dong, Y. Ding, Z. Wang, and D. Wu. Friction and wear properties of 3d carbon/silicon carbide composites prepared by liquid silicon infiltration. *Tribol. Lett.*, 37(2) (2009).
67. Z. Stadler, K. Krnel, and T. Kosmač. Friction and wear of sintered metallic brake linings on a C/C-SiC composite brake disc. *Wear*, 265 (2008).
68. S. Fouget, M. Rollin, R. Pailler, and X. Bourrat. Tribological behaviour of composites made of carbon fibers and ceramic matrix in the Si-C-system. *Wear*, 264 (2008).
69. Y. H. Zhang, Z. C. Xiao, J. P. Wang, R. P. Xing, Z. G. Peng, J. M. Su, and Z. H. Jin. Effect of pyrocarbon content on thermal and frictional properties in C/C preforms of C/C-SiC composites. *Wear*, 269 (2010).
70. S. Fan, L. Zhang, L. Cheng, G. Tian, and S. Yang. Effect of braking pressure and braking speed on the tribological properties of C/SiC aircraft brake materials. *Compos. Sci. Technol.*, 70(6) (2010).
71. Y. Cai, S. Fan, H. Liu, L. Zhang, L. Cheng, B. Dong, and J. Jiang. Microstructures and improved wear resistance of 3D needled C/SiC composites with graphite filler. *Compos. Sci. Technol.*, 69(14) (2009).
72. S. Fan, L. Zhang, L. Cheng, J. Zhang, S. Yang, and H. Liu. Wear mechanisms of the C/SiC brake materials. *Tribol. Int.*, 44(1) (2011).
73. R. Weiß. Carbon/carbons and their industrial applications, in *Ceramic matrix composites*, edited by W. Krenkel (Wiley-VCH, Weinheim, Germany, 2008).
74. H. Abu El-Hija, B. Heidenreich, W. Krenkel, and S. Hugel. Development of C/C-SiC brake pads for high-performance elevators. *Int. J. Appl. Ceram. Technol.*, 2(2) (2005)
75. R. Renz and W. Krenkel. C/C-SiC composites for high performance emergency brake systems, Proceedings of the 9th European Conference on Composite Materials (ECCM-9), Brighton, UK, 2000.

Index

Abbreviations, 210–211
Ablated material morphology, 378
Ablation, of C/C materials, 353–388
Ablation morphologies, 363–365
Ablation rate, 367
Ablation resistance, 367
Ablation solver algorithms, 373
Ablation tests, 358, 359
Ablative C/C composites, variety of, 355–356
Ablative conditions, C/C and C/phenolic composites in, 357
Ablative material design, 355
ABO_4 coatings, 260
Accelerated mission tests (AMTs), 596
Acheson process, 149
Acoustic emission (AE), 349. *See also* AE parameters
  for CMCs, 571–590
  future directions for, 588–589
  lifetime prediction and, 586–588
  principles and practice, 572–574
Acoustic emission acquisition, approaches to, 574, 575–580, 580–584
Acoustic emission events, location of, 576
Acoustic emission monitoring, 576, 577
  event-based, 575–580, 582
  high-temperature testing and, 584–586
Acoustic emission signal analysis
  supervised pattern recognition method in, 584
  unsupervised clustering methodology in, 583–584
  using pattern recognition techniques, 580–584
Acoustic emission systems, 575
Acoustic emission tensile tests, 572
Adherence, of environmental barrier coatings, 435
Adhesive joining, 553
Adhesives, 552
Adhesive technology, 553
AE parameters, 349. *See also* Acoustic emission entries
Aerodynamic performance, improving supersonic, 192–193
Aeroengine CMC components, designing, 425
Aeroengines, 405, 593
Aeronautics applications, 200–202
Aero-propulsion, SiC/SiC composites for, 217–235
Aero-propulsion gas turbine engines, ceramic components for, 218
Aerospace technology advancement, 236
AESTUS (Astrium ST/EADS IW), 193–195
Afterburner components, 598–599
Aircraft brake design, safety in, 648
Aircraft braking, 647–650
Air cross-linking, 629. *See also* Cross-links
Air plasma spraying (APS), 442–443
Aitken's relaxation methods, 505
Alumina-based composites, 240
Alumina-mullite matrix, 249. *See also* Mullite-alumina composite strength
Alumina surface, 438
Alumina-YAG matrices, 249–250, 253
Aluminosilicate matrices, 249
Aluminosilicate matrix materials, 250
Aluminosilicate matrix systems, 253
Aluminum, in SA3 fibers, 629
Anisotropic damage, simplest, 472
Anisotropic damage theory, 467, 468–481
Anisotropic framework, tools for implementing, 479–481
Anisotropic interphases, 52
Anisotropy, 309
Anti-ablation components, of oxidation protective coatings, 461
Anti-ablation layer, 460
Anti-oxidation coatings, preparation methods for, 454–456
Anti-oxidation methods, for C/C composites, 452
Antioxidation protection, for carbon brakes, 654–655
Applications. *See also* Aeronautics applications; CMC applications; Divertor structural component applications; Friction applications; High energy laser (HEL) applications; Industrial applications; Nuclear applications; Space applications
  atmospheric reentry, 354, 356–359
  for C/SiC and C/C-SiC materials, 191–209
  for friction systems, 202–207
  one-step-cycle, 92
  for oxide–oxide composites, 261–263
  tribological, 655–657
Architecture, 624, 630
Arc image furnace, 378
ArcjoinT technology, 554–555
Ariane 5, 193, 195
ARIES fusion reactor, 611, 614–616
Arrhenius plots, 119
Aspa, Yvan, xvii, 353
ASTM (American Society for Testing and Materials) standards, 562
Astrium, 196, 197, 198
Asymmetrical four-point bending test, 562
ATK-COIC composites, 252, 253–256
Atmospheric reentry, 353–354
Atmospheric reentry application, 354, 356–359
Atmospheric reentry materials, 355–356
Atomic displacement damage, 390
Atomic level, of carbon fibers, 28
Autoclave technique, 163–164
Autoclaving, 244
Automotive brake disks (discs), 203–204. *See also* Brake disk entries
Automotive components, crash behavior of, 206–207

Backside impact damage, 413, 414
Balance equations, 371

Ceramic Matrix Composites: Materials, Modeling and Technology, First Edition. Edited by Narottam P. Bansal and Jacques Lamon.

Ballistic impact gun, 406
Bansal, Narottam, xvi
Baranger, E., xvii, 467
Barrier coatings, 554
Barrier layer, 460
Basic structural units (BSUs), 28, 29, 30
Bayesian fragmentation model, 533–535
Behavior loop, 505–506
Behavior–microstructure relationships, 520–547
Behavior models, 628
Bending behavior, 70–71
Bending deformation, 106
Bending force, 80
Bending tests, cantilever, 70, 71
Benioff law, assessing CMC lifetime with, 588
Bias-extension tests, 68, 69
Biaxial tensile machine, 70
Bilayered specimens, flexural test on, 57
Binocular magnifier (BM), 359
Blanket component, 614–616
BN (boron nitride) coatings, 225, 240
BN interphases, 56, 338–339, 398
  debonding of, 344
  improving, 344
  SiC-fiber-reinforced SiC matrix composites with, 341–344
BN oxidation, 341, 342
Boisse, Philippe, xvii, 65
Bolt/rivet system, 552
Bond coatings, 435
Bonding, 40, 41. *See also* Debond entries
Bonding strength, of oxidation protective coatings, 461
Bond stress relaxation (BSR) test, 222
BoraSiC® coating, 123, 124
Borides, 123
Boron, in SA3 fibers, 630
Boron carbide oxidation reaction, 494
Boron nitride (BN). *See* BN entries; iBN fibers; Silicon-doped boron nitride (BN); Sylramic™-iBN SiC fiber
Boron oxide evacuation, 496
Borosilicate glass, 456
Bouncing-back velocity, impact velocity and, 418
Braiding angle, 633
Braiding textures, 630
Brake configuration, multidisk, 649
Brake disk development, 664
Brake disk manufacturing, 661–662
Brake disks (discs), 118, 126, 127, 203–204. *See also* Disk (disc) brakes
  carbon/ceramic, 666
  CFRP preforms for, 163, 662
  for emergency brake systems, 666–667
  for racing cars, 655–656
Brake disk temperatures, 664, 665
Brake materials, properties of, 664
Brake pads, for high-rise lifts, 667
Brakes
  antioxidation protection for, 654–655
  carbon/carbon materials for, 653–654
  mechanical properties of C/C materials for, 652
Brake systems, high performance, 666–667
Braking, aircraft, 647–650
Braking efficiency, unbalanced, 657
BraSiC™ alloys, 561
Brazilian test, 57
Brazing, 553–554, 561
Brazing alloys, 552, 553–554, 560, 561
Bridging, elastic, 282
Brittle failure stress criterion, 482
Brittle fracture(s), 289
Brush-like joining technology, 559
BSAS/BSAS coatings, 438–440
BSAS (barium strontium aluminosilicate) coating, 433, 434, 435, 436–437, 438–440, 444
  hot pressed, 446
  plasma-sprayed, 446
Bulk ceramics, 174–175
Bulk diffusion, 381
Bulk protection systems, for carbon/carbon composites, 119–122
Bundle models, 13–14
Bundle-unit fractures, 101
BWR (boiling water reactor) applications, 622–623

CAD (computer-aided design) models, 67
Calcium magnesium aluminosilicate (CMAS), 421–422. *See also* CMAS entries
Camphene, 246
Cantilever bending test, 70, 71
Cantilever support, effects of, 422
Capillary infiltration, 169
Capillary pressure, 246
Carbon (C), free, 19–21. *See also* C/C entries; C/SiC entries; Graphene entries; Graphite entries; Pyrocarbon entries
Carbon-based CMC, 554. *See also* Ceramic matrix composites (CMCs)
Carbon-based coatings, 224
Carbon-based interphase, 401–402
Carbon brakes, antioxidation protection for, 654–655
Carbon/carbon brake disks (discs), 118
Carbon/carbon brakes, 126, 127. *See also* C/C carbon disk brake market
Carbon/carbon chains, 133
Carbon/carbon (C/C) composites, xv, 379. *See also* C/C entries
  ablation of, 353–355, 383
  ablation tests performed on, 358
  in ablative conditions, 357
  in aircraft braking, 650
  anti-oxidation methods for, 452, 453–454
  behavior of, 624
  bulk protection systems for, 119–122
  characteristics of commercial, 626
  compressive fracture behavior of, 106
  electrical properties of, 115–116
  dimensional changes for, 400
  ductility of, 400
  for fission reactor applications, 627
  fracture behavior of, 97
  for friction applications, 647–657
  for fusion applications, 624–627
  for high-temperature furnaces, 129–134
  industrial applications for, 87–146
  interfacial bonding with, 454
  manufacture of, 87–97
  multilayer-coated, 125
  for nuclear applications, 560–561, 610
  oxidation behavior of, 119, 120
  processing for nuclear applications, 624–627
  radiation effects on, 389, 399–401
  shear strength tests on, 562
  specific heat of, 117
  stability with, 453–454
  strengths of, 97–109
  thermal conductivity of, 110–115
  thermal expansion of, 116–117, 453
  thermal properties of, 109–118
  thermal shock resistance of, 118
  thermomechanical properties of, 625–627
  3D reinforced, 106–107
  unidirectional, 89, 96
Carbon/carbon crucibles, 139
Carbon/carbon fixtures, 134–137
Carbon/carbon furnace chamber, 130
Carbon/carbon grades, properties of, 120
Carbon/carbon (C/C) materials. *See also* C/C friction materials
  ablation of, 353–388
  for brakes, 653–654

characteristics and properties of, 113, 114, 115
demand for, 138
future directions of, 118
limitations and trends in, 657
maximum application temperature of, 134
mechanical properties for brakes, 652
oxidation protection of, 118–126
for racing car brake disks, 655–656
in the solar energy market, 137–140
thermal conductivity of, 140
tribological behavior of, 651–655
Carbon/carbon (C/C) preforms, 149, 160, 166, 169, 189, 658
Carbon/carbon screws, 130, 131
Carbon/carbon systems, load-carrying, 134
Carbon/carbon thermal insulation, 139
Carbon/carbon ventilators, 133
Carbon/ceramic brake disks, 666, 668
Carbon ceramic brakes, 203, 659, 665–666
Carbon/ceramic materials, for friction applications, 657–668
Carbon coatings, 240, 260, 261
Carbon fabrics, 89, 90
Carbon fiber(s), 27–39, 74, 75, 161
advantages of, 27
atomic level of, 28
graphitic microstructures in, 399–400
graphitized, 147
hierarchical levels of, 28–32
increasing demand for, 36–37
information concerning, 36–37
lower nanometer range of, 28–31
nonlinear stress–strain curve of, 35
PAN-based, 30, 31, 88
pre-impregnated, 88, 92
production routes for, 27
properties of, 89, 181
stiffness and strength of, 32–36
structure of, 28–32
10-μm range of, 32
types of, 88
upper nanometer range of, 31–32
Young's modulus of, 33
Carbon fiber bundles, *in situ* XRD of, 36
Carbon fiber composites, radiation effects in, 399–401
Carbon-fiber-reinforced carbon (C/C) materials, 147. *See also* Carbon/carbon entries
Carbon fiber-reinforced ceramic matrix composites (CMCs), 286–288
Carbon-fiber-reinforced plastic (CFRP) preforms, 149, 150, 161, 171, 172, 209
for brake disks (discs), 163, 662
manufacture of, 158–159, 162–166
pyrolysis of, 166–168
Carbon-fiber-reinforced polymers (CFRPs), 90–91. *See also* CFRP materials
Carbon-fiber-reinforced SiC composites, 148, 337–338
Carbon fiber reinforcements, 81–82, 88–90
Carbon fiber removal, stressed-oxidation degradation modeling in, 345–346
Carbon fiber seal, 638
Carbon interphase oxidation, 339
Carbon interphases, SiC-fiber-reinforced SiC matrix composites with, 338–341
Carbon interphase system, 347. *See also* Interphase carbon
Carbonization, 27, 126
Carbon materials, 111, 112, 115, 394
Carbon matrices, 92–94, 625
Carbon matrix systems, 90–94
Carbon nanotubes (CNTs), 27, 112
Carbon oxidation, 652
Carbon phase reactivity, 384
Carbon/phenolic composites, 353–354
Carbons, thermophysical properties of monolithic, 110
Cars. *See* Automotive entries; Passenger cars; Racing car brake disks
Casalegno, Valentina, xvii, 551
Catastrophic fracture(s), 289
Cauchy stress tensor, 74
CBS (carbon-boron-silicon-containing) coating, 337
C/C carbon disk brake market, 653. *See also* Carbon/carbon entries
C/C costs, 657
C/C friction materials, 202–203
C/C limitations/trends, 657
C/C manufacturing cycle, 88
C-coupon testing, event-based AE of, 580, 582
C/C parameters, 87–88
C/C-SiC components, 170
C/C–SiC composites, 147–216
C/C-SiC handling device, 208–209
C/C-SiC materials, 199–200. *See also* C/C–SiC composites
applications for, 191–209
composition and microstructure of, 176–178
mechanical properties of, 178–187
MI-derived, 160–161
oxidation behavior of, 190–191
properties of, 174–191
thermal properties of, 187–189
tribological behavior of, 191
C/C-SiC sliding elements, 128–129, 207
C/C testing, in SRM facility, 362–365
C/C tribological applications, 655–657
CEA, 631–633
CEA sandwich design, 638–641
Celsian phase, 436
Ceramic-based composites, ultrahigh temperature, 273–292
Ceramic brake pads, 205, 206
Ceramic cladding, 637
Ceramic coatings, 457–459
Ceramic components, for aero-propulsion gas turbine engines, 218
Ceramic composites, 74–76, 236, 255
Ceramic fiber–matrix composites (CFMCs), 236
Ceramic fibers, coatings onto, 242
Ceramic materials, mechanical properties of, 65
Ceramic matrix composites (CMCs), 23, 65, 74, 75, 147, 148, 175, 217–218, 220, 223, 224–233, 236, 401. *See also* CMC entries; Continuous-fiber-reinforced ceramic matrix composites (CFCCs); SiC/SiC CMCs
acoustic emission for, 571–590
adhesive joining of, 553
advantages of, 295–296
in aeroengines, 405
behavior at interfaces and interphases of, 40–64
brazing of, 553–554, 561
carbon-based, 554
carbon fiber-reinforced, 286–288
continuous-fiber-reinforced, 3
creep-recovery behavior of, 304–308
damage modeling for, 467–468
event-based AE monitoring of, 575–580
fission reactor applications for, 616–623
FOD factors in, 425–426
FOD responses for, 424–425
foreign object damage in, 405–429
for friction applications, 647–671
fusion applications for, 610–616
integration and joining of, 551–567
microporous oxide–oxide, 250
microstructure–behavior relationships for, 520–547
nuclear applications for, 609–646
oxide–oxide, 236, 237–238, 248–250, 261–263, 295–296

Ceramic matrix composites (CMCs) (*Continued*)
  for pin cladding or structural materials, 621–623
  porous oxide–oxide, 248–250
  radiation effects in, 389, 394–401
  reinforcement of, 3–26
  role of interfacial domain in, 41–49
  silicon nitrides *vs.*, 424–425
  topics related to, xvi
Ceramic particles, 457
Ceramics, 3, 174–175, 392–394
C-fiber bundles, 92. *See also* Carbon fiber entries
C fiber embedding, 151
C fiber reinforcement, for brakes, 653–654
CFRP materials, 175, 662. *See also* Carbon-fiber-reinforced plastic (CFRP) preforms; Carbon-fiber-reinforced polymers (CFRPs)
C_f/[Si-B-C] composites, 585, 586
Chain-of-element model, in fragmentation modeling, 531
Chain-of-segment model, in fragmentation modeling, 531–532
Characteristic strength, 538–539
Characteristic volume, 538
Chemical bonding, at an interface, 40, 41. *See also* Bond- entries
Chemical compatibility, of environmental barrier coatings, 435–436
Chemical fuses, 496
Chemical vapor deposition (CVD), xv, 149, 219, 242, 248, 443, 454. *See also* CVD entries
Chemical vapor infiltration (CVI), xv, 41, 75, 149, 151–156, 173, 174, 176–177, 219, 248, 356. *See also* CVI entries
  conductive matrices through, 112
Choi, Sung R., xvii, 405
Cladding, CMCs for, 621–623, 627–628
Cladding closure process, 640
Cladding designs, 637–638
Closure effects, 467, 468, 471
Clustering, 583–584
Clustering methodology, applied to AE signals, 583–584
Cluzel, C., xvii, 467
CMAS composition model, 445–447. *See also* Calcium magnesium aluminosilicate (CMAS)
CMAS melting point, 445
CMAS/salt, 421–422
CMC adhesive joining, 553. *See also* Ceramic matrix composites (CMCs)
CMC afterburner components, 598–599
CMC application designs, 641
CMC applications, for gas turbines, 593–608
CMC brake disks (discs), 126, 127, 203–204
CMC brazing, 553–554
CMC candidate turbine components, 431
CMC C-coupon testing, event-based AE of, 580, 582
CMC/coating interface, 441
CMC coatings, 438–441
CMC components, 442, 580–581, 607
CMC damage mechanisms, 571
CMC damage models, 471–472, 472–481
CMC degradation, 405
CMC developments, for military engines, 594–600
CMC-elevated temperature stress-rupture, event-based AE of, 579
CMC exhaust components, Safran R&D on, 605–606
CMC fabrication routes, 219–220
CMC failure behavior, 238
CMC foreign object damage, 409–422
CMC hot section components, 430
CMC impact damage, 412–413
CMC integration, 551–552, 552–553, 555–558
CMC jet vanes, 202
CMC joining, 551–552
  dismantling of, 563
  exotic techniques for, 555–558
  future directions for, 562–563
  mechanical interlocking in, 558–559
  mechanical tests on, 561–562
  for nuclear applications, 560–561
  for ultrastable structures, 561
  via glass and glass-ceramic materials, 557, 561
CMC lifetime, Benioff law to assess, 588
CMC materials, 184–185, 406–407
CMC material systems, 416, 431
CMC mechanical joining, 552
CMC mixer nozzle, for helicopter demonstration, 598
CMC mixer nozzle technology, 261–262
CMC processing, xv
CMC program demonstrations, on industrial gas turbines, 606–607
CMC properties, 217
CMC research and development
  for combustion chambers, 601–602
  for commercial engines, 600–607
  for exhaust sections/components, 603, 605–606
  for turbine components, 602–603
CMC specifications, for divertor component applications, 613
CMC stresses, 230, 231
CMC stress–strain behavior, event-based AE of, 576–579
CMC targets, configuration and support of, 407–408
Coating/CMC interface, 441
Coating processes, 443
Coating requirements, 219
Coatings, 240. *See also* Interface coating entries; Oxidation protective coatings; Oxidation-resistant coating systems
  ABO_4, 260
  barrier, 554
  bond, 435
  BoraSiC®, 123, 124
  boron nitride, 240
  BSAS, 433, 434, 435, 436–437, 438–440
  carbon, 240, 260, 261
  carbon-based, 224
  ceramic, 457–459
  onto ceramic fibers, 242
  CMC, 438–441
  composite, 460
  CVD-SiC, 122–123, 124
  discrete, 242
  double-layer, 460
  ductile polymeric, 424
  EB-PVD, 443
  fugitive, 56, 260–261
  glass, 456
  gradient, 459
  impact damage of, 424
  lacquer, 424
  layered, 459
  metal, 456–457
  monazite, 242, 258–260
  mullite, 249, 430, 433, 434, 437–438
  multilayer, 122–124, 437, 459, 460
  nanoparticle-toughened, 458–459
  nanowire-toughened, 458
  oxide, 56
  porous, 56, 260
  powder immersion reaction assisted, 121–122
  "SiCralee," 170
  whisker-toughened, 458
  zirconia, 260, 261
Coatings research, 437

Coating temperature, 147
Coating work, for Si-based ceramics, 430
Coefficient of friction (COF, CoF), 191, 207, 650, 651, 652, 657. *See also* CoF stability
Coefficient of moisture expansion (CME), 561
Coefficient of thermal expansion (CTE), 116–117, 147, 187–189, 436–437, 441, 457, 551–552, 561, 661, 664
  of environmental barrier coatings, 433
Coefficient of variation, 14
CoF stability, 660, 664, 665, 666. *See also* Coefficient of friction (COF, CoF)
Cold-grip experiments, 508
Collaborative programs, 600
Collision cascade, 391
Combustion chambers, CMC research and development for, 601–602
Commercial C/C composites, characteristics of, 626. *See also* Carbon/carbon entries
Commercial engines, CMC research and development for, 600–607
Compaction, 415
Components manufacturing/testing, 599
Composite approach, 273
Composite behavior, probability-based general model of, 536
Composite coating, 460
Composite design, 60, 237–240, 536
Composite joints, 557
Composite laws, 380–383
Composite materials. *See also* Composites; C/SiC materials
  damage monitoring of, 581–583
  in nuclear industry, 609, 610
  radiation effects on, 392
  thermostructural, 647
Composite mechanical behavior, 521–526
Composite reinforcements, draping in woven, 76
Composite reinforcement forming, using semidiscrete approach, 77
Composites. *See also* Composite materials; C/SiC composites; Ultrahigh temperature ceramic-based composites
  alumina-based, 240
  choice of interphase, 630–631
  densification of, 631–635
  hermetic, 635, 636, 637
  with interface coatings, 257–261
  Nextel-based, 254, 259–260
  one-dimensional, 521
  oxidation protective coatings for ultrahigh-temperature, 452–463
  properties of, 250–257
  silica-free porous, 250
  UCSB, 255, 256
  ultimate strength of, 252
  uses of, 36–37
Composite specimens, structural phenomena in, 543
Composite systems, 337–344
Composite versatility, 546
Compression, 313–317
Compression behavior, of PAN and MPP fibers, 35
Compression creep, 313–317
Compression creep rates, 314
Compression creep strain, 316
Compression double-notch shear test, 562
Compressive fracture behavior, of carbon/carbon composites, 106
Compressive strength, 106–108, 252
Compressive tests method, 104–105
Computational approach, for woven composites, 539–542
Computed microtomography (CMT), 359, 360, 361
Computer technology, 571, 573
Concorde supersonic aircraft program, 650
Conductive matrices, 110–112
Cone cracking, 415, 416–417
Conservation equations, 368
Consolidation methods, for oxide–oxide composites, 247–248
Constant stress, filaments under, 21–22
Constituent (constitutive) properties, 526–531
Contact yield pressure, 419
Continuous FE approaches, 76–77
Continuous fiber embedment, 175
Continuous-fiber-reinforced ceramic matrix composites (CFCCs), 3, 101, 236, 520
  microstructure–behavior relationships for, 520–547
  tensile stress–strain behavior of, 521–524
Continuous-fiber-reinforced ceramics, 3
Continuous fiber reinforcements, 74
Continuous Lower Energy, Emissions, and Noise (CLEEN) Program, 604
Continuous mechanical models, 77
Continuous refractory fibers, 23
Control rod failure modes, 620
Control rods, 618, 619, 620, 635
Cook/Gordon mechanism, 43, 44–45
Cooling channels, 126
Cost benefits, 607
Costs, for carbon/ceramic brake disks, 668
C/phenolic composites, in ablative conditions, 357
C/phenolic resin composites, 355, 356
Crack arrest, 530
Crack deflection, 50, 51, 237, 239, 240, 257–260, 282, 324, 325
  degradation of, 253
  at interfaces, 41–44
  material elastic properties and, 45–49
Crack deflection mechanisms, 43
Crack density, of minicomposites, 488
Crack density indicator, 488
Crack deviation, 54
Crack formation, 349
Crack-free oxide fiber, 246
Crack growth, 327–328
  slow, 496, 544, 545
  stressed-oxidation degradation modeling in unbridged, 347–348
  subcritical, 16–23, 297–298, 301
Crack growth parameters, 298, 301
Crack growth rate, 21, 330
Cracking, of oxidation protective coatings, 460
Crack initiation, at interfaces, 41
Crack morphologies, 443
Crack networks, 482
Crack opening, residual, 489. *See also* Crack openings
Crack opening/closing effects, 123
Crack opening indicator, 488–489, 490, 498
Crack openings, 482, 499. *See also* Crack opening
Crack reduction, 459
Cracks. *See also* Macrocrack entries; Matrix crack entries; Microcrack entries; Multiple cracking
  macroscopic, 513
  transverse, 522, 523, 540
  vertical, 437, 440
Crack spacing, 538
Crack tips, stress states at, 43, 44
Crack-velocity formulation, 301
Crash behavior, of automotive components, 206–207
Creep, 103–104, 307, 317–321
Creep behavior, 103–104, 218–219, 221–222, 224, 259
Creep lifetime, 300, 301, 306
Creep loading, 304–308, 330
Creep/oxidation/fatigue performance, of Nextel 720/A system, 254–255

Creep rates, 299–300, 306, 311
  secondary, 300, 309–311, 319
  in steam, 319
Creep-recovery behavior, 304–308
Creep resistance, 223, 254, 263–264
Creep run-out, 259–260, 299–300, 304
Creep rupture, 254
Creep strain(s), 299, 309, 310, 319, 326
Creep strain recovery, 304–308
Creep stresses, 298, 312
Creep tests, 312
Critical elements, 542
Critical fiber, 11
Critical fiber-dictated failure mode, 18
Cross-links, 36. *See also* Air cross-linking
Cross-ply (CP) lamination, 98, 99
Cross sections, of impact damage, 413–416
C/R testing, 359, 362
Crucibles, 139
Crystalline mullite coating, 437
Crystallite orientation, Young's modulus and, 34
Crystallite size, 30, 36
Crystals, deformation of, 35
C/SiC brake development, 664
C/SiC brake disks (discs), 203, 661, 662–663, 664–666
C/SiC components, 174
C/SiC composites, 74, 75, 147–216, 337–338
  anti-oxidation methods for, 452, 453–454
  interfacial bonding with, 454
  multilayer strategies for, 339–341
  stability with, 453–454
  stressed-oxidation degradation modeling in, 345–346
  thermal expansion with, 453
C/SiC development, history and milestones of, 659–663
C/SiC devices, advantages of, 207–208
C/SiC materials. *See also* SICARBON®
  applications for, 191–209
  composition and microstructure of, 176–178
  for frictional applications, 669
  future challenges for, 668
  gradient, 661
  for industrial applications, 209, 666–667
  industrial manufacture of, 668
  LSI-based, 170
  manufacturing methods for, 149–174
  manufacturing methods comparison for, 173–174
  mechanical properties of, 178–187
  MI-derived, 160–161
  modified, 660
  oxidation behavior of, 190–191
  PIP-derived, 158
  process time for, 173–174
  properties of, 174–191, 664
  thermal conductivity of, 660
  thermal properties of, 187–189
  tribological behavior of, 191
  wear rates of, 660–661
C/SiC nozzle flaps, 200–202
C/SiC parameters, 87–88
CT specimen, 513, 514
Cumulative oxygen concentration, 496–497
Cumulative slip length, 485
CVD mullite coatings, 249, 437. *See also* Chemical vapor deposition (CVD)
CVD polycrystalline cubic SiC, radiation effects on, 392
CVD process, 631
CVD SiC, radiation-resistant, 394
CVD-SiC coatings, 122–123, 124
CVI densification, 94, 95. *See also* Chemical vapor infiltration (CVI)
CVI impregnation, 95, 96
CVI materials, 178, 180, 181
CVI matrix, 633
CVI-MI approach, 228, 229, 231. *See also* Matrix infiltration entries; Melt infiltration entries
CVI-PIP hybrid approach, 227, 228, 229, 231. *See also* Polymer infiltration and pyrolysis (PIP)
CVI-processed 2.5D composites, 356
CVI processes, 92–94, 151–154, 625, 635, 636
  advantages and disadvantages of, 156
  for composite properties, 631–635, 641
CVI processing, 227
CVI-pyC fiber coating, 153. *See also* Pyrocarbon entries
CVI reactor, 94
CVI SiC composites, 346, 631–635
CVI SiC matrix composites, 339, 395–397
CVI SiC/SiC composites, xv, 41, 52
Cyclic creep loading, 304–308
Cyclic creep-recovery tests, 304, 306, 307–308
Cyclic fatigue, 321–324, 329–330. *See also* Fatigue cycle(s)
Cyclic fatigue lifetime test, 498
Cyclic fatigue tests, 497, 586
Cyclic loading, 326, 327, 328, 330, 491
  damage during, 109
  influence on cumulative slip length, 485
Cyclic loading case, 496

Damage, coupling plasticity with, 470–471
Damage detection, 571
Damaged SiC/SiC composites, 638
Damage evolution laws, 468, 470, 475–477, 484–487
  macroscopic, 487
Damage field predictions, 507, 510–511
Damage-insensitive behavior, 522
Damage kinematics, 469, 472, 473, 475–477
Damage localization control, 513–515
Damage macromodel, 490
Damage mechanisms, 476, 477, 478, 479
  identifying, 584, 585–586
Damage modeling
  based on anisotropic damage theory, 468–481
  for ceramic matrix composites, 467–468
  general notations for, 468–469
  for structure computations, 467–519
  thermodynamic framework for, 469–471
Damage models, 521
  CMC, 471–472, 472–481
  mechanical, 513
Damage modes, 522, 545
Damage monitoring, of composite materials, 581–583
Damage phenomena, features of, 524
Damage-sensitive behavior, 522
Damage tolerance, 520, 528–530
Damage tolerance analysis, 510–512
Damage variables, 471
David, Patrick, xvii, 87
Dead molds, 127
Debond cracking, 50
Debond energy, 60
Debonding, 41, 42, 46, 47, 55, 99, 530
  of BN interphase, 344
  of fiber bundles, 105
  influence of, 50–51
Debonding crack topography, 50
Debonding criteria, 44–45
Debonding potential approach, 60
Debonding potentials, 46, 47
Debonding stress, interfacial, 100
Decrimping, of tows, 69–70
Defect clusters, 391
Defect families, 539
Defect migration/evolution, 391

Defects, 393, 394
Deflected cracks, influence of, 49–51
Deflection criteria, 44–45
Deformation, 541–542. *See also* Deformations
of crystals, 35
Deformation-based analysis, 16
Deformations, 536. *See also* Deformation
Degradation. *See also* EBC degradation entries; Environment-degradation mechanisms; Fiber degradation; Strength degradation entries
of CMCs, 405
of environmental barrier coatings, 431
of mechanical properties, 350
stressed-oxidation, 337–338, 339–341, 343–344, 345–349
time–stress–temperature–environment, 334
via radiation, 401–402
of yarn, 502
Degradation mechanism, 345
Degradation mechanism simulation, 497–502
Degradation sources, 447–448
Delamination, 159, 167–168, 595, 596
Delayed fiber failure(s), oxygen and, 19
Delay-effect method, 513, 514
DEMO fusion reactor, 560
Dense matrix composites, 238
Densification, 155, 156, 247
CVI, 94, 95
matrix, 330
Densification processes, 288, 625, 631–636
Densified carbon/carbon, 90
Density, effects of, 101
Design, for stress-oxidation degradation, 345–349
Design tool simulation, 209
Deuterium–tritium (D–T) fusion reaction, 610–611
Diborides, 273–290
DiCarlo, James A., xvii, 217
Diffusion, 392, 494
Diffusion coefficient, 494
Diffusion-controlled regime, 382
Diffusion model, 490, 492
Diffusion–reaction competition, 383
Diffusion/reaction micromodels, 482
Diffusion/reaction model, 497, 498
Dimensional changes, for C/C composites, 400
Dimensional stability, 396
Directed metal oxidation (DIMOX), 248
Directed vapor deposition (DVD), 443
Discrete coatings, 242
Discretization, wireframe, 492–493
Disilicates, 441
Disk (disc) brakes, thermomechanical properties of, 651. *See also* Brake disks (discs)
Disk materials, 654
Dislocation–interstitial bias, 393
Distributed damage mechanism, 237
Divertor component, 612–613
Divertor structural component applications, 613
Dorn equation, 104
Double-dome forming, simulation of, 77, 78, 79
Double-layer coating, 460
Double-notch shear (DNS) test, 317–318, 319
Double plug healing mechanism, 501
Draping, in woven composite reinforcements, 76
DREAM (DRastically EAsy Maintenance) fusion reactor, 611, 614–616
Dry fiber preforms, 150, 151
Dual-mode ramjets (DMRs), 196
Ductile metals, 639
Ductile polymeric coating, 424
Ductility, 400, 641
Dynamic test methods, 33

EBC-coated CMC components, 442. *See also* Environmental barrier coatings (EBCs)
EBC degradation, 431, 447–448
EBC degradation categories, 444
EBC development, 448
EBC lifing, 448
EBC lifing method, 430–431, 442–448
EBC requirements, 432–437
EBC research and development, 448, 606–607
EBC spallation, 447
EBC test rigs, 444
EB-PVD coating, 443
Economy, in aircraft braking, 648
Effective reactivity, 380–383
Eigenvalue problems, 479–480
Elastic bridging, 282
Elastic characteristics, of SiC-based fibers, 7
Elastic damageable behavior, 521
Elastic energy, of undamaged materials, 471
Elastic energy release, 586–587
Elasticity properties, 256–257. *See also* Elastic properties
Elastic modulus ratio, 46
Elastic properties, crack deflection and, 45–49
Elastic strain, 475
Elastic unwrinkling model, 34
Electrical conductivity, of ultrahigh temperature ceramic-based composites, 280–282
Electrical properties, of carbon/carbon composites, 115–116
Electrical resistivity (ER), 130, 132, 280–282, 409
Electron beam physical vapor deposition (EB-PVD), 442–443
Electron diffraction, 28
Electron fluxes, 354
Electrophoretic deposition (EPD), 242, 247
Elevated temperatures, 100–101, 102–103
Elevator emergency brakes, 205, 206
Emergency brakes, elevator, 205, 206
Emergency brake systems, brake disks for, 666–667
Energetic radiation, 389
Energy-absorbing phenomena, 50
Energy balance in impact, 417–418
Energy function, 481
Energy potential, 469–470, 471, 472–475
Energy release acceleration, detection of, 587–588
Energy security, 401
Engine components, fabrication routes for, 219–220
Engineering materials, CMC integration to, 551–552
Engineering notation, 469
Engineering structures, lifetimes of, 467
Enhanced thermal cyclic life, 435
Environmental barrier coatings (EBCs), 126, 233. *See also* EBC entries
evolution of, 437–442
FOD response for, 422–424
processing, testing, and lifing, 442–448
for SiC/SiC composites, 430–451
validating, 443–444
Environmental coatings, 438–441, 442
Environmental concerns, technology improvements for, 430
Environmental effects. *See also* Radiation environments
of ablation of C/C materials, 353–388
in FOD, 421–422
on mechanical behavior, 296–330
on oxide–oxide composites, 295–333
Environment-degradation mechanisms, 334–337
Equilibrated 4D C/C composites, 355, 356. *See also* Carbon/carbon entries
Equivalent fiber length, 535

Erosion, mechanical, 360
Erosion resistance, improving in oxidation protective coatings, 460–461
Eulerian transpose, 468
Eutectics, 552, 553
Event-based AE monitoring. *See also* Acoustic emission entries
  of ceramic matrix composites, 575–580
  of CMC C-coupon testing, 580, 582
  of CMC-elevated temperature stress-rupture, 579
  of CMC stress–strain behavior, 576–579
Evolution laws, for matrix consumption, 494
Exclusion zones, 540
Exhaust sections/components, CMC research and development for, 603, 605–606
EXPERT experiment, 193, 194
Extensional waves, 573
Extreme temperatures, 353
Extreme value theory, 532, 534

FAA CLEEN Program, 604
Fabrication routes, for SiC/SiC engine components, 219–220
Failure, of fibers, 13–14
Failure behavior, of C/C composites, 97
Failure characteristics, 7, 8, 12
Failure criteria, 506–507
Failure modes, 18, 503, 620
Failure of interface, 44
Failure probability, 541
Failure probability equations, 536
Failure strains, of filaments, 10
Failure strengths, 6–8
Fast fracture data, 23
Fast-spectrum reactors, 621–622
Fatigue, 482, 486
Fatigue analysis, 496
Fatigue behavior, 108–109, 478
Fatigue cycle(s), 325, 326, 327, 488. *See also* Cyclic fatigue entries
Fatigue damage evolution law, 484–487
Fatigue damage law, 490
Fatigue fracture, 108
Fatigue inelastic strain evolution law, 487–488
Fatigue loading, 321–324, 470, 487, 489–490
Fatigue macromodeling, 497–498
Fatigue mechanism, 484
Fatigue modeling, 482–490
Fatigue performance, of Nextel 720/A system, 254–255
Fatigue resistance, 21, 254
Fatigue tests, lifetime prediction during, 586–588
FE approaches, 76–77
Felt susceptors, 139–140
Ferraris, Monica, xvii, 551
Fiber(s), 65, 240–241
  bonding to matrix, 41
  choices of, 624
  continuous refractory, 23
  creep characteristics of, 254
  critical, 11
  distance between neighboring, 344
  first-generation, 4
  fracture surface of, 19
  higher temperature capable, 263–264
  high modulus, 147, 150
  high tenacity, 149
  intermediate modulus, 149–150
  non-oxide, 4, 16
  oxide, 3, 4
  pitch-based reinforcing, 99–100
  reinforcing against fracture, 40
  SiC-based, 3
  stress on, 50
  strongly bonded, 347–348
  subcritical cracking of, 496–497
  third-generation, 4
  treated, 50, 52
  ultra-high modulus, 147, 150
  woven, 344
Fiber architectures, 225, 232, 240–241
Fiber bundle morphology, 377
Fiber bundles, 12, 16, 105, 107
Fiber bundle tests, 33
Fiber characteristics, 521
Fiber coater, 242, 243
Fiber/coating bonds, 55
Fiber coatings, 154
  advances in, 224–225
  CVI-pyC, 153
  oxide, 240
  porous, 56
Fiber decoupling, 339
Fiber degradation, 101, 253, 346–347
Fiber-direction (F1) microcracking mechanisms, 476–477
Fiber equilibrium local equation, 483
Fiber embedment, continuous, 175
Fiber failure(s), 13–14, 33, 237, 252, 304, 343, 347–348, 542
  oxygen and delayed, 19
Fiber failure stress, 6
Fiber flaw growth, stressed-oxidation degradation modeling in, 346–347
Fiber fracture(s), 11, 312
Fiber grain size, 221–222
Fiber interactions, 15, 16, 525
Fiber interfacial coating system, 232
Fiber irradiation (FIR), 348
Fiber length, equivalent, 535
Fiber lifetime studies, 496
Fiber load–carrying capacity, 542–543
Fiber–matrix bonding (FMB), 161, 166–167, 175, 624–625
Fiber/matrix interface(s), 97, 98–101, 109
  weak, 238–239
Fiber/matrix interfacial domain, 40, 51–56, 530
  role in CMCs, 41–49
Fiber–matrix interfacial sliding properties, radiation effects on, 398–399
Fiber–matrix reactions, 290
Fiber/matrix sliding, 49–50
Fiber modulus, 251
Fiber orientation, 185, 186
Fiber overloading, 13
Fiber preforms, 89–90, 93
Fiber pull-out, 261
Fiber/PyC interface, 47–48. *See also* Pyrocarbon (PyC) fiber coatings
Fiber-reinforced ceramic matrix composites (CMCs), behavior at interfaces and interphases of, 40–64
Fiber-reinforced composites, xv–xvi
  microstructure–behavior relationships for, 520–547
Fiber-reinforced materials, 181
Fiber-reinforced plastics (FRPs), 158, 192
Fiber-reinforced SiC-based composites, stress-environmental effects on, 334–352
Fiber reinforcement, constraining effect of, 241
Fiber relaxation (FR), 348
Fiber rotation, 73
Fiber seals, porous carbon, 638
Fiber strength, as a function of gauge length, 9
Fiber stress redistribution, 14
Fiber stress-rupture (SR), 348
Fiber structure, oxide matrices and, 243
Fiber surface flaw formation and/or growth, 336–337
Fiber-to-fiber bonding, 343
Fiber volatilization, 336, 337
Fiber volume fraction, 286–288
Fiber weakening, 347
Fickian regime, 494
Field-assisted sintering (SPS), 247–248

Filament failure strains, 10
Filaments, 21–22, 23, 242. *See also* Multifilament tows; SiC-based filaments/tows; Single filaments
Filament strengths, 14–15
Filament strengths histogram, 10
Filament–tow relations, 14–16
Filament winding, 633
Filament-winding process, 165–166
Film-boiling method, 94
Final failure, 506. *See also* Ultimate failure
First-generation environmental coatings, 438–441
First-generation fibers, 4
First wall component, 614–616
Fishnet algorithm, 76
Fission reactor applications, 616–623, 627, 641
Fission reactors, 609, 610, 616
Fixed-point method, 504–505
Flame holder components, 598–599
Flaw density functions, 528
Flaw populations, 539
Flaws, fracture-inducing, 541
Flaw severity, 528
Flaw strength distributions, 528, 534
Flaw strengths, 9, 10
Flexural strength(s), 287–288, 289. *See also* Flexure strengths
  temperature dependence of, 102–103
  of ultrahigh temperature ceramic-based composites, 284–285
Flexural test on bilayered specimens, 57
Flexural waves, 573
Flexure strengths, 252. *See also* Flexural strength(s)
Flow–material interaction, 355
Flux balance equations, 495, 496
FOD-and-CMAS/salt condition, 421–422. *See also* Foreign object damage (FOD)
FOD behavior, 421
FOD modeling, 419, 425–426
FOD responses, 422–424, 424–425, 426
FOD testing, 406, 408
Force–strain diagram, 33, 34
Force–strain relation, nonlinear, 11
Foreign object damage (FOD), 405, 528, 529
  in ceramic matrix composites, 405–429
  in CMCs, 409–422
  effects in, 420–422
  factors in CMCs, 425–426
Foreign object impact, 446–447
Formed ZrO_2 coating, 459
Forming process, 65
4D C/C testing, in SRM facility, 362–365
4D composites, 355, 356
Fracture(s)
  bundle-unit, 101
  determining tensile, 101–102
  premature, 107
  reinforcing fibers against, 40
  single-fiber, 101
Fracture behavior, of C/C composites, 97
Fracture-inducing flaws, 533, 541
Fracture mechanics model, 42
Fracture signals, 586
Fracture strain enhancement, 109
Fracture strains, 108
Fracture strength, of matrices, 527–528
Fracture stress, 101
Fracture surfaces, 32, 252, 301–304, 306, 307–308, 312, 313–315, 317, 319–321, 324, 328, 329, 343
  of fibers, 19
  planar, 308
  stress-rupture, 340
Fracture toughness, 6, 74, 287, 288
  enhanced, 50
  of ultrahigh temperature ceramic-based composites, 282–283
Fracture toughness plots, 283
Fracture toughness values, 55
Fracture zone, 510
Fragmentation modeling, 531–535, 544
Fragmentation process, 532
Fragmentation strength predictions, 534
Fragment dichotomy model, 532–533, 536
Fragment failure, 528
Fragment strength, 532–533
Free carbon, 19–21
Free energy, 469
Freeze forming/freeze drying/freeze gelation, 246–247
Frenkel defects, 390, 391
Friable graded insulation (FGI), 607
Frictional bridging mechanism, 282
Friction applications
  carbon/carbon composites for, 647–657
  carbon/ceramic materials for, 657–668
  CMCs for, 647–671
  C/SiC materials for, 669
Friction pads, 205, 207
Friction rings, 204
Friction system applications, 202–207
Frontal impact damage, 411–413
Fu, Qiangang, xvii, 452
Fuel-pin concept, 621
Fugitive coatings, 56, 260–261
Fukushima Daiichi accident, 622
Full CVI method, 227–228, 231. *See also* Chemical vapor infiltration (CVI)
Full frequency spectrum waveforms, digital capture of, 573
Full PIP approach, 227, 228, 229, 231. *See also* Polymer infiltration and pyrolysis (PIP)
Furnaces. *See also* High-temperature furnaces
  arc image, 378
  high-performance vacuum, 129–130
  inert gas, 131, 132
  oxidation, 362
Fusion applications, 610–616, 624–627
Fusion Power Reactor (FPR), 610–611
Fusion reactor concepts, 611
Fusion reactors, 609–610

Gas-cooled fast reactors (GFRs), 616, 617. *See also* GFR entries
  CMCs for structural materials in, 621–623
Gas mole flux, 371
Gas-phase-derived carbon matrices, 92–94
Gas transport, 371–372
Gas turbine engines, ceramic components for, 218
Gas Turbine-Modular Helium Reactor (GT-MTR), 618, 619
Gas turbines, 593–608
Gas velocity, 443
Gauge length, fiber strength as a function of, 9
Gaussian probability density function, 9
GE CMC shrouds, for gas turbines, 607
General Electric (GE) Aviation demonstration, 601
General Electric evaluations, 600
General Electric F110 engine, 595–596
General Electric next-generation engines, 602
General notations, for damage modeling, 468–469
Generation IV Nuclear Energy Systems Initiative, 616
Generation IV systems, 617
Genet, M., xvii, 467
GFR applications, 621–623. *See also* Gas-cooled fast reactors (GFRs)
GFR cladding, 638–639, 641
Glass-ceramic method, 557, 561
Glass coatings, 456
Glass/glass-ceramics, as CMC joining materials, 557, 561

Glass sealing layers, 124–126
Glassy phases, 439, 441
Glassy reaction zone, 435, 438
Global solar energy market, 137–140
Godin, Nathalie, xvii, 571
Gradient coatings, 459
Gradient C/SiC materials, 661
Grain/boundary sliding, 222
Grain size, 241, 282, 284
Graphene layers, 36
Graphene planes, 33–34. *See also* Graphitic planes
Graphene sheets, 28, 30
Graphite, 379–380, 394
Graphite radiation damage, 394
Graphitic microstructures, in carbon fibers, 399–400
Graphitic planes, orientation of, 48. *See also* Graphene planes
Graphitization, 96–97, 112, 189
Graphitized C fibers, 147
Green bodies, 150, 245, 246
Green–Naghdi's (GN) frame, 73
Guo, Shuqi, xvii, 273

Hafnium carbide (HfC), 460. *See also* Hf- entries
Hafnium diborides (HfB_2), 273–290
Halogenides, 122
Hamilton–Jacobi equation, 368–370
Hardening function, 478
Hatta, Hiroshi, xvii, 87
Healing mechanisms, double plug, 501
Healing plug, 482
Healing process modeling, 490–496
Healing scenario, 492–494
Heat flux, 611, 612, 613
Heating elements, 130, 132
Heat transfer, via radiation, 172
Heat treatment temperature (HTT), 27, 30, 31, 32, 36, 95–97, 99, 101
Heidenreich, Bernhard, xvii, 147
Helicopter demonstration, CMC mixer nozzle for, 598
Helium permeation testing, 639–640
Helium production, 630
Hemispherical forming, simulation of, 80
Henegar–Jones mechanism map, 348
Hermetic composites, 635, 636, 637
Hertz–Knudsen–Langmuir relationship, 370
Heterogeneous mass transfer constant, 368
Hexacelsian phase, 436
Hexagonal-tube fuel-assembly body, 620–621
HfB_2-based ceramic matrix composites, 286–288. *See also* Hafnium entries
HfB_2-based composites, 273–290
HfC-SiC multilayer coatings, 459
High ductility, 641
High energy laser (HEL) applications, 240–241
Higher temperature capable fibers, 263–264
Higher temperature SiC/SiC components, microstructural design guidelines and service issues for, 232–233
High flux plasma jet ablation, 378
Highly oriented polycrystalline graphite (HOPG), 394
High modulus (HM) C-fibers, 88
High modulus (HM) fibers, 147, 150
High performance brake systems, 666–667
High performance components, manufacture of, 209
High-performance vacuum furnaces, 129–130
High-rise lifts, brake pads for, 667
High strengths, 6
High temperature behavior, of Young's modulus, 35–36
High-temperature brazing alloys, 554
High-temperature electrolysis (HTE), 618
High-temperature furnaces, carbon/carbon composites for, 129–134
High temperature glue, brazing alloys as, 553
High-temperature materials, 87
High-temperature mechanical tests, 633–635
High-temperature metal treatment applications, 207–209
High temperature processing, 289
High temperature range, 336
High temperatures, 16–23, 221
High-temperature SiC/SiC components, 218–219
High-temperature structural applications, CMC components for, 580–581
High-temperature testing, acoustic emission monitoring and, 584–586
High Temperature Test Reactor (HTTR), 624
High-temperature thermal treatments, 223
High-temperature thermal cycling, 593
High-temperature treatments, 112
High tenacity (HT) fibers, 149
High tensile (HT) C-fibers, 88
High tensile Young's modulus, 33
Hi-Nicalon™ fiber bundles, rupture time of, 19
Hi-Nicalon™ fibers, 16, 17, 18
  free carbon in, 21
Hi-Nicalon™ Type S CMC, 223–224
Hi-Nicalon™ Type S fiber process, 4
Hi-Nicalon™ Type S (HNS, HNLS) fibers, 220–224, 395–397, 627, 628–629, 632, 634. *See also* HNS entries; Nicalon™ entries
  free carbon in, 21
HM 7 (Herakles) experiment, 193, 195
HNS (NOX grade) fiber, 221. *See also* Hi-Nicalon™ entries
HNS reinforcement, 641
Hold times, tension–tension fatigue with, 324–330
Homogeneous temperature field, lifetime predictions and, 507–508
Hooke's operator, 468, 469
Hot-grip testing, 507
Hot isostatic pressing, 247–248
Hot pressed BSAS, 446. *See also* BSAS (barium strontium aluminosilicate) coating
Hot (isostatic) pressed SiC [H(i)PSiC], 149
Hot pressed Y_2SiO_5, 446
Hot pressing (HP), 247–248, 274–275, 288–289
  reactive, 276
Hot section components, 430
HSCT-EPM program, 438
Hybrid CVI-PIP approach, 227, 228, 229, 231. *See also* Chemical vapor infiltration (CVI); Polymer infiltration and pyrolysis (PIP)
Hybrid structures, 551
Hyperelastic models, 82
Hypoelastic behavior, 77
Hypoelastic models, 73, 82
Hysteresis, unload/reload, 576, 577
Hysteresis loops, 57–59

iBN CMCs, 230. *See also* Boron entries; Ceramic matrix composites (CMCs)
iBN fibers, 223, 224, 226. *See also* Sylramic™-iBN SiC fiber
iBN SiC fiber, 224
Immobile defects, 393
Impact damage, 405
  in CMCs, 412–413
  of coatings, 424
  cross sections of, 413–416
Impact damage assessments, 408–409
Impact damage morphology, 410–417
Impact damage size, 419–420

Impact energy balance, 417–418
Impact force, prediction of, 417–420
Impact gun, 406
Impacting kinetic energy, 418
Impact response, of environmental barrier coatings, 422–424
Impact testing, results of, 422, 423
Impact velocity, 410, 411, 412, 413, 415, 419
  bouncing-back velocity and, 418
Imperfect load sharing, 14
Implicit functions, 368, 370–371
Indentation testing, 257–258
Industrial applications, 87–146, 209, 666–667
Industrial gas turbines, CMC program demonstrations on, 606–607
Industrial manufacture, of C/SiC materials, 668
Inelastic behaviors, 477–478
Inelastic strains, 470–471
Inert gas furnaces, 131, 132
Infiltration, of textile preform, 75. *See also* Melt infiltration (MI)
Infiltration depth, 154
Infiltration processes, 227–228. *See also* Capillary infiltration; Chemical vapor infiltration (CVI); Liquid silicon infiltration (LSI) process; Matrix infiltration entries; Nonreactive melt infiltration (NRMI); Polymer infiltration entries; Pressure infiltration; RMI fabrication route; Vacuum-assisted polymer (VAP) infiltration; Vacuum infiltration
In-plane deformation, *in situ* measurement of, 35
In-plane shear behavior, 68–69, 74
In-plane shear force, 80
In-plane tensile strength, 217, 218
In situ Raman scattering, 35
In situ reaction, 456
In situ XRD, 35, 36. *See also* X-ray diffraction (XRD)
Interface behavior, predicting, 47
Interface characteristics, 60
Interface coating processing, 242. *See also* Interfacial coating concepts
Interface coatings, 240, 257–261
Interface control, 238–240, 257
Interface cracks, 530
  mechanical behavior and, 530–531
Interface crack size, 50
Interface debond stresses, 48, 100
Interface engineering, 563
Interface failure(s), 44
Interface optimization, 530
Interface properties, 41, 56–60
Interfaces
  coating/CMC, 441
  crack deflection at, 41–44
  crack initiation at, 41
  fiber/matrix, 97, 98–101, 109
  fiber-reinforced ceramic matrix composites and, 40–64
  functions of, 60
  strengthened, 51–56
  weak, 56
Interface selection, 60
Interface strength, 43, 530–531. *See also* Interfacial shear strength; Interfacial tensile strength
  at elevated temperatures, 100–101
  at room temperature, 98–100
Interfacial bonding, with C/C and C/SiC composites, 454
Interfacial coating concepts, 261
Interfacial domain characteristics, criteria for, 60–61
Interfacial domains
  control of, 51–52
  fiber/matrix, 40, 51–56, 530
  guidelines to designing, 60
  role in CMCs, 41–49
Interfacial domain strength, features associated with, 53
Interfacial fiber coatings, advances in, 224–225
Interfacial fracture energy, 42
Interfacial shear strength, 56, 57–60, 624
Interfacial shear stresses, 52, 57, 59, 60, 490
Interfacial sliding, 398–399
Interfacial stress, 50. *See also* Interfacial shear stresses
Interfacial tensile strength, 56, 57, 60
Intergranular cracking, 282
Interior oxidation, 334–335, 336–337
Interlaminar failure(s), 317–321
Interlaminar shear, creep in, 317–321
Interlaminar shear strength (ILSS), 318–321, 654
Interlayers, 555
Interlocking, in CMC joints, 558–559
Intermediate modulus (IM) C-fibers, 88
Intermediate modulus (IM) fibers, 149–150
Intermediate temperature range, 335–336
Internal oxidation modes, 339
International Nuclear Energy Research Initiative (I-NERI), 620
Interphase carbon, 360
Interphase composites, choice of, 630–631
Interphase oxidation, 339
Interphase recession, stressed-oxidation degradation modeling in, 346–347
Interphase region, 338–340
Interphase removal (IR), 348
Interphases. *See also* Pyrocarbon (PyC) interphases
  anisotropic, 52
  carbon-based, 401–402
  defined, 40
  fiber-reinforced ceramic matrix composites and, 40–64
  functions of, 60
  multilayered, 47–48, 631
  in SiC/SiC composites, 398–399
  strengthened, 51–56
  weak, 56
Interphase thickness, 635–636
Interphase volatilization, 337
Intertow matrix, 522
Ion thrusters, 354
Iridium (Ir), 460
Irradiation, toughness and, 394
Irreversible material loss, 353
Isothermal conditions, 371–372
Isothermal CVI, 92, 93. *See also* Chemical vapor infiltration (CVI)
Isothermal/isobaric CVI (I-CVI), 154–156, 173, 174
Isotropic pitches, 91
ITAR (International Traffic in Arms Regulations) agreement, 593
ITER (International Thermonuclear Experimental Reactor), 560, 610–611, 624
ITER divertor, 612, 613
IXV experiment, 193, 194

Japan Atomic Energy Research Institute (JAERI), 624
Jefferson, George, xvii, 236
Jet vanes, 202
Joining, 552–553, 557. *See also* CMC joining; Micro-wave-assisted joining method
Joining materials, 552
Joining technology, 551–552
Joints, 557–558
Jones–Henegar mechanism map, 348

Kagawa, Yutaka, xvii, 273
Katoh, Yutai, xvii, 389
Keller, Kristin A., xvii, 236
Kerans, Ronald J., xvii, 236
Kinematic models, 76
Kinetic data, 368

Kinetic energy, of impact, 418
Kinetic energy transfer, 390
Knudsen regime, 494
Krenkel, Walter, xvii, 647

Lachaud, Jean, xvii, 353
Lacquer coating, 424
Ladevèze, Pierre, xvii, 467
Laminates, 163–164, 167–168
Lamination, cross-ply, 98, 99
Lamon, Jacques, xvi, xviii, 3, 40, 520
Lap tests, 562
Larson–Miller (LM) approach, 229, 230–231
Laser-assisted joining, 557
Laser communication terminals (LCTs), 199
Laser profilometry, 359
Layered coatings, 459
Lead-cooled fast reactors (LFRs), 616, 617
Lee, Kang N., xviii, 430
Legendre–Fenchel transform, 474
Length scales, 526–531
Life cycle cost benefits, 607
Lifetime
  maximum, 500
  temperature *vs.*, 501
  tow, 545
Lifetime control mechanisms, 542–544
Lifetime modeling, 467–519
Lifetime predictions, 497–503, 506–512, 544–545
  damage tolerance analysis and, 512
  during fatigue tests, 586–588
  model *vs.* experiment comparison, 509
Lifing, 444–448
  EBC, 448
Lifing method, 430–431, 442–448
Lifing method development, 448
Linear variable differential transformer (LVDT), 410
Link failure, 533–535
Liquid composite molding (LCM), 65, 75
Liquid phase formers, 278
Liquid phase impregnation (LPI), 625
Liquid phase reaction method, 454–455
Liquid phase sintered (LPS) SiC (LPSSiC), 149, 392
Liquid precursors, 245
Liquid silicon infiltration (LSI) process, 161, 209, 554, 555, 658–659. *See also* LSI entries
  advantages of, 173
Load-carrying ability, impacts on, 345
Load-carrying capacity, 522
Load-carrying carbon/carbon systems, 134
Load-deflection behavior, 237
Load displacement curves, 57, 59–60, 410–411
Load effects, on static fatigue, 16–23
Loading
  creep, 330
  quasi-static, 479
  static, 330
  unixial, 35
Loading conditions, 59
Loading frequency, 322, 323, 324
Loading rate, 297
Loads, mechanical, 647
Load sharing, 13, 50, 535, 536
  fiber interactions by, 15
  imperfect, 14
Load–strain curves, 12, 71
Load train, 6
Load transfers, 55
Local behavior calculation, 503–506
Local loop, 504–505
Local stress, 345
Locking angles, 69
Lock-in thermography, 562
Longitudinal tows, 525, 544
Long-term thermal exposure, 253
Low activation materials (LAMs), 611–612
Low-cost raw materials, 173
Lower nanometer range, of carbon fibers, 28–31
"Low-met" materials, 665
Low stress damage detection, 571
Low temperature range, 335
Low toughness, 6
LSI-based C/SiC materials, 170. *See also* Liquid silicon infiltration (LSI) process
LSI materials, 177, 178, 181, 184, 185, 190
LSI process time, 171
LWR (light water reactor) applications, 623, 627–628
LWR pin cladding, US designs for, 637–638

Macrocrack density indicator, 490. *See also* Macroscopic cracks
Macrocrack formation, 349
Macrocrack opening indicator, 490
Macro-lifetime models, 468
Macromodels, 490, 497–498. *See also* Macroscale models
Macro-multiphysic models, 468
Macropores, 355, 522, 529
Macroscale models, 368
Macro-scale prediction, 482–483
Macroscopic cracks, 513
Macroscopic damage evolution law, 487
Macroscopic effects, on materials, 392
Macroscopic flaws, 528–530
Macroscopic identification, 489–490
Macroscopic models, 481–482
Macroscopic roughness, 375–378
Macroscopic scale, 66
Macrotests, 500
Mass flux, 368
Mass loss, in ablation, 375–378
Mass transfer constants, heterogeneous, 368
Master curve, 46–47
Material elastic properties, crack deflection and, 45–49
Material–flow interaction, 355
Material functions, for quasi-static loading, 479
Material hardening function, 478
Material loss, irreversible, 353
Material optimization, 383
Materials, 392, 393. *See also* Carbon materials
Materials observation, 365–383
Materials tests, 356–359
Material surface morphology, 362–363, 364
Matrices. *See also* SiC matrices
  alumina-mullite, 249
  alumina-YAG, 249–250. 253
  aluminosilicate, 249
  bonding of fibers to, 41
  carbon, 625
  dense, 238
  fracture strength of, 527–528
  multilayered, 155, 584
  oxygen inhibited, 341
  porous, 237–238
  self-healing, 55–56, 467, 468, 491
  "self-sealing," 155
Matrix buildup, 156, 157
Matrix characteristics, 521
Matrix-coated tows, 525
Matrix consumption, evolution laws for, 494
Matrix crack deflection, 61
Matrix crack density, 576
Matrix cracking, 335, 337, 467, 522–523, 529–530, 531, 580, 581
  degree of, 345
  types of, 585–586
Matrix cracking saturation, 533
Matrix cracking strengths (MCSs), 218, 223, 225, 232
Matrix cracking stress, 348–349
Matrix crack regions, impacts on, 345

Matrix cracks, 41, 42, 53, 55, 59, 237, 341, 342, 343, 347–348, 493, 523
Matrix crack spacing, 538
Matrix damage, 521, 522, 523, 541–542
Matrix decomposition, 155, 156
Matrix densification, 308, 330
Matrix density, 253, 264
Matrix-dominated properties, improvements in, 264
Matrix–fiber reactions, 290
Matrix flaws, 537
Matrix fragmentation modeling, 544
Matrix fragmentation stochastic model, 531–535
Matrix infiltration (MI), 149, 243–247. *See also* Melt infiltration entries; MI materials
Matrix infiltration approaches, 228, 231. *See also* CVI-MI approach
Matrix interfacial crack patterns, 53
Matrix materials, 249, 250
Matrix microcracks, 55
Matrix porosity, 238, 250, 252
Matrix precursors, 90–94
  thermoplastics as, 91–92
  thermosetting resins as, 91
Matrix processes, 232
Matrix sintering, 238
Matrix strength, 538
Matrix systems, 90–94
Maximum application temperature, of carbon/carbon materials, 134
Maximum force, 14
Maximum lifetime, 500
Maximum stress, 325, 329
Mazerat, Stéphane, xviii, 3
Mechanical behavior
  of composites, 521–526
  environmental effects on, 296–330
  experimental analysis of, 67–72
  at high temperatures, 16–23
  interface cracks and, 530–531
  modeling, 73–74
Mechanical bonding, at an interface, 40
Mechanical damage model, 513
Mechanical erosion, 360
Mechanical interlocking, in CMC joints, 558–559
Mechanical joining, 552–553
Mechanical loads, 647
Mechanical macro-modeling, 468
Mechanical models, continuous, 77
Mechanical properties
  of C/C materials for brakes, 652
  of C/SiC and C/C-SiC materials, 178–187
  degradation of, 350
  of oxide–oxide composites, 251–253
  of ultrahigh temperature ceramic-based composites, 282–285
Mechanical tensile tests, 635
Mechanical tests, 67–72, 599
Mechanism analysis, for lifetime modeling, 498–503
Melt-infiltrated (MI) SiC/SiC CMC Sylramic™ fibers, 406, 407, 408, 410, 411, 412, 413, 415–416, 418
  FOD in, 420
Melt infiltration (MI), 248, 289, 395. *See also* Matrix infiltration entries; MI materials
Melt infiltration process, 160–174, 177
Melting ranges, for brazing alloys, 553
Mesophase pitch as precursor (MPP) approach, 27. *See also* MPP fibers
Mesophase pitches, 91–92
Mesoscopic scale, 66
Metal coatings, 456–457
Metal oxidation processing, 248
Metals, 134–137, 639
Metal treatment applications, high-temperature, 207–209
Methane formation, 134
Methanization rate, 138–139
Microbuckling, 107–108
Microbuckling stress, 106, 108
Microcomposite models, 482, 483–484
Microcomposites, 527
Microcrack density, 55
Microcrack formation, 349
Microcracking, 523–524
Microcracking mechanism modeling, 467, 471
Microcracking mechanisms, 475–476, 476–477, 479
Microcracks, 238, 479
Micromechanical models, 537–539
Micromodels, 482, 483–484. *See also* Microscale models
Micro-multiphysic models, 468
Microporous oxide/oxide CMCs, 250
Microscale models, 368
Microscopic scale, 66, 67
Microstructural design guidelines, for higher temperature SiC/SiC components, 232–233
Microstructure–behavior relationships, 520–547
Microstructures
  graphitic, 399–400
  SiC/SiC, 217, 229–231
  tailored, 160
  thermal spray coating, 443
Micro-wave-assisted joining method, 557
Migrating vacancies, 393
Military engines, 594–600
MI materials, 181, 183. *See also* Matrix infiltration entries; Melt infiltration entries; SiC fiber-reinforced MI matrix composites
Minicomposite behavior, 484, 486
Minicomposites, 50, 55, 527
  crack density of, 488
  stress–strain behavior in, 536–537
  ultimate failure in, 535
Misorientation angle, 31–32
Mixed matrix systems, 138
Mixer nozzle, for helicopter demonstration, 598
Mixer nozzle technology, 261–262
Mixtures law, 521
Modal acoustic emission (MAE), 573
Model development, 350
Models. *See also* Healing process modeling; Lifetime modeling; Models; Multiscale modeling
  of receding ablative surfaces, 368–372
  for stress-oxidation degradation, 345–349
Modeling strategies, multiscale, 369, 481–503
Models. *See also* Damage model- entries; Diffusion model
  behavior, 628
  CMAS composition, 445–447
  diffusion/reaction, 497
  for fatigue, 482–490
  FOD, 419, 425–426
  macro- and micro-multiphysic, 468
  macro-lifetime, 468
  macroscopic, 481–482
  microcomposite, 482, 483–484
  micromechanical, 537–539
  multiaxial elemental strength, 541
  physicochemistry in, 384
  plasticity, 478
  probabilistic, 546
  shear fracture, 106
  silica volatility, 432, 433, 444–445
  of stress state, 544
  of stress–strain behavior, 531–539
  of transverse intra-yarn crack opening, 482–490
  Weibull, 527, 528
Modified C/SiC materials, 660
Molten salt reactors (MSRs), 616, 617
Monazite coatings, 242, 258–260
Monazites, 56, 257
Monolithic carbons, thermophysical properties of, 110

Monolithic SiC, 659
Monosilicates, 441, 445
Monte Carlo random walk (MC/RW) algorithm, 372, 373
Monte Carlo simulation method, 534–535
Monte Carlo simulations, in fragmentation modeling, 531
Morscher, Gregory N., xviii, 334, 571
MPP fibers, 29, 30, 31, 32, 35. *See also* Mesophase pitch as precursor (MPP) approach
Mullite-alumina composite strength, 253. *See also* Alumina-mullite matrix
Mullite + BSAS mixture, 439. *See also* BSAS (barium strontium aluminosilicate) coating
Mullite coatings, 249, 430, 433, 434, 437–438
Multiaxial elemental strength model, 541
Multidisk brake configuration, 649
Multifilament tows, 11–16
Multilayer-coated carbon/carbon, 125
Multilayer coatings, 122–124, 437, 459, 460
Multilayered (PyC/SiC) fiber coatings, 41
Multilayered interphases, 47–48, 631
Multilayered matrices, 155, 584
Multilayer strategies, for C/SiC composites, 339–341
Multiphase inlay ceramic coatings, 457–458
Multiple cracking, 536, 537, 539–541, 545
Multiscale materials, 65–67
Multiscale modeling, of oxidation/damage coupling and self-healing effects, 481–503
Multiscale modeling strategies, 369

N720/alumina, 410, 411
N720/alumina composites, 295, 296–304, 308–313. *See also* Nextel™ entries
N720/alumina oxide/oxide, 422
N720/aluminosilicate, 413–415
Nanoparticle-toughened coatings, 458–459
Nanopowder infiltration and transient eutectic phase (NITE) method, 555–556, 560–561, 628, 641
  associated composites and, 635–636
Nanopowder infiltration and transient eutectic-phase (NITE) SiC/SiC screws, 552
Nanopowders, 635
Nanowire-toughened coatings, 458
NASA (National Aeronautics and Space Administration), 224, 225, 227, 228, 233, 261–262
NASA advancements, in SiC/SiC materials/processes, 220–231
NASA UEET Program, 601, 603
"Needle forest" situation, 374
Neighboring fibers, distance between, 344
Neutron fluxes, 354
Neutron irradiation, 400
Newton–Raphson method, 505
Newton–Raphson solver, 505
Nextel™ 720/alumina (N720/A) composite, environmental effects on mechanical behavior of, 296–330. *See also* N720/alumin- entries
Nextel™ 720/alumina–mullite (N720/AM) composite, environmental effects on mechanical behavior of, 296–330
Nextel™ 720/aluminosilicate composite, 295
Nextel™ 720/A system, creep/oxidation/fatigue performance of, 254–255
Nextel™ 720-based composites, 254
Nextel™ 720 oxide fibers, 406–407, 410
Nextel™-based composites, 259–260
Nextel™ fibers, 240, 241
Nextel™ oxide fibers, 4
Next-generation environmental coatings, 442
Nicalon™ fiber bundles, 12
Nicalon™-fiber-reinforced CVI SiC composites, 346
Nicalon™ fibers, 8, 20, 21. *See also* Hi-Nicalon™ entries
Nicalon™ filaments, flaw strengths for, 10
Nicalon™ SiC-based single filaments, strains-to-failure for, 15
Nicalon™ Type S fibers, 406, 407. *See also* Hi-Nicalon™ Type S (HNS, HNLS) fibers
Nondestructive tests, 562
Nonlinear force–strain relation, 11
Nonlinear stress–strain curves, of carbon fibers, 35
Non-oxide fibers, 4, 16
Non-pure SiC, radiation effects on, 392
Nonreactive melt infiltration (NRMI), 219. *See also* NRMI entries; RMI fabrication route
Norgett–Robinson–Torrens (NRT) model, 390, 391
Normal distribution, 9–10
Normalized strength, 410
Notations, for damage modeling, 468–469
Notch sensitivity, 255
NRMI CMCs, 229, 230, 231. *See also* Ceramic matrix composites (CMCs); Nonreactive melt infiltration (NRMI)
NRMI fabrication route, 223, 225, 226
Nuclear applications
  CMC joining for, 560–561
  for CMCs, 609–646
  hermetic composite design for, 637
  processing of C/C composites for, 624–627
  processing of SiC/SiC composites for, 627–641
Nuclear industry, 609, 610, 622
Nuclear Reaction Analysis (NRA), 630
Nuclear reactors, SiC/SiC components in, 610
Nuclear transmutation, 390
Numerical approaches, 372
Numerical results, of roughness features, 378–380

Objective derivatives, 73
Octets, 355
Off-axis compressive tests method, 104–105
Off-axis properties, of oxide–oxide composites, 256–257
One-dimensional (1D) composites, 521, 535, 536–537
One-step-cycle applications, 92
Open-hole plate, 513, 515
Open structures, impact damage morphology and, 411
Optical microscope (OM), 359
Optical systems, 199
Organic matrix composites, 75
Original equipment manufacturers (OEMs), 430, 448
Orthogonal microcracking (MO) mechanisms, 475–476, 479
Orthonormal frames, 73
Outer glass sealing layers, 124–126
Outer multilayer coatings, 122–124
Overablation, 354, 367
Overloading, of fibers, 13
Oxidation, 336–337
  of BN, 341, 342
  of carbon, 652
  of C/SiC and C/C-SiC materials, 190–191
  directed metal, 248
  at high temperatures, 16
  interior, 334–335, 336–337

methods to prevent, 337–338
protection against, 457–458
of protection tiles, 379
of SiC matrices, 336
in TBC failure, 444
Oxidation/damage coupling, multiscale modeling of, 481–503
Oxidation embrittlement (OE), 336–337, 348
Oxidation furnace, 2D C/R testing in, 362
Oxidation mechanisms, for SiC composites, 335
Oxidation performance, of Nextel 720/A system, 254–255
Oxidation protection, of carbon/carbon materials, 118–126
Oxidation protective coatings, 452–463
Oxidation rates, 121, 123, 190
Oxidation reactions, 345, 494
Oxidation-resistant coating systems, 456–459
Oxidation-resistant temperature, 460
Oxidation stability, 147
Oxidation tests, 337, 366, 367
of 2D C/R, 362
of 3D C/C, 359–361
Oxide, total volume of, 492, 493
Oxide-based fibrous composites, processing of, 241–248
Oxide-based porous coatings, 56
Oxide coatings, 56
Oxide composite stability, 236
Oxide fiber coating, 240
Oxide fibers, 3, 4, 246
Oxide layers, 20
Oxide matrices, fiber structure and, 243
Oxide–oxide CMCs, 236, 237–238, 248–250, 261–263, 295–296, 324, 330. *See also* Ceramic matrix composites (CMCs)
creep and recovery behavior of, 304–308
environmental effects on mechanical behavior of, 296–330
interlaminar shear and, 317–321
Oxide–oxide composites, 236–272
applications for, 261–263
consolidation methods for, 247–248
durability of, 254–255
environmental effects on, 295–333
future for, 263–264
matrix densification and, 308
mechanical properties of, 251–253
off-axis properties of, 256–257
physical characteristics of, 250–251
porous matrix, 262, 263
reinforcement of, 240–241
technology development for, 261–263
tough, 264
with weak oxide–oxide phase boundaries, 257–261
Oxide-oxide materials, 262
Oxide–oxide phase boundaries, weak, 257–260
Oxide-oxide TPS systems, 263. *See also* Thermal protection systems (TPSs)
Oxide plug, role of, 499
Oxide scale growth, 346
Oxide volatilization flux, 495
Oxygen, delayed fiber failure and, 19
Oxygen barrier layer, 459
Oxygen concentration, 490, 496–497
Oxygen concentration evolution, 500
Oxygen concentration indicator, 482
Oxygen diffusion, 453, 494
Oxygen diffusivity, 124, 126
Oxygen-free SiC fiber, 4
Oxygen inhibited matrix, 341

Pack cementation and capillary infiltration (PCI), 169
Pack cementation method, 454, 458
PAN-based carbon fibers, 30, 31, 88
PAN fibers, 29–30, 32, 35. *See also* Polyacrylonitrile (PAN)
Parallelepipedic macropores, 355
Parallel yarn ablation, 360
Paris law, 496
Passenger cars, C/SiC brake disks for, 664–666
Passive oxidation, 190
Pattern clustering, steps in, 583–584
Pattern recognition techniques, acoustic emission signal analysis using, 580–584
Peak stress, 45, 255
Peak-to-valley roughness, 374–375
Pellet/cladding interaction, 638
Peterlik, Herwig, xviii, 27
Phase boundaries, weak oxide–oxide, 257–260
Phase stability, of environmental barrier coatings, 436–437
Phenolic resin, 355, 356
Phenolics, 91
Phosphate glass coating, 456
Physicochemistry, in models, 384
Picture frame test, 68–69
Pin cladding, 621–623, 637–638
PIP-derived C/SiC materials, 158. *See also* Polymer infiltration and pyrolysis (PIP)
PIP materials, 178, 181
PIP process, 158–160
advantages and disadvantages of, 160
Pitch-based reinforcing fibers, 99–100
Pitches, 91–92
Pitch impregnation, 625
Planar fracture(s), 304, 327
Planar fracture surface, 308
Planar interface construction (PLIC), 372, 373, 374
Plasma arc jet, 378
Plasma-facing components (PFCs), 609, 611–616
Plasma jet ablation, high flux, 378
Plasma jet tests, 359–361, 362
Plasma-sprayed BSAS, 446. *See also* BSAS (barium strontium aluminosilicate) coating
Plasma spraying, 437, 442–443, 446, 447, 455, 459
Plasticity, coupling damage with, 470–471
Plasticity model, 478
Plug truncation, 493
Plug volume, evolution of, 495
Point defects, 393, 394
Point stress failure criterion, 255
Poisson's ratio, 278, 289
Poisson–Weibull model, in fragmentation modeling, 531
Polyacrylonitrile (PAN), 27. *See also* PAN entries
Polyacrylonitrile (PAN)-based carbon fibers, 399
Polycarbosilanes, 157
Polycarbosiloxane precursors, 157–158
Polycrystalline graphite, 379–380
Polycrystalline YAG fibers, 264
Polymeric coatings, ductile, 424
Polymer infiltration, 158, 159
Polymer infiltration and pyrolysis (PIP), 149, 157–160, 174, 176–177, 227, 289, 395. *See also* PIP entries
Polymer precursors, 91, 94
Polymer reinforcement, 87
Polysilicon, 137
Pore-filling, 95, 96
Pore structures, 154
Porous alumina surface, 438
Porous carbon fiber seal, 638
Porous C/C components, 554
Porous coatings, 260
Porous fiber coatings, 56
Porous matrices, 237–238
Porous matrix composites, 238
Porous matrix composite systems, 248–250
Porous matrix oxide–oxide CMCs, 261, 317–321. *See also* Ceramic matrix composites (CMCs)

Porous matrix oxide–oxide composites, 262, 263
Porous oxide–oxide CMCs, 248–250
Porsche Ceramic Composite Brake (PCCB), 203, 204
Postimpact residual strength testing, 409
Postimpact strength, 420–421, 409–410, 424
Powder immersion reaction assisted coating (PIRAC), 121–122
Power-law crack-velocity formulation, 301
Power law form, of time to failure, 22
Power laws, 587, 588
Pratt & Whitney F100 engine, 596, 597
Pratt & Whitney PYBBN vectoring nozzle, 597
Preceramic-polymer joints, 558
Prediction capabilities, 503–515
Preforming simulation, 76–81
Preforms, dry fiber, 150, 151. *See also* Carbon-fiber-reinforced plastic (CFRP) preforms; C/C preforms
Preloading effects, in FOD, 421
Premature fracture, 107
Prepreg (pre-impregnated) C-fibers, 88, 92
Prepreg layers, 172
Prepreg MI SiC/SiC, 410, 411, 413, 415–416, 418. *See also* Matrix infiltration entries; Melt infiltration entries
Prepreg processing, 243–244
Prepregs, 158–159
Pressure gradient CVI, 92, 93–94. *See also* Chemical vapor infiltration (CVI)
Pressure infiltration, 244–245
Pressureless sintering (PS), 247, 273–274, 277–278, 288–289
Pressurized melt infiltration (PMI), 169–170
Primary knock-on atom (PKA), 390, 391
Probabilistic models, 546
Probability-based general model, of composite behavior, 536
Processing methods, for oxide-based fibrous composites, 241–248
Production routes, for carbon fibers, 27
Projectile materials, 406–407
Projectiles, 408, 410–411
Propagation stages, 496
Propeller brakes, 204–205
Protection tiles, oxidation of, 379
PTAH-SOCAR technology, 196, 200
Push-in/push-out tests, 57, 59
Push-out curve, 57, 58
Push-out tests, 98–99, 100, 106, 257–258
PWR (pressurized water reactor) applications, 622–623
PyC interlayers, 396, 398
Pyrocarbon-based tailored interface, 61. *See also* Pyrolytic carbon (PyC)
Pyrocarbon (PyC) fiber coatings, 41. *See also* Fiber/PyC interface
Pyrocarbon (PyC) interphases, 47, 50, 52, 55, 631, 635, 641
Pyrolized C/phenolic resin composites, 355, 356
Pyrolysis, 159–160, 166–168. *See also* PIP entries; Polymer infiltration and pyrolysis (PIP)
Pyrolysis behavior, 91
Pyrolysis process times, 171–172
Pyrolytic carbon (PyC), 93, 398. *See also* Pyrocarbon entries
Pyrolytic graphite, 394

Quasi-Newton method, 481
Quasi-secant operator, 505
Quasi-static loading, identified material functions for, 479

Racing car brake disks, C/C composites for, 655–656
Radial gripping, at an interface, 40
Radiation, energetic, 389
Radiation-cooled structures, 196
Radiation damage, graphite, 394
Radiation damage evolution, in SiC, 392–393
Radiation damage theory, 389–392
Radiation defect production, 390–391
Radiation degradation, 401–402
Radiation effects, 389–404
  in carbon materials, 394
  on ceramics, 392–394
  on fiber–matrix interfacial sliding properties, 398–399
  in SiC, 393–394
Radiation enhanced diffusion, 392
Radiation environments, extended service life in, 401
Radiation heat transfer, 172
Radiation-resistant CVD SiC, 394. *See also* Chemical vapor deposition (CVD)
Radiation-resistant material, 393
Raman measurements, 30
Raman scattering, *in situ*, 35
Rapid CVI methods, 92, 94
Rare earth silicate coating volatility, 433, 434
Rare earth (RE) silicates, 441, 445
Rate constitutive models, 73
Raw materials, low-cost, 173
Reaction bonded SiC (RBSiC), 149
Reaction bonding, 248
Reaction constant, 494
Reaction-controlled regime, 382
Reaction–diffusion competition, 383
Reaction/diffusion micromodels, 482
Reaction/diffusion model, 497, 498
Reaction rates, 367
Reactive agents, solid, 278
Reactive hot pressing (RHP), 276
Reactive phase, 366
Reactive sintering (RS), 273–274, 276–277, 288–289
Reactive spark plasma sintering (RSPS), 276
Reactivity, 380–383, 384, 554
Receding ablative surfaces, modeling, 368–372
Recession, 444
Recession rates, 366, 432
Recrystallized SiC (RSiC), 149
Redensification/recarbonization cycles, 95
Refractory fibers, continuous, 23
Refractory metals, 639
Reinforcement, 3–26
Reinforcing architecture, creating, 624
Relaxation domains, 539–540
Relaxation methods, 505
Representative unit cell (RUC), 66, 67
Research and development (R&D), 607
  in aero-engine field, 593
  on CMCs for commercial engines, 600–607
  EBC, 448, 606–607
  in gas-turbine field, 593
  on oxidation protective coatings, 460–461
Research reactors, 610
Residual crack opening, 489
Residual strains, 477–478, 484
Residual strength, 108–109, 409–410, 421–422
Resin transfer molding (RTM), 158. *See also* RTM process
Response surfaces, 502
Reynolds number (*Re*), 381, 383
Rig testing, 599, 600, 605
Ring cracking, 413
RMI fabrication route, 225. *See also* Nonreactive melt infiltration (NRMI)
R'Mili, Mohamed, xviii, 3
Robust cladding closure process, 640
Rocket jet vanes, 202
Rocket motor nozzles/divergents, 354

Rocket nozzle tests, 365–366
Room temperature, interfacial strength at, 98–100
Room temperature fatigue resistance, 254
Rough-interface formalism, 57
Roughness features, 359–365, 378–380
RTM process, 162–163. *See also* Resin transfer molding (RTM)
Ruggles-Wrenn, Marina B., xviii, 295
Runge–Kutta 45 algorithm, 496
Rupture criteria, 478
Rupture tests, 349
Rupture time, of Hi-Nicalon^{TM} fiber bundles, 19
Rupture time predictions, 542–545

SA3 fiber surface, 50. *See also* Tyranno^{TM}-SA3 SiC fiber
SA3 reinforcement, 641
Safety, in aircraft brake design, 648
Safran demonstrations, 601–602, 605–606
Safran/Herakles developments, 594, 599
Salt impregnation, 119–121
Sandwich design, 638–641
Sapphire fibers, 240, 241
SAS (strontium aluminum silicate) coating, 436
Satellite structures, 199–200
Satellite thrusters, 195–196
Sauder, Cédric, xviii, 609
Scale effects, 545
Scale formation, 337
Scale parameter, 36
Scales, in woven fabrics, 65–66
Scanning electron microscope (SEM), 20, 32, 33, 359, 360, 361
Scatter, 526, 527, 545
Scattering angle, 28
Scherrer's formula, 28
Screw/nut system, 552
Seal coat, 493, 502, 503
Seal coat healing, 500–501
Sealing layers, glass, 124–126
Seal layer, 460
Secondary creep rates, 300, 309–311, 319
Second-generation environmental coatings, 441
Self-healing, 460, 490, 493
Self-healing capacity, 123
Self-healing effects, multiscale modeling of, 481–503
Self-healing layer, 123
Self-healing matrix, 55–56, 467, 468, 491
Self-interstitial atom (SIA), 390, 391. *See also* SIA clusters
"Self-sealing" matrices, 155, 191
Semidiscrete approach, composite reinforcement forming using, 77
Service issues, for higher temperature SiC/SiC components, 232–233
Service process, volatility during, 453
SFR applications, 620–621. *See also* Sodium-cooled fast reactors (SFRs)
Sharp Edge Flight Experiment (SHEFEX), 193, 195
Shear angles, 77
Shear behavior, 68–69, 74, 104
Shear curves, 69, 70
Shear failure, 317–321
Shear forces, in-plane, 80
Shear fracture model, 106
Shear modulus, 278–279, 289
Shear strength, 104–106
  interfacial, 56, 57–60, 624
  interlaminar, 654
Shear strength tests, on joined CMC, 561–562
Shear stress(es), 318, 483
  crack initiation and, 41
  interface, 52, 59, 60, 490
  mean, 485
Shear stress–strain curves, 105
Shell finite elements, 80
Sherwood number (*Sh*), 374–376, 378, 381, 382
Short fiber-based C/SiC materials, 181–182, 188
Short-fiber reinforcements, 662
Short fibers, 285–288
Shrouds, for gas turbines, 607
SIA clusters, 392–393. *See also* Self-interstitial atom (SIA)
Si-based ceramics, coating work for, 430
SiC (silicon carbide), 657–658
  electrical resistivity and, 282
  monolithic, 659
  radiation damage evolution in, 392–393
  radiation effects on, 392, 393–394
  radiation tolerance of, 402
SICARBON®, 197, 198, 199
SiC-based CMCs, 334. *See also* Ceramic matrix cmposites (CMCs)
SiC-based composites, fiber-reinforced, 334–352
SiC-based fiber(s), 3. *See also* SiC fiber(s)
  advancements in, 220–224
  elastic and failure characteristics of, 7
  failure characteristics measured on tows for, 12
  fatigue resistance of, 21
  oxygen-free, 4
  physical properties and composition of, 5
SiC-based fiber bundles, delayed failure of, 19
SiC-based filaments/tows, 3–26
SiC-based matrices, advances in, 227–229
SiC-based single filaments, 15, 17
SiC/BN/SiC composites, stressed-oxidation degradation modeling in, 347–348
SiC CMC metal liner hybrid design, 638. *See also* Ceramic matrix cmposites (CMCs)
SiC component manufacturing, 149
SiC composite oxidation mechanisms, 335
SiC composites, 75, 337–338
SiC/C/SiC composites, 346–347, 389, 394–398
SiC deposition, 154
SiC fiber(s), 74–75, 219–220, 641. *See also* SiC-based fiber entries
  key properties of, 221
  selecting, 628–630
  in SiC/SiC components, 220
  temperature capability of, 222–223
SiC fiber architectures, advances in, 225–227
SiC fiber composites, stressed-oxidation degradation modeling in, 347
SiC fiber-reinforced MI matrix composites, 578–579
SiC fiber-reinforced SiC matrix composites, 338–341, 341–344. *See also* SiC/SiC entries
SiC fiber reinforcements, 81–82
$SiC_f/PyC/SiC_m$ composites, 47
SiC_f/[Si-B-C] composite, 585
SiC grain size, 282, 284
SiC materials, properties of, 181
SiC matrices, 149, 154, 157, 336
SiC matrix production, 657–658
SiC nanowire network, 458
SiCN matrices, 157
SiC oxidation, 341
SiC particles, 275, 280, 282
"SiCralee" coatings, 170
SiC/SiC CMC material systems, 431. *See also* Ceramic matrix composites (CMCs)
SiC/SiC CMCs, 217, 218, 226, 227, 229
SiC/SiC CMC Sylramic^{TM} fibers, 406, 407, 421–422
SiC/SiC components, 225
  high-temperature, 218–219
  in nuclear reactors, 610

SiC/SiC components (*Continued*)
  service issues for, 232–233
  SiC fiber in, 220
SiC/SiC composite properties, densification processes for, 631–636
SiC/SiC composites, 41, 42, 56, 60–61, 74, 530. *See also* CVI SiC/SiC composites
  for aero-propulsion, 217–235
  damaged, 638
  densification of, 631
  environmental barrier coatings for, 430–451
  features observed on, 53
  improved, 222
  interphase in, 398–399
  joining, 552
  joints for, 557–558
  materials and processes for, 233
  for nuclear applications, 560–561, 642
  processing for nuclear applications, 627–641
SiC/SiC engine components, 217, 219–220
SiC/SiC fiber matrix constituents, requirements of, 218
SiC/SiC materials/processes, NASA advancements in, 220–231
SiC/SiC microstructures, 217
SiC/SiC microstructural design methods, advances in, 229–231
SiC/SiC minicomposites, 55
SiC/SiC screws, 552
SiC volatility, 433, 434
SiC volatility data, 432
SiC-whisker-toughened SiC, 458
σ_0 estimates, 7
Silica activity, 438. *See also* SiO_2 entries
Silica-based composites, 263
Silica-free matrix materials, 249
Silica-free porous composites, 250
Silica glass sealings, 126
Silica volatility, 431–432
Silica volatility model, 432, 433, 444–445
Silicides, 457
Silicon, properties of, 181
Silicon carbide–based ceramics, 555. *See also* C/SiC composites; SiC entries
Silicon carbide fiber-reinforced ZrB_2-based ceramic matrix composites, 288
Silicon-doped boron nitride (BN), 219
Siliconization, 161, 168–173, 658
Silicon nitrides, 424–425
Silicon treatments, 121, 128, 134
Simplified marching-cube (SMC) discretization, 372, 373, 374
Simulations
  of double-dome forming, 77, 78, 79
  of hemispherical forming, 80
  of preforming, 76–81
Simulation tools, 209
Si/mullite + BSAS/BSAS EBC, 440–441, 444. *See also* BSAS (barium strontium aluminosilicate) coating
Single-fiber fracture, 101
Single fiber tests, 33
Single-filament properties, determination of, 14–16
Single filaments, 6–10, 15
Sintered SiC (SSiC), 149
Sintering, 133, 249. *See also* Spark plasma sintering (SPS) method
  pressureless, 247, 273–274, 277–278, 288–289
  reactive, 273–274, 276–277, 288–289
  spark plasma, 273–274, 275–276, 279, 288
Sintering approach, 222–223. *See also* Matrix sintering
SiO_2 scale formation, 337. *See also* Silica entries
SiO_2 volatilization, 432
SiSiC matrix, 169
SITE process, 636
Skin–core structure, 32
Sliding elements, 128–129
Sliding surface topography, 50
Slip length, 485
Slow crack growth (SCG), 496, 544, 545
  in filaments under constant stress, 21–22
  SEM micrographs of, 20
Slow crack growth mechanism, 19–21
Slow crack growth parameters, 298, 301, 329
Slurries, 243, 244–245, 246. *See also* Solution/sol/slurry coating techniques
Slurry method, 455
Small-angle X-ray scattering (SAXS), 31
S–N curves, 108–109
Snecma jet engine applications, 594–595
Snecma M53 engine, 594
Snecma M88 engine, 594–595
Sodium-cooled fast reactors (SFRs), 616, 617
  CMCs for structural materials in, 618, 620–621
Soft structures, impact damage morphology and, 411
Solar energy market, C/C materials in, 137–140
Solar Turbines field evaluations, 606–607
Sol–gel method, 455
Solid/liquid rocket motor nozzles/divergents, 354
Solid oxide reaction products, 337
Solid reactive agents, 278
Solid state displacement reactions, 557–558
Solution/sol/slurry coating techniques, 242
Solver acceleration, 505–506
Sound, speed of, 577
Sound travel, in CMCs, 573–574
Sound waves, synchronized, 575
Space applications, for C/SiC and C/C-SiC materials, 192–200
Space propulsion, 193–199
Spallation, 422–424, 444, 447
Spark plasma sintering (SPS), 273–274, 275–276, 279, 288, 556
  reactive, 276
Specific heat, of carbon/carbon composites, 117
Speed of sound, 577
Spriet, Patrick, xviii, 593
SRM facilities, 362–365
SRM ground experiments, 359
Stability, with C/C and C/SiC composites, 453–454
Stable failure(s), 13
Standardization research, 562–563
Standard normalized tests, 628
State, laws of, 469
Static failure, 503
Static failure predictions, 513–515
Static fatigue, 16–23
Static fatigue constants, 18, 19
Static fatigue failure, 503
Static fatigue lifetime test, 498
Static fatigue loading, 507
Static fatigue tests, 498, 586–588
Static loading, 330, 486–487, 488
Static test methods, 33
Statistical strength distributions, 6
Steady morphologies, 374–375
Steady-state creep, 103–104
Steady-state morphology, 376, 378
Steam cycling, 443
Steam effects, 254, 299–304, 311, 312, 314, 319, 325
Steam environment, 296, 307
Stiffness, 32–36, 70–71
Stochastic models, of matrix fragmentation, 531–535
Stochastic structural features, 543–544
Strain, residual, 484
Strain data, 6

Strain increment, 74
Strain measurement, 16
Strain rates, 326
Strain recovery, 306
Strains, residual, 477–478
Strain–stress curves, 14
Strain–time curves, 304
Strain(s)-to-failure, 6, 9, 50, 525
Strength. *See also* Compressive strength; Flexural strength; Interfacial strength; Shear strength; Tensile (tension) strength
  of carbon fibers, 32–36
  of C/C composites, 97–109
  characteristic, 538–539
  of filaments, 14–15
  interface, 530–531
  normalized, 410
  postimpact, 409–410, 420–421, 424
  residual, 108–109, 409–410
  of tows, 14
Strength data, 6
Strength degradation, 410, 420, 421, 425, 512
  in cantilever support, 422
  at high temperatures, 16
Strength degradation mechanism, 109
Strength degradation patterns, 510
Strength distributions, 6, 534
Strength equation, 535
Strength tests, on joined CMC, 561–562
Stress(es)
  on fiber, 50
  filaments under constant, 21–22
  fracture, 101
  local, 345
  microbuckling, 106, 108
Stress-based approaches, in fragmentation modeling, 531
Stress concentration, 347
Stress concentration factor, 512
Stress condition tests, 349
Stress criteria, 477
Stress distribution, 486
Stressed-oxidation degradation, 337–338, 339–341, 343–344
  modeling and design for, 345–349
Stressed-oxidation degradation modeling, 345–349
  in C/SiC composites, 345–346
  in SiC/BN/SiC composites, 347–348
  in SiC/C/SiC composites, 346–347
  in SiC fiber composites, 347
Stressed-oxidation mechanisms, 337, 350
Stress-environmental effects, on fiber-reinforced SiC-based composites, 334–352
Stress rate, 297
Stress redistribution, 14
Stress-rupture (SR), of fibers, 348
Stress–rupture behavior, 300, 311
Stress-rupture experiments, 346–347
Stress-rupture fracture surfaces, 340
Stress-rupture results, 344
Stress–rupture tests, 229, 337, 579, 581
Stress–rupture time behavior, 16, 18
Stress–rupture time data, 545
Stress–rupture time diagrams, 16–19
Stress–rupture time relationship, 22
Stress sensors, universal, 35
Stress state(s), 43, 44, 49, 51
Stress state modeling, 544
Stress–strain behavior, 309
  event-based AE of, 576–579
  modeling of, 531–539
  in 1D composites and minicomposites, 536–537
  tensile, 521–524, 536
Stress–strain curves, 14, 33, 104–106, 296, 297, 521–522, 523
  nonlinear, 35
  tensile, 54, 55, 59
Stress/strain law, 483–484
Stress–strain plot, 251, 252
Stress–strain response, 480, 520
Strongly bonded fibers, 347–348
Structural components, modeling and design of, 264
Structural features, stochastic, 543–544
Structural materials, CMCs for, 621–623
Structural performance, at high temperatures, 221
Structural phenomena, in composite specimens, 543
Subcomponents, 170, 588
Subcritical crack growth, 16–23, 297–298, 301
Subcritical cracking, of fibers, 496–497
Subelement feasibility tests, 599, 600
Supercritical fluid technique, 455
Supercritical-water reactors (SCWRs), 616, 617
Supersonic aerodynamic performance, improving, 192–193
Super Sylramic™-iBN fiber, 223
Supervised classification technique, 584, 586
Supervised pattern recognition method, applied to AE signals, 584
Surface gas transfer, 371–372
Surface impregnation, oxidation behavior and, 190
Surface morphology, 362–363, 364
Surface recession, 334–335, 336
Surface recession equation, 368
Surface recession velocity, 370
Surface roughening, 354
Surface roughness, 50
Surface roughness scales, 359
Sustained creep, 307
Sylramic™ composites, 344
Sylramic™ fiber(s), 4, 6, 223, 406, 407, 421–422
Sylramic™-iBN fiber, 223, 225
Sylramic™-iBN SiC fiber, 56. *See also* Super Sylramic™-iBN fiber
Symmetric operators, computing square roots of, 480
Synchronized sound waves, 575

Tactical air cycles (TACs), 596
Tailored interfaces, pyrocarbon-based, 61
Tailored microstructures, 160
TAURO fusion reactor, 611, 614–616
TBC failure, 444. *See also* Thermal barrier coatings (TBCs)
TBC spallation, 444
Technology development, for oxide–oxide composites, 261–263
Technology improvements, for environmental concerns, 430
Telescope tubes, 199
Temperature(s), 353, 501, 664, 665. *See also* Coating temperature; Elevated temperatures; Maximum application temperature; Oxidation-resistant temperature
Temperature effects, 16–23, 420
Temperature fields, homogeneous, 507–508
Temperature gradient, 508–510
Temperature gradient cycling, 444
Temperature ranges, 335–336
10-μm range, of carbon fibers, 32
Tensile (tension) strength, 6, 36, 101–104, 301
  in-plane, 217, 218
  interfacial, 56, 57, 60
  temperature dependence of, 102–103
  thru-thickness, 217, 218
  ultimate, 251–252
Tensile behavior, 11, 69–70
  analysis of, 57
  of tows, 13
Tensile creep, 298–304, 308–313
Tensile creep and recovery, 304–308
Tensile creep rates, 314
Tensile creep strain, 316
Tensile fatigue, S–N curves for, 108–109
Tensile fracture, determining, 101–102
Tensile internal forces, 80

Tensile preloading, 421
Tensile properties, at elevated temperatures, 102–103
Tensile stresses, crack initiation and, 41
Tensile stress–strain behavior, 296–298, 309, 310, 521–524, 536
Tensile stress–strain curves, 54, 55, 59
Tensile tests, 191, 572, 635
Tension–tension fatigue, 321–324, 324–330
Tensor notation, 468
Textile composite reinforcements, 65–74
Textile materials, measuring stiffness of, 70–71
Textile preform, infiltration of, 75
Textile reinforcements, 65–84
  mechanical behavior of, 67–72
  unbalanced, 81
Thébault, Jacques Georges, xviii, 647
Thermal barrier coatings (TBCs), 233, 444, 445
Thermal conductivity, 188–189, 394, 396–398, 400, 624, 625, 627, 636, 637, 654
  of carbon/carbon composites, 110–115
  of carbon/carbon materials, 140
  of C/SiC materials, 660
  evolution of, 112–115
  of ultrahigh temperature ceramic-based composites, 279–280
Thermal conductivity plots, 281
Thermal creep, 633–634
Thermal cyclic life, enhanced, 435
Thermal cycling, 435, 438, 440, 593
Thermal expansion
  of carbon/carbon composites, 116–117
  of environmental barrier coatings, 433
  with C/C and C/SiC composites, 453
Thermal expansion mismatch, 458
Thermal exposure, long-term, 253
Thermal fatigue testing, 599
Thermal gradient CVI, 92, 93, 154–156, 174. *See also* Chemical vapor infiltration (CVI)
Thermal insulation, 139
Thermally grown oxide (TGO), 444
Thermal performance, 236
Thermal properties, 109–118, 187–189
Thermal protection materials, 383
Thermal protection systems (TPSs), 147, 148, 192–193, 236, 263, 353–354, 379
Thermal resistivity, 394
Thermal shock, 189, 191, 202
Thermal shock resistance, 125
  of carbon/carbon composites, 118
  of oxidation protective coatings, 460
Thermal shock resistance evolution, 627
Thermal spray coating microstructure, 443
Thermal stability, 189, 190
Thermal treatments, of metals, 134–137
Thermochemical capability, 640
Thermodynamical bond, 552
Thermodynamics, in damage modeling, 469–471
Thermodynamic stability, 554
Thermoelastic properties, 256–257
Thermography, 562
Thermomechanical properties, 625–627, 651
Thermomechanical stability, 557
Thermophysical properties, of monolithic carbons, 110
Thermoplastics, as matrix precursors, 91–92
Thermosetting resins, as matrix precursors, 91
Thermostructural composite materials, 647
Thermostructural composites, tribological applications for, 668–669
Third-generation fibers, 4
3D C/C composites, 355, 359
3D C/C testing, under plasma jet and oxidation, 359–361
3D fiber preforms, 89–90, 150
3D geometrical models, 67
Three-dimensional (3D) strength properties, 218
3D-orthogonal architectures, 226–227
3D-reinforced carbon/carbons, compressive strength of, 106–107
3D-woven fabrics, 152
Three-node shell finite element, 80
Three-point bend testing, 257–258
Thrust vector control (TVC) systems, 202
Thru-thickness properties, 225–227
Thru-thickness tensile strength, 217, 218
Time-dependent rupture, 347
Time derivatives, 486
Time–space approximations, 504
Time–stress–temperature–environment degradation, 334
Time–temperature performance, 250
Time to failure, 22, 325, 588
Tokamak armor, 354–355
Tokamak system, 610
Tolerance analysis, 506–512
Tough ceramic composites, 237
Toughness, 255, 394
Tough oxide–oxide composites, 264
Tow-based testing methods, 14–16
Tow lifetimes, 545
Tows. *See also* SiC-based filaments/tows
  choices of, 624
  decrimping of, 69–70
  longitudinal, 544
  matrix-coated, 525
  multifilament, 11–16
  static fatigue of, 22–23
  stress–rupture time behavior of, 18
  ultimate failures of, 525
Tow strength, 12
Tow tensile behavior, fiber properties and, 13
Tow testing technique, 16
Tow thickness, 540
Tow ultimate strength, 14
Trace operator, 469
Traction energy, 474
Traditional acoustic emission (TAE), 572–573
Transient-liquid-phase bonding (TLPB) method, 555
Transient regime, 382
Transition layer, 460
Transmission electron microscopy (TEM), 28, 29–31
Transverse bend test, 57
Transverse compaction behavior, 72
Transverse cracks, 75, 497, 522, 523, 540
Transverse intra-yarn crack opening, modeling of, 482–490
Transverse matrix cracks, 484
Transverse tensile test, 57
Transverse tows, multiple cracking in, 539–541
Treated fibers, 50, 52
Tribological applications, 655–657, 668–669
Tribological behavior, 651–653
  of C/C materials, 651–655
  of C/SiC and C/C-SiC materials, 191
Triplex cladding, 637
Truncations, 525
Turbine component demonstrations, on military engines, 599–600
Turbine components, CMC research and development for, 602–603
Turbine technology, 603
Turbostatic structure, 28
2D braiding, 633
2D C/R testing, in plasma jet and in oxidation furnace, 362
2D CVI composites, 634. *See also* Chemical vapor infiltration (CVI)
2D fabric-based materials, 185–187
2D fabrics, 75, 152–154

2D fiber architecture, 225–226, 227
2D woven composites, matrix damage in, 541–542
2.5D architecture, 75, 76
2.5D composites, 356, 633
2.5D C/C testing, in SRM facility, 363–365
2.5D fiber preforms, 89–90, 150
Two-step nonlinear force–strain relation, 11
Tyranno™ family fibers, 4, 6, 16
Tyranno™-SA3 SiC fiber, 223, 395, 627, 628–630, 634
  aluminum in, 629
  boron in, 630
  free carbon in, 21

UCSB composites, 255, 256
Ultimate failure, 524–526. *See also* Final failure
  of a longitudinal tow, 525
  in 1D composites and minicomposites, 535
  of a tow, 525
Ultimate strength, of tows, 14
Ultimate tensile strength (UTS), 251–252, 296–297
Ultraefficient engine technology (UEET), 441
Ultra Efficient Engine Technology (UEET) Program, 601, 603
Ultra-high modulus (UHM) fibers, 147, 150
Ultrahigh temperature ceramic-based composites, 273 292
  elastic properties of, 278–279
  electrical conductivity of, 280–282
  fabrication methods for, 273–278
  flexural strength of, 284–285
  fracture toughness of, 282–283
  mechanical properties of, 282–285
  oxidation protective coatings for, 452–463
  with particulates, 273–285
  physical properties of, 278–282
  with short fibers, 285–288
  thermal conductivity of, 279–280
Ultrahigh-temperature composites, 452
Ultrasonic testing, 571
Ultrastable structures, CMC joining for, 561
Unbalanced textile reinforcement, 81
Unbridged crack growth, stressed-oxidation degradation modeling in, 347–348
Undamaged materials, elastic energy of, 471
Unidirectional (UD) C/C composites, 89, 96
Unidirectional (UD) fiber layers, 185–187
Uniform strain model, 34
Uniform stress model, 34
UniGrid™ system, 136, 137
United States FAA CLEEN Program, 604. *See also* US entries
Unit woven cell, architecture and geometry of, 67
Universal stress sensor, 35
Unixial loading, 35
Unload/reload hysteresis, 576, 577
Unstable failure(s), 13–14
Unsupervised clustering methodology, applied to AE signals, 583–584
Upper nanometer range, of carbon fibers, 31–32
US Air Force, CMC developments for, 595–597. *See also* United States FAA CLEEN Program
US military engines, CMC developments for, 595–598
US Navy, CMC developments for, 597–598

V 22 Osprey brakes, 657
Vacancies, migrating, 393
Vacuum-assisted polymer (VAP) infiltration, 158
Vacuum furnaces, high-performance, 129–130
Vacuum infiltration, vibration-assisted, 245–246
Variation coefficient, 14
Ventilated C/SiC brake disks, 662–663
Vertical cracks, 437, 440
Very-high-temperature reactors (VHTRs), 616, 617, 618–620
VHTR applications, 618–620, 641
VHTR control rods, 635
Vibration-assisted vacuum infiltration, 245–246
Vignoles, Gerard L., xviii, 353
Vinci engine, 195, 197
Virtual testing, 539–542
Viscoelastic behavior, 36
Viscoplasticity, 470–471
Viscous sliding (VS), 348
Void swelling, 393
Void volume, 242, 250
Volatility, 433, 453
Volatility data, 432
Volatilization, 336, 337
Volume-of-fluid (VOF) techniques, 372, 374
Vulcain nozzle extensions, 193, 196

Wang, Yiguang, xviii, 452
Warm pressing, 164–165
Warp and weft directions, 78, 80
Water jet cutting, 127
Water vapor, volatility in, 433
Water vapor stability, of environmental barrier coatings, 433–434
Waveforms, 575, 576–577, 579
Wave guides, 584
Wave velocity, 584
Weakest-link law/rule, 375–376, 383
Weak fiber–matrix interfaces, 238–239
Weak interfaces, 56
Weak interphases, 56
Weak oxide interface coatings, 257–260
Wear effect, 485, 487
Wear rates, of C/SiC materials, 660–661
Weft and warp directions, 78, 80
Weibull distribution, in fragmentation modeling, 531
Weibull distribution function, 9, 10
Weibull equation, 22
Weibull model, 6–8, 527, 528
Weibull modulus, 36, 525–526, 527, 536, 537
Weibull plots, 8, 9
Weibull statistical parameters, determination of, 8–9
Weibull theory, 36
Weiss, Roland, xviii, 87
Wet filament winding, 165–166
Wettability, 557
Wettability theory, 554
Whisker-toughened coatings, 458
Wireframe discretization, 492–493
Wood-based products, joints for, 558–559
Wound highly porous oxide (WHIPOX) ceramic composite, 249
Wovan fabric, 241
Woven composite reinforcements, draping in, 76
Woven composites, 539–542
Woven fabric reinforcement internal work, 77–78
Woven fibers, 344

Xenotimes, 257
X-ray computed microtomography (CMT), 359, 360, 361
X-ray diffraction (XRD), 28, 29–31, 35, 36

YAG (yttrium aluminum garnet) fibers, 240–241, 264
Yarn ablation, 360
Yarn damage, 507
Yarn degradation, 502

Yarn matrix damage, 510–511
Yarn prepreg process, 92
Yarns, 65. *See also* Transverse intra-yarn crack opening
Young's modulus, 27, 185, 189, 278–279, 289, 624
  crystallite orientation and, 34
  deflection criteria and, 45
  high temperature behavior of, 35–36
  high tensile, 33
  of carbon fibers, 33
Ytterbium (Yb) silicates, 445, 446
Yttria-stabilized zirconia (YSZ), 438, 440, 446
Yttrium (Y) silicates, 445, 459

Zircalloy cladding, 638
Zirconia, 249
Zirconia coatings, 260, 261. *See also* ZrO_2 coating
Zirconium diborides (ZrB_2), 273–290
ZrB_2-based ceramic matrix composites, 286–288
ZrB_2 composites, 273–290
ZrC-SiC multilayer coating, 459
ZrO_2 coating, 459. *See also* Zirconia coatings

www.ingramcontent.com/pod-product-compliance
Lightning Source LLC
LaVergne TN
LVHW080057220926
848484LV00007BA/1808